PRINCIPLES OF HUMAN ANATOMY

Principles of
Human Anatomy
FIFTH EDITION

Gerard J. Tortora
BERGEN COMMUNITY COLLEGE

1817

Harper & Row, Publishers, New York

Cambridge, Philadelphia, San Francisco,

London, Mexico City, São Paulo, Singapore, Sydney

Sponsoring Editor: Elizabeth A. Dollinger
Project Editor: Thomas R. Farrell
Text Design: Caliber Design Planning, Inc.
Cover Design: John Keithley and Associates, based on text design
Cover Photo: Schematic of human circulatory system. © Steve Allen, Peter Arnold, Inc.
Text Art: Vantage Art, Inc.
Photo Research: Mira Schachne
Production Manager: Kewal K. Sharma
Compositor: Arcata Graphics/Kingsport
Printer and Binder: Arcata Graphics/Kingsport

PRINCIPLES OF HUMAN ANATOMY, Fifth Edition

Library of Congress Cataloging-in-Publication Data

Tortora, Gerard J.
    Principles of human anatomy.

    Bibliography: p.
    Includes indexes.
    1. Human physiology. 2. Anatomy, Human. I. Title.
QP34.5.T68 1989   612   88–6232
ISBN 0-06-046685-5

89 90 91 9 8 7 6 5 4 3 2

The fifth edition of Principles of Human Anatomy *is dedicated to Geraldine C. Tortora, my wife, whose love, support, and assistance have contributed so much to my success as an author.*

Contents in Brief

Contents in Detail ix
Preface xix
Note to the Student xxiii

1 **An Introduction to the Human Body** 1

2 **Cells** 31

3 **Tissues** 63

4 **The Integumentary System** 92

5 **Osseous Tissue** 113

6 **The Skeletal System: The Axial Skeleton** 131

7 **The Skeletal System: The Appendicular Skeleton** 164

8 **Articulations** 185

9 **Muscle Tissue** 223

10 **The Muscular System** 245

11 **Surface Anatomy** 311

12 **The Cardiovascular System: Blood** 327

13 **The Cardiovascular System: The Heart** 344

14 **The Cardiovascular System: Blood Vessels** 369

15 **The Lymphatic System** 415

16 **Nervous Tissue** 443

17 **The Spinal Cord and the Spinal Nerves** 459

18 **The Brain and the Cranial Nerves** 486

19 **The Autonomic Nervous System** 533

20 **Sensory and Motor Systems** 549

21 **The Endocrine System** 590

22 **The Respiratory System** 616

23 **The Digestive System** 644

24 **The Urinary System** 687

25 **The Reproductive Systems** 708

26 Developmental Anatomy 751

Appendix A: Symbols and Abbreviations A-1
Appendix B: Eponyms Used in This Text B-1
Appendix C: Answers to Self-Quizzes C-1
Selected Readings SR-1
Glossary of Terms G-1
Index I-1

Contents in Detail

Preface—xix
Note to the Student—xxiii

1 AN INTRODUCTION TO THE HUMAN BODY—1

Anatomy and Physiology Defined—2
Levels of Structural Organization—2
Life Processes—4
Structural Plan—4
Anatomical Position and Regional Names—4
Directional Terms—4
Planes and Sections—12
Body Cavities—13
Adominopelvic Regions—16
Abdominopelvic Quadrants—16
Medical Imaging—19
Conventional Radiography—19
Computed Tomography (CT) Scanning—19
Dynamic Spatial Reconstructor (DSR)—22
Magnetic Resonance Imaging (MRI)—23
Ultrasound (US)—24
Positron Emission Tomography (PET)—25
Digital Subtraction Angiography (DSA)—25
Measuring the Human Body—25
STUDY OUTLINE—27
REVIEW QUESTIONS—28
SELF-QUIZ—29

2 CELLS—31

Generalized Animal Cell—32
Plasma (Cell) Membrane—32
Chemistry and Structure—32
Functions—33
Movement of Materials Across Plasma
    Membranes—35
        Passive Processes 35—Active Processes 36
Cytoplasm—38
Organelles—38
Nucleus—38
Ribosomes—40

Endoplasmic Reticulum (ER)—40
Golgi Complex—40
Mitochondria—42
Lysosomes—44
Peroxisomes—45
The Cytoskeleton—46
Centrosome and Centrioles—47
Flagella and Cilia—47
Cell Inclusions—47
Extracellular Materials—48
Normal Cell Division—48
Somatic Cell Division—49
        Mitosis 49—Cytokinesis 52
Reproductive Cell Division—52
        Meiosis 53
Abnormal Cell Division: Cancer (CA)—54
Definition—54
Spread—55
Types—56
Possible Causes—56
Treatment—57
Cells and Aging—57
KEY MEDICAL TERMS ASSOCIATED WITH CELLS—58
STUDY OUTLINE—59
REVIEW QUESTIONS—61
SELF-QUIZ—61

3 TISSUES—63

Types of Tissues—64
Epithelial Tissue—64
Covering and Lining Epithelium—65
        Arrangement of Layers 65—Cell Shapes 65—Cell
        Junctions 65—Classification 67—Simple
        Epithelium 67—Stratified Epithelium 67—
        Pseudostratified Epithelium 74
Glandular Epithelium—74
        Structural Classification of Exocrine
        Glands 74—Functional Classification of
        Exocrine Glands 75
Connective Tissue—75
Classification—76
Embryonic Connective Tissue—76

Adult Connective Tissue—76
    Connective Tissue Proper 76—Cartilage 84—
    Osseous Tissue (Bone) 84—Vascular Tissue
    (Blood) 84
Membranes—85
Mucous Membranes—85
Serous Membranes—85
Cutaneous Membrane—85
Synovial Membranes—85
Muscle Tissue—85
Nervous Tissue—87
STUDY OUTLINE—88
REVIEW QUESTIONS—89
SELF-QUIZ—90

4 THE INTEGUMENTARY SYSTEM—92

Skin—93
Functions—93
Structure—93
Epidermis—93
Dermis—96
Skin Color—96
Epidermal Ridges and Grooves—97
Blood Supply—98
Skin and Temperature Regulation—98
Skin Wound Healing—98
Epidermal Wound Healing—98
Deep Wound Healing—99
Epidermal Derivatives—99
Hair—99
    Development and Distribution 100—
    Structure 100—Color 103—Growth and
    Replacement 103
Glands—104
    Sebaceous (Oil) Glands 104—Sudoriferous (Sweat)
    Glands 104—Ceruminous Glands 104
Nails—105
Aging and the Integumentary System—105
Developmental Anatomy of the Integumentary
    System—105
Applications to Health—106
Acne—106
Systemic Lupus Erythematosus (SLE)—106
Psoriasis—106
Decubitus Ulcers—107
Sunburn—107
Skin Cancer—107
Burns—107
    Classification 107—Treatment 109
KEY MEDICAL TERMS ASSOCIATED WITH THE INTEGUMENTARY
SYSTEM—109
STUDY OUTLINE—110

REVIEW QUESTIONS—111
SELF-QUIZ—112

5 OSSEOUS TISSUE—113

Functions—114
Histology—114
Compact Bone—116
Spongy Bone—118
Ossification—118
Intramembranous Ossification—118
Endochondral Ossification—118
Bone Growth—119
Bone Replacement—121
Blood and Nerve Supply—122
Aging and the Skeletal System—123
Developmental Anatomy of the Skeletal System—123
Applications to Health—123
Osteoporosis—123
Vitamin Deficiencies—124
    Rickets 124—Osteomalacia 125
Paget's Disease—125
Osteomyelitis—125
Fractures—125
    Types 125—Fracture Repair 125
KEY MEDICAL TERMS ASSOCIATED WITH OSSEOUS TISSUE—127
STUDY OUTLINE—127
REVIEW QUESTIONS—128
SELF-QUIZ—129

6 THE SKELETAL SYSTEM: THE AXIAL SKELETON—131

Types of Bones—132
Surface Markings—132
Divisions of the Skeletal System—132
Skull—133
Sutures—135
Fontanels—135
Cranial Bones—135
    Frontal Bone 135—Parietal Bones 135—Temporal
    Bones 139—Occipital Bone 140—Sphenoid
    Bone 140—Ethmoid Bone 142
Cranial Fossae—143
Facial Bones—143
    Nasal Bones 143—Maxillae 144—Paranasal
    Sinuses 144—Zygomatic Bones 145—
    Mandible 145—Lacrimal Bones 146—Palatine
    Bones 146—Inferior Nasal Conchae 146—
    Vomer 146
Foramina—146
Hyoid Bone—146

Vertebral Column—148
Divisions—148
Normal Curves—148
Typical Vertebra—148
Cervical Region—151
Thoracic Region—151
Lumbar Region—151
Sacrum and Coccyx—156
Thorax—157
Sternum—157
Ribs—158
Applications to Health—159
Herniated (Slipped) Disc—159
Abnormal Curves—160
Spina Bifida—160
Fractures of the Vertebral Column—160
STUDY OUTLINE—161
REVIEW QUESTIONS—161
SELF-QUIZ—162

7 THE SKELETAL SYSTEM: THE APPENDICULAR SKELETON—164

Pectoral (Shoulder) Girdle—165
Clavicle—166
Scapula—166
Upper Extremity—169
Humerus—169
Ulna and Radius—170
Carpals, Metacarpals, and Phalanges—170
Pelvic (Hip) Girdle—170
Lower Extremity—174
Femur—175
Patella—175
Tibia and Fibula—176
Tarsals, Metatarsals, and Phalanges—179
Arches of the Foot—180
Female and Male Skeletons—183
STUDY OUTLINE—183
REVIEW QUESTIONS—184
SELF-QUIZ—184

8 ARTICULATIONS—185

Classification—186
Functional—186
Structural—186
Fibrous Joints—186
Suture—186
Syndesmosis—186
Gomphosis—186
Cartilaginous Joints—187
Synchondrosis—187

Symphysis—187
Synovial Joints—187
Structure—187
Movements—189
    *Gliding 189—Angular 189—Rotation 189—
    Circumduction 189—Special 189*
Types—192
    *Gliding 193—Hinge 193—Pivot 193—
    Ellipsoidal 193—Saddle 193—Ball-and-
    Socket 196*
Summary of Joints—196
Selected Articulations of the Body—197
Applications to Health—218
Rheumatism—218
Arthritis—218
    *Rheumatoid Arthritis (RA) 218—
    Osteoarthritis 218—Gouty Arthritis 218*
Bursitis—219
Dislocation—219
Sprain and Strain—219
KEY MEDICAL TERMS ASSOCIATED WITH ARTICULATIONS—219
STUDY OUTLINE—219
REVIEW QUESTIONS—220
SELF-QUIZ—221

9 MUSCLE TISSUE—223

Characteristics—224
Functions—224
Types—224
Skeletal Muscle Tissue—224
Connective Tissue Components—224
Nerve and Blood Supply—226
Histology—226
Contraction—230
Sliding-Filament Theory—230
Neuromuscular Junction (Motor End-Plate)—230
Motor Unit—231
Mechanism—233
Muscle Length and Force of Contraction—233
All-or-None Principle—233
Muscle Tone—233
Types of Skeletal Muscle Fibers—234
Cardiac Muscle Tissue—235
Smooth Muscle Tissue—237
Aging and Muscle Tissue—238
Developmental Anatomy of the Muscular System—238
Applications to Health—239
Fibrosis—239
Fibrositis—239
''Charley Horse''—240
Muscular Dystrophies—240
Myasthenia Gravis (MG)—240
Abnormal Contractions—240

KEY MEDICAL TERMS ASSOCIATED WITH THE MUSCULAR
SYSTEM—240
STUDY OUTLINE—241
REVIEW QUESTIONS—242
SELF-QUIZ—243

10. THE MUSCULAR SYSTEM—245

How Skeletal Muscles Produce Movement—246
Origin and Insertion—246
Lever Systems and Leverage—246
Arrangement of Fasciculi—248
Group Actions—248
Naming Skeletal Muscles—248
Principal Skeletal Muscles—248
Intramuscular (IM) Injections—307
STUDY OUTLINE—308
REVIEW QUESTIONS—308
SELF-QUIZ—309

11 SURFACE ANATOMY—311

Head—312
Neck—312
Trunk—312
Upper Extremity—315
Lower Extremity—317
STUDY OUTLINE—325
REVIEW QUESTIONS—325
SELF-QUIZ—325

12 THE CARDIOVASCULAR SYSTEM: BLOOD—327

Physical Characteristics—328
Functions—328
Components—329
Formed Elements—329
    Origin 332—Erythrocytes 332—Leucocytes 334—
    Thrombocytes 336
Plasma—337
Applications to Health—338
Anemia—338
    Nutritional Anemia 338—Pernicious Anemia 338—
    Hemorrhagic Anemia 338—Hemolytic
    Anemia 338—Aplastic Anemia 339—Sickle-Cell
    Anemia (SCA) 339
Polycythemia—339
Infectious Mononucleosis (IM)—339
Chronic Epstein-Barr Virus (CEBV) Syndrome—340
Leukemia—340
KEY MEDICAL TERMS ASSOCIATED WITH BLOOD—340
STUDY OUTLINE—341

REVIEW QUESTIONS—342
SELF-QUIZ—342

13 THE CARDIOVASCULAR SYSTEM: THE HEART—344

Location—345
Pericardium—345
Heart Wall—345
Chambers of the Heart—347
Great Vessels of the Heart—347
Valves of the Heart—348
Atrioventricular (AV) Valves—348
Semilunar Valves—351
Skeleton of the Heart—351
Surface Projection—353
Conduction System—354
Electrocardiogram (ECG)—354
Cardiac Cycle—356
Autonomic Control—356
Blood Supply—357
Artificial Heart—358
Risk Factors in Heart Disease—359
Developmental Anatomy of the Heart—360
Applications to Health—360
Coronary Artery Disease (CAD)—360
    Atherosclerosis 360—Coronary Artery Spasm 362
Congenital Defects—362
Arrhythmias—364
    Heart Block 364—Flutter and Fibrillation 365—
    Ventricular Premature Contraction (VPC) 365
Congestive Heart Failure (CHF)—365
Cor Pulmonale (CP)—365
KEY MEDICAL TERMS ASSOCIATED WITH THE HEART—365
STUDY OUTLINE—366
REVIEW QUESTIONS—367
SELF-QUIZ—367

14 THE CARDIOVASCULAR SYSTEM: BLOOD VESSELS—369

Arteries—370
Elastic (Conducting) Arteries—370
Muscular (Distributing) Arteries—370
Anastomoses—370
Arterioles—371
Capillaries—371
Venules—373
Veins—373
Blood Reservoirs—374
Circulatory Routes—375
Systemic Circulation—375
Pulmonary Circulation—405
Hepatic Portal Circulation—405

Fetal Circulation—406
Aging and the Cardiovascular System—409
**Developmental Anatomy of Blood and Blood
    Vessels—409**
Applications to Health—409
Hypertension—409
Aneurysm—410
Coronary Artery Disease (CAD)—410
Deep-Venous Thrombosis (DVT)—411
Key Medical Terms Associated with Blood Vessels—411
Study Outline—411
Review Questions—412
Self-Quiz—413

15   THE LYMPHATIC SYSTEM—415

Lymphatic Vessels—416
Lymphatic Tissue—418
Lymph Nodes—418
Tonsils—421
Spleen—421
Thymus Gland—424
Lymph Circulation—424
Route—424
    *Thoracic (Left Lymphatic) Duct 425—Right
    Lymphatic Duct—425*
Maintenance—427
Principal Groups of Lymph Nodes—427
**Developmental Anatomy of the Lymphatic
    System—435**
Applications to Health—435
Acquired Immune Deficiency Syndrome (AIDS)—435
    *HIV: Structure and Pathogenesis 435—
    Symptoms 437—HIV Outside the Body 437—
    Classification 437—Transmission 437—Drugs and
    Vaccines Against HIV 437—Prevention of
    Transmission 438*
Autoimmune Diseases—438
Severe Combined Immunodeficiency (SCID)—438
Hypersensitivity (Allergy)—439
Tissue Rejection—439
Hodgkin's Disease (HD)—439
Key Medical Terms Associated with the Lymphatic
System—440
Study Outline—440
Review Questions—441
Self-Quiz—441

16   NERVOUS TISSUE—443

Organization—444
Histology—445
Neuroglia—445

Neurons—445
    *Structure 445—Structural Variation 448—
    Classification 450*
Nerve Impulse—451
Regeneration—454
Organization of Neurons—455
Study Outline—457
Review Questions—457
Self-Quiz—458

17   THE SPINAL CORD AND THE
        SPINAL NERVES—459

Grouping of Neural Tissue—460
Spinal Cord—460
Protection and Coverings—460
    Vertebral Canal 460—Meninges 460
General Features—461
Structure in Cross Section—465
Functions—466
    Impulse Conduction 466—Reflex Center 467
Spinal Nerves—469
Names—469
Composition and Coverings—469
Distribution—469
    *Branches 469—Plexuses 470—Intercostal
    (Thoracic) Nerves 474*
Dermatomes—478
Applications to Health—480
Spinal Cord Injury—480
Peripheral Nerve Damage and Repair—480
    *Chromatolysis 480—Wallerian Degeneration 480—
    Retrograde Degeneration 481—Regeneration 481*
Neuritis—481
Sciatica—481
Shingles—482
Study Outline—482
Review Questions—483
Self-Quiz—483

18   THE BRAIN AND THE CRANIAL
        NERVES—486

Brain—487
Principal Parts—487
Protection and Coverings—488
Cerebrospinal Fluid (CSF)—488
Blood Supply—492
Brain Stem—492
    Medulla Oblongata 492—Pons 496—Midbrain 497
Diencephalon—497
    Thalamus 497—Hypothalamus 497

Cerebrum—501
  *Lobes 502—White Matter 504—Basal Ganglia
  (Cerebral Nuclei) 504—Limbic System 506—
  Functional Areas of Cerebral Cortex 506—
  Electroencephalogram (EEG)—508*
Brain Lateralization (Split-Brain Concept)—508
Cerebellum—509
  Structure 509—Functions 509
Cranial Nerves—511
Olfactory (I)—511
Optic (II)—512
Oculomotor (III)—513
Trochlear (IV)—513
Trigeminal (V)—514
Abducens (VI)—515
Facial (VII)—516
Vestibulocochlear (VIII)—516
Glossopharyngeal (IX)—516
Vagus (X)—517
Accessory (XI)—517
Hypoglossal (XII)—519
Aging and the Nervous System—521
Developmental Anatomy of the Nervous System—521
Applications to Health—521
Cerebrovascular Accident (CVA)—521
Transient Ischemic Attack (TIA)—521
Brain Tumors—522
Poliomyelitis—524
Cerebral Palsy (CP)—524
Parkinson's Disease (PD)—524
Multiple Sclerosis (MS)—525
Epilepsy—526
Dyslexia—526
Tay-Sachs Disease—526
Headache—526
Trigeminal Neuralgia (Tic Douloureux)—527
Reye's Syndrome (RS)—528
Alzheimer's Disease (AD)—528
KEY MEDICAL TERMS ASSOCIATED WITH THE CENTRAL NERVOUS
SYSTEM—528
STUDY OUTLINE—529
REVIEW QUESTIONS—530
SELF-QUIZ—531

19 THE AUTONOMIC NERVOUS SYSTEM—533

**Somatic Efferent and Autonomic Nervous
  Systems—534**
Structure of the Autonomic Nervous System—534
Visceral Efferent Pathways—534
  *Preganglionic Neurons 535—Autonomic
  Ganglia 535—Postganglionic Neurons 537*
Sympathetic Division—538
Parasympathetic Division—540

Physiology of the Autonomic Nervous System—541
Neurotransmitters—541
Receptors—542
Activities—542
Visceral Autonomic Reflexes—544
Control by Higher Centers—545
Biofeedback—545
Meditation—546
STUDY OUTLINE—546
REVIEW QUESTIONS—547
SELF-QUIZ—547

20 SENSORY AND MOTOR SYSTEMS—549

Sensations—550
Definition—550
Characteristics—550
Classification of Receptors—551
  *Location 551—Stimulus Detected 551—Simplicity
  or Complexity 551*
General Senses—551
Cutaneous Sensations—551
  *Tactile Sensations 551—Thermoreceptive
  Sensations 553—Pain Sensations 553*
Proprioceptive Sensations—555
  Receptors 555
Levels of Sensation—556
Sensory Pathways—556
Somatosensory Cortex—556
Posterior Column Pathway—557
Spinothalamic Pathway—557
Cerebellar Tracts—558
Special Senses—558
Olfactory Sensations—559
  *Structure of Receptors 559—Olfactory
  Pathway 560*
Gustatory Sensations—560
  *Structure of Receptors 560—Gustatory
  Pathway 561*
Visual Sensations—561
  *Accessory Structures of Eye 562—Structure of
  Eyeball 563—Visual Pathway 568*
Auditory Sensations and Equilibrium—568
  *External (Outer) Ear 570—Middle Ear 570—
  Internal (Inner) Ear 573—Auditory Pathway 575—
  Mechanism of Equilibrium 576*
Motor Pathways—579
Linkage of Sensory Input and Motor Responses—579
Motor Cortex—580
Pyramidal Pathways—580
Extrapyramidal Pathways—581
Developmental Anatomy of the Eye and Ear—581
Applications to Health—583
Cataract—583

Glaucoma—583
Conjunctivitis (Pinkeye)—583
Trachoma—584
Deafness—584
Labyrinthine Disease—584
Ménière's Syndrome—584
Vertigo—584
Otitis Media—584
Motion Sickness—584
KEY MEDICAL TERMS ASSOCIATED WITH SENSORY
STRUCTURES—585
STUDY OUTLINE—585
REVIEW QUESTIONS—587
SELF-QUIZ—587

21   THE ENDOCRINE SYSTEM—590

Endocrine Glands—591
Receptors—591
Feedback Control—592
Pituitary (Hypophysis)—593
Adenohypophysis—593
Neurohypophysis—596
Thyroid—597
Parathyroids—600
Adrenals (Suprarenals)—601
Adrenal Cortex—602
Adrenal Medulla—605
Pancreas—606
Ovaries and Testes—608
Pineal (Epiphysis Cerebri)—608
Thymus—609
Aging and the Endocrine System—610
Developmental Anatomy of the Endocrine
   System—610
Other Endocrine Tissues—610
KEY MEDICAL TERMS ASSOCIATED WITH THE ENDOCRINE
SYSTEM—611
STUDY OUTLINE—612
REVIEW QUESTIONS—613
SELF-QUIZ—614

22   THE RESPIRATORY SYSTEM—616

Organs—618
Nose—618
Pharynx—620
Larynx—620
Trachea—624
Bronchi—626
Lungs—628
   Gross Anatomy 628—Lobes and Fissures 628—
   Lobules 628—Alveolar–Capillary (Respiratory)
   Membrane 630—Blood and Nerve Supply 630

Nervous Control of Respiration—634
Medullary Rhythmicity Area—634
Pneumotaxic Area—635
Apneustic Area—635
Cortical Influences—635
Aging and the Respiratory System—635
Developmental Anatomy of the Respiratory
   System—635
Applications to Health—636
Bronchogenic Carcinoma (Lung Cancer)—636
Bronchial Asthma—636
Bronchitis—636
Emphysema—637
Pneumonia—637
Tuberculosis (TB)—637
Respiratory Distress Syndrome (RDS) of the
   Newborn—637
Respiratory Failure—638
Sudden Infant Death Syndrome (SIDS)—638
Coryza (Common Cold) and Influenza (Flu)—639
Pulmonary Embolism (PE)—639
Pulmonary Edema—639
Carbon Monoxide (CO) Poisoning—639
Smoke Inhalation Injury—639
KEY MEDICAL TERMS ASSOCIATED WITH THE RESPIRATORY
SYSTEM—640
STUDY OUTLINE—640
REVIEW QUESTIONS—641
SELF-QUIZ—642

23   THE DIGESTIVE SYSTEM—644

Digestive Processes—645
Organization—645
General Histology—645
   Mucosa 645—Submucosa 646—Muscularis 646—
   Serosa 646
Peritoneum—646
Mouth (Oral Cavity)—648
Tongue—649
Salivary Glands—651
Teeth—652
   Dental Terminology 656—Dentitions 656—Blood
   and Nerve Supply 657
Pharynx—657
Esophagus—657
Histology—657
Activities—657
Blood and Nerve Supply—659
Stomach—659
Anatomy—659
Histology—659
Activities—662
Blood and Nerve Supply—663

Pancreas—663
Anatomy—663
Histology—663
Activities—663
Blood and Nerve Supply—663
Liver—665
Anatomy—665
Histology—665
Activities—665
Blood and Nerve Supply—667
Gallbladder (GB)—670
Histology—670
Activities—670
Blood and Nerve Supply—671
Small Intestine—671
Anatomy—671
Histology—671
Activities—672
Blood and Nerve Supply—674
Large Intestine—674
Anatomy—675
Histology—675
Activities—675
Blood and Nerve Supply—679
Aging and the Digestive System—679
Developmental Anatomy of the Digestive System—679
Applications to Health—679
Dental Caries—679
Periodontal Disease—681
Peritonitis—681
Peptic Ulcers—681
Appendicitis—682
Tumors—682
Diverticulitis—682
Cirrhosis—682
Hepatitis—682
Gallstones—683
Anorexia Nervosa—683
Bulimia—683
KEY MEDICAL TERMS ASSOCIATED WITH THE DIGESTIVE SYSTEM—683
STUDY OUTLINE—684
REVIEW QUESTIONS—685
SELF-QUIZ—685

24   THE URINARY SYSTEM—687

Kidneys—688
External Anatomy—688
Internal Anatomy—688
Nephron—692
Blood and Nerve Supply—693
Juxtaglomerular Apparatus (JGA)—696
Physiology—697
Hemodialysis Therapy—697

Ureters—698
Structure—698
Histology—699
Physiology—700
Blood and Nerve Supply—700
Urinary Bladder—700
Structure—700
Histology—700
Physiology—701
Blood and Nerve Supply—701
Urethra—701
Histology—701
Physiology—702
Aging and the Urinary System—702
Developmental Anatomy of the Urinary System—702
Applications to Health—702
Renal Calculi (Kidney Stones)—702
Gout—703
Glomerulonephritis (Bright's Disease)—704
Pyelitis and Pyelonephritis—704
Cystitis—704
Nephrosis—704
Polycystic Disease—704
Renal Failure—704
Urinary Tract Infections (UTIs)—705
KEY MEDICAL TERMS ASSOCIATED WITH THE URINARY SYSTEM—705
STUDY OUTLINE—705
REVIEW QUESTIONS—706
SELF-QUIZ—706

25   THE REPRODUCTIVE SYSTEMS—708

Male Reproductive System—709
Scrotum—709
Testes—709
    Spermatogenesis 712—Spermatozoa 714
Ducts—714
    Ducts of the Testis 714—Epididymis 715—Ductus (Vas) Deferens 715—Ejaculatory Duct 716—Urethra 716
Accessory Sex Glands—716
Semen (Seminal Fluid)—718
Penis—719
Female Reproductive System—722
Ovaries—722
    Oogenesis—722
Uterine (Fallopian) Tubes—726
Uterus—727
Menstrual Cycle—728
    Hormonal Control 729—Menstrual Phase (Menstruation) 729—Preovulatory Phase 731—Ovulation 731—Postovulatory Phase 731

Vagina—732
Vulva—732
Perineum—733
Mammary Glands—733
Sexual Intercourse—735
Male Sexual Act—736
 Erection 736—Lubrication 736—Orgasm 736
Female Sexual Act—737
 Erection 737—Lubrication 737—Orgasm
 (Climax) 737
Birth Control (BC)—737
Removal of Gonads and Uterus—737
Sterilization—737
Contraception—737
 Natural 737—Mechanical 737—
 Chemical 738
Aging and the Reproductive Systems—739
**Developmental Anatomy of the Reproductive
 Systems—741**
Applications to Health—741
Sexually Transmitted Diseases (STDs)—741
 Gonorrhea 741—Syphilis 742—Genital
 Herpes 742—Trichomoniasis 743—
 Chlamydia 743
Male Disorders—743
 Testicular Cancer 743—Prostate 743—Sexual
 Functional Abnormalities 743
Female Disorders—744
 Menstrual Abnormalities 744—Toxic Shock
 Syndrome (TSS) 744—Ovarian Cysts 745—
 Endometriosis 745—Infertility 745—Disorders
 Involving the Breasts 745—Cervical
 Cancer 745—Pelvic Inflammatory Disease
 (PID) 746
KEY MEDICAL TERMS ASSOCIATED WITH THE REPRODUCTIVE
SYSTEMS—746
STUDY OUTLINE—746
REVIEW QUESTIONS—748
SELF-QUIZ—749

26 DEVELOPMENTAL ANATOMY—751

Development During Pregnancy—752
Fertilization and Implantation—752
 Fertilization 752—Formation of the Morula 753—
 Development of the Blastocyst 753—
 Implantation 753
External Human Fertilization—754
Embryo Transfer—754
Gamete Intrafallopian Transfer (GIFT) and Transvaginal
 Oocyte Retrieval—754
Embryonic Development—756
Beginnings of Organ Systems—756
 Embryonic Membranes 758—Placenta and
 Umbilical Cord 759
Fetal Growth—761
Gestation—761
Prenatal Diagnostic Techniques—765
Amniocentesis—765
Chorionic Villi Sampling (CVS)—765
Parturition and Labor—765
KEY MEDICAL TERMS ASSOCIATED WITH DEVELOPMENTAL
ANATOMY—767
STUDY OUTLINE—768
REVIEW QUESTIONS—768
SELF-QUIZ—769

*Appendix A: Symbols and
 Abbreviations—A-1*
*Appendix B: Eponyms Used in This
 Text—B-1*
*Appendix C: Answers to Self-
 Quizzes—C-1*
Selected Readings—SR-1
Glossary of Terms—G-1
Index—I-1

Preface

AUDIENCE

Designed for the introductory course in human anatomy, the fifth edition of *Principles of Human Anatomy* assumes no previous study of the human body. The text is geared to students in health-oriented, medical, and biological programs. Among the students specifically served by this text are those pursuing careers as nurses, medical office assistants, physicians' assistants, medical laboratory technologists, radiological technologists, respiratory therapists, dental hygienists, physical therapists, surgical assistants and technologists, diagnostic medical sonographers, cytotechnologists and histologic technologists, electroencephalographic (EEG) technologists, emergency medical technicians–paramedics, nuclear medicine technologists, morticians, and medical record keepers. However, because of the scope of the book, *Principles of Human Anatomy,* Fifth Edition, is also useful for students in the biological sciences, science technology, liberal arts, physical education, and premedical, predental, and prechiropractic programs.

OBJECTIVES

The objectives of *Principles of Human Anatomy* remain unchanged in this fifth edition. Because human anatomy is such a large and complex body of knowledge to present in an introductory course, the first objective is to concentrate on data and unified concepts that contribute to a basic understanding of the structure of the human body. Data unessential to this objective have been minimized. The second objective is to present essential technical vocabulary and important, but difficult, concepts at a reading level that can be handled by the average student. Easy-to-comprehend explanations of terms and step-by-step development of concepts are presented to meet this objective.

THEMES

The fifth edition of this textbook departs from the approaches of most other anatomy texts in that somewhat more emphasis is given to physiology, disorders, and clinical applications. The basic content of the book is anatomy, but *because structure and function are so inseparably related,* it makes little sense to present students with anatomical detail without relating anatomy to some function. The discussion of function gives the students a better understanding of anatomical concepts.

An understanding of structure and function may be enhanced by considering variations from the normal, or the defects and disorders observed in clinical situations. In turn, such conditions are better understood once the context of normal anatomy has been established. Disorders are treated under the special heading **Applications to Health** at the ends of chapters and **Clinical Applications** integrated throughout the text.

ORGANIZATION

The book is organized by systems rather than by regions. Chapter 1 introduces the levels of structural organization, organ systems, structural plan, anatomical position, regional names, directional terms, planes and sections, cavities of the human body and units of measurement. The chapter has been improved by the inclusion of two exhibits—one that describes the various subdivisions of anatomy and another that summarizes representative structures found in the abdominopelvic regions. There is also a new section on medical imaging that includes conventional radiography, computed tomography (CT) scanning, dynamic spatial reconstruction (DSR), magnetic resonance imaging (MRI), ultrasound (US), positron emission tomography (PET), and digital subtraction angiography (DSA). Several new color photos of cadavers that show viscera in body cavities have also been added.

Chapter 2, which deals with the cellular level of organization, describes a generalized animal cell to demonstrate the basic structural features and functions of cells. It includes updated material on the structure, chemistry, and functions of plasma membranes; receptor-mediated endocytosis; lysosomes; extracellular materials; and cells and cancer. New to the chapter are discussions on the microtrabecular lattice and reproductive cell division. Several new scanning electron micrographs (SEMs) have also been added. The tissue level or organization is presented in Chapter 3 through descriptions of the structure, functions, and locations of the principal kinds of tissues. New to the chapter are sections dealing with regeneration of epithelial tissue, fibronectins, and Marfan's syndrome. The discussions of muscle tissue and nervous tissue have been placed after the section on membranes. There is a new summary exhibit on tissues and several new color photomicrographs have been added as replacements.

The discussion of the organ and organ-system levels of organization begins with Chapter 4 on the integumentary system. This chapter includes new material on the histology of the epidermis, epidermal and deep wound healing, skin grafts (SGs), types of skin cancer, and suntanning salons. As in all the chapters dealing with organ systems, there is some reference to physiology, disorders, and relevant medical terminology. Chapters dealing with organ systems also contain discussions of the effects of aging and developmental anatomy.

In Chapter 5, there are revised sections on the histology of bone and osteoporosis. Chapter 6, on the axial skeleton, has been updated to include a new section on temporomandibular joint (TMJ) syndrome and a new summary exhibit on cranial bones. Numerous color photographs of bones have also been added. Many new color photographs of bones have also been added to Chapter 7, which covers the appendicular skeleton. New to Chapter 8 on articulations are new photographs of synovial joints.

The muscular system is analyzed through a study of muscle tissue and the locations and actions of the principal skeletal muscles. Chapter 9 contains new sections on recruitment of motor units and muscle length and force of contraction. Among the revised areas are types of skeletal muscle fibers, electromyography (EMG), and the histology of cardiac and smooth muscle tissue. New color photographs of tendons have been added. Chapter 10, on skeletal muscles, has been revised extensively. First, there are new exhibits on muscles of the soft palate; muscles that move the wrist, hand, and fingers; intrinsic muscles of the hand; and intrinsic muscles of the foot. Also, every muscle exhibit now contains an overview section that focuses on the muscles under consideration. Another striking change is the use of a new technique for illustrating skeletal muscles. Several new color photographs and cross-sectional illustrations have also been added.

As in the previous edition, the chapter on surface anatomy has been placed immediately following the muscular system, rather than at the end of the book. However, because of the way Chapter 11 is structured, that is, separate exhibits of various regions of the body with accompanying art, surface anatomy can be studied *anywhere* in the book. The approach used to study surface anatomy is left to the option of the instructor and student. Many new surface anatomy features have been added.

The student is next introduced to the cardiovascular and lymphatic systems. Chapter 12, on blood, contains newly added material on obtaining blood samples and chronic Epstein–Barr virus (EBV) syndrome. New color photomicrographs of blood have also been added. The sections on the histology of neutrophils and plasmapheresis have been revised. Chapter 13, on the heart, contains new sections on the skeleton of the heart, artificial pacemaker, anastomoses of the heart, and coronary artery bypass grafting (CABG). Diagrams of the skeleton of the heart and heart anomalies are also new. The sections dealing with the pericardium and atherosclerosis have been revised. New to Chapter 14, on blood vessels, are color photomicrographs of the histology of blood vessels and color photographs of the gross anatomy of blood vessels, valves in veins, and fetal circulation. The sections dealing with the branches of the aorta and hypertension have been revised. Chapter 15, on the lymphatic system, contains new sections on severe combined immunodeficiency (SCID) and Hodgkin's disease (HD) and new color photographs on the gross anatomy of the spleen and thoracic duct. The section on the organization of lymphatic tissue, histology of the thymus, hypersensitivity, and acquired immune deficiency syndrome (AIDS) have been expanded and revised.

The next major area of emphasis is the nervous system. Students are introduced to the structure and function of nervous tissue, the spinal cord and spinal nerves, the brain and cranial nerves, and the autonomic nervous system. Chapter 16, on nervous tissue, contains a new section on the effects of crack on nerve impulse conduction and a revised section on regeneration of nervous tissue. Chapter 17, on the spinal cord, contains new line art and color photographs of the gross anatomy of the brachial plexus and lumbar plexus. The section dealing with the blood–brain barrier (BBB), cerebellar structures, cerebrovascular accidents (CVAs), Parkinson's disease (PD), dyslexia, and Alzheimer's disease (AD) have been revised and there is a new section on transient ischemic attacks (TIAs) and new line art on the origin of cranial nerves in Chapter 18, which deals with the brain. Chapter 19 now contains an expanded exhibit on the activities of the autonomic nervous system.

Sensory structures are considered next. New to Chapter 20 are sections on the histology of the lens, vertigo, the linkage of sensory input and motor responses, and new line art on sensory pathways. Among the topics revised are how acupuncture works, the histology of the olfactory epithelium and retina, and auditory pathways. Attention is then turned to the endocrine system in Chapter 21. The sections on receptors and the vascular and nerve supply to the pancreas have been revised and new line art has been added regarding the histology of the thyroid, parathyroids, and pancreas. Several new color photographs of endocrine disorders have also been added.

Chapter 22, on the respiratory system, contains a new section on smoke inhalation injury and line art on the histology of the alveolar–capillary (respiratory) membrane. The topics of innervation of the pharynx, vascular supply of the larynx, vascular and nerve supply of the lungs, and the common cold have been revised.

New topics added to Chapter 23, on the digestive system, include root canal therapy and occult blood plus new line art on the histology of the general organization of the gastrointestinal tract, stomach, small intestine, and large intestine, several color histology photomicrographs, and color gross anatomy photographs of the liver. The sections on the greater omentum and hemorrhoids have been revised.

Chapter 24, on the urinary system, has been expanded by new discussions of renal failure and urinary tract infections (UTIs). A new color photograph on the gross anatomy of the urinary bladder has also been added. The sections on the histology of the endothelial–capsular membrane and urethra have been revised.

Chapter 25, on the reproductive systems, has further been strengthened by the addition of material on the menstrual cycle; mifepristone (RU486), a birth control drug; chlamydia; and testicular cancer. New color photographs illustrating the gross anatomy of the male internal reproductive structures, penis, and mammary glands have also been added. The sections dealing with spermatogenesis, oogenesis, histology of the endometrium, vulva, and mammography have been revised.

Finally, Chapter 26, on developmental anatomy, now contains new sections on the successful separation of Siamese twins, fetal surgery, electronic fetal monitoring (EFM), and the stress of parturition. The discussions of gestation and Down's syndrome (DS) have been revised.

SPECIAL FEATURES

As in previous editions, the book contains numerous learning aids. Users of the book have cited the pedagogical aids as one of the book's many strengths. All the tested and successful learning aids of previous editions have been kept in the fifth edition, and several new ones have been added. These special features include the following:

1. **Student Objectives.** Each chapter opens with a comprehensive list of student objectives. Each objective describes a knowledge or skill students should acquire while studying the chapter. (See **Note to the Student** for an explanation of how the objectives can be used.)
2. **Chapter Outline.** This is new to the fifth edition. Each chapter now contains an outline of its contents to help students overview the sequence of topics.
3. **Study Outline.** A study outline at the end of each chapter provides a brief summary of major topics. This section consolidates the essential points covered in the chapter so that students can recall and relate the points to one another. Page numbers have been added to the outlines so that topics within chapters can be located more easily.
4. **Review Questions.** Review questions at the end of each chapter provide a check to see if the objectives stated at the beginning of the chapter have been met. After answering the questions, students should reread the objectives to determine whether they have met the goals.
5. **Key Medical Terms.** Glossaries of selected medical terms appear at the end of appropriate chapters; these listings are entitled ''Key Medical Terms.'' All glossa-

ries have been revised for the fifth edition. In addition, phonetic pronunciations have been added to the medical terminology lists in the fifth edition.

6. **Self-Quizzes.** New to the fifth edition is a self-quiz at the end of each chapter. These quizzes are designed to evaluate for an understanding of principles and concepts discussed in the chapter. Answers to the questions are provided at the back of the book in Appendix C.
7. **Exhibits.** Health-science students are generally expected to learn a great deal about the anatomy of certain organ systems, specifically, skeletal muscles, articulations, blood vessels, lymph nodes, and nerves. To avoid interrupting the discussion of concepts and to organize the data, anatomical details have been presented in tabular form in exhibits, most of which are accompanied by illustrations. This method allows a clearer statement of general concepts in the narrative and organizes the specific details to be learned. New summary exhibits have been added to the fifth edition.
8. **Applications to Health.** Abnormalities of structure or function are grouped at the end of appropriate chapters in sections entitled ''Applications to Health.'' These sections provide a review of normal body processes as well as demonstrations of the importance of the study of anatomy and physiology to a career in any of the health fields. All disorders have been updated, and many new ones have been added.
9. **Phonetic Pronunciations.** Throughout the text, phonetic pronunciations are provided in parentheses for selected terms. These pronunciations are given at the point where the terms are introduced and are repeated in the ''Glossary of Terms.'' The **Note to the Student** explains the pronunciation key.
10. **Clinical Applications.** Throughout the text, clinical applications are integrated within the narrative. Many new ones have been added.
11. **Line Art.** The anatomically accurate line drawings in the book are large so that details are easily seen. In the fifth edition, many new illustrations have been added and full color is used *throughout* to differentiate structures and regions. The line art in Chapter 10 has been completely redone using new techniques.
12. **Photographs.** The photographs amplify the narrative and the line drawings. Numerous photomicrographs (in full color), newly added scanning electron micrographs, and transmission electron micrographs enhance the histological discussions. Color photographs of specimens clarify gross anatomy discussions and have been added throughout. Photographs of regional dissections, most in full color, have also been added to the fifth edition.
13. **Appendixes.** Appendix A, ''Symbols and Abbreviations,'' is a list of several medical symbols and an alphabetical list of commonly encountered medical abbreviations. The abbreviations list has been greatly ex-

panded. Appendix B, "Eponyms Used in This Text," is an alphabetical list of commonly encountered eponyms and the corresponding current terminology. Eponyms are cited in the text in parentheses immediately following the preferred current terms and cross referenced in the index. Appendix C, "Answers to Self-Quizzes," contains the correct responses to the self-quiz tests for each chapter.

14. **Selected Readings.** The revised bibliography lists current references for instructors and students.

15. **Glossary.** The comprehensive glossary of terms has been updated for the fifth edition.

SUPPLEMENTARY MATERIALS

The following supplementary items are available to accompany the fifth edition of *Principles of Human Anatomy*.

1. **Instructor's Manual.** Each chapter in the complimentary *Instructor's Manual,* prepared by Gerard J. Tortora, consists of a chapter overview, a list of instructional concepts relating to the chapter, and lists of audiovisual materials relating to the chapter topic.

2. **Test Bank.** A complimentary test bank, prepared by Gerard J. Tortora, contains a variable number of multiple-choice questions for each chapter in the textbook. The test bank is also available on computer disks (Harper Test) and may be secured from the publisher.

3. **Laboratory Manual.** Written by Patricia J. Donnelly, this laboratory manual uses the cat as the dissection animal. Each muscle dissection is accompanied by a photo, line illustration, and simple step-by-step instructions. Frequent references are made to comparable human organ systems.

4. **The Anatomy Coloring Book.** This popular study aid, written by Wynn Kapit and Larry Elson, helps students memorize structures of the body through coloring in line drawings.

5. **Software Based on The Anatomy Coloring Book.** Available for use on the Apple Macintosh with Hypercard. This interactive software program allows the student to focus on individual anatomical parts of each graphic representation in the coloring book. Audio pronunciations are also included.

6. **Biosource Software.** For the Apple II, II+, IIe, III, and IBM PC: Neuromuscular Concepts, Skeletal Muscular Anatomy and Physiology, and Brain. Each set of three diskettes consists of five tutorials, a series of 10-question, multiple-choice, quizzes, and a 50-question review exam.

7. **Body Language.** For the Apple II, II+, IIe, III, and IBM PC: Skeletal System, Muscular System, Respiratory System, Cardiovascular System, Nervous System, Digestive System, Urinary System, and Reproductive System. Disks are user-friendly and interactive. Each system disk gives students practice in identifying key structures with the aid of effective illustrations. Pronunciation and spelling drills are also included.

8. **Transparencies.** One hundred twenty-five full-color transparencies are available. Illustrations from the text that are often shown and discussed in class are the subjects for the transparencies, and special care has been taken to make them clear and usable as overhead projections.

ACKNOWLEDGMENTS

I wish to thank the following people for their contributions to the fifth edition of *Principles of Human Anatomy:* Marian C. Diamond, University of California, Berkeley; Gail P. Greenwald, University of Iowa School of Medicine; Jackson E. Jeffrey, Virginia Commonwealth University; Michael C. Kennedy, Hahnemann University; Andrew J. Kuntzman, Wright State University School of Medicine; Lynn Romrell, University of Florida; and John B. Schumann, Gannon University.

Continuing thanks go to the many instructors and students who were kind enough to send me their suggestions for improvement and those people who worked on the four previous editions. Gratitude is also extended for the contributions of the individuals and organizations whose names appear with the photographs in the text. Special thanks to Andrew J. Kuntzman of Wright State University School of Medicine for providing numerous excellent color photomicrographs and Dr. Michael C. Kennedy of Hahnemann Medical College for his help in reviewing the illustration program. Finally, for typing drafts of the manuscript and numerous other duties associated with the task of putting together a textbook, thanks to Geraldine C. Tortora.

As the acknowledgments indicate, the participation of many individuals of diverse talent and expertise is required in the production of a textbook of this scope and complexity. For this reason, readers and users of the fifth edition are invited to send their reactions and suggestions to me so that plans can be formulated for subsequent editions.

Gerard J. Tortora
Natural Sciences and Mathematics S229
Bergen Community College
400 Paramus Road
Paramus, NJ 07652

Note to the Student

At the beginning of each chapter is a listing of **Student Objectives.** Before you read the chapter, please read the objectives carefully. Each objective is a statement of a skill or knowledge that you should acquire. To meet these objectives, you will have to perform several activities. Obviously, you must read the chapter carefully. If there are sections of the chapter that you do not understand after one reading, you should reread those sections before continuing. In conjunction with your reading, pay particular attention to the figures and exhibits; they have been carefully coordinated with the textural narrative. Also, at the beginning of each chapter is a **Chapter Outline,** designed to provide you with an overview of the sequence of topics discussed in each chapter.

At the end of each chapter are several other learning guides that you may find useful. The **Study Outline** is a concise summary of important topics discussed in the chapter. This section is designed to consolidate the essential points covered in the chapter, so that you may recall and relate them to one another. Page numbers have been added to the outline so that you can locate topics in the chapter more early. The **Review Questions** are a series of questions designed specifically to help you master the objectives. After you have answered the review questions, you should return to the beginning of the chapter and reread the objectives to determine whether you have achieved the goals. **Key Medical Terms** appear in some chapters. This is a listing of terms, with phonetic pronunciations and definitions designed to build your medical vocabulary. New to the fifth edition are **Self-Quizzes** for each chapter. The questions are designed to evaluate for an understanding of principles and concepts discussed in the chapter. Answers to the questions are provided in Appendix C.

As a further aid, we have included pronunciations for many terms that may be new to you. These appear in parentheses immediately following the new words, and they are repeated in the glossary of terms at the back of the book. (Of course, since there will always be some conflict among medical personnel and dictionaries about pronunciation, you will come across variations in different sources.) Look at the words carefully and say them out loud several times. Learning to pronounce a new word will help you remember it and make it a useful part of your medical vocabulary. Take a few minutes now to read the following pronunciation key, so it will be familiar as you encounter new words. The key is repeated at the beginning of the glossary of terms.

PRONUNCIATION KEY

1. The strongest accented syllable appears in capital letters, for example, bilateral (bī-LAT-er-al) and diagnosis (dī-ag-NŌ-sis).
2. If there is a secondary accent, it is noted by a single quote mark('), for example, constitution (kon'-sti-TOO-shun) and physiology (fiz'-ē-OL-ō-jē). Any additional secondary accents are also noted by a single quote mark, for example, decarboxylation (dē'-kar-bok'-si-LĀ-shun).
3. Vowels marked with a line above the letter are pronounced with the long sound as in the following common words.

    ā as in *māke*
    ē as in *bē*
    ī as in *īvy*
    ō as in *pōle*

4. Vowels not so marked are pronounced with the short sound as in the following words.

    e as in *bet*
    i as in *sip*
    o as in *not*
    u as in *bud*

5. Other phonetic symbols are used to indicate the following sounds:

    a as in *above*
    oo as in *sue*
    yoo as in *cute*
    oy as in *oil*

1 An Introduction to the Human Body

STUDENT OBJECTIVES

1. Define anatomy, with its subdivisions, and physiology.
2. Define each of the following levels of structural organization that make up the human body: chemical, cellular, tissue, organ, system, and organismic.
3. Identify the principal systems of the human body, list representative organs of each system, and describe the function of each system.
4. List and define several important life processes of humans.
5. Define the anatomical position and compare common and anatomical terms used to describe various regions of the human body.
6. Define several directional terms used in association with the human body.
7. Define the common anatomical planes that may be passed through the human body and distinguish a cross section, frontal section, and midsagittal section.
8. List by name and location the principal body cavities and the organs contained within them.
9. Describe how the abdominopelvic cavity is divided into nine regions and quadrants.
10. Contrast the principles employed in conventional radiography, computed tomography (CT) scanning, dynamic spatial reconstructing (DSR), magnetic resonance imaging (MRI), ultrasound (US), positron emission tomography (PET), and digital subtraction angiography (DSA) in the diagnosis of disease.
11. Define the common metric units of length, mass, and volume, and their U.S. equivalents, which are used in measuring the human body.

CHAPTER OUTLINE

■ Anatomy and Physiology Defined
■ Levels of Structural Organization
■ Life Processes
■ Structural Plan
■ Anatomical Position and Regional Names
■ Directional Terms
■ Planes and Sections
■ Body Cavities
■ Abdominopelvic Regions
■ Abdominopelvic Quadrants
■ Medical Imaging
Conventional Radiography
Computed Tomography (CT) Scanning
Dynamic Spatial Reconstructor (DSR)
Magnetic Resonance Imaging (MRI)
Ultrasound (US)
Positron Emission Tomography (PET)
Digital Subtraction Angiography (DSA)
■ Measuring the Human Body

You are about to begin a study of the human body in order to learn how your body is organized and how it functions. The study of the human body involves many branches of science. Each contributes to a comprehensive understanding of how your body normally works and what happens when it is injured, diseased, or placed under stress.

ANATOMY AND PHYSIOLOGY DEFINED

Two branches of science that will help you understand your body parts and functions are anatomy and physiology. *Anatomy* (a-NAT-ō-m̄e; *anatome* = to dissect) refers to the study of *structure* and the relation among structures. Anatomy is a broad science, and the study of structure becomes more meaningful when specific aspects of the science are considered. Several subdivisions of anatomy are described in Exhibit 1-1.

Whereas anatomy and its branches deal with structures

EXHIBIT 1-1

Subdivisions of Anatomy

SUBDIVISION	DESCRIPTION
Surface Anatomy	Study of the form (morphology) and markings of the surface of the body.
Gross (Macroscopic) Anatomy	Study of structures that can be examined without the use of a micropscope.
Systemic (Systematic) Anatomy	Study of specific systems of the body such as the nervous system or respiratory system.
Regional Anatomy	Study of a specific region of the body such as the head or chest.
Developmental Anatomy	Study of development from the fertilized egg to adult form.
Embryology (em'-brē-OL-ō- jē; *logos* = study of)	Study of development from the fertilized egg through the eighth week in utero.
Pathological (path'-ō-LOJ-i-kal; *patho* = disease) Anatomy	Study of structural changes associated with disease.
Histology (hiss-TOL-ō-jē; *histio* = tissue)	Microscopic study of the structure of tissues.
Cytology (sī-TOL-ō-jē; *cyto* = cell)	Microscopic study of the structure of cells.
Radiographic (rā'-dē-ō-GRAF-ik; *radio* = ray; *graph* = to write) Anatomy	Study of the structure of the body that includes the use of several imaging techniques, such as x rays.

of the body, *physiology* (fiz'-ē-OL-ō-jē) deals with *functions* of the body parts, that is, how the body parts work. However, the two sciences cannot be separated completely. Since each part of the body is designed to carry out a particular set of functions, structure and function are studied together. The structure of a part often determines the functions it will perform and, in turn, body functions often influence the size, shape, and health of the structures.

LEVELS OF STRUCTURAL ORGANIZATION

The human body consists of several levels of structural organization that are associated with one another in various ways. The lowest level of organization, the *chemical level*, includes all chemical substances essential for maintaining life. The chemicals are made up of atoms joined together in various ways (Figure 1-1).

The chemicals, in turn, are put together to form the next higher level of organization: the *cellular level. Cells* are the basic structural and functional units of an organism. Among the many kinds of cells in your body are muscle cells, nerve cells, and blood cells. Figure 1-1 shows several isolated cells from the lining of the stomach. Each has a different structure, and each performs a different function.

The next higher level of structural organization is the *tissue level. Tissues* are made up of groups of similarly specialized cells and their intercellular material (substance between cells) that have a similar embryological origin and perform certain special functions. When the cells shown in Figure 1-1 are joined together, they form a tissue called epithelium, which lines the stomach. Each cell in the tissue has a specific function. Mucous cells produce mucus, a secretion that lubricates food as it passes through the stomach. Parietal cells produce acid in the stomach. Zymogenic cells produce enzymes needed to digest proteins. Other examples of tissues in your body are muscle tissue, connective tissue, and nervous tissue.

In many places in the body, different kinds of tissues are joined together to form an even higher level of organization: the *organ level. Organs* are structures of definite form and function composed of two or more different tissues. Organs usually have a recognizable shape. Examples of organs are the heart, liver, lungs, brain, and stomach. Figure 1-1 shows three of the tissues that make up the stomach. The serosa is a layer of connective tissue and epithelium around the outside that protects the stomach and reduces friction when the stomach moves and rubs against other organs. The muscle tissue layers of the stomach contract to mix food and pass it on to the next digestive organ. The epithelial tissue layer lining the stomach produces mucus, acid, and enzymes.

The next higher level of structural organization in the body is the *system level. A system* consists of an association

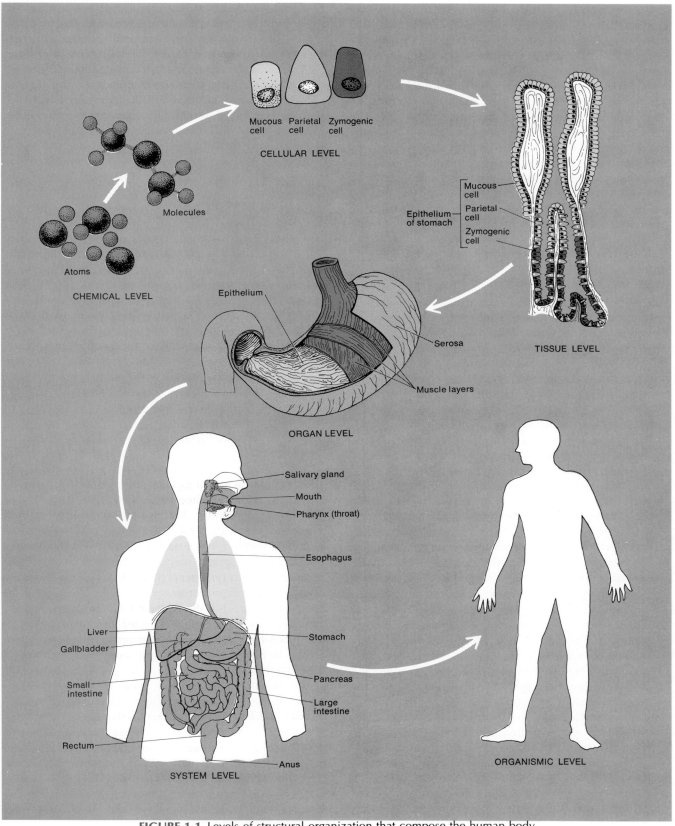

FIGURE 1-1 Levels of structural organization that compose the human body.

of organs that have a common function. The digestive system, which functions in the breakdown of food, is composed of the mouth, saliva-producing glands called salivary glands, pharynx (throat), esophagus, stomach, small intestine, large intestine, rectum, liver, gallbladder, and pancreas.

The highest level of organization is the *organismic level.* All the parts of the body functioning with one another constitute the total *organism*—one living individual.

In the chapters that follow, you will examine the anatomy and physiology of the major body systems. Exhibit 1-2 illustrates these systems, their representative organs, and their general functions. Unless otherwise stated, male and female bodies are similar. The systems are presented in the exhibit in the order in which they are discussed in later chapters.

LIFE PROCESSES

All living forms have certain characteristics that distinguish them from nonliving things. Following are several of the more important life processes of humans:

1. *Metabolism.* Metabolism is the sum of all the chemical processes that occur in the body. One phase of metabolism, called *catabolism,* provides us with energy needed to sustain life. The other phase of metabolism, called *anabolism,* uses the energy from catabolism to make various substances that form the body's structural and functional components. Among the many processes contributing to metabolism are: *ingestion,* the taking in of foods; *digestion,* the breaking down of foods into simpler forms that can be used by cells; *absorption,* the uptake of substances by cells; *assimilation,* the buildup of absorbed substances into different materials required by cells; *respiration,* the generation of energy, usually in the presence of oxygen with the release of carbon dioxide; *secretion,* the production and release of a useful substance by cells; and *excretion,* the elimination of wastes produced as a result of metabolism.
2. *Excitability.* Excitability is our ability to sense changes within and around us. We do this by responding to stimuli (changes in the environment) such as light, pressure, heat, noises, chemicals, and pain. We constantly respond to our environment to make adjustments that maintain health.
3. *Conductivity.* Conductivity refers to the ability to carry the effect of a stimulus from one part of a cell to another. This characteristic is highly developed in nerve cells and is developed to a great extent in muscle fibers (cells).
4. *Contractility.* Contractility is the capacity of cells or parts of cells to undergo shortening and change form. Muscle fibers exhibit a high degree of contractility.
5. *Growth.* Growth refers to an increase in size. It involves an increase in the number of cells or an increase in the size of existing cells.
6. *Differentiation.* Differentiation is the basic mechanism whereby unspecialized cells change to specialized cells. Specialized cells have structural and functional characteristics that differ from cells from which they originated. Through differentiation, a fertilized egg normally develops into an embryo, fetus, infant, child, and adult, each of which consists of a variety of diversified cells.
7. *Reproduction.* Reproduction refers to either the formation of new cells for growth, repair, or replacement, or the production of a new individual. Through reproduction, life is transmitted from one generation to the next.

STRUCTURAL PLAN

The human body has certain general *anatomical characteristics* that will help you to understand its overall structural plan. For example, humans have a *backbone* (*vertebral column*), a characteristic that places them in a large group of organisms called *vertebrates.* Another characteristic of the body's organization is that it basically resembles a *tube within a tube* construction. The outer tube is formed by the body wall; the inner tube is the gastrointestinal tract. Moreover, humans are for the most part *bilaterally symmetrical:* essentially, the left and right sides of the body are mirror images.

ANATOMICAL POSITION AND REGIONAL NAMES

In all anatomical texts and charts, descriptions of any region or part of the human body assume that the body is in a specific position, called the *anatomical position,* so that directional terms are clear and any part can be related to any other part. In the anatomical position, the subject is standing erect (upright position) facing the observer, the upper extremities (limbs) are placed at the sides, the palms of the hands are turned forward, and the feet are flat on the floor (Figure 1-2). Once the body is in the anatomical position, it is easier to visualize and understand how it is organized into various regions. The common and anatomical terms, in parentheses, of the principal body regions are also presented in Figure 1-2.

DIRECTIONAL TERMS

In order to explain exactly where various body structures are located relative to each other, anatomists use certain *directional terms.* Such terms are precise and avoid the use of unnecessary words. If you want to point out the sternum (breastbone) to someone who knows where the clavicle (collarbone) is, you can say that the sternum is inferior (farther away from the head) and medial (toward the middle of the body) to the clavicle. As you can see,

EXHIBIT 1-2

Principal Systems of Human Body: Representative Organs and Functions

1. Integumentary

Definition: The skin and structures derived from it, such as hair, nails, and sweat and oil glands.

Function: Helps regulate body temperature, protects the body, eliminates wastes, synthesizes vitamin D, and receives certain stimuli such as temperature, pressure, and pain.

2. Skeletal

Defintion: All the bones of the body, their associated cartilages, and the joints of the body.

Function: Supports and protects the body, provides leverage, produces blood cells, and stores minerals.

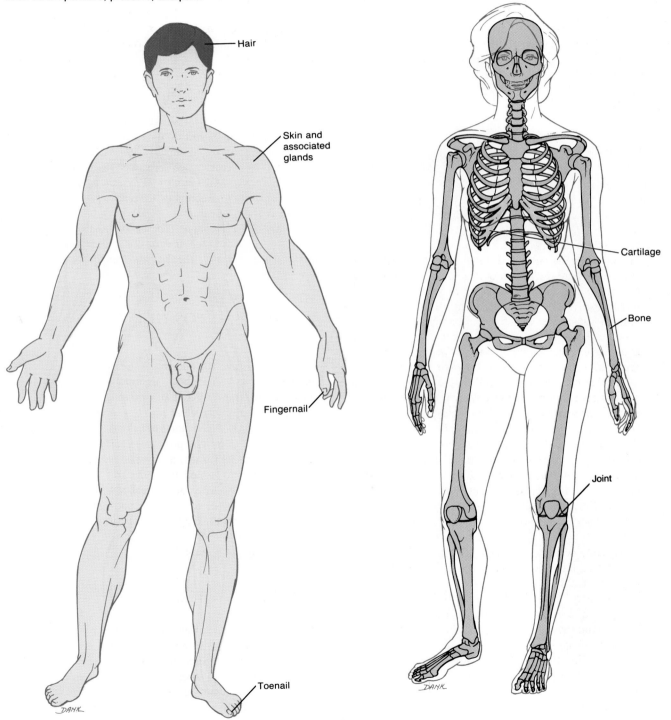

3. Muscular
Defintion: All the muscle tissue of the body, including skeletal (shown in the illustration), visceral, and cardiac.
Function: Participates in bringing about movement, maintains posture, and produces heat.

4. Cardiovascular
Definition: Blood, heart, and blood vessels.
Function: Distributes oxygen and nutrients to cells, carries carbon dioxide and wastes from cells, maintains the acid–base balance of the body, protects against disease, prevents hemorrhage by forming blood clots, and helps regulate body temperature.

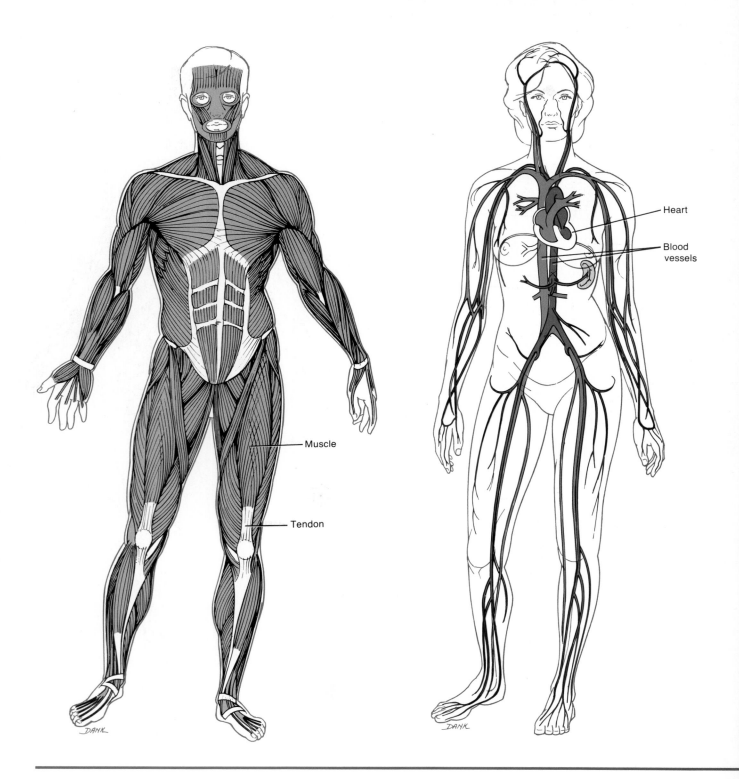

EXHIBIT 1-2 (*Continued*)

5. Lymphatic

Definition: Lymph, lymph vessels, and structures or organs containing lymphatic tissue (large numbers of white blood cells called lymphocytes) such as lymph nodes, the spleen, thymus gland, and tonsils.

Function: Returns proteins and plasma to the cardiovascular system, transports fats from the gastrointestinal tract to the cardiovascular system, filters body fluid, produces white blood cells, and protects against disease.

6. Nervous

Definition: Brain, spinal cord, nerves, and sense organs, such as the eye and ear.

Function: Regulates body activities through nerve impulses.

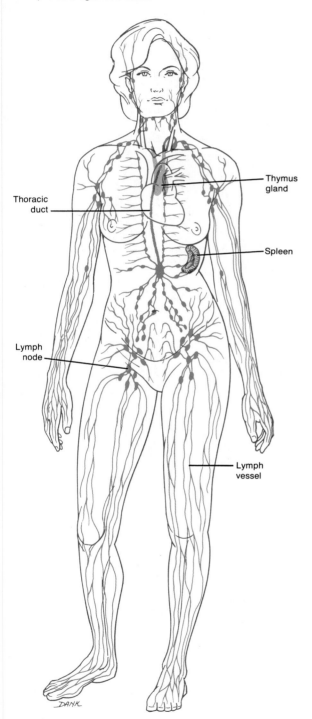

Thoracic duct

Thymus gland

Spleen

Lymph node

Lymph vessel

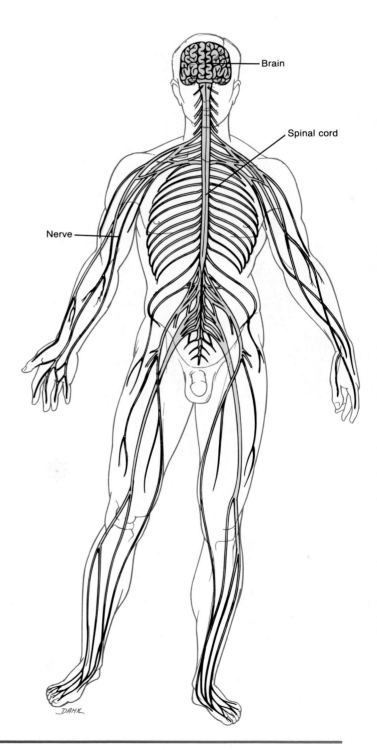

Brain

Spinal cord

Nerve

7. Endocrine
Definition: All glands that produce hormones.
Function: Regulates body activities through hormones transported by the cardiovascular system.

8. Respiratory
Definition: The lungs and a series of associated passageways leading into and out of them.
Function: Supplies oxygen, eliminates carbon dioxide, and helps regulate the acid–base balance of the body.

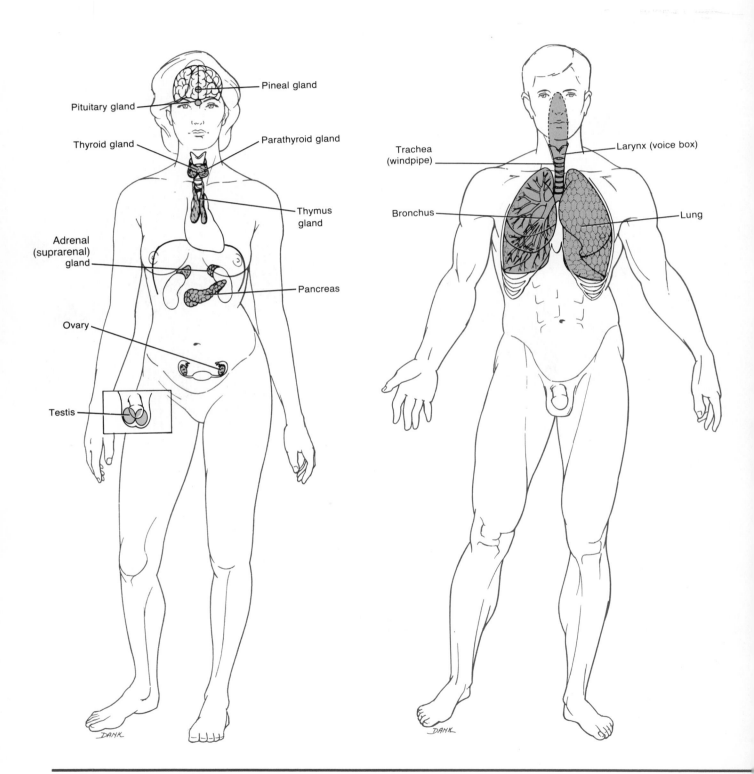

Pineal gland

Pituitary gland

Thyroid gland

Parathyroid gland

Thymus gland

Adrenal (suprarenal) gland

Pancreas

Ovary

Testis

Trachea (windpipe)

Larynx (voice box)

Bronchus

Lung

EXHIBIT 1-2 (*Continued*)

9. Digestive

Definition: A long tube called the gastrointestinal (GI) tract and associated organs such as the salivary glands, liver, gallbladder, and pancreas.

Function: Performs the physical and chemical breakdown of food for use by cells and eliminates solid wastes.

10. Urinary

Definition: Organs that produce, collect, and eliminate urine.

Function: Regulates the chemical composition of blood, eliminates wastes, regulates fluid and electrolyte balance and volume, and helps maintain the acid–base balance of the body.

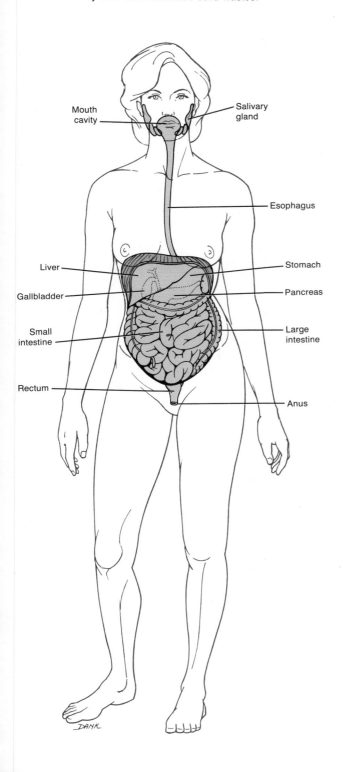

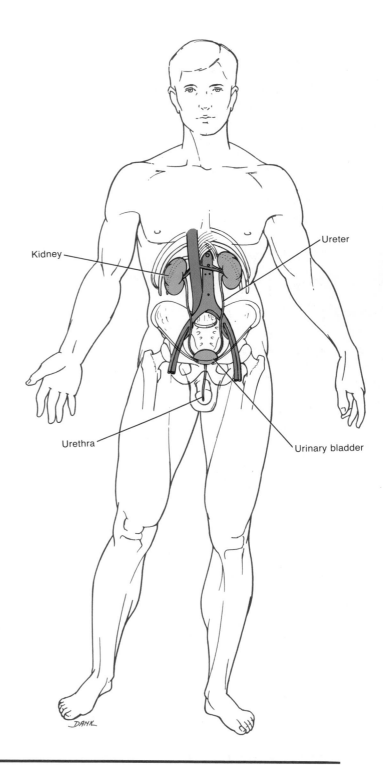

11. Reproductive

Definition: Organs (testes and ovaries) that produce reproductive cells (sperm and ova) and other organs that transport and store reproductive cells.
Function: Reproduces the organism.

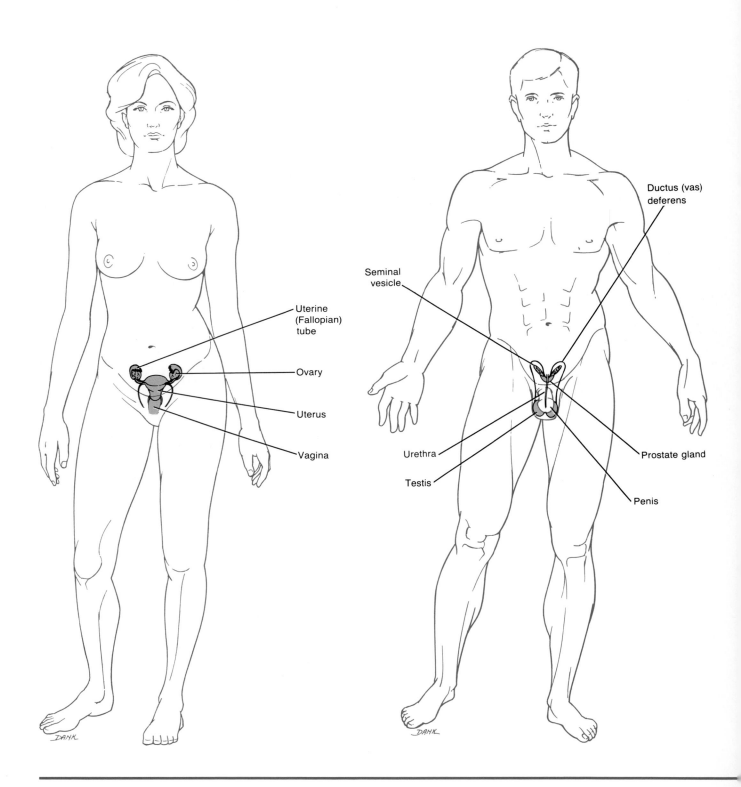

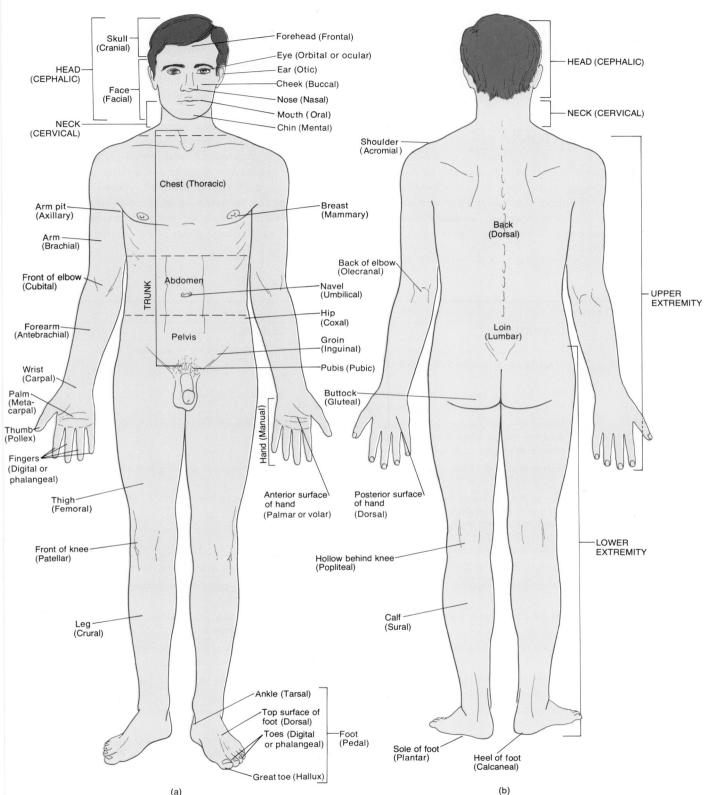

FIGURE 1-2 Anatomical position. The common names and anatomical terms, in parentheses, are indicated for many of the regions of the body. (a) Anterior view. (b) Posterior view.

using the terms *inferior* and *medial* avoids a great deal of complicated description. Many directional terms are defined in Exhibit 1-3. The parts of the body used in the examples are labeled in Figure 1-3. Studying the exhibit and the figure together should make clear to you many of the directional relations among various body parts.

PLANES AND SECTIONS

The structural plan of the human body may also be described with respect to *planes* (imaginary flat surfaces) that pass through it. Several of the commonly used planes are illustrated in Figure 1-4. A *midsagittal (median) plane* is a vertical plane that passes through the midline of the body and divides the body or organs into *equal* right and left sides. A *parasagittal (para = near) plane* is a vertical plane that does not pass through the midline of the body and divides the body or organs into *unequal* left and right portions. A *frontal (coronal; kō-RŌ-nal) plane* is a plane at a right angle to a midsagittal (or parasagittal) plane that divides the body or organs into anterior and posterior portions. Finally, a *horizontal (transverse) plane* is a plane that is parallel to the ground, that is, at a right angle to the midsagittal, parasagittal, and frontal planes. It divides the body or organs into superior and inferior portions.

When you study a body structure, you will often view it in *section,* that is, you look at the flat surface resulting from a cut made through the three-dimensional organ. It is important to know the plane of the section so that you

EXHIBIT 1-3

Directional Terms*

TERM	DEFINITION	EXAMPLE
Superior (Cephalic or **Cranial)**	Toward the head or the upper part of a structure; generally refers to structures in the trunk.	The heart is superior to the liver.
Inferior (Caudad)	Away from the head or toward the lower part of a structure; generally refers to structures in the trunk.	The stomach is inferior to the lungs
Anterior (Ventral)	Nearer to or at the front of the body. In the **prone position,** the body lies anterior side down. In the **supine position,** the body lies anterior side up.	The sternum is anterior to the heart.
Posterior (Dorsal)	Nearer to or at the back of the body.	The esophagus is posterior to the trachea.
Medial	Nearer the midline of the body or a structure. The **midline** is an imaginary vertical line that divides the body into equal left and right sides. The *anterior midline* is on the anterior surface of the body and the *posterior midline* is on the posterior surface.	The ulna is on the medial side of the forearm.
Lateral	Farther from the midline of the body or a structure.	The lungs are lateral to the heart.
Intermediate	Between two structures.	The ring finger is intermediate between the little and middle fingers.
Ipsilateral	On the same side of the body.	The gallbladder and ascending colon of the large intestine are ipsilateral.
Contralateral	On the opposite side of the body.	The ascending and descending colons of the large intestine are contralateral.
Proximal	Nearer the attachment of an extremity to the trunk or a structure; nearer to the point of origin.	The humerus is proximal to the radius.
Distal	Farther from the attachment of an extremity to the trunk or a structure; farther from the point of origin.	The phalanges are distal to the carpals.
Superficial	Toward or on the surface of the body.	The muscles of the thoracic wall are superficial to the viscera in the thoracic cavity (see Figure 1-7c).
Deep	Away from the surface of the body.	The ribs are deep to the skin of the chest (see Figure 1-7c).
Parietal	Pertaining to the outer wall of a body cavity.	The parietal pleura forms the outer layer of the pleural sacs that surround the lungs (see Figure 1-7c).
Visceral	Pertaining to the covering of an organ (viscus).	The visceral pleura forms the inner layer of the pleural sacs and covers the external surface of the lungs (see Figure 1-7c).

* Study this exhibit with Figures 1-3 and 1-7c in order to visualize the examples given.

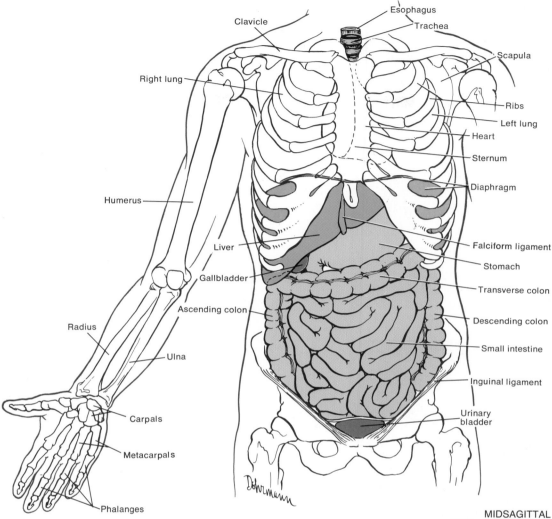

FIGURE 1-3 Anatomical and directional terms. By studying
Exhibit 1-3 with this figure, you should gain an understanding
of the meanings of the terms *superior, inferior, anterior, posterior, medial, lateral, intermediate, ipsilateral, contralateral,
proximal,* and *distal.*

can understand the anatomical relation of one part to another.
Figure 1-5 indicates how three different sections—a *cross
section,* a *frontal section,* and a *midsagittal section*—are
made through different parts of the brain.

BODY CAVITIES

Spaces within the body that contain internal organs are
called *body cavities.* Specific cavities may be distinguished
if the body is divided into right and left halves. Figure 1-6
shows the two principal body cavities. The *dorsal body
cavity* is located near the posterior (dorsal) surface of the
body. It is further subdivided into a *cranial cavity,* which
is a bony cavity formed by the cranial (skull) bones and
contains the brain, and a *vertebral (spinal) canal,* which
is a bony cavity formed by the vertebrae of the backbone

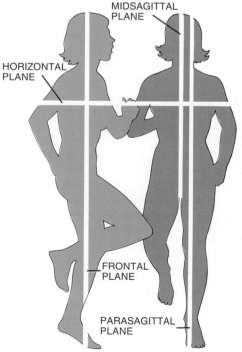

FIGURE 1-4 Planes of the human body.

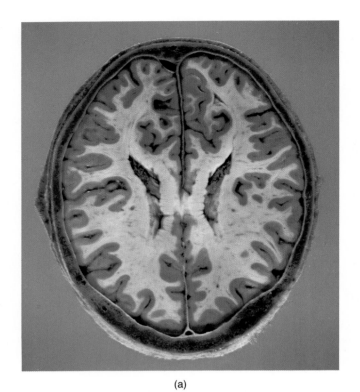

(a)

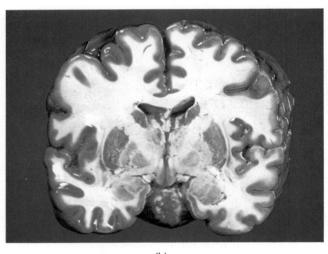

(b)

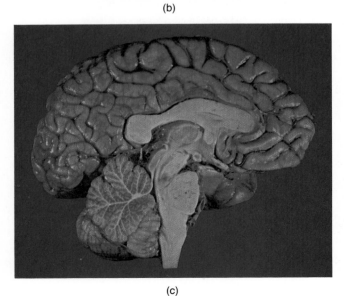

(c)

FIGURE 1-5 Sections through different parts of the brain. (a) Cross section (horizontal or transverse). (Courtesy of Stephen A. Kieffer and E. Robert Heitzman, *An Atlas of Cross-Sectional Anatomy*, Harper & Row, Publishers, Inc., New York, 1979.) (b) Frontal section. (Courtesy of C. Yokochi and J. W. Rohen, *Photographic Anatomy of the Human Body*, 2nd ed., 1979, IGAKU-SHOIN, Ltd., Tokyo, New York.) (c) Midsagittal section. (Courtesy of C. Yokochi and J. W. Rohen, *Photographic Anatomy of the Human Body*, 2nd ed., 1979, IGAKU-SHOIN, Ltd., Tokyo, New York.)

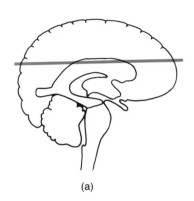

(a)

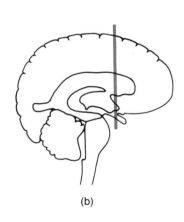

(b)

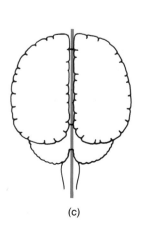

(c)

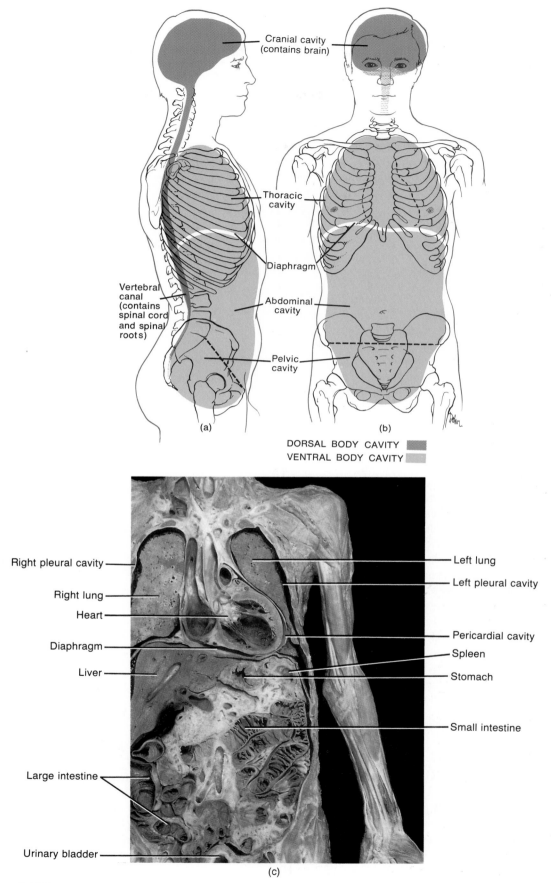

DORSAL BODY CAVITY

VENTRAL BODY CAVITY

Cranial cavity (contains brain)

Thoracic cavity

Diaphragm

Vertebral canal (contains spinal cord and spinal roots)

Abdominal cavity

Pelvic cavity

(a)

(b)

Right pleural cavity

Right lung

Heart

Diaphragm

Liver

Large intestine

Urinary bladder

Left lung

Left pleural cavity

Pericardial cavity

Spleen

Stomach

Small intestine

(c)

FIGURE 1-6 Body cavities. (a) Location of the dorsal and ventral body cavities seen in right lateral view. (b) Subdivisions of the ventral body cavity seen in anterior view. (c) Photograph of a frontal portion of the chest, abdomen, and pelvis. (Courtesy of J. A. Gosling, P. F. Harris, et al., *Atlas of Human Anatomy*, Gower Medical Publishing Ltd., 1985.)

and contains the spinal cord and the beginnings of spinal nerves.

The other principal body cavity is the *ventral body cavity.* This cavity is located on the anterior (ventral) aspect of the body. Its walls are composed of skin, connective tissue, bone, muscles, and serous membranes. The organs inside the ventral body cavity are called *viscera* (VIS-er-a). Like the dorsal body cavity, the ventral body cavity has two principal subdivisions—an upper portion, called the *thoracic* (thō-RAS-ik) *cavity* (or chest cavity), and a lower portion, called the *abdominopelvic* (ab-dom'-i-nō-PEL-vik) *cavity.* The anatomical landmark that divides the ventral body cavity into the thoracic and abdominopelvic cavities is the muscular diaphragm.

The thoracic cavity contains several divisions. There are two *pleural cavities,* one surrounding each lung (Figure 1-7). Each pleural cavity is a small potential space (not a natural space) between the parietal pleura, a membrane that lines the pleural cavities, and the visceral pleura, a membrane that covers the lungs. The pleural cavities contain a small amount of fluid. The *mediastinum* (mē'-dē-as-TĪ-num) is a broad, median partition, actually a mass of tissue, between the pleurae of the lungs that extends from the sternum to the vertebral column (Figure 1-7a,b). The mediastinum includes all of the structures in the thoracic cavity, except the lungs themselves. Among the structures in the mediastinum are the heart, thymus gland, trachea, esophagus, and many large blood and lymphatic vessels. The *pericardial* (per'-i-KAR-dē-al; *peri* = around; *cardi* = heart) *cavity* is a small potential space between the visceral pericardium and parietal pericardium, the membranes covering the heart (Figure 1-7b). It also contains a fluid.

The abdominopelvic cavity, as the name suggests, is divided into two portions, although no wall separates them (see Figure 1-6). The upper portion, the *abdominal cavity,* contains the stomach, spleen, liver, gallbladder, pancreas, small intestine, and most of the large intestine. The lower portion, the *pelvic cavity,* contains the urinary bladder, sigmoid colon, rectum, and the internal male or female reproductive organs. One way to mark the division between the abdominal and pelvic cavities is to draw an imaginary line from the symphysis pubis (anterior joint between hip-bones) to the superior border of the sacrum (sacral promontory).

ABDOMINOPELVIC REGIONS

To describe the location of organs easily, the abdominopelvic cavity may be divided into the *nine regions* shown in Figure 1-8. Note which organs and parts of organs are in the various regions by examining Figure 1-8b–f and Exhibit 1-4. Although some parts of the body shown in the figure and described in the exhibit may be unfamiliar to you at this point, they will be discussed in detail in later chapters.

EXHIBIT 1-4

Representative Structures Found in the Abdominopelvic Regions

REGION	REPRESENTATIVE STRUCTURES
Epigastric (ep-i-GAS-trik; *epi* = above; *gaster* = stomach)	Left lobe and medial part of right lobe of liver, pyloric part and lesser curvature of stomach, superior and descending portions of duodenum, body and superior part of head of pancreas, and the two adrenal (suprarenal) glands.
Right Hypochondriac (hī-pō-KON-drē-ak; *hypo* = under; *chondro* = cartilage)	Right lobe of liver, gallbladder, and upper superior third of right kidney.
Left Hypochondriac	Body and fundus of stomach, spleen, left colic (splenic) flexure, superior two-thirds of left kidney, and tail of pancreas.
Umbilical (um-BIL-i-kul)	Middle portion of transverse colon, inferior part of duodenum, jejunum, ileum, hilar regions of both kidneys, and bifurcations (branching) of abdominal aorta and inferior vena cava.
Right Lumbar (*lumbus* = loin)	Superior part of cecum, ascending colon, right colic (hepatic) flexure, inferior lateral portion of right kidney, and small intestine.
Left Lumbar	Descending colon, inferior third of left kidney, and small intestine.
Hypogastric (Pubic)	Urinary bladder when full, small intestine, and part of sigmoid colon.
Right Iliac (Inguinal) (IL-ē-ak; iliac refers to superior part of hipbone)	Lower end of cecum, appendix, and small intestine.
Left Iliac (Inguinal)	Junction of descending and sigmoid parts of colon and small intestine.

ABDOMINOPELVIC QUADRANTS

The abdominopelvic cavity may be divided more simply into *quadrants* (*quad* = four). These are shown in Figure 1-9. In this method, frequently used by clinical personnel, a horizontal line and a vertical line are passed through the umbilicus. These two lines divide the abdomen into a *right upper quadrant* (*RUQ*), *left upper quadrant* (*LUQ*), *right lower quadrant* (*RLQ*), and *left lower quadrant* (*LLQ*). Whereas the nine-region division is more widely used for anatomical studies, the quadrant division is more commonly used for locating the site of an abdominopelvic pain, tumor, or other abnormality.

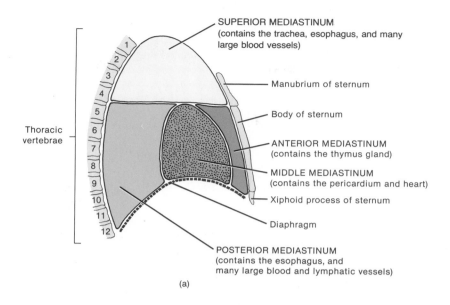

SUPERIOR MEDIASTINUM
(contains the trachea, esophagus, and many
large blood vessels)

Manubrium of sternum

Body of sternum

ANTERIOR MEDIASTINUM
(contains the thymus gland)

MIDDLE MEDIASTINUM
(contains the pericardium and heart)

Xiphoid process of sternum

Diaphragm

POSTERIOR MEDIASTINUM
(contains the esophagus, and
many large blood and lymphatic vessels)

Thoracic
vertebrae

(a)

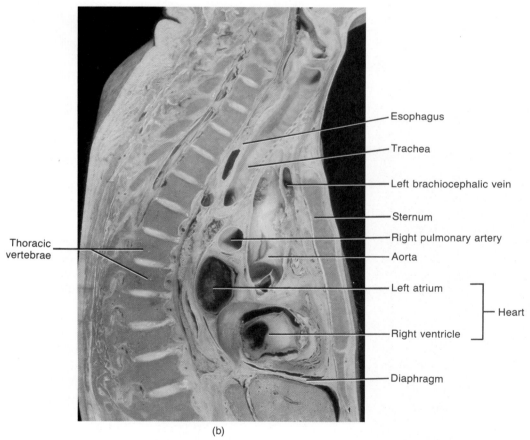

Esophagus

Trachea

Left brachiocephalic vein

Sternum

Right pulmonary artery

Aorta

Left atrium

Right ventricle

Heart

Diaphragm

Thoracic
vertebrae

(b)

FIGURE 1-7 Mediastinum. (a) Diagram of subdivisions of the mediastinum seen in right
lateral view. (b) Photograph of a near midline sagittal section through the thorax showing
some mediastinal structures. (Courtesy of J. A. Gosling, P. F. Harris, et al., *Atlas of Human
Anatomy*, Gower Medical Publishing Ltd., 1985.) (c) Mediastinum seen in cross section of
the thorax. Some of the structures shown and labeled may be unfamiliar to you now. However,
they are discussed in detail in later chapters. (d) Photograph of a cross section of the right
side of the thorax, emphasizing the pleurae and pleural cavity. (Courtesy of J. A. Gosling,
P. F. Harris, et al., *Atlas of Human Anatomy*, Gower Medical Publishing Ltd., 1985.)

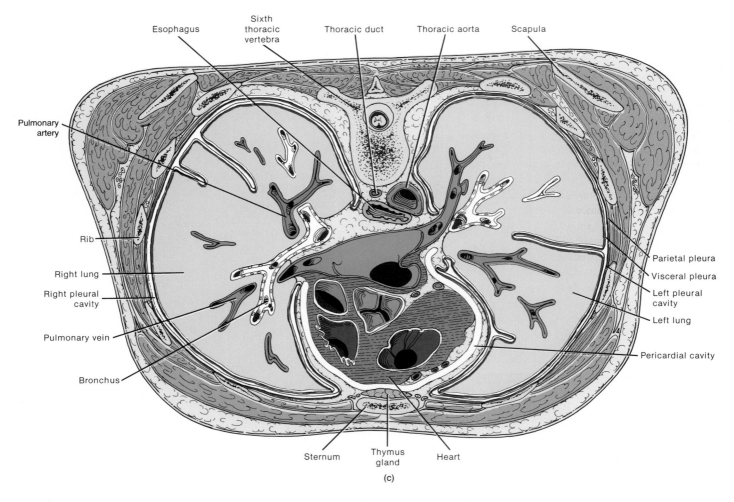

Esophagus

Sixth thoracic vertebra

Thoracic duct

Thoracic aorta

Scapula

Pulmonary artery

Rib

Right lung

Right pleural cavity

Pulmonary vein

Bronchus

Parietal pleura

Visceral pleura

Left pleural cavity

Left lung

Pericardial cavity

Sternum

Thymus gland

Heart

(c)

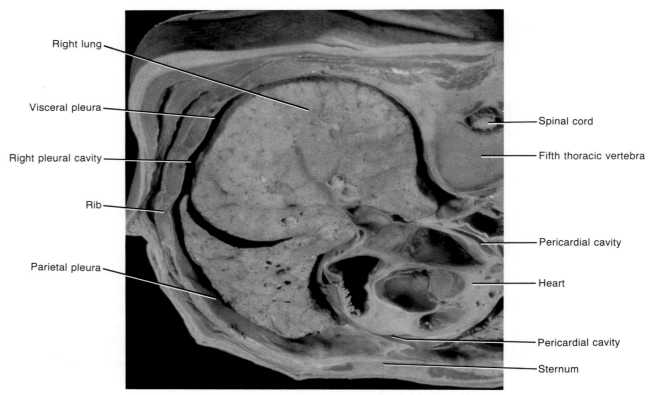

Right lung

Visceral pleura

Right pleural cavity

Rib

Parietal pleura

Spinal cord

Fifth thoracic vertebra

Pericardial cavity

Heart

Pericardial cavity

Sternum

(d)

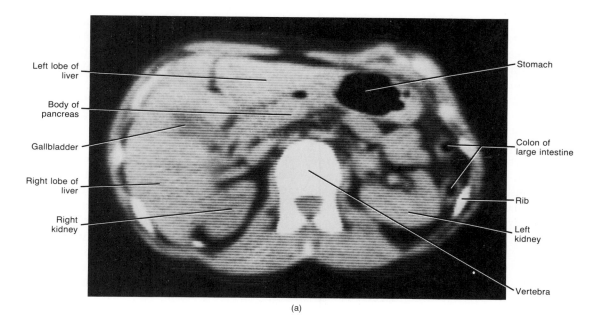

(a)

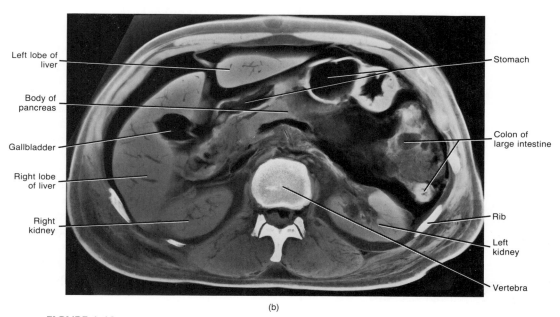

(b)

FIGURE 1-10 Radiographic anatomy. (a) Roentgenogram of the abdomen. (b) CT scan of the abdomen. Note the greater detail in the CT scan. (Courtesy of Stephen A. Kieffer and E. Robert Heitzman, *An Atlas of Cross-Sectional Anatomy*, Harper & Row, Publishers, Inc., New York, 1979.)

action, replay, high-speed, and slow-motion viewing. The level of radiation exposure of the DSR is about double that for a chest roentgenogram.

The DSR has been designed to provide three-dimensional imaging of the heart, lungs, and circulation. It can be used to measure the volumes and movements of the heart and lungs and other internal organs, to detect cancer and heart defects, and to measure tissue damage following a heart attack, stroke, or other disease.

MAGNETIC RESONANCE IMAGING (MRI)

New in the arsenal for diagnosing disease is *magnetic resonance imaging (MRI)*, formerly called *nuclear magnetic resonance, (NMR)*. MRI focuses on the nuclei of atoms of a single element in a tissue and determines their response to an external force such as magnetism. In most studies to date, MRI of hydrogen nuclei has been popular because of the body's large water content. The part of the body to be studied, ranging from a finger to the entire body, is placed in the scanner, exposing the nuclei to a uniform magnetic field (Figure 1-11a). The part of the body to be examined is placed in the center of the magnet. The image produced shows the density and energy loss (voltage) of the nuclei of a particular element and somewhat resembles a CT scan (Figure 1-11b). The images, which indicate a biochemical blueprint of cellular activity, can be displayed in color and as two- or three-dimensional images. The procedure typically takes about 30 minutes. MRI is not indicated for pregnant women, individuals who have artificial pacemakers or metal joints, or persons dependent on life-support equipment that has metal components.

The diagnostic advantage of MRI is that in addition to providing images of diseased organs and tissues, it also provides information about what chemicals are present in the organ and tissue. Such an analysis can indicate a particular disease is in progress, even before symptoms occur.

MRI also offers the advantages of being noninvasive (not involving puncture or incision of the skin or insertion of an instrument or foreign material into the body), not utilizing radiation, and gathering biochemical information without time-consuming chemical analyses.

MRI studies have confirmed the findings of other diagnostic techniques such as CT scans and, in some cases, have added more detail. Thus, MRI has proved useful in identifying existing pathologies. Scientists hope that MRI can also be used to perform a "biopsy" on tumors without an operation, assess mental disorders, measure blood flow, study the evolution of hematomas and infarctions in the brain following a stroke, identify the potential for developing a stroke, assess treatment for conditions such as heart disease and stroke, monitor the progress and treatment of a disease, study the effects of toxic drugs on tissues, measure intracellular pH, study metabolism, and determine how donor organs from cadavers may function after transplantation.

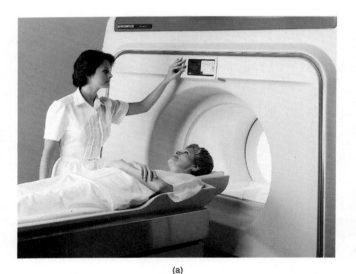

(a)

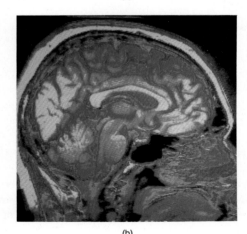

(b)

FIGURE 1-11 Magnetic resonance imaging (MRI). (a) MRI scanner. (b) Image of the human brain. (Courtesy of Technicare Corporation, Cleveland, Ohio.)

ULTRASOUND (US)

When high-frequency sound waves (sound waves that cannot be heard by the human ear) travel forward, they continue to move until they make contact with an object. Then, a certain amount of the sound bounces back. This is the principle of *ultrasound* (*US*). Submarines use the principle to detect the presence of other vessels and mines and to determine the depth of the ocean floor.

In medical practice, high-frequency sound waves are generated by a hand-held instrument called a *transducer* that is moved over the part of the body to be examined. The returning sound waves are also detected by the transducer. Since normal and abnormal body tissues have different densities, they reflect sound waves differently. This information is translated into an image that appears on a screen. The image, called a *sonogram,* can be photographed for future reference. Some images are still, two-dimensional cross sections; others, such as the heart or fetus, can be moving images.

Ultrasound may be used to study most abdominal organs, especially in diagnosing gallstones; pelvic organs; blood flow in arteries and veins (Doppler ultrasound); the heart (echocardiography); and a developing fetus (fetal ultrasound).

Among the advantages of ultrasound are the following: there is no exposure to radiation, it is noninvasive, it is painless, it does not require any dyes that might cause nausea or allergic reactions, and it is quick and readily available. Numerous experiments and years of experience have uncovered no clear evidence that US produces any harmful effects in humans. However, since ultrasound does not penetrate bone and air-filled spaces, it is not useful in diagnosing abnormalities in the skull, lungs, or intestines. Obesity and scars also interfere with the efficiency of the procedure.

POSITRON EMISSION TOMOGRAPHY (PET)

In *radioisotopic scanning* the chemical activity of various tissues of the body can be recorded. Physicians can diagnose certain diseases by charting the course of an injected radioisotope through the body and by noting how much of the substance concentrates and where. The branch of medicine that is concerned with the use of radioisotopes in the diagnosis of disease and therapy is called *nuclear medicine.*

Although radioisotope scans have been used in diagnosis since the 1950s, their value was limited by distorted images of the distribution of a radioisotope. Within the past few years more precise localization of radioisotopes in the body has been made possible by new computer imaging techniques similar to those used in CT scanning. The result is a sophisticated version of radioisotope scanning called *positron emission tomography (PET)*.

The principle behind PET scanning is as follows. Short-lived radioisotopes such as ^{11}C, ^{13}N, or ^{15}O are produced and incorporated into a solution that can be injected into the body. As the radioisotope circulates through the body, it emits positively charged electronlike particles called *positrons*. Positrons collide with negatively charged electrons in body tissues, causing their annihilation and the release of gamma rays. Gamma rays are bundles of energy similar to high-energy x rays. Once released, gamma rays travel in opposite directions and are detected and recorded by PET receptors. A computer then takes the information and constructs a colored *PET scan* that shows where the radioisotopes are being used in the body (Figure 1-12).

PET provides information that cannot be obtained by any other technique by sensing function rather than structure. Using PET, physicians can study the effects of drugs in body organs, measure blood flow through organs such as the brain and heart, diagnose coronary artery disease, identify the extent of damage as a result of strokes or heart attacks, and detect cancers and measure the effects of treatment. PET studies with schizophrenic and manic–depressive patients have revealed that schizophrenics tend to use less glucose in certain brain areas, whereas manic—depressive patients use more glucose during manic phases. PET technology is also used to study the chemical changes that occur during epileptic seizures and in persons suffering from senile dementia. As impressive as PET is in the diagnosis of disease, it is also being used by scientists to probe the healthy brain. For example, by detecting and recording changes in glucose consumption in the brain, scientists can identify which specific areas of the brain are involved in specific sensory and motor activities.

DIGITAL SUBTRACTION ANGIOGRAPHY (DSA)

A procedure that provides physicians with a much clearer look at diseased arteries is called *digital subtraction angiography (DSA)*. The procedure employs a computer technique that compares an x-ray image of the same region of the body before and after a contrast substance containing iodine has been introduced intravenously. Any tissue or blood vessels that show up in the first image can be subtracted (erased) from the second image, leaving an unobstructed view of an artery. DSA figuratively lifts an artery out of the body so that it can be studied in isolation. DSA is frequently used to study the blood vessels of the brain and heart. DSA also helps in diagnosing lesions in the carotid arteries leading to the brain, a potential cause of strokes, and in evaluating patients before surgery and after coronary artery bypass grafting (CABG) and certain transplant operations.

MEASURING THE HUMAN BODY

An important aspect of describing the body and understanding how it works is *measurement*—what the dimensions of an organ are, how much it weighs, how long it takes for a physiological event to occur. Such measurements also have clinical importance, for example, in determining how much of a given medication should be administered. As you will see, measurements involving time, weight, temperature, size, length, and volume are a routine part of your studies in a medical science program.

Whenever you come across a measurement in the text, the measurement will be given in metric units. The metric system is standardly used in sciences. To help you compare the metric unit to a familiar unit, the approximate U.S. equivalent will also be given in parenthesis directly after the metric unit. For example, you might be told that the length of a particular part of the body is 2.54 cm (1 in.).

As a first step in helping you understand the correlation between the metric system and the U.S. system of measurement, three exhibits have been prepared. Exhibit 1-5 contains metric units of length and some U.S. equivalents. Exhibit 1-6 contains metric units of mass and some U.S. equivalents. Exhibit 1-7 contains metric units of volume and some U.S. equivalents. Carefully examine the exhibits. Even if you do not learn all the metric units and their equivalents at this point, you can refer back to the exhibits later.

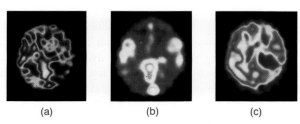

FIGURE 1-12 Positron emission tomography (PET) scans. (a) Oxygen metabolism in the brain. (b) Blood volume in the brain. (c) Blood flow through the brain. (Courtesy of Dr. Michel M. Ter-Rogossian, Washington University, School of Medicine.)

EXHIBIT 1-5

Metric Units of Length and Some U.S. Equivalents

METRIC UNIT	MEANING OF PREFIX	METRIC EQUIVALENT	U.S. EQUIVALENT
1 Kilometer (km)	kilo = 1,000	1,000 m = 10^3 m	3,280.84 ft = 0.62 mi 1 mi = 1.61 km
1 Hectometer (hm)	hecto = 100	100 = 10^2 m	328 ft
1 Dekameter (dam)	deka = 10	10 m = 10^0 m	32.8 ft
1 Meter (m)		Standard unit of length	39.37 in. = 3.28 ft = 1.09 yd
1 Decimater (dm)	deci = 1/10	0.1 m = 10^{-1} m	3.94 in.
1 Centimeter (cm)	centi = 1/100	0.01 m = 10^{-2} m	0.394 in. 1 in. = 2.54 cm
1 Millimeter (mm)	milli = 1/1,000	0.001 m = 1/10 cm = 10^{-3} m	0.0394 in.
1 Micrometer (μm) [Formerly Micron (μ)]	micro = 1/1,000,000	0.000,001 m = 1/10,000 cm = 10^{-6} m	3.94×10^{-5} in.
1 Nanometer (nm) [Formerly Millimicron (mμ)]	nano = 1/1,000,000,000	0.000,000,001 m = 1/10,000,000 cm = 10^{-9} m	3.94×10^{-8} in.

EXHIBIT 1-6

Metric Units of Mass and Some U.S. Equivalents

METRIC UNIT	METRIC EQUIVALENT	U.S. EQUIVALENT
1 Kilogram (kg)	1,000 g	2.205 lb 1 ton = 907 kg
1 Hectogram (hg)	100 g	
1 Dekagram (dag)	10 g	0.0353 oz
1 Gram (g)	1 g	1 lb = 453.6 g 1 oz = 28.35 g
1 Decigram (dg)	0.1 g	
1 Centigram (cg)	0.01 g	
1 Milligram (mg)	0.001 g	0.015 gr
1 Microgram (μg)	0.000,001 g	
1 Nanogram (ng)	0.000,000,001 g	
1 Picogram (pg)	0.000,000,000,001 g	

EXHIBIT 1-7

Metric Units of Volume and Some U.S. Equivalents

METRIC UNIT (ABBREVIATION)	METRIC EQUIVALENT	U.S. EQUIVALENT
1 Liter (l)	1,000 ml	33.81 fl oz or 1.057 qr; 946 ml = 1 qt
1 Milliliter (ml)	0.001 L	0.0338 fl oz; 30 ml = 1 fl oz 5 ml = 1 teaspoon
1 Cubic Centimeter (cm^3)	0.999972 ml	0.0338 fl oz

STUDY OUTLINE

Anatomy and Physiology Defined (p. 2)

1. Anatomy is the study of structure and the relation among structures.
2. Subdivisions of anatomy include surface anatomy (form and markings of surface features), gross anatomy (macroscopic), systemic, or systematic, anatomy (systems), regional anatomy (regions), developmental anatomy (development from fertilization to adulthood), embryology (development from fertilized egg through eighth week in utero), pathological anatomy (disease), histology (tissues), cytology (cells), and radiographic anatomy (x rays).
3. Physiology is the study of how body structures function.

Levels of Structural Organization (p. 2)

1. The human body consists of several levels of structural organization; among these are the chemical, cellular, tissue, organ, system, and organismic levels.
2. Cells are the basic structural and functional units of an organism.
3. Tissues consist of groups of similarly specialized cells and their intercellular material that perform certain special functions.
4. Organs are structures of definite form that are composed of two or more different tissues and have specific functions.
5. Systems consist of associations of organs that have a common function.
6. The human organism is a collection of structurally and functionally integrated systems.
7. The systems of the human body are the integumentary, skeletal, muscular, nervous, endocrine, cardiovascular, lymphatic, respiratory, digestive, urinary, and reproductive systems (see Exhibit 1-2).

Life Processes (p. 4)

1. All living forms have certain characteristics that distinguish them from nonliving things.
2. Among the life processes in humans are metabolism, excitability, conductivity, contractility, growth, differentiation, and reproduction.

Structural Plan (p. 4)

1. The human body has certain general characteristics.
2. Among the characteristics are a backbone, a tube within a tube organization, and bilateral symmetry.

Anatomical Position and Anatomical Names (p. 4)

1. When in the anatomical position, the subject stands erect facing the observer, the upper extremities are placed at the sides, the palms of the hands are turned forward, and the feet are flat on the floor.
2. Regional names are terms given to specific regions of the body for reference. Examples of regional names include cranial (skull), thoracic (chest), brachial (arm), patellar (knee), cephalic (head), and gluteal (buttock).

Directional Terms (p. 4)

1. Directional terms indicate the relation of one part of the body to another.
2. Commonly used directional terms are superior (toward the head or upper part of a structure), inferior (away from the head or toward the lower part of a structure), anterior (near or at the front of the body), posterior (near or at the back of the body), medial (nearer the midline of the body or a structure), lateral (farther from the midline of the body or a structure), intermediate (between a medial and lateral structure), ipsilateral (on the same side of the body), contralateral (or on the opposite side of the body), proximal (nearer the attachment of an extremity to the trunk or a structure), distal (farther from the attachment of an extremity to the trunk or a structure), superficial (toward or on the surface of the body), deep (away from the surface of the body), parietal (pertaining to the outer wall of a body cavity), and visceral (pertaining to the covering of an organ).

Planes and Sections (p. 12)

1. Planes are imaginary flat surfaces that are used to divide the body or organs into definite areas. A midsagittal (median) plane is a vertical plane through the midline of the body that divides the body or organs into equal right and left sides; a parasagittal plane is a plane that does not pass through the midline of the body and divides the body or organs into unequal right and left sides; a frontal (coronal) plane is a plane at a right angle to a midsagittal (or parasagittal) plane that divides the body or organs into anterior and posterior portions; and a horizontal (transverse) plane is a plane parallel to the ground and at a right angle to the midsagittal, parasagittal, and frontal planes that divides the body or organs into superior and inferior portions.
2. Sections are flat surfaces resulting from cuts through body structures. They are named according to the plane on which the cut is made and include cross sections, frontal sections, and midsagittal sections.

Body Cavities (p. 13)

1. Spaces in the body that contain internal organs are called cavities.
2. The dorsal and ventral cavities are the two principal body cavities. The dorsal cavity contains the brain and spinal cord. The organs of the ventral cavity are collectively called viscera.
3. The dorsal cavity is subdivided into the cranial cavity, which contains the brain, and the vertebral, or spinal canal, which contains the spinal cord and beginnings of spinal nerves.
4. The ventral body cavity is subdivided by the diaphragm into an upper thoracic cavity and a lower abdominopelvic cavity.
5. The thoracic cavity contains two pleural cavities and a mediastinum, which includes the pericardial cavity.
6. The mediastinum is a broad, median partition, actually a mass of tissue between the pleurae of the lungs that extends from the sternum to the vertebral column; it contains all contents of the thoracic cavity, except the lungs.
7. The abdominopelvic cavity is divided into a superior abdominal and an inferior pelvic cavity by an imaginary line extending from the symphysis pubis to the sacral promontory.
8. Viscera of the abdominal cavity include the stomach, spleen, pancreas, liver, gallbladder, small intestine, and most of the large intestine.
9. Viscera of the pelvic cavity include the urinary bladder, sigmoid colon, rectum, and internal female and male reproductive structures.

Abdominopelvic Regions (p. 16)

1. To describe the location of organs easily, the abdominopelvic cavity may be divided into nine regions by drawing four imaginary lines: subcostal (superior horizontal), transtubercular (inferior horizontal), and right and left midclavicular (both vertical lines).
2. The names of the nine abdominopelvic regions are epigastric, right hypochondriac, left hypochondriac, umbilical, right lumbar, left lumbar, hypogastric (pubic), right iliac (inguinal), and left iliac (inguinal).

Abdominopelvic Quadrants (p. 16)

1. To locate the site of an abdominopelvic abnormality in clinical studies, the abdominopelvic cavity may be divided into quadrants by passing imaginary horizontal and vertical lines through the umbilicus.
2. The names of the abdominopelvic quadrants are right upper quadrant (RUQ), left upper quadrant (LUQ), right lower quadrant (RLQ), and left lower quadrant (LLQ).

Medical Imaging (p. 19)

Conventional Radiography (p. 19)

1. Conventional radiography uses a single barrage of x rays.
2. The photographic two-dimensional image produced is called a roentgenogram.
3. Conventional radiography has several diagnostic drawbacks, including overlapping of organs and tissues and inability to differentiate subtle differences in tissue density.

Computed Tomography (CT) Scanning (p. 19)

1. Computed tomography (CT) scanning combines the principles of x-ray and advanced computer technologies.
2. The image produced, called a CT scan, provides a very accurate cross-sectional picture of any area of the body.

Dynamic Spatial Reconstructor (DSR) (p. 22)

1. The DSR is a highly sophisticated x-ray machine that can produce moving, three-dimensional images of different organs of the body.

2. The DSR was designed to provide imaging of the heart, lungs, and circulation.

Magnetic Resonance Imaging (MRI) (p. 23)

1. Magnetic resonance imaging (MRI) is based on the reaction of atomic nuclei to magnetism.
2. MRI can identify existing pathologies, evaluate drug therapy, measure metabolism, and assess the potential for certain diseases.

Ultrasound (US) (p. 24)

1. Ultrasound (US) is based on high-frequency sound waves that are reflected and are translated into images.
2. Applications of US include diagnosing diseases of abdominal and pelvic organs, analyzing blood flow, assessing heart disorders, and evaluating fetal development.

Positron Emission Tomography (PET) (p. 25)

1. Positron emission tomography (PET) is a form of radioisotope scanning based on the emission of positrons from radioisotopes.
2. PET not only helps in the diagnosis of disease but also analyzes a healthy brain.

Digital Subtraction Angiography (DSA) (p. 25)

1. This procedure compares a blood vessel before and after a contrast medium is introduced.
2. It is frequently used to study blood vessels of the heart.

Measuring the Human Body (p. 25)

1. Various kinds of measurements are important in understanding the human body.
2. Examples of such measurements include organ size, body weight, and amount of medication to be administered.
3. Measurements in this book are given in metric units followed by the approximate U.S. equivalents in parentheses.
4. Metric units of length may be reviewed in Exhibit 1-5, metric units of mass in Exhibit 1-6, and metric units of volume in Exhibit 1-7.

REVIEW QUESTIONS

1. Define anatomy. List and define the various subdivisions of anatomy. Define physiology.
2. Construct a diagram to illustrate the levels of structural organization that characterize the human body. Be sure to define each level.
3. Using Exhibit 1-2 as a guide, outline the function of each system of the body, and list several organs that compose each system.
4. List and define the important life processes in humans.
5. What does bilateral symmetry mean? Why is the body considered to be a tube within a tube? What is a vertebrate?
6. Define the anatomical position. Why is the anatomical position used?
7. Review Figure 1-2. See if you can locate each region on your own body and give both its common and its anatomical name.

8. What is a directional term? Why are these terms important? Can you use each of the directional terms listed in Exhibit 1-3 in a complete sentence?
9. Name the various planes that may be passed through the body. Describe how each plane divides the body.
10. What is meant by the phrase "a part of the body has been sectioned"?
11. Define a body cavity. List the body cavities discussed, and tell which major organs are located in each. What landmarks separate the various body cavities from one another?
12. What is the mediastinum?
13. Name and locate the nine regions of the abdominopelvic area. List the organs, or parts of organs, in each.
14. Describe how the abdominopelvic cavity is divided into quadrants and name each quadrant.
15. Why is an autopsy performed? Describe the basic procedure.

16. Explain the principle of computed tomography (CT) scanning. Contrast it with conventional radiography in terms of principle and diagnostic value.

17. Explain the principle and clinical applications of the dynamic spatial reconstructor (DSR), magnetic resonance imaging (MRI), ultrasound (US), positron emission tomography (PET), and digital subtraction angiography (DSA).

18. Convert the following lengths:
    a. A bacterial cell measures 100 μm in length. Give its length in nanometers.
    b. How many meters are in 1 mi?
    c. If a road sign reads 35 km/hour, what would your speedometer have to read (in miles) to obey the sign?
    d. How many millimeters are in 1 km? In 1 in?
    e. A person's arm measures 2 ft in length. How many centimeters is this?
    f. Convert 0.40 m to millimeters.
    g. How many millimeters are in 5 in?
    h. If you ran 295.2 ft, how many meters would you have run?
    i. If the distance to the moon is 239,000 mi, what is it in meters?

19. Solve the following conversions of mass.
    a. Calculate the milligrams in 0.4 kg and in 1 lb.
    b. If a bottle contains 1.42 g, how many centigrams does it contain?
    c. The indicated dosage of a certain drug is 50 μg. How many milligrams is this?
    d. If you weigh 110 lb, how many kilograms do you weight?
    e. How many centigrams are in 1 g?

20. Convert the following volumes.
    a. If you excrete 1,200 ml of urine in a day, how many liters is this?
    b. How many milliliters are in 2 l?
    c. Convert 2 pt to milliliters.
    d. If you remove 15 cm³ of blood from a patient, how many milliliters have you removed?

SELF-QUIZ

1. Complete the following table relating common terms to anatomical terms.

COMMON TERM	ANATOMICAL TERM
a.	Axillary
b. Arm	
c.	Cephalic
d. Chest	
e.	Neck

Choose the one best answer to these questions.

___ 2. One milliliter (ml) is equal to:
A. about 1 oz; B. ¹/₁₀₀ liter; C. 0.1 liter; D. ¹/₁₀₀₀ liter; E. about ¹/₁₀₀ liter.

___ 3. Which of the following statements about the function of the respiratory system is *not* true?
A. it supplies oxygen; B. it eliminates carbon dioxide; C. it helps regulate acid–base balance in the body; D. it filters blood; E. it includes the lungs.

___ 4. Which is most inferiorly located?
A. abdomen; B. pelvic cavity; C. mediastinum; D. diaphragm; E. pleural cavity.

___ 5. The spleen, tonsils, and thymus are all organs in which system?
A. nervous; B. lymphatic; C. cardiovascular; D. digestive; E. endocrine.

___ 6. The system responsible for providing support, protection, and leverage, storage of minerals, and production of blood cells is
A. nervous; B. lymphatic; C. cardiovascular; D. digestive; E. endocrine.

___ 7. In the anatomical position
A. the body is in a supine position; B. the body is in a prone position; C. the palmar surface of the hand is anterior; D. the dorsal surface of the hand is anterior; E. the body is prone and the palmar surface of the hand is anterior.

___ 8. The general anatomical characteristics of the human body include:
(1) a backbone (vertebral column).
(2) a tube-within-a-tube construction.
(3) bilateral symmetry.
A. (1) only; B. (2) only; C. (3) only; D. (1) and (2); E. all of the above.

___ 9. Which word describes the location of the stomach with reference to the pancreas?
A. anterior; B. distal; C. dorsal; D. proximal; E. sagittal.

___ 10. Any part of the body that is away from the midline is said to be
A. medial; B. lateral; C. distal; D. superior; E. posterior.

___ 11. Where does the knee joint lie in reference to the leg bones?
A. anterior; B. medial; C. inferior; D. lateral; E. proximal.

___ 12. The microscopic examination of tissues is referred to as
A. embryology; B. physiology; C. histology; D. gross anatomy; E. surface anatomy.

___ 13. The visceral pleura
A. is the same as the parietal pleura; B. forms around the heart as a layer called the pericardium; C. is so called because it is part of the internal tissue of a lung; D. is the membrane closely adhering to the outer surface of a lung; E. lines the abdomen.

___ 14. Which of the following statements is/are true of a plane that divides the body into superior and inferior parts?
(1) This plane must pass through the umbilicus (navel).
(2) This plane is a transverse (horizontal) plane.
(3) This plane is a midsagittal (median) plane.
(4) This plane is called a coronal (frontal) plane.
A. (1) only; B. (2) only; C. (3) only; D. (4) only; E. (1) and (3).

___ **15.** Which of the following is/are true?
    (1) The thoracic cavity is the same as the pleural cavity.
    (2) The diaphragm divides the abdominopelvic cavity into abdominal and pelvic portions.
    (3) The organs of importance in the ventral body cavity are the brain and spinal cord.
    A. (1) only; B. (2) only; C. (3) only; D. (1) and (2); E. none of the statements is true.

___ **16.** You would *not* look in the pelvic cavity to find the
    A. rectum; B. sigmoid colon; C. thymus gland; D. uterus; E. urinary bladder.

Complete the following.

17. From superior to inferior, the three abdominal regions on the right side are right hypochondriac, right lumbar, and _____.

18. When you lie face down in a pool, as if to do the "deadman's float," you are lying on your _____ surface.

19. Since the stomach and the spleen are both located on the left side of the abdomen, they could be described as _____ -lateral.

20. A _____ plane divides the brain into equal left and right sides.

2 Cells

STUDENT OBJECTIVES

1. Define a cell and list its generalized parts.
2. Explain the chemistry, structure, and functions of the plasma membrane.
3. Describe how materials move across plasma membranes by diffusion, facilitated diffusion, osmosis, filtration, active transport, and endocytosis.
4. Describe the chemical composition and functions of cytoplasm.
5. Describe the structure and functions of the following cellular structures: nucleus, ribosomes, endoplasmic reticulum (ER), Golgi complex, mitochondria, lysosomes, peroxisomes, cytoskeleton, centrioles, cilia, and flagella.
6. Define a cell inclusion and give several examples.
7. Define extracellular material and give several examples.
8. Describe cancer as a homeostatic imbalance of cells.
9. Explain the relation between aging and cells.
10. Define key medical terms associated with cells.

CHAPTER OUTLINE

■ **Generalized Animal Cell**
■ **Plasma (Cell) Membrane**
Chemistry and Structure
Functions
Movement of Materials Across Plasma Membranes
  Passive Processes
  Active Processes
■ **Cytoplasm**
■ **Organelles**
Nucleus
Ribosomes
Endoplasmic Reticulum (ER)
Golgi Complex
Mitochondria
Lysosomes
Peroxisomes
The Cytoskeleton
Centrosome and Centrioles
Flagella and Cilia
■ **Cell Inclusions**
■ **Extracellular Materials**
■ **Normal Cell Division**
Somatic Cell Division
  Mitosis
  Cytokinesis
Reproductive Cell Division
  Meiosis
■ **Abnormal Cell Division: Cancer (CA)**
Definition
Spread
Types
Possible Causes
Treatment
■ **Cells and Aging**
■ **Key Medical Terms Associated with Cells**

The study of the body at the cellular level of organization is important because activities essential to life occur in cells and disease processes originate there. A *cell* may be defined as the basic, living, structural and functional unit of the body and, in fact, of all organisms. *Cytology* (sī-TOL-ō-jē; *cyt* = cell; *logos* = study of) is the branch of science concerned with the study of cells. This chapter concentrates on the structure, functions, and reproduction of cells.

A series of illustrations accompanies each cell structure that you study. A diagram of a generalized animal cell shows the location of the structure within the cell. An electron micrograph shows the actual appearance of the structure.* A diagram of the electron micrograph clarifies some of the small details by exaggerating their outlines. Finally, an enlarged diagram of the structure shows its details.

CLINICAL APPLICATION

Despite the enormous advantages provided by electron microscopy in helping scientists understand detailed cellular structure, one drawback is the fact that specimens must be killed, since they are placed in a vacuum. This problem is now overcome by a procedure called *microtomography* (mī-krō-tō-MOG-ra-fē), in which the principles of electron microscopy and computed tomography (CT) scanning are combined. Microtomography produces high-magnification, three-dimensional images of *living* cells.

The potential applications of microtomography include studying how normal and cancerous cells move, grow, and reproduce. It could also be used to follow the changes that occur as an embryo progresses through its developmental sequences. The technique should enable scientists to observe the microscopic effects of drugs on living cells and how cancer-causing substances alter living cells.

GENERALIZED ANIMAL CELL

A *generalized animal cell* is a composite of many different cells in the body. Examine the generalized cell illustrated in Figure 2-1, but keep in mind that no such single cell actually exists.

For convenience, we can divide the generalized cell into four principal parts:

1. *Plasma (cell) membrane.* The outer, limiting membrane separating the cell's internal parts from the extracellular fluid and external environment.

2. *Cytoplasm.* The substance that surrounds organelles and is located between the nucleus and the plasma membrane.

3. *Organelles.* Permanent structures with characteristic morphology that are highly specialized for specific cellular activities.

4. *Inclusions.* The secretions and storage products of cells.

Extracellular materials, which are substances external to the cell surface, will also be examined in connection with cells.

PLASMA (CELL) MEMBRANE

The exceedingly thin structure that separates one cell from other cells and from the external environment is called the *plasma (cell) membrane* (Figure 2-2a). The membrane measures from 4.5 nm at the phospholipid bilayer regions up to 10 nm in regions where membrane proteins are present. Formerly, dimensions of cells and parts of cells were given in angstroms (Å). One angstrom equals 0.1 nm.

CHEMISTRY AND STRUCTURE

Plasma membranes consist primarily of phospholipids (lipids that contain phosphorus), the most abundant chemicals, and proteins. Other chemicals in lesser amounts include cholesterol (a lipid), glycolipids (combinations of carbohydrates and lipids), and carbohydrates called oligosaccharides. Studies of membrane structure suggest a concept regarding the arrangement of the various molecules referred to as the *fluid mosaic model* (Figure 2-2b).

The phospholipid molecules are arranged in two parallel rows, forming a *phospholipid bilayer.* A phospholipid molecule consists of a polar, phosphate-containing "head" that mixes with water (hydrophilic) and nonpolar fatty acid "tails" that do not mix with water (hydrophobic). The molecules are oriented in the bilayer so that "heads" face outward on either side and the "tails" face each other in the membrane's interior. The phospholipid bilayer is soft and flexible since the phospholipid molecules can move sideways and exchange places in their own row; movement of phospholipid molecules between rows rarely occurs. The bilayer is also self-sealing; if a needle is pushed through it and pulled out, the puncture site will seal automatically. The phospholipid bilayer forms the basic framework of plasma membranes.

The plasma membrane proteins (PMPs) are classified into two categories: integral and peripheral. *Integral proteins* are embedded in the phospholipid bilayer among the fatty acid "tails." Some of the integral proteins lie at or near the inner and outer membrane surfaces; others penetrate the membrane completely. Since the phospholipid bilayer is somewhat fluid and flexible and the integral proteins have been observed moving from one location to another in the membrane, the relation has been compared to icebergs

---

* An *electron micrograph* (*EM*) is a photograph taken with an electron microscope. Some electron microscopes can magnify objects up to 1,000,000 times. In comparison, the light microscope that you probably use in your laboratory magnifies objects up to 1,000 times their size.

membrane uses energy supplied by ATP. In fact, a typical body cell probably expends up to 40 percent of its ATP for active transport. Although the molecular events involved in active transport are not completely understood, integral proteins in the plasma membrane do assume a role.

As just noted, glucose can be transported across cell membranes via facilitated diffusion from areas of high to low concentration. Glucose can also be moved by the cells lining the gastrointestinal tract from the cavity of the tract into the blood, even though blood concentration of glucose is higher. This movement involves active transport. One proposed mechanism is that glucose (or other substance) enters a channel in an integral membrane protein. When the glucose molecule makes contact with an active site in the channel, the energy from ATP induces a change in the membrane protein that expels the glucose on the opposite side of the membrane. Active transport is also an important process in maintaining the concentrations of some ions inside body cells and other ions outside body cells.

● **Endocytosis** Large molecules and particles pass through plasma membranes by a process called **endocytosis,** in which a segment of the plasma membrane surrounds the substance, encloses it, and brings it into the cell. The export of substances from the cell by the reverse process is called **exocytosis,** a very important mechanism for secretory cells. There are three basic kinds of endocytosis: phagocytosis, pinocytosis, and receptor-mediated endocytosis.

In **phagocytosis** (fag'-ō-sī-TŌ-sis), or "cell eating," projections of cytoplasm, called **pseudopodia** (soo'-dō-PŌ-dē-a), engulf large solid particles external to the cell (Figure 2-3a,b). Once the particle is surrounded, the membrane folds inwardly, forming a membrane sac around the particle. This newly formed sac, called a **phagocytic vesicle,** breaks off from the plasma membrane, and the solid material inside the vesicle is digested by enzymes provided by lysosomes. Indigestible particles and cell products are removed from the cell by exocytosis. Phagocytosis is important because molecules and particles of material that would normally be restricted from crossing the plasma membrane because of their large size can be brought into or removed from the cell. The phagocytic white blood cells of the body constitute a vital defense mechanism that helps protect us against disease. Through phagocytosis, the white blood cells engulf and destroy bacteria and other foreign substances.

In **pinocytosis** (pi'-nō-sī-TŌ-sis), or "cell drinking," the engulfed material consists of an extracellular liquid rather than a solid (Figure 2-3c). Moreover, no cytoplasmic pro-

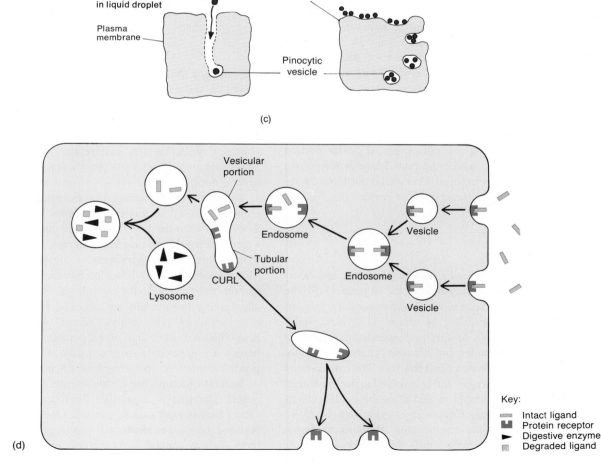

(c)

(d)

jections are formed. Instead, a minute droplet of liquid is attracted to the surface of the membrane. The membrane folds inwardly, forms a pinocytic vesicle that surrounds the liquid, and detaches from the rest of the intact membrane. Few cells are capable of phagocytosis, but many cells carry on pinocytosis. Examples include endothelial (lining) cells that make up the wall of blood vessels, especially capillaries, and cells in the kidneys and urinary bladder.

Receptor-mediated endocytosis is a highly selective process in which cells can take up large molecules or particles. The interstitial fluid that bathes cells contains a large number of chemicals, most of which are in concentrations lower than in the cells themselves. These chemicals, called ***ligands,*** serve a variety of functions, all of which are related to homeostasis. For example, some of the substances are nutrients, like amino acids, iron, and vitamins, needed for various chemical reactions that sustain life. Some are hormones that deliver messages to cells so that cells can carry on specific physiological responses. Other substances are waste products or poisonous materials that certain cells have the ability to break down so that they do not interfere with the normal functioning of cells.

Receptor-mediated endocytosis is accomplished as follows. The plasma membrane contains protein receptors that have binding sites for ligands (Figure 2-3c). The binding between receptor and ligand causes the plasma membrane to fold inward, forming a ***vesicle*** around the ligand. As a vesicle moves inward from the plasma membrane, it fuses with another vesicle to form a larger structure called an ***endosome.*** Each endosome develops into an even larger structure called a ***CURL*** (compartment of uncoupling of receptor and ligand). Within the CURL, the ligands separate from the receptors. The ligands move into one part of the CURL (vesicular portion) and the receptors accumulate in another portion of the CURL (tubular portion). Ligands have different destinations in the cell. However, in most instances following this separation, the vesicular portion fuses with a lysosome where the ligands are broken down by powerful digestive enzymes. The tubular portion recycles receptors to the plasma membrane for reuse.

CYTOPLASM

The substance inside the cell's plasma membrane and external to the nucleus is called ***cytoplasm*** (SĪ-tō-plazm') (Figure 2-4a,b). It is the substance in which various cellular components are found. Physically, cytoplasm is a thick, semi-transparent, elastic fluid containing suspended particles and a series of minute tubules and filaments that form a cytoskeleton (which will be described shortly). The cytoskeleton provides support and shape, and is involved in the movement of structures in the cytoplasm and even the entire cell, as occurs in phagocytosis. Chemically, cytoplasm is 75–90 percent water plus solid components. Proteins, carbohydrates, lipids, and inorganic substances compose the bulk of the solid components. The inorganic substances and most carbohydrates are soluble in water and are present as a true solution. Many organic compounds, however, are found as colloids—large molecules that remain suspended in solution. Since the particles of a colloid bear electrical charges that repel each other, they remain suspended and separated from each other.

Functionally, cytoplasm is the substance in which some chemical reactions occur. The cytoplasm receives raw materials from the external environment and converts them into usable energy by decomposition reactions. Cytoplasm is also the site where new substances are synthesized for cellular use. It packages chemicals for transport to other parts of the cell or other cells of the body and facilitates the excretion of waste materials.

ORGANELLES

Despite the numerous chemical activities occurring simultaneously in the cell, there is little interference of one reaction with another. This is because the cell has a system of compartmentalization provided by structures called ***organelles.*** These structures are specialized portions of the cell with characteristic shape that assume specific roles in growth, maintenance, repair, and control. The number and types of organelles vary among different cells, depending on their functions.

NUCLEUS

The ***nucleus*** (NOO-klē-us) is generally a spherical or oval organelle and is the largest structure in the cell (Figure 2-4a–c). It contains hereditary factors of the cell, called genes, which control cellular structure and direct many cellular activities. Most body cells contain a single nucleus, although some, such as mature red blood cells, do not. Skeletal muscle fibers (cells) and a few other cells contain several nuclei.

The nucleus is separated from the cytoplasm by a double membrane called the ***nuclear membrane*** or ***envelope*** (Figure 2-4c). Between the two layers of the nuclear membrane is a space called the ***perinuclear cisterna.*** Each nuclear membrane resembles the structure of the plasma membrane. Minute ***pores*** in the nuclear membrane allow the nucleus to communicate with the cytoplasm. Substances entering and leaving the nucleus are believed to pass through the tiny pores. At many points the two nuclear membranes fuse with each other and, at the points of fusion, the membranes are permeable to practically all dissolved or suspended substances, including newly synthesized ribosomes.

Several structures are visible internal to the nuclear membrane. The first is a gel-like fluid that fills the nucleus called ***karyolymph*** (***nucleoplasm***). One or more spherical bodies called the ***nucleoli*** may also be present. These structures do not contain a membrane and are composed of

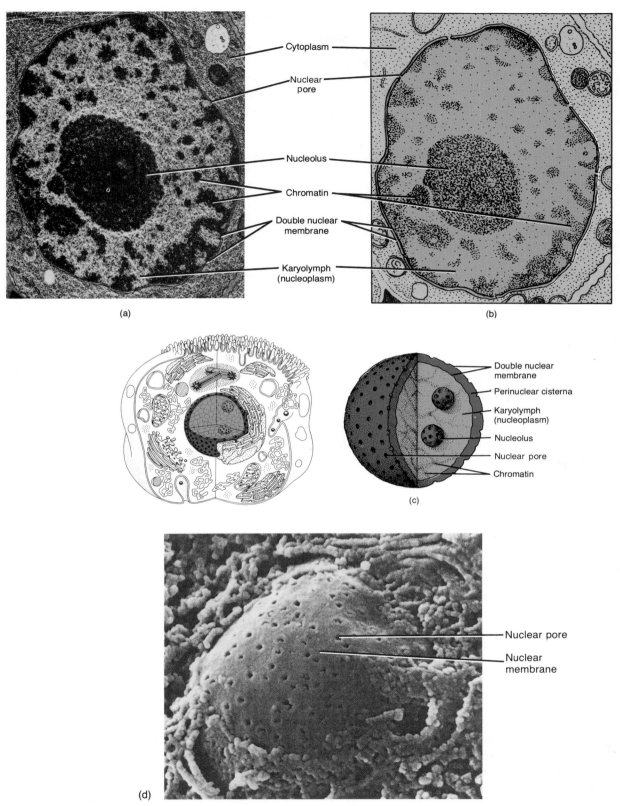

FIGURE 2-4 Cytoplasm and nucleus. (a) Electron micrograph of cytoplasm and the nucleus at a magnification of 31,600×. (Copyright © Dr. Myron C. Ledbetter, Biophoto Associates/ Photo Researchers.) (b) Diagram of the electron micrograph. (c) Diagram of a nucleus with two nucleoli. (d) Scanning electron micrograph at a magnification of 11,200×. (Courtesy of T. Fujita, *SEM Atlas of Cells and Tissues,* Igaku-Shoin Ltd.)

protein, DNA, and RNA. Nucleoli disperse and disappear during cell division and reform once new cells are formed. Nucleoli are the sites of the synthesis of a type of RNA called ribosomal RNA. The RNA is stored in nucleoli and, as you will see shortly, assumes a function in protein synthesis. Finally, there is the **genetic material** consisting principally of DNA. When the cell is not reproducing, the genetic material appears as a threadlike mass called **chromatin.** Prior to cellular reproduction the chromatin shortens and coils into rod-shaped bodies called **chromosomes.**

Chromosomes consist of DNA and proteins called **histones** organized in a very specific way. Elementary subunits of chromosomal structure are called **nucleosomes.** A nucleosome consists of different histone proteins associated with a relatively fixed length of DNA (about 200 nitrogenous base pairs). Present evidence indicates that the DNA is wrapped around the histones in some way. One type of histone maintains adjacent nucleosomes in a helical coil **(solenoid).** Nucleosomes are separated by stretches of DNA only so that the arrangement resembles beads on a string. The functional significance of nucleosome structure is still speculative. It is believed that histones may facilitate changes in chromosomal structures that expose activated genes (DNA) to perform a specific task in the cell.

RIBOSOMES

Ribosomes (RĪ-bo-sōms) are tiny granules, 25 nm at their largest dimension, that are composed of a type of RNA called ribosomal RNA (rRNA) and a number of specific ribosomal proteins. The rRNA is manufactured by DNA in the nucleolus. Ribosomes were so named because of their high content of rRNA. Structurally, a ribosome consists of two subunits, one about half the size of the other. Scientists have provided three-dimensional models of the structure of ribosomes (Figure 2-5e). Functionally, ribosomes are the sites of protein synthesis: they receive genetic instructions and use them to produce proteins. Amino acids are joined one at a time on ribosomes into a protein chain. The completed chain then folds into a protein molecule that can serve as part of the cell's structure or as an enzyme, for example. The mechanism of ribosomal function is discussed later in the chapter.

Some ribosomes, called **free ribosomes,** are scattered in the cytoplasm; they have no attachments to other parts of the cell. The free ribosomes occur singly or in clusters, and they are primarily concerned with synthesizing proteins for use inside the cell. Other ribosomes are attached to a cellular structure called the endoplasmic reticulum (ER). These ribosomes are concerned with the synthesis of proteins for export from the cell.

ENDOPLASMIC RETICULUM (ER)

Within the cytoplasm, there is a system of pairs of parallel membranes enclosing narrow channels, called **cisternae,** of varying shapes. This system is known as the **endoplasmic reticulum** (en'-dō-PLAS-mik re-TIK-yoo-lum) or **ER** (Figure 2-5a–d). The channels are continuous with the nuclear membrane.

On the basis of its association with ribosomes, ER is divided into two types. **Granular (rough) ER** is studded with ribosomes; **agranular (smooth) ER** is free of ribosomes. Agranular ER is synthesized from granular ER.

Numerous functions related to homeostasis are attributed to the ER. It contributes to the mechanical support and distribution of the cytoplasm. The ER is also involved in the intracellular exchange of materials with the cytoplasm and provides a surface area for chemical reactions. Various products are transported from one portion of the cell to another via the ER, so the ER is considered an intracellular transportation system. The ER also serves as a storage area for synthesized molecules. And, together with a cellular structure called the Golgi complex, the ER assumes a role in the synthesis and packaging of molecules.

GOLGI COMPLEX

A cytoplasmic structure called the **Golgi** (GOL-jē) **complex** is generally near the nucleus. In cells with high secretory activity, the Golgi complex is extensive. It consists of four to eight flattened membranous sacs, stacked upon each other like a pile of dishes with expanded areas at their ends. Like those of the ER, the stacked elements are called **cisternae** (Figure 2-6). On the basis of function, cisternae are designated as cis, medial, and trans (described shortly).

The principal function of the Golgi complex is to process, sort, and deliver proteins to various parts of the cell. Proteins synthesized at ribosomes associated with granular ER are transported into the granular ER (Figure 2-7). While still in the granular ER, sugar molecules are added to the proteins if needed (glycoproteins). Next, the proteins become surrounded by a vesicle formed by a piece of the granular ER membrane; the vesicle buds off and fuses with the cis cisterna of the Golgi complex. The cis cisternae are the ones closest to the agranular ER. As a result of this fusion, the proteins enter the Golgi complex. Once inside the Golgi complex, the proteins are transported by other vesicles formed by the Golgi complex from the cis cisternae to the medial cisternae to the trans cisternae, the ones farthest from the agranular ER. As the proteins pass in succession through the Golgi cisternae, they become modified in various ways depending on their function and destination. The proteins are sorted and packaged (in vesicles) in the trans cisternae. Some vesicles become **secretory granules,** which move toward the surface of the cell where the protein is released. The contents of the granules are discharged into the extracellular space, and the granule membrane is incorporated into the plasma membrane. Cells of the gastrointestinal tract that secrete protein enzymes utilize this mechanism. The secretory granule prevents "digestion" of the cytoplasm of the cells by the enzymes as it moves toward the cell surface. Other vesicles that pinch off from the Golgi complex are loaded with special digestive enzymes

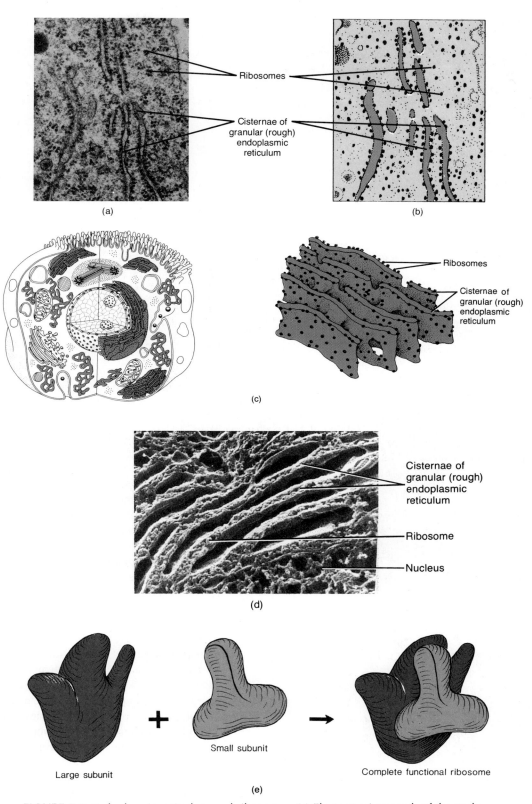

FIGURE 2-5 Endoplasmic reticulum and ribosomes. (a) Electron micrograph of the endoplasmic reticulum and ribosomes at a magnification of 76,000×. (Copyright © Dr. Myron C. Ledbetter, Biophoto Associates/Photo Researchers.) (b) Diagram of the electron micrograph. (c) Diagram of the endoplasmic reticulum and ribosomes. See if you can find the agranular (smooth) endoplasmic reticulum in Figure 3-9a. (d) Scanning electron micrograph at a magnification of 42,000×. (Courtesy of T. Fujita, *SEM Atlas of Cells and Tissues,* Igaku-Shoin Ltd.) (e) Diagram of the three-dimensional structure of ribosomes.

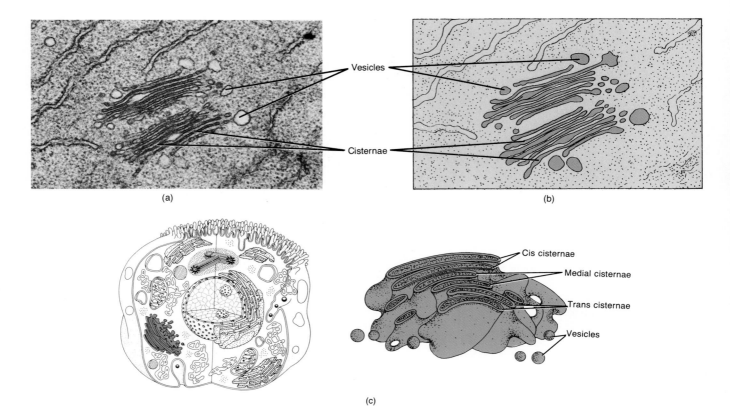

(a)

(b)

Vesicles

Cisternae

Cis cisternae

Medial cisternae

Trans cisternae

Vesicles

(c)

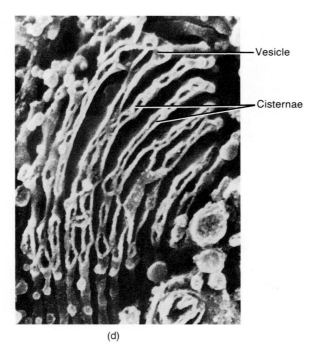

Vesicle

Cisternae

(d)

FIGURE 2-6 Golgi complex. (a) Electron micrograph of the two Golgi complexes at a magnification of 78,000×. (Copyright © Dr. Myron C. Ledbetter, Biphoto Associates/Photo Researchers.) (b) Diagram of the electron micrograph. (c) Diagram of a Golgi complex. (d) Scanning electron micrograph at a magnification of 55,200×. (Courtesy of T. Fujita, *SEM Atlas of Cells and Tissues*, Igaku-Shoin Ltd.)

and remain within the cell. They become cellular structures called lysosomes.

The Golgi complex is also associated with lipid secretion. Lipids synthesized by the agranular ER pass through the ER into the Golgi complex. Packaged lipids are discharged at the surface of the cell. In the course of moving through the cytoplasm, the vesicle may release lipids into the cytoplasm before being discharged from the cell. These appear in the cytoplasm as lipid droplets. Among the lipids secreted in this manner are steroids.

MITOCHONDRIA

Small, spherical, rod-shaped, or filamentous structures called **mitochondria** (mī'-tō-KON-drē-a) appear throughout the cytoplasm. Because of their function in generating energy, they are referred to as "powerhouses" of the cell. When sectioned and viewed under an electron microscope, each reveals an elaborate internal organization (Figure 2-8). A mitochondrion consists of two membranes, each of which is similar in structure to the plasma membrane. The outer mitochondrial membrane is smooth, but the inner membrane is arranged in a series of folds called **cristae**. The center of a mitochondrion is called the **matrix**.

Because of the nature and arrangement of the cristae, the inner membrane provides an enormous surface area for chemical reactions. Enzymes involved in energy-releasing reactions that form ATP are located on the cristae. Active cells, such as muscle, liver, and kidney tubule cells,

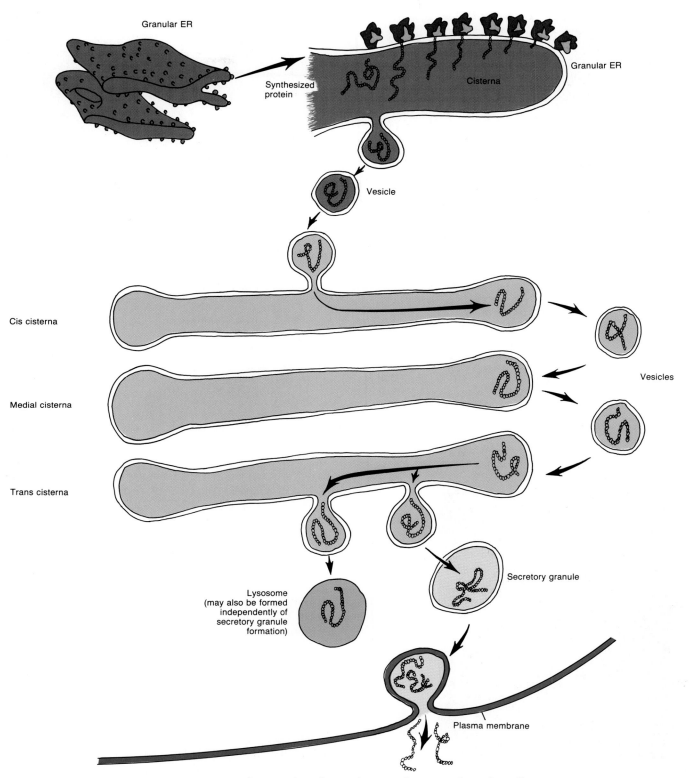

FIGURE 2-7 Packaging of synthesized protein for export from the cell.

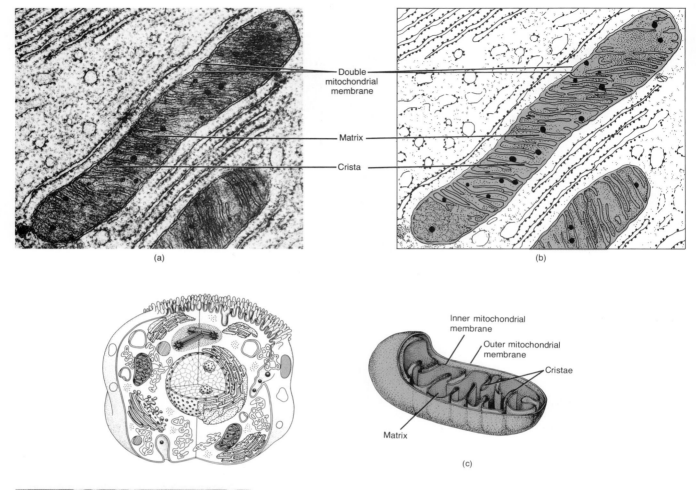

(a)

(b)

Double
mitochondrial
membrane

Matrix

Crista

Inner mitochondrial
membrane

Outer mitochondrial
membrane

Cristae

Matrix

(c)

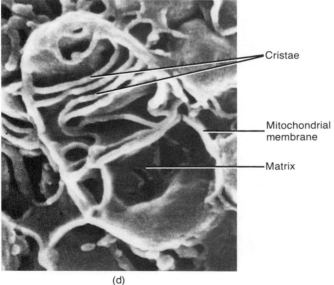

Cristae

Mitochondrial
membrane

Matrix

(d)

FIGURE 2-8 Mitochondria. (a) Electron micrograph of an entire mitochondrion (above) and a portion of another (below). (Courtesy of Lester V. Bergman & Associates, Inc.) (b) Diagram of the electron micrograph. (c) Diagram of a mitochondrion. (d) Scanning electron micrograph of an isolated mitochondrion at a magnification of 38,400×. (Courtesy of T. Fujita, *SEM Atlas of Cells and Tissues,* Igaku-Shoin Ltd.)

have a large number of mitochondria because of their high energy expenditure.

Mitochondria are self-replicative; that is, they can divide to form new ones. The replication process is controlled by DNA that is incorporated into the mitochondrial structure. Self-replication usually occurs in response to increased cellular need for ATP.

LYSOSOMES

When viewed under the electron microscope, *lysosomes* (LĪ-so-sōms; *lysis* = dissolution; *soma* = body) appear as membrane-enclosed spheres (Figure 2-9). They are formed from Golgi complexes and have a single membrane. They contain powerful digestive enzymes capable of breaking down many kinds of molecules. You will see in Chapter 18 that Tay-Sachs disease results from a deficiency of a lysosomal enzyme. These enzymes are also capable of digesting bacteria and other substances that enter the cell in phagocytic vesicles. White blood cells, which ingest bacteria by phagocytosis, contain large numbers of lysosomes.

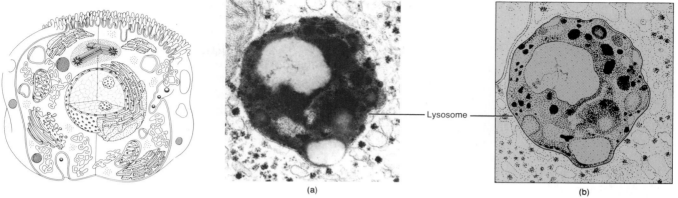

FIGURE 2-9 Lysosome. (a) Electron micrograph of a lysosome at a magnification of 55,000×. (Courtesy of F. Van Hoof, Université Catholique de Louvain.) (b) Diagram of the electron micrograph.

Lysosomal enzymes are believed to be synthesized on the granular ER and then transported to the Golgi complex, probably in an inactive form to prevent the unwanted digestion of structures through which they pass. Lysosomes develop as buds that pinch off the ends of Golgi cisternae. When first formed, the lysosome is referred to as a *primary lysosome,* meaning that it contains enzymes but is not yet engaged in digestive activity. In order to participate in digestion, a primary lysosome must fuse with a membrane-bound vacuole or other organelles containing food particles. A *vacuole* is a membrane-bound organelle that, in animal cells, frequently functions in temporary storage or transportation. As a result of this fusion, the primary lysosome becomes known as a *secondary lysosome,* that is, a lysosome engaged in digestive activity. In the digestive process, the contents of the vacuole are broken down into smaller and smaller components. Eventually, the products of digestion are small enough to pass out of the lysosome into the cytoplasm of the cell to be recycled in the synthesis of various molecules needed by the cell.

Lysosomes function in intracellular digestion in a number of ways. For example, during phagocytosis lysosomal enzymes digest the solid material contained in phagocytic vesicles. A similar process occurs in pinocytosis and receptor-mediated endocytosis. Lysosomes also use their enzymes to recycle the cells own molecules. A lysosome can engulf another organelle, digest it, and return the digested components to the cytoplasm for reuse. In this way, worn-out cellular structures are continually renewed. The process by which worn-out organelles are digested is called *autophagy* (aw-TOF-a-jē; *auto* = self; *phagio* = to eat). A human liver cell can recycle about half its contents in a week. Under normal conditions, the lysosomal membrane is impermeable to the passage of enzymes into cytoplasm. This prevents digestion of the cytoplasm. However, under certain conditions, *autolysis* or self-destruction by lysosomes occurs. For example, during human embryological development, the fingers and toes are webbed. The normal development of individual fingers and toes requires the selective removal of the webbed cells between the digits. This involves autolysis, in which lysosomal enzymes digest the tissue between the digits. It is actually a programmed destruction of cells. Because of this activity, lysosomes are sometimes referred to as "suicide packets."

Lysosomes also function in extracellular digestion. Prior to fertilization, the head of a sperm cell releases lysosomal enzymes capable of digesting a barrier around the egg so that the sperm cell can penetrate it. Once this is accomplished, fertilization can take place. Also, release of lysosomal enzymes from a cell may be the process responsible for bone removal. In bone reshaping, especially during the growth process, special bone-destroying cells, called osteoclasts, secrete extracellular enzymes that dissolve bone. Bone tissue cultures given excess amounts of vitamin A remove bone apparently through an activation process involving lysosomes.

CLINICAL APPLICATION

Animals overfed on vitamin A interestingly develop spontaneous fractures, suggesting greatly increased lysosomal activity. On the other hand, cortisone and hydrocortisone, steroid hormones produced by the adrenal gland, have a stabilizing effect on lysosomal membranes. The steroid hormones are well known for their anti-inflammatory properties, which suggests that they reduce destructive cellular activity by lysosomes.

PEROXISOMES

Organelles similar in structure to lysosomes, but smaller, are called *peroxisomes* (pe-ROKS-i-sōms). They are abundant in liver cells and contain several enzymes related to the metabolism of hydrogen peroxide (H_2O_2), a substance that is toxic to body cells. One of the enzymes in peroxisomes, called *catalase*, immediately breaks down H_2O_2 into water and oxygen:

$$H_2O_2 \xrightarrow{\text{catalase}} H_2O + O_2$$

Hydrogen   Water   Oxygen
peroxide

THE CYTOSKELETON

Cytoplasm has a complex internal structure, consisting of a series of exceedingly small microfilaments, microtubules, and intermediate filaments, together referred to as the ***cytoskeleton*** (see Figure 2-1).

Microfilaments are rodlike structures that are 6 nm in diameter. They are of variable length and may occur in bundles, randomly scattered throughout the cytoplasm, or arranged in a meshwork, depending on the type of cell in which they are found. Microfilaments consist of a protein called ***actin.*** In muscle tissue, actin microfilaments (thin myofilaments) and myosin microfilaments (thick myofilaments) are involved in the contraction of muscle fibers (cells). This mechanism is described in Chapter 9. In nonmuscle cells, microfilaments help provide support and shape and assist in the movement of entire cells (phagocytes and cells of developing embryos) and movements within cells (secretion, phagocytosis, pinocytosis).

Microtubules are relatively straight, slender, cylindrical structures that range in diameter from 18 to 30 nm, usually averaging about 24 nm. They consist of a protein called ***tubulin.*** Microtubules dispersed in the cytoplasm, together with microfilaments, help provide support and shape for cells. Microtubules may also form conducting channels through which various substances can move throughout the cytoplasm. This mechanism has been studied extensively in nerve cells. Microtubules also assist in the movement of pseudopodia that are characteristic of phagocytes. As you will see shortly, microtubules form the structure of flagella and cilia (cellular appendages involved in motility), centrioles (organelles that may direct the assembly of microtubules), and the mitotic spindle (structures that are involved in cell division).

Intermediate filaments range between 8 and 12 nm in diameter. There are at least five classes of intermediate filaments, each based on their protein composition. Examples include myosin myofilaments in muscle fibers (Chapter 9) and tonofilaments of epithelial cells (Chapter 3). Intermediate filaments are usually dispersed throughout cytoplasm. In some cells, they are individual filaments; in others, they are bundled together. Although the functions of intermediate filaments are not completely understood, it appears that they provide structural reinforcement in some cells and assist in contraction of others.

Some investigators believe that the microfilaments, microtubules, and intermediate filaments, as well as other cytoplasmic components, are held together by a three-dimensional meshwork of fine filaments called ***microtrabeculae*** (mī'-krō-tra-BEK-yoo-lē) that are about 10–15 nm thick.

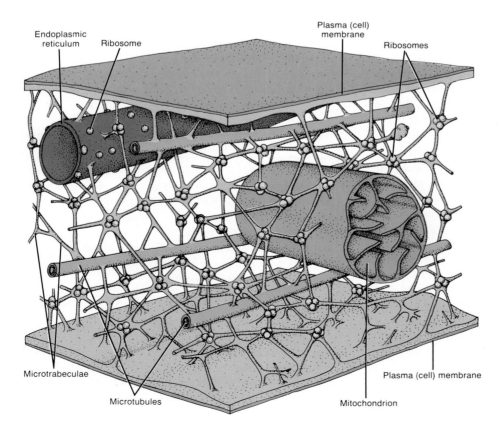

Endoplasmic reticulum   Ribosome   Plasma (cell) membrane   Ribosomes

Microtrabeculae   Microtubules   Mitochondrion   Plasma (cell) membrane

FIGURE 2-10 Microtrabecular lattice.

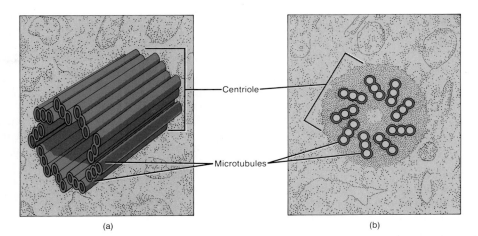

FIGURE 2-11 Centrosome and centrioles. (a) Diagram of a centriole in longitudinal view. (b) Diagram of a centriole in cross section. (c) Electron micrograph of a centriole in cross section at a magnification of 67,000×. (Courtesy of Biophoto Associates/Photo Researchers.)

Together, all the microtrabeculae constitute the **microtrabecular lattice** (Figure 2-10). This lattice is believed to provide organization for chemical reactions that occur within the cytoplasm and to assist in the transport of substances through the cytoplasm.

CENTROSOME AND CENTRIOLES

A dense area of cytoplasm, generally spherical and located near the nucleus, is called the **centrosome (centrosphere).** Within the centrosome is a pair of cylindrical structures: the **centrioles** (Figure 2-11). Each centriole is composed of nine triplet clusters of microtubules arranged in a circular pattern. Centrioles lack the two central single microtubules found in flagella and cilia. The two centrioles are situated so that the long axis of one is at right angles to the long axis of the other. Centrioles assume a role in cell reproduction by serving as centers about which microtubules involved in chromosome movement are organized. This role will be described shortly as part of cell division. Certain cells, such as most mature nerve cells, do not have a centrosome and so do not reproduce. This is why they cannot be replaced if they are destroyed. Like mitochondria, centrioles contain DNA that controls their self-replication.

FLAGELLA AND CILIA

Some body cells possess projections for moving the entire cell or for moving substances along the surface of the cell. These projections contain cytoplasm and are bounded by the plasma membrane. If the projections are few (typically occurring singly or in pairs) and long in proportion to the size of the cell, they are called **flagella** (fla-JEL-a). The only example of a flagellum in the human body is the tail of a sperm cell, used for locomotion (see Figure 25-5b). If the projections are numerous and short, resembling many hairs, they are called **cilia** (SIL-ē-a). In humans, ciliated cells of the respiratory tract move mucus that has trapped foreign particles over the surface of the tissue (see Figure 22-5b). Electron microscopy has revealed no fundamental

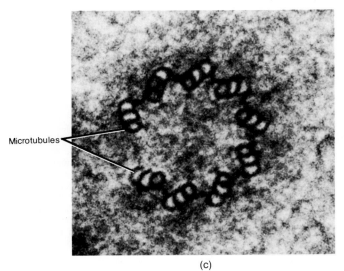

(c)

structural difference between cilia and flagella. Both consist of nine pairs of microtubules that form a ring around two single microtubules in the center.

CELL INCLUSIONS

Cell inclusions are a large and diverse group of chemical substances produced by cells, some of which have recognizable shapes. These products are principally organic and may appear or disappear at various times in the life of the cell. Examples include melanin, glycogen, and lipids. **Melanin** is a pigment stored in certain cells of the skin, hair, and eyes. It protects the body by screening out harmful ultraviolet rays from the sun. **Glycogen** is a polysaccharide that is stored in the liver, skeletal muscle fibers (cells), and the vaginal mucosa. When the body requires quick energy, liver cells can break down the glycogen into glucose and release it. **Lipids,** which are stored in adipocytes (fat cells), may be decomposed for producing energy.

The major parts of the cell and their functions are summarized in Exhibit 2-1.

EXHIBIT 2-1

Cell Parts and Their Functions

PART	FUNCTIONS
PLASMA MEMBRANE	Protects cellular contents; makes contact with other cells; provides receptors for hormones, enzymes, and antibodies; mediates the entrance and exit of materials.
CYTOPLASM	Serves as the ground substance in which chemical reactions occur.
ORGANELLES	
Nucleus	Contains genes and controls cellular activities.
Ribosomes	Sites of protein synthesis.
Endoplasmic Reticulum (ER)	Contributes to mechanical support; conducts intracellular nerve impulses in muscle fibers (cells); facilitates intracellular exchange of materials with cytoplasm; provides a surface area for chemical reactions; provides a pathway for transporting chemicals; serves as a storage area; together with Golgi complex synthesizes and packages molecules for export.
Golgi Complex	Packages synthesized proteins for secretion in conjunction with endoplasmic reticulum; forms lysosomes; secretes lipids; synthesizes carbohydrates; combines carbohydrates with proteins to form glycoproteins for secretion.
Mitochondria	Sites of production of ATP.
Lysosomes	Digest substances and foreign microbes; may be involved in bone removal.
Peroxisomes	Contain several enzymes, such as catalase, related to hydrogen peroxide metabolism.
Microfilaments	Form part of cytoskeleton; involved in muscle fiber (cell) contraction; provide support and shape; assist in cellular and intracellular movement.
Microtubules	Form part of cytoskeleton; provide support and shape; form intracellular conducting channels; assist in cellular movement; form the structure of flagella, cilia, centrioles, and spindle fibers.
Intermediate Filaments	Form part of cytoskeleton; probably provide structural reinforcement in some cells.
Centrioles	Help organize mitotic spindle during cell division.
Flagella and Cilia	Allow movement of entire cell (flagella) or movement of particles along surface of cell (cilia).
INCLUSIONS	Melanin (pigment in skin, hair, eyes) screens out ultraviolet rays; glycogen (stored glucose) can be decomposed to provide energy; lipids (stored in fat cells) can be decomposed to produce energy.

EXTRACELLULAR MATERIALS

The substances that lie outside cells are called *extracellular materials*. They include body fluids, which provide a medium for dissolving, mixing, and transporting substances. Among the body fluids are interstitial fluid, the fluid that fills the microscopic spaces (interstitial spaces), and plasma, the liquid portion of blood. Extracellular materials also include secreted inclusions like mucus and special substances that form the matrix (substance between cells) in which some cells are embedded.

The matrix materials are produced by certain cells and deposited outside their plasma membranes. The matrix supports cells, binds them together, and gives strength and elasticity to the tissue. Some matrix materials are *amorphous* (*a* = without, *morpho* = shape); they have no specific shape. These include hyaluronic acid, chondroitin sulfate, dermatan sulfate, and keratan sulfate. *Hyaluronic* (hī'-a-loo-RON-ik) *acid* is a viscous, fluidlike substance that binds cells together, lubricates joints, and maintains the shape of the eyeballs. *Chondroitin* (kon-DROY-tin) *sulfate* is a jellylike substance that provides support and adhesiveness in cartilage, bone, and blood vessels. *Dermatan sulfate* is found in the skin, tendons, and heart valves, and *keratan sulfate* is found in the cornea of the eye and bone.

Other matrix materials are *fibrous,* or threadlike. Fibrous materials provide strength and support for tissues. Among these are *collagenous* (*kolla* = glue) *fibers* consisting of the protein *collagen*. These fibers are found in all types of connective tissue, especially in bones, cartilage, tendons, and ligaments. *Reticular* (*rete* = net) *fibers* consisting of the protein collagen and a coating of glycoprotein, form a network around fat cells, nerve fibers, muscle fibers (cells), and blood vessels. They also are found in close association with the basal lamina of most epithelia and form the framework or stroma for many soft organs of the body such as the spleen. *Elastic fibers,* consisting of the protein *elastin*, give elasticity to skin and to tissues forming the walls of blood vessels.

NORMAL CELL DIVISION

Most of the cell activities mentioned thus far maintain the life of the cell on a day-to-day basis. However, cells become damaged, diseased, or worn out and then die. New cells must be produced as replacements and for growth. In addition, sperm and egg cells must be produced by cell division.

Cell division is the process by which cells reproduce themselves. It consists of a nuclear division and a cytoplasmic division. Because nuclear division can be of two types, two kinds of cell division are recognized. In the first kind of division, often called *somatic cell division,* a single starting cell called a *parent cell* duplicates itself and the result is two identical cells called *daughter cells*. This process consists of a nuclear division called *mitosis* and a cytoplasmic division called *cytokinesis.* The process ensures that each daughter cell has the same *number* and *kind* of chromosomes as the original parent cell. After the process is complete, the two daughter cells have the same hereditary material and genetic potential as the parent cell. This kind of cell division results in an increase in the number of body cells. In a 24-hour period, the average adult loses billions of cells from different parts of the body. Obviously, these cells must be replaced. Cells that have a short life span—the cells of the outer layer of skin, the cornea of the eye, the gastrointestinal tract—are continually being replaced. Mitosis and cytokinesis are the means by which dead or injured cells are replaced and new cells are added for body growth.

The second type of cell division is called *reproductive cell division* and is the mechanism by which sperm and egg cells are produced, cells required to form a new organism. The process consists of a nuclear division called *meiosis* plus *cytokinesis.* We shall first discuss somatic cell division.

SOMATIC CELL DIVISION

When a cell reproduces, it must replicate (produce duplicates of) its chromosomes so that its hereditary traits may be passed on to succeeding generations of cells. A *chromosome* is a highly coiled DNA molecule that is partly covered by protein. The protein causes changes in the length and thickness of the chromosome. Hereditary information is contained in the DNA portion of the chromosome in units called *genes.* Humans have about 100,000 functional genes.

Before taking a look at the relation between chromosomes and cell division, it is necessary to examine briefly the structure of DNA, the basic component of chromosomes.

A molecule of DNA is a chain composed of repeating units called *nucleotides.* Each nucleotide of DNA consists of three basic parts (Figure 2-12a): (1) It contains one of four possible *nitrogen bases,* which are ring-shaped structures containing atoms of C, H, O, and N. The nitrogen bases found in DNA are named adenine, thymine, cytosine, and guanine. (2) It contains a sugar called *deoxyribose.* (3) It has a phosphoric acid called the *phosphate group.* The nucleotides are named according to the nitrogen base that is present. Thus, a nucleotide containing thymine is called a *thymine nucleotide,* one containing adenine is called an *adenine nucleotide,* and so on.

The chemical composition of the DNA molecule was known before 1900, but it was not until 1953 that a model of the organization of the chemicals was constructed. This model was proposed by J. D. Watson and F. H. C. Crick on the basis of data from many investigations. Figure 2-12b shows the following structural characteristics of the DNA molecule: (1) The molecule consists of two strands with crossbars. The strands twist about each other in the form of a *double helix* so that the shape resembles a twisted ladder. (2) The uprights of the DNA ladder consist of alternating phosphate groups and the deoxyribose portions of the nucleotides. (3) The rungs of the ladder contain paired nitrogen bases. As shown, adenine always pairs off with thymine, and cytosine always pairs off with guanine. It is estimated that the nucleus of a human diploid cell contains about three billion nitrogen base pairs.

When a cell is between divisions it is said to be in *interphase (metabolic phase).* It is during this stage that the replication (synthesis) of chromosomes occurs and the RNA and protein needed to produce structures required for doubling all cellular components are manufactured.

The period of interphase during which chromosomes are replicated is referred to as the S (for synthesis) *period.* When DNA replicates, its helical structure partially uncoils. Those portions of DNA that remain coiled stain darker than the uncoiled portions. This unequal distribution of stain causes the DNA to appear as a granular mass called *chromatin* (see Figure 2-13a). During uncoiling, DNA separates at the points where the nitrogen bases are connected. Each exposed nitrogen base then picks up a complementary nitrogen base (with associated sugar and phosphate group) from the cytoplasm of the cell. This uncoiling and complementary base pairing continues until each of the two original DNA strands is matched and joined with two newly formed DNA strands. The original DNA molecule has become two DNA molecules.

The S period is preceded by a G_1 (for gap or growth) *period,* during which cells are engaged in growth, metabolism, and the production of substances required for division. Following chromosomal replication in the S period, there is another G period called the G_2 *period.* During this gap period there is activity similar to that occurring in the G_1 period. Since the G periods are times during which there are no events related to chromosomal replication, they are thought of as gaps in the divisional cycle (see Figure 2-4).

A microscopic view of a cell during interphase shows a clearly defined nuclear membrane, nucleoli, karyolymph, chromatin, and a pair of centrioles. Once a cell completes its replication of DNA and its production of RNA and proteins during interphase, mitosis begins.

Mitosis

The events that take place during mitosis and cytokinesis are plainly visible under a microscope after the cells have been stained in the laboratory.

The process called *mitosis* is the distribution of the two sets of chromosomes into two separate and equal nuclei following the replication of the chromosomes of the parent nucleus. For convenience, biologists divide the process into four stages: prophase, metaphase, anaphase, and telophase. These are arbitrary classifications. Mitosis is actually a continuous process, one stage merging imperceptibly into the next.

● *Prophase* During *prophase* (Figure 2-13b), the first stage of mitosis, chromatin shortens and coils into chromo-

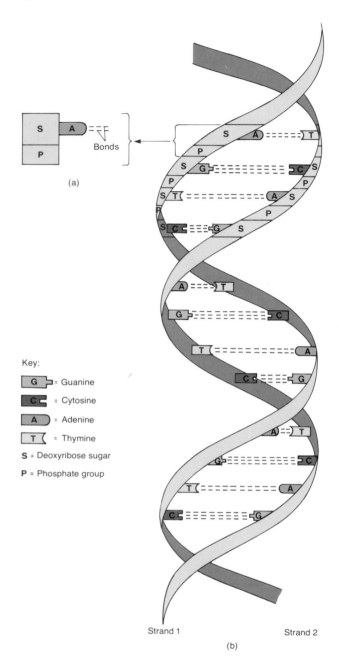

Key:

G = Guanine

C = Cytosine

A = Adenine

T = Thymine

S = Deoxyribose sugar

P = Phosphate group

(a)

Bonds

Strand 1   Strand 2

(b)

(c)

FIGURE 2-12 DNA molecule. (a) Adenine nucleotide. (b) Portion of an assembled DNA molecule. (c) Space-filling (three-dimensional) model to show relative size and location of atoms. (Courtesy of Ealing Corp.)

somes. The nucleoli become less distinct and the nuclear membrane disappears. Each prophase "chromosome" is actually composed of a pair of structures called *chromatids*. A chromatid is a complete chromosome consisting of a double-stranded DNA molecule and each is attached to its chromatid pair by a small spherical body called a *centromere*. During prophase, the chromatid pairs assemble near the center of the cell in a region called the *equatorial plane region (equator)* of the cell.

Also during prophase, the paired centrioles separate and each pair moves to an opposite pole (end) of the cell. Between the centrioles, a series of microtubules are organized into two groups of fibers. The *continuous (interpolar) microtubules* originate from the vicinity of each pair of cen-

trioles and grow toward each other. Thus, they extend from one pole of the cell to another. As they grow toward each other, the second group of microtubules develops. These are called *chromosomal microtubules* and apparently grow out of the centromeres; they extend from a centromere to a pole of the cell. Together, the continuous and chromosomal microtubules constitute the *mitotic spindle* and, with the centrioles, are referred to as the *mitotic apparatus*.

● *Metaphase* During *metaphase* (Figure 2-13c), the second stage of mitosis, the centromeres of the chromatid pairs line up on the equatorial plane of the cell. The centromeres of each chromatid pair form a chromosomal microtubule that attaches the centromere to a pole of the cell.

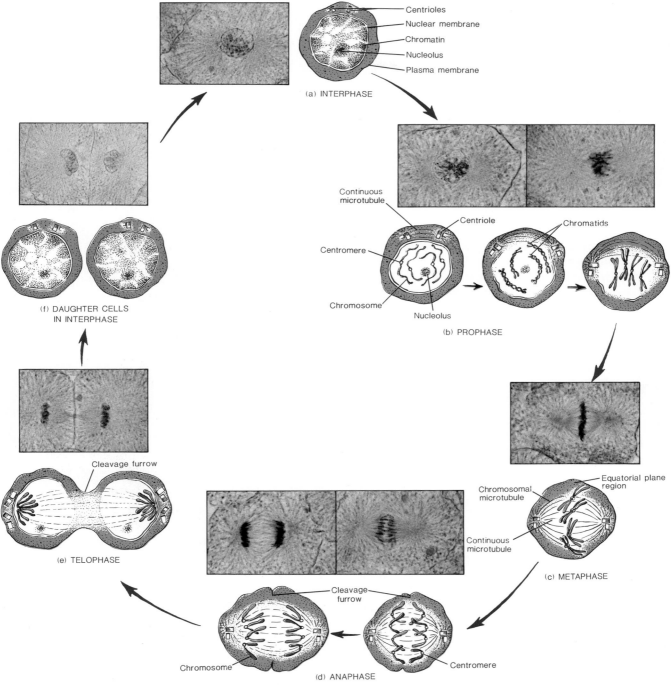

FIGURE 2-13 Cell division: mitosis and cytokinesis. Photomicrographs and diagrammatic representations of the various stages of cell division in whitefish eggs. Read the sequence starting at (a), and move clockwise until you complete the cycle. (Photographs by Carolina Biological Supply Company.)

● **Anaphase** The third stage of mitosis, *anaphase* (Figure 2-13d), is characterized by the division of the centromeres and the movement of complete identical sets of chromatids, now called chromosomes, to opposite poles of the cell. During this movement, the centromeres attached to the chromosomal microtubules seem to drag the trailing parts of the chromosomes toward opposite poles. Although several theories have been proposed, the mechanism by which the chromosomes move to opposite poles is not completely understood.

● **Telophase** Telophase (Figure 2-13e), the final stage of mitosis, consists of a series of events nearly the reverse of prophase. By this time, two identical sets of chromosomes have reached opposite poles. As telophase progresses, new nuclear membranes begin to enclose the chromosomes, the chromosomes start to assume their chromatin form, nucleoli reappear, and the mitotic spindles disappear. The centrioles also replicate so that each cell has two centriole pairs. The formation of two nuclei identical to those of cells in interphase terminates telophase. A mitotic cycle has thus been completed (Figure 2-13f).

● **Time Required**   The time required for mitosis varies with the kind of cell, its location, and the influence of factors such as temperature. Furthermore, the different stages of mitosis are not equal in duration. However, in order to give you some idea of the length of a cell cycle, mammalian cells in culture have been studied and often have the following time intervals: The G_1 period is highly variable, ranging from almost nonexistent in rapidly dividing cells to days, weeks, or years. However, it typically takes about 8–10 hours. The S period takes about 6–8 hours,

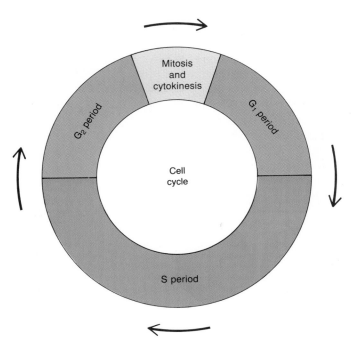

FIGURE 2-14 Various phases and periods in a cell cycle. Relative amounts of time are indicated by the size of the compartments.

the G_2 period about 4–6 hours, and mitosis and cytokinesis about 30–45 minutes. Within the mitosis and cytokinesis time interval, prophase takes longest and anaphase is shortest. As you can see, mitosis and cytokinesis represent only a small part of the life cycle of a cell. Together, the various phases of the cell cycle require about 18–24 hours in many cultured mammalian cells.

Cytokinesis

Division of the cytoplasm, a process called *cytokinesis* (sī'-tō-ki-NĒ-sis), often begins during late anaphase and terminates at the same time as telophase. Cytokinesis begins with the formation of a *cleavage furrow* that extends around the cell's equatorial plane region. The furrow progresses inward, resembling a constricting ring, and cuts completely through the cell to form two separate portions of cytoplasm (Figure 2-13d–f).

If we consider the cell cycle in its entirety, the sequence of events can be summarized as follows: G_1 period → S period → G_2 period → mitosis → cytokinesis (Figure 2-14).

A summary of the events that occur during the interphase and cell division is presented in Exhibit 2-2.

REPRODUCTIVE CELL DIVISION

In sexual reproduction, each new organism is produced by the union and fusion of two different sex cells, one produced by each parent. The sex cells, called *gametes,* are the ovum produced in the female gonads (ovaries) and the sperm produced in the male gonads (testes). The union

EXHIBIT 2-2

Summary of Events Associated with Interphase and Cell Division

PERIOD OR STAGE	ACTIVITY
INTERPHASE	Cell is between divisions.
G_1 Period	Cell engages in growth, metabolism, and production of substances required for division; no chromosomal replication.
S Period	Chromosomal replication occurs.
G_2 Period	Same as for G_1 period.
CELL DIVISION	Single parent cell produces two identical daughter cells.
Prophase	Chromatin shortens and coils into chromosomes (chromatids), nucleoli and nuclear membrane become less distinct, centrioles separate and move to opposite poles of cell, and mitotic spindle forms.
Metaphase	Centromeres of chromatid pairs line up on equatorial plane of cell and form chromosomal microtubules that attach centromeres to poles of cell.
Anaphase	Centromeres divide and identical sets of chromosomes move to opposite poles of cells.
Telophase	Nuclear membrane reappears and encloses chromosomes, chromosomes resume chromatin form, nucleoli reappear, mitotic spindle disappears, and centrioles duplicate.
Cytokinesis	Cleavage furrow forms around equatorial plane of cell, progresses inward, and separates cytoplasm into two separate and equal portions.

and fusion of gametes is called *fertilization* and the cell thus produced is known as a *zygote*. The zygote contains a mixture of chromosomes (DNA) from the two parents and, through its repeated mitotic division, develops into a new organism.

Gametes differ from all other body cells (somatic cells) with respect to the number of chromosomes in their nuclei. Somatic cells, such as brain cells, stomach cells, kidney cells, and all other uninucleated somatic cells, contain 46 chromosomes in their nuclei. Some somatic cells, such as skeletal muscle fibers (cells), are multinucleated and thus contain more than 46 chromosomes. However, since most somatic cells are uninucleated, these are the cells to which we shall refer in the following discussion. Of the 46 chromosomes, 23 are a complete set that contain one copy of all the genes necessary for carrying out the activities of the cell. In a sense, the other 23 chromosomes are a duplicate set. The symbol *n* is used to designate the number of different chromosomes within the nucleus. Since somatic cells contain two sets of chromosomes, they are referred to as *diploid* (DIP-loyd; *di* = two) *cells,* symbolized as *2n*. In a diploid cell, two chromosomes that belong to a pair are called *homologous* (hō-MOL-ō-gus) *chromosomes,* or *homologues*. In human diploid cells, 22 of the 23 pairs of chromosomes are morphologically similar and are called *autosomes*. The other pair is called *sex chromosomes,* designated as X and Y. In females, the homologous pair of sex chromosomes consists of two X chromosomes; in males, the pair consists of an X and a Y chromosome.

If gametes had the same number of chromosomes as somatic cells, the zygote formed from their fusion would have double the number. The somatic cells of the resulting individual would have twice the number of chromosomes (*4n*) as the somatic cells of the parents, and with every succeeding generation, the number of chromosomes would double. The chromosome number does not double with each generation because of a special nuclear division called *meiosis*. Meiosis occurs only in the development of gametes and it results in the production of cells that contain only 23 chromosomes. Thus, gametes are *haploid* (HAP-loyd) *cells,* meaning "one-half," and are symbolized as *n*.

Meiosis

The formation of haploid sperm cells in the testes of the male consists of several phases and is called spermatogenesis. One of the phases involves meiosis. The formation of haploid ova (eggs) in the ovaries of the female also involves several phases and is referred to as oogenesis. It too involves meiosis. Both spermatogenesis and oogenesis are discussed in detail in Chapter 25. At this point, we shall examine only the essentials of meiosis.

Meiosis occurs in two successive nuclear divisions referred to as *reduction division (meiosis I)* and *equatorial division (meiosis II)*. During the interphase that precedes reduction division of meiosis, the chromosomes replicate themselves. During the S period, this replication is similar to that in interphase preceding the mitosis of somatic cell division. Once chromosomal replication is complete, reduction division begins. It consists of four phases referred to as prophase I, metaphase I, anaphase I, and telophase I (Figure 2-15).

Prophase I is an extended phase in which the chromosomes shorten and thicken, the nuclear membrane and nucleoli disappear, the centrioles replicate, and the mitotic spindle appears. Unlike the prophase of mitosis, however, a unique event occurs in prophase I of meiosis. The chromosomes line up along the equatorial plane region in homologous pairs. The pairing is called *synapsis.* The four chromatids of each homologous pair are referred to as a *tetrad.* Another unique event of meiosis occurs within a tetrad. Portions of one chromatid may be exchanged with portions of another, a process called *crossing-over* (Figure 2-16). This process, among others, permits an exchange of genes among chromatids so that subsequent daughter cells produced are unlike each other genetically and unlike the parent cell that produced them. This phenomenon accounts for part of the great genetic variation among humans and other organisms that form gametes by meiosis. In metaphase I, the paired chromosomes line up along the equatorial plane of the cell, with one member of each pair on either side. Recall that there is no pairing of homologous chromosomes during the metaphase of mitosis. The centromeres of each chromatid pair form chromosomal microtubules that attach the centromeres to opposite poles of the cell. Anaphase I is characterized by separation of the members of each homologous pair, with one member of each pair moving to an opposite pole of the cell. During anaphase I, unlike mitotic anaphase, the centromeres do not split and the paired chromatids, held by a centromere, remain together. Telophase I and cytokinesis are similar to telophase and cytokinesis of mitosis. The net effect of reduction division is that each resulting daughter cell contains the haploid number of chromosomes; each cell contains only one member of each pair of the original homologous chromosomes in the starting parent cell.

The interphase between reduction division and equatorial division is either brief or lacking altogether. It does differ from the interphase preceding reduction division in that there is no replication of DNA between the reduction and equatorial divisions.

The equatorial division of meiosis consists of four phases referred to as prophase II, metaphase II, anaphase II, and telophase II. These phases are essentially similar to those that occur during mitosis since the centromeres divide and chromatids separate and move toward opposite poles of the cell.

In reviewing the overall process, not that during reduction division we start with a parent cell with the diploid number and end up with two daughter cells, each with haploid number. During equatorial division, each haploid cell formed during reduction division divides and the net result

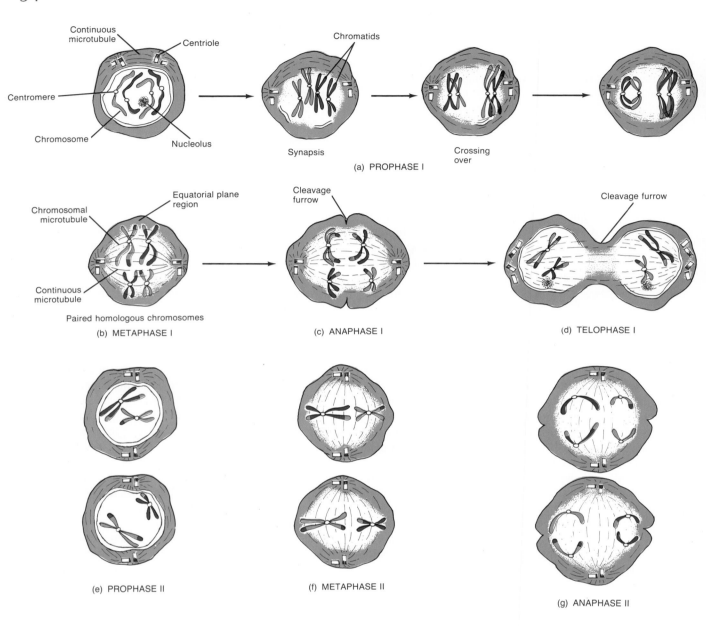

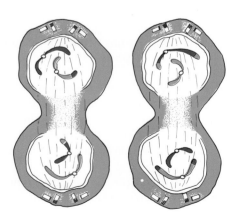

FIGURE 2-15 Meiosis. See text for details.

is four haploid cells. As you will see later, all four haploid cells develop into sperm cells in the testes of the male, but only one of the haploid cells has the potential to develop into an ovum in the female. The other three become structures called polar bodies that do not function as gametes.

A very simplified comparison of mitosis and meiosis is illustrated in Figure 2-17.

ABNORMAL CELL DIVISION: CANCER (CA)

DEFINITION

By the time adolescence is completed, cells of the body have reached an equilibrium in which the rate of new cell production equals the number of cells that die. In most tissues, only a very small percentage of the cells retain

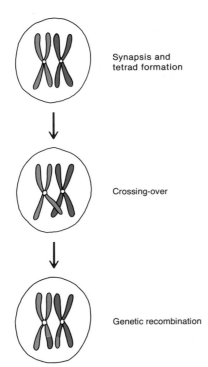

FIGURE 2-16 Crossing-over within a tetrad resulting in genetic recombination.

the ability to undergo an indefinite sustained series of mitotic divisions. Also, as a general rule, after cells have undergone differentiation, their ability to reproduce is greatly reduced and ultimately stops. Many such cells are locked into the G_1 period of the cell cycle and do not divide at all. After the first few years of life, all nerve cells of the brain and spinal cord and skeletal muscle and cardiac muscle fibers (cells) lose their reproductive capability. Any cell capable

of mitotic division can undergo transformation into a cancer cell. Thus, cancers can only arise from those tissues that have active cells capable of division or cells that can revert to an active mitotic state, such as liver cells (hepatocytes) or small lymphocytes (one type of white blood cell).

When cells in some area of the body duplicate without control, the excess of tissue that develops is called a ***tumor, growth,*** or ***neoplasm.*** The study of tumors is called ***oncology*** (*onco* = swelling or mass; *logos* = study of) and a physician who specializes in this field is called an ***oncologist.*** Tumors may be cancerous and sometimes fatal or they may be quite harmless. A cancerous growth is called a ***malignant tumor,*** or ***malignancy.*** A noncancerous growth is called a ***benign growth.*** Benign tumors are composed of cells that do not spread to other parts of the body, but they may be removed if they interfere with a normal body function or are disfiguring.

SPREAD

Cells of malignant growths duplicate continuously and very often quickly and without control. The majority of cancer patients are not killed by the ***primary tumor*** that develops but by secondary infections of bacteria and viruses due to lowered resistance as a result of ***metastasis*** (me-TAS-ta-sis), the spread of the disease to other parts of the body. One of the unique properties of a malignant tumor is its ability to metastasize. Tumor cells secrete a protein called ***autocrine motility factor*** (***AMF***) that enables them to metastasize. Metastatic groups of cells are more difficult to detect and eliminate than primary tumors.

In the process of metastasis, there is an initial invasion of the malignant cells into surrounding tissues. As the cancer grows, it expands and begins to compete with normal tissues

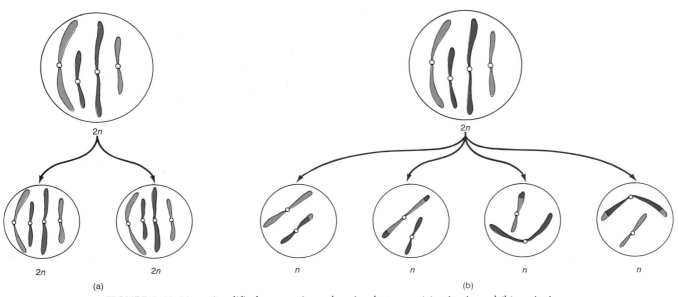

FIGURE 2-17 Very simplified comparison showing between (a) mitosis and (b) meiosis.

for space and nutrients. Eventually, the normal tissue atrophies and dies. The invasiveness of the malignant cells may be related to mechanical pressure of the growing tumor, motility of the malignant cells, and enzymes produced by the malignant cells. Also, malignant cells lack what is called **contact inhibition.** When nonmalignant cells of the body divide and migrate (e.g., skin cells that multiply to heal a superficial cut), their further migration is inhibited by contact on all sides with other skin cells. Unfortunately, malignant cells do not conform to the rules of contact inhibition; they have the ability to invade healthy body tissues with very few restrictions.

Following invasion, some of the malignant cells may detach from the primary tumor and invade a body cavity (abdominal or thoracic) or enter the blood or lymph. This latter condition can lead to widespread metastasis. In the next step in metastasis, those malignant cells that survive in the blood or lymph invade adjacent body tissues and establish **secondary tumors.** It is believed that some of the invading cells involved in metastasis have properties different from those of the primary tumor that enhance metastasis. These include appropriate mechanical, enzymatic, and surface properties. In the final stage of metastasis, the secondary tumors become vascularized; that is, they take on new networks of blood vessels that provide nutrients for their further growth. Any new tissue, whether it results from repair, normal growth, or tumors, requires a blood supply. A protein that serves as a chemical trigger for blood vessel growth is called **angiogenin,** which has been isolated from human colon-tumors. In all stages of metastasis, the malignant cells resist the antitumor defenses of the body. The pain associated with cancer develops when the growth puts pressure on nerves or blocks a passageway so that secretions build up pressure.

TYPES

At present, cancers are classified by their microscopic appearance and the body site from which they arise. At least 100 different cancers have been identified in this way. If finer details of appearance are taken into consideration, the number can be increased to 200 or more. The name of the cancer is derived from the type of tissue in which it develops. **Carcinoma** (*carc* = cancer; *oma* = tumor) refers to a malignant tumor consisting of epithelial cells. A tumor that develops from a gland is called an **adenosarcoma** (*adeno* =gland). **Sarcoma** is a general term for any cancer arising from connective tissue. **Osteogenic sarcomas** (*osteo* = bone; *genic* = origin), the most frequent type of childhood cancer, destroy normal bone tissue and eventually spread to other areas of the body. **Myelomas** (*myelos* = marrow) are malignant tumors, usually occurring in middle-aged and older people, that interfere with the blood-cell-producing function of bone marrow and cause anemia. **Chondrosarcomas** (*chondro* = cartilage) are cancerous growths of cartilage.

POSSIBLE CAUSES

What triggers a perfectly normal cell to lose control and become abnormal? Scientists are uncertain. First, there are environmental agents: substances in the air we breathe, the water we drink, the food we eat. A chemical or other environmental agent that produces cancer is called a **carcinogen.** The World Health Organization estimates that carcinogens may be associated with 60–90 percent of all human cancer. Examples of carcinogens are the hydrocarbons found in cigarette tar. Ninety percent of all lung cancer patients are smokers. Another environmental factor is radiation. Ultraviolet (UV) light from the sun, for example, may cause genetic mutations in exposed skin cells and lead to cancer, especially among light-skinned people.

Viruses are a second cause of cancer, at least in animals. These agents are tiny packages of nucleic acids, either DNA or RNA, that are capable of infecting cells and converting them to virus producers. With over 100 separate viruses identified as carcinogens in many species and tissues of animals, it is also probable that at least some cancers in humans are due to viruses. For example, *human T-cell leukemia-lymphoma virus-1* (*HTLV*-1) is strongly associated with *leukemia,* a malignant disease of blood-forming tissues, and *lymphoma,* a cancer of lymphoid tissue. A variant of HTLV-1, known as *human immunodeficiency virus* (HIV), is the causative agent of acquired immune deficiency syndrome (AIDS). The *Epstein-Barr virus* (*EBV*), the causative agent of infectious mononucleosis, has been linked as the causative agent of several human cancers—*Burkitt's lymphoma* (a cancer of white blood cells called B cells), *nasopharyngeal carcinoma* (common in Chinese males), and *Hodgkin's disease* (a cancer of the lymphatic system). The *hepatitis B virus* (*HBV*) has been associated with cancer of the liver. Also, *type 2 herpes simplex virus,* the causative agent of genital herpes, has been implicated in cancer of the cervix of the uterus, and the *papilloma virus,* a virus that causes warts, has been associated with cancer of the cervix, vagina, vulva, penis, and colon.

A great deal of cancer research is now directed toward studying **oncogenes** (ONG-kō-jēnz), genes that have the ability to transform a normal cell into a cancerous cell. Oncogenes are derived from normal genes that control growth and development, called **proto-oncogenes,** that undergo some change that either causes them to produce an abnormal product or disrupts their control so that they are expressed inappropriately, making their products in excessive amounts or at the wrong time. It is believed that some oncogenes cause extra production of growth factors, chemicals that stimulate cell growth. Other oncogenes may cause changes in surface receptors, causing them to send signals as if they are being activated by growth factors. As a result, the growth pattern of the cell becomes abnormal. Oncogenes are genes that can cause cancer when they are inappropriately activated.

Every human cell contains oncogenes. In fact, oncogenes apparently carry out normal cellular functions until a malignant change occurs. It appears that some proto-oncogenes are activated to oncogenes by various types of mutations in which the DNA of the proto-oncogenes is altered. Such mutations are induced by carcinogens. Other proto-oncogenes are activated by viruses. Some oncogenes can also be activated by a rearrangement of a cell's chromosomes in which segments of DNA are exchanged. This rearrangement is sufficient to activate oncogenes by placing them near genes that enhance their activity. Such a mechanism occurs in Burkitt's lymphoma. Malignant tumors of the colon and rectum and one type of lung cancer are linked to oncogenes.

Researchers have also determined that some cancers are not caused by oncogenes but may be caused by genes called **anti-oncogenes.** These genes can cause cancer when they are inappropriately inactivated. The prototype for a cancer caused by an anti-oncogene is a rare, inherited childhood cancer of the eye called retinoblastoma.

Currently, scientists are also trying to establish a relation between stress and cancer. Some believe that stress may play a role not only in the development but also in the metastasis of cancer. There is also great interest in determining the effects of alterations of the immune system and nutrition in the development of cancer.

In 1982, the National Research Council issued a series of guidelines related to cancer and diet. The main recommendations were to (1) reduce fat intake from 40 percent of total calories to 30 percent; (2) increase consumption of fiber, fruits, and vegetables; (3) increase intake of complex carbohydrates (e.g., potatoes and pasta); and (4) reduce consumption of salted, smoked, and pickled foods and simple carbohydrates (refined sugars).

TREATMENT

Treating cancer is difficult because it is not a single disease and because all the cells in a single population (tumor) do not behave in the same way. The same cancer may contain a diverse population of cells by the time it reaches a clinically detectable size. Although tumor cells look alike when stained and viewed under the microscope, they do not necessarily behave in the same manner in the body. For example, some metastasize and others do not. Some divide and others do not. Some are sensitive to drugs and some are resistant. As a consequence of differences in drug resistance, a single chemotherapeutic drug may destroy susceptible cells but permit resistant cells to proliferate. This is probably one of the reasons that combination chemotherapy is usually more successful. In addition to chemotherapy, radiation therapy, surgery, and hyperthermia (abnormally high temperatures), and immunotherapy (bolstering the body's own defenses) may be used alone or in combination.

Scientists are moving closer to developing a vaccine for cancer. What happens in cancer is that the immune system fails to protect the body. Accordingly, the goal of a cancer vaccine is to stimulate the immune system into marshaling a successful attack against the cancer cells.

There has been considerable debate over the use of **Laetrile** (LĀ-e-tril) in the treatment of human cancer. Laetrile is a naturally derived substance prepared from apricot pits. In response to public pressure, the National Cancer Institute (NCI) and Food and Drug Administration (FDA) conducted a clinical trial to determine the effectiveness of Laetrile in the treatment of advanced cancer. The results of the trial were published in 1982 and indicated that Laetrile does not work and is a toxic drug.

CELLS AND AGING

Aging is a progressive failure of the body's homeostatic adaptive responses. It is a general response that produces observable changes in structure and function and increased vulnerability to environmental stress and disease. Disease and aging probably accelerate each other. The specialized branch of medicine that deals with the medical problems and care of elderly persons is called **geriatrics** (jer'ē-AT-riks; *geras* = old age; *iatrike* = surgery, medicine).

The obvious characteristics of aging are well known: graying and loss of hair, loss of teeth, wrinkling of skin, decreased muscle mass, and increased fat deposits. The physiological signs of aging are gradual deterioration in function and capacity to respond to environmental stress. Thus, basic kidney and digestive metabolic rates decrease, as does the ability to respond effectively to changes in temperature, diet, and oxygen supply in order to maintain a constant internal environment. These manifestations of aging are related to a net decrease in the number of cells in the body (thousands of brain cells are lost each day) and to the dysfunctioning of the cells that remain.

The extracellular components of tissues also change with age. Collagen fibers, responsible for the strength in tendons, increase in number and change in quality with aging. These changes in the collagen of arterial walls are as much responsible for their loss of extensibility as are the deposits associated with atherosclerosis, the deposition of fatty materials in arterial walls. Elastin, another extracellular component, is responsible for the elasticity of blood vessels and skin. It thickens, fragments, and acquires a greater affinity for calcium with age—changes that may also be associated with the development of atherosclerosis.

Several kinds of cells in the body—heart cells, skeletal muscle fibers (cells), nerve cells—are incapable of replacement. Experiments have proved that many other cell types are limited when it comes to cell division. Cells grown outside the body divide only a certain number of times and then stop. The number of divisions correlates with the donor's age and with the normal life span of the different species from which the cells are obtained—strong evidence for the hypothesis that cessation of mitosis is a normal, genetically programmed event. According to this view, an "aging" gene is part of the genetic blueprint at birth, and it turns on at a preprogrammed time, slowing down or halting processes vital to life.

Another theory of aging is the free radical theory. Free radicals, or oxygen radicals, are oxygen molecules that bear free electrons, are highly reactive, and can easily tie up and weaken proteins. As a result, cells grow rigid as nutrients are excluded and wastes are locked in. Such effects are exhibited as wrinkled skin, stiff joints, and hardened arteries. Free radicals may also cause damage to DNA. Among the factors that produce free radicals are pollution, radiation, and certain foods we eat. Other substances in the diet, such as vitamin E, vitamin C, beta-carotene, and selenium, are antioxidants and inhibit free radical formation.

Recently, it has been learned that glucose, the most abundant sugar in the body, may play a role in the aging process. According to one hypothesis, glucose is added, haphazardly, to proteins, forming irreversible cross-links between adjacent protein molecules. As a person ages, more cross-links are formed and this probably contributes to the stiffening and loss of elasticity that occurs in aging tissues.

Whereas some theories of aging explain the process at the cellular level, others concentrate on regulatory mechanisms operating within the entire organism. For example, one such theory holds that the immune system, which manufactures antibodies against foreign invaders, turns on its own cells. This autoimmune response might be caused by changes in the surfaces of cells, causing antibodies to attack the body's own cells. As surface changes in cells increase, the autoimmune response intensifies, producing the well-known signs of aging. Another organismic theory suggests that aging is programmed in the pituitary gland, a gland that produces and stores hormones and is attached to the undersurface of the brain. Supposedly, at a set time in life, the gland releases a hormone that triggers age-associated disruptions.

The effects of aging on the various body systems are discussed in their respective chapters.

KEY MEDICAL TERMS ASSOCIATED WITH CELLS

NOTE TO THE STUDENT

Each chapter in this text that discusses a major system of the body is followed by a glossary of *key medical terms*. Both normal and pathological conditions of the system are included in these glossaries. You should familiarize yourself with the terms, since they will play an essential role in your medical vocabulary.

Some of these disorders, as well as disorders discussed in the text, are referred to as local or systemic. A *local disease* is one that affects one part or a limited area of the body. A *systemic disease* affects either the entire body or several parts.

The science that deals with why, when, and where diseases occur and how they are transmitted in a human community is known as *epidemiology* (ep'-ide'-me-OL-o-je; *epidemios* = prevalent; *logos* = study of). The science that deals with the effects and uses of drugs in the treatment of disease is called *pharmacology* (far'-ma-KOL-o-je; *pharmakon* = medicine; *logos* = study of).

Atrophy (AT-rō-fē; *a* = without; *tropho* = nourish)   A decrease in the size of cells with subsequent decrease in the size of the affected tissue or organ; wasting away.

Biopsy (BĪ-op-sē; *bio* = life; *opsis* = vision)   The removal and microscopic examination of tissue from the living body for diagnosis.

Deterioration   (de-te'-rē-ō-RĀ-shun;   *deterior* = worse or poorer)   The process or state of growing worse; disintegration or wearing away.

Dysplasia (dis-PLĀ-zē-a; *dys* = abnormal; *plas* =to grow)   Alteration in the size, shape, and organization of cells due to chronic irritation or inflammation; may progress to neoplasia (tumor formation, usually malignant) or revert to normal if the stress is removed.

Hyperplasia (hī'-per-PLĀ-zē-a; *hyper* = over)   Increase in the number of cells due to an increase in the frequency of cell division.

Hypertrophy (hī-PER-trō-fē)   Increase in the size of cells without cell division.

Insidious (in-SID-ē-us)   Hidden, not apparent, as a disease that does not exhibit distinct symptoms of its arrival.

Metaplasia (met'-a-PLĀ-zē-a; *meta* = change)   The transformation of one cell into another.

Metastasis (me-TAS-ta-sis; *stasis* = standing still)   The transfer of disease from one part of the body to another that is not directly connected with it.

Necrosis (ne-KRŌ-sis; *necros* = death; *osis* = condition)   Death of a group of cells.

Neoplasm (NĒ-ō-plazm; *neo* = new)   Any abnormal formation or growth, usually a malignant tumor.

Progeny (PROJ-e-nē; *progignere* = to bring forth)   Offspring or descendants.

Senescence (se-NES-ens)   The process of growing old.

STUDY OUTLINE

Generalized Animal Cell (p. 32)

1. A cell is the basic, living, structural and functional unit of the body.
2. A generalized cell is a composite that represents various cells of the body.
3. Cytology is the science concerned with the study of cells.
4. The principal parts of a cell are the plasma (cell) membrane, cytoplasm, organelles, and inclusions. Extracellular materials are manufactured by the cell and deposited outside the plasma membrane.

Plasma (Cell) Membrane (p. 32)
Chemistry and Structure (p. 32)

1. The plasma (cell) membrane surrounds the cell and separates it from other cells and the external environment.
2. It is composed primarily of phospholipids and proteins. According to the fluid mosaic model, the membrane consists of a phospholipid bilayer with integral and peripheral proteins.

Functions (p. 33)

1. Functionally, the plasma membrane facilitates contact with other cells, provides receptors, and regulates the passage of materials.
2. The membrane's selectively permeable nature restricts the passage of certain substances. Substances can pass through the membrane depending on their molecular size, lipid solubility, electrical charges, and the presence of carriers.

Movement of Materials Across Plasma Membranes (p. 35)

1. Passive (physical) processes involve the kinetic energy of individual molecules.
2. Diffusion is the net movement of molecules or ions from an area of higher concentration to an area of lower concentration until an equilibrium is reached.
3. In facilitated diffusion, certain molecules, such as glucose, combine with a carrier to become soluble in the phospholipid portion of the membrane.
4. Osmosis is the movement of water through a selectively permeable membrane from an area of higher water concentration to an area of lower water concentration.
5. In an isotonic solution, red blood cells maintain their normal shape; in a hypotonic solution, they undergo hemolysis; in a hypertonic solution, they undergo crenation.
6. Filtration is the movement of water and dissolved substances across a selectively permeable membrane by pressure.
7. Active (physiological) processes involve the use of ATP by the cell.
8. Active transport is the movement of ions across a cell membrane from lower to higher concentration.
9. Endocytosis is the movement of substances through plasma membranes in which the membrane surrounds the substance, encloses it, and brings it into the cell.
10. Phagocytosis is the ingestion of solid particles by pseudopodia. It is an important process used by white blood cells to destroy bacteria that enter the body.
11. Pinocytosis is the ingestion of a liquid by the plasma membrane. In this process, the liquid becomes surrounded by a vacuole.
12. Receptor-mediated endocytosis is the selective uptake of large molecules and particles by cells.

Cytoplasm (p. 38)

1. Cytoplasm is the substance inside the cell between the plasma membrane and nucleus that contains organelles and inclusions.
2. It is composed mostly of water plus proteins, carbohydrates, lipids, and inorganic substances. The chemicals in cytoplasm are either in solution or in a colloid (suspended) form.
3. Functionally, cytoplasm is the medium in which chemical reactions occur.

Organelles (p. 38)

1. Organelles are specialized portions of the cell with characteristic morphology that carry on specific activities.
2. They assume specific roles in cellular growth, maintenance, repair, and control.

Nucleus (p. 38)

1. Usually the largest organelle, the nucleus controls cellular activities and contains the genetic information.
2. Most body cells have a single nucleus; some (red blood cells) have none, while others (skeletal muscle fibers) have several.
3. The parts of the nucleus include the double nuclear membrane, karyolymph, nucleoli, and genetic material (DNA), comprising the chromosomes.
4. Chromosomes consist of DNA and histones and of subunits called nucleosomes.

Ribosomes (p. 40)

1. Ribosomes are granular structures consisting of ribosomal RNA and ribosomal proteins.
2. They occur free (singly or in clusters) or in conjunction with endoplasmic reticulum.
3. Functionally, ribosomes are the sites of protein synthesis.

Endoplasmic Reticulum (ER) (p. 40)

1. The ER is a network of parallel membranes continuous with the plasma membrane and nuclear membrane.
2. Granular or rough ER has ribosomes attached to it. Agranular or smooth ER does not contain ribosomes.
3. The ER provides mechanical support, conducts intracellular nerve impulses in muscle fiber (cells), exchanges materials with cytoplasm, transports substances intracellularly, stores synthesized molecules, and helps export chemicals from the cell.

Golgi Complex (p. 40)

1. The Golgi complex consists of four to eight stacked, flattened, membranous sacs (cisternae) referred to as cis, medial, and trans.
2. The principal function of the Golgi complex is to process, sort, and deliver proteins within the cell. It also secretes proteins.

Mitochondria (p. 42)

1. Mitochondria consist of a smooth outer membrane and a folded

inner membrane surrounding the interior matrix. The inner folds are called cristae.

2. The mitochondria are called "powerhouses of the cell" because ATP is produced in them.

Lysosomes (p. 44)

1. Lysosomes are spherical structures that contain digestive enzymes. They are formed from Golgi complexes.
2. They are found in large numbers in white blood cells, which carry on phagocytosis.
3. If the cell is injured, lysosomes release enzymes and digest the cell (autolysis).
4. Lysosomes may be involved in bone removal.

Peroxisomes (p. 45)

1. Peroxisomes are similar to lysosomes, but smaller.
2. They contain enzymes (e.g., catalase) involved in the metabolism of hydrogen peroxide.

The Cytoskeleton (p. 46)

1. Together, microfilaments, microtubules, and intermediate filaments form the cytoskeleton.
2. Microfilaments are rodlike structures consisting of the protein actin or myosin. They are involved in muscular contraction, support, and movement.
3. Microtubules are cylindrical structures consisting of the protein tubulin. They support, provide movement, and form the structure of flagella, cilia, centrioles, and the mitotic spindle.
4. Intermediate filaments appear to provide structural reinforcement in some cells.

Centrosome and Centrioles (p. 47)

1. The dense area of cytoplasm containing the centrioles is called a centrosome.
2. Centrioles are paired cylinders arranged at right angles to one another. They assume an important role in cell reproduction by helping to organize the mitotic spindle.

Flagella and Cilia (p. 47)

1. These cellular projections have the same basic structure and are used in movement.
2. If projections are few (typically occurring singly or in pairs) and long, they are called flagella. If they are numerous and hairlike, they are called cilia.
3. The flagellum on a sperm cell moves the entire cell. The cilia on cells of the respiratory tract move foreign matter trapped in mucus along the cell surfaces toward the throat for elimination.

Cell Inclusions (p. 47)

1. Cell inclusions are chemical substances produced by cells. They are usually organic and may have recognizable shapes.
2. Examples are melanin, glycogen, and lipids.

Extracellular Materials (p. 48)

1. These are all the substances that lie outside the plasma membrane.
2. They provide support and a medium for the diffusion of nutrients and wastes.

3. Some, like hyaluronic acid and chondroitin sulfate, are amorphous. Others, like collagenous, reticular, and elastic fibers, are fibrous.

Normal Cell Division (p. 48)

1. Cell division is the process by which cells reproduce themselves. It consists of nuclear division and cytoplasmic division (cytokinesis).
2. Cell division that results in an increase in body cells is called somatic cell division and involves a nuclear division called mitosis and cytokinesis.
3. Cell division that results in the production of sperm and eggs is called reproductive cell division and consists of a nuclear division called meiosis and cytokinesis.

Somatic Cell Division (p. 49)

1. Prior to mitosis and cytokinesis, the DNA molecules, or chromosomes, replicate themselves so the same chromosomal complement can be passed on to future generations of cells.
2. A cell carrying on every life process except division is said to be in interphase (metabolic phase).
3. Mitosis is the distribution of two sets of chromosomes into separate and equal nuclei following their replication.
4. It consists of prophase, metaphase, anaphase, and telophase.
5. Cytokinesis usually begins in late anaphase and terminates in telophase.
6. A cleavage furrow forms at the cell's equatorial plane and progresses inward, cutting through the cell to form two separate portions of cytoplasms.

Reproductive Cell Division (p. 52)

1. Gametes contain the haploid (n) chromosome number and uninucleated somatic cell contain the diploid ($2n$) chromosome number.
2. Meiosis is one of the processes that produces haploid gametes. It consists of two successive nuclear divisions called reduction division and equatorial division.
3. During reduction division, homologous chromosomes undergo synapsis and crossing-over; the net result is two haploid daughter cells.
4. During equatorial division, the two haploid daughter cells undergo mitosis and the net result is four haploid cells.

Abnormal Cell Division: Cancer (CA) (p. 54)

1. Cancerous tumors are referred to as malignant; noncancerous tumors are called benign; the study of tumors is called oncology.
2. The spread of cancer from its primary site is called metastasis.
3. Carcinogens include environmental agents and viruses.
4. Treating cancer is difficult because all the cells in a single population do not behave the same way.

Cells and Aging (p. 57)

1. Aging is a progressive failure of the body's homeostatic adaptive responses.
2. Many theories of aging have been proposed, including genetically programmed cessation of cell division and excessive immune responses, but none successfully answers all the experimental objections.
3. All the various body systems exhibit definitive and sometimes exclusive changes with aging.

REVIEW QUESTIONS

1. Define a cell. What are the four principal portions of a cell? What is meant by a generalized cell?
2. Discuss the chemistry and structure of the plasma membrane with respect to the fluid mosaic model.
3. How do integral and peripheral membrane proteins differ in function?
4. Describe the various functions of the plasma membrane. What determines selective permeability?
5. Define and give an example of each of the following: diffusion, facilitated diffusion, osmosis, filtration, active transport, phagocytosis, pinocytosis, and receptor-mediated endocytosis.
6. Compare the effect on red blood cells of an isotonic, hypertonic, and hypotonic solution.
7. Discuss the chemical composition and physical nature of cytoplasm. What is its function?
8. What is an organelle? By means of a labeled diagram, indicate the parts of a generalized animal cell.
9. Describe the structure and functions of the nucleus of a cell. What are nucleosomes?
10. Discuss the distribution of ribosomes. What is their function?
11. Distinguish between granular (rough) and agranular (smooth) endoplasmic reticulum (ER). What are the functions of ER?
12. Describe the structure and functions of the Golgi complex.
13. Why are mitochondria referred to as "powerhouses of the cell"?
14. List and describe the various functions of lysosomes.
15. What is the importance of peroxisomes?
16. Contrast the structure and functions of microfilaments, microtubules, and intermediate filaments. What is the microtrabecular lattice?
17. Describe the structure and function of centrioles.
18. How are cilia and flagella distinguished on the basis of structure and function?
19. Define a cell inclusion. Provide examples and indicate their functions.
20. What are extracellular materials? Give examples and the functions of each.
21. Distinguish between the two types of cell division. Why is each important?
22. Define interphase. Explain what happens during the G_1, S, and G_2 periods of interphase.
23. Describe the principal events of each stage of mitosis.
24. Distinguish between haploid and diploid cells.
25. Define meiosis and contrast the principal events of reduction division and equatorial division.
26. What is a tumor? Distinguish between malignant and benign tumors. Describe the principal types of malignant tumors.
27. Define metastasis. What factors contribute to metastasis?
28. Discuss several possible causes of cancer.
29. What are some of the problems with respect to treating cancer?
30. What is aging? List some of the characteristics of aging.
31. Briefly describe the various theories regarding aging.
32. Refer to the glossary of key medical terms associated with cells. Be sure that you can define each term.

SELF-QUIZ

Choose the one best answer to these questions.

___ 1. A membrane that permits the passage of only certain materials is described as
A. impermeable; B. freely permeable; C. selectively permeable; D. endocytic; E. macromolecular.

___ 2. As a result of cell division, each daughter cell has
A. half as many chromosomes as its parent cell; B. twice as many chromosomes as its parent cell; C. exactly the same number of chromosomes as its parent cell; D. one-quarter as many chromosomes as its parent cell; E. three-quarters as many chromosomes as its parent cell.

___ 3. Two cells of the same kind have close to the same concentration of constituents in their cytoplasm. Cell A is placed in solution A, a 2 percent glucose solution. Cell B is placed in solution B, a 5 percent glucose solution. Which of the following is correct?
A. cell A will swell more than cell B; B. cell B will swell more than cell A; C. both cells will increase in size, at the same rate; D. both cells will decrease in size, at the same rate; E. neither cell will change size or shape.

___ 4. An extracellular material that binds cells together, lubricates joints, and maintains the shape of the eyeballs is
A. collagen; B. chondroitin sulfate; C. hyaluronic acid; D. reticulin; E. elastin.

___ 5. The important difference between meiosis and mitosis is that

A. in meiosis the number of chromosomes in the resulting cells is reduced by one-half; B. fewer sperm cells are produced, as compared with the number of egg cells; C. meiosis occurs only in somatic cells; D. abnormalities seldom occur in mitosis but are fairly common in meiosis; E. egg cells are produced by meiosis, sperm cells by mitosis.

___ 6. The net movement of water molecules across a selectively permeable membrane from a region of high concentration of water to a region of low concentration of water is called
A. osmosis; B. diffusion; C. filtration; D. dialysis; E. endocytosis.

___ 7. Which of the following statements is/are true of living cells placed into a hypertonic solution?
(1) Initially, the cells will take in water molecules from the surrounding solution.
(2) Initially, the cells will swell up and burst.
(3) Initially, solute (e.g., dissolved sugar, salt) molecules will leak out of the cell through the cell membrane.
(4) Initially, the cells will lose water to the surrounding solution.
A. (1) only; B. (2) only; C. (3) only; D. (4) only; E. (1) and (2).

___ 8. Cytokinesis is the division of the cytoplasm of a cell. A similar equal division of the nucleus is a process called
A. synapsis; B. pinocytosis; C. involution; D. diplopia; E. mitosis.

9. Match the following:

___ a. site of direction of cellular activities by means of genes located here

___ b. cytoplasmic sites of protein synthesis; may occur attached to ER or scattered freely in cytoplasm

___ c. system of membranous channels providing pathways for transport within the cell and surface areas for chemical reactions; may be granular or agranular

___ d. stacks of cisternae; involved in packaging and secretion of proteins and lipids and synthesis of carbohydrates

___ e. called "suicide packets" since they may release enzymes that lead to autolysis of the cell

___ f. similar to lysosomes, but smaller; contain the enzyme catalase that breaks down hydrogen peroxide

___ g. cristae-containing structures, called "powerhouses of the cell" since ATP production occurs here

___ h. form part of cytoskeleton; involved with cell movement and contraction

___ i. part of cytoskeleton, provide support and give shape to cell; form flagellae, cilia, centrioles, and spindle fibers

___ j. help organize spindle fibers used in cell division

A. centrioles
B. cilia
C. endoplasmic reticulum (ER)
D. flagella
E. Golgi complex
F. lysosomes
G. microfilaments
H. microtubules
I. mitochondria
J. nucleus
K. peroxisomes
L. ribosomes

___ k. long, hairlike structures that help move entire cell, for example, sperm cell

___ l. short, hairlike structures that move particles over cell surface

Complete the following.

10. Phagocytosis is the process of cell (eating? drinking?).

11. The plasma membrane is composed of two main chemical components: a bilayer of _____ embedded with protein.

12. In the formation of mature sperm and egg cells, the nuclear division is known as _____.

13. A solution that is hypertonic to red blood cells contains (more? fewer?) solute particles and (more? fewer?) water molecules than blood.

14. By the time that cells complete meiosis, they will contain (46? 23?) chromosomes, the ($2n$? n?) number.

15. Match the following:

___ a. chromosomes move toward opposite poles of the cell

___ b. nuclear membrane disappears; chromatin thickens into distinct chromosomes (chromatids); centrioles move to opposite end of cell, enabling microtubules to form mitotic spindles

___ c. The series of events is essentially the reverse of prophase; cytokinesis occurs

___ d. chromatids line up on the equatorial plane and are attached to spindle fibers

___ e. cell is not involved in cell division; it is in a metabolic phase that follows telophase

A. anaphase
B. interphase
C. metaphase
D. prophase
E. telophase

3 Tissues

CHAPTER OUTLINE
■ **Types of Tissues**
■ **Epithelial Tissue**
Covering and Lining Epithelium
  Arrangement of Layers
  Cell Shapes
  Cell Junctions
  Classification
  Simple Epithelium
  Stratified Epithelium
  Pseudostratified Epithelium
Glandular Epithelium
  Structural Classification of Exocrine Glands
  Functional Classification of Exocrine Glands
■ **Connective Tissue**
Classification
Embryonic Connective Tissue
Adult Connective Tissue
  Connective Tissue Proper
  Cartilage
  Osseous Tissue (Bone)
  Vascular Tissue (Blood)
■ **Membranes**
Mucous Membranes
Serous Membranes
Cutaneous Membrane
Synovial Membranes
■ **Muscle Tissue**
■ **Nervous Tissue**

STUDENT OBJECTIVES

1. Define a tissue and classify the tissues of the body into four major types.
2. Discuss the distinguishing characteristics of epithelial tissue.
3. Compare the layering arrangements and cell shapes of covering and lining epithelium.
4. Describe the various types of cell junctions between epithelial cells.
5. List the structure, location, and function for the following types of epithelium: simple squamous, simple cuboidal, simple columnar (nonciliated and ciliated), stratified squamous, stratified cuboidal, stratified columnar, transitional, and pseudostratified.
6. Define a gland and distinguish between exocrine and endocrine glands.
7. Classify exocrine glands according to structural complexity and function and give an example of each.
8. Identify the distinguishing characteristics of connective tissue.
9. Discuss the intercellular substance, fibers, and cells that constitute connective tissue.
10. List the structure, function, and location of loose (areolar) connective tissue, adipose tissue, dense, elastic, and reticular connective tissue, cartilage, osseous (bone) tissue, and vascular (blood) tissue.
11. Define an epithelial membrane and list the location and function of mucous, serous, cutaneous, and synovial membranes.
12. Contrast the three types of muscle tissue with regard to structure and location.
13. Describe the structural features and functions of nervous tissue.

Cells are highly organized units, but they do not function in isolation. They work together in a group of similar cells called a tissue.

TYPES OF TISSUES

A *tissue* is a group of similar cells and their intercellular substance that have a similar embryological origin and function. The science that deals with the study of tissues is called *histology* (hiss-TOL-ō-jē; *histio* = tissue; *logos* = study of). The various tissues of the body are classified into four principal types according to their function and structure:

1. *Epithelial* (ep'-i-THĒ-lē-al) *tissue,* which covers body surfaces or tissues, lines body cavities and ducts, and forms glands.
2. *Connective tissue,* which protects and supports the body and its organs and binds organs together.
3. *Muscular tissue,* which is responsible for movement.
4. *Nervous tissue,* which initiates and transmits nerve impulses that coordinate body activities.

Epithelial tissue and connective tissue, except for bone and blood, will be discussed in detail in this chapter. The general features of bone tissue and blood will be introduced here, but their detailed discussion occurs later in the book. Similarly, the detailed discussion of muscle tissue and nervous tissue will be postponed.

EPITHELIAL TISSUE

Epithelial tissues perform many activities in the body, ranging from protection of underlying tissues against microbial invasion, drying out, and harmful environmental factors to secretion. *Epithelial tissue,* or more simply *epithelium,* may be divided into two subtypes: (1) *covering and lining epithelium* and (2) *glandular epithelium.* Covering and lining epithelium forms the outer covering of external body surfaces and the outer covering of some internal organs. It lines body cavities and the interiors of the respiratory and gastrointestinal tracts, blood vessels, and ducts. It makes up, along with nervous tissue, the parts of the sense organs that are receptive to the stimuli that produce sensations such as smell, hearing, vision, and touch. In addition, it is the tissue from which gametes (sperm and eggs) develop. Glandular epithelium constitutes the secreting portion of glands.

Both types of epithelium consist largely or entirely of closely packed cells with little or no intercellular material between adjacent cells. (Such intercellular material is also called the matrix.) The points of attachment between adjacent plasma membranes of epithelial cells are called *cell junctions* (described shortly). They not only provide cell-to-cell attachments but also inhibit the movement of materials into certain cells and provide channels for communication between other cells. Epithelial cells are arranged in continuous sheets that may be either single or multilayered. Nerves may extend through these sheets, but blood vessels do not. They are *avascular* (*a* = without; *vascular* = blood vessels). The vessels that supply nutrients and remove wastes are located in underlying connective tissue.

Both types of epithelium overlie and adhere firmly to the connective tissue, which holds the epithelium in position and prevents it from being torn. The attachment between the epithelium and the connective tissue is a thin extracellular layer called the *basement membrane.* With few exceptions, epithelial cells secrete along their basal surfaces a material consisting of a special type of collagen and glycoproteins. This structure averages 50–80 nm in thickness and is referred to as the *basal lamina.* Frequently, the basal lamina is reinforced by an underlying *reticular lamina* consisting of reticular fibers and glycoproteins. This lamina is produced by cells in the underlying connective tissue. The combination of the basal lamina and reticular lamina constitutes the basement membrane.

Within a tissue, most normal cells remain in place, anchored to basement membranes and connective tissues. In an adult, a few cells, such as phagocytes, routinely move through the intercellular material (matrix) during an infection. In an embryo, certain cells migrate extensively as part of the growth and development process. Various activities of cells depend on a variety of *adhesion proteins* found in intercellular material and blood. These proteins interact with receptors on plasma membranes called *integrins.* Many adhesion proteins contain tripeptides (arginine, glycine, and aspartic acid) as the recognition sites for integrins. Among the adhesion proteins are fibronectin, vitronectin, osteopontin, collagens, thrombospondin, and fibrinogen. Together, adhesion proteins and their integrins function in anchoring cells in position, providing traction for the movement of cells, differentiation, positioning of cells, and possibly growth.

Some tissues of the body are so highly differentiated (specialized) that they have lost their capacity for mitosis. Examples are muscle tissue and nervous tissue. Other tissues, such as epithelium, which are subjected to a certain amount of wear and tear and injury, have a continuous capacity for renewal because they contain stem cells that are capable of renewing the tissue. Examining cells that are sloughed off provides the basis for the Pap smear, a test for precancer and cancer diagnosis of the uterus (Chapter 25).

CLINICAL APPLICATION

The *ability of epithelial tissue to regenerate* is of clinical significance following injury and surgery. On the basis of regenerative capacity, epithelial cells (as well as other body cells) are distinguished into two principal types:

labile cells and stable cells. *Labile* (LĀ-bīl) *cells* continue to regenerate throughout life under normal conditions. Examples include epidermal cells of the epidermis of the skin and epidermal cells of mucous membranes of the gastrointestinal (GI), urinary, and respiratory tracts. *Stable cells* cease regeneration at about the time of puberty but may resume regeneration under special conditions, such as injury. Epithelial cells of the liver, pancreas, kidneys, and thyroid glands are examples.

There are also cells of the body called *permanent cells* that cease to regenerate after birth and remain that way throughout life. An example is a neuron (nerve cell).

COVERING AND LINING EPITHELIUM

Arrangement of Layers

Covering and lining epithelium is arranged in several different ways related to location and function. If the epithelium is specialized for absorption or filtration and is in an area that has minimal wear and tear, the cells of the tissue are arranged in a single layer. Such an arrangement is called **simple epithelium.** If the epithelium is found in an area with a high degree of wear and tear, then the cells are stacked in several layers. This tissue is referred to as **stratified epithelium.** A third, less common arrangement of epithelium is called **pseudostratified.** Like simple epithelium, pseudostratified epithelium has only one layer of cells. However, some of the cells do not reach the surface—an arrangement that gives the tissue a multilayered, or stratified, appearance. The cells in pseudostratified epithelium that do reach the surface either secrete mucus or contain cilia that move mucus and foreign particles for eventual elimination from the body.

Cell Shapes

In addition to classifying covering and lining epithelium according to the number of its layers, we may also categorize it by cell shape. The cells may be flat, cubelike, or columnar or a combination of shapes. **Squamous** (SKWĀ-mus) cells are flattened and scalelike. They are attached to each other like a mosaic. **Cuboidal** cells are usually cube-shaped in cross section. They sometimes appear as hexagons. **Columnar** cells are tall and cylindrical, appearing as rectangles set on end. **Transitional** cells often have a combination of shapes and are found where there is a great degree of distention or expansion in the body. Transitional cells in the bottom layer of an epithelial tissue may range in shape from cuboidal to columnar. In the intermediate layer, they may be cuboidal or polyhedral. In the superficial layer, they may range from cuboidal to squamous, depending on how much they are pulled out of shape during certain body functions.

Cell Junctions

Epithelial cells, unlike connective tissue cells, are tightly joined to form a close functional unit. The points of attachment between adjacent plasma membranes of epithelial cells (and a few other types of cells as well) are called **cell junctions (junctional complexes).** In addition to providing cell-to-cell attachments, cell junctions prevent the movement of materials between certain cells and provide channels for communication between other cells. Four categories of cell junctions are recognized: (1) tight junction (zonula occludens), (2) intermediate junction (zonula adherens), (3) desmosome (macula adherens), and (4) gap junction (nexus). These cell junctions are illustrated in Figure 3-1.

● **Tight Junction (Zonula occludens)** A **tight junction,** or **zonula occludens** (*zonula* = small zone; *occludens* = occlude), is typically located along the lateral surfaces of adjacent plasma membranes, closest to the luminal (free) surface. In a tight junction, the outer layers of adjacent plasma membranes are fused together at various points. The junction completely encircles each cell that is joined and occludes the intercellular spaces. Tight junctions are common in epithelial cells lining the small intestine. Here, they prevent the movement of substances across epithelial cells via the intercellular route. Some substances, however, can selectively enter the small intestinal cells through microvilli by absorption.

● **Intermediate Junction (Zonula adherens)** An **intermediate junction,** or **zonula adherens,** like the tight junction, completely encircles each cell that is joined. The intermediate junction is located just below the tight junction between small intestinal epithelial cells. In an intermediate junction, the adjacent plasma membranes are separated by an intercellular space measuring about 20 nm. The inner surfaces of the neighboring plasma membranes contain a moderately dense area of cytoplasmic material (not as dense as the plaques in desmosomes) in which actin-containing **microfilaments** are embedded. Collectively, the microfilaments constitute the **terminal web.** These filaments are smaller in diameter than the tonofilaments of desmosomes. Intermediate junctions are believed to form firm connections between adjacent cells.

● **Desmosome (Macula adherens)** One of the most frequent types of cell junction is called a **desmosome** (*desmos* = bond; *soma* = body), or **macula adherens** (*macula* = spot; *adhaereo* = to stick). Desmosomes are numerous in stratified epithelium, especially the epidermis of the skin. They are also found in the epithelial lining of the small intestine. Desmosomes are scattered over adjacent cell surfaces somewhat like spot welds and form firm intercellular attachments between cells. The adjacent plasma membranes at a desmosome appear thickened because there is a dense proteinaceous **plaque** of cytoplasmic material on the inner surfaces of the neighboring membranes. Cytoplasmic fila-

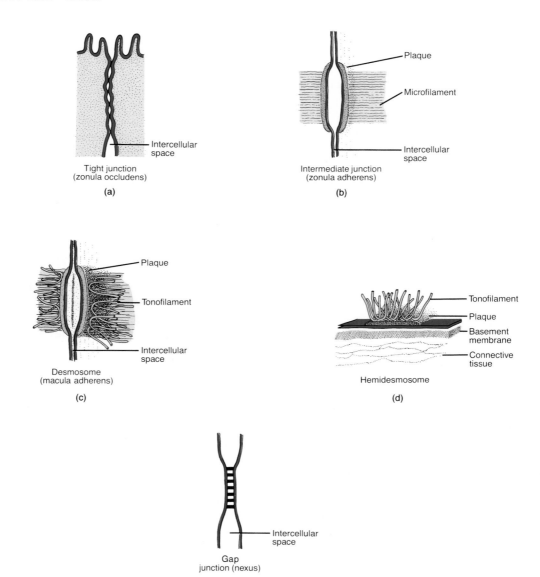

FIGURE 3-1 Cell junctions.

ments called **tonofilaments** converge in the region of the plaques and make a U-turn back into the cytoplasm. The plasma membranes of the adjacent cells at a desmosome are separated by an intercellular space measuring about 22–24 nm. In addition to serving as firm sites of adhesion between adjacent cells, desmosomes also help to support the entire epithelial tissue.

In the basal cells of epithelium, the plasma membranes are in contact with the underlying basement membrane rather than adjacent plasma membranes. Here, there are structures that look like half a desmosome, called **hemidesmosomes** (*hemi* = half). Hemidesmosomes are assumed to provide anchorage for the basal epithelial cell plasma membranes to the basement membrane.

● **Gap Junction (Nexus)** In a **gap junction,** or **nexus** (*nexus* = bond), the outer layers of adjacent plasma membranes approach each other, leaving a gap of about 2 nm between them. The gap is bridged by structures that link the cells together. These structures may be minute channels formed by membrane proteins that enable adjacent cells to communicate. It is postulated that ions and small molecules pass rapidly through gap junctions from one cell to the next, resulting in spontaneous impulse conduction. Gap junctions are found between smooth and cardiac muscle fibers (cells), as well as between epithelial cells. The presence of gap junctions between muscle fibers presumably enables the cells to conduct the impulses rapidly and contract in an orderly and coordinated manner.

Classification

Considering arrangement of layers and cell shapes in combination, we may classify covering and lining epithelium as follows:

Simple
1. Squamous
2. Cuboidal
3. Columnar

Statified
1. Squamous
2. Cuboidal
3. Columnar
4. Transitional

Pseudostratified

Each of the epithelial tissues described in the following sections is illustrated in Exhibit 3-1.

Simple Epithelium

● *Simple Squamous Epithelium* This type of simple epithelium consists of a single layer of flat, scalelike cells. Its surface resembles a tiled floor. The nucleus of each cell is centrally located and oval or spherical. Since simple squamous epithelium has only one layer of cells, it is highly adapted to diffusion, osmosis, and filtration. Thus, it lines the air sacs of the lungs, where gases (oxygen and carbon dioxide) are exchanged between air spaces and blood. It is present in the part of the kidney that filters blood. It also lines the inner surfaces of the membranous labyrinth and tympanic membrane of the ear. Simple squamous epithelium is found in body parts that have little wear and tear.

Simple squamous epithelium that lines the heart, blood vessels, and lymph vessels and forms the walls of capillaries is known as *endothelium*. Simple squamous epithelium that forms the epithelial layer of serous membranes is called *mesothelium*. Serous membranes line the thoracic and abdominopelvic cavities and cover the viscera within them.

● *Simple Cuboidal Epithelium* Viewed from above, the cells of simple cuboidal epithelium appear as closely fitted polygons. The cuboidal nature of the cells is obvious only when the tissue is sectioned at right angles. Like simple squamous epithelium, these cells possess a central nucleus, which is usually round. Simple cuboidal epithelium covers the surface of the ovaries, lines the anterior surface of the capsule of the lens of the eye, and forms the pigmented epithelium of the retina of the eye. In the kidneys, where it forms the kidney tubules and contains microvilli, it functions in water reabsorption. It also lines the smaller ducts of some glands and the secreting units of glands, such as the thyroid.

Simple cuboidal epithelium performs the functions of secretion and absorption. *Secretion,* usually a function of epithelium, is the production and release by cells of a fluid that may contain a variety of substances such as mucus, perspiration, or enzymes. *Absorption* is the intake of fluids or other substances by cells of the skin or mucous membranes.

● *Simple Columnar Epithelium* The surface view of simple columnar epithelium is similar to that of simple cuboidal tissue. When sectioned at right angles to the surface, however, the cells appear somewhat rectangular. The nuclei, usually located near the bases of the cells, are commonly oval.

The luminal surfaces (surfaces adjacent to the lumen, or cavity of a hollow organ, vessel, or duct) of simple columnar epithelial cells are modified in several ways, depending on location and function. Simple columnar epithelium lines the gastrointestinal tract from the cardia of the stomach to the anus, the gallbladder, and excretory ducts of many glands. In such sites, the cells protect the underlying tissues. Many of them are also modified to aid in food-related activities. For example, in the small intestine especially, the plasma membranes of the cells are folded into *microvilli* (see Figure 2-1). The microvilli arrangement increases the surface area of the plasma membrane and thereby allows larger amounts of digested nutrients and fluids to be absorbed into the body.

Interspersed among the typical columnar cells of the intestine are other modified columnar cells called *goblet cells*. These cells, which secrete mucus, are so named because the mucus accumulates in the upper half of the cell, causing the area to bulge out. The whole cell resembles a goblet or wine glass. The secreted mucus serves as a lubricant between the food and walls of the gastrointestinal tract.

A third modification of columnar epithelium is found in cells with hairlike processes called *cilia*. In a few portions of the upper respiratory tract, ciliated columnar cells are interspersed with goblet cells. Mucus secreted by the goblet cells forms a film over the respiratory surface. This film traps foreign particles that are inhaled. The cilia wave in unison and move the mucus, with any foreign particles, toward the throat, where it can be swallowed or eliminated. Ciliated columnar epithelium is also found in the uterus and uterine (Fallopian) tubes of the female reproductive system, some paranasal sinuses, and the central canal of the spinal cord.

Stratified Epithelium

In contrast to simple epithelium, stratified epithelium consists of at least two layers of cells. Thus, it is more durable and can protect underlying tissues from the external environment and from wear and tear. Some stratified epithelial cells also produce secretions. The name of the specific kind of stratified epithelium depends on the shape of the *surface* cells.

EXHIBIT 3-1

Epithelial Tissues

COVERING AND LINING EPITHELIUM

Simple squamous (250 ×)

Description: Single layer of flat, scalelike cells; large, centrally located nuclei.

Location: Lines air sacs of lungs, glomerular (Bowman's) capsule of kidneys, and inner surfaces of the membranous labyrinth and tympanic membrane of ear. Called endothelium when it lines heart, blood and lymphatic vessels, and forms capillaries. Called mesothelium when it lines the ventral body cavity and covers viscera as part of a serous membrane.

Function: Filtration, absorption and secretion in serous membranes.

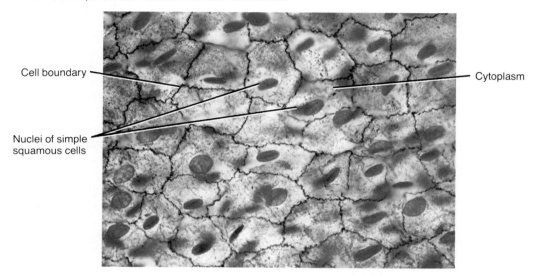

Cell boundary

Cytoplasm

Nuclei of simple squamous cells

Simple cuboidal (450 ×)

Description: Single layer of cube-shaped cells; centrally located nuclei.

Location: Covers surface of ovary, lines anterior surface of capsule of the lens of eyes, forms pigmented epithelium of retina of eye, and lines kidney tubules and smaller ducts of many glands.

Function: Secretion and absorption.

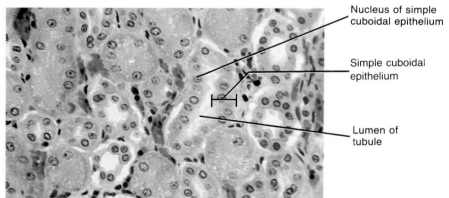

Nucleus of simple cuboidal epithelium

Simple cuboidal epithelium

Lumen of tubule

Simple columnar (nonciliated) (600 ×)
Description: Single layer of nonciliated rectangular cells; contains goblet cells in some locations; nuclei at bases of cells.
Location: Lines the gastrointestinal tract from the cardia of the stomach to the anus, excretory ducts of many glands, and gallbladder.
Function: Secretion and absorption.

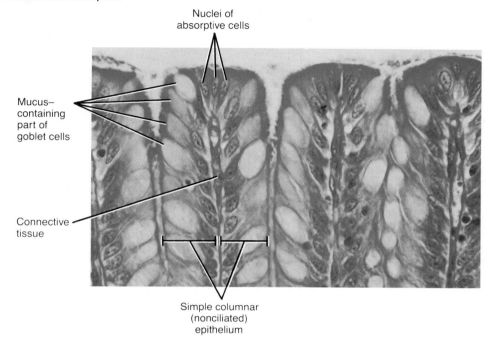

Nuclei of
absorptive cells

Mucus–
containing
part of
goblet cells

Connective
tissue

Simple columnar
(nonciliated)
epithelium

Simple columnar (ciliated) (400 ×)
Description: Single layer of ciliated columnar cells; contains goblet cells in some locations; nuclei at bases of cells.
Location: Lines a few portions of upper respiratory tract, uterine (Fallopian) tubes, uterus, some paranasal sinuses, and central canal of spinal cord.
Function: Moves mucus by ciliary action.

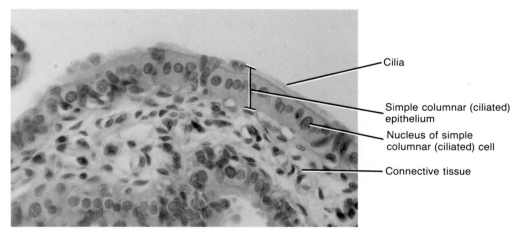

Cilia

Simple columnar (ciliated)
epithelium

Nucleus of simple
columnar (ciliated) cell

Connective tissue

Stratified squamous (185 ×)
Description: Several layers of cells; cuboidal to columnar shape in deep layers; scalelike shape in superficial layers; basal cells replace surface cells as they are lost.
Location: Nonkeratinized variety lines wet surfaces such as lining of the mouth, tongue, esophagus, part of epiglottis, and vagina; keratinized variety forms outer layer of skin.
Function: Protection

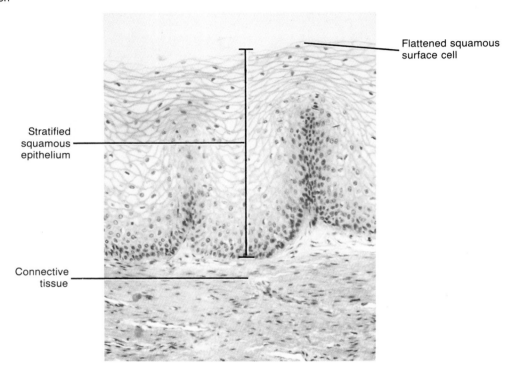

Flattened squamous surface cell

Stratified squamous epithelium

Connective tissue

Stratified cuboidal (185 ×)
Description: Two or more layers of cells in which the surface cells are cube shaped.
Location: Ducts of adult sweat glands, fornix of conjunctiva of eye, cavernous urethra of male urogenital system, pharynx, and epiglottis.
Function: Protection.

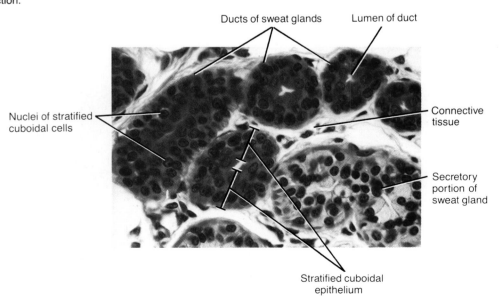

Ducts of sweat glands   Lumen of duct

Nuclei of stratified cuboidal cells

Connective tissue

Secretory portion of sweat gland

Stratified cuboidal epithelium

Stratified columnar (600 ×)
Description: Several layers of polyhedral cells; columnar cells only in superficial layer.
Location: Lines part of male urethra, large excretory ducts of some glands, and small areas in anal mucous membrane.
Function: Protection and secretion.

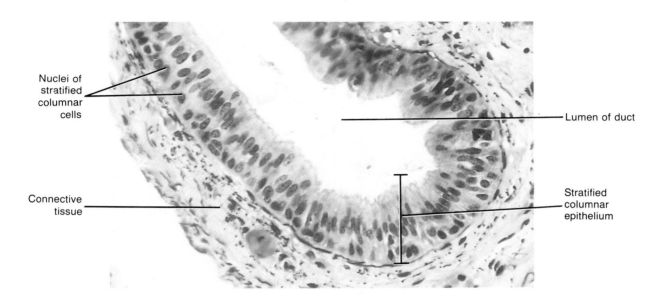

Nuclei of stratified columnar cells

Connective tissue

Lumen of duct

Stratified columnar epithelium

Stratified transitional (185 ×)
Description: Resembles nonkeratinized stratified squamous tissue, except that superficial cells are larger and have a rounded free surface.
Location: Lines urinary bladder and portions of ureters and urethra.
Function: Permits distention.

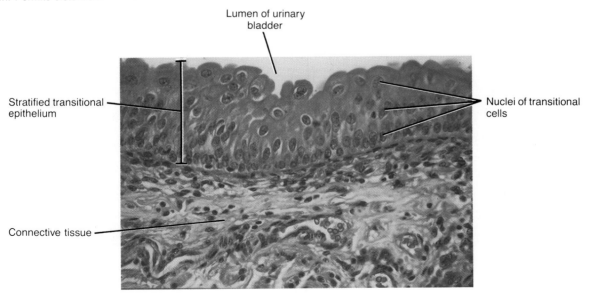

Lumen of urinary bladder

Stratified transitional epithelium

Connective tissue

Nuclei of transitional cells

EXHIBIT 3-1 (*Continued*)

Pseudostratified (500 ×)

Description: Not a true stratified tissue; nuclei of cells at different levels; all cells attached to basement membrane, but not all reach surface.

Location: Lines larger excretory ducts of many large glands, epididymis, male urethra, and auditory (Eustachian) tubes; ciliated variety with goblet cells lines most of the upper respiratory tract and some ducts of male reproductive system.

Function: Secretion and movement of mucus and sperm cells by ciliary action.

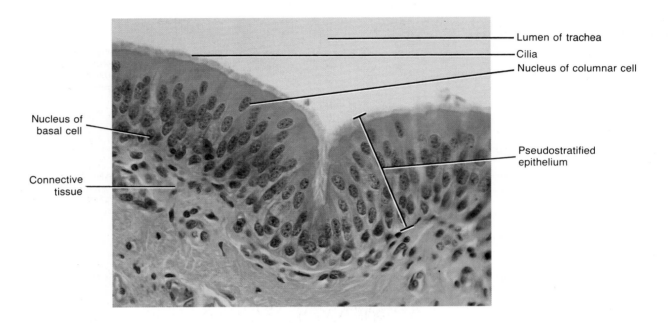

Lumen of trachea
Cilia
Nucleus of columnar cell
Nucleus of basal cell
Connective tissue
Pseudostratified epithelium

GLANDULAR EPITHELIUM

Exocrine gland (300 ×)

Description: Secretes products into ducts.

Location: Sweat, oil, wax, and mammary glands of the skin; digestive glands such as salivary glands that secrete into mouth cavity.

Function: Produces mucus, perspiration, oil, wax, milk, or digestive enzymes.

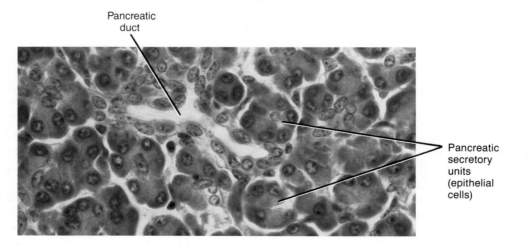

Pancreatic duct
Pancreatic secretory units (epithelial cells)

Endocrine gland (180×)
Description: Secretes hormones into blood.
Location: Pituitary at base of brain, thyroid and parathyroids near larynx, adrenals above kidneys, pancreas below stomach, ovaries in pelvic cavity, testes in scrotum, pineal at base of brain, and thymus in thoracic cavity.
Function: Produces hormones that regulate various body activities.

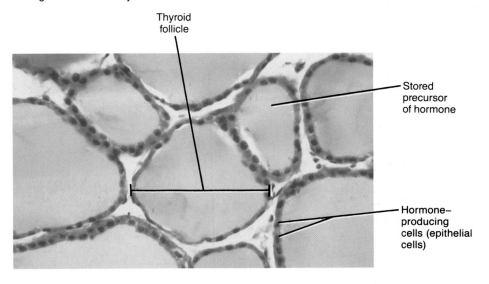

Thyroid
follicle

Stored
precursor
of hormone

Hormone-
producing
cells (epithelial
cells)

Photomicrographs of simple cuboidal, simple columnar (ciliated), stratified squamous, stratified columnar, and pseudo stratified courtesy Andrew Kuntzman.
All other photomicrographs copyright © 1983 by Michael H. Ross. Used by permission.

● *Stratified Squamous Epithelium*   In the more superficial layers of this type of epithelium, the cells are flat, whereas in the deep layers, cells vary in shape from cuboidal to columnar. The basal, or bottom, cells continually replicate by cell division. As new cells grow, they push the surface cells outward. The basal cells continually shift upward and outward. As they move farther from the deep layer and their blood supply, they become dehydrated, shrink, and become harder. At the surface, the cells are rubbed off. New cells continually emerge, are sloughed off, and replaced.

One form of stratified squamous epithelium is called *nonkeratinized stratified squamous epithelium.* This tissue is found on wet surfaces that are subjected to considerable wear and tear—the lining of the mouth, the tongue, the esophagus, and the vagina. Another form of stratified squamous epithelium is called *keratinized stratified squamous epithelium.* The surface cells of this type form a tough layer of material containing keratin. *Keratin* is a protein that is waterproof and resistant to friction and helps resist bacterial invasion. The outer layer of skin consists of keratinized tissue.

● *Stratified Cuboidal Epithelium*   This relatively rare type of epithelium is found in the ducts of the sweat glands of adults, fornix of the conjunctiva of the eye, cavernous urethra of the male urogenital system, pharynx, and epiglottis. It sometimes consists of more than two layers of cells. Its function is mainly protective.

● *Stratified Columnar Epithelium*   Like stratified cuboidal epithelium, this type of tissue is also uncommon in the body. Usually, the basal layer or layers consist of shortened, irregularly polyhedral cells. Only the superficial cells are columnar in form. This kind of epithelium lines part of the male urethra, some larger excretory ducts such as lactiferous (milk) ducts in the mammary glands, and small areas in the anal mucous membrane. It functions in protection and secretion.

● *Transitional Epithelium*   This kind of epithelium is very much like nonkeratinized stratified squamous epithelium. The distinction is that cells of the outer layer in transitional epithelium tend to be large and rounded rather than flat. This feature allows the tissue to be stretched (distended) without the outer cells breaking apart from one another. When stretched, they are drawn out into squamous-like cells. Because of this arrangement, transitional epithelium lines hollow structures that are subjected to expansion from within, such as the urinary bladder, and parts of the ureters and urethra. Its function is to help prevent a rupture of the organ.

Pseudostratified Epithelium

The third category of covering and lining epithelium consists of columnar cells and is called pseudostratified epithelium. The nuclei of the cells are at varying depths. Even though all the cells are attached to the basement membrane in a single layer, some do not reach the surface. This feature gives the impression of a multilayered tissue, the reason for the designation *pseudo*stratified epithelium. It lines the larger excretory ducts of many glands, epididymis, parts of the male urethra, and the auditory (Eustachian) tubes, the tubes that connect the middle ear cavity and upper part of the throat. Pseudostratified epithelium may be ciliated and may contain goblet cells. It lines most of the upper repiratory tract and certain ducts of the male reproductive system.

GLANDULAR EPITHELIUM

The function of glandular epithelium is secretion, accomplished by glandular cells that lie in clusters deep to the covering and lining epithelium. A *gland* may consist of one cell or a group of highly specialized epithelial cells that secrete substances into ducts, onto a surface, or into the blood. The production of such substances always requires active work by the cells and results in an expenditure of energy.

All glands of the body are classified as exocrine or endocrine according to whether they secrete substances into ducts (or directly onto a free surface) or into the blood. *Exocrine glands* secrete their products into ducts (tubes) that empty at the surface of covering and lining epithelium or directly onto a free surface. The product of an exocrine gland may be released at the skin surface or into the lumen of a hollow organ. The secretions of exocrine glands include mucus, perspiration, oil, wax, and digestive enzymes. Examples of exocrine glands are sweat glands, which eliminate perspiration to cool the skin, salivary glands, which secrete a digestive enzyme, and goblet cells, which produce mucus. *Endocrine glands* are ductless and secrete their products into the blood. The secretions of endocrine glands are always hormones, chemicals that regulate various physiological activities. The pituitary, thyroid, and adrenal (suprarenal) glands are examples of endocrine glands.

Structural Classification of Exocrine Glands

Exocrine glands are classified into two structural types: unicellular and multicellular. *Unicellular glands* are single celled. A good example of a unicellular gland is a goblet cell (see Exhibit 3-1; simple columnar, nonciliated). Goblet cells are found in the epithelial lining of the digestive and respiratory systems. They produce mucus to lubricate the free surfaces of these membranes.

Multicellular glands occur in several different forms (Figure 3-2). If the secretory portions of a gland are tubular, it is referred to as a *tubular gland.* If they are flasklike, it is called an *acinar* (AS-i-nar) *gland.* If the gland contains both tubular and flasklike secretory portions, it is called a

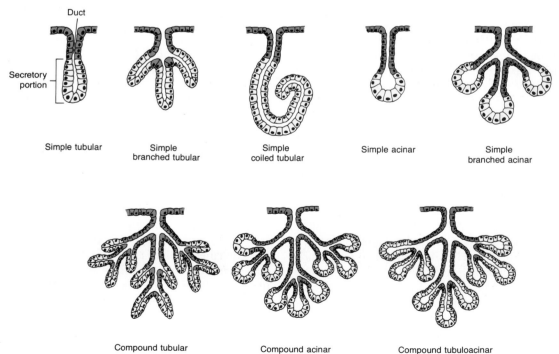

FIGURE 3-2 Structural types of multicellular exocrine glands. The secretory portions of the glands are indicated in gold. The blue areas represent the ducts of the glands.

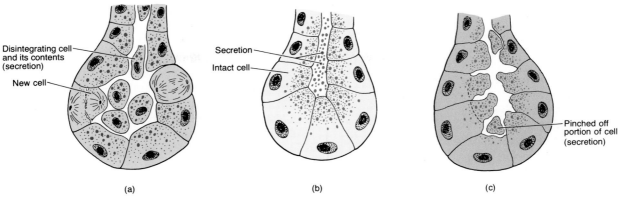

Disintegrating cell and its contents (secretion)

New cell

Secretion

Intact cell

Pinched off portion of cell (secretion)

(a) (b) (c)

FIGURE 3-3 Functional classification of multicellular exocrine glands. (a) Holocrine gland. (b) Merocrine gland. (c) Apocrine gland.

tubuloacinar gland. Furthermore, if the duct of the gland does not branch, it is referred to as a *simple gland;* if the duct does branch, it is called a *compound gland.* By combining the shape of the secretory portion with the degree of branching of the duct, we arrive at the following structural classification for exocrine glands:

I. *Unicellular.* Single-celled gland that secretes mucus. Example: goblet cell of the digestive and respiratory systems.

II. *Multicellular.* Many-celled glands.
  A. *Simple.* Single, nonbranched duct.
    1. *Tubular.* The secretory portion is straight and tubular. Example: intestinal glands.
    2. *Branched tubular.* The secretory portion is branched and tubular. Examples: gastric and uterine glands.
    3. *Coiled tubular.* The secretory portion is coiled. Example: eccrine sudoriferous (sweat) glands.
    4. *Acinar.* The secretory portion is flasklike. Example: seminal vesicle glands.
    5. *Branched acinar.* The secretory portion is branched and flasklike. Example: sebaceous (oil) glands.
  B. *Compound.* Branched duct.
    1. *Tubular.* The secretory portion is tubular. Examples: bulbourethral (Cowper's) glands, testes, and liver.
    2. *Acinar.* The secretory portion is flasklike. Examples: salivary glands (sublingual and submandibular).
    3. *Tubuloacinar.* The secretory portion is both tubular and flasklike. Examples: salivary glands (parotid) and pancreas.

Functional Classification of Exocrine Glands

The functional classification of exocrine glands is based on whether a secretion is a product of a cell or consists of entire or partial glandular cells themselves. The three recognized categories are holocrine, merocrine, and apocrine glands. *Holocrine glands* accumulate a secretory product in their cytoplasm. The cell then dies and is discharged with its contents as the glandular secretion (Figure 3-3a). The discharged cell is replaced by a new cell. One example of a holocrine gland is a sebaceous (oil) gland of the skin. *Merocrine (eccrine) glands* simply form the secretory product and discharge it from the cell (Figure 3-3b). Most exocrine glands of the body are merocrine. Examples of merocrine glands are the salivary glands and pancreas. *Apocrine glands* accumulate their secretory product at the apical (outer) margin of the secreting cell. That portion of the cell pinches off from the rest of the cell to form the secretion (Figure 3-3c). The remaining part of the cell repairs itself and repeats the process. An example of an apocrine gland is the large, modified sweat gland found in the axilla.

CONNECTIVE TISSUE

The most abundant tissue in the body is *connective tissue.* This binding and supporting tissue usually is highly vascular and thus has a rich blood supply. An exception is cartilage, which is avascular. The connective tissue cells are usually widely scattered, rather than closely packed, and there is considerable intercellular material (matrix). In contrast to epithelium, connective tissues do not occur on free surfaces, such as the surfaces of a body cavity or the external surface of the body. The general functions of connective tissues are protection, support, binding together various organs, and separating structures such as skeletal muscles.

The intercellular substance in a connective tissue largely determines the tissue's qualities. This substance is nonliving and may consist of fluid, semifluid, gel-like, or fibrous material. In cartilage, the intercellular material is firm but pliable. In bone, it is considerably harder and not pliable. The cells of connective tissue produce the intercellular substances. The cells may also store fat, ingest bacteria and cell debris, form anticoagulants, or give rise to antibodies that protect against disease.

CLASSIFICATION

Connective tissues may be classified in several ways. We shall classify them as follows:

  I. *Embryonic connective tissue*
    A. Mesenchyme
    B. Mucous connective tissue
 II. *Adult connective tissue*
    A. Connective tissue proper
      1. Loose (areolar) connective tissue
      2. Adipose tissue
      3. Dense (collagenous) connective tissue
      4. Elastic connective tissue
      5. Reticular connective tissue
    B. Cartilage
      1. Hyaline cartilage
      2. Fibrocartilage
      3. Elastic cartilage
    C. Osseous (bone) tissue
    D. Vascular (blood) tissue

Each of the connective tissues described in the following sections is illustrated in Exhibit 3-2.

EMBRYONIC CONNECTIVE TISSUE

Connective tissue that is present primarily in the embryo or fetus is called ***embryonic connective tissue.*** The term *embryo* refers to a developing human from fertilization through the first two months of pregnancy; a *fetus* refers to a developing human from the third month of pregnancy to birth.

One example of embryonic connective tissue found almost exclusively in the embryo is ***mesenchyme*** (MEZ-en-kīm)—the tissue from which all other connective tissues eventually arise. Mesenchyme is located beneath the skin and along the developing bones of the embryo. Some mesenchymal cells are scattered irregularly throughout adult connective tissue, most frequently around blood vessels. Here mesenchymal cells differentiate into fibroblasts that assist in wound healing.

Another kind of embryonic connective tissue is ***mucous connective tissue (Wharton's jelly),*** found primarily in the fetus. This tissue is located in the umbilical cord of the fetus, where it supports the wall of the cord.

ADULT CONNECTIVE TISSUE

Adult connective tissue is connective tissue that exists in the newborn and that does not change after birth. It is subdivided into several kinds.

Connective Tissue Proper

Connective tissue that has a more or less fluid intercellular material and a fibroblast as the typical cell is termed ***connec-*** ***tive tissue proper.*** Five examples of such tissues may be distinguished.

● ***Loose (Areolar) Connective Tissue***   Loose or areolar (a-RĒ-ō-lar) connective tissue is one of the most widely distributed connective tissues in the body. Structurally, it consists of fibers and several kinds of cells embedded in a semifluid intercellular substance. The term ''loose'' refers to the loosely woven arrangement of fibers in the intercellular substance. The fibers are neither abundant nor arranged to prevent stretching.

The intercellular substance consists primarily of hyaluronic acid, chondroitin sulfate, dermatan sulfate, and keratan sulfate. It is secreted by connective tissue cells, mostly fibroblasts. The intercellular substance normally facilitates the passage of nutrients from the blood vessels of the connective tissue into adjacent cells and tissues, although the thick consistency of this acid may impede the movement of some drugs. But if an enzyme called ***hyaluronidase*** is injected into the tissue, the intercellular substance changes to a watery consistency. This feature is of clinical importance because the reduced viscosity hastens the absorption and diffusion of injected drugs and fluids through the tissue and thus can lessen tension and pain. Some bacteria, white blood cells, and sperm cells produce hyaluronidase.

The three types of fibers embedded between the cells of loose connective tissue are collagenous, elastic, and reticular fibers. ***Collagenous fibers*** are very tough and resistant to a pulling force, yet are somewhat flexible because they are usually wavy. These fibers often occur in bundles. They are composed of many minute fibers called fibrils lying parallel to one another. The bundle arrangement affords a great deal of strength. Chemically, collagenous fibers consist of the protein collagen. ***Elastic fibers,*** by contrast, are smaller than collagenous fibers and freely branch and rejoin one another. Elastic fibers consist of a protein called elastin. These fibers also provide strength and have great elasticity, up to 50 percent of their length. ***Reticular fibers*** also consist of collagen, plus some glycoprotein. They are very thin fibers that form branching networks. Like collagenous fibers, reticular fibers provide support and strength and also form the ***stroma*** (framework) of many soft organs.

The cells in loose connective tissue are numerous and varied. Most are ***fibroblasts***—large, flat cells with branching processes. If the tissue is injured, fibroblasts are believed to form collagenous fibers, elastic fibers, and the intercellular substance. When a fibroblast becomes relatively inactive, it is sometimes referred to as a ***fibrocyte.***

Other cells found in loose connective tissue are called ***macrophages*** (MAK-rō-fā-jez; *macro* = large; *phagein* = to eat), which are derived from white blood cells called monocytes. They are irregular in form with short branching projections and are capable of engulfing bacteria and cellular debris by phagocytosis. Thus, they provide a vital defense for the body.

A third kind of cell in loose connective tissue is a

EXHIBIT 3-2

Connective Tissues

EMBRYONIC

Mesenchymal (180×)

Description: Consists of highly branched mesenchymal cells embedded in a fluid substance.

Location: Under skin and along developing bones of embryo; some mesenchymal cells found in adult connective tissue, especially along blood vessels.

Function: Forms all other kinds of connective tissue.

Blood vessels

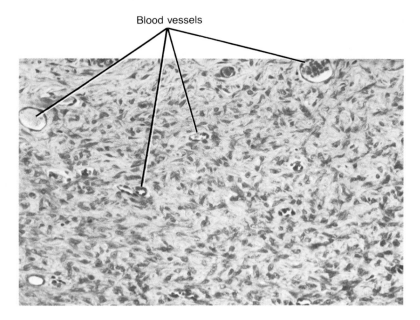

Mucous (320×)

Description: Consists of flattened or spindle-shaped cells embedded in a mucuslike substance containing fine collagenous fibers.

Location: Umbilical cord of fetus.

Function: Support.

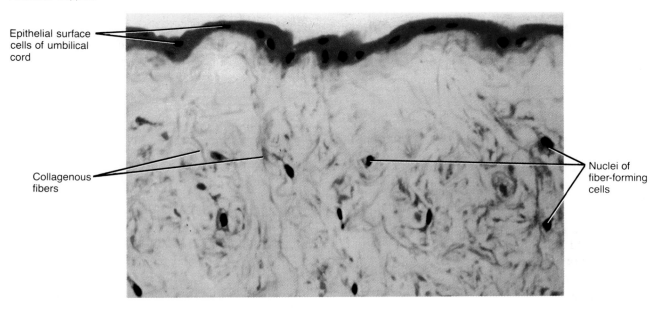

Epithelial surface cells of umbilical cord

Collagenous fibers

Nuclei of fiber-forming cells

ADULT

Loose or areolar (180 ×)

Description: Consists of fibers (collagenous, elastic, and reticular) and several kinds of cells (fibroblasts, macrophages, plasma cells, and mast cells) embedded in a semifluid ground substance.
Location: Subcutaneous layer of skin, mucous membranes, blood vessels, nerves, and body organs.
Function: Strength, elasticity, and support.

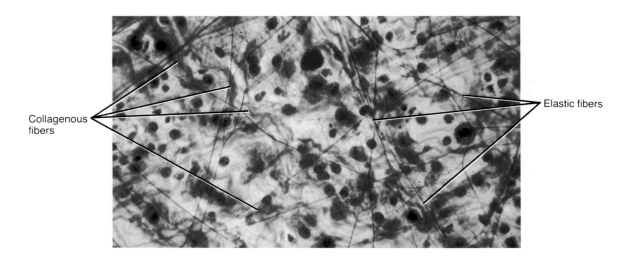

Collagenous fibers

Elastic fibers

Adipose (250 ×)

Description: Consists of adipocytes, "signet ring"-shaped cells with peripheral nuclei, that are specialized for fat storage.
Location: Subcutaneous layer of skin, around heart and kidneys, marrow of long bones, and padding around joints.
Function: Reduces heat loss through skin, serves as an energy reserve, supports, and protects.

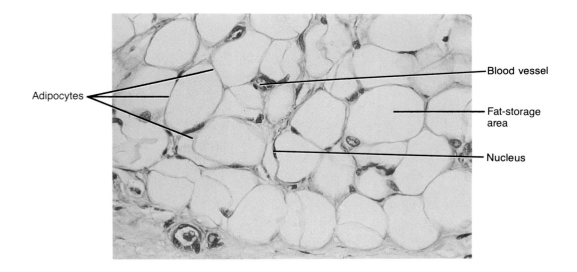

Adipocytes

Blood vessel

Fat-storage area

Nucleus

Dense or collagenous (250 ×)
Description: Consists of predominantly collagenous fibers arranged in bundles; fibroblasts present in rows between bundles.
Location: Forms tendons, ligaments, aponeuroses, membranes around various organs, and fasciae.
Function: Provides strong attachment between various structures.

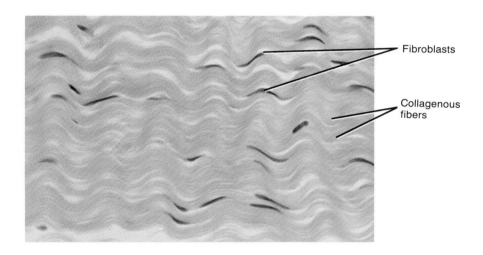

Elastic (180 ×)
Description: Consists of predominantly freely branching elastic fibers; fibroblasts present in spaces between fibers.
Location: Lung tissue, wall of arteries, trachea, bronchial tubes, true vocal cords, and ligamenta flava of vertebrae.
Function: Allows stretching of various organs.

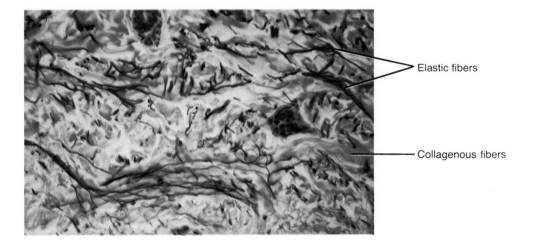

Reticular (250 ×)
Description: Consists of a network of interlacing reticular fibers with thin, flat cells wrapped around fibers.
Location: Liver, spleen, lymph nodes, basal lamina underlying epithelia, and surrounding blood vessels and muscle.
Function: Forms stroma of organs; binds together smooth muscle tissue cells.

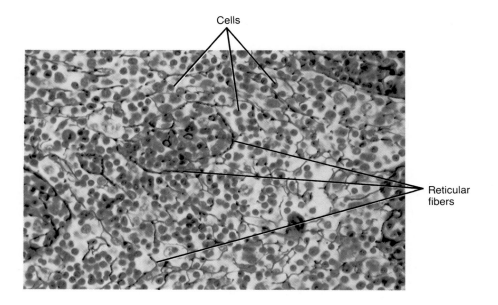

Cells

Reticular fibers

Hyaline cartilage (350 ×)
Description: Also called gristle; appears as a bluish white, glossy mass; contains numerous chondrocytes; is the most abundant type of cartilage.
Location: Ends of long bones, ends of ribs, nose, parts of larynx, trachea, bronchi, bonchial tubes, and embryonic skeleton.
Function: Provides movement at joints, flexibility, and support.

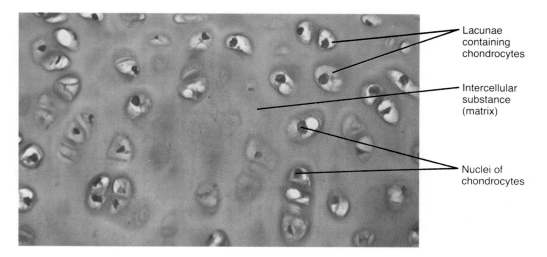

Lacunae containing chondrocytes

Intercellular substance (matrix)

Nuclei of chondrocytes

Fibrocartilage (180 ×)
Description: Consists of chondrocytes scattered among bundles of collagenous fibers.
Location: Symphysis pubis, intervertebral discs, and menisci of knee.
Function: Support and fusion.

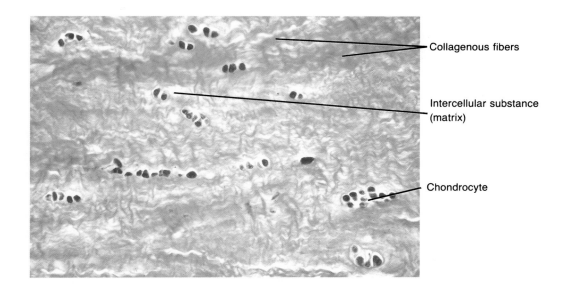

Collagenous fibers

Intercellular substance
(matrix)

Chondrocyte

Elastic cartilage (350 ×)
Description: Consists of chondrocytes located in a threadlike network of elastic fibers.
Location: Epiglottis of larynx, external ear, and auditory (Eustachian) tubes.
Function: Gives support and maintains shape.

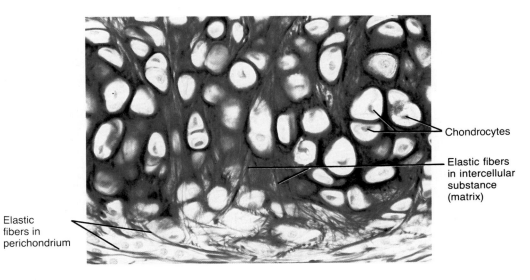

Chondrocytes

Elastic fibers
in intercellular
substance
(matrix)

Elastic
fibers in
perichondrium

EXHIBIT 3-2 (*Continued*)

Osseous (bone) (150×)

Description: Compact bone consists of osteons (Haversian systems) that contain lamellae, lacunae, osteocytes, canaliculi, and central (Haversian) canals.

Location: Both compact and spongy bone comprise the various bones of the body.

Function: Support, protection, movement, storage, and blood cell production.

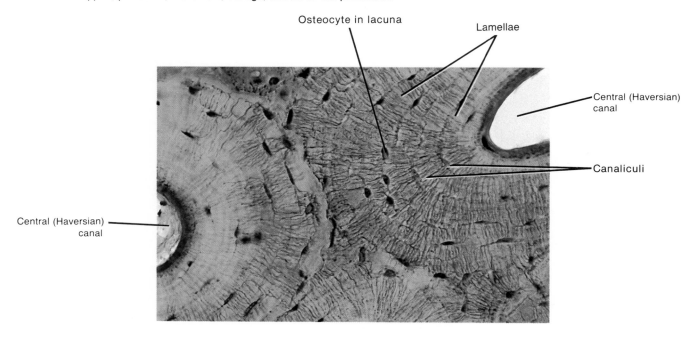

Vascular (blood) (1800×)

Description: Consists of plasma (intercellular substance) and formed elements (erythrocytes, leucocytes, and thrombocytes).

Location: Within blood vessels (arteries, arterioles, capillaries, venules, and veins).

Function: Erythrocytes transport oxygen and carbon dioxide, leucocytes carry on phagocytosis and are involved in allergic reactions and immunity, and thrombocytes are essential to the clotting of blood.

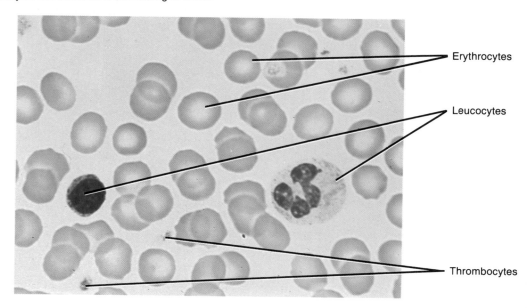

plasma cell. It is small and either round or irregular and develops from a type of white blood cell called a lymphocyte (B cell). Plasma cells give rise to antibodies and accordingly provide a defensive mechanism through immunity. They are found in many places of the body, but most are found in connective tissue, especially that of the gastrointestinal tract and the mammary glands.

Another cell in loose connective tissue is a *mast cell.* The mast cell is found in abundance along blood vessels. It forms heparin, an anticoagulant that prevents blood from clotting in the vessels. Mast cells are also believed to produce histamine and serotonin, chemicals that dilate small blood vessels.

Other cells in loose connective tissue include *adipocytes* (*fat cells*) and *leucocytes* (*white blood cells*).

Loose connective tissue is continuous throughout the body. It is present in all mucous membranes and around all blood vessels and nerves. And it occurs around body organs and in the papillary (upper) region of the dermis of the skin. Combined with adipose tissue, it forms the *subcutaneous* (sub'-kyoo-TĀ-ne-us; *sub* = under; *cut* = skin) *layer*—the layer of tissue that attaches the skin to underlying tissues and organs. The subcutaneous layer is also referred to as the *superficial fascia* (FASH-ē-a) or *hypodermis.*

CLINICAL APPLICATION

Marfan (mar-FAN) *syndrome* is an inherited disorder that affects about 1 person in 10,000. It results in abnormalities of connective tissue, especially in the skeleton, eyes, and cardiovascular system. Victims of Marfan Syndrome tend to be tall, with disproportionately long arms, an unusually long lower half of the body, very long fingers and toes, an especially elongated thumb, an overly curved backbone, a malformed breastbone that either curves outward or inward, a backward curve of the legs, flat feet, and loose joints. Other common, visible signs of Marfan syndrome include leanness, small muscle mass, crowded teeth, and nearsightedness. Among the cardiovascular problems, the most serious ones are a weakened area in the aorta (main artery that emerges from the heart), weakened heart valves, and heart murmurs. Flo Hyman, the best American female volleyball player ever, died of Marfan syndrome in January 1986.

● *Adipose Tissue* Adipose tissue is basically a form of loose connective tissue in which the cells, called *adipocytes,* are specialized for fat storage. Adipocytes are derived from fibroblasts and the cells have the shape of a "signet ring" because the cytoplasm and nucleus are pushed to the edge of the cell by a large droplet of fat. Adipose tissue is found wherever loose connective tissue is located. Specifically, it is in the subcutaneous layer below the skin, around the kidneys, at the base and on the surface of the heart, in the marrow of long bones, as a padding around joints,

and behind the eyeball in the orbit. Adipose tissue is a poor conductor of heat and therefore reduces heat loss through the skin. It is also a major energy reserve and generally supports and protects various organs.

CLINICAL APPLICATION

A surgical procedure, called *suction lipectomy* (*lipo* = fat; *ectomy* = removal of) involves suctioning out small amounts of fat from certain areas of the body. The technique can be used in a number of areas of the body in which fat accumulates, such as the inner and outer thighs, buttocks, the area behind the knee, underside of the arms, breasts, and abdomen. However, suction lipectomy is not a treatment for obesity, since it will not result in a permanent reduction in fat in the area.

● *Dense* (*Collagenous*) *Connective Tissue* Dense (collagenous) connective tissue is characterized by a closer packing of fibers than in loose connective tissue. The fibers can be either irregularly arranged or regularly arranged. In areas of the body where tensions are exerted in various directions, the fiber bundles are interwoven and without regular orientation. Such dense connective tissue is referred to as *irregularly arranged* and occurs in sheets. It forms most fasciae, the reticular (lower) region of the dermis of the skin, the periosteum of bone, the perichondrium of cartilage, and the *membrane* (*fibrous*) *capsules* around organs, such as the kidneys, liver, testes, and lymph nodes.

In other areas of the body, dense connective tissue is adapted for tension in one direction, and the fibers have an orderly, parallel arrangement. Such dense connective tissue is known as *regularly arranged.* The most common variety of dense regularly arranged connective tissue has a predominance of collagenous (white) fibers arranged in bundles. Fibroblasts are placed in rows between the bundles. The tissue is silvery white, tough, yet somewhat pliable. Because of its great strength, it is the principal component of *tendons,* which attach muscles to bones; *aponeuroses* (ap'-ō-noo-RŌ-sēz), which are sheetlike tendons connecting one muscle with another or with bone; and many *ligaments* (collagenous ligaments), which hold bones together at joints.

CLINICAL APPLICATION

Scientists have developed a carbon fiber implant for *reconstructing severely torn ligaments and tendons.* The implant consists of carbon fibers coated with a plastic called polylactic acid. The coated fibers are sewn in and around torn ligaments and tendons to reinforce them and to provide a scaffolding around which the body's own collagenous fibers grow. Within 2 weeks, the polylactic acid is absorbed by the body and the carbon fibers eventually fracture. By this time, the fibers are completely clad in collagen produced by fibroblasts.

● **Elastic Connective Tissue** Unlike collagenous connective tissue, elastic connective tissue has a predominance of freely branching elastic fibers. These fibers give the unstained tissue a yellowish color. Fibroblasts are present only in the spaces between the fibers. Elastic connective tissue can be stretched and will snap back into shape. It is a component of the walls of elastic arteries, trachea, bronchial tubes to the lungs, and the lungs themselves. Elastic connective tissue provides stretch and strength, allowing structures to perform their functions efficiently. Yellow elastic ligaments, as contrasted with collagenous ligaments, are composed mostly of elastic fibers; they form the ligamenta flava of the vertebrae (ligaments between successive vertebrae), the suspensory ligament of the penis, and the true vocal cords.

● **Reticular Connective Tissue** Reticular connective tissue consists of interlacing reticular fibers. It helps form a delicate supporting stroma (framework) for many organs, including the liver, spleen, and lymph nodes. It is also found in the basal lamina underlying epithelia and around blood vessels and muscle. Reticular connective tissue also helps bind together the fibers (cells) of smooth muscle tissue.

Cartilage

Cartilage is capable of enduring considerably more stress than the tissues just discussed. Unlike other connective tissues, cartilage has no blood vessels (except for the perichondrium) or nerves of its own. **Cartilage** consists of a dense network of collagenous fibers and elastic fibers firmly embedded in chondroitin sulfate, a jellylike substance. Whereas the strength of cartilage is due to its collagenous fibers, its resilience (ability to assume its original shape after deformation) is due to chondroitin sulfate. The cells of mature cartilage, called **chondrocytes** (KON-drō-sīts), occur singly or in groups within spaces called **lacunae** (la-KOO-nē) in the intercellular substance. The surface of cartilage is surrounded by irregularly arranged dense connective tissue called the **perichondrium** (per'-i-KON-drē-um; *peri* = around; *chondro* = cartilage). Three kinds of cartilage are recognized; hyaline cartilage, fibrocartilage, and elastic cartilage (Exhibit 3-2).

● **Hyaline Cartilage** This cartilage, also called gristle, appears in the body as a bluish-white, shiny substance. The collagenous fibers, although present, are not visible with ordinary staining techniques, and the prominent chondrocytes are found in lacunae. Hyaline cartilage is the most abundant kind of cartilage in the body. It is found at joints over the ends of the long bones (where it is called **articular cartilage**) and forms the **costal cartilages** at the ventral ends of the ribs. Hyaline cartilage also helps to form the nose, larynx, trachea, bronchi, and bronchial tubes leading to the lungs. Most of the embryonic skeleton consists of hyaline cartilage. Hyaline cartilage affords flexibility and support.

● **Fibrocartilage** Chondrocytes scattered through many bundles of visible collagenous fibers are found in this type of cartilage. Fibrocartilage is found at the symphysis pubis, the point where the coxal (hip) bones fuse anteriorly at the midline. It is also found in the intervertebral discs between vertebrae and the menisci of the knee. This tissue combines strength and rigidity.

● **Elastic Cartilage** In this tissue, chondrocytes are located in a threadlike network of elastic fibers. Elastic cartilage provides strength and maintains the shape of certain organs—the epiglottis of the larynx, the external part of the ear (pinna), and the auditory (Eustachian) tubes.

Osseous Tissue (Bone)

Together, cartilage, joints, and **osseous** (OS-ē-us) **tissue** (**bone**) comprise the skeletal system. Mature bone cells are called **osteocytes.** The intercellular substance consists of mineral salts, primarily calcium phosphate and calcium carbonate, and collagenous fibers. The salts are responsible for the hardness of bone.

Bone tissue is classified as either compact (dense) or spongy (cancellous), depending on how the intercellular substance and cells are organized. At this point, we shall discuss compact bone only. The basic unit of compact bone is called an **osteon (Haversian system).** Each osteon consists of **lamellae,** concentric rings of hard, intercellular substance; **lacunae,** small spaces between lamellae that contain osteocytes; **canaliculi,** radiating minute canals that provide numerous routes so that nutrients can reach osteocytes and wastes can be removed from them; and a **central (Haversian) canal** that contains blood vessels and nerves. Note that whereas osseous tissue is vascular, cartilage is not. Also, the lacunae of osseous tissue are interconnected by canaliculi; those of cartilage are not.

Functionally, the skeletal system supports soft tissues, protects delicate structures, works with skeletal muscles to produce movement, stores calcium and phosphorus, and houses red marrow, which produces several kinds of blood cells.

The details of compact and spongy bone are discussed in Chapter 5.

Vascular Tissue (Blood)

Vascular tissue (**blood**) is a liquid connective tissue that consists of an intercellular substance called plasma and formed elements (cells and cell-like structures). **Plasma** is a straw-colored liquid that consists mostly of water plus some dissolved substances (nutrients, enzymes, hormones, respiratory gases, and ions). The formed elements are erythrocytes (red blood cells), leucocytes (white blood cells), and thrombocytes (platelets). The fibers characteristic of all connective tissues are present in blood only when it is clotted.

Erythrocytes function in transporting oxygen to body

cells and removing carbon dioxide from them. *Leucocytes* are involved in phagocytosis, immunity, and allergic reactions. *Thrombocytes* function in blood clotting.

The details of blood are considered in Chapter 12.

MEMBRANES

The combination of an epithelial layer and an underlying connective tissue layer constitutes an *epithelial membrane.* The principal epithelial membranes of the body are mucous membranes, serous membranes, and the cutaneous membrane, or skin. Another kind of membrane, a *synovial membrane,* does not contain epithelium.

MUCOUS MEMBRANES

A *mucous membrane,* or *mucosa,* lines a body cavity that opens directly to the exterior. Mucous membranes line the entire gastrointestinal, respiratory, excretory, and reproductive tracts (see Figure 23-2) and consist of a lining layer of epithelium and an underlying layer of connective tissue. In addition, most mucous membranes also contain a layer of smooth muscle called the muscularis mucosae.

The epithelial layer of a mucous membrane secretes mucus, which prevents the cavities from drying out. It also traps particles in the respiratory passageways and lubricates food as it moves through the gastrointestinal tract. In addition, the epithelial layer is responsible for the secretion of digestive enzymes and the absorption of food.

The connective tissue layer of a mucous membrane is called the *lamina propria.* The lamina propria is so named because it belongs to the mucous membrane (*proprius =* one's own). It binds the epithelium to the underlying structures and allows some flexibility of the membrane. It also holds the blood vessels in place, protects underlying muscles from abrasion or puncture, provides the epithelium covering it with oxygen and nutrients, and removes wastes.

The *muscularis mucosae* contains smooth muscle fibers and separates the mucosa from the submucosa beneath. Its role in the gastrointestinal tract is considered in Chapter 23.

SEROUS MEMBRANES

A *serous membrane,* or *serosa,* lines a body cavity that does not open directly to the exterior, and it covers the organs that lie within the cavity. Serous membranes consist of thin layers of loose connective tissue covered by a layer of mesothelium, and they consist of two portions. The part attached to the cavity wall is called the *parietal* (pa-RĪ-e-tal) *portion.* The part that covers the organs inside these cavities is the *visceral portion.* The serous membrane lining the thoracic cavity and covering the lungs is called the *pleura* (see Figure 1-7). The membrane lining the heart cavity and covering the heart is the *pericardium* (*cardio* = heart). The serous membrane lining the abdominal cavity and covering the abdominal organs and some pelvic organs is called the *peritoneum.*

The epithelial layer of a serous membrane secretes a lubricating fluid that allows the organs to glide easily against one another or against the walls of the cavities. The connective tissue layer of the serous membrane consists of a relatively thin layer of loose connective tissue.

CUTANEOUS MEMBRANE

The *cutaneous membrane,* or skin, constitutes an organ of the integumentary system and is discussed in Chapter 4.

SYNOVIAL MEMBRANES

Synovial membranes line the cavities of the joints (see Figure 8-1a). Like serous membranes, they line structures that do not open to the exterior. Unlike mucous, serous, and cutaneous membranes, they do not contain epithelium and are therefore not epithelial membranes. They are composed of loose connective tissue with elastic fibers and varying amounts of fat. Synovial membranes secrete *synovial fluid,* which lubricates the ends of bones as they move at joints and nourishes the articular cartilage covering the bones that form the joints.

MUSCLE TISSUE

Muscle tissue consists of fibers (cells) that are highly specialized for contraction. As a result of this characteristic, muscle tissue provides motion, maintenance of posture, and heat production. On the basis of certain structural and functional characteristics, muscle tissue is classified into three types: skeletal, cardiac, and smooth (Exhibit 3-3).

Skeletal muscle tissue is named for its location—attached to bones. It is also *striated;* that is, the fibers (cells) contain alternating light and dark bands (striations) that are perpendicular to the long axes of the fibers. The striations are visible under a microscope. Skeletal muscle tissue is also *voluntary* because it can be made to contract by conscious control. A single skeletal muscle fiber is cylindrical, and the fibers are parallel to each other in a tissue. Each muscle fiber contains a plasma membrane, the *sarcolemma,* surrounding the cytoplasm, or *sarcoplasm.* Skeletal muscle fibers are multinucleate (they have more than one nucleus), and the nuclei lie close to the sarcolemma. The contractile elements of skeletal muscle fibers are proteins called *myofilaments.* They contain wide, transverse, dark bands and narrow light ones that give the fibers the striated appearances.

Cardiac muscle tissue forms the bulk of the wall of the heart. Like skeletal muscle tissue, it is striated. However, unlike skeletal muscle tissue, it is usually involuntary; its contraction is usually not under conscious control. Cardiac muscle fibers are roughly quadrangular and branch to form networks throughout the tissue. The fibers usually have only one nucleus that is centrally located. Cardiac muscle

EXHIBIT 3-3

Muscle Tissue

Skeletal muscle tissue (800 ×)
Description: Cylindrical, striated fibers with several peripheral nuclei; voluntary.
Location: Attached to bones.
Function: Motion, posture, heat production.

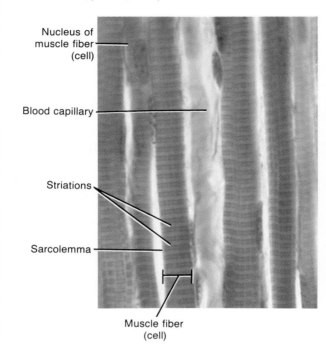

Nucleus of muscle fiber (cell)

Blood capillary

Striations

Sarcolemma

Muscle fiber (cell)

Cardiac muscle tissue (400×*)*
Description: Quadrangular, branching, striated fibers with one centrally located nucleus; contains intercalated discs; usually involuntary.
Location: Heart wall.
Function: Motion (contraction of heart).

Intercalated disc

Muscle fiber (cell)

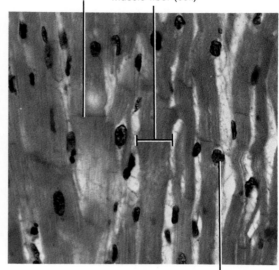

Nucleus of muscle fiber (cell)

Smooth muscle tissue (840 ×)
Description: Spindle-shaped, nonstriated fibers with one centrally located nucleus; usually involuntary.
Location: Walls of hollow internal structures such as blood vessels, stomach, intestines, and urinary bladder.
Function: Motion (constriction of blood vessels, propulsion of foods through gastrointestinal tract, contraction of gall bladder).

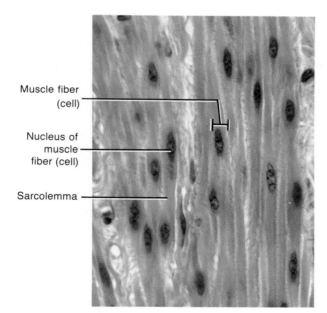

Muscle fiber (cell)

Nucleus of muscle fiber (cell)

Sarcolemma

Photomicrographs courtesy of Andrew Kuntzman.

fibers are separated from each other by transverse thickenings of the sarcolemma called **intercalated discs.** These are unique to cardiac muscle and serve to strengthen the tissue and aid nerve impulse conduction.

Smooth muscle tissue is located in the walls of hollow internal structures such as blood vessels, the stomach, intestines, and urinary bladder. Smooth muscle fibers are usually involuntary and they are **nonstriated (smooth).** Each smooth muscle fiber is thickest in the midregion, with either end tapering to a point, and contains a single, centrally located nucleus.

A more detailed discussion of muscle tissue is considered in Chapter 9.

NERVOUS TISSUE

Despite the tremendous complexity of the nervous system, it consists of only two principal kinds of cells; neurons and neuroglia. *Neurons,* or nerve cells, are highly specialized cells that are capable of picking up stimuli; converting stimuli to nerve impulses; and conducting nerve impulses to other neurons, muscle fibers, or glands. Neurons are the structural and functional units of the nervous system.

Most consist of three basic portions: cell body and two kinds of processes called dendrites and axons (Exhibit 3-4). The *cell body (perikaryon)* contains the nucleus and other organelles. *Dendrites* are highly branched processes of the cell body that conduct nerve impulses toward the cell body. *Axons* are single, long processes of the cell body that conduct nerve impulses away from the cell body.

Neuroglia are cells that protect and support neurons. They are of clinical interest because they are frequently the sites of tumors of the nervous system.

The detailed structure and function of neurons and neuroglia are considered in Chapter 16.

A summary of the principal tissues discussed in this chapter is presented in Exhibit 3-5.

EXHIBIT 3-4

Nervous Tissue (640 ×)

Description: Neurons (nerve cells) consist of a cell body and processes extending from the cell body called dendrites or axons.
Location: Nervous system.
Function: Pick up stimuli, convert stimuli to nerve impulses, and conduct nerve impulses to other neurons, muscle fibers, or glands.

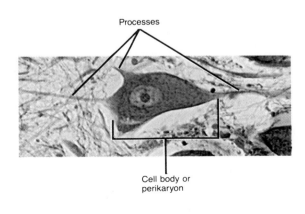

Processes

Cell body or perikaryon

EXHIBIT 3-5

Summary of Tissues

TISSUE	COMMENT
EPITHELIUM	
Covering and Lining	Forms outer covering of body and some viscera; lines body cavities, some viscera, blood vessels, and ducts; makes up parts of sense organs; and produces gametes.
Simple	Single layer of cells.
Squamous	Flat, scalelike cells.
Cuboidal	Cube-shaped cells.
Columnar	Rectangular-shaped cells.
Stratified	Two or more layers of cells.
Squamous	Flat, scalelike cells.
Cuboidal	Cube-shaped cells.
Columnar	Rectangular-shaped cells.
Transitional	Cells variable in shape.
Pseudostratified	Single layer of cells that appears to be stratified.
Glandular	Forms secretory portions of glands.
Exocrine	Secrete products into ducts.
Endocrine	Secrete hormones into the blood.
CONNECTIVE	
Embryonic	Present primarily in embryo and fetus.
Mesenchyme	Embryonic tissue from which all other connective tissues develop.
Mucous	Fetal tissue found in umbilical cord.
Adult	Found in newborn and does not change after birth.
Connective Tissue Proper	Intercellular substance is fluid and fibroblast is the typical cell.
Loose (Areolar)	Most abundant connective tissue.
Adipose	Specialized for fat storage.
Dense (Collagenous)	Forms tendons and ligaments.
Elastic	Provides stretch and strength.
Reticular	Forms stroma (framework) of many organs.
Cartilage	Has no blood or nervous supply.
Hyaline	Also called gristle; found at ends of long bones and ribs.
Fibrocartilage	Found in symphysis pubis and intervertebral discs.
Elastic	Found in larynx and ear.
Osseous (Bone)	Contains very rigid intercellular substance and is classified as compact or spongy.
Vascular (Blood)	Liquid connective tissue consisting of plasma and formed elements (erythrocytes, leucocytes, and thrombocytes).
MUSCULAR	Highly specialized for contraction.
Skeletal	Found attached to bones, striated, voluntary.
Cardiac	Found in the heart, striated, involuntary.
Smooth	Found in viscera and blood vessels, nonstriated, involuntary.
NERVOUS	
Neurons	Specialized for detecting stimuli, converting them into nerve impulses, and conducting nerve impulses.
Neuroglia	Protect and support neurons.

STUDY OUTLINE

Types of Tissues (p. 64)

1. A tissue is a group of similar cells and their intercellular substance that have a similar embryological origin and are specialized for a particular function.
2. Depending on their function and structure, the various tissues of the body are classified into four principal types: epithelial, connective, muscular, and nervous.

Epithelial Tissue (p. 64)

1. Epithelium has many cells, little intercellular material, and no blood vessels (avascular). It is attached to connective tissue by a basement membrane. It can replace itself.
2. The subtypes of epithelium include covering and lining epithelium and glandular epithelium.
3. Cell junctions found between epithelial cells include tight junctions, intermediate junctions, desmosomes, and gap junctions.

Covering and Lining Epithelium (p. 65)

1. Layers are arranged as simple (one layer), stratified (several layers), and pseudostratified (one layer that appears as several); cell shapes include squamous (flat), cuboidal (cubelike), columnar (rectangular), and transitional (variable).
2. Simple squamous epithelium is adapted for diffusion and filtration and is found in lungs and kidneys. Endothelium lines the heart and blood vessels. Mesothelium lines the thoracic and abdominopelvic cavities and covers the organs within them.
3. Simple cuboidal epithelium is adapted for secretion and absorption. It is found covering ovaries, in kidneys and eyes, and lining some glandular ducts.
4. Nonciliated simple columnar epithelium lines most of the gastrointestinal tract. Goblet cells secrete mucus. In a few portions of the respiratory tract, the cells are ciliated to move foreign particles trapped in mucus out of the body.
5. Stratified squamous epithelium is protective. It lines the upper gastrointestinal tract and vagina and forms the outer layer of skin.
6. Stratified cuboidal epithelium is found in adult sweat glands, pharynx, epiglottis, and portions of the urethra.
7. Stratified columnar epithelium protects and secretes. It is found in the male urethra and large excretory ducts.
8. Transitional epithelium lines the urinary bladder and is capable of stretching.
9. Pseudostratified epithelium has only one layer but gives the appearance of many. It lines larger excretory ducts, parts of urethra, auditory (Eustachian) tubes, and most upper respiratory structures, where it protects and secretes.

Glandular Epithelium (p. 74)

1. A gland is a single cell or a mass of epithelial cells adapted for secretion.
2. Exocrine glands (sweat, oil, and digestive glands) secrete into ducts or directly onto a free surface.
3. Structural classification includes unicellular and multicellular glands; multicellular glands are further classified as tubular, acinar, tubuloacinar, simple, and compound.
4. Functional classification includes holocrine, merocrine, and apocrine glands.
5. Endocrine glands secrete hormones into the blood.

Connective Tissue (p. 75)

1. Connective tissue is the most abundant body tissue. It has few cells, an extensive intercellular substance, and a rich blood supply (vascular), except for cartilage. It does not occur on free surfaces.
2. The intercellular substance determines the tissue's qualities.
3. Connective tissue protects, supports, and binds organs together.
4. Connective tissue is classified into two principal types: embryonic and adult.

Embryonic Connective Tissue (p. 76)

1. Mesenchyme forms all other connective tissues.
2. Mucous connective tissue is found in the umbilical cord of the fetus, where it gives support.

Adult Connective Tissue (p. 76)

1. Adult connective tissue is connective tissue that exists in the newborn and does not change after birth. It is subdivided into several kinds: connective tissue proper, cartilage, bone tissue, and vascular tissue.
2. Connective tissue proper has a more or less fluid intercellular material, and a typical cell is the fibroblast. Five examples of such tissues may be distinguished.
3. Loose (areolar) connective tissue is one of the most widely distributed connective tissues in the body. Its intercellular substance (hyaluronic acid) contains fibers (collagenous, elastic, and reticular) and various cells (fibroblasts, macrophages, plasma, and mast). Loose connective tissue is found in all mucous membranes, around body organs, and in the subcutaneous layer.
4. Adipose tissue is a form of loose connective tissue in which the cells, called adipocytes, are specialized for fat storage. It is found in the subcutaneous layer and around various organs.
5. Dense (collagenous) connective tissue has a close packing of fibers (regularly or irregularly arranged). It is found as a component of fascia, membranes of organs, tendons, ligaments, and aponeuroses.
6. Elastic connective tissue has a predominance of freely braching elastic fibers that give it a yellow color. It is found in elastic arteries, trachea, bronchial tubes, and true vocal cords.
7. Reticular connective tissue consists of interlacing reticular fibers and forms the stroma of the liver, spleen, and lymph nodes.
8. Cartilage has a jellylike matrix containing collagenous and elastic fibers and chondrocytes.
9. Hyaline cartilage is found in the embryonic skeleton, at the ends of bones, in the nose, and in respiratory structures. It is flexible, allows movement, and provides support.
10. Fibrocartilage connects the pelvic bones and the vertebrae. It provides strength.

11. Elastic cartilage maintains the shape of organs such as the larynx, auditory (Eustachian) tubes, and external ear.
12. Osseous tissue (bone) consists of mineral salts and collagenous fibers that contribute to the hardness of bone and cells called osteocytes. It supports, protects, helps provide movement, stores minerals, and houses red marrow.
13. Vascular tissue (blood) consists of plasma and formed elements (erythrocytes, leucocytes, and thrombocytes). Functionally, its cells transport, carry on phagocytosis, participate in allergic reactions, provide immunity, and bring about blood clotting.

Membranes (p. 85)
1. An epithelial membrane is an epithelial layer overlying a connective tissue layer. Examples are mucous, serous, and cutaneous membranes.
2. Mucous membranes line cavities that open to the exterior, such as the gastrointestinal tract.
3. Serous membranes (pleura, pericardium, peritoneum) line closed cavities and cover the organs in the cavities. These membranes consist of parietal and visceral portions.
4. The cutaneous membrane is the skin.
5. Synovial membranes line joint cavities and do not contain epithelium.

Muscle Tissue (p. 85)
1. Muscle tissue is modified for contraction and thus provides motion, maintenance of posture, and heat production.
2. Skeletal muscle tissue is attached to bones, is striated, and is voluntary.
3. Cardiac muscle tissue forms most of the heart wall, is striated, and is usually involuntary.
4. Smooth muscle tissue is found in the walls of hollow internal structures (blood vessels and viscera), is nonstriated, and is usually involuntary.

Nervous Tissue (p. 87)
1. The nervous system is composed of neurons (nerve cells) and neuroglia (protective and supporting cells).
2. Most neurons consist of a cell body and two types of processes called dendrites and axons.
3. Neurons are specialized to pick up stimuli, convert stimuli into nerve impulses, and conduct nerve impulses.

REVIEW QUESTIONS

1. Define a tissue. What are the four basic kinds of human tissue?
2. Distinguish covering and lining epithelium from glandular epithelium. What characteristics are common to all epithelium?
3. Describe the origin and composition of the basement membrane.
4. Describe the various layering arrangements and cell shapes of epithelium.
5. Define a cell junction. List several functions of cell junctions.
6. Compare tight junctions, intermediate junctions, desmosomes, hemidesmosomes, and gap junctions on the basis of structure, location, and function.
7. How is epithelium classified? List the various types.
8. For each of the following kinds of epithelium, briefly describe the microscopic appearance, location in the body, and functions: simple squamous, simple cuboidal, simple columnar (nonciliated and ciliated), stratified squamous, stratified cuboidal, stratified columnar, transitional, and pseudostratified.
9. Define the following terms: endothelium, mesothelium, secretion, absorption, goblet cell, and keratin.
10. What is a gland? Distinguish between endocrine and exocrine glands.
11. Describe the classification of exocrine glands according to structure and function and give at least one example of each.
12. Enumerate the ways in which connective tissue differs from epithelium.
13. How are connective tissues classified? List the various types.
14. How are embryonic connective tissue and adult connective tissue distinguished?
15. Describe the following connective tissues with regard to microscopic appearance, location in the body, and function: loose (areolar), adipose, dense (collagenous), elastic, reticular, hyaline cartilage, fibrocartilage, elastic cartilage, osseous tissue (bone), and vascular tissue (blood).
16. Define the following terms: hyaluronic acid, collagenous fiber, elastic fiber, reticular fiber, fibroblast, macrophage, plasma cell, mast cell, melanoctye, adipocyte, chondrocyte, lacuna, osteocyte, lamella, and canaliculi.
17. Define the following kinds of membranes: mucous, serous, cutaneous, and synovial. Where is each located in the body? What are their functions?
18. Describe the histology of muscle tissue. How is it classified? What are its functions?
19. Distinguish between neurons and neuroglia. Describe the structure and function of neurons.
20. Following are some descriptions of various tissues of the body. For each description, name the tissue described.
    a. An epithelium that permits distention (stretching).
    b. A single layer of flat cells concerned with filtration and absorption.
    c. Forms all other kinds of connective tissue.
    d. Specialized for fat storage.
    e. An epithelium with waterproofing qualities.
    f. Forms the framework of many organs.
    g. Produces perspiration, wax, oil, or digestive enzymes.
    h. Cartilage that shapes the external ear.
    i. Contains goblet cells and lines the intestine.
    j. Most widely distributed connective tissue.
    k. Forms tendons, ligaments, and aponeuroses.
    l. Specialized for the secretion of hormones.
    m. Provides support in the umbilical cord.
    n. Lines kidney tubules and is specialized for absorption and secretion.
    o. Permits extensibility of lung tissue.
    p. Stores red marrow, protects, supports.
    q. Nonstriated, usually involuntary muscle tissue.
    r. Consists of granulocytes and agranulocytes.
    s. Composed of a cell body, dendrites, and axon.

SELF-QUIZ

Choose the one best answer to these questions.

_____ 1. A surgeon performing abdominal surgery will pass through the skin, then subcutaneous tissue, then muscle, to reach the _____ membrane lining the inside wall of the abdomen: A. parietal pleura; B. parietal pericardium; C. parietal peritoneum; D. visceral pleura; E. visceral pericardium; F. visceral peritoneum.

_____ 2. Which statement about connective tissue is false? A. cells are very closely packed together; B. connective tissue has an abundant blood supply; C. intercellular substance is present in large amounts; D. it is the most abundant tissue in the body; E. it does not cover or line body surfaces.

_____ 3. Modified columnar cells that are unicellular glands secreting mucus are known as A. cilia; B. microvilli; C. goblet cells; D. branched tubular glands; E. basal cells.

_____ 4. A group of similar cells and their intercellular substance operating together to perform a specialized activity is called a(n) A. organ; B. tissue; C. system; D. organ system; E. organism.

_____ 5. Which statement best describes covering and lining epithelium? A. it is always arranged in a single layer of cells; B. it contains large amounts of intercellular substance; C. it has an abundant blood supply; D. its free surface is exposed to the exterior of the body or to the interior of a hollow structure; E. its cells are widely scattered.

_____ 6. Which statement best describes connective tissue? A. usually contains a large amount of intercellular substance; B. always arranged in a single layer of cells; C. primarily concerned with secretion; D. usually lines a body cavity; E. is avascular.

_____ 7. A gland
A. is either exocrine or endocrine; B. may be single celled or muticellular; C. with a duct system may be classed as simple or compound; D. forms and produces secretions; E. is described by all of the above.

_____ 8. Mucous membranes
A. line cavities of the body that are not open to the outside; B. secrete a thin watery serous fluid; C. cover the outside of such organs as the kidney and stomach; D. are found lining the respiratory and urinary passages; E. are described by none of the above.

_____ 9. Which of the following statements are correct?
(1) Simple squamous epithelium lines blood vessels.
(2) Endothelium is composed of cuboidal cells.
(3) Ciliated epithelium is found only in the respiratory system.
(4) Stratified epithelium is found on the surface of skin.
(5) Transitional epithelium is found in the urinary tract.
A. (1), (2), and (3); B. (1), (3), and (5); C. (1), (4), and (5); D. (1), (2), (3), and (4); E. (2), (4), and (5).

_____ 10. Which type of cell junction assumes a role in nerve impulse conduction between muscle fibers?
A. desmosome; B. intermediate junction; C. tight junction; D. hemidesmosome; E. gap junction.

_____ 11. Since the seminal vesicles contain single, nonbranched ducts and flasklike secretory portions, they are classified as A. compound tubular; B. simple branched acinar; C. simple tubular; D. compound tubuloacinar; E. simple acinar.

_____ 12. Which of the following is involuntary and striated? A. skeletal muscle tissue; B. cardiac muscle tissue; C. smooth muscle tissue; D. visceral muscle tissue; E. neural tissue.

_____ 13. Which tissue is characterized by the presence of cell bodies, dendrites, and axons?
A. muscle; B. vascular; C. nervous; D. epithelial; E. osseous.

14. Match the following:

_____ a. lines the inner surface of the stomach and intestine

_____ b. lines urinary tract, as in bladder, permitting distension

_____ c. lines mouth; present on outer surface of skin

_____ d. single layer of cube-shaped cells; found in kidney tubules and ducts of some glands

_____ e. lines air sacs of lungs where thin cells are required for diffusion of gases into blood

_____ f. not a true stratified; all cells on basement membrane, but some do not reach surface of tissue

_____ g. derived from lymphocyte, gives rise to antibodies and so is helpful in defense

_____ h. phagocytic cell; engulfs bacteria and cleans up debris; important during infection

_____ i. believed to form collagenous and elastic fibers in injured tissue

_____ j. abundant along walls of blood vessels; believed to produce heparin, an anticoagulant, as well as histamine, which dilates blood vessels

_____ k. tissue forming most of heart wall

_____ l. contains lacunae and chondrocytes

_____ m. forms fasciae, tendons, and dermis of skin

_____ n. stores fat and provides insulation

A. cardiac muscle
B. stratified transitional
C. fibroblast
D. pseudostratified
E. dense connective tissue
F. simple columnar
G. macrophage
H. stratified squamous
I. adipose
J. simple cuboidal
K. plasma cell
L. simple squamous
M. mast cell
N. cartilage

Complete the following.

15. Bone tissue is also known as _____ tissue. Compact bone consists of concentric rings, or (lamellae? canaliculi?) with bone cells, called (chondrocytes? osteocytes?), located in tiny spaces called _____.

16. The portion of serous membranes that covers organs (viscera) is called the _____ layer; that portion lining the cavity is named the _____ layer.

17. The embryonic tissue from which all other connective tissues arise is called _____.

18. The structure that attaches epithelium to underlying connective tissue is called the _____.

19. Blood, or _____ tissue, consists of a fluid called _____ containing three types of formed elements.

20. A gland that secretes its product into a duct is referred to as an _____ gland.

4 The Integumentary System

STUDENT OBJECTIVES

1. Define the integumentary system.
2. List the various layers of the epidermis and describe their structure and functions.
3. Describe the composition and function of the dermis.
4. Explain the basis for skin color.
5. Describe the blood supply of the skin.
6. Outline the steps involved in epidermal wound healing and deep wound healing.
7. Compare the structure, distribution, and functions of hair, skin glands, and nails.
8. Describe the effects of aging on the integumentary system.
9. Describe the development of the epidermis, its derivatives, and the dermis.
10. Describe the causes and effects for the following skin disorders: acne, systemic lupus erythematosus (SLE), psoriasis, decubitus ulcers, sunburn, skin cancer, and burns.
11. Define key medical terms associated with the integumentary system.

CHAPTER OUTLINE

■ **Skin**
Functions
Structure
Epidermis
Dermis
Skin Color
Epidermal Ridges and Grooves
Blood Supply
Skin and Temperature Regulation
■ **Skin Wound Healing**
Epidermal Wound Healing
Deep Wound Healing
■ **Epidermal Derivatives**
Hair
   Development and Distribution
   Structure
   Color
   Growth and Replacement
Glands
   Sebaceous (Oil) Glands
   Sudoriferous (Sweat) Glands
   Ceruminous Glands
Nails
■ **Aging and the Integumentary System**
■ **Developmental Anatomy of the Integumentary System**
■ **Applications to Health**
Acne
Systemic Lupus Erythematosus (SLE)
Psoriasis
Decubitus Ulcers
Sunburn
Skin Cancer
Burns
   Classification
   Treatment
■ **Key Medical Terms Associated with the Integumentary System**

An aggregation of tissues that performs a specific function is an *organ*. The next higher level of organization is a *system*—a group of organs operating together to perform specialized functions. The skin and its derivatives, such as hair, nails, glands, and several specialized receptors, constitute the *integumentary* (in'-teg-yoo-MEN-tar-ē) *system* of the body.

The developmental anatomy of the integumentary system is considered at the end of the chapter.

SKIN

The *skin* is an organ because it consists of tissues structurally joined together to perform specific activities. It is one of the larger organs of the body in terms of surface area. For the average adult, the skin occupies a surface area of approximately 19,355 sq cm (3,000 sq in.). The skin is not just a simple thin covering that keeps the body together and gives it protection. The skin is quite complex in structure and performs several functions essential for survival. *Dermatology* (der'-ma-TOL-ō-jē; *dermato* = skin; *logos* = study of) is the medical specialty that deals with the diagnosis and treatment of skin disorders.

FUNCTIONS

The numerous functions of the skin are as follows:

1. *Maintenance of body temperature.* In response to high environmental temperature or strenuous exercise, the production of perspiration by sudoriferous (sweat) glands helps to lower body temperature back to normal. This is described in detail later in the chapter.
2. *Protection.* The skin covers the body and provides a physical barrier that protects underlying tissues from physical abrasion, bacterial invasion, dehydration, and ultraviolet (UV) radiation.
3. *Perception of stimuli.* The skin contains numerous nerve endings and receptors that detect stimuli related to temperature, touch, pressure, and pain (Chapter 20).
4. *Excretion.* Not only does perspiration assume a role in helping to maintain normal body temperature, it also assists in the excretion of small amounts of water, salts, and several organic compounds.
5. *Synthesis of vitamin D.* The term **vitamin D** actually refers to a group of closely related compounds synthesized naturally from a precursor molecule present in the skin upon exposure to ultraviolet (UV) radiation. In the skin, the precursor substance, 7-dehydrocholesterol, is converted to cholecalciferol (vitamin D_3) in the presence of UV radiation. In the liver, cholecalciferol is converted to 25-hydroxycholecalciferol. Then, in the kidneys, this substance is changed into 1,25-dihydroxycalciferol (calcitriol), the most active form of vitamin D that stimulates the absorption of calcium and phosphorus from dietary foods. In subsequent chapters, when we speak of vitamin D, we are really referring to 1,25-dihydroxycalciferol. Vitamin D is actually a hormone, since it is produced in one location in the body, transported by the blood, and then exerts its effect in another location.
6. *Immunity.* Certain cells of the epidermis assume a role in bolstering immunity, your ability to fight disease by producing antibodies.

STRUCTURE

Structurally, the skin consists of two principal parts (Figure 4-1). The outer, thinner portion, which is composed of *epithelium,* is called the **epidermis.** The epidermis is cemented to the inner, thicker, *connective tissue* part called the *dermis.* Thick skin has a relatively thick epidermis, whereas thin skin has a relatively thin epidermis. Beneath the dermis is a *subcutaneous (SC) layer.* This layer, also called the *superficial fascia* or *hypodermis,* consists of areolar and adipose tissues. Fibers from the dermis extend down into the subcutaneous layer and anchor the skin to it. The subcutaneous layer, in turn, is firmly attached to underlying tissues and organs.

EPIDERMIS

The *epidermis* is composed of stratified squamous epithelium and contains four distinct types of cells (see Figure 4-3). The most numerous is known as a *keratinocyte,* a cell that undergoes keratinization. This process will be described shortly. The functions of these cells are to produce keratin, which helps waterproof and protect the skin and underlying tissues, and to participate in immunity. A second type of cell in the epidermis is called a *melanocyte.* It is located at the base of the epidermis and its role is to produce melanin, one of the pigments responsible for skin color. The third and fourth types of cells in the epidermis are called *nonpigmented granular dendrocytes,* two distinct cell types, formerly known as *Langerhans cells* and *Granstein cells.* These cells differ both in their sensitivity to damage by ultraviolet (UV) radiation and their functions in immunity. Langerhans cells are a small population of dendrocytes that arise from bone marrow and migrate to the epidermis and other stratified squamous epithelial tissue in the body. These cells are sensitive to UV radiation and lie above the basal layer of keratinocytes. Langerhans cells interact with cells called helper T cells to assist in the immune response. Granstein cells are dendrocytes that are more resistant to UV radiation and interact with cells called suppressor T cells to assist in the immune response.

The keratinocytes of the epidermis are organized into four or five cell layers, depending on location in the body (see Figures 4-1 and 4-2). Where exposure to friction is greatest, such as in the palms and soles, the epidermis

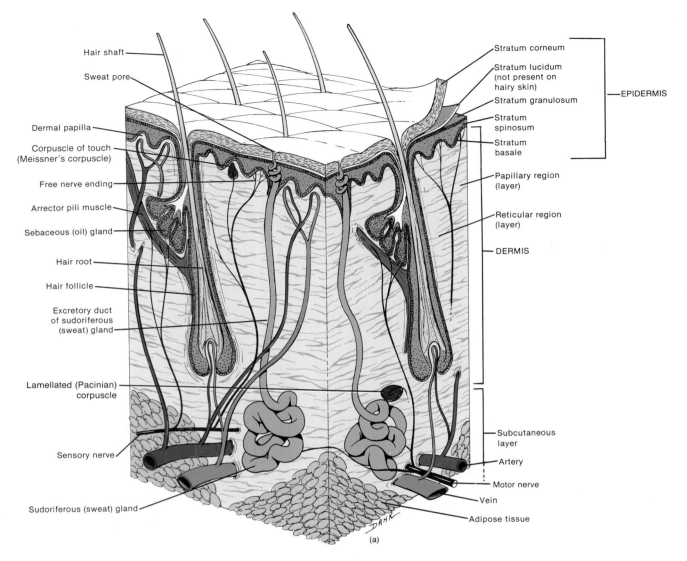

Hair shaft

Sweat pore

Dermal papilla

Corpuscle of touch
(Meissner's corpuscle)

Free nerve ending

Arrector pili muscle

Sebaceous (oil) gland

Hair root

Hair follicle

Excretory duct
of sudoriferous
(sweat) gland

Lamellated (Pacinian)
corpuscle

Sensory nerve

Sudoriferous (sweat) gland

Stratum corneum

Stratum lucidum
(not present on
hairy skin)

Stratum granulosum

Stratum
spinosum

Stratum
basale

EPIDERMIS

Papillary region
(layer)

Reticular region
(layer)

DERMIS

Subcutaneous
layer

Artery

Motor nerve

Vein

Adipose tissue

(a)

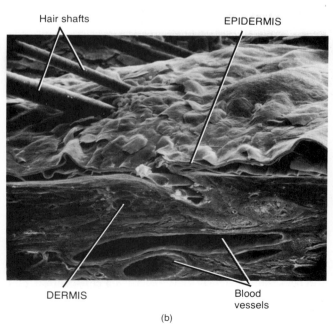

Hair shafts

EPIDERMIS

DERMIS

Blood
vessels

(b)

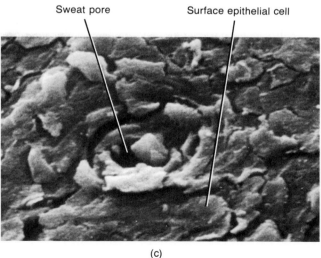

Sweat pore

Surface epithelial cell

(c)

FIGURE 4-1 Skin. (a) Structure of the skin and underlying subcutaneous layer. (b) Scanning electron micrograph of the skin and several hairs at a magnification of 260×. (Courtesy of Richard K. Kessel and Randy H. Kardon, *Tissues and Organs: A Text-Atlas of Scanning Electron Microscopy.* Copyright © 1979 by Scientific American, Inc.) (c) Scanning electron micrograph of a sweat pore at a magnification of 110×. (Courtesy of Brain/Science Photo Library/Photo Researchers.)

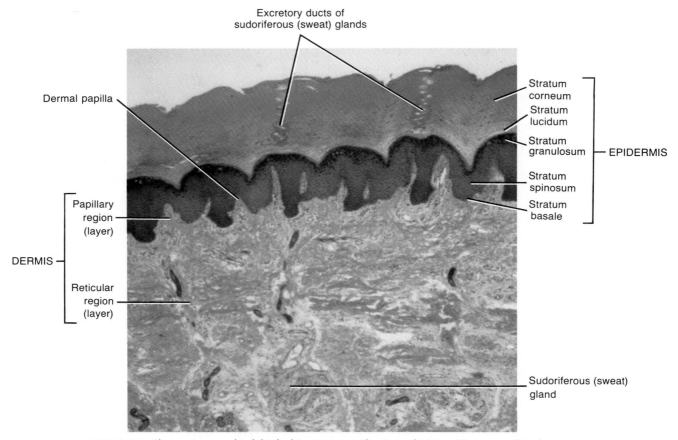

Excretory ducts of
sudoriferous (sweat) glands

Dermal papilla

Stratum
corneum

Stratum
lucidum

Stratum
granulosum

Stratum
spinosum

Stratum
basale

EPIDERMIS

Papillary
region
(layer)

DERMIS

Reticular
region
(layer)

Sudoriferous (sweat)
gland

FIGURE 4-2 Photomicrograph of thick skin at a magnification of 155×. (Courtesy of Andrew Kuntzman.)

has five layers. In all other parts it has four layers. The names of the five layers from the deepest to the most superficial are as follows:

1. **Stratum basale.** This single layer of cuboidal to columnar cells is capable of continued cell division. As these cells multiply, they push up toward the surface and become part of the layers to be described next. Their nuclei degenerate, and the cells die. Eventually, the cells are shed from the top layer of the epidermis. The stratum basale is sometimes referred to as the **stratum germinativum** (jer'-mi-na-TĒ-vum) to indicate its role in germinating new cells. The stratum basale of hairless skin contains nerve endings sensitive to touch called **tactile (Merkel's) discs.**

2. **Stratum spinosum.** This layer of the epidermis contains eight to ten rows of polyhedral (many-sided) cells that fit closely together. The surfaces of these cells contain spinelike projections (*spinosum* = prickly) that join the cells together.

3. **Stratum granulosum.** The third layer of the epidermis consists of three to five rows of flattened cells that contain darkly staining granules of a substance called **keratohyalin** (ker'-a-tō-HĪ-a-lin). This compound is involved in the first step of keratin formation. **Keratin** is a waterproofing protein found in the top layer of the epidermis.

The nuclei of the cells in the stratum granulosum are in various stages of degeneration. As these nuclei break down, the cells are no longer capable of carrying out vital metabolic reactions and die.

4. **Stratum lucidum.** This layer is normally found only in the thick skin of the palms and soles and is absent in thin skin. It consists of three to five rows of clear, flat, dead cells that contain droplets of a substance called **eleidin** (el-Ē-i-din). The layer is so named because eleidin is translucent (*lucidum* = clear). Eleidin is formed from keratohyalin and is eventually transformed to keratin.

5. **Stratum corneum.** This layer consists of 25 to 30 rows of flat, dead cells completely filled with keratin. These cells are continuously shed and replaced. The stratum corneum serves as an effective barrier against light and heat waves, bacteria, and many chemicals.

In the process of **keratinization,** newly formed cells produced in the basal layers are pushed up to more superficial layers. As the cells move toward the surface, the cytoplasm, nucleus, and other organelles are replaced by keratin and the cells die. Eventually, the keratinized cells are sloughed off and replaced by underlying cells that, in turn, become keratinized. The whole process by which a cell forms in the basal layers, rises to the surface, becomes keratinized, and sloughs off takes about 2 weeks.

DERMIS

The second principal part of the skin, the **dermis,** is composed of connective tissue containing collagenous and elastic fibers (see Figure 4-1). The dermis is very thick in the palms and soles and very thin in the eyelids, penis, and scrotum. It also tends to be thicker on the dorsal aspects of the body than the ventral and thicker on the lateral aspects of extremities than medial aspects. Numerous blood vessels, nerves, glands, and hair follicles are embedded in the dermis.

The upper region of the dermis, about one-fifth of the thickness of the total layer, is named the **papillary region** or **layer.** It consists of loose connective tissue containing fine elastic fibers. Its surface area is greatly increased by small, fingerlike projections called **dermal papillae** (pa-PIL-ē). These structures project into the epidermis, and many contain loops of capillaries. Some dermal papillae also contain **corpuscles of touch,** also called **Meissner's** (MĪS-nerz) **corpuscles.** As the name implies, these nerve endings are sensitive to touch.

The remaining portion of the dermis is called the **reticular region** or **layer.** It consists of dense, irregularly arranged connective tissue containing interlacing bundles of collagenous and coarse elastic fibers. It is named the reticular (*rete* = net) region because the bundles of collagenous fibers interlace in a netlike manner. Spaces between the fibers are occupied by a small quantity of adipose tissue, hair follicles, nerves, oil glands, and the ducts of sweat glands. Varying thicknesses of the reticular region, among other factors, are responsible for differences in the thickness of the skin.

The combination of collagenous and elastic fibers in the reticular region provides the skin with strength, extensibility, and elasticity. (Extensibility is the ability to stretch; elasticity is the ability to return to original shape after extension or contraction.) The ability of the skin to stretch can readily be seen during conditions of pregnancy, obesity, and edema. The small tears that occur during extreme stretching are initially red and remain visible afterward as silvery white streaks called **striae** (STRĪ-ē).

The reticular region is attached to underlying organs, such as bone and muscle, by the subcutaneous layer. The subcutaneous layer also contains nerve endings called **lamellated** or **Pacinian** (pa-SIN-ē-an) **corpuscles** that are sensitive to pressure (see Figure 20-1).

CLINICAL APPLICATION

The collagenous fibers in the dermis run in all directions, but in particular regions of the body they tend to run more in one direction than another. The predominant direction of the underlying collagenous fibers is indicated in the skin by **lines of cleavage (tension lines).** The lines are especially evident on the palmar surfaces of the fingers where they run parallel to the long axis of the digit. Lines of cleavage are of particular interest to a surgeon because an incision running parallel to the collagen fibers will heal with only a fine scar. An incision made across the rows of fibers disrupts the collagen, and the wound tends to gape open and heal in a broad, thick scar.

SKIN COLOR

The color of the skin is due to melanin, a pigment in the epidermis; carotene, a pigment mostly in the dermis; and blood in capillaries in the dermis. The amount of **melanin** varies the skin color from pale yellow to black. This pigment is found primarily in the basale and spinosum layers. Melanin is synthesized in cells called **melanocytes** (MEL-a-nō-sīts), located either just beneath or between cells of the stratum basale. Melanocytes are produced from **melanoblasts** (MEL-a-nō-blasts), precursor cells that vary in number from 800 to 2000 cells per cubic millimeter (mm^3). Maximum numbers of melanoblasts are found in mucous membranes, the penis, face, and extremities. Since the number of melanocytes is about the same in all races, differences in skin color are due to the amount of pigment the melanocytes produce and disperse. An inherited inability of an individual in any race to produce melanin results in **albinism** (AL-bin-izm). The pigment is absent in the hair and eyes as well as the skin. An individual affected with albinism is called an **albino** (al-BĪ-no). The partial or complete loss of melanocytes from areas of skin produces patchy, white spots and the condition is called **vitiligo** (vit-i-LĪ-gō). In some people, melanin tends to form in patches called **freckles.**

Melanocytes synthesize melanin from the amino acid *tyrosine* in the presence of an enzyme called **tyrosinase.** Exposure to ultraviolet radiation increases the enzymatic activity of melanocytes and leads to increased melanin production. The cell bodies of melanocytes send out long processes between epidermal cells (Figure 4-3). Upon contact with the processes, epidermal cells take up the melanin by phagocytosis. When the skin is again exposed to ultraviolet radiation, both the amount and the darkness of melanin increase, tanning and further protecting the body against radiation. Thus, melanin serves a vital protective function. Melanocyte-stimulating hormone (MSH) produced by the anterior pituitary gland causes increased melanin synthesis and distribution through the epidermis.

CLINICAL APPLICATION

Overexposure of the skin to the ultraviolet light of the sun may lead to skin cancer. Among the most lethal skin cancers is **malignant melanoma** (*melano* = dark-colored; *oma* = tumor), cancer of the melanocytes. Fortunately, most skin cancers involve basal and squamous

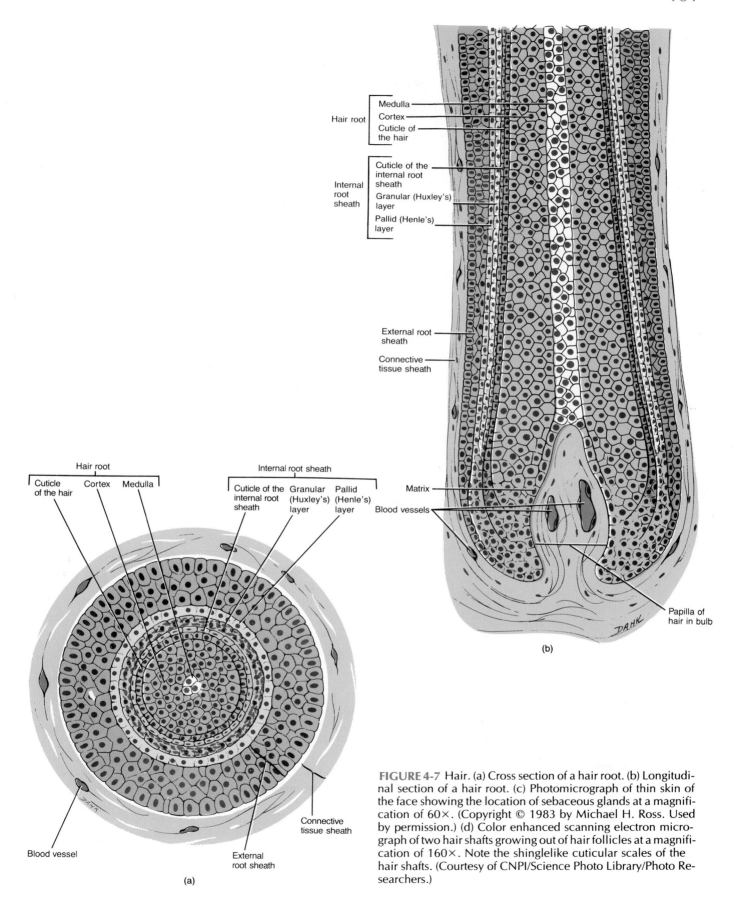

Hair root
Medulla
Cortex
Cuticle of the hair

Internal root sheath
Cuticle of the internal root sheath
Granular (Huxley's) layer
Pallid (Henle's) layer

External root sheath

Connective tissue sheath

Matrix

Blood vessels

Papilla of hair in bulb

(b)

Hair root
Cuticle of the hair
Cortex
Medulla

Internal root sheath
Cuticle of the internal root sheath
Granular (Huxley's) layer
Pallid (Henle's) layer

Blood vessel

External root sheath

Connective tissue sheath

(a)

FIGURE 4-7 Hair. (a) Cross section of a hair root. (b) Longitudinal section of a hair root. (c) Photomicrograph of thin skin of the face showing the location of sebaceous glands at a magnification of 60×. (Copyright © 1983 by Michael H. Ross. Used by permission.) (d) Color enhanced scanning electron micrograph of two hair shafts growing out of hair follicles at a magnification of 160×. Note the shinglelike cuticular scales of the hair shafts. (Courtesy of CNPI/Science Photo Library/Photo Researchers.)

Opening of hair follicle
containing sebum from
sebaceous (oil) gland

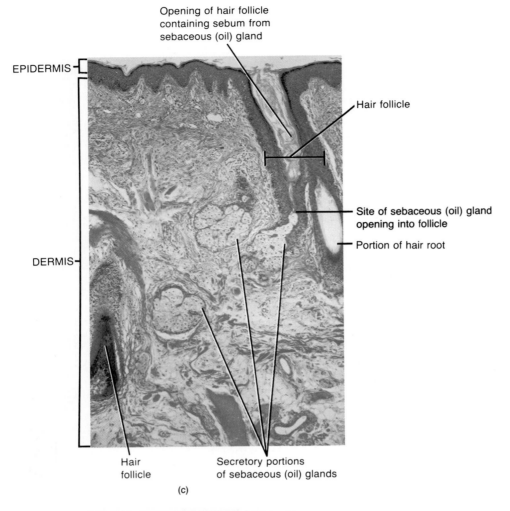

EPIDERMIS

Hair follicle

Site of sebaceous (oil) gland
opening into follicle

Portion of hair root

DERMIS

Hair
follicle

Secretory portions
of sebaceous (oil) glands

(c)

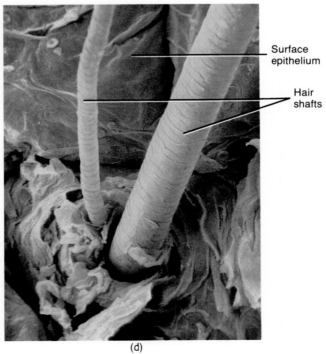

Surface
epithelium

Hair
shafts

(d)

shingles. The **root** is the portion below the surface that penetrates into the dermis and even into the subcutaneous layer and, like the shaft, contains a medulla, cortex, and cuticle.

Surrounding the root is the **hair follicle,** which is made up of an external zone of epithelium (the external root sheath) and an internal zone of epithelium (the internal root sheath). The **external root sheath** is a downward continuation of the basale and spinosum layers of the epidermis. Near the surface it contains all the epidermal layers. As it descends, it does not exhibit the superficial epidermal layers. At the bottom of the hair follicle, the external root sheath contains only the stratum basale. The **internal root sheath** is formed from proliferating cells of the matrix (described shortly) and takes the form of a cellular tubular sheath deep to the external root sheath. The internal root sheath extends only partway up the follicle and consists of (1) an inner layer, the **cuticle of the internal root sheath,** which is a single layer of flattened cells with atrophied nuclei, (2) a middle **granular (Huxley's) layer,** which is one to three layers of flattened nucleated cells, and (3) an outer **pallid (Henle's) layer,** which is a single layer of cuboidal cells with flattened nuclei.

The base of each follicle is enlarged into an onion-shaped structure, the **bulb.** This structure contains an indentation, the **papilla of the hair,** filled with loose connective tissue. The papilla of the hair contains many blood vessels and provides nourishment for the growing hair. The bulb also contains a region of cells called the **matrix,** a germinal layer. The cells of the matrix produce new hairs by cell division when older hairs are shed. This replacement occurs within the same follicle.

CLINICAL APPLICATION

A substance that removes superfluous hair is called a **depilatory.** It dissolves the protein in the hair shaft, turning it into a gelatinous mass that can be wiped away. Since the hair root is not affected, regrowth of the hair occurs. In **electrolysis,** the hair bulb is destroyed by an electric current so that the hair cannot regrow.

Sebaceous (oil) glands and a bundle of smooth muscle are also associated with hair. Details of the sebaceous glands will be discussed shortly. The smooth muscle is called **arrector pili;** it extends from the dermis of the skin to the side of the hair follicle (Figure 4-1a). In its normal position hair is arranged at an angle to the surface of the skin. The arrectores pilorum muscles contract under stresses of fright, cold, and emotions and pull the hairs into a vertical position. This contraction results in ''goosebumps'' or ''gooseflesh'' because the skin around the shaft forms slight elevations.

Around each hair follicle are nerve endings, called **root hair plexuses,** that are sensitive to touch (see Figure 20-1). They respond if a hair shaft is moved.

Color

The color of hair is due primarily to melanin. It is formed by melanocytes distributed in the matrix of the bulb of the follicle. There are two basic classes of melanin: eumelanin (brown-black) and pheomelanin (yellow to reddish). One distinguishing characteristic is that pheomelanin has a higher content of sulfur. The many variations in hair color are combinations of different amounts of the pigments. Light-colored hair has a predominance of pheomelanin, as does red hair. Dark-colored hair contains mostly eumelanin. Graying of hair is the loss of pigment believed to be the result of a progressive inability of the melanocytes to make tyrosinase, the enzyme necessary for the synthesis of melanin. White hair results from air in the medullary shaft.

Growth and Replacement

Hair replacement occurs according to a cyclic pattern, alternating between growing and resting periods. During the growing phase of the cycle, the cells of the matrix are active. They increase in number by cell division and are pushed upward and eventually die, a situation similar to epidermal growth. The product is a hair, which is essentially dead protein tissue. Scalp hair grows about 1 mm (0.04 in.) every 3 days.

During the resting phase, the matrix becomes inactive and undergoes atrophy. At this point the root of the hair detaches from the matrix and the hair slowly moves up the follicle. It may remain there for some time until pulled out, shed, or pushed up by a replacing hair.

Either before or after the hair comes out of the follicle, proliferation of cells by the external root sheath forms a new matrix. The new matrix undergoes cell division, forming a new hair that grows up the follicle and replaces the old hair. You might be interested to know that shaving or cutting the hair has no effect on its growth.

The cycle of hair growth varies in different parts of the body. In the scalp, each hair grows steadily and continuously for 2–6 years; growth then stops, and after 3 months, the hair is shed. After another 3 months of a resting phase, a new hair starts to grow from the same follicle. Eyebrows, by contrast, have a growing phase of only about 10 weeks. It is for this reason that these hairs do not grow very long.

Normal hair loss in an adult scalp is about 70–100 hairs per day. Both the rate of growth and the replacement cycle may be altered by illness, diet, and other factors. For example, high fever, major illness, major surgery, blood loss, or severe emotional stress may increase the rate of shedding. Rapid weight-loss diets involving severe restriction of calories or protein also increase hair loss. An increase in the rate of shedding can also occur for 3–4 months after childbirth. Certain drugs and radiation therapy are also factors in increasing hair loss.

CLINICAL APPLICATION

The age of onset, degree of thinning, and ultimate hair pattern associated with **common baldness (male-pattern baldness)** are determined by male hormones, called androgens, and heredity. Androgens are also involved in promoting normal sexual development. In recent years, minoxidil, a potent vasodilator drug used to treat high blood pressure, has been applied topically to stimulate hair regrowth in some persons with common baldness. Since minoxidil increases blood flow in the skin, it is believed to stimulate new hair growth by preventing the progressive senescence of germinal cells of the hair matrix.

GLANDS

Three kinds of glands associated with the skin are sebaceous glands, sudoriferous glands, and ceruminous glands.

Sebaceous (Oil) Glands

Sebaceous (se-BĀ-shus) or **oil glands,** with few exceptions, are connected to hair follicles (see Figure 4-7c). The secreting portions of the glands lie in the dermis, and those glands associated with hairs open into the necks of hair follicles. Sebaceous glands not associated with hair follicles open directly onto the surface of the skin (lips, glans penis, labia minora, and tarsal glands of the eyelids). Sebaceous glands are simple branched acinar glands. Absent in the palms and soles, they vary in size and shape in other regions of the body. For example, they are small in most areas of the trunk and extremities, but large in the skin of the breasts, face, neck, and upper chest.

The sebaceous glands secrete an oily substance called **sebum** (SĒ-bum), a mixture of fats, cholesterol, proteins, and inorganic salts. Sebum helps keep hair from drying and becoming brittle, forms a protective film that prevents excessive evaporation of water from the skin, keeps the skin soft and pliable, and inhibits the growth of certain bacteria.

CLINICAL APPLICATION

When sebaceous glands of the face become enlarged because of accumulated sebum, **blackheads** develop. Since sebum is nutritive to certain bacteria, **pimples** or **boils** often result. The color of blackheads is due to melanin and oxidized oil, not dirt.

Sudoriferous (Sweat) Glands

Sudoriferous (soo'-dor-IF-er-us; *sudor* = sweat; *ferre* = to bear) or **sweat glands** are divided into two principal types on the basis of structure and location. **Apocrine sweat glands** are simple, branched tubular glands. Their distribution is limited primarily to the skin of the axilla, pubic region, and pigmented areas (areolae) of the breasts. The secretory portion of apocrine sweat glands is located in the dermis or subcutaneous layer and the excretory duct opens into hair follicles. Apocrine sweat glands begin to function at puberty and produce a more viscous secretion than eccrine sweat glands.

Eccrine sweat glands are much more common than apocrine sweat glands and are simple, coiled tubular glands. They are distributed throughout the skin except for the margins of the lips, nail beds of the fingers and toes, glans penis, glans clitoris, labia minora, and eardrums. Eccrine sweat glands are most numerous in the skin of the palms and the soles; their density can be as high as 3,000 per square inch in the palms. The secretory portion of eccrine sweat glands is located in the subcutaneous layer and the excretory duct projects upward through the dermis and epidermis to terminate at a pore at the surface of the epidermis (see Figure 4-1). Eccrine sweat glands function throughout life and produce a secretion that is more watery than that of apocrine sweat glands.

Perspiration, or **sweat,** is the substance produced by sudoriferous glands. It is a mixture of water, salts (mostly NaCl), urea, uric acid, animo acids, ammonia, sugar, lactic acid, and ascorbic acid. Its principal function is to help maintain body temperature. It also helps eliminate wastes.

Since the mammary glands are actually modified sudoriferous glands, they could be discussed here. However, because of their relation to the reproductive system, they will be considered in Chapter 25.

Ceruminous Glands

In certain parts of the skin, sudoriferous glands are modified as **ceruminous** (se-ROO-mi-nus) **glands.** Such modified glands are simple, coiled tubular glands present in the external auditory meatus (canal). The secretory portions of ceruminous glands lie in the submucosa, deep to sebaceous glands, and the excretory ducts open either directly onto the surface of the external auditory meatus or into sebaceous ducts. The combined secretion of the ceruminous and sebaceous glands is called **cerumen** (*cera* = wax). Cerumen, together with hairs in the external auditory meatus, provides a sticky barrier that prevents the entrance of foreign bodies.

CLINICAL APPLICATION

Some people produce an abnormal amount of cerumen, or earwax, in the external auditory canal. It then becomes impacted and prevents sound waves from reaching the tympanic membrane. The treatment for **impacted cerumen** is usually periodic ear irrigation or removal of wax with a blunt instrument by trained medical personnel.

NAILS

Hard, keratinized cells of the epidermis are referred to as *nails*. The cells form a clear, solid covering over the dorsal surfaces of the terminal portions of the fingers and toes. Each nail (Figure 4-8) consists of a nail body, a free edge, and a nail root. The *nail body* is the portion of the nail that is visible, the *free edge* is the part that may project beyond the distal end of the digit, and the *nail root* is the portion that is hidden in the proximal nail groove. Most of the nail body is pink because of the underlying vascular tissue. The whitish semilunar area of the proximal end of the body is called the *lunula* (LOO-nyoo-la). It appears whitish because the vascular tissue underneath does not show through due to the thickened stratum basale in the area.

The fold of skin that extends around the proximal and lateral borders of the nail is known as the *nail fold,* and the epidermis beneath the nail constitutes the *nail bed.* The furrow between the two is the *nail groove.*

The *eponychium* (ep'-ō-NIK-ē-um) or *cuticle* is a narrow band of epidermis that extends from the margin of the nail wall (lateral border), adhering to it. It occupies the proximal border of the nail and consists of stratum corneum. The thickened area of stratum corneum below the free edge of the nail is referred to as the *hyponychium* (hī'-pō-NIK-ē-um).

The epithelium of the proximal part of the nail bed is known as the *nail matrix.* Its function is to bring about the growth of nails. Essentially, growth occurs by the transformation of superficial cells of the matrix into nail cells. In the process, the outer, harder layer is pushed forward over the stratum germinativum. The average growth in the length of fingernails is about 1 mm (0.04 in.) per week. The growth rate is somewhat slower in toenails. For some reason, the longer the digit, the faster the nail grows. Adding supplements such as gelatin to an otherwise healthy diet has no effect on making nails grow faster or stronger.

Functionally, nails help us to grasp and manipulate small objects in various ways.

AGING AND THE INTEGUMENTARY SYSTEM

Although skin is constantly aging, the pronounced effects do not occur until a person reaches the late forties. Collagen fibers stiffen, break apart, and form into a shapeless, matted tangle. Elastic fibers thicken into clumps and fray, and as a result, the skin wrinkles. Fibroblasts, which produce both collagenous and elastic fibers, decrease in number, and macrophages become less efficient phagocytes. In addition, there is a loss of subcutaneous fat; atrophy of sebaceous (oil) glands, producing dry and broken skin that is susceptible to infection; a decrease in the number of functioning melanocytes, resulting in gray hair and atypical skin pigmentation; and an increase in the size of some melanocytes that produces pigmented blotching. Aged skin also becomes susceptible to pathological conditions such as senile pruritus (itching), decubitus ulcers (bedsores), and herpes zoster (shingles). Prolonged exposure to the ultraviolet (UV) rays of sunlight accelerates the aging of skin.

DEVELOPMENTAL ANATOMY OF THE INTEGUMENTARY SYSTEM

In this and subsequent chapters, the developmental anatomy of the systems of the body will be discussed at the end of each appropriate chapter. Since the principal features of embryonic development are not treated in detail until Chapter 26, it will be necessary to explain a few terms at this point so that you can follow the development of organ systems.

As part of the early development of a fertilized egg, a portion of the developing embryo differentiates into three layers of tissue called primary germ layers. On the basis of position, the primary germ layers are referred to as *ectoderm, mesoderm,* and *endoderm.* They are the embryonic tissues from which all tissues and organs of the body will eventually develop (see Exhibit 26-1).

The *epidermis* is derived from the *ectoderm.* At the

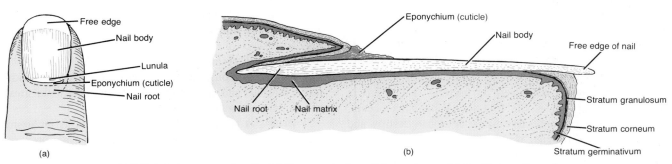

FIGURE 4-8 Structure of nails. (a) Fingernail viewed from above. (b) Sagittal section of a fingernail and nail bed.

beginning of the second month, the ectoderm consists of simple cuboidal epithelium. These cells become flattened and are known as the **periderm.** By the fourth month, all layers of the epidermis are formed and each layer assumes its characteristic structure.

Nails are developed during the third month. Initially, they consist of a thick layer of epithelium called the **primary nail field.** The nail itself is keratinized epithelium and grows forward from its base. It is not until the ninth month that the nails actually reach the tips of the digits.

Hair follicles develop between the third and fourth months as downgrowths of the stratum basale of the epidermis into the dermis below. The downgrowths soon differentiate into the bulb, papilla of the hair, beginnings of the epithelial portions of sebaceous glands, and other structures associated with hair follicles. By the fifth or sixth month, the follicles produce **lanugo** (delicate fetal hair), first on the head and then on other parts of the body. The lanugo is usually shed prior to birth.

The epithelial (secretory) portions of *sebaceous* (*oil*) *glands* develop from the sides of the hair follicles and remain connected to the follicles.

The epithelial portions of *sudoriferous* (*sweat*) *glands* are also derived from downgrowths of the stratum basale of the epidermis into the dermis. They appear during the fourth month on the palms and soles and a little later in other regions. The connective tissue and blood vessels associated with the glands develop from **mesoderm.**

The *dermis* is derived from wandering **mesenchymal (mesodermal) cells.** The mesenchyme becomes arranged in a zone beneath the ectoderm and there undergoes changes into the connective tissues that form the dermis.

APPLICATIONS TO HEALTH

ACNE

Acne is an inflammation of sebaceous (oil) glands and usually begins at puberty. At puberty, the sebaceous glands, under the influence of androgens (male hormones) grow in size and increase production of sebum. Although testosterone, a male hormone, appears to be the most potent circulating androgen for sebaceous cell stimulation, adrenal and ovarian androgens can stimulate sebaceous secretions as well.

Acne occurs predominantly in sebaceous follicles. The four basic types of acne lesions in order of increasing severity are comedones, papules, pustules, and cysts. The sebaceous follicles are rapidly colonized by bacteria that thrive in the lipid-rich sebum of the follicles. When this occurs, the cyst or sac of connective tissue cells can destroy and displace epidermal cells, resulting in permanent scarring. This type of acne, called *cystic acne,* may be treated successfully with a drug called Accutane, a synthetic form of vitamin A. Care must be taken to avoid squeezing, pinching, or scratching the lesions.

SYSTEMIC LUPUS ERYTHEMATOSUS (SLE)

Systemic lupus erythematosus (er-i'-them-a-TŌ-sus). **SLE,** or **lupus** is an autoimmune, inflammatory, connective tissue disease occurring mostly in young women in their reproductive years. An **autoimmune disease** is one in which the body attacks its own tissues, failing to differentiate between what is foreign and what is not. In SLE, damage to blood vessel walls results in the release of chemicals that mediate the inflammatory response. The blood vessel damage can be associated with virtually every body system.

The cause of SLE is not known and its onset may be abrupt or gradual. It is not contagious and is thought to be hereditary. There seems to be a strong incidence of other connective tissue disorders—especially rheumatoid arthritis (RA) and rheumatic fever—in relatives of SLE victims. The disease may be triggered by medication, such as penicillin, sulfa, or tetracycline, exposure to excessive sunlight, injury, emotional upset, infection, or other stress. These triggering factors, once recognized, are to be avoided by the patient in the future.

Symptoms include painful joints, low-grade fever, fatigue, mouth ulcers, photosensitivity, rapid loss of large amounts of scalp hair, and sometimes an eruption across the bridge of the nose and cheeks called a "butterfly rash." Other skin lesions may occur with blistering and ulceration. The erosive nature of some of the SLE skin lesions are thought to resemble the damage inflicted by the bite of a wolf, thus the term **lupus.** The most serious complications of the disease involve inflammation of the kidneys, liver, spleen, lungs, heart, and the central nervous system.

PSORIASIS

Psoriasis (sō-RĪ-a-sis) is a chronic, occasionally acute, noncontagious, relapsing skin disease. It is characterized by distinct, reddish, slightly raised plaques or papules (small, round skin elevations) covered with scales. Itching is seldom severe, and the lesions heal without scarring. Psoriasis ordinarily involves the scalp, the elbows and knees, the back, and the buttocks. Occasionally, the disease is generalized.

Studies have traced the complex cause of psoriasis to an abnormally high rate of mitosis in epidermal cells that may be related to a substance carried in the blood, a defect in the immune system, or a virus. Triggering factors such as trauma, infections, certain drugs (beta blockers and lithium), seasonal and hormonal changes, and emotional stress can initiate and intensify the skin eruptions.

Psoriasis is treated by a number of methods including steroid ointments and creams, natural sunlight, tar preparations, retinoids (chemicals similar to vitamin A), and PUVA, a therapy that combines psoralen (a chemical that increases the skin's reaction to light) and artificial ultraviolet (UV) light.

DECUBITUS ULCERS

Decubitus (dē-KYOO-bi-tus) *ulcers,* also known as *bedsores, pressure sores,* or *trophic ulcers,* are caused by a constant deficiency of blood to tissues overlying a bony projection that has been subjected to prolonged pressure against an object such as a bed, cast, or splint. The deficiency results in tissue ulceration. Small breaks in the epidermis become infected, and the sensitive subcutaneous and deeper tissues are damaged. Eventually the tissue is destroyed.

Decubitus ulcers are seen most frequently in patients who are bedridden for long periods of time. The most common areas involved are the skin over the sacrum, heels, ankles, buttocks, and other large bony projections. The chief causes are pressure from infrequent turning of the patient, trauma and maceration of the skin, and malnutrition. Maceration of the skin often follows soaking of bed and clothing by perspiration, urine, or feces.

SUNBURN

Sunburn is injury to the skin as a result of acute, prolonged exposure to the ultraviolet (UV) rays of sunlight. The damage to skin cells caused by sunburn is due to inhibition of DNA and RNA synthesis, which leads to cell death. There can also be damage to blood vessels as well as other structures in the dermis. Overexposure over a period of years results in a leathery skin texture, wrinkles, skin folds, sagging skin, warty growths called keratoses, freckling, a yellow discoloration due to abnormal elastic tissue, premature aging of the skin, and skin cancer.

SKIN CANCER

Excessive sun exposure can result in *skin cancer,* the most common cancer in Caucasians. However, everyone, regardless of skin pigmentation, is a potential victim of skin cancer if exposure to sunlight is sufficiently intense and continuous. Natural skin pigment can never give complete protection.

The three most common forms of skin cancer are basal cell carcinoma (BCC), squamous cell carcinoma (SCC), and malignant melanoma. *Basal cell carcinomas* (*BCCs*) account for over 75 percent of all skin cancers. The tumors arise from the epidermis and rarely metastasize. They are believed to be caused by years of chronic sun exposure. *Squamous cell carcinomas* (*SCCs*) also arise from the epidermis and, although less common than BCCs, they have a variable tendency to metastasize. Most SCCs arise from pre-existing lesions on sun-exposed skin. *Malignant melanomas* arise from melanocytes and are the leading cause of death from all diseases arising in the skin since they metastasize rapidly. Fortunately, malignant melanomas are the least common of the skin cancers. The principal cause is chronic sun exposure. A clinical trial is underway to test an antimelanoma vaccine made from human melanoma cells. The vaccine apparently sensitizes the patients so that they are more receptive to other anticancer substances, such as interleukin-2, or chemotherapy. The antimelanoma vaccine is designed to stop continued tumor growth rather than prevent tumor formation.

The treatment of most skin cancers consists of curettage, excision, cryosurgery, and radiation therapy.

Among the risk factors for skin cancer are:

1. *Skin type.* Persons with fair skin who never tan but always burn are at high risk.
2. *Geographic location.* Areas with many days of sunlight per year and high-altitude locations show high incidences of skin cancer.
3. *Age.* Older people are more prone to skin cancer because of greater exposure to sunlight.
4. *Personal habits.* Individuals engaged in outdoor occupations have a higher risk.
5. *Immunologic status.* Persons who are immunosuppressed have a higher incidence of skin cancer.

In recent years, **suntanning salons** have become very popular, and this has become a concern to physicians. Many salons claim to use ''safe'' wavelengths of ultraviolet (UV) light, that is, longer or A portions of the UV spectrum (UVA). According to some medical authorities, the radiation emitted by lamps in many suntanning parlors is potentially dangerous. Among the harmful effects are increased risk for skin cancer, premature aging of the skin, suppression of the immune system, and retinal damage or cataracts.

If you must be in direct sunlight for long periods of time, use a suitable sunscreen. One of the best agents for protection against overexposure to ultraviolet rays of the sun is para-aminobenzoic acid (PABA). The alcohol preparations of PABA are best because the active ingredient binds to the stratum corneum of the skin.

BURNS

Tissues may be damaged by thermal (heat), electrical, radioactive, or chemical agents, all of which can destroy the proteins in the exposed cells and cause cell injury or death. Such damage is a *burn.* The injury to tissues directly or indirectly in contact with the damaging agent, such as the skin or the linings of the respiratory and gastrointestinal tracts, is the local effect of a burn. Generally, however, the systemic effects of a burn are a greater threat to life than the local effects. The systemic effects of a burn may include (1) a large loss of water, plasma, and plasma proteins, which causes shock; (2) bacterial infection; (3) reduced circulation of blood; and (4) decreased production of urine.

Classification

A *first-degree burn* involves only the surface epithelium. It is characterized by mild pain, erythema (redness), dry skin, slight edema, and no blisters. Skin functions remain intact. The pain of a first-degree burn may be lessened by flushing the affected area with cold water. Generally, a first-degree burn will heal in about 2–3 days and may be accompanied by flaking or peeling. A typical sunburn is an example of a first-degree burn.

A *second-degree burn* involves the deeper layers of the epidermis or the upper levels of the dermis and skin functions are lost. In a superficial second-degree burn, the deeper layers of the epidermis are injured and there is characteristic redness blister formation, marked edema, and pain. Blisters beneath or within the epidermis are called *bullae* (BYOOL-ē), meaning bubbles. Such an injury usually heals within 7–10 days with only mild scarring. In a deep second-degree burn, there is destruction of the epidermis as well as the upper levels of the dermis. Epidermal derivatives, such as hair follicles, sebaceous glands, and sweat glands are usually not injured. If there is no infection, deep second-degree burns heal without grafting in about 3–4 weeks. Scarring may result.

First- and second-degree burns are collectively referred to as *partial-thickness burns*. A *third-degree burn* or *full-thickness burn* involves destruction of the epidermis, dermis, and epidermal derivatives and skin functions are lost. Such burns vary in appearance from marble-white to mahogany colored to charred, dry wounds. There is marked edema and such a burn is usually not painful to the touch due to destruction of nerve endings. Regeneration is slow and much granulation tissue forms before being covered by epithelium. Even if skin grafting is quickly begun, third-degree burns quickly contract and produce scarring.

A fairly accurate method for estimating the amount of surface area affected by a burn is to apply the *Lund-Browder method.* This method estimates the extent by measuring the areas affected against the percentage of total surface area for body parts shown in Figure 4-9. For example, if the anterior of the head and neck of an adult is affected, the burn covers 4½ percent of the body surface. Because the proportions of the body change with growth, the percentages vary for different ages. Thus, the extent of burn damage can be made fairly accurately for any age group.

A somewhat less accurate, although easy to apply, means for estimating the extent of a burn is the "rule of nines."

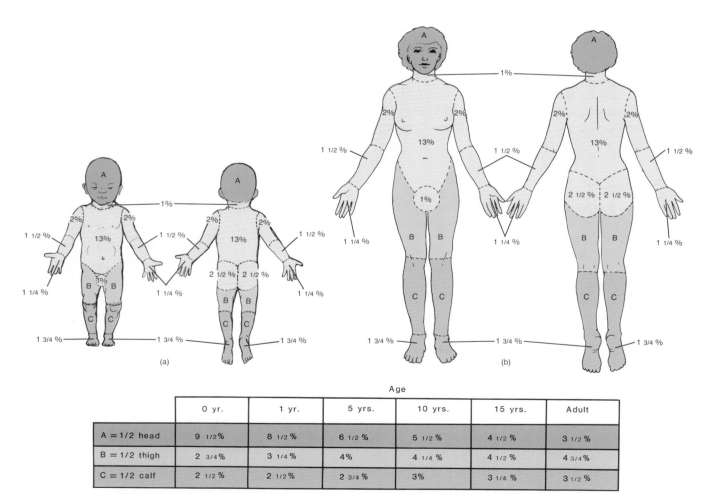

	Age					
	0 yr.	1 yr.	5 yrs.	10 yrs.	15 yrs.	Adult
A = 1/2 head	9 1/2 %	8 1/2 %	6 1/2 %	5 1/2 %	4 1/2 %	3 1/2 %
B = 1/2 thigh	2 3/4 %	3 1/4 %	4%	4 1/4 %	4 1/2 %	4 3/4 %
C = 1/2 calf	2 1/2 %	2 1/2 %	2 3/4 %	3%	3 1/4 %	3 1/2 %

FIGURE 4-9 The Lund-Browder method for estimating the extent of burns. Relative proportions of various body regions in (a) a young child compared to (b) an adult.

1. If the anterior and posterior surfaces of the head and neck are affected, the burn covers 9 percent of the body surface.
2. The anterior and posterior surfaces of each shoulder, arm, forearm, and hand constitute 9 percent of the body surface.
3. The anterior and posterior surfaces of the trunk, including the buttocks, constitute 36 percent.
4. The anterior and posterior surfaces of each foot, leg, and thigh as far up as the buttocks total 18 percent.
5. The perineum (per-i-NĒ-um) consists of 1 percent. The perineum includes the anal and urogenital regions.

Treatment

A severely burned individual should be moved as quickly as possible to a hospital. Treatment may then include:

1. Cleansing the burn wounds thoroughly.
2. Removing all dead tissue (debridement) so antibacterial

agents can directly contact the wound surface and thereby prevent infection.
3. Replacing lost body fluids and electrolytes (ions).
4. Covering wounds with temporary protection as soon as possible.
5. Removing a thin layer of skin from another part of the burn victim's body and transplanting it to the injured area, a procedure called a skin graft (SG). When a victim is so extensively burned that there is not enough healthy skin left intact for removal and grafting onto burned areas, a new source of skin grafts is needed. A new procedure makes this possible. Human skin cells taken from a small skin-biopsy sample of the burn victim can be grown in the laboratory to produce thin sheets of skin that can be used to cover burn areas. Once the sheets adhere to burn wounds, they generate a permanent new skin to cover the burned areas. The obvious advantage to this procedure is that there is no limit to the amount of new epithelium that can be grown from a very small skin sample.

KEY MEDICAL TERMS ASSOCIATED WITH THE INTEGUMENTARY SYSTEM

Abrasion (a-BRĀ-shun; *ab* = away; *rasion* = scraped)   A portion of the skin that has been scraped away.

Anhidrosis (an'-hī-DRŌ-sis; *an* = without; *hidr* = sweating; *osis* = condition)   A rare genetic condition characterized by inability to sweat.

Antiperspirant (an'-tī-PER-spi-rant'; *anti* = against; *perspirare* = to breathe through)   An agent that inhibits or prevents perspiration and usually contains aluminum compounds as the active ingredient.

Athlete's (ATH-lēts) **foot**   A superficial fungus infection of the skin of the foot.

Callus (KAL-lus)   An area of hardened and thickened skin that is usually seen in palms and soles and is due to pressure and friction.

Carbuncle (KAR-bung-kl; *carbunculus* = little coal)   A hard, round, deep, painful inflammation of the subcutaneous tissue that causes necrosis (death) and pus formation (abscess).

Chickenpox   Highly contagious disease that initiates in the respiratory system and is caused by a varicella-zoster virus and characterized by vesicular eruptions on the skin that fill with pus, rupture, and form a scab before healing. Also called *varicella* (var'-i-SEL-a). Shingles is caused by the latent chickenpox virus.

Cold sore (KŌLD sor)   A lesion, usually in oral mucous membrane, caused by type 1 herpes simplex virus (HSV), transmitted by oral or respiratory routes. Triggering factors include ultraviolet (UV) radiation, hormonal changes, and emotional stress. Also called a **fever blister.**

Comedo (KOM-ē-dō; *comedo* = to eat up)   A collection of sebaceous material and dead cells in the hair follicle and excretory duct of the sebaceous (oil) gland. Usually found over the face, chest, and back, and more commonly during adolescence. Also called **blackhead** or **whitehead.**

Contusion (kon-TOO-shun; *contundere* = to bruise)   Condition in which tissue below the skin is damaged, but the skin is not broken.

Corn (KORN)   A painful conical thickening of the skin found principally over toe joints and between the toes. It may be hard or soft, depending on the location. Hard corns are usually found over toe joints, and soft corns are usually found between the fourth and fifth toes.

Cyst (SIST; *cyst* = sac containing fluid)   A sac with a distinct connective tissue wall, containing a fluid or other material.

Deodorant (dē-Ō-dor-ant; *de* = from; *odorare* = perfume)   An agent that masks offensive odors; usually contains aluminum compounds as the active ingredient. Body odor is not caused by perspiration itself as much as by the activities of bacteria that grow in perspiration.

Dermabrasion (der-ma-BRA-shun; *derm* = skin)   Removal of acne, scars, tattoos, or moles by sandpaper or a high-speed brush.

Dermatome (DER-ma-tōm; *tomy* = cut)   An instrument for excising areas of skin to be used for grafting.

Detritus (de-TRI-tus; *deterere* = to rub away)   Particulate matter produced by or remaining after the wearing away or disintegration of a substance or tissue; scales, crusts, and loosened skin.

Eczema (EK-ze-ma; *ekzein* = to boil out)   An acute or chronic superficial inflammation of the skin, characterized by redness, oozing, crusting, and scaling. Also called **chronic dermatitis.**

Erythema (er'-e-THĒ-ma; *erythema* = redness)   Redness of the skin caused by an engorgement of capillaries in lower layers of the skin. Erythema occurs with any skin injury, infection, or inflammation.

Furuncle (FYOOR-ung-kul)   A boil; an abscess resulting from infection of a hair follicle.

German measles   Highly contagious disease that initiates in the respiratory system and is caused by the rubella virus and characterized by a rash of small red spots on the skin. Also called **rubella** (roo-BEL-a).

Hives (HĪVZ)   Condition of the skin marked by reddened elevated patches that are often itchy. Also called **urticaria.**

Hypodermic (hī'-pō-DER-mik; *hypo* = under)   Relating to the area beneath the skin. Also called **subcutaneous.**

Impetigo (im'-pe-TĪ-go)   Superficial skin infection caused by staphylococci or streptococci; most common in children.

Intradermal (in'-tra-DER-mal; *intra* = within)   Within the skin. Also called **intracutaneous.**

Keratosis (ker'-a-TŌ-sis; *kera* = horn)   Formation of a hardened growth of tissue.

Laceration (las'-er-Ā-shun; *lacerare* = to tear)   Wound or irregular tear of the skin.

Measles   Highly contagious disease that initiates in the respiratory system and is caused by the measles virus and characterized by a papular rash on the skin. Also called **rubeola** (roo-BĒ-ō-la).

Nevus (NE-vus)   A round, pigmented, flat, or raised skin area that may be present at birth or develop later. Varying in color from yellow-brown to black. Also called **mole** or **birthmark.**

Nodule (NOD-yool; *nodulus* = little knot)   A large cluster of cells raised above the skin but extending deep into the tissues.

Papule (PAP-yool)   A small, round skin elevation varying in size from a pinpoint to that of a split pea. One example is a pimple.

Polyp (POL-ip)   A tumor on a stem found especially on mucous membranes.

Pruritus (proo-RĪ-tus; *pruire* = to itch)   Itching, one of the most common dermatological disorders. It may be caused by skin disorders (infections), systemic disorders (cancer, kidney failure), or psychogenic factors (emotional stress).

Pustule (PUS-tyool)   A small, round elevation of the skin containing pus.

Subcutaneous (sub'-kyoo-TĀ-nē-us; *sub* = under)   Beneath the skin; also called **hypodermis.**

Topical (TOP-i-kal)   Pertaining to a definite area; local. Also in reference to a medication, applied to the surface rather than ingested or injected.

Wart (WORT)   Mass produced by uncontrolled growth of epithelial skin cells; caused by a virus (papovavirus). Most warts are noncancerous.

STUDY OUTLINE

Skin (p. 93)

1. The skin and its derivatives (hair, glands, and nails) constitute the integumentary system.
2. The skin is one of the larger organs of the body. It performs the functions of maintaining body temperature; protection; receiving stimuli; excretion of water, salts, and several organic compounds; and synethesis of vitamin D.
3. The principal parts of the skin are the outer epidermis and inner dermis. The dermis overlies the subcutaneous layer.
4. The epidermal layers, from deepest to most superficial, are the stratum basale, spinosum, granulosum, lucidum, and corneum. The basale undergoes continuous cell division and produces all other layers. Epidermal cells include keratinocytes, melanocytes, and nonpigmented granular dendrocytes (Langerhans and Granstein cells).
5. The dermis consists of a papillary region and a reticular region. The papillary region is loose connective tissue containing blood vessels, nerves, hair follicles, dermal papillae, and corpuscles of touch (Meissner's). The reticular region is dense, irregularly arranged connective tissue containing adipose tissue, hair follicles, nerves, sebaceous (oil) glands, and ducts of sudoriferous (sweat) glands.
6. Lines of cleavage indicate the direction of collagenous fiber bundles in the dermis and are considered during surgery.
7. The color of the skin is due to melanin, carotene, and blood in capillaries in the dermis.
8. Epidermal ridges increase friction for better grasping ability and provide the basis for fingerprints and footprints.
9. The epidermis is avascular; the dermis is abundantly supplied by the cutaneous plexus and papillary plexus.
10. In response to nerve impulses from the brain, sudoriferous (sweat) glands increase their output of perspiration, skin blood vessels dilate, metabolic rate decreases, and muscle tone decreases. These responses decrease body temperature.

Skin Wound Healing (p. 98)
Epidermal Wound Healing (p. 98)

1. In an epidermal wound, the central portion of the wound usually extends deep down to the dermis, whereas the wound edges usually involve only superficial damage to the epidermal cells.
2. Epidermal wounds are repaired by enlargement and migration of basal cells, contact inhibition, and division of migrating and stationary basal cells.

Deep Wound Healing (p. 99)

1. During the inflammatory phase, a blood clot unites the wound edges, epithelial cells migrate across the wound, vasodilation and increased permeability of blood vessels deliver phagocytes, and fibroblasts form.
2. During the migratory phase, epithelial cells beneath the scab bridge the wound, fibroblasts begin to synthesize scar tissue, and damaged blood vessels begin to regrow.
3. During the proliferative phase, the events of the migratory phase intensify and the open wound tissue is called granulation tissue.
4. During the maturation phase, the scab sloughs off, the epidermis is restored to normal thickness, collagenous fibers become more organized, fibroblasts begin to disappear, and blood vessels are restored to normal.

Epidermal Derivatives (p. 99)

1. Epidermal derivatives are structures developed from the embryonic epidermis.
2. Among the epidermal derivatives are hair, skin glands (sebaceous, sudoriferous, and ceruminous), and nails.

Hair (p. 99)

1. Hairs are epidermal growths that function in protection.

2. Hair consists of a shaft above the surface, a root that penetrates the dermis and subcutaneous layer, and a hair follicle.

3. Associated with hairs are sebaceous (oil) glands, arrectores pilorum muscles, and root hair plexuses.

4. Hair color is due to combinations of various amounts of two basic classes of melanin: eumelanin (brown-black) and pheomelanin (yellow to reddish). Graying is due to the loss of melanin.

5. New hairs develop from cell division of the matrix in the bulb; hair replacement and growth occurs in a cyclic pattern. "Male-pattern" baldness is caused by androgens and heredity.

Glands (p. 104)

1. Sebaceous (oil) glands are usually connected to hair follicles; they are absent in the palms and soles. Sebaceous glands produce sebum, which moistens hairs and waterproofs the skin. Enlarged sebaceous glands may produce blackheads, pimples, and boils.

2. Sudoriferous (sweat) glands are divided into apocrine and eccrine. Apocrine sweat glands are limited in distribution to the skin of the axilla, pubis, and areolae; their ducts open into hair follicles. Eccrine sweat glands have an extensive distribution; their ducts terminate at pores at the surface of the epidermis. Sudoriferous glands produce perspiration, which carries small amounts of wastes to the surface and assists in maintaining body temperature.

3. Ceruminous glands are modified sudoriferous glands that secrete cerumen. They are found in the external auditory meatus.

Nails (p. 105)

1. Nails are hard, keratinized epidermal cells over the dorsal surfaces of the terminal portions of the fingers and toes.

2. The principal parts of a nail are the body, free edge, root, lunula, eponychium, hyponychium, and matrix. Cell division of the matrix cells produces new nails.

Aging and the Integumentary System (p. 105)

1. Most effects of aging occur when an individual reaches the late forties.

2. Among the effects of aging are wrinkling, loss of subcutaneous fat, atrophy of sebaceous glands, and decrease in the number of melanocytes.

Developmental Anatomy of the Integumentary System (p. 105)

1. The epidermis is derived from ectoderm. Hair, nails, and skin glands are epidermal derivatives.

2. The dermis is derived from wandering mesodermal cells.

Applications to Health (p. 106)

1. Acne is an inflammation of sebaceous glands.

2. Systemic lupus erythematosus (SLE) is an autoimmune disease of connective tissue.

3. Psoriasis is a chronic skin disease characterized by reddish, raised plaques or papules.

4. Decubitus ulcers are caused by a chronic deficiency of blood to tissues subjected to prolonged pressure.

5. Sunburn is a skin injury resulting from prolonged exposure to the ultraviolet (UV) rays of sunlight.

6. Skin cancer can be caused by excessive exposure to sunlight.

7. Tissue damage that destroys protein is called a burn. Depending on the depth of damage, skin burns are classified as first-degree and second-degree (partial-thickness) and third-degree (full-thickness). One method employed for determining the extent of a burn is the Lund-Browder method; another is the "rule of nines." Burn treatment may include cleansing the wound, removing dead tissue, replacing lost body fluids, covering wounds with temporary protection, and skin grafting.

REVIEW QUESTIONS

1. What is the integumentary system?

2. List the principal functions of the skin.

3. Compare the structure of epidermis and dermis. What is the subcutaneous layer?

4. List and describe the epidermal layers from the deepest outward. What is the importance of each layer? Describe the various cells that comprise the epidermis.

5. Contrast the structural differences between the papillary and reticular regions of the dermis.

6. What are lines of cleavage? What is their importance during surgery?

7. Explain the factors that produce skin color. What is an albino?

8. Describe how melanin is synthesized and distributed to epidermal cells.

9. How are epidermal ridges formed? Why are they important?

10. List the receptors in the epidermis, dermis, and subcutaneous layer, and indicate the location and role of each.

11. Describe the blood supply of the skin.

12. Outline the steps involved in epidermal wound healing and deep wound healing.

13. Describe the development and distribution of hair.

14. Describe the structure of a hair. How are hairs moistened? What produces "goosebumps" or "gooseflesh"?

15. Contrast the locations and functions of sebaceous (oil) glands, sudoriferous (sweat) glands, and ceruminous glands.

16. Distinguish between apocrine and eccrine sweat glands.

17. From what layer of the skin do nails form? Describe the principal parts of a nail.

18. Describe the effects of aging on the integumentary system.

19. Describe the origin of the epidermis, its derivatives, and the dermis.

20. Define each of the following disorders of the integumentary system; acne, systemic lupus erythematosus (SLE), psoriasis, decubitus ulcers, sunburn, and skin cancer.

21. Define a burn. Classify burns according to degree.

22. Explain how the Lund-Browder method is used to estimate the extent of a burn. What is the "rule of nines"? How are burns treated?

23. Refer to the glossary of key medical terms associated with the integumentary system. Be sure that you can define each term.

SELF-QUIZ

Choose the one best answer to these questions.

___ 1. Which of the following statements about the function of skin is *not* true?
A. it helps regulate body temperature; B. it protects against bacterial invasion and dehydration; C. it absorbs water and salts; D. it participates in the synthesis of vitamin D; E. it detects stimuli related to temperature and pain.

___ 2. The layer of the dermis that is in direct contact with the epidermis is the
A. papillary region; B. stratum corneum; C. stratum basale; D. reticular region; E. stratum granulosum.

___ 3. In that part of a course in anatomy dealing with the integumentary system, one is concerned with
A. mucous membranes; B. the viscera of the abdominal and thoracic cavities; C. the skin and related structures; D. bones and joints; E. lymphatic tissues.

___ 4. The layer of the skin from which new epidermal cells are derived is the
A. stratum corneum; B. stratum basale; C. stratum lucidum; D. stratum granulosum; E. reticular layer.

___ 5. "Goosebumps" occur as a result of
A. contraction of arrector pili muscles; B. secretion of sebum; C. contraction of elastic fibers in the bulb of the hair follicle; D. contraction of dermal papillae; E. secretion of perspiration.

6. Match the following:

___ a. deepest layer of epithelium, consisting of a single layer of cuboidal to columnar cells
___ b. eight to ten rows of polyhedral cells that contain spinelike projections
___ c. third layer of epidermis, consisting of three to five rows of flattened, degenerating cells
___ d. flat, dead cells that are clear due to presence of eleiden; normally found only in thick skin of palms and soles
___ e. most superficial layer of skin, consisting of 25 to 30 rows of flat, dead cells

A. basale
B. corneum
C. granulosum
D. lucidum
E. spinosum

Complete the following:

7. This type of cell produces the pigments that give skin its color: _____.

8. The epidermis is derived from the (ecto-? meso-? endo-?) derm. All its layers are formed by the (second? fourth?) month. The dermis arises from the _____ -derm.

9. The _____ is the part of a hair follicle where cells undergo mitosis permitting growth of a new hair.

10. Present in fingerlike projections known as dermal _____ are corpuscles of touch (Meissner's corpuscles), sense receptors sensitive to (pressure? touch?).

11. The outer layer of skin is named the _____. It is composed of (connective tissue? epithelium?).

12. The inner portion of skin, called the _____, is made of (connective tissue? epithelium?).

___ ___ ___ 13. Arrange the parts of a hair from superficial to deep:
A. shaft; B. bulb; C. root

___ ___ ___ 14. Arrange the layers of a hair from external to internal:
A. medulla; B. cuticle; C. cortex

15. State whether the following descriptions refer to sebaceous, sudoriferous, or ceruminous glands:
A. sweat glands: _____
B. simple branched acinar glands leading directly to hair follicle: secrete sebum that keeps hair and skin from drying out: _____
C. line the external auditory meatus: secrete ear wax: _____

Circle T (true) or F (false) for the following:

T F 16. Eccrine sweat glands are more numerous than apocrine sweat glands and are especially dense on palms and soles.

T F 17. The dermis consists of two regions; the papillary region is most superficial, and the reticular region is deeper.

T F 18. The internal root sheath is a downward continuation of the epidermis.

T F 19. Both epidermis and dermis contain blood vessels (are vascular).

T F 20. In deep wound healing, the correct sequence of phases for repair is inflammatory, migratory, maturation, proliferation.

5 Osseous Tissue

CHAPTER OUTLINE
- **Functions**
- **Histology**

Compact Bone

Spongy Bone
- **Ossification**

Intramembranous Ossification

Endochondral Ossification
- **Bone Growth**
- **Bone Replacement**
- **Blood and Nerve Supply**
- **Aging and the Skeletal System**
- **Developmental Anatomy of the Skeletal System**
- **Applications to Health**

Osteoporosis

Vitamin Deficiencies

  Rickets

  Osteomalacia

Paget's Disease

Osteomyelitis

Fractures

  Types

  Fracture Repair
- **Key Medical Terms Associated with Osseous Tissue**

STUDENT OBJECTIVES

1. Discuss the components and functions of the skeletal system.
2. List and describe the gross features of a long bone.
3. Describe the histological features of compact and spongy bone tissue.
4. Contrast the steps involved in intramembranous and endochondral ossification.
5. Describe the processes of bone construction and destruction involved in bone remodeling.
6. Describe the conditions necessary for normal bone growth and replacement.
7. Describe the blood and nerve supply of bone tissue.
8. Explain the effects of aging on the skeletal system.
9. Describe the development of the skeletal system.
10. Define the following: osteoporosis, rickets, osteomalacia, Paget's disease, and osteomyelitis.
11. Define a fracture (Fx) describe several common kinds of fractures, and describe the sequence of events involved in fracture repair.
12. Define key medical terms associated with osseous tissue.

Without the skeletal system we would be unable to perform movements, such as walking or grasping. The slightest jar to the head or chest could damage the brain or heart. It would even be impossible to chew food. The framework of bones and cartilage that protects our organs and allows us to move is called the **skeletal system.** The specialized branch of medicine that deals with the preservation and restoration of the skeletal system, articulations (joints), and associated structures is called **orthopedics** (or′-tho-PĒ-diks; *ortho* = correct or straighten; *pais* = child).

The developmental anatomy of the skeletal system is considered at the end of the chapter.

FUNCTIONS

The skeletal system performs several basic functions.

1. **Support.** The skeleton provides a framework for the body and, as such, it supports soft tissues and provides a point of attachment for many muscles.
2. **Protection.** Many internal organs are protected from injury by the skeleton. For example, the brain is protected by the cranial bones, the spinal cord by the vertebrae, the heart and lungs by the rib cage, and internal reproductive organs by the pelvic bones.
3. **Movement.** Bones serve as levers to which muscles are attached. When the muscles contract, the bones acting as levers produce movement.
4. **Mineral storage.** Bones store several minerals that can be distributed to other parts of the body on demand. The principal stored minerals are calcium and phosphorus.
5. **Blood cell production.** Red marrow in certain bones is capable of producing blood cells, a process called **hematopoiesis** (hēm′-a-tō-poy-Ē-sis) or **hemopoiesis. Red marrow** consists of blood cells in immature stages, fat cells, and macrophages. Red marrow produces red blood cells, some white blood cells, and platelets.

HISTOLOGY

Structurally, the skeletal system consists of two types of connective tissue: cartilage and bone. We described the microscopic structure of cartilage in Chapter 3. Here, our attention will be directed to a detailed discussion of the microscopic structure of bone tissue.

Like other connective tissues, **bone,** or **osseous** (OS-ē-us) **tissue,** contains a great deal of intercellular substance surrounding widely separated cells. Four types of cells are characteristic of bone tissue: osteoprogenitor (osteogenic), osteoblasts, osteocytes, and osteoclasts. **Osteoprogenitor** (os′-tē-ō-prō-JEN-i-tor; *osteo* = bone; *pro* = precursor; *gen* = to produce) **cells** are stem cells derived from mesenchyme. They possess mitotic potential and have the ability to differentiate into osteoblasts (described shortly). Osteo-progenitor cells are found in the inner portion of the membrane (periosteum) around a bone, the membrane (endosteum) that lines the medullary cavity, and in canals in bone (perforating and central) that contain blood vessels. **Osteoblasts** (OS-tē-ō-blasts′; *blast* = germ or bud) do not have mitotic potential and are associated with bone formation. They secrete some of the organic components and mineral salts involved in bone formation. Osteoblasts are found on the surfaces of bone. **Osteocytes** (OS-te-ō-sīts; *cyte* = cell) or mature bone cells are the principal cells of bone tissue. Like osteoblasts, osteocytes have no mitotic potential. Osteocytes are actually osteoblasts that become isolated right in the bony intercellular substance that they deposit around themselves and whose morphology changes. Whereas osteoblasts initially form bone, osteocytes maintain daily cellular activities of bone tissue. **Osteoclasts** (OS-tē-o-clasts′; *clast* = to break) develop from circulating monocytes (one type of white blood cell). They are found around the surfaces of bone and function in bone degradation. This is important in the development, growth, maintenance, and repair of bone. Unlike other connective tissues, the intercellular substance of bone contains abundant mineral salts, primarily calcium phosphate ($Ca_3(PO_4)_2 \cdot (OH)_2$) and some calcium carbonate ($CaCO_3$). In addition, there are small amounts of magnesium hydroxide, fluoride, and sulfate. As these salts are deposited in the framework formed by the collagenous fibers of the intercellular substance, the tissue hardens, that is, becomes ossified. The mineral salts compose 67 percent of the weight of bone, and the collagenous fibers make up the remaining 33 percent.

The microscopic structure of bone may be analyzed by first considering the anatomy of a long bone such as the humerus (arm bone) shown in Figure 5-1a. A typical long bone consists of the following parts:

1. **Diaphysis** (dī-AF-i-sis). The shaft or long, main portion of the bone.
2. **Epiphyses** (ē-PIF-i-sēz). The extremities or ends of the bone (singular is **epiphysis**).
3. **Metaphysis** (me-TAF-i-sis). The region in a mature bone where the diaphysis joins the epiphysis. In a growing bone, it is the region adjacent to or near the epiphyseal plate where calcified cartilage is reinforced and then replaced by bone (described later in the chapter).
4. **Articular cartilage.** A thin layer of hyaline cartilage covering the epiphysis where the bone forms a joint with another bone.
5. **Periosteum** (per′-ē-OS-tē-um). A dense, white, fibrous covering around the surface of the bone not covered by articular cartilage. The periosteum (*peri* = around; *osteo* = bone) consists of two layers. The outer **fibrous layer** is composed of connective tissue containing blood vessels, lymphatic vessels, and nerves that pass into the bone. The inner **osteogenic** (os′-tē-ō-JEN-ik) **layer** contains elastic fibers, blood vessels, and osteoprogenitor cells and osteoblasts. The periosteum is essential for

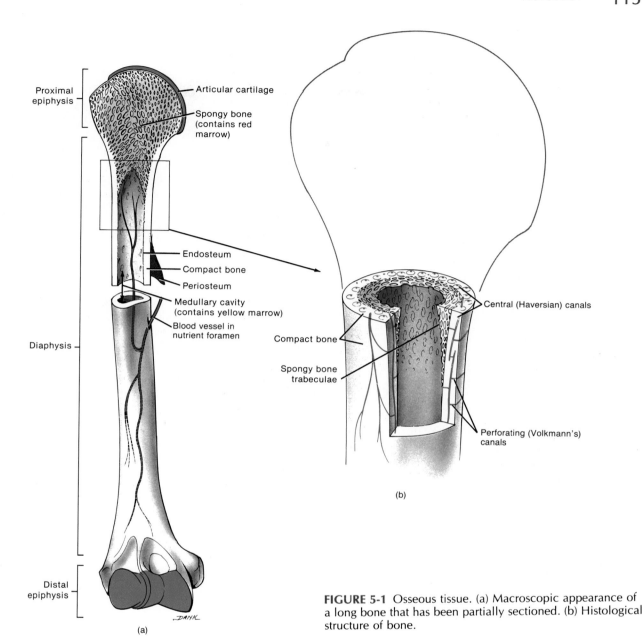

FIGURE 5-1 Osseous tissue. (a) Macroscopic appearance of a long bone that has been partially sectioned. (b) Histological structure of bone.

bone growth, repair, and nutrition. It also serves as a point of attachment for ligaments and tendons.

6. *Medullary* (MED-yoo-lar'-ē) or *marrow cavity.* The space within the diaphysis that contains the fatty *yellow marrow* in adults. Yellow marrow consists primarily of fat cells and a few scattered blood cells. Thus, yellow marrow functions in fat storage.

7. *Endosteum* (end-OS-tē-um). A layer of osteoprogenitor cells and osteoblasts that lines the medullary cavity and also contains scattered osteoclasts (cells that assume a role in the removal of bone).

Bone is not a completely solid, homogeneous substance. In fact, all bone has some spaces between its hard components. The spaces provide channels for blood vessels that supply bone cells with nutrients. The spaces also make bones lighter. Depending on the size and distribution of the spaces, the regions of a bone may be categorized as spongy or compact (Figure 5-2; see Figure 5-1 also).

Compact (*dense*) bone tissue contains few spaces. It is deposited in a layer over the spongy bone tissue. The layer of compact bone is thicker in the diaphysis than the epiphyses. Compact bone tissue provides protection and support and helps the long bones resist the stress of weight placed on them. *Spongy* (*cancellous*) bone tissue by contrast contains many large spaces filled with red marrow. It makes up most of the bone tissue of short, flat, and irregularly shaped bones and most of the epiphyses of long bones. Spongy bone tissue also provides a storage area for some red marrow.

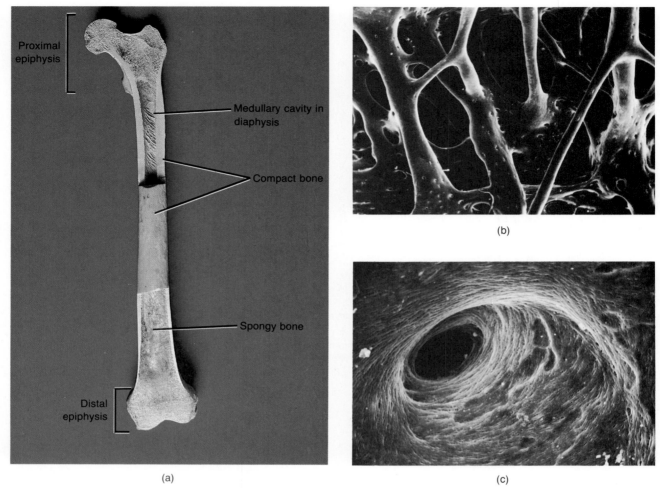

(a)

(b)

(c)

FIGURE 5-2 Spongy and compact bone. (a) Photograph of a section through the femur to illustrate the positional and structural differences between spongy and compact bone. (Courtesy of Lester V. Bergman & Associates, Inc.) (b) Scanning electron micrograph of spongy bone trabeculae at a magnification of 25×. (c) Scanning electron micrograph of a view inside a central (Haversian) canal of compact bone at a magnification of 250×. (Electron micrographs courtesy of Fisher Scientific Company and S.T.E.M. Laboratories, Inc. Copyright © 1975.)

COMPACT BONE

We can compare the differences between spongy and compact bone tissues by looking at the highly magnified sections in Figure 5-3. One main difference is that adult compact bone has a concentric-ring structure, whereas spongy bone does not. Blood vessels and nerves from the periosteum penetrate the compact bone through *perforating (Volkmann's) canals*. The blood vessels of these canals connect with blood vessels and nerves of the medullary cavity and those of the *central (Haversian) canals*. The central canals run longitudinally through the bone. Around the canals are *concentric lamellae* (la-MEL-ē)—rings of hard, calcified, intercellular substance. Between the lamellae are small spaces called *lacunae* (la-KOO-nē), which contain osteocytes. *Osteocytes* as noted earlier, are mature osteoblasts that no longer produce new bone tissue and function to

support daily cellular activities of bone tissue. Radiating in all directions from the lucunae are minute canals called *canaliculi* (kan'-a-LIK-yoo-lī), which contain slender processes of osteocytes. The canaliculi connect with other lacunae and, eventually, with the central canals. Thus, an intricate network is formed throughout the bone. This branching network of canaliculi provides numerous routes so that nutrients can reach the osteocytes and wastes can be removed. Each central canal, with its surrounding lamellae, lacunae, osteocytes, and canaliculi, is called an *osteon (Haversian system)*. Osteons are characteristic of adult bone. The areas between osteons contain *interstitial lamellae*. These also possess lacunae with osteocytes and canaliculi, but their lamellae are usually not connected to the osteons. Interstitial lamellae are fragments of older osteons that have been partially destroyed during bone replacement.

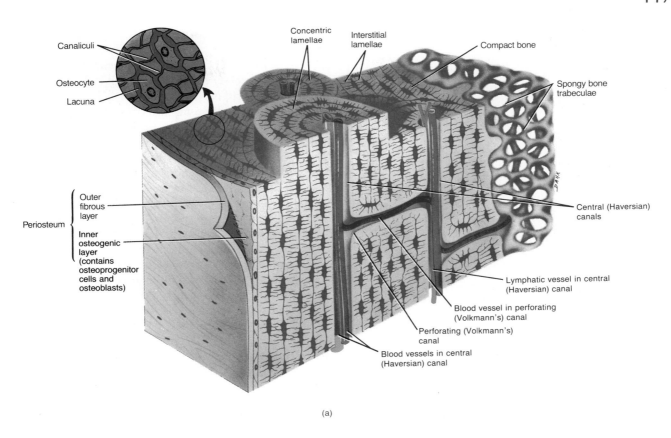

(a)

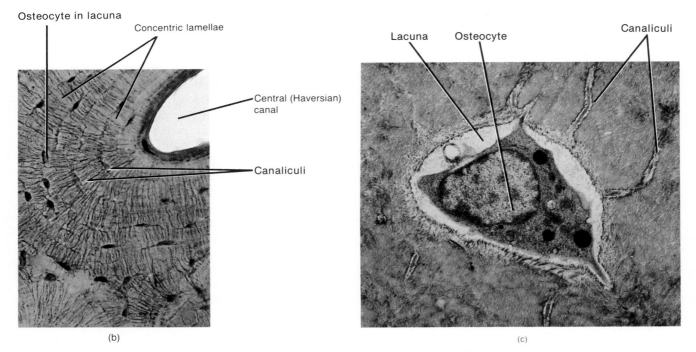

(b)

(c)

FIGURE 5-3 Histology of osseous tissue. (a) Enlarged aspect of several osteons (Haversian systems) in compact bone. (b) Photomicrograph of portions of several osteons (*Haversian systems*) with interstitial lamellae between at a magnification of 150×. (Courtesy of Carolina Biological Supply Company.) Histology of osseous tissue. (c) Electron micrograph of an osteocyte at a magnification of 10,000×. (Courtesy of Biophoto Associates.)

SPONGY BONE

In contrast to compact bone, spongy bone does not contain true osteons. It consists of an irregular latticework of thin plates of bone called *trabeculae* (tra-BEK-yoo-lē). See Figure 5-2b. The spaces between the trabeculae of some bones are filled with red marrow. The cells of red marrow are responsible for producing blood cells. Within the trabeculae lie lacunae, which contain osteocytes. Blood vessels from the periosteum penetrate through to the spongy bone, and osteocytes in the trabeculae are nourished directly from the blood circulating through the marrow cavities.

Most people think of all bone as a very hard, rigid material. Yet the bones of an infant are generally more pliable than those of an adult. The final shape and hardness of adult bones require many years to develop and depend on a complex series of chemical changes. Let us now see how bones are formed and how they grow.

OSSIFICATION

The process by which bone forms in the body is called *ossification* (os'-i-fi-KĀ-shun) or *osteogenesis*. The "skeleton" of a human embryo is composed of fibrous membranes and hyaline cartilage. Both are shaped like bones and provide the medium for ossification. Ossification begins around the sixth or seventh week of embryonic life and continues throughout adulthood. Two kinds of bone formation occur. The first is called intramembranous (in'-tra-MEM-bra-nus) ossification. This term refers to the formation of bone directly on or within the fibrous membranes (*intra* = within; *membranous* = membrane). The second kind, endochondral (en'-dō-KON-dral) ossification, refers to the formation of bone in cartilage (*endo* = within; *chondro* = cartilage). These two kinds of ossification do *not* lead to differences in the structure of mature bones. They simply indicate different methods of bone formation. Both mechanisms involve the replacement of a preexisting connective tissue with bone.

The first stage in the development of bone is the migration of mesenchymal (embryonic connective tissue) cells into the area where bone formation is about to begin. These cells increase in number and size and become osteoprogenitor cells. In some skeletal structures where capillaries are lacking they become chondroblasts; in others where capillaries are present they become osteoblasts. The *chondroblasts* are responsible for cartilage formation. Osteoblasts form bone tissue by intramembranous or endochondral ossification.

INTRAMEMBRANOUS OSSIFICATION

Of the two types of bone formation, the simpler and more direct is *intramembranous ossification.* Most of the surface skull bones and the clavicles (collarbones) are formed in this way (Figure 5-4). The essentials of this process are as follows.

Osteoblasts formed from osteoprogenitor cells cluster in a fibrous membrane. The site of such a cluster is called a *center of ossification.* The osteoblasts then secrete intercellular substances partly composed of collagenous fibers that form a framework, or matrix, in which calcium salts are quickly deposited. The deposition of calcium salts is called *calcification.* When a cluster of osteoblasts is completely surrounded by the calcified matrix, it is called a *trabecula.* As trabeculae form in nearby ossification centers, they fuse into the open latticework characteristic of spongy bone. With the formation of successive layers of bone, some osteoblasts become trapped in the lacunae. The entrapped osteoblasts lose their ability to form bone and are called osteocytes. The spaces between the trabeculae fill with red marrow. The original connective tissue that surrounds the growing mass of bone then becomes the periosteum. The ossified area has now become true spongy bone. Eventually, the surface layers of the spongy bone will be reconstructed into compact bone. Much of this newly formed bone will be destroyed and reformed so the bone may reach its final adult size and shape.

ENDOCHONDRAL OSSIFICATION

The replacement of cartilage by bone is called *endochondral (intracartilaginous) ossification.* Most bones of the body, including bones of the base of the skull, are formed in this way (Figure 5-4). Since this type of ossification is best observed in a long bone, we shall study the process in the tibia, or shinbone (Figure 5-5).

Early in embryonic life, a cartilage model or template of the future bone is laid down. This model is covered by a membrane called the *perichondrium* (per-i-KON-drē-um). Midway along the shaft of this model, a blood vessel penetrates the perichondrium, stimulating the osteoprogenitor cells in the internal layer of the perichondrium to enlarge and become osteoblasts. The cells begin to form a collar of compact bone around the middle of the diaphysis of the cartilage model. Once the perichondrium starts to form bone, it is called the *periosteum.* Simultaneously with the appearance of the bone collar and the penetration of blood vessels, changes occur in the cartilage in the center of the diaphysis. In this area, called the *primary ossification center,* cartilage cells hypertrophy (increase in size)—probably because they accumulate glycogen for energy and produce enzymes to catalyze future chemical reactions. These cells burst, resulting in a change in extracellular pH to a more alkaline pH causing the intercellular substance to become *calcified;* that is, minerals are deposited within it. Once the cartilage becomes calcified, nutritive materials required by the cartilage cells can no longer diffuse through the intercellular substance, and this may cause the cartilage cells to die. Then the intercellular substance begins to degen-

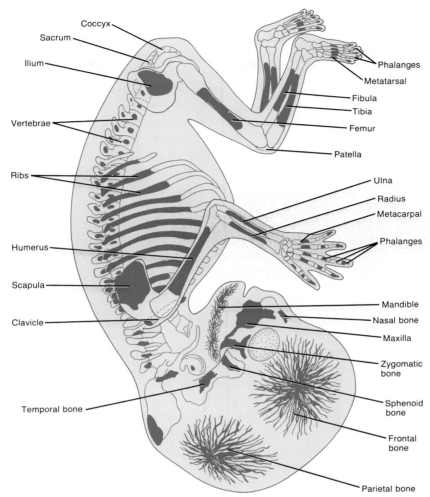

FIGURE 5-4 Comparison between bones that develop by intramembranous (red) and endochondral (blue) ossification as seen in a 10-week old human embryo.

erate, leaving large cavities in the cartilage model. Blood vessels grow along the spaces where cartilage cells were previously located and enlarge the cavities further. Gradually, these spaces in the middle of the shaft join with each other, and the marrow cavity is formed. Osteoprogenitor cells that may develop into osteoblasts are carried into the cartilage with the invading blood vessels.

As these developmental changes are occurring, the osteoblasts of the periosteum deposit successive layers of bone on the outer surface so that the collar thickens, becoming thickest in the diaphysis. The cartilage model continues to grow at its ends, steadily increasing in length. Eventually, blood vessels enter the epiphyses, and *secondary ossification centers* appear in the epiphyses and lay down spongy bone. In the tibia, one secondary ossification center develops in the proximal epiphysis soon after birth. The other center develops in the distal epiphysis during the child's second year.

After the two secondary ossification centers have formed, bone tissue completely replaces cartilage, except in two regions. Cartilage continues to cover the articular surfaces of the epiphyses, where it is called *articular cartilage.* It also remains as a region between the epiphysis and diaphysis, where it is called the *epiphyseal plate* (metaphysis of a growing bone).

BONE GROWTH

In order to understand how a bone grows in length, you will need to know some of the details of the structure of the epiphyseal plate.

The epiphyseal (ep'-i-FIZ-ē-al) plate consists of four zones (Figure 5-6). The *zone of reserve cartilage* is adjacent to the epiphysis and consists of small chondrocytes that are scattered irregularly throughout the intercellular matrix. The cells of this zone do not function in bone growth; they anchor the epiphyseal plate to the bone of the epiphysis.

The *zone of proliferating cartilage* consists of slightly larger chondrocytes arranged like stacks of coins. The func-

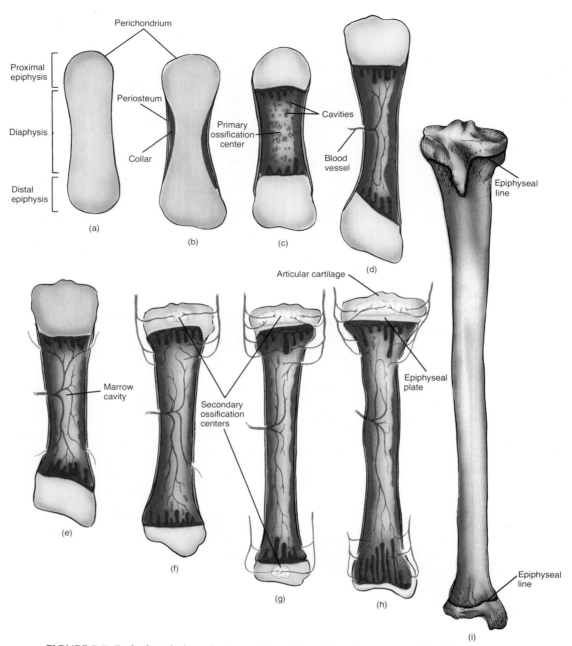

FIGURE 5-5 Endochondral ossification of the tibia. (a) Cartilage model. (b) Collar formation. (c) Development of primary ossification center. (d) Entrance of blood vessels. (e) Marrow-cavity formation. (f) Thickening and lengthening of collar. (g) Formation of secondary ossification centers. (h) Remains of cartilage as the articular cartilage and epiphyseal lines.

tion of this zone is to make new chondrocytes by cell division to replace those that die at the diaphyseal surface of the epiphyseal plate.

The *zone of hypertrophic* (hī-per-TRŌF-ik) *cartilage* consists of even larger chondrocytes that are also arranged in columns, with the more mature cells closer to the diaphysis. The lengthwise expansion of the epiphyseal plate is the result of cellular proliferation of the zone of proliferating cartilage and maturation of the cells in the zone of hypertrophic cartilage.

The *zone of calcified matrix* is only a few cells thick and consists mostly of dead cells because the intercellular matrix around them has calcified. The calcified matrix is taken up by osteoclasts, and the area is invaded by osteoblasts and capillaries from the bone in the diaphysis. These cells lay down bone on the calcified cartilage that persists. As a result, the diaphyseal border of the epiphyseal plate is firmly cemented to the bone of the diaphysis.

The region between the diaphysis and epiphysis of a bone where the calcified matrix is replaced by bone is called the *metaphysis* (me-TAF-i-sis). The activity of the epiphyseal plate is the only mechanism by which the diaphysis can increase in length. Unlike cartilage, which can grow by both interstitial and appositional growth, bone can grow only by appositional growth.

The epiphyseal plate allows the diaphysis of the bone to increase in length until early adulthood. The rate of growth is controlled by hormones such as growth hormone (GH) produced by the pituitary gland. As the child grows, cartilage cells are produced by mitosis on the epiphyseal side of the plate. Cartilage cells are then destroyed and the cartilage is replaced by bone on the diaphyseal side of the plate. In this way, the thickness of the epiphyseal plate remains fairly constant, but the bone on the diaphyseal side increases in length.

Growth in diameter occurs along with growth in length. In this process, the bone lining the marrow cavity is destroyed so that the cavity increases in diameter. At the same time, osteoblasts from the periosteum add new osseous tissue around the outer surface of the bone. Initially, diaphyseal and epiphyseal ossifications produce only spongy bone. Later, by reconstruction, the outer region of spongy bone is reorganized into compact bone.

The epiphyseal cartilage cells stop dividing and the cartilage is eventually replaced by bone. The newly formed bony structure is called the *epiphyseal line,* a remnant of the once active epiphyseal plate. With the appearance of the epiphyseal line, appositional bone growth stops. The clavicle is the last bone to stop growing. Ossification of most bones is usually completed by age 25. In general, bone growth in females is completed before that in males.

CLINICAL APPLICATION

Osteogenic sarcoma is a malignant bone tumor that primarily affects osteoblasts. It predominantly affects young people between the ages of 10 and 25 years and occurs slightly more often in males than females. Tumors often occur in the metaphyses of long bones such as the femur (thighbone), tibia (shinbone), and humerus (arm bone). Left untreated, osteogenic sarcoma metastasizes and leads to death quite rapidly. Metastases occur most frequently in the lungs. Treatment consists of multidrug chemotherapy following amputation or resection.

BONE REPLACEMENT

Bones undergoing either intramembranous or endochondral ossification are continually remodeled from the time that initial calcification occurs until the final structure appears. *Remodeling* is the replacement of old bone tissue by new bone tissue. Compact bone is formed by the transformation of spongy bone. The diameter of a long bone is increased by the destruction of the bone closest to the marrow cavity and the construction of new bone around the outside of

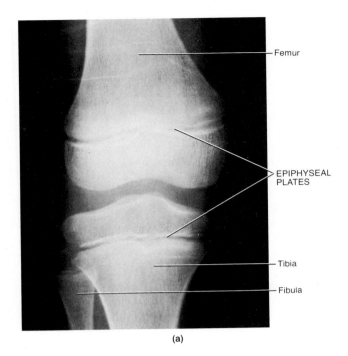

(a)

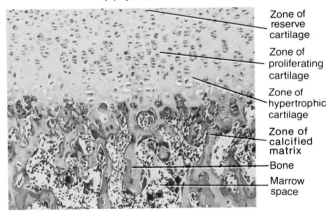

(b)

FIGURE 5-6 Epiphyseal plate. (a) Anteroposterior projection showing epiphyseal plates in the femur and tibia of a 10-year-old child. (Courtesy of Arthur Provost, R.T.) (b) Photomicrograph of epiphysis of a long bone showing the epiphyseal plate at a magnification of 160×. Copyright © 1983 by Michael H. Ross. Used by permission.

the diaphysis. However, even after bones have reached their adult shapes and sizes, old bone is perpetually destroyed and new bone tissue is formed in its place. Bone is never metabolically at rest; it constantly remodels and reappropriates its matrix and minerals along lines of mechanical stress.

Bone shares with skin the feature of replacing itself throughout adult life. Remodeling takes place at different rates in various body regions. The distal portion of the femur (thighbone) is replaced about every 4 months. By contrast, bone in certain areas of the shaft will not be completely replaced during the individual's life. Remodeling allows worn or injured bone to be removed and replaced with new tissue. It also allows bone to serve as the body's storage area for calcium. Many other tissues in the body need calcium in order to perform their functions. For example, a nerve cell needs calcium for nerve impulse conduction, muscle needs calcium to contract, and blood needs calcium to clot. The blood continually trades off calcium with the bones, removing calcium when it and other tissues are not receiving enough of this element and resupplying the bones with dietary calcium to keep them from losing too much bone mass.

The cells believed to be responsible for the resorption (loss of a substance through a physiological or pathological process) of bone tissue are osteoclasts. In the healthy adult, a delicate homeostasis is maintained between the action of the osteoclasts in removing calcium and the action of the bone-making osteoblasts in depositing calcium. Should too much new tissue be formed, the bones become abnormally thick and heavy. If too much calcium is deposited in the bone, the surplus may form thick bumps, or spurs, on the bone that interfere with movement at joints. A loss of too much tissue or calcium weakens the bones and allows them to break easily or to become very flexible. As you will see later, a greatly accelerated remodeling process results in a condition called Paget's disease.

In the process of resorption, it is believed that osteoclasts send out projections that secrete protein-digesting enzymes released from lysosomes and several acids (lactic and citric). The enzymes may function by digesting the collagen and other organic substances, while the acids may cause the bone salts (minerals) to dissolve. It is also presumed that the osteoclastic projections may phagocytose whole fragments of collagen and bone salts. Magnesium deficiency inhibits the activity of osteoclasts.

Normal bone growth in the young and bone replacement in the adult depend on several factors. First, sufficient quantities of calcium and phosphorus, components of the primary salt that makes bone hard, must be included in the diet. Manganese may also be important in bone growth. It has been shown that manganese deficiency significantly inhibits laying down new bone tissue.

Second, the individual must obtain sufficient amounts of vitamins, particularly vitamin D, which participates in the absorption of calcium from the gastrointestinal tract into the blood, calcium removal from bone, and kidney reabsorption of calcium that might otherwise be lost in urine.

Third, the body must manufacture the proper amounts of the hormones responsible for bone tissue activity (Chapter 21). Growth hormone (GH), secreted by the pituitary gland, is responsible for the general growth of bones. Too much or too little of this hormone during childhood makes the adult abnormally tall or short. Other hormones specialize in regulating the osteoclasts. Calcitonin (CT), produced by the thyroid gland, inhibits osteoclastic activity and accelerates calcium absorption by bones, while parathormone (PTH), synthesized by the parathyroid glands, increases the number and activity of osteoclasts. PTH also releases calcium and phosphate from bones into blood, transports calcium from urine into blood, and transports phosphate from blood into urine. And still others, especially the sex hormones, aid osteoblastic activity and thus promote the growth of new bone. The sex hormones act as a double-edged sword. They aid in the growth of new bone, but they also bring about the degeneration of all the cartilage cells in epiphyseal plates. Because of the sex hormones, the typical adolescent experiences a spurt of growth during puberty, when sex hormone levels start to increase. The individual then quickly completes the growth process as the epiphyseal cartilage disappears. Premature puberty can actually prevent one from reaching an average adult height because of the simultaneous premature degeneration of the plates.

BLOOD AND NERVE SUPPLY

Bone is richly supplied with blood, and blood vessels are especially abundant in portions of bone containing red bone marrow. Blood vessels pass into bones from the periosteum. Although the blood supply to bones varies according to their shape, here we shall consider the blood supply to a long bone only.

Because the artery to the diaphysis of a long bone is usually the largest, it is referred to as the *nutrient artery*. The artery first enters the bone during the early development and passes through an opening in the diaphysis, the *nutrient foramen* (see Figure 5-1a), which leads into a *nutrient canal*. On entering the medullary cavity, the nutrient artery divides into a proximal and a distal branch, each of which supplies most of the marrow, inner portion of compact bone of the diaphysis, and metaphysis. As the nutrient artery passes through compact bone on its way to the medullary cavity, it sends branches into the central (Haversian) canals. Branches of the *articular arteries* enter many vascular foramina in the epiphysis and supply the marrow, metaphysis, and bony tissue of the epiphysis. *Periosteal arteries,* accompanied by nerves, enter the diaphysis at numerous points through perforating (Volkmann's) canals (see Figure 5-3a). These blood vessels run in the central canals and supply the outer part of the compact bone of the diaphysis. Veins accompany the several types of arteries; the principal veins leave the bone by numerous vascular foramina at the epiphyses of the bone.

Although the nerve supply to bone is not extensive, some nerves (vasomotor) accompany blood vessels and some (sensory) occur in the periosteum. The nerves of the periosteum are primarily concerned with pain, as might be associated with a fracture or tumor.

AGING AND THE SKELETAL SYSTEM

There are two principal effects of aging on the skeletal system. The first effect is the loss of calcium from bones. This loss usually begins after age 40 in females and continues thereafter until as much as 30 percent of the calcium in bones is lost by age 70. In males, calcium loss typically does not begin until after age 60. The loss of calcium from bones is one of the factors related to a condition called osteoporosis, which will be described shortly.

The second principal effect of aging on the skeletal system is a decrease in the rate of protein formation that results in a decreased ability to produce the organic portion of bone matrix. As a consequence, bone matrix accumulates a lesser proportion of organic matrix and a greater proportion of inorganic matrix. In some elderly individuals, this process can cause their bones to become quite brittle and more susceptible to fracture.

DEVELOPMENTAL ANATOMY OF THE SKELETAL SYSTEM

Bone forms about the sixth or seventh week by either of two processes, *intramembranous ossification* or *endochondral ossification*. Both processes begin when *mesenchymal (mesodermal) cells* migrate into the area where bone formation will occur. In some skeletal structures, the mesenchymal cells develop into *chondroblasts* that form *cartilage*. In other skeletal structures, the mesenchymal cells develop into *osteoblasts* that form *bone tissue* by intramembranous or endochondral ossification. (The details of ossification have already been discussed earlier in the chapter.)

Discussion of the development of the skeletal system provides us with an excellent opportunity to trace the development of the extremities. The *extremities* make their appearance about the fifth week as small elevations at the sides of the trunk called *limb buds* (Figure 5-7). They consist of masses of general *mesoderm* surrounded by *ectoderm*. At this point, a mesenchymal skeleton exists in the limbs; some of the masses of mesoderm surrounding the developing bones will develop into the skeletal muscles of the extremities. By the sixth week, the limb buds develop a constriction around the middle portion. The constriction produces distal segments of the upper buds called *hand plates* and distal segments of the lower buds called *foot plates*. These plates represent the beginnings of the *hands* and *feet*, respectively. At this stage of limb development, a cartilaginous skeleton is present. By the seventh week, the *arm*, *forearm*, and *hand* are evident in the upper limb bud, and the *thigh*, *leg*, and *foot* appear in the lower limb bud. Endochondral ossification has begun. By the eighth week, the upper limb bud is appropriately called the *upper extremity* as the *shoulder*, *elbow*, and *wrist* areas become

apparent, and the lower limb bud is referred to as the *lower extremity* with the appearance of the *knee* and *ankle* areas.

The *notochord* is a flexible rod of tissue that lies in a position where the future vertebral column will develop (see Figure 9-10a). As the vertebrae develop, the notochord becomes surrounded by the developing vertebral bodies and the notochord eventually disappears, except for remnants that persist as the *nucleus pulposus* of the intervertebral discs (see Figure 6-19).

APPLICATIONS TO HEALTH

Some bone disorders result from deficiencies in vitamins or minerals or from too much or too little of the hormones that regulate bone homeostasis. Infections and tumors are also responsible for certain bone disorders.

OSTEOPOROSIS

Osteoporosis (os'-tē-ō-pō-RŌ-sis) is an age-related disorder characterized by decreased bone mass and increased susceptibility to fractures. It primarily affects middle-aged and elderly people—white women more than men and whites more than blacks. Between puberty and the middle years, sex hormones maintain osseous tissue by stimulating the osteoblasts to form new bone. Women produce smaller amounts of sex hormones after menopause, and both men and women produce smaller amounts during old age. As a result, the osteoblasts become less active and there is a decrease in bone mass. Osteoporosis can also occur during nursing and pregnancy and in individuals exposed to prolonged treatment with cortisone or high levels of thyroid hormones. The first symptom of osteoporosis occurs when bone mass is so depleted that the skeleton can no longer withstand the mechanical stresses of everyday living and compression fractures result. Osteoporosis is responsible for shrinkage of the backbone and height loss, hunched backs, hip fractures, and considerable pain. Osteoporosis affects the entire skeletal system, especially the vertebral bodies, ribs, proximal femur (hip), humerus, and distal radius.

Among the factors implicated in osteoporosis besides race and sex, are (1) body build (short females are at greater risk since they have less total bone mass), (2) weight (thin females are at greater risk since adipose tissue is a great source of estrone, an estrogen that retards bone loss), (3) smoking (decreases blood estrogen levels), (4) calcium deficiency and malabsorption, (5) vitamin D deficiency, (6) exercise (sedentary people are more likely to develop bone loss), and (7) certain drugs (alcohol, some diuretics, cortisone, and tetracycline promote bone loss). Estrogen replacement therapy (ERT) and exercise are prescribed to treat osteoporosis. Studies indicate that calcium by itself has little effect on preventing the progression of osteoporosis. Calcium is helpful when taken with estrogen.

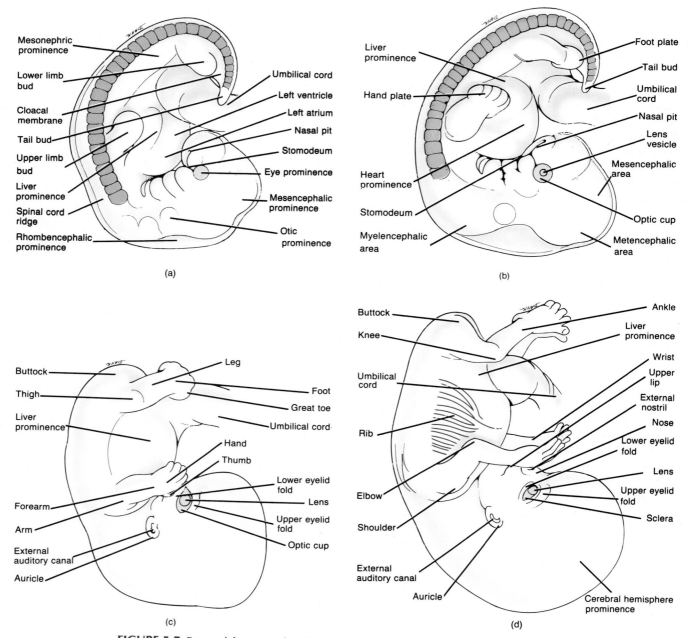

FIGURE 5-7 External features of a developing embryo at various stages of development. (a) Fifth week. (b) Sixth week. (c) Seventh week. (d) Eighth week. Many of the labeled structures are discussed in later chapters.

VITAMIN DEFICIENCIES

Rickets

A deficiency of vitamin D in children results in a condition called *rickets.* It is characterized by an inability of the body to transport calcium and phosphorus from the gastrointestinal tract into the blood for utilization by bones. As a result, epiphyseal cartilage cells cease to degenerate and new cartilage continues to be produced. Epiphyseal cartilage thus becomes wider than normal. At the same time, the soft matrix laid down by the osteoblasts in the diaphysis fails to calcify. As a result, the bones stay soft. When the child walks, the weight of the body causes the bones in the legs to bow. Malformations of the head, chest, and pelvis also occur.

The cure and prevention of rickets consist of adding generous amounts of calcium, phosphorus, and vitamin D to the diet. Exposing the skin to the ultraviolet rays of sunlight also aids the body in manufacturing additional vitamin D.

Osteomalacia

A deficiency of vitamin D in adults causes the bones to give up excessive amounts of calcium and phosphorus. This loss, called **demineralization,** is especially heavy in the bones of the pelvis, legs, and spine. Demineralization caused by vitamin D deficiency is called **osteomalacia** (os'-tē-ō-ma-LĀ-shē-a; *malacia* = softness). After the bones demineralize, the weight of the body produces a bowing of the leg bones, shortening of the backbone, and flattening of the pelvic bones. Osteomalacia mainly affects women who live on poor cereal diets devoid of milk, are seldom exposed to the sun, and have repeated pregnancies that deplete the body of calcium. The condition responds to the same treatment as rickets; if the disease is severe enough and threatens life, large doses of vitamin D are given.

Osteomalacia may also result from failure to absorb fat (steatorrhea) because vitamin D is soluble in fats and calcium combines with fats. As a result, vitamin D and calcium remain with the unabsorbed fat and are lost in the feces.

PAGET'S DISEASE

Paget's disease is characterized by a greatly accelerated remodeling process in which osteoclastic resorption is massive and new bone formation by osteoblasts is extensive. As a result, there is an irregular thickening and softening of the bones and greatly increased vascularity. It rarely occurs in individuals under 50. The cause, or **etiology** (ē'-tē-OL-ō-jē), of the disease is unknown.

OSTEOMYELITIS

The term **osteomyelitis** (os'-tē-ō-mī-i-LĪ-tis) includes all infectious diseases of bone. These diseases may be localized or widespread and may also involve the periosteum, marrow, and cartilage. Various microorganisms may give rise to bone infection, but the most frequent are bacteria known as *Staphylococcus aureus,* commonly called "staph." These bacteria may reach the bone by various means: the bloodstream, an injury such as a fracture, or an infection such as a sinus infection or a tooth abscess. Antibiotics have been effective in treating the disease and in preventing it from spreading through extensive areas of bone, but complete recovery is usually a prolonged and painful process.

FRACTURES

In simplest terms, a **fracture (Fx)** is any break in a bone. Usually, the fracture is restored to normal position by manipulation without surgery. This procedure of setting a fracture is called **closed reduction.** In other cases, the fracture must be exposed by surgery before the break is rejoined. This procedure is known as **open reduction.**

Types

Although fractures may be classified in several different ways, the following scheme is useful (Figure 5-8):

1. **Partial (incomplete).** A fracture in which the break across the bone is incomplete.
2. **Complete.** A fracture in which the break across the bone is complete, so that the bone is broken into two pieces.
3. **Closed (simple).** A fracture in which the bone does not break through the skin.
4. **Open (compound).** A fracture in which the broken ends of the bone protrude through the skin.
5. **Comminuted** (KOM-i-nyoo'-ted). A fracture in which the bone is splintered at the site of impact and smaller fragments of bone are found between the two main fragments.
6. **Greenstick.** A partial fracture in which one side of the bone is broken and the other side bends; occurs only in children.
7. **Spiral.** A fracture in which the bone is usually twisted apart.
8. **Transverse.** A fracture at right angles to the long axis of the bone.
9. **Impacted.** A fracture in which one fragment is firmly driven into the other.
10. **Pott's.** A fracture of the distal end of the fibula, with serious injury of the distal tibial articulation.
11. **Colles'** (KOL-ēz). A fracture of the distal end of the radius in which the distal fragment is displaced posteriorly.
12. **Displaced.** A fracture in which the anatomical alignment of the bone fragments is not preserved.
13. **Nondisplaced.** A fracture in which the anatomical alignment of the bone fragments is preserved.
14. **Stress.** A partial fracture resulting from inability to withstand repeated stress due to a change in training, harder surfaces, longer distances, and greater speed. About 25 percent of all stress fractures involve the fibula, typically the distal third.
15. **Pathologic.** A fracture due to weakening of a bone caused by disease processes such as neoplasia, osteomyelitis, osteoporosis, or osteomalacia.

Fracture Repair

Unlike the skin, which may repair itself within days, or muscle, which may mend in weeks, a bone sometimes requires months to heal. A fractured femur, for example, may take 6 months to heal. Sufficient calcium to strengthen and harden new bone is deposited only gradually. Bone cells also grow and reproduce slowly. Moreover, the blood supply to bone is decreased, which helps to explain the difficulty in the healing of an infected bone.

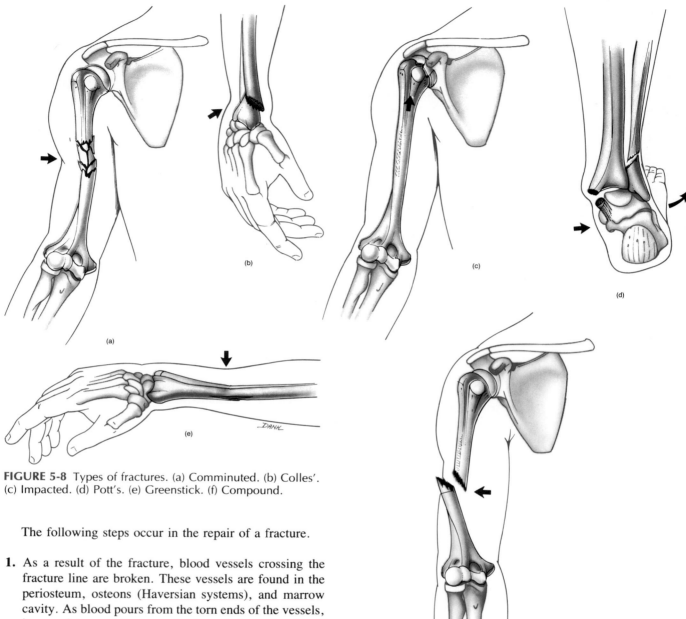

FIGURE 5-8 Types of fractures. (a) Comminuted. (b) Colles'. (c) Impacted. (d) Pott's. (e) Greenstick. (f) Compound.

The following steps occur in the repair of a fracture.

1. As a result of the fracture, blood vessels crossing the fracture line are broken. These vessels are found in the periosteum, osteons (Haversian systems), and marrow cavity. As blood pours from the torn ends of the vessels, it coagulates and forms a clot in and about the site of the fracture. This clot, called a *fracture hematoma* (hē-ma-TŌ-ma), usually occurs 6–8 hours after the injury. Since the circulation of blood ceases when the fracture hematoma forms, bone cells and periosteal cells at the fracture line die. The hematoma serves as a focus for subsequent cellular invasion.

2. A growth of new bone tissue—a *callus*—develops in and around the fractured area. It forms a bridge between separated areas of bone. The part of the callus that forms from the osteogenic cells of the torn periosteum and develops around the outside of the fracture is called an *external callus.* The part of the callus that forms from the osteogenic cells of the endosteum and develops between the two ends of bone fragments and between the two marrow cavities is called the *internal callus.*

Approximately 48 hours after a fracture occurs, the cells that ultimately repair the fracture become actively mitotic. These cells come from the osteogenic layer of the periosteum, the endosteum of the marrow cavity, and the bone marrow. As a result of their accelerated mitotic activity, the cells of the three regions grow toward the fracture. During the first week following the fracture, the cells of the endosteum and bone marrow form new trabeculae in the marrow cavity near the line of fracture. This is the internal callus. During the next few days, osteogenic cells of the periosteum form a collar around each bone fragment. The collar, or external callus, is replaced by trabeculae. The trabeculae of the calli are joined to living and dead portions of the original bone fragments.

3. The final phase of fracture repair is the *remodeling* of the calli. Dead portions of the original fragments are gradually resorbed. Compact bone replaces spongy bone around the periphery of the fracture. In some cases, the healing is so complete that the fracture line is undetectable, even by x ray. However, a thickened area on the surface of the bone usually remains as evidence of the fracture site.

CLINICAL APPLICATION

In the past when a fracture failed to unite, the patient could either wait, hoping that nature and time would solve the problem, or choose surgery. Now there is another alternative, called *pulsating electromagnetic fields* (*PEMFs*), that involves electrotherapy to stimulate bone repair.

Essentially, the fracture is exposed to a weak electrical current generated from coils that are fastened around the cast. Osteoblasts adjacent to the fracture site become more active metabolically in response to the stimulation of the alteration of electrical charges on the surfaces of their membranes. The increased activity apparently causes an acceleration of calcification, vascularization, and endochondral ossification, resulting in acceleration of fracture repair. It was noted earlier that parathyroid hormone (PTH) increases osteoclastic activity and thus stimulates bone destruction. One hypothesis suggests that electricity also heals fractures by keeping PTH from acting on osteoclasts, thus increasing bone formation and repair.

Although the original application of PEMFs was to treat improperly healing fractures, research is now underway to determine if electrical stimulation can be used to regenerate limbs, stop the growth of tumor cells in humans, reduce postoperative swelling, and improve the healing of tendons and ligaments.

KEY MEDICAL TERMS ASSOCIATED WITH OSSEOUS TISSUE

Achondroplasia (a-kon'-drō-PLĀ-zē-a; *a* = without; *chondro* = cartilage; *plasia* = growth) Imperfect ossification within cartilage of long bones during fetal life; also called **fetal rickets.**

Craniotomy (krā'-nē-OT-ō-mē; *cranium* = skull; *tome* = a cutting) Any surgery that requires cutting through the bones surrounding the brain.

Necrosis (ne-KRŌ-sis; *necros* = death; *osis* = condition) Death of tissues or organs; in the case of bone, results from deprivation of blood supply resulting from fracture, extensive removal of periosteum in surgery, exposure to radioactive substances, or other causes.

Osteitis (os'-tē-Ī-tis; *osteo* = bone) Inflammation or infection of bone.

Osteoarthritis (os'-tē-ō-ar-THRĪ-tis; *arthro* = joint) A degenerative condition of bone and also the joint.

Osteoblastoma (os'-tē-ō-blas-TŌ-ma; *oma* = tumor) A benign tumor of osteoblasts.

Osteochondroma (os'-tē-ō-kon-DRŌ-ma; *chondro* = cartilage) A benign tumor of bone and cartilage.

Osteoma (os'-tē-Ō-ma) A benign bone tumor.

Osteosarcoma (os'-tē-ō-sar-KŌ-ma; *sarcoma* = connective tissue tumor) A malignant tumor composed of osseous tissue.

Pott's (POTS) **disease** Inflammation of the backbone, caused by the microorganism that produces tuberculosis.

STUDY OUTLINE

Functions (p. 114)
1. The skeletal system consists of all bones attached at joints and cartilage between joints.
2. The functions of the skeletal system include support, protection, leverage, mineral storage, and blood cell production.

Histology (p. 114)
1. Osseous tissue consists of widely separated cells surrounded by large amounts of intercellular substance. The four principal types of cells are osteoprogenitor cells, osteoblasts, osteocytes, and osteoclasts. The intercellular substance contains collagenous fibers and abundant hydroxyapatites (mineral salts).
2. Parts of a typical long bone are the diaphysis (shaft), epiphyses (ends), metaphysis, articular cartilage, periosteum, medullary (marrow) cavity, and endosteum.

3. Compact (dense) bone consists of osteons (Haversian systems) with little space between them. Compact bone lies over spongy bone and composes most of the bone tissue of the diaphysis. Functionally, compact bone protects, supports, and resists stress.
4. Spongy (cancellous) bone consists of trabeculae surrounding many red marrow-filled spaces. It forms most of the structure of short, flat, and irregular bones, and the epiphyses of long bones. Functionally, spongy bone stores some red marrow and provides some support.

Ossification: Bone Formation (p. 118)
1. Bone forms by a process called ossification (osteogenesis), which begins when mesenchymal cells become transformed into osteoprogenitor cells, which differentiate into osteoblasts.

2. The process begins during the sixth or seventh week of embryonic life and continues throughout adulthood. The two types of ossification, intramembranous and endochondral, involve the replacement of a preexisting connective tissue with bone.

3. Intramembranous ossification occurs within fibrous membranes of the embryo and the adult.

4. Endochondral ossification occurs within a cartilage model. The primary ossification center of a long bone is in the diaphysis. Cartilage degenerates, leaving cavities that merge to form the marrow cavity. Osteoblasts lay down bone. Next, ossification occurs in the epiphyses, where bone replaces cartilage, except for the epiphyseal plate.

Bone Growth (p. 119)

1. The anatomical zones of the epiphyseal plate are the zones of reserve cartilage, proliferating cartilage, hypertrophic cartilage, and calcified matrix.

2. Because of the activity of the epiphyseal plate, the diaphysis of a bone increases in length by appositional growth.

3. Bone grows in diameter as a result of the addition of new bone tissue by periosteal osteoblasts around the outer surface of the bone.

Bone Replacement (p. 121)

1. The homeostasis of bone growth and development depends on a balance between bone formation and resorption.

2. Old bone is constantly destroyed by osteoclasts, while new bone is constructed by osteoblasts. This process is called remodeling.

3. Normal growth depends on calcium, phosphorus, and vitamins, especially vitamin D, and is controlled by hormones that are responsible for bone mineralization and resorption.

Blood and Nerve Supply (p. 122)

1. Long bones are supplied by nutrient, articular, and periosteal arteries; veins accompany the arteries.

2. The nerve supply to bones consists of vasomotor and sensory nerves.

Aging and the Skeletal System (p. 123)

1. The principal effect of aging is a loss of calcium from bones, which may result in osteoporosis.

2. Another effect is a decreased production of organic matrix, which makes bones more susceptible to fracture.

Developmental Anatomy of the Skeletal System (p. 123)

1. Bone forms from mesoderm by intramembranous or endochondral ossification.

2. Extremities develop from limb buds, which consist of mesoderm and ectoderm.

Applications to Health (p. 123)

1. Osteoporosis is a decrease in the amount and strength of bone tissue due to decreases in hormone output.

2. Rickets is a vitamin D deficiency in children in which the body does not absorb calcium and phosphorus. The bones soften and bend under the body's weight.

3. Demineralization caused by vitamin D deficiency in adults results in osteomalacia.

4. Paget's disease is the irregular thickening and softening of bones, related to a greatly accelerated remodeling process.

5. Osteomyelitis is a term for the infectious diseases of bones, marrow, and periosteum. It is frequently caused by "staph" bacteria.

6. A fracture (Fx) is any break in a bone.

7. The types of fractures include partial, complete, simple, compound, comminuted, greenstick, spiral, transverse, impacted, Pott's, Colles', displaced, nondisplaced, and stress.

8. Fracture repair consists of forming a fracture hematoma, forming a callus, and remodeling.

9. Treatment by pulsating electromagnetic fields (PEMFs) has provided dramatic results in healing fractures that would otherwise not have mended properly. Its application for limb regeneration and stopping the growth of tumor cells is being investigated.

REVIEW QUESTIONS

1. Define the skeletal system. What are its five principal functions?

2. Why is osseous tissue considered a connective tissue? What is its composition? Describe the cells present and the intercellular substance.

3. Diagram the parts of a long bone and list the functions of each part.

4. Distinguish between compact and spongy bone in terms of microscopic appearance, location, and function.

5. Diagram the microscopic appearance of compact bone and indicate the functions of the various components.

6. What is meant by ossification? Describe the initial events of ossification.

7. Outline the major events involved in intramembranous and endochondral ossification and explain the principal differences.

8. Describe the histology of the various zones of the epiphyseal plate. How does the plate grow? What is the significance of the epiphyseal line?

9. Define remodeling. How does balance between osteoblast activity and osteoclast activity control the replacement of bone.

10. Describe the blood and nerve supplies of a long bone.

11. Explain the effects of aging on the skeletal system.

12. Describe the development of the skeletal system.

13. What are the principal symptoms of osteoporosis, Paget's disease, and osteomyelitis? What is the etiology of each?

14. Distinguish between rickets and osteomalacia. What do the two diseases have in common?

15. What is a fracture (Fx)? Distinguish several principal kinds. Outline the three basic steps involved in fracture repair.

16. Refer to the glossary of key medical terms associated with the osseous tissue. Be sure that you can define each term.

EXHIBIT 6-2

Divisions of the Skeletal System

REGION OF THE SKELETON	NUMBER OF BONES
AXIAL SKELETON	
Skull	
Cranium	8
Face	14
Hyoid	1
Auditory ossicles[*]	6
(3 in each ear)	
Vertebral column	26
Thorax	
Sternum	1
Ribs	24
	80
APPENDICULAR SKELETON	
Pectoral (shoulder) girdles	
Clavicle	2
Scapula	2
Upper extremities	
Humerus	2
Ulna	2
Radius	2
Carpals	16
Metacarpals	10
Phalanges	28
Pelvic (hip) girdle	
Coxal (pelvic or hip) bone	2
Lower extremities	
Femur	2
Fibula	2
Tibia	2
Patella	2
Tarsals	14
Metatarsals	10
Phalanges	28
	126
	Total = 206

[*] Although the auditory ossicles are not considered part of the axial or appendicular skeleton, but rather as a separate group of bones, they are placed with the axial skeleton for convenience. They are discussed in detail in Chapter 20.

bone, sphenoid bone, and ethmoid bone. There are 14 *facial bones:* nasal bones (2), maxillae (2), zygomatic bones (2), mandible, lacrimal bones (2), palatine bones (2), inferior nasal conchae (2), and vomer. Be sure you can locate all the skull bones in the various views of the skull (Figure 6-2).

SUTURES

A *suture* (SOO-chur), meaning seam or stitch, is an immovable joint found only between skull bones. Very little connective tissue is found between the bones of a suture. Four prominent skull sutures are:

1. *Coronal suture* between the frontal bone and the two parietal bones.
2. *Sagittal suture* between the two parietal bones.

3. *Lambdoidal* (lam-DOY-dal) *suture* between the parietal bones and the occipital bone.
4. *Squamosal* (skwa-MŌ-sal) *suture* between the parietal bones and the temporal bones.

Refer to Figures 6-2 and 6-3 for the locations of these sutures. Several other sutures are also shown. Their names are descriptive of the bones they connect. For example, the frontonasal suture is between the frontal bone and the nasal bones. These sutures are indicated in Figures 6-2 through 6-5.

FONTANELS

The "skeleton" of a newly formed embryo consists of cartilage or fibrous membrane structures shaped like bones. Gradually, the cartilage or fibrous membrane is replaced by bone. At birth, membrane-filled spaces called *fontanels* (fon'-ta-NELZ; = little fountains) are found between cranial bones (Figure 6-3). These "soft spots" are areas where intramembranous ossification will eventually replace the membrane-filled spaces with bone. Physicians find the fontanels helpful in determining the position of the infant's head prior to delivery. Fontanels also allow the skull to be compressed as it passes through the birth canal during delivery. Finally, fontanels permit rapid growth of the brain during infancy. Although an infant may have many fontanels at birth, the form and location of six are fairly constant.

The *anterior (frontal) fontanel* is located between the angles of the two parietal bones and the two segments of the frontal bone. This fontanel is roughly diamond shaped, and it is the largest of the six fontanels. It usually closes 18–24 months after birth.

The *posterior (occipital;* ok-SIP-i-tal) *fontanel* is situated between the two parietal bones and the occipital bones. This diamond-shaped fontanel is considerably smaller than the anterior fontanel. It generally closes about 2 months after birth.

The *anterolateral (sphenoidal;* sfē-NOY-dal) *fontanels* are paired. One is located on each side of the skull at the junction of the frontal, parietal, temporal, and sphenoid bones. These fontanels are quite small and irregular in shape. They normally close about 3 months after birth.

The *posterolateral (mastoid) fontanels* are also paired. One is situated on each side of the skull at the junction of the parietal, occipital, and temporal bones. These fontanels are irregularly shaped. They begin to close 1 or 2 months after birth, but closure is not generally complete until 12 months.

CRANIAL BONES

Frontal Bone

The *frontal bone* forms the forehead (the anterior part of the cranium), the roofs of the *orbits* (eye sockets), and most of the anterior part of the cranial floor. Soon after

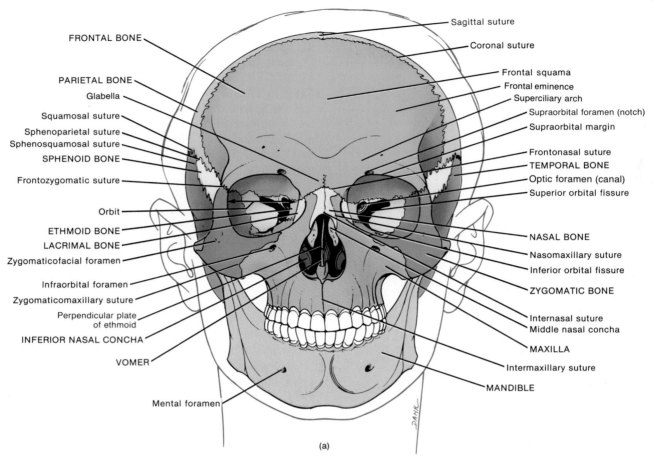

FRONTAL BONE

PARIETAL BONE
Glabella
Squamosal suture
Sphenoparietal suture
Sphenosquamosal suture
SPHENOID BONE
Frontozygomatic suture

Orbit
ETHMOID BONE
LACRIMAL BONE
Zygomaticofacial foramen

Infraorbital foramen
Zygomaticomaxillary suture
Perpendicular plate
of ethmoid
INFERIOR NASAL CONCHA

VOMER

Mental foramen

Sagittal suture
Coronal suture

Frontal squama
Frontal eminence
Superciliary arch
Supraorbital foramen (notch)
Supraorbital margin
Frontonasal suture
TEMPORAL BONE
Optic foramen (canal)
Superior orbital fissure

NASAL BONE
Nasomaxillary suture
Inferior orbital fissure
ZYGOMATIC BONE

Internasal suture
Middle nasal concha
MAXILLA
Intermaxillary suture
MANDIBLE

(a)

FIGURE 6-2 Skull. (a) Anterior view.

birth, the left and right parts of the frontal bone are united by a suture. The suture usually disappears by age 6. If, however, the suture persists throughout life, it is referred to as the *metopic suture.*

If you examine the anterior and lateral views of the skull in Figure 6-2, you will note the *frontal squama* (SKWĀ-ma), or *vertical plate* (*squam* = scale). This scale-like plate, which corresponds to the forehead, gradually slopes down from the coronal suture, then turns abruptly downward. It projects slightly above its lower edge on either side of the midline to form the *frontal eminences.* Inferior to each eminence is a horizontal ridge, the *superciliary arch,* caused by the projection of the frontal sinuses posterior to the eyebrow.

Between the eminences and the arches just superior to the nose is a flattened area, the *glabella.* A thickening of the frontal bone is called the *supraorbital margin.* From this margin the frontal bone extends posteriorly to form the roof of the orbit and part of the floor of the cranial cavity.

CLINICAL APPLICATION

Just above the supraorbital margin is a relatively sharp ridge that overlies the frontal sinus. A blow to the ridge frequently lacerates the skin over it, resulting in bleeding. Bruising of the skin over the ridge causes tissue fluid and blood to accumulate in the surrounding connective tissue and gravitate into the upper eyelid. The resulting swelling and discoloration is called a *black eye.*

Within the supraorbital margin, slightly medial to its midpoint, is a hole (frequently a notch) called the *supraorbital foramen* (*notch).* The supraorbital nerve and artery pass through this foramen. The *frontal sinuses* lie deep to the frontal squama. These mucus-lined cavities act as sound chambers that give the voice resonance.

Parietal Bones

The two *parietal bones* (pa-RĪ-e-tal; *paries* = wall) form the greater portion of the sides and roof of the cranial cavity. The external surface contains two slight ridges that may be observed by looking at the lateral view of the skull in Figure 6-2. These are the *superior temporal line* and a less conspicuous *inferior temporal line* below it. The internal surfaces of the bones contain many eminences and depressions that accommodate the blood vessels supplying the outer meninx (covering) of the brain called the *dura mater.*

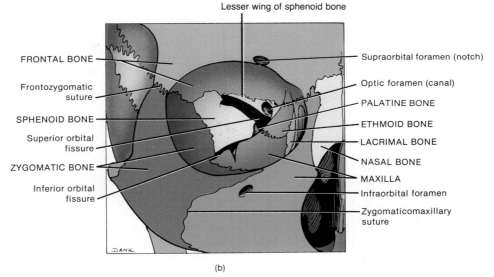

Lesser wing of sphenoid bone

FRONTAL BONE

Frontozygomatic suture

SPHENOID BONE

Superior orbital fissure

ZYGOMATIC BONE

Inferior orbital fissure

Supraorbital foramen (notch)

Optic foramen (canal)

PALATINE BONE

ETHMOID BONE

LACRIMAL BONE

NASAL BONE

MAXILLA

Infraorbital foramen

Zygomaticomaxillary suture

(b)

FIGURE 6-2 (*Continued*)

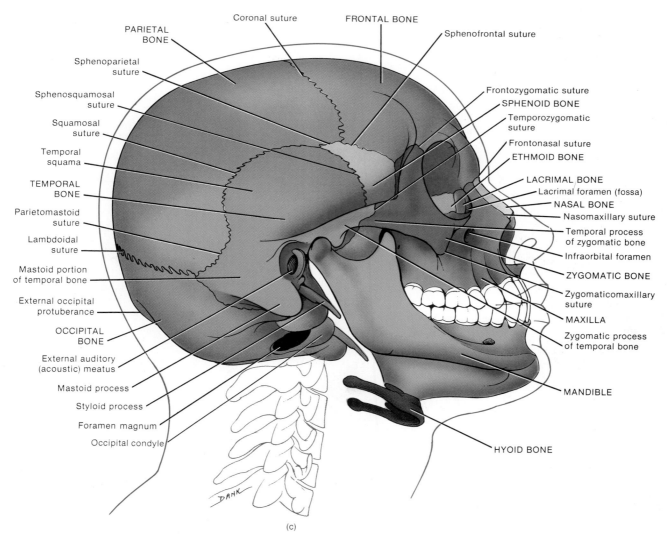

Coronal suture

FRONTAL BONE

PARIETAL BONE

Sphenofrontal suture

Sphenoparietal suture

Sphenosquamosal suture

Squamosal suture

Temporal squama

TEMPORAL BONE

Parietomastoid suture

Lambdoidal suture

Mastoid portion of temporal bone

External occipital protuberance

OCCIPITAL BONE

External auditory (acoustic) meatus

Mastoid process

Styloid process

Foramen magnum

Occipital condyle

Frontozygomatic suture

SPHENOID BONE

Temporozygomatic suture

Frontonasal suture

ETHMOID BONE

LACRIMAL BONE

Lacrimal foramen (fossa)

NASAL BONE

Nasomaxillary suture

Temporal process of zygomatic bone

Infraorbital foramen

ZYGOMATIC BONE

Zygomaticomaxillary suture

MAXILLA

Zygomatic process of temporal bone

MANDIBLE

HYOID BONE

(c)

FIGURE 6-2 (*Continued*) Skull. (c) Right lateral view. Although the hyoid bone is not part of the skull, it is included in the illustration for reference.

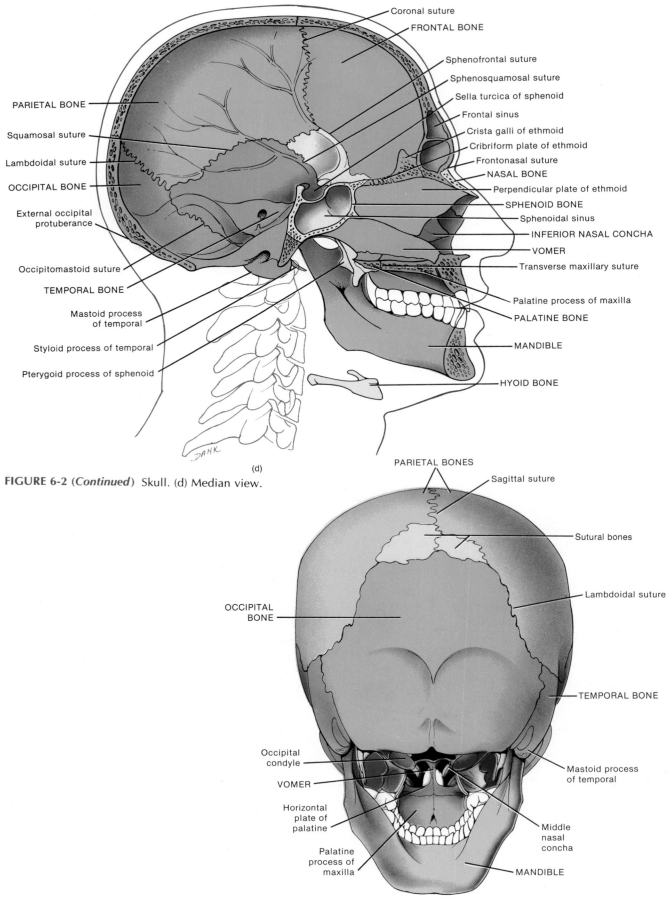

Coronal suture
FRONTAL BONE
Sphenofrontal suture
Sphenosquamosal suture
Sella turcica of sphenoid
Frontal sinus
Crista galli of ethmoid
Cribriform plate of ethmoid
Frontonasal suture
NASAL BONE
Perpendicular plate of ethmoid
SPHENOID BONE
Sphenoidal sinus
INFERIOR NASAL CONCHA
VOMER
Transverse maxillary suture
Palatine process of maxilla
PALATINE BONE
MANDIBLE
HYOID BONE

PARIETAL BONE
Squamosal suture
Lambdoidal suture
OCCIPITAL BONE
External occipital protuberance
Occipitomastoid suture
TEMPORAL BONE
Mastoid process of temporal
Styloid process of temporal
Pterygoid process of sphenoid

(d)

FIGURE 6-2 (Continued) Skull. (d) Median view.

PARIETAL BONES
Sagittal suture
Sutural bones
Lambdoidal suture
TEMPORAL BONE
Mastoid process of temporal
Middle nasal concha
MANDIBLE

OCCIPITAL BONE
Occipital condyle
VOMER
Horizontal plate of palatine
Palatine process of maxilla

FIGURE 6-2 (Continued) Skull. (e) Posterior view.

(e)

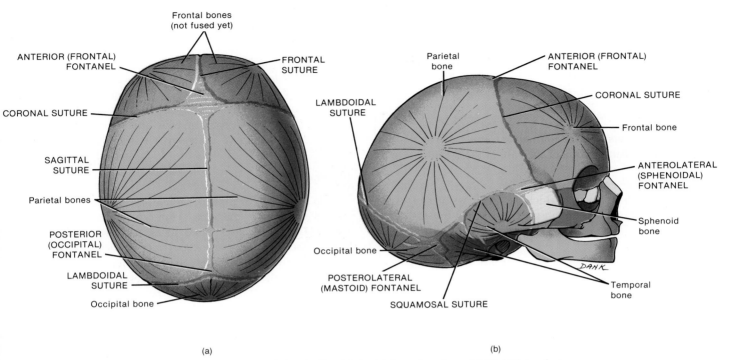

FIGURE 6-3 Fontanels of the skull at birth. (a) Superior view. (b) Right lateral view.

Temporal Bones

The two *temporal* (*tempora* = temples) *bones* form the inferior sides of the cranium and part of the cranial floor.

In the lateral view of the skull in Figure 6-2, note the *temporal squama*—a thin, large, expanded area that forms the anterior and superior part of the temple. Projecting from the inferior portion of the temporal squama is the *zygomatic process,* which articulates with the temporal process of the zygomatic bone. The zygomatic process of the temporal bone and the temporal process of the zygomatic bone constitute the *zygomatic arch.*

At the floor of the cranial cavity, shown in Figure 6-5, is the *petrous portion* of the temporal bone. This portion is triangular and located at the base of the skull between the sphenoid and occipital bones. The petrous portion contains the internal ear in which are located the structures involved in hearing and equilibrium (balance). It also contains the *carotid foramen (canal)* through which the internal carotid artery passes (see Figure 6-4). Posterior to the carotid foramen and anterior to the occipital bone in the *jugular foramen (fossa)* through which the internal jugular vein and the glossopharyngeal (IX) nerve, vagus (X) nerve, and accessory (XI) nerve pass. (As you will see later, the Roman numerals associated with cranial nerves indicate the order in which the nerves arise from the brain, from front to back.)

Between the squamous and petrous portions is a socket called the *mandibular fossa.* Anterior to the mandibular fossa is a rounded eminence, the *articular tubercle.* The mandibular fossa and articular tubercle articulate with the condylar process of the mandible (lower jawbone) to form the temporomandibular joint (TMJ). The mandibular fossa and articular tubercle are seen best in Figure 6-4.

In the lateral view of the skull in Figure 6-2, you will see the *mastoid portion* of the temporal bone, located posterior and inferior to the external auditory meatus, or ear canal. In the adult, the portion of the bone contains a number of *mastoid air "cells."* These air spaces are separated from the brain only by thin bony partitions.

CLINICAL APPLICATION

If *mastoiditis,* the inflammation of the mastoid air cells, occurs, the infection may spread to the brain or its outer covering. These bony cells do not drain as do the paranasal sinuses.

The *mastoid process* is a rounded projection of the temporal bone posterior to the external auditory meatus. It serves as a point of attachment for several neck muscles. Near the posterior border of the mastoid process is the *mastoid foramen* through which a vein (emissary) to the transverse sinus and a small branch of the occipital artery to the dura mater pass. The *external auditory (acoustic) meatus* is the canal in the temporal bone that leads to the middle ear. The *internal auditory (acoustic) meatus* is superior to the jugular foramen (see Figure 6-5a). It transmits the facial (VII) and vestibulocochlear (VIII) nerves and the internal auditory artery. The *styloid process* projects downward from the undersurface of the temporal bone and serves as a point of attachment for muscles and ligaments of the tongue and neck. Between the styloid process and the mastoid process is the *stylomastoid foramen,* which transmits the facial (VII) nerve and stylomastoid artery (see Figure 6-4).

Occipital Bone

The *occipital* (ok-SIP-i-tal) *bone* forms the posterior part and a prominent portion of the base of the cranium (Figure 6-4).

The *foramen magnum* is a large hole in the inferior part of the bone through which the medulla oblongata (part of the brain) and its membranes, the spinal portion of the accessory (XI) nerve, the vertebral and spinal arteries, and the meninges pass.

The *occipital condyles* are oval processes with convex surfaces, one on either side of the foramen magnum, that articulate (form a joint) with depressions on the first cervical vertebra. Extending laterally from the posterior portion of

the condyles are quadrilateral plates of bone called the *jugular processes.* At the base of the condyles is the *hypoglossal canal (fossa)* through which the hypoglossal (XII) nerve passes (see Figure 6-5).

The *external occipital protuberance* is a prominent projection on the posterior surface of the bone just superior to the foramen magnum. You can feel this structure as a definite bump on the back of your head, just above your neck. The protuberance is also visible in Figure 6-2d.

Sphenoid Bone

The *sphenoid* (SFĒ-noyd) *bone* is situated at the middle part of the base of the skull (Figure 6-5). *Spheno* means wedge. This bone is referred to as the keystone of the cranial floor because it articulates with all the other cranial bones. If you view the floor of the cranium from above, you will note that the sphenoid articulates with the temporal bones anteriorly and the occipital bone posteriorly. It lies posterior and slightly superior to the nasal cavities and

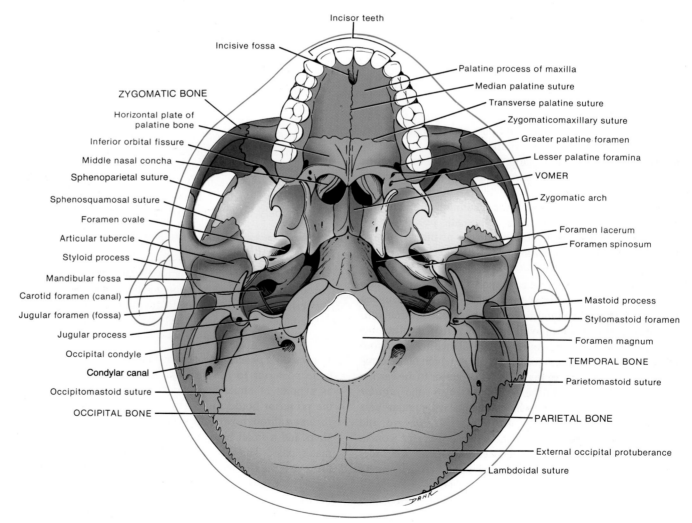

FIGURE 6-4 Skull in inferior view.

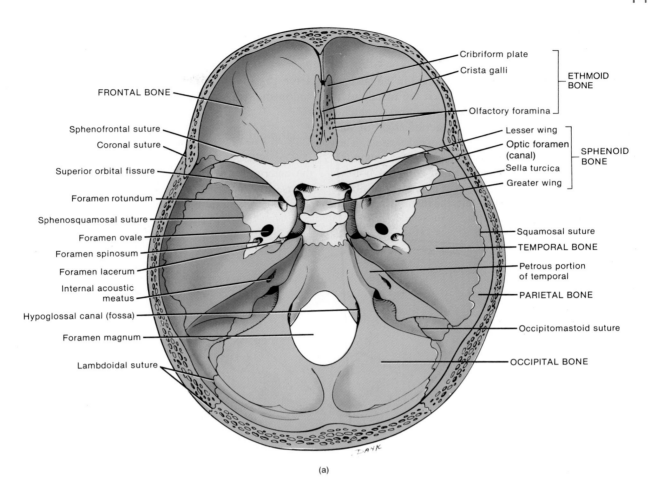

FRONTAL BONE

Sphenofrontal suture
Coronal suture

Superior orbital fissure

Foramen rotundum
Sphenosquamosal suture
Foramen ovale
Foramen spinosum
Foramen lacerum
Internal acoustic meatus
Hypoglossal canal (fossa)
Foramen magnum

Lambdoidal suture

Cribriform plate
Crista galli
Olfactory foramina
ETHMOID BONE

Lesser wing
Optic foramen (canal)
Sella turcica
Greater wing
SPHENOID BONE

Squamosal suture
TEMPORAL BONE
Petrous portion of temporal
PARIETAL BONE

Occipitomastoid suture

OCCIPITAL BONE

(a)

forms part of the floor and sidewalls of the orbit (eye socket). The shape of the sphenoid is frequently described as a bat with outstretched wings.

The *body* of the sphenoid is the cubelike central portion between the ethmoid and occipital bones. It contains the *sphenoidal sinuses,* which drain into the nasal cavity (see Figure 6-8). On the superior surface of the sphenoid body is a depression called the *sella turcica* (SEL-a TUR-sika = Turk's saddle). This depression houses the pituitary gland.

The *greater wings* of the sphenoid are lateral projections from the body and form the anterolateral floor of the cranium. The greater wings also form part of the lateral wall of the skull just anterior to the temporal bone. The posterior portion of each greater wing contains a triangular projection, the *sphenoidal spine,* that fits into the angle between the squama and petrous portion of the temporal bone.

The *lesser wings* are anterior and superior to the greater wings. They form part of the floor of the cranium and the posterior part of the orbit.

Between the body and lesser wing, you can locate the *optic foramen (canal)* through which the optic (II) nerve and ophthalmic artery pass. Lateral to the body between the greater and lesser wings is a somewhat triangular slit called the *superior orbital fissure.* It is an opening for the oculomotor (III) nerve, trochlear (IV) nerve, ophthalmic

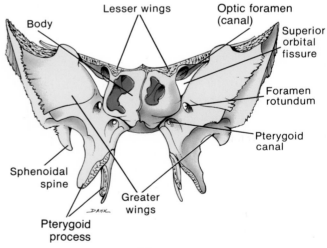

Body
Lesser wings
Optic foramen (canal)
Superior orbital fissure
Foramen rotundum
Pterygoid canal
Sphenoidal spine
Greater wings
Pterygoid process

(b)

FIGURE 6-5 Sphenoid bone. (a) Viewed in the floor of the cranium from above. (b) Anterior view.

branch of the trigeminal (V) nerve, and abducens (VI) nerve. This fissure may also be seen in the anterior view of the skull in Figure 6-2.

On the inferior part of the sphenoid bone you can see the *pterygoid* (TER-i-goyd) *processes.* These structures project inferiorly from the points where the body and greater wings unite. The pterygoid processes form part of the lateral walls of the nasal cavities.

At the base of the lateral pterygoid process in the greater wing is the *foramen ovale* through which the mandibular branch of the trigeminal (V) nerve passes. Another foramen, the *foramen spinosum,* lies at the posterior angle of the sphenoid and transmits the middle meningeal vessels. The *foramen lacerum* is bounded anteriorly by the sphenoid bone and medially by the sphenoid and occipital bones. Although the foramen is covered in part by a layer of fibrocartilage in living subjects, it transmits the meningeal branch

of the ascending pharyngeal artery. A final foramen associated with the sphenoid bone is the *foramen rotundum* through which the maxillary branch of the trigeminal (V) nerve passes. It is located at the junction of the anterior and medial parts of the sphenoid bone.

Ethmoid Bone

The *ethmoid bone* is a light, spongy bone located in the anterior part of the floor of the cranium between the orbits. It is anterior to the sphenoid and posterior to the nasal bones (Figure 6-6). The ethmoid bone forms part of

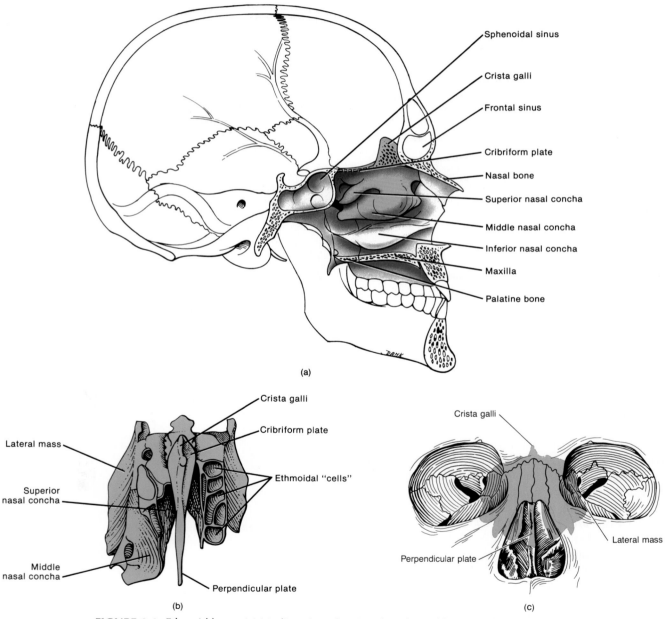

FIGURE 6-6 Ethmoid bone. (a) Median view showing the ethmoid bone on the inner aspect of the left part of the skull. (b) Anterior view. A frontal section has been made through the left side of the bone to expose the ethmoidal "cells." (c) Highly diagrammatic representation showing the approximate position of the ethmoid bone in the skull in anterior view.

the anterior portion of the cranial floor, the medial wall of the orbits, the superior portions of the nasal septum, or partition, and most of the sidewalls of the nasal roof. The ethmoid is the principal supporting structure of the nasal cavities.

Its *lateral masses (labyrinths)* compose most of the wall between the nasal cavities and the orbits. They contain several air spaces, or "cells," ranging in number from 3 to 18. It is from these "cells" that the bone derives its name (*ethmos* = sieve). The ethmoid "cells" together form the *ethmoidal sinuses.* The sinuses are shown in Figure 6-8. The *perpendicular plate* forms the superior portion of the nasal septum (see Figure 6-7). The *cribriform (horizontal) plate,* lies in the anterior floor of the cranium and forms the roof of the nasal cavity. The cribriform plate contains the *olfactory foramina* through which the olfactory (I) nerves pass. These nerves function in smell. Projecting upward from the horizontal plate is a triangular process called the *crista galli* (= cock's comb). This structure serves as a point of attachment for the membranes (meninges) that cover the brain.

The labyrinths contain two thin, scroll-shaped bones on either side of the nasal septum. These are called the *superior nasal concha* (KONG-ka; *concha* = shell) and the *middle nasal concha.* The conchae allow for the efficient circulation and filtration of inhaled air before it passes into the trachea, bronchi, and lungs by causing the air to whirl around.

CRANIAL FOSSAE

If you examine Figure 6-5a, you can see that the floor of the cranium contains three distinct levels, from anterior to posterior, called *cranial fossae.* The fossae contain depressions for the various brain convolutions, grooves for cranial blood vessels, and numerous foramina. The highest level, the *anterior cranial fossa,* is formed largely by the portion of the frontal bone that constitutes the roof of the orbits and nasal cavity, the crista galli and cribriform plate of the ethmoid bone, and the lesser wings and part of the body of the sphenoid bone. This fossa houses the frontal lobes of the cerebral hemispheres. The next fossa is below the level of the anterior cranial fossa and is called the *middle cranial fossa.* It is shaped like a butterfly and has a small median portion and two expanded lateral portions. The median portion is formed by part of the body of the sphenoid bone, and the lateral portions are formed by the greater wings of the sphenoid bone, the temporal squama, and the parietal bone. The middle cranial fossa cradles the temporal lobes of the cerebral hemispheres. The last fossa, at the lowest level, is the *posterior cranial fossa,* the largest of the fossae. It is formed largely by the occipital bone and the petrous and mastoid portions of the temporal bone. It is a very deep fossa that accommodates the cerebellum, pons, and medulla.

FACIAL BONES

Nasal Bones

The paired *nasal bones* are small, oblong bones that meet at the middle and superior part of the face (see Figures 6-2 and 6-7). Their fusion forms part of the bridge of the nose. The inferior portion of the nose, indeed the major portion, consists of cartilage.

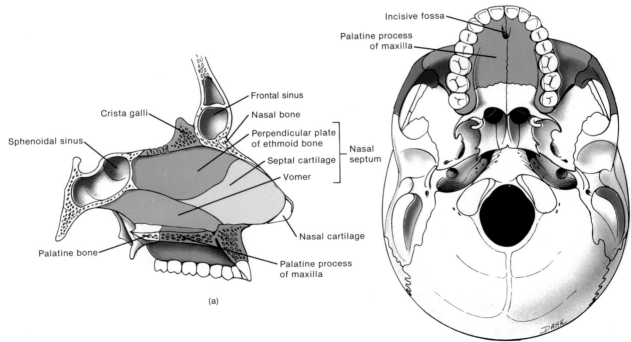

FIGURE 6-7 Maxillae. (a) Median view of the left maxilla. (b) Inferior view of the skull, showing the maxillae.

Maxillae

The paired *maxillae* (mak-SIL-ē) unite to form the upper jawbone (Figure 6-7) and articulate with every bone of the face except the mandible, or lower jawbone. They form part of the floors of the orbits, part of the roof of the mouth (most of the hard palate), and part of the lateral walls and floor of the nasal cavity.

Each maxillary bone contains a *maxillary sinus* that empties into the nasal cavity (see Figure 6-8). The *alveolar* (al-VĒ-ō-lar) *process* (*alveolus* = hollow) contains the *alveoli* (bony sockets) into which the maxillary (upper) teeth are set. The *palatine process* is a horizontal projection of the maxilla that forms the anterior three-fourths of the hard palate, or anterior portion of the roof of the oral cavity. The two portions of the maxillary bones unite, and the fusion is normally completed before birth.

CLINICAL APPLICATION

If the palatine processes of the maxillary bones do not unite before birth, a condition called *cleft palate* results. The condition may also involve incomplete fusion of the horizontal plates of the palatine bones (see Figure 6-4). Another form of this condition, called *cleft lip,* involves a split in the upper lip. Cleft lip is often associated with cleft palate. Depending on the extent and position of the cleft, speech and swallowing may be affected. A surgical procedure can sometimes improve the condition.

The *infraorbital foramen,* which can be seen in the anterior view of the skull in Figure 6-2, is an opening in the maxilla inferior to the orbit. The infraorbital nerve and artery are transmitted through this opening. Another prominent fossa in the maxilla is the *incisive fossa* just posterior to the incisor teeth. Through it pass branches of the greater palatine vessels and the nasopalatine nerve. A final fossa associated with the maxilla and sphenoid bone is the *inferior orbital fissure.* It is located between the greater wing of the sphenoid and the maxilla (see Figure 6-4). It transmits the maxillary branch of the trigeminal (V) nerve, the infraorbital vessels, and the zygomatic nerve.

Paranasal Sinuses

Although they are not cranial or facial bones, this is an appropriate point to discuss paired cavities, called *paranasal sinuses,* that are are located in certain cranial and facial bones near the nasal cavity (Figure 6-8). The paranasal sinuses are lined with mucous membranes that are continuous with the lining of the nasal cavity. Cranial bones contain-

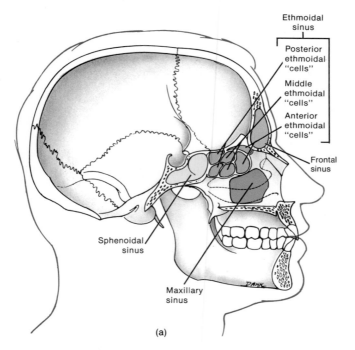

FIGURE 6-8 Paranasal sinuses. (a) Median view showing the location of the sinuses. (b) Median view showing the openings of the paranasal sinuses into the nasal cavity.

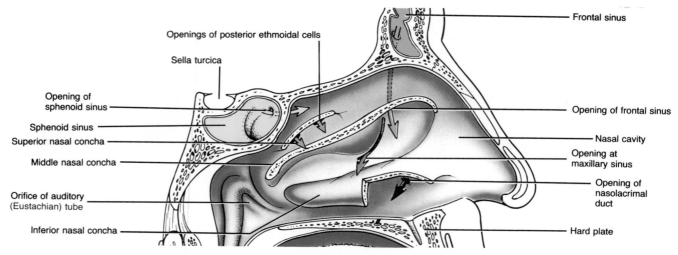

(b)

ing paranasal sinuses are the frontal, sphenoid, ethmoid, and maxillae. (The paranasal sinuses were described in the discussion of each of these bones.) Besides producing mucus, the paranasal sinuses lighten the skull bones and serve as resonant chambers for sound.

CLINICAL APPLICATION

Secretions produced by the mucous membranes of the paranasal sinuses drain into the nasal cavity. An inflammation of the membranes due to an allergic reaction or infection is called *sinusitis*. If the membranes swell enough to block drainage into the nasal cavity, fluid pressure builds up in the paranasal sinuses and a sinus headache results.

Zygomatic Bones

The two *zygomatic bones* (*malars*), commonly referred to as the cheekbones, form the prominences of the cheeks and part of the outer wall and floor of the orbits (Figure 6-2b).

The *temporal process* of the zygomatic bone projects posteriorly and articulates with the zygomatic process of the temporal bone. These two processes form the *zygomatic arch* (Figure 6-4). A foramen associated with the zygomatic bone is the *zygomaticofacial foramen* near the center of the bone (see Figure 6-2). It transmits the zygomaticofacial nerve and vessels.

Mandible

The *mandible,* or lower jawbone, is the largest, strongest facial bone (Figure 6-9). It is the only movable skull bone.

In the lateral view you can see that the mandible consists of a curved, horizontal portion called the *body* and two perpendicular portions called the *rami*. The *angle* of the mandible is the area where each ramus meets the body. Each ramus has a *condylar* (KON-di-lar) *process* that articulates with the mandibular fossa and articular tubercle of the temporal bone to form the temporomandibular joint (TMJ). It also has a *coronoid* (KOR-ō-noyd) *process* to which the temporalis muscle attaches. The depression between the coronoid and condylar processes is called the *mandibular notch*. The *alveolar process* is an arch containing the *alveoli* (sockets) for the mandibular (lower) teeth.

The *mental foramen* (*mentum* = chin) is approximately below the first molar tooth. The mental nerve and vessels pass through this opening. It is through this foramen that dentists sometimes reach the nerve when injecting anesthetics. Another foramen associated with the mandible is the *mandibular foramen* on the medial surface of the ramus, another site frequently used by dentists to inject anesthetics. It transmits the inferior alveolar nerve and vessels. The mandibular foramen is the beginning of the *mandibular canal,* which runs forward in the ramus deep to the roots of the teeth. The canal carries branches of the inferior alveolar nerve and vessels to the teeth. Parts of these nerves and vessels emerge through the mental foramen.

CLINICAL APPLICATION

Opening of the mouth very wide, as in a very large yawn, may cause displacement of the condylar process of the mandible from the mandibular fossa of the temporal bone, thereby producing a *dislocation of the jaw*. The displacement is usually anterior and bilateral and the patient is unable to close the mouth.

Another problem associated with the temporomandibular joint (TMJ) is called *TMJ syndrome*. It is characterized by dull pain around the ear, tenderness of the jaw muscles, clicking or popping noise when opening

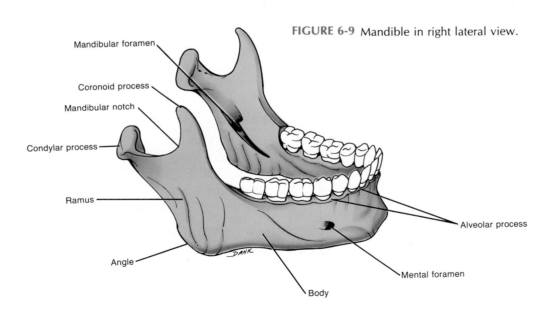

FIGURE 6-9 Mandible in right lateral view.

Mandibular foramen

Coronoid process

Mandibular notch

Condylar process

Ramus

Angle

Alveolar process

Mental foramen

Body

or closing the mouth, limited or abnormal opening of the mouth, headache, tooth sensitivity, and abnormal wearing of the teeth. TMJ syndrome might be caused by improperly aligned teeth, grinding or clenching the teeth, trauma to the jaw, or arthritis. Treatment may consist of application of heat or ice, keeping lips together and teeth apart, a soft diet, taking aspirin, muscle retraining, use of an occlusal splint, adjusting or reshaping the teeth, orthodontic treatment, or surgery.

Lacrimal Bones

The paired *lacrimal bones* (LAK-ri-mal; *lacrima* = tear) are thin bones roughly resembling a fingernail in size and shape. They are the smallest bones of the face. These bones are posterior and lateral to the nasal bones in the medial wall of the orbit. They can be seen in the anterior and lateral views of the skull in Figure 6-2.

The lacrimal bones form a part of the medial wall of the orbit. They also contain the *lacrimal fossae* through which the tear ducts pass into the nasal cavity (see Figure 6-2).

Palatine Bones

The two *palatine* (PAL-a-tīn) *bones* are L shaped and form the posterior portion of the hard palate, part of the floor and lateral wall of the nasal cavity, and a small portion of the floors of the orbits. The posterior portion of the hard palate, which separates the nasal cavity from the oral cavity, is formed by the *horizontal plates* of the palatine bones. These can be seen in Figure 6-4.

Two foramina associated with the palatine bones are the greater and lesser palatine foramina (Figure 6-4). The *greater palatine foramen,* at the posterior angle of the hard palate, transmits the greater palatine nerve and descending palatine vessels. The *lesser palatine foramina,* usually two or more on each side, are posterior to the greater palatine foramina. They transmit the lesser palatine nerve and artery.

Inferior Nasal Conchae

Refer to the views of the skull in Figures 6-2a and 6-6a. The two *inferior nasal conchae* (KONG-kē) are scroll-like bones that form a part of the lateral wall of the nasal cavity and project into the nasal cavity inferior to the superior and middle nasal conchae of the ethmoid bone. They serve the same function as the superior and middle nasal conchae, that is, they allow for the circulation and filtration of air before it passes into the lungs. The inferior nasal conchae are separate bones and not part of the ethmoid.

Vomer

The *vomer* (= plowshare) is a roughly triangular bone that forms the inferior and posterior part of the nasal septum.

It is clearly seen in the anterior view of the skull in Figure 6-2a and the inferior view in Figure 6-4.

The inferior border of the vomer articulates with the cartilage septum that divides the nose into a right and left nostril. Its superior border articulates with the perpendicular plate of the ethmoid bone. The structures that form the *nasal septum,* or partition, are the perpendicular plate of the ethmoid, septal cartilage, vomer, and parts of the palatine bones and maxillae (Figure 6-7a).

CLINICAL APPLICATION

A *deviated nasal septum (DNS)* is deflected laterally from the midline of the nose. The deviation usually occurs at the junction of bone with the septal cartilage. If the deviation is severe, it may entirely block the nasal passageway. Even though the blockage may not be complete, infection and inflammation may develop and cause nasal congestion, blockage of the paranasal sinus openings, and chronic sinusitis.

A summary of bones of the skull is presented in Exhibit 6-3.

FORAMINA

Some foramina (singular = foramen) of the skull were mentioned along with the descriptions of the cranial and facial bones with which they are associated. As preparation for studying other systems of the body, especially the nervous and cardiovascular systems, these foramina, as well as some additional ones, and the structures passing through them are listed in Exhibit 6-4. For your convenience and for future reference, the foramina are listed alphabetically.

HYOID BONE

The single *hyoid bone* (*hyoedes* = U shaped) is a unique component of the axial skeleton because it does not articulate with any other bone. Rather, it is suspended from the styloid process of the temporal bone by ligaments and muscles.

EXHIBIT 6-3

Summary of Bones of the Skull[*]

CRANIAL BONES	FACIAL BONES
Frontal (1)	Nasal (2)
Parietal (2)	Maxillae (2)
Temporal (2)	Zygomatic (2)
Occipital (1)	Mandible (1)
Sphenoid (1)	Lacrimal (2)
Ethmoid (1)	Palatine (2)
	Inferior nasal conchae (2)
	Vomer (1)

[*] The numbers in parentheses indicate how many of each bone are present.

EXHIBIT 6-4

Summary of Foramina of the Skull

FORAMEN	LOCATION	STRUCTURES PASSING THROUGH
Carotid Figure 6-4)	Petrous portion of temporal.	Internal carotid artery.
Greater palatine (Figure 6-4)	Posterior angle of hard palate.	Greater palatine nerve and greater palatine vessels.
Hypoglossal (Figure 6-5)	Superior to base of occipital condyles.	Hypoglossal (XII) nerve and branch of ascending pharyngeal artery.
Incisive (Figure 6-7)	Posterior to incisor teeth.	Branches of greater palatine vessels and nasopalatine nerve.
Inferior orbital (Figure 6-4)	Between greater wing of sphenoid and maxilla.	Maxillary branch of trigeminal (V) nerve, zygomatic nerve, and infraorbital vessels.
Infraorbital (Figure 6-2a)	Inferior to orbit in maxilla.	Infraorbital nerve and artery.
Jugular (Figure 6-4)	Posterior to carotid canal between petrous portion of temporal and occipital.	Internal jugular vein, glossopharyngeal (IX) nerve, vagus nerve (X), and accessory (XI) nerve.
Lacerum (Figure 6-5)	Bounded anteriorly by sphenoid, posteriorly by petrous portion of temporal, and medially by the sphenoid and occipital.	Branch of ascending pharyngeal artery.
Lacrimal (Figure 6-2c)	Lacrimal bone.	Lacrimal (tear) duct.
Lesser palatine (Figure 6-4)	Posterior to greater palatine foramen.	Lesser palatine nerves and artery.
Magnum (Figure 6-4)	Occipital bone.	Medulla oblongata and its membranes, accessory (XI) nerve, and the vertebral and spinal arteries and meninges.
Mandibular (Figure 6-9)	Medial surface of ramus of mandible.	Inferior alveolar nerve and vessels.
Mastoid	Posterior border of mastoid process of temporal bone.	Emissary vein to transverse sinus and branch of occipital artery to dura mater.
Mental (Figure 6-9)	Inferior to second premolar tooth in mandible.	Mental nerve and vessels.
Olfactory (Figure 6-5)	Cribriform plate of ethmoid.	Olfactory (I) nerve.
Optic (Figure 6-5)	Between upper and lower portions of small wing of sphenoid.	Optic (II) nerve and ophthalmic artery.
Ovale (Figure 6-5)	Greater wing of sphenoid.	Mandibular branch of trigeminal (V) nerve.
Rotundum (Figure 6-5)	Junction of anterior and medial parts of sphenoid.	Maxillary branch of trigeminal (V) nerve.
Spinosum (Figure 6-5)	Posterior angle of sphenoid.	Middle meningeal vessels.
Stylomastoid (Figure 6-4)	Between styloid and mastoid processes of temporal.	Facial (VII) nerve and stylomastoid artery.
Superior orbital (Figure 6-5)	Between greater and lesser wings of sphenoid.	Oculomotor (III) nerve, trochlear (IV) nerve, ophthalmic branch of trigeminal (V) nerve, and abducens (VI) nerve.
Supraorbital (Figure 6-2a)	Supraorbital margin of orbit.	Supraorbital nerve and artery.
Zygomaticofacial (Figure 6-2a)	Zygomatic bone.	Zgyomaticofacial nerve and vessels.

The hyoid is located in the neck between the mandible and larynx. It supports the tongue and provides attachment for some of its muscles. It also provides attachment for muscles of the neck and pharynx. Refer to the median and lateral views of the skull in Figure 6-2c,d to see the position of the hyoid bone.

The hyoid consists of a horizontal *body* and paired projections called the *lesser cornu* (*cornu* = horn) and the *greater cornu* (Figure 6-10). Muscles and ligaments attach to these paired projections.

The hyoid bone is frequently fractured during strangulation. As a result, it is carefully examined in an autopsy when strangulation is suspected.

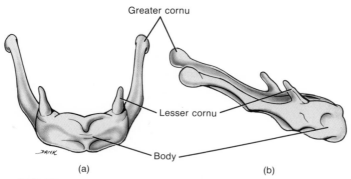

FIGURE 6-10 Hyoid bone. (a) Anterior view. (b) Right lateral view.

VERTEBRAL COLUMN

DIVISIONS

The *vertebral column* (*spine*) together with the sternum and ribs, constitutes the skeleton of the *trunk* of the body. The vertebral column makes up about two-fifths of the total height of the body and is composed of a series of bones called *vertebrae.* In an average adult male, the column measures about 71 cm (28 in.) in length; in an average adult female it measures about 61 cm (24 in.) in length. In effect, the vertebral column is a strong, flexible rod that rotates and moves anteriorly, posteriorly, and laterally. It encloses and protects the spinal cord, supports the head, and serves as a point of attachment for the ribs and the muscles of the back. Between vertebrae are openings called *intervertebral foramina.* The nerves that connect the spinal cord to various parts of the body pass through these openings.

The adult vertebral column typically contains 26 vertebrae (Figure 6-11a,b). These are distributed as follows: 7 *cervical vertebrae* (*cervix* = neck) in the neck region; 12 *thoracic vertebrae* (*thorax* = chest) posterior to the thoracic cavity; 5 *lumbar vertebrae* (*lumbus* = loin) supporting the lower back; 5 *sacral vertebrae* fused into one bone called the *sacrum;* and usually 4 *coccygeal* (kok-SIJ-ē-al) *vertebrae* fused into one or two bones called the *coccyx* (KOK-six). Prior to the fusion of the sacral and coccygeal vertebrae, the total number of vertebrae is 33.

Between adjacent vertebrae from the first vertebra (axis) to the sacrum are fibrocartilaginous *intervertebral discs.* Each disc is composed of an outer fibrous ring consisting of fibrocartilage called the *annulus fibrosus* and an inner soft, pulpy, highly elastic structure called the *nucleus pulposus* (Figure 6-11c). The discs form strong joints, permit various movements of the vertebral column, and absorb vertical shock. Under compression, they flatten, broaden, and bulge from their intervertebral spaces (Figure 6-11c).

NORMAL CURVES

When viewed from the side with the subject facing to the right, the vertebral column shows four *normal curves* (Figure 6-11b), two of which are convex, (and two of which are concave). The curves of the column, like the curves in a long bone, are important because they increase its strength. The curves also help maintain balance in the upright position, absorb shocks from walking, and help protect the column from fracture.

In the fetus, there is only a single anteriorly concave curve. At approximately the third postnatal month, when an infant begins to hold its head erect, the *cervical curve* develops. Later, when the child stands and walks, the *lumbar curve* develops. The cervical and lumbar curves are anteriorly convex. Because they are modifications of the fetal positions, they are called *secondary curves.* The other two curves, the *thoracic curve* and the *sacral curve,* are anteriorly concave. Since they retain the anterior concavity of the fetus, they are referred to as *primary curves.*

CLINICAL APPLICATION

In recent years, *gravity inversion* has been used by some people to decompress the backbone by using gravity and the body's own weight. This may be accomplished by hanging upside down by the ankles in gravity inversion boots suspended from a horizontal bar. Despite the popularity of this method of traction and exercise, there are some potentially harmful effects. For example, it has been found that some healthy young individuals may experience increases in general blood pressure, pulse rate, intraocular pressure, and blood pressure in the eyes. Moreover, gravity inversion is not recommended for people with glaucoma, high blood pressure, heart disease, hiatal hernias, or disorders of the vertebral column.

TYPICAL VERTEBRA

Although there are variations in size, shape, and detail in the vertebrae in different regions of the column, all the vertebrae are basically similar in structure (Figure 6-12). A typical vertebra consists of the following components.

1. The *body* is the thick, disc-shaped anterior portion that is the weight-bearing part of a vertebra. Its superior

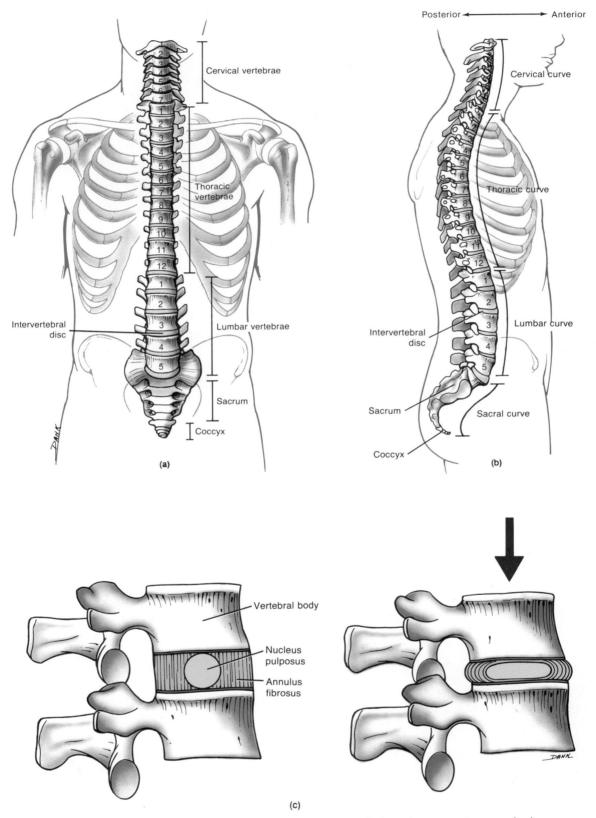

FIGURE 6-11 Vertebral column. (a) Anterior view. (b) Right lateral view. (c) Intervertebral disc in its normal position (left) and under compression (right). The relative size of the disc has been enlarged for emphasis. A "window" has been cut in the annulus fibrosus so that the nucleus pulposus can be seen.

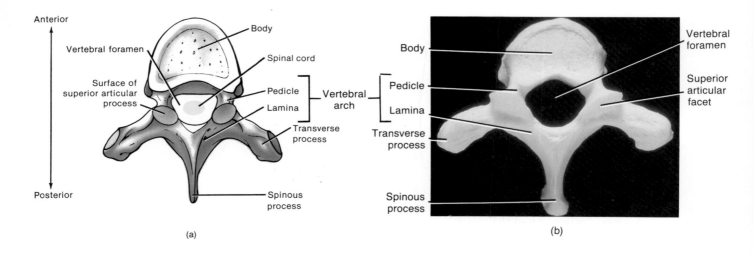

(a)

(b)

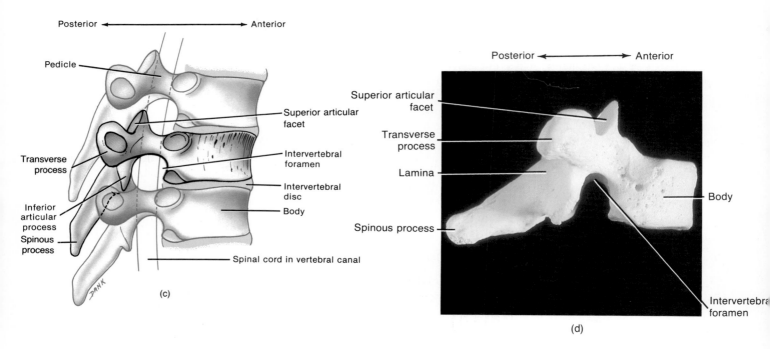

(c)

(d)

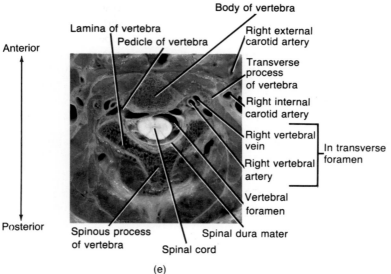

(e)

FIGURE 6-12 Typical vertebra. (a) Diagram of superior view. (b) Photograph of superior view. (Courtesy of J. A. Gosling, P. F. Harris, et al., *Atlas of Human Anatomy,* Gower Medical Publishing Ltd., 1985.) (c) Diagram of right lateral view. (d) Photograph of right lateral view. (Courtesy of J. A. Gosling, P. F. Harris, et al., *Atlas of Human Anatomy,* Gower Medical Publishing Ltd., 1985.) (e) Photograph of a cross section through the upper cervical region of the vertebral column to show the relationship of a vertebra to surrounding structures. (Courtesy of Stephen A. Kieffer and E. Robert Heitzman, *An Atlas of Cross-Sectional Anatomy,* Harper & Row Publishers, Inc., New York, 1979.)

and inferior surfaces are roughened for the attachment of intervertebral discs. The anterior and lateral surfaces contain nutrient foramina for blood vessels.

2. The *vertebral (neural) arch* extends posteriorly from the body of the vertebra. With the body of the vertebra, it surrounds the spinal cord. It is formed by two short, thick processes, the *pedicles* (PED-i-kuls), which project posteriorly from the body to unite with the laminae. The *laminae* (LAM-i-nē) are the flat parts that join to form the posterior portion of the vertebral arch. The space that lies between the vertebral arch and body contains the spinal cord. This space is known as the *vertebral foramen.* The vertebral foramina of all vertebrae together form the *vertebral (spinal) canal.* The pedicles are notched superiorly and inferiorly in such a way that, when they are arranged in the column, there is an opening between vertebrae on each side of the column. This opening, the *intervertebral foramen,* permits the passage of a single spinal nerve.

3. Seven *processes* arise from the vertebral arch. At the point where a lamina and pedicle join, a *transverse process* extends laterally on each side. A single *spinous process (spine)* projects posteriorly and inferiorly from the junction of the laminae. These three processes serve as points of muscular attachment. The remaining four processes form joints with other vertebrae. The two *superior articular processes* of a vertebra articulate with the vertebra immediately superior to them. The two *inferior articular processes* of a vertebra articulate with the vertebra inferior to them. The articulating surfaces of the articular processes are referred to as *facets.*

CERVICAL REGION

When viewed from above, it can be seen that the bodies of *cervical vertebrae* are smaller than those of thoracic vertebrae (Figure 6-13). The vertebral arches, however, are larger. The spinous processes of the second through sixth cervical vertebrae are often *bifid,* that is, with a cleft. All cervical vertebrae have three foramina: the vertebral foramen and two transverse foramina. Each cervical transverse process contains a *transverse foramen* through which the vertebral artery and its accompanying vein and nerve fibers pass.

The first two cervical vertebrae differ considerably from the others. The first cervical vertebra (C1), the *atlas,* is named for its support of the head. Essentially, the atlas is a ring of bone with *anterior* and *posterior arches* and large *lateral masses.* It lacks a body and a spinous process. The superior surfaces of the lateral masses, called *superior articular facets,* are concave and articulate with the occipital condyles of the occipital bone. This articulation permits the movement seen when nodding the head. The inferior surfaces of the lateral masses, the *inferior articular facets,* articulate with the second cervical vertebra. The transverse processes and transverse foramina of the atlas are quite large.

The second cervical vertebra (C2), the *axis,* does have a body. A peglike process called the *dens* projects up through the ring of the atlas. The dens makes a pivot on which the atlas and head rotate. This arrangement permits side-to-side rotation of the head.

CLINICAL APPLICATION

In various instances of trauma, the dens of the axis may be driven into the medulla oblongata of the brain. This injury is the usual cause of fatality in *whiplash injuries* that result in death.

The third through sixth cervical vertebrae (C3–C6) correspond to the structural pattern of the typical cervical vertebra previously described.

The seventh cervical vertebra (C7), called the *vertebra prominens,* is somewhat different. It is marked by a large, nonbifid spinous process that may be seen and felt at the base of the neck.

THORACIC REGION

Viewing a typical *thoracic vertebra* from above, you can see that it is considerably larger and stronger than a vertebra of the cervical region (Figure 6-14). In addition, the spinous process on each vertebra is long, laterally flattened, and directed inferiorly. Thoracic vertebrae also have longer and heavier transverse processes than cervical vertebrae.

Except for the eleventh and twelfth thoracic vertebrae, the transverse processes have *facets* for articulating with the tubercles of the ribs. The bodies of thoracic vertebrae also have whole facets or half-facets, called *demifacets,* for articulation with the heads of the ribs. The first thoracic vertebra (T1) has, on either side of its body, a superior whole facet and an inferior demifacet. The superior facet articulates with the first rib, and the inferior demifacet, together with the superior demifacet of the second thoracic vertebra (T2), forms a facet for articulation with the second rib. The second through eighth thoracic vertebrae (T2–T8) have two demifacets on each side, a larger superior demifacet and a smaller inferior demifacet. When the vertebrae are articulated, they form whole facets for the heads of the ribs. The ninth thoracic vertebra (T9) has a single superior demifacet on either side of its body. The tenth through twelfth thoracic vertebrae (T10–T12) have whole facets on either side of their bodies.

LUMBAR REGION

The *lumbar vertebrae* (L1–L5) are the largest and strongest in the column (Figure 6-15). Their various projections are short and thick. The superior articular processes are directed medially instead of superiorly. The inferior articular processes are directed laterally instead of inferiorly. The spinous

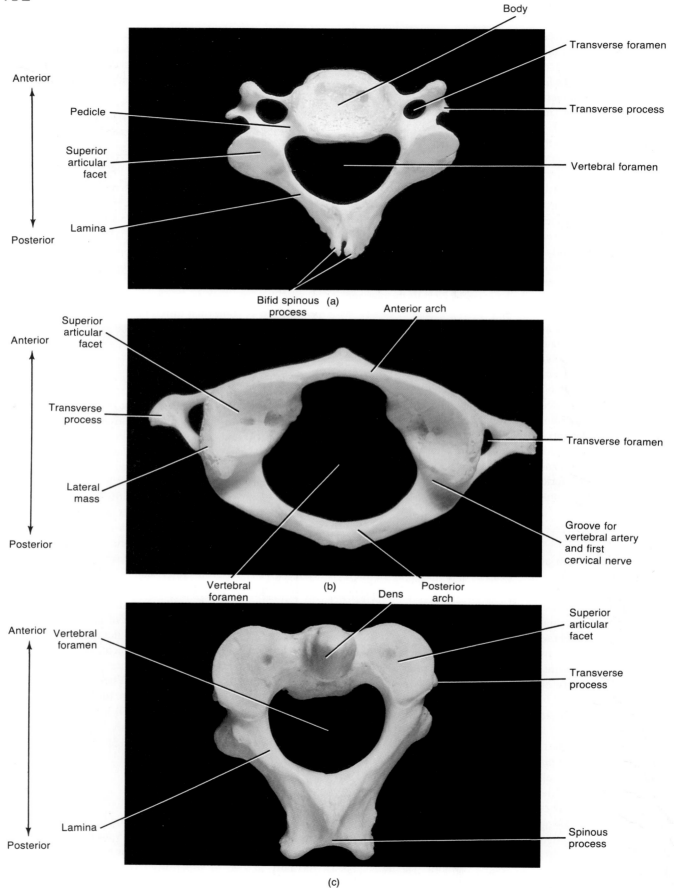

Anterior

Posterior

Body

Transverse foramen

Pedicle

Transverse process

Superior articular facet

Vertebral foramen

Lamina

Bifid spinous process   (a)

Anterior arch

Superior articular facet

Anterior

Transverse process

Transverse foramen

Lateral mass

Posterior

Groove for vertebral artery and first cervical nerve

Vertebral foramen   (b)   Posterior arch

Dens

Superior articular facet

Anterior   Vertebral foramen

Transverse process

Lamina

Spinous process

Posterior

(c)

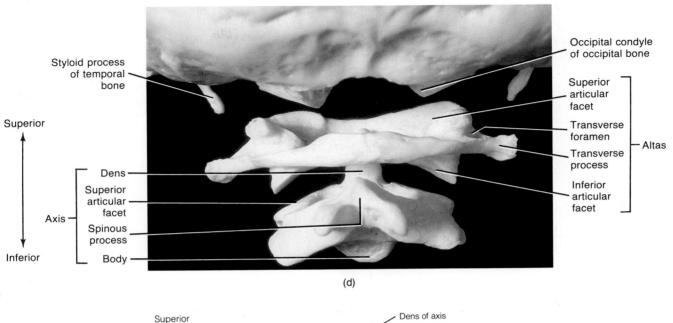

Occipital condyle
of occipital bone

Styloid process
of temporal
bone

Superior
articular
facet

Transverse
foramen

Altas

Transverse
process

Superior

Inferior
articular
facet

Dens

Superior
articular
facet

Axis

Spinous
process

Inferior

Body

(d)

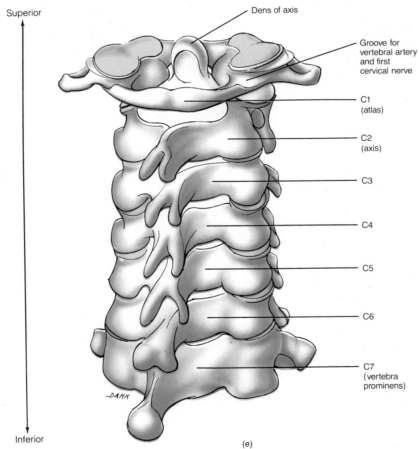

Superior

Dens of axis

Groove for
vertebral artery
and first
cervical nerve

C1
(atlas)

C2
(axis)

C3

C4

C5

C6

C7
(vertebra
prominens)

Inferior

(e)

FIGURE 6-13 Cervical vertebrae. (a) Photograph of a superior view of a typical cervical vertebra. (b) Photograph of a superior view of the atlas. (c) Photograph of a superior view of the axis. (d) Photograph of the atlas and axis in posterior view. (e) Diagram of cervical vertebrae articulated in posterior view. (Photographs courtesy of J. A. Gosling, P. F. Harris,

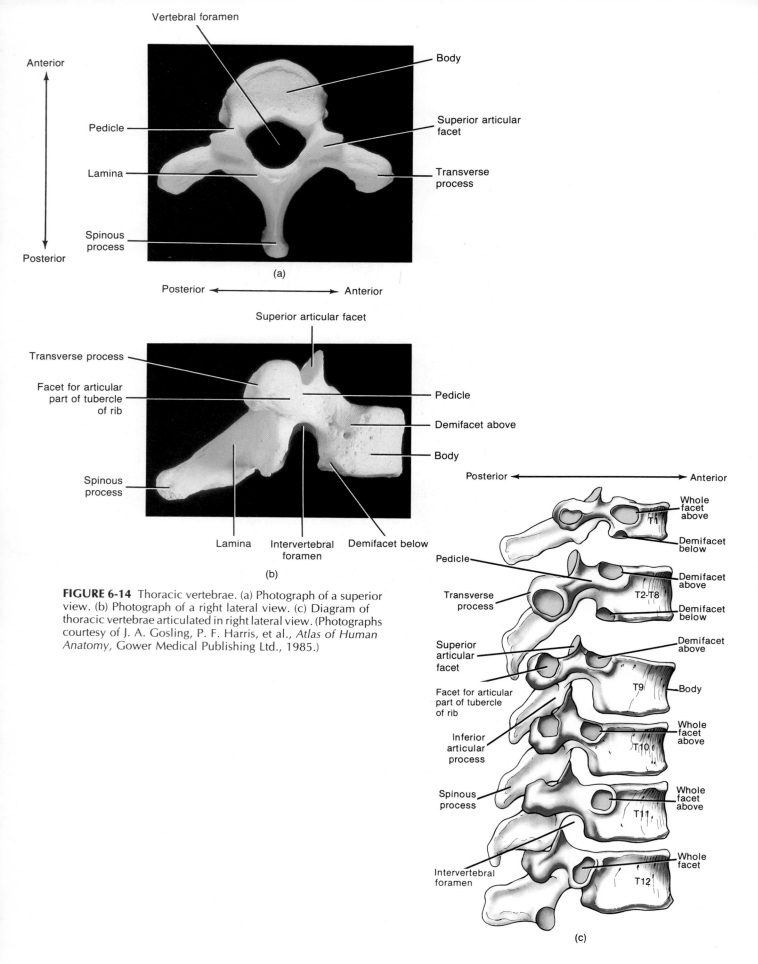

FIGURE 6-14 Thoracic vertebrae. (a) Photograph of a superior view. (b) Photograph of a right lateral view. (c) Diagram of thoracic vertebrae articulated in right lateral view. (Photographs courtesy of J. A. Gosling, P. F. Harris, et al., *Atlas of Human Anatomy,* Gower Medical Publishing Ltd., 1985.)

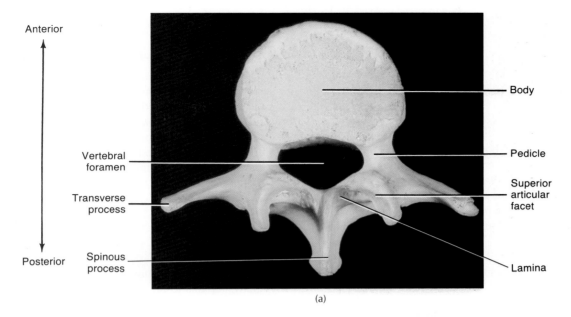

Anterior

Posterior

Body

Pedicle

Superior
articular
facet

Vertebral
foramen

Transverse
process

Spinous
process

Lamina

(a)

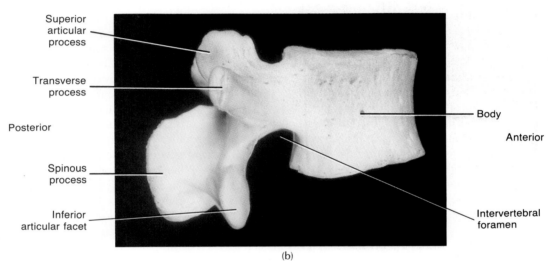

Superior
articular
process

Transverse
process

Posterior

Spinous
process

Inferior
articular facet

Body

Anterior

Intervertebral
foramen

(b)

Posterior ← → Anterior

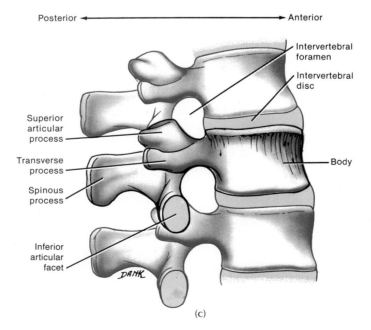

Intervertebral
foramen

Intervertebral
disc

Superior
articular
process

Transverse
process

Spinous
process

Inferior
articular
facet

Body

DANK

(c)

FIGURE 6-15 Lumbar vertebrae. (a) Photograph in superior view. (b) Photograph in right lateral view. (c) Diagram of lumbar vertebrae articulated in right lateral view. (Photographs courtesy of J. A. Gosling, P. F. Harris, et al., *Atlas of Human Anatomy*, Gower Medical Publishing Ltd., 1985.)

155

processes are quadrilateral in shape, thick and broad, and project nearly straight posteriorly. The spinous processes are well adapted for the attachment of the large back muscles.

SACRUM AND COCCYX

The **sacrum** is a triangular bone formed by the union of five sacral vertebrae. These are indicated in Figure 6-16 as S1–S5. Fusion occurs between 16 and 18 years and is usually completed by the mid-twenties. The sacrum serves as a strong foundation for the pelvic girdle. It is positioned at the posterior portion of the pelvic cavity between the two coxal (hip) bones.

The concave anterior side of the sacrum faces the pelvic cavity. It is smooth and contains four **transverse lines (ridges)** that mark the joining of the sacral vertebral bodies. At the ends of these lines are four pairs of **anterior sacral (pelvic) foramina.**

The convex, posterior surface of the sacrum is irregular. It contains a **median sacral crest,** the fused spinous processes of the upper sacral vertebrae; a **lateral sacral crest,** the transverse processes of the sacral vertebrae; and four pairs of **posterior sacral (dorsal) foramina.** These foramina communicate with the anterior sacral foramina through which nerves and blood vessels pass. The **sacral canal** is a continuation of the vertebral canal. The laminae of the fifth sacral vertebra, and sometimes the fourth, fail to meet. This leaves an inferior entrance to the vertebral canal called the **sacral hiatus** (hī-Ā-tus). On either side of the sacral hiatus are the **sacral cornua,** the inferior articular processes of the fifth sacral vertebra. They are connected by ligaments to the coccygeal cornua of the coccyx.

CLINICAL APPLICATION

Anesthetic agents that act on the sacral and coccygeal nerves are sometimes injected through the sacral hiatus, a procedure called **caudal anesthesia.** Since the sacral hiatus is between the sacral cornua, the cornua are important bony landmarks for locating the hiatus. Anesthetic agents may also be injected through the posterior sacral (dorsal) foramina.

The superior border of the sacrum exhibits an anteriorly projecting border, the **sacral promontory** (PROM-on-tō-rē). It is an obstetrical landmark for measurements of the pelvis. An imaginary line running from the superior surface of the symphysis pubis to the sacral promontory separates the abdominal and pelvic cavities. Laterally, the sacrum has a large **auricular surface** for articulating with the ilium of the coxal (hip) bone. Posterior to the auricular surface is a roughened surface, the **sacral tuberosity,** which contains depressions for the attachment of ligaments. The sacral tuberosity unites with the coxal bone to form is the sacroiliac joint. The **superior articular processes** of the sacrum articulate with the fifth lumbar vertebra.

The **coccyx** is also triangular in shape and is formed by the fusion of the coccygeal vertebrae, usually the last four. These are indicated in Figure 6-16 as Co1–Co4. Fusion generally occurs between 20 and 30 years. The dorsal surface of the body of the coccyx contains two long **coccygeal cornua** that are connected by ligaments to the sacral cornua. The coccygeal cornua are the pedicles and superior articular processes of the first coccygeal vertebra. On the lateral

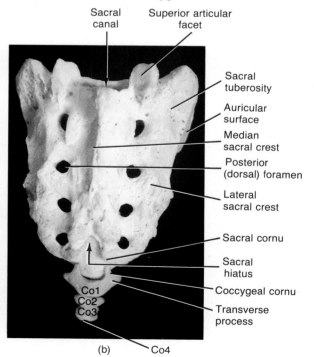

FIGURE 6-16 Sacrum and coccyx. (a) Anterior view. (b) Posterior view. (Photographs copyright © 1987 by Michael H. Ross. Used by permission.)

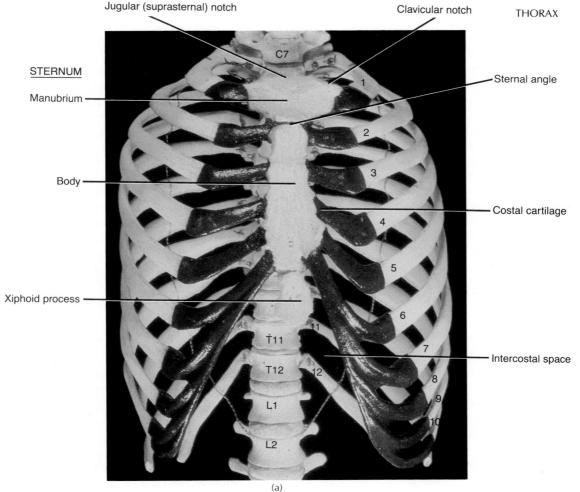

STERNUM

Jugular (suprasternal) notch

Clavicular notch

Sternal angle

Manubrium

C7

1

2

3

Body

Costal cartilage

4

5

Xiphoid process

6

11

T11

7

Intercostal space

12

T12

8

9

L1

10

L2

(a)

FIGURE 6-17 Skeleton of the thorax. (a) Photograph of an anterior view. (b) Diagram of a right lateral view of the sternum. (Photograph courtesy of J. A. Gosling, P. F. Harris, et al., *Atlas of Human Anatomy,* Gower Medical Publishing Ltd., 1985.)

surfaces of the body of the coccyx are a series of *transverse processes,* the first pair being the largest. The coccyx articulates superiorly with the sacrum.

THORAX

Anatomically, the term *thorax* refers to the chest. The skeletal portion of the thorax is a bony cage formed by the sternum, costal cartilage, ribs, and the bodies of the thoracic vertebrae (Figure 6-17a).

The thoracic cage is roughly cone shaped, the narrow portion being superior and the broad portion inferior. It is flattened from front to back. The thoracic cage encloses and protects the organs in the thoracic cavity and upper abdomen. It also provides support for the bones of the shoulder girdle and upper extremities.

STERNUM

The *sternum,* or breastbone, is a flat, narrow bone measuring about 15 cm (6 in.) in length. It is located in the median line of the anterior thoracic wall.

The sternum (Figure 6-17a,b) consists of three basic portions: the *manubrium* (ma-NOO-brē-um), the superior

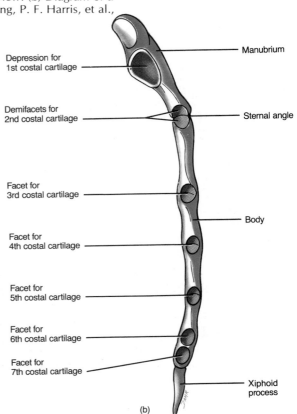

Depression for 1st costal cartilage

Manubrium

Demifacets for 2nd costal cartilage

Sternal angle

Facet for 3rd costal cartilage

Body

Facet for 4th costal cartilage

Facet for 5th costal cartilage

Facet for 6th costal cartilage

Facet for 7th costal cartilage

Xiphoid process

(b)

portion; the *body,* the middle, largest portion; and the *xiphoid* (ZĪ-foyd) *process,* the inferior, smallest portion. The junction of the manubrium and body forms the *sternal angle* (see Figure 11-7). The manubrium has a depression on its superior surface called the *jugular (suprasternal) notch.* On each side of the jugular notch are *clavicular notches* that articulate with the medial ends of the clavicles. The manubrium also articulates with the first and second ribs. The body of the sternum articulates directly or indirectly with the second through tenth ribs. The xiphoid process has no ribs attached to it but provides attachment for some abdominal muscles. The xiphoid process consists of hyaline cartilage during infancy and childhood and does not ossify completely until about age 40. If the hands of a rescuer are mispositioned during cardiopulmonary resuscitation (CPR), there is the danger of fracturing the ossified xiphoid process, separating it from the body, and driving it into the liver.

CLINICAL APPLICATION

Since the sternum possesses red bone marrow throughout life and because it is readily accessible, it is a common site for *marrow biopsy.* Under a local anesthetic, a wide-bore needle is introduced into the marrow cavity of the sternum for aspiration of a sample of red bone marrow. This procedure is called a *sternal puncture.*

The sternum may also be split in the midsagittal plane to allow surgeons access to mediastinal structures such as the thymus gland, heart, and great vessels of the heart.

RIBS

Twelve pairs of *ribs* make up the sides of the thoracic cavity (Figure 6-17). The ribs increase in length from the first through seventh. Then they decrease in length to the twelfth rib. Each rib articulates posteriorly with its corresponding thoracic vertebra. In order to number the ribs anteriorly, count downward from the second costal cartilage, which articulates at the angle of the sternal angle.

The first through seventh ribs have a direct anterior attachment to the sternum by a strip of hyaline cartilage, called *costal cartilage* (*costa* = rib). These ribs are called *true (vertebrosternal) ribs.* The remaining five pairs of ribs are referred to as *false ribs* because their costal cartilages do not attach directly to the sternum. The cartilages of the eighth, ninth, and tenth ribs attach to each other and then to the cartilage of the seventh rib. These false ribs are called *vertebrochondral ribs.* The eleventh and twelfth false ribs are designated as *floating (vertebral) ribs* because their anterior ends do not attach even indirectly to the sternum. They attach only posteriorly to the thoracic vertebrae.

Although there is some variation in rib structure, we shall examine the parts of a typical (third through ninth) rib when viewed from the right side and from behind (Figure

6-18). The *head* of a typical rib is a projection at the posterior end of the rib. It is wedge shaped and consists of one or two *facets* that articulate with facets on the bodies of adjacent thoracic vertebrae. The facets on the head of a rib are separated by a horizontal *interarticular crest.* The inferior facet on the head of a rib is larger than the superior facet. The *neck* is a constricted portion just lateral to the head. A knoblike structure on the posterior surface where the neck joins the body is called a *tubercle.* It consists of a *nonarticular part* that affords attachment to the ligament of the tubercle and an *articular part* that articulates with the facet of a transverse process of the inferior of the two vertebrae to which the head of the rib is connected. The *body (shaft)* is the main part of the rib. A short distance beyond the tubercle, there is an abrupt change in the curvature of the shaft. This point is called the *costal angle.* The inner surface of the rib has a *costal groove* that protects blood vessels and a small nerve.

CLINICAL APPLICATION

Rib fractures usually result from direct blows, most commonly from steering wheel impact, falls, and crushing injuries to the chest. Ribs tend to break at the weakest point, that is, the site of greatest curvature just anterior to the costal angle. In children, the ribs are highly elastic and fractures are less frequent than in adults. Since the first two ribs are protected by the clavicle and pectoralis major muscle, and the last two ribs are mobile, they are the least commonly injured. The middle ribs are the ones most commonly fractured. In some cases, fractured ribs may cause damage to the heart, great vessels of the heart, lungs, trachea, bronchi, esophagus, spleen, and liver.

The posterior portion of the rib is connected to a thoracic vertebra by its head and articular part of a tubercle. The facet of the head fits into a facet on the body of a vertebra, and the articular part of the tubercle articulates with the facet of the transverse process of the vertebra. Each of the second through ninth ribs articulates with the bodies of two adjacent vertebrae. The first, tenth, eleventh, and twelfth ribs articulate with only one vertebra each. On the eleventh and twelfth ribs, there is no articulation between the tubercles and the transverse processes of their corresponding vertebrae.

Spaces between ribs, called *intercostal spaces,* are occupied by intercostal muscles, blood vessels, and nerves.

CLINICAL APPLICATION

Surgical access to the lungs or structures in the mediastinum is commonly undertaken through an intercostal space. Special rib retractors are used to create a wide separation between ribs. The costal cartilages are sufficiently elastic to permit considerable bending.

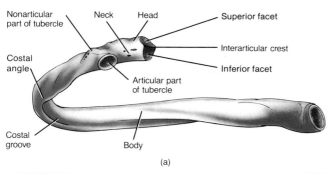

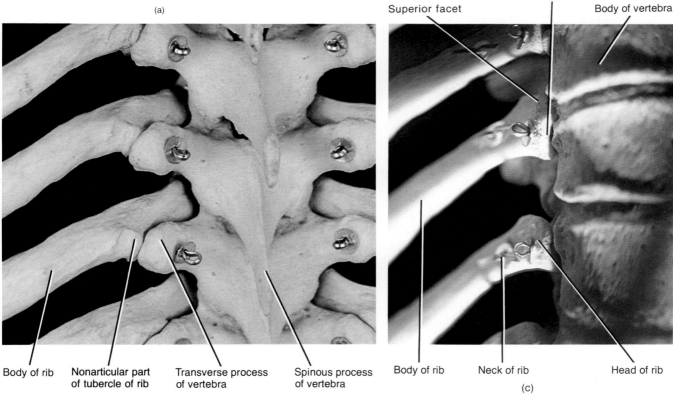

FIGURE 6-18 Typical rib. (a) A left rib viewed from below and behind. (b) Photograph of several ribs articulating with their respective vertebrae in posterior view. (Courtesy of Matt Iacobino.) (c) Photograph of several ribs articulating with their respective vertebrae in anterior view. (Courtesy of Matt Iacobino.)

APPLICATIONS TO HEALTH

HERNIATED (SLIPPED) DISC

In their function as shock absorbers, intervertebral discs are subject to compressional forces. The discs between the fourth and fifth lumbar vertebrae and between the fifth lumbar vertebra and sacrum usually are subject to more forces than other discs. If the anterior and posterior ligaments of the discs become injured or weakened, the pressure developed in the nucleus pulposus may be great enough to rupture the surrounding fibrocartilage. If this occurs, the nucleus pulposus may protrude posteriorly or into one of the adjacent vertebral bodies (herniate). This condition is called a ***herniated (slipped) disc*** or ***herniated intervertebral disc (HIVD).***

Most often the nucleus pulposus slips posteriorly toward the spinal cord and spinal nerves (Figure 6-19). This movement exerts pressure on the spinal nerves, causing considerable, sometimes very acute, pain. If the roots of the sciatic nerve, which passes from the spinal cord to the foot, are pressured, the pain radiates down the back of the thigh, through the calf, and occasionally into the foot. If pressure is exerted on the spinal cord itself, nervous tissue may be destroyed.

Traction, bed rest, and analgesia usually relieve the pain. If such treatment is ineffective, surgical decompression of the spinal nerves or removal of some of the nucleus pulposus may be necessary to relieve pain. Removal may involve conventional surgery or ***chemonucleolysis,*** the use of a proteolytic enzyme called chymopapain (Chymodiactin), the active ingredient in meat tenderizer. This enzyme is injected into a herniated disc where it dissolves the nucleus pulposus, thus relieving pressure on spinal nerves and pain. Although chemonucleolysis provides good short-term results, long-term results are poor. Another procedure, called ***percutaneous lumbar discectomy with aspiration probe,*** can be used on selected patients. This procedure involves the use of an automated needlelike probe (nucleosome)

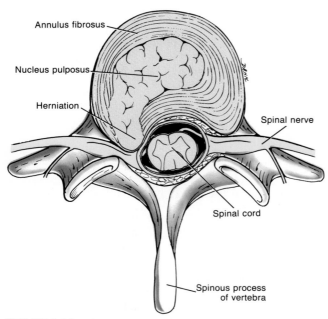

FIGURE 6-19 Superior view of a herniated (slipped) disc.

that cuts away and removes disc material. The probe is inserted through the skin and guided by special x-ray equipment with an image projected on a TV screen. The nucleus pulposus is aspirated through the probe.

ABNORMAL CURVES

As a result of various conditions, the normal curves of the vertebral column may become exaggerated or the column may acquire a lateral bend, resulting in *abnormal curves* of the spine.

Scoliosis (skō-lē-Ō-sis; *scolio* = bent) is a lateral bending of the vertebral column, usually in the thoracic region. This is the most common of the curvatures. It may be congenital, due to an absence of the lateral half of a vertebra (hemivertebra). It may also be acquired from a persistent severe sciatica. Poliomyelitis may cause scoliosis by the paralysis of muscles on one side of the body, which produces a lateral deviation of the trunk toward the unaffected side. Poor posture is also a contributing factor. Scoliosis can also result from one leg being shorter than the other.

Until recently, a child with progressive scoliosis faced either years of treatment with a brace or major corrective surgery. Several research teams are experimenting with using electrical stimulation of muscles to limit the progression of scoliosis and even to reduce curvatures in some cases. The skeletal muscles on the convex side of the spinal curve are electrically stimulated.

Kyphosis (kī-FŌ-sis; *kypho* = hunchback) is an exaggeration of the thoracic curve of the vertebral column. In tuberculosis of the spine, vertebral bodies may partially collapse, causing an acute angular bending of the vertebral column.

In the elderly, degeneration of the intervertebral discs leads to kyphosis. Kyphosis may also be caused by rickets and poor posture. The term "round-shouldered" is an expression for mild kyphosis. Electrical muscle stimulation is being studied to assess its effects on kyphosis as well as on scoliosis.

Lordosis (lor-DŌ-sis; *lordo* = swayback) is an exaggeration of the lumbar curve of the vertebral column. It may result from increased weight of abdominal contents as in pregnancy or extreme obesity. Other causes include poor posture, rickets, and tuberculosis of the spine.

SPINA BIFIDA

Spina bifida (SPĪ-na BIF-i-da) is a congenital defect of the vertebral column in which laminae fail to unite at the midline. The lumbar vertebrae are involved in about 50 percent of all cases. In less serious cases, the defect is small and the area is covered with skin. Only a dimple or tuft of hair may mark the site. In such cases, symptoms are mild, with perhaps intermittent urinary problems. An operation is not ordinarily required. Larger defects in the vertebral arches with protrusion of the membranes (meninges) around the spinal cord or the spinal cord itself produce serious problems, such as partial or complete paralysis, partial or complete loss of urinary bladder control, and the absence of reflexes. Spina bifida may be diagnosed prenatally by a test of the mother's blood, sonography, and/or amniocentesis.

FRACTURES OF THE VERTEBRAL COLUMN

Fractures of the vertebral column most commonly involves T5, T6, and T9–L2. They usually result from a flexion–compression type of injury such as might be sustained in landing on the feet or buttocks after a fall from a height or having a heavy weight fall on the shoulders. The forceful compression wedges the involved vertebrae. If, in addition to compression, there is forceful forward movement, one vertebra may displace forward on its adjacent vertebra below, with either dislocation or fracture of the articular facets between the two (fracture dislocation) and with rupture of the interspinous ligaments.

The cervical vertebrae may be fractured or, more commonly, dislodged by a fall on the head with acute flexion of the neck, as might happen on diving into shallow water. Dislocation may even result from the sudden forward jerk that may occur when an automobile or airplane crashes ("whiplash"). The relatively horizontal intervertebral facets of the cervical vertebrae allow dislocation to take place without their being fractured, whereas the relatively vertical thoracic and lumbar intervertebral facets nearly always fracture in forward dislocation of the thoracolumbar region. Spinal nerve damage may occur as a result of fractures of the vertebral column.

STUDY OUTLINE

Types of Bones (p. 132)
1. On the basis of shape, bones are classified as long, short, flat, or irregular.
2. Sutural (Wormian) bones are found between the sutures of certain cranial bones. Sesamoid bones develop in tendons or ligaments.

Surface Markings (p. 132)
1. Surface markings are structural features visible on the surfaces of bones.
2. Each marking is structured for a specific function—joint formation, muscle attachment, or passage of nerves and blood vessels.
3. Terms that describe markings include fissure, foramen, meatus, fossa, process, condyle, head, facet, tuberosity, crest, and spine.

Divisions of the Skeletal System (p. 132)
1. The axial skeleton consists of bones arranged along the longitudinal axis. The parts of the axial skeleton are the skull, hyoid bone, auditory ossicles, vertebral column, sternum, and ribs.
2. The appendicular skeleton consists of the bones of the girdles and the upper and lower extremities. The parts of the appendicular skeleton are the pectoral (shoulder) girdles, bones of the upper extremities, pelvic (hip) girdle, and bones of the lower extremities.

Skull (p. 133)
1. The skull consists of the cranium and the face. It is composed of 22 bones.
2. Sutures are immovable joints between bones of the skull. Examples are coronal, sagittal, lambdoidal, and squamosal sutures.
3. Fontanels are membrane-filled spaces between the cranial bones of fetuses and infants. The major fontanels are the anterior, posterior, anterolaterals, and posterolaterals.
4. The 8 cranial bones include the frontal, parietal (2), temporal (2), occipital, sphenoid, and ethmoid.
5. The 14 facial bones are the nasal (2), maxillae (2), zygomatic (2), mandible, lacrimal (2), palatine (2), inferior nasal conchae (2), and vomer.
6. Paranasal sinuses are cavities in bones of the skull that communicate with the nasal cavity. They are lined by mucous membranes. The cranial bones containing the paranasal sinuses are the frontal, sphenoid, ethmoid, and maxillae.
7. The foramina of the skull bones provide passages for nerves and blood vessels.

Hyoid Bone (p. 146)
1. The hyoid bone is a U-shaped bone that does not articulate with any other bone.
2. It supports the tongue and provides attachment for some of its muscles as well as some neck muscles and muscles of the pharynx.

Vertebral Column (p. 148)
1. The vertebral column, sternum, and ribs constitute the skeleton of the trunk.
2. The bones of the adult vertebral column are the cervical vertebrae (7), thoracic vertebrae (12), lumbar vertebrae (5), the sacrum (5, fused), and the coccyx (4, fused).
3. The vertebral column contains normal primary curves (thoracic and sacral) and normal secondary curves (cervical and lumbar). These curves give strength, support, and balance.
4. The vertebrae are similar in structure, each consisting of a body, vertebral arch, and seven processes. Vertebrae in the different regions of the column vary in size, shape, and detail.

Thorax (p. 157)
1. The thoracic skeleton consists of the sternum, ribs and costal cartilages, and thoracic vertebrae.
2. The thoracic cage protects vital organs in the chest area and upper abdomen.

Applications to Health (p. 159)
1. Protrusion of the nucleus pulposus of an intervertebral disc posteriorly or into an adjacent vertebral body is called a herniated (slipped) disc.
2. Exaggeration of a normal curve or lateral bending of the vertebral column results in an abnormal curve. Examples include scoliosis, kyphosis, and lordosis.
3. The imperfect union of the vertebral laminae at the midline, a congenital defect, is referred to as spina bifida.
4. Fractures of the vertebral column most often involve T5, T6, and T9–L2.

REVIEW QUESTIONS

1. What are the four principal types of bones? Give an example of each. Distinguish between a sutural (Wormian) and a sesamoid bone.
2. What are surface markings? Describe and give an example of each.
3. Distinguish between the axial and appendicular skeletons. What subdivisions and bones are contained in each?
4. What are the bones that compose the skull? The cranium? The face?
5. Define a suture. What are the four prominent sutures of the skull? Where are they located?
6. What is a fontanel? Describe the location of the six fairly constant fontanels.
7. What is a paranasal sinus? What cranial bones contain paranasal sinuses?
8. Define the following: "black eye," mastoiditis, cleft palate, cleft lip, sinusitis, dislocation of the jaw, TMJ syndrome, deviated nasal septum (DNS), gravity inversion, caudal anesthesia, and sternal puncture.

7 The Skeletal System: The Appendicular Skeleton

STUDENT OBJECTIVES

1. Identify the bones of the pectoral (shoulder) girdle and their major markings.
2. Indentify the upper extremity, its component bones, and their markings.
3. Identify the components of the pelvic (hip) girdle and their principal markings.
4. Identify the lower extremity, its component bones, and their markings.
5. Define the structural features and importance of the arches of the foot.
6. Compare the principal structural differences between female and male skeletons, especially those that pertain to the pelvis.

CHAPTER OUTLINE

■ **Pectoral (Shoulder) Girdle**
Clavicle
Scapula
■ **Upper Extremity**
Humerus
Ulna and Radius
Carpals, Metacarpals, and Phalanges
■ **Pelvic (Hip) Girdle**
■ **Lower Extremity**
Femur
Patella
Tibia and Fibula
Tarsals, Metatarsals, and Phalanges
Arches of the Foot
■ **Female and Male Skeletons**

This chapter discusses the bones of the appendicular skeleton, that is, the bones of the pectoral (shoulder) and pelvic (hip) girdles and extremities. The differences between female and male skeletons are also compared.

PECTORAL (SHOULDER) GIRDLE

The *pectoral* (PEK-tō-ral) or *shoulder girdles* attach the bones of the upper extremities to the axial skeleton (Figure 7-1). Each of the two pectoral girdles consists of two bones:

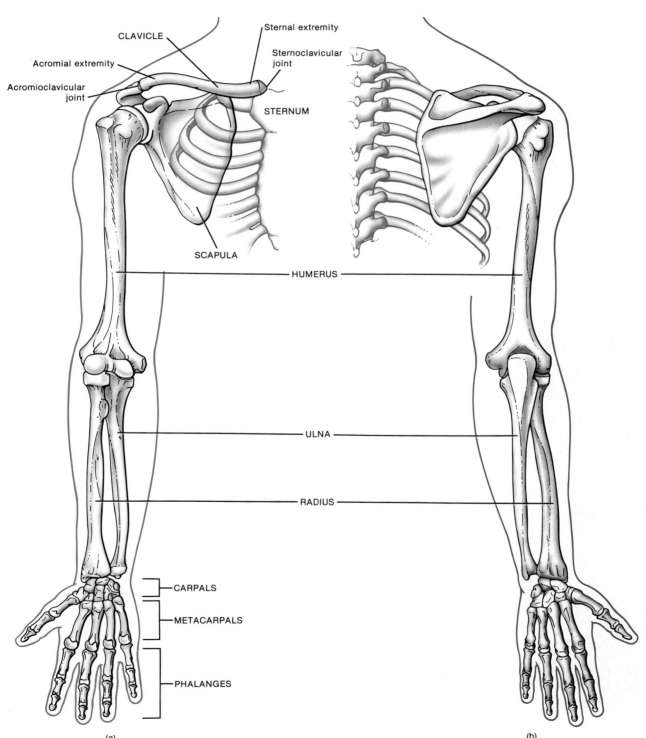

FIGURE 7-1 Right pectoral (shoulder) girdle and upper extremity. (a) Anterior view. (b) Posterior view.

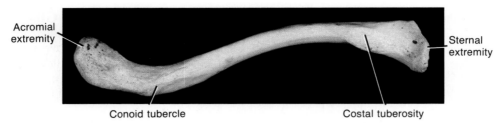

FIGURE 7-2 Photograph of right clavicle viewed from below. (Copyright © 1987 by Michael H. Ross. Used by permission.)

a clavicle and a scapula. The clavicle is the anterior component and articulates with the sternum at the sternoclavicular joint. The posterior component, the scapula, which is positioned freely by complex muscle attachments, articulates with the clavicle and humerus. The pectoral girdles have no articulation with the vertebral column. Although the shoulder joints are not very stable, they are freely movable and thus allow movement in many directions.

CLAVICLE

The *clavicles* (KLAV-i-kuls), or collarbones, are long, slender bones with a double curvature (Figure 7-2). The medial one-third of the clavicle is convex anteriorly, whereas the lateral one-third is concave anteriorly. The clavicles lie horizontally in the superior and anterior part of the thorax superior to the first rib.

The medial end of the clavicle, the *sternal extremity,* is rounded and articulates with the sternum. The broad, flat, lateral end, the *acromial* (a-KRŌ-mē-al) *extremity,* articulates with the acromion of the scapula. This joint is called the *acromioclavicular joint.* (Refer to Figure 7-1

for a view of these articulations.) The *conoid tubercle* on the inferior surface of the lateral end of the bone serves as a point of attachment for a ligament. The *costal tuberosity* on the inferior surface of the medial end also serves as a point of attachment for a ligament.

CLINICAL APPLICATION

Because of its position, the clavicle transmits forces from the upper extremity to the trunk. If such forces are excessive, as in falling on one's outstretched arm, a *fractured clavicle* may result. In fact, it is the most frequently broken bone in the body.

SCAPULA

The *scapulae* (SCAP-yoo-lē), or shoulder blades, are large, triangular, flat bones situated in the posterior part of the thorax between the levels of the second and seventh ribs (Figure 7-3). Their medial borders are located about 5 cm (2 in.) from the vertebral column.

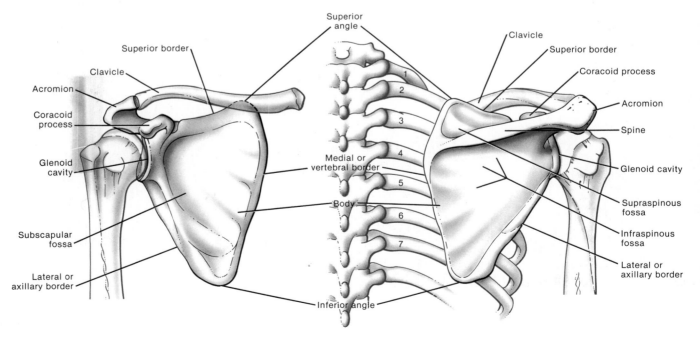

(a)   (b)

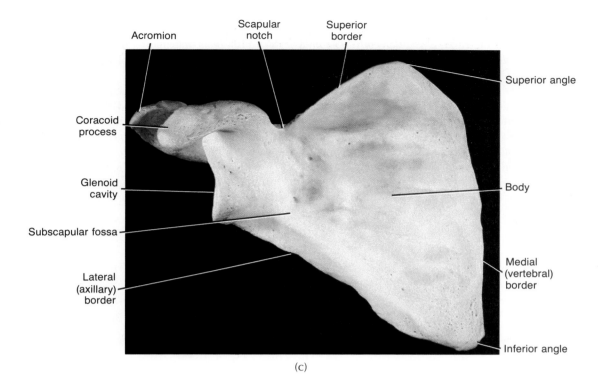

(c)

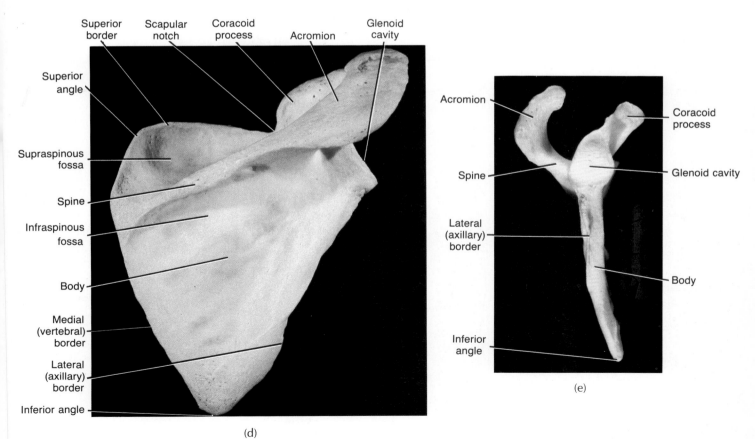

(d)

(e)

FIGURE 7-3 Right scapula. Diagrams of (a) anterior view and (b) posterior view. Photographs in (c) anterior view, (d) posterior view, and (e) lateral border view. (Copyright © 1987 by Michael H. Ross. Used by permission.)

A sharp ridge, the *spine,* runs diagonally across the posterior surface of the flattened, triangular *body.* The end of the spine projects as a flattened, expanded process called the *acromion* (a-KRŌ-mē-on), easily felt as the high point of the shoulder. This process articulates with the clavicle. Inferior to the acromion is a depression called the *glenoid cavity.* This cavity articulates with the head of the humerus to form the shoulder joint.

The thin edge of the body near the vertebral column is the *medial (vertebral) border.* The thick edge closer to the arm is the *lateral (axillary) border.* The medial and lateral borders join at the *inferior angle.* The superior edge of the scapular body, called the *superior border,* joins the vertebral border at the *superior angle.* The *scapular notch* is a prominent indentation along the superior border near the coracoid process; the notch permits passage of the supra-scapular nerve.

At the lateral end of the superior border is a projection of the anterior surface called the *coracoid* (KOR-a-koyd) *process* to which muscles attach. Above and below the spine are two fossae: the *supraspinous* (soo'-pra-SPĪ-nus) *fossa* and the *infraspinous fossa,* respectively. Both serve as surfaces of attachment for shoulder muscles. On the ventral (costal) surface is a lightly hollowed-out area called the *subscapular fossa,* also a surface of attachment for shoulder muscles.

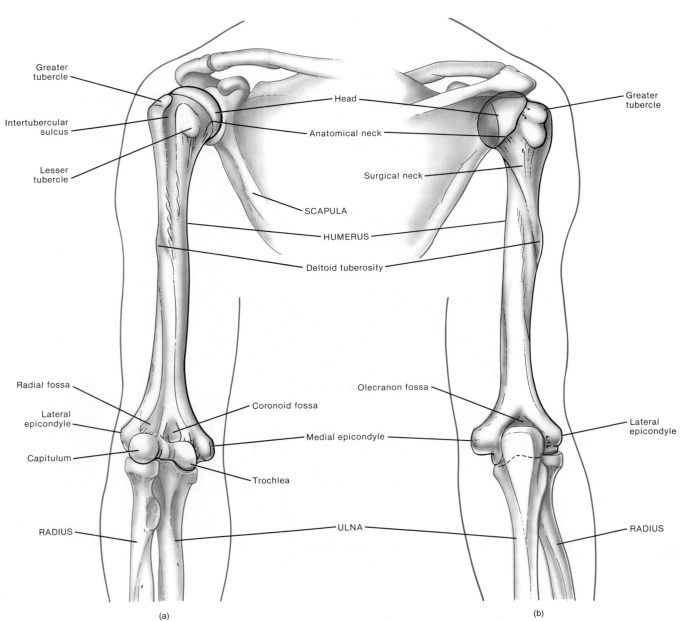

(a)   (b)

FIGURE 7-4 Right humerus. Diagrams of (a) anterior view and (b) posterior view.

UPPER EXTREMITY

The *upper extremities* consist of 60 bones. The skeleton of the right upper extremity is shown in Figure 7-1. Each upper extremity includes a humerus in the arm, ulna and radius in the forearm, carpals (wrist bones), metacarpals (palm bones), and phalanges in the fingers of the hand.

HUMERUS

The *humerus* (HYOO-mer-us), or arm bone, is the longest and largest bone of the upper extremity (Figure 7-4). It articulates proximally with the scapula and distally at the elbow with both ulna and radius.

The proximal end of the humerus consists of *a head* that articulates with the glenoid cavity of the scapula. It also has an *anatomical neck,* which is an oblique groove just distal to the head. The *greater tubercle* is a lateral projection distal to the neck. It is the most laterally palpable bony landmark of the shoulder region. The *lesser tubercle* is an anterior projection. Between these tubercles runs an *intertubercular sulcus (bicipital groove).* The *surgical neck* is a constricted portion just distal to the tubercles. It is so named because of its liability to fracture.

The *body (shaft)* of the humerus is cylindrical at its proximal end. It gradually becomes triangular and is flattened and broad at its distal end. Along the middle portion of the shaft, there is a roughened, V-shaped area called the *deltoid tuberosity.* This area serves as a point of attachment for the deltoid muscle.

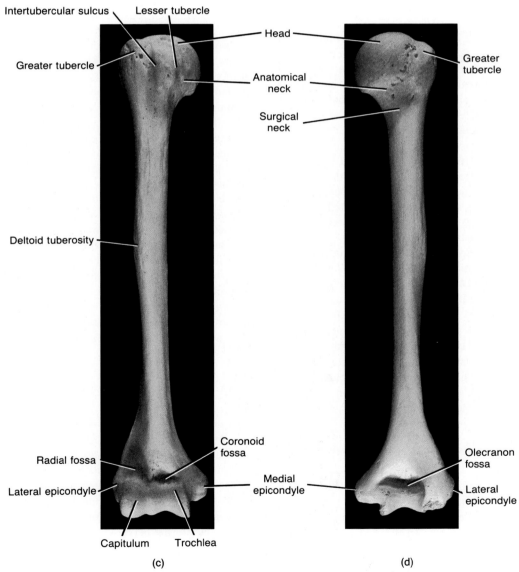

FIGURE 7-4 (*Continued*) Photographs in (c) anterior view and (d) posterior view. (Copyright © 1987 by Michael H. Ross. Used by permission.)

The following parts are found at the distal end of the humerus. The *capitulum* (ka-PIT-yoo-lum) is a rounded knob that articulates with the head of the radius. The *radial fossa* is a depression that receives the head of the radius when the forearm is flexed. The *trochlea* (TRŌK-lē-a) is a pulleylike surface that articulates with the ulna. The *coronoid fossa* is an anterior depression that receives part of the ulna when the forearm is flexed. The *olecranon* (ō-LEK-ra-non) *fossa* is a posterior depression that receives the olecranon of the ulna when the forearm is extended. The *medial epicondyle* and *lateral epicondyle* are rough projections on either side of the distal end to which most muscles of the forearm are attached. The ulnar nerve lies on the posterior surface of the medial epicondyle and may easily be rolled between the finger and the medial epicondyle.

ULNA AND RADIUS

The *ulna* is the medial bone of the forearm (Figure 7-5). In other words, it is located at the little finger side. The proximal end of the ulna presents an *olecranon* (*olecranon process*), which forms the prominence of the elbow. The *coronoid process* is an anterior projection that, together with the olecranon, receives the trochlea of the humerus. The *trochlear* (*semilunar*) *notch* is a curved area between the olecranon and the coronoid process. The trochlea of the humerus fits into this notch. The *radial notch* is a depression located laterally and inferiorly to the trochlear notch. It receives the head of the radius. The distal end of the ulna consists of a *head* that is separated from the wrist by a fibrocartilage disc. A *styloid process* is on the posterior side of the distal end.

The *radius* is the lateral bone of the forearm, that is, it is situated on the thumb side. The proximal end of the radius has a disc-shaped *head* that articulates with the capitulum of the humerus and radial notch of the ulna. It also has a raised, roughened area on the medial side called the *radial tuberosity.* This is a point of attachment for the biceps brachii muscle. The shaft of the radius widens distally to form a concave inferior surface that articulates with two bones of the wrist called the lunate and scaphoid bones. Also at the distal end is a *styloid process* on the lateral side and a medial, concave *ulnar notch* for articulation with the distal end of the ulna.

CLINICAL APPLICATION

When one falls on the outstretched arm, the radius bears the brunt of forces transmitted through the hand. If a fracture occurs in such a fall, it is usually a transverse break about 3 cm (1 in.) from the distal end of the bone. In this type of injury, called a *Colles' fracture,* the hand is displaced backward and upward (see Figure 5-8b).

CARPALS, METACARPALS, AND PHALANGES

The *carpus* (or wrist) consists of eight small bones, the *carpals,* united to each other by ligaments (Figure 7-6). The bones are arranged in two transverse rows, with four bones in each row and they are named for their shapes. In the anatomical position, the proximal row of carpals, from the lateral to medial position, consists of the *scaphoid* (resembles a boat), *lunate* (resembles a crescent moon in its anteroposterior aspect), *triquetral* (has three articular surfaces), and *pisiform* (peashaped). In about 70 percent of cases involving carpal fractures, only the scaphoid is involved. The distal row of carpals, from the lateral to medial position, consists of the *trapezium* (four sided), *trapezoid* (also four sided), *capitate* (its rounded projection, the head, articulates with the lunate), and *hamate* (named for a large hood-shaped projection on its anterior surface).

The five bones of the *metacarpus* constitute the palm of the hand. Each metacarpal bone consists of a proximal *base,* a *shaft,* and a distal *head.* The metacarpal bones are numbered I to V, starting with the lateral bone. The bases articulate with the distal row of carpal bones and with one another. The heads articulate with the proximal phalanges of the fingers. The heads of the metacarpals are commonly called the "knuckles" and are readily visible when the fist is clenched.

The *phalanges* (fa-LAN-jēz), or bones of the fingers, number 14 in each hand. A single bone of the finger (or toe) is referred to as a *phalanx* (FĀ-lanks). Each phalanx consists of a proximal *base,* a *shaft,* and a distal *head.* There are two phalanges in the first digit, called the *pollex* (*thumb*), and three phalanges in each of the remaining four digits. These digits, moving medially from the thumb, are commonly referred to as the index finger, middle finger, ring finger, and little finger. The first row of phalanges, the *proximal row,* articulates with the metacarpal bones and second row of phalanges. The second row of phalanges, the *middle row,* articulates with the proximal row and the third row. The third row of phalanges, the *distal row,* articulates with the middle row. The thumb has no middle phalanx.

PELVIC (HIP) GIRDLE

The *pelvic (hip) girdle* consists of the two *coxal* (KOK-sal) *bones,* or *hip bones* (Figure 7-7). The pelvic girdle provides a strong and stable support for the lower extremities on which the weight of the body is carried. The coxal bones are united to each other anteriorly at the symphysis (SIM-fi-sis) pubis. They unite posteriorly to the sacrum.

Together with the sacrum and coccyx, the two coxal (hip) bones of the pelvic girdle form the basinlike structure called the *pelvis.* The pelvis is divided into a greater pelvis and a lesser pelvis by an oblique plane that passes through sacral promontory (posterior), iliopectineal lines (laterally), and symphysis pubis (anteriorly). The circumference of this oblique plane is called the **brim of the pelvis.** The **greater (false) pelvis** represents the expanded portion situated superior to the brim of the pelvis. The greater pelvis consists laterally of the superior portions of the ilia and posteriorly of the superior portion of the sacrum. There is

no bony component in the anterior aspect of the greater pelvis. Rather, the front is formed by the walls of the abdomen.

The **lesser (true) pelvis** is inferior and posterior to the brim of the pelvis. It is formed by the inferior portions of

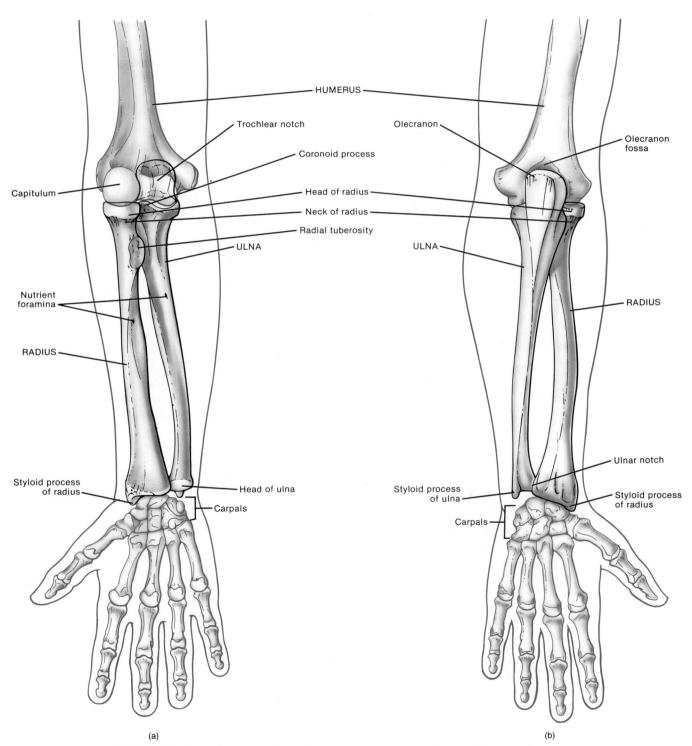

(a)   (b)

FIGURE 7-5 Right ulna and radius. Diagrams of (a) anterior view and (b) posterior view in relation to humerus and hand.

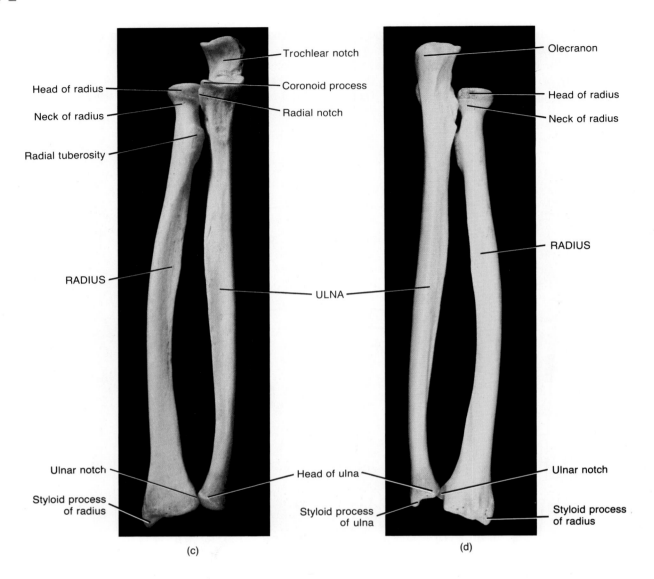

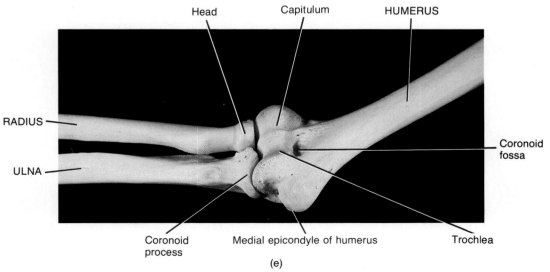

FIGURE 7-5 (*Continued*) Photographs in (c) anterior view, (d) posterior view, and (e) medial view of the right elbow. (Copyright © 1987 by Michael H. Ross. Used by permission.)

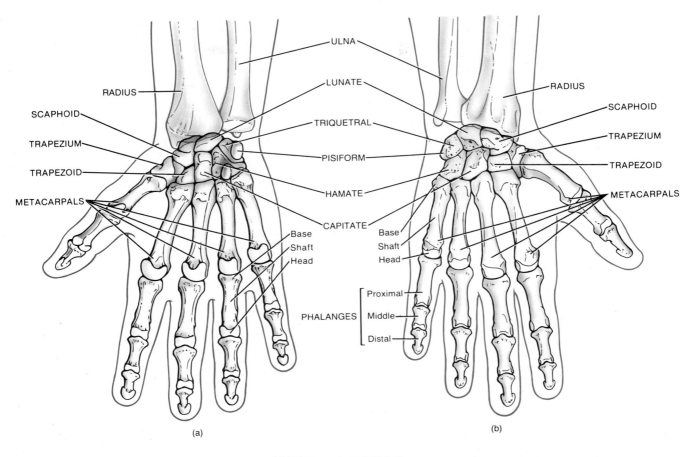

(a)

(b)

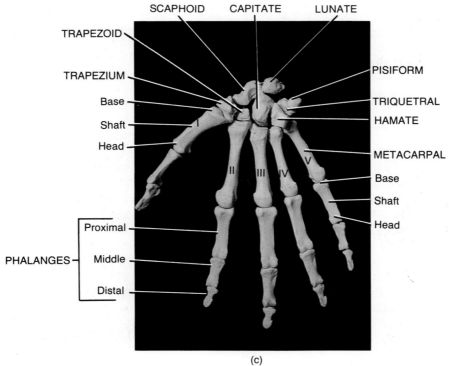

(c)

FIGURE 7-6 Right wrist and hand. Diagram of (a) anterior view and (b) posterior view in relation to the ulna and radius. (c) Photograph in anterior view. (Courtesy of J. A. Gosling et al., *Atlas of Human Anatomy with Integrated Text.* Copyright © 1985 by Gower Medical Publishing Ltd.)

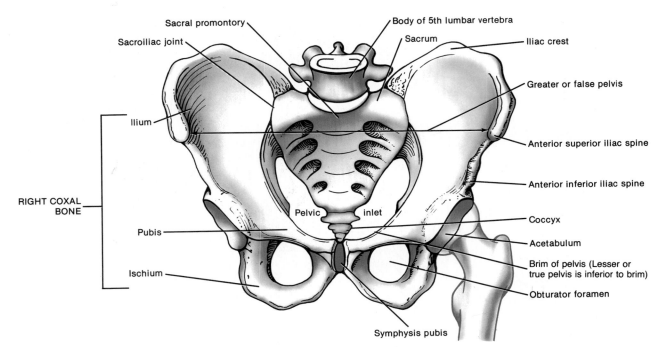

FIGURE 7-7 Pelvic (hip) girdle of a female in anterior view.

the ilia and sacrum, the coccyx, and the pubes. The lesser pelvis contains a superior opening called the **pelvic inlet** and an inferior opening called the **pelvic outlet.**

CLINICAL APPLICATION

Pelvimetry is the measurement of the size of the inlet and outlet of the birth canal. Measurement of the pelvic cavity in females is important to the physician, because the fetus must pass through the narrower opening of the lesser pelvis at birth.

Each of the two **coxal (hip) bones** of a newborn consists of three components: a superior **ilium,** an inferior and anterior **pubis,** and an inferior and posterior **ischium** (Figure 7-8). Eventually, the three separate bones fuse into one. The area of fusion is a deep, lateral fossa called the **acetabulum** (as'-e-TAB-yoo-lum). Although the adult coxae are both single bones, it is common to discuss the bones as if they still consisted of three portions.

The ilium is the largest of the three subdivisions of the coxal bone. Its superior border, the **iliac crest,** ends anteriorly in the **anterior superior iliac spine.** Posteriorly, the iliac crest ends in the **posterior superior iliac spine.** The **posterior inferior iliac spine** is just inferior. The spines serve as points of attachment for muscles of the abdominal wall. Slightly inferior to the posterior inferior iliac spine is the **greater sciatic** (sī-AT-ik) **notch.** The internal surface of the ilium seen from the medial side is the **iliac fossa.** It is a concavity where the iliacus muscle attaches. Posterior

to this fossa are the **iliac tuberosity,** a point of attachment for the sacroiliac ligament, and the **auricular surface,** which articulates with the sacrum. The other conspicuous markings of the ilium are three arched lines on its gluteal (buttock) surface called the **posterior gluteal line,** the **anterior gluteal line,** and the **inferior gluteal line.** The gluteal muscles attach to the ilium between these lines.

The ischium is the inferior, posterior portion of the coxal bone. It contains a prominent **ischial spine,** a **lesser sciatic notch** below the spine, and an **ischial tuberosity.** The rest of the ischium, the **ramus,** joins with the pubis and together they surround the **obturator** (OB-too-rā'-ter) **foramen.**

The pubis is the anterior and inferior part of the coxal bone. It consists of a **superior ramus,** an **inferior ramus,** and a **body** that contributes to the formation of the symphysis pubis.

The **symphysis pubis** is the joint between the two coxal bones (see Figure 7-7). It consists of fibrocartilage. The **acetabulum** is the fossa formed by the ilium, ischium, and pubis. It is the socket for the head of the femur. Two-fifths of the acetabulum is formed by the ilium, two-fifths by the ischium, and one-fifth by the pubis. On the inferior portion of the acetabulum is the **acetabular notch.**

LOWER EXTREMITY

The **lower extremities** are composed of 60 bones (Figure 7-9). Each extremity includes the femur in the thigh, patella (kneecap), fibula and tibia in the leg, tarsals (ankle bones), metatarsals, and phalanges in the toes.

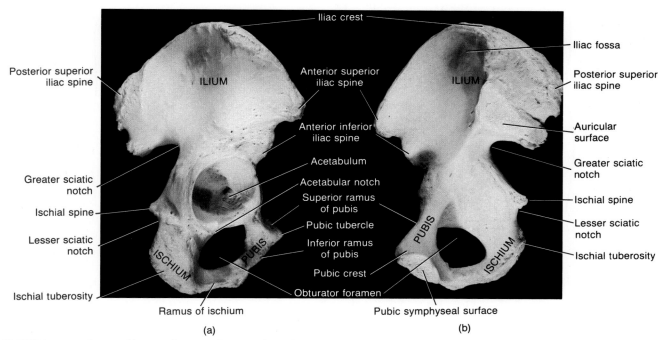

FIGURE 7-8 Right coxal bone. Photographs in (a) lateral view and (b) medial view. (c) The lines of fusion of the ilium, ischium, and pubis are not always visible in an adult bone. (Copyright © 1987 by Michael H. Ross. Used by permission.)

FEMUR

The *femur,* or thighbone, is the longest and heaviest bone in the body (Figure 7-10). Its proximal end articulates with the coxal bone. Its distal end articulates with the tibia. The body (shaft) of the femur bows medially so that it approaches the femur of the opposite thigh. As a result of this convergence, the knee joints are brought nearer to the body's line of gravity. The degree of convergence is greater in the female because the female pelvis is broader.

The proximal end of the femur consists of a rounded *head* that articulates with the acetabulum of the coxal bone. The *neck* of the femur is a constricted region distal to the head. A fairly common fracture in the elderly occurs at the neck of the femur. Apparently, the neck becomes so weak that it fails to support the body. The *greater trochanter* (trō-KAN-ter) and *lesser trochanter* are projections that serve as points of attachment for some of the thigh and buttock muscles. The greater trochanter is the prominence felt and seen anterior to the hollow on the side of the hip. Between the trochanters on the anterior surface is a narrow *intertrochanteric line.* Between the trochanters on the posterior surface is an *intertrochanteric crest.*

The body (shaft) of the femur contains a rough vertical ridge on its posterior surface called the *linea aspera.* This ridge serves for the attachment of several thigh muscles.

The distal end of the femur is expanded and includes the *medial condyle* and *lateral condyle.* These articulate with the tibia. Superior to the condyles are the *medial*

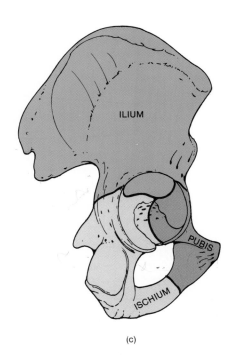

(c)

epicondyle and *lateral epicondyle.* A depressed area between the condyles on the posterior surface is called the *intercondylar* (in'-ter-KON-di-lar) *fossa.* The *patellar surface* is located between the condyles on the anterior surface.

PATELLA

The *patella,* or kneecap, is a small, triangular bone anterior to the knee joint (Figure 7-11). It is a sesamoid bone that develops in the tendon of the quadriceps femoris muscle.

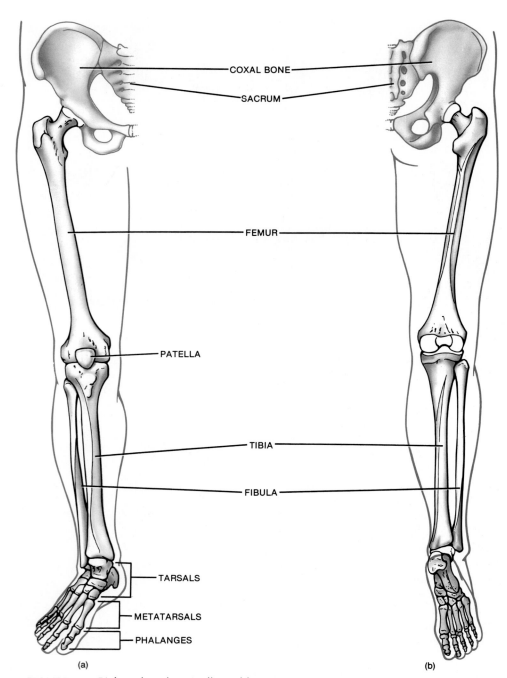

FIGURE 7–9 Right pelvic (hip) girdle and lower extremity. (a) Anterior view. (b) Posterior view.

The broad superior end of the patella is called the **base.** The pointed inferior end is the **apex.** The posterior surface contains two **articular facets,** one for the medial condyle and the other for the lateral condyle of the femur.

TIBIA AND FIBULA

The **tibia,** or shinbone, is the larger, medial bone of the leg (Figure 7-12). It bears the major portion of the weight of the leg. The tibia articulates at its proximal end with

the femur and fibula, and at its distal end with the fibula of the leg and talus bone of the ankle.

The proximal end of the tibia is expanded into a **lateral condyle** and a **medial condyle.** These articulate with the condyles of the femur. The inferior surface of the lateral condyle articulates with the head of the fibula. The slightly concave condyles are separated by an upward projection called the **intercondylar eminence.** The **tibial tuberosity** on the anterior surface is a point of attachment for the patellar ligament.

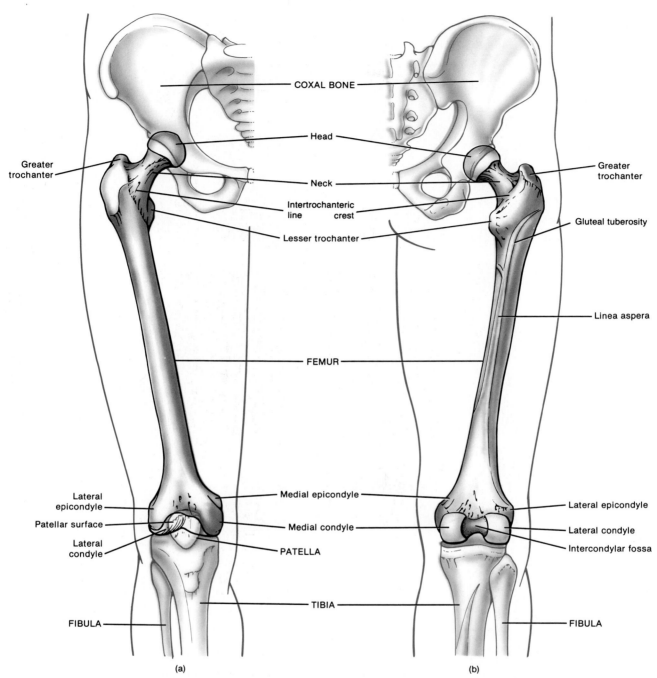

FIGURE 7-10 Right femur. Diagrams of (a) anterior view and (b) posterior view in relation to the coxal bone, patella, tibia, and fibula.

The medial surface of the distal end of the tibia forms the *medial malleolus* (mal-LĒ-ō-lus). This structure articulates with the talus bone of the ankle and forms the prominence that can be felt on the medial surface of your ankle. The *fibular notch* articulates with the fibula.

CLINICAL APPLICATION

Shinsplints (*tibia stress syndrome*) refers to soreness or pain along the tibia, probably caused by inflammation of the periosteum (periostitis) brought on by the repeated tugging of the muscles and tendons attached to the periosteum. The condition may occur as a result of walking or running up and down hills or by vigorous activity of the legs following a period of relative inactivity. Rest usually alleviates the pain. Patients who do not respond to rest may be given local injections of cortisonelike steroid drugs or may have to undergo minor surgery to release pressure in the soft tissues around the bone.

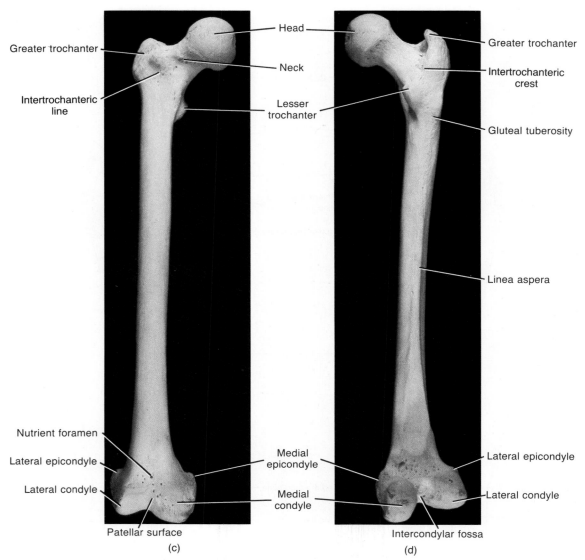

Greater trochanter

Head

Intertrochanteric line

Neck

Lesser trochanter

Greater trochanter

Intertrochanteric crest

Gluteal tuberosity

Linea aspera

Nutrient foramen

Lateral epicondyle

Lateral condyle

Medial epicondyle

Medial condyle

Patellar surface

Lateral epicondyle

Lateral condyle

Intercondylar fossa

(c)

(d)

FIGURE 7-10 (*Continued*) Photographs in (c) anterior view and (d) posterior view. (Copyright © 1987 by Michael H. Ross. Used by permission.)

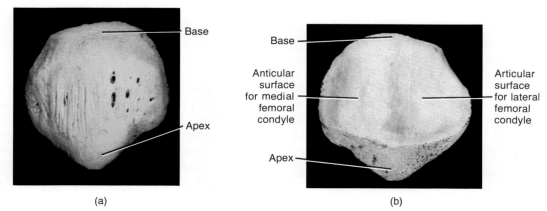

Base

Apex

Base

Anticular surface for medial femoral condyle

Articular surface for lateral femoral condyle

Apex

(a)

(b)

FIGURE 7-11   Right patella. Photographs in (a) anterior view and (b) posterior view. (Copyright © 1987 by Michael H. Ross. Used by permission.)

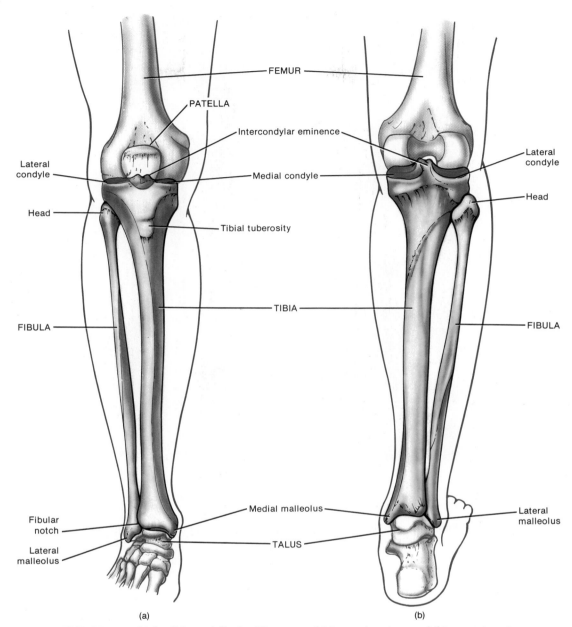

FIGURE 7-12 Right tibia and fibula. Diagrams of (a) anterior view and (b) posterior view. Photographs on pg. 180 in (c) anterior view and (d) posterior view. (Copyright © 1987 by Michael H. Ross. Used by permission.)

The *fibula* is parallel and lateral to the tibia. It is considerably smaller than the tibia. The *head* of the fibula, the proximal end, articulates with the inferior surface of the lateral condyle of the tibia below the level of the knee joint. The distal end has a projection called the *lateral malleolus* that articulates with the talus bone of the ankle. This forms the prominence on the lateral surface of the ankle. The inferior portion of the fibula also articulates with the tibia at the fibular notch. A fracture of the lower end of the fibula with injury to the tibial articulation is called a *Pott's fracture.*

TARSALS, METATARSALS, AND PHALANGES

The *tarsus* is a collective designation for the seven bones of the ankle called *tarsals* (Figure 7-13). The term *tarsos* pertains to a broad, flat surface. The *talus* and *calcaneus* (kal-KĀ-nē-us) are located on the posterior part of the foot. The anterior part contains the *cuboid, navicular,* and three *cuneiform bones* called the *first (medial), second (intermediate),* and *third (lateral) cuneiform.* The talus, the uppermost tarsal bone, is the only bone of the foot that articulates with the fibula and tibia. It is surrounded on one side by the medial malleolus of the tibia and on the other side by

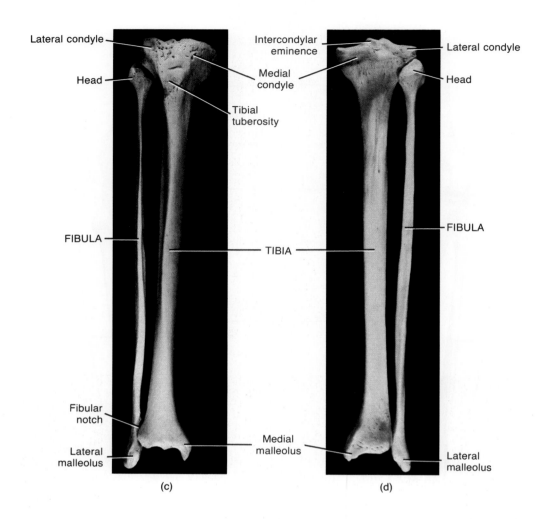

(c)   (d)

the lateral malleolus of the fibula. During walking, the talus intially bears the entire weight of the body. About half the weight is then transmitted to the calcaneus. The remainder is transmitted to the other tarsal bones. The calcaneus, or heel bone, is the largest and strongest tarsal bone.

The **metatarsus** consists of five metatarsal bones numbered I to V from the medial to lateral position. Like the metacarpals of the palm of the hand, each metatarsal consists of a proximal **base,** a **shaft,** and a distal **head.** The metatarsals articulate proximally with the first, second, and third cuneiform bones and with the cuboid. Distally, they articulate with the proximal row of phalanges. The first metatarsal is thicker than the others because it bears more weight.

The **phalanges** of the foot resemble those of the hand both in number and arrangement. Each also consists of a proximal **base,** a middle **shaft,** and a distal **head.** The **hallux** (great or big toe) has two large, heavy phalanges called proximal and distal phalanges. The other four toes each have three phalanges—proximal, middle, and distal.

ARCHES OF THE FOOT

The bones of the foot are arranged in two **arches** (Figure 7-14). These arches enable the foot to support the weight

of the body and provide leverage while walking. The arches are not rigid. They yield as weight is applied and spring back when the weight is lifted.

The **longitudinal arch** has two parts. Both consist of tarsal and metatarsal bones arranged to form an arch from the anterior to the posterior part of the foot. The **medial** (inner) part of the longitudinal arch originates at the calcaneus. It rises to the talus and descends through the navicular, the three cuneiforms, and the three medial metatarsals. The talus is the keystone of this arch. The **lateral** (outer) part of the longitudinal arch also begins at the calcaneus. It rises at the cuboid and descends to the two lateral metatarsals. The cuboid is the keystone of this arch.

The **transverse arch** is formed by the calcaneus, navicular, cuboid, and the posterior parts of the five metatarsals.

CLINICAL APPLICATION

The bones composing the arches are held in position by ligaments and tendons. If these ligaments and tendons are weakened, the height of the medial longitudinal arch may decrease or "fall." The result is **flatfoot.**

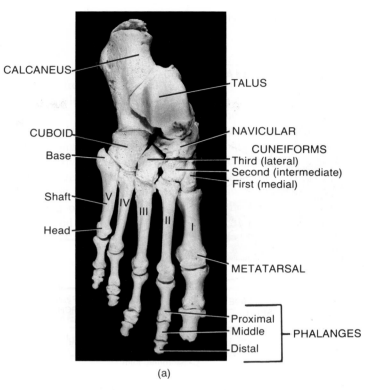

(a)

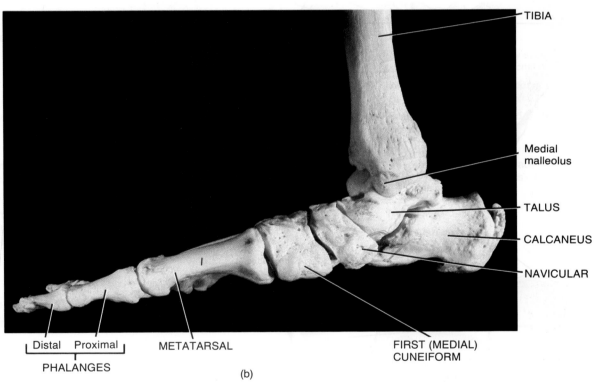

(b)

FIGURE 7-13 Right foot. Photographs in (a) superior view and (b) medial view. (Copyright © 1987 by Michael H. Ross. Used by permission.)

Lower Extremity (p. 174)

1. The bones of each lower extremity include the femur, tibia, fibula, tarsals, metatarsals, and phalanges.
2. The bones of the foot are arranged in two arches, the longitudinal arch and the transverse arch, to provide support and leverage.

Female and Male Skeletons (p. 183)

1. The female pelvis is adapted for pregnancy and childbirth. Differences in pelvic structure are listed in Exhibit 7-1.
2. Male bones are generally larger and heavier than female bones and have more prominent markings for muscle attachment.

REVIEW QUESTIONS

1. What is the pectoral (shoulder) girdle? Why is it important?
2. What are the bones of the upper extremity? What is a Colles' fracture?
3. What is the pelvic (hip) girdle? Why is it important?
4. What are the bones of the lower extremity? What is a Pott's fracture?
5. What is pelvimetry? What is its clinical importance?
6. In what ways do the upper extremity and lower extremity differ structurally?
7. Describe the structure of the longitudinal and transverse arches of the foot. What is the function of an arch?
8. How do flatfeet, clawfeet, and bunions arise?
9. What are the principal structural differences between typical female and male skeletons? Use Exhibit 7-1 as a guide in formulating your response.

SELF-QUIZ

Choose the one best answer to these questions.

___ 1. Which structures are on the posterior surface of the upper extremity (in the anatomical position)?
A. radial fossa and radial notch; B. trochlea and capitulum; C. coronoid process and coronoid fossa; D. olecranon process and olecranon fossa; E. lesser tubercle and intertubercular sulcus.

___ 2. Choose the false statement:
A. the capitulum articulates with the head of the radius; B. the medial and lateral epicondyles are located at the distal ends of the tibia and fibula; C. the coronoid fossa articulates with the ulna when the ulna is flexed; D. the trochlea articulates with the trochlear notch of the ulna; E. the ulna is the medial bone of the forearm.

___ 3. The bones of the pectoral girdle
A. articulate with the sternum anteriorly and the vertebrae posteriorly; B. include both clavicles and scapulae and the upper part of the sternum; C. are considered to be part of the axial skeleton; D. both articulate with the head of the humerus, forming the shoulder joint; E. are described by none of the above.

___ 4. Which of the following pairs of terms is correctly matched or related?
A. clavicle—breastbone; B. shoulder blade—scapula; C. sternum—wrist; D. coronoid bone—collar bone; E. pelvis—backbone.

___ 5. The anatomical name for the socket that the humerus fits into is the
A. acetabulum; B. subscapular fossa; C. glenoid cavity; D. sella turcica; E. iliac fossa.

___ 6. Which two structures (anatomical features) are directly anterior/posterior to each other?
A. the two tubercles of the humerus; B. the coronoid and olecranon fossae of the humerus; C. the two epicondyles of the humerus; D. the anatomical and surgical necks of the humerus; E. the trochlea and capitulum of the humerus.

___ 7. The radius
A. has a tuberosity on its distal end; B. has an acromial process at its proximal end; C. is longer and larger than the ulna; D. has a styloid process on its distal end; E. and the ulna are called carpal bones.

___ 8. The pelvic inlet
A. is narrower than the pelvic outlet, in both sexes; B. lies above the "false pelvis"; C. is formed by the symphysis pubis, the sacrum, and the ala of each ilium; D. is the space bounded by the iliac crests, sacrum, and symphysis pubis; E. is described by none of the above.

___ 9. The greater trochanter is a large bony prominence located
A. on the humerus; B. on the proximal part of the femur; C. near the tuberosity of the tibia; D. at the base of the skull; E. near the medial (vertebral) border of the scapula.

___ 10. Which of the following is a part of the femur?
A. obturator foramen; B. trochlea; C. tibial tuberosity; D. medial malleolus; E. none of the above.

___ 11. Which of the following statements regarding the male pelvis is *not* true?
A. the bones are heavier and rougher than in the female; B. the male pelvis is narrow and deep; C. the male pubic arch is wider than in the female; D. the cavity of the true pelvis is small in the male; E. none of the above.

Arrange the answers in correct sequence

___ ___ ___ 12. According to size of the bones, from largest to smallest:
A. femur
B. ulna
C. humerus

___ ___ ___ 13. From proximal to distal:
A. phalanges
B. metacarpals
C. carpals

Complete the following.

14. Wrist bones are called _____. There are (5? 7? 8? 14?) of them in each wrist.
15. The bones that constitute the palm of the hand are called _____ bones. There are _____ in each hand. The one on the thumb side is numbered (I? V?).

8 Articulations

CHAPTER OUTLINE

■ **Classification**
Functional
Structural
■ **Fibrous Joints**
Suture
Syndesmosis
Gomphosis
■ **Cartilaginous Joints**
Synchondrosis
Symphysis
■ **Synovial Joints**
Structure
Movements
   Gliding
   Angular
   Rotation
   Circumduction
   Special
Types
   Gliding
   Hinge
   Pivot
   Ellipsoidal
   Saddle
   Ball-and-Socket
Summary of Joints
■ **Selected Articulations of the Body**
■ **Applications to Health**
Rheumatism
Arthritis
   Rheumatoid Arthritis (RA)
   Osteoarthritis
   Gouty Arthritis
Bursitis
Dislocation
Sprain and Strain
■ **Key Medical Terms Associated with Articulations**

Bones are too rigid to bend without damage. Fortunately, the skeletal system consists of many separate bones, most of which are held together at joints by flexible connective tissue. All movements that change the positions of the bony parts of the body occur at joints. You can understand the importance of joints if you imagine how a cast over the knee joint prevents flexing the leg or how a splint on a finger limits the ability to manipulate small objects.

An *articulation (joint)* is a point of contact between bones, between cartilage and bones, or between teeth and bones. The scientific study of joints is referred to as *arthrology* (ar-THROL-ō-jē; *arthro* = joint; *logos* = study of). The joint's structure determines how it functions. Some joints permit no movement, others permit slight movement, and still others afford considerable movement. In general, the closer the fit at the point of contact, the stronger the joint. At tightly fitted joints, however, movement is restricted. The looser the fit, the greater the movement. Unfortunately, loosely fitted joints are prone to dislocation. Movement at joints is also determined by the structure of the articulating bones, the flexibility of the connective tissue that binds the bones together, and the position of ligaments, muscles, and tendons.

CLASSIFICATION

FUNCTIONAL

The functional classification of joints takes into account the degree of movement they permit. Functionally, joints are classified as *synarthroses* (sin'-ar-THRŌ-sēz), which are immovable joints; *amphiarthroses* (am'-fē-ar-THRŌ-sēz), which are slightly movable joints; and *diarthroses* (dī-ar-THRŌ-sēz), which are freely movable joints.

STRUCTURAL

The structural classification of joints is based on the presence or absence of a synovial (joint) cavity (a space between the articulating bones) and the kind of connective tissue that binds the bones together. Structurally, joints are classified as *fibrous,* in which there is no synovial cavity and the bones are held together by fibrous connective tissue; *cartilaginous,* in which there is no synovial cavity and the bones are held together by cartilage; and *synovial,* in which there is a synovial cavity and the bones forming the joint are united by a surrounding articular capsule and frequently by accessory ligaments (described in detail later). We shall discuss the joints of the body based on their structural classification, but with reference to their functional classification as well.

FIBROUS JOINTS

Fibrous joints lack a synovial cavity, and the articulating bones are held very closely together by fibrous connective tissue. They permit little or no movement. The three types of fibrous joints are (1) sutures, (2) syndesmoses, and (3) gomphoses.

SUTURE

Sutures (SOO-cherz) are found between bones of the skull. In a suture, the bones are united by a thin layer of dense fibrous connective tissue. Based on the form of the margins of the bones, several types of sutures can be distinguished. In a *serrated suture,* the margins of the bones are serrated like the teeth of a saw. An example is the sagittal suture between the two parietal bones (see Figure 6-2a,e). In a *squamous suture,* the margin of one bone overlaps that of the adjacent bone. The squamosal suture between the parietal and temporal bones is an example (see Figure 6-2c). In a *plane suture,* the even, fairly regular margins of the adjacent bones are brought together. An example is the intermaxillary suture between maxillary bones (see Figure 6-2a). Since sutures are immovable, they are functionally classified as synarthroses.

Some sutures, present during growth, are replaced by bone in the adult. In this case they are called *synostoses* (sin'-os-TŌ-sēz), or bony joints—joints in which there is a complete fusion of bone across the suture line. An example is the frontal suture between the left and right sides of the frontal bone, which begins to fuse during infancy (see Figure 6-3a). Synostoses are also functionally classified as synarthroses.

SYNDESMOSIS

A *syndesmosis* (sin'-dez-MŌ-sis) is a fibrous joint in which the uniting fibrous connective tissue is present in a much greater amount than in a suture, but the fit between the bones is not quite as tight. The fibrous connective tissue forms an interosseous membrane or ligament. A syndesmosis is slightly movable because the bones are separated more than in a suture and some flexibility is permitted by the interosseous membrane or ligament. Syndesmoses are functionally classified as amphiarthrotic and typically permit slight movement. Examples of syndesmoses include the distal articulation of the tibia and fibula (see Figure 7-12) and the articulations between the shafts of the ulna and radius (see Figure 7-5).

GOMPHOSIS

A *gomphosis* (gom-FŌ-sis) is a type of fibrous joint in which a cone-shaped peg fits into a socket. The intervening substance is the periodontal ligament. A gomphosis is functionally classified as synarthrotic. Examples are the articulations of the roots of the teeth with the alveoli (sockets) of the maxillae and mandible.

CARTILAGINOUS JOINTS

Another joint that has no synovial cavity is a ***cartilaginous joint.*** Here the articulating bones are tightly connected by cartilage. Like fibrous joints, they allow little or no movement. The two types of cartilaginous joints are (1) synchondroses and (2) symphyses.

SYNCHONDROSIS

A ***synchondrosis*** (sin'-kon-DRŌ-sis) is a cartilaginous joint in which the connecting material is hyaline cartilage. The most common type of synchondrosis is the epiphyseal plate (see Figure 5-6). Such a joint is found between the epiphysis and diaphysis of a growing bone and is immovable. Thus, it is synarthrotic. Since the hyaline cartilage is eventually replaced by bone when growth ceases, the joint is temporary. It is replaced by a synostosis. Another example of a synchondrosis is the joint between the first rib and the sternum. The cartilage in this joint undergoes ossification during adult life.

SYMPHYSIS

A ***symphysis*** (SIM-fi-sis) is a cartilaginous joint in which the connecting material is a broad, flat disc of fibrocartilage. This joint is found between bodies of vertebrae (see Figure 6-11). A portion of the intervertebral disc is cartilaginous material. The symphysis pubis between the anterior surfaces of the coxal bones is another example (see Figure 7-7). These joints are slightly movable, or amphiarthrotic.

FIGURE 8-1 Synovial joint. (a) Diagram of a generalized synovial joint in frontal section. (b) Photograph of the internal structure of the right knee joint in frontal section. (Courtesy of C. Yokochi and J. W. Rohen, *Photographic Anatomy of the Human Body*, 2nd ed., 1979, IGAKU-SHOIN, Ltd., Tokyo, New York.)

SYNOVIAL JOINTS

STRUCTURE

A joint in which there is a space between articulating bones is called a ***synovial*** (si-NŌ-vē-al) ***joint.*** The space is called a ***synovial (joint) cavity*** (Figure 8-1). Because of this cavity and because of the arrangement of the articular capsule and accessory ligaments, synovial joints are freely movable. Thus, synovial joints are functionally classified as diarthrotic.

Synovial joints are also characterized by the presence of ***articular cartilage.*** Articular cartilage covers the surfaces of the articulating bones but does not bind the bones together. The articular cartilage of synovial joints is hyaline cartilage.

Synovial joints are surrounded by a sleevelike ***articular capsule*** that encloses the synovial cavity and unites the articulating bones. The articular capsule is composed of two layers. The outer layer, the ***fibrous capsule,*** consists of dense connective (collagenous) tissue. It is attached to the periosteum of the articulating bones at a variable distance from the edge of the articular cartilage. The flexibility of the fibrous capsule permits movement at a joint, whereas its great tensile strength resists dislocation. The fibers of some fibrous capsules are arranged in parallel bundles and are therefore highly adapted to resist recurrent strain. Such fibers are called ***ligaments*** and are given special names. The strength of the ligaments is one of the principal factors in holding bone to bone.

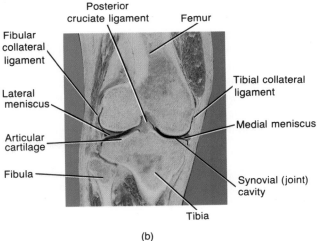

(b)

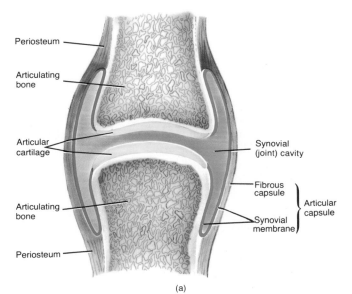

(a)

The inner layer of the articular capsule is formed by a *synovial membrane.* The synovial membrane is composed of loose connective tissue with elastic fibers and a variable amount of adipose tissue. It secretes *synovial fluid,* which lubricates the joint and provides nourishment for the articular cartilage. Synovial fluid also contains phagocytic cells that remove microbes and debris resulting from wear and tear in the joint. Synovial fluid consists of hyaluronic acid and an interstitial fluid formed from blood plasma and is similar in appearance and consistency to uncooked egg white. When there is no joint movement, the fluid is quite viscous, but as movement increases, the fluid becomes less viscous. The amount of synovial fluid varies in different joints of the body, ranging from a thin, viscous layer to about 3.5 ml (about 0.1 oz) of free fluid in a large joint such as the knee. The amount present in each joint is sufficient only to form a thin film over the surfaces within an articular capsule.

CLINICAL APPLICATION

One interesting feature of some synovial joints is their ability to produce a *cracking sound* when pulled apart. When a synovial joint is first pulled on, there is little separation between the opposing articular surfaces. However, as the pull continues, a negative pressure develops in the synovial fluid, driving out carbon dioxide (CO_2). As a result, a bubble of gas forms in the fluid. Then, the opposing articular surfaces abruptly separate, until limited by the articular capsule. Once the surfaces are separated, the pressure within the joint exceeds that in the bubble and the bubble collapses, causing the crackling noise. The collapse of the large bubble creates a series of smaller bubbles, which gradually go back into solution. Until the small bubbles disappear and the gas is completely dissolved, the joint cannot be cracked again. Synovial joints whose surfaces are more congruent, such as those between phalanges and metacarpals, crack more easily than joints whose surfaces are less congruent.

Many synovial joints also contain *accessory ligaments,* which are called extracapsular ligaments and intracapsular ligaments. *Extracapsular ligaments* are outside the articular capsule. An example is the fibular collateral ligament of the knee joint (see Figure 8-14). *Intracapsular ligaments* occur within the articular capsule but are excluded from the synovial cavity by reflections of the synovial membrane. Examples are the cruciate ligaments of the knee joint (see Figure 8-14).

Inside some synovial joints, there are pads of fibrocartilage that lie between the articular surfaces of the bones and are attached by their margins to the fibrous capsule. These pads are called *articular discs (menisci).* The discs usually subdivide the synovial cavity into two separate spaces. Articular discs allow two bones of different shapes to fit tightly; they modify the shape of the joint surfaces of the articulating bones. Articular discs also help maintain the stability of the joint and direct the flow of synovial fluid to areas of greatest friction.

CLINICAL APPLICATION

A tearing of articular discs in the knee, commonly called *torn cartilage,* occurs frequently among athletes. Such damaged cartilage requires surgical removal (meniscectomy) or it will begin to wear and cause arthritis. At one time, knee joint surgery for torn cartilage necessitated cutting through layers of healthy tissue and removing much, if not all, of the cartilage. This procedure is usually painful and expensive and does not always provide full recovery.

These problems have been overcome by *arthroscopy (arthroscopic surgery).* The arthroscope is a device, several inches long, that resembles a pencil and is designed like a telescope with a series of lenses aligned one above the other. Its optical fibers give off light, like a lamp on a miner's hard hat. It is inserted into the knee joint through an incision as small as one-quarter inch. A second small incision is made to insert a tube through which a salt solution is injected into the joint. Another small incision is used for insertion of an instrument that shaves off and reshapes the damaged cartilage and then suctions out the shaved cartilage along with the salt solution. Some orthopedic surgeons attach a lightweight television camera to the arthroscope so that the image from inside the knee can be projected onto a screen. Since arthroscopy requires only small incisions, recovery is usually speedy and there is very little pain or discomfort. Although arthroscopy is used mostly to remove torn cartilage, it can also be used for other types of knee surgery, for surgery on other joints of the body to obtain tissue samples, for diagnosing certain pathologies, deciding whether surgery is needed, and for planning surgical procedures.

The various movements of the body create friction between moving parts. To reduce this friction, saclike structures called *bursae* are situated in the body tissues. These sacs resemble joints in that their walls consist of connective tissue lined by a synovial membrane. They are also filled with a fluid similar to synovial fluid. Bursae are located between the skin and bone in places where skin rubs over bone. They are also found between tendons and bones, muscles and bones, and ligaments and bones. As fluid-filled sacs, they cushion the movement of one part of the body over another. An inflammation of a bursa is called *bursitis.*

The articular surfaces of synovial joints are kept in contact with each other by several factors. One factor is the fit of the articulating bones. This interlocking is very obvious at the hip joint, where the head of the femur articulates with the acetabulum of the coxal bone. Another factor is the strength of the joint ligaments. This is especially important in the hip joint. A third factor is the tension of the muscles around the joint. For example, the fibrous capsule of the knee joint is formed principally from tendinous expansions by muscles acting on the joint.

MOVEMENTS

The movements permitted at synovial joints are limited by several factors. The most important is the **structure of the articulating bones,** that is, the precise manner in which the articulating bones fit with respect to each other. A second factor is the **tension of ligaments.** The different components of a fibrous capsule are tense only when the joint is in certain positions. Tense ligaments not only restrict the range of movement but also direct the movement of the articulating bones with respect to each other. In the knee joint, for example, the major ligaments are lax when the knee is bent, but tense when the knee is straightened. Also, when the knee is straightened, the surfaces of the articulating bones are in fullest contact with each other. A third factor that restricts movement at a synovial joint is **muscle tension,** which reinforces the restraint placed on a joint by ligaments. A good example of the effect of muscle tension on a joint is seen at the hip joint. When the thigh is raised with the knee straight, the movement is restricted by the tension of the hamstring muscles on the posterior surface of the thigh. But if the knee is bent, the tension on the hamstring muscles is lessened and the thigh can be raised further. Finally, in a few joints, the **apposition of soft parts** may limit mobility. For example, during bending at the elbow, the anterior surface of the forearm is pressed against the anterior surface of the arm.

Following is a description of the specific movements that occur at synovial joints.

Gliding

A **gliding movement** is the simplest kind that can occur at a joint. One surface moves back-and-forth and from side-to-side over another surface without angular or rotary motion. Some joints that glide are those between the carpals and between the tarsals. The heads and tubercles of ribs glide on the bodies and transverse processes of vertebrae.

Angular

Angular movements increase or decrease the angle between bones. Among the angular movements are flexion, extension, abduction, and adduction (Figure 8-2). **Flexion** involves a decrease in the angle between the surfaces of the articulating bones. Examples of flexion include bending the head forward (the joint is between the occipital bone and the atlas), bending the elbow, and bending the knee.

Extension involves an increase in the angle between the surfaces of the articulating bones. Extension restores a body part to its anatomical position after it has been flexed. Examples of extension are returning the head to the anatomical position after flexion, straightening the arm after flexion, and straightening the leg after flexion. Continuation of extension beyond the anatomical position, as in bending the head backward, is called **hyperextension.**

Abduction usually means movement of a bone *away from* the midline of the body. An example of abduction is moving the arm upward and away from the body until it is held straight out at right angles to the chest. With the fingers and toes, however, the midline of the body is not used as the line of reference. Abduction of the fingers (not the thumb) is a movement away from an imaginary line drawn through the middle finger; in other words, it is spreading the fingers. In abduction of the thumb, the thumb moves away from the plane of the palm at a right angle to the palm. Abduction of the toes is relative to an imaginary line drawn through the second toe.

Adduction is usually movement of a part *toward* the midline of the body. An example of adduction is returning the arm to the side after abduction. As in abduction, adduction of the fingers (not the thumb) is relative to the middle finger, and adduction of the toes is relative to the second toe. In adduction of the thumb, the thumb moves toward the plane of the palm at a right angle to the palm.

Rotation

Rotation is the movement of a bone around its own longitudinal axis. During rotation, no other motion is permitted. In **medial rotation,** the anterior surface of a bone or extremity moves toward the midline. In **lateral rotation,** the anterior surface moves away from the midline. We rotate the atlas around the dens of the axis when we shake the head from side to side. Rotation of the humerus turns the anterior surface of the bone either medially or laterally. (Figure 8-3a).

Circumduction

Circumduction is a movement in which the distal end of a bone moves in a circle while the proximal end remains stable. The bone describes a cone in the air. Circumduction typically involves flexion, abduction, adduction, extension, and rotation. It involves a 360° rotation. An example is moving the outstretched arm in a circle to wind up to pitch a ball (Figure 8-3b).

Special

Special movements are those found only at the joints indicated in Figure 8-4. **Inversion** is the movement of the sole of the foot inward (medially). **Eversion** is the movement of the sole outward (laterally). **Dorsiflexion** involves bend-

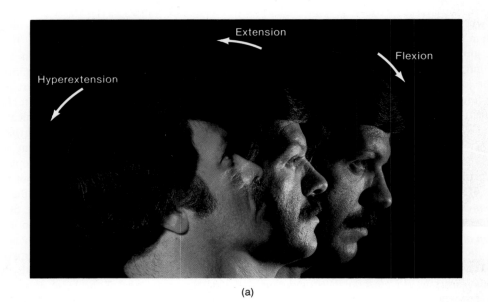

(a)

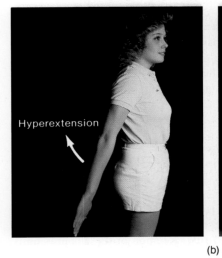

(b)

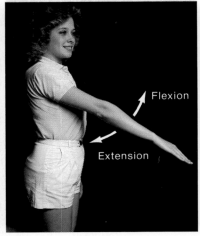

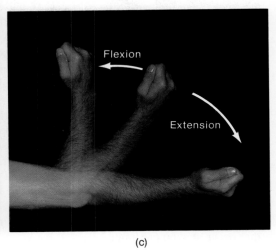

(c)

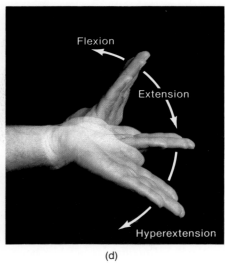

(d)

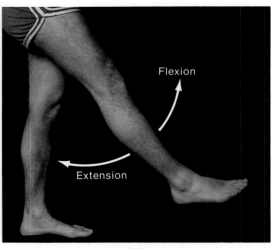

(e)

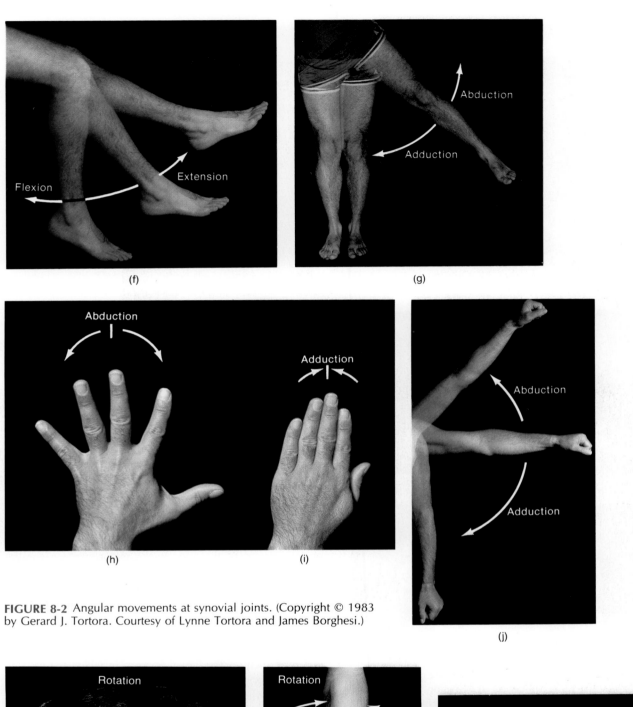

FIGURE 8-2 Angular movements at synovial joints. (Copyright © 1983 by Gerard J. Tortora. Courtesy of Lynne Tortora and James Borghesi.)

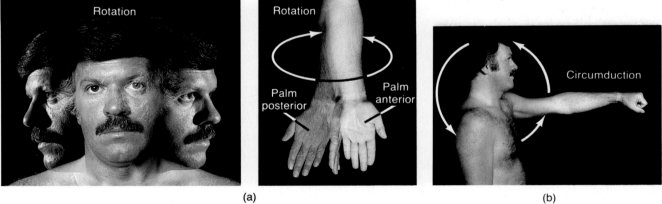

FIGURE 8-3 Rotation and circumduction. (a) Rotation at the atlantoaxial joint (left) and rotation of the humerus (right). (b) Circumduction of the humerus at the shoulder joint. (Copyright © 1983 by Gerard J. Tortora.)

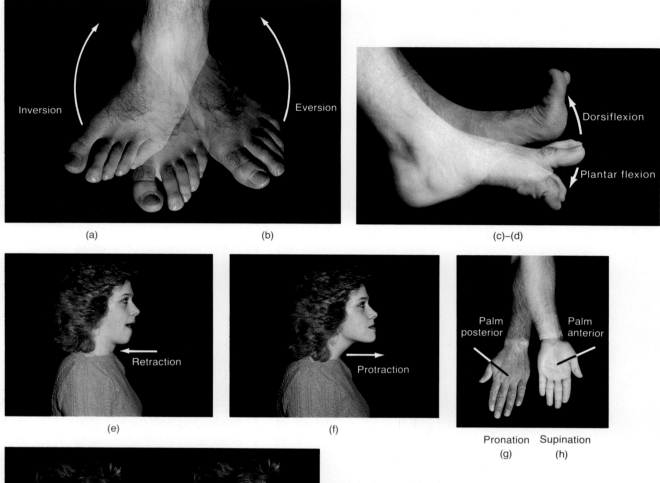

FIGURE 8-4 Special movements. (a) Inversion. (b) Eversion. (c) Dorsiflexion. (d) Plantar flexion. (e) Retraction. (f) Protraction. (g) Pronation. (h) Supination. (i) Elevation. (j) Depression. (Copyright © 1983 by Gerard J. Tortora. Courtesy of Lynne Tortora and James Borghesi.)

ing the foot in the direction of the dorsum (upper surface). *Plantar flexion* involves bending the foot in the direction of the plantar surface (sole).

Protraction is the movement of the mandible or shoulder girdle forward on a plane parallel to the ground. Thrusting the jaw outward is protraction of the mandible. Bringing your arms forward until the elbows touch requires protraction of the shoulder girdle. *Retraction* is the movement of a protracted part of the body backward on a plane parallel to the ground. Pulling the lower jaw back in line with the upper jaw is retraction to the mandible.

Supination is a movement of the forearm in which the palm of the hand is turned anterior or superior. To demonstrate supination, flex your forearm at the elbow to prevent rotation of the humerus in the shoulder joint. *Pronation* is a movement of the flexed forearm in which the palm is turned posterior or inferior.

Elevation is an upward movement of a part of the body. You elevate your mandible when you close your mouth. *Depression* is a downward movement of a part of the body. You depress your mandible when you open your mouth. The shoulders can also be elevated and depressed.

A summary of movements that occur at synovial joints is presented in Exhibit 8-1.

TYPES

Though all synovial joints are similar in structure, variations exist in the shape of the articulating surfaces. Accordingly, synovial joints are divided into six subtypes: gliding, hinge, pivot, ellipsoidal, saddle, and ball-and-socket joints.

EXHIBIT 8-1

Summary of Movements at Synovial Joints

MOVEMENT	DEFINITION
GLIDING	One surface moves back-and-forth and from side-to-side over another surface without angular or rotary motion.
ANGULAR	There is an increase or decrease at the angle between bones.
Flexion	Involves a decrease in the angle between the surfaces of articulating bones.
Extension	Involves an increase in the angle between the surfaces of articulating bones.
Hyperextension	Continuation of extension beyond the anatomical position.
Abduction	Movement of a bone away from the midline.
Adduction	Movement of a bone toward the midline.
ROTATION	Movement of a bone around its longitudinal axis; may be medial or lateral.
CIRCUMDUCTION	A movement in which the distal end of a bone moves in a circle while the proximal end remains stable.
SPECIAL	Occurs at specific joints.
Inversion	Movement of the sole of the foot inward.
Eversion	Movement of the sole of the foot outward.
Dorsiflexion	Bending the foot in the direction of the dorsum (upper surface).
Plantar flexion	Bending the foot in the direction of the plantar surface (sole).
Protraction	Movement of the mandible or shoulder girdle forward on a plane parallel to the ground.
Retraction	Movement of a protracted part backward on a plane parallel to the ground.
Supination	Movement of the forearm in which the palm is turned anterior or superior.
Pronation	Movement of the flexed forearm in which the palm is turned posterior or inferior.
Elevation	Movement of a part of the body upward.
Depression	Movement of a part of the body downward.

Gliding

The articulating surfaces of bones in **gliding joints** or **arthrodia** (ar-THRŌ-dē-a) are usually flat. Only side-to-side and back-and-forth movements are permitted (Figure 8-5a). Twisting and rotation are inhibited at gliding joints, generally because ligaments or adjacent bones restrict the range of movement. Since gliding joints do not move around an axis, they are referred to as **nonaxial**. Examples are the joints between carpal bones, tarsal bones, sternum and clavicle, and scapula and clavicle.

Hinge

A **hinge** or **ginglymus** (JIN-gli-mus) **joint** is one in which the convex surface of one bone fits into the concave surface of another bone. Movement is primarily in a single plane, and the joint is therefore known as **monaxial** or **uniaxial** (Figure 8-5b). The motion is similar to that of a hinged door. Movement is flexion and extension. Examples of hinge joints are the elbow, ankle, and interphalangeal joints. The movement allowed by a hinge joint is illustrated by flexion and extension at the elbow (see Figure 8-2c).

Pivot

In a **pivot** or **trochoid** (TRŌ-koyd) **joint,** a rounded, pointed, or conical surface of one bone articulates within a ring formed partly by another bone and partly by a ligament. The primary movement permitted is rotation, and the joint is therefore **monaxial** (Figure 8-5c). Examples include the joint between the atlas and axis (atlantoaxial) and between the proximal ends of the radius and ulna. Movement at a pivot joint is illustrated by supination and pronation of the palms and rotation of the head from side-to-side (see Figure 8-3a).

Ellipsoidal

In an **ellipsoidal** or **condyloid** (KON-di-loyd) **joint,** an oval-shaped condyle of one bone fits into an elliptical cavity of another bone. Since the joint permits side-to-side and back-and-forth movements, it is **biaxial** (Figure 8-5d). The joint at the wrist between the radius and carpals is ellipsoidal. The movement permitted by such a joint is illustrated when you flex and extend (see Figure 8-2d) and abduct and adduct the wrist.

Saddle

In a **saddle** or **sellaris** (sel-A-ris) **joint,** the articular surfaces of both bones are saddle shaped, that is, concave in one direction and convex in the other. Essentially, the saddle joint is a modified ellipsoidal joint in which the movement is somewhat freer. Movements at a saddle joint are side-to-side and back-and-forth. Thus, the joint is **biaxial** (Figure 8-5e). The joint between the trapezium of the carpus and metacarpal of the thumb is an example of a saddle joint.

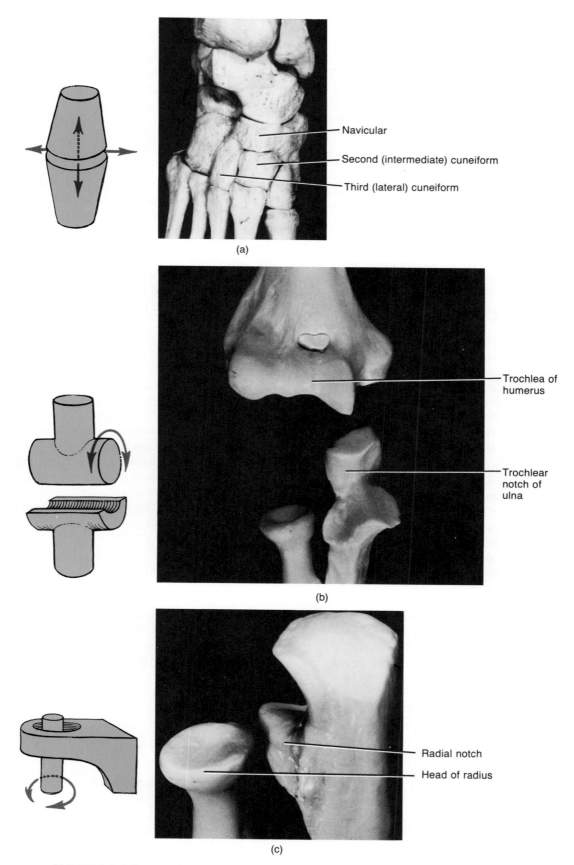

(a)

Navicular

Second (intermediate) cuneiform

Third (lateral) cuneiform

(b)

Trochlea of humerus

Trochlear notch of ulna

(c)

Radial notch

Head of radius

FIGURE 8-5 Subtypes of synovial joints. For each subtype shown, there is a simplified diagram and a photograph of the actual joint. (a) Gliding joint between the navicular and second and third cuneiforms of the tarsus. (b) Hinge joint at the elbow between the trochlea of the humerus and trochlear notch of the ulna. (c) Pivot joint between the head of the radius and the radial notch of the ulna. (d) Ellipsoidal joint at the wrist between the distal end of the

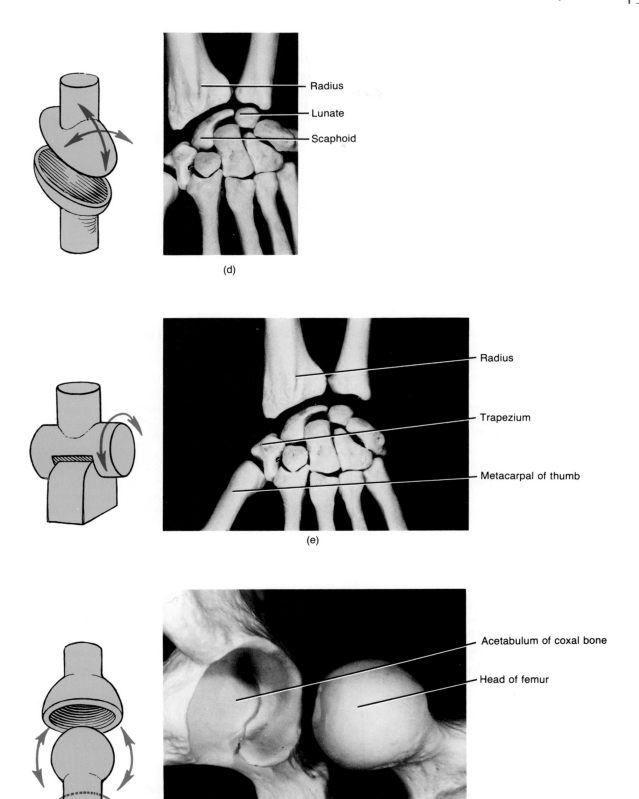

(d)

Radius
Lunate
Scaphoid

(e)

Radius
Trapezium
Metacarpal of thumb

(f)

Acetabulum of coxal bone
Head of femur

radius and the scaphoid and lunate bones of the carpus. (e) Saddle joint between the trapezium of the carpus and the metacarpal of the thumb. (f) Ball-and-socket joint between the head of the femur and the acetabulum of the coxal bone. (Copyright © 1985 by Michael H. Ross. Used by permission.)

Ball-and-Socket

A **ball-and-socket** or **spheroid** (SFĒ-royd) **joint** consists of a ball-like surface of one bone fitted into a cuplike depression of another bone. Such a joint permits **triaxial** movement, or movement in three planes of motion: flexion–extension, abduction–adduction, and rotation (Figure 8-5f). Examples of ball-and-socket joints are the shoulder joint and hip joint. The range of movement at a ball-and-socket joint is illustrated by circumduction of the arm (Figure 8-3b).

SUMMARY OF JOINTS

The summary of joints presented in Exhibit 8-2 is based on the anatomy of the joints. If we rearrange the types of joints into a classification based on movement, we arrive at the following:

Synarthroses: immovable joints
1. **Suture**
2. **Synchondrosis**
3. **Gomphosis**

EXHIBIT 8-2

Summary of Joints

TYPE	DESCRIPTION	MOVEMENT	EXAMPLES
FIBROUS	No synovial (joint) cavity; bones held together by a thin layer of fibrous tissue or dense fibrous tissue.		
Suture	Found only between bones of the skull; articulating bones separated by a thin layer of fibrous tissue.	None (synarthrotic).	Lambdoidal suture between occipital and parietal bones.
Syndesmosis	Articulating bones united by dense fibrous tissue.	Slight (amphiarthrotic).	Distal ends of tibia and fibula.
Gomphosis	Cone-shaped peg fits into a socket; articulating bones separated by periodontal ligament.	None (synarthrotic).	Roots of teeth in alveoli (sockets).
CARTILAGINOUS	No synovial cavity, articulating bones united by cartilage.		
Synchondrosis	Connecting material is hyaline cartilage.	None (synarthrotic).	Temporary joint between the diaphysis and epiphyses of a long bone and permanent joint between true ribs and sternum.
Symphysis	Connecting material is a broad, flat disc of fibrocartilage.	Slight (amphiarthrotic).	Intervertebral joints and symphysis pubis.
SYNOVIAL	Synovial cavity and articular cartilage present; articular capsule composed of an outer fibrous capsule and an inner synovial membrane; may contain accessory ligaments, articular discs (menisci), and bursae.	Freely movable (diarthrotic).	
Gliding	Articulating surfaces usually flat.	Nonaxial.	Intercarpal and intertarsal joints.
Hinge	Spoollike surface fits into a concave surface.	Monaxial (flexion–extension).	Elbow, ankle, and interphalangeal joints.
Pivot	Rounded, pointed, or concave surface fits into a ring formed partly by bone and partly by a ligament.	Monaxial (rotation).	Atlantoaxial and radioulnar joints.
Ellipsoidal	Oval-shaped condyle fits into an elliptical cavity.	Biaxial (flexion–extension, abduction–adduction).	Radiocarpal joint.
Saddle	Articular surfaces concave in one direction and convex in opposite direction.	Biaxial (flexion–extension, abduction–adduction).	Carpometacarpal joint of thumb.
Ball-and-socket	Ball-like surface fits into a cuplike depression.	Triaxial (flexion–extension, abduction–adduction, rotation).	Shoulder and hip joints.

Amphiarthroses: slightly movable joints
1. *Syndesmosis*
2. *Symphysis*

Diarthroses: freely movable joints
1. *Gliding*
2. *Hinge*
3. *Pivot*
4. *Ellipsoidal*
5. *Saddle*
6. *Ball-and-socket*

SELECTED ARTICULATIONS OF THE BODY

We shall now examine in some detail selected articulations of the body. In order to simplify your learning efforts, a series of exhibits have been prepared. Each exhibit considers a specific articulation and contains (1) a definition, that is, a description of the bones that form the joint; (2) the type of joint (its structural classification); and (3) its anatomical components—a description of the major connecting ligaments, articular disc, articular capsule, and other distinguishing features of the joint. Each exhibit also refers you to an illustration of the joint.

EXHIBIT 8-3

Temporomandibular Joint (TMJ) (Figure 8-6)

DEFINITION	Joint formed by mandibular condyle of the mandible and mandibular fossa and articular tubercle of temporal bone. The temporomandibular joint (TMJ) is the only movable joint between skull bones; all other skull joints are sutures and therefore immovable.
TYPE OF JOINT	Synovial; combined hinge (ginglymus) and gliding (arthrodial) type.
ANATOMICAL COMPONENTS	1. **Articular disc (meniscus).** Fibrocartilage disc that separates the synovial cavity into a superior and inferior compartment, each of which has a synovial membrane. 2. **Articular capsule.*** Thin, fairly loose envelope around the circumference of the joint. 3. **Lateral ligament.** Two short bands on the lateral surface of the articular capsule that extend inferiorly and posteriorly from the lower border and tubercle of the zygomatic arch to the lateral and posterior aspect of the condylar process of the mandible. It is covered by the parotid gland and helps prevent displacement of the mandible. 4. **Sphenomandibular ligament.** Thin band that extends inferiorly and anteriorly from the spine of the sphenoid bone to the ramus of the mandible. 5. **Stylomandibular ligament.** Thickened band of deep cervical fascia that extends from the styloid process of the temporal bone to the inferior and posterior border of the ramus of the mandible. The ligament separates the parotid from the submandibular gland.

* Also referred to as the **capsular ligament.**

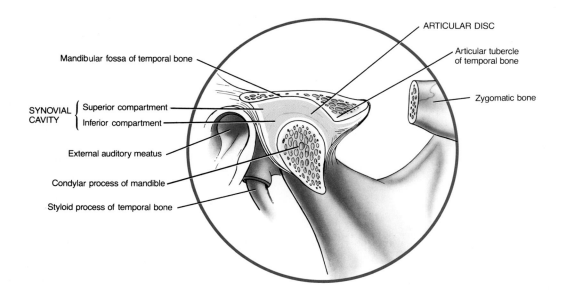

(a)

FIGURE 8-6 Temporomandibular joint (TMJ). (a) Sagittal section through the TMJ.

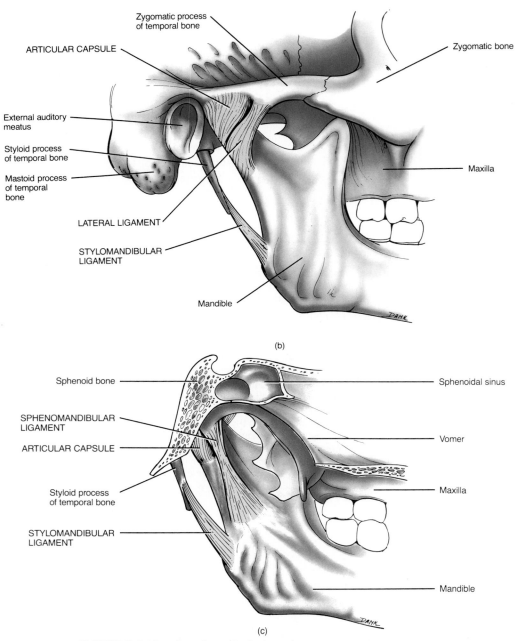

(b)

(c)

FIGURE 8-6 (*Continued*) (b) Right lateral view. (c) Median view.

EXHIBIT 8-4

Atlanto-occipital Joints (Figure 8-7)

DEFINITION	Joints formed by the superior articular surfaces of the atlas and the occipital condyles of the occipital bone.
TYPE OF JOINT	Synovial; ellipsoidal (condyloid) type.
ANATOMICAL COMPONENTS	1. **Articular capsules.** Thin, loose envelopes that surround the occipital condyles and attach them to the articular processes of the atlas.
	2. **Anterior atlanto-occipital ligament.** Broad, thick ligament that extends from the anterior surface of the foramen magnum of the occipital bone to the anterior arch of the atlas.
	3. **Posterior atlanto-occipital ligament.** Broad, thin ligament that extends from the posterior surface of the foramen magnum of the occipital bone to the posterior arch of the atlas. The inferior portion of the ligament is sometimes ossified.
	4. **Lateral atlanto-occipital ligaments.** Thickened portions of the articular portions of the articular capsules that also contain fibrous tissue bundles and extend from the jugular process of the occipital bone to the transverse process of the atlas.

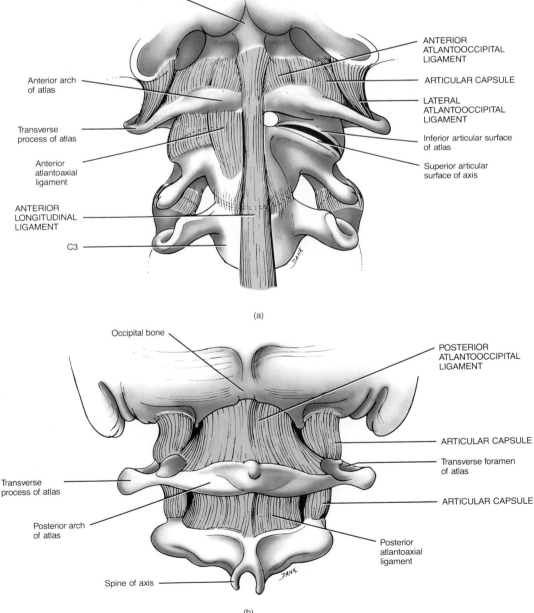

(a)

(b)

FIGURE 8-7 Atlantooccipital joints. (a) Anterior view. (b) Posterior view.

EXHIBIT 8-5

Intervertebral Joints (Figure 8-8)

DEFINITION	Joints formed between (1) vertebral bodies and (2) vertebral arches.
TYPE OF JOINT	Joints between vertebral bodies—cartilaginous (fibrocartilage), symphysis type. Joints between vertebral arches—synovial, gliding (arthrodial) type.
ANATOMICAL COMPONENTS OF JOINTS BETWEEN VERTEBRAL BODIES	1. **Anterior longitudinal ligament.** Broad, strong band that extends along the anterior surfaces of the vertebral bodies from the axis to the sacrum. It is firmly attached to the intervertebral discs. 2. **Posterior longitudinal ligament.** Extends along the posterior surfaces of the vertebral bodies within the vertebral canal from the axis to the sacrum. The free surface of the ligament is separated from the spinal dura mater by loose (areolar) tissue. 3. **Intervertebral discs.** Unite with adjacent surfaces of the vertebral bodies from the axis to the sacrum. Each disc is composed of a peripheral **anulus fibrosus** consisting of fibrous tissue and fibrocartilage and a central **nucleus pulposus** composed of a soft, pulpy, highly elastic substance.
ANATOMICAL COMPONENTS OF JOINTS BETWEEN VERTEBRAL ARCHES	1. **Articular capsules.** Thin, loose ligaments attached to the margins of the articular processes of adjacent vertebrae. 2. **Ligamenta flava.** Contain elastic tissue and connect the laminae of adjacent vertebrae from the axis to the first segment of the sacrum. 3. **Supraspinous ligament.** Strong fibrous cord that connects the spinous processes from the seventh cervical vertebra to the sacrum. 4. **Ligamentum nuchae.** Fibrous ligament that represents an enlargement of the supraspinous ligament in the neck. It extends from the external occipital protuberance of the occipital bone to the spinous process of the seventh cervical vertebra. 5. **Interspinous ligaments.** Relatively weak bands that run between adjacent spinous processes. 6. **Intertransverse ligaments.** Bands between transverse processes that are readily apparent in the lumbar region.

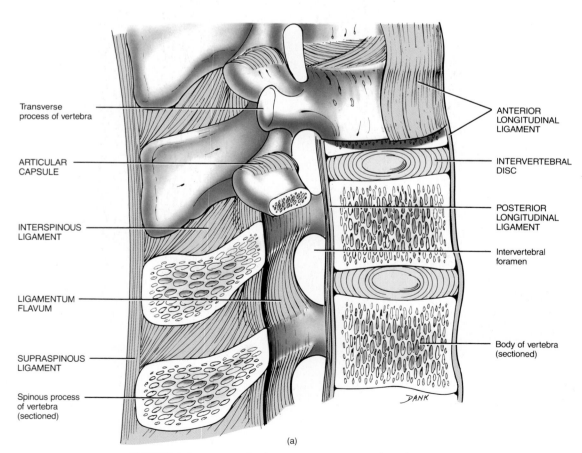

FIGURE 8-8 Intervertebral joints. (a) Diagram of median view.

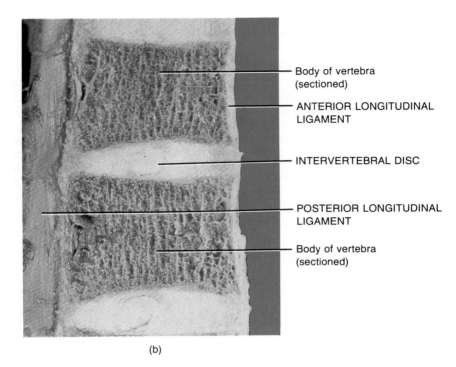

— Body of vertebra
(sectioned)

— ANTERIOR LONGITUDINAL
LIGAMENT

— INTERVERTEBRAL DISC

— POSTERIOR LONGITUDINAL
LIGAMENT

— Body of vertebra
(sectioned)

(b)

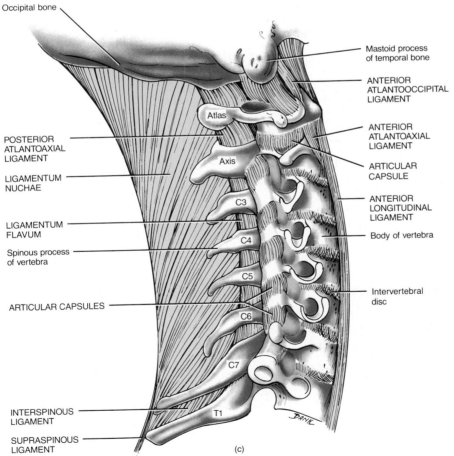

Occipital bone

Mastoid process
of temporal bone

ANTERIOR
ATLANTOOCCIPITAL
LIGAMENT

Atlas

ANTERIOR
ATLANTOAXIAL
LIGAMENT

POSTERIOR
ATLANTOAXIAL
LIGAMENT

Axis

ARTICULAR
CAPSULE

LIGAMENTUM
NUCHAE

ANTERIOR
LONGITUDINAL
LIGAMENT

C3

LIGAMENTUM
FLAVUM

Body of vertebra

C4

Spinous process
of vertebra

C5

ARTICULAR CAPSULES

Intervertebral
disc

C6

C7

INTERSPINOUS
LIGAMENT

T1

SUPRASPINOUS
LIGAMENT

(c)

FIGURE 8-8 (*Continued*) Intervertebral joints. (b) Photograph of an intervertebral disc between two adjacent vertebral bodies. (Courtesy of C. Yokochi and J. W. Rohen, *Photographic Anatomy of the Human Body,* 2nd ed., 1979, IGAKU-SHOIN, Ltd., Tokoyo, New York.) (c) Diagram of right lateral view.

EXHIBIT 8-6

Lumbosacral Joint (Figure 8-9)

DEFINITION	Joint formed by the body of the fifth lumbar vertebra and the superior surface of the first sacral vertebra of the sacrum.
TYPE OF JOINT	Joint between the bodies of the fifth lumbar vertebra and the first sacral vertebra—cartilaginous joint, symphysis type. Joint between the articular processes—synovial joint, gliding (arthrodial) type.
ANATOMICAL COMPONENTS	1. The lumbosacral joint is similar to the joints between typical vertebrae, united by an **intervertebral disc, anterior** and **posterior longitudinal ligaments, ligamenta flava, interspinous** and **supraspinous ligaments,** and **articular capsules** between articular processes. The lumbosacral joint also contains an iliolumbar ligament. 2. **Iliolumbar ligament.** Strong ligament that connects the transverse process of the fifth lumbar vertebra with the base of the sacrum and the iliac crest.

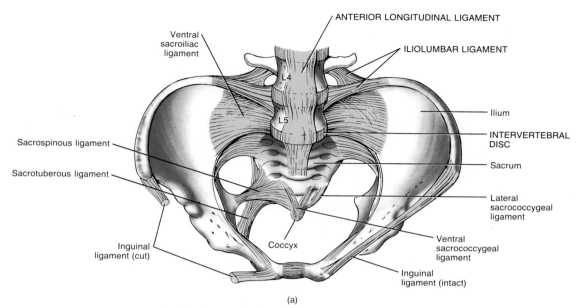

FIGURE 8–9 Lumbosacral joint. (a) Anterior view.

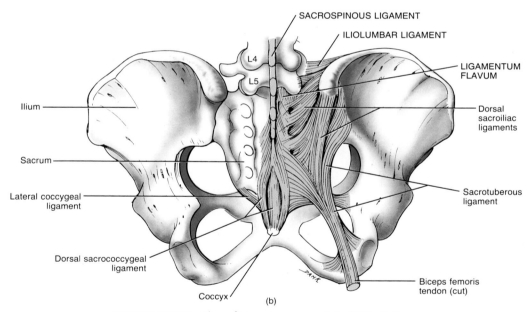

FIGURE 8-9 (*Continued*) Lumbosacral joint. (b) Posterior view.

EXHIBIT 8-7

Humeroscapular or Glenohumeral (Shoulder) Joint (Figure 8-10)

DEFINITION	Joint formed by the head of the humerus and the glenoid cavity of the scapula.
TYPE OF JOINT	Synovial joint, ball-and-socket (spheroid) type.

ANATOMICAL COMPONENTS	1. **Articular capsule.** Loose sac that completely envelopes the joint, extending from the circumference of the glenoid cavity to the anatomical neck of the humerus.
	2. **Coracohumeral ligament.** Strong, broad ligament that extends from the coracoid process of the scapula to the greater tubercle of the humerus.
	3. **Glenohumeral ligaments.** Three thickenings of the articular capsule over the ventral surface of the joint.
	4. **Transverse humeral ligament.** Narrow sheet extending from the greater tubercle to the lesser tubercle of the humerus.
	5. **Glenoid labrum.** Narrow rim of fibrocartilage around the edge of the glenoid cavity.
	6. Among the **bursae** associated with the shoulder joint are:
	   a. **Subscapular bursa** between the tendon of the subscapularis muscle and the underlying joint capsule.
	   b. **Subdeltoid bursa** between the deltoid muscle and joint capsule.
	   c. **Subacromial bursa** between the acromion and joint capsule.
	   d. **Subcoracoid bursa** either lies between the coracoid process and joint capsule or appears as an extension from the subacromial bursa.

CLINICAL APPLICATION

Dislocations of the shoulder joint are based on the relationship of the displaced head of the humerus to the glenoid cavity. Anterior displacement of the humeral head accounts for almost all shoulder dislocations.

Separation of the shoulder refers to dislocation of the acromioclavicular joint.

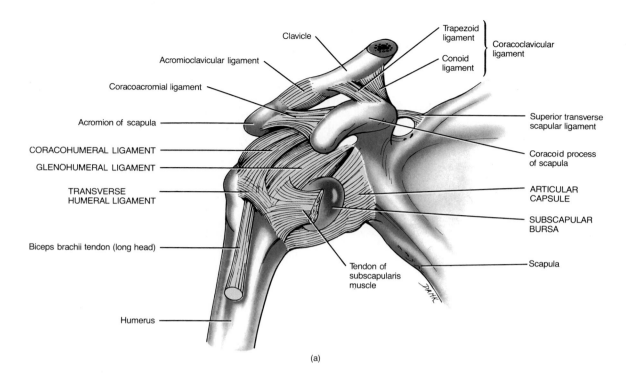

(a)

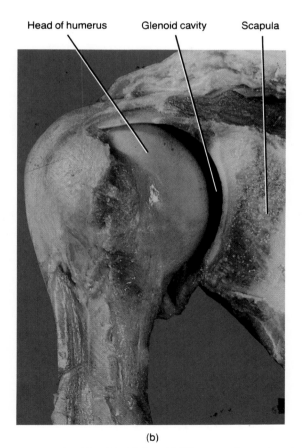

(b)

FIGURE 8-10 Humeroscapular (shoulder) joint. (a) Diagram of anterior view. (b) Photograph of anterior view. (Courtesy of C. Yokochi and J. W. Rohen, *Photographic Anatomy of the Human Body*, 2nd ed., 1979, IGAKU-SHOIN, Ltd., Tokyo, New York.)

EXHIBIT 8-8

Elbow Joint (Figure 8-11)

DEFINITION	Joint formed by the trochlea of the humerus, the trochlear notch of the ulna, and the head of the radius.
TYPE OF JOINT	Synovial joint, hinge (ginglymus) type.
ANATOMICAL COMPONENTS	1. **Articular capsule.** The anterior part covers the anterior part of the joint from the radial and coronoid fossae of the humerus to the coronoid process of the ulna and the annular ligament of the radius. The posterior part extends from the capitulum, olecranon fossa, and lateral epicondyle of the humerus to the annular ligament of the radius, the olecranon of the ulna, and the ulna posterior to the radial notch. 2. **Ulnar collateral ligament.** Thick, triangular ligament that extends from the medial epicondyle of the humerus to the coronoid process and olecranon of the ulna. 3. **Radial collateral ligament.** Strong, triangular ligament that extends from the lateral epicondyle of the humerus to the annular ligament of the radius and the radial notch of the ulna.

CLINICAL APPLICATION

Tennis elbow most commonly refers to pain at or near the lateral epicondyle of the humerus, usually caused by an improperly executed backhand. In this maneuver, the extensor muscles, especially the extensor carpi radialis brevis muscle, strain or sprain. Also, the extensors rub and roll over the lateral epicondyle and head of the radius, resulting in pain.

Little-league elbow typically develops as a result of a heavy pitching schedule, especially among youngsters.

The disorder occurs mostly on the medial side of the elbow, in which there is some degree of involvement of the medial epicondylar epiphysis that may cause it to enlarge, fragment, or separate. Disorders on the lateral side of the elbow are more severe and chronic and are caused by the repeated jamming of the radial head against the capitulum of the humerus, which results in necrosis of the head of the radius, osteochondritis, and osteoarthritis.

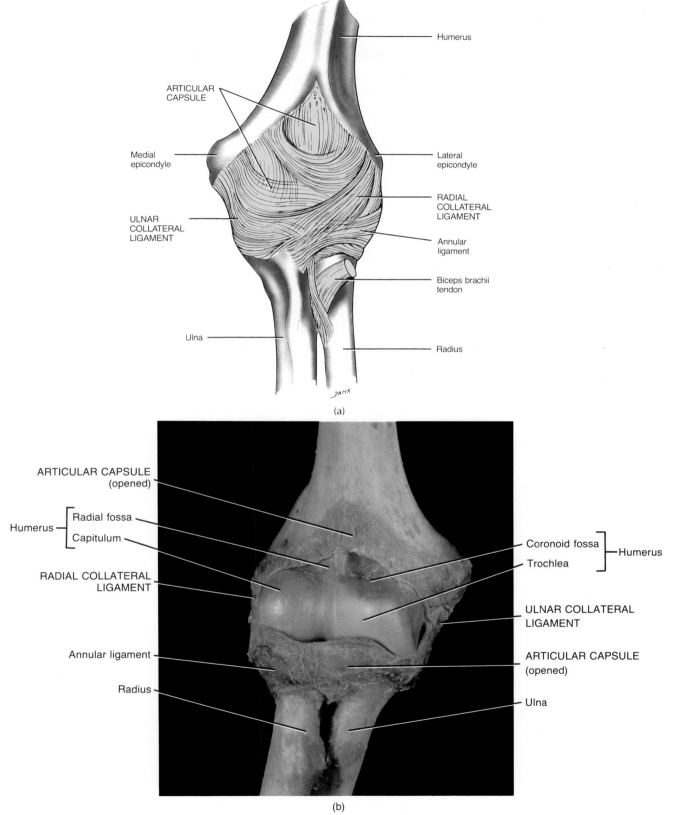

FIGURE 8-11 Elbow joint. Anterior view. (a) Diagram. (b) Photograph. (Courtesy of J. A. Gosling, P. F. Harris, et al., *Atlas of Human Anatomy,* Gower Medical Publishing Ltd., 1985.)

EXHIBIT 8-9

Radiocarpal (Wrist) Joint (Figure 8-12)

DEFINITION	Joint formed by the distal end of the radius, the distal surface of the articular disc separating the carpal and distal radioulnar joint, and the scaphoid, lunate, and triquetral carpal bones.
TYPE OF JOINT	Synovial joint, ellipsoidal (condyloid) type.
ANATOMICAL COMPONENTS	1. **Articular capsule.** Extends from the styloid processes of the ulna and radius to the proximal row of carpal bones. 2. **Palmar radiocarpal ligament.** Thick, strong ligament that extends from the anterior border of the distal end of the radius to the proximal row of carpals. Some longer fibers extend to the capitate in the distal row. 3. **Dorsal radiocarpal ligament.** Extends from the posterior border of the distal end of the radius to the proximal row of carpals, especially the triquetral. 4. **Ulnar collateral ligament.** Extends from the styloid process of the ulna to the triquetral and pisiform bones and the transverse carpal ligament. 5. **Radial collateral ligament.** Extends from the styloid process of the radius to the scaphoid and pisiform bones.

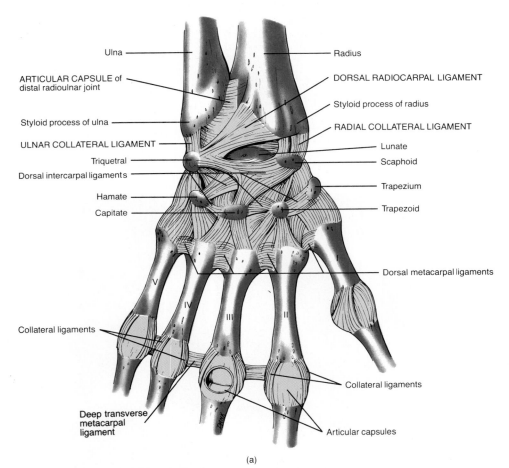

(a)

FIGURE 8-12 Radiocarpal (wrist) joint. (a) Dorsal view.

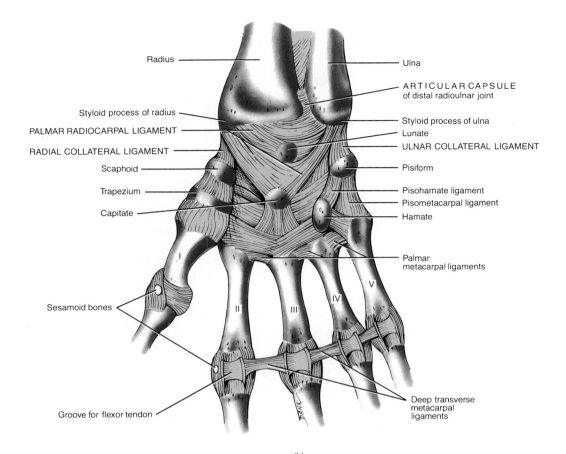

Radius

Ulna

ARTICULAR CAPSULE
of distal radioulnar joint

Styloid process of radius

Styloid process of ulna

PALMAR RADIOCARPAL LIGAMENT

Lunate

RADIAL COLLATERAL LIGAMENT

ULNAR COLLATERAL LIGAMENT

Scaphoid

Pisiform

Trapezium

Pisohamate ligament

Capitate

Pisometacarpal ligament

Hamate

Palmar
metacarpal ligaments

Sesamoid bones

Deep transverse
metacarpal
ligaments

Groove for flexor tendon

(b)

FIGURE 8-12 (*Continued*) Radiocorpal (wrist) joint. (b) Palmar view.

EXHIBIT 8-10

Coxal (Hip) Joint (Figure 8-13)

DEFINITION	Joint formed by the head of the femur and the acetabulum of the coxal bone.
TYPE OF JOINT	Synovial, ball-and-socket (spheroid) type.
ANATOMICAL COMPONENTS	1. **Articular capsule.** Extends from the rim of the acetabulum to the neck of the femur. One of the strongest ligaments of the body, the capsule consists of circular and longitudinal fibers. The circular fibers, called the **zona orbicularis,** form a collar around the neck of the femur. The longitudinal fibers are reinforced by accessory ligaments known as the iliofemoral ligament, pubofemoral ligament, and ischiofemoral ligament. 2. **Iliofemoral ligament.** Thickened portion of the articular capsule that extends from the anterior inferior iliac spine of the coxal bone to the intertrochanteric line of the femur. 3. **Pubofemoral ligament.** Thickened portion of the articular capsule that extends from the pubic part of the rim of the acetabulum to the neck of the femur. 3. **Ischiofemoral ligament.** Thickened portion of the articular capsule that extends from the ischial wall of the acetabulum to the neck of the femur. 5. **Ligament of the head of the femur (capitate ligament).** Flat, triangular band that extends from the fossa of the acetabulum to the head of the femur. 6. **Acetabular labrum.** Fibrocartilage rim attached to the margin of the acetabulum. 7. **Transverse ligament of the acetabulum.** Strong ligament that crosses over the acetabular notch, converting it to a foremen. It supports part of the acetabular labrum and is connected with the ligament of the head of the femur and the articular capsule.

CLINICAL APPLICATION

Dislocation of the hip among adults is quite rare because of (1) the stability of the ball-and-socket joint, (2) the strong, tough, articular capsule, (3) the strength of the intracapsular ligaments, and (4) the extensive musculature over the joint.

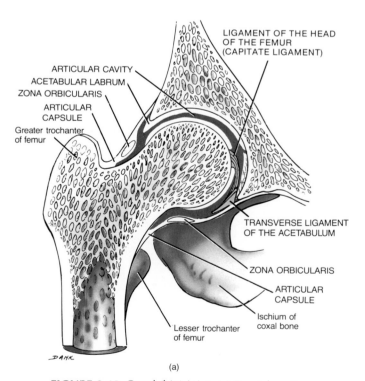

(a)

FIGURE 8-13 Coxal (hip) joint. (a) Frontal section.

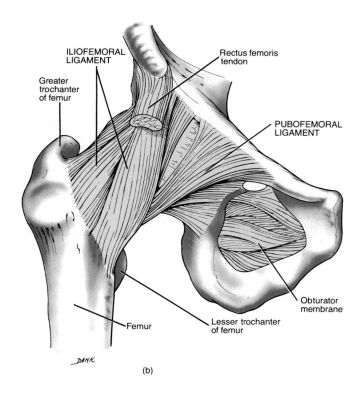

(b)

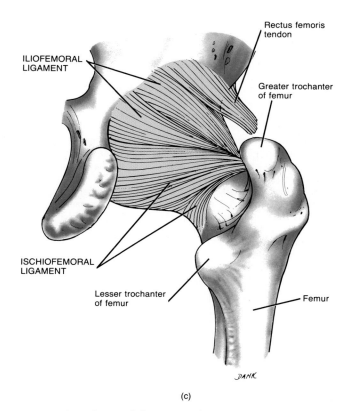

(c)

FIGURE 8-13 (*Continued*) Coxal (hip) joint. (b) Anterior view. (c) Posterior view.

EXHIBIT 8-11

Tibiofemoral (Knee) Joint (Figure 8-14)

DEFINITION	The largest joint of the body, actually consisting of three joints: (1) an intermediate patellofemoral joint between the patella and the patellar surface of the femur; (2) a lateral tibiofemoral joint between the lateral condyle of the femur, lateral meniscus, and lateral condyle of the tibia; and (3) a medial tibiofemoral joint between the medial condyle of the femur, medial meniscus, and medial condyle of the tibia.
TYPE OF JOINT	Patellofemoral joint—partly synovial, gliding (arthrodial) type. Lateral and medial tibiofemoral joints—synovial, hinge (ginglymus) type.
ANATOMICAL COMPONENTS	1. **Articular capsule.** No complete, independent capsule unites the bones. The ligamentous sheath surrounding the joint consists mostly of muscle tendons or expansions of them. There are, however, some capsular fibers connecting the articulating bones.

    2. **Medial and lateral patellar retinacula.** Fused tendons of insertion of the quadriceps femoris muscle and the fascia lata that strengthen the anterior surface of the joint.

    3. **Patellar ligament.** Central portion of the common tendon of insertion of the quadriceps femoris muscle that extends from the patella to the tibial tuberosity. This ligament also strengthens the anterior surface of the joint. The posterior surface of the ligament is separated from the synovial membrane of the joint by an **infrapatellar fat pad.**

    4. **Oblique popliteal ligament.** Broad, flat ligament that extends from the intercondylar fossa of the femur to the head of the tibia. The tendon of the semimembranous muscle is superficial to the ligament and passes from the medial condyle of the tibia to the lateral condyle of the femur. The ligament and tendon afford strength for the posterior surface of the joint.

    5. **Arcuate popliteal ligament.** Extends from the lateral condyle of the femur to the styloid process of the head of the fibula. It strengthens the lower lateral part of the posterior surface of the joint.

    6. **Tibial collateral ligament.** Broad, flat ligament on the medial surface of the joint that extends from the medial condyle of the femur to the medial condyle of the tibia. The ligament is crossed by tendons of the sartorius, gracilis, and semitendinosus muscles, all of which strengthen the medial aspect of the joint.

    7. **Fibular collateral ligament.** Strong, rounded ligament on the lateral surface of the joint that extends from the lateral condyle of the femur to the lateral side of the head of the fibula. The ligament is covered by the tendon of the biceps femoris muscle. The tendon of the popliteal muscle is deep to the tendon.

    8. **Intraarticular ligaments.** Ligaments within the capsule that connect the tibia and femur.

        a. **Anterior cruciate ligament.** Extends posteriorly and laterally from the area anterior to the intercondylar eminence of the tibia to the posterior part of the medial surface of the lateral condyle of the femur. This ligament is stretched or torn in about 70 percent of all serious knee injuries. Both O. J. Simpson and Gale Sayers had their careers in professional football ended by torn anterior cruciate ligaments.

        b. **Posterior cruciate ligament.** Extends anteriorly and medially from the posterior intercondylar fossa of the tibia and lateral meniscus to the anterior part of the medial surface of the medial condyle of the femur.

    9. **Articular discs.** Fibrocartilage discs between the tibial and femoral condyles. They help to compensate for the incongruence of the articulating bones.

        a. **Medial meniscus.** Semicircular piece of fibrocartilage (C shaped). Its anterior end is attached to the anterior intercondylar fossa of the tibia, in front of the anterior cruciate ligament. Its posterior end is attached to the posterior intercondylar fossa of the tibia between the attachments of the posterior cruciate ligament and lateral meniscus.

        b. **Lateral meniscus.** Nearly circular piece of fibrocartilage (approaches an incomplete O in shape). Its anterior end is attached anterior to the intercondylar eminence of the tibia and lateral and posterior to the anterior cruciate ligament. Its posterior end is attached posterior to the intercondylar eminence of the tibia and anterior to the posterior end of the medial meniscus. The medial and lateral menisci are connected to each other by the **transverse ligament** and to the margins of the head of the tibia by the **coronary ligaments.**

  10. The principal **bursae** of the knee include the following:

        a. **Anterior bursae:** (1) between the patella and skin **(prepatellar bursa),** (2) between upper part of tibia and patellar ligament **(infrapatellar bursa),** (3) between lower part of tibial tuberosity and skin, and (4) between lower part of femur and deep surface of quadriceps femoris muscle **(suprapatellar bursa).**

        b. **Medial bursae:** (1) between medial head of gastrocnemius muscle and the articular capsule, (2) superficial to the tibial collateral ligament between the ligament and tendons of the sartorius, gracilis, and semitendinosus muscles, (3) deep to the tibial collateral ligament between the ligament and the tendon of the semimembranosus muscle, (4) between the tendon of the semimembranosus muscle and the head of the tibia, and (5) between the tendons of the semimembranosus and semitendinosus muscles.

        c. **Lateral bursae:** (1) between the lateral head of the gastrocnemius muscle and articular capsule, (2) between the tendon of the biceps femoris muscle and fibular collateral ligament, (3) between the tendon of the popliteal muscle and fibular collateral ligament, and (4) between the lateral condyle of the femur and the popliteal muscle.

CLINICAL APPLICATION

The most common type of **knee injury** in football is rupture of the tibial collateral ligament, often associated with tearing of the anterior cruciate ligament and medial meniscus (torn cartilage). It is caused by a blow to the lateral side of the knee. When a knee is examined for such an injury, the three C's are kept in mind: collateral ligament, cruciate ligament, and cartilage.

Runner's knee (patellofemoral syndrome) refers to an aching pain around the front of a runner's knee that may result from excessive pronation of the foot or tight hamstrings. The pain typically occurs after a person has been sitting for a while, especially after exercise. A common cause of runner's knee is constantly walking, running, or jogging on the same side of the road. Since roads are high in the middle and slope down on the sides, the slope stresses the knee that is closer to the center of the road.

A **swollen knee** may occur immediately or be delayed. The immediate swelling is due to escape of blood from rupture of the anterior cruciate ligament, torn menisci, fractures, or collateral ligament sprains. Delayed swelling is due to an excessive production of synovial fluid as a result of conditions that irritate the synovial membrane.

A **dislocated knee** refers to the displacement of the tibia relative to the femur. Accordingly, such dislocations are classified as anterior, posterior, medial, lateral, or rotatory. The most common type is anterior dislocation, resulting from hyperextension of the knee. A frequent consequence of a dislocated knee is damage to the popliteal artery.

In November 1987, a group of orthopedic surgeons at the Hospital of the University of Pennsylvania performed the **first transplant of an entire human knee.** The surgery was necessitated by a potentially malignant tumor on the knee of a 32-year-old female recipient. The donor was an 18-year-old male. Once the tumor was removed, the patient's knee joint was removed and replaced with the donor's joint. The donor knee joint, about 41 cm (16 in.) long, consisted of the lower portion of a femur that was connected to the recipient's femur by a rod; the upper portion of the tibia and head of the fibula that were connected by a metal plate; the patella; medial and lateral menisci; intracapsular and extracapsular ligaments; and certain tendons. The recipient's own muscles, nerves, and blood vessels were used. Since rejection occurs in only about 3 to 5 percent of bone transplant cases, no immunosupressive drugs were needed. Although the recipient will not be permitted to participate in strenuous exercise and contact sports, she is expected to have near-normal use of her leg.

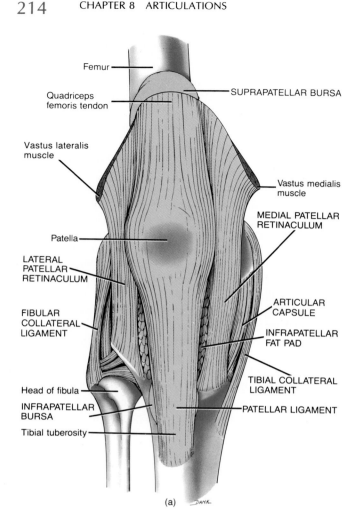

(a)

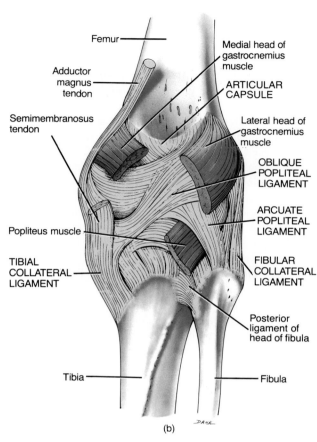

(b)

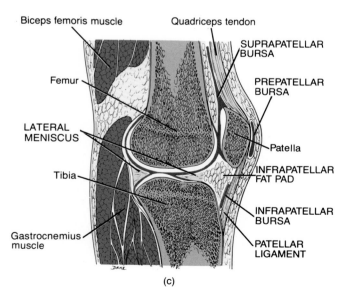

(c)

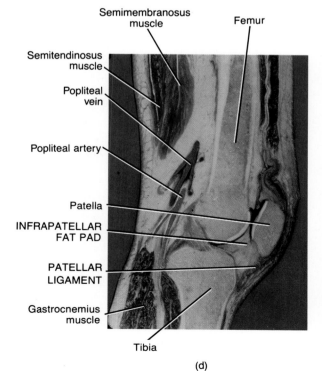

(d)

FIGURE 8-14 Tibiofemoral (knee) joint. (a) Diagram of anterior view. (b) Diagram of posterior view. (c) Diagram of sagittal section. (d) Photograph of sagittal section. (Courtesy of C. Yokochi and J. W. Rohen, *Photographic Anatomy of the Human Body,* 1969, IGAKU-SHOIN, Ltd., Tokyo, New York.)

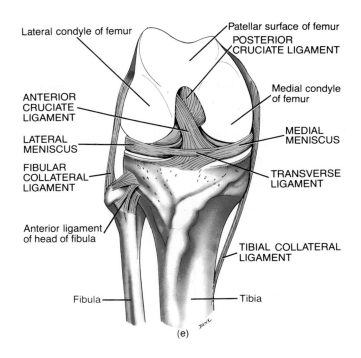

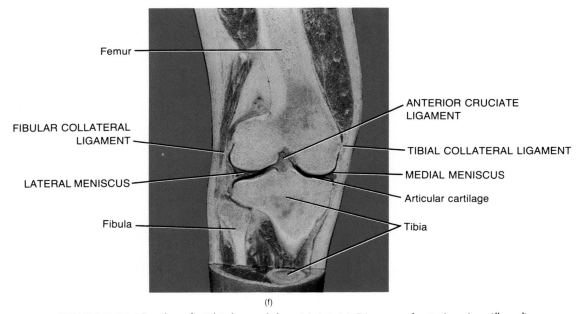

FIGURE 8–14 (*Continued*) Tibiofemoral (knee) joint. (e) Diagram of anterior view (flexed). (f) Photograph of frontal section. (Courtesy of C. Yokochi and J. W. Rohen, *Photographic Anatomy of the Human Body*, 2nd ed., 1979, IGAKU-SHOIN, Ltd., Tokyo, New York.)

EXHIBIT 8-12

Talocrural (Ankle) Joints (Figure 8-15)

DEFINITION	Joints between (1) the distal end of the tibia and its medial malleolus and the talus and (2) the lateral malleolus of the fibula and the talus.
TYPE OF JOINT	Both joints—synovial, hinge (ginglymus) type.
ANATOMICAL COMPONENTS	1. **Articular capsule.** Surrounds the joint and extends from the borders of the tibia and malleoli to the talus. 2. **Deltoid (medial) ligament.** Strong, triangular ligament that extends from the medial malleolus of the tibia to the navicular, calcaneus, and talus. 3. **Anterior talofibular ligament.** Extends from the anterior margin of the lateral malleolus of the fibula to the talus. 4. **Posterior talofibular ligament.** Extends from the posterior margin of the lateral malleolus of the fibula to the talus. 5. **Calcaneofibular ligament.** Extends from the apex of the lateral malleolus of the fibula to the talus. Together, the anterior talofibular, posterior talofibular, and calcaneofibular ligaments are referred to as the **lateral ligament.**

CLINICAL APPLICATION

The most common **ankle sprain** is a lateral sprain. A person is more likely to turn the foot inward rather than outward due to normal adduction that occurs during running and the fact that the lateral ligament is weaker than the deltoid (medial) ligament. Most lateral ankle sprains involve the anterior talofibular ligament and frequently occur on irregular surfaces.

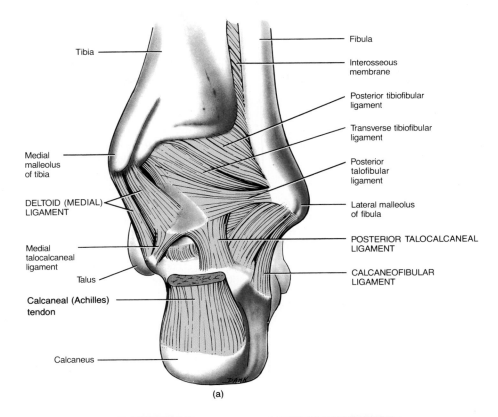

(a)

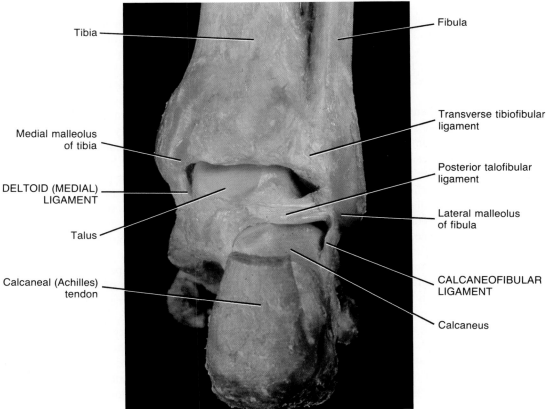

(b)

FIGURE 8-15 Talocrural (ankle) joints. Posterior view. (a) Diagram. (b) Photograph. (Courtesy of J. A. Gosling, P. F. Harris, et al., *Atlas of Human Anatomy*, Gower Medical Publishing Ltd., 1985.)

APPLICATIONS TO HEALTH

RHEUMATISM

Rheumatism (*rheumat* = subject to flux) refers to any painful state of the supporting structures of the body—its bones, ligaments, joints, tendons, or muscles. Arthritis is a form of rheumatism in which the joints have become inflamed.

ARTHRITIS

The term **arthritis** refers to many different diseases, the most common of which are rheumatoid arthritis (RA), osteoarthritis, and gouty arthritis. All are characterized by inflammation in one or more joints. Inflammation, pain, and stiffness may also be present in adjacent parts of the body, such as the muscles near the joint. The chronic pain that accompanies various arthritic conditions may be related to the patient's inability to produce endorphins, naturally produced painkillers. The causes of arthritis are unknown.

Rheumatoid Arthritis (RA)

Rheumatoid (ROO-ma-toyd) **arthritis** (**RA**) is the most common inflammatory form of arthritis. It is an autoimmune disease in which the body attacks its own cartilage and joint linings. It is characterized by inflammation of the joint, swelling, pain, and a loss of function. Usually this form occurs bilaterally—if your left knee is affected, your right knee may also be affected, although usually not to the same degree.

The primary symptom of rheumatoid arthritis is inflammation of the synovial membrane. If untreated, the following sequential pathology may occur. The membrane thickens and synovial fluid accumulates. The resulting pressure causes pain and tenderness. The membrane then produces an abnormal granulation tissue called **pannus,** which adheres to the surface of the articular cartilage. The pannus formation sometimes erodes the cartilage completely. When the cartilage is destroyed, fibrous tissue joins the exposed bone ends. The tissue ossifies and fuses the joint so that it is immovable—the ultimate crippling effect of rheumatoid arthritis. Most cases do not progress to this stage, but the range of motion of the joint is greatly inhibited by the severe inflammation and swelling.

Damaged joints may be surgically replaced, either partly or entirely, with artificial joints. The artificial parts are inserted after removal of the diseased portion of the articulating bone and its cartilage. The new metal or plastic joint is fixed in place with a special acrylic element. When freshly mixed in the operating room, it hardens as strong as bone in minutes. These new parts function nearly as well as a normal joint—and much better than a diseased joint.

Osteoarthritis

A degenerative joint disease far more common than rheumatoid arthritis, and usually less damaging, is **osteoarthritis** (os'-tē-ō-ar-THRĪ-tis). It apparently results from a combination of aging, irritation of the joints, and wear and abrasion.

Degenerative joint disease is a noninflammatory, progressive disorder of movable joints, particularly weight-bearing joints. It is characterized by the deterioration of articular cartilage and by formation of new bone in the subchondral areas and at the margins of the joint. The cartilage slowly degenerates, and as the bone ends become exposed, small bumps, or **spurs,** of new osseous tissue are deposited on them. These spurs decrease the space of the joint cavity and restrict joint movement. Unlike rheumatoid arthritis, osteoarthritis usually affects only the articular cartilage. The synovial membrane is rarely destroyed, and other tissues are unaffected.

Gouty Arthritis

Uric acid is a waste product produced during the metabolism of nucleic acids. Normally, all the acid is quickly excreted in the urine. In fact, it gives urine its name. The person who suffers from **gout** either produces excessive amounts of uric acid or is not able to excrete normal amounts. The result is a buildup of uric acid in the blood. This excess acid then reacts with sodium to form a salt called sodium urate. Crystals of this salt are deposited in soft tissues. Typical sites are the kidneys and the cartilage of the ears and joints.

In **gouty** (GOW-tē) **arthritis,** sodium urate crystals are deposited in the soft tissues of the joints. The crystals irritate the cartilage, causing inflammation, swelling, and acute pain. Eventually, the crystals destroy all the joint tissues. If the disorder is not treated, the ends of the articulating bones fuse and the joint becomes immovable.

Gouty arthritis occurs primarily in males of any age. It is believed to be the cause of 2–5 percent of all chronic joint diseases. Numerous studies indicate that gouty arthritis is sometimes caused by an abnormal gene. As a result of this gene, the body manufactures unusually large amounts of uric acid. Diet and environmental factors such as stress and climate are also suspected causes of gouty arthritis.

Although other forms of arthritis cannot be treated with complete success, the treatment of gouty arthritis with the use of various drugs has been quite effective. A chemical called colchicine has been utilized periodically since the sixth century to relieve pain, swelling, and tissue destruction that occur during attacks of gouty arthritis. This chemical is derived from the variety of crocus plant from which the spice saffron is obtained. Other drugs, which either inhibit uric acid production or assist in the elimination of excess uric acid by the kidneys, are used to prevent further attacks. The drug allopurinol is used for the treatment of gouty arthritis because it prevents the formation of uric acid without interfering with nucleic acid synthesis.

BURSITIS

An acute chronic inflammation of a bursa is called **bursitis.** The condition may be caused by trauma, by an acute or chronic infection (including syphilis and tuberculosis), or by rheumatoid arthritis. Repeated excessive friction often results in a bursitis with local inflammation and the accumulation of fluid. Bunions are frequently associated with a friction bursitis over the head of the first metatarsal bone. Symptoms include pain, swelling, tenderness, and the limitation of motion involving the inflamed bursa. The prepatellar or subcutaneous infrapatellar bursa may become inflamed in individuals who spend a great deal of time kneeling. This bursitis is usually called *"housemaid's knee"* (*"carpet layer's knee"*).

Aspiration of the knee joint may be necessary to relieve pressure, to evacuate blood, or to obtain a fluid sample for laboratory studies. In this procedure, the needle is inserted somewhat proximal and lateral to the patella through the tendinous part of the vastus lateralis muscle and is directed toward the middle of the joint. Through the same route, the joint cavity may be anesthetized or irrigated, and steroids may be administered in the treatment of knee pathologies.

DISLOCATION

A **dislocation** or **luxation** (luks-Ā-shun) is the displacement of a bone from a joint with tearing of ligaments, tendons, and articular capsules. a partial or incomplete dislocation is called a **subluxation.** The most common dislocations are those involving a finger or shoulder. Those of the mandible, elbow, knee, or hip are less common. Symptoms include loss of motion, temporary paralysis of the involved joint, pain, swelling, and occasional shock. A dislocation is usually caused by a blow or fall, although unusual physical effort may lead to this condition.

SPRAIN AND STRAIN

A **sprain** is the forcible wrenching or twisting of a joint with partial rupture or other injury to its attachment without luxation. It occurs when the attachments are stressed beyond their normal capacity. There may be damage to the associated blood vessels, muscles, tendons, ligaments, or nerves. A sprain is more serious than a **strain,** which is the overstretching of a muscle. Severe sprains may be so painful that the joint cannot be moved. There is considerable swelling, with reddish to blue discoloration due to hemorrhage from ruptured blood vessels. The ankle joint is most often sprained; the low back area is another frequent location for sprains.

DMSO stands for **dimethyl sulfoxide** (Rimso-50). It is a clear, colorless, and practically odorless liquid that has been used as an industrial solvent since the 1940s. Although the FDA has approved the use of a 50 percent solution of DMSO for the treatment of interstitial cystitis, a specific urinary bladder disorder, a 90 percent solution (approved for veterinary use) and a 99 percent solution (used as an industrial solvent) are also being used as an anti-inflammatory, without FDA approval, to treat human disorders such as chronic arthritis, sprains, and strains.

The FDA points to the potential for side effects of DMSO since the drug is rapidly absorbed from the surface of the skin into the bloodstream. Among the side effects are allergic reactions (erythema and itching), headache, nausea, diarrhea, burning on urination, and disturbances in vision. Users also experience an oysterlike taste in their mouths and a garliclike body odor for up to 72 hours after administration. Lens opacities (cloudiness) have also been reported in laboratory animals. Studies of the effectiveness of DMSO for human disorders are now underway.

KEY MEDICAL TERMS ASSOCIATED WITH ARTICULATIONS

Ankylosis (ang'-ki-LŌ-sis; *ankyle* = stiff joint; *osis* = condition) Severe or complete loss of movement at a joint.

Arthralgia (ar-THRAL-jē-a; *arth* = joint; *algia* = pain) Pain in a joint.

Arthrosis (ar-THRŌ-sis) Refers to an articulation; also a disease of a joint.

Bursectomy (bur-SEK-tō-mē; *ectomy* = removal of) Removal of a bursa.

Chondritis (kon-DRĪ-tis; *chondro* = cartilage) Inflammation of cartilage.

Rheumatology (roo'-ma-TOL-ō-jē; *rheumat* = subject to flux) The medical specialty devoted to arthritis.

Synovitis (sin'-ō-VĪ-tis; *synov* = joint) Inflammation of a synovial membrane in a joint.

STUDY OUTLINE

Classification (p. 186)

1. An articulation (joint) is a point of contact between two or more bones.

2. Functional classification of joints is based on the degree of movement permitted. Joints may be synarthroses, amphiarthroses, or diarthroses.

3. Structural classification is based on the presence of a synovial (joint) cavity and type of connecting tissue. Structurally, joints are classified as fibrous, cartilaginous, or synovial.

Fibrous Joints (p. 186)

1. Bones held by fibrous connective tissue, with no synovial cavity, are fibrous joints.
2. These joints include immovable sutures (found in the skull), slightly movable syndesmoses (such as the tibiofibular articulation), and immovable gomphoses (roots of teeth in alveoli of mandible and maxilla).

Cartilaginous Joints (p. 187)

1. Bones held together by cartilage, with no synovial cavity, are cartilaginous joints.
2. These joints include immovable synchondroses united by hyaline cartilage (temporary cartilage between diaphysis and epiphyses) and partially movable symphyses united by fibrocartilage (the symphysis pubis).

Synovial Joints (p. 187)

1. Synovial joints contain a synovial cavity, articular cartilage, and a synovial membrane; some also contain ligaments, articular discs, and bursae.
2. All synovial joints are freely movable.
3. Movements at synovial joints are limited by the structure of articulating bones, tension of ligaments, muscle tension, and apposition of soft parts.
4. Types of movements at synovial joints include gliding movements, angular movements, rotation, circumduction, and special movements.
5. Types of synovial joints include gliding joints (between wrist bones), hinge joints (elbow), pivot joints (radioulnar), ellipsoidal joints (radiocarpal), saddle joints (carpometacarpal), and ball-and-socket joints (shoulder and hip).
6. A joint may be described according to the number of planes of movement it allows as nonaxial, monaxial, biaxial, or triaxial.

Selected Articulations of the Body (p. 197)

1. Temporomandibular joint (TMJ) is between the mandible and temporal bone.
2. Atlanto-occipitals are between the atlas and occipital bones.
3. Intervertebrals are between vertebral bodies and between vertebral arches.
4. Lumbosacral is between the fifth lumbar vertebra and the first sacral vertebra of the sacrum.
5. Humeroscapular (glenohumeral) is between the humerus and scapula.
6. Elbow is between the humerus and ulna and between the humerus and radius.
7. Radiocarpal is between the radius and scaphoid, lunate, and triquetral carpal bones.
8. Coxal is between the femur and coxal (hip) bone.
9. Tibiofemoral is between the patella and femur and between the femur and tibia.
10. Talocrural is between the tibia and talus and between the fibula and talus.

Applications to Health (p. 218)

1. Rheumatism is a painful state of supporting body structures such as bones, ligaments, tendons, joints, and muscles.
2. Arthritis refers to several disorders characterized by inflammation of joints, often accompanied by stiffness of adjacent structures.
3. Rheumatoid arthritis (RA) refers to inflammation of a joint accompanied by pain, swelling, and loss of function.
4. Osteoarthritis is a degenerative joint disease characterized by deterioration of articular cartilage and spur formation.
5. Gouty arthritis is a condition in which sodium urate crystals are deposited in the soft tissues of joints and eventually destroy the tissues.
6. Bursitis is an acute or chronic inflammation of bursae.
7. A dislocation, or luxation, is a displacement of a bone from its joint; a partial dislocation is called subluxation.
8. A sprain is the forcible wrenching or twisting of a joint with partial rupture to its attachments without dislocation, while a strain is the stretching of a muscle.

REVIEW QUESTIONS

1. Define an articulation. What factors determine the degree of movement at joints?
2. Distinguish among the three kinds of joints on the basis of structure and function. List the subtypes. Be sure to include degree of movement and specific examples.
3. Explain the components of a synovial joint. Indicate the relation of ligaments and tendons to the strength of the joint and restrictions on movement.
4. Explain how the articulating bones in a synovial joint are held together.
5. What is an accessory ligament? Define the two principal types.
6. What is an articular disc? Why are they important?
7. Describe the principle and importance of anthroscopy.
8. What are bursae? What is their function?
9. Define the following principal movements: gliding, angular, rotation, circumduction, and special. Name a joint where each occurs.
10. Have another person assume the anatomical position and execute each of the movements at joints discussed in the text. Reverse roles, and see if you can execute the same movements.
11. Contrast nonaxial, monaxial, biaxial, and triaxial planes of movement. Give examples of each, and name a joint at which each occurs.
12. For each joint of the body discussed in Exhibits 8-3–8-12, be sure that you can name the bones that form the joint, identify the joint by type, and list the anatomical components of the joint.
13. Distinguish between rheumatoid arthritis (RA), osteoarthritis, and gouty arthritis with respect to causes and symptoms.
14. Define bursitis. How is it caused?
15. Define dislocation. What are the symptoms of dislocation?
16. Distinguish between a sprain and a strain.
17. Refer to the glossary of key medical terms associated with articulations. Be sure that you can define each term.

1. It serves as a storehouse for water and particularly for fat. Much of the fat of an overweight person is in the superficial fascia.
2. It forms a layer of insulation protecting the body from loss of heat.
3. It provides mechanical protection from blows.
4. It provides a pathway for nerves and vessels.

Deep fascia is by far the most extensive of the three types. It is a dense connective tissue composed of a superficial layer (external investing layer) and a deep layer (internal investing layer). The deep fascia lines the body wall and extremities and holds muscles together, separating them into functioning groups. Functionally, deep fascia allows free movement of muscles, carries nerves and blood vessels, fills spaces between muscles, and sometimes provides the origin for muscles.

Subserous (visceral) fascia is located between the internal investing layer of deep fascia and a serous membrane. It is composed of loose connective tissue. It forms the fibrous layer of serous membranes, covers and supports the viscera, and attaches the parietal layer of serous membranes to the internal surface of the body wall.

CLINICAL APPLICATION

Lines of fusion of fascial sheets are essentially avascular and for this reason they are often sites for *surgical incisions*. Surgeons also prefer fascial junctional areas for anchoring of sutures because of the strength of fasciae and the strong union that results from wound healing.

Skeletal muscles are further protected, strengthened, and attached to other structures by several other connective tissue coverings (Figure 9-1). The entire muscle is usually wrapped with a substantial quantity of fibrous connective tissue called the *epimysium* (ep'-i-MĪZ-ē-um). Bundles of fibers (cells) called *fasciculi* (fa-SIK-yoo-lī) or *fascicles* (FAS-i-kuls) are covered by a fibrous connective tissue called the *perimysium* (per'-i-MĪZ-ē-um). *Endomysium* (en'-dō-MĪZ-ē-um) is a fibrous connective tissue that penetrates into the interior of each fascicle and separates the muscle fibers. All three connective tissue coverings are extensions of deep fascia.

Epimysium, perimysium, and endomysium are all continuous with the connective tissue that attaches the muscle

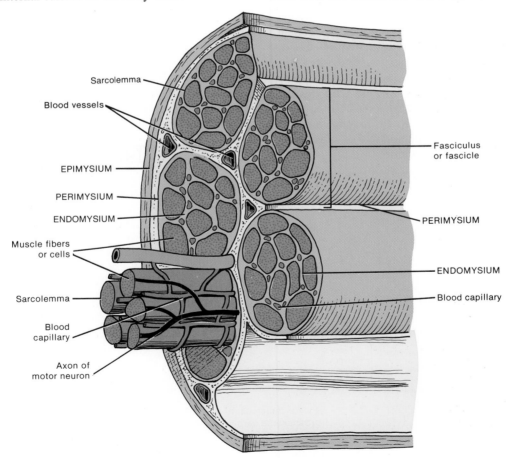

FIGURE 9-1 Relationships of connective tissue to skeletal muscle. Shown is a cross section and longitudinal section of a portion of a skeletal muscle indicating the relative positions of the epimysium, perimysium, and endomysium. Compare this figure with the photomicrograph in Figure 9-4b.

to another structure, such as bone or other muscle. All three elements may be extended beyond the muscle fibers (cells) as a *tendon*—a cord of connective tissue that attaches a muscle to the periosteum of a bone. When the connective tissue elements extend as a broad, flat layer, the tendon is called an *aponeurosis*. This structure also attaches to the coverings of a bone or another muscle. An example of an aponeurosis is the galea aponeurotica (see Figure 10-4). When a muscle contracts, the tendon and its corresponding bone or muscle are pulled toward the contracting muscle. In this way skeletal muscles produce movement. Certain tendons, especially those of the wrist and ankle, are enclosed by tubes of fibrous connective tissue called *tendon (synovial) sheaths*. They are similar in structure to bursae. The inner layer of a tendon sheath, the *visceral layer,* is applied to the surface of the tendon. The outer layer is known as the *parietal layer* (Figure 9-2). Between the layers is a cavity that contains a film of synovial fluid. Tendon sheaths permit tendons to slide back-and-forth more easily.

CLINICAL APPLICATION

Tendinitis or *tenosynovitis* (ten'-ō-sin-ō-VĪ-tis) frequently occurs as inflammation involving the tendon sheaths and synovial membrane surrounding certain joints. The wrists, shoulders, elbows (tennis elbow), finger joints (trigger finger), ankles, and associated tendons are most often affected. The affected sheaths may become visibly swollen because of fluid accumulation or they may remain dry. Local tenderness is variable, and there may be disabling pain with movement of the body part. The condition often follows some form of trauma, strain, or excessive exercise.

NERVE AND BLOOD SUPPLY

Skeletal muscles are well supplied with nerves and blood vessels. The innervation and vascularization are directly related to contraction, the chief characteristic of muscle. For a skeletal muscle fiber (cell) to contract, it must first be stimulated by an impulse from a nerve cell. Muscle contraction also requires a good deal of energy and therefore large amounts of nutrients and oxygen. Moreover, the waste products of these energy-producing reactions must be eliminated. Thus, prolonged muscle action depends on a rich blood supply to deliver nutrients and oxygen and remove wastes.

Generally, an artery and one or two veins accompany each nerve that penetrates a skeletal muscle. The larger branches of the blood vessels accompany the nerve branches through the connective tissue of the muscle (Figure 9-3). Microscopic blood vessels called capillaries are distributed within the endomysium. Each muscle fiber is thus in close contact with one or more capillaries. Each skeletal muscle fiber usually makes contact with a portion of a nerve cell called a synaptic end-bulb.

HISTOLOGY

When a typical skeletal muscle is teased apart and viewed microscopically, it can be seen to consist of thousands of elongated, cylindrical cells called *muscle fibers* or *myofibers* (see Exhibit 3-3). These fibers lie parallel to one another and range from 10 to 100 μm in diameter. Some fibers may reach lengths of 30 cm (12 in.) or more. Each muscle fiber is enveloped by a plasma membrane called the *sarco-*

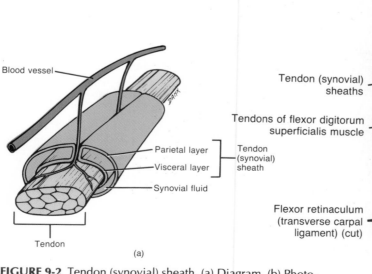

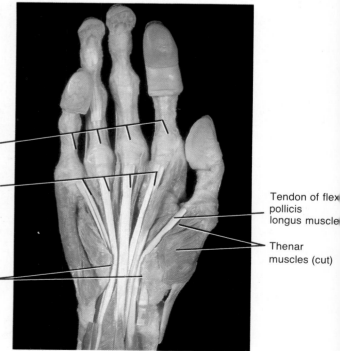

FIGURE 9-2 Tendon (synovial) sheath. (a) Diagram. (b) Photograph. (Courtesy of J. A. Gosling, P. F. Harris, et al., *Atlas of Human Anatomy,* Gower Medical Publishing Ltd., 1985.)

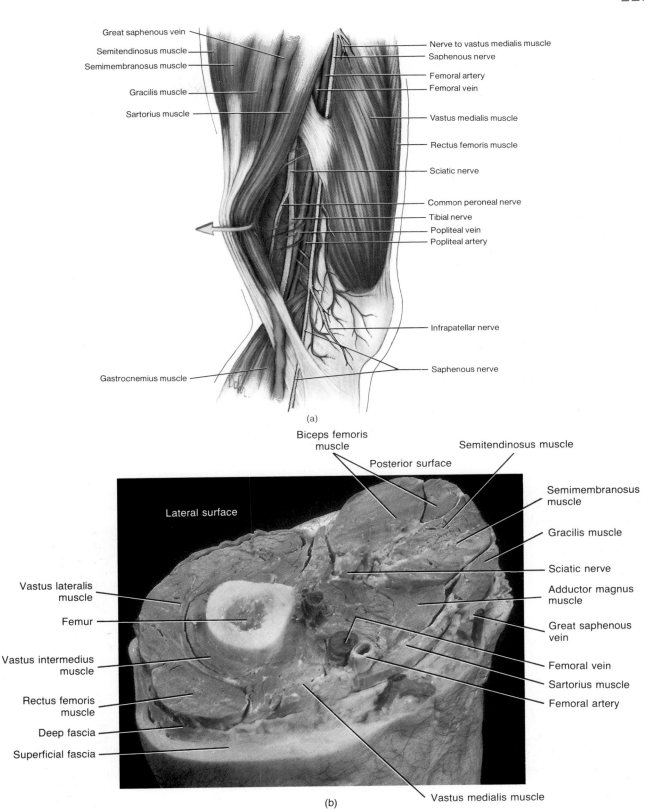

FIGURE 9-3 Relation of blood vessels and nerves to skeletal muscles. (a) Diagram of the left thigh and knee in medial view. (b) Photograph of a cross section through the midportion of the thigh. (Courtesy of J. A. Gosling, P. F. Harris, et al., *Atlas of Human Anatomy*, Gower Medical Publishing Ltd., 1985.)

lemma (*sarco* = flesh; *lemma* = sheath). The sarcolemma surrounds a quantity of cytoplasm called *sarcoplasm.* Within the sarcoplasm of a muscle fiber and lying close to the sarcolemma are many nuclei. Skeletal muscle fibers are thus multinucleate. The sarcoplasm also contains myofibrils (described shortly), special high-energy molecules, enzymes, and *sarcoplasmic reticulum* (sar'-kō-PLAZ-mik re-TIK-yoo-lum), a network of membrane-enclosed tubules comparable to smooth endoplasmic reticulum (Figure 9-4a). Dilated sacs of sarcoplasmic reticulum, called *terminal cisterns,* form ringlike channels around myofibrils. Run-

ning transversely through the fiber and perpendicularly to the sarcoplasmic reticulum are *transverse tubules (T tubules).* The turbules are extensions of the sarcolemma that open to the outside of the fiber. A *triad* consists of a transverse tubule and the segments of sarcoplasmic reticulum (terminal cisterns) on either side.

A highly magnified view of skeletal muscle fibers reveals that they are composed of cylindrical structures, about 1 or 2 μm in diameter, called *myofibrils* (Figure 9-4a,b). Myofibrils, ranging in number from several hundred to several thousand, run longitudinally through the muscle fiber

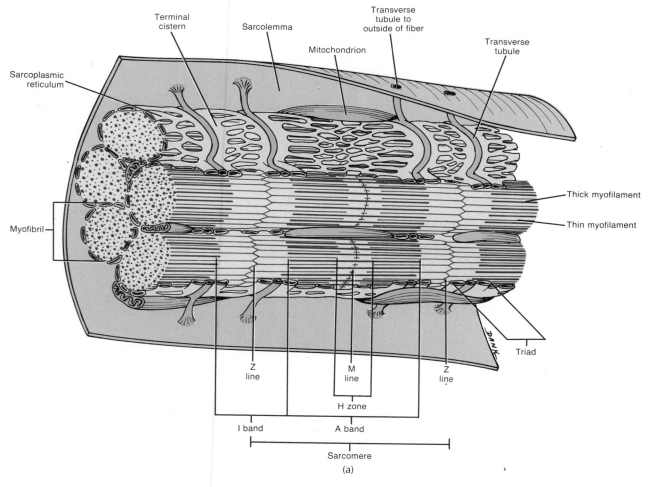

FIGURE 9-4 Histology of skeletal muscle tissue. (a) Enlarged aspect of several myofibrils of a muscle fiber based on an electron micrograph. (b) Enlarged aspect of a sarcomere showing thin and thick myofilaments. (c) Electron micrograph of several sarcomeres at a magnification of 35,000X. (Courtesy of D. E. Kelly, from *Introduction to the Musculoskeletal System* by Cornelius Rosse and D. Kay Clawson, Harper & Row, Publishers, Inc., New York, 1970.)

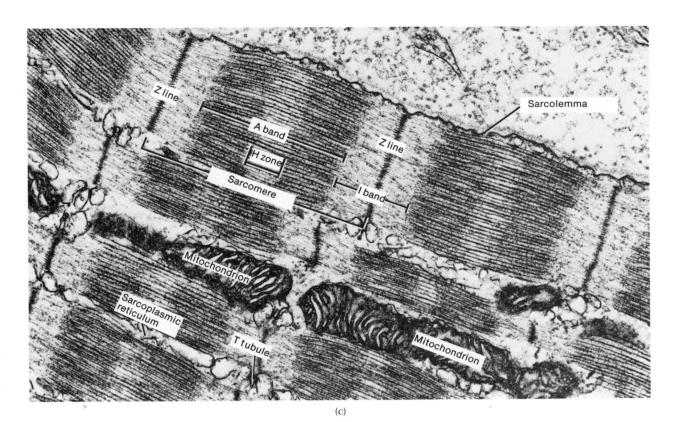

(c)

and consist of two kinds of even smaller structures called *myofilaments.* The *thin myofilaments* are about 5 nm in diameter. The *thick myofilaments* are about 16 mm in diameter.

The myofilaments of a myofibril do not extend the entire length of a muscle fiber—they are arranged in compartments called *sarcomeres.* Sarcomeres are separated from one another by narrow zones of dense material called *Z lines.* Within a sarcomere, certain areas can be distinguished (Figure 9-4b). A dark dense area, called the *anisotropic* or *A band,* represents the length of thick myofilaments. The sides of the A band are darkened by the overlapping of thick and thin myofilaments. The length of darkening depends on the extent of overlapping. As you will see later, the greater the degree of contraction, the greater the overlapping of thick and thin myofilaments. A light-colored, less dense area called the *isotropic band,* or *I band,* is composed of thin myofilaments only. This combination of alternating dark A bands and light I bands gives the muscle fiber its striated (striped) appearance. A narrow *H zone* contains thick myofilaments only. In the center of the H zone is the *M line,* a series of fine threads that appear to connect the middle parts of adjacent thick myofilaments.

Thin myofilaments are anchored to the Z lines and project in both directions. They are composed mostly of the protein *actin.* The actin molecules are arranged in two single strands that entwine helically and give the thin myofilaments their characteristic shape (Figure 9-5a). Each actin molecule contains a *myosin-binding site* that interacts with a cross bridge

of a myosin molecule (described shortly). Besides actin, the thin myofilaments contain two other protein molecules, *tropomyosin* and *troponin,* that are involved in the regulation of muscle contractions. Tropomyosin is arranged in strands that are loosely attached to the actin helices. Troponin is located at regular intervals on the surface of tropomyosin and is made up of three subunits: troponin I, which binds to actin, troponin C, which binds to calcium ions, and troponin T, which binds to tropomyosin. Together, tropomyosin and troponin are referred to as the *tropomyosin–troponin complex.*

Thick myofilaments overlap the free ends of the thin myofilaments and occupy the A band region of a sarcomere. These myofilaments are composed mostly of the protein *myosin.* A myosin molecule is shaped like a golf club. The tails (handles of the golf club) are arranged parallel to each other forming the shaft of the thick myofilament. The heads of the golf clubs project outward from the shaft and are arranged spirally on the surface of the shaft. The projecting heads are referred to as *cross bridges* and contain an *actin-binding site* and an *ATP-binding site* (Figure 9-5b).

CLINICAL APPLICATION

Steroids are chemical substances derived from cholesterol. Most steroids are hormones. At the IX Pan American Games in Caracas, Venezuela, held in 1983, worldwide attention was focused on the use of *muscle-building*

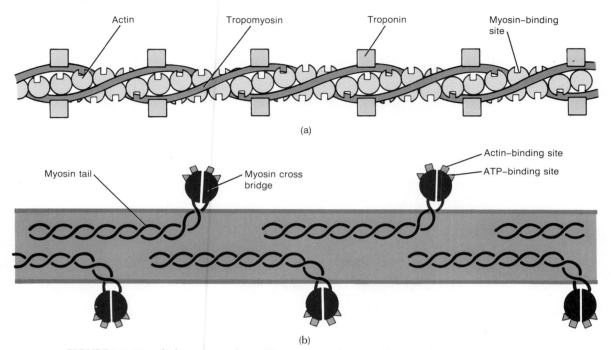

FIGURE 9-5 Detailed structure of myofilaments. (a) Thin myofilament. (b) Thick myofilament.

anabolic steroids by amateur athletes. These steroids, a derivative of the hormone testosterone, were used by the athletes supposedly to build muscle proteins and therefore increase strength and endurance during athletic events. However, physicians point out that use of anabolic steroids may cause a number of side effects, including liver cancer, kidney damage, increased risk of heart disease, muscle spasm, increased irritability and aggressive behavior, psychotic symptoms (including hallucinations), manic episodes, major depression, and mood swings. In females additional side effects include sterility, the development of facial hair and deepening of the voice, atrophy of the breasts and uterus, enlargement of the clitoris, and irregularities of menstruation. In males, additional side effects include testicular atrophy, excessive development of breast glands and diminished hormone secretion and sperm production by the testes.

CONTRACTION

SLIDING-FILAMENT THEORY

During muscle contraction, thin myofilaments slide inward toward the H zone. The sarcomere shortens, but the lengths of the thin and thick myofilaments do not change. The myosin cross bridges of the thick myofilaments connect with portions of actin of the thin myofilaments. The myosin cross bridges move like the oars of a boat on the surface of the thin myofilaments, and the thin and thick myofilaments slide past each other. As the thin myofilaments move past

the thick myofilaments, the H zone narrows and even disappears when the thin myofilaments meet at the center of the sarcomere (Figure 9-6). In fact, the myosin cross bridges may pull the thin myofilaments of each sarcomere so far inward that their ends overlap. As the thin myofilaments slide inward, the Z lines are drawn toward each other and the sarcomere is shortened. The sliding of myofilaments and shortening of sarcomeres cause the shortening of the muscle fibers. All these events associated with the movement of myofilaments are known as the *sliding-filament theory* of muscle contraction.

NEUROMUSCULAR JUNCTION (MOTOR END-PLATE)

For a skeletal muscle fiber to contract, a stimulus must be applied to it. The stimulus is delivered by a nerve cell, or *neuron.* A neuron has a threadlike process called a fiber, or axon, that may run 91 cm (3 ft) or more to a muscle. A bundle of such fibers from many different neurons composes a nerve. A neuron that stimulates muscle tissue is called a *motor neuron.*

On entering a skeletal muscle, the axon of a motor neuron branches into *axon terminals* (*telodendria*) that come into close approximation with a portion of the sarcolemma of a muscle fiber. The term *neuromuscular function, myoneural function (MNJ),* or *motor end-plate* refers to the axon terminal of a motor neuron together with the portion of the sarcolemma of a muscle fiber in close approximation with the axon terminal (Figure 9-7). Close examination of a neuromuscular junction reveals that the distal ends of

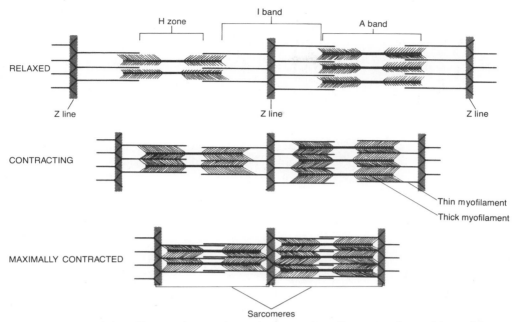

FIGURE 9-6 Sliding-filament theory of muscle contraction. Shown are the positions of the various parts of two sarcomeres in relaxed, contracting, and maximally contracted states. Note the movement of the thin myofilaments and the relative sizes of the I band and H zones.

the axon terminals are expanded into bulblike structures called *synaptic end-bulbs.* The bulbs contain membrane-enclosed sacs, the *synaptic vesicles,* that store chemicals called *neurotransmitters.* These chemicals determine whether a nerve impulse is passed on to a muscle (or gland or another nerve cell). The invaginated area of the sarcolemma under the axon terminal is referred to as a *synaptic gutter (trough),* and the space between the axon terminal and sarcolemma is known as a *synaptic cleft.* There are numerous folds of the sarcolemma along the synaptic gutter, called *subneural clefts,* which greatly increase the surface area of the synaptic gutter.

When a nerve impulse (nerve action potential) reaches an axon terminal, it initiates a sequence of reactions that liberate neurotransmitter molecules from synaptic vesicles (and possibly the cytoplasm as well). The neurotransmitter released at neuromuscular junctions is *acetylcholine* (as'-ē-til-KŌ-lēn), or *ACh.* Upon its release, ACh diffuses across the synaptic cleft and combines with receptors on the sarcolemma of the muscle fiber. This combination alters the permeability of the sarcolemma and ultimately results in the development of a muscle action potential that travels along the sarcolemma, thus initiating the events leading to contraction.

MOTOR UNIT

A motor neuron, together with all the muscle fibers it stimulates, is referred to as a *motor unit.* A single motor neuron may innervate about 150 muscle fibers, depending on the region of the body. This means that stimulation of one neuron will tend to cause the simultaneous contraction of about 150 muscle fibers. In addition, all the muscle fibers of a motor unit that are sufficiently stimulated will contract and relax together. Muscles that control precise movements, such as the extrinsic (external) eye muscles, have fewer than 10 muscle fibers to each motor unit. Muscles of the body that are responsible for gross movements, such as the biceps brachii and gastrocnemius, may have as many as 2000 muscle fibers in each motor unit.

Stimulation of a motor neuron produces a contraction in all the muscle fibers in a particular motor unit. Accordingly, the total tension in a muscle can be varied by adjusting the number of motor units that are activated. The process of increasing the number of active motor units is called *recruitment* and is determined by the needs of the body at a given time. The various motor neurons to a given muscle fire asynchronously; that is, while some are excited, others are inhibited. This means that while some motor units are active, others are inactive; all the motor units are not contracting at the same time. This pattern of firing of motor neurons prevents fatigue while maintaining contraction by allowing a brief rest for the inactive units. The alternating motor units relieve one another so smoothly that the contraction can be sustained for long periods. It also helps maintain a state of partial contraction in a muscle, a phenomenon called muscle tone (described later in the chapter). Also, recruitment is one factor responsible for producing smooth movements during a muscle contraction, rather than a series of jerky movements.

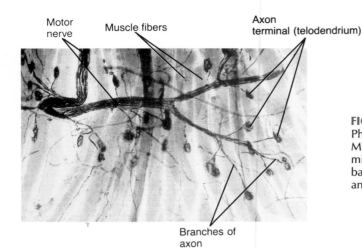

Motor nerve

Muscle fibers

Axon terminal (telodendrium)

Branches of axon

(a)

FIGURE 9-7 Neuromuscular junction (motor end plate) (a) Photomicrograph at a magnification of 400×. (© 1983 by Michael H. Ross. Used by permission.) (b) Scanning electron micrograph at a magnification of 1,650× (FUJIJA.) (c) Diagram based on a photomicrograph. (d) Enlarged aspect based on an electron micrograph.

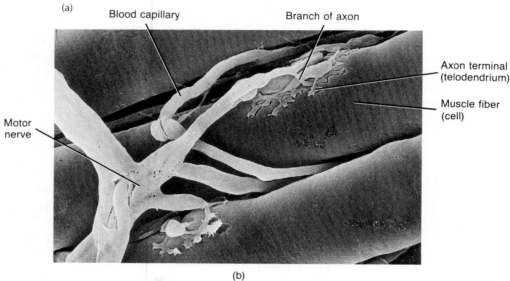

Blood capillary

Branch of axon

Axon terminal (telodendrium)

Muscle fiber (cell)

Motor nerve

(b)

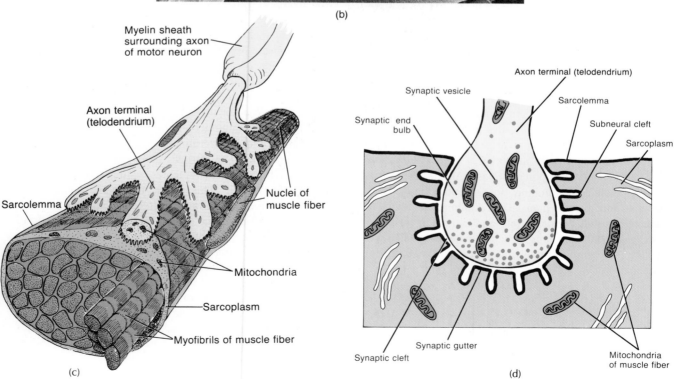

Myelin sheath surrounding axon of motor neuron

Axon terminal (telodendrium)

Sarcolemma

Nuclei of muscle fiber

Mitochondria

Sarcoplasm

Myofibrils of muscle fiber

(c)

Axon terminal (telodendrium)

Synaptic vesicle

Sarcolemma

Synaptic end bulb

Subneural cleft

Sarcoplasm

Synaptic cleft

Synaptic gutter

Mitochondria of muscle fiber

(d)

MECHANISM

Rather than discuss the details of the mechanism of muscle contraction, the process is outlined in Exhibit 9-1. By following the step-by-step summary, you should understand the basics of muscle contraction.

CLINICAL APPLICATION

Following death, certain chemical changes occur in muscle tissue that affect the status of the muscles. Because of a lack of ATP, myosin cross bridges remain attached to the actin myofilaments, thus preventing relaxation. The resulting condition, in which muscles are in a state of partial contraction, is called *rigor mortis* (rigidity of death). The time elapsing between death and the onset of rigor mortis varies greatly among individuals. Those who have had long, wasting illnesses undergo rigor mortis more quickly.

EXHIBIT 9-1

Summary of Events Involved in Contraction and Relaxation of a Skeletal Muscle Fiber

1. A nerve impulse (nerve action potential) causes synaptic vesicles in motor axon terminals to release acetylcholine (ACh).
2. Acetylcholine diffuses across the synaptic cleft within the neuromuscular junction and initiates a muscle action potential that spreads over the surface of the sarcolemma.
3. The muscle action potential enters the transverse tubules and sarcoplasmic reticulum and stimulates the sarcoplasmic reticulum to release calcium ions from storage into the sarcoplasm.
4. Calcium ions combine with troponin, causing the tropomyosin–troponin complex to move, thus exposing the myosin-binding sites on actin.
5. The muscle action potential also causes ATP → ADP + P. The released energy activates myosin cross bridges, which combine with the exposed myosin-binding sites on actin, and causes the myosin cross bridges to move toward the H zone. This movement results in the sliding of the thin myofilaments past thick myofilaments.
6. The sliding draws the Z lines toward each other, the sarcomere shortens, the muscle fibers contract, and the muscle contracts.
7. Acetylcholine is inactivated by acetylcholinesterase (AChE), thus terminating the muscle action potential. As a result, ACh no longer has any effect on the neuromuscular junction.
8. Once the muscle action potential is inhibited, calcium ions are actively transported back into the sarcoplasmic reticulum by calsequestrin and calcium ATPase, using energy from ATP breakdown.
9. The low calcium concentration in the sarcoplasm permits the tropomyosin–troponin complex to reattach to actin. As a result, myosin-binding sites of actin become covered, myosin cross bridges separate from actin, ADP is resynthesized into ATP (which reattaches to the ATP-binding site of the myosin cross bridge), and the thin myofilaments return to their relaxed position.
10. Sarcomeres return to their resting lengths, muscle fibers relax, and the muscle relaxes.

MUSCLE LENGTH AND FORCE OF CONTRACTION

Having already considered the sliding-filament theory of muscle contraction, we can now examine the relation between muscle length and force of contraction (tension). As you already know, a skeletal muscle fiber contracts when myosin cross birdges of thick myofilaments connect with portions of thin myofilaments within a sarcomere. As it turns out, a muscle fiber develops its greatest tension when there is maximum overlap between thick and thin myofilaments. At this length, the optimal length, the maximum number of myosin cross bridges make contact with thin myofilaments to bring about the greatest force of contraction. As a muscle fiber is stretched, fewer and fewer myosin cross bridges make contact with thin myofilaments and the force of contraction progressively decreases. In fact, if a muscle fiber is stretched to 175 percent its optimal length, no myosin cross bridges attach to thin myofilaments and no contraction occurs. At lengths less than the optimum length, the force of contraction also decreases. This is because extreme shortening of sarcomeres causes thin myofilaments to overlap and thick myofilaments to crumple as they run into Z lines, resulting in fewer myosin cross bridge contacts with thin myofilaments. In general, changes in resting muscle fiber length above or below the optimum length rarely exceed 30 percent.

ALL-OR-NONE PRINCIPLE

The weakest stimulus from a neuron that can still initiate a contraction is called a *threshold (liminal) stimulus.* A stimulus of lesser intensity, or one that cannot initiate contraction, is referred to as a *subthreshold (subliminal) stimulus.* According to the *all-or-none principle,* once a threshold stimulus is applied, individual muscle fibers of a motor unit will contact to their fullest extent or will not contact at all, provided conditions remain constant. In other words, *muscle fibers* do not partly contract. The principle does not mean that the entire muscle must be either fully relaxed or fully contracted because, of the many motor units that comprise the entire muscle, some are contracting and some are relaxing. Thus, the muscle as a whole can have graded contractions. The strength of contraction may be decreased by fatigue, lack of nutrients, or lack of oxygen.

MUSCLE TONE

A muscle may be in a state of partial contraction even though contraction of the muscle fibers is always complete. At any given time, some cells in a muscle are contracted while others are relaxed. This contraction tightens a muscle, but there may not be enough fibers contracting at the time to produce movement. Recruitment (asynchronous firing) allows the contraction to be sustained for long periods.

A sustained partial contraction of portions of a skeletal muscle in response to activation of stretch receptors results in *muscle tone.* Tone is essential for maintaining posture. For example, when the muscles in the back of the neck are in tonic contraction, they keep the head in the anatomical position and prevent it from slumping forward onto the chest, but they do not apply enough force to pull the head back into hyperextension. The degree of tone in a skeletal muscle is monitored by receptors in the muscle called *muscle spindles.* They provide feedback information on tone to the spinal cord and brain so that adjustments can be made (Chapter 20).

CLINICAL APPLICATION

The term *flaccid* (FLAK-sid or FLAS-sid) is applied to muscles with less than normal tone. Such a loss of tone may be the result of damage or disease of the nerve that conducts a constant flow of impulses to the muscle. If the muscle does not receive impulses for an extended period of time, it may progress from flaccidity to *atrophy* (AT-rō-fē), which is a state of wasting away. Individual muscle fibers decrease in size due to a progressive loss of myofibrils. Muscles may also become flaccid and atrophied if they are not used. Bedridden individuals and people with casts may experience atrophy because the flow of impulses to the inactive muscle is greatly reduced. If the nerve supply to a muscle is cut, it will undergo complete atrophy. In about 6 months to 2 years, the muscle will be one-quarter its original size and the muscle fibers will be replaced by fibrous tissue. The transition to fibrous tissue, when complete, cannot be reversed. Battery-operated transcutaneous muscle stimulations (TMSs) are used to maintain the strength of atrophied muscles immobilized by a cast by stimulating the muscles to contract. TMSs are also used to prevent atrophy of muscles whose motor nerves have been temporarily damaged by trauma or stroke. Some athletes even use TMSs to strengthen certain muscles.

Muscular hypertrophy (hī-PER-trō-fē) is the reverse of atrophy. It refers to an increase in the diameters of muscle fibers due to the production of more myofibrils, mitochondria, sarcoplasmic reticulum, nutrients (glycogen and triglycerides), and energy-supplying molecules (ATP and phosphocreatine). Hypertrophy results from exercises that build muscles. Hypertrophic muscles are capable of more forceful contractions. Weak muscular activity does not produce significant hypertrophy. It results from very forceful muscular activity or repetitive muscular activity at moderate levels. There is evidence that the number of muscle fibers does not increase after birth. During childhood, the increase in the *size* of muscle fibers appear to be at least partially under the control of growth hormone (GH) produced by the anterior pituitary gland. A further increase in the size of muscle fibers appears to be due to the hormone testosterone, produced by the testes. The influence of testosterone probably accounts for the generally larger muscles in males than females. More forceful muscular contractions, as in weight lifting, also contribute to generally larger muscles in males.

TYPES OF SKELETAL MUSCLE FIBERS

All skeletal muscle fibers are not alike in structure or function. For example, skeletal muscle fibers vary in color depending on their content of *myoglobin,* a reddish pigment similar to hemoglobin in blood. Myoglobin stores oxygen until needed by mitochondria, the organelles in which ATP generation occurs. Skeletal muscle fibers that have a high myoglobin content are referred to as *red muscle fibers.* Conversely, skeletal muscle fibers that have a low content of myoglobin are called *white muscle fibers.* Red muscle fibers are smaller in diameter than white muscle fibers, and red muscle fibers have more mitochondria and more blood capillaries. White muscle fibers have a more extensive sarcoplasmic reticulum than red muscle fibers.

Skeletal muscle fibers contract with different velocities, depending on their ability to split ATP. Faster contracting fibers have greater ability to split ATP. In addition, skeletal muscle fibers vary with respect to the metabolic processes they use to generate ATP. They also differ in terms of the onset of fatigue.

On the basis of various structural and functional characteristics, skeletal muscle fibers are classified into three types:

1. *Slow-twitch red fibers.* These fibers contain large amounts of myoglobin, many mitochondria, and many blood capillaries. Slow-twitch red fibers have a high capacity to generate ATP by oxidative metabolic processes. Such fibers also split ATP at a slow rate and, as a result, contraction velocity is slow. Slow-twitch red fibers are very resistant to fatigue.
2. *Fast-twitch red fibers.* These fibers contain very large amounts of myoglobin, very many mitochondria, and very many blood capillaries. Fast-twitch red fibers have a very high capacity for generating ATP by oxidative metabolic processes. Such fibers also split ATP at a very rapid rate and, as a result, contraction velocity is fast. Fast-twitch red fibers are resistant to fatigue, but not quite as much as slow-twitch red fibers.
3. *Fast-twitch white fibers.* These fibers have a low content of myoglobin, relatively few mitochondria, and relatively few blood capillaries. They do, however, contain large amounts of glycogen. Fast-twitch white fibers are geared to generate ATP by anaerobic metabolic processes, which are not able to supply skeletal muscle fibers continuously with sufficient ATP. Accordingly, these fibers fatigue easily, but they split ATP at a fast rate so that contraction velocity is fast.

Most skeletal muscles of the body are a mixture of all three types of skeletal muscle fibers, but their proportion varies depending on the usual action of the muscle. For example, postural muscles of the neck, back, and legs have a higher proportion of slow-twitch red fibers. Muscles of the shoulders and arms are not constantly active but are used intermittently, usually for short periods of time, to produce large amounts of tension such as in lifting and throwing. These muscles have a higher proportion of fast-twitch white fibers. Leg muscles not only support the body but are also used for walking and running. Such muscles have higher proportions of slow-twitch red fibers and fast-twitch white fibers.

Even though most skeletal muscles are a mixture of all three types of skeletal muscle fibers, all the skeletal muscle fibers of any one motor unit are all the same. In addition, the different skeletal muscle fibers in a muscle may be used in various ways, depending on need. For example, if only a weak contraction is needed to perform a task, only slow-twitch red fibers are activated by their motor units. If a stronger contraction is needed, the motor units of fast-twitch red fibers are activated. And, if a maximal contraction is required, motor units of fast-twitch white fibers are activated as well. Activation of various motor units is determined in the brain and spinal cord.

Although the number of the different skeletal muscle fibers does not change, the characteristics of those present can be altered. Various types of exercises can bring about changes in the fibers in a skeletal muscle. Endurance-type exercises, such as running or swimming, cause a gradual transformation of fast-twitch white fibers into fast-twitch red fibers. The transformed muscle fibers show a slight increase in diameter, mitochondria, blood capillaries, and strength. Endurance exercises result in cardiovascular and respiratory changes that cause skeletal muscles to receive better supplies of oxygen and carbohydrates but do not contribute to muscle mass. On the other hand, exercises that require great strength for short periods of time, such as weight lifting, produce an increase in the size and strength of fast-twitch white fibers. The increase in size is due to increased synthesis of thin and thick myofilaments. The overall result is that the person develops large muscles.

CLINICAL APPLICATION

Whenever an action potential passes over a skeletal muscle, a small portion of the action potential also spreads to the skin. The recording and study of electrical changes that occur in muscle tissue is called *electromyography* or *EMG* (*electro* = electricity; *myo* = muscle; *graph* = to write). The record of the electrical changes is known as an *electromyogram.* To record electromyograms, needles or wire electrodes are inserted directly into muscles or electrodes are placed on the skin. Electromyograms may be used to determine abnormalities in muscles, to ascertain which muscles control various movements, to reveal our command of individual motor units, and as part of biofeedback studies.

CARDIAC MUSCLE TISSUE

The principal constituent of the heart wall is *cardiac muscle tissue.* Although it is striated in appearance like skeletal muscle, it is involuntary. The fibers of cardiac muscle tissue are roughly quadrangular and usually have only a single centrally located nucleus (see Exhibit 3-3). Skeletal muscle fibers contain several nuclei that are peripherally located. The thin sarcolemma of cardiac muscle fibers is similar to that of skeletal muscle, but the sarcoplasm is more abundant and the mitochondria are larger and more numerous. Cardiac muscle fibers have the same arrangement of actin and myosin and the same bands, zones, and lines as skeletal muscle fibers (Figure 9-8). Myofilaments in cardiac muscle fibers are not arranged in discrete myofibrils as in skeletal muscle. The transverse tubules of mammalian cardiac muscle are larger than those of skeletal muscle and are located at the Z lines rather than at the A–I band junctions as in skeletal muscle fibers. The sarcoplasmic reticulum of cardiac muscle is less well developed than that in skeletal muscle.

Note in Exhibit 3-3 that cardiac muscle fibers branch and interconnect with each other. Recall that skeletal muscle fibers are arranged in parallel fashion. Cardiac muscle fibers are also physically connected to each other by cell junctions called *gap junctions,* which permit a direct conduction of muscle action potentials from one fiber to another (see Figure 3-1e). Cardiac muscle fibers form two separate networks. The muscular walls and partition of the upper chambers of the heart (atria) compose one network. The muscular walls and partition of the lower chambers of the heart (ventricles) compose the other network. Each fiber in a network is separated from the next fiber by an irregular transverse thickening of the sarcolemma called an *intercalated* (in-TER-ka-lāt-ed) *disc.* These discs strengthen cardiac muscle tissue and aid in the conduction of muscle action potentials from one muscle fiber to another. When a single fiber of either network is stimulated, all the fibers in the network become stimulated as well. Thus, each network contracts as a functional unit. As you will see in Chapter 13, when the fibers of the atria contract as a unit, blood moves into the ventricles. Then, when the ventricular fibers contract as a unit, blood is pumped into arteries.

Under normal resting conditions, cardiac muscle tissue contracts and relaxes rapidly, continuously, and rhythmically about 75 times a minute without stopping. This is a major physiological difference between cardiac and skeletal muscle tissue. Accordingly, cardiac muscle tissue requires a constant supply of oxygen. Energy generation occurs in large, numerous mitochondria. Another difference is the source of stimulation. Skeletal muscle tissue ordinarily contracts only when stimulated by a nerve impulse. In contrast, cardiac muscle tissue can contract without extrinsic (outside) nerve stimulation. Its source of stimulation is a conducting tissue of specialized intrinsic (internal) muscle within the heart. Nerve stimulation merely causes the conducting tissue to increase or decrease its rate of discharge. Some types of smooth muscle fibers and nerve cells in the brain and spinal cord also possess spontaneous, rhythmical, self-excitation, a phenomenon referred to as *auto-rhythmicity.*

Another difference between cardiac and skeletal muscle tissue is that cardiac muscle tissue remains contracted (depo-

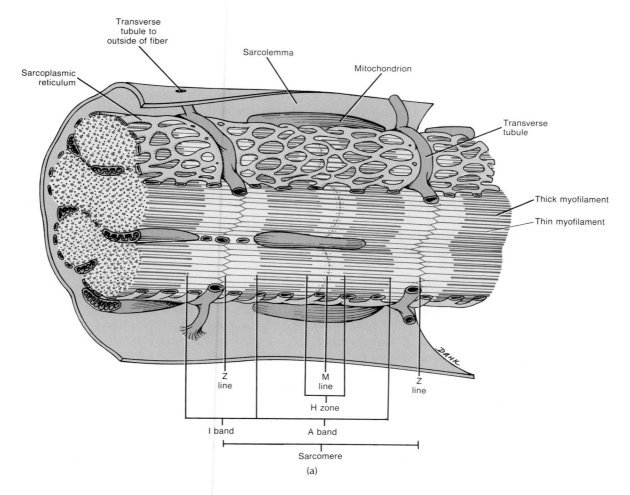

Transverse
tubule to
outside of fiber

Sarcolemma

Mitochondrion

Sarcoplasmic
reticulum

Transverse
tubule

Thick myofilament

Thin myofilament

DANK

Z
line

M
line

Z
line

H zone

I band

A band

Sarcomere

(a)

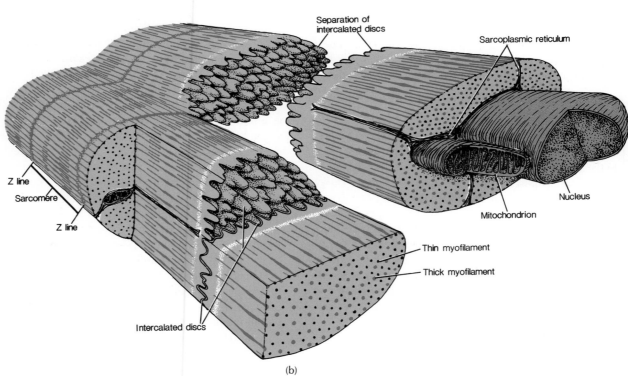

Separation of
intercalated discs

Sarcoplasmic reticulum

Z line

Sarcomere

Z line

Nucleus

Mitochondrion

Thin myofilament

Thick myofilament

Intercalated discs

(b)

FIGURE 9-8 Histology of cardiac muscle tissue. (a) Diagram based on an electron micrograph showing several myofibrils. (b) Diagram based on an electron micrograph showing intercalated discs and related structures.

larized) 10–15 times longer than skeletal muscle tissue. Cardiac muscle tissue also has an extra-long refractory period that allows time for the heart to relax between beats.

SMOOTH MUSCLE TISSUE

Like cardiac muscle tissue, *smooth muscle tissue* is usually involuntarily. However, it is nonstriated. Smooth muscle fibers are considerably smaller than skeletal muscle fibers. A single fiber of smooth muscle tissue is about 5–10 μm in diameter and 30–200 μm long. Each fiber is thickest in the midregion and tapers at each end. Within the fiber is a single, oval, centrally located nucleus (Figure 9-9 and see Exhibit 3-3). The sarcoplasm of smooth muscle fibers contains thick myofilaments (longer than those of skeletal muscle fibers) and thin myofilaments, but not arranged as orderly as in striated muscle. In smooth muscle fibers, there are 10–15 thin myofilaments for each thick myofilament in the regions of myofilament overlap; in skeletal muscle fibers, the ratio is 2:1. Smooth muscle fibers also contain *intermediate filaments.* Since the various myofilaments have no regular pattern of organization and since there are no A or I bands of sarcomeres, smooth muscle fibers have no characteristic striations—thus the name smooth.

Intermediate filaments are attached to structures called *dense bodies,* which have characteristics similar to Z lines in striated muscle fibers. Some dense bodies are dispersed throughout the cytoplasm; others are attached to the sarcolemma. Bundles of intermediate filaments stretch from one dense body to another (Figure 9-9b). The sliding filament mechanism involving thick and thin myofilaments during contraction generates tension that is transmitted to intermediate filaments. These, in turn, pull on the dense bodies attached to the sarcolemma and there is a lengthwise shortening of the muscle fiber. Note in Figure 9-9b that shortening of the muscle fiber produces a bubblelike expansion of the sarcolemma. Evidence suggests that a smooth muscle fiber contracts in a corkscrewlike manner; the fiber twists in a helix as it shortens and rotates in the opposite direction as it lengthens.

Smooth muscle fibers also possess a less well-developed sarcoplasmic reticulum than skeletal muscle fibers. Small subsurface vesicles termed *caveolae* open onto the surface of the smooth muscle fiber and are thought to function in a way similar to the transverse tubules of striated fibers, namely, to carry muscle action potentials into the fibers. Tubules of the sarcoplasmic reticulum are associated with longitudinal rows of caveolae.

Two kinds of smooth muscle tissue, visceral and multiunit, are recognized (Figure 9-9a). The more common type is called *visceral (single-unit) muscle tissue.* It is found in wraparound sheets that form part of the walls of small arteries and veins and hollow viscera such as the stomach, intestines, uterus, and urinary bladder. The terms *smooth muscle tissue* and *visceral muscle tissue* are sometimes used interchangeably. The fibers in visceral muscle tissue are tightly bound together to form a continuous network. They contain gap junctions to facilitate muscle action potential conduction between fibers. When a neuron stimulates one fiber, the muscle action potential travels over the other fibers so that contraction occurs in a wave over many adjacent fibers. Whereas skeletal muscle fibers contract as individual units, visceral muscle cells contract in sequence as the impulse spreads from one cell to another.

The second kind of smooth muscle tissue, *multiunit smooth muscle tissue,* consists of individual fibers each with its own motor-nerve endings. Whereas stimulation

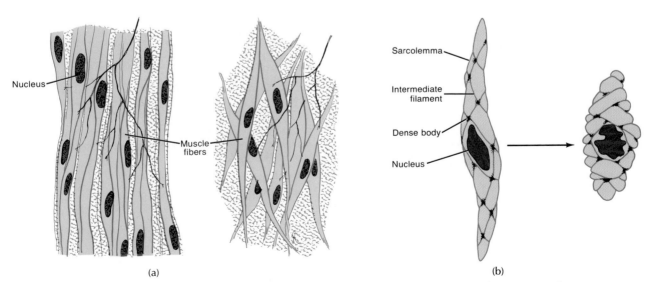

FIGURE 9-9 Histology of smooth muscle tissue. (a) Diagram of visceral (single-unit) smooth muscle tissue (left) and multiunit smooth muscle tissue (right). (b) Details of a fiber before contraction (left) and after contraction (right).

of a single visceral muscle fiber causes contraction of many adjacent fibers, stimulation of a single multiunit fiber causes contraction of only that fiber. In this respect, multiunit muscle tissue is like skeletal muscle tissue. Multiunit smooth muscle tissue is found in the walls of large arteries, large airways to the lungs, in the arrector pili muscles that attach to hair follicles, and in the intrinsic (internal) muscles of the eye, such as the iris.

Although the principles of contraction are essentially the same in smooth and striated muscle tissue, smooth muscle tissue exhibits several important physiological differences. First of all, the duration of contraction and relaxation of smooth muscle fibers is about 5–500 times longer than in skeletal muscle fibers. Smooth muscle tissue can also undergo sustained, long-term tone, which is important in the gastrointestinal tract where the walls of the tract maintain a steady pressure on the contents of the tract, in the walls of blood vessels called arterioles that maintain a steady pressure on blood, and in the wall of the urinary bladder that maintains a steady pressure on urine.

Some smooth muscle fibers contract in response to nerve impulses from the autonomic (involuntary) nervous system. Thus, smooth muscle is normally not under voluntary control. Other smooth muscle fibers contract in response to hormones or local factors such as pH, oxygen and carbon dioxide levels, temperature, and ion concentrations.

Finally, unlike skeletal muscle fibers, smooth muscle fibers can stretch considerably without developing tension. When smooth muscle fibers are stretched, they initially develop increased tension. However, there is an almost immediate decrease in the tension. This phenomenon is referred to as *stress-relaxation* and is important because it permits smooth muscle to accommodate great changes in size while still retaining the ability to contract effectively. Thus, the smooth muscle in the wall of hollow organs such as the stomach, intestines, and urinary bladder can stretch as the viscera distend, while the pressure within them remains the same.

A summary of the principal characteristics of the three types of muscle tissue is presented in Exhibit 9-2.

AGING AND MUSCLE TISSUE

Beginning at about 30 years of age, there is a progressive loss of skeletal muscle mass that is largely replaced by fat. Accompanying the loss of muscle mass, there is a decrease in maximal strength and a diminishing of muscle reflexes.

DEVELOPMENTAL ANATOMY OF THE MUSCULAR SYSTEM

In this brief discussion of the development of the muscular system, we shall concentrate mostly on skeletal muscles. Except for the muscles of the iris of the eyes and the arrector pili muscles attached to hairs, all muscles of the body are derived from *mesoderm*. As the mesoderm develops, a portion of it becomes arranged in dense columns on either side of the developing nervous system. These columns of mesoderm undergo segmentation into a series of blocks of cells called *somites* (Figure 9-10a). The first pair of somites appears on the twentieth day. Eventually, 44 pairs of somites are formed by the thirtieth day.

With the exception of the skeletal muscles of the head and extremities, *skeletal muscles* develop from the *mesoderm of somites*. Since there are very few somites in the head region of the embryo, most of the skeletal muscles there develop from the *general mesoderm* in the head region.

EXHIBIT 9-2

Summary of the Principal Characteristics of Muscle Tissue

CHARACTERISTIC	SKELETAL MUSCLE	CARDIAC MUSCLE	SMOOTH MUSCLE
Location	Primarily attached to bones.	Heart.	Walls of hollow viscera, blood vessels, iris, arrector pili.
Microscopic appearance	Striated, multinucleated, unbranched fibers.	Striated uninucleated, branched fibers with intercalated discs.	Nonstriated (smooth) uninucleated, spindle-shaped fibers.
Nervous control	Voluntary.	Involuntary.	Involuntary.
Sarcomeres	Yes.	Yes.	No.
Transverse tubules	Yes.	Yes.	No.
Gap junctions	No.	Yes.	Yes, in visceral smooth muscle.
Cell size	Large.	Large.	Small.
Source of calcium	Sarcoplasmic reticulum.	Sarcoplasmic reticulum and extracellular fluids.	Sarcoplasmic reticulum and extracellular fluids.
Speed of contraction	Fast.	Moderate.	Slow.

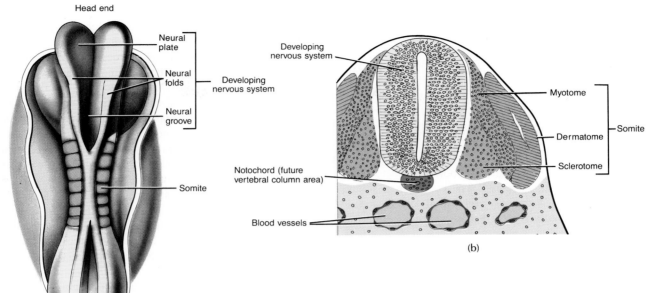

FIGURE 9-10 Development of the muscular system. (a) Dorsal aspect of an embryo indicating the location of somites. (b) Cross section of a somite.

The skeletal muscles of the limbs develop from masses of general mesoderm around developing bones in embryonic limb buds (origins of future extremities: see Figure 5-7).

The cells of a somite are differentiated into three regions: (1) *myotome,* which forms most of the skeletal muscles; (2) *dermatome,* which forms the connective tissues, including the dermis, under the epidermis; and (3) *sclerotome,* which gives rise to the vertebrae (Figure 9-10b).

In the development of a skeletal muscle from a myotome of a somite, certain patterns emerge. For example, a myotome may split longitudinally into two or more portions. This represents the manner in which the trapezius muscle forms. Other myotomes split into two or more layers. Such a pattern represents how the external oblique, internal oblique, and transversus abdominis muscles develop. In other instances, several myotomes fuse to form a single muscle. Such an example is the rectus abdominis muscle. Muscles may also migrate, wholly or in part, from their sites of origin. The latissimus dorsi, for example, originates from the cervical myotomes but extends all the way down to the thoracic and lumbar vertebrae and hip bone. Another trend that may be observed is a change in the direction of muscle fibers. Initially, muscle fibers in a myotome are parallel to the long axis of the embryo. However, nearly all developed skeletal muscles do not have fibers that are parallel to the long axis. One example is the external oblique.

Finally, *fasciae, ligaments,* and *aponeuroses* may form as a result of degeneration of all or parts of myotomes.

Smooth muscle develops from **mesodermal cells** that migrate to and envelop the developing gastrointestinal tract and viscera.

Cardiac muscle develops from **mesodermal cells** that migrate to and envelop the developing heart while it is still in the form of primitive heart tubes (see Figure 13-10).

APPLICATIONS TO HEALTH

Disorders of the muscular system are related to disruptions of homeostasis. The disorders may involve a lack of nutrients, accumulation of toxic products, disease, injury, disuse, or faulty nervous connections (innervations).

FIBROSIS

The formation of fibrous (containing fibers) connective tissue in locations where it normally does not exist is called *fibrosis.* Skeletal and cardiac muscle fibers cannot undergo mitosis, and dead muscle fibers are normally replaced with fibrous connective tissue. Fibrosis, then, is often a consequence of muscle injury or degeneration.

FIBROSITIS

Fibrositis is an inflammation of fibrous tissue. If it occurs in the lumbar region, it is termed *lumbago* (lum-BĀ-gō). Fibrositis is a common condition characterized by pain, stiffness, or soreness of fibrous tissue, especially in the muscle coverings. It is not destructive or progressive. It may persist for years or spontaneously disappear. Attacks of fibrositis may follow an injury, repeated muscular strain, or prolonged muscular tension.

"CHARLEY HORSE"

Fibromyositis refers to a group of symptoms that include pain, tenderness, and stiffness of the joints, muscles, or adjacent structures. These symptoms ordinarily occur in various combinations. When major muscles, especially in the lower extremity are involved, such as the quadriceps femoris muscle of the thigh, it is usually called *"charley horse."* It results from a contusion (bruising) and tearing of muscle fibers, that may produce a hematoma (collection of blood). "Charley horse" is characterized by the sudden onset of pain and is aggravated by motion. In some cases, local muscle spasms are noted. The condition is relieved with heat, massage, and rest and completely disappears, although occasionally it may become chronic or recur at frequent intervals.

MUSCULAR DYSTROPHIES

Muscular dystrophies (*dystrophy* = degeneration) are muscle-destroying diseases. The diseases are characterized by degeneration of individual muscle fibers, which leads to a progressive atrophy of the skeletal muscle. Usually, the voluntary skeletal muscles are weakened equally on both sides of the body, whereas the internal muscles, such as the diaphragm, are not affected. Histologically, the changes that occur include the variation in muscle fiber size, degeneration of fibers, and deposition of fat. Muscular dystrophies are classified by mode of inheritance, age of onset, and clinical features. The most common form of muscular dystrophy is called *Duchenne* (du-SHĀN) *muscular dystrophy* (*DMD*). The gene responsible for DMD has been identified and its DNA code has been worked out. Hopefully, this information will lead to replacement therapy and a halt to muscle loss.

The cause of muscular dystrophies has been variously attributed to a genetic defect, faulty metabolism of potassium, protein deficiency, and inability of the body to utilize creatine.

There is no specific drug therapy for this disorder. Treatment involves attempts to prolong ambulation by muscle-strengthening exercises, corrective surgical measures, and appropriate braces. Patients are encouraged to keep physically active.

MYASTHENIA GRAVIS (MG)

Myasthenia (mī-as-THĒ-nē-a) *gravis* (*MG*) is a weakness of skeletal muscles. It is caused by an abnormality at the neuromuscular junction that prevents muscle fibers from contracting. Recall that motor neurons stimulate skeletal muscle fibers to contract by releasing acetylcholine (ACh). Myasthenia gravis is an autoimmune disorder caused by antibodies directed against ACh receptors of the muscle fiber sarcolemma. The antibodies bind to the receptors and hinder the attachment of ACh to the receptors. As the disease progresses, more neuromuscular junctions become affected. The muscle becomes increasingly weaker and may eventually cease to function altogether.

Myasthenia gravis is more common in females, occurring most frequently between the age of 20 and 50. The muscles of the face and neck are most apt to be involved. Initial symptoms include a weakness of the eye muscles and difficulty in swallowing. Later, the individual has difficulty chewing and talking. Eventually, the muscles of the limbs may become involved. Death may result from paralysis of the respiratory muscles, but usually the disorder does not progress to this stage.

Anticholinesterase drugs such as neostigmine and pyridostigmine, derivatives of physostigmine, have been the primary treatment for the disease. They act as inhibitors of acetylcholinesterase, thus raising the level of ACh to bind with available receptors. More recently, steroid drugs, such as prednisone, have been used with great success to reduce antibody levels. Immunosuppressant drugs are also used to decrease the production of antibodies that interfere with normal muscle contraction. Another treatment involves *plasmapheresis,* a procedure that separates blood cells from the plasma that contains the unwanted antibodies. The blood cells are then mixed with a plasma substitute and pumped back into the individual. In some individuals, surgical removal of the thymus gland is indicated (thymectomy).

ABNORMAL CONTRACTIONS

One kind of abnormal contraction of a muscle is *spasm,* a sudden, involuntary contraction of short duration. A *cramp* is a painful spasmodic contraction of a muscle. It is an involuntary, complete tetanic contraction. *Convulsions* are violent, involuntary tetanic contractions of an entire group of muscles. Convulsions occur when motor neurons are stimulated by fever, poisons, hysteria, or changes in body chemistry due to withdrawal of certain drugs. The stimulated neurons send many bursts of seemingly disordered impulses to muscle fibers. *Fibrillation* is the uncoordinated contraction of individual muscle fibers preventing the smooth contraction of the muscle. A *tic* is a spasmodic twitching made involuntarily by muscles that are ordinarily under voluntary control. Twitching of the eyelid and face muscles are examples. In general, tics are of psychological origin.

KEY MEDICAL TERMS ASSOCIATED WITH THE MUSCULAR SYSTEM

Gangrene (GANG-rēn; *gangraena* = an eating sore)   Death of a soft tissue, such as muscle, that results from interruption of its blood supply. It is caused by various species of *Clostridium,* bacteria that live anaerobically in the soil.

Myalgia (mī-AL-jē-a; *algia* = painful condition)   Pain in or associated with muscles.

Myoma (mī-Ō-ma; *oma* = tumor) A tumor consisting of muscle tissue.

Myomalacia (mī'-ō-ma-LĀ-shē-a; *malaco* = soft) Softening of a muscle.

Myopathy (mī-OP-a-thē; *pathos* = disease) Any disease of muscle tissue.

Myosclerosis, (mī'-ō-skle-RŌ-sis; *scler* = hard) Hardening of a muscle.

Myositis (mī-ō-SĪ-tis; *itis* = inflammation of) Inflammation of muscle fibers (cells).

Myospasm (MĪ-ō-spazm) Spasm of a muscle.

Myotonia (mī-ō-TO-nē-a; *tonia* = tension) Increased muscular excitability and contractility with decreased power of relaxation; tonic spasm of the muscle.

Paralysis (pa-RAL-a-sis; *para* = beyond; *lyein* = to loosen) Loss or impairment of motor (muscular) function resulting from a lesion of nervous or muscular origin.

Trichinosis (trik'-i-NŌ-sis) A myositis caused by the parasitic worm *Trichinella spiralis,* which may be found in the muscles of humans, rats, and pigs. People contract the disease by eating insufficiently cooked infected pork.

Volkmann's contracture (FŌLK-manz kon-TRAK-tur; *contra* = against) Permanent contraction of a muscle due to replacement of destroyed muscle fibers (cells) with fibrous tissue that lacks ability to stretch. Destruction of muscle fibers may occur from interference with circulation caused by a tight bandage, a piece of elastic, or a cast.

Wryneck or **torticollis** (RĪ-neck; *tortus* = twisted; *collum* = neck Contracted state of several superficial and deep muscles of the neck that produces twisting of the neck and an unnatural position of the head; one of the common triggering factors is vigorous activity.

STUDY OUTLINE

Characteristics (p. 224)
1. Excitability is the property of receiving and responding to stimuli.
2. Contractility is the ability to shorten and thicken (contract).
3. Extensibility is the ability to be stretched (extended).
4. Elasticity is the ability to return to original shape after contraction or extension.

Functions (p. 224)
1. Through contraction, muscle tissue performs three important functions.
2. These functions are motion, maintenance of posture, and heat production.

Types (p. 224)
1. Skeletal muscle tissue is primarily attached to bones. It is striated and voluntary.
2. Cardiac muscle tissue forms the wall of the heart. It is striated and involuntary.
3. Visceral muscle tissue is located in viscera. It is nonstriated (smooth) and involuntary.

Skeletal Muscle Tissue (p. 224)
Connective Tissue Components (*p. 224*)
1. The term fascia is applied to a sheet or broad band of fibrous connective tissue underneath the skin or around muscles and organs of the body.
2. Other connective tissue components are epimysium, covering the entire muscle; perimysium, covering fasciculi; and endomysium, covering fibers; all are extensions of deep fascia.
3. Tendons and aponeuroses are extensions of connective tissue beyond muscle cells that attach the muscle to bone or other muscle.

Nerve and Blood Supply (*p. 226*)
1. Nerves convey impulses for muscular contraction.
2. Blood provides nutrients and oxygen for contraction.

Histology (*p. 226*)
1. Skeletal muscle consists of fibers (cells) covered by a sarcolemma. The fibers contain sarcoplasm, nuclei, sarcoplasmic reticulum, and transverse tubules.
2. Each fiber contains myofibrils that consist of thin and thick myofilaments. The myofilaments are compartmentalized into sarcomeres.
3. Thin myofilaments are composed of actin, tropomyosin, and troponin; thick myofilaments consist mostly of myosin.
4. Projecting myosin heads are called cross bridges and contain actin- and ATP-binding sites.

Contraction (p. 230)
Sliding-Filament Theory (*p. 230*)
1. A muscle action potential travels over the sarcolemma and enters the transverse tubules and sarcoplasmic reticulum.
2. The action potential leads to the release of calcium ions from the sarcoplasmic reticulum, triggering the contractile process.
3. Actual contraction is brought about when the thin myofilaments of a sarcomere slide toward each other.

Neuromuscular Junction (Motor End-Plate) (*p. 230*)
1. A motor neuron transmits a nerve impulse to a skeletal muscle for contraction.
2. A neuromuscular junction (motor end-plate) refers to an axon terminal of a motor neuron and the portion of the muscle fiber sarcolemma in close approximation with it.

Motor Unit (*p. 231*)
1. A motor neuron and the muscle fibers it stimulates form a motor unit.
2. A single motor unit may innervate as few as 10 or as many as 2000 muscle fibers.

Mechanism (*p. 233*)
1. When a nerve impulse (nerve action potential) reaches an axon terminal, the synaptic vesicles of the terminal release acetylcholine (ACh), which initiates a muscle action potential in the muscle fiber sarcolemma. The action potential then travels into the transverse tubules and sarcoplasmic reticulum.
2. The transmitted action potential releases calcium ions that combine with troponin, causing it to pull on tropomyosin, thus exposing myosin-binding sites on actin.
3. The energy released from the breakdown of ATP causes myosin cross bridges to attach to actin, and their movement results in the sliding of thin myofilaments.

Muscle Length and Force of Contraction (p. 233)

1. A muscle fiber develops its greatest tension when there is maximum overlap between thick and thin myofilaments (optimum length).
2. As a fiber is stretched or shortened, force of contraction continuously decreases.

All-or-None Principle (p. 233)

1. The weakest stimulus capable of causing contraction is a liminal (threshold) stimulus.
2. A stimulus not capable of inducing contraction is a subliminal (subthreshold) stimulus.
3. Muscle fibers of a motor unit contract to their fullest extent or not a all.

Muscle Tone (p. 233)

1. A sustained partial contraction of portions of a skeletal muscle results in muscle tone.
2. Tone is essential for maintaining posture.
3. Flaccidity is a condition of less than normal tone. Atrophy is a wasting away or decrease in size; hypertrophy is an enlargement or overgrowth.

Types of Skeletal Muscle Fibers (p. 234)

1. On the basis of structure and function, skeletal muscle fibers are classified as slow-twitch red, fast-twitch red, and fast-twitch white.
2. Most skeletal muscles contain a mixture of all three fiber types, their proportions varying with the usual action of the muscle.
3. Various exercises can modify the types of skeletal muscle fibers.

Cardiac Muscle Tissue (p. 235)

1. This muscle is found only in the heart. It is striated and involuntary.
2. The fibers are quadrangular and usually contain a single centrally placed nucleus.
3. Compared to skeletal muscle tissue, cardiac muscle tissue has more sarcoplasm, more mitochondria, less well-developed sarcoplasmic reticulum, and larger transverse tubules located at Z lines rather than at A–I band junctions. Myofilaments are not arranged in discrete myofibrils.
4. The fibers branch freely and are connected via gap junctions.
5. Intercalated discs provide strength and aid in conduction of muscle action potential.
6. Unlike skeletal muscle tissue, cardiac muscle tissue contracts and relaxes rapidly, continuously, and rhythmically. Energy is supplied by glycogen and fat in large, numerous mitochondria.

7. Cardiac muscle tissue can contract without extrinsic stimulation and can remain contracted longer than skeletal muscle tissue.
8. Cardiac muscle tissue has a long refractory period, which prevents tetanus.

Smooth Muscle Tissue (p. 237)

1. Smooth muscle is nonstriated and involuntary.
2. Smooth muscle fibers contain intermediate filaments, dense bodies (function as Z lines), and caveolae (function as transverse tubules).
3. Visceral (single-unit) smooth muscle is found in the walls of viscera. The fibers are arranged in a network.
4. Multiunit smooth muscle is found in blood vessels and the eye. The fibers operate singly rather than as a unit.
5. The duration of contraction and relaxation of smooth muscle is longer than in skeletal muscle.
6. Smooth muscle fibers contract in response to nerve impulses, hormones, and local factors.
7. Smooth muscle fibers can stretch considerably without developing tension.

Aging and Muscle Tissue (p. 238)

1. At about 30 years of age, there is a progressive loss of skeletal muscle, which is replaced by fat.
2. There is also a decrease in muscle strength and diminished muscle reflexes.

Developmental Anatomy of the Muscular System (p. 238)

1. With few exceptions, muscles develop from mesoderm.
2. Skeletal muscles of the head and extremities develop from general musoderm; the remainder of the skeletal muscles develop from the mesoderm of somites.

Applications to Health (p. 239)

1. Fibrosis is the formation of fibrous tissue where it normally does not exist; it frequently occurs in damaged muscle tissue.
2. Fibrositis is an inflammation of fibrous tissue. If it occurs in the lumbar region, it is called lumbago.
3. "Charley horse" refers to pain, tenderness, and stiffness of joints, muscles, and related structure in major muscles, especially in the lower extremity, such as the quadriceps femoris muscle in the thigh.
4. Muscular dystrophy is a hereditary disease of muscles characterized by degeneration of individual muscle fibers.
5. Myasthenia gravis (MG) is a disease characterized by great muscular weakness and fatigability resulting from improper neuromuscular transmission.
6. Abnormal contractions include spasms, cramps, convulsions, fibrillations, and tics.

REVIEW QUESTIONS

1. How is the skeletal system related to the muscular system? What are the four characteristics of muscle tissue?
2. What are the three basic functions of the muscular system?
3. How can the three types of muscle tissue be distinguished on the basis of location, microscopic appearance, and nervous control?
4. What is fascia? What are the three different types of fascia and where are they found in the body?
5. Define epimysium, perimysium, endomysium, tendon, and aponeurosis. Describe the nerve and blood supply to a skeletal muscle.
6. Describe the microscopic structure of skeletal muscle tissue.

7. What are some problems associated with muscle-building anabolic steroids.
8. In considering the contraction of skeletal muscle tissue, describe the following: neuromuscular junction (motor endplate), motor unit, role of calcium, sources of energy, and sliding-filament theory.
9. How does muscle length relate to force of contraction?
10. What is the all-or-none principle? Relate it to a threshold and subthreshold stimulus.
11. What is muscle tone? Why is it inportant? Distinguish between atrophy and hypertrophy.
12. Compare slow-twitch red, fast-twitch red, and fast-twitch white fibers with respect to structure and function.
13. Compare skeletal, cardiac, and smooth muscle with regard to differences in structure and physiology.
14. Describe the effects of aging on muscle tissue.
15. Describe how the muscular system develops.
16. Define fibrosis, fibrositis, and ''charley horse.''
17. What is muscular dystrophy?
18. What is myasthenia gravis (MG)? In this disease, why do the muscles not contract normally?
19. Define each of the following abnormal muscular contractions: spasm, cramp, convulsion, fibrillation, and tic.
20. Refer to the glossary of key medical terms associated with the muscular system. Be sure that you can define each term.

SELF-QUIZ

1. Contrast the kinds of fascia by indicating which of the following are characteristics of superficial fascia (S) or deep fascia (D).
 ___ a. immediately under the skin
 ___ b. composed of dense connective tissue
 ___ c. contains much fat and so serves as insulator of the body
 ___ d. also called subcutaneous layer
 ___ e. forms epimysium, perimysium, and endomysium that holds muscles into functional groups

Complete the following:

2. Each muscle fiber (also known as a _____) is surrounded by a membrane called the _____ and contains cytoplasm called _____.
3. A T tubule, along with sarcoplasmic reticulum (terminal cisterns) on either side, is called a _____. The hundreds or thousands of myofibrils in a skeletal muscle fiber each consist of bundles of thick and thin _____.
4. To describe a sarcomere during contraction, complete each statement with one of the following: lengthens, shortens, or stays the same length.
 a. The sarcomere _____.
 b. Each thick myofilament (A band) _____.
 c. Each thin myofilament _____.
 d. The I band _____.
 e. The H zone _____.
5. Smooth muscle fibers are (cylinder? spindle?) shaped with (several nuclei? one nucleus?) per cell. They (do? do not?) contain actin and myosin. However, because of the irregular arrangement of these filaments, smooth muscle tissue appears (striated? nonstriated or ''smooth''?). In general, smooth muscle contracts and relaxes more (rapidly? slowly?) than skeletal muscle does, and smooth muscle holds the contraction for a (shorter? longer?) period of time than skeletal muscle does.
6. Which of the three types of muscle tissue develop from mesoderm? _____. Part of the mesoderm forms columns on either side of the developing nervous system. This tissue segments into blocks of tissue called _____.
7. Match the muscle types listed at the right with descriptions below:
 ___ a. involuntary muscle found in blood vessels and intestine   A. cardiac
 ___ b. involuntary striated muscle   B. skeletal
 ___ c. striated voluntary muscle attached to bones   C. smooth

8. Match the following:
 ___ a. invagination of deep fascia that surrounds muscle fibers and bundles and holds muscles into functional groups   A. aponeurosis
 ___ b. cord of dense connective tissue that attaches muscle into periosteum   B. endomysium, perimysium, epimysium
 ___ c. similar in function to tendon but consists of a broad, flat layer of dense connective tissue   C. tendon
 ___ d. tube of fibrous connective tissue lined with synovial membrane that permits tendons to slide easily, as in wrist and ankle   D. tendon sheath
9. Arrange the answers in correct sequence:
 ___ ___ ___ a. From superficial to deep:
    A. superficial fascia
    B. subserous fascia
    C. deep fascia
 ___ ___ ___ b. According to the amount of muscle tissue they surround, from most to least:
    A. perimysium
    B. endomysium
    C. epimysium
 ___ ___ ___ c. From largest to smallest:
    A. myofibril
    B. myofilament
    C. muscle fiber

Choose the one best answer to these questions:
 ___ 10. All of the following molecules are parts of thin filaments except A. actin; B. myosin; C. tropomyosin; D. troponin.
 ___ 11. Choose the one statement that is false:
    A. A band refers to the anisotropic band; B. the A band is darker than the I band; C. thick myofilaments reach the Z line in relaxed muscle; D. the H zone contains thick myofilaments, but not thin ones; E. thick myofilaments are made of myosin.
 ___ 12. The ability of muscle tissue to receive and respond to a stimulus is referred to as
    A. contractility; B. excitability; C. elasticity; D. extensibility; E. conductivity.

—— **13.** Which of the following is *not* performed by muscles?
A. motion; B. excretion; C. maintenance of posture; D. heat production.

—— **14.** Choose the false statement about cardiac muscle:
A. cardiac muscle has a long refractory period; B. cardiac muscle has one centrally located nucleus per fiber; C. cardiac muscle cells are called cardiac muscle fibers; D. cardiac muscle fibers are separated by intercalated discs; E. cardiac muscle fibers are spindle shaped with no striations.

—— **15.** A motor unit is defined as a
A. Nerve and a muscle; B. single neuron and a single muscle fiber; C. neuron and the muscle fibers it supplies; D. single muscle fiber and the nerves that innervate it; E. muscle and the efferent and afferent nerves that innervate it.

—— **16.** The all-or-none principle applied to muscle means
A. individual motor units contract in a graded fashion, that is, from maximum to minimum; B. individual motor units have only one type of contraction; C. muscles contract with minimal stimuli; D. muscles contract by summation of stimuli; E. the whole muscle contracts rather than single motor units.

—— **17.** Skeletal muscle tissue
A. has the ability to contract rhythmically by itself; B. is composed of long, spindlelike cells, each containing a single nucleus; C. has the ability to contract when stimulated; D. is present in the walls of arteries; E. is not described by any of the above.

10 The Muscular System

CHAPTER OUTLINE

■ **How Skeletal Muscles Produce Movement**
Origin and Insertion
Lever Systems and Leverage
Arrangement of Fasciculi
Group Actions
■ **Naming Skeletal Muscles**
■ **Principal Skeletal Muscles**
■ **Intramuscular (IM) Injections**

STUDENT OBJECTIVES

1. Describe the relation between bones and skeletal muscles in producing body movements.
2. Define a lever and fulcrum and compare the three classes of levers on the basis of placement of the fulcrum, effort, and resistance.
3. Identify the various arrangements of muscle fibers in a skeletal muscle and relate the arrangements to the strength of contraction and range of movement.
4. Discuss most body movements as activities of groups of muscles by explaining the roles of the prime mover, antagonist, synergist, and fixator.
5. Define the criteria employed in naming skeletal muscles.
6. Identify the principal skeletal muscles in different regions of the body by name, origin, insertion, action, and innervation.
7. Discuss the administration of drugs by intramuscular (IM) injection.

The term *muscle tissue* refers to all the contractile tissues of the body: skeletal, cardiac, and smooth muscle. The *muscular system,* however, refers to the *skeletal* muscle system: the skeletal muscle tissue and connective tissues that make up individual muscle organs, such as the biceps brachii muscle. Cardiac muscle tissue is located in the heart and is therefore considered part of the cardiovascular system. Smooth muscle tissue of the intestine is part of the digestive system, whereas smooth muscle tissue of the urinary bladder is part of the urinary system. In this chapter, we discuss only the muscular system. We shall see how skeletal muscles produce movement, and we shall describe the principal skeletal muscles.

HOW SKELETAL MUSCLES PRODUCE MOVEMENT

ORIGIN AND INSERTION

Skeletal muscles produce movements by exerting force on tendons, which in turn pull on bones. Most muscles cross at least one joint and are attached to the articulating bones that form the joint (Figure 10-1). When such a muscle contracts, it draws one articulating bone toward the other.

The two articulating bones usually do not move equally in response to the contraction. One is held nearly in its original position because other muscles contract to pull it in the opposite direction or because its structure makes it less movable. Ordinarily, the attachment of a muscle tendon to the stationary bone is called the *origin.* The attachment of the other muscle tendon to the movable bone is the *insertion.* A good analogy is a spring on a door. The part of the spring attached to the door represents the insertion; the part attached to the frame is the origin. The fleshy portion of the muscle between the tendons of the origin and insertion is called the *belly (gaster).* The origin is usually proximal and the insertion distal, especially in the extremities. In addition, muscles that move a body part generally do not cover the moving part. Figure 10-1a shows that, although contraction of the biceps brachii muscle moves the forearm, the belly of the muscle lies over the humerus.

LEVER SYSTEMS AND LEVERAGE

In producing a body movement, bones act as levers and joints function as fulcrums of these levers. A *lever* may be defined as a rigid rod that moves about on some fixed point called a *fulcrum.* A fulcrum may be symbolized as △. A lever is acted on at two different points by two

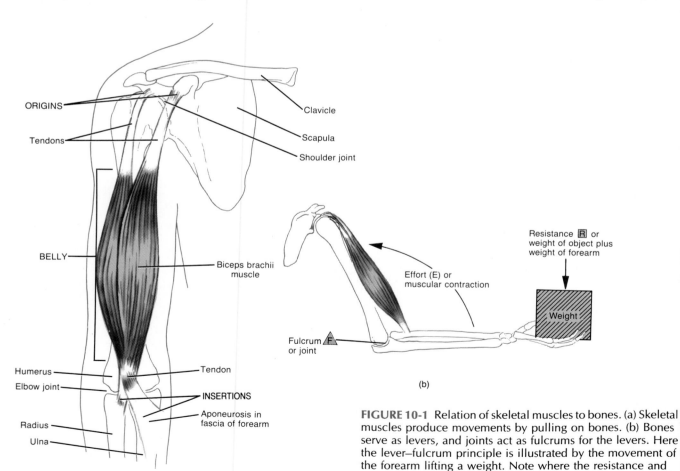

(a)

(b)

FIGURE 10-1 Relation of skeletal muscles to bones. (a) Skeletal muscles produce movements by pulling on bones. (b) Bones serve as levers, and joints act as fulcrums for the levers. Here the lever–fulcrum principle is illustrated by the movement of the forearm lifting a weight. Note where the resistance and effort are applied in this example.

different forces: the *resistance* `R` and the *effort* (E). The resistance may be regarded as a force to be overcome, whereas the effort is the force exerted to overcome the resistance. The resistance may be the weight of a part of the body part that is to be moved. The muscular effort (contraction) is applied to the bone at the insertion of the muscle and produces motion. Consider the biceps brachii flexing the forearm at the elbow as a weight is lifted (Figure 10-1b). When the forearm is raised, the elbow is the fulcrum. The weight of the forearm plus the weight in the hand is the resistance. The shortening of the biceps brachii pulling the forearm up is the effort.

Levers are categorized into three types according to the positions of the fulcrum, the effort, and the resistance.

1. In *first-class levers*, the fulcrum is between the effort and resistance (Figure 10-2a). An example of a first-class lever is a seesaw. There are not many first-class levers in the body. One example is the head resting on the vertebral column. When the head is raised, the facial portion of the skull is the resistance. The joint between the atlas and occipital bone (atlanto-occipital joint) is the fulcrum. The contraction of the muscles of the back is the effort.
2. *Second-class levers* have the fulcrum at one end, the effort at the opposite end, and the resistance between them (Figure 10-2b). They operate like a wheelbarrow. Most authorities agree that there are very few examples of second-class levers in the body. One example is raising the body on the toes. The body is the resistance, the ball of the foot is the fulcrum, and the contraction of the calf muscles to pull the heel upward is the effort.
3. *Third-class levers* consist of the fulcrum at one end, the resistance at the opposite end, and the effort between them (Figure 10-2c). They are the most common levers in the body. An example is flexing the forearm at the elbow. As we have seen, the weight of the forearm is the resistance, the contraction of the biceps brachii is the effort, and the elbow joint is the fulcrum.

Leverage, the mechanical advantage gained by a lever, is largely responsible for a muscle's strength and range of movement. Consider strength first. Suppose we have two muscles of the same strength crossing and acting on a joint. Assume also that one is attached farther from the joint and one is nearer. The muscle attached farther will produce the more powerful movement. Thus, strength of movement depends on the placement of muscle attachments.

In considering range of movement, again assume that we have two muscles of the same strength crossing and acting on a joint and that one is attached farther from the joint than the other. The muscle inserting closer to the joint will produce the greater range of movement. Thus, range of movement also depends on the placement of muscle attachments. Since strength increases with distance from the joint and range of movement decreases, maximal strength and maximal range are incompatible; strength and range very inversely.

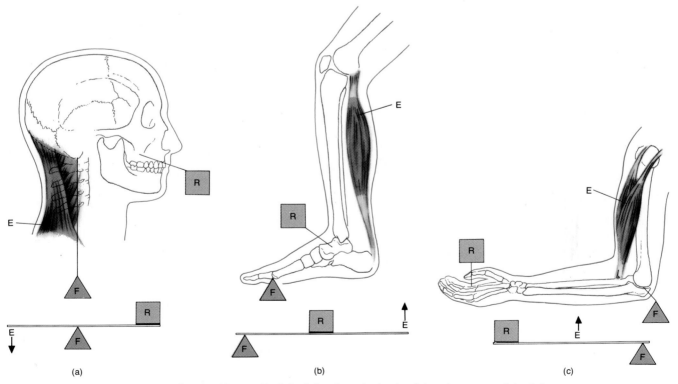

(a)   (b)   (c)

FIGURE 10-2 Classes of levers. Each is defined on the basis of the placement of the fulcrum, effort, and resistance. (a) First-class lever. (b) Second-class lever. (c) Third-class lever.

ARRANGEMENT OF FASCICULI

Recall from Chapter 9 that skeletal muscle fibers (cells) are arranged within the muscle in bundles called fasciculi (fascicles). The muscle fibers are arranged in a parallel fashion within each bundle, but the arrangement of the fasciculi with respect to the tendons may take one of four characteristic patterns.

The first pattern is called *parallel*. The fasciculi are parallel with the longitudinal axis and terminate at either end in flat tendons. The muscle is typically quadrilateral in shape. An example is the stylohyoid muscle (see Figure 10-7). In a modification of the parallel arrangement, called *fusiform*, the fasciculi are nearly parallel with the longitudinal axis and terminate at either end in flat tendons, but the muscle tapers toward the tendons, where the diameter is less than that of the belly. An example is the biceps brachii muscle (see Figure 10-18).

The second distinct pattern is called *convergent*. A broad origin of fasciculi converges to a narrow, restricted insertion. Such a pattern gives the muscle a triangular shape. An example is the deltoid muscle (see Figure 10-17).

The third distinct pattern is referred to as *pennate*. The fasciculi are short in relation to the entire length of the muscle and the tendon extends nearly the entire length of the muscle. The fasciculi are directed obliquely toward the tendon like the plumes of a feather. If the fasciculi are arranged on only one side of a tendon, as in the extensor digitorum longus muscle, the muscle is referred to as *unipennate* (see Figure 10-24). If the fasciculi are arranged on both sides of a centrally positioned tendon, as in the rectus femoris muscle, the muscle is referred to as *bipennate* (see Figure 10-23).

The final distinct pattern is referred to as *circular*. The fasciculi are arranged in a circular pattern and enclose an orifice. An example is the orbicularis oris muscle (see Figure 10-4).

Fascicular arrangement is correlated with the power of a muscle and range of movement. When a muscle fiber contracts, it shortens to a length just slightly greater than half of its resting length. Thus, the longer the fibers in a muscle, the greater the range of movement it can produce. By contrast, the strength of a muscle depends on the total number of fibers it contains, since a short fiber can contract as forcefully as a long one. Because a given muscle can contain either a small number of long fibers or a large number of short fibers, fascicular arrangement represents a compromise between power and range of movement. Pennate muscles, for example, have a large number of fasciculi distributed over their tendons, giving them greater power, but a smaller range of movement. Parallel muscles, on the other hand, have comparatively few fasciculi that extend the length of the muscle. Thus, they have a greater range of movement but less power.

GROUP ACTIONS

Most movements are coordinated by several skeletal muscles acting in groups rather than individually and most skeletal muscles are arranged in opposing pairs at joints, that is, flexors–extensors, abductors–adductors, and so on. Consider flexing the forearm at the elbow, for example. A muscle that causes a desired action is referred to as the *prime mover (agonist)*. In this instance, the biceps brachii is the prime mover (see Figure 10-18). Simultaneously with the contraction of the biceps brachii, another muscle, called the *antagonist*, is relaxing. In this movement, the triceps brachii serves as the antagonist (see Figure 10-18). The antagonist has an effect opposite to that of the prime mover; that is, the antagonist relaxes and yields to the movement of the prime mover. You should not assume, however, that the biceps brachii is always the prime mover and the triceps brachii is always the antagonist. For example, when extending the forearm at the elbow, the triceps brachii serves as the prime mover and the biceps brachii functions as the antagonist; their roles are reversed. Note that if the prime mover and antagonist contracted simultaneously with equal force, there would be no movement.

In addition to prime movers and antagonists, most movements also involve muscles called *synergists*, which serve to steady a movement, thus preventing unwanted movements and helping the prime mover function more efficiently. For example, flex your hand at the wrist and then make a fist. Note how difficult this is to do. Now, extend your hand at the wrist and then make a fist. Note how much easier it is to clench your fist. In this case, the extensor muscles of the wrist act as synergists in cooperation with the flexor muscles of the fingers acting as prime movers. The extensor muscles of the fingers serve as antagonists (see Figure 10-19).

Some muscles in a group also act as *fixators*, which stabilize the origin of the prime mover so that the prime mover can act more efficiently. For example, the scapula is a freely movable bone in the pectoral (shoulder) girdle that serves as a firm origin for several muscles that move the arm. However, for the scapula to do this, it must be held steady. This is accomplished by fixator muscles that hold the scapula firmly against the back of the chest. In abduction of the arm, the deltoid muscle serves as the prime mover, whereas fixators (pectoralis minor, rhomboideus major, rhomboideus minor, trapezius, subclavius, and serratus anterior muscles) hold the scapula firmly (see Figure 10-16). These fixators stabilize the scapula that serves as the attachment site for the origin of the deltoid muscle while the insertion of the muscle pulls on the humerus to abduct the arm. Under different conditions and depending on the movement and which point is fixed, many muscles act, at various times, as prime movers, antagonists, synergists, or fixators.

NAMING SKELETAL MUSCLES

The names of most of the nearly 700 skeletal muscles are based on several types of characteristics. Learning the terms used to indicate specific characteristics will help you remember the names of muscles.

EXHIBIT 10-4

Muscles That Move the Eyeballs—Extrinsic Muscles (Figure 10-6)

OVERVIEW Muscles associated with the eyeball are of two principal types: extrinsic and intrinsic. ***Extrinsic muscles*** originate outside the eyeball and are inserted on its outer surface (sclera). ***Intrinsic muscles*** originate and insert entirely within the eyeball.

Movements of the eyeballs are controlled by three pairs of extrinsic muscles. The four rectus muscles move the eyeball in the direction indicated by their respective names—superior, inferior, lateral, and medial. The two oblique muscles—superior and inferior—rotate the eyeball on its axis. The extrinsic muscles of the eyeballs are among the fastest contracting and precisely controlled skeletal muscles of the body.

MUSCLE	ORIGIN	INSERTION	ACTION	INNERVATION
Superior rectus (*superior* = above; *rectus* = in this case, muscle fibers running parallel to long axis of eyeball)	Tendinous ring attached to bony orbit around optic foramen.	Superior and central part of eyeball.	Rolls eyeball upward.	Oculomotor (III) nerve.
Inferior rectus (*inferior* = below)	Same as above.	Inferior and central part of eyeball.	Rolls eyeball downward.	Oculomotor (III) nerve
Lateral rectus	Same as above.	Lateral side of eyeball.	Rolls eyeball laterally.	Abducens (VI) nerve.
Medial rectus	Same as above.	Medial side of eyeball.	Rolls eyeball medially.	Oculomotor (III) nerve.
Superior oblique (*oblique* = in this case, muscle fibers running diagonally to long axis of eyeball)	Same as above.	Eyeball between superior and lateral recti.	Rotates eyeball on its axis; directs cornea downward and laterally; note that it moves through a ring of fibro-cartilaginous tissue called the trochlea (*trochlea* = pulley).	Trochlear (IV) nerve.
Inferior oblique	Maxilla (front of orbital cavity).	Eyeball between inferior and lateral recti.	Rotates eyeball on its axis; directs cornea upward and laterally.	Oculomotor (III) nerve.

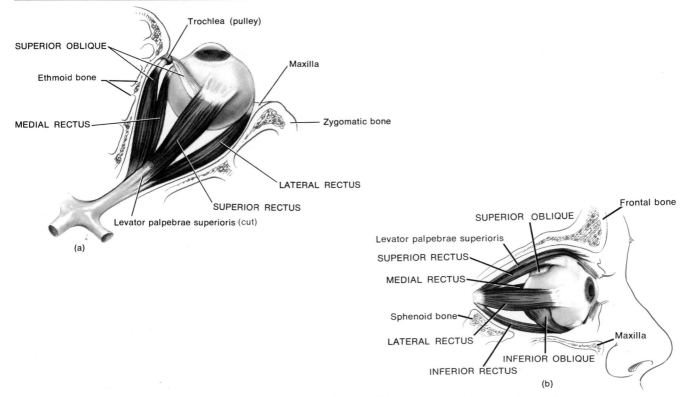

FIGURE 10-6 Extrinsic muscles of the eyeball. (a) Diagram of the right eyeball seen from above. (b) Diagram of lateral view of the right eyeball.

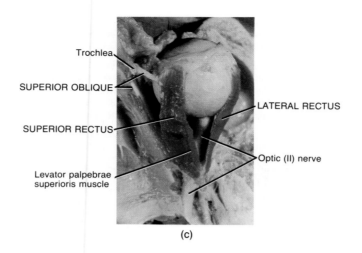

(c)

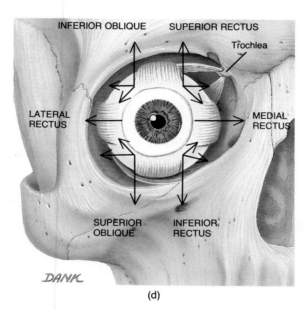

(d)

FIGURE 10-6 (*Continued*) (c) Photograph of superior view of the right eyeball. (Courtesy of C. Yokochi and J. W. Rohen, *Photographic Anatomy of the Human Body,* 2nd ed., 1979, IGAKU-SHOIN, Ltd., Tokyo, New York.) (d) Movements of the right eyeball in response to contraction of the various extrinsic muscles.

EXHIBIT 10-5

Muscles That Move the Tongue—Extrinsic Muscles (Figure 10-7)

OVERVIEW The tongue is divided into lateral halves by a median fibrous septum. The septum extends throughout the length of the tongue and is attached inferiorly to the hyoid bone. Like the muscles of the eyeball, muscles of the tongue are of two principal types—extrinsic and intrinsic. *Extrinsic muscles* originate outside the tongue and insert into it. *Intrinsic muscles* originate and insert within the tongue. The extrinsic and intrinsic muscles of the tongue are arranged in both lateral halves of the tongue.

MUSCLE	ORIGIN	INSERTION	ACTION	INNERVATION
Genioglossus (*geneion* = chin; *glossus* = tongue)	Mandible.	Undersurface of tongue and hyoid bone.	Depresses tongue and thrusts it forward (protraction).	Hypoglossal (XII) nerve.
Styloglossus (*stylo* = stake or pole; styloid process of temporal bone)	Styloid process of temporal bone.	Side and undersurface of tongue.	Elevates tongue and draws it backward (retraction).	Hypoglossal (XII) nerve.
Palatoglossus (*palato* = palate)	Anterior surface of soft palate.	Side of tongue.	Elevates posterior portion of tongue and draws soft palate down on tongue.	Pharyngeal plexus.
Hyoglossus	Body of hyoid bone.	Side of tongue.	Depresses tongue and draws down its sides.	Hypoglossal (XII) nerve.

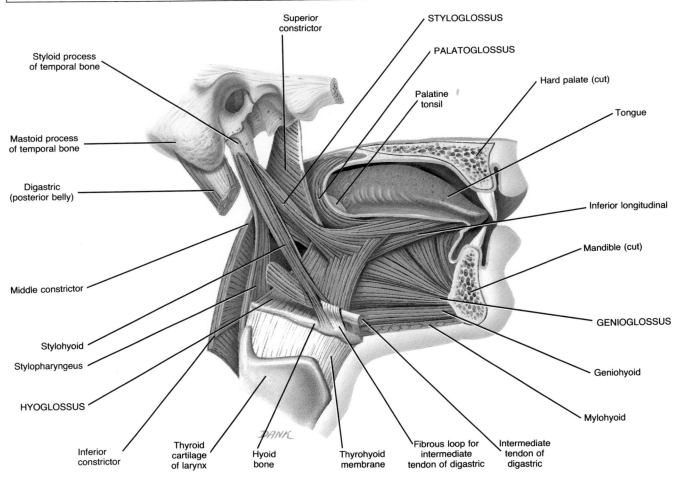

FIGURE 10-7 Muscles that move the tongue as viewed from the right side.

EXHIBIT 10-6

Muscles of the Soft Palate (Figure 10-8)

OVERVIEW The soft palate is mainly muscular in structure and attaches anteriorly to the hard palate and blends posteriorly with the pharynx. The soft palate forms the posterior aspect of the partition that separates the oral (mouth) cavity from the pharynx. Hanging down from the free edge of the soft palate is a nipplelike structure, the uvula. During swallowing, the muscles of the soft palate tighten and elevate it, thus preventing food from entering the nasal cavities.

MUSCLE	ORIGIN	INSERTION	ACTION	INNERVATION
Levator veli palatini (*levator* = raises; *velum* = veil; *palato* = palate; see also Figure 10-9)	Petrous portion of temporal bone and medial wall of auditory (Eustachian) tube.	Blends with corresponding muscle of opposite side.	Elevates soft palate during swallowing.	Pharnyngeal plexus.
Tensor veli palatini (*tensor* = makes tense; see also Figure 10-9)	Medial pterygoid plate of sphenoid bone, spine of sphenoid, lateral wall of auditory (Eustachian) tube.	Palatine aponeurosis and palatine bone.	Tenses (tightens) soft palate during swallowing.	Mandibular branch of trigeminal (V) nerve.
Musculus uvulae (*uvulae* = uvula)	Posterior border of hard palate and palatine aponeurosis.	Uvula.	Tenses (tightens) and raises uvula.	Pharyngeal plexus.
Palatoglossus (*palato* = palate; *glossus* = tongue) (see Figure 10-7)	Anterior surface of soft palate.	Side of tongue.	Elevates posterior portion of tongue and draws soft palate down on tongue.	Pharyngeal plexus.
Palatopharyngeus (*pharyngo* = pharynx)	Posterior border of hard palate and palatine aponeurosis.	Posterior border of thyroid cartilage and lateral and posterior walls of pharynx.	Elevates larynx and pharynx and helps close nasopharynx during swallowing.	Pharyngeal plexus.

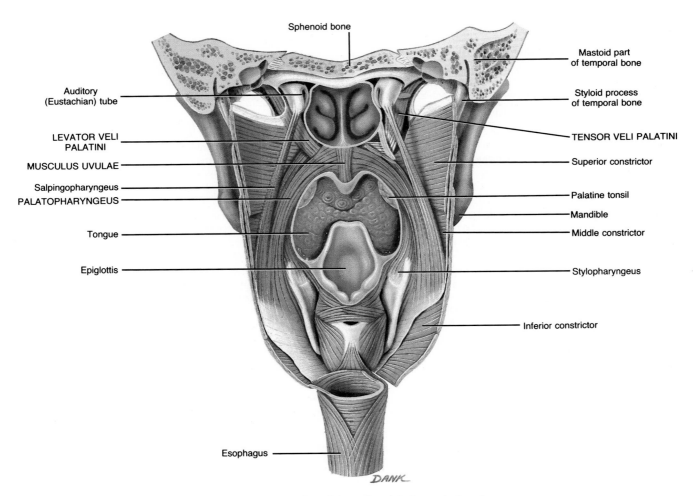

FIGURE 10-8 Muscles of the soft palate in posterior view.

EXHIBIT 10-7

Muscles of the Pharynx (Figure 10-9)

OVERVIEW The pharynx (throat) is a funnel-shaped, muscular tube located posterior to the nasal cavities, mouth, and larynx (voice box). The muscles are arranged in two layers—an outer circular layer and an inner longitudinal layer. The **circular layer** is composed of the three constrictors, each overlapping the one above. The remaining muscles constitute the **longitudinal layer.**

MUSCLE	ORIGIN	INSERTION	ACTION	INNERVATION
CIRCULAR LAYER				
Inferior constrictor (*inferior* = below; *constrictor* = decreases diameter of a lumen)	Cricoid and thyroid cartilages of larynx.	Posterior median raphe of pharynx.	Constricts inferior portion of pharynx to propel a bolus into esophagus.	Pharyngeal plexus.
Middle constrictor	Greater and lesser cornua of hyoid bone and stylohyoid ligament.	Posterior median raphe of pharynx.	Constricts middle portion of pharynx to propel a bolus into esophagus.	Pharyngeal plexus.
Superior constrictor (*superior* = above)	Pterygoid process, pterygomandibular raphe, and mylohyoid line of mandible.	Posterior median raphe of pharynx.	Constricts superior portion of pharynx to propel a bolus into esophagus.	Pharyngeal plexus.
LONGITUDINAL LAYER				
Stylopharyngeus (*stylo* = stake or pole; styloid process of temporal bone; *pharyngo* = pharynx) (see also Figure 10-7)	Medial side of base of styloid process.	Lateral aspects of pharynx and thyroid cartilage.	Elevates larynx and dilates pharynx to help bolus descend.	Glossopharyngeal (IX) nerve.
Salpingopharyngeus (*salping* = pertaining to the auditory or uterine tube) (see Figure 10-8)	Inferior portion of auditory (Eustachian) tube.	Posterior fibers of palatopharyngeus muscle.	Elevates superior portion of lateral wall of pharynx during swallowing and opens orifice of auditory (Eustachian) tube.	Pharyngeal plexus.
Palatopharyngeus (*palato* = palate) (see Figure 10-8)	Soft palate.	Posterior border of thyroid cartilage and lateral and posterior wall of pharynx.	Elevates larynx and pharynx and helps close nasopharynx during swallowing.	Pharyngeal plexus.

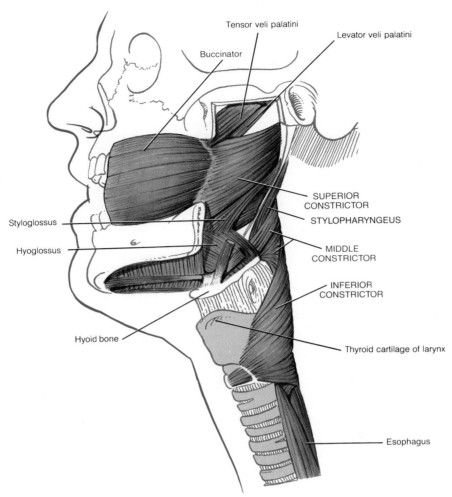

Tensor veli palatini

Levator veli palatini

Buccinator

SUPERIOR
CONSTRICTOR

STYLOPHARYNGEUS

Styloglossus

Hyoglossus

MIDDLE
CONSTRICTOR

INFERIOR
CONSTRICTOR

Hyoid bone

Thyroid cartilage of larynx

Esophagus

FIGURE 10-9 Muscles of the pharynx. Left lateral view.

EXHIBIT 10-8

Muscles of the Floor of the Oral Cavity (Figure 10-10)

OVERVIEW As a group, these muscles are referred to as **_suprahyoid muscles._** They lie superior to the hyoid bone and all insert into it. The digastric muscle consists of an anterior belly and a posterior belly united by an intermediate tendon that is held in position by a fibrous loop (see also Figure 10-7).

MUSCLE	ORIGIN	INSERTION	ACTION	INNERVATION
Diagastric (*di* = two; *gaster* = belly)	Anterior belly from inner side of lower border of mandible; posterior belly from mastoid process of temporal bone.	Body of hyoid bone via an intermediate tendon.	Elevates hyoid bone and depresses mandible as in opening the mouth.	Anterior belly from mandibular division of trigeminal (V) nerve; posterior belly from facial (VII) nerve.
Stylohyoid (*stylo* = stake or pole, styloid process of temporal bone; *hyoedes* = U-shaped, pertaining to hyoid bone; see also Figure 10-7)	Styloid process of temporal bone.	Body of hyoid bone.	Elevates hyoid bone and draws it posteriorly.	Facial (VII) nerve.
Mylohyoid	Inner surface of mandible.	Body of hyoid bone.	Elevates hyoid bone and floor of mouth and depresses mandible.	Mandibular division of trigeminal (V) nerve.
Geniohyoid (*geneion* = chin) (see Figure 10-7)	Inner surface of mandible.	Body of hyoid bone.	Elevates hyoid bone, draws hyoid bone and tongue anteriorly, depresses mandible.	Cervical nerve C1.

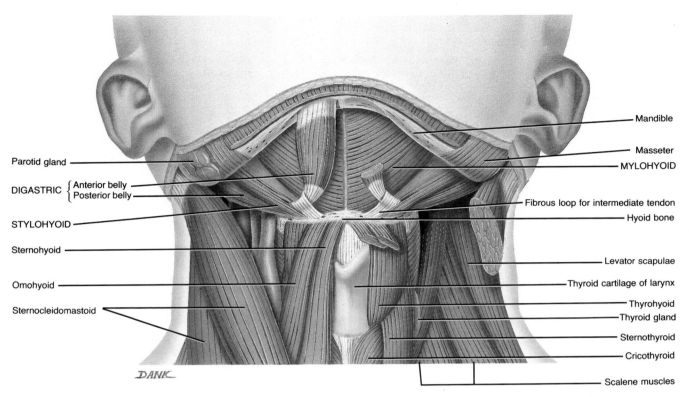

Parotid gland

DIGASTRIC { Anterior belly
            Posterior belly

STYLOHYOID

Sternohyoid

Omohyoid

Sternocleidomastoid

Mandible

Masseter

MYLOHYOID

Fibrous loop for intermediate tendon

Hyoid bone

Levator scapulae

Thyroid cartilage of larynx

Thyrohyoid

Thyroid gland

Sternothyroid

Cricothyroid

Scalene muscles

DANK

FIGURE 10-10 Muscles of the floor of the oral cavity. Superficial muscles are shown on the left side of the illustration, deep muscles are shown on the right side of the illustration.

EXHIBIT 10-9

Muscles of the Larynx (Figure 10-11)

OVERVIEW The muscles of the larynx, like those of the eyeballs and tongue, are grouped into extrinsic and intrinsic. The extrinsic muscles of the larynx marked with * are together referred to as *infrahyoid (strap) muscles.* They lie inferior to the hyoid bone. The omohyoid muscle, like the digastric muscle, is composed of two bellies and an intermediate tendon. In this case, however, the two bellies are referred to as superior and inferior, rather than anterior and posterior.

MUSCLE	ORIGIN	INSERTION	ACTION	INNERVATION
EXTRINSIC				
Omohyoid* (*omo* = relationship to the shoulder, *hyoedes* = U-shaped; pertaining to hyoid bone)	Superior border of scapula and superior transverse ligament.	Body of hyoid bone.	Depresses hyoid bone.	Branches of ansa cervicalis nerve (C1–C3).
Sternohyoid* (*sterno* = sternum)	Medial end of clavicle and manubrium of sternum.	Body of hyoid bone.	Depresses hyoid bone.	Branches of ansa cervicalis nerve (C1–C3).
Sternothyroid* (*thyro* = thyroid gland)	Manubrium of sternum.	Thyroid cartilage of larynx.	Depresses thyroid cartilage.	Branches of ansa cervicalis nerve (C1–C3).
Thyrohyoid*	Thyroid cartilage of larynx.	Greater cornu of hyoid bone.	Elevates thyroid cartilage and depresses hyoid bone.	Branches of ansa cervicalis nerve (C1–C2) and descending hypoglossal (XII) nerve.
Stylopharyngeus	See Exhibit 10-7.			
Palatopharyngeus	See Exhibit 10-7.			
Inferior constrictor	See Exhibit 10-7.			
Middle constrictor	See Exhibit 10-7.			
INTRINSIC				
Cricothyroid (*crico* = cricoid cartilage of larynx)	Anterior and lateral portion of cricoid cartilage of larynx.	Anterior border of inferior cornu of thyroid cartilage of larynx and posterior part of inferior border of lamina of thyroid cartilage.	Produces tension and elongation of vocal folds.	External laryngeal branch of vagus (X) nerve.
Posterior cricoarytenoid (*arytaina* = shaped like a jug.	Posterior surface of cricoid cartilage.	Posterior surface of muscular process of arytenoid cartilage of larynx.	Opens glottis.	Recurrent laryngeal branch of vagus (X) nerve.
Lateral cricoarytenoid	Superior border of cricoid cartilage.	Anterior surface of muscular process of arytenoid cartilage.	Closes glottis.	Recurrent laryngeal branch of vagus (X) nerve.
Arytenoid	Posterior surface and lateral border of one arytenoid cartilage.	Corresponding parts of opposite arytenoid cartilage.	Closes glottis.	Recurrent laryngeal branch of vagus (X) nerve.
Thyroarytenoid	Inferior portion of angle of thyroid cartilage and middle of cricothyroid ligament.	Base and anterior surface of arytenoid cartilage.	Shortens and relaxes vocal folds.	Recurrent laryngeal branch of vagus (X) nerve.

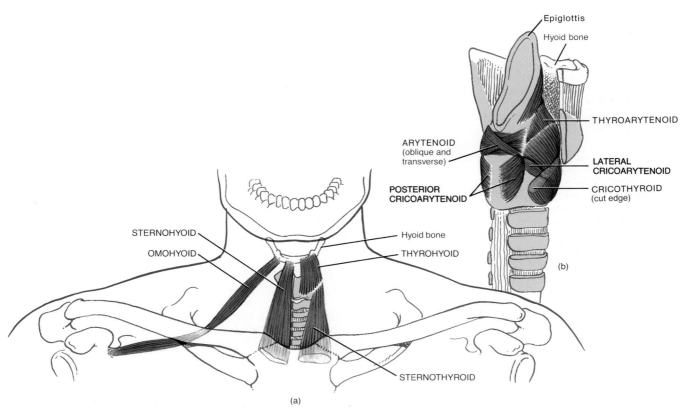

FIGURE 10-11 Muscles of the larynx. (a) Anterior view. (b) Right posterolateral view.

EXHIBIT 10-10

Muscles That Move the Head

OVERVIEW The cervical region is divided by the sternocleidomastoid muscle into two principal triangles—anterior and posterior. The **anterior triangle** is bordered superiorly by the mandible, inferiorly by the sternum, medially by the cervical midline, and laterally by the anterior border of the sternocleidomastoid muscle (see Figure 11-5). The **posterior triangle** is bordered inferiorly by the clavicle, anteriorly by the posterior border of the sternocleiodomastoid muscle, and posteriorly by the anterior border of the trapezius muscle (see Figure 11-5). Subsidiary triangles exist within the two principal triangles.

MUSCLE	ORIGIN	INSERTION	ACTION	INNERVATION
Sternocleidomastoid (*sternum* = breastbone; *cleido* = clavicle; *mastoid* = mastoid process of temporal bone) (see Figure 10-16)	Sterum and clavicle.	Mastoid process of temporal bone.	Contraction of both muscles flex the cervical part of the vertebral column, draw the head forward, and elevate chin; contraction of one muscle rotates face toward side opposite contracting muscle.	Accessory (XI) nerve; cervical nerves C2–C3.
Semispinalis capitis (*semi* = half; *spine* = spinous process; *caput* = head) (see Figure 10-21)	Articular process of seventh cervical vertebra and transverse processes of first six thoracic vertebrae.	Occipital bone.	Both muscles extend head; contraction of one muscle rotates face toward same side as contracting muscle.	Dorsal rami of spinal nerves.
Splenius capitis (*splenion* = bandage) (see Figure 10-21)	Ligamentum nuchae and spines of seventh cervical vertebra and first four thoracic vertebrae.	Occipital bone and mastoid process of temporal bone.	Both muscles extend head; contraction of one rotates it to same side as contracting muscle.	Dorsal rami of middle and lower cervical nerves.
Longissimus capitis (*longissimus* = longest) (see Figure 10-21)	Transverse processes of last four cervical vertebrae.	Mastoid process of temporal bone.	Extends head and rotates face toward side opposite contracting muscle.	Dorsal rami of middle and lower cervical nerves.

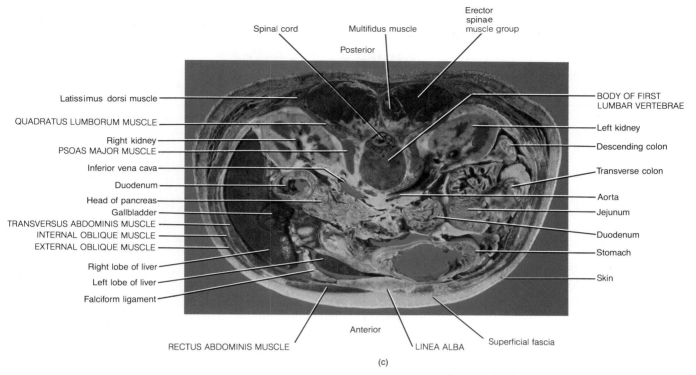

Spinal cord
Multifidus muscle
Erector spinae muscle group
Posterior

Latissimus dorsi muscle
QUADRATUS LUMBORUM MUSCLE
Right kidney
PSOAS MAJOR MUSCLE
Inferior vena cava
Duodenum
Head of pancreas
Gallbladder
TRANSVERSUS ABDOMINIS MUSCLE
INTERNAL OBLIQUE MUSCLE
EXTERNAL OBLIQUE MUSCLE
Right lobe of liver
Left lobe of liver
Falciform ligament

BODY OF FIRST LUMBAR VERTEBRAE
Left kidney
Descending colon
Transverse colon
Aorta
Jejunum
Duodenum
Stomach
Skin

RECTUS ABDOMINIS MUSCLE
Anterior
LINEA ALBA
Superficial fascia

(c)

FIGURE 10-12 (*Continued*) Muscles of the abdominal wall. (c) Photograph of a cross section through the abdomen showing the musculature and related viscera. (Courtesy of Stephen A. Kieffer and E. Robert Heitzman, *An Atlas of Cross-Sectional Anatomy*, Harper & Row, Publishers, Inc., New York, 1979.) The surface anatomy of the anterior abdominal muscle is illustrated in Figure 11-8.

EXHIBIT 10-12

Muscles Used in Breathing (Figure 10-13)

OVERVIEW The muscles described here are attached to the ribs and by their contraction and relaxation alter the size of the thoracic cavity during normal breathing. In forced breathing, other muscles are involved as well. Essentially, inspiration occurs when the thoracic cavity increases in size. Expiration occurs when the thoracic cavity decreases in size.

MUSCLE	ORIGIN	INSERTION	ACTION	INNERVATION
Diaphragm (*dia* = across; *phragma* = wall)	Xiphoid process, costal cartilages of last six ribs, and lumbar vertebrae.	Central tendon.	Forms floor of thoracic cavity; pulls central tendon downward during inspiration and thus increases vertical length of thorax.	Phrenic nerve.
External intercostals (*external* = closer to surface; *inter* = between; *costa* = rib)	Inferior border of rib above.	Superior boarder of rib below.	May elevate ribs during inspiration and thus increase lateral and anteroposterior dimensions of thorax.	Intercostal nerves.
Internal Intercostals (*internal* = farther from surface)	Superior border of rib below.	Inferior border of rib above.	May draw adjacent ribs together during forced expiration and thus decrease lateral and anteroposterior dimensions of thorax.	Intercostal nerves.

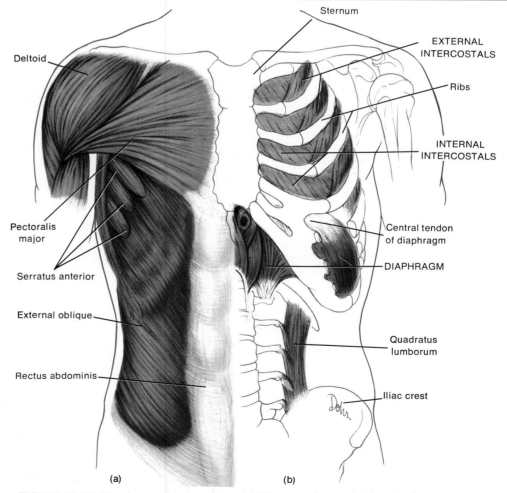

FIGURE 10-13 Muscles used in breathing. (a) Diagram of superficial view. (b) Diagram of deep view.

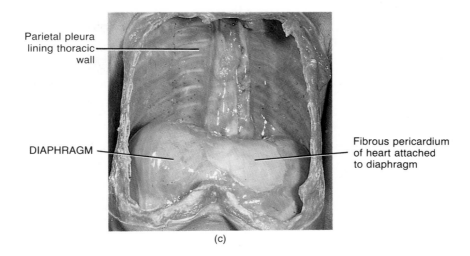

Parietal pleura lining thoracic wall

DIAPHRAGM

Fibrous pericardium of heart attached to diaphragm

(c)

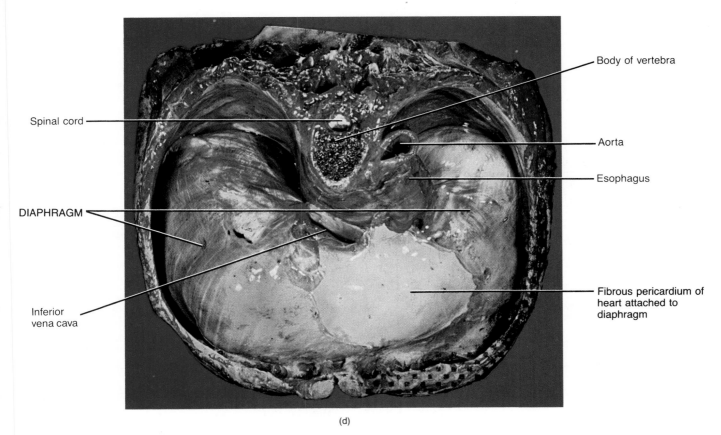

Spinal cord

DIAPHRAGM

Inferior vena cava

Body of vertebra

Aorta

Esophagus

Fibrous pericardium of heart attached to diaphragm

(d)

FIGURE 10-13 (*Continued*) Muscles used in breathing. (c) Photograph of anterior view of diaphragm (Courtesy of C. Yokochi and J. W. Rohen, *Photographic Anatomy of the Human Body,* 2nd ed., Igaku-Shoin, Ltd.) Muscles used in breathing. (d) Photograph of superior view of diaphragm. (Courtesy of C. Yokochi and J. W. Rohen, *Photographic Anatomy of the Human Body,* 1969, IGAKU-SHOIN, Ltd., Tokyo, New York.)

EXHIBIT 10-13

Muscles of the Pelvic Floor (Figure 10-14)

OVERVIEW The muscles of the pelvic floor, together with the fascia covering their external and internal surfaces, are referred to as the ***pelvic diaphragm.*** The diaphragm is funnel shaped and forms the floor of the abdominopelvic cavity where it supports the pelvic viscera. It is pierced by the anal canal and urethra in both sexes and also by the vagina in females.

MUSCLE	ORIGIN	INSERTION	ACTION	INNERVATION
Levator ani (*levator* = raises; *ani* = anus)	This muscle is divisible into two parts, the pubococcygeus muscle and the iliococcygeus muscle.			
Pubococcygeus (*pubo* = pubis; *coccygeus* = coccyx)	Pubis.	Coccyx, urethra, anal canal, central tendon of perineum, and anococcygeal raphe.	Supports and slightly raises pelvic floor, resists increased intra-abdominal pressure, and draws anus toward pubis and constricts it.	Sacral nerves S3–S4 or S4 and perineal branch of pudendal nerve.
Iliococcygeus (*ilio* = ilium)	Ischial spine.	Coccyx.	Supports and slightly raises pelvic floor, resists increased intra-abdominal pressure, and draws anus toward pubis and constricts it.	Sacral nerves S3-S4 or S4 and perineal branch of pudendal nerve.
Cocygeus	Ischial spine.	Lower sacrum and upper coccyx.	Supports and slightly raises pelvic floor, resists intra-abdominal pressure, and pulls coccyx forward following defecation or parturition.	Sacral nerve S3 or S4.

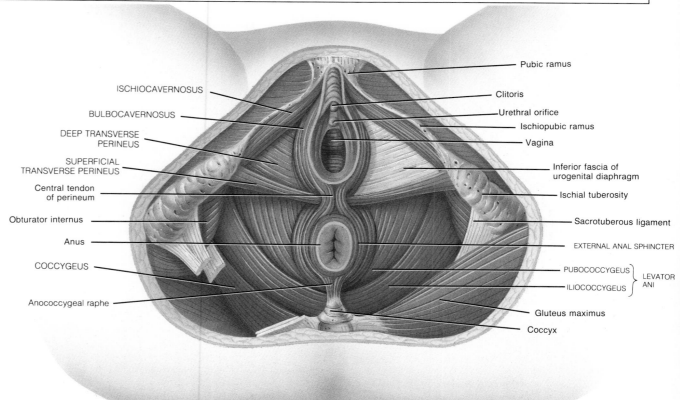

FIGURE 10-14 Muscles of the pelvic floor seen in the female perineum.

EXHIBIT 10-14

Muscles of the Perineum (Figure 10-15; see also Figure 10-14)

OVERVIEW The *perineum* is the entire outlet of the pelvis. It is a diamond-shaped area at the lower end of the trunk between the thighs and buttocks. It is bordered anteriorly by the symphysis pubis, laterally by the ischial tuberosities, and posteriorly by the coccyx. A transverse line drawn between the ischial tuberosities divides the perineum into an anterior *urogenital triangle* that contains the external genitals and a posterior *anal triangle* that contains the anus (see Figure 25-21).

The deep transverse perineus, the urethral sphincter, and a fibrous membrane constitute the *urogenital diaphragm.* It surrounds the urogenital ducts and helps strengthen the pelvic floor.

MUSCLE	ORIGIN	INSERTION	ACTION	INNERVATION
Superficial transverse perineus (*superficial* = closer to surface; *transverse* = across, *perineus* = perineum)	Ischial tuberosity.	Central tendon of perineum.	Helps stabilize the central tendon of the perineum.	Perineal branch of pudendal nerve.
Bulbocavernosus (*bulbus* = bulb; *caverna* = hollow place)	Central tendon of perineum.	Inferior fascia of urogenital diaphragm, corpus spongiosum of penis, and deep fascia on dorsum of penis in male; pubic arch and root and dorsum of clitoris in female.	Helps expel last drops of urine during micturition, helps propel semen along urethra, and may assist in erection of the penis in male; decreases vaginal orifice and assists in erection of clitoris in female.	Perineal branch of pudendal nerve.
Ischiocavernosus (*ischion* = hip)	Ischial tuberosity and ischial and pubic rami.	Corpus cavernosum of penis in male and clitoris in female	May maintain erection of penis in male and clitoris in female.	Perineal branch of pudendal nerve.
Deep transverse perineus (*deep* = farther from surface)	Ischial rami.	Central tendon of perineum.	Helps expel last drops of urine and semen in male and urine in female.	Perineal branch of pudendal nerve.
Urethral sphincter (*sphincter* = circular muscle that decreases size of an opening; *urethrae* = urethra)	Ischial and pubic rami.	Median raphe in male and vaginal wall in female.	Helps expel last drops of urine and semen in male and urine in female.	Perineal branch of pudendal nerve.
External anal sphincter	Anococcygeal raphe.	Central tendon of perineum.	Keeps anal canal and orifice closed.	Sacral nerve S4 and inferior rectal branch of pudendal nerve.

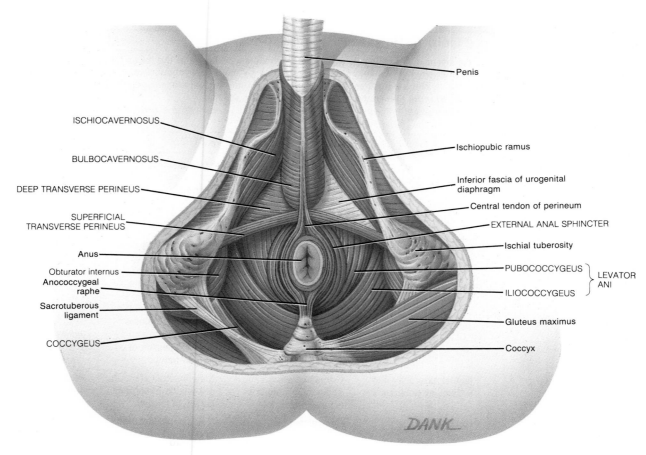

FIGURE 10-15 Muscles of the male perineum.

EXHIBIT 10-15

Muscles That Move the Pectoral (Shoulder) Girdle (Figure 10-16)

OVERVIEW Muscles that move the pectoral (shoulder) girdle originate on the axial skeleton and insert on the clavicle or scapula. The muscles can be distinguished into *anterior* and *posterior* groups. The principal action of the muscles is to stabilize the scapula so that it can function as a stable point of origin for most of the muscles that move the humerus (arm).

MUSCLE	ORIGIN	INSERTION	ACTION	INNERVATION
ANTERIOR MUSCLES **Subclavius** (*sub* = under; *clavius* = clavicle)	First rib.	Clavicle.	Depresses clavicle.	Nerve to subclavius.
Pectoralis minor (*pectus* = breast, chest, thorax; *minor* = lesser)	Third through fifth ribs.	Coracoid process of scapula.	Depresses and moves scapula anteriorly and elevates third through fifth ribs during forced inspiration when scapula is fixed.	Medial pectoral nerve.
Serratus anterior (*serratus* = saw-toothed; *anterior* = front)	Upper eight or nine ribs.	Vertebral border and inferior angle of scapula.	Rotates scapula upward and laterally and elevates ribs when scapula is fixed.	Long thoracic nerve.
POSTERIOR MUSCLES **Trapezius** (*trapezoides* = trapezoid-shaped)	Occipital bone, ligamentum nuchae, and spines of seventh cervical and all thoracic vertebrae.	Clavicle and acromion and spine of scapula.	Elevates clavicle, adducts scapula, rotates scapula upward, elevates or depresses scapula, and extends head.	Accessory (XI) nerve and cervical nerves C3–C4.
Levator scapulae (*levator* = raises; *scapulae* = scapula)	Upper four or five cervical vertebrae.	Superior vertebral border of scapula.	Elevates scapula and slightly rotates it downward.	Dorsal scapula nerve and cervical nerves C3–C5.
Rhomboideus major (*rhomboides* = rhomboid or diamond-shaped)	Spines of second to fifth thoracic vertebrae.	Vertebral border of scapula below spine.	Adducts scapula and slightly rotates it downward.	Dorsal scapular nerve.
Rhomboideus minor	Spines of seventh cervical and first thoracic vertebrae.	Vertebral border of scapula below spine.	Adducts scapula and slightly rotates it downward.	Dorsal scapular nerve.

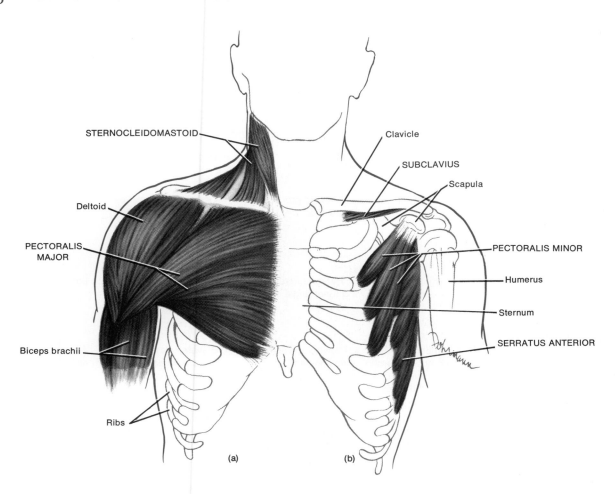

STERNOCLEIDOMASTOID

Deltoid

PECTORALIS
MAJOR

Biceps brachii

Ribs

(a)

Clavicle

SUBCLAVIUS

Scapula

PECTORALIS MINOR

Humerus

Sternum

SERRATUS ANTERIOR

(b)

FIGURE 10-16 Muscles that move the pectoral (shoulder) girdle. (a) Anterior superficial view. (b) Anterior deep view.

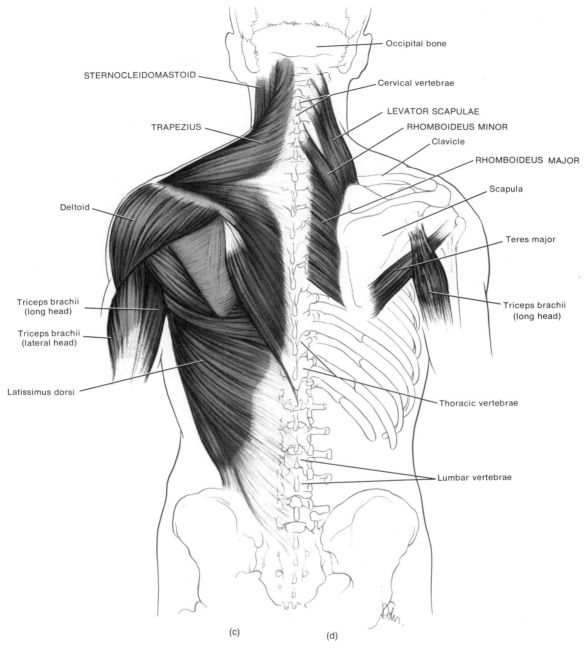

FIGURE 10-16 (*Continued*) Muscles that move the pectoral (shoulder) girdle. (c) Posterior superficial view. (d) Posterior deep view.

EXHIBIT 10-16

Muscles That Move the Arm (Humerus) (Figure 10-17)

OVERVIEW Of the nine muscles that cross the shoulder joint, only two of them (pectoralis major and latissimus dorsi) do not originate on the scapula. These two muscles are thus designated as *axial muscles,* since they originate on the axial skeleton. The remaining seven muscles, the *scapular muscles,* arise from the scapula.

The strength and stability of the shoulder joint are not provided by the shape of the articulating bones or its ligaments. Instead, four deep muscles of the shoulder and their tendons—subscapularis, supraspinatus, infraspinatus, and teres minor—strengthen and stabilize the shoulder joint. The muscles and their tendons are so arranged as to form a nearly complete circle around the joint. This arrangement is referred to as the *rotator (musculotendinous) cuff* and is a common site of injury to baseball pitchers, especially tearing of the supraspinatus muscle.

After you have studied the muscles in this exhibit, arrange them according to the following actions: flexion, extension, abduction, adduction, medial rotation, and lateral rotation. (The same muscle can be used more than once).

MUSCLE	ORIGIN	INSERTION	ACTION	INNERVATION
AXIAL MUSCLES				
Pectoralis major (see also Figure 10-16a)	Clavicle, sternum, cartilages of second to sixth ribs.	Greater tubercle and intertubercular sulcus of humerus.	Flexes, adducts, and rotates arm medially.	Medial and lateral pectoral nerves.
Latissimus dorsi (*latissimus* = widest; *dorsum* = back)	Spines of lower six thoracic vertebrae, lumbar vertebrae, crests of sacrum and ilium, lower four ribs.	Intertubercular sulcus of humerus.	Extends, adducts, and rotates arm medially; draws arm downward and backward.	Thoracodorsal nerve.
SCAPULAR MUSCLES				
Deltoid (*delta* = triangular)	Acromial extremity of clavicle and acromion and spine of scapula.	Deltoid tuberosity of humerus.	Abducts, flexes, extends, and medially and laterally rotates arm.	Axillary nerve.
Subscapularis (*sub* = below; *scapularis* = scapula)	Subscapular fossa of scapula.	Lesser tubercle of humerus.	Rotates arm medially.	Upper and lower subscapular nerves.
Supraspinatus (*supra* = above; *spinatus* = spine of scapula)	Fossa superior to spine of scapula.	Greater tubercle of humerus.	Assists deltoid muscle in abducting arm.	Suprascapular nerve.
Infraspinatus (*infra* = below)	Fossa inferior to spine of scapula.	Greater tubercle of humerus.	Rotates arm laterally; adducts arm.	Suprascapular nerve.
Teres major (*teres* = long and round)	Inferior angle of scapula.	Intertubercular sulcus of humerus.	Extends arm and assists in adduction and medial rotation of arm.	Lower subscapular nerve.
Teres minor	Interior lateral border of scapula.	Greater tubercle of humerus.	Rotates arm laterally and extends and adducts arm.	Axillary nerve.
Coracobrachialis (*coraco* = coracoid process)	Coracoid process of scapula.	Middle of medial surface of shaft of humerus.	Flexes and adducts arm.	Musculocutaneous nerve.

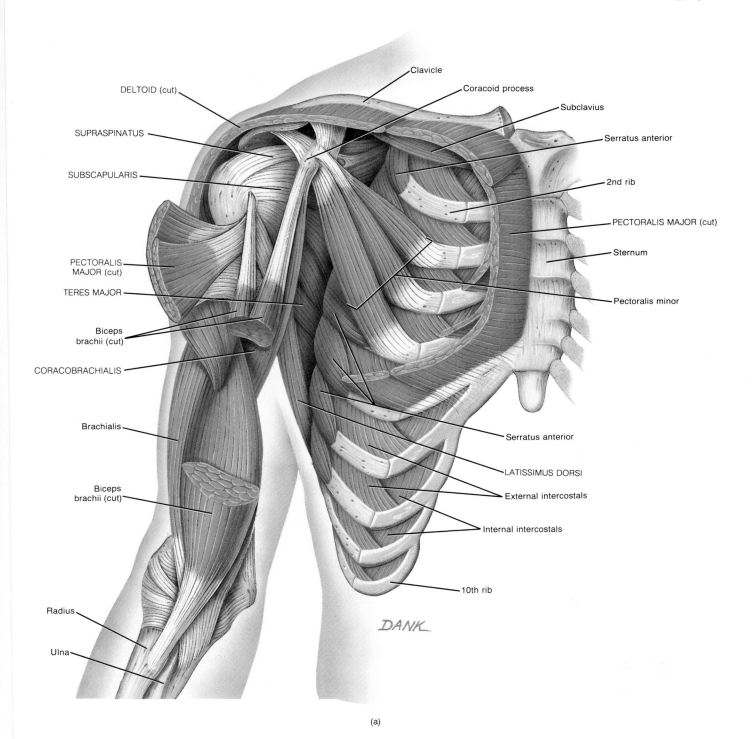

DELTOID (cut)

SUPRASPINATUS

SUBSCAPULARIS

PECTORALIS
MAJOR (cut)

TERES MAJOR

Biceps
brachii (cut)

CORACOBRACHIALIS

Brachialis

Biceps
brachii (cut)

Radius

Ulna

Clavicle

Coracoid process

Subclavius

Serratus anterior

2nd rib

PECTORALIS MAJOR (cut)

Sternum

Pectoralis minor

Serratus anterior

LATISSIMUS DORSI

External intercostals

Internal intercostals

10th rib

DANK

(a)

FIGURE 10-17 Muscles that move the arm (humerus). (a) Anterior deep view.

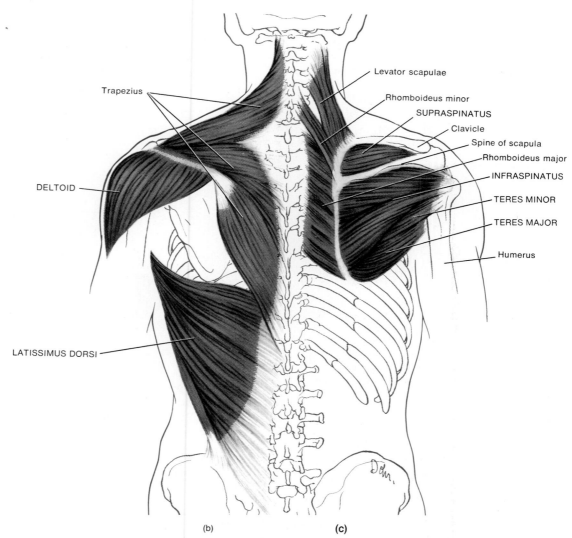

FIGURE 10-17 (*Continued*) Muscles that move the arm (humerus). (b) Posterior superficial view. (c) Posterior deep view.

EXHIBIT 10-17

Muscles That Move the Forearm (Radius and Ulna) (Figure 10-18)

OVERVIEW The muscles that move the forearm (radius and ulna) are conveniently divided into *flexors* and *extensors*. Recall that the elbow joint is a hinge joint, capable only of flexion and extension under normal conditions. Whereas the biceps brachii, brachialis, and brachioradialis are flexors of the elbow joint, the triceps brachii and anconeus are extensors.

MUSCLE	ORIGIN	INSERTION	ACTION	INNERVATION
FLEXORS				
Biceps brachii (*biceps* = two heads of origin; *brachion* = arm)	Long head originates from tubercle above glenoid cavity; short head originates from coracoid process of scapula.	Radial tuberosity and bicipital aponeurosis.	Flexes and supinates forearm; flexes arm.	Musculocutaneous nerve.
Brachialis	Distal, anterior surface of humerus.	Tuberosity and coronoid process of ulna.	Flexes forearm.	Musculocutaneous and radial nerves.
Brachioradialis (*radialis* = radius) (see also Figure 10-19)	Supracondyloid ridge of humerus.	Superior to styloid process of radius.	Flexes forearm; semisupinates and semipronates forearm.	Radial nerve.
EXTENSORS				
Triceps brachii (*triceps* = three heads of origin)	Long head originates from infraglenoid tuberosity of scapula; lateral head originates from lateral and posterior surface of humerus superior to radial groove; medial head originates from posterior surface of humerus inferior to radial groove.	Olecranon of ulna.	Extends forearm; extends arm.	Radial nerve.
Anconeus (*anconeal* = pertaining to elbow) (see Figure 10-19)	Lateral epicondyle of humerus.	Olecranon and superior portion of shaft of ulna.	Extends forearm.	Radial nerve.

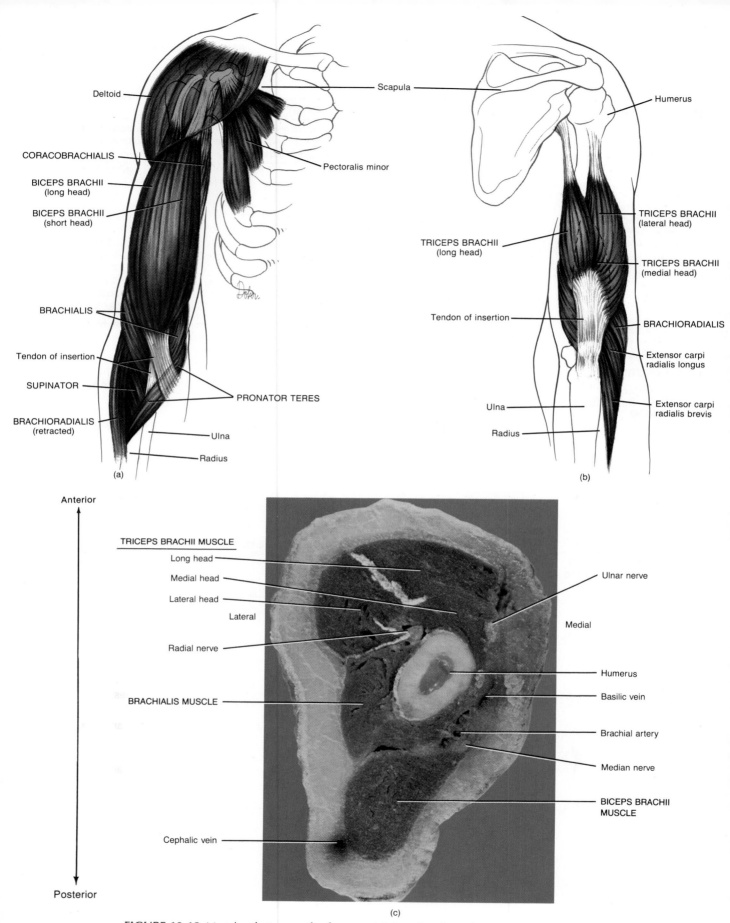

FIGURE 10-18 Muscles that move the forearm. (a) Anterior view. (b) Posterior view. (c) Photograph of a cross section of the arm showing the musculature and related structures. (Courtesy of Stephen A. Kieffer and E. Robert Heitzman, *An Atlas of Cross-Sectional Anatomy,* Harper & Row, Publishers, Inc., New York, 1979.) The surface anatomy of the muscles that move the forearm is illustrated in Figures 11-10 and 11-11.

EXHIBIT 10-18

Muscles That Move the Wrist, Hand, and Fingers (Figure 10-19)

OVERVIEW Muscles that move the wrist, hand, and fingers are many and varied. However, as you will see, their names for the most part give some indication of their origin, insertion, or action. On the basis of location and function, the muscles are divided into two principal groups—anterior and posterior. The **anterior muscles** function as flexors and pronators. They originate mostly on the humerus and typically insert on the carpals, metacarpals, and phalanges. The bellies of these muscles form the bulk of the proximal forearm. The **posterior muscles** function as extensors and supinators. The majority of these muscles arise on the humerus and insert on the metacarpals and phalanges. Each of the two principal groups is also divided into superficial and deep muscles.

The tendons of the muscles of the forearm that attach to the wrist or continue into the hand, along with blood vessels and nerves, are held close to bones by strong fascial structures. The tendons are also surrounded by tendon sheaths. At the wrist, the deep fascia is thickened into fibrous bands called retinacula. The **flexor retinaculum (transverse carpal ligament)** is located over the palmar surface of the carpal bones. Through it pass the long flexor tendons of the digits and wrist and the median nerve. The **extensor retinaculum (dorsal carpal ligament)** is located over the dorsal surface of the carpal bones. Through it pass the extensor tendons of the wrist and digits.

After you have studied the muscles in this exhibit, arrange them according to the following actions: flexion, extension, abduction, adduction, supination, and pronation. (The same muscles can be used more than once.)

MUSCLE	ORIGIN	INSERTION	ACTION	INNERVATION
ANTERIOR GROUP (Flexors and pronators)				
Superficial				
Pronator teres (*pronation* = turning palm downward or posteriorly	Medial epicondyle of humerus and coronoid process of ulna.	Midlateral surface of radius.	Pronates and flexes forearm.	Median nerve.
Flexor carpi radialis (*flexor* = decreases angle at joint; *carpus* = wrist; *radialis* = radius)	Medial epicondyle of humerus.	Second and third metacarpals.	Flexes and abducts wrist.	Median nerve.
Palmaris longus (*palma* = palm; *longus* = long)	Medial epicondyle of humerus.	Flexor retinaculum.	Flexes wrist.	Median nerve.
Flexor carpi ulnaris (*ulnaris* = ulna)	Medial epicondyle of humerus and upper posterior border of ulna.	Pisiform, hamate, and fifth metacarpal.	Flexes and adducts wrist.	Ulnar nerve.
Flexor digitorum superficialis (*digit* = finger or toe; *superficialis* = closer to surface)	Medial epicondyle of humerus, coronoid process of ulna, and oblique line of radius.	Middle phalanges.	Flexes middle phalanges of each finger.	Median nerve.
Deep				
Flexor digitorum profundus (*profundus* = deep)	Anterior medial surface of body of ulnar.	Bases of distal phalanges.	Flexes distal phalanges of each finger.	Median and ulnar nerves.
Flexor pollicis longus (*pollex* = thumb)	Anterior surface of radius and interosseous membrane.	Base of distal phalanx of thumb.	Flexes thumb.	Median nerve.
Pronator quadratus (*quadratus* = squared, foursided)	Distal portion of shaft of ulna.	Distal portion of shaft of radius.	Pronates forearm.	Median nerve.

EXHIBIT 10-18 (*Continued*)

MUSCLE	ORIGIN	INSERTION	ACTION	INNERVATION
POSTERIOR GROUP (Extensors and supinators)				
Superficial				
Brachioradialis	See Exhibit 10-17.			
Extensor carpi radialis longus (*extensor* = increases angle at joint)	Lateral epicondyle of humerus.	Second metacarpal.	Extends and abducts wrist.	Radial nerve.
Extensor carpi radialis brevis (brevis = short)	Lateral epicondyle of humerus.	Third metacarpal.	Extends wrist.	Radial nerve.
Extensor digitorum	Lateral epicondyle of humerus.	Second through fifth phalanges.	Extends phalanges.	Radial nerve.
Extensor digiti minimi (*minimi* = little finger)	Tendon of extensor digitorum.	Tendon of extensor digitorum on fifth phalanx.	Extends little finger.	Deep radial nerve.
Extensor carpi ulnaris	Lateral epicondyle of humerus and posterior border of ulna.	Fifth metacarpal.	Extends and adducts wrist.	Deep radial nerve.
Anconeus	See Exhibit 10-17.			
Deep				
Supinator (*supination* = turning palm upward or anteriorly	Lateral epicondyle of humerus and ridge of ulna.	Lateral surface of proximal one-third of radius.	Supinates forearm.	Deep radial nerve.
Abductor pollicis longus (*abductor* = moves part away from midline	Posterior surface of middle of radius and ulna and interosseous membrane.	First metacarpal.	Extends and abducts wrist.	Deep radial nerve.
Extensor pollicis brevis	Posterior surface of middle of radius and interosseous membrane.	Base of proximal phalanx of thumb.	Extends thumb and abducts wrist.	Deep radial nerve.
Extensor pollicis longus	Posterior surface of middle of ulna and interosseous membrane.	Base of distal phalanx of thumb.	Extends thumb and abducts wrist.	Deep radial nerve.
Extensor indicis (*indicis* = index)	Posterior surface of ulna.	Tendon of extensor digitorum of index finger.	Extends index finger.	Deep radial nerve.

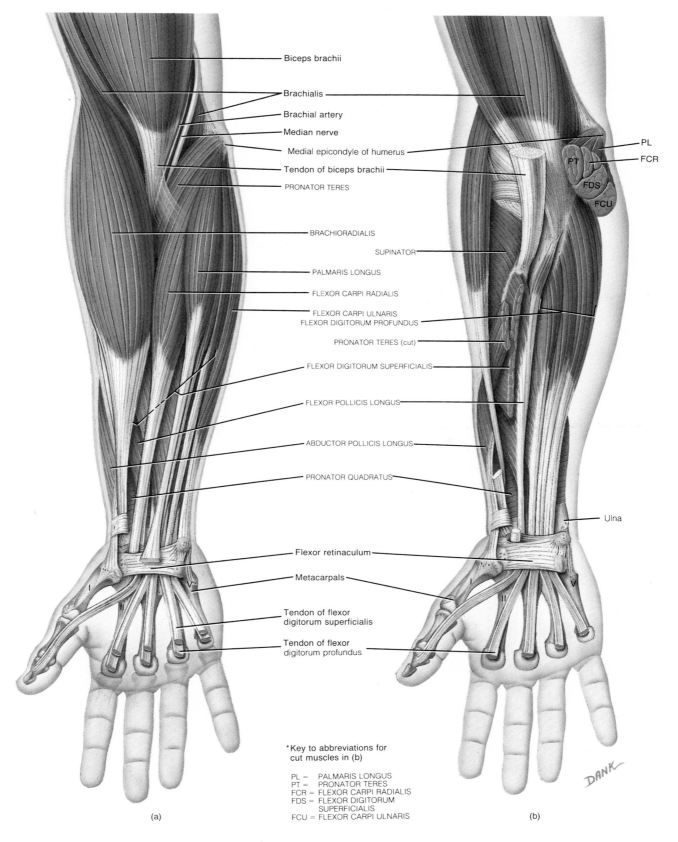

Biceps brachii

Brachialis

Brachial artery

Median nerve

Medial epicondyle of humerus

Tendon of biceps brachii

PRONATOR TERES

PL

FCR

PT

FDS

FCU

BRACHIORADIALIS

SUPINATOR

PALMARIS LONGUS

FLEXOR CARPI RADIALIS

FLEXOR CARPI ULNARIS
FLEXOR DIGITORUM PROFUNDUS

PRONATOR TERES (cut)

FLEXOR DIGITORUM SUPERFICIALIS

FLEXOR POLLICIS LONGUS

ABDUCTOR POLLICIS LONGUS

PRONATOR QUADRATUS

Ulna

Flexor retinaculum

Metacarpals

Tendon of flexor
digitorum superficialis

Tendon of flexor
digitorum profundus

*Key to abbreviations for
cut muscles in (b)

PL =   PALMARIS LONGUS
PT =   PRONATOR TERES
FCR = FLEXOR CARPI RADIALIS
FDS = FLEXOR DIGITORUM
          SUPERFICIALIS
FCU = FLEXOR CARPI ULNARIS

(a)

(b)

FIGURE 10-19 Muscles that move the wrist, hand, and fingers. (a) Superficial anterior view. (b) Deep anterior view.

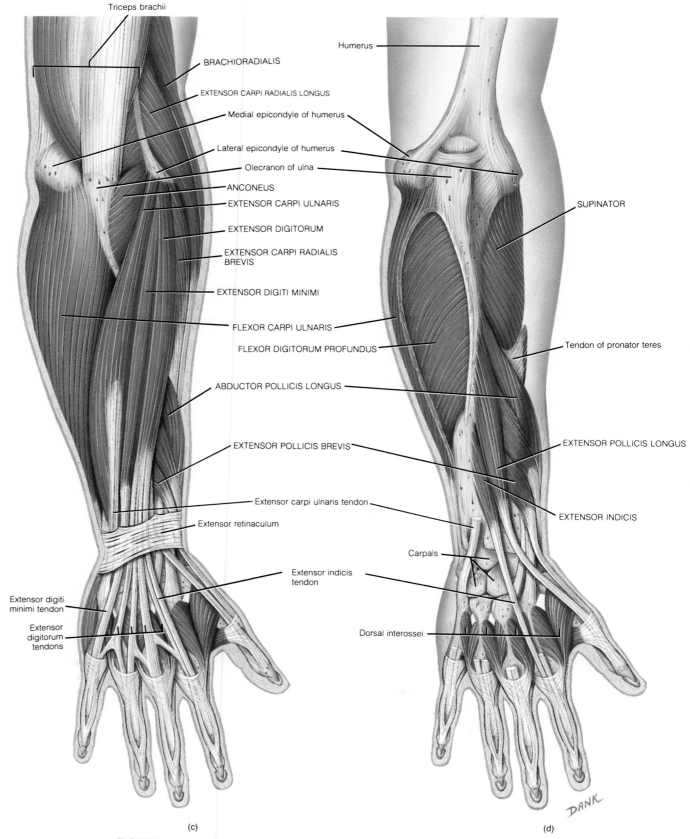

Triceps brachii

BRACHIORADIALIS

EXTENSOR CARPI RADIALIS LONGUS

Medial epicondyle of humerus

Lateral epicondyle of humerus

Olecranon of ulna

ANCONEUS

EXTENSOR CARPI ULNARIS

EXTENSOR DIGITORUM

EXTENSOR CARPI RADIALIS BREVIS

EXTENSOR DIGITI MINIMI

FLEXOR CARPI ULNARIS

FLEXOR DIGITORUM PROFUNDUS

ABDUCTOR POLLICIS LONGUS

EXTENSOR POLLICIS BREVIS

Extensor carpi ulnaris tendon

Extensor retinaculum

Extensor indicis tendon

Extensor digiti minimi tendon

Extensor digitorum tendons

Humerus

SUPINATOR

Tendon of pronator teres

EXTENSOR POLLICIS LONGUS

EXTENSOR INDICIS

Carpals

Dorsal interossei

DANK

(c)   (d)

FIGURE 10-19 (Continued) (c) Superficial posterior view. (d) Deep posterior view.

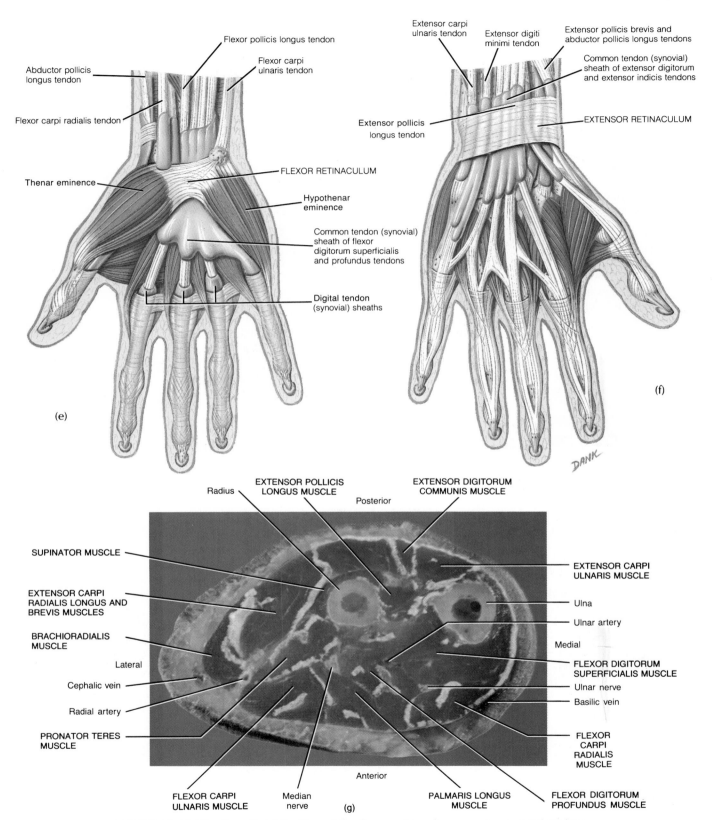

FIGURE 10-19 (*Continued*) (e) Position of the flexor retinaculum seen in an anterior view of the right wrist and hand. (f) Position of the extensor retinaculum seen in a posterior view of the right wrist and hand. (g) Photograph of a cross section of the forearm showing the musculature and related structures. (Courtesy of Stephen A. Kieffer and E. Robert Heitzman, *An Atlas of Cross-Sectional Anatomy*, Harper & Row, Publishers, Inc., New York, 1979.)

EXHIBIT 10-19

Intrinsic Muscles of the Hand (Figure 10-20)

OVERVIEW Several of the muscles discussed in Exhibit 10-18 help move the digits in various ways. In addition, there are muscles in the palmar surface of the hand called *intrinsic muscles* that also help move the digits. Such muscles are so named because their origins and insertions are both within the hands. These muscles assist in the intricate and precise movements that are characteristic of the human hand.

The intrinsic muscles of the hand are divided into three principal groups—thenar, hypothenar, and intermediate. The four *thenar* (THĒ-nar) *muscles* act on the thumb and form the *thenar eminence* (see Figure 11-13). The four *hypothenar* (HĪ-pō-thē'-nar) *muscles* act on the little finger and form the *hypothenar eminence* (see Figure 11-13). The eleven *intermediate (midpalmar) muscles* act on all the digits, except the thumb.

MUSCLE	ORIGIN	INSERTION	ACTION	INNERVATION
THENAR MUSCLES				
Abductor pollicis brevis (*abductor* = moves part away from midline; *pollex* = thumb; *brevis* = short)	Flexor retinaculum, scaphoid, and trapezium.	Proximal phalanx of thumb.	Abducts thumb.	Median nerve.
Opponens pollicis (*opponens* = opposes)	Flexor retinaculum and trapezium.	Metacarpal of thumb.	Draws thumb across palm to meet little finger (opposition).	Median nerve.
Flexor pollicis brevis (*flexor* = decreases angle at joint)	Flexor retinaculum, trapezium, and first metacarpal.	Proximal phalanx of thumb.	Flexes and adducts thumb.	Median and ulnar nerves.
Adductor pollicis (*adductor* = moves part toward midline)	Capitate and second and third metacarpals.	Proximal phalanx of thumb.	Adducts thumb.	Ulnar nerve.
HYPOTHENAR MUSCLES				
Palmaris brevis* (*palma* = palm)	Flexor retinaculum and palmar aponeurosis.	Skin on ulnar border of palm of hand.	Draws skin toward middle of palm as in clenching fist.	Ulnar nerve.
Abductor digiti minimi (*digit* = finger or toe; *minimi* = little finger)	Pisiform and tendon of flexor carpi ulnaris.	Proximal phalanx of little finger.	Abducts little finger.	Ulnar nerve.
Flexor digiti minimi brevis	Flexor retinaculum and hamate.	Proximal phalanx of little finger.	Flexes little finger.	Ulnar nerve.
Opponens digiti minimi	Flexor retinaculum and hamate.	Metacarpal of little finger.	Draws little finger across palm to meet thumb.	Ulnar nerve.
INTERMEDIATE (MIDPALMAR) MUSCLES				
Lumbricals (four muscles)	Tendons of flexor digitorum profundus.	Tendons of extensor digitorum.	Extend interphalangeal joints.	Median and ulnar nerves.
Dorsal interossei (four muscles; *dorsal* = back surface; *inter* = between; *ossei* = bones)	Adjacent sides of metacarpals.	Proximal phalanx of second, third, and fourth fingers.	Abduct fingers from middle finger and flex fingers at metacarpophalangeal joints.	Ulnar nerve.
Palmar interossei (three muscles)	Medial side of second metacarpal and lateral sides of fourth and fifth metacarpals.	Proximal phalanx of same finger.	Adduct fingers toward middle finger.	Ulnar nerve.

* Not illustrated

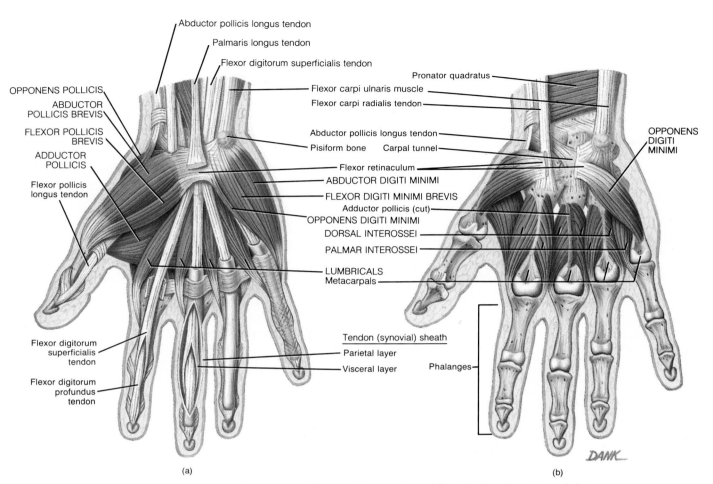

Abductor pollicis longus tendon

Palmaris longus tendon

Flexor digitorum superficialis tendon

OPPONENS POLLICIS

ABDUCTOR POLLICIS BREVIS

FLEXOR POLLICIS BREVIS

ADDUCTOR POLLICIS

Flexor pollicis longus tendon

Flexor digitorum superficialis tendon

Flexor digitorum profundus tendon

Pronator quadratus

Flexor carpi ulnaris muscle

Flexor carpi radialis tendon

Abductor pollicis longus tendon

Pisiform bone   Carpal tunnel

Flexor retinaculum

ABDUCTOR DIGITI MINIMI

FLEXOR DIGITI MINIMI BREVIS

Adductor pollicis (cut)

OPPONENS DIGITI MINIMI

DORSAL INTEROSSEI

PALMAR INTEROSSEI

LUMBRICALS

Metacarpals

Tendon (synovial) sheath

Parietal layer

Visceral layer

OPPONENS DIGITI MINIMI

Phalanges

DANK

(a)   (b)

FIGURE 10-20 Palmar view of the intrinsic muscles of the right hand. (a) Superficial. (b) Deep.

EXHIBIT 10-20

Muscles That Move the Vertebral Column (Figure 10-21)

OVERVIEW The muscles that move the vertebral column are quite complex because they have multiple origins and insertions and there is considerable overlapping among them. One way to group the muscles is on the basis of the general direction of the muscle bundles and their approximate lengths. For example, the **splenius muscles** arise from the midline and run laterally and superiorly to their insertions. The **erector spinae (sacrospinalis) muscle** arises from either the midline or more laterally but, usually, runs almost longitudinally, with neither a marked outward nor inward direction as it is traced superiorly. The **transversospinalis muscles** arise laterally but run toward the midline as they are traced superiorly. Deep to these three muscle groups are small **segmental muscles** that run between spinous processes or transverse processes of vertebrae. Since the scalene muscles also assist in moving the vertebral column, they are included in this exhibit.

Note in Exhibit 10-11 that the rectus abdominis and quadratus lumborum muscles also assume a role in moving the vertebral column.

MUSCLE	ORIGIN	INSERTION	ACTION	INNERVATION
SPLENIUS MUSCLES				
Splenius capitis (*splenium* = bandage; *caput* = head)	Ligamentum nuchae and spinous processes of seventh cervical vertebra and first three or four thoracic vertebrae.	Occipital bone and mastoid process of temporal bone.	Acting together, they extend the head and neck; acting singly, each laterally flexes and rotates head to same side.	Dorsal rami of middle cervical nerves.
Splenius cervicis (*cervix* = neck)	Spinous processes of third through sixth thoracic vertebrae.	Transverse processes of first two to four cervical vertebrae.	Acting together, they extend the head and neck; acting singly, each laterally flexes and rotates head to same side.	Dorsal rami of lower cervical nerves.
ERECTOR SPINAE (SACROSPINALIS)	This is the largest muscular mass of the back and consists of three groupings—iliocostalis, longissimus, and spinalis. These groups, in turn, consist of a series of overlapping muscles. The iliocostalis group is laterally placed, the longissimus group is intermediate in placement, and the spinalis group is medially placed.			
Iliocostalis (lateral) group				
Iliocostalis lumborum (*ilium* = flank; *costa* = rib)	Iliac crest.	Lower six ribs.	Extends lumbar region of vertebral column.	Dorsal rami of lumbar nerves.
Iliocostalis thoracis (*thorax* = chest)	Lower six ribs.	Upper six ribs.	Maintains erect position of spine.	Dorsal rami of thoracic (intercoastal) nerves.
Longissimus (intermediate) group				
Longissimus thoracis (*longissimus* = longest)	Transverse processes of lumbar vertebrae.	Transverse processes of all thoracic and upper lumbar vertebrae and ninth and tenth ribs.	Extends thoracic region of vertebral column.	Dorsal rami of spinal nerves.
Longissimus cervicis	Transverse processes of fourth and fifth thoracic vertebrae.	Transverse processes of second to sixth cervical vertebrae.	Extends cervical region of vertebral column.	Dorsal rami of spinal nerves.
Longissimus capitis	Transverse processes of upper four thoracic vertebrae.	Mastoid process of temporal bone.	Extends head and rotates it to opposite side.	Dorsal rami of middle and lower cervical nerves.
Spinalis (medial) group				
Spinalis thoracis (*spinalis* = vertebral column)	Spinous processes of upper lumbar and lower thoracic vertebrae.	Spinous processes of upper thoracic vertebrae.	Extends vertebral column.	Dorsal rami or spinal nerves.

EXHIBIT 10-20 (Continued)

MUSCLE	ORIGIN	INSERTION	ACTION	INNERVATION
Spinalis cervicis	Ligamentum nuchae and spinous process of seventh cervical vertebra.	Spinous process of axis.	Extends vertebral column.	Dorsal rami of spinal nerves.
Spinalis capitis	Arises with semispinalis thoracis.	Inserts with spinalis thoracis.	Extends vertebral column.	Dorsal rami of spinal nerves
TRANSVERSOSPINALIS MUSCLES				
Semispinalis thoracis (*semi* = partially or one-half)	Transverse processes of sixth to tenth thoracic vertebrae.	Spinous processes of first four thoracic and last two cervical vertebrae.	Extends vertebral column and rotates it to opposite side.	Dorsal rami of thoracic and cervical spinal nerves.
Semispinalis cervicis	Transverse processes of first five or six thoracic vertebrae.	Spinous processes of first to fifth cervical vertebrae.	Extends vertebral column and rotates it to opposite side.	Dorsal rami of thoracic and cervical spinal nerves.
Semispinalis capitis	Transverse processes of first six or seven thoracic vertebrae and seventh cervical vertebra and articular processes of fourth, fifth, and sixth cervical vertebrae.	Occipital bone.	Extends vertebral column and rotates it to opposite side.	Dorsal rami of cervical nerves.
Multifidus (*multi* = many; *findere* = to split)	Sacrum, ilium, transverse processes of lumbar, thoracic, and lower four cervical vertebrae.	Spinous process of a higher vertebra.	Extends vertebral column and rotates it to opposite side.	Dorsal rami of spinal nerves.
Rotatores (*rotate* = turn on an axis)	Transverse processes of all vertebrae.	Spinous process of vertebra above the one of origin.	Extend vertebral column and rotate it to opposite side.	Dorsal rami of spinal nerves.
SEGMENTAL MUSCLES **Interspinales** (*inter* = between)	Superior surface of all spinous processes.	Inferior surface of spinous process of vertebra above the one of origin.	Extend vertebral column.	Dorsal rami of spinal nerves.
Intertransversarii (*inter* = between)	Transverse processes of all vertebrae.	Transverse process of vertebra above the one of origin.	Abduct vertebral column.	Dorsal and ventral rami of spinal nerves.
SCALENE MUSCLES				
Anterior scalene (*anterior* = front; *skalenos* = uneven)	Transverse processes of third through sixth cervical vertebrae.	First rib.	Flexes and rotates neck and assists in inspiration.	Ventral rami of fifth and sixth cervical nerves.
Middle scalene	Transverse processes of last six cervical vertebrae.	First rib.	Flexes and rotates neck and assists in inspiration.	Ventral rami of third through eighth cervical nerves.
Posterior scalene	Transverse processes of fourth through sixth cervical vertebrae.	Second or third rib.	Flexes and rotates neck and assists in inspiration.	Ventral rami of last three cervical nerves.

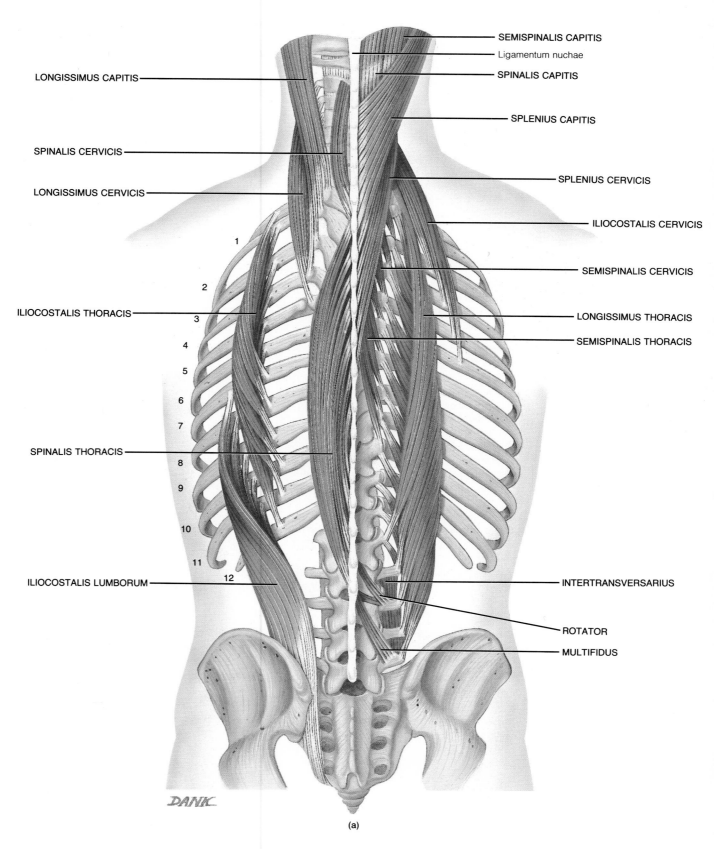

LONGISSIMUS CAPITIS

SPINALIS CERVICIS

LONGISSIMUS CERVICIS

ILIOCOSTALIS THORACIS

SPINALIS THORACIS

ILIOCOSTALIS LUMBORUM

SEMISPINALIS CAPITIS

Ligamentum nuchae

SPINALIS CAPITIS

SPLENIUS CAPITIS

SPLENIUS CERVICIS

ILIOCOSTALIS CERVICIS

SEMISPINALIS CERVICIS

LONGISSIMUS THORACIS

SEMISPINALIS THORACIS

INTERTRANSVERSARIUS

ROTATOR

MULTIFIDUS

1
2
3
4
5
6
7
8
9
10
11
12

DANK

(a)

FIGURE 10-21 Muscles that move the vertebral column. (a) Posterior view of deep muscles of the neck and back.

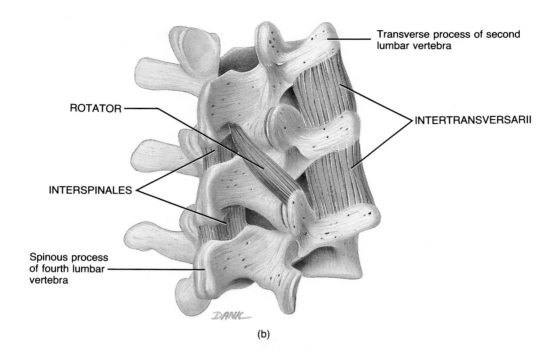

Transverse process of second lumbar vertebra

INTERTRANSVERSARII

ROTATOR

INTERSPINALES

Spinous process of fourth lumbar vertebra

(b)

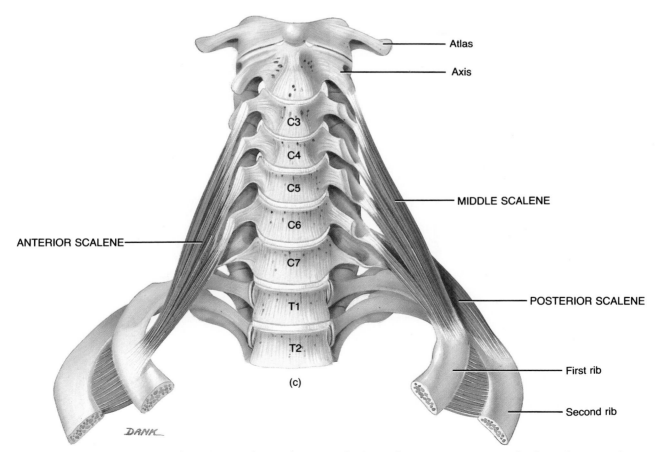

Atlas

Axis

C3

C4

C5

C6

C7

T1

T2

MIDDLE SCALENE

ANTERIOR SCALENE

POSTERIOR SCALENE

First rib

Second rib

(c)

FIGURE 10-21 (Continued) (b) Posterolateral view of several intervertebral muscles. (c) Anterior view of right scalene muscles.

EXHIBIT 10-21

Muscles That Move the Thigh (Femur) (Figure 10-22)

OVERVIEW As you will see, muscles of the lower extremities are larger and more powerful than those of the upper extremities since lower extremity muscles function in stability, locomotion, and maintenance of posture. Upper extremity muscles are characterized by versatility of movement. In addition, muscles of the lower extremities frequently cross two joints and act equally on both.

The majority of muscles that act on the thigh (femur) originate on the pelvic (hip) girdle and insert on the femur. The anterior muscles are the psoas major and iliacus, together referred to as the iliopsoas muscle. The remaining muscles (except for the pectineus, adductors, and tensor fasciae latae) are posterior muscles. Technically, the pectineus and adductors are components of the medial compartment of the thigh, but they are included in this exhibit because they act on the thigh. The tensor fasciae latae muscle is laterally placed. The *fascia lata* is a deep fascia of the thigh that encircles the entire thigh. It is well developed laterally where, together with the tendons of the gluteus maximus and tensor fasciae latae muscles, forms a structure called the *iliotibial tract.* The tract inserts into the lateral condyle of the tibia.

After you have studied the muscles in this exhibit, arrange them according to the following actions: flexion, extension, abduction, adduction, medial rotation, and lateral rotation. (The same muscles can be used more than once).

MUSCLE	ORIGIN	INSERTION	ACTION	INNERVATION
Psoas major (*psoa* = muscle of loin)	Transverse processes and bodies of lumbar vertebrae.	Lesser trochanter of femur.	Flexes and rotates thigh laterally; flexes vertebral column.	Lumbar nerves L2–L3.
Iliacus (*iliac* = ilum)	Iliac fossa.	Tendon of psoas major.	Flexes and rotates thigh laterally.	Femoral nerve.
Gluteus maximus (*glutos* = buttock; *maximus* = largest)	Iliac crest, sacrum, coccyx, and aponeurosis of sacrospinalis.	Iliotibial tract of fascia lata and gluteal tuberosity of femur.	Extends, abducts, adducts, and rotates thigh laterally.	Inferior gluteal nerve.
Gluteus medius (*media* = middle)	Ilium.	Greater trochanter of femur.	Abducts and rotates thigh medially.	Superior gluteal nerve.
Gluteus minimus (*minimus* = smallest)	Ilium.	Greater trochanter of femur.	Abducts and rotates thigh medially.	Superior gluteal nerve.
Tensor fasciae latae (*tensor* = makes tense; *fascia* = band; *latus* = wide)	Iliac crest.	Tibia by way of the iliotibial tract.	Flexes and abducts thigh.	Superior gluteal nerve.
Piriformis (*pirum* = pear; *forma* = shape)	Sacrum.	Greater trochanter of femur.	Rotates thigh laterally and abducts it.	Sacral nerves S2 or S1–S2.
Obturator Internus (*obturator* = obturator foramen; *internus* = inside)	Inner surface of obturator foramen, pubis, and ischium.	Greater trochanter of femur.	Rotates thigh laterally and abducts it.	Nerve to obturator internus.
Obturator externus (*externus* = outside)	Outer surface of obturator membrane.	Trochanteric fossa of femur.	Rotates thigh laterally.	Obturator nerve.
Superior gemellus (*superior* = above; *gemellus* = twins)	Ischial spine.	Greater trochanter of femur.	Rotates thigh laterally and abducts it.	Nerve to obturator internus.
Inferior gemellus (*inferior* = below)	Ischial tuberosity.	Greater trochanter of femur.	Rotates thigh laterally and abducts it.	Nerve to quadratus femoris.
Quadratus femoris (*quad* = four; *femoris* = femur)	Ischial tuberosity.	Small tubercle on posterior femur.	Laterally rotates and adducts thigh.	Nerve to quadratus femoris.
Adductor longus (*adductor* = moves part closer to midline; *longus* = long)	Pubic crest and symphysis pubis.	Linea aspera of femur.	Adducts, laterally rotates, and flexes thigh.	Obturator nerve.
Adductor brevis (*brevis* = short)	Inferior ramus of pubis.	Linea aspera of femur.	Adducts, laterally rotates, and flexes thigh.	Obturator nerve.

EXHIBIT 10-21 (Continued)

MUSCLE	ORIGIN	INSERTION	ACTION	INNERVATION
Adductor magnus (*magnus* = large)	Inferior ramus of pubis and ischium to ischial tuberosity.	Linea aspera of femur.	Adducts, flexes, laterally rotates, and extends thigh (anterior part flexes, posterior part extends).	Obturator and sciatic nerves.
Pectineus (*pecten* = comb-shaped)	Fascia of pubis.	Pectineal line of femur.	Flexes and adducts thigh.	Femoral nerve.

CLINICAL APPLICATION

A common sports injury is a **"pulled groin."** Very simply this means there is a strain, stretching, and probably some tearing away of the tendinous origins of the adductor muscles that move the thigh.

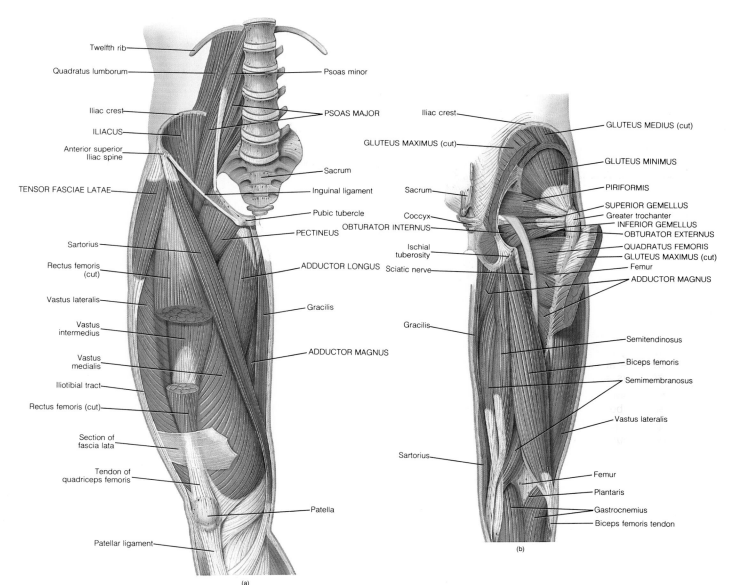

FIGURE 10-22 Muscles that move the thigh (femur). (a) Anterior superficial view. (b) Posterior deep view.

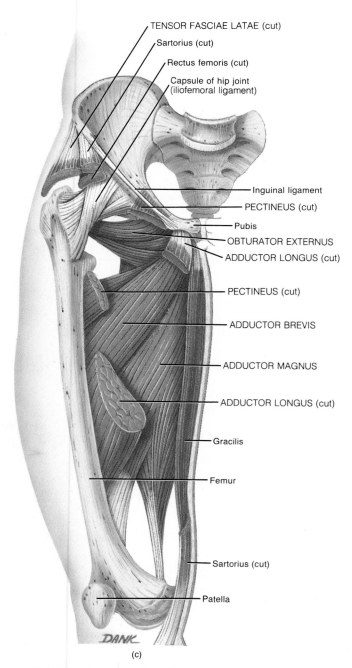

TENSOR FASCIAE LATAE (cut)

Sartorius (cut)

Rectus femoris (cut)

Capsule of hip joint
(iliofemoral ligament)

Inguinal ligament

PECTINEUS (cut)

Pubis

OBTURATOR EXTERNUS

ADDUCTOR LONGUS (cut)

PECTINEUS (cut)

ADDUCTOR BREVIS

ADDUCTOR MAGNUS

ADDUCTOR LONGUS (cut)

Gracilis

Femur

Sartorius (cut)

Patella

DANK

(c)

FIGURE 10-22 (*Continued*) (c) Anterior deep view.

EXHIBIT 10-22

Muscles That Act on the Leg (Tibia and Fibula) (Figure 10-23)

OVERVIEW   The muscles that act on the leg (tibia and fibula) originate in the hip and thigh and are separated into compartments by deep fascia. The **medial (adductor) compartment** is so named because its muscles adduct the thigh. It is innervated by the obturator nerve. As noted earlier, the adductor magnus, adductor longus, adductor brevis, and pectineus muscles, components of the medial compartment, are included in Exhibit 10-21 because they act on the femur. The gracilis, the other muscle in the medial compartment, not only adducts the thigh but also flexes the leg. For this reason, it is included in this exhibit.

The **anterior (extensor) compartment** is so designated because its muscles act mainly to extend the leg. It is composed of the quadriceps femoris and sartorius muscles and is innervated by the femoral nerve. The quadriceps femoris muscle is a composite muscle that includes four distinct parts, usually described as four separate muscles (rectus femoris, vastus lateralis, vastus intermedius, and vastus medialis). The common tendon for the four muscles is known as the **patellar ligament** and attaches to the tibial tuberosity.

The **posterior (flexor) compartment** is so named because its muscles flex the leg. It is innervated by branches of the sciatic nerve. Included are the hamstrings (biceps femoris, semitendinosus, and semimembranosus). The hamstrings are so named because their tendons are long and stringlike in the popliteal area. The **popliteal fossa** is a diamond-shaped space on the posterior aspect of the knee bordered laterally by the biceps femoris and medially by the semitendinosus and semimembranosus muscles (see also Figure 11-15b).

MUSCLE	ORIGIN	INSERTION	ACTION	INNERVATION
MEDIAL (ADDUCTOR) COMPARTMENT				
Adductor magnus **Adductor longus** **Adductor brevis** **Pectineus**	See Exhibit 10-21			
Gracilis (*gracilis* = slender)	Symphysis pubis and pubic arch.	Medial surface of body of tibia.	Adducts thigh and flexes leg.	Obturator nerve.
ANTERIOR (EXTENSOR) COMPARTMENT				
Quadriceps femoris (*quadriceps* = four heads of origin; *femoris* = femur)				
Rectus femoris (*rectus* = fibers parallel to midline)	Anterior inferior iliac spine.	Upper border of patella.	All four heads extend leg; rectus portion alone also flexes thigh.	Femoral nerve.
Vastus lateralis (*vastus* = large; *lateralis* = lateral)	Greater trochanter and linea aspera of femur.			Femoral nerve.
Vastus medialis (*medialis* = medial)	Linea aspera of femur.	Tibial tuberosity through patellar ligament (tendon of quadriceps).		Femoral nerve.
Vastus intermedius (*intermedius* = middle)	Anterior and lateral surfaces of body of femur.			Femoral nerve.
Sartorius (*sartor* = tailor; refers to cross-legged position of tailors)	Anterior superior spine of ilium.	Medial surface of body of tibia.	Flexes and medially rotates leg; flexes thigh and rotates it laterally, thus crossing leg.	Femoral nerve.

EXHIBIT 10-22 (Continued)

MUSCLE	ORIGIN	INSERTION	ACTION	INNERVATION
POSTERIOR (FLEXOR) COMPARTMENT				
Hamstrings	A collective designation for three separate muscles.			
Biceps femoris (*biceps* = two heads of origin)	Long head arises from ischial tuberosity; short head arises from linea aspera of femur.	Head of fibula and lateral condyle of tibia.	Flexes and laterally rotates leg and extends thigh.	Tibial nerve from sciatic nerve.
Semitendinosus (*semi* = half; *tendo* = tendon)	Ischial tuberosity.	Proximal part of medial surface of body of tibia.	Flexes and medially rotates leg and extends thigh.	Tibial nerve from sciatic nerve.
Semimembranosus (*membran* = membrane)	Ischial tuberosity.	Medial condyle of tibia.	Flexes and medially rotates leg and extends thigh.	Tibial nerve from sciatic nerve.

CLINICAL APPLICATION

"Pulled hamstrings" are common sports injuries in individuals who run very hard. Sometimes the violent muscular exertion required to perform a feat tears off part of the tendinous origins of the hamstrings from the ischial tuberosity. This is usually accompanied by a contusion (bruising) and tearing of some of the muscle fibers and rupture of blood vessels producing a hematoma (collection of blood).

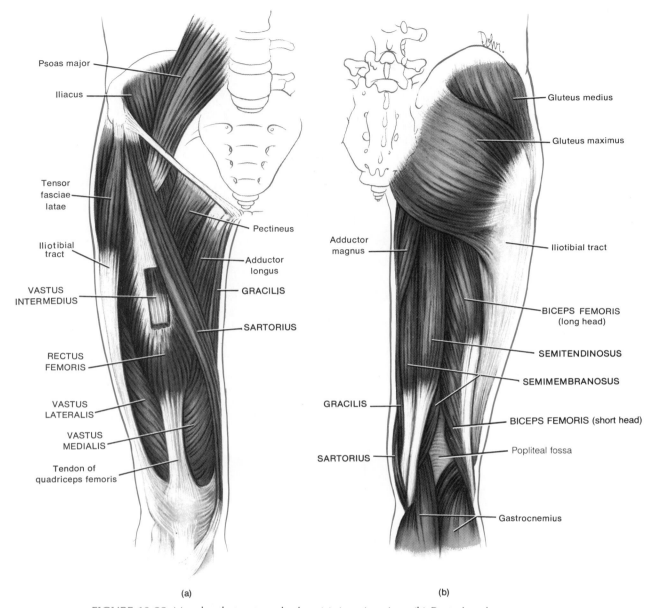

Psoas major

Iliacus

Tensor
fasciae
latae

Iliotibial
tract

VASTUS
INTERMEDIUS

RECTUS
FEMORIS

VASTUS
LATERALIS

VASTUS
MEDIALIS

Tendon of
quadriceps femoris

Pectineus

Adductor
longus

GRACILIS

SARTORIUS

(a)

Gluteus medius

Gluteus maximus

Iliotibial tract

Adductor
magnus

BICEPS FEMORIS
(long head)

SEMITENDINOSUS

SEMIMEMBRANOSUS

GRACILIS

BICEPS FEMORIS (short head)

SARTORIUS

Popliteal fossa

Gastrocnemius

(b)

FIGURE 10-23 Muscles that act on the leg. (a) Anterior view. (b) Posterior view.

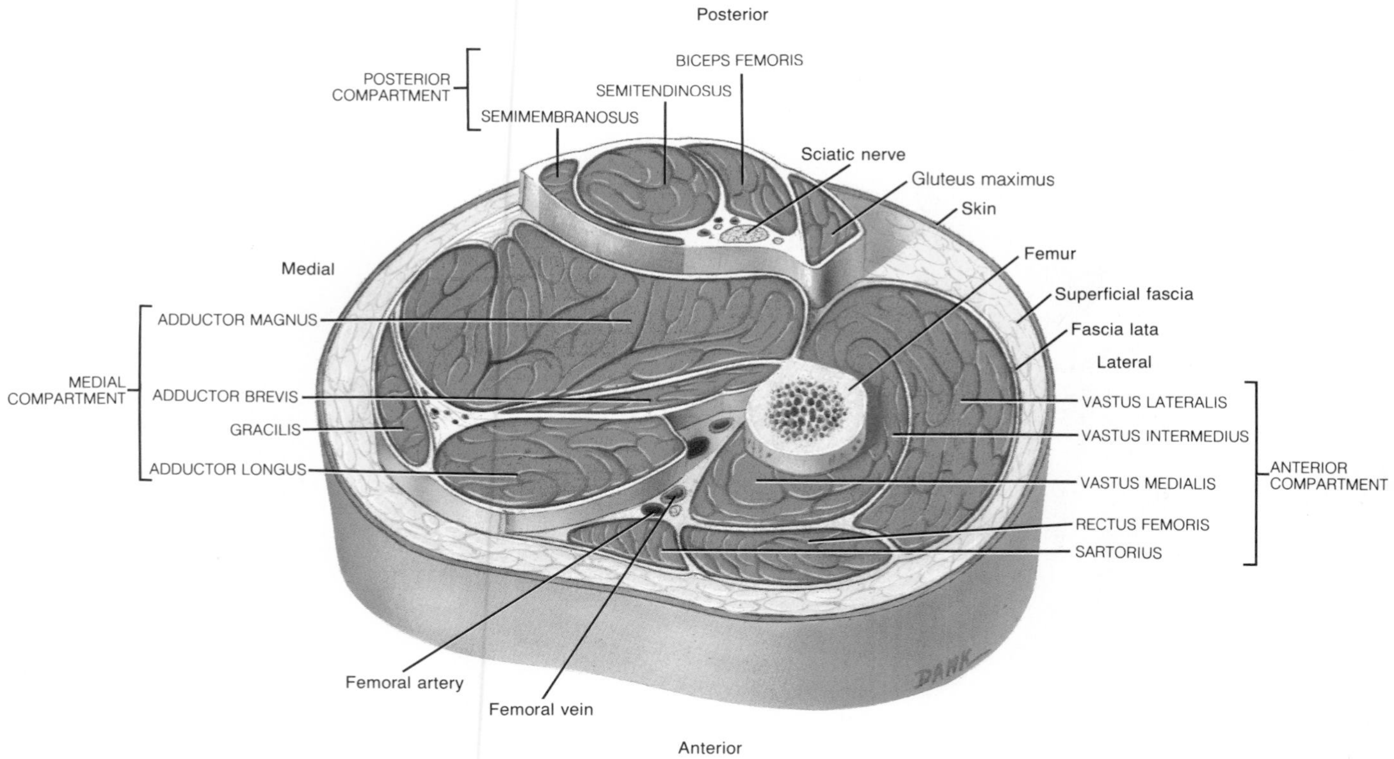

FIGURE 10-23 (*Continued*) (c) Diagram of a cross section of the thigh showing the musculature and related structure. (d) Photograph of a cross section of the thigh showing the musculature and related structures. (Courtesy of Stephen A. Kieffer and E. Robert Heitzman, *An Atlas of Cross-Sectional Anatomy,* Harper & Row, Publishers, Inc., New York, 1979.) The surface anatomy of the muscles that act on the leg is illustrated in Figure 11-14.

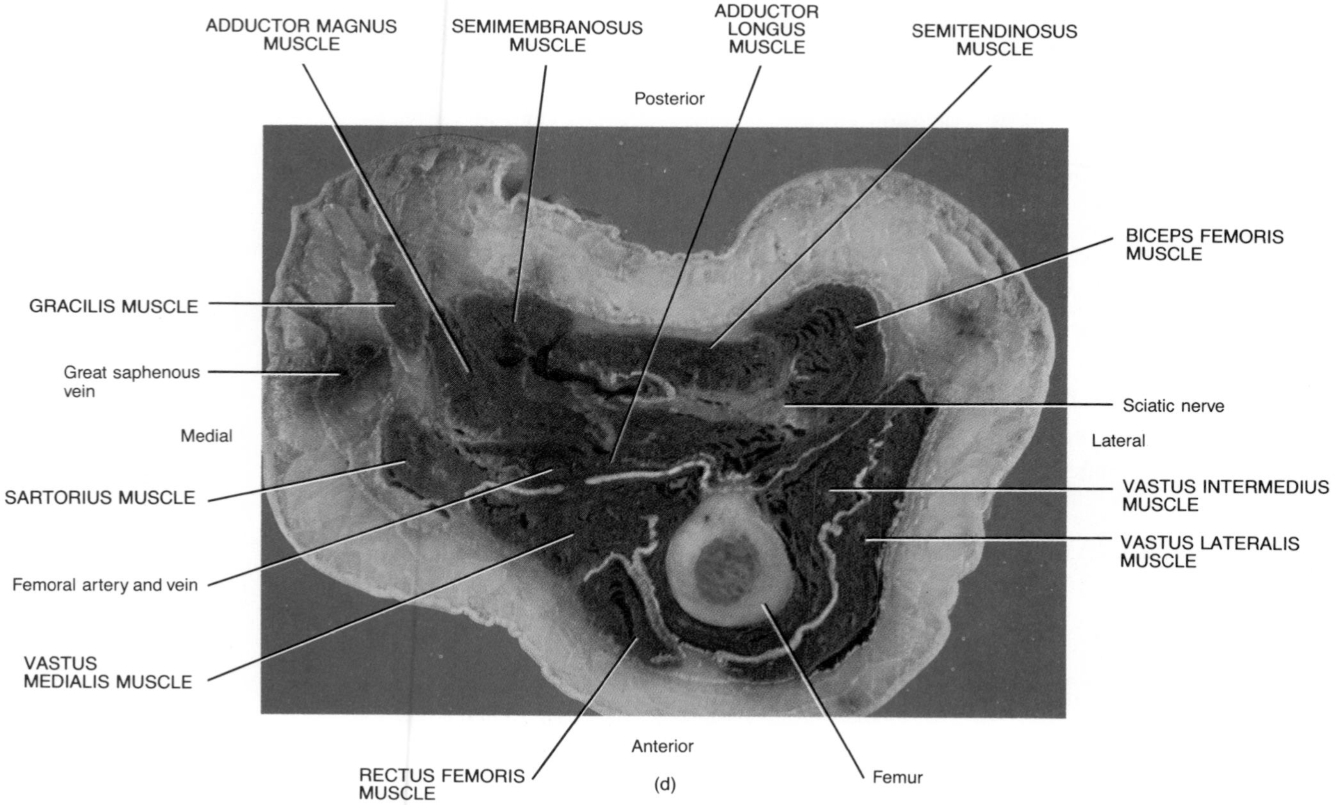

EXHIBIT 10-23

Muscles That Move the Foot and Toes (Figure 10-24)

OVERVIEW The musculature of the leg, like that of the thigh, can be distinguished into three compartments by deep fascia. In addition, all the muscles in the respective compartments are innervated by the same nerve. The ***anterior compartment*** consists of muscles that dorsiflex the foot and are innervated by the deep peroneal nerve. In a situation analogous to the wrist, the tendons of the muscles of the anterior compartment are held firmly to the ankle by thickenings of deep fascia called the ***superior extensor retinaculum*** (***transverse ligament of the ankle***) and ***inferior extensor retinaculum*** (***cruciate ligament of the ankle***).

The ***lateral (peroneal) compartment*** contains two muscles that plantar flex and evert the foot. They are supplied by the superficial peroneal nerve.

The ***posterior compartment*** consists of muscles that are divisible into superficial and deep groups. All are innervated by the tibial nerve. All three superficial muscles share a common tendon of insertion, the calcaneal (Achilles) tendon that inserts into the calcaneal bone of the ankle. The superficial muscles are plantar flexors of the foot. Of the four deep muscles, three plantar flex the foot.

MUSCLE	ORIGIN	INSERTION	ACTION	INNERVATION
ANTERIOR COMPARTMENT				
Tibialis anterior (*tibialis* = tibia; *anterior* = front)	Lateral condyle and body of tibia and interosseous membrane.	First metatarsal and first (medial) cuneiform.	Dorsiflexes and inverts foot.	Deep peroneal nerve.
Extensor hallucis longus (*extensor* = increases angle at joint; *hallucis* = hallux or great toe; *longus* = long)	Anterior surface of fibula and interosseous membrane.	Distal phalanx of great toe.	Dorsiflexes and inverts foot and extends great toe.	Deep peroneal nerve.
Extensor digitorum longus (*digitorum* = finger or toe)	Lateral condyle of tibia, anterior surface of fibula, and interosseous membrane.	Middle and distal phalanges of four outer toes.	Dorsiflexes and everts foot and extends toes.	Deep peroneal nerve.
Peroneus tertius (*perone* = fibula; *tertius* = third)	Distal third of fibula and interosseous membrane.	Fifth metatarsal.	Dorsiflexes and everts foot.	Deep peroneal nerve.
LATERAL (PERONEAL) COMPARTMENT				
Peroneus longus	Head and body of fibula and lateral condyle of tibia.	First metatarsal and first cuneiform.	Plantar flexes and everts foot.	Superficial peroneal nerve.
Peroneus brevis (*brevis* = short)	Body of fibula.	Fifth metatarsal.	Plantar flexes and everts foot.	Superficial peroneal nerve.
POSTERIOR COMPARTMENT				
Superficial				
Gastrochemius (*gaster* = belly; *kneme* = leg)	Lateral and medial condyles of femur and capsule of knee.	Calcaneus by way of calcaneal (Achilles) tendon.	Plantar flexes foot and flexes leg.	Tibial nerve.
Soleus (*soleus* = sole of foot)	Head of fibula and medial border of tibia.	Calcaneus by way of calcaneal (Achilles) tendon.	Plantar flexes foot.	Tibial nerve.
Plantaris (*plantar* = sole of foot)	Femur above lateral condyle.	Calcaneus by way of calcaneal (Achilles) tendon.	Plantar flexes foot.	Tibial nerve.
Deep				
Popliteus (*poples* = posterior surface of knee)	Lateal condyle of femur.	Proximal tibia.	Flexes and medially rotates leg.	Tibial nerve.
Flexor hallucis longus (*flexor* = decreases angle at joint)	Lower two-thirds of fibula.	Distal phalanx of great toe.	Plantar flexes and everts foot and flexes great toe.	Tibial nerve.
Flexor digitorum longus	Posterior surface.	Distal phalanges of four outer toes.	Plantar flexes and inverts foot and flexes toes.	Tibial nerve.

EXHIBIT 10-23 (*Continued*)

MUSCLE	ORIGIN	INSERTION	ACTION	INNERVATION
Tibialis posterior (*posterior* = back)	Tibia, fibula, and interosseous membrane.	Second, third, and fourth metatarsals; navicular; third (lateral) cuneiform; and cuboid.	Plantar flexes and inverts foot.	Tibial nerve.

CLINICAL APPLICATION

The calcaneal tendon, the **strongest tendon of the body,** is able to withstand a 1,000-pound force without tearing.

Despite this, however, the calcaneal tendon ruptures more frequently than any other tendon.

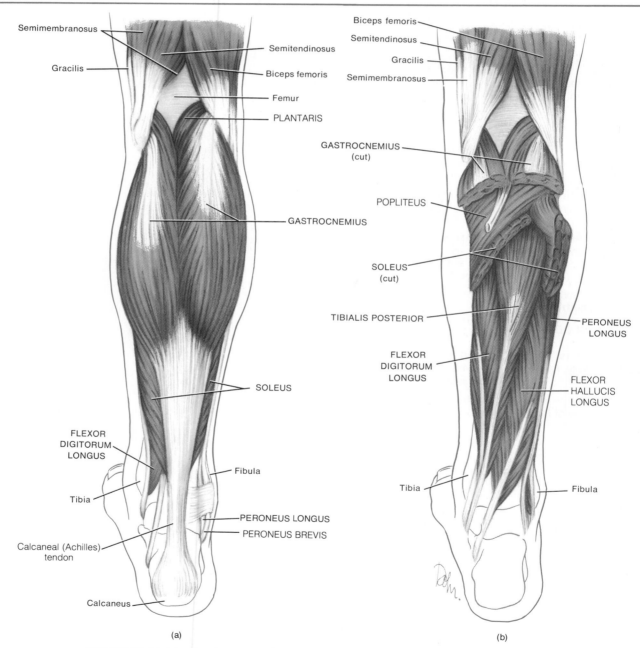

(a)   (b)

FIGURE 10-24 Muscles that move the foot and toes. (a) Superficial posterior view. (b) Deep posterior view.

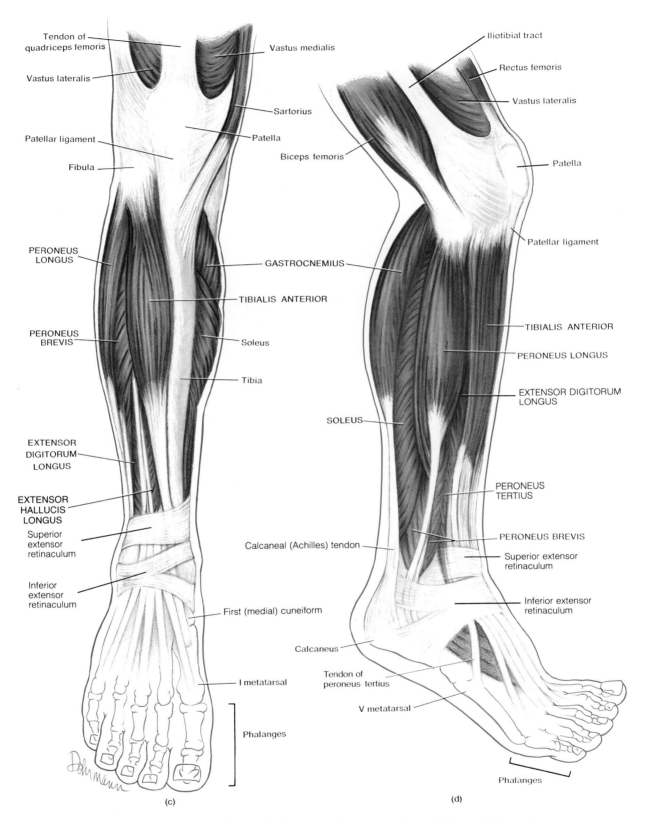

Tendon of quadriceps femoris

Vastus lateralis

Patellar ligament

Fibula

PERONEUS LONGUS

PERONEUS BREVIS

EXTENSOR DIGITORUM LONGUS

EXTENSOR HALLUCIS LONGUS

Superior extensor retinaculum

Inferior extensor retinaculum

Vastus medialis

Sartorius

Patella

GASTROCNEMIUS

TIBIALIS ANTERIOR

Soleus

Tibia

First (medial) cuneiform

I metatarsal

Phalanges

(c)

Iliotibial tract

Rectus femoris

Vastus lateralis

Patella

Biceps femoris

Patellar ligament

TIBIALIS ANTERIOR

PERONEUS LONGUS

EXTENSOR DIGITORUM LONGUS

SOLEUS

PERONEUS TERTIUS

PERONEUS BREVIS

Superior extensor retinaculum

Calcaneal (Achilles) tendon

Inferior extensor retinaculum

Calcaneus

Tendon of peroneus tertius

V metatarsal

Phalanges

(d)

FIGURE 10-24 (*Continued*) Muscles that move the foot and toes. (c) Superficial anterior view. (d) Superficial lateral view.

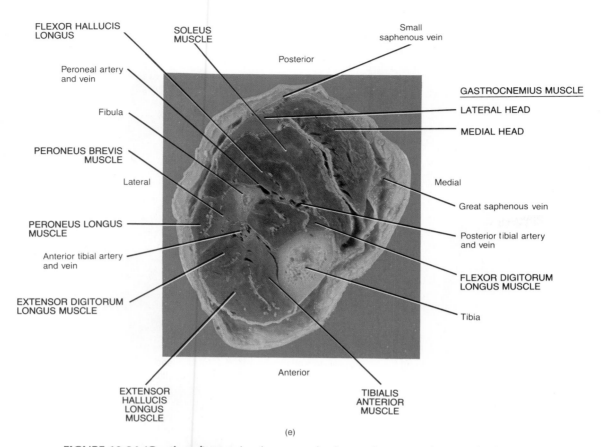

FLEXOR HALLUCIS LONGUS

SOLEUS MUSCLE

Small saphenous vein

Posterior

Peroneal artery and vein

Fibula

GASTROCNEMIUS MUSCLE

LATERAL HEAD

MEDIAL HEAD

PERONEUS BREVIS MUSCLE

Lateral

Medial

Great saphenous vein

PERONEUS LONGUS MUSCLE

Posterior tibial artery and vein

Anterior tibial artery and vein

FLEXOR DIGITORUM LONGUS MUSCLE

EXTENSOR DIGITORUM LONGUS MUSCLE

Tibia

EXTENSOR HALLUCIS LONGUS MUSCLE

Anterior

TIBIALIS ANTERIOR MUSCLE

(e)

FIGURE 10-24 (*Continued*) Muscles that move the foot and toes. (e) Photograph of a cross section of the leg showing the musculature and related structures. (Courtesy of Stephen A. Kieffer and E. Robert Heitzman, *An Atlas of Cross-Sectional Anatomy,* Harper & Row, Publishers, Inc., New York, 1979.) The surface anatomy of the muscles that move the foot and toes is illustrated in Figures 11-15 and 11-16.

EXHIBIT 10-24

Intrinsic Muscles of the Foot (Figure 10-25)

OVERVIEW The intrinsic muscles of the foot are, for the most part, comparable to those in the hand. Where the muscles of the hand are specialized for precise and intricate movements, those of the foot are limited to support and locomotion. The deep fascia of the foot forms the **plantar aponeurosis (fascia)** that extends from the calcaneus to the phalanges. The aponeurosis supports the longitudinal arch of the foot and transmits the flexor tendons of the foot.

The intrinsic musculature of the foot is divided into two groups—dorsal and plantar. There is only one **dorsal muscle.** The **plantar muscles** are arranged in four layers, the most superficial layer being referred to as the first layer.

MUSCLE	ORIGIN	INSERTION	ACTION	INNERVATION
DORSAL MUSCLE				
Extensor digitorum brevis[*] (*extensor* = increases angle at joint; *digit* = finger or toe; *brevis* = short)	Calcaneus.	Tendon of extensor digitorum longus and proximal phalanx of great toe.	Extends first through fourth toes.	Deep peroneal nerve.
PLANTAR MUSCLES **First (superficial) layer**				
Abductor hallucis (*abductor* = moves part away from midline; *hallucis* = hallux or great toe)	Calcaneus and plantar aponeurosis.	Proximal phalanx of great toe.	Abducts great toe and flexes metatarsophalangeal joint.	Medial plantar nerve.
Flexor digitorum brevis (*flexor* = decreases angle at joint)	Calcaneus and plantar aponeurosis.	Middle phalanx of second through fifth toes.	Flexes second through fifth toes.	Medial plantar nerve.
Abductor digiti minimi (*minimi* = small toe)	Calcaneus and plantar aponeurosis.	Proximal phalanx of small toe.	Abducts and flexes small toe.	Lateral plantar nerve.
Second Layer **Quadratus plantae** (*quad* = four; *planta* = sole of foot)	Calcaneus.	Tendons of flexor digitorum longus.	Flexes second through fifth toes.	Lateral plantar nerve.
Lumbricales (four muscles)	Tendons of flexor digitorum longus.	Tendons of extensor digitorum longus.	Extend second through fifth toes.	Medial and lateral plantar nerves.
Third Layer **Flexor hallucis brevis**	Cuboid and third (lateral) cuneiform.	Proximal phalanx of great toe.	Flexes great toe.	Medial plantar nerve.
Adductor hallucis (*adduct* = moves part closer to midline)	Second through fourth metatarsals and ligaments of metatarsophalangeal joints.	Proximal phalanx of great toe.	Adducts and flexes great toe.	Lateral plantar nerve.
Flexor digiti minimi brevis	Fifth metatarsal.	Proximal phalanx of small toe.	Flexes small toe.	Lateral plantar nerve.
Fourth (deep) Layer **Dorsal interossei** (four muscles; *dorsal* = back surface; *inter* = between; *ossei* = bone)	Adjacent side of metatarsals.	Proximal phalanges, both sides of second toe, lateral side of third and fourth toes.	Abduct toes and flex proximal phalanges.	Lateral plantar nerve.
Plantar interossei	Third, fourth, and fifth metatarsals.	Proximal phalanges of same toes.	Adduct third, fourth, and fifth toes and flex proximal phalanges.	Lateral plantar nerve.

[*] Not illustrated.

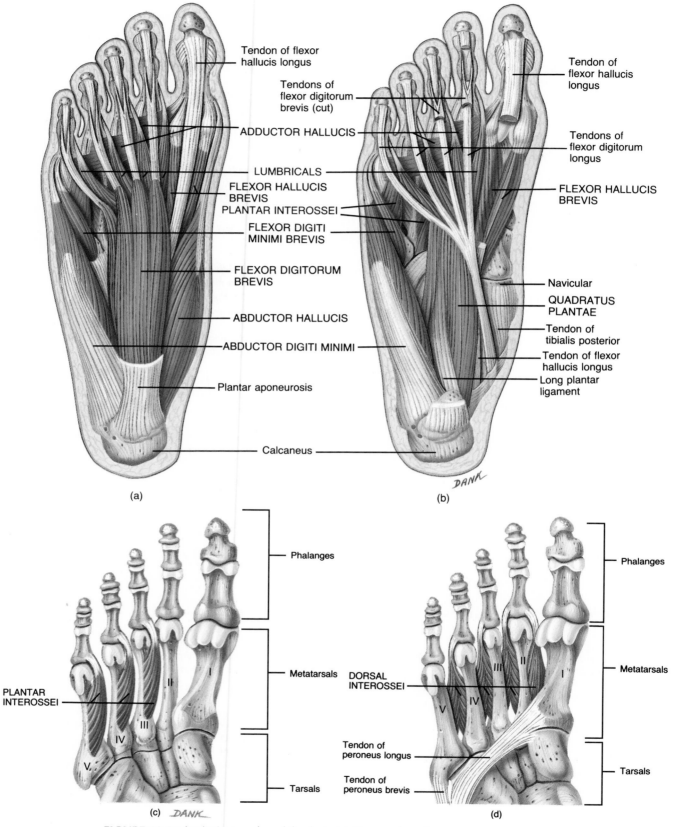

FIGURE 10-25 Intrinsic muscles of the foot. (a) Plantar view showing some superficial and deeper muscles. (b) Plantar view showing deeper muscles. (c) Plantar view showing plantar interossei. (d) Plantar view showing dorsal interossei.

INTRAMUSCULAR (IM) INJECTIONS

An *intramuscular (IM) injection* penetrates the skin and subcutaneous tissue to enter the muscle itself. Intramuscular injections are preferred when prompt absorption is desired, when larger doses than can be given cutaneously are indicated, or when the drug is too irritating to give subcutaneously. The common sites for intramuscular injections include the buttock, lateral side of the thigh, and the deltoid region of the arm. Muscles in these areas, especially the gluteal muscles in the buttock, are fairly thick. Because of the large number of muscle fibers and extensive fascia, the drug has a large surface area for absorption. Absorption is further promoted by the extensive blood supply to muscles. Ideally, intramuscular injections should be given deep within the muscle and away from major nerves and blood vessels.

For many intramuscular injections, the preferred site is the *gluteus medius muscle* of the buttock (Figure 10-26a). The buttock is divided into quadrants, and the upper outer quadrant is used as the injection site. The iliac crest serves as a landmark for this quadrant. The spot for injection in an adult is usually about 5–7½ cm (2–3 in.) below the iliac crest. The upper outer quadrant is chosen because the muscle in this area is quite thick and has few nerves. Injection in this area thus reduces the chance of injury to the sciatic nerve, which can cause paralysis of the lower extremity. The probability of injecting the drug into a blood vessel is also remote in this area. After the needle is inserted into the gluteus medius muscle, the plunger is pulled up for a few seconds. If the syringe fills with blood, the needle is in a blood vessel and a different injection site on the opposite buttock is chosen.

Injections given in the lateral side of the thigh are inserted into the midportion of the *vastus lateralis muscle* (Figure 10-26b). This site is determined by using the knee and greater trochanter of the femur as landmarks. The midportion of the muscle is located by measuring a handbreadth above the knee and a handbreadth below the greater trochanter.

The *deltoid* injection is given in the midportion of the muscle about two to three fingerbreadths below the acromion of the scapula and lateral to the axilla (Figure 10-26c).

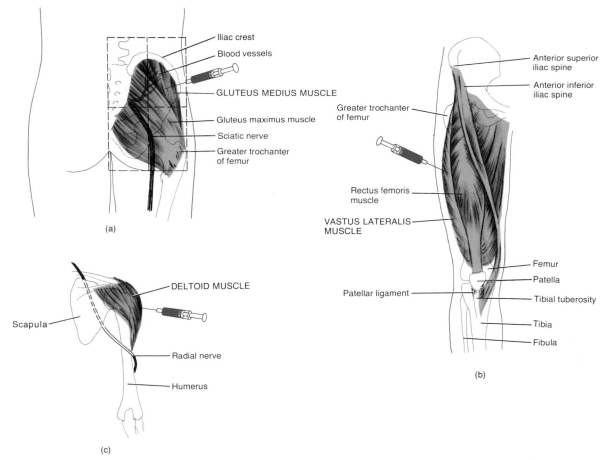

FIGURE 10-26 Intramuscular injections. Shown are the three common sites for intramuscular injections. (a) Buttock. (b) Lateral surface of the thigh. (c) Deltoid region of the arm.

STUDY OUTLINE

How Skeletal Muscles Produce Movement (p. 246)

1. Skeletal muscles produce movement by pulling on bones.
2. The attachment to the stationary bone is the origin. The attachment to the movable bone is the insertion.
3. Bones serve as levers and joints serve as fulcrums. The lever is acted on by two different forces: resistance and effort.
4. Levers are categorized into three types—first class, second class, and third class—according to the position of the fulcrum, effort, and resistance on the lever.
5. Fascicular arrangements include parallel, convergent, pennate, and circular. Fascicular arrangement is correlated with the power of a muscle and the range of movement.
6. The prime mover produces the desired action. The antagonist produces an opposite action. The synergist assists the prime mover by reducing unnecessary movement. The fixator stabilizes the origin of the prime mover so that it can act more efficiently.

Naming Skeletal Muscles (p. 248)

Skeletal muscles are named on the basis of distinctive criteria: direction of fibers, location, size, number of origins (or heads), shape, origin and insertion, and action.

Principal Skeletal Muscles (p. 249)

The principal skeletal muscles of the body are grouped according to region in Exhibits 10-2–10-24.

Intramuscular (IM) Injections (p. 307)

1. Advantages of intramuscular injections are prompt absorption, use of larger doses than can be given cutaneously, and minimal irritation.
2. Common sites for intramuscular injections are the buttock, lateral side of the thigh, and deltoid region of the arm.

REVIEW QUESTIONS

1. What is meant by the muscular system? Explain fully.
2. Using the terms origin, insertion, and belly in your discussion, describe how skeletal muscles produce body movements by pulling on bones.
3. What is a lever? Fulcrum? Apply these terms to the body, and indicate the nature of the forces that act on levers. Describe the three classes of levers, and provide one example for each in the body.
4. Describe the various arrangments of fasciculi. How is fascicular arrangement correlated with the strength of a muscle and its range of movement?
5. Define the role of the agonist, antagonist, synergist, and fixator in producing body movements.
6. Select at random several muscles presented in Exhibits 10-2–10-24 and see if you can determine the criterion or criteria employed for naming each. In addition, refer to the prefixes, suffixes, roots, and definitions in each exhibit as a guide. Select as many muscles as you wish, as long as you feel you understand the concept involved.
7. What muscles would you use to do the following: (a) frown, (b) pout, (c) show surprise, (d) show your upper teeth, (e) pucker your lips, (f) squint, (g) blow up a balloon, (h) smile?
8. What are the principal muscles that move the mandible? Give the function of each.
9. What would happen if you lost tone in the masseter and temporalis muscles?
10. What extrinsic muscles move the eyeball? In which direction does each muscle move the eyeball?
11. Describe the action of each of the muscles acting on the tongue.
12. What tongue, facial, and mandibular muscles would you use when chewing food?
13. What muscles tense and elevate the soft palate?
14. What muscles constrict the pharynx? What muscle dilates the pharynx? Distinguish the circular and longitudinal layers.
15. Describe the muscles involved, and their actions, in moving the hyoid bone.
16. Describe the actions of the extrinsic and intrinsic muscles of the larynx.
17. What muscles are responsible for moving the head, and how do they move the head?
18. What muscles would you use to signify ''yes'' and ''no'' by moving your head?
19. Describe the composition of the anterolateral and posterior abdominal wall.
20. What muscles accomplish compression of the abdominal wall?
21. What are the principal muscles involved in breathing? What are their actions?
22. Describe the actions of the muscles of the pelvic floor. What is the pelvic diaphragm?
23. Describe the actions of the muscles of the perineum. What is the urogenital diaphragm?
24. In what directions is the pectoral (shoulder) girdle drawn? What muscles accomplish these movements?
25. What muscles are used to raise your shoulders, lower your shoulders, join your hands behind your back, and join your hands in front of your chest?
26. What movements are possible at the shoulder joint? What muscles accomplish these movements?
27. Distinguish axial and scapular muscles involved in moving the arm.
28. What muscles move the arm? In which directions do these movements occur?
29. Organize the muscles that move the forearm into flexors and extensors. What muscles move the forearm and what actions are used when striking a match?
30. Discuss the various movements possible at the wrist, hand, and fingers. What muscles accomplish these movements?
31. How many muscles and actions of the wrist, hand, and fingers used when writing can you list?

32. What is the flexor retinaculum? Extensor retinaculum?
33. What muscles form the thenar eminence? What are their functions?
34. What muscles form the hypothenar eminence? What are their functions?
35. Discuss the various muscles and movements of the vertebral column. How are the muscles grouped?
36. What muscles accomplish movements of the thigh? What actions are produced by these muscles? What is the iliotibial tract?
37. Organize the muscles that act on the leg into medial, anterior, and posterior compartments. What is the popliteal fossa?
38. What muscles act at the knee joint? What kinds of movements do these muscles perform?

39. Determine the muscles and their actions listed in Exhibit 10-22 that you would use to climb a ladder to a diving board, dive into the water, swim the length of a pool, and then sit at pool side.
40. Name the muscles that plantar flex, evert, pronate, and dorsiflex the foot. What is the superior extensor retinaculum? Inferior extensor retinaculum?
41. How are the intrinsic muscles of the foot organized?
42. In which directions are the toes moved? What muscles bring about these movements?
43. What are the advantages of intramuscular (IM) injections?
44. Describe how you would locate the sites for an intramuscular (IM) injection in the buttock, lateral side of the thigh, and deltoid region of the arm.

SELF-QUIZ

Complete the following:

1. In flexion, your forearm serves as a rigid rod, or _____, that moves about a fixed point, called a _____ (your elbow joint, in this case).
2. The attachment of a muscle to the bone that moves is called the _____ end of the muscle.
3. Hyperextend your head as if to look at the sky. The weight of your face and jaw serves as the ((E)? ⚠? Ⓡ?), while your neck muscles provide the ((E)? ⚠? Ⓡ?). The fulcrum is the joint between the atlas and _____ bone. This is an example of a _____-class lever.
4. A muscle that contracts to cause the desired action is called the _____. Muscles that assist or cooperate with the muscle that causes the desired action are known as _____.
5. Name the skeletal muscle that:
   a. produces a frowning expression: _____
   b. allows you to blow air out of your mouth and produce a sucking action: _____
6. Essentially, all the muscles controlling facial expression receive nerve impulses via the _____ nerve, which is cranial nerve (III? V? VII?).
7. Two large muscles help you to close your mouth forcefully, as in chewing. Both of these act by (lowering the maxilla? elevating the mandible?). The _____ covers your temple and the _____ covers the ramus of the mandible.
8. All the rectus muscles move the eyeball in the direction (opposite to? the same as?) that given in the muscle name. For example, the superior rectus moves the eyeball (superiorly? inferiorly?).
9. Contraction of the diaphragm flattens the dome, causing the size of the thorax to (increase? decrease?), as occurs during (inspiration? expiration?).
10. The deep transverse perineus, urethral sphincter, and a fibrous membrane constitute the _____ diaphragm.
11. Of the nine muscles that cross the shoulder, only two of them, the pectoralis major and _____, do not originate on the scapula.
12. The majority of muscles that act on the thigh originate in the pelvic (hip) girdle and insert on the _____.
13. Whereas the _____ muscles act on the thumb, the _____ muscles act on the little finger.

14. The deep fascia organized into fibrous bands through which flexor tendons of the digits and wrist pass is the _____ (transverse carpal ligament).
15. Examine your own forearm, palm, and fingers. There is more muscle mass on the (anterior? posterior?) surface. You therefore have more muscles that can (flex? extend?) your wrist and fingers.
16. Of the muscles that move the vertebral column, the deepest group is the _____ muscles.
17. Quadriceps femoris is the main muscle mass on the (anterior? posterior?) surface of the thigh. The name of this muscle mass indicates it has _____ heads of origin. All converge to insert on the _____ bone by means of the patellar ligament. All therefore cause (flexion? extension?) of the leg. Feel their contraction as you extend your own leg. The hamstrings are (synergists? antagonists?) of the quadriceps. Hamstrings are located on the (anterior? posterior?) surface of the thigh.
18. Of the muscles that move the foot and toes, those of the posterior compartment are innervated by the _____ nerve.
19. All three of the hamstring muscles originate on the _____ and insert just below the knee. The actions they cause are (flexion? extension?) of the thigh and (flexion? extension?) of the leg.
20. The deep fascia of the foot forms the _____ aponeurosis that supports the longitudinal arch and transmits the flexor tendons of the foot.

Choose the one best answer to these questions:

___ 21. Which generalization concerning movement by skeletal muscles is *not* true?
   A. muscles produce movements by pulling on bones; B. during contractions the two articulating bones move equally; C. the tendon attachment to the more stationary bone is the origin; D. bones serve as levers and joints as fulcrums of the levers; E. in the extremities, the insertion is usually distal.

___ 22. The muscle
   A. that is used to elevate the scapula and extend the head is the trapezius; B. that is considered part of the abdominal muscle group is the latissimus dorsi; C. that

is found on the lateral and posterior part of the thigh is the sartorius; D. that is considered one of the hamstring muscles is the rectus femoris; E. that is used to extend the elbow and abduct the arm is the triceps brachii.

___ 23. The action of the pectoralis major muscle is to
A. abduct the arm and rotate the arm laterally; B. adduct the arm and rotate the arm medially; C. adduct the arm and rotate the arm laterally; D. abduct the arm and rotate the arm medially; E. abduct and raise the arm.

___ 24. What muscle is located on the posterior surface of the arm?
A. biceps brachii; B. biceps femoris; C. pronator teres; D. vastus lateralis; E. triceps brachii.

___ 25. Of these muscles, which one adducts the thigh?
A. gluteus medius; B. superior gemellus; C. inferior gemellus; D. pectineus; E. obturator internus.

26. For each of the following, indicate the type of clue that each part of the name gives:

___ a. rectus abdominis   A. action
___ b. gluteus maximus   B. direction of fibers
___ c. biceps brachii   C. location
___ d. sternocleidomastoid   D. number of heads or origins
___ e. adductor longus   E. points of attachment of origin and insertion
                                 F. size or shape

27. Match the following:

___ a. located superior to the hyoid (suprahyoid), these form the floor of the oral cavity

___ b. "strap muscles" that are infrahyoid, these muscles cover the anterior of the larynx and trachea; these permit you to elevate your thyroid cartilage ("Adam's apple") during swallowing to prevent food from entering your larynx (try it)

___ c. forming part of the wall of the pharynx, these muscles elevate the larynx and pharynx during swallowing; and they close off the nasopharynx (so that food does not back up into the nasal cavity) during swallowing.

___ d. three muscles that squeeze the pharynx to propel a bolus of food into the esophagus during swallowing

___ e. muscles permitting tongue movements

A. styloglossus, hyoglossus, and genioglossus
B. superior, middle, and inferior constrictors
C. stylopharyngeus, salpingopharyngeus, and palatopharyngeus
D. diagastric, stylohyoid, mylohyoid, and geniohyoid
E. thyrohyoid, sternohyoid, omohyoid, and sternothyroid

11 Surface Anatomy

CHAPTER OUTLINE
- Head
- Neck
- Trunk
- Upper Extremity
- Lower Extremity

STUDENT OBJECTIVES

1. Define surface anatomy.
2. Identify the principal regions of the human body.
3. Describe the surface anatomy features of the head.
4. Describe the surface anatomy features of the neck.
5. Describe the surface anatomy features of the trunk.
6. Describe the surface anatomy features of the upper extremities.
7. Describe the surface anatomy features of the lower extremities.

In Chapter 1, several branches of anatomy were defined and their importance to an understanding of the structure of the body was noted. Now that you have a fairly good idea how the body is organized, we can take a closer look at surface anatomy. Very simply, **surface anatomy** is the study of the form and markings of the surface of the body. A knowledge of surface anatomy will help you identify certain superficial structures through visual inspection or palpation through the skin. **Palpation** means to feel with the hand.

To introduce you to surface anatomy, we shall consider some key features of each of the principal regions of the body. These regions are outlined as follows and may be reviewed in Figure 1-2.

Head (cephalic region, or caput)
  Cranium (skull, or brain case)
  Face (facies)

Neck (collum)

Trunk
  Back (dorsum)
  Chest (thorax)
  Abdomen (venter)
  Pelvis

Upper extremity
  Armpit (axilla)
  Shoulder (acromial, or omos)
  Arm (brachium)
  Elbow (cubitus)
  Forearm (antebrachium)
  Wrist (carpus)
  Hand (manus)
    1. Metacarpus (dorsum and palm)
    2. Fingers (phalanges, or digits)

Lower extremity
  Buttocks (gluteal region)
  Thigh (femoral region)
  Knee (genu)
  Leg (crus)
  Ankle (tarsus)

Foot (pes)
    1. Metatarsus (dorsum and sole)
    2. Toes (phalanges, or digits)

HEAD

The **head (cephalic region, or caput)** is divided into the cranium and face. The **cranium (skull, or brain case)** consists of a **frontal region** (front of skull, or sinciput, that includes the forehead), **parietal region** (crown of skull, or vertex), **temporal region** (side of skull, or tempora), and **occipital region** (base of skull, or occiput).

The **face (facies)** is subdivided into an **orbital** or **ocular region** that includes the eyeballs (bulbi oculorum), eyebrows (supercilia), and eyelids (palpebrae); **nasal region** (nose, or nasus); **infraorbital region** (inferior to orbit); **oral region** (mouth); **mental region** (anterior portion of mandible); **buccal region** (cheek); **parotid-masseteric region** (external to the parotid gland and masseter muscle); **zygomatic region** (inferolateral to orbit); and **auricular region** (ear).

Examine Exhibit 11-1. It contains illustrations of various surface features of the head with an accompanying description of each of the features.

NECK

The **neck (collum)** is divided into an **anterior cervical region**, two **lateral cervical regions**, and a **posterior region (nucha)**.

The most prominent structure in the midline of the anterior cervical region is the **thyroid cartilage (Adam's apple)** of the larynx. Just superior to it, the **hyoid bone** can be palpated. Inferior to the thyroid cartilage, the **cricoid cartilage** of the larynx can be felt.

A major portion of the lateral cervical regions is formed by the **sternocleidomastoid muscles.** Each muscle extends from the mastoid process of the temporal bone, felt as a bump behind the auricle of the ear, to the sternum and clavicle. Each sternocleidomastoid muscle divides its portion of the neck into an anterior triangle and a posterior (lateral) triangle. The **anterior triangle** is bordered superiorly by the mandible, inferiorly by the sternum, medially by the cervical midline, and laterally by the anterior border of the sternocleidomastoid muscle. The **posterior (lateral) triangle** is bordered inferiorly by the clavicle, anteriorly by the posterior border of the sternocleidomastoid muscle, and posteriorly by the anterior border of the trapezius muscle.

A muscle that extends downward and outward from the base of the skull and occupies a portion of the lateral cervical region is the **trapezius muscle.**

A "stiff neck" is frequently associated with an inflammation of this muscle. A very prominent vein that runs along the lateral surface of the neck is the **external jugular vein.** It is readily seen if you are angry or if your collar is too tight. The locations of a few surface features of the neck are shown in Exhibit 11-2.

TRUNK

The **trunk** is divided into the back, chest, abdomen, and pelvis.

One of the most striking surface features of the **back (dorsum)** is the **vertebral spines,** the posteriorly pointed projections of the vertebrae. A very prominent vertebral spine is the **vertebra prominens** of the seventh cervical vertebra. It is easily seen when the head is flexed. Another easily identifiable surface landmark of the back is the **scapula.** In fact, several parts of the scapula (axillary border,

EXHIBIT 11-1

Surface Anatomy of the Head

HEAD (Figure 11-1)

1. **Zygomatic region.** Inferolateral to orbit.
2. **Orbital (ocular) region.** Includes eyeball, eyelids, and eyebrows.
3. **Infraorbital region.** Inferior to orbit.
4. **Nasal region.** Nose.
5. **Buccal region.** Cheek.
6. **Oral region.** Mouth.
7. **Mental region.** Anterior portion of mandible.
8. **Parotid-masseteric region.** External to parotid gland and masseter muscle.
9. **Occipital region.** Base of skull (occiput).
10. **Auricular region.** Ear.
11. **Parietal region.** Crown of skull (vertex).
12. **Temporal region.** Side of skull (tempora).
13. **Frontal region.** Front of skull (sinciput).

EYE (Figure 11-2)

1. **Pupil.** Opening of center of iris of eyeball for light transmission.
2. **Iris.** Circular pigmented muscular membrane behind cornea.
3. **Sclera.** "White" of eye; a coat of fibrous tissue that covers entire eyeball except for cornea.
4. **Conjunctiva.** Membrane that covers exposed surface of eyeball and lines eyelids.
5. **Palpebrae (eyelids).** Folds of skin and muscle lined on their inner aspect by conjunctiva.
6. **Palpebral fissure.** Space between eyelids when they are open.
7. **Medial commissure.** Site of union of upper and lower eyelids near nose.
8. **Lateral commissure.** Site of union of upper and lower eyelids away from nose.
9. **Lacrimal caruncle.** Fleshy, yellowish projection of medial commissure that contains modified sudoriferous (sweat) and sebaceous (oil) glands.
10. **Eyelashes.** Hairs on margins of eyelids, usually arranged in two or three rows.
11. **Supercilla (eyebrows).** Several rows of hairs superior to upper eyelids.

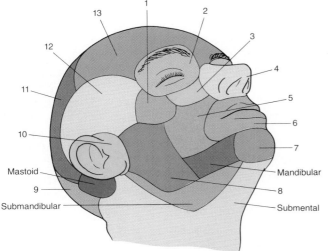

FIGURE 11-1 Regions of the head in anterolateral view.

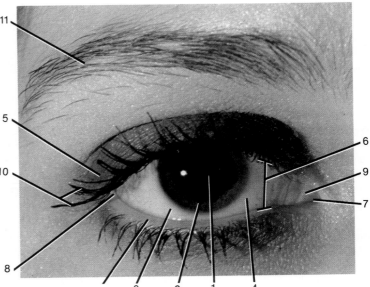

FIGURE 11-2 Surface anatomy of the right eye in anterior view. (Copyright © 1982 by Gerard J. Tortora. Courtesy of Lynne Tortora.)

EXHIBIT 11-1 (Continued)

EAR (Figure 11-3)

1. **Pinna.** Portion of external ear not contained in head; also called auricle or trumpet.
2. **Tragus.** Cartilaginous projection anterior to external opening to ear.
3. **Antitragus.** Cartilaginous projection opposite tragus.
4. **Concha.** Hollow of pinna.
5. **Helix.** Superior and posterior free margin of pinna.
6. **Antihelix.** Semicircular ridge posterior and superior to concha.
7. **Triangular fossa.** Depression in superior portion of antihelix.
8. **Lobule.** Inferior portion of pinna devoid of cartilage.
9. **External auditory canal (meatus).** Canal extending from external ear to eardrum.

NOSE AND LIPS (Figure 11-4)

1. **Root.** Superior attachment of nose at forehead located between eyes.
2. **Apex.** Tip of nose.
3. **Dorsum nasi.** Rounded anterior border connecting root and apex; in profile, may be straight, convex, concave, or wavy.
4. **Nasofacial angle.** Point at which side of nose blends with tissues of face.
5. **Ala.** Convex flared portion of inferior lateral surface; unites with cheek.
6. **External naris.** External opening into nose.
7. **Bridge.** Superior portion of dorsum nasi, superficial to nasal bones.
8. **Philtrum.** Vertical groove in medial portion of upper lip.
9. **Lips.** Upper and lower fleshy borders of oral cavity.

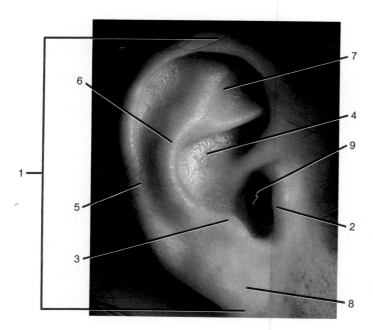

FIGURE 11-3 Surface anatomy of the right ear in lateral view. (Copyright © 1982 by Gerard J. Tortora. (Courtesy of James Borghesi.)

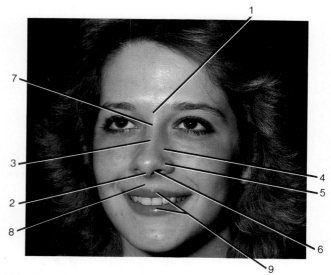

FIGURE 11-4 Surface anatomy of the nose and lips in anterior view. (Copyright © 1982 by Gerard J. Tortora. (Courtesy of Lynne Tortora.)

vertebral border, inferior angle, spine, and acromion) may also be seen or palpated. In lean individuals the *ribs* may also be seen. The vertebral border of the scapula crosses ribs 2 to 7. Among the superficial muscles of the back that can be seen are the *latissimus dorsi, erector spinae (sacrospinalis), infraspinatus, trapezius,* and *teres major.* These, as well as the other surface features of the back, are described in Exhibit 11-3.

The *chest (thorax)* presents a number of anatomical landmarks. At its superior region are the *clavicles.* The *sternum* lies in the midline of the chest and is divisible into a superior manubrium, a middle body, and an inferior xiphoid process. Its superior border attaches to the clavicles. Between the medial ends of the clavicles, there is a depression on the superior surface of the sternum called the *jugular (suprasternal) notch.* The trachea can be palpated posterior to the jugular notch. The *sternal angle* is formed by a junction

line between the manubrium and body of the sternum and is palpable under the skin. It locates the costal cartilage of the second rib and is the most reliable surface landmark of the chest. At or inferior to the sternal angle and slightly to the right, the trachea bifurcates (divides in two) into the left and right primary bronchi. The inferior portion of the sternum, the *xiphoid process,* may be palpated. Also visible or palpable are the *ribs.* The apex of the lung rises as high as the neck of the first rib, superior to the clavicle. The *costal margins,* the inferior edges of the costal cartilages of ribs 7 through 10, can also be seen or palpated. The margins are very near the iliac crests, sometimes only 1–2 cm away. Among the more prominent superficial chest muscles are the *pectoralis major* and *serratus anterior.* These and other surface features of the chest are described in Exhibit 11-3. The surface features of the heart are shown in Figure 13-5.

EXHIBIT 11-2

Surface Anatomy of the Neck (Figure 11-5)

1. **Anterior triangle of the neck.** Bordered superiorly by mandible, inferiorly by sternum, medially by cervical midline, and laterally by anterior border of sternocleidomastoid muscle.
2. **Posterior triangle of neck.** Bordered inferiorly by clavicle, anteriorly by posterior border of sternocleidomastoid muscle, and posteriorly by anterior border of trapezius muscle.
3. **Trapezius muscle.** Occupies a portion of lateral surface of neck; helps raise, lower, and draw shoulders backward and extends head.
4. **Sternocleidomastoid muscle.** Forms major portion of lateral surface of neck; flexes head and rotates it to opposite side.
5. **Cricoid cartilage.** Inferior laryngeal cartilage that attaches larynx to trachea; can be palpated by running your fingertip down from your chin over the thyroid cartilage (after you pass the cricoid cartilage, your fingertip sinks in).
6. **Thyroid cartilage (Adam's apple).** Triangular laryngeal cartilage in the midline of the anterior cervical region.
7. **Hyoid bone.** Lies just superior to the thyroid cartilage; it is the first resistant structure palpated in the midline below the chin.

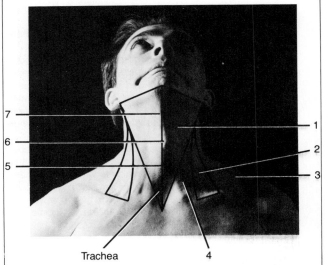

FIGURE 11-5 Surface anatomy of the neck in anterior view. (Courtesy of Carroll H. Weiss, Camera M.D. Studios, Inc.)

The **abdomen (venter)** and **pelvis** have already been discussed in terms of their nine regions or four quadrants (see Figures 1-8 and 1-9). External features of the abdomen and pelvis include the **umbilicus, linea alba, external oblique muscle, rectus abdominis muscle, tendinous intersections,** and **symphysis pubis.** The surface features of the abdomen and pelvis are described in Exhibit 11-3.

UPPER EXTREMITY

The **upper extremity** consists of the armpit (axilla), shoulder, arm, elbow, forearm, wrist, and hand.

At the **shoulder (acromial or omos)** moving laterally along the top of the clavicle, it is possible to palpate a slight elevation at the lateral end of the clavicle, the **acromioclavicular joint.** Less than 2.5 cm (1 in.) distal to this joint, one can also feel the **acromion** of the scapula, which forms the tip of the shoulder. The rounded prominence of the shoulder is formed by the **deltoid muscle,** a frequent site for intramuscular injections (Exhibit 11-4).

Most of the anterior surface of the **arm (brachium)** is occupied by the **biceps brachii muscle,** whereas most of the posterior surface is occupied by the **triceps brachii muscle.** The **brachial artery,** which provides the main arterial supply to the arm, begins at the lower border of the teres major muscle as the continuation of the axillary artery and terminates at the level of the neck of the radius by dividing into the ulnar and radial arteries. The radial artery is palpable throughout its length (Exhibit 11-4).

At the **elbow (cubitus)** it is possible to locate three bony protuberances. The **medial** and **lateral epicondyles** of the humerus form visible eminences on the dorsum of the elbow. The **olecranon** of the ulna forms the large eminence in the middle of the dorsum of the elbow and lies between and slightly superior to the epicondyles when the forearm is extended. The **ulnar nerve** can be palpated in a groove behind the medial epicondyle. The triangular space of the anterior region of the elbow is the **cubital fossa.** The **median cubital vein** usually crosses the cubital fossa obliquely. This vein is the one frequently selected for removal of blood for diagnosis, transfusions, and intravenous therapy (Exhibit 11–4).

One of the most prominent landmarks of the **forearm (antebrachium)** is the **styloid process** of the ulna. It may be seen as a protuberance on the medial side of the wrist. The ulna is the medial bone of the forearm, and the radius is the lateral bone of the forearm. On the lateral side of the upper forearm is the **brachioradialis muscle.** Next to it is the **flexor carpi radialis muscle.** On the medial side of the upper forearm is the **flexor carpi ulnaris muscle** (Exhibit 11-4).

At the **wrist (carpus),** several structures may be palpated or seen. On the anterior surface are **wrist creases,** designated as proximal, middle, and distal, where the skin is firmly attached to the underlying deep fascia. On the anterior surface of the wrist, it is possible to see the **tendon of the palmaris longus muscle** by making a fist. Next to this tendon, as you move toward the thumb, you can feel the **tendon of the flexor carpi radialis muscle.** If you continue toward the thumb, you can palpate the **radial artery** just medial to the styloid process of the radius. This artery is frequently used to take the pulse. The **pisiform bone,** the medial bone of the proximal carpals, can be palpated as a projection distal to the styloid process of the ulna. By bending the thumb backward, two prominent tendons may be located along the posterior surface of the wrist. The one closer to the styloid process of the radius is the **tendon of the extensor pollicus brevis muscle.** The one closer to the styloid process of the ulna is the **tendon of the extensor pollicus longus muscle.** The depression between these two

EXHIBIT 11-3

Surface Anatomy of the Trunk

BACK (Figure 11-6)

1. **Vertebral spines.** Posteriorly pointed projections of vertebrae. The spine of C2 is the first bony prominence encountered when the finger is drawn downward in the midline; the spine of C7 is the upper of the two prominences found at the base of the neck; the spine of T1 is the lower prominence at the base of the neck; the spine of T3 is about at the same level as the spine of the scapula; the spine of T7 is about opposite the inferior angle of the scapula; a line passing through the highest points of the iliac crests, called the supracristal line, passes through the spine of L4.
2. **Scapula.** Shoulder blade (label line is on vertebral border).
3. **Latissimus dorsi muscle.** Broad muscle of back that helps draw shoulders backward and downward.
4. **Erector spinae (sacrospinalis) muscle.** Parallel to vertebral column on either side of midline between the twelfth rib and iliac crest; moves vertebral column in various directions.
5. **Infraspinatus muscle.** Located inferior to spine of scapula; helps laterally rotate humerus.
6. **Trapezius muscle.** Extends from cervical and thoracic vertebrae to spine of scapula and lateral end of clavicle; occupies portion of lateral surface of neck (see Figure 11-5); helps raise, lower, and draw shoulders backward and extend head.
7. **Teres major muscle.** Located inferior to infraspinatus; helps extend, adduct, and medially rotate humerus.
8. **Posterior axillary fold.** Formed by the latissimus dorsi and teres major muscles; can be palpated between the finger and thumb.

CHEST (Figure 11-7)

1. **Clavicle.** Collarbone.
2. **Jugular (suprasternal) notch of sternum.** Depression on superior border of manubrium of sternum between medial ends of clavicles; trachea can be palpated in the notch.
3. **Manubrium of sternum.** Superior portion of sternum at the same levels as the bodies of the third and fourth thoracic vertebrae and anterior to the arch of the aorta.
4. **Sternal angle of sternum.** Formed by junction between manubrium and body of sternum, located about 4 cm (1½ in.) below jugular notch.
5. **Body of sternum.** Midportion of sternum anterior to heart and the vertebral bodies of T5–T8.
6. **Xiphoid process of sternum.** Inferior portion of sternum, between the seventh costal cartilages.
7. **Costal margin.** Inferior edges of costal cartilages of ribs 7 through 10. The first costal cartilage lies below the medial end of the clavicle; the seventh costal cartilage is the lowest to articulate directly with the sternum; the tenth costal cartilage forms the lowest part of the costal margin, when viewed anteriorly.
8. **Serratus anterior muscle.** Inferior and lateral to pectoralis major muscle; helps rotate scapula superiorly and elevate ribs; also illustrated in Exhibit 11-3, abdomen.

   Ribs. Form bony cage of thoracic cavity (not illustrated here, see Figure 6-17a).

   Mammary glands. Accessory organs of the female reproductive system located inside the breasts. They overlie the pectoralis major muscle (two-thirds) and serratus anterior muscle (one-third). After puberty, they enlarge to their hemispherical shape, and in young adult females, they extend from the second through sixth ribs and from the lateral margin of the sternum to the midaxillary line (see Figure 25-22).

9. **Nipples.** Superficial to fourth intercostal space or fifth rib about 10 cm (4 in.) from the midline in males and most females. The position of the nipples in females is variable depending on the size and pendulousness of the breasts. The right dome of the diaphragm is just inferior to the right nipple, the left dome is about 2–3cm (1 in.) inferior to the left nipple, and the central tendon is at the level of the junction of the body of the sternum and xiphoid process.
10. **Anterior axillary fold.** Formed by the lateral border of the pectoralis major muscle; can be palpated between the fingers and thumb.
11. **Pectoralis major muscle.** Principal upper chest muscle; flexes, adducts, and medially rotates humerus.

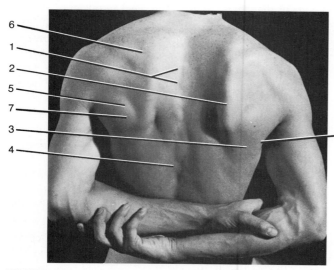

FIGURE 11-6 Surface anatomy of the back in posterior view. (Courtesy of Victor B. Eichler, copyright © 1980.)

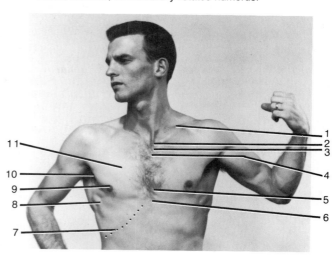

FIGURE 11-7 Surface anatomy of the chest in anterior view. (Courtesy of Vincent P. Destro, Mayo Foundation.)

EXHIBIT 11-3 *(Continued)*

ABDOMEN AND PELVIS (Figure 11-8)

1. **Umbilicus.** Also called navel; previous site of attachment of umbilical cord in fetus. It is level with the intervertebral disc between the bodies of L3 and L4. The abdominal aorta bifurcates into the right and left common iliac arteries anterior to the body of vertebra L4. The inferior vena cava lies to the right of the abdominal aorta and is wider; it arises anterior to the body of vertebra L5.
2. **External oblique muscle.** Located inferior to serratus anterior; helps compress abdomen and bend vertebral column laterally.
3. **Rectus abdominis muscle.** Located just lateral to midline of abdomen; helps compress abdomen and flex vertebral column.
4. **Linea alba.** Flat, tendinous raphe forming a furrow along midline between rectus abdominis muscles. The furrow extends from the xiphoid process to the symphysis pubis. It is broad above the umbilicus and narrow below it.
5. **Tendinous intersection.** Fibrous band that runs transversely or obliquely across the rectus abdominis muscles.
6. **McBurney's point.** An important landmark at the junction of the lateral and middle thirds of a line joining the umbilicus and anterior superior iliac spine. An oblique incision through McBur-

ney's point is made for appendectomy. Pressure of the finger on McBurney's point produces tenderness in acute appendicitis. **Symphysis pubis.** Anterior joint of hipbones; palpated as a firm resistance in the midline at the inferior portion of the anterior abdominal wall (see Figure 7-7).

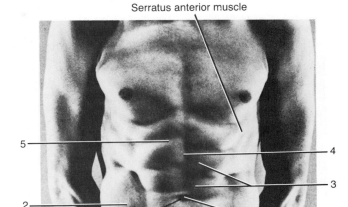

Serratus anterior muscle

Anterior superior iliac spine

FIGURE 11-8 Surface anatomy of the abdomen in anterior view. (Courtesy of R. D. Lockhart, *Living Anatomy,* Faber & Faber, London, 1974.)

tendons is known as the *"anatomical snuffbox."* By palpating the depression, you can feel the radial artery (Exhibit 11-4).

On the dorsum of the *hand (manus)* the distal ends of the second through fifth metacarpal bones are commonly referred to as the *"knuckles."* The term also includes the joints between metacarpals and phalanges and the joints between the phalanges of the fingers. The *dorsal venous arch,* the superficial veins on the dorsum of the hand, can be displayed by compressing the blood vessels at the wrist for a few moments as the hand is opened and closed. The *tendon of the extensor digiti minimi muscle* can be seen in line with phalanx V (little finger), and the *tendons of the extensor digitorum muscle* can be seen in line with phalanges II, III, and IV. By examining the palm of the hand, it is possible to see a number of *skin creases,* as well as the location of the joints of the fingers. The skin creases of the palm are collectively called *palmar flexion creases.* The skin creases on the anterior surface of the fingers are called *digital flexion creases.* Also on the palm are the *thenar eminence,* a rounded contour formed by thumb muscles, and the *hypothenar eminence,* a rounded contour formed by little finger muscles (Exhibit 11-4).

LOWER EXTREMITY

The *lower extremity* consists of the buttocks, thigh, knee, leg, ankle, and foot.

The outline of the superior border of the *buttock (gluteal region)* is formed by the *iliac crest.* The *posterior superior iliac spine,* the posterior termination of the iliac crest, lies deep to a dimple (skin depression) about 4 cm (1.5 in.) lateral to the midline. Most of the prominence of the buttocks is formed by the *gluteus maximus* and *gluteus medius muscles.* The depression that separates the buttocks is the *gluteal cleft,* whereas the inferior limit of the buttock formed by the inferior margin of the gluteus maximums muscle is called the *gluteal fold.* The bony prominence in each buttock is the *ischial tuberosity* of the hipbone. This structure bears the weight of your body when you are seated. About 20 cm (8 in.) below the highest portion of the iliac crest, the *greater trochanter* of the femur can be felt on the lateral side of the thigh. The iliac crest and greater trochanter are useful landmarks when giving an intramuscular injection in the gluteal muscle (Exhibit 11-5).

Among the prominent superficial muscles on the anterior surface of the *thigh (femoral region)* are the *sartorius* and three of the four components of the *quadriceps femoris (vastus lateralis, vastus medialis,* and *rectus femoris).* The vastus lateralis is frequently used as an injection site for diabetics when administering insulin. The superficial medial thigh muscles include the *adductor magnus, adductor brevis, adductor longus, gracilis, obturator externus,* and *pectineus.* The superficial posterior thigh muscles are the *hamstrings (semitendinosus, semimembranosus,* and *biceps femoris)* (Exhibit 11-5).

On the anterior surface of the *knee (genu),* the *patella,*

EXHIBIT 11-4

Surface Anatomy of the Upper Extremity

SHOULDER (Figure 11-9)

1. **Acromion.** Expanded end of spine of scapula; forms tip of shoulder; clearly visible in some individuals and can be palpated about 2.5 cm (1 in.) distal to acromioclavicular joint.
2. **Deltoid muscle.** Triangular muscle that forms rounded prominence of shoulder; primary action is to abduct arm.

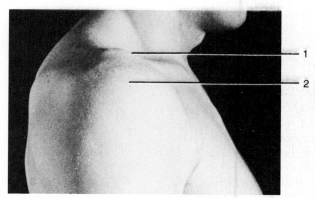

FIGURE 11-9 Surface anatomy of the right shoulder in lateral view. (Courtesy of Victor B. Eichler, copyright © 1980.)

ARM AND ELBOW (Figure 11-10)

1. **Biceps brachii muscle.** Forms bulk of anterior surface of arm; helps flex forearm.
2. **Triceps brachii muscle.** Forms bulk of posterior surface of arm; helps extend forearm.
3. **Medial epicondyle.** Medial projection at distal end of humerus.
4. **Lateral epicondyle.** Lateral projection at distal end of humerus.
5. **Olecranon.** Projection of proximal end of ulna; forms elbow.
6. **Cubital fossa.** Triangular space in anterior region of elbow; contains tendon of biceps brachii muscle, brachial artery and its terminal branches (radial and ulnar arteries), and parts of median and radial nerves.
7. **Median cubital vein** (not illustrated here; see Figure 14-14a). Crosses cubital fossa obliquely.
8. **Brachial artery** (not illustrated here; see Figure 14-9a). Passes posterior to coracobrachialis muscle and then medial to biceps brachii muscle. It enters the middle of the cubital fossa and passes under the bicipital aponeurosis, which separates it from the median cubital vein. The artery is frequently used to take blood pressure.
9. **Bicipital aponeurosis.** An aponeurotic band that inserts the biceps brachii muscle into the deep fascia in the medial aspect of the forearm. It can be felt when the muscle contracts.

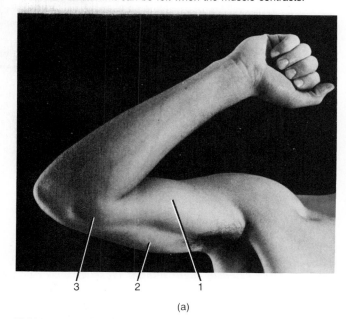

(a)

FIGURE 11-10 Surface anatomy of the arm and elbow. (a) Medial view of right upper extremity. (Courtesy of Victor B. Eichler, copyright © 1980.)

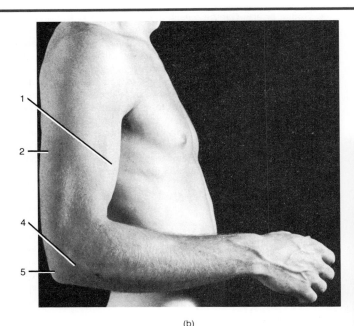

(b)

FIGURE 11-10 (*Continued*) Surface anatomy of the arm and elbow. (b) Lateral view of right upper extremity. (Courtesy of Victor B. Eichler, copyright © 1980.) (c) Anterior view of right arm. (Courtesy of Vincent P. Destro, Mayo Foundation.)

FOREARM (Figure 11-11)

1. **Styloid process of ulna.** Projection of distal end of head of ulna at medial side of wrist.
2. **Flexor carpi radialis muscle.** Located along midportion of forearm; helps flex wrist.
3. **Flexor carpi ulnaris muscle.** Located at medial aspect of forearm; helps flex wrist.
4. **Brachioradialis muscle.** Located at superior and lateral aspect of forearm; helps flex forearm. See Figure 11-10c.

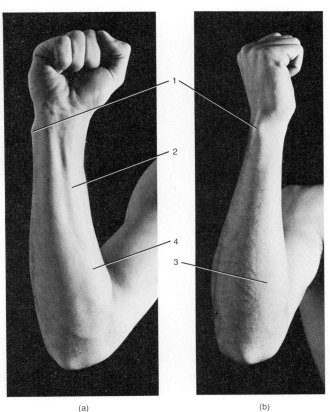

(a)   (b)

FIGURE 11-11 Surface anatomy of the right forearm. (a) Anterior view. (b) Medial view. (Courtesy of Victor B. Eichler, copyright © 1980.)

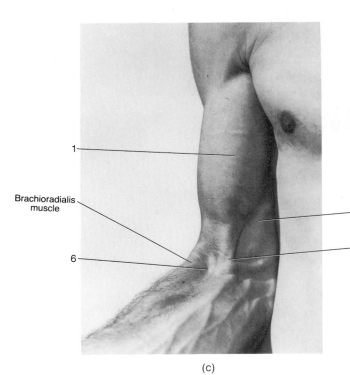

Brachioradialis muscle

(c)

EXHIBIT 11-4 (*Continued*)

WRIST (Figure 11-12)

1. **Wrist creases.** Three more or less constant lines on anterior aspect of wrist (proximal, middle, and distal) where skin is firmly attached to underlying deep fascia.
2. **Tendon of palmaris longus muscle.** By making a fist, the tendon can be seen on anterior surface of wrist nearer ulna; muscle helps to flex wrist (this muscle is not present in about 13 percent of the population).
3. **Tendon of flexor carpi radialis muscle.** Tendon on anterior surface of wrist lateral to tendon of palmaris longus.

4. **Radial artery.** Can be palpated just medial to styloid process of radius; frequently used to take pulse.
5. **Pisiform bone.** Medial bone of proximal carpals; easily palpated as a projection distal to styloid process of ulna.
6. **Tendon of extensor pollicus brevis muscle.** Tendon closer to styloid process of radius along posterior surface of wrist, best seen when thumb is bent backward; muscle extends thumb.
7. **Tendon of extensor pollicus longus muscle.** Tendon closer to styloid process of ulna along posterior surface of wrist, best seen when thumb is bent backward; muscle extends thumb.
8. **"Anatomical snuffbox."** Depression between tendons of extensor pollicus brevis and extensor pollicus longus muscles; radial artery can be palpated in the depression.

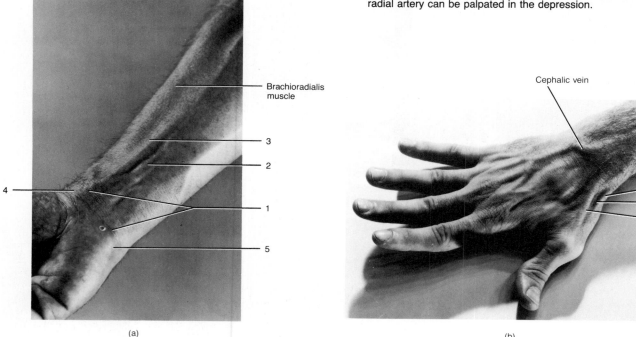

(a)   (b)

FIGURE 11-12 Surface anatomy of the right forearm. (a) Anterior view. (b) Dorsum. (Courtesy of Vincent P. Destro, Mayo Foundation.)

hormones from endocrine glands to the cells; enzymes to various cells.

2. *It regulates:* pH through buffers; normal body temperature through the heat-absorbing and coolant properties of its water content; the water content of cells, principally through dissolved sodium ions.

3. *It protects against:* blood loss through the clotting mechanism; toxins; and foreign microbes.

COMPONENTS

Blood is composed of two portions: formed elements (cells and cell-like structures) and plasma (liquid containing dissolved substances). The formed elements compose about 45 percent of the volume of blood; plasma constitutes about 55 percent (Figure 12-1).

FORMED ELEMENTS

In clinical practice, the most common classification of the *formed elements* of the blood is the following (Figure 12-2):

Erythrocytes (red blood cells)

Leucocytes (white blood cells)
 Granular leucocytes (granulocytes)
  Neutrophils
  Eosinophils
  Basophils
 Agranular leucocytes (agranulocytes)
  Lymphocytes
  Monocytes

Thrombocytes (platelets)

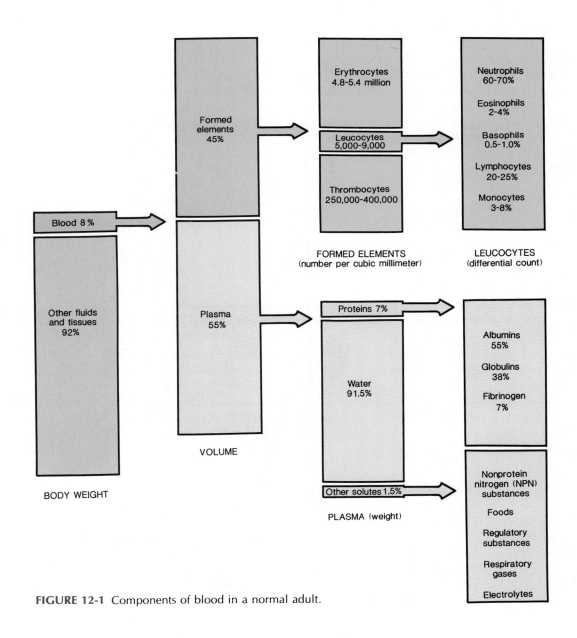

FIGURE 12-1 Components of blood in a normal adult.

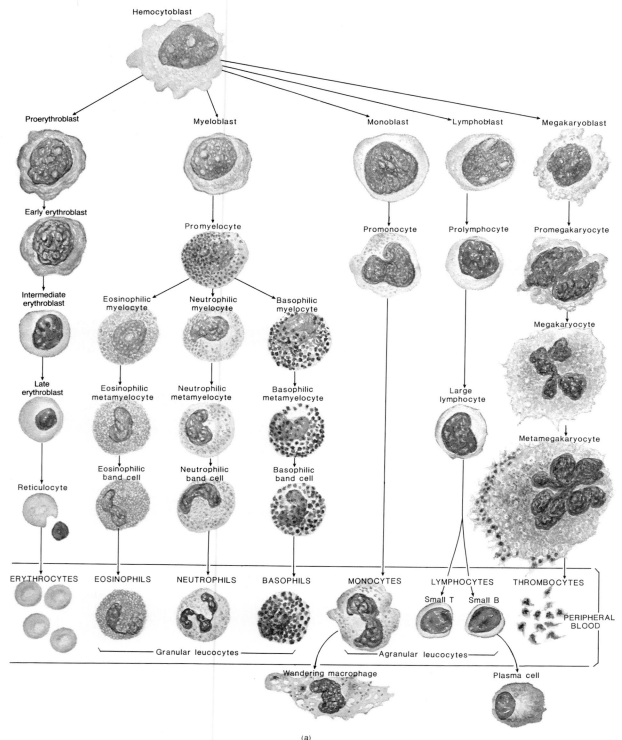

FIGURE 12-2 Blood cells. (a) Diagram of origin, development, and structure.

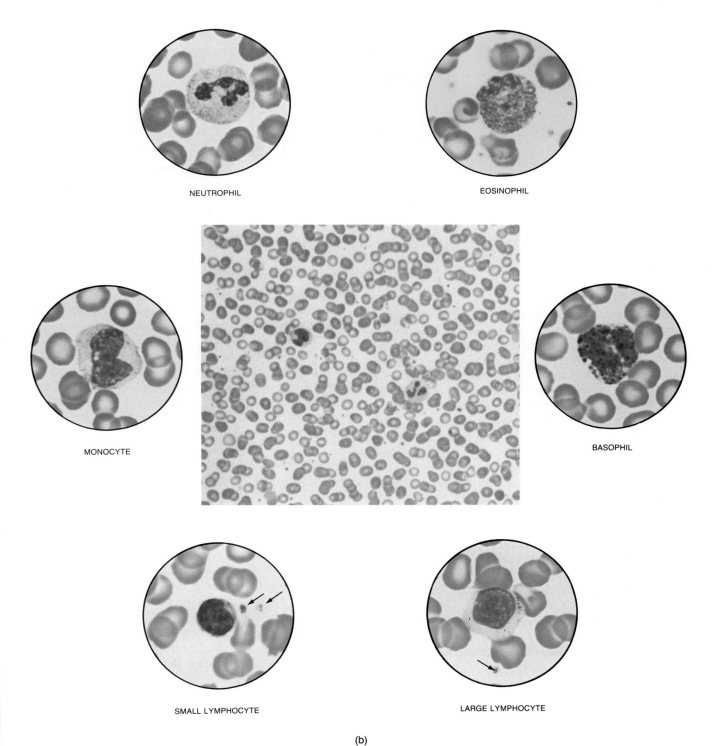

NEUTROPHIL

EOSINOPHIL

MONOCYTE

BASOPHIL

SMALL LYMPHOCYTE

LARGE LYMPHOCYTE

(b)

FIGURE 12-2 (*Continued*) Blood cells. (b) Photomicrograph of blood cells. Shown in the center is a blood smear at a magnification of 400×. The three larger cells with darker-staining nuclei are white blood cells; the more numerous cells are red blood cells, which lack nuclei; and the specklike objects are platelets. The principal types of white blood cells are shown in the large circles at a magnification of 1,800×. These white blood cells are surrounded by red blood cells. Note the platelets, indicated by the arrows. (Courtesy of Michael H. Ross and Edward J. Reith, *Histology: A Text and Atlas*. Copyright © 1985 by Michael H. Ross and Edward J. Reith, Harper & Row, Publishers, Inc., New York.)

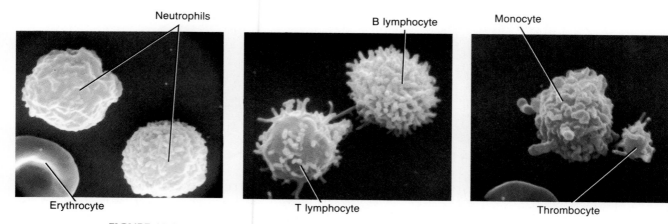

FIGURE 12-2 (*Continued*) Blood cells. (c) Scanning electron micrograph of several different blood cells. (From *Tissues and Organs: A Text-Atlas of Scanning Electron Microscopy* by Richard G. Kessel and Randy H. Kardin. Copyright © 1979 W. H. Freeman and Company.)

Origin

The process by which blood cells are formed is called **hemopoiesis** (hē-mō-poy-Ē-sis), or **hematopoiesis**. During embryonic and fetal life, there are no clear-cut centers for blood-cell production. The yolk sac, liver, spleen, thymus gland, lymph nodes, and bone marrow all participate at various times in producing the formed elements. In the adult, however, we can pinpoint the production process to the red bone marrow in the proximal epiphyses of the humerus and femur, sternum, ribs, vertebrae, and pelvis and lymphoid tissue. Red blood cells, granular leucocytes, and platelets are produced in red bone marrow (myeloid tissue). Agranular leucocytes arise from both myeloid tissue and from lymphoid tissue—spleen, tonsils, lymph nodes. Undifferentiated mesenchymal cells in red bone marrow are transformed into **hemocytoblasts** (hē'-mō-SĪ-tō-blasts), immature cells that are eventually capable of developing into mature blood cells (Figure 12-2a). The hemocytoblasts undergo differentiation into five types of cells from which the major types of blood cells develop.

1. **Proerythroblasts (rubriblasts)** form mature red blood cells.
2. **Myeloblasts** form mature neutrophils, eosinophils, and basophils.
3. **Megakaryoblasts** form mature thrombocytes (platelets).
4. **Lymphoblasts** form lymphocytes.
5. **Monoblasts** form monocytes.

Erythrocytes

● **Structure** Microscopically, **erythrocytes** (e-RITH-rō-sīts) or **red blood cells (RBCs)** appear as biconcave discs when viewed externally in profile. Mature red blood cells average about 8 μm in diameter and are quite simple in structure. They lack a nucleus and can neither reproduce nor carry on extensive metabolic activities. Their plasma membrane is selectively permeable and consists of protein (stromatin) and lipids (lecithin and cholesterol). The membrane encloses cytoplasm and a red pigment called **hemoglobin**. Hemoglobin, which constitutes about 33 percent of the cell weight, is responsible for the red color of blood. Normal values for hemoglobin are 14–20 g/100 ml in infants, 12–15 g/100 ml in adult females, and 14–16.5 g/100 ml in adult males. Certain proteins (antigens) on the surfaces of red blood cells are responsible for the various blood groups. The ABO and Rh groups are examples.

● **Functions** The hemoglobin in erythrocytes combines with oxygen to form oxyhemoglobin and with carbon dioxide to form carbaminohemoglobin and then transports them through blood vessels. The hemoglobin molecule consists of a protein called globin and a pigment called heme, which contains iron (Figure 12-3). As the erythrocytes pass through the lungs, each of the four iron atoms in the hemoglobin molecules combines with a molecule of oxygen. The oxygen is transported in this state to other tissues of the body. In the tissues, the iron–oxygen reaction reverses, and the oxygen is released to diffuse into the interstitial fluid. On the return trip, the globin portion combines with a molecule of carbon dioxide from the interstitial fluid to form carbaminohemoglobin. This complex is transported to the lungs, where the carbon dioxide is released and then exhaled. Although some carbon dioxide is transported by hemoglobin in this manner, the greater portion is transported in blood plasma.

Red blood cells are highly specialized for their transport function. They contain a large number of hemoglobin molecules in order to increase their oxygen-carrying capacity. One estimate is 280 million molecules of hemoglobin per erythrocyte. The biconcave shape of a red blood cell has a much greater surface area than, say, a sphere or a cube. The erythrocyte thus presents a large surface area for the diffusion of gas molecules that pass through the membrane to combine with hemoglobin.

FIGURE 12-3 Diagram of a hemoglobin molecule. The globin (protein) portions of the molecule are indicated in green and blue and the four heme (iron-containing) portions are in the center of each globin molecule.

CLINICAL APPLICATION

A **blood substitute** called *Fluosol-DA* is a slippery, white liquid that has a very high capacity for oxygen. Since its only function is to transport oxygen, it is actually a hemoglobin substitute. It may eventually turn out to be useful in providing oxygen to tissues that are oxygen deficient as a result of various disorders. In clinical trials, Fluosol-DA is being used to oxygenate the heart during procedures in which blood flow to the heart is blocked, to oxygenate tumors to make them more responsive to chemotherapy or radiotherapy, to reperfuse tissues following clot removal while not causing further damage, and to protect heart muscle during a heart attack. Some researchers believe that Fluosol-DA may be useful in keeping donor organs oxygenated while they are readied for transplantation and treating cerebral ischemia. Use of Fluosol-DA is limited by the fact that it must be kept frozen until just before use and recipients must receive oxygen by mask. Nevertheless, patients would be free of the risks of transfusion reactions and post-transfusion AIDS and hepatitis.

● **Life Span and Number**   The plasma membrane of a red blood cell becomes fragile and the cell is nonfunctional in about 120 days. The plasma membranes of wornout red blood cells are removed from circulation by macrophages (phagocytes) in the spleen, liver, and bone marrow. The

hemoglobin is broken down into hemosiderin, an iron-containing pigment; bilirubin, a non-iron-containing pigment; and globin, a protein. The hemosiderin is stored or used in bone marrow to produce new hemoglobin for new red blood cells. Bilirubin is secreted by the liver into bile, and globin is metabolized by the liver.

A healthy male has about 5.4 million red blood cells per cubic millimeter (mm^3) of blood, and a healthy female has about 4.8 million. The higher value in the male is because of his higher rate of metabolism. To maintain normal quantities of erythrocytes, the body must produce new mature cells at the astonishing rate of 2 million per second. In the adult, production takes place in the red bone marrow in the spongy bone of the cranium, ribs, sternum, bodies of vertebrae, and proximal epiphyses of the humerus and femur.

● **Production**   The process by which erythrocytes are formed is called **erythropoiesis** (e-rith'-rō-poy-Ē-sis). Erythropoiesis starts in red bone marrow with the transformation of a hemocytoblast into a proerythroblast (see Figure 12-2a). The **proerythroblast (rubriblast)** gives rise to an **early erythroblast (prorubricyte),** which then develops into an **intermediate erythroblast (rubricyte),** the first cell in the sequence that begins to synthesize hemoglobin. The intermediate erythroblast next develops into a **late erythroblast (metarubricyte),** in which hemoglobin synthesis is at a maximum. In the next stage, the late erythroblast develops into a **reticulocyte,** a cell in the developmental sequence that contains about 34 percent hemoglobin, retains some endoplasmic reticulum, and loses its nucleus by extrusion. Reticulocytes pass from bone marrow into the blood stream by squeezing between the endothelial cells of blood capillaries. If a reticulocyte attempts to cross the endothelial barrier before it has lost its nucleus, the nucleus is forced out since it is too rigid to pass through. Reticulocytes generally become **erythrocytes,** or mature red blood cells, within 1–2 days after their release from bone marrow. The normal proportion of reticulocytes in blood is between 0.5 and 1.5 percent. Aged erythrocytes are destroyed by fixed phagocytic macrophages in the liver and spleen. The hemoglobin molecules are split apart, the iron is reused, and the rest of the molecule is converted into other substances for reuse or elimination.

Normally erythropoiesis and red blood cell destruction proceed at the same pace. But if the body suddenly needs more erythrocytes or if erythropoiesis is not keeping up with red blood cell destruction, a homeostatic mechanism steps up erythrocyte production.

CLINICAL APPLICATION

The rate of erythropoiesis is measured by a procedure called a **reticulocyte** (re-TIK-yoo-lō-sīt) **count.** Some reticulocytes are normally released into the bloodstream before they become mature red blood cells. If the number of reticulocytes in a sample of blood is less than 0.5 percent of the number of mature red blood cells in the sample, erythropoiesis is occurring too slowly. A low reticulocyte count might confirm a diagnosis of anemia, or it might indicate a kidney disease that prevents the kidney cells from producing erythropoietin. If the reticulocytes number more than 1.5 percent of the mature red blood cells, erythropoiesis is abnormally rapid. Any number of problems may be responsible for a high reticulocyte count. These include oxygen deficiency and uncontrolled red blood cell production caused by a cancer in the bone marrow.

Another test with important clinical applications is the hematocrit. *Hematocrit (Hct)* is the percentage of blood that is made up of red blood cells. It is determined by centrifuging blood and noting the ratio of red blood cells to whole blood. The average hematocrit for males is 40–54 percent. The average hematocrit for females is 38–47 percent. Anemic blood may have a hematocrit of 15 percent; polycythemic blood (an abnormal increase in the number of functional erythrocytes) may have a hematocrit of 65 percent. Athletes, however, may have higher-than-normal hematocrits, reflecting constant physical activity rather than a pathological condition.

Leucocytes

● *Structure and Types*   Unlike red blood cells, *leucocytes* (LOO-kō-sīts), or *white blood cells (WBCs),* have nuclei and do not contain hemoglobin (see Figure 12-2). Leucocytes fall into two major groups. The first group is the *granular leucocytes.* They develop from red bone marrow, have large granules in their cytoplasm, and possess lobed nuclei. The three kinds of granular leucocytes are *neutrophils,* or *polymorphs* (10–12 μm in diameter), *eosinophils* (10–12 μm in diameter), and *basophils* (8–10 μm in diameter). The nuclei of neutrophils have two to six lobes, connected by very thin strands. As the cells age, the extent of lobulation increases. Fine, evenly distributed granules are found in the cytoplasm. Eosinophils contain nuclei that are usually bilobed, with the lobes connected by a thin strand or thick isthmus. The cytoplasm is packed with large, uniform-sized granules that do not cover or obscure the nucleus. The nuclei of basophils are bilobed or irregular in shape, often in the form of a letter S. The cytoplasmic granules are round, variable in size, and commonly obscure the nucleus.

The second principal group of leucocytes is the *agranular leucocytes.* They develop from lymphoid and myeloid tissue and no cytoplasmic granules can be seen under a light microscope because of their small size. The two kinds of agranular leucocytes are *lymphocytes* (7–15 μm in diameter) and *monocytes* (14–19 μm in diameter). The nuclei of lymphocytes are darkly staining, round, or slightly indented. The cytoplasm forms a narrow rim around the nucleus.

The nuclei of monocytes are usually indented or kidney shaped and the cytoplasm has a foamy appearance.

Just as red blood cells have surface proteins, so do white blood cells and all other nucleated cells in the body. These proteins, called *HLA (human leucocyte associated) antigens,* are unique for each person (except for identical twins) and can be used to identify a tissue. If an incompatible tissue is transplanted, it is rejected by the recipient as foreign due, in part, to differences in donor and recipient HLA antigens. The HLA antigens are used to type tissues to help prevent rejection.

● *Functions*   The skin and mucous membranes of the body are continuously exposed to microbes and their toxins. Some of these microbes are capable of invading deeper tissues to cause disease, and once they enter the body, the general function of leucocytes is to combat them by phagocytosis or antibody production. Neutrophils and monocytes are actively *phagocytotic:* they can ingest bacteria and dispose of dead matter (see Figure 2-3b). Neutrophils (NOO-trō-fils) are the most active leucocytes in responding to tissue destruction by bacteria. In addition to carrying on phagocytosis, they release the enzyme lysozyme, which destroys certain bacteria. Apparently, monocytes (MON-ō-sīts) take longer to reach the site of infection than do neutrophils, but once they arrive, they do so in large numbers and destroy more microbes. Monocytes that have migrated to infected tissues and differentiate into phagocytes are called *wandering macrophages.* They clean up cellular debris following an infection.

A number of different chemicals in inflamed tissue cause phagocytes to migrate toward the tissue. This phenomenon is called *chemotaxis.* Among the substances that provide stimuli for chemotaxis are toxins produced by microbes and degenerative products of damaged tissues.

Most leucocytes possess, to some degree, the ability to migrate through the minute spaces between the cells that form the walls of capillaries and through connective and epithelial tissue. This movement, like that of amoebas, is called *diapedesis* (dī'-a-pe-DĒ-sis). First, part of the cell membrane projects outward. Then the cytoplasm and nucleus flow into the projection. Finally, the rest of the membrane snaps up into place. Another projection is made, and so on, until the cell has migrated to its destination. Neutrophils also contain *defensins,* amino acids that exhibit a broad range of antibiotic activity against bacteria, fungi, and viruses.

Eosinophils (ē'-ō-SIN-ō-fils) are believed to release substances that combat the effects of histamine and other mediators of inflammation in allergic reactions. Eosinophils leave the capillaries, enter the tissue fluid, and phagocytize antigen–antibody complexes. Eosinophils are also effective against certain parasitic worms. Thus, a high eosinophil count frequently indicates an allergic condition or a parasitic infection.

Basophils (BĀ-sō-fils) are also believed to be involved

in allergic reactions. Basophils leave the capillaries, enter the tissues, and liberate heparin, histamine, and serotonin. These substances intensify the overall inflammatory reaction and are involved in hypersensitivity (allergic) reactions.

Lymphocytes (LIM-fō-sīts) are involved in the production of antibodies. *Antibodies* (AN-ti-bod'-ēz) are special proteins that inactivate antigens. *Antigens* (AN-ti-jens) are substances that will stimulate the production of antibodies and that are capable of reacting specifically with antibodies. Most antigens are proteins, and most are not synthesized by the body. Many of the proteins that make up the cell structures and enzymes of bacteria are antigens. The toxins released by bacteria are also antigens. The branch of science that deals with the responses of the body when challenged by antigens is referred to as *immunology* (im'-yoo-NOL-ō-jē; *immunis* = free). When antigens enter the body, they react chemically with substances in the lymphocytes and stimulate some lymphocytes, called B cells, to become *plasma cells* (Figure 12-4). The plasma cells then produce antibodies, globulin-type proteins that attach to antigens much as enzymes attach to substrates. Like enzymes, a specific antibody will generally attach only to a certain antigen. However, unlike enzymes, which enhance the reactivity of the substrate, antibodies "cover" their antigens so the antigens cannot come in contact with other chemicals in the body. In this way, bacterial poisons can be sealed up and rendered harmless. The bacteria themselves are destroyed by the antibodies. This process is called the *antigen–antibody response*. Phagocytes in tissues destroy the antigen–antibody complexes.

CLINICAL APPLICATION

Multiple myeloma is a malignant disorder of plasma cells in bone marrow. The symptoms are caused by the growing tumor cell mass itself and antibodies produced by the malignant cells. Among the effects are pain and osteoporosis due to multiple osteolytic (bone-dissolving) lesions in bones such as the sternum, ribs, backbone, clavicles, skull, pelvis, and proximal extremities. Other effects include hypercalcemia (excessive blood calcium), anemia, leucopenia (abnormally low level of white blood cells), thrombocytopenia (abnormally low level of thrombocytes), spinal cord compression, enlarged liver, kidney damage, and increased susceptibility to infection.

Other lymphocytes are called *T cells.* One group of T cells, the *killer T cells,* are activated by certain antigens and react by destroying them directly or indirectly by recruiting other lymphocytes and macrophages. T cells are especially effective against bacteria, viruses, fungi, transplanted cells, and cancer cells.

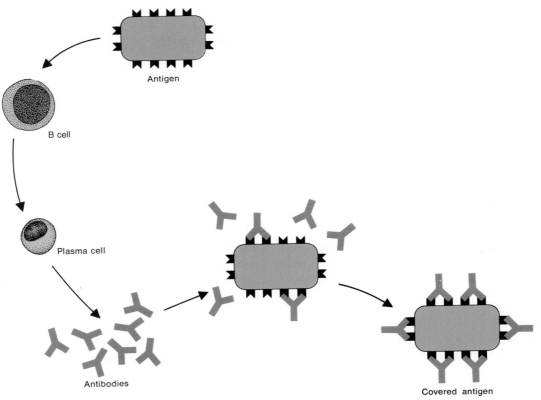

FIGURE 12-4 Antigen–antibody response. An antigen entering the body stimulates a B cell to develop into an antibody-producing plasma cell. The antibodies attach to the antigen, cover it, and render it harmless.

The antigen–antibody response helps us combat infection and gives us immunity to some diseases. It is also responsible for blood types, allergies, and the body's rejection of organs transplanted from an individual with a different genetic makeup.

An increase in the number of white cells present in the blood typically indicates a state of inflammation or infection. Because each type of white cell plays a different role, determining the percentage of each type in the blood assists in diagnosing the condition. A *differential blood count (DIF)* is the number of each kind of white cell in 100 white blood cells. A normal differential blood count falls within the following percentages:

Neutrophils	60–70%
Eosinophils	2–4%
Basophils	0.5–1%
Lymphocytes	20–25%
Monocytes	3–8%
	100%

Particular attention is paid to the neutrophils in a differential blood count. More often than not, a high neutrophil count indicates response to invading bacteria. An increase in the number of monocytes generally indicates a chronic infection. Eosinophils and basophils are elevated during allergic reactions. High lymphocyte counts indicate antigen–antibody reactions.

● *Life Span and Number* Bacteria exist everywhere in the environment and have continuous access to the body through the mouth, nose, and pores of the skin. Furthermore, many cells, especially those of epithelial tissue, age and die, and their remains must be disposed of daily. Even when the body is healthy, the leucocytes actively ingest bacteria and debris. However, a leucocyte can phagocytize only a certain number of substances before they interfere with the leucocyte's normal metabolic activities and bring on its death. Consequently, the life span of most leucocytes is very short. In a healthy body, some white blood cells can live up to several months, but most live only a few days. During a period of infection they may live only a few hours.

Leucocytes are far less numerous than red blood cells, averaging from 5,000 to 9,000 cells per cubic millimeter of blood. Red blood cells therefore outnumber white blood cells about 700 to 1. The term *leucocytosis* (loo'-kō-sī-TŌ-sis) refers to an increase in the number of white blood cells. If the increase exceeds 10,000, a pathological condition is usually indicated. An abnormally low level of white blood cells (below 5,000/mm³) is termed *leucopenia* (loo-kō-PĒ-nē-a).

● *Production* Granular leucocytes are produced in red bone marrow (myeloid tissue); agranular leucocytes are produced in both myeloid and lymphoid tissue. The developmental sequences for the five types of leucocytes are shown in Figure 12-2.

CLINICAL APPLICATION

Bone marrow, as indicated earlier, contains hemocytoblasts that differentiate into erythrocytes, leucocytes, and platelets. Among the leucocytes are cells that combat infections and cause tissue rejection. Until recently, the use of *bone marrow transplantation,* the transfer of bone marrow from a donor to a recipient, required that the marrow of the donor be obtained from a relative and closely matched to that of the recipient. In cases where the match is not identical, T cells in donor marrow, which are part of the immune system, recognize the recipient's tissues as foreign and attack them. The principal tissues rejected in this way are in the skin, liver, and gastrointestinal tract. Several recent advances now permit bone marrow transplantation in which the tissue match between donor and recipient is not even close.

Bone marrow transplantation has been used to treat aplastic anemia, certain types of leukemia, and severe combined immunodeficiency disease (SCID), an inherited deficiency of infection-fighting blood cells. The technique is also being expanded to treat other kinds of leukemia, non-Hodgkin's lymphoma, thalassemia, multiple myeloma, sickle-cell anemia, and hemolytic anemia.

Thrombocytes

● *Structure* In addition to the immature cell types that develop into erythrocytes and leucocytes, hemocytoblasts differentiate into still another kind of cell, called a megakaryoblast (see Figure 12-2a). Megakaryoblasts are ultimately transformed into megakaryocytes, large cells that shed fragments of cytoplasm. Each fragment becomes enclosed by a piece of the cell membrane and is called a *thrombocyte* (THROM-bō-sīt) or *platelet.* Platelets are round or oval discs without a nucleus. They average from 2 to 4 μm in diameter.

● *Function* Platelets prevent fluid loss by initiating a chain of reactions that results in blood clotting. This mechanism is described shortly.

● *Life Span and Number* Platelets have a short life span, probably only 5–9 days. Between 250,000 and 400,000 platelets appear in each cubic millimeter of blood.

● *Production* Platelets are produced in red bone marrow according to the developmental sequence shown in Figure 12-2a.

A summary of the formed elements in blood is presented in Exhibit 12-2.

EXHIBIT 12-2

Summary of the Formed Elements in Blood

FORMED ELEMENT	NUMBER	DIAMETER (μm)	LIFE SPAN	FUNCTION
Erythrocyte (red blood cell)	4.8 million/mm³ in females 5.4 million/mm³ in males	8	120 days	Transports oxygen and carbon dioxide.
Leucocyte (white blood cell)	5,000–9,000/mm³		A few hours to a few days*	
Granular				
Neutrophil	60–70% of total	10–12		Phagocytosis.
Eosinophil	2–4% of total	10–12		Combats the effects of histamine in allergic reactions, phagocytizes antigen–antibody complexes, and destroys certain parasitic worms.
Basophil	0.5–1% of total	8–10		Liberates heparin, histamine, and serotonin in allergic reactions that intensify the overall inflammatory response.
Agranular				
Lymphocyte	20–25% of total	7–15		Immunity (antigen–antibody reactions).
Monocyte	3–8% of total	14–19		Phagocytosis.
Thrombocyte (platelet)	250,000–400,000/mm³	2–4	5–9 days	Blood clotting.

* Some lymphocytes, called T and B memory cells, can live throughout life once they are formed. Most white blood cells, however, have life spans ranging from a few hours to a few days.

CLINICAL APPLICATION

The most commonly ordered hematology test is the ***complete blood count (CBC).*** It generally includes determination of hemoglobin, hematocrit, red blood cell count, white blood cell count, differential blood count, and comments about red blood cell, white blood cell, and platelet morphology.

PLASMA

When the formed elements are removed from blood, a straw-colored liquid called *plasma* is left. Exhibit 12-3 outlines the chemical composition of plasma. Some of the proteins in plasma are also found elsewhere in the body, but in blood they are called *plasma proteins.* **Albumins,** which constitute 55 percent of plasma proteins, are largely responsible for blood's viscosity. The concentration of albumins is about four times higher in plasma than in interstitial fluid. Along with the electrolytes (ions), albumins also help regulate blood volume by preventing the water in the blood from diffusing into the interstitial fluid. Recall that water moves by osmosis from an area of high water (low solute) concentration to an area of low water (high solute) concentration. *Globulins,* which comprise 38 percent of plasma proteins, are antibody proteins released by plasma cells. Gamma globulin is especially well known because it is able to form an antigen–antibody complex with the proteins of the hepati-

tis and measles viruses and the tetanus bacterium, among others. ***Fibrinogen*** makes up about 7 percent of plasma proteins and takes part in the blood-clotting mechanism along with the platelets.

CLINICAL APPLICATION

Plasmapheresis (plaz'-ma-fe-RĒ-sis; *aphairesis* = removal), also called ***therapeutic plasma exchange (TPE),*** refers to a procedure in which blood is withdrawn from the body, its components are selectively separated, the undesirable component causing disease is removed, and the remainder is returned to the body. Among the substances removed are toxins, metabolic substances, and antibodies. Blood is withdrawn by a needle or catheter, mixed with an anticoagulant, and pumped through a separator where red blood cells, white blood cells, platelets, and plasma are separated by spinning the sample. The process usually takes 3–5 hours and no more than 15 percent of the patient's total blood volume is allowed outside the body at any one time.

The ability of plasmapheresis to remove antibodies and other immunologically active substances has made the procedure useful for neurological conditions in which autoimmunity is believed to play a role. About half of plasmapheresis procedures are done on patients with neurological disorders. Among the neurological disorders for which plasmapheresis benefits patients are myasthenia gravis, Eaton-Lambert syndrome, chronic inflammatory demyelinating polyneuropathy, and Guillain-Barré syndrome. Unfortunately, the procedure shows no benefit in amyotrophic lateral sclerosis (ALS). Plasmapheresis may also be used in the treatment of aplastic anemia, hemolytic anemia, multiple myeloma, thrombocytemia, post-transfusion Rh incompatibility, life-threatening sickle-cell crisis, kidney diseases, systemic lupus erythematosus (SLE), rheumatoid arthritis (RA), and certain drug overdoses. Because of some potentially serious complications of apheresis (hypovolemia, shock, congestive heart failure, pulmonary edema, muscle twitching, thrombosis), the procedure is used discriminately and for short periods of time when more conservative therapy is unsuccessful.

APPLICATIONS TO HEALTH

ANEMIA

Anemia is a sign, not a diagnosis. Many kinds of anemia exist, all characterized by insufficient erythrocytes or hemoglobin. These conditions lead to fatigue and intolerance to cold, both of which are related to lack of oxygen needed for energy and heat production, and to paleness, which is due to low hemoglobin content.

Nutritional Anemia

Nutritional anemia arises from an inadequate diet, one that provides insufficient amounts of iron, the necessary amino acids, or vitamin B_{12}.

Pernicious Anemia

Pernicious anemia is an insufficient production of erythrocytes because of lack of vitamin B_{12} resulting from an inability of the body to produce intrinsic factor (IF). Intrinsic factor is a substance produced by stomach mucosal cells and is required for absorption of vitamin B_{12} by the lining of the small intestine.

Hemorrhagic Anemia

An excessive loss of erythrocytes through bleeding is called *hemorrhagic anemia.* Common causes are large wounds, stomach ulcers, and heavy menstrual bleeding. If bleeding is extraordinarily heavy, the anemia is termed acute. Excessive blood loss can be fatal. Slow, prolonged bleeding is apt to produce a chronic anemia; the chief symptom is fatigue.

Hemolytic Anemia

If erythrocyte plasma membranes rupture prematurely, the cells remain as ''ghosts'' and their hemoglobin pours

EXHIBIT 12-3

Chemical Composition and Description of Substances in Plasma

CONSTITUENT	DESCRIPTION
WATER	Liquid portion of blood; constitutes about 91.5 percent of plasma. Ninety percent of water derived from absorption from gastrointestinal tract; 10 percent from cellular respiration. Acts as solvent and suspending medium for solid components of blood and absorbs heat.
SOLUTES	Constitute about 8.5 percent of plasma.
Proteins Albumins	Smallest plasma proteins. Produced by liver and provide blood with viscosity, a factor related to maintenance and regulation of blood pressure. Also exert considerable osmotic pressure to maintain water balance between blood and tissues and regulate blood volume.
Globulins	Protein group to which antibodies belong. Gamma globulins attack measles, hepatitis, and polio viruses and tetanus bacterium.
Fibrinogen	Produced by liver. Plays essential role in clotting.
Nonprotein nitrogen (NPN) substances	Contain nitrogen but are not proteins. Include urea, uric acid, creatine, creatinine, and ammonium salts. Represent breakdown products of protein metabolism and are carried by blood to organs of excretion.
Food substances	Products of digestion passed into blood for distribution to all body cells. Include amino acids (from proteins), glucose (from carbohydrates), fatty acids, glycerides, and glycerol (from fats).
Regulatory substances	Enzymes, produced by body cells, to catalyze chemical reactions. Hormones, produced by endocrine glands, to regulate growth and development in body.
Respiratory gases	Oxygen (O_2) and carbon dioxide (CO_2). Whereas oxygen is more closely associated with hemoglobin of red blood cells, carbon dioxide is more closely associated with plasma.
Electrolytes	Inorganic salts of plasma. Cations include Na^+, K^+, Ca^{2+}, Mg^{2+}; anions include Cl^-, PO_4^{3-}, SO_4^{2-}, HCO_3^-. Help maintain osmotic pressure, normal pH, physiological balance between tissues and blood.

out into the plasma. A characteristic sign of this condition, called *hemolytic anemia,* is distortions in the shapes of erythrocytes that are progessing toward hemolysis. There may also be a sharp increase in the number of reticulocytes, since the destruction of red blood cells stimulates erythropoiesis.

The premature destruction of red cells may result from inherent defects, such as hemoglobin defects, abnormal red cell enzymes, or defects of the red cell membrane.

Agents that may cause hemolytic anemia are parasites, toxins, and antibodies from incompatible blood (Rh⁻ mother and Rh⁺ fetus, for instance). Hemolytic disease of newborn (HDN) or erythroblastosis fetalis is an example of a hemolytic anemia.

The term **thalassemia** (thal'-a-SĒ-mē-a) represents a group of hereditary hemolytic anemias resulting from a defect in the synthesis of hemoglobin, which produces extremely thin and fragile erythrocytes. It occurs primarily in populations from countries bordering the Mediterranean Sea. Treatment generally consists of blood transfusions.

Aplastic Anemia

Destruction or inhibition of the red bone marrow results in **aplastic anemia.** Typically, the marrow is replaced by fatty tissue, fibrous tissue, or tumor cells. Toxins, gamma radiation, and certain medications are causes. Many of the medications inhibit the enzymes involved in hemopoiesis. Bone marrow transplants can now be done with a reasonable hope of success in patients with aplastic anemia. They are done early, before the victim is sensitized by transfusions. Immunosuppressive drugs are given for several days before the transplant and with decreasing frequency afterward.

Sickle-Cell Anemia (SCA)

The erythrocytes of a person with **sickle-cell anemia (SCA)** manufacture an abnormal kind of hemoglobin. When an erythrocyte gives up its oxygen to the interstitial fluid, its hemoglobin tends to lose its integrity in places of low oxygen tension and forms long, stiff, rodlike structures that bend the erythrocyte into a sickle shape (Figure 12-5). The sickled cells rupture easily. Even though erythropoiesis is stimulated by the loss of the cells, it cannot keep pace with the hemolysis. The individual consequently suffers from a hemolytic anemia that reduces the amount of oxygen that can be supplied to the tissues. Prolonged

oxygen reduction may eventually cause extensive tissue necrosis, pain, and organ damage. Sickled cells also increase blood viscosity, which leads to occlusion of small blood vessels. Furthermore, because of the shape of the sickled cells, they tend to get stuck in blood vessels and can cut off the blood supply to an organ altogether.

Sickle-cell anemia is characterized by several symptoms. In young children, hand-foot syndrome is present, in which there is swelling and pain in the wrists and feet. Older patients experience pain in the back and extremities without swelling and abdominal pain. Complications include neurological disorders (meningitis, seizures, stroke), impaired pulmonary function, orthopedic abnormalities (femoral head necrosis, osteomyelitis), genitourinary tract disorders (involuntary urination, blood in urine, kidney failure), ocular disturbances (hemorrhage, detached retina, blindness), and obstetric complications (convulsions, coma, infection).

Sickle-cell anemia is inherited. The gene responsible for the tendency of the erythrocytes to sickle during hypoxia also seems to prevent erythrocytes from rupturing during a malarial crisis. The gene also alters the permeability of the plasma membranes of sickled cells, causing potassium to leak out. Low levels of potassium kill the malarial parasites that infect sickled cells. Sickle-cell genes are found primarily among populations, or descendants of populations, that live in the malaria belt around the world, including parts of Mediterranean Europe and subtropical Africa and Asia. A person with only one of the sickling genes is said to have sickle-cell trait. Such an individual has a high resistance to malaria—a factor that may have tremendous survival value—but does not develop the anemia. Only people who inherit a sickling gene from both parents get sickle-cell anemia.

Treatment consists of administration of analgesics to relieve pain, antibiotics to counter infections, and transfusion therapy.

POLYCYTHEMIA

The term **polycythemia** (pol'-ē-sī-THĒ-mē-a) refers to an abnormal increase in the number of red blood cells. The hematocrit (Hct) is an important factor in diagnosing the condition. When the hematocrit increases, especially above 55, it is very suggestive of polycythemia. There is an increased blood viscosity associated with the elevated hematocrit. The increased viscosity causes a rise in blood pressure and contributes to thrombosis and hemorrhage. The thrombosis results from too many red blood cells piling up as they try to enter smaller vessels. The hemorrhage is due to widespread hyperemia (unusually large amount of blood in an organ part).

INFECTIOUS MONONUCLEOSIS (IM)

Infectious mononucleosis (IM) is a contagious disease primarily affecting lymphoid tissue throughout the body. It

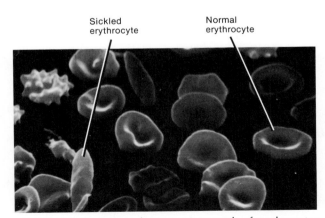

Sickled erythrocyte   Normal erythrocyte

FIGURE 12-5 Scanning electron micrograph of erythrocytes in sickle-cell anemia at a magnification of about 2,000×. (Courtesy of Fisher Scientific Company and S.T.E.M. Laboratories, Inc. Copyright © 1975.)

is caused by the *Epstein-Barr virus* (*EBV*), the same agent that has been linked to Burkitt's lymphoma, nasopharyngeal carcinoma, and Hodgkin's disease. It occurs mainly in children and young adults, with the peak incidence at 15–20 years of age. The virus most commonly enters the body through intimate oral contact, multiplies in lymphoid tissues, and spreads into the blood where it infects and multiplies in B lymphocytes, the primary host cells. As a result of this infection, the B lymphocytes become enlarged and abnormal in appearance and resemble monocytes, the primary reason for which the disease receives its name mononucleosis. Infectious mononucleosis is characterized by an elevated white blood cell count with an abnormally high percentage of lymphocytes. Symptoms include sore throat, lymphadenopathy (enlarged and tender lymph nodes), fever, brilliant red throat and soft palate, stiff neck, cough, and malaise. The spleen may also enlarge. Secondary complications involving the liver, eye, heart, kidneys, and nervous system may develop. There is no cure for infectious mononucleosis, and treatment consists of watching for and treating complications. Usually, the disease runs its course in a few weeks, and the individual generally suffers no permanent ill effects.

CHRONIC EPSTEIN-BARR VIRUS (CEBV) SYNDROME

The Epstein-Barr virus (EBV) has recently been implicated in causing a disease called **chronic Epstein-Barr virus (CEBV) syndrome**. This disease, unheard of several years ago, is characterized by sore throat, extreme fatigue, headache, muscular aches and weakness, fever and chills, night sweats, swollen glands, swelling of the eyelids, inability to concentrate, allergy, memory loss, and depression. Females in high-stress occupations seem to be especially vulnerable. It is suspected that the disease might be caused by a new strain of the EBV or some factor that suppresses immunity and allows the EBV to become reactivated. For the most part, there is no treatment for the disease, but patients are advised to get plenty of rest.

LEUKEMIA

Clinically, **leukemia** is classified on the basis of the duration and character of the disease, that is, acute or chronic. In its simplest terms, acute leukemia refers to a malignant disease of blood-forming tissues characterized by uncontrolled production and accumulation of immature leucocytes and many of the cells fail to reach maturity. In chronic leukemia, there is an accumulation of mature leucocytes in the bloodstream because they do not die at the end of their normal life span. The *human T-cell leukemia-lymphoma virus-1* (*HTLV-1*) is strongly associated with leukemia. Leukemia is also classified according to the identity and site of origins of the predominant cell involved, such as myelocytic (myelogenous, myeloblastic, granulocytic), lymphocytic (lymphogenous, lymphatic), and monocytic.

In acute leukemia, the anemia and bleeding problems commonly seen result from the crowding out of normal bone marrow cells by the overproduction of immature cells, preventing normal production of red blood cells and platelets. One cause of death from acute leukemia is internal hemorrhaging, especially cerebral hemorrhage that destroys the vital centers in the brain. Perhaps the most frequent cause of death is uncontrolled infection due to lack of mature or normal white blood cells. The abnormal accumulation of immature leucocytes may be reduced by using x rays and antileukemic drugs. Partial or complete remissions may be induced, with some lasting as long as 15 years.

KEY MEDICAL TERMS ASSOCIATED WITH BLOOD

Autologous (aw-TOL-o-gus; *auto* = self) **transfusion** (trans-FYOO-zhun)   Donating one's own blood before elective surgery to ensure an abundant supply and reduce transfusion complications such as those associated with AIDS and hepatitis. It is not indicated for individuals who cannot donate 500 ml of blood in a short period of time and who have preexisting anemia, bleeding disorders, infections, unstable blood pressure, and malignancy involving bone marrow.

Citrated (SIT-rāt-ed) **whole blood**   Whole blood protected from coagulation by CPD (citrate phosphate dextrose) or a similar compound.

Cyanosis (sī'-a-NŌ-sis; *cyano* = blue)   Slightly bluish, dark purple skin coloration due to oxygen deficiency in systemic blood.

Direct (immediate) transfusion (*trans* = through)   Transfer of blood directly from one person to another without exposing the blood to air.

Exchange transfusion   Removing blood from the recipient while simultaneously replacing it with donor blood. This method is used for hemolytic disease of newborn (HDN) and poisoning.

Gamma globulin (GLOB-yoo-lin)   Solution of globulins from nonhuman blood consisting of antibodies that react with specific pathogens, such as measles, epidemic hepatitis, tetanus, and possibly poliomyelitis viruses. It is prepared by injecting the specific virus into animals, removing blood from the animals after antibodies have accumulated, isolating antibodies, and injecting them into a human for short-term immunity.

Hemochromatosis (hē'-mō-krō'-ma-TŌ-sis; *heme* = iron; *chroma* = color)   Disorder of iron metabolism characterized by excess deposits of iron in tissues, especially the liver and pancreas, that result in bronze coloration of the skin, cirrhosis, diabetes, mellitus, and bone and joint abnormalities.

Hemorrhage (HEM-or-ij; *rrhage* = bursting forth)   Bleeding, either internal (from blood vessels into tissues) or external (from blood vessels directly to the surface of the body).

Indirect (mediate) transfusion   Transfer of blood from a donor to a container and then to the recipient, permitting blood to be stored for an emergency. The blood may be separated into its components so that a patient will receive only a needed part.

Platelet (PLĀT-let) **concentrates** A preparation of platelets obtained from freshly drawn whole blood and used for transfusions in platelet-deficiency disorders such as hemophilia.

Reciprocal (re-CIP-rō-cal) **transfusion** Transfer of blood from a person who has recovered from a contagious infection into the vessels of a patient suffering with the same infection. An equal amount of blood is returned from the patient to the well person. This method allows the patient to receive antibody-bearing lymphocytes from the recovered person.

Septicemia (sep'-ti-SĒ-mē-a; *sep* = decay; *emia* = condition of blood) Toxins or disease-causing bacteria in the blood. Also called "blood poisoning."

Thrombocytopenia (throm'-bō-sī'-tō-PĒ-nē-a; *thrombo* = clot; *penia* = poverty) Very low platelet count that results in a tendency to bleed from capillaries.

Transfusion (trans-FYOO-zhun) Transfer of whole blood, blood components (red blood cells only or plasma only), or bone marrow directly into the bloodstream.

Venesection (vēn'-e-SEK-shun; *veno* = vein) Opening of a vein for withdrawal of blood. Although *phlebotomy* (fle-BŌT-ō-mē; *phlebo* = vein; *tome* = to cut) is a synonym for venesection, in clinical practice, phlebotomy refers to therapeutic bloodletting, such as might be done to remove a pint of blood to lower the viscosity of blood of a patient with polycythemia.

Whole blood Blood containing all formed elements, plasma, and plasma solutes in natural concentrations.

STUDY OUTLINE

Physical Characteristics (p. 328)

1. The cardiovascular system consists of blood, the heart, and blood vessels. The lymphatic system consists of lymph, lymph vessels, and lymph glands.
2. Physical characteristics of blood include viscosity, 4.5–5.5; temperature, 38°C (100.4°F); pH, 7.35–7.45; and a salt (NaCl) concentration of 0.90 percent. Blood constitutes about 8 percent of body weight.

Functions (p. 328)

1. Blood transports oxygen, carbon dioxide, nutrients, wastes, hormones, and enzymes.
2. It helps to regulate pH, body temperature, and water content of cells.
3. It prevents blood loss through clotting and combats toxins and microbes through special combat-unit cells.

Components (p. 329)

1. The formed elements in blood include erythrocytes (red blood cells), leucocytes (white blood cells), and thrombocytes (platelets).
2. Blood cells are formed by a process called hemopoiesis.
3. Red bone marrow (myeloid tissue) is responsible for producing red blood cells, granular leucocytes, and platelets; lymphoid tissue and myeloid tissue produce agranular leucocytes.

Erythrocytes (p. 332)

1. Erythrocytes are biconcave discs without nuclei and containing hemoglobin.
2. The function of red blood cells is to transport oxygen and carbon dioxide.
3. Red blood cells live about 120 days. A healthy male has about 5.4 million/mm^3 of blood; a healthy female, about 4.8 million/mm^3.
4. Erythrocyte formation, called erythropoiesis, occurs in adult red marrow of certain bones.
5. A reticulocyte count is a diagnostic test that indicates the rate of erythropoiesis.
6. A hematocrit (Hct) measures the percentage of red blood cells in whole blood.

Leucocytes (p. 334)

1. Leucocytes are nucleated cells. Two principal types are granular (neutrophils, eosinophils, basophils) and agranular (lymphocytes and monocytes).
2. The general function of leucocytes is to combat inflammation and infection. Neutrophils and monocytes (wandering macrophages) do so through phagocytosis.
3. Eosinophils combat the effects of histamine in allergic reactions, phagocytize antigen–antibody complexes, and combat parasitic worms; basophils liberate heparin, histamine, and serotonin in allergic reactions that intensify the inflammatory response.
4. Lymphocytes, in response to the presence of foreign substances called antigens, differentiate into tissue plasma cells that produce antibodies. Antibodies attach to the antigens and render them harmless. This antigen–antibody response combats infection and provides immunity.
5. A differential blood count (DIF) is a diagnostic test in which white blood cells are enumerated.
6. White blood cells usually live for only a few hours or a few days. Normal blood contains 5,000–9,000/mm^3.

Thrombocytes (p. 336)

1. Thrombocytes are disc-shaped structures without nuclei.
2. They are formed from megakaryocytes and are involved in clotting.
3. Normal blood contains 250,000–400,000/mm^3.

Plasma (p. 337)

1. The liquid portion of blood, called plasma, consists of 91.5 percent water and 8.5 percent solutes.
2. Principal solutes include proteins (albumins, globulins, fibrinogen), nonprotein nitrogen (NPN) substances, food, enzymes and hormones, respiratory gases, and electrolytes.

Applications to Health (p. 338)

1. Anemia is a decreased erythrocyte count or hemoglobin deficiency. Kinds of anemia include nutritional, pernicious, hemorrhagic, hemolytic, aplastic, and sickle-cell anemia (SCA).
2. Polycythemia is an abnormal increase in the number of erythrocytes.

3. Infectious mononucleosis (IM) is a contagious disease that primarily affects lymphoid tissue. It is characterized by an elevated white blood cell count, with an abnormally high percentage of lymphocytes. The cause is the Epstein-Barr virus (EBV).
4. Chronic Epstein-Barr virus (CEBV) syndrome is characterized by sore throat, extreme fatigue, muscular aches, fever and chills, and neurological defects.
5. Leukemia is a malignant disease of blood-forming tissues characterized by the uncontrolled production of white blood cells that interferes with normal clotting and vital body activities.

REVIEW QUESTIONS

1. How are blood, interstitial fluid, and lymph related?
2. List the principal physical characteristics of blood.
3. List the functions of blood and their relation to other systems of the body.
4. Describe the origin of blood cells.
5. Describe the microscopic appearance of erythrocytes. What is the essential function of erythrocytes?
6. What is a reticulocyte count? What is its diagnostic significance?
7. Define hematocrit (Hct). Compare the hematocrits of anemic and polycythemic blood.
8. Describe the classification of leucocytes. What are their functions?
9. What is a differential blood count (DIF)? What is its significance?
10. Describe the antigen–antibody response. How is it protective?
11. Describe the structure and function of thrombocytes.
12. Compare erythrocytes, leucocytes, and thrombocytes with respect to size, number per mm^3, and life span.
13. What are the major constituents in plasma? What do they do?
14. Define anemia. Contrast the causes of nutritional, pernicious, hemorrhagic, hemolytic, aplastic, and sickle-cell anemia (SCA).
15. What is infectious mononucleosis (IM)?
16. Describe the symptoms of chronic Epstein-Barr virus (CEBV) syndrome.
17. What is leukemia? What are the causes of some of its symptoms?
18. Refer to the glossary of key medical terms associated with blood. Be sure that you can define each term.

SELF-QUIZ

1. Match the following:
___ a. constitute the largest percentage (about 70 percent) of leucocytes
___ b. 20–25 percent of leucocytes
___ c. involved in immunity; form plasma cells for antibody production
___ d. involved in allergic response; release serotonin, heparin, and histamine
___ e. involved in allergic response; release antihistamines
___ f. develop into wandering macrophages that clean up sites of infection
___ g. important in phagocytosis (two answers)
___ h. classified as granular leucocytes (three answers)

A. basophils
B. eosinophils
C. lymphocytes
D. monocytes
E. neutrophils

Complete the following.

2. Blood is a connective tissue that consists of about _____ percent intercellular material and about _____ percent cells or formed elements. The intercellular material is a liquid named _____. The process of blood formation is called _____.
3. Arrange, in order, these stages in erythropoiesis. Write letters on the lines provided. _____ _____ _____ _____
A. late erythroblast formed; B. reticulocyte loses nucleus; C. erythroblast starts to synthesize hemoglobin; D. erythrocyte moves out of marrow into general circulation.
4. During infection, it is likely that a person's leucocyte count will (increase? decrease?) A count of (4,000? 8,000? 15,000?) leucocytes/mm^3 blood is most likely. This condition is known as (leucocytosis? leucopenia?). In order to confirm changes in counts of specific types of leucocytes, a _____ count may be taken.

Choose the one best answer to these questions.

___ 5. Choose the false statement about neutrophils:
A. they are actively phagocytic; B. they are the most abundant type of leucocyte; C. their number decreases during most infections; D. an increase in their number would be a form of leucocytosis; E. they are granular leucocytes.
___ 6. The normal red blood cell count in healthy males is _____RBCs/mm^3.
A. 5.4 million; B. 2 million; C. 0.5 million; D. 250,000; E. 8,000.
___ 7. All of the following types of formed elements are produced in bone marrow *except*
A. neutrophils; B. basophils; C. platelets; D. erythrocytes; E. lymphocytes.
___ 8. Megakaryocytes are involved in the formation of
A. red blood cells; B. basophils; C. lymphocytes; D. platelets; E. neutrophils.
___ 9. Choose the false statement about blood:
A. blood is thicker than water. B. it normally has a pH of 6.5–7.0; C. the male human body normally contains about 5–6 liters of it; D. it normally consists of more plasma than cells; E. it constitutes about 8 percent of total body weight.

___ **10.** Cells that form mature red blood cells are
A. proerythroblasts; B. lymphoblasts; C. myeloblasts; D. megakaryoblasts; E. monoblasts.

___ **11.** Which of the following statements about the function of blood is *not* true?
A. it transports gases, nutrients, and wastes; B. it plays a role in the regulation of temperature; C. it transports tissue cells; D. it regulates the water content of cells; E. it regulates pH.

___ **12.** The parent (stem) cell of all blood cells is the
A. myeloblast; B. megakaryoblast; C. hemocytoblast; D. monoblast; E. proerythroblast.

___ **13.** The ability of white blood cells to crawl through capillaries and reach an injured or infected body tissue is
A. diapedesis; B. phagocytosis; C. leucocytosis; D. leucopenia; E. pinocytosis.

___ **14.** Which of these values is *not* considered normal for blood?
A. viscosity of 4.5–5.5; B. temperature of 98.6°F; C. pH of 7.35–7.45; D. volume of 4–6 liters; E. salt concentration of 0.90 percent.

___ **15.** Which of the following numbers of white blood cells per cubic millimeter of blood represents leucopenia?
A. 12,000; B. 16,000; C. 4,000; D. 10,000; E. 8,000.

___ **16.** Erythrocytes
A. are nonnucleated, biconcave discs; B. are usually called white blood corpuscles; C. are formed in the lungs; D. contain fibroblasts; E. are phagocytes.

___ **17.** The percentage of formed elements in the blood, by volume, is called the
A. red count; B. white count; C. hematocrit; D. serum volume; E. Cushny determination.

___ **18.** One function of the platelets is to aid in the
A. transport of carbon dioxide; B. production of vitamin K; C. utilization of calcium and phosphorus; D. destruction of bacteria; E. clotting of blood.

Circle T (true) or F (false) for the following:

T F **19.** The liquid portion of blood that contains the clotting proteins is called plasma.

T F **20.** The three kinds of granular leucocytes are neutrophils, monocytes, and eosinophils.

T F **21.** The pigment hemoglobin is resposible for the color of blood and the carriage of respiratory gases.

T F **22.** The process by which blood cells are formed is called fibrinolysis.

T F **23.** A high lymphocyte count may indicate that an allergic reaction is taking place in the body.

T F **24.** A count of 150,000 thrombocytes per cubic millimeter of blood would probably interfere with the blood's ability to clot.

13 The Cardiovascular System: The Heart

CHAPTER OUTLINE
- ■ **Location**
- ■ **Pericardium**
- ■ **Heart Wall**
- ■ **Chambers of the Heart**
- ■ **Great Vessels of the Heart**
- ■ **Valves of the Heart**
- Atrioventricular (AV) Valves
- Semilunar Valves
- Skeleton of the Heart
- Surface Projection
- ■ **Conduction System**
- ■ **Electrocardiogram (ECG)**
- ■ **Cardiac Cycle**
- ■ **Autonomic Control**
- ■ **Blood Supply**
- ■ **Artificial Heart**
- ■ **Risk Factors in Heart Disease**
- ■ **Developmental Anatomy of the Heart**
- ■ **Applications to Health**
- Coronary Artery Disease (CAD)
  - *Atherosclerosis*
  - *Coronary Artery Spasm*
- Congenital Defects
- Arrhythmias
  - *Heart Block*
  - *Flutter and Fibrillation*
  - *Ventricular Premature Contraction (VPC)*
- Congestive Heart Failure (CHF)
- Cor Pulmonale (CP)
- ■ **Key Medical Terms Associated with the Heart**

STUDENT OBJECTIVES

1. Describe the location of the heart and identify its borders.
2. Describe the structure of the pericardium and heart wall.
3. Identify and describe the chambers, great vessels, valves, and cardiac skeleton of the heart.
4. Describe the surface anatomy features of the heart.
5. Explain the structural and functional features of the conduction system of the heart and describe an electrocardiogram (ECG) and explain its significance.
6. Contrast the effects of sympathetic and parasympathetic stimulation of the heart.
7. Discuss the route of blood in coronary (cardiac) circulation.
8. List the risk factors involved in heart disease.
9. Describe the development of the heart.
10. Describe how atherosclerosis and coronary artery spasm contribute to coronary artery disease (CAD).
11. Define coarctation of the aorta, patent ducts arteriosus, septal defects, valvular stenosis, tetralogy of Fallot, atrioventricular (AV) block, atrial flutter, atrial fibrillation, ventricular fibrillation (VF), congestive heart failure (CHF), and cor pulmonale (CP).
12. Define key medical terms associated with the heart.

The **heart** is the center of the cardiovascular system. Whereas the term *cardio* refers to the heart, the term *vascular* refers to blood vessels (or an abundant blood supply). The heart is a hollow, muscular organ that weighs about 342 grams (11 oz) and beats over 100,000 times a day to pump 3,784 liters (1,000 gallons) of blood per day through over 60,000 miles of blood vessels. The blood vessels form a network of tubes that carry blood from the heart to the tissues of the body and then return it to the heart. The specific aspects of the heart that we shall consider are its location, covering, wall and chambers, great vessels, valves, surface anatomy, conduction system, cardiac cycle (heartbeat), and autonomic control. Several disorders related to the heart will also be considered.

The study of the normal heart and disease associated with it is known as **cardiology** (kar-dē-OL-ō-jē).

The developmental anatomy of the heart is considered at the end of the chapter.

LOCATION

The heart is situated between the lungs in the mediastinum. Recall that the mediastinum is the space, actually the mass of tissue, between the lungs that extends from the sternum to the vertebral column. Specifically, the heart is in the middle mediastinum (see Figure 1-7). About two-thirds of its mass lies to the left of the body's midline (Figure 13-1 and see also Figure 22-7e). The heart is shaped like a blunt cone about the size of your closed fist—12 cm (5 in.) long, 9 cm (3.5 in.) wide at its broadest point, and 6 cm (2.5 in.) thick.

Its pointed end, the **apex**, is formed by the tip of the left ventricle, projects inferiorly, anteriorly, and to the left, and lies superior to the central tendon of the diaphragm. Anteriorly, the apex is in the fifth intercostal space.

The **left border** is formed almost entirely by the left ventricle, although the left atrium forms part of the upper end of the border. The left border of the heart is indicated approximately as a curved line drawn from the left fifth costochondral junction to the left second intercostal space, about 1 cm (0.8 in.) from the sternum.

The **superior border,** where the great vessels enter and leave the heart, is represented by a line joining the second left intercostal space to the third right costal cartilage.

The **base** of the heart projects superiorly, posteriorly, and to the right. It is formed by the atria, mostly the left atrium. It lies opposite the fifth to ninth thoracic vertebrae. Anteriorly, it lies just inferior to the second rib.

The **right border** is formed by the right atrium and corresponds to a curved line drawn from the xiphisternal articulation on the right side of the sternum to about the middle of the right third costal cartilage.

The **inferior border** is formed by the right ventricle and slightly by the left ventricle. It is represented by a line passing from the inferior end of the right border through the xiphisternal joint to the apex of the heart.

The **sternocostal (anterior) surface** is formed mainly by the right ventricle, right atrium, and left ventricle, whereas the **diaphragmatic (inferior) surface** is formed by the left and right ventricles, mostly the left.

PERICARDIUM

The heart is enclosed and held in place by the **pericardium,** an ingenious structure designed to confine the heart to its position in the mediastinum, yet allow it sufficient freedom of movement so that it can contract vigorously and rapidly when the need arises.

The pericardium consists of two portions referred to as the fibrous pericardium and the serous pericardium (Figure 13-2a). The outer **fibrous pericardium** consists of very heavy fibrous connective tissue. The fibrous pericardium resembles a bag that rests on the diaphragm with its open end fused to the connective tissues of the great vessels entering and leaving the heart. The fibrous pericardium prevents overdistension of the heart, provides a tough protective membrane around the heart, and anchors the heart in the mediastinum. The inner **serous pericardium** is a thinner, more delicate membrane that forms a double layer around the heart (Figure 13-2b). The outer **parietal layer** of the serous pericardium is directly beneath the fibrous pericardium. The inner **visceral layer** of the serous pericardium, also called the **epicardium,** is beneath the parietal layer, attached to the myocardium (muscle) of the heart. Between the parietal and visceral layers of the serous pericardium is a thin film of serous fluid that holds the two layers together much like a thin film of water does to two microscope slides. The serous fluid, known as **pericardial fluid,** is an ultrafiltrate of plasma and prevents friction between the membranes as the heart moves. There are up to 150 ml of pericardial fluid. The space occupied by the pericardial fluid is a potential space (not an actual space), called the **pericardial cavity.**

An inflammation of the pericardium is known as **pericarditis.** Pericarditis with a buildup of pericardial fluid or extensive bleeding into the pericardium, if untreated, is a life-threatening condition. Since the pericardium cannot stretch to accommodate the excessive fluid or blood buildup, the heart is subjected to compression. This compression is known as **cardiac tamponade** (tam'-pon-ĀD) and can result in cardiac failure.

HEART WALL

The wall of the heart (Figure 13-2a) is divided into three layers: the epicardium (external layer), myocardium (middle layer), and endocardium (inner layer). The **epicardium** (also called the serous pericardium) is the thin, transparent outer layer of the wall. It is composed of serous tissue and mesothelium.

The **myocardium,** which is cardiac muscle tissue, constitutes the bulk of the heart. Cardiac muscle fibers (cells)

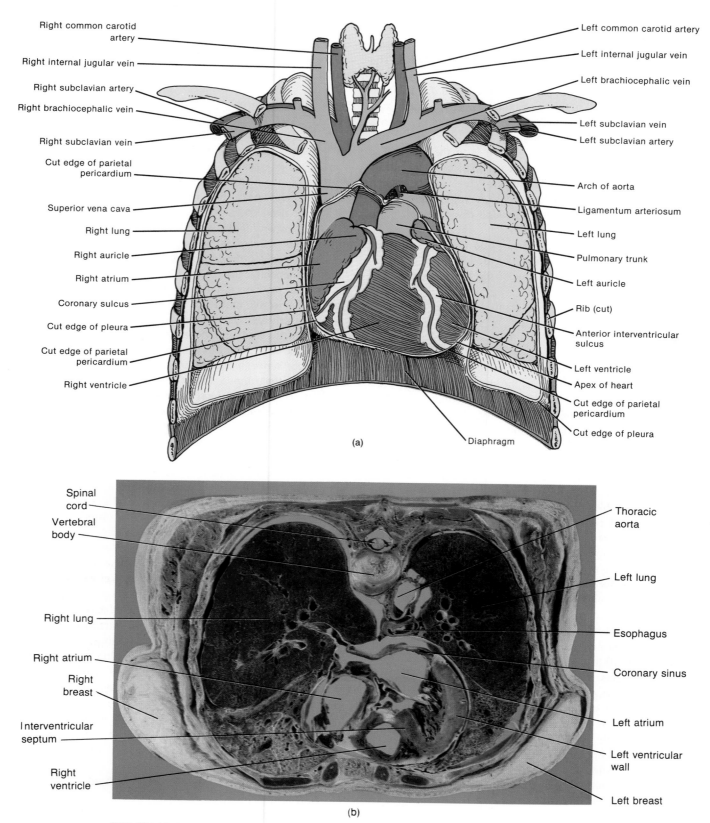

FIGURE 13-1 Position of the heart and associated blood vessels in the thoracic cavity. (a) Diagram. In this and subsequent illustrations, vessels that carry oxygenated blood are colored red; vessels that carry deoxygenated blood are colored blue. (b) Photograph of a cross section through the lower portion of the thoracic cavity. (Courtesy of Stephen A. Kieffer and E. Robert Heitzman, *An Atlas of Cross-Sectional Anatomy*, Harper & Row, Publishers, Inc., New York, 1979.)

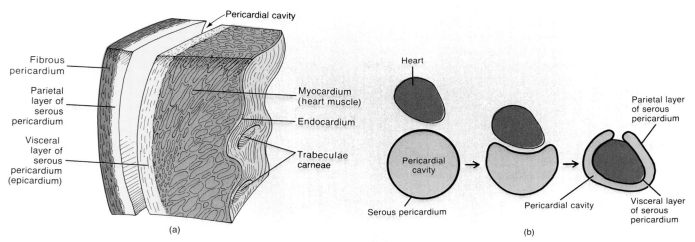

FIGURE 13-2 Pericardium and heart wall. (a) Structure of the pericardium and heart wall. (b) Relation of the serous pericardium to the heart.

are involuntary, striated, and branched, and the tissue is arranged in interlacing bundles of fibers. The myocardium is responsible for the contraction of the heart.

The **endocardium** is a thin layer of endothelium overlying a thin layer of connective tissue. It lines the inside of the myocardium and covers the valves of the heart and the tendons that hold them open. It is continuous with the endothelial lining of the large blood vessels of the heart.

Inflammation of the epicardium, myocardium, and endocardium is referred to as **epicarditis, myocarditis,** and **endocarditis,** respectively.

CHAMBERS OF THE HEART

The interior of the heart is divided into four cavities called **chambers** that receive circulating blood (Figure 13-3). The two superior chambers are called the right and left **atria.** Each atrium has an appendage called an **auricle** (OR-i-kul; *auris* = ear), so named because its shape resembles a dog's ear. The auricle increases the atrium's surface area. The lining of the atria is smooth, except for their anterior walls and the lining of the auricles, which contain projecting muscle bundles that are parallel to one another and resemble the teeth of a comb: the **pectinate** (PEK-ti-nāt) **muscles.** These bundles give the lining of the auricles a ridged appearance.

The atria are separated by a partition called the **interatrial septum.** A prominent feature of this septum is an oval depression, the **fossa ovalis,** which corresponds to the site of the foramen ovale, an opening in the interatrial septum of the fetal heart. The fossa ovalis faces the opening of the inferior vena cava and is located in the septal wall of the atrium (see Figure 14-19).

The two inferior chambers are the right and left **ventricles.** They are separated by an **interventricular septum.**

The muscle tissue of the atria and ventricles is separated by connective tissue that also forms the valves. This "cardiac skeleton" effectively divides the myocardium into two sepa-rate muscle masses. Externally, a groove known as the **coronary sulcus** (SUL-kus) separates the atria from the ventricles. It encircles the heart and houses the coronary sinus and circumflex branch of the left coronary artery. The **anterior interventricular sulcus** and **posterior interventricular sulcus** separate the right and left ventricles externally. The sulci contain coronary blood vessels and a variable amount of fat (Figure 13-3a–c).

GREAT VESSELS OF THE HEART

The right atrium receives blood from all parts of the body except the lungs. It receives the blood through three veins. In general, the **superior vena cava** (**SVC**) brings blood from parts of the body superior to the heart; in general, the **inferior vena cava** (**IVC**) brings blood from parts of the body inferior to the heart; and the **coronary sinus** drains blood from most of the vessels supplying the wall of the heart (Figure 13-3c–e). The right atrium then delivers the blood into the right ventricle, which pumps it into the **pulmonary trunk.** The pulmonary trunk divides into a **right** and **left pulmonary artery,** each of which carries blood to the lungs. In the lungs, the blood releases its carbon dioxide and takes on oxygen. Blood returns to the heart via four **pulmonary veins** that empty into the left atrium. The blood then passes into the left ventricle, which pumps the blood into the **ascending aorta.** From here the blood is passed into the **coronary arteries, arch of the aorta, thoracic aorta,** and **abdominal aorta.** These blood vessels and their branches transport the blood to all body parts.

During fetal life, there is a temporary blood vessel, called the ductus arteriosus, that connects the pulmonary trunk with the aorta (see Figure 14-19). Its purpose is to redirect blood so that only a small volume enters the nonfunctioning fetal lungs. The ductus arteriosus normally closes shortly after birth, leaving a remnant known as the **ligamentum arteriosum.**

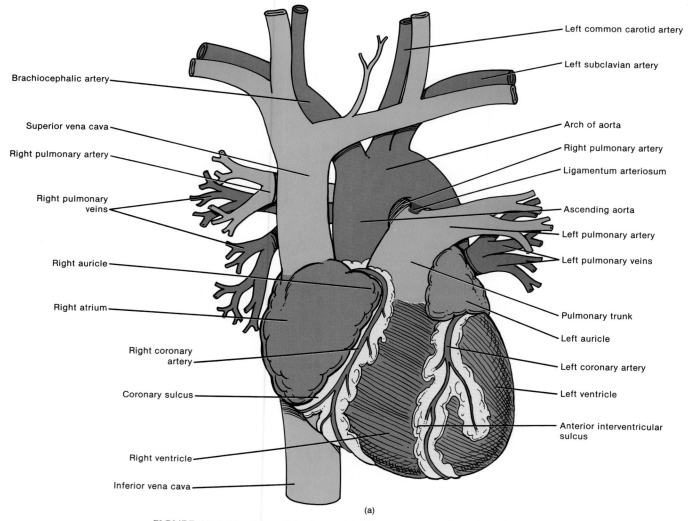

Brachiocephalic artery

Superior vena cava

Right pulmonary artery

Right pulmonary veins

Right auricle

Right atrium

Right coronary artery

Coronary sulcus

Right ventricle

Inferior vena cava

Left common carotid artery

Left subclavian artery

Arch of aorta

Right pulmonary artery

Ligamentum arteriosum

Ascending aorta

Left pulmonary artery

Left pulmonary veins

Pulmonary trunk

Left auricle

Left coronary artery

Left ventricle

Anterior interventricular sulcus

(a)

FIGURE 13-3 Structure of the heart. (a) Diagram of anterior external view.

The thickness of the four chambers varies according to function (Figure 13-3e). The atria are thin-walled because they need only enough cardiac muscle tissue to deliver the blood into the ventricles with the aid of gravity and a reduced pressure created by the expanding ventricles. The right ventricle has a thicker layer of myocardium than the atria, since it must send blood to the lungs and back around to the left atrium. The left ventricle has the thickest wall, since it must pump blood at high pressure through literally thousands of miles of vessels in the head, trunk, and extremities.

VALVES OF THE HEART

As each chamber of the heart contracts, it pushes a portion of blood into a ventricle or out of the heart through an artery. In order to keep the blood from flowing backward, the heart has structures composed of dense connective tissue called *valves*.

ATRIOVENTRICULAR (AV) VALVES

Atrioventricular (AV) valves lie between the atria and ventricles (Figure 13-3e,f). The right atrioventricular valve between the right atrium and right ventricle is also called the *tricuspid valve* because it consists of three cusps (flaps). These cusps are fibrous tissues that grow out of the walls of the heart and are covered with endocardium. The pointed ends of the cusps project into the ventricle. Cords called *chordae tendinea* (KOR-dē TEN-di-nē) connect the pointed ends and undersurfaces to small conical projections—the *papillary muscles* (muscular columns)—located on the inner surface of the ventricles. The irregular surface of ridges and folds of the myocardium in the ventricles is known as the *trabeculae carnea* (tra-BEK-yoo-lē KAR-nē). The chordae tendineae and their papillary muscles keep the cusps pointing in the direction of the blood flow. The atrioventricular valve between the left atrium and left ventricle is called the *bicuspid (mitral) valve.* It is also known as the left atrioventricular (AV) valve. It has two cusps that work in the same way as the cusps of the tricuspid valve. Its cusps

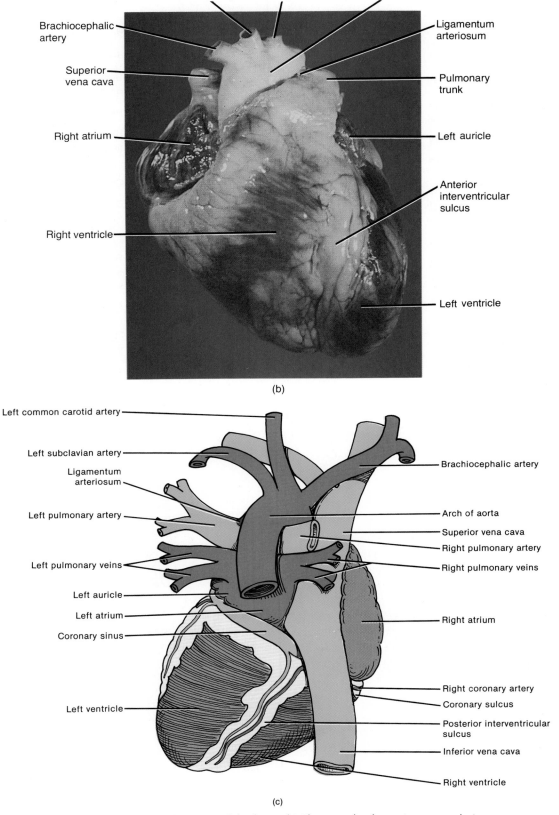

(b)

(c)

FIGURE 13-3 (*Continued*) Structure of the heart (b) Photograph of anterior external view. (Courtesy of C. Yokochi and J. W. Rohen, *Photographic Anatomy of the Human Body,* 2nd ed., Igaku-Shoin, Ltd.) (c) Diagram of posterior external view.

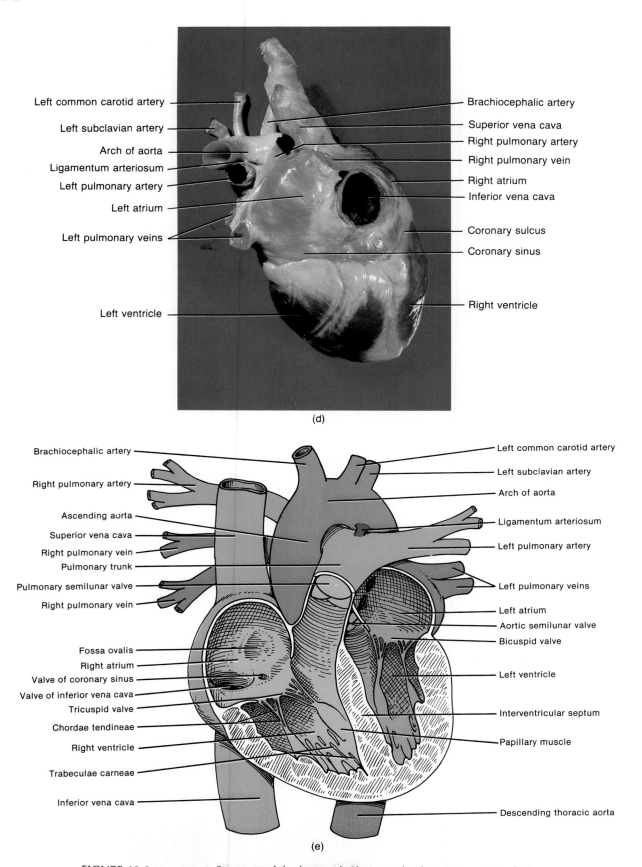

(d)

(e)

FIGURE 13-3 (*Continued*) Structure of the heart. (d) Photograph of posterior external view. (Courtesy of C. Yokochi and J. W. Rohen, *Photographic Anatomy of the Human Body,* 2nd ed., 1979, IGAKU-SHOIN, Ltd., Tokyo, New York.) (e) Diagram of anterior internal view.

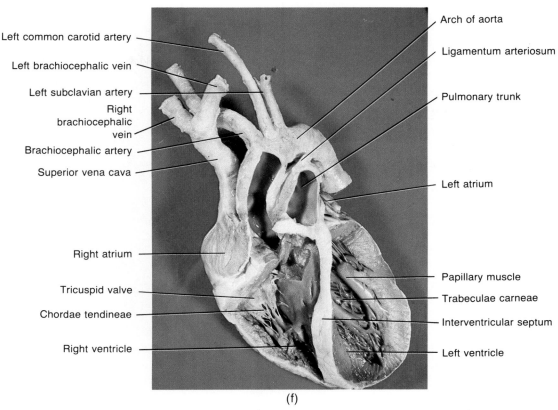

Left common carotid artery

Left brachiocephalic vein

Left subclavian artery

Right brachiocephalic vein

Brachiocephalic artery

Superior vena cava

Right atrium

Tricuspid valve

Chordae tendineae

Right ventricle

Arch of aorta

Ligamentum arteriosum

Pulmonary trunk

Left atrium

Papillary muscle

Trabeculae carneae

Interventricular septum

Left ventricle

(f)

FIGURE 13-3 (***Continued***) Structure of the heart. (f) Photograph of anterior internal view. (Courtesy of C. Yokochi and J. W. Rohen, *Photographic Anatomy of the Human Body,* 2nd ed., 1979, IGAKU-SHOIN, Ltd., Tokyo, New York.)

are also attached by way of the chordae tendineae to papillary muscles.

In order for blood to pass from an atrium to a ventricle, the atrioventricular (AV) valve opens down as the papillary muscles relax and the chordae tendineae slacken (Figure 13-4a). When the ventricle pumps blood out of the heart into an artery, any blood driven back toward the atrium is pushed between the flaps and ventricle wall (Figure 13-4b). This action drives the cusps upward until their edges meet and close the opening. At the same time, contraction of the papillary muscles and tightening of the chordae tendineae help prevent the valve from swinging upward into the atrium.

SEMILUNAR VALVES

Both arteries that leave the heart have a valve that prevents blood from flowing back into the heart. These are the ***semilunar valves*** (see Figure 13-3e,f). The ***pulmonary semilunar valve*** lies in the openings where the pulmonary trunk leaves the right ventricle. The ***aortic semilunar valve*** is situated at the opening between the left ventricle and the aorta.

Both valves consist of three semilunar (half-moon or crescent-shaped) cusps. Each cusp is attached by its convex margin to the artery wall. The free borders of the cusps curve outward and project into the opening inside the blood

vessel. Like the atrioventricular valves, the semilunar valves permit blood to flow in one direction only—in this case, the flow is from the ventricles into the arteries.

SKELETON OF THE HEART

In addition to cardiac muscle tissue, the heart wall also consists of fibrous connective tissue that forms the ***skeleton of the heart.*** The skeleton forms the foundation to which the heart valves attach, serves as an attachment for cardiac muscle fibers, and acts as an electrical insulator between atria and ventricles. Essentially, the skeleton consists of fibrous connective tissue rings that surround the valves of the heart, fuse with one another, and merge with the interventricular septum. The components of the skeleton of the heart are as follows (Figure 13-4e):

1. Four ***fibrous rings*** that support the four valves of the heart and are fused to each other: ***right atrioventricular fibrous ring, left atrioventricular fibrous ring, pulmonary fibrous ring,*** and ***aortic fibrous ring.***
2. ***Right fibrous trigone,*** a larger triangular mass that is formed by the fusion of the fibrous connective tissue of the left atrioventricular, aortic, and right atrioventricular fibrous rings.
3. ***Left fibrous trigone,*** a smaller mass that is formed by

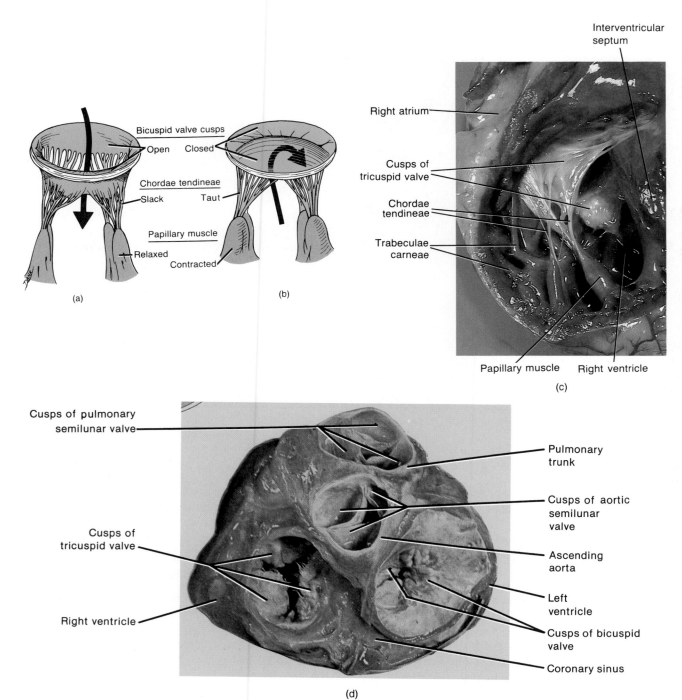

FIGURE 13-4 Atrioventricular (AV) valves. (a) Bicuspid valve open. (b) Bicuspid valve closed. The tricuspid valve operates in a similar manner. (c) Photograph of the tricuspid valve. (d) Photograph of the valves of the heart in superior view in which the atria have been removed. (Photographs courtesy of C. Yokochi and J. W. Rohen, *Photographic Anatomy of the Human Body,* 2nd ed., Igaku-Shoin, Ltd.)

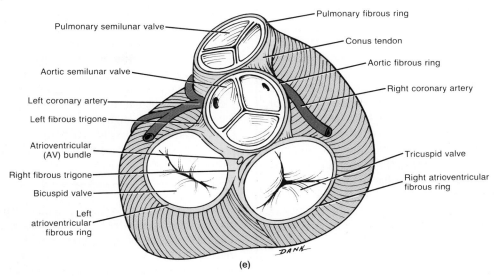

FIGURE 13-4 (*Continued*) Atrioventricular (AV) valves. (e) Diagram of the heart in superior view in which the atria have been removed. The various components of the heart muscle are illustrated in black.

the fusion of the fibrous connective tissue of the left atrioventricular and aortic fibrous rings.

4. ***Conus tendon,*** a mass that is formed by the fusion of the fibrous connective tissue of the pulmonary and aortic fibrous rings.

SURFACE PROJECTION

The location of the valves of the heart may be identified by surface projection (Figure 13-5). The pulmonary and aortic semilunar valves are represented on the surface by a line about 2.5 cm (1 in.) in length. The pulmonary semilunar valve lies horizontally behind the inner end of the left third costal cartilage and the adjoining part of the sternum. The aortic semilunar valve is placed obliquely behind the left side of the sternum, at the level of the third intercostal space. The tricuspid valve lies behind the sternum, extending from the midline at the level of the fourth costal cartilage down toward the right sixth chondrosternal junction. The bicuspid valve lies behind the left side of the sternum obliquely at the level of the fourth costal cartilage. It is represented by a line about 3 cm in length.

Although heart sounds are produced in part by the closure of valves, they are not necessarily heard best over these valves. Each sound tends to be clearest in a slightly different location closest to the surface of the body (Figure 13-5).

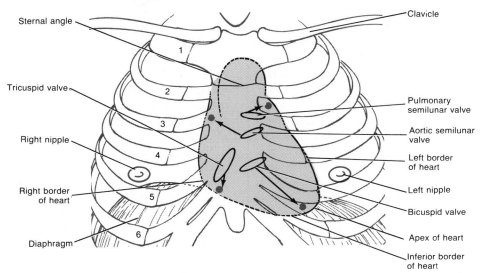

FIGURE 13-5 Surface projection of the heart. The red dots indicate where heart sounds caused by the respective valves are best heard.

CLINICAL APPLICATION

Heart sounds provide valuable information about the valves. Unusual sounds are called **murmurs**. Some murmurs are caused by the noise made by a little blood passing back up into an atrium because of improper closure of an atrioventricular valve. Frequently, a portion of a normal mitral valve is pushed back too far (prolapsed) during contraction due to expansion of the cusps and elongation of the chordae tendineae. Such a condition is called **mitral valve prolapse (MVP),** an inherited disorder resulting from defective collagen. Or, a murmur might be caused by a diseased mitral valve due to rheumatic fever (**valvular heart disease**). In the absence of any complications, many murmurs have no clinical significance.

CONDUCTION SYSTEM

The heart is innervated by the autonomic nervous system (Chapter 19), but the autonomic neurons only increase or decrease the time it takes to complete a cardiac cycle (heartbeat); that is, they do not initiate contraction. The chamber walls can go on contracting and relaxing, contracting and relaxing, without any direct stimulus from the nervous system. This action is possible because the heart has an intrinsic regulating system called the **conduction system.** The conduction system is composed of specialized muscle tissue that generates and distributes the electrical impulses that stimulate the cardiac muscle fibers to contract. These tissues are the sinoatrial (sinuatrial) node, the atrioventricular (AV) node, the atrioventricular (AV) bundle (of His), the bundle branches, and the conduction myofibers (Purkinje fibers). The cells of the conduction system develop during embryological life from certain cardiac muscle fibers.

The **sinoatrial (sinuatrial) node,** also known as the **SA node,** or **pacemaker,** is a compact mass of cells located in the right atrial wall inferior to the opening of the superior vena cava (Figure 13-6a). The SA node initiates each cardiac cycle and thereby sets the basic pace for the heart rate. The SA node spontaneously depolarizes and generates action potentials faster than other components of the conduction system and myocardium. As a result, nerve impulses from the SA node spread to other areas of the conduction system and myocardium and stimulate them so frequently that they are not able to generate action potentials at their own inherent rates. Thus, the faster rate of discharge of the SA node sets the rhythm for the rest of the heart, hence its common name pacemaker. The rate set by the SA node may be altered by nerve impulses from the autonomic nervous system or by certain blood-borne chemicals such as thyroid hormones and epinephrine.

CLINICAL APPLICATION

Sinus node dysfunction refers to a disorder of impulse formation within the SA node and impulse conduction out of the node. The condition is characterized by dizziness, lightheadedness, and/or syncope (fainting). Less commonly, mental changes, stroke, congestive heart failure (CHF), and angina (chest pain) may occur. Sinus node dysfunction may be caused by certain heart diseases, autonomic nervous system dysfunction, and certain drugs.

Once an action potential is initiated by the SA node, the impulse spreads out over both atria, causing them to contract, and at the same time depolarizing the **atrioventricular (AV) node.** Because of its location near the inferior portion of the interatrial septum, the AV node is one of the last portions of the atria to be depolarized.

From the AV node, a tract of conduction fibers called the **atrioventricular (AV) bundle (bundle of His)** runs through the cardiac skeleton to the top of the interventricular spetum. It then continues down both sides of the septum as the **right** and **left bundle branches.** The atrioventricular bundle distributes the action potential over the medial surfaces of the ventricles. Actual contraction of the ventricles is stimulated by **conduction myofibers (Purkinje fibers)** that emerge from the bundle branches and pass into the cells of the myocardium.

ELECTROCARDIOGRAM (ECG)

In a normal heartbeat, the two atria contract simultaneously while the two ventricles relax. Then, when the two ventricles contract, the two atria relax. The term **systole** (SIS-tō-lē) refers to the phase of contraction; **diastole** (dī-AS-tō-lē) is the phase of relaxation. A **cardiac cycle,** or complete heartbeat, consists of the systole and diastole of both atria plus the systole and diastole of both ventricles.

Impulse transmission through the conduction system generates electrical currents that can be detected on the body's surface. A recording of the electrical changes that accompany the cardiac cycle is called an **electrocardiogram (ECG or EKG).** The instrument used to record the changes is an **electrocardiograph.**

Each portion of the cardiac cycle produces a different electrical impulse. These impulses are transmitted from the electrodes to a recording pen that graphs the impulses as a series of up-and-down waves called **deflection waves.** In a typical record (Figure 13-6b), three clearly recognizable waves accompany each cardiac cycle. The first, called the **P wave,** is a small upward wave. It indicates atrial depolarization—the spread of an impulse from the SA node through the muscle of the two atria. A fraction of a second after

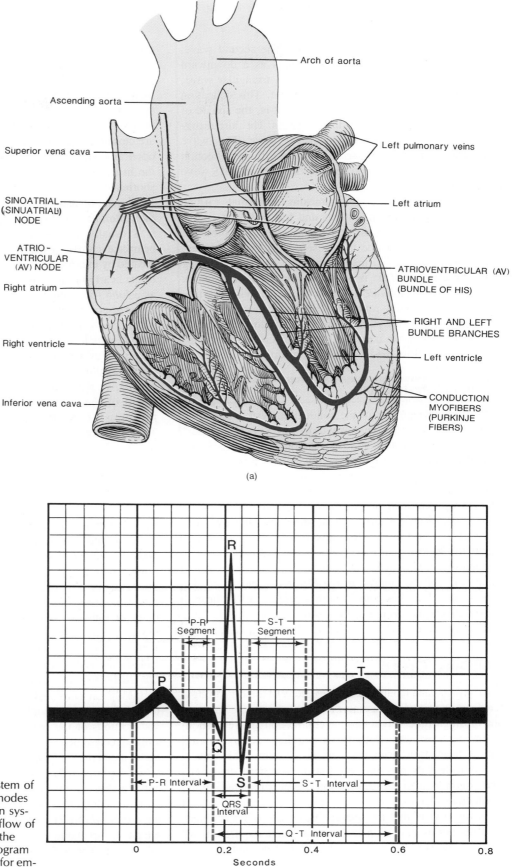

Arch of aorta

Ascending aorta

Superior vena cava

SINOATRIAL (SINUATRIAL) NODE

ATRIO-VENTRICULAR (AV) NODE

Right atrium

Right ventricle

Inferior vena cava

Left pulmonary veins

Left atrium

ATRIOVENTRICULAR (AV) BUNDLE (BUNDLE OF HIS)

RIGHT AND LEFT BUNDLE BRANCHES

Left ventricle

CONDUCTION MYOFIBERS (PURKINJE FIBERS)

(a)

R

P-R Segment

S-T Segment

P

T

Q

S

P-R Interval

QRS Interval

S-T Interval

Q-T Interval

0   0.2   0.4   0.6   0.8

Seconds

(b)

FIGURE 13-6 Conduction system of the heart. (a) Location of the nodes and bundles of the conduction system. The arrows indicate the flow of electricity (impulses) through the atria. (b) Normal electrocardiogram of a single heartbeat, enlarged for emphasis.

the P wave begins, the atria contract. The second wave, called the **QRS wave (complex)**, begins as a downward deflection, continues as a large, upright, triangular wave, and ends as a downward wave at its base. This deflection represents ventricular depolarization, that is, the spread of the electrical impulse through the ventricles. The third recognizable deflection is a dome-shaped **T wave**. This wave indicates ventricular repolarization. There is no deflection to show atrial repolarization because the stronger QRS wave masks this event.

The ECG is useful in diagnosing abnormal cardiac rhythms and conduction patterns and following the course of recovery from a heart attack. It can also detect the presence of fetal life or determine the presence of several fetuses. For some people, ECG monitoring may be done by using a **Holter monitor**. This is an ambulatory system in which the monitoring hardware is worn by an individual while going about everyday routines. The Holter monitor is used especially to detect rhythm disorders in the conduction system. It is also used to correlate rhythm disorders and symptoms and to follow the effectiveness of drugs.

CLINICAL APPLICATION

When major elements of the conduction system are disrupted, an irregular heart rhythm may occur. In one type of rhythm disturbance, the ventricles fail to receive atrial impulses, causing the ventricles and atria to beat independently of each other. In patients with such a condition, normal heart rhythm can be restored and maintained with an **artificial pacemaker,** a device that sends out small electrical charges that stimulate the heart. It consists of three basic parts: *pulse generator,* which contains the battery cells and produces the impulse; a *lead,* which is a flexible wire connected to the pulse generator that delivers the impulse to the electrode; and an *electrode,* which makes contact with a portion of the heart and delivers the charge to the heart.

CARDIAC CYCLE

Two phenomena control the movement of blood through the heart as part of its cardiac cycle (heartbeat): the opening and closing of the valves and the contraction and relaxation of the myocardium. Both these activities occur without direct stimulation from the nervous system. The valves are controlled by pressure changes in each heart chamber. The contraction of the cardiac muscle is stimulated by its conduction system.

If we assume that the average heart rate (HR) is about 75 times per minute, then each cardiac cycle requires about 0.8 sec. During the first 0.1 sec, the atria contract and the ventricles relax. The atrioventricular valves are open, and the semilunar valves are closed. For the next 0.3 sec, the atria are relaxing and the ventricles are contracting. During the first part of this period, all valves are closed;

during the second part, the semilunar valves are open. The last 0.4 sec of the cycle is the relaxation, or quiescent, period and all chambers are in diastole. In a complete cycle, then, the atria are in systole 0.1 sec and in diastole 0.7 sec; the ventricles are in systole 0.3 sec and in diastole 0.5 sec. For the first part of the quiescent period, all valves are closed; during the latter part, the atrioventricular valves open and blood starts draining into the ventricles. When the heart beats faster than normal, the quiescent period is shortened accordingly.

The act of listening to sounds within the body is called **auscultation** (aws-kul-TĀ-shun; *auscultare* = to listen) and it is usually done with a stethoscope. The sound of the heartbeat comes primarily from turbulence in blood flow created by the closure of the valves, not from the contraction of the heart muscle. The first sound, which can be described as a *lubb* ($\overline{oo}$) sound, is a long, booming sound. The lubb is the sound created by the closure of the atrioventricular valves soon after ventricular systole begins. The second sound, which is heard as a short, sharp sound, can be described as a *dupp* (ŭ) sound. Dupp is the sound created as the semilunar valves close toward the end of ventricular systole. A pause between the second sound and the first sound of the next cycle is about two times longer than the pause between the first and second sound of each cycle. Thus, the cardiac cycle can be heard as a lubb, dupp, pause; lubb, dupp, pause; lubb, dupp, pause.

AUTONOMIC CONTROL

The sinoatrial (SA) node initiates contraction and, left to itself, would set an unvarying heart rate. However, the body's need for blood supply varies under different conditions. There are several regulatory mechanisms stimulated by factors such as chemicals present in the body, temperature, emotional state, and age. The most important control of heart rate and strength of contraction is the effect of the autonomic nervous system.

Within the medulla of the brain is a group of neurons called the **cardioacceleratory center (CAC).** Arising from this center are sympathetic fibers that travel down a tract in the spinal cord and then pass outward in the *cardiac (accelerator) nerves* and innervate the SA node, AV node, and portions of the myocardium (Figure 13-7). When the cardioacceleratory center is stimulated, nerve impulses travel along the sympathetic fibers and cause them to release norepinephrine (NE), which increases the rate of heartbeat and the strength of contraction.

The medulla also contains a group of neurons that form the **cardioinhibitory center (CIC).** Arising from this center are parasympathetic fibers that reach the heart via the **vagus (X) nerve.** These fibers innervate the SA node and AV node. When this center is stimulated, nerve impulses transmitted along the parasympathetic fibers cause the release of acetylcholine (ACh), which decreases the rate of heartbeat and strength of contraction.

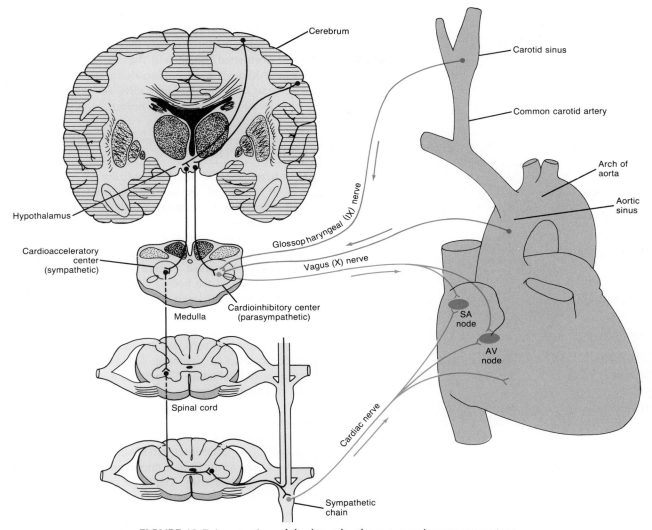

FIGURE 13-7 Innervation of the heart by the autonomic nervous system.

The autonomic control of the heart is therefore the result of opposing sympathetic (stimulatory) and parasympathetic (inhibitory) influences. Sensory impulses from receptors in different parts of the cardiovascular system act on the centers so that a balance between stimulation and inhibition is maintained. Nerve cells capable of responding to changes in blood pressure are called **baroreceptors (pressoreceptors)**. These receptors affect the rate of heartbeat.

Heart rate may also be influenced by certain chemicals (epinephrine, sodium, potassium), temperature, emotions, sex, and age.

BLOOD SUPPLY

The wall of the heart, like any other tissue, has its own blood vessels. Nutrients could not possibly diffuse through all the layers of cells that make up the heart tissue. The flow of blood through the numerous vessels that pierce

the myocardium is called **coronary (cardiac) circulation** (Figure 13-8). The term coronary refers to the fact that the blood vessels of the heart resemble a crown.

The vessels that serve the myocardium include the *left coronary artery,* which originates as a branch of the ascending aorta. This artery runs under the left atrium and divides into the anterior interventricular and circumflex branches. The *anterior interventricular branch* follows the anterior interventricular sulcus and supplies oxygenated blood to the walls of both ventricles. The *circumflex branch* distributes oxygenated blood to the walls of the left ventricle and left atrium.

The *right coronary artery* also originates as a branch of the ascending aorta. It runs under the right atrium and divides into the posterior interventricular and marginal branches. The *posterior interventricular branch* follows the posterior interventricular sulcus and supplies the walls of the two ventricles with oxygenated blood. The *marginal branch* transports oxygenated blood to the myocardium of

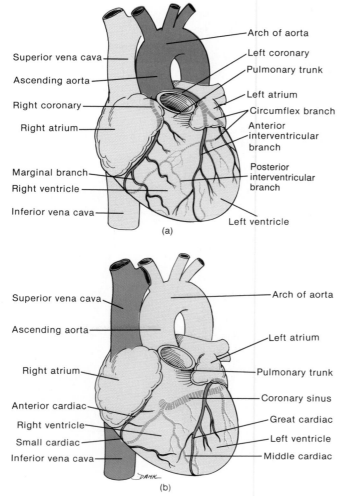

Superior vena cava

Ascending aorta

Right coronary

Right atrium

Marginal branch

Right ventricle

Inferior vena cava

Arch of aorta

Left coronary

Pulmonary trunk

Left atrium

Circumflex branch

Anterior interventricular branch

Posterior interventricular branch

Left ventricle

(a)

Superior vena cava

Ascending aorta

Right atrium

Anterior cardiac

Right ventricle

Small cardiac

Inferior vena cava

Arch of aorta

Left atrium

Pulmonary trunk

Coronary sinus

Great cardiac

Left ventricle

Middle cardiac

(b)

FIGURE 13-8 Coronary (cardiac) circulation. (a) Anterior view of arterial distribution. (b) Anterior view of venous drainage.

the right ventricle. The left ventricle receives the most abundant blood supply because of the enormous work it must do.

As blood passes through the coronary system of the heart, it delivers oxygen and nutrients and collects carbon dioxide and wastes. Most of the deoxygenated blood, which carries the carbon dioxide and wastes, is collected by a large vein, the *coronary sinus,* which empties into the right atrium. A vascular sinus is a vein with a thin wall that has no smooth muscle to alter its diameter. The principal tributaries of the coronary sinus are the *great cardiac vein,* which drains the anterior aspect of the heart, and the *middle cardiac vein,* which drains the posterior aspect of the heart.

Most parts of the body receive branches from more than one artery, and where two or more arteries supply the same region, they usually connect with each other. This is referred to as **anastomosis.** Anastomoses between arteries provide collateral circulation (alternate routes) for blood to reach a particular organ or tissue. The myocardium contains nu-

merous anastomoses either connecting branches of one coronary artery or between branches of different coronary arteries. When a major coronary blood vessel is about 90 percent obstructed, blood will flow through the collateral vessels. Although most collaterals in the heart are quite small, heart muscle can remain alive as long as it receives as little as 10–15 percent of its normal supply.

CLINICAL APPLICATION

Most heart problems result from faulty coronary circulation. If a reduced oxygen supply weakens the fibers (cells) but does not actually kill them, the condition is called *ischemia* (is-KĒ-mē-a). *Angina pectoris* (an-JĪ-na, or AN-ji-na, PEK-tō-ris), meaning "chest pain," results from ischemia of the myocardium. (Pain impulses originating from most viscera are referred to an area on the surface of the body.) Common causes of angina include stress, strenuous exertion after a heavy meal, atherosclerosis, coronary artery spasm, hypertension, anemia, thyroid disease, and aortic stenosis. The symptoms of angina pectoris include chest pain, accompanied by constriction or tightness, labored breathing, and a sensation of foreboding. Sometimes weakness, dizziness, and perspiration occur.

A much more serious problem is *myocardial infarction* (in-FARK-shun) or *MI,* commonly called a heart attack. *Infarction* means the death of an area of tissue because of an interrupted blood supply. Myocardial infarction may result from a thrombus or embolus in one of the coronary arteries. The tissue distal to the obstruction dies and is replaced by noncontractile scar tissue. Thus, the heart muscle loses at least some of its strength. The aftereffects depend partly on the size and location of the infarcted, or dead, area.

It has recently been learned that under some conditions most tissue damage occurs when blood flow is restored (reprofusion) to an area, not when it is interrupted. When blood supply is interrupted, the affected tissue produces an enzyme called *xanthinoxidase.* When blood is restored, the enzyme reacts with oxygen in the blood to produce *superoxide free radicals,* which cause the tissue damage. Attempts are now underway to restore oxygen-starved tissue by preventing the formation of superoxide free radicals using drugs such as superoxide dismutase and allopurinol (also used to treat gout).

ARTIFICIAL HEART

On December 2, 1982, Dr. Barney B. Clark, a 61-year-old retired dentist, made medical history by becoming the first human to receive an **artificial heart.** The 7½-hour operation was performed by Dr. William DeVries, then a surgeon at the University of Utah Medical Center.

The artificial heart, known as Jarvik-7 after its inventor, Dr. Robert Jarvik, consists of an aluminum base and a pair of rigid plastic chambers that serve as ventricles (Figure 13-9). Each chamber contains a flexible diaphragm driven by compressed air that forces blood past mechanical valves into the major blood vessels. The surgery essentially consisted of removing the ventricles from Dr. Clark's heart but leaving the atria intact. Then Dacron connectors were sutured onto the atria, aorta, and pulmonary trunk. The artificial heart was then snapped into position via the connectors. Essentially, the artificial heart is designed to pump sufficient blood to sustain only moderate activity.

Recently, the artificial heart has come under some criticism because of its possible role in causing several disabling or fatal strokes in its recipients. While the use of permanent artificial hearts has slowed down dramatically, more and more emphasis is now being placed on implanting artificial hearts as bridges to transplants. That is, artificial hearts are used as temporary mechanical support devices until a human heart transplantation can be performed. Research is continuing and clinical trials are being undertaken in the hopes of developing a safer, more efficient and practical artificial heart.

CLINICAL APPLICATION

Teams of surgeons at two hospitals in Baltimore performed a *three-way transplant* in May 1987. At the University of Maryland Hospital, physicians removed the heart and lungs from an accident victim who had been declared brain dead. The organs were then transported to Johns Hopkins Medical Center and transplanted into a person whose lungs were destroyed by cystic fibrosis (CF) but whose heart was healthy. The healthy heart was removed and transplanted into another person who was in need of a heart transplant. The procedure, referred to as a "domino donor" organ exchange, marked the first time in U.S. medical history that a healthy human heart was taken from a living person and transplanted into another human.

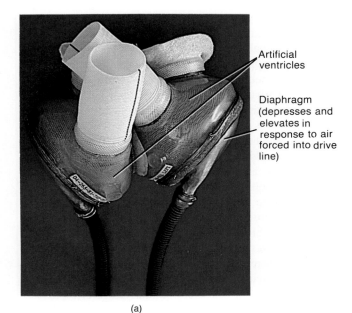

(a)

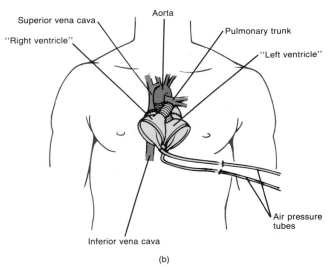

(b)

FIGURE 13-9 Artificial heart. (a) Photograph of an artificial heart prior to implantation. (Courtesy of Bradley Nelson, University of Utah.) (b) Diagram of an implanted heart.

RISK FACTORS IN HEART DISEASE

It is estimated that one in every five persons who reaches age 60 will have a myocardial infarction (heart attack). One in every four persons between 30 and 60 has the potential to be stricken. Heart disease is epidemic in this country, despite the fact that some of the causes can be foreseen and prevented. The results of research indicate that people who develop combinations of certain risk factors eventually have heart attacks. *Risk factors* are characteristics, symptoms, or signs present in a person free of disease that are statistically associated with an excessive rate of development of a disease. Among the major risk factors in heart disease are:

1. High blood cholesterol level.
2. High blood pressure.
3. Cigarette smoking.
4. Obesity.
5. Lack of regular exercise.
6. Diabetes mellitus.
7. Genetic predisposition (family history of heart disease at an early age).
8. Sex.
9. Age.
10. Fibrinogen level.

The first five risk factors all contribute to increasing the heart's work load. High blood cholesterol is discussed shortly, and hypertension is discussed in the next chapter. Cigarette smoking, through the effects of nicotine, stimulates the adrenal gland to oversecrete aldosterone, epinephrine, and norepinephrine (NE)—powerful vasoconstrictors. Overweight people develop miles of extra capillaries to nourish fat tissue. The heart has to work harder to pump the blood through more vessels. Without exercise, venous return gets less help from contracting skeletal muscles. In addition, regular exercise strengthens the smooth muscle of blood vessels and enables them to assist general circulation. Exercise also increases cardiac efficiency and output. In diabetes mellitus, fat metabolism dominates glucose metabolism. As a result, cholesterol levels get progressively higher and result in plaque formation, a situation that may lead to high blood pressure. High blood pressure drives fat into the vessel wall, encouraging atherosclerosis. Up to age 50, there is a 10–15 year lag in the extent of heart disease in females compared to males; after that, the rates of disease in both sexes are similar. Regardless of sex, the incidence of heart disease increases with age. Regarding fibrinogen, the higher the level, the greater the risk of heart disease (more so in males than females). Fibrinogen enhances blood clot formation.

DEVELOPMENTAL ANATOMY OF THE HEART

The *heart,* a derivative of **mesoderm,** begins to develop before the end of the third week. It begins its development in the ventral region of the embryo beneath the foregut (see Figure 23-20). The first step is the formation of a pair of tubes, the **endothelial (endocardial) tubes,** from mesodermal cells (Figure 13-10). These tubes then unite to form a common tube, the **primitive heart tube.** Next, the primitive heart tube develops into five regions: **ventricle, bulbus cordis, atrium, sinus venosus,** and **truncus arteriosus.** Since the bulbus cordis and ventricle subsequently grow more rapidly than the others, and the heart grows more rapidly than its superior and inferior attachments, the heart assumes a U shape and later an S shape. The flextures of the heart reorient the regions so that the atrium and sinus venosus eventually come to lie superior to the bulbus cordis, ventricle, and truncus arteriosus.

At about the seventh week a partition, the **interatrial septum,** forms in the atrial region, dividing it into a *right* and *left atrium.* The opening in the partition is the *foramen ovale,* which normally closes at birth and later forms a depression called the *fossa ovalis.* An **interventricular septum** also develops and partitions the ventricular region into a *right* and *left ventricle.* The bulbus cordis and truncus arteriosus divide into two vessels, the *aorta* (arising from the left ventricle) and the *pulmonary trunk* (arising from the right ventricle). Recall that the ductus arteriosus is a temporary vessel between the aorta and pulmonary trunk until birth. The great veins of the heart, *superior vena*

cava and *inferior vena cava,* develop from the venous end of the primitive heart tube.

APPLICATIONS TO HEALTH

CORONARY ARTERY DISEASE (CAD)

Coronary artery disease (CAD) is a condition in which the heart muscle receives an inadequate amount of blood because of an interruption of its blood supply. It is presently the leading cause of death in the United States. Depending on the degree of interruption, symptoms can range from a mild chest pain to a full-scale heart attack. Generally, symptoms manifest themselves when there is about a 75 percent narrowing of coronary artery lumina. The underlying causes of CAD are many and varied. Two of the principal ones are atherosclerosis and coronary artery spasm. Another is a thrombus or an embolus in a coronary artery.

Atherosclerosis

Atherosclerosis (ath'-er-ō-skle-RŌ-sis) is a process in which fatty substances, especially cholesterol (CH) and triglycerides (ingested fats), are deposited in the walls of medium-sized and large arteries in response to certain stimuli. It is believed that the first event in atherosclerosis is damage to the endothelial lining of the artery. Contributing factors include high blood pressure (hypertension), carbon monoxide in cigarettes, diabetes mellitus, and a high cholesterol level. Following endothelial damage, monocytes begin to stick to the smooth inner endothelial lining of the tunica interna (inner coat) of the artery. Then, they slip between endothelial cells, penetrate the endothelium, and develop into macrophages. Here, the macrophages begin to take up cholesterol from low-density lipoproteins (described shortly). At the same time, smooth muscle fibers (cells) in the tunica media (middle coat) of the artery also ingest cholesterol. The more cholesterol there is in the blood, the faster the process occurs. The accumulated cholesterol and cells form a lesion called an **atherosclerotic plaque** (Figure 13-11). The plaque looks like a pearly gray or yellow mound of tissue. As it grows, it may obstruct blood flow in the affected artery and damage the tissues the artery supplies. An additional danger is that the plaque provides a roughened surface that causes blood platelets to release *platelet-derived growth factor (PDGF)*, a hormone that promotes the division of smooth muscle fibers. In fact, macrophages and endothelial cells also produce PDGF. This further complicates the process since PDGF causes the lesion to grow larger. Platelets in the area of the atherosclerotic plaque also release clot-forming chemicals. Thus, a thrombus may form. If the clot breaks off and forms an embolus, it may obstruct small arteries and capillaries quite a distance from the site of formation. Among the therapies used to reduce cholesterol levels are diet, exercise, and drugs (niacin, cholestyramine, colestipol, and lovastatin).

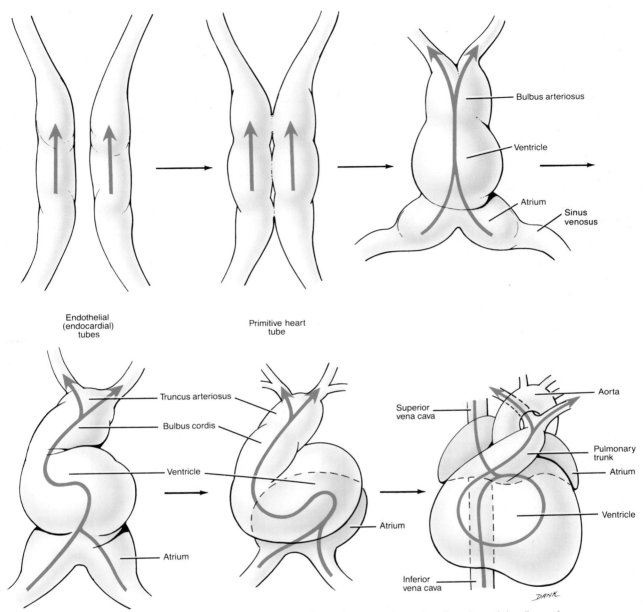

FIGURE 13-10 Development of the heart. The red arrows show the direction of the flow of blood from the venous to the arterial end.

The treatment of CAD varies with the nature and urgency of the symptoms. Among the possible treatments are drug therapy, coronary artery bypass grafting (CABG), percutaneous transluminal coronary angioplasty (PTCA), and laser angioplasty. The two most commonly used drugs are ***nitroglycerin*** and ***beta blockers***. Antithrombotic drugs that suppress the activity of platelets are being tested on a trial basis.

Coronary artery bypass grafting (CABG) is the most direct way of increasing the blood supply to the heart. In a typical operation, a portion of a blood vessel is removed from another part of the body. At one time, the saphenous vein from the leg was frequently used. Now, however,

the internal mammary artery from the chest is used. Research has shown that obstructions reform in grafted veins more readily than arteries. Once the heart is exposed, circulation is maintained by a heart–lung machine. Next, a clamp is placed across the aorta above the openings of the coronary arteries. This stops the blood supply to the heart. A cold saline solution containing potassium is then flushed into the heart through a small catheter inserted into the aorta. The saline solution ensures that the myocardium will maintain its natural state, and the potassium and low temperature of the solution discourage any tendency for the heart to start beating prematurely. A segment of the grafted blood vessel is then sutured between the aorta and the unblocked

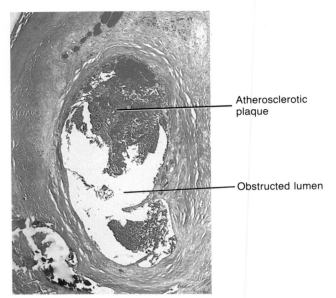

FIGURE 13-11 Photomicrograph of a cross section of an artery partially obstructed by an atherosclerotic plaque. (Courtesy of Sklar/Photo Researchers.)

portion of the coronary artery, distal to the obstruction (Figure 13-12). If more than one artery is clogged, additional bypasses may be made. Once the repair is completed, the aortic clamp is removed and, as blood at normal temperature flows into the coronary arteries, the cold saline solution and potassium leave. This normally causes the heart to start beating again. If it does not, a mild shock from a defibrillator is applied from a pair of paddlelike electrodes.

Although coronary bypass grafting increases blood flow and decreases angina, it tends not to increase survival rates. Whether treated with surgery or medication, patients live about equally as long. The unfavorable conditions that initially led to the surgery will in time frequently clog the new blood vessels, or other poor aspects of the circulatory system will take their toll in increased blood pressure, heart enlargement, or stroke.

CAD may also be treated by a procedure referred to as *percutaneous transluminal coronary angioplasty* (*PTCA*). Like coronary artery bypass grafting, it is an attempt to increase the blood supply to the heart muscle. In this procedure, a balloon catheter is inserted into an artery of an arm or leg and gently guided through the arterial system under x-ray observation. Once in place in an obstructed coronary artery, the balloon is inflated for a few seconds, thus compressing the obstructing material against the vessel wall and permitting increased blood flow. A newly approved laser system called *laser angioplasty,* is used as an adjunct to balloon dilation of blood vessels. After a laser vaporizes atherosclerotic plaques and makes a channel through the blood vessel obstruction, the balloon catheter is inserted, and the balloon is inflated. A modification of PTCA, called *percutaneous balloon valvuloplasty,* is used to treat faulty heart valves.

Coronary Artery Spasm

Atherosclerosis results in a fixed obstruction to blood flow. Obstruction can also be caused by *coronary artery spasm,* a condition in which the smooth muscle of a coronary artery undergoes a sudden contraction, resulting in vasoconstriction. Coronary artery spasm typically occurs in individuals with atherosclerosis and may result in chest pain during rest (variant angina), chest pain during exertion (typical angina), heart attacks, and sudden death. Although the causes of coronary artery spasm are unknown, several factors are receiving attention. These include smoking, stress, and a vasoconstrictor chemical released by platelets. There is considerable interest in aspirin and other drugs that might inhibit the vasoconstrictor chemical. Coronary artery spasm is treated with long-acting nitrates (nitroglycerin) and calcium-blocking agents.

CONGENITAL DEFECTS

A defect that exists at birth, and usually before, is called a *congenital defect.* Many defects are not serious and may go unnoticed for a lifetime. Others heal themselves. But some are life-threatening and must be mended by surgical techniques ranging from simple suturing to replacement of malfunctioning parts with synthetic materials.

One congenital defect that can occur is *coarctation* (kō'-ark-TĀ-shun) *of the aorta* (Figure 13-13a). In this condition, a segment of the aorta is too narrow. As a result, the flow of oxygenated blood to the body is reduced, the left ventricle is forced to pump harder, and high blood pressure develops. The condition may be corrected by insertion of a synthetic graft.

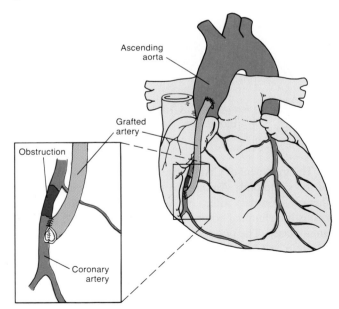

FIGURE 13-12 Coronary artery bypass grafting (CABG). The internal mammary artery is removed from the patient's chest. At a point distal to the obstruction, the grafted artery is sutured to the coronary artery.

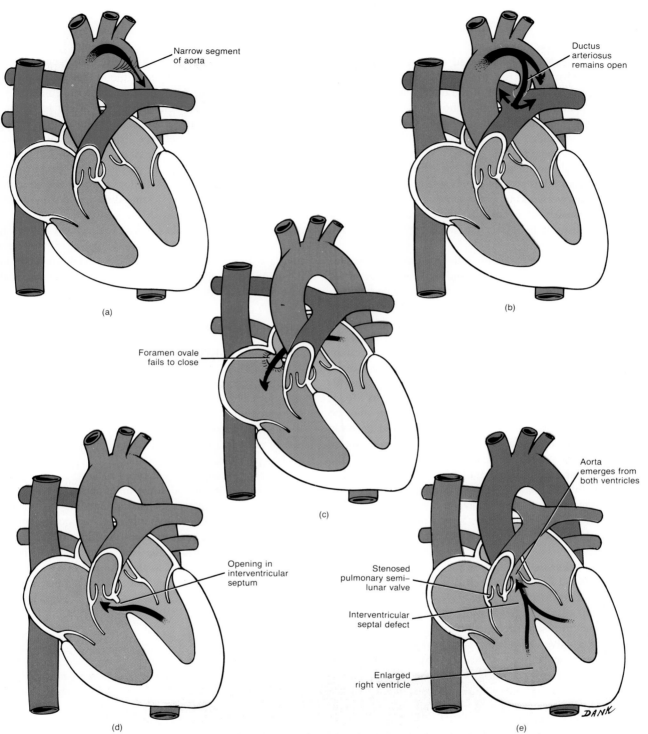

FIGURE 13-13 Some common congenital heart defects. (a) Coarctation of the aorta. (b) Patent ductus arteriosus. (c) Interatrial septal defect. (d) Interventricular septal defect. (e) Tetralogy of Fallot.

Another common congenital defect is **patent ductus arteriosus** (Figure 13-13b). The ductus arteriosus between the aorta and the pulmonary trunk normally closes shortly after birth in response to increased oxygenation of blood flowing through the ductus arteriosus. Only a few hours after birth, the muscular wall of the ductus arteriosus constricts and, within a week, constriction is sufficient to stop all blood flow. During the second month, the ductus arteriosus becomes occluded by the formation of fibrous tissue in its lumen. In some babies, the ductus arteriosus remains open. As a result, aortic blood flows into the lower-pressure pulmonary trunk, thus increasing the pulmonary trunk blood pressure and overworking both ventricles and the heart. Treatment consists of a surgical procedure in which the ductus arteriosus is ligated or cut so that the cut ends are tied off.

A **septal defect** is an opening in the septum that separates the interior of the heart into a left and right side. **Interatrial septal defect** is failure of the fetal foramen ovale between the two atria to close after birth (Figure 13-13c). Because pressure in the right atrium is low, interatrial septal defect generally allows a good deal of blood to flow from the left atrium to the right without going through systemic circulation. This defect, an example of left-to-right shunt, overloads the pulmonary circulation, produces fatigue, and increases respiratory infections. If it occurs early in life, it inhibits growth because the systemic circulation may be deprived of a considerable portion of the blood destined for the organs and tissues of the body. **Interventricular septal defect** is caused by an incomplete closure of the interventricular septum (Figure 13-13d). Due to the higher pressure in the left ventricle, blood is shunted directly from the left ventricle into the right ventricle. This is another example of a left-to-right shunt. Septal openings can now be sewn shut or covered with synthetic patches inserted via a catheter.

Valvular stenosis is a narrowing of one of the valves regulating blood flow in the heart. *Stenosis* means narrowing. Stenosis may occur in the valve itself, most commonly in the mitral valve from rheumatic heart disease or in the aortic valve from sclerosis or rheumatic fever. Or it may occur near a valve. All stenoses are serious because they place a severe work load on the heart by making it work harder to push the blood through the abnormally narrow valve openings. As a result of mitral stenosis, blood pressure is increased. Angina pectoris and heart failure may accompany this disorder. Most stenosed valves are totally replaced with artificial valves.

Tetralogy of Fallot (tet-RAL-ō-jē of fal-Ō) is a combination of four defects: an interventricular septal defect, an aorta that emerges from both ventricles instead of from the left ventricle only, a stenosed pulmonary semilunar valve, and an enlarged right ventricle (Figure 13-13e). The condition is an example of a right-to-left shunt, in which blood is shunted from the right ventricle to the left ventricle without going through pulmonary circulation. Because there is stenosis of the pulmonary semilunar valve, the increased right ventricular pressure forces deoxygenated blood from the right ventricle to enter the left ventricle through the interventricular septum. As a result, deoxygenated blood gets mixed with the oxygenated blood that is pumped into systemic circulation. Also, because the aorta emerges from the right ventricle and the pulmonary artery is stenosed, very little blood gets to the lungs, and pulmonary circulation is bypassed almost completely. The insufficient amount of oxygenated blood in systemic circulation results in *cyanosis* (sī-a-NŌ-sis), a blue or dark purple discoloration of skin. For this reason tetralogy of Fallot is one of the conditions that causes a "blue baby." Lung disorders and suffocation also result in cyanosis.

It is possible to correct cases of tetralogy of Fallot when the patient is of proper age and condition. Open-heart operations are performed in which the narrowed pulmonary valve is cut open and the interventricular septal defect is sealed with a Dacron patch.

ARRHYTHMIAS

Arrhythmia (a-RITH-mē-a) is a general term that refers to an abnormality or irregularity in the heart rhythm. More physicians are using the term **dysrhythmia** since it implies an abnormal rhythm, whereas arrhythmia implies no rhythm. An arrhythmia results when there is a disturbance in the conduction system of the heart, either due to faulty production of electrical impulses or faulty conduction of impulses as they pass through the system.

There are many different types of arrhythmias that can occur, some normal and some quite serious. Arrhythmias may be caused by factors such as caffeine, nicotine, alcohol, anxiety, certain drugs, hyperthyroidism, potassium deficiency, and certain heart diseases. Serious arrhythmias can result in cardiac arrest if the heart cannot supply its own oxygen demands, as well as those of the rest of the body. They can be controlled, and the normal heart rhythm can be reestablished, if they are detected and treated early enough.

Heart Block

One serious arrhythmia is called a **heart block**. Perhaps the most common blockage is in the atrioventricular (AV) node, which conducts impulses from the atria to the ventricles. This disturbance is called **atrioventricular (AV) block**. It usually indicates a myocardial infarction, atherosclerosis, rheumatic heart disease, diphtheria, or syphilis. With complete AV block, patients may have vertigo, unconsciousness, or convulsions. These symptoms result from a decreased cardiac output with diminished cerebral blood flow and cerebral hypoxia or lack of sufficient oxygen.

Among the causes of AV block are excessive stimulation by the vagus (X) nerves (which depresses conductivity of the junctional fibers), destruction of the AV bundle as a result of coronary infarct, atherosclerosis, myocarditis, or depression caused by various drugs.

Flutter and Fibrillation

Two other abnormal rhythms are *flutter* and *fibrillation.* In *atrial flutter* the atrial rhythm averages between 240 and 360 beats per minute. The condition is essentially rapid atrial contractions accompanied by a second-degree AV block. It generally indicates severe damage to heart muscle. *Atrial fibrillation* is asynchronous contraction of the atrial muscles that causes the atria to contract irregularly and still faster. Atrial flutter and fibrillation occur in myocardial infarction, acute and chronic rheumatic heart disease, and hyperthyroidism. Atrial fibrillation results in complete uncoordination of atrial contraction so that atrial pumping ceases altogether. When the muscle fibrillates, the muscle fibers of the atrium quiver individually instead of contracting together. The quivering cancels out the pumping of the atrium. In a strong heart, atrial fibrillation reduces the pumping effectiveness of the heart by only 25–30 percent.

Ventricular fibrillation (*VF*), is another abnormality that almost always indicates imminent cardiac arrest and death unless corrected quickly. It is characterized by asychronous, haphazard, ventricular muscle contractions. The rate may be rapid or slow. The impulse travels to the different parts of the ventricles at different rates. Thus, part of the ventricle may be contracting while other parts are still unstimulated. Ventricular contraction becomes ineffective and circulatory failure and death occur. Ventricular fibrillation may be caused by coronary occlusion. It sometimes occurs during surgical procedures on the heart or pericardium. It may be the cause of death in electrocution. It is possible to pass a very strong electrical current through the ventricles for a short interval of time in order to stop ventricular fibrillation. This is accomplished by placing electrodes on two sides of the heart and is called *defibrillation.*

Ventricular Premature Contraction (VPC)

Another form of arrhythmia arises when a small region of the heart outside the pacemaker becomes more excitable than normal, causing an occasional abnormal impulse to be generated between normal impulses. The region from which the abnormal impulse is generated is called an *ectopic focus.* As a wave of depolarization spreads outward from the ectopic focus, it causes a *ventricular premature contraction* (*VPC*), also called *premature ventricular contraction*

(*PVC*). The contraction occurs early in diastole before the SA node is normally scheduled to discharge its impulses. A person with such a ventricular premature contraction might feel a thump in the chest. The contractions may be relatively benign and may be caused by emotional stress, excessive intake of stimulants such as caffeine or nicotine, and lack of sleep. In other cases, the contractions may indicate an underlying pathology.

CONGESTIVE HEART FAILURE (CHF)

Congestive heart failure (*CHF*) may be defined as a chronic or acute state that results when the heart is not capable of supplying the oxygen demands of the body. Symptoms and signs of CHF include fatigue, peripheral and pulmonary edema, and visceral congestion. These symptoms are produced by diminished blood flow to the various tissues of the body and by accumulation of excess blood in the various organs because the heart is unable to pump out the blood returned to it by the great veins. Both diminished blood flow and accumulation of excess blood in the organs occur together, but certain symptoms result from congestion, while others are produced by poor tissue nutrition. CHF is usually caused by coronary artery disease in which atherosclerosis leads to myocardial ischemia.

COR PULMONALE (CP)

Cor pulmonale (kor pul-mōn-ALE; *cor* = heart; *pulmon* = lung), or *CP,* refers to right ventricular hypertrophy from disorders that bring about hypertension (high blood pressure) in the pulmonary circulation. Certain disorders such as those involving the respiratory center in the brain, lung diseases, diseases involving airways in the lungs, deformities of the thoracic cage, and neuromuscular disorders can result in hypoxia in the lungs. As a result, vasoconstriction of pulmonary arteries and arterioles occurs and blood flow to the lungs is decreased. Also, any conditions such as tumors, aneurysms, and diseased blood vessels in the lungs can result in decreased blood flow. The reduction in blood flow and increased resistance in the lungs bring about pulmonary hypertension. Ultimately, this causes increased pressure in the pulmonary arteries and pulmonary trunk, resulting in right ventricular hypertrophy.

Individuals with CP may exhibit fatigue, weakness, dyspnea, right upper quadrant (RUQ) pain due to an enlarged liver, warm and cyanotic extremities, cyanosis of the face, distended neck veins, chest pain, and edema.

KEY MEDICAL TERMS ASSOCIATED WITH THE HEART

Angiocardiography (an'-jē-ō-kar'-dē-OG-ra-fē; *angio* = vessel; *cardio* = heart; *graph* = writing) X-ray examination of the heart and great blood vessels after injection of a radiopaque dye into the blood stream.

Aortic insufficiency (ā-OR-tik in'-su-FISH-en-sē) An improper

closure of the aortic semilunar valve that permits a backflow of blood.

Cardiac arrest (KAR-dē-ak a-REST) Complete stoppage of the heartbeat.

Cardiomegaly (kar'-dē-ō-MEG-a-lē; *mega* = large) Heart enlargement. Long-distance runners and weight lifters frequently develop heart enlargement as a natural adaptation to increased work load produced by regular exercise.

Commissurotomy (kom'-i-shur-OT-ō-mē) An operation that is performed to widen the opening in a heart valve that has become narrowed by scar tissue.

Compliance (kom-PLĪ-ans) The passive or diastolic stiffness properties of the left ventricle. A hypertrophied or fibrosed heart with a stiff wall, for example, has decreased compliance. Also, the stiffness properties of the lung.

Constrictive pericarditis (kon-STRIK-tiv per'-i-kar-DĪ-tis) A shrinking and thickening of the pericardium that prevents heart muscle from expanding and contracting normally.

Incompetent valve (in-KOM-pe-tent VALV) Any valve that does not close properly thus permitting a backflow of blood; also called *valvular insufficiency.*

Palpitation (pal'-pi-TĀ-shun) A fluttering of the heart or abnormal rate or rhythm of the heart.

Pancarditis (pan'-kar-DĪ-tis; *pan* = all) Inflammation of the whole heart including the inner layer (endocardium), heart muscle (myocardium), and outer sac (paricardium).

Paroxysmal tachycardia (par'-ok-SIZ-mal tak'-e-KAR-dē-ā) A period of rapid heartbeats that begins and ends suddenly.

Stokes-Adams syndrome Sudden attacks of unconsciousness, sometimes with convulsions, that may accompany heart block.

STUDY OUTLINE

Location (p. 345)
1. The heart is situated between the lungs in the mediastinum.
2. About two-thirds of its mass is to the left of the midline.

Pericardium (p. 345)
1. The pericardium consists of an outer fibrous layer and an inner serous pericardium.
2. The serous pericardium is composed of a parietal and visceral layer.
3. Between the parietal and visceral layers of the serous pericardium is the pericardial cavity, a potential space filled with pericardial fluid that prevents friction between the two membranes.

Wall; Chambers; Vessels; and Valves (p. 345)
1. The wall of the heart has three layers: epicardium, myocardium, and endocardium.
2. The chambers include two upper atria and two lower ventricles.
3. The blood flows through the heart from the superior and inferior venae cavae and the coronary sinus to the right atrium, through the tricuspid valve to the right ventricle, through the pulmonary trunk to the lungs, through the pulmonary veins into the left atrium, through the bicuspid valve to the left ventricle, and out through the aorta.
4. Valves prevent backflow of blood in the heart.
5. Atrioventricular (AV) valves, between the atria and their ventricles, are the tricuspid valve on the right side of the heart and the bicuspid (mitral) valve on the left.
6. The chordae tendineae and their muscles stop blood from backflowing into the atria.
7. The two arteries that leave the heart both have a semilunar valve.
8. The skeleton of the heart consists of fibrous connective tissue around the valves of the heart and fused with each other; the skeleton anchors the valves and cardiac fibers and forms an electrical insulator between atria and ventricles.

Conduction System (p. 354)
1. The conduction system consists of tissue specialized for impulse conduction.
2. Components of this system are the sinoatrial (SA) node (pacemaker), atrioventricular (AV) node, atrioventricular (AV) bundle (bundle of His), bundle branches, and conduction myofibers (Purkinje fibers).

Electrocardiogram (p. 354)
1. A cardiac cycle consists of the systole (contraction) and diastole (relaxation) of both atria plus the systole and diastole of both ventricles followed by a short pause.
2. The record of electrical changes during each cardiac cycle is referred to as an electrocardiogram (ECG).
3. A normal ECG consists of a P wave (spread of impulse from SA node over atria), QRS wave (spread of impulse through ventricles), and T wave (ventricular repolarization).
4. The ECG is invaluable in diagnosing abnormal cardiac rhythms and conduction patterns, detecting the presence of fetal life, determining the presence of several fetuses, and following the course of recovery from a heart attack.
5. An artificial pacemaker may be used to restore an abnormal cardiac rhythm.

Cardiac Cycle (p. 356)
1. With an average heartbeat of 75/min, a complete cardiac cycle requires 0.8 sec.
2. The first heart sound (lubb) represents the closing of the atrioventricular valves. The second sound (dupp) represents the closing of semilunar valves.
3. A peculiar sound is called a murmur.

Autonomic Control (p. 356)
1. Heart rate and strength of contraction may be increased by sympathetic stimulation from the cardioacceleratory center (CAC) in the medulla and decreased by parasympathetic stimulation from the cardioinhibitory center (CIC) in the medulla.
2. Pressoreceptors are nerve cells that respond to changes in blood pressure. They act on the cardiac centers in the medulla.
3. Other influences on heart rate include chemicals (epinephrine, sodium, potassium), temperature, emotion, sex, and age.

Blood Supply (p. 357)
1. The coronary (cardiac) circulation delivers oxygenated blood to the myocardium and removes carbon dioxide from it.
2. Deoxygenated blood returns to the right atrium via the coronary sinus.
3. Complications of this system are angina pectoris and myocardial infarction (MI).

Artificial Heart (p. 358)
1. The Jarvik artificial heart consists of an aluminum base and rigid plastic chambers that serve as ventricles.

2. When it is implanted, the recipient's atria are left intact.

Risk Factors in Heart Disease (p. 359)

1. Risk factors in heart disease include high blood cholesterol, high blood pressure, cigarette smoking, obesity, lack of regular exercise, diabetes mellitus, genetic predisposition, sex, age, and fibrinogen level.

Developmental Anatomy of the Heart (p. 360)

1. The heart develops from mesoderm.
2. The endothelial tubes develop into the four-chambered heart and great vessels of the heart.

Applications to Health (p. 360)

1. Coronary artery disease (CAD) refers to a condition in which the myocardium receives inadequate blood due to atherosclerosis, coronary artery spasm, or thrombi or emboli; it may be treated with drugs or surgically.
2. Congenital heart defects include coarctation of the aorta; patent ductus arteriosus, septal defects, valvular stenosis, and tetralogy of Fallot.
3. Arrhythmias result in heart block, flutter, and fibrillation.
4. Congestive heart failure (CHF) results when the heart cannot supply the oxygen demands of the body.
5. Cor pulmonale (CP) refers to right ventricular hypertrophy secondary to pulmonary hypertension.

REVIEW QUESTIONS

1. Describe the location of the heart in the mediastinum and identify the various borders and surfaces of the heart.
2. Distinguish the subdivisions of the pericardium. What is the purpose of this structure?
3. Compare the three layers of the heart wall according to composition, location, and function.
4. Define atria and ventricles. What vessels enter or exit the atria and ventricles?
5. Describe the principal valves in the heart and how they operate.
6. Describe the components of the skeleton of the heart.
7. Describe how the valves of the heart may be identified by surface projection.
8. Describe the structure and function of the heart's conducting system.
9. Define and label the deflection waves of a normal electrocardiogram (ECG). Explain why the ECG is an important diagnostic tool.
10. Define a cardiac cycle and describe its timing. What causes heart sounds?
11. Distinguish between the cardioacceleratory center (CAC) and cardioinhibitory center (CIC) with respect to the regulation of heart rate.
12. Describe the route of blood in coronary (cardiac) circulation. Distinguish between angina pectoris and myocardial infarction (MI).
13. Explain how an artificial heart is implanted and how it operates.
14. Describe how the heart develops.
15. Describe the risk factors involved in heart disease.
16. What is coronary artery disease (CAD)? What factors contribute to it?
17. Define each of the following congenital heart defects: coarctation of the aorta, patent ductus arteriosus, septal defects, valvular stenosis, and tetralogy of Fallot.
18. Define an arrhythmia. Give several examples.
19. Describe the symptoms of congestive heart failure (CHF) and cor pulmonale (CP).
20. Refer to the glossary of key medical terms associated with the heart. Be sure that you can define each term.

SELF-QUIZ

1. Match the following:
   ___ a. also called the mitral valve
   ___ b. prevents backflow of blood from right ventricle to right atrium
   ___ c. prevents backflow of blood from pulmonary trunk to right ventricle
   ___ d. prevents backflow of blood into left atrium
   ___ e. have half-moon-shaped leaflets or cusps (two answers)
   ___ f. also called atrioventricular (AV) valve (two answers)

   A. aortic semilunar
   B. bicuspid
   C. pulmonary semilunar
   D. tricuspid

Complete the following.

2. The heart is derived from _____-derm. The heart begins to develop during the (third? fifth? seventh?) week. Its initial formation consists of two endothelial tubes that unite to form the _____ tube.
3. The first heart sound (lubb? dupp?) is created by turbulence of blood at the (opening? closing?) of the _____ valves.
4. Oxygenated blood returns from the lungs to the heart by means of the pulmonary _____. They empty into the _____ of the heart.
5. The pointed part of the heart, called the _____, lies in the _____ intercostal space. The base of the heart is located just inferior to your _____.
6. The heart is divided into superior chambers called _____ and inferior chambers called _____.
7. Blood from all parts of the body except the lungs flows into a chamber of the heart named the _____. Blood from superior body parts enters the heart via the vein called the _____. Blood from vessels supplying heart tissue returns to the right atrium via the vessel named the _____.

Choose the one best answer to these questions.

___ 8. The structure that serves the heart in much the same way the pleura serves the lungs is the
   A. mesentery; B. myocardium; C. diaphragm; D. endocardium; E. pericardium

___ **9.** The right atrium of the heart
(1) contains the pacemaker.
(2) receives blood from the superior and inferior vena cavae.
(3) receives blood directly from the lungs.
(4) empties into the aorta.
A. (1) only; B. (2) only; C. (3) only; D. (4) only; E. (1) and (2).

___ **10.** Stimulation of the vagus (X) nerve
A. increases conductivity of the heart tissue; B. increases heart rate; C. decreases excitability of the heart; D. decreases the inhibitory mechanisms of the heart; E. has no effect on the heart.

___ **11.** Which of these vessels carries oxygenated blood?
A. superior vena cava; B. coronary sinus; C. inferior vena cava; D. pulmonary vein; E. pulmonary artery.

___ **12.** Which vessel is *not* associated with the right side of the heart?
A. superior vena cava; B. pulmonary trunk; C. inferior vena cava; D. aorta; E. coronary sinus.

___ **13.** Which sequences correctly represent the conduction of an impulse through the heart?
A. SA node, AV node, AV bundle, bundle branches;
B. SA node, AV bundle, AV node, bundle branches;
C. AV node, SA node, AV bundle, bundle branches;
D. SA node, bundle branches, AV node, AV bundle.

___ **14.** Heart sounds are produced by the
A. contraction of the myocardium; B. closure of the AV and semilunar valves; C. the flow of blood in the atria; D. the flow of blood in the ventricles; E. expansion and recoil of the aorta.

___ **15.** The groove that separates the atria from the ventricles is the
A. coronary sulcus; B. interatrial septum; C. interventricular sulcus; D. interventricular septum; E. fossa ovalis.

___ **16.** Choose the false statement about heart structure:
A. the heart chamber with the thickest wall is the left ventricle; B. the apex of the heart is more superior in location than the base; C. the heart has four chambers; D. the left ventricle forms the apex and most of the left border of the heart; E. pectinate muscles are found in atrial walls.

___ **17.** Which of the following structures are located in ventricles?
A. trabeculae carneae; B. fossa ovalis; C. ligamentum arteriosum; D. pectinate muscles; E. orifice of coronary sinus.

Arrange the answers in correct sequence.

— — — — — **18.** From most superficial to deepest:
A. epicardium (visceral pericardium)
B. myocardium
C. parietal pericardium, fibrous layer
D. pericardial space containing pericardial fluid
E. parietal pericardium, serous layer

— — — — — **19.** Route of a red blood cell now in the right atrium:
A. left atrium
B. left ventricle
C. right ventricle
D. pulmonary artery
E. pulmonary vein

14 The Cardiovascular System: Blood Vessels

STUDENT OBJECTIVES

1. Contrast the structure and function of arteries, arterioles, capillaries, venules, and veins.
2. Identify the principal arteries and veins of systemic circulation.
3. Identify the major blood vessels of pulmonary circulation.
4. Trace the route of blood involved in hepatic portal circulation and explain its importance.
5. Contrast fetal and adult circulation.
6. Describe the effects of aging on the cardiovascular system.
7. Describe the development of blood vessels and blood.
8. List the causes and symptoms of hypertension, aneurysms, and deep-venous thrombosis (DVT).
9. Define key medical terms associated with blood vessels.

CHAPTER OUTLINE

■ **Arteries**
Elastic (Conducting) Arteries
Muscular (Distributing) Arteries
Anastomoses
■ **Arterioles**
■ **Capillaries**
■ **Venules**
■ **Veins**
■ **Blood Reservoirs**
■ **Circulatory Routes**
Systemic Circulation
Pulmonary Circulation
Hepatic Portal Circulation
Fetal Circulation
■ **Aging and the Cardiovascular System**
■ **Developmental Anatomy of Blood and Blood Vessels**
■ **Applications to Health**
Hypertension
Aneurysm
Coronary Artery Disease (CAD)
Deep-Venous Thrombosis (DVT)
■ **Key Medical Terms Associated with Blood Vessels**

Blood vessels form an extensive network of tubes that carry blood away from the heart, transport it to the tissues of the body, and then return it to the heart. *Arteries* are vessels that carry blood from the heart to the tissues. Large, elastic arteries leave the heart and divide into medium-sized, muscular arteries that branch out into the various regions of the body. Medium-sized arteries divide into small arteries, which, in turn, divide into still smaller arteries called *arterioles*. As the arterioles enter a tissue, they branch into countless microscopic vessels called *capillaries*. Through the walls of capillaries, substances are exchanged between the blood and body tissues. Before leaving the tissue, groups of capillaries reunite to form small veins called *venules*. These, in turn, merge to form progressively larger tubes called veins. *Veins*, then, are blood vessels that convey blood from the tissues back to the heart. Since blood vessels require oxygen and nutrients just like other tissues of the body, they also have blood vessels in their own walls called *vasa vasorum*.

The developmental anatomy of blood vessels and blood is considered later in the chapter.

ARTERIES

Arteries have walls constructed of three coats, or tunics, and a hollow core, called a *lumen*, through which blood flows (Figures 14-1 and 14-4). The inner coat of an arterial wall is the *tunica interna (intima)*. It is composed of a lining of endothelium (simple squamous epithelium) that is in contact with the blood and a layer of elastic tissue called the internal elastic membrane. The middle coat, or *tunica media*, is usually the thickest layer. It consists of elastic fibers and smooth muscle. The outer coat, the *tunica externa (adventitia)*, is composed principally of elastic and collagenous fibers. An external elastic membrane may separate the tunica externa from the tunica media.

As a result of the structure of the middle coat (tunica media) especially, arteries have two major properties: elasticity and contractility. When the ventricles of the heart contract and eject blood into the large arteries, they expand to accommodate the extra blood. Then, as the ventricles relax, the elastic recoil of the arteries forces the blood onward. The contractility of an artery comes from its smooth muscle, which is arranged longitudinally and in rings around the lumen somewhat like a doughnut and is innervated by sympathetic branches of the autonomic nervous system. When there is sympathetic stimulation, the smooth muscle contracts, squeezes the wall around the lumen, and narrows the vessel. Such a decrease in the size of the lumen is called *vasoconstriction*. Conversely, when sympathetic stimulation is removed, the smooth muscle fibers relax and the size of the arterial lumen increases. This increase is called *vasodilation* and is usually due to the inhibition of vasoconstriction.

The contractility of arteries also serves a function in stopping bleeding. The blood flowing through an artery is under a great deal of pressure. Thus, great quantities of blood can quickly be lost from a broken artery. When an artery is cut, its wall constricts so that blood does not escape quite so rapidly. However, there is a limit to how much vasoconstriction can help.

ELASTIC (CONDUCTING) ARTERIES

Large arteries are referred to as *elastic (conducting) arteries*. They include the aorta and brachiocephalic, common carotid, subclavian, vertebral, and common iliac arteries. The wall of elastic arteries is relatively thin in proportion to their diameter, and their tunica media contains more elastic fibers and less smooth muscle. As the heart alternately contracts and relaxes, the rate of blood flow tends to be intermittent. When the heart contracts and forces blood into the aorta, the wall of the elastic arteries stretches to accommodate the surge of blood and stores the pressure energy. During relaxation of the heart, the wall of the elastic arteries recoils, moving the blood forward in a more continuous flow. Elastic arteries are called conducting arteries because they *conduct* blood from the heart to medium-sized muscular arteries.

MUSCULAR (DISTRIBUTING) ARTERIES

Medium-sized arteries are called *muscular (distributing) arteries*. They include the axillary, brachial, radial, intercostal, splenic, mesenteric, femoral, popliteal, and tibial arteries. Their tunica media contains more smooth muscle than elastic fibers, and they are capable of greater vasoconstriction and vasodilation to adjust the volume of blood to suit the needs of the structure supplied. The wall of muscular arteries is relatively thick, mainly due to the large amounts of smooth muscle. Muscular arteries are called distributing arteries because they *distribute* blood to various parts of the body.

ANASTOMOSES

Most parts of the body receive branches from more than one artery. In such areas the distal ends of the vessels unite. The junction of two or more vessels supplying the same body region is called an *anastomosis* (a-nas-tō-MŌ-sis). Anastomoses may also occur between the origins of veins and between arterioles and venules. Anastomoses between arteries provide alternate routes by which blood can reach a tissue or organ. Thus, if a vessel is occluded by disease, injury, or surgery, circulation to a part of the body is not necessarily stopped. The alternate route of blood to a body part through an anastomosis is known as *collateral circulation*. An alternate blood route may also be supplied by nonanastomosing vessels that supply the same region of the body.

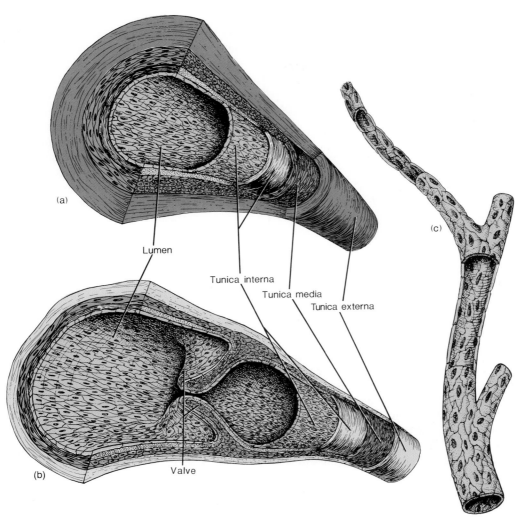

FIGURE 14-1 Comparative structure of (a) an artery, (b) a vein, and (c) a capillary. The relative size of the capillary is enlarged.

Arteries that do not anastomose are known as ***end arteries.*** Occlusion of an end artery interrupts the blood supply to a whole segment of an organ, producing necrosis (death) of that segment.

ARTERIOLES

An ***arteriole*** is a very small, almost microscopic artery that delivers blood to capillaries. Arterioles closer to the arteries from which they branch have a tunica interna like that of arteries, a tunica media composed of smooth muscle and very few elastic fibers, and a tunica externa composed mostly of elastic and collagenous fibers (see Figure 14-4c). As arterioles get smaller in size, the tunics change character so that arterioles closest to capillaries consist of little more than a layer of endothelium surrounded by a few scattered smooth muscle fibers (Figure 14-2).

Arterioles play a key role in regulating blood flow from arteries into capillaries. The smooth muscle of arterioles, like that of arteries, is subject to vasoconstriction and vasodilation. During vasoconstriction, blood flow into capillaries is restricted; during vasodilation, the flow is significantly increased. The relation of arterioles and blood flow will be considered shortly.

CAPILLARIES

Capillaries are microscopic vessels that usually connect arterioles and venules (see Figure 14-1). They are found near almost every cell in the body. The distribution of capillaries in the body varies with the activity of the tissue. For example, in places where activity is higher, such as the liver, kidneys, lungs, muscles, and nervous system, there are rich capillary supplies. In areas where activity is

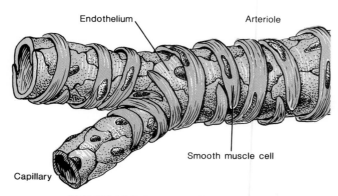

FIGURE 14-2 Structure of an arteriole.

lower, such as tendons and ligaments, the capillary supply is not as extensive. The epidermis, cornea of the eye, and cartilage are devoid of capillaries.

The primary function of capillaries is to permit the exchange of nutrients and wastes between the blood and tissue cells. The structure of the capillaries is admirably suited to this purpose. Capillary walls are composed of only a single layer of cells (endothelium) and a basement membrane. They have no tunica media or tunica externa. Thus, a substance in the blood must pass through the plasma membrane of just one cell to reach tissue cells. This vital exchange of materials occurs only through capillary walls—the thick walls of arteries and veins present too great a barrier.

Although capillaries pass directly from arterioles to venules in some places in the body, in other places they form extensive branching networks. These networks increase the surface area for diffusion and thereby allow a rapid exchange of large quantities of materials. In most tissues, blood normally flows through only a small portion of the capillary network when metabolic needs are low. But, when a tissue becomes active, the entire capillary network fills with blood.

The flow of blood through capillaries is regulated by vessels with smooth muscle in their walls. A *metarteriole* (*met* = beyond) is a vessel that emerges from an arteriole, passes through the capillary network, and empties into a venule (Figure 14-3). The proximal portions of metarterioles are surrounded by scattered smooth muscle fibers whose contraction and relaxation help regulate the amount and force of blood. The distal portion of a metarteriole has no smooth muscle fibers and is called a *thoroughfare channel*. It serves as a low-resistance channel that increases blood flow. *True capillaries* emerge from arterioles or metarterioles and are not on the direct flow route from arteriole to venule. At their sites of origin, there is a ring of smooth muscle fibers called a *precapillary sphincter* that controls the flow of blood entering a true capillary.

Autoregulation refers to a local, automatic adjustment of blood flow in a given region of the body in response to the particular needs of the tissue. In most body tissues, oxygen is the principal stimulus for autoregulation. A suggested mechanism for autoregulation is as follows. In re-

sponse to low oxygen supplies, the cells in the immediate area produce and release *vasodilator substances*. Such substances are thought to include potassium ions, hydrogen ions, carbon dioxide, lactic acid, and adenosine. Once released, the vasodilator substances produce a local dilation of arterioles and relaxation of precapillary sphincters. The result is an increased flow of blood into the tissue, which restores oxygen levels to normal. The autoregulation mechanism is important in meeting the nutritional demands of active tissues, such as muscle tissue, where the demand might increase as much as tenfold.

Some capillaries of the body, such as those found in muscle tissue and other locations, are referred to as *continuous capillaries*. These capillaries are so named because the cytoplasm of the endothelial cells is continuous when viewed in cross section through a microscope; the cytoplasm appears as an uninterrupted ring, except for the endothelial junction. Other capillaries of the body are referred to as *fenestrated capillaries*. They differ from continuous capillaries in that their endothelial cells have numerous fenestrae (pores) where the cytoplasm is absent. The fenestrae range from 70 to 100 nm in diameter and are closed by a thin diaphragm, except in the capillaries in the kidneys where they are assumed to be open. Fenestrated capillaries are also found in the villi of the small intestine, choroid plexuses of the ventricles in the brain, ciliary processes of the eyes, and endocrine glands.

Microscopic blood vessels in certain parts of the body, such as the liver, are termed *sinusoids*. They are wider than capillaries and more tortuous. Also, instead of the usual endothelial lining, sinusoids contain spaces between endothelial cells and the basal lamina is incomplete or absent. In addition, sinusoids contain specialized lining cells that are adapted to the function of the tissue. For example, in the liver, sinusoids contain phagocytic cells called *stellate reticuloendothelial (Kupffer's) cells*. Like capillaries, sinusoids convey blood from arterioles to venules. Other regions

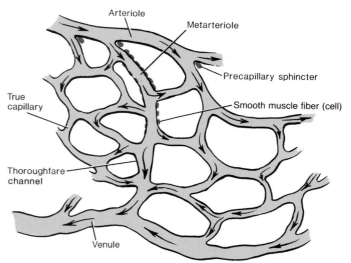

FIGURE 14-3 Details of a capillary network.

containing sinusoids include the spleen, adenohypophysis, parathyroid glands, adrenal cortex, and bone marrow.

VENULES

When several capillaries unite, they form small veins called *venules* (see Figure 14-4b). Venules collect blood from capillaries and drain it into veins. The venules closest to the capillaries consist of a tunica interna of endothelium and a tunica externa of connective tissue. As the venules approach the veins, they also contain the tunica media characteristic of veins.

VEINS

Veins are composed of essentially the same three coats as arteries, but there are variations in their relative thicknesses. The tunica interna of veins is extremely thin compared to that of their accompanying arteries. In addition, the tunica media of veins is much thinner than that of accompanying arteries, and the tunica externa is thicker in veins (Figures 14-1 and 14-4). Despite these differences, veins are still distensible enough to adapt to variations in the volume and pressure of blood passing through them.

By the time the blood leaves the capillaries and moves into the veins, it has lost a great deal of pressure. The difference in pressure can be observed in the blood flow from a cut vessel; blood leaves a cut vein in an even flow rather than in the rapid spurts characteristic of arteries. Most of the structural differences between arteries and veins reflect this pressure difference. For example, the walls of veins are not as strong as those of arteries. The low pressure in veins, however, has its disadvantages. When you stand, the pressure pushing blood up the veins in your lower extremities is barely enough to balance the force of gravity pushing it back down. For this reason, many veins, especially those in the limbs, contain valves that prevent backflow (Figure 14-5). Normal valves ensure the flow of blood toward the heart.

CLINICAL APPLICATION

In people with weak venous valves, gravity forces large quantities of blood back down into distal parts of the vein. This pressure overloads the vein and pushes its wall outward. After repeated overloading, the walls lose their elasticity and become stretched and flabby. Such dilated and tortuous veins caused by incompetent valves are called *varicose veins* (*VVs*). They may be due to heredity, mechanical factors (prolonged standing and pregnancy), or aging. Because a varicosed wall is not able to exert a firm resistance against the blood, blood tends to accumulate in the pouched-out area of the vein, causing it to swell and forcing fluid into the surrounding tissue. Veins close to the surface of the legs are highly susceptible to varicosities. Veins that lie deeper are not as vulnerable because surrounding skeletal muscles prevent their walls from overstretching.

Varicose veins may be treated by several methods, depending on the severity of the condition. These include: (1) frequent periods of rest with elevation of the lower extremities; (2) external pressure with elastic stockings or bandages; (3) a procedure known as sclerotherapy, an intravenous injection of sclerosing chemicals that col-

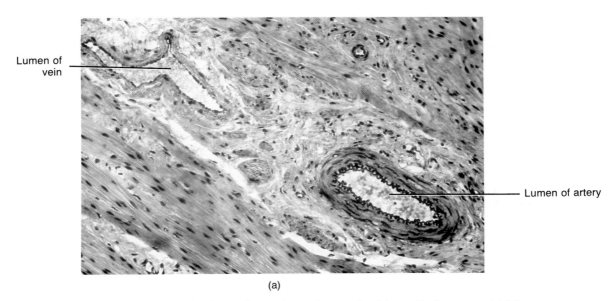

Lumen of vein

Lumen of artery

(a)

FIGURE 14-4 Histology of blood vessels. (a) Photomicrograph of the wall of an artery (right) and its accompanying vein (left) at a magnification of 250×. Note that the lumen of the vein is larger than that of the artery, but the wall of the vein is thinner and the vein frequently appears collapsed (flattened). (Courtesy Andrew Kuntzman.)

Lumen of venule

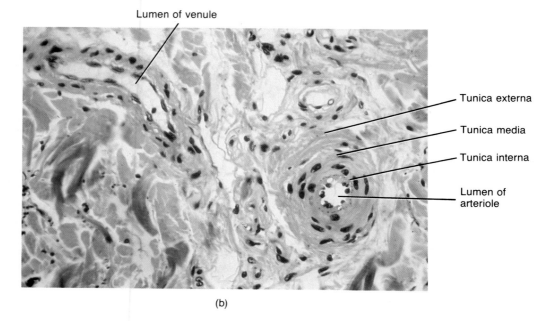

Tunica externa

Tunica media

Tunica interna

Lumen of
arteriole

(b)

Venule

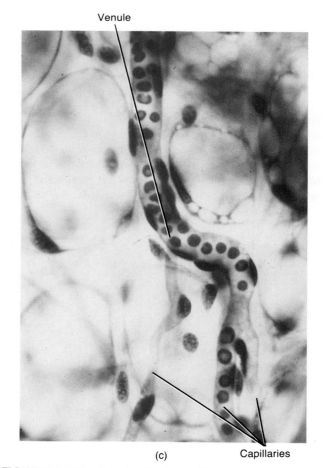

lapses the veins and prevents blood from flowing into them, thus eliminating the purple-blue discoloration; and (4) surgery, in which the saphenous veins with the incompetent valves are ligated and removed (''stripping'').

A mild form of varicose veins in the legs is referred to as **spider burst veins,** a condition that is most commonly treated by sclerotherapy.

A *vascular (venous) sinus* is a vein with a thin endothelial wall that has no smooth muscle to alter its diameter. Surrounding tissue replaces the tunica media and tunica externa to provide support. Intracranial vascular sinuses, which are supported by the dura mater, return cerebrospinal fluid and deoxygenated blood from the brain to the heart. Another example of a vascular sinus is the coronary sinus of the heart.

BLOOD RESERVOIRS

The volume of blood in various parts of the cardiovascular system varies considerably. Veins, venules, and venous sinuses contain about 59 percent of the blood in the system, arteries about 13 percent, pulmonary vessels about 12 percent, the heart about 9 percent, and arterioles and capillaries about 7 percent. Since systemic veins contain so much of the blood, they are referred to as *blood reservoirs.* They serve as storage depots for blood, which can be moved quickly to other parts of the body if the need arises. When there is increased muscular activity, the vasomotor center sends increasing sympathetic impulses to veins that serve as blood reservoirs. The result is vasoconstriction, which permits the distribution of blood from venous reservoirs to skeletal muscles, where it is needed most. A similar mechanism operates in cases of hemorrhage, when blood volume and pressure decrease. Vasoconstriction of veins

Capillaries

(c)

FIGURE 14-4 (Continued) Histology of blood vessels. (b) Photomicrograph of a venule and an ateriole at a magnification of 250×. (Courtesy Andrew Kuntzman.) (c) Photomicrograph of several capillaries uniting to from a venule. Note the red blood cells in single file in the capillary. (Courtesy of Michael H. Ross and Edward J. Reith, *Histology: A Text and Atlas.* Copyright © 1985 by Michael H. Ross and Edward J. Reith, Harper & Row, Publishers, Inc., New York.)

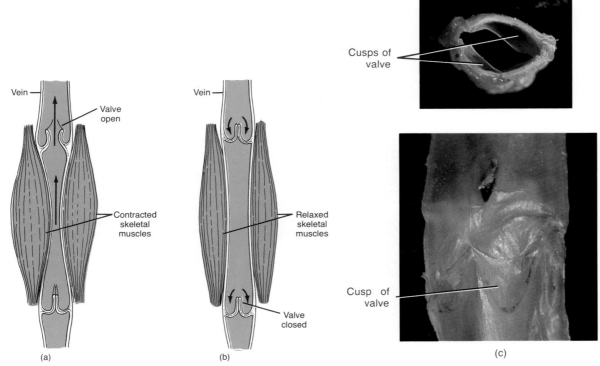

FIGURE 14-5 Role of skeletal muscle contractions and venous valves in returning blood to the heart. (a) When skeletal muscles contract, the valves open, and blood is forced toward the heart. (b) When skeletal muscles relax, the valves close to prevent the backflowing of blood from the heart. (c) Photograph of a one-way valve in a vein (cross section above, longitudinal section below). (Courtesy of J. A. Gosling et al., *Atlas of Human Anatomy with Integrated Text.* Copyright © 1985 by Gower Medical Publishing Ltd.)

in venous reservoirs helps compensate for the blood loss. Among the principal blood reservoirs are the veins of the abdominal organs (especially the liver and spleen) and the veins of the skin.

CIRCULATORY ROUTES

Arteries, arterioles, capillaries, venules, and veins are organized into definite routes that circulate blood throughout the body. We can now look at the basic routes the blood takes as it is transported through its vessels.

Figure 14-6 shows a number of basic *circulatory routes* through which the blood travels. *Systemic circulation* includes all the oxygenated blood that leaves the left ventricle through the aorta and the deoxygenated blood that returns to the right atrium after traveling to all the organs, including the nutrient arteries to the lungs. Two of the many subdivisions of the systemic circulation are the *coronary (cardiac) circulation* (see Figure 13-8), which supplies the myocardium of the heart, and the *hepatic portal circulation,* which runs from the gastrointestinal tract to the liver (see Figure 14-18). Blood leaving the aorta and traveling through the systemic arteries is a bright red color. As it moves through capillaries, it loses its oxygen and takes on carbon dioxide, so that blood in systemic veins is a dark red color.

When blood returns to the heart from the systemic route, it goes out of the right ventricle through the *pulmonary circulation* to the lungs (see Figure 14-17). In the lungs, it loses its carbon dioxide and takes on oxygen. It is now bright red again. It returns to the left atrium of the heart and reenters the systemic circulation.

Another major route—*fetal circulation*—exists only in the fetus and contains special structures that allow the developing fetus to exchange materials with its mother (see Figure 14-19).

Cerebral circulation (cerebral arterial circle, or circle of Willis) is discussed in Exhibit 14-3.

SYSTEMIC CIRCULATION

The flow of blood from the left ventricle to all parts of the body and back to the right atrium is called the *systemic circulation.* The purpose of systemic circulation is to carry oxygen and nutrients to body tissues and to remove carbon dioxide and other wastes from the tissues. All systemic arteries branch from the *aorta,* which arises from the left ventricle of the heart.

As the aorta emerges from the left ventricle, it passes upward and deep to the pulmonary trunk. At this point, it is called the *ascending aorta.* The ascending aorta gives off two coronary branches to the heart muscle. Then it

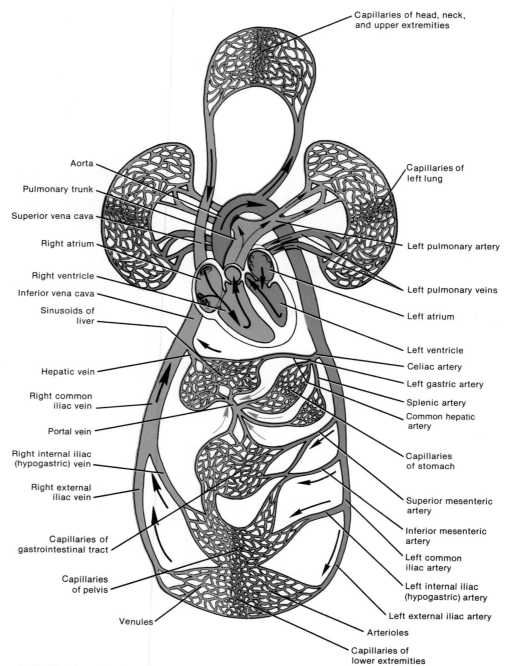

Capillaries of head, neck, and upper extremities

Aorta

Pulmonary trunk

Superior vena cava

Right atrium

Right ventricle

Inferior vena cava

Sinusoids of liver

Hepatic vein

Right common iliac vein

Portal vein

Right internal iliac (hypogastric) vein

Right external iliac vein

Capillaries of gastrointestinal tract

Capillaries of pelvis

Venules

Capillaries of left lung

Left pulmonary artery

Left pulmonary veins

Left atrium

Left ventricle

Celiac artery

Left gastric artery

Splenic artery

Common hepatic artery

Capillaries of stomach

Superior mesenteric artery

Inferior mesenteric artery

Left common iliac artery

Left internal iliac (hypogastric) artery

Left external iliac artery

Arterioles

Capillaries of lower extremities

FIGURE 14-6 Circulatory routes. Systemic circulation is indicated by heavy black arrows, pulmonary circulation by thin black arrows, and hepatic portal circulation by thin colored arrows. Refer to Figure 13-8 for the details of coronary (cardiac) circulation and Figure 14-19 for the details of fetal circulation.

turns to the left, forming the **arch of the aorta** before descending to the level of the fourth thoracic vertebra as the **descending aorta.** The descending aorta lies close to the vertebral bodies, passes through the diaphragm, and divides at the level of the fourth lumbar vertebra into two **common iliac arteries,** which carry blood to the lower extremities. The section of the descending aorta between the arch of aorta and the diaphragm is referred to as the **thoracic aorta.** The section between the diaphragm and the common iliac arteries is termed the **abdominal aorta.**

Each section of the aorta gives off arteries that continue to branch into distributing arteries leading to organs and finally into the arterioles and capillaries that service the tissues.

Blood is returned to the heart through the systemic veins. All the veins of the systemic circulation flow into either the **superior** or **inferior vena cava** or the **coronary sinus.** They in turn empty into the right atrium. The principal arteries and veins of systemic circulation are described and illustrated in Exhibits 14-1–14-12 and Figures 14-7–14-16.

EXHIBIT 14-1

Aorta and Its Branches (Figure 14-7)

DIVISION OF AORTA	ARTERIAL BRANCH	REGION SUPPLIED
Ascending aorta	Right and left coronary	Heart.
Arch of aorta	Brachiocephalic — Right common carotid	Right side of head and neck.
	Brachiocephalic — Right subclavian	Right upper extremity.
	Left common carotid	Left side of head and neck.
	Left subclavian	Left upper extremity.
Thoracic aorta	Intercostals	Intercostal and chest muscles and pleurae.
	Superior phrenics	Posterior and superior surfaces of diaphragm.
	Bronchials	Bronchi of lungs.
	Esophageals	Esophagus.
Abdominal aorta	Inferior phrenics	Inferior surface of diaphragm.
	Celiac — Common hepatic	Liver.
	Celiac — Left gastric	Stomach and esophagus.
	Celiac — Splenic	Spleen, pancreas, and stomach.
	Superior mesenteric	Small intestine, cecum, and ascending and transverse colons.
	Suprarenals	Adrenal (suprarenal) glands.
	Renals	Kidneys.
	Gonadals — Testicular	Testes.
	Gonadals or Ovarians	Ovaries.
	Inferior mesenteric	Transverse, descending, and sigmoid colons and rectum.
	Common iliacs — External iliacs	Lower extremities.
	Common iliacs — Internal iliacs (hypogastrics)	Uterus, prostate gland, muscles of buttocks, and urinary bladder.

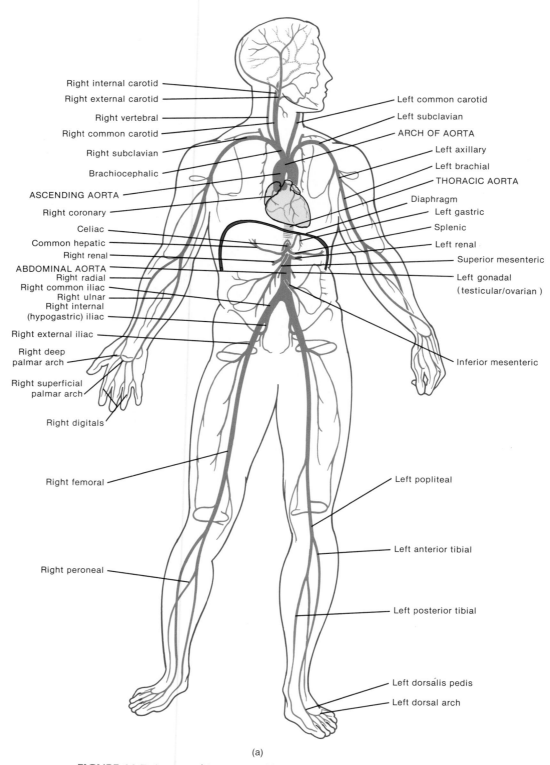

(a)

FIGURE 14-7 Aorta and its principal branches in anterior view. (a) Diagram.

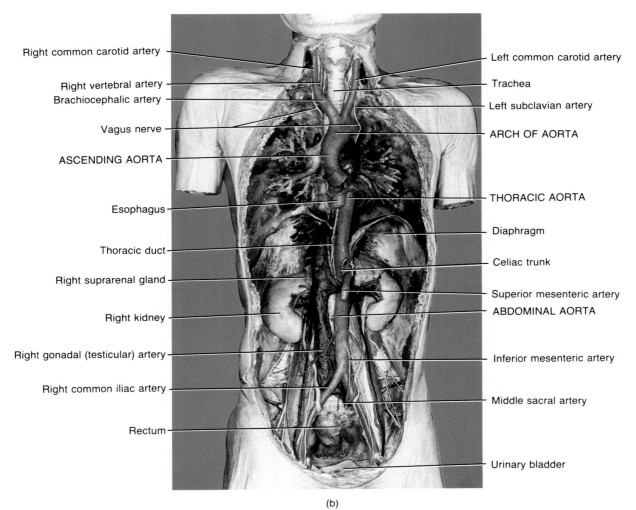

Right common carotid artery

Right vertebral artery

Brachiocephalic artery

Vagus nerve

ASCENDING AORTA

Esophagus

Thoracic duct

Right suprarenal gland

Right kidney

Right gonadal (testicular) artery

Right common iliac artery

Rectum

Left common carotid artery

Trachea

Left subclavian artery

ARCH OF AORTA

THORACIC AORTA

Diaphragm

Celiac trunk

Superior mesenteric artery

ABDOMINAL AORTA

Inferior mesenteric artery

Middle sacral artery

Urinary bladder

(b)

FIGURE 14-7 (*Continued*) Aorta and its principal branches in anterior view. (b) Photograph. (Courtesy of C. Yokochi and J. W. Rohen, *Photographic Anatomy of the Human Body,* 2nd ed., Igaku-Shoin, Ltd.)

EXHIBIT 14-2

Ascending Aorta (Figure 14-8)

BRANCH	DESCRIPTION AND REGION SUPPLIED
Coronary arteries	Right and left branches arise from ascending aorta just superior to aortic semilunar valve. They form crown around heart, giving off branches to atrial and ventricular myocardium.

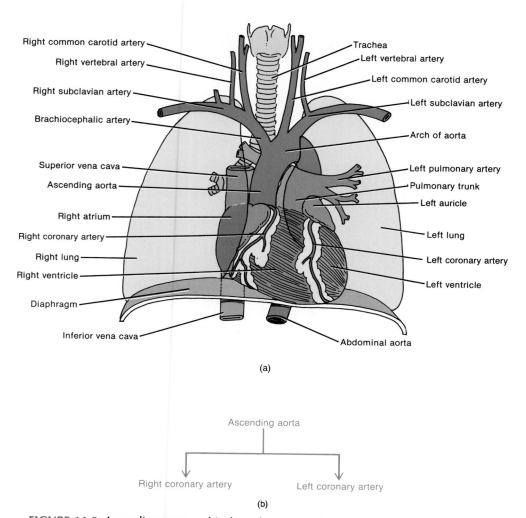

(a)

(b)

FIGURE 14-8 Ascending aorta and its branches in anterior view. (a) Diagram. (b) Scheme of distribution.

EXHIBIT 14-3

Arch of Aorta (Figure 14-9)

BRANCH	DESCRIPTION AND REGION SUPPLIED
Brachiocephalic	**Brachiocephalic artery** is first branch off arch of aorta. It divides to form right subclavian artery and right common carotid artery. **Right subclavian artery** extends from brachiocephalic to first rib and then passes into armpit (axilla) and supplies arm, forearm, and hand. Continuation of right subclavian into axilla is called **axillary artery.*** From here, it continues into arm as **brachial artery.** At bend of elbow, brachial artery divides into medial **ulnar** and lateral **radial arteries.** These vessels pass down to palm, one on each side of forearm. In palm, branches of two arteries anastomose to form two palmar arches—**superficial palmar arch** and **deep palmar arch.** From these arches arise **digital arteries,** which supply fingers and thumb. Before passing into axilla, right subclavian gives off major branch to brain called **vertebral artery.** Right vertebral artery passes through foramina of transverse processes of cervical vertebrae and enters skull through foramen magnum to reach undersurface of brain. Here it unites with left vertebral artery to form **basilar artery.** **Right common carotid artery** passes upward in neck. At upper level of larynx, it divides into **right external** and **right internal carotid arteries.** External carotid supplies right side of thyroid gland, tongue, throat, face, ear, scalp, and dura mater. Internal carotid supplies brain, right eye, and right sides of forehead and nose. Anastomoses of left and right internal carotids along with basilar artery form a somewhat hexagonal arrangement of blood vessels at base of brain near sella turcica called **cerebral arterial circle (circle of Willis).** From this anastomosis arise arteries supplying brain. Essentially, cerebral arterial circle is formed by union of **anterior cerebral arteries** (branches of internal carotids) and **posterior cerebral arteries** (branches of basilar artery). Posterior cerebral arteries are connected with internal carotids by **posterior communicating arteries.** Anterior cerebral arteries are connected by **anterior communicating arteries.** The **internal carotid arteries** are also considered as part of cerebral arterial circle. The function of the cerebral arterial circle is to equalize blood pressure to brain and provide alternate routes for blood to brain, should arteries become damaged.
Left common carotid	**Left common carotid** is second branch off arch of aorta (see Figure 14-8). Corresponding to right common carotid, it divides into basically same branches with same names, except that arteries are now labeled "left" instead of "right."
Left subclavian	**Left subclavian artery** is third branch off arch of aorta (see Figure 14-8). It distributes blood to left vertebral artery and vessels of left upper extremity. Arteries branching from left subclavian are named like those of right subclavian.

* The right subclavian artery is a good example of the practice of giving the same vessel different names as it passes through different regions.

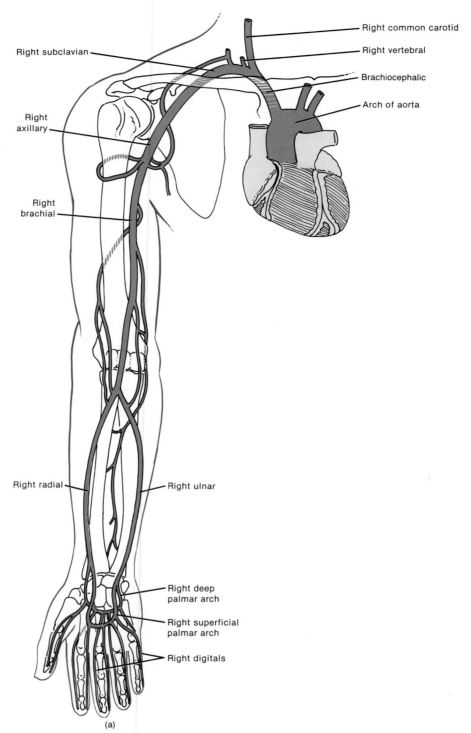

Right subclavian

Right common carotid

Right vertebral

Brachiocephalic

Arch of aorta

Right
axillary

Right
brachial

Right radial

Right ulnar

Right deep
palmar arch

Right superficial
palmar arch

Right digitals

(a)

FIGURE 14-9 Arch of the aorta and its branches. (a) Diagram of anterior view of the principal arteries of the right upper extremity. Arch of the aorta and its branches. (b) Photograph of the principal arteries of the right upper extremity. (Courtesy of C. Yokochi and J. W. Rohen, *Photographic Anatomy of the Human Body,* 2nd ed., 1979, IGAKU-SHOIN, Ltd., Tokyo, New York.) (c) Right lateral view of the principal arteries of the neck and head.

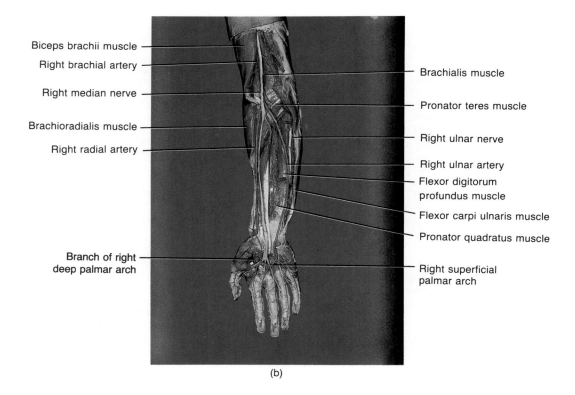

Biceps brachii muscle

Right brachial artery

Right median nerve

Brachioradialis muscle

Right radial artery

Brachialis muscle

Pronator teres muscle

Right ulnar nerve

Right ulnar artery

Flexor digitorum profundus muscle

Flexor carpi ulnaris muscle

Pronator quadratus muscle

Branch of right deep palmar arch

Right superficial palmar arch

(b)

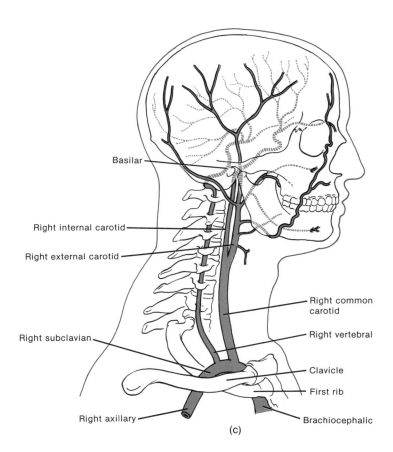

Basilar

Right internal carotid

Right external carotid

Right subclavian

Right axillary

Right common carotid

Right vertebral

Clavicle

First rib

Brachiocephalic

(c)

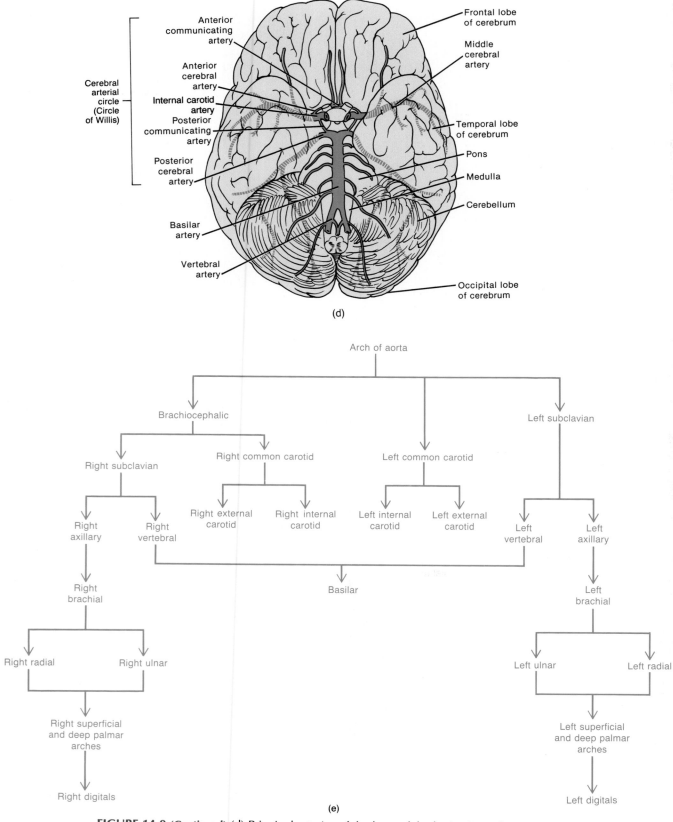

(d)

(e)

FIGURE 14-9 (*Continued*) (d) Principal arteries of the base of the brain. Note the arteries that comprise the cerebral arterial circle (circle of Willis). Arch of the aorta and its branches. (e) Scheme of distribution.

EXHIBIT 14-4

Thoracic Aorta (Figure 14-10)

BRANCH	DESCRIPTION AND REGION SUPPLIED
	Thoracic aorta runs from fourth to twelfth thoracic vertebra. Along its course, it sends off numerous small arteries to viscera and skeletal muscles of the chest. Branches of an artery that supply viscera are called **visceral branches.** Those that supply body wall structures are called **parietal branches.**
VISCERAL Pericardial	Several minute **pericardial arteries** supply blood to dorsal aspect of pericardium.
Bronchial	One right and two left **bronchial arteries** supply the bronchial tubes, visceral pleurae, bronchial lymph nodes, and esophagus.
Esophageal	Four or five **esophageal arteries** supply the esophagus.
Mediastinal	Numerous small **mediastinal arteries** supply blood to structures in the posterior mediastinum.
PARIETAL Posterior intercostal	Nine pairs of **posterior intercostal arteries** supply the intercostal, pectoral, and abdominal muscles; overlying subcutaneous tissue and skin; mammary glands; and vertebral canal and its contents.
Subcostal	The left and right **subcostal arteries** have a distribution similar to that of the posterior intercostals.
Superior phrenic	Small **superior phrenic arteries** supply the posterior surface of the diaphragm.

EXHIBIT 14-5

Abdominal Aorta (Figure 14-10)

BRANCH	DESCRIPTION AND REGION SUPPLIED
VISCERAL Celiac	**Celiac artery (trunk)** is first visceral aortic branch below diaphragm. It has three branches: (1) **common hepatic artery,** (2) **left gastric artery,** and (3) **splenic artery.** The common hepatic artery has three main branches: (1) **hepatic artery proper,** a continuation of the common hepatic artery, which supplies the liver and gallbladder; (2) **right gastric artery,** which supplies the stomach and duodenum; and (3) **gastroduodenal artery,** which supplies the stomach, duodenum, and pancreas. The left gastric artery supplies the stomach and its **esophageal branch** supplies the esophagus. The splenic artery supplies the spleen and has three main branches: (1) **pancreatic arteries,** which supply the pancreas; (2) **left gastroepiploic artery,** which supplies the stomach; and (3) **short gastric arteries,** which supply the stomach.
Superior mesenteric	The **superior mesenteric artery** has several principal branches: (1) **inferior pancreaticoduodenal artery,** which supplies the pancreas and duodenum; (2) **jejunal** and **ileal arteries,** which supply the jejunum and ileum; (3) **ileocolic artery,** which supplies the ileum and ascending colon; (4) **right colic artery,** which supplies the ascending colon; and (5) **middle colic artery,** which supplies the transverse colon.
Suprarenals	Right and left **suprarenal arteries** supply blood to adrenal (suprarenal) glands. The glands are also supplied by branches of the renal and inferior phrenic arteries.
Renals	Right and left **renal arteries** carry blood to kidneys and adrenal (suprarenal) glands.
Gonadals (testiculars or ovarians)	Right and left **testicular arteries** extend into scrotum and terminate in testes; right and left **ovarian arteries** are distributed to ovaries.
Inferior mesenteric	The principal branches of the **inferior mesenteric artery** are: (1) **left colic artery,** which supplies the transverse and descending colons; (2) **sigmoid arteries,** which supply the descending and sigmoid colons; and (3) **superior rectal artery,** which supplies the rectum.
PARIETAL Inferior phrenics	**Inferior phrenic arteries** are distributed to undersurface of diaphragm and adrenal (suprarenal) glands.
Lumbars	**Lumbar arteries** supply spinal cord and its meninges and muscles and skin of lumbar region of back.
Middle sacral	**Middle sacral artery** supplies sacrum, coccyx, and rectum.

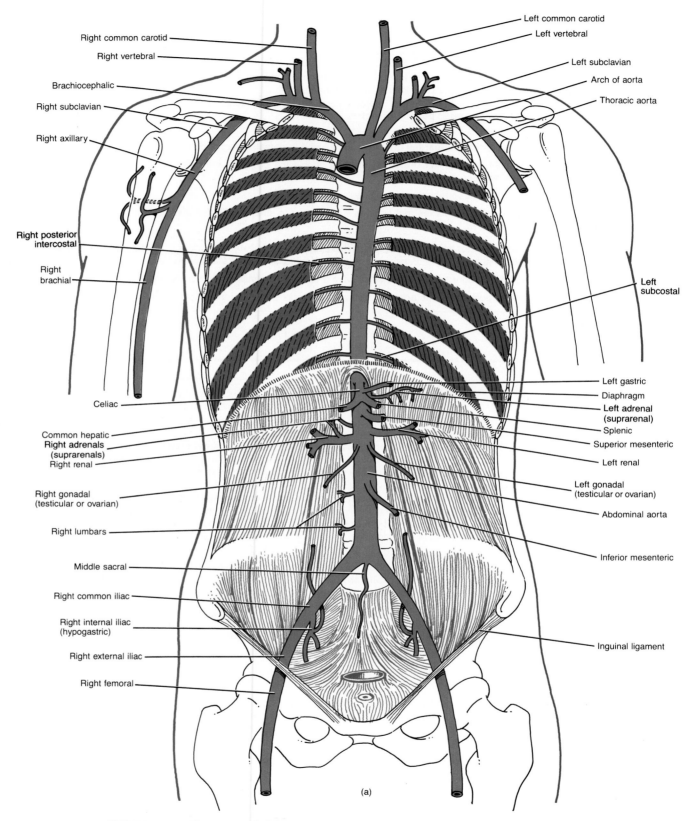

FIGURE 14-10 Thoracic and abdominal aorta and their principal branches. (a) Diagram of thoracic and abdominal aorta in anterior view.

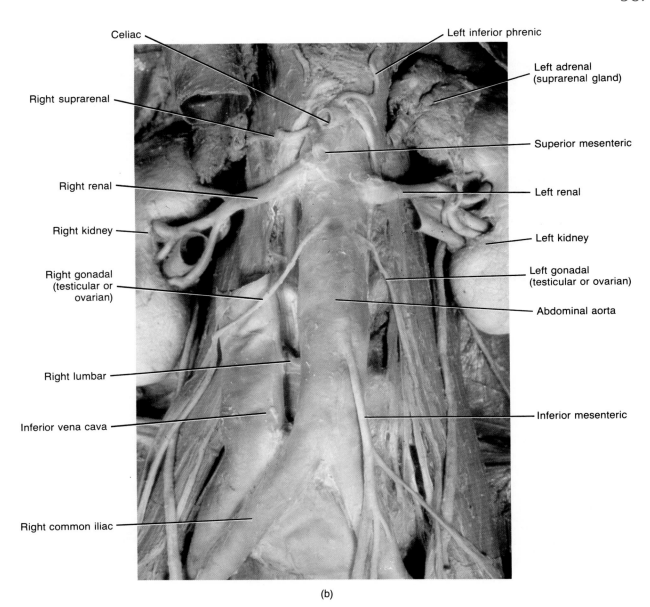

Celiac

Left inferior phrenic

Right suprarenal

Left adrenal (suprarenal gland)

Right renal

Superior mesenteric

Right kidney

Left renal

Right gonadal (testicular or ovarian)

Left kidney

Right lumbar

Left gonadal (testicular or ovarian)

Inferior vena cava

Abdominal aorta

Right common iliac

Inferior mesenteric

(b)

FIGURE 14-10 (*Continued*) (b) Photograph of abdominal aorta. (Courtesy of J. A. Gosling, P. F. Harris, et al., *Atlas of Human Anatomy,* Gower Medical Publishing Ltd., 1985.)

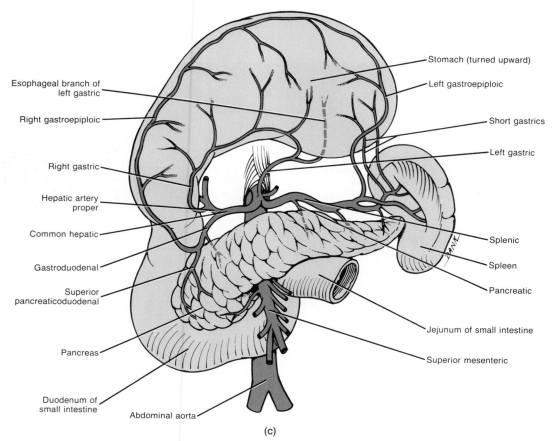

Stomach (turned upward)

Esophageal branch of left gastric

Left gastroepiploic

Right gastroepiploic

Short gastrics

Left gastric

Right gastric

Hepatic artery proper

Common hepatic

Gastroduodenal

Splenic

Spleen

Pancreatic

Superior pancreaticoduodenal

Pancreas

Jejunum of small intestine

Superior mesenteric

Duodenum of small intestine

Abdominal aorta

(c)

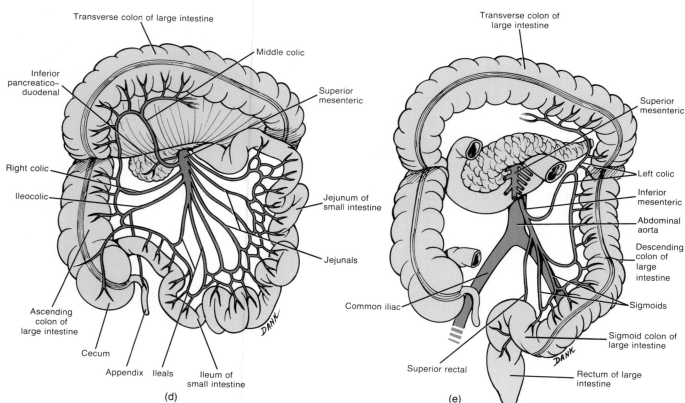

Transverse colon of large intestine

Middle colic

Inferior pancreaticoduodenal

Superior mesenteric

Right colic

Ileocolic

Jejunum of small intestine

Jejunals

Ascending colon of large intestine

Cecum

Appendix

Ileals

Ileum of small intestine

(d)

Transverse colon of large intestine

Superior mesenteric

Left colic

Inferior mesenteric

Abdominal aorta

Descending colon of large intestine

Sigmoids

Common iliac

Sigmoid colon of large intestine

Superior rectal

Rectum of large intestine

(e)

FIGURE 14-10 (*Continued*) (c) Diagram of principal branches of common hepatic, left gastric, and splenic arteries in anterior view. (d) Diagram of principal branches of superior mesenteric artery in anterior view. (e) Diagram of principal branches of inferior mesenteric artery in anterior view.

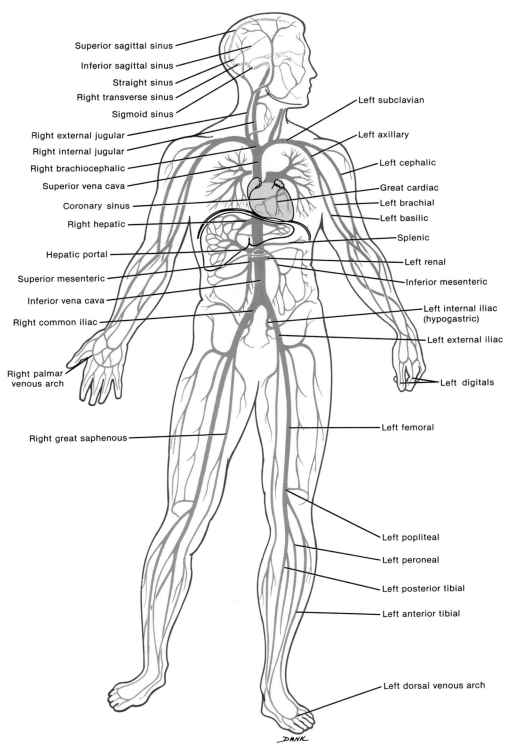

Superior sagittal sinus
Inferior sagittal sinus
Straight sinus
Right transverse sinus
Sigmoid sinus
Right external jugular
Right internal jugular
Right brachiocephalic
Superior vena cava
Coronary sinus
Right hepatic
Hepatic portal
Superior mesenteric
Inferior vena cava
Right common iliac
Right palmar venous arch
Right great saphenous

Left subclavian
Left axillary
Left cephalic
Great cardiac
Left brachial
Left basilic
Splenic
Left renal
Inferior mesenteric
Left internal iliac (hypogastric)
Left external iliac
Left digitals
Left femoral
Left popliteal
Left peroneal
Left posterior tibial
Left anterior tibial
Left dorsal venous arch

DANK

FIGURE 14-12 Principal veins in anterior view.

EXHIBIT 14-8

Veins of Head and Neck (Figure 14-13)

VEIN	DESCRIPTION AND REGION DRAINED
Internal jugulars	Right and left **internal jugular veins** receive blood from face and neck. They arise as continuation of **sigmoid sinuses** at base of skull. Intracranial vascular sinuses are located between layers of dura mater and receive blood from brain. Other sinuses that drain into internal jugular include **superior sagittal sinus, inferior sagittal sinus, straight sinus,** and **transverse (lateral) sinuses** (see also Figure 18-13d). Internal jugulars descend on either side of neck and pass behind clavicles, where they join with right and left subclavian veins. Unions of internal jugulars and subclavians form right and left brachiocephalic veins. From here blood flows into superior vena cava.
External jugulars	Right and left **external jugular veins** run down neck along outside of internal jugulars. They drain blood from parotid (salivary) glands, facial muscles, scalp, and other superficial structures into subclavian veins.

CLINICAL APPLICATION

In cases of heart failure, the venous pressure in the right atrium may rise. In such instances, the pressure in the column of blood in the external jugular vein rises so that, even with the patient at rest and sitting in a chair, the external jugular vein will be visibly distended. Temporary distention of the vein is often seen in healthy adults when the intrathoracic pressure is raised in singing, coughing, and physical exertion.

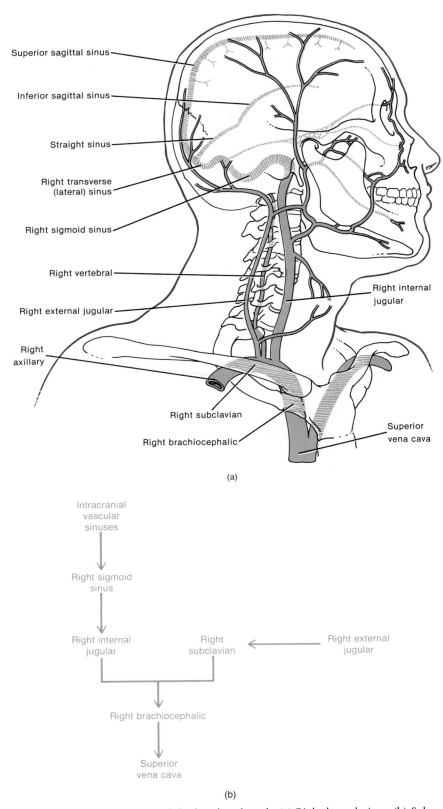

FIGURE 14-13 Principal veins of the head and neck. (a) Right lateral view. (b) Scheme of drainage.

EXHIBIT 14-9

Veins of Upper Extremities (Figure 14-14)

VEIN	DESCRIPTION AND REGION DRAINED
	Blood from each upper extremity is returned to the heart by superficial and deep veins. Both sets of veins contain valves. **Superficial veins** are located just below the skin and are often visible. They anastomose extensively with each other and deep veins. **Deep veins** are located deep in the body. They usually accompany arteries, and many have the same names as corresponding arteries and are thus commonly called venae comitantes.
SUPERFICIAL **Cephalics**	**Cephalic vein** of each upper extremity begins in the medial part of **dorsal venous arch** and winds upward around radial border of forearm. In front of elbow, it is connected to basilic vein by the **median cubital vein.** Just below elbow, cephalic vein unites with **accessory cephalic vein** to form cephalic vein of upper extremity. Ultimately, cephalic vein empties into axillary vein.
Basilics	**Basilic vein** of each upper extremity originates in the ulnar part of **dorsal venous arch.** It extends along posterior surface of ulna to point below elbow where it receives **median cubital vein.** If a vein must be punctured for an injection, transfusion, or removal of a blood sample, median cubitals are preferred. The median cubital vein joins the basilic vein to form the axillary vein.
Median antebrachials	**Median antebrachial veins** drain **palmar venous arch,** ascend on ulnar side of anterior forearm, and end in median cubital veins.
DEEP **Radials**	**Radial veins** receive **dorsal metacarpal veins.**
Ulnars	**Ulnar veins** receive tributaries from **palmar venous arch.** Radial and ulnar veins unite in bend of elbow to form brachial veins.
Brachials	Located on either side of brachial artery, **brachial veins** join into axillary veins.
Axillaries	**Axillary veins** are a continuation of brachials and basilics. Axillaries end at first rib, where they become subclavians.
Subclavians	Right and left **subclavian veins** unite with internal jugulars to form brachiocephalic veins. Thoracic duct of lymphatic system delivers lymph into left subclavian vein at junction with internal jugular. Right lymphatic duct delivers lymph into right subclavian vein at corresponding junction.

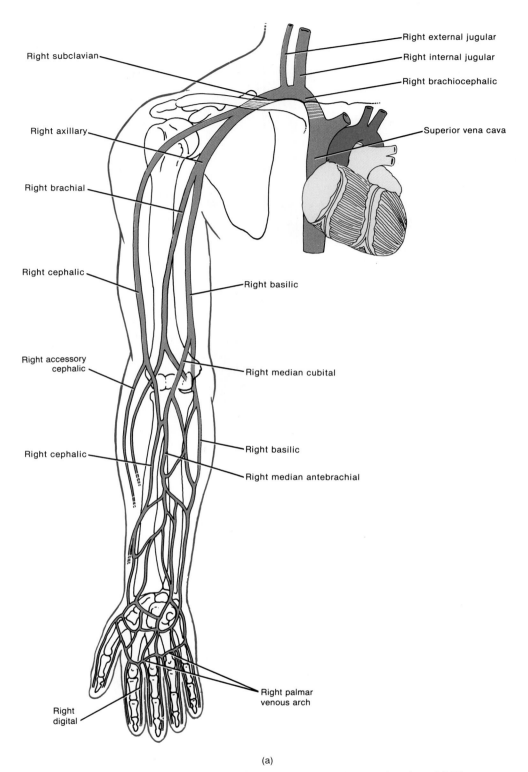

Right subclavian

Right axillary

Right brachial

Right cephalic

Right accessory cephalic

Right cephalic

Right external jugular

Right internal jugular

Right brachiocephalic

Superior vena cava

Right basilic

Right median cubital

Right basilic

Right median antebrachial

Right palmar venous arch

Right digital

(a)

FIGURE 14-14 Principal veins of the right upper extremity in anterior view. (a) Diagram.

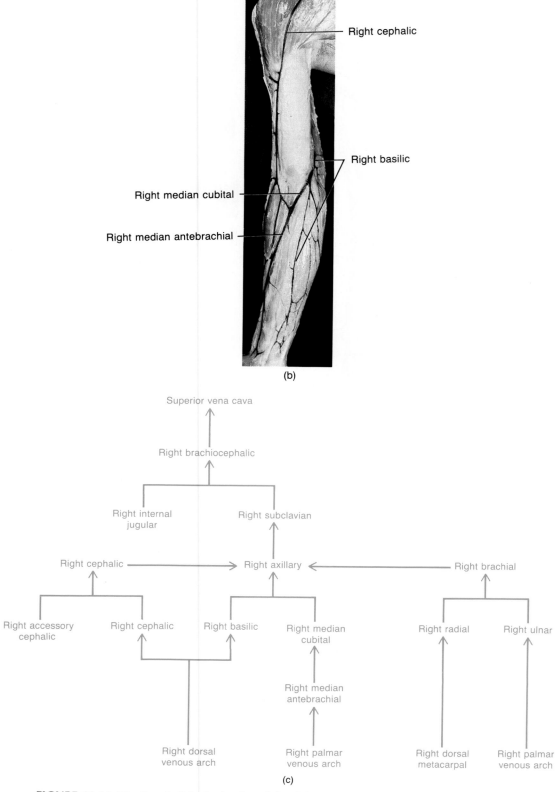

(b)

(c)

FIGURE 14-14 (*Continued*) Principal veins of the right upper extremity in anterior view. (b) Photograph. (Courtesy of C. Yokochi and J. W. Rohen, *Photographic Anatomy of the Human Body*, 2nd ed., 1979, IGAKU-SHOIN, Ltd., Tokyo, New York.) (c) Scheme of drainage.

EXHIBIT 14-10

Veins of Thorax (Figure 14-15)

VEIN	DESCRIPTION AND REGION DRAINED
Brachiocephalic	Right and left **brachiocephalic veins,** formed by union of subclavians and internal jugulars, drain blood from head, neck, upper extremities, mammary glands, and upper thorax. Brachiocephalics unite to form superior vena cava.
Azygos veins	**Azygos veins,** besides collecting blood from thorax, may serve as bypass for inferior vena cava that drains blood from lower body. Several small veins directly link azygos veins with inferior vena cava. Large veins that drain lower extremities and abdomen dump blood into azygos. If inferior vena cava or hepatic portal vein becomes obstructed, azygos veins can return blood from lower body to superior vena cava.
Azygos	**Azygos vein** lies in front of vertebral column, slightly right of midline. It begins as continuation of right ascending lumbar vein. It connects with inferior vena cava, right common iliac, and lumbar veins. Azygos receives blood from **right intercostal veins** that drain chest muscles; from hemiazygos and accessory hemiazygos veins; from several **esophageal, mediastinal,** and **pericardial veins;** and from right **bronchial vein.** Vein ascends to fourth thoracic vertebra, arches over right lung, and empties into superior vena cava.
Hemiazygos	**Hemiazygos vein** is in front of vertebral column and slightly left of midline. It begins as continuation of left ascending lumbar vein. It receives blood from lower four or five **intercostal veins** and some **esophageal** and **mediastinal veins.** At level of ninth thoracic vertebra, it joins azygos vein.
Accessory hemiazygos	**Accessory hemiazygos vein** is also in front and to left of vertebral column. It receives blood from three or four **intercostal veins** and left **bronchial vein.** It joins azygos at level of eighth thoracic vertebra.

EXHIBIT 14-11

Veins of Abdomen and Pelvis (Figure 14-15)

VEIN	DESCRIPTION AND REGION DRAINED
Inferior vena cava	**Inferior vena cava** is the largest vein of the body. It is formed by union of two common iliac veins that drain lower extremities and abdomen. Inferior vena cava extends upward through abdomen and thorax to right atrium. Numerous small veins enter the inferior vena cava. Most carry return flow from branches of abdominal aorta, and names correspond to names of arteries.
Common iliacs	**Common iliac veins** are formed by union of internal (hypogastric) and external iliac veins and represent distal continuation of inferior vena cava at its bifurcation.
Internal iliacs	Tributaries of **internal iliac (hypogastric) veins** basically correspond to branches of internal iliac arteries. Internal iliacs drain gluteal muscles, medial side of thigh, urinary bladder, rectum, prostate gland, ductus (vas) deferens, uterus, and vagina.
External iliacs	**External iliac veins** are continuation of femoral veins and receive blood from lower extremities and inferior part of anterior abdominal wall.
Renals	**Renal veins** drain kidneys.
Gonadals (testiculars or ovarians)	**Testicular veins** drain testes (left testicular vein empties into left renal vein); **ovarian veins** drain ovaries (left ovarian vein empties into left renal vein).
Suprarenals	**Suprarenal veins** drain adrenal (suprarenal) glands (left suprarenal vein empties into left renal vein).
Inferior phrenics	**Inferior phrenic veins** drain diaphragm (left inferior phrenic vein sends tributary to left renal vein).
Hepatics	**Hepatic veins** drain liver.
Lumbars	A series of parallel **lumbar veins** drain blood from both sides of posterior abdominal wall. Lumbars connect at right angles with right and left **ascending lumbar veins,** which form origin of corresponding azygos or hemiazygos vein. Lumbars drain blood into ascending lumbars and then run to inferior vena cava, where they release remainder of flow.

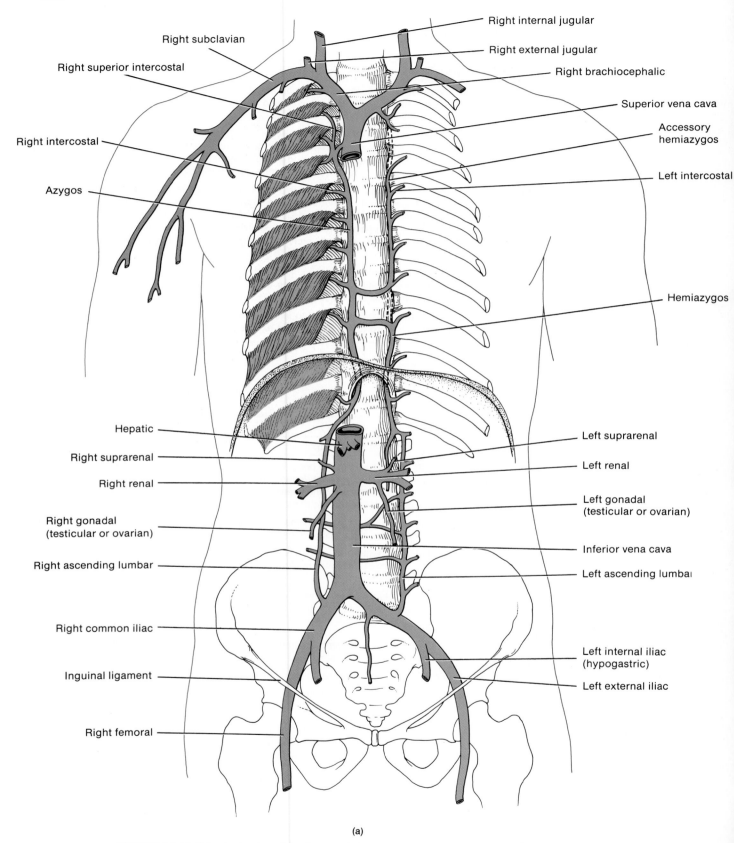

FIGURE 14-15 Principal veins of the thorax, abdomen, and pelvis in anterior view. (a) Diagram.

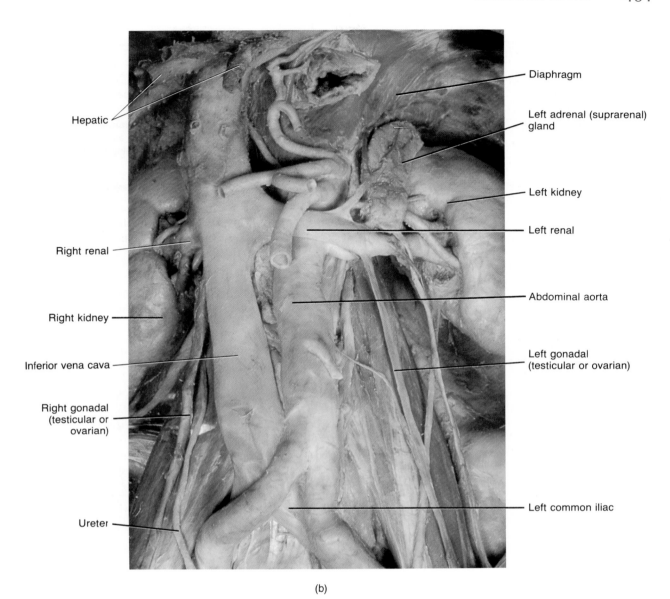

Hepatic

Right renal

Right kidney

Inferior vena cava

Right gonadal
(testicular or
ovarian)

Ureter

Diaphragm

Left adrenal (suprarenal)
gland

Left kidney

Left renal

Abdominal aorta

Left gonadal
(testicular or ovarian)

Left common iliac

(b)

FIGURE 14-15 (*Continued*) (b) Photograph of beginning of inferior vena cava. (Courtesy of J. A. Gosling, P. F. Harris, et al., *Atlas of Human Anatomy,* Gower Medical Publishing Ltd., 1985.)

EXHIBIT 14-12

Veins of Lower Extremities (Figure 14-16)

VEIN	DESCRIPTION AND REGION DRAINED
	Blood from each lower extremity is returned by superficial set and deep set of veins.
SUPERFICIAL VEINS Great saphenous	**Great saphenous vein,** longest vein in body, begins at medial end of **dorsal venous arch** of foot. It passes in front of medial malleolus and then upward along medial aspect of leg and thigh. It receives tributaries from superficial tissues and connects with deep veins as well. It empties into femoral vein in groin.

CLINICAL APPLICATION

The great saphenous vein is very constant in its position anterior to the medial malleolus. It is frequently used for prolonged administration of intravenous fluids, which is particularly important in very young babies and in patients of any age who are in shock and whose veins are collapsed. It and the small saphenous vein are subject to varicosity.

VEIN	DESCRIPTION AND REGION DRAINED
Small saphenous	**Small saphenous vein** begins at lateral end of dorsal venous arch of foot. It passes behind lateral malleolus and ascends under skin of back of leg. It receives blood from foot and posterior portion of leg. It empties into popliteal vein behind knee.
DEEP VEINS Posterior tibial	**Posterior tibial vein** is formed by union of **medial** and **lateral plantar veins** behind medial malleolus. It ascends deep in muscle at back of leg, receives blood from **peroneal vein,** and unites with anterior tibial vein just below knee.
Anterior tibial	**Anterior tibial vein** is upward continuation of **dorsalis pedis veins** in foot. It runs between tibia and fibula and unites with posterior tibial to form popliteal vein.
Popliteal	**Popliteal vein,** just behind knee, receives blood from anterior and posterior tibials and small saphenous vein.
Femoral	**Femoral vein** is upward continuation of popliteal just above knee. Femorals run up posterior of thighs and drain deep structures of thighs. After receiving great saphenous veins in groin, they continue as right and left external iliac veins.

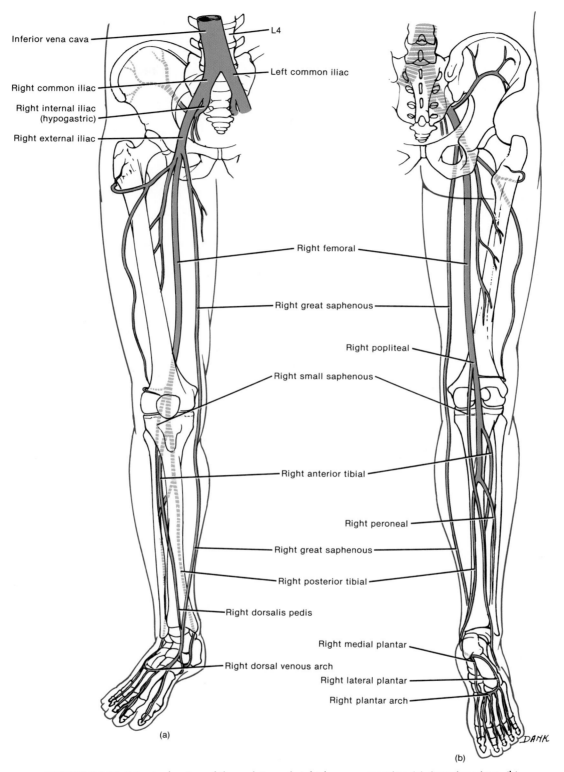

FIGURE 14-16 Principal veins of the pelvis and right lower extremity. (a) Anterior view. (b) Posterior view.

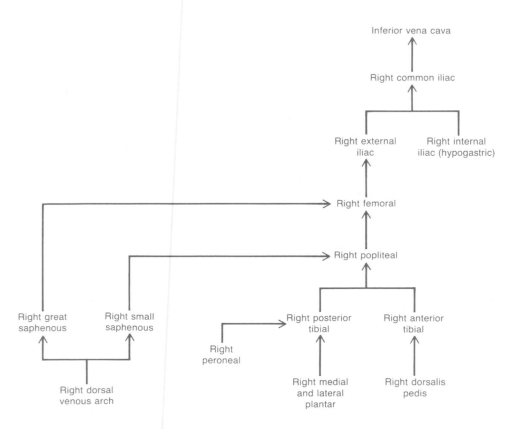

FIGURE 14-16 (*Continued*) Principal veins of the pelvis and right lower extremity. (c) Scheme of drainage.

PULMONARY CIRCULATION

The flow of deoxygenated blood from the right ventricle to the lungs and the return of oxygenated blood from the lungs to the left atrium is called ***pulmonary circulation*** (Figure 14-17). The ***pulmonary trunk*** emerges from the right ventricle and passes upward, backward, and to the left. It then divides into two branches: the ***right pulmonary artery*** runs to the right lung; the ***left pulmonary artery*** goes to the left lung. On entering the lungs, the branches divide and subdivide until ultimately they form capillaries around the alveoli in the lungs. Carbon dioxide is passed from the blood to the alveoli to be breathed out of the lungs. Oxygen breathed in by the lungs is passed from the alveoli into the blood. The capillaries unite, venules and veins are formed, and eventually two ***pulmonary veins*** exit from each lung and transport the oxygenated blood to the left atrium. The pulmonary veins are the only postnatal (after birth) veins that carry oxygenated blood. Contractions of the left ventricle then send the blood into the systemic circulation.

HEPATIC PORTAL CIRCULATION

Blood enters the liver from two sources. The hepatic artery delivers oxygenated blood from the systemic circulation; the hepatic portal vein delivers deoxygenated blood from the gastrointestinal tract. ***Hepatic portal circulation*** involves the flow of venous blood from the gastrointestinal tract to the liver before returning to the heart (Figure 14-18). Hepatic portal blood is rich with substances absorbed from the gastrointestinal tract. The liver monitors these substances before they pass into the general circulation. For example, the liver stores nutrients such as glucose. It also modifies other digested substances so they may be used by cells, detoxifies harmful substances that have been absorbed by the gastrointestinal tract, and destroys bacteria by phagocytosis.

The hepatic portal system includes veins that drain blood from the pancreas, spleen, stomach, intestines, and gallbladder and transport it to the hepatic portal vein of the liver. The ***hepatic portal vein*** is formed by the union of the superior mesenteric and splenic veins. The ***superior mesenteric vein*** drains blood from the small intestine and portions

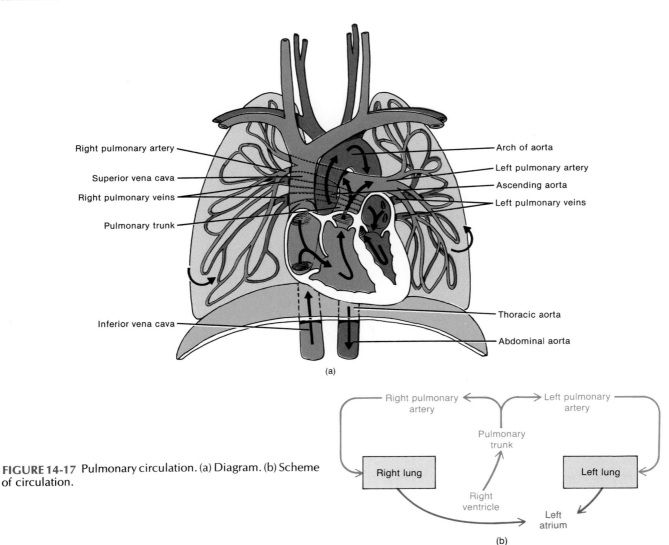

FIGURE 14-17 Pulmonary circulation. (a) Diagram. (b) Scheme of circulation.

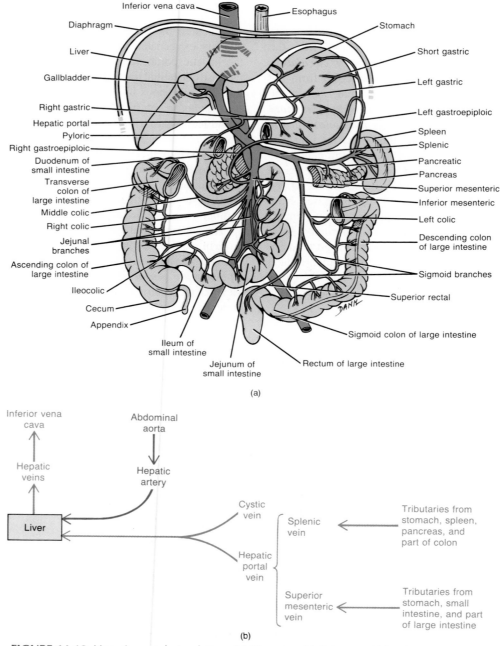

FIGURE 14-18 Hepatic portal circulation. (a) Diagram. (b) Scheme of blood flow through the liver, including arterial circulation. Deoxygenated blood is indicated in blue; oxygenated blood in red.

of the large intestine and stomach. The *splenic vein* drains the spleen and receives tributaries from the stomach, pancreas, and portions of the colon. The tributaries from the stomach are the *gastric, pyloric,* and *gastroepiploic veins.* The *pancreatic veins* come from the pancreas, and the *inferior mesenteric veins* come from the portions of the colon. Before the hepatic portal vein enters the liver, it receives the *cystic vein* from the gallbladder and other veins. Ultimately, deoxygenated blood leaves the liver through the *hepatic veins,* which enter the inferior vena cava.

FETAL CIRCULATION

The circulatory system of a fetus, called *fetal circulation,* differs from an adult's because the lungs, kidneys, and gastrointestinal tract of a fetus are nonfunctional. The fetus derives its oxygen and nutrients from the maternal blood and eliminates its carbon dioxide and wastes into the maternal blood (Figure 14-19).

The exchange of materials between fetal and maternal circulation occurs through a structure called the *placenta* (pla-SEN-ta). It is attached to the umbilicus (navel) of the

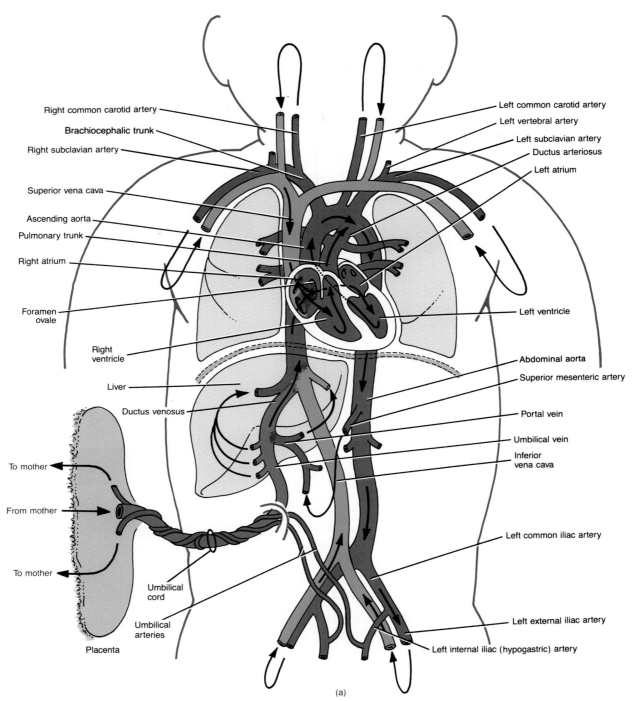

FIGURE 14-19 Fetal circulation. (a) Diagram. The various colors represent different degrees of oxygenation of blood ranging from greatest (red) to intermediate (two shades of purple) to least (blue).

fetus by the umbilical (um-BIL-i-kal) cord, and it communicates with the mother through countless small blood vessels that emerge from the uterine wall. The umbilical cord contains blood vessels that branch into capillaries in the placenta. Wastes from the fetal blood diffuse out of the capillaries, into spaces containing maternal blood (intervillous spaces) in the placenta, and finally into the mother's uterine

blood vessels. Nutrients travel the opposite route—from the maternal blood vessels to the intervillous spaces to the fetal capillaries. Normally, there is no mixing of maternal and fetal blood since all exchanges occur through capillaries.

Blood passes from the fetus to the placenta via two *umbilical arteries*. These branches of the internal iliac (hypogastric) arteries are included in the umbilical cord. At

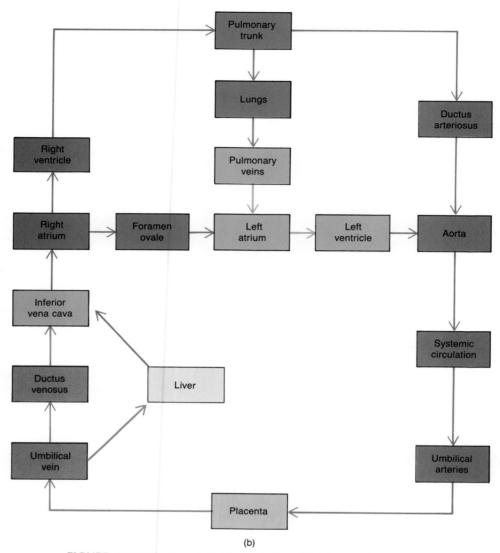

(b)

FIGURE 14-19 (*Continued*) Fetal circulation. (b) Scheme of circulation.

the placenta, the blood picks up oxygen and nutrients and eliminates carbon dioxide and wastes. The oxygenated blood returns from the placenta to the fetus via a single *umbilical vein.* This vein ascends to the liver of the fetus, where it divides into two branches. Some blood flows through the branch that joins the hepatic portal vein and enters the liver. Although the fetal liver manufactures red blood cells, it does not funciton in digestion. Therefore, most of the blood flows into the second branch, the *ductus venosus* (DUK-tus ve-NŌ-sus). The ductus venosus eventually passes its blood to the inferior vena cava, bypassing the liver.

In general, circulation through other portions of the fetus is not unlike postnatal circulation. Deoxygenated blood returning from the lower regions is mingled with oxygenated blood from the ductus venosus in the inferior vena cava. This mixed blood then enters the right atrium. The circula-

tion of blood through the upper portion of the fetus is also similar to postnatal flow. Deoxygenated blood returning from the upper regions of the fetus is collected by the superior vena cava, and it also passes into the right atrium.

Most of the blood does not pass through the right ventricle to the lungs, as it does in postnatal circulation, since the fetal lungs do not operate. In the fetus, an opening called the *foramen ovale* (fō-RĀ-men ō-VAL-ē) exists in the septum between the right and left atria. A valve in the inferior vena cava directs about one-third of the blood through the foramen ovale so that it may be sent directly into the systemic circulation. The blood that does descend into the right ventricle is pumped into the pulmonary trunk, but little of this blood actually reaches the lungs. Most blood in the pulmonary trunk is sent through the *ductus arteriosus* (ar-tē-rē-Ō-sus). This small vessel connecting the pulmonary trunk with the aorta enables most blood to bypass the fetal lungs.

The blood in the aorta is carried to all parts of the fetus through its systemic branches. When the common iliac arteries branch into the external and internal iliacs, part of the blood flows into the internal iliacs. It then goes to the umbilical arteries and back to the placenta for another exchange of materials. The only vessel that carries fully oxygenated blood is the umbilical vein.

At birth, when pulmonary, renal, digestive, and liver functions are established, the special structures of fetal circulation are no longer needed and the following changes occur.

1. The umbilical arteries atrophy to become the **medial umbilical ligaments.**
2. The umbilical vein becomes the **round ligament** of the liver.
3. The placenta is delivered by the mother as the *"afterbirth."*
4. The ductus venosus becomes the **ligamentum venosum,** a fibrous cord in the liver.
5. The foramen ovale normally closes shortly after birth to become the *fossa ovalis,* a depression in the interatrial septum.
6. The ductus arteriosus closes, atrophies, and becomes the **ligamentum arteriosum.**

Anatomical defects resulting from failure of these changes to occur are described in Chapter 13.

AGING AND THE CARDIOVASCULAR SYSTEM

General changes associated with aging and the cardiovascular system include loss of extensibility of the aorta, reduction in cardiac muscle fiber (cell) size, progressive loss of cardiac muscular strength, a reduced output of blood by the heart, and an increase in blood pressure. There is an increase in the incidence of coronary artery disease (CAD), the major cause of heart disease and death in older Americans. Congestive heart failure (CHF), a set of symptoms associated with impaired pumping performance of the heart, also occurs. Changes in blood vessels such as hardening of the arteries and cholesterol deposits in arteries that serve brain tissue reduce nourishment to the brain and result in the malfunction or death of brain cells.

DEVELOPMENTAL ANATOMY OF BLOOD AND BLOOD VESSELS

Since the human egg and yolk sac have little yolk to nourish the developing embryo, blood and blood vessel formation starts as early as 15–16 days. The development begins in the *mesoderm* of the yolk sac, chorion, and body stalk.

Blood vessels develop from isolated masses and cords of mesenchyme in the mesoderm called **blood islands** (Fig-

ure 14-20). Spaces soon appear in the islands and become the lumens of the blood vessels. Some of the mesenchymal cells immediately around the spaces give rise to the *endothelial lining of the blood vessels.* Mesenchyme around the endothelium forms the *tunics* (intima, media, externa) of the larger blood vessels. Growth and fusion of blood islands form an extensive network of blood vessels throughout the embryo.

Blood plasma and *blood cells* are produced by the endothelial cells and appear in the blood vessels of the yolk sac and allantois quite early. Blood formation in the embryo itself begins at about the second month in the liver and spleen, a little later in bone marrow, and much later in lymph nodes.

APPLICATIONS TO HEALTH

HYPERTENSION

Hypertension, or high blood pressure, is the most common disease affecting the heart and blood vessels. Statistics indicate that hypertension afflicts one out of every five American adults. Although there is some disagreement as to what defines hypertension, a strong consensus has emerged suggesting that a blood pressure of 120/80 is normal and desirable in a healthy adult. A reading of 140/90 is generally regarded as the threshold of hypertension, while higher

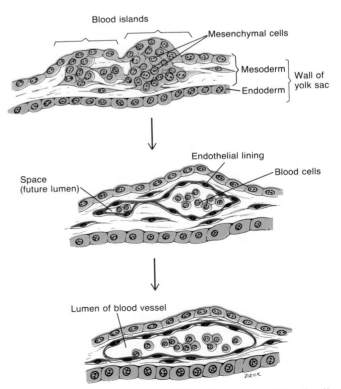

FIGURE 14-20 Development of blood vessels and blood cells from blood islands.

values, especially those over 160/95, are classified as dangerously hypertensive.

Primary hypertension (*essential hypertension*) is a persistently elevated blood pressure that cannot be attributed to any particular organic cause. Approximately 85 percent of all hypertension cases fit this definition. The other 15 percent are *secondary hypertension*. Secondary hypertension has an identifiable underlying cause such as atherosclerosis, kidney disease, and adrenal hypersecretion. Atherosclerosis increases blood pressure by reducing the elasticity of the arterial wall and narrowing the lumen through which the blood can flow. Kidney disease and obstruction of blood flow may cause the kidneys to release renin into the blood. This enzyme catalyzes the formation of angiotensin II from a plasma protein. Angiotensin II is a powerful blood-vessel constrictor—and the most potent agent known for raising blood pressure. It also stimulates aldosterone release. Aldosteronism, the hypersecretion of aldosterone, may also cause an increase in blood pressure. Aldosterone is the adrenal cortex hormone that promotes the retention of salt and water by the kidneys. It thus tends to increase plasma volume. *Pheochromocytoma* (fē-ō-krō-mō-sī-TŌ-ma) is a tumor of the adrenal medulla. It produces and releases into the blood large quantities of norepinephrine and epinephrine. These hormones also raise blood pressure. Epinephrine causes an increase in heart rate and norepinephrine causes vasoconstriction.

High blood pressure is of considerable concern because of the harm it can do to the heart, brain, and kidneys if it remains uncontrolled. The heart is most commonly affected by high blood pressure. When pressure is high, the heart uses more energy in pumping. Because of the increased effort, the heart muscle thickens and the heart becomes enlarged. The heart also needs more oxygen. If it cannot meet the demands put on it, angina pectoris or even myocardial infarction may develop. Hypertension is also a factor in the development of atherosclerosis. Continued high blood pressure may produce a cerebral vascular accident (CVA), or stroke. In this case, severe strain has been imposed on the cerebral arteries that supply the brain. These arteries are usually less protected by surrounding tissues than are the major arteries in other parts of the body. These weakened cerebral arteries may finally rupture, and a brain hemorrhage follows.

The kidneys are also prime targets of hypertension. The principal site of damage is in the arterioles that supply them. The continual high blood pressure pushing against the walls of the arterioles causes them to thicken, thus narrowing the lumen. The blood supply to the kidneys is thereby gradually reduced. In response, the kidneys may secrete renin, which raises the blood pressure even higher and complicates the problem. The reduced blood flow to the kidney cells may eventually lead to the death of the cells.

Medical science cannot cure hypertension. However, almost all cases of hypertension, whether mild or severe, can be controlled in a number of ways. The overweight person with hypertension will usually be placed on a reducing diet, because blood pressure often falls with weight loss. Treatment often involves the restriction of sodium intake. Sodium restriction curbs fluid retention by the body and also tends to reduce blood volume. Evidence suggests that less fat and more potassium and calcium may also lead to a reduction in blood pressure whereas magnesium deficiency may cause hypertension. Since nicotine is a vasoconstrictor, it elevates blood pressure. Stopping smoking may help to decrease blood pressure. Since alcohol raises blood pressure, lowering intake can help reduce it. As indicated earlier, exercise can help to reduce hypertension. In recent years, some physicians have advocated relaxation techniques (yoga, meditation, and biofeedback) to treat hypertension. For those individuals who require medication, a number of drugs are available. Many people can be treated with diuretics, which eliminate large amounts of water and sodium, thus decreasing blood volume and reducing blood pressure. Vasodilators are often used in combination with diuretics. They relax the smooth muscle in arterial walls, causing vasodilation and thus lowering blood pressure. Beta blockers are also used to lower blood pressure, often in combination with diuretics.

ANEURYSM

An *aneurysm* (AN-yoo-rizm) is a thin, weakened section of the wall of an artery or a vein that bulges outward forming a balloonlike sac of the blood vessel. Common causes of aneurysms include atherosclerosis, syphilis, congenital blood vessel defects, and trauma. If an aneurysm goes untreated, it grows larger and larger until the blood vessel wall becomes so thin that it may burst, causing massive hemorrhage with shock, severe pain, stroke, or death, depending on which vessel is involved. Even an unruptured aneurysm can lead to damage by interrupting blood flow or putting pressure on adjacent blood vessels, organs, or bones.

Surgical repair of an aneurysm consists of temporarily clamping the damaged artery above and below the aneurysm and then opening the aneurysm. A graft, usually of Dacron, is then sutured to healthy segments of the artery to reestablish normal blood flow.

CORONARY ARTERY DISEASE (CAD)

In Chapter 13 it was indicated that the common causes of heart disease are related to inadequate coronary blood supply, anatomical disorders, and arrhythmias. *Coronary artery disease* (*CAD*) is a condition in which the heart muscle receives inadequate blood because of an interruption of blood supply. Depending on the degree of interruption, symptoms can range from a mild chest pain to a full-scale heart attack. The underlying causes of CAD are many and varied. Two of the principal ones are atherosclerosis and coronary artery spasm, both of which were discussed in detail in Chapter 13.

DEEP-VENOUS THROMBOSIS (DVT)

Venous thrombosis, the presence of a thrombus in a vein, typically occurs in deep veins of the lower extremities. In such cases, the condition is referred to as *deep-venous thrombosis (DVT)*. The two most serious complications of DVT are pulmonary embolism, in which the thrombus dislodges and finds its way into the pulmonary arterial blood flow, and postphlebitic syndrome, which consists of edema, pain, and skin changes due to destruction of venous valves. Treatment consists of intravenous heparin therapy and elevation of the extremity, fibrinolytic therapy (streptokinase or urokinase), and, in rare cases, thrombectomy.

KEY MEDICAL TERMS ASSOCIATED WITH BLOOD VESSELS

Aortography (ā'-or-TOG-ra-fē) X-ray examination of the aorta and its main branches after injection of a radiopaque dye.

Arteritis (ar'-te-RĪ-tis; *itis* = inflammation of) Inflammation of an artery, probably due to an autoimmune response.

Claudication (klaw'-di-KĀ-shun) Pain and lameness or limping caused by defective circulation of the blood in the vessels of the limbs.

Compensation (kom'-pen-SĀ-shun) A change in the circulatory system made to compensate for some abnormality; an adjustment of size of heart or rate of heartbeat made to counterbalance a defect in structure or function; often used specifically to describe the maintenance of adequate circulation in spite of the presence of heart disease.

Coronary enderterectomy (KOR-o-na-rē end'-ar-ter-EK-tō-mē) The removal of the obstructing area within the lumen of the vessel.

Hypercholesteremia (hī-per-kō-les'-ter-Ē-mē-a; *hyper* = over; *heme* = blood) An excess of cholesterol in the blood.

Hypotension (hī'-pō-TEN-shun; *hypo* = below; *tension* = pressure) Low blood pressure; most commonly used to describe an acute drop in blood pressure, as occurs in circulatory shock.

Normotensive (nor'-mō-TEN-siv) Characterized by normal blood pressure.

Occlusion (o-KLOO-zhun) The closure or obstruction of the lumen of a structure such as a blood vessel.

Percussion (per-KUSH-un) Tapping a part of the body as an aid in diagnosing the condition of parts of the body by the sound obtained.

Phlebitis (fle-BĪ-tis; *phleb* = vein) Inflammation of a vein often in a leg.

Postural hypotension (POS-chur-al hī'-pō-TEN-shun) Lowering of systemic blood pressure with the assumption of an erect or semierect posture; it is usually a sign of a disease. May be caused by cardiovascular or neurogenic factors.

Raynaud's (rā-NOZ) **disease** A vascular disorder, primarily of females, characterized by bilateral attacks of ischemia, usually of the fingers and toes, in which the skin becomes pale and exhibits burning and pain; it is brought on by cold or emotional stimuli. If the condition is secondary to another disorder, it is called **Raynaud's phenomenon.**

Shunt A passage between two blood vessels or between the two sides of the heart.

Syncope (SIN-kō-pē) A temporary cessation of consciousness; a faint. One cause might be insufficient blood supply to the brain.

Thrombectomy (throm-BEK-tō-mē; *thrombo* = clot) An operation to remove a blood clot from a blood vessel.

Thrombophlebitis (throm'-bō-fle-BĪ-tis) Inflammation of a vein with clot formation.

STUDY OUTLINE

Arteries (p. 370)

1. Arteries carry blood away from the heart. Their wall consists of a tunica interna, tunica media (which maintains elasticity and contractility), and tunica externa.
2. Large arteries are referred to as elastic (conducting) arteries and medium-sized arteries are called muscular (distributing) arteries.
3. Many arteries anastomose—the distal ends of two or more vessels unite. An alternate blood route from an anastomosis is called collateral circulation. Arteries that do not anastomose are called end arteries.

Arterioles (p. 371)

1. Arterioles are small arteries that deliver blood to capillaries.
2. Through constriction and dilation they assume a key role in regulating blood flow from arteries into capillaries.

Capillaries (p. 371)

1. Capillaries are microscopic blood vessels through which materials are exchanged between blood and tissue cells; some capillaries are continuous, others are fenestrated.
2. Capillaries branch to form an extensive capillary network throughout the tissue. This network increases the surface area, allowing a rapid exchange of large quantities of materials.
3. Precapillary sphincters regulate blood flow through capillaries.
4. In response to low levels of oxygen, cells produce vasodilator substances that cause dilation of arterioles and relaxation of precapillary sphincters, a phenomenon called autoregulation.
5. Microscopic blood vessels in the liver are called sinusoids.

Venules (p. 373)

1. Venules are small vessels that continue from capillaries and merge to form veins.
2. They drain blood from capillaries into veins.

Veins (p. 373)

1. Veins consist of the same three tunics as arteries but have less elastic tissue and smooth muscle.
2. They contain valves to prevent backflow of blood.
3. Weak valves can lead to varicose veins (VVs) or hemorrhoids.
4. Vascular (venous) sinuses are veins with very thin walls.

Blood Reservoirs (p. 374)

1. Systemic veins are collectively called blood reservoirs.
2. They store blood that, through vasoconstriction, can move to other parts of the body if the need arises.
3. The principal reservoirs are the veins of the abdominal organs (liver and spleen) and skin.

Circulatory Routes (p. 375)

1. The largest circulatory route is the systemic circulation.
2. Two of the many subdivisions of the systemic circulation are coronary (cardiac) circulation and hepatic portal circulation.
3. Other routes include the cerebral, pulmonary, and fetal circulation.

Systemic Circulation (p. 375)

1. The systemic circulation takes oxygenated blood from the left ventricle through the aorta to all parts of the body including lung tissue.
2. The aorta is divided into the ascending aorta, the arch of the aorta, and the descending aorta. Each section gives off arteries that branch to supply the whole body.
3. Blood is returned to the heart through the systemic veins. All the veins of the systemic circulation flow into either the superior or inferior vena cava or the coronary sinus. They in turn empty into the right atrium.

Pulmonary Circulation (p. 405)

1. The pulmonary circulation takes deoxygenated blood from the right ventricle to the lungs and returns oxygenated blood from the lungs to the left atrium.
2. It allows blood to be oxygenated for systemic circulation.

Hepatic Portal Circulation (p. 405)

1. The hepatic portal circulation collects blood from the veins of the pancreas, spleen, stomach, intestines, and gallbladder and directs it into the hepatic portal vein of the liver.
2. This circulation enables the liver to utilize nutrients and detoxify harmful substances in the blood.

Fetal Circulation (p. 406)

1. The fetal circulation involves the exchange of materials between fetus and mother.
2. The fetus derives its oxygen and nutrients and eliminates its carbon dioxide and wastes through the maternal blood supply by means of a structure called the placenta.
3. At birth, when pulmonary, digestive, and liver functions are established, the special structures of fetal circulation are no longer needed.

Aging and the Cardiovascular System (p. 409)

1. General changes include loss of elasticity of blood vessels, reduction in cardiac muscle size, and reduced cardiac output.
2. The incidence of coronary artery disease (CAD), congestive heart failure (CHF), and atherosclerosis increases with age.

Developmental Anatomy of Blood and Blood Vessels (p. 409)

1. Blood vessels develop from isolated masses of mesenchyme in mesoderm called blood islands.
2. Blood is produced by the endothelium of blood vessels.

Applications to Health (p. 409)

1. Hypertension, or high blood pressure, is classified as primary and secondary.
2. An aneurysm is a thin, weakened section of the wall of an artery or vein that bulges outward forming a balloonlike sac.
3. Deep-venous thrombosis (DVT) refers to a blood clot in a deep vein, especially in the lower extremities.

REVIEW QUESTIONS

1. Describe the structural and functional differences among arteries, arterioles, capillaries, venules, and veins.
2. Discuss the importance of the elasticity and contractility of arteries.
3. Distinguish between elastic (conducting) and muscular (distributing) arteries in terms of location, histology, and function.
4. What is an anastomosis? What is collateral circulation?
5. Describe how capillaries are structurally adapted for exchanging materials with body cells. Distinguish between true, continuous, and fenestrated capillaries.
6. What are blood reservoirs? Why are they important?
7. What is meant by a circulatory route? Define systemic circulation.
8. Diagram the major divisions of the aorta, their principal arterial branches, and the regions supplied.
9. Trace a drop of blood from the arch of the aorta through its systemic circulatory route to the tip of the big toe on your left foot and back to the heart again. Remember that the major branches of the arch are the brachiocephalic artery, left common carotid artery, and left subclavian artery. Be sure also to indicate which veins return the blood to the heart.
10. What is the cerebral arterial circle (circle of Willis)? Why is it important?
11. What major organs are supplied by branches of the thoracic aorta? How is blood returned from these organs to the heart?
12. What organs are supplied by the celiac, superior mesenteric, renal, inferior mesenteric, inferior phrenic, and middle sacral arteries? How is blood returned to the heart?
13. Trace a drop of blood from the brachiocephalic artery into the digits of the right upper extremity and back again to the right atrium.
14. What are the three major groups of systemic veins?
15. Define pulmonary circulation. Prepare a diagram to indicate the route. What is the purpose of the route?
16. What is hepatic portal circulation? Describe the route by means of a diagram. Why is this route significant?
17. Discuss in detail the anatomy and physiology of fetal circulation. Be sure to indicate the function of the umbilical arteries, umbilical vein, ductus venosus, foramen ovale, and ductus arteriosus.
18. Describe the effects of aging on the cardiovascular system.
19. Describe the development of blood vessels and blood.

20. Compare the causes of primary and secondary hypertension. How does hypertension affect the body? How is hypertension treated?

21. What is an aneurysm? Why is an aneurysm a serious problem?

22. What is deep-venous thrombosis (DVT)?

23. Refer to the glossary of key medical terms associated with blood vessels. Be sure that you can define each term.

SELF-QUIZ

Complete the following.

1. For each of the following pairs of fetal vessels, circle the vessel with higher oxygen content. A. umbilical artery/umbilical vein; B. ductus arteriosus/ductus venosus; C. femoral artery/femoral vein; D. pulmonary artery/pulmonary vein; E. aorta/thoracic portion of inferior vena cava.

2. Virtually all blood from gastrointestinal organs, as well as blood from the spleen, empties into veins that lead to the single _____ vein. This vessel enters the undersurface of the liver.

3. The aorta ends at about the level of the _____ vertebra by dividing into right and left _____.

4. Besides supplying the arms, the subclavian arteries each send a branch that ascends the neck through foramina in cervical vertebrae. These are the _____ arteries. They join at the base of the brain to form the _____ artery.

5. The major artery from which all systemic arteries branch is the _____. It exits from the chamber of the heart known as the (right? left?) ventricle. The first arteries to branch off this artery are the _____ arteries.

6. Draw arrows next to each factor listed below to indicate whether it increases (↑) or decreases (↓) with normal aging.
  ___ A. size and strength of cardiac muscle fibers (cells)
  ___ B. cardiac output (CO)
  ___ C. deposits of cholesterol in blood vessels supplying brain
  ___ D. blood pressure

7. Name the three main vessels that empty venous blood into the right atrium of the heart:
  A. _____ B. _____ C. _____

8. Blood from all the cranial vascular sinuses eventually drains into the _____ veins, which descend in the neck.

9. In order for blood to flow from the left brachial veins to the left brachial artery, which of these must it pass through? (a brachial capillary? the heart and a lung?)

10. In order for blood to flow from the left brachial vein to the right arm, it must pass through the heart? (Yes? No?) Both sides of the heart, that is, right and left? (Yes? No?) One lung (Yes? No?)

Choose *all* correct answers to questions 11 and 12.

___ **11.** In the most direct route from the left leg to the left arm of an adult, blood must pass through all these structures: A. inferior vena cava; B. brachiocephalic artery; C. capillaries in lung; D. hepatic portal vein; E. left subclavian artery; F. right ventricle of heart; G. left external iliac vein.

___ **12.** In the most direct route from the fetal right ventricle to the fetal left leg, blood must pass through all these structures: A. aorta; B. umbilical artery; C. lung; D. ductus arterio-sus; E. ductus venosus; F. left ventricle; G. left common iliac artery.

Choose the one best answer to these questions.

___ **13.** Choose the false statement:
A. in order for blood to pass from a vein to an artery, it must pass through chambers of the heart; B. in its passage from an artery to a vein a red blood cell must ordinarily travel through a capillary; C. the wall of the femoral artery is thicker than the wall of the femoral vein; D. most of the smooth muscle in arteries is in the tunica interna.

___ **14.** Choose the false statement:
A. arteries contain valves, but veins do not; B. decrease in the size of the lumen of a blood vessel by contraction of a smooth muscle is called vasoconstriction; C. end arteries are vessels that do not anastomose; D. sinusoids are wider and more tortuous (winding) than capillaries.

___ **15.** A small vessel connecting the pulmonary trunk with the aorta and bypassing the fetal lungs is the
A. foramen ovale; B. ductus venosus; C. ductus arteriosus; D. fossa ovalis; E. vasa vasorum.

___ **16.** Which statement best describes arteries?
A. all carry oxygenated blood to the heart; B. all contain valves to prevent the backflow of blood; C. all carry blood away from the heart; D. only large arteries are lined with endothelium; E. all branch from the descending aorta.

___ **17.** Which statement is *not* true of veins?
A. they have less elastic tissue and smooth muscle than arteries; B. they contain more fibrous tissue than arteries; C. most veins in the extremities have valves; D. they always carry deoxygenated blood; E. all empty into the inferior vena cava.

___ **18.** All these vessels are in the leg or foot *except* the
A. saphenous vein; B. azygos vein; C. peroneal artery; D. dorsalis pedis artery; E. popliteal artery.

___ **19.** Which vessel returns blood to the heart from systemic circulation?
A. pulmonary artery; B. pulmonary vein; C. superior vena cava; D. aorta; E. subclavian artery.

___ **20.** In fetal circulation, blood passes from the right atrium to the left atrium through the
A. ductus venosus; B. ductus arteriosus; C. umbilical vein; D. foramen ovale; E. umbilical artery.

___ **21.** Which of the following are involved in pulmonary circulation?
A. superior vena cava, right atrium, and left ventricle; B. inferior vena cava, right atrium, left ventricle; C. right ventricle, pulmonary trunk, and left atrium; D. left ventricle, aorta, and inferior vena cava; E. superior vena cava, right atrium, right ventricle.

—— **22.** In fetal circulation, the blood containing the highest amount of oxygen is found in the
A. umbilical arteries; B. ductus venosus; C. aorta; D. umbilical vein; E. ductus arteriosus.

—— **23.** Which of the following is/are true?
(1) The left common carotid artery branches off the brachiocephalic artery.
(2) The right subclavian artery branches off the brachiocephalic artery.
(3) The right common carotid artery branches off the brachiocephalic artery.
A. (1) only; B. (2) only; C. (3) only; D. (2) and (3); E. none of the above.

—— **24.** Which of the following arteries do(es) *not* arise from the arch of the aorta?
A. brachiocephalic; B. left common carotid; C. right common carotid; D. left subclavian; E. all of the above *do* arise from the arch of the aorta.

—— **25.** Blood is supplied to the pelvic viscera by way of the
A. inferior vena cava; B. anterior tibial artery; C. external iliac artery; D. internal iliac artery; E. femoral artery.

—— **26.** The superior vena cava
(1) is one of three veins carrying blood to the right atrium.
(2) is formed by joining of the right and left brachiocephalic veins.
(3) is joined by the azygos vein in the upper posterior part of the thoracic cavity.

A. (1) only; B. (2) only; C. (3) only; D. all of the above; E. (1) and (2).

—— **27.** The vessel that brings blood from the liver and other abdominal organs to the heart is
A. the portal vein; B. the hepatic vein; C. the abdominal aorta; D. the inferior vena cava; E. none of the above.

—— **28.** From which parts of the body does blood come, draining into the right brachiocephalic vein?
A. head, right arm and shoulder, upper right thoracic region; B. head and neck, both arms and shoulders; C. right side of head and neck; D. right side of the body; E. right side of head and neck, right arm and shoulder, upper right thoracic region.

Arrange the answers in correct sequence.

—— —— —— —— **29.** Route of a drop of blood from the right side of the heart to the left side of the heart:
A. pulmonary artery
B. arterioles
C. capillaries
D. venules and veins

—— —— —— —— —— **30.** Route of a drop of blood from small intestine to heart:
A. superior mesenteric vein
B. hepatic portal vein
C. small vessels within the liver
D. hepatic vein
E. inferior vena cava

15 The Lymphatic System

STUDENT OBJECTIVES

1. Describe the components of the lymphatic system and list their functions.
2. Describe the structure and origin of lymphatics and contrast them with veins.
3. Describe the histological aspects of lymph nodes and explain their functions.
4. Trace the general plan of lymph circulation from lymphatics into the thoracic duct or right lymphatic duct.
5. Describe the principal lymph nodes of the head and neck, extremities, and trunk, their location, and the areas they drain.
6. Describe the development of the lymphatic system.
7. Describe the clinical symptoms of the following disorders: acquired immune deficiency syndrome (AIDS), autoimmune diseases, severe combined immunodeficiency (SCID), hypersensitivity (allergy), and Hodgkin's disease (HD).
8. Define key medical terms associated with the lymphatic system.

CHAPTER OUTLINE

■ **Lymphatic Vessels**
■ **Lymphatic Tissue**
Lymph Nodes
Tonsils
Spleen
Thymus Gland
■ **Lymph Circulation**
Route
 Thoracic (Left Lymphatic) Duct
 Right Lymphatic Duct
Maintenance
■ **Principal Groups of Lymph Nodes**
■ **Developmental Anatomy of the Lymphatic System**
■ **Applications to Health**
Acquired Immune Deficiency Syndrome (AIDS)
 HIV: Structure and Pathogenesis
 Symptoms
 HIV Outside the Body
 Classification
 Transmission
 Drugs and Vaccines Against HIV
 Prevention of Transmission
Autoimmune Diseases
Severe Combined Immunodeficiency (SCID)
Hypersensitivity (Allergy)
Tissue Rejection
Hodgkin's Disease (HD)
■ **Key Medical Terms Associated with the Lymphatic System**

The **lymphatic** (lim-FAT-ik) **system** consists of a pale yellow fluid called lymph, vessels that transport lymph called lymphatics, and a number of structures and organs that contain lymphatic (lymphoid) tissue (see Figure 15-2b). **Lymphatic tissue** is a specialized form of reticular connective tissue that contains large numbers of lymphocytes. The stroma (framework) of lymphatic tissue is a meshwork of reticular fibers and reticular cells (fibroblasts and fixed macrophages). One exception to this is the thymus gland, which has a stroma that is somewhat different, composed of epithelioreticular tissue (discussed later in the chapter).

Lymphatic tissue occurs in the body in various ways. Lymphatic tissue not enclosed by a capsule is referred to as **diffuse lymphatic tissue.** This is the simplest form of lymphatic tissue and is found in the lamina propria (connective tissue) of mucous membranes of the gastrointestinal tract, respiratory passageways, urinary tract, and reproductive tract. It is also normally found in small amounts in the stroma of almost every organ of the body.

Lymphatic nodules also do not have capsules and are oval-shaped concentrations of lymphatic tissue that usually consist of a central, lighter-staining region consisting of large lymphocytes (**germinal center**) and a peripheral, darker-staining region of small lymphocytes (**cortex**). Most lymphatic nodules are solitary, small, and discrete. Such nodules are found randomly in the lamina propria of mucous membranes of the gastrointestinal tract, respiratory passageways, urinary tract, and reproductive tract. Some lymphatic nodules occur in multiple, large aggregations in specific parts of the body. Among these are the tonsils in the pharyngeal region and aggregated lymphatic follicles (Peyer's patches) in the ileum of the small intestine (Chapter 23). Aggregations of lymphatic nodules also occur in the appendix.

Lymphatic organs of the body—the lymph nodes, spleen, and thymus gland—all contain lymphatic tissue enclosed by a connective tissue capsule. Since bone marrow produces lymphocytes, it is also a component of the lymphatic system.

The lymphatic system has several functions. Lymphatics drain from tissue spaces a protein-containing fluid (interstitial fluid) that escapes from blood capillaries. The proteins, which cannot be directly reabsorbed by blood vessels, are returned to the cardiovascular system by lymphatics. Lymphatics also transport fats from the gastrointestinal tract to the blood. Lymphatic tissue also functions in surveillance and defense; that is, lymphocytes, with the aid of macrophages, protect the body from foreign cells, microbes, and cancer cells. Lymphocytes recognize foreign cells and substances, microbes, and cancer cells and respond to them in two general ways. Some lymphocytes (T cells) destroy them directly or indirectly by releasing various substances. Other lymphocytes (B cells) differentiate into plasma cells that secrete antibodies against foreign substances to help eliminate them. Overall, the lymphatic system concentrates foreign substances in certain lymphatic organs, circulates lymphocytes through the organs to make contact with the foreign substances, and destroys the foreign substances and eliminates them from the body.

The developmental anatomy of the lymphatic system is considered later in the chapter.

LYMPHATIC VESSELS

Lymphatic vessels originate as microscopic vessels in spaces between cells called **lymph capillaries** (Figure 15-1a). Lymph capillaries may occur singly or in extensive plexuses. They originate throughout the body, but not in avascular tissue, the central nervous system, splenic pulp, and bone marrow. They are slightly larger and more permeable in one direction only than blood capillaries.

Lymph capillaries also differ from blood capillaries in that they end blindly; blood capillaries have an arterial and a venous end. In addition, lymph capillaries are structurally adapted to ensure the return of proteins to the cardiovascular system when they leak out of blood capillaries. Close examination of lymph capillaries reveals that the endothelial cells making up the capillary wall overlap each other, forming pores (Figure 15-1b). This overlapping arrangement permits fluid to flow easily into the capillary but prevents the flow of fluid out of the capillary, much like a one-way valve would operate. Note also that the outer surfaces of the endothelial cells of the capillary wall are attached to the surrounding tissue by structures called **anchoring filaments.** During edema, there is an excessive accumulation of fluid in the tissue, causing tissue swelling. This swelling produces a pull on the anchoring filaments, opening the pores even more so that more fluid can flow into the lymph capillary.

Just as blood capillaries converge to form venules and veins, lymph capillaries unite to form larger and larger lymph vessels called **lymphatics** (Figure 15-2). Lymphatics resemble veins in structure but have thinner walls and more valves and contain lymph nodes at various intervals along their length. Lymphatics of the skin travel in loose subcutaneous tissue and generally follow veins. Lymphatics of the viscera generally follow arteries, forming plexuses around them. Ultimately, lymphatics deliver lymph into two main channels—the thoracic duct and the right lymphatic duct. These will be described shortly.

CLINICAL APPLICATION

Lymphangiography (lim-fan'-jē-OG-ra-fē) is the x-ray examination of lymphatic vessels and lymph organs after they are filled with a radiopaque substance. Such an x ray is called a **lymphangiogram** (lim-FAN-jē-ō-gram). Lymphagiograms are useful in detecting edema and carcinomas, and in locating lymph nodes for surgical or radiotherapeutic treatment.

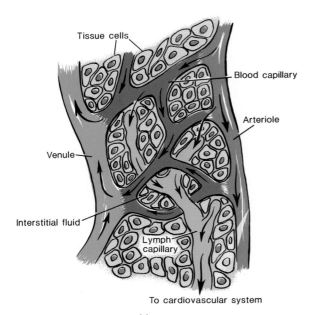

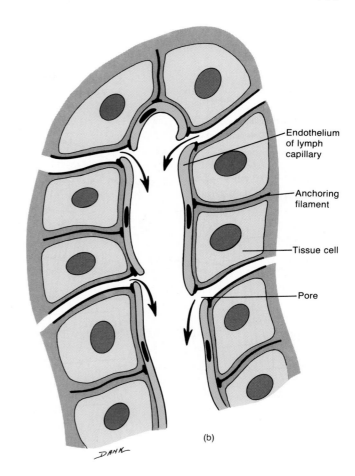

FIGURE 15-1 Lymph capillaries. (a) Relationship of lymph capillaries to tissue cells and blood capillaries. (b) Details of a lymph capillary.

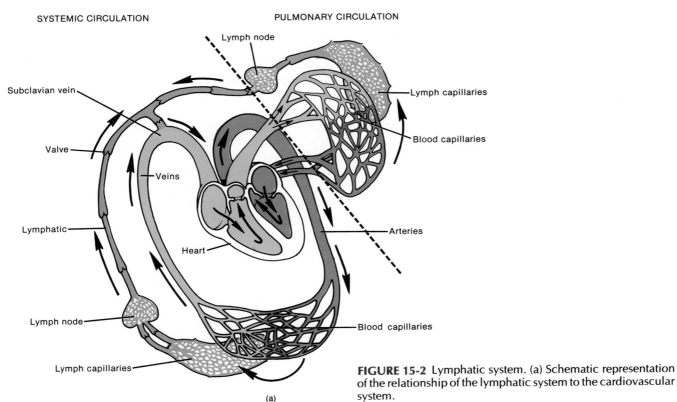

FIGURE 15-2 Lymphatic system. (a) Schematic representation of the relationship of the lymphatic system to the cardiovascular system.

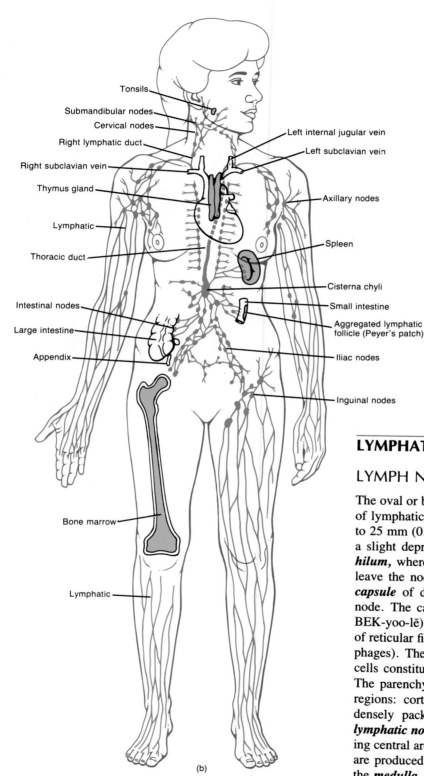

Tonsils
Submandibular nodes
Cervical nodes
Right lymphatic duct
Right subclavian vein
Thymus gland
Lymphatic
Thoracic duct
Intestinal nodes
Large intestine
Appendix
Bone marrow
Lymphatic

Left internal jugular vein
Left subclavian vein
Axillary nodes
Spleen
Cisterna chyli
Small intestine
Aggregated lymphatic follicle (Peyer's patch)
Iliac nodes
Inguinal nodes

(b)

(c)

FIGURE 15-2 (*Continued*) Lymphatic system. (b) Location of the principal components of the lymphatic system. (c) The light gold area indicates those portions of the body drained by the right lymphatic duct. All other areas of the body are drained by the thoracic duct.

LYMPHATIC TISSUE

LYMPH NODES

The oval or bean-shaped structures located along the length of lymphatics are called **lymph nodes**. They range from 1 to 25 mm (0.04 to 1 in.) in length. A lymph node contains a slight depression on one side called a **hilus** (HĪ-lus) or **hilum,** where blood vessels and efferent lymphatic vessels leave the node (Figure 15-3). Each node is covered by a **capsule** of dense connective tissue that extends into the node. The capsular extensions are called **trabeculae** (tra-BEK-yoo-lē). Internal to the capsule is a supporting network of reticular fibers and reticular cells (fibroblasts and macrophages). The capsule, trabeculae, and reticular fibers and cells constitute the stroma (framework) of a lymph node. The parenchyma of a lymph node is specialized into two regions: cortex and medulla. The outer **cortex** contains densely packed lymphocytes arranged in masses called **lymphatic nodules.** The nodules often contain lighter-staining central areas, the **germinal centers,** where lymphocytes are produced. The inner region of a lymph node is called the **medulla.** In the medulla, the lymphocytes are arranged in strands called **medullary cords.** These cords also contain macrophages and plasma cells.

The circulation of lymph through a node involves afferent (to convey toward a center) lymphatic vessels, sinuses in the node, and efferent (to convey away from a center) lymphatic vessels. **Afferent lymphatic vessels** enter the convex surface of the node at several points. They contain valves

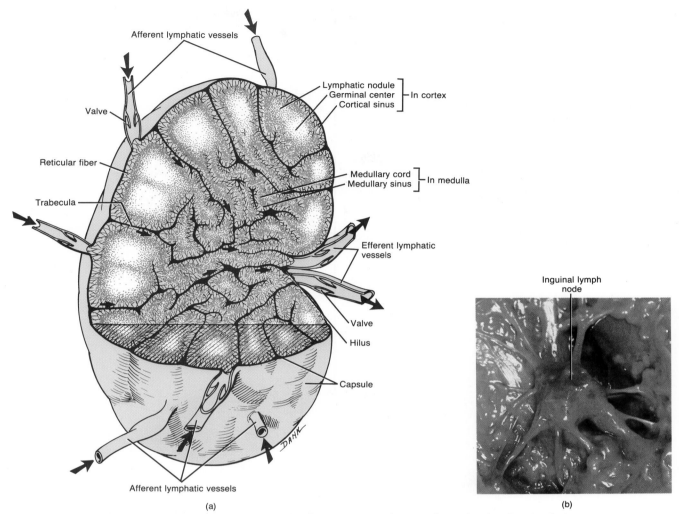

Afferent lymphatic vessels

Valve

Reticular fiber

Trabecula

Lymphatic nodule
Germinal center — In cortex
Cortical sinus

Medullary cord
Medullary sinus — In medulla

Efferent lymphatic vessels

Valve

Hilus

Capsule

Afferent lymphatic vessels

(a)

Inguinal lymph node

(b)

FIGURE 15-3 Structure of a lymph node. (a) Diagram showing the path taken by circulating lymph. (b) Photograph of an inguinal lymph node. (Courtesy of C. Yokochi and J. W. Rohen, *Photographic Anatomy of the Human Body,* 2nd ed., 1979, IGAKU-SHOIN, Ltd., Tokyo, New York.)

that open toward the node so that the lymph is directed *inward*. Once inside the node, the lymph enters the sinuses, which are a series of irregular channels. Lymph from the afferent lymphatic vessels enters the **cortical sinuses** just inside the capsule. From here it circulates to the **medullary sinuses** between the medullary cords. From these sinuses the lymph usually circulates into one or two **efferent lymphatic vessels,** located at the hilus of the lymph node. Efferent lymphatic vessels are wider than the afferent vessels and contain valves that open away from the node to convey lymph *out* of the node.

Lymph nodes are scattered throughout the body, usually in groups (see Figure 15-2b). Typically, these groups are arranged in two sets: **superficial** and **deep**.

Lymph passing from tissue spaces through lymphatics on its way back to the cardiovascular system is filtered through lymph nodes. As lymph passes through the nodes, it is filtered of foreign substances. These substances are trapped by the reticular fibers within the node. Then, macrophages destroy the foreign substances by phagocytosis, T

cells may destroy them by releasing various products, and/or B cells may develop into plasma cells that produce antibodies that destroy them. Lymph nodes also produce lymphocytes, some of which can circulate to other parts of the body.

Histological features of lymphatics and lymph nodes are shown in Figure 15-4.

CLINICAL APPLICATION

Knowledge of the location of the lymph nodes and the direction of lymph flow is important in the diagnosis and prognosis of the spread of cancer by **metastasis**. Cancer cells usually spread by way of the lymphatic system and produce aggregates of tumor cells where they lodge. Such secondary tumor sites are predictable by the direction of lymph flow from the organ primarily involved.

Valve

Wall of
lymphatic

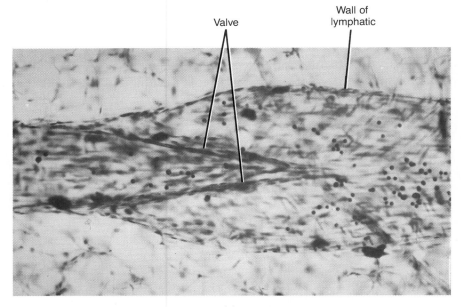

(a)

Blood vessel

Hilus

Cortical
sinus

Medulla

Cortex

Lymph
nodules

Medullary
sinus

Cortical
sinus

Trabecula

Capsule

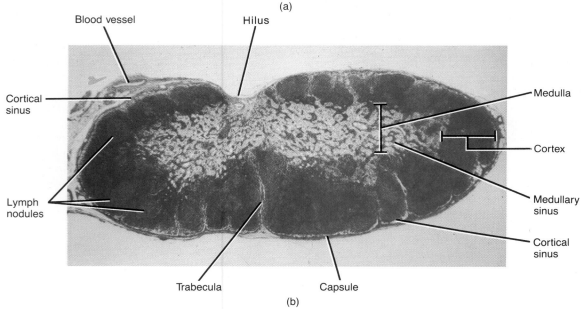

(b)

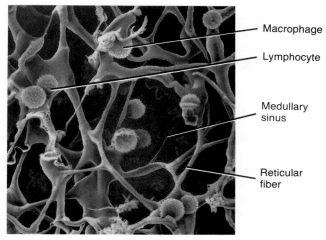

Macrophage

Lymphocyte

Medullary
sinus

Reticular
fiber

(c)

FIGURE 15-4 Histology of lymphatics and lymph nodes. (a) Photomicrograph of a lymphatic at a magnification of 180×. (b) Photomicrograph of a lymph node at a magnification of 25×. (Photomicrographs copyright © 1983 by Michael H. Ross. Used by permission.) Histology of lymphatics and lymph nodes. (c) Illustration based on a scanning electron micrograph of a portion of a medullary sinus of a lymph node at a magnification of 1,000×. (Courtesy of Leroy, Biocosmos/Science Photo Library, Photo Researchers.)

TONSILS

Tonsils are multiple aggregations of large lymphatic nodules embedded in a mucous membrane. The tonsils are arranged in a ring at the junction of the oral cavity and pharnyx. The single *pharyngeal* (fa-RIN-jē-al) *tonsil* or *adenoid* is embedded in the posterior wall of the nasopharynx (see Figure 22-2a). The paired *palatine* (PAL-a-tīn) *tonsils* are situated in the tonsillar fossae between the pharyngopalatine and glossopalatine arches (see Figure 23-4). These are the ones commonly removed by a tonsillectomy. The paired *lingual* (LIN-gwal) *tonsils* are located at the base of the tongue and may also have to be removed by a tonsillectomy (see Figure 20-9c).

The tonsils are situated strategically to protect against invasion of foreign substances. Functionally, the tonsils produce lymphocytes and antibodies.

SPLEEN

The oval *spleen* is the largest mass of lymphatic tissue in the body, measuring about 12 cm (5 in.) in length. It is situated in the left hypochondriac region between the fundus of the stomach and diaphragm (see Figure 1-8d). Its *visceral surface* (Figure 15-5a) contains the contours of the organs adjacent to it—the gastric impression (stomach), renal impression (left kidney), and colic impression (left flexure of colon). The *diaphragmatic* (dī-a-fra-MAT-ik) *surface* is smooth and convex and conforms to the concave surface of the diaphragm to which it is adjacent (Figure 15-5b).

The spleen is surrounded by a capsule of dense connective tissue and scattered smooth muscle fibers (Figure 15-5c).

The capsule, in turn, is covered by a serous membrane, the peritoneum. Like lymph nodes, the spleen contains a hilus, trabeculae, and reticular fibers and cells. The capsule, trabeculae, reticular fibers, and reticular cells constitute the stroma of the spleen.

The parenchyma of the spleen consists of two different kinds of tissue called white pulp and red pulp (Figure 15-6). *White pulp* is essentially lymphatic tissue, mostly lymphocytes, arranged around arteries called central arteries. In various areas, the lymphocytes are thickened into lymphatic nodules referred to as *splenic nodules* (*Malpighian corpuscles*). The *red pulp* consists of *venous sinuses* filled with blood and cords of splenic tissue called *splenic* (*Billroth's*) *cords*. Veins are closely associated with the red pulp. Splenic cords consist of erythrocytes, macrophages, lymphocytes, plasma cells, and granulocytes.

The splenic artery and vein and the efferent lymphatics pass through the hilus. Since the spleen has no afferent lymphatic vessels or lymph sinuses, it does not filter lymph. One key splenic function related to immunity is the production of B cells, which develop into antibody-producing plasma cells. The spleen also phagocytizes bacteria and worn-out and damaged red blood cells and platelets. In addition, the spleen stores and releases blood in case of demand, such as during hemorrhage. Sympathetic impulses cause the smooth muscle of the capsule of the spleen to contract. During early fetal development, the spleen participates in blood cell formation.

About 10 percent of the population has *accessory spleens.* They are most commonly found near the hilus of the primary spleen or embedded in the tail of the pancreas. In general, accessory spleens are about 1 cm (0.5 in.) or less in diameter.

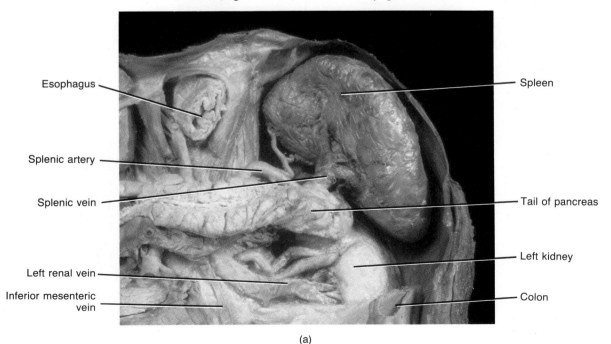

Esophagus

Splenic artery

Splenic vein

Left renal vein

Inferior mesenteric vein

Spleen

Tail of pancreas

Left kidney

Colon

(a)

FIGURE 15-5 Gross structure of the spleen. (a) Photograph of the spleen in relation to abdominal viscera.

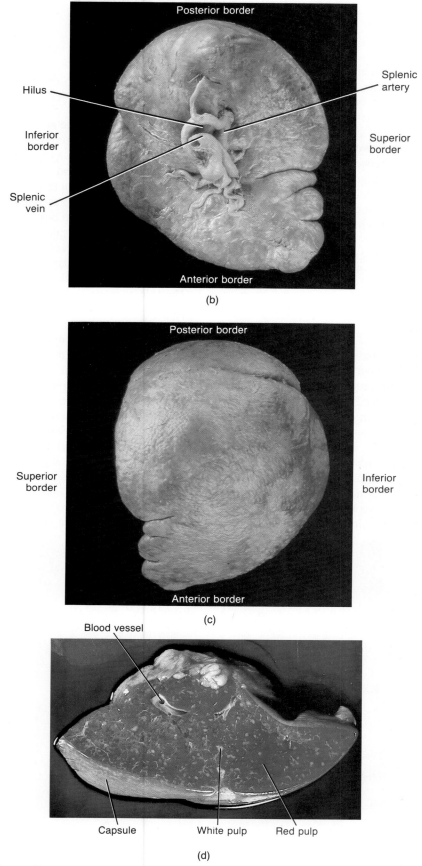

(b)

(c)

(d)

Figure 15-5 (Continued) (b) Photograph of visceral surface. (c) Photograph of diaphragmatic surface. (Photographs courtesy of J. A. Gosling, P. F. Harris, et al., *Atlas of Human Anatomy,* Gower Medical Publishing Ltd., 1985.) (d) Photograph of a section through the spleen showing white and red pulp. (Courtesy of C. Yokochi and J. W. Rohen, *Photographic Anatomy of the Human Body,* 2nd ed., 1979, IGAKU-SHOIN, Ltd., Tokyo, New York.)

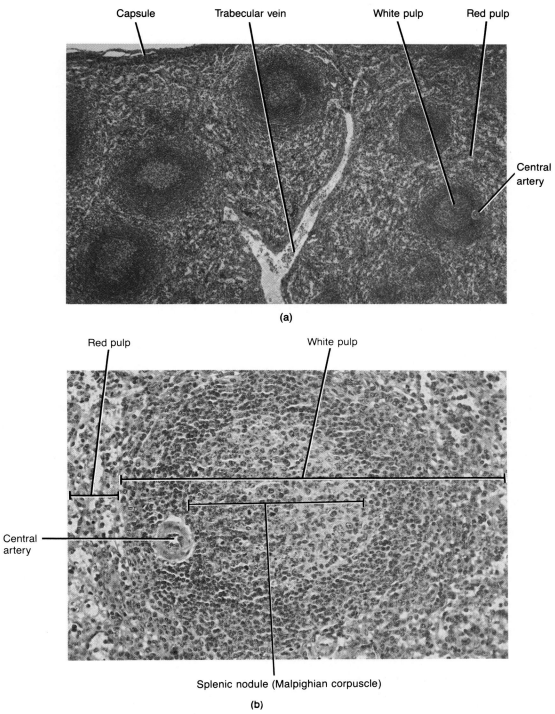

Capsule   Trabecular vein   White pulp   Red pulp

Central artery

(a)

Red pulp   White pulp

Central artery

Splenic nodule (Malpighian corpuscle)

(b)

FIGURE 15-6 Histology of the spleen. (a) Photomicrograph of a portion of the spleen at a magnification of 60×. (b) Photomicrograph of an enlarged aspect of white pulp at a magnification of 150×. (© 1983 by Michael H. Ross. Used by permission.)

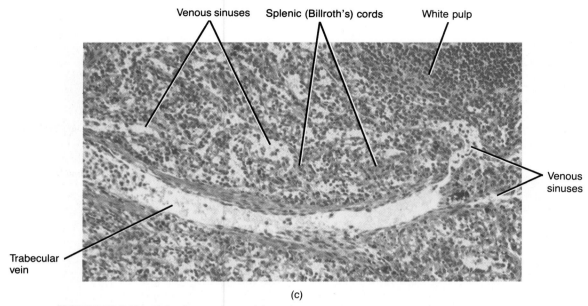

(c)

FIGURE 15-6 (*Continued*) Histology of the spleen. (c) Photomicrograph of an enlarged aspect of red pulp at a magnification of 150×. (Photomicrographs copyright © 1983 by Michael H. Ross. Used by permission.)

CLINICAL APPLICATION

The spleen is the most frequently damaged organ in cases of abdominal trauma, particularly those involving severe blows over the lower left chest or upper abdomen that fracture the protecting ribs. Such a crushing injury may **rupture the spleen,** which causes severe intraperitoneal hemorrhage and shock. Prompt removal of the spleen, called a **splenectomy,** is needed to prevent the patient from bleeding to death. The functions of the spleen are then assumed by other structures, particularly bone marrow.

THYMUS GLAND

Usually, a bilobed lymphatic organ, the **thymus gland** is located in the superior mediastinum, posterior to the sternum and between the lungs (Figure 15-7a). The two **thymic lobes** are held in close proximity by an enveloping layer of connective tissue. Each lobe is enclosed by a connective tissue **capsule.** The capsule gives off extensions into the lobes called **trabeculae,** which divide the lobes into **lobules.** Each lobule consists of a deeply staining peripheral **cortex** and a lighter-staining central **medulla.** The cortex is composed almost entirely of small, medium, and large tightly packed lymphocytes held in place by reticular tissue fibers (Figure 15-7b-c). Since the reticular tissue of the thymus gland differs in origin and structure from that usually found in other lymphatic organs, it is referred to as **epithelioreticular** supporting tissue. The medulla consists mostly of epithelial cells and more widely scattered lymphocytes, and its

reticulum is more cellular than fibrous. In addition, the medulla contains characteristic **thymic (Hassall's) corpuscles,** concentric layers of epithelial cells. Their significance is unknown.

The thymus gland is conspicuous in the infant and it reaches its maximum size of about 40 grams during puberty. After puberty, much of the thymic tissue is replaced by fat and connective tissue. By the time the person reaches maturity, the gland has atrophied but still continues to be functional.

Its role in immunity is to help produce T cells that destroy invading microbes directly or indirectly by producing various substances.

LYMPH CIRCULATION

ROUTE

When plasma is filtered by blood capillaries, it passes into the interstitial spaces; it is then known as interstitial fluid. Fluid movement between blood capillaries and body cells depends on hydrostatic and osmotic pressures. When this fluid passes from interstitial spaces into lymph capillaries, it is called lymph. Lymph from lymph capillaries flows into lymphatics that run toward lymph nodes. At the nodes, afferent vessels penetrate the capsules at numerous points, and the lymph passes through the sinuses of the nodes. Efferent vessels from the nodes either run with afferent vessels into another node of the same group or pass on to another group of nodes. From the most proximal group of each chain of nodes, the efferent vessels unite to form

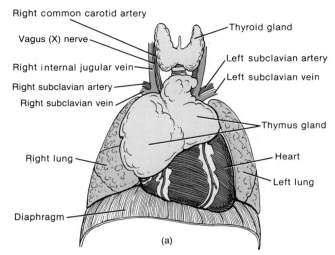

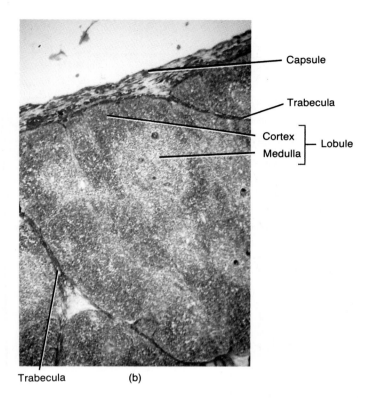

FIGURE 15-7 Thymus gland. (a) Location of the thymus gland in a young child. (b) Photomicrograph of several lobules at a magnification of 40×. (c) Photomicrograph of an enlarged aspect of thymic corpuscles at a magnification of 600×. (Photomicrographs courtesy Andrew Kuntzman.)

lymph trunks. The principal trunks are the *lumbar, intestinal, bronchomediastinal, subclavian,* and *jugular trunks* (Figure 15-8a).

Thoracic (Left Lymphatic) Duct

The principal trunks pass their lymph into two main channels, the thoracic duct and the right lymphatic duct. The *thoracic (left lymphatic) duct* is about 38–45 cm (15–18 in.) in length and begins as a dilation in front of the second lumbar vertebra called the *cisterna chyli* (sis-TER-na KĪ-lē). (Figure 15-8b) The thoracic duct is the main collecting duct of the lymphatic system and receives lymph from the left side of the head, neck, and chest, the left upper extremity, and the entire body below the ribs (see Figure 15-2c).

The cisterna chyli receives lymph from the right and left lumbar trunks and from the intestinal trunk. The lumbar trunks drain lymph from the lower extremities, wall and viscera of the pelvis, kidneys, adrenals (suprarenals) and the deep lymphatics from most of the abdominal wall. The intestinal trunk drains lymph from the stomach, intestines, pancreas, spleen, and visceral surface of the liver. In the neck, the thoracic duct also receives lymph from the left jugular, left subclavian, and left bronchomediastinal trunks. The left jugular trunk drains lymph from the left side of the head and neck; the left subclavian trunk drains lymph from the upper left extremity; and the left bronchomediastinal trunk drains lymph from the left side of the head and neck; the left subclavian trunk drains lymph from the upper left extremity; and the left bronchomediastinal trunk drains lymph from the left side of the deeper parts of the anterior thoracic wall, upper part of the anterior abdominal wall, anterior part of the diaphragm, left lung, and left side of the heart.

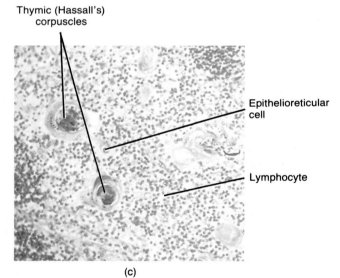

Right Lymphatic Duct

The *right lymphatic duct* is about 1.25 cm (0.5 in.) long and drains lymph from the upper right side of the body (see Figure 15-2c). The right lymphatic duct collects lymph from its trunks as follows (Figure 15-8). It receives lymph from the right jugular trunk, which drains the right side of the head and neck, from the right subclavian trunk, which drains the right upper extremity, and from the right bronchomediastinal trunk, which drains the right side of the thorax, right lung, right side of the heart, and part of the convex surface of the liver.

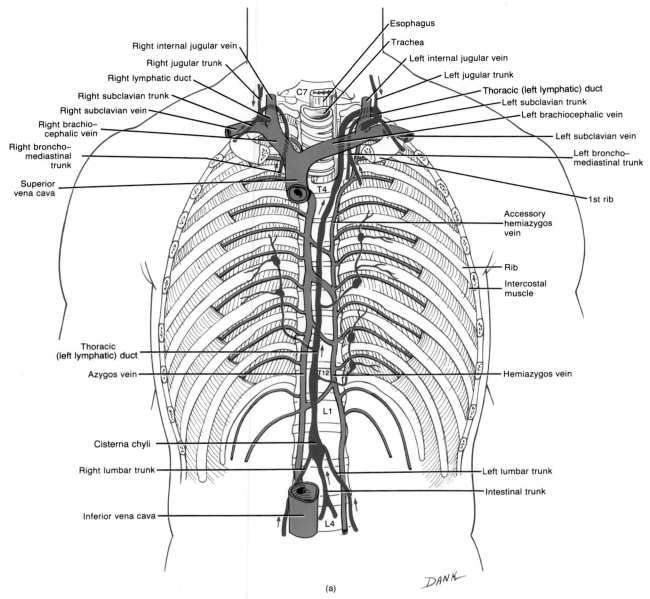

FIGURE 15-8 Scheme of lymphatic circulation. (a) Relation of lymph trunks to the thoracic duct and right lymphatic duct. See also Figure 15-2b.

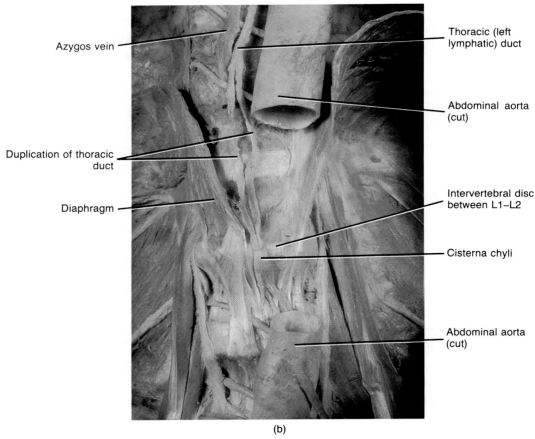

Azygos vein

Thoracic (left lymphatic) duct

Duplication of thoracic duct

Abdominal aorta (cut)

Diaphragm

Intervertebral disc between L1–L2

Cisterna chyli

Abdominal aorta (cut)

(b)

FIGURE 15-8 *(Continued)* (b) Photograph of the origin of the cisterna chyli. Note the duplication of the thoracic duct after it emerges from the cisterna chyli. The single duct is apparent at the top of the photograph. (Courtesy of J. A. Gosling, P. F. Harris, et al., *Atlas of Human Anatomy*, Gower Medical Publishing Ltd., 1985.)

Ultimately, the thoracic duct empties all its lymph into the junction of the left internal jugular vein and left subclavian vein, and the right lymphatic duct empties all its lymph into the junction of the right internal jugular vein and right subclavian vein. Thus, lymph is drained back into the blood and the cycle repeats itself continuously.

MAINTENANCE

The flow of lymph from tissue spaces to the large lymphatic ducts to the subclavian veins is maintained primarily by the milking action of skeletal muscles. Skeletal muscle contractions compress lymph vessels and force lymph toward the subclavian veins. Lymph vessels, like veins, contain valves, and the valves ensure the movement of lymph toward the subclavian veins (see Figure 15-4a).

Another factor that maintains lymph flow is respiratory movements. These movements create a pressure gradient between the two ends of the lymphatic system. Lymph flows from the abdominal region, where the pressure is higher, toward the thoracic region, where it is lower.

CLINICAL APPLICATION

Edema, an excessive accumulation of interstitial fluid in tissue spaces, may be caused by an obstruction, such as an infected node or a blockage of vessels, in the pathway between the lymphatic capillaries and the subclavian veins. Another cause is excessive lymph formation and increased permeability of blood capillary walls. A rise in capillary blood pressure, in which interstitial fluid is formed faster than it is passed into lymphatics, also may result in edema.

PRINCIPAL GROUPS OF LYMPH NODES

Lymph nodes usually appear in groups and typically are arranged in two sets: *superficial* and *deep.*

Exhibits 15-1–15-5 list the principal groups of lymph nodes of the body by region and the general areas of the body they drain.

EXHIBIT 15-1

Principal Lymph Nodes of the Head and Neck (Figure 15-9)

LYMPH NODES	LOCATION AND AREAS DRAINED
LYMPH NODES OF THE HEAD	
Occipital	Near trapezius and semispinalis capitis muscles. They drain the occipital portion of scalp and upper neck.
Retroauricular	Behind ear. They drain skin of ear and posterior parietal region of scalp.
Preauricular	Anterior to tragus. They drain pinna and temporal region of the scalp.
Parotid	Embedded in and below parotid gland. They drain root of nose, eyelids, anterior temporal region, external auditory meatus, tympanic cavity, nasopharynx, and posterior portions of nasal cavity.
Facial	Consist of three groups: infraorbital, buccal, and mandibular.
Infraorbital	Below the orbit. They drain eyelids and conjunctiva.
Buccal	At angle of mouth. They drain the skin and mucous membrane of nose and cheek.
Mandibular	Over mandible. They drain the skin and mucous membrane of nose and cheek.
LYMPH NODES OF THE NECK	
Submandibular	Along inferior border of mandible. They drain chin, lips, nose, nasal cavity, cheeks, gums, lower surface of palate, and anterior portion of tongue.
Submental	Between digastric muscles. They drain chin, lower lip, cheeks, tip of tongue, and floor of mouth.
Superficial cervical	Along external jugular vein. They drain lower part of ear and parotid region.
Deep cervical	Largest group of nodes in neck, consisting of numerous large nodes forming a chain extending from base of skull to root of neck. They are arbitrarily divided into superior deep cervical nodes and inferior deep cervical nodes.
Superior deep cervical	Under sternocleidomastoid muscle. They drain posterior head and neck, pinna, tongue, larynx, esophagus, thyroid gland, nasopharynx, nasal cavity, palate, and tonsils.
Inferior deep cervical	Near subclavian vein. They drain posterior scalp and neck, superficial pectoral region, and part of arm.

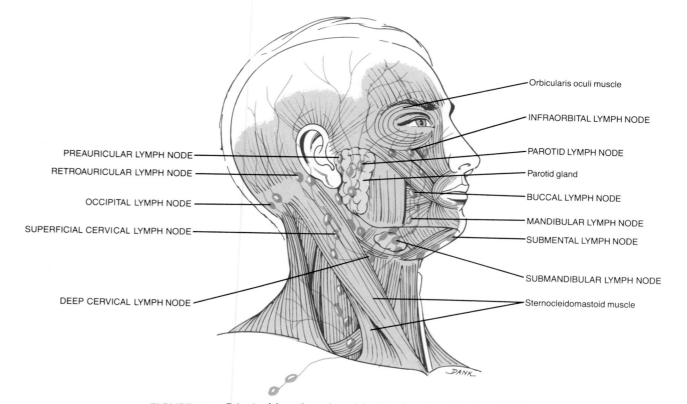

FIGURE 15-9 Principal lymph nodes of the head and neck in lateral view.

EXHIBIT 15-2

Principal Lymph Nodes of the Upper Extremities (Figure 15-10)

LYMPH NODES	LOCATION AND AREAS DRAINED
Supratrochlear	Above medial epicondyle of humerus. They drain medial fingers, palm, and forearm.
Deltopectoral	Below clavicle. They drain lymphatic vessels on radial side of upper extremity.
Axillary	Most deep lymph nodes of the upper extremities are in the axilla and are called the axillary nodes. They are large in size and may be grouped as follows.
Lateral	Medial and posterior aspects of axillary artery. They drain most of whole upper extremity. Since infection or malignancy of upper extremity may cause tenderness and swelling in axilla, the axillary nodes, expecially the lateral group, are clinically important since they filter lymph from much of upper extremity.
Pectoral (anterior)	Along inferior border of the pectoralis minor muscle. They drain skin and muscles of anterior and lateral thoracic walls and central and lateral portions of mammary gland.
Subscapular (posterior)	Along subscapular artery. They drain skin and muscles of posterior part of neck and thoracic wall.
Central (intermediate)	Base of axilla embedded in adipose tissue. They drain lateral, pectoral (anterior), and subscapular (posterior) nodes.
Subclavicular (medial)	Posterior and superior to pectoralis minor muscle. They drain deltopectoral nodes.

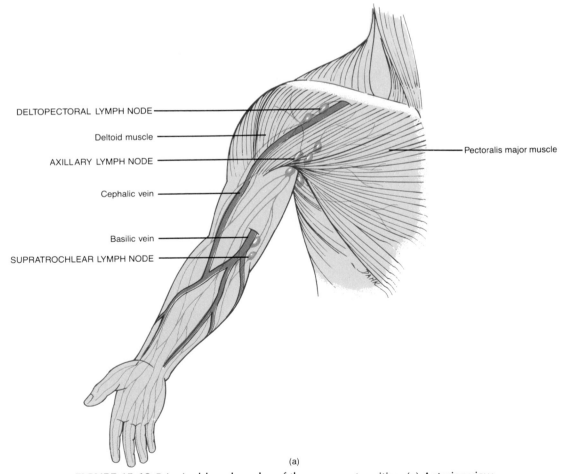

(a)

FIGURE 15-10 Principal lymph nodes of the upper extremities. (a) Anterior view.

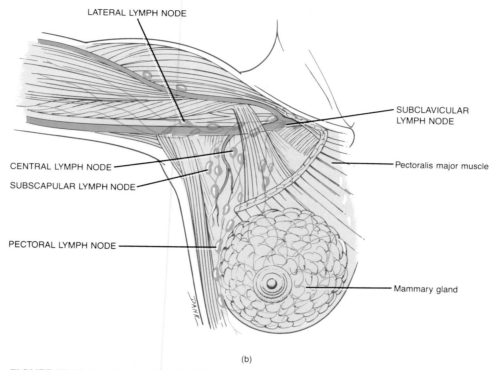

(b)

FIGURE 15-10 (*Continued*) Principal lymph nodes of the upper extremities. (b) Anterior view.

EXHIBIT 15-3

Principal Lymph Nodes of the Lower Extremities (Figure 15-11)

LYMPH NODES	LOCATION AND AREAS DRAINED
Popliteal	In adipose tissue in popliteal fossa. They drain knee and portions of leg and foot, especially heel.
Superficial inguinal	Parallel to saphenous vein. They drain anterior and lateral abdominal wall to level of umbilicus, gluteal region, external genitals, perineal region, and entire superficial lymphatics of lower extremity.
Deep inguinal	Medial to femoral vein. They drain deep lymphatics of lower extremity, penis, and clitoris.

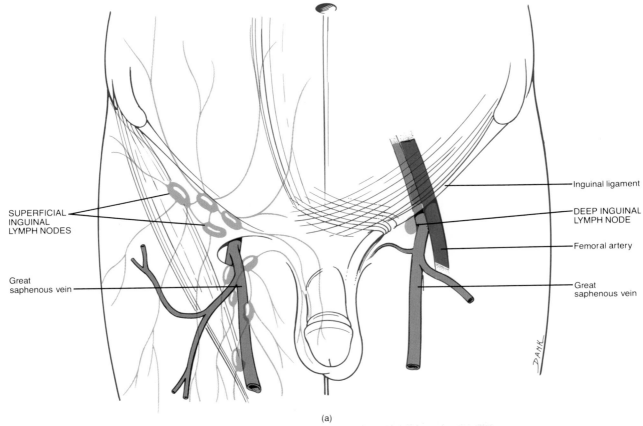

(a)

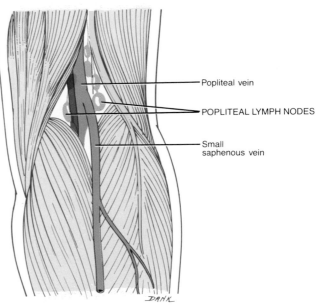

(b)

FIGURE 15-11 Principal lymph nodes of the lower extremities. (a) Anterior view. Principal lymph nodes of the lower extremities. (b) Posterior view.

EXHIBIT 15-4

Principal Lymph Nodes of the Abdomen and Pelvis (Figure 15-12)

LYMPH NODES	LOCATION AND AREAS DRAINED
	Lymph nodes of the abdomen and pelvis are divided into **parietal lymph nodes** that are retroperitoneal (behind the parietal peritoneum) and in close association with larger blood vessels and **visceral lymph nodes** found in association with visceral arteries.
PARIETAL **External iliac**	Arranged about external iliac vessels. They drain the deep lymphatics of abdominal wall below umbilicus, adductor region of thigh, urinary bladder, prostate gland, ductus (vas) deferens, seminal vesicles, prostatic and membranous urethra, uterine (Fallopian) tubes, uterus, and vagina.
Common iliac	Arranged along course of common iliac vessels. They drain pelvic viscera.
Internal iliac	Near internal iliac artery. They drain pelvic viscera, perineum, gluteal region, and posterior surface of thigh.
Sacral	In hollow of sacrum. They drain rectum, prostate gland, and posterior pelvic wall.
Lumbar	From aortic bifurcation to diaphragm; arranged around aorta and designated as **right lateral aortic nodes, left lateral aortic nodes, preaortic nodes,** and **retroaortic nodes.** They drain the efferents from testes, ovaries, uterine (Fallopian) tubes, uterus, kidneys, adrenal (suprarenal) glands, abdominal surface of diaphragm, and lateral abdominal wall.

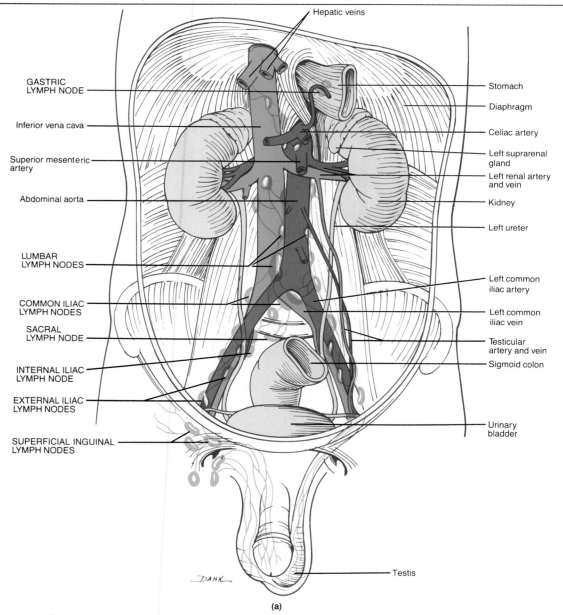

(a)

FIGURE 15-12 Principal lymph nodes of the abdomen and pelvis. (a) Anterior view. See also Figure 15-2b.

LYMPH NODES	LOCATION AND AREAS DRAINED
VISCERAL **Celiac**	Consist of three groups of nodes: gastric, hepatic, and pancreaticosplenic.
Gastric	Lie along lesser and greater curvatures of stomach. They drain lesser curvature of stomach, inferior, anterior, and posterior aspects of stomach, and esophagus.
Hepatic	Along the hepatic artery. They drain stomach, duodenum, liver, gallbladder, and pancreas.
Pancreaticosplenic	Along splenic artery. They drain stomach, spleen, and pancreas.
Superior mesenteric	These nodes are divided into mesenteric, ileocolic, and mesocolic groups.
Mesenteric	Along superior mesenteric artery. They drain jejunum and all parts of ileum, except for terminal portion.
Ileocolic	Along ileocolic artery. They drain terminal portion of ileum, appendix, cecum, and ascending colon.
Mesocolic	Between layers of transverse mesocolon. They drain descending iliac and sigmoid parts of colon.
Inferior mesenteric	Near left colic, sigmoid, and superior rectal arteries. They drain descending, iliac, and sigmoid parts of colon, superior part of rectum, and superior anal canal.

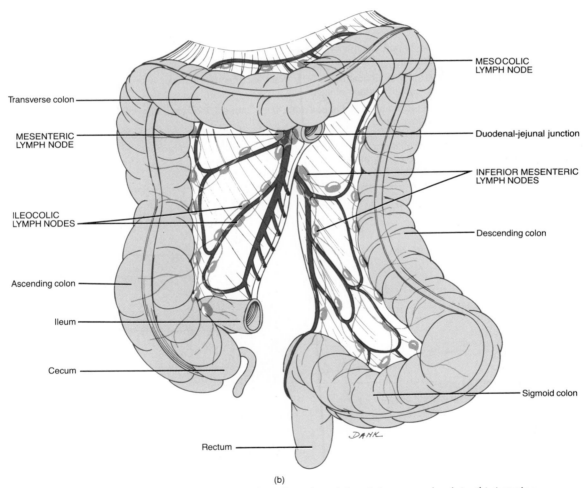

(b)

FIGURE 15-12 (Continued) Principal lymph nodes of the abdomen and pelvis. (b) Anterior view.

433

EXHIBIT 15-5

Principal Lymph Nodes of the Thorax (Figure 15-13)

LYMPH NODES	LOCATION AND AREAS DRAINED
	Lymph nodes of the thorax are divided into **parietal lymph nodes,** which drain the wall of the thorax, and **visceral lymph nodes,** which drain the viscera.
PARIETAL **Sternal (parasternal)**	Alongside internal thoracic artery. They drain central and lateral parts of mammary gland, deeper structures of anterior abdominal wall above umbilicus, diaphragmatic surface of liver, and deeper parts of anterior portion of thoracic wall.
Intercostal	Near heads of ribs at posterior parts of intercostal spaces. They drain posterolateral aspect of thoracic wall.
Phrenic (diaphragmatic)	Located on thoracic aspect of the diaphragm and divisible into three sets called anterior phrenic, middle phrenic, and posterior phrenic.
Anterior phrenic	Behind base of xiphoid process. They drain convex surface of liver, diaphragm, and anterior abdominal wall.
Middle phrenic	Close to phrenic nerves where they pierce diaphragm. They drain middle part of diaphragm and convex surface of liver.
Posterior phrenic	Back of diaphragm near aorta. They drain posterior part of diaphragm.
VISCERAL **Anterior mediastinal**	Anterior part of superior mediastinum anterior to arch of aorta. They drian thymus gland and pericardium.
Posterior mediastinal	Posterior to pericardium. They drain esophagus, posterior aspect of the pericardium, diaphragm, and convex surface of liver.
Tracheobronchial	The tracheobronchial nodes drain lungs, bronchi, thoracic part of trachea, and heart and are divisible into four groups:
Tracheal	On either side of trachea.
Bronchial	At inferior part of trachea and in angle between two bronchi.
Bronchopulmonary	In the hilus of each lung.
Pulmonary	Within lungs on larger bronchial branches.

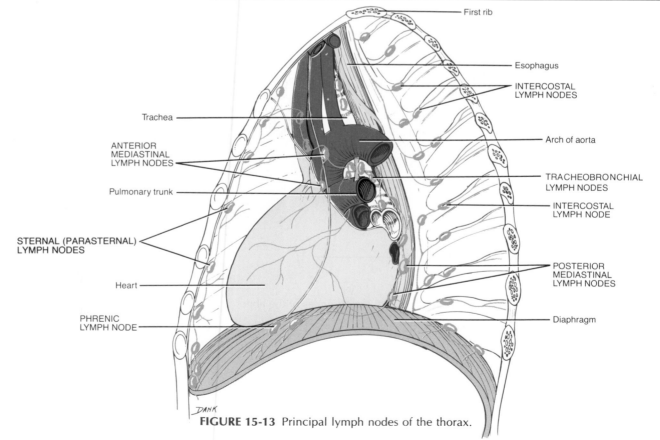

FIGURE 15-13 Principal lymph nodes of the thorax.

DEVELOPMENTAL ANATOMY OF THE LYMPHATIC SYSTEM

The lymphatic system begins its development by the end of the fifth week. *Lymphatic vessels* develop from **lymph sacs** that arise from developing veins. Thus, the lymphatic system is also derived from **mesoderm.**

The first lymph sacs to appear are the paired **jugular lymph sacs** at the junction of the internal jugular and subclavian veins (Figure 15-14). From the jugular lymph sacs, capillary plexuses spread to the *thorax, upper extremities, neck,* and *head.* Some of the plexuses enlarge and form lymphatics in their respective regions. Each jugular lymph sac retains at least one connection with its jugular vein, the left one developing into the superior portion of the thoracic duct (left lymphatic duct). The next lymph sac to appear is the unpaired **retroperitoneal lymph sac** at the root of the mesentery of the intestine. It develops from the primitive vena cava and mesonephric (primitive kidney) veins. Capillary plexuses and lymphatics spread from the retroperitoneal lymph sac to the *abdominal viscera* and *diaphragm.* The sac establishes connections with the cisterna chyli but loses its connections with neighboring veins.

At about the time the retroperitoneal lymph sac is developing, another lymph sac, the **cisterna chyli,** develops below the diaphragm on the posterior abdominal wall. It gives rise to the inferior portion of the *thoracic duct* and the *cisterna chyli* of the thoracic duct. Like the retroperitoneal lymph sac, the cisterna chyli also loses its connections with surrounding veins.

The last of the lymph sacs, the paired **posterior lymph sacs,** develop from the iliac veins at their union with the posterior cardinal veins. The posterior lymph sacs produce capillary plexuses and lymphatics of the *abdominal wall, pelvic region,* and *lower extremity.* The posterior lymph sacs join the cisterna chyli and lose their connections with adjacent veins.

With the exception of the anterior part of the sac from which the cisterna chyli develops, all lymph sacs become invaded by **mesenchymal cells** and are converted into groups of *lymph nodes.*

APPLICATIONS TO HEALTH

ACQUIRED IMMUNE DEFICIENCY SYNDROME (AIDS)

Never before has science been confronted with an epidemic in which the primary disease only lowers the victim's immunity, and then a second unrelated disease produces the symptoms that may result in death. The primary disease is called **acquired immune deficiency syndrome (AIDS).**

AIDS surfaced in 1981 with reports from New York and California that a few homosexual males died of a rare form of cancer called Kaposi's sarcoma (KS) and suffered from problems with their immune systems. As more and

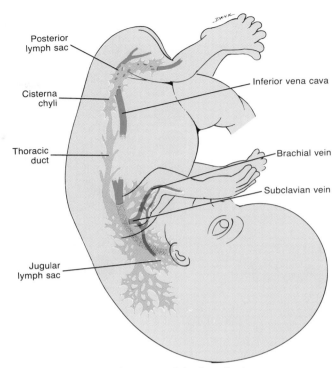

FIGURE 15-14 Development of the lymphatic system.

more cases appeared, the disease was also identified among intravenous drug users, who died of a previously rare type of pneumonia known as *Pneumocystis carinii* pneumonia (PCP) and also experienced immune system problems. A group of scientists in France first isolated the virus that causes the disease in 1983. The virus is called **human immunodeficiency virus (HIV).** At present, there are two distinct classes of AIDS virus: HIV-1 and HIV-2.

Among adult AIDS patients, 93 percent are males. It is estimated that by 1991, there will be over 270,000 cases of AIDS and nearly 179,000 deaths unless current trends are changed. All carriers of the virus, whether they have the disease or not, are assumed to be infected for life and capable of transmitting the virus to others. The risk of developing AIDS increases yearly after infection with the virus; the longer a person is infected, the greater the chances of developing AIDS.

HIV: Structure and Pathogenesis

Viruses consist of a core of DNA or RNA surrounded by a protein coat (capsid). Some viruses also contain an envelope (outer layer) composed of a double layer of lipid penetrated by proteins (Figure 15-15). Outside a living host cell, a virus has no metabolic functions and is unable to replicate. However, once a virus makes contact with a host cell, the viral nucleic acid enters the host cell. Once inside, the viral nucleic acid uses the host cell's enzymes, ribosomes, nutrients, and other resources to make copies of itself and new protein coats and envelopes. As these components accumulate, they are assembled into a large number

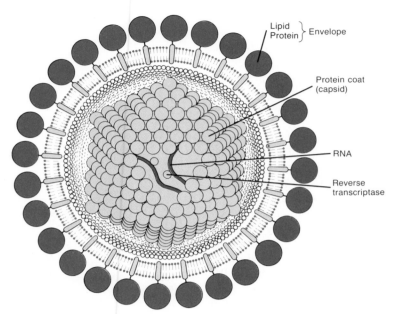

Lipid } Envelope
Protein

Protein coat
(capsid)

RNA

Reverse
transcriptase

FIGURE 15-15 Structure of the human immunodeficiency virus (HIV), the causative agent of AIDS.

of viruses that then leave the host cell to infect other cells. As viruses go through their cycle of replication, they can damage or kill host cells in various ways. These include shutting down the synthesis of host cell proteins, RNA, and DNA; inhibiting cell division; damaging DNA; rupturing lysosomes that can lead to autolysis; and inducing toxic effects (viral proteins). Moreover, the body's own defenses, attacking the infected cells, can kill the cell as well as the viruses it harbors. Not only can the AIDS virus damage and kill host cells, it can also lie dormant for years before causing its effects, attacking more different types of host cells than imagined, and it has more complex means of destroying the immune system than suspected.

Soon after infection with the AIDS virus, the host develops antibodies against several proteins in the virus. The presence of these antibodies in blood is used as a basis for diagnosing AIDS. Normally, antibodies so developed are protective, but in the case of the AIDS virus, this is not necessarily so. In fact, one of the most remarkable aspects of the AIDS virus is its ability to produce a carrier state in a high proportion of infected people, even though antibodies are present.

The reason the AIDS virus lowers the body's immune system is that the virus primarily attacks helper T cells (lymphocytes). As a consequence, several key roles of helper T cells in the immune response are inhibited. For example, helper T cells cooperate with B cells (lymphocytes) to amplify antibody production. The reduction of the number of helper T cells inhibits antibody production by descendants of B cells against the AIDS virus and other microbes. Also, helper T cells secrete interleukin 2 that stimulates proliferation of killer T cells. With the loss of helper T cells, fewer

killer T cells are available to destroy various antigens. In AIDS, the ratio of helper T cells to suppressor T cells, which is normally 2:1, is reversed. Since T4 cells orchestrate a large portion of our immune defense, their destruction leads to collapse of the immune system and susceptibility to opportunistic infections such as Kaposi's sarcoma (KS) and *Pneumocystis carinii* pneumonia (PCP). Kaposi's sarcoma (KS) is a deadly form of skin cancer prevalent in equatorial Africa but previously almost unknown in the United States. It arises from endothelial cells of blood vessels and produces painless, purple or brownish lesions that resemble bruises on the skin or on the inside of the mouth, nose, or rectum. *Pneumocystis carinii* pneumonia (PCP) is a rare form of pneumonia caused by the protozoan *Pneumocystis carinii*. It results in shortness of breath, persistent dry cough, sharp chest pains, and difficulty in breathing. AIDS victims are also subject to a form of herpes that attacks the central nervous system, cryptococcal meningitis, toxoplasmosis, oral and esophageal candidiasis, cryptosporidiosis, a severe form of arthritis called Reiter's syndrome, and psoriasis. The AIDS virus also attacks macrophages; brain cells; endothelial cells that line various organs, body cavities, and blood vessels; and possibly colon cells. Heart and liver cells are also possible target cells. The virus might kill these cells or multiply in them, thus spreading the virus.

As more information becomes available about the AIDS virus, it is becoming clear that it is more complex in structure than originally thought, it is capable of attacking more different types of host cells than imagined, and it has more complex means of destroying the immune system than suspected.

Symptoms

As a result of cell destruction and viral multiplication, several symptoms are associated with AIDS. Symptoms of AIDS may develop for months or years and include a general feeling of discomfort or uneasiness (malaise); low-grade fever or night sweats; coughing; shortness of breath; sore throat; extreme fatigue; muscle aches; unexplained weight loss; enlarged lymph nodes in the neck, armpits, and groin; and blue-violet or brownish spots on the skin, usually the lower extremities.

HIV Outside the Body

Outside the body, the AIDS virus is fragile and can be eliminated in a number of ways. Dishwashing and clothes-washing by exposing the virus to 133°F (56°C) for 10 minutes will kill HIV. Chemicals such as hydrogen peroxide (H_2O_2), rubbing alcohol, Lysol, household bleach, and germicidal skin cleaner (such as Betadine and Hibiclens) are also very effective against the virus. Standard clorination is also sufficient to kill HIV in swimming pools and hot tubs.

Classification

The Centers for Disease Control (CDC) in Atlanta have established the following classification system for AIDS.

Group I. Acute infection. Includes patients with transient signs and symptoms that appear at or shortly after initial infection with HIV. Individuals show a mononucleosis-like syndrome with or without aseptic meningitis.

Group II. Asymptomatic infection. Patients in this group show no signs or symptoms of HIV infection and must have had no previous signs or symptoms that would have led to their classification in groups III or IV.

Group III. Persistent generalized lymphadenopathy. Included here are patients who exhibit palpable lymphadenopathy (lymph node enlargement of 1 cm or greater) at two or more sites for more than 3 months in the absence of a disease other than AIDS.

Group IV. Other disease. The clinical symptoms of patients in this group are arranged into one or more subgroups (A through E).

Subgroup A. Constitutional disease. Patients exhibit one or more of the following: fever persisting for more than 1 month, involuntary weight loss greater than 10 percent, or diarrhea persisting for more than 1 month in the absence of a disease other than AIDS.

Subgroup B. Neurologic disease. Patients exhibit one or more of the following: dementia, myelopathy, or peripheral neuropathy in the absence of a disease other than AIDS.

Subgroup C. Secondary infectious diseases. Patients have an infectious disease associated with AIDS.

Category C-1. Includes patients with diseases such as *Pneumocystis carinii* pneumonia (PCP), toxoplasmosis, histoplasmosis, and herpes simplex virus infection.

Category C-2. Includes patients with diseases such as oral hairy leucoplakia, tuberculosis, or oral candidiasis.

Subgroup D. Secondary cancers. Includes patients with Kaposi's sarcoma (KS), non-Hodgkin's lymphoma, or lymphoma of the brain.

Subgroup E. Other conditions. Patients in this group have diseases not classified above, such as lymphoid interstitial pneumonitis.

Transmission

Although HIV has been isolated from a number of body fluids, such as blood, semen, vaginal fluid, tears, and saliva, it appears that the fluids that provide sufficient virus for transmission seem to be limited to blood, semen, and vaginal secretions. It is assumed that the presence of lymphocytes in these fluids is important, or even essential, for concentrating the viruses. The sites best suited for establishment of infection after exposure appear to be the cardiovascular system, open wounds of the skin, the penis, vagina, and rectum.

HIV is effectively transmitted by sexual contact between males, from males to females, and from females to males through vaginal or anal intercourse with infected persons. HIV is also effectively transmitted through exchanges of blood, such as by contaminated hypodermic needles and needle stick, open wound, or mucous membrane exposure in health-care workers. The virus is also transmitted from infected mothers to their infants before or during birth or through breast-feeding. It does not appear that individuals become infected as a result of routine, nonintimate contacts.

No evidence exists that AIDS can be spread through *normal* kissing (some experts feel that prolonged, vigorous, wet deep kissing could theoretically transmit the virus if it results in breaks or tears in the lining of the mouth or if preexisting sores are present). Also, although mosquitoes can carry the AIDS virus for several days, there is no evidence that the virus can multiply inside mosquitoes or that they are capable of transmitting the disease. It also appears that health-care personnel who take proper routine barrier precautions (gloves, masks, safety glasses) are not at risk.

Drugs and Vaccines Against HIV

Medical scientists are mounting what is probably the greatest concentrated effort ever to find a cure for a single virus disease. Antiviral therapy is aimed at disrupting various points in the viral life cycle. Some research centers on preventing the attachment of the virus to the host cell plasma membrane. Other research is designed to counterattack the virus once it gets inside a host cell and takes over the machinery of the host cell. Still other research is aimed at inhibiting assembly of viruses within the host cell and their subsequent release. Most drugs now used against the AIDS virus are directed against a key viral enzyme called reverse transcriptase. In the cytoplasm of a host cell, reverse transcriptase creates a double strand of DNA using viral RNA as a template. The viral DNA is then incorporated into the host cell's DNA. There it lies dormant until some mechanism turns it on. Once it is turned on, it directs the synthesis of new viruses.

The drug most widely tried against the AIDS virus and given approval by the Food and Drug Administration (FDA) is zidovudine (Retrovir), formerly known as azidothymidine (AZT). The action of this drug is to inhibit the action of reverse transcriptase, thus preventing the virus from making DNA from RNA. Clinical improvements among patients taking zidovudine are weight gain, increased energy, and neurological improvements (reversal of loss of mental function and dementia). Side effects include severe bone marrow damage, anemia, and suppression of the immune system. Moreover, patients must take the medication every 4 hours and are required to receive weekly or bimonthly transfusions. Dideoxycytidine (DDC), a drug similar in action to zidovudine but with fewer toxic effects, is now in early human trials as is phosphonoformate. A drug called ribavirin (Virazole) appears to prevent the synthesis of viral proteins. It is not yet approved for AIDS patients in the United States. Alpha interferon is believed to attach at the final stage of virus production. It is being tried in trials both alone and in combination with other drugs. Other experimental drugs being considered are granulocyte-macrophage colony-stimulating factor (GM-MCS) that boosts the number of white blood cells, AL-721, which are thought to hamper viral attachment to cells, and fusidic acid, whose mechanism of action is unknown.

In addition to drug development and testing, considerable emphasis is being given to development of a vaccine against AIDS. The purpose of a vaccine is to stimulate the production of antibodies against the virus that will kill the virus directly and to bolster immune defenses after the virus has invaded the body. The principal experimental vaccines under study make use of various subunits of the AIDS virus. In a process that is largely trial and error, researchers have selected different elements of the virus (proteins from the envelope or coat) that are believed to be most likely to produce the broadest range of antibodies. Although some scientists propose using the entire killed AIDS virus in a vaccine to stimulate antibody production, there is some concern that some of the viruses might remain alive and thus cause disease.

Once a vaccine is developed, it is tested on laboratory animals to see which antibodies are produced. Then, potentially useful vaccines are used with small numbers of humans to determine safety and the range of antibodies produced. This phase took place late in 1987 in the United States using a vaccine called Vax-Syn HIV-1. If the vaccine still

seems promising, the next step is to involve large numbers of patients to see if it actually protects against AIDS. This process takes many years. Finally, once safety and effectiveness are established, the vaccine will be widely distributed.

Prevention of Transmission

At the present time, there are no drugs or vaccines to prevent AIDS. The only means of prevention is to stop transmission of the virus. Sexual transmission of the AIDS virus can be prevented if infected persons do not have vaginal, oral, or anal intercourse with susceptible persons, or, if during intercourse, effective barrier techniques (condoms and spermicides) are used. Infection from donated blood and blood products can be prevented by testing blood for evidence of the AIDS virus. AIDS transmitted by needles and syringes could be avoided if the use of intravenous drugs is stopped or if unsterilized injection paraphernalia are not used. Mother-to-infant infections could be avoided if infected females would not become pregnant. If these measures are to be effective, they must be part of an overall program involving education, counseling, screening individuals at high risk, tracing contacts, and modifying behavior.

Until there is effective drug therapy or an effective vaccine, preventing the spread of AIDS must rely on education and safer sexual practices.

AUTOIMMUNE DISEASES

Under normal conditions, the body's immune mechanism is able to recognize its own tissues and chemicals. It normally does not produce T cells or B cells against its own substances. Such recognition of self is called *immunologic tolerance.* Although the mechanism of tolerance is not completely understood, it is believed that suppressor T cells may inhibit the differentiation of B cells into antibody-producing plasma cells, or inhibit helper T cells that cooperate with B cells to amplify antibody production.

At times, however, immunologic tolerance breaks down and the body has difficulty in discriminating between its own antigens and foreign antigens. This loss of immunologic tolerance leads to an *autoimmune disease (autoimmunity).* Such diseases are immunologic responses mediated by antibodies against a person's own tissue antigens. Among human autoimmune diseases are rheumatoid arthritis (RA), systemic lupus erythematosus (SLE), thyroiditis, rheumatic fever, glomerulonephritis, encephalomyelitis, hemolytic and pernicious anemias, Addison's disease, Grave's disease, possibly some forms of diabetes, myasthenia gravis, and multiple sclerosis (MS).

SEVERE COMBINED IMMUNODEFICIENCY (SCID)

Severe combined immunodeficiency (SCID) is a rare immunodeficiency disease in which both B cells and T cells are missing or inactive in providing immunity. Perhaps the most famous patient with SCID was David, the "bubble boy," who lived in a sterile plastic chamber for all but 15 days of his life; he died on February 22, 1984, at age 12. He was the oldest untreated survivor of SCID.

David was placed in the sterile chamber shortly after birth to protect him from microbes that his body could not fight. In an effort to correct his disorder, David underwent a bone marrow transplant from his older sister, who was the most closely but still not perfectly matched donor available. Eighty days after the transplant and still in a germ-free environment, David developed some of the symptoms of infectious mononucleosis (IM), a condition caused by the Epstein-Barr virus (EBV). David was brought out of isolation for easier treatment, with the hope that the transplant would provide the same protection as his sterile plastic chamber. Unfortunately, the transplant was not successful, and David died about 4 months after it was performed. It was discovered that what killed David was not the immediate failure of the transplant but cancer. His B cells had proliferated as a result of the EBV. This is a very clear demonstration of a virus causing a cancer in humans.

HYPERSENSITIVITY (ALLERGY)

A person who is overly reactive to an antigen is said to be *hypersensitive* (*allergic*). Whenever an allergic reaction occurs, there is tissue injury. The antigens that induce an allergic reaction are called *allergens*. Almost any substance can be an allergen for some individual. Common allergens include certain foods (milk, peanuts, eggs), antibiotics such as penicillin, cosmetics, chemicals in plants such as poison ivy, pollens, dust, molds, substances released by insects (wasp, hornet, and bee stings), iodine-containing dyes used in certain x rays, and even microbes.

There are four basic types of hypersensitivity reactions: type I (anaphylaxis), type II (cytotoxic), type III (immune complex), and type IV (cell mediated). The first three involve antibodies; the last involves T cells. Here we shall consider only types I and IV.

Type I (*anaphylaxis*) *reactions* occur within a few minutes after a person sensitized to an allergen is reexposed to it. *Anaphylaxis* (an'-a-fi-LAK-sis) literally means "against protection" and results from the interaction of humoral antibodies (IgE) with mast cells and basophils. In response to certain allergens, some people produce IgE antibodies that bind to the surfaces of mast cells and basophils. This binding is what causes a person to be allergic to the allergen. Whereas basophils circulate in blood, mast cells are especially numerous in connective tissue of the skin, respiratory system, and endothelium of blood vessels. In response to the attachment of IgE antibodies to basophils and mast cells, the cells release chemicals called mediators of anaphylaxis, among which are histamine and prostaglandins (PGs). Collectively, the mediators increase blood capillary permeability, increase smooth muscle contraction, and increase mucus secretion. As a result, a person may experi-

ence edema and redness, along with other inflammatory responses; difficulty in breathing from constricted bronchial tubes; and a "runny" nose from excess mucus secretion.

Some anaphylactic reactions, such as hay fever, bronchial asthma, hives, eczema, swelling of the lips or tongue, abdominal cramps, and diarrhea are referred to as localized (affecting one part or a limited area). Other anaphylactic reactions are considered systemic (affecting several parts or the entire body). An example is acute anaphylaxis (anaphylactic shock), which may produce life-threatening systemic effects such as circulatory shock and asphyxia (oxygen starvation) and can be fatal within a few minutes. The effects of histamine may be counteracted by the administration of epinephrine.

Type IV (*cell-mediated*) *reactions* involve T cells and often are not apparent for a day or more. Type IV reactions occur when allergens that bind to tissue cells are phagocytized by macrophages and presented to receptors on the surfaces of T cells. This results in the proliferation of T cells, which respond by destroying the allergens. An example of a type IV reaction that involves the skin is the familiar skin test for tuberculosis.

TISSUE REJECTION

Transplantation involves the replacement of an injured or diseased tissue or organ. Usually, the body recognizes the proteins in the transplanted tissue or organ as foreign and produces antibodies against them. This phenomenon is known as *tissue rejection*. Rejection can be somewhat reduced by matching donor and recipient HLA antigens and by administering drugs that inhibit the body's ability to form antibodies. Recall from Chapter 12 that white blood cells and other nucleated cells have surface antigens called HLA antigens. These are unique for each person, except identical twins. The more closely matched the HLA antigens between donor and recipient, the less the likelihood of tissue rejection.

Until recently, *immunosuppressive drugs* suppressed not only the recipient's immune rejection of the donor organ but also the immune response to all antigens as well. This causes patients to become very susceptible to infectious diseases. A drug called cyclosporine, derived from a fungus, has largely overcome this problem with regard to kidney, heart, and liver transplants. A selective immunosuppressive drug, it inhibits T cells that are responsible for tissue rejection but has only a minimal effect on B cells. Thus, rejection is avoided and resistance against disease is still maintained

HODGKIN'S DISEASE (HD)

Hodgkin's disease (*HD*) is a form of cancer, usually arising in lymph nodes, the cause of which is unknown. It may, however, arise from a combination of genetic predisposition, disturbance of the immune system, and an infectious agent (Epstein-Barr virus). The histological diagnosis of the disease is made by the presence of large, malignant, multinucleate cells in the affected lymph node called Reed-Sternberg cells. The disease is initially characterized by a painless, nontender lymph node, most commonly in the neck, but occasionally in the axilla, inguinal, or femoral region. About one-quarter to one-third of patients also have an unexplained and persistent fever and/or night sweats. Fatigue and weight loss are also associated complaints, as is pruritus (itching). Treatment consists of radiation therapy, chemotherapy, and combinations of the two. Hodgkin's disease is considered to be a curable malignancy.

KEY MEDICAL TERMS ASSOCIATED WITH THE LYMPHATIC SYSTEM

Adenitis (ad'-e-NĪ-tis; *adeno* = gland; *itis* = inflammation of)   Enlarged, tender, and inflamed lymph nodes resulting from an infection.

Elephantiasis (el'-e-fan-TĪ-a-sis)   Great enlargement of a limb (especially lower limbs) and/or scrotum resulting from obstruction of lymph glands or vessels by a parasitic worm.

Hypersplenism (hī'-per-SPLĒN-izm; *hyper* = over)   Abnormal splenic activity involving highly increased blood cell destruction.

Lymphadenectomy (lim-fad'-e-NEK-tō-mē; *ectomy* = removal)   Removal of a lymph node.

Lymphadenopathy (lim-fad'-e-NOP-a-thē; *patho* = disease)   Enlarged, sometimes tender lymph glands.

Lymphangioma (lim-fan'-jē-Ō-ma; *angio* = vessel; *oma* = tumor).   A benign tumor of the lymph vessels.

Lymphangitis (lim'-fan-JĪ-tis)   Inflammation of the lymphatic vessels.

Lymphedema (lim'-fe-DĒ-ma; *edema* = swelling)   Accumulation of lymph fluid producing subcutaneous tissue swelling.

Lymphoma (lim'-FŌ-ma)   Any tumor composed of lymph tissue.

Lymphostasis (lim-FŌS-tā-sis; *stasis* = halt)   A lymph flow stoppage.

Splenomegaly (splē'-nō-MEG-a-lē; *mega* = large)   Enlarged spleen.

STUDY OUTLINE

Lymphatic Vessels (p. 416)
1. The lymphatic system consists of lymph, lymphatic vessels, and structures and organs that contain lymphatic tissue (specialized reticular tissue containing large numbers of lymphocytes).
2. Among the lymphatic tissue-containing components of the lymphatic system are diffuse lymphatic tissue, lymphatic nodules, and lymphatic organs (lymph nodes, spleen, and thymus gland).
3. Lymphatic vessels begin as blind-ended lymph capillaries in tissue spaces between cells.
4. Lymph capillaries merge to form larger vessels, called lymphatics, which ultimately converge into the thoracic duct or right lymphatic duct.
5. Lymphatics have thinner walls and more valves than veins.

Lymphatic Tissue (p. 418)
1. Lymph nodes are oval structures located along lymphatics.
2. Lymph enters nodes through afferent lymphatic vessels and exits through efferent lymphatic vessels.
3. Lymph passing through the nodes is filtered. Lymph nodes also produce lymphocytes.
4. Tonsils are multiple aggregations or large lymphatic nodules embedded in mucous membranes. They include the pharnygeal, palatine, and lingual tonsils.
5. The spleen is the largest mass of lymphatic tissue in the body and functions in production of lymphocytes and antibodies, phagocytosis of bacteria and worn-out red blood cells, and storage of blood.
6. The thymus gland functions in immunity by producing T cells.

Lymph Circulation (p. 424)
1. The passage of lymph is from interstitial fluid, to lymph capillaries, to lymphatics, to lymph trunks, to the thoracic duct or right lymphatic duct, to the subclavian veins.
2. Lymph flows as a result of skeletal muscle contractions and respiratory movements. It is also aided by valves in the lymphatics.

Principal Groups of Lymph Nodes (p. 427)
1. Lymph nodes are scattered throughout the body as superficial and deep groups.
2. The principal groups of lymph nodes are found in the head and neck, upper extremities, lower extremities, abdomen and pelvis, and thorax.

Developmental Anatomy (p. 435)
1. Lymphatic vessels develop from lymph sacs, which develop from veins. Thus, they are derived from mesoderm.
2. Lymph nodes develop from lymph sacs that become invaded by mesenchymal cells.

Applications to Health (p. 435)
1. Acquired immune deficiency syndrome (AIDS) lowers the body's immunity by decreasing the number of helper T cells and reversing the ratio of helper T cells to suppressor T cells. AIDS victims frequently develop Kaposi's sarcoma (KS) and *Pneumocystis carinii* pneumonia (PCP).
2. Autoimmune diseases result when the body does not recognize "self" antigens and produces antibodies against them. Several human autoimmune diseases are rheumatoid arthritis (RA), systemic lupus erythematosus (SLE), rheumatic fever, hemolytic and pernicious anemias, myasthenia gravis, and multiple sclerosis (MS).
3. Severe combined immunodeficiency (SCID) is an immunodeficiency disease in which both B cells and T cells are missing or inactive in providing immunity.
4. Hypersensitivity is overreactivity to an antigen. Localized ana-

responses only in involuntary muscles and glands, it is usually considered to be involuntary.

With few exceptions, the viscera receive nerve fibers from the two divisions of the autonomic nervous system: the *sympathetic division* and the *parasympathetic division.* In general, the fibers of one division stimulate or increase an organ's activity, while the fibers from the other inhibit or decrease activity (see Chatper 19).

HISTOLOGY

Despite the organizational complexity of the nervous system, it consists of only two principal kinds of cells: neurons and neuroglia. Neurons make up the nervous tissue that forms the structural and functional portion of the system. They are highly specialized for nerve impulse conduction and for all special functions attributed to the nervous system: thinking, controlling muscle activity, regulating glands. Neuroglia serve as a special supporting and protective component of the nervous system.

NEUROGLIA

The cells of the nervous system that perform the functions of support and protection are called *neuroglia* (noo-ROG-lē-a; *neuro* = nerve; *glia* = glue) or *glial cells* (Figure 16-2). Neuroglia are generally smaller than neurons and outnumber them by 5–10 times. Many of the glial cells form a supporting network by twining around nerve cells or lining certain structures in the brain and spinal cord. Others bind nervous tissue to supporting structures and attach the neurons to their blood vessels. A few types of glial cells also serve specialized functions. For example, some produce a phospholipid covering, called a myelin sheath, around nerve fibers in the central nervous system, which increases the speed of nerve impulse conduction and insulates the fibers. Certain small glial cells are phagocytic; they protect the central nervous system from disease by engulfing invading microbes and clearing away debris. Neuroglia are of clinical interest because they are a common source of tumors (gliomas) of the nervous system. It is estimated that gliomas account for 40–45 percent of brain tumors. Unfortunately, gliomas are very invasive.

Exhibit 16-1 lists the neuroglial cells and summarizes their functions.

NEURONS

Nerve cells, called *neurons,* are responsible for conducting nerve impulses from one part of the body to another. They are the structural and functional units of the nervous system.

Structure

Most neurons consist of three distinct portions: (1) cell body, (2) dendrites, and (3) axon (Figure 16-3a). The *cell body, soma,* or *perikaryon* (per'-i-KAR-ē-on) contains a well-defined nucleus and nucleolus surrounded by a granular cytoplasm. Within the cytoplasm are typical organelles such as lysosomes, mitochondria, and Golgi complexes. Many neurons also contain cytoplasmic inclusions such as *lipofuscin* pigment that occurs as clumps of yellowish-brown granules. Lipofuscin may be a by-product of lysosomal activity. Although its significance is unknown, lipofusin is related to aging; the amount of pigment increases with age. Also located in the cytoplasm are structures characteristic of neurons: chromatophilic substance and neurofibrils. The *chromatophilic substance* (*Nissl bodies*) is an orderly arrangement of granular (rough) endoplasmic reticulum whose

EXHIBIT 16-1

Neuroglia of Central Nervous System

TYPE	DESCRIPTION	FUNCTION
Astrocytes (*astro* = star; *cyte* = cell)	Star-shaped cells with numerous processes. **Protoplasmic astrocytes** are found in the gray matter of the CNS, and **fibrous astrocytes** are found in the white matter of the CNS.	Twine around nerve cells to form supporting network in CNS; attach neurons to their blood vessles; help form blood–brain barrier (Chapter 17).
Oligodendrocytes (*oligo* = few; *dendro* = tree)	Resemble astrocytes in some ways, but processes are fewer and shorter.	Give support by forming semirigid connective tissue rows between neurons in CNS; produce a phospholipid myelin sheath around axons of neurons of CNS.
Microglia (*micro* = small; *glia* = glue)	Small cells with few processes; derived from monocytes; normally stationary but may migrate to site of injury; also called brain macrophages.	Engulf and destroy microbes and cellular debris; may migrate to area of injured nervous tissue and function as small macrophages.
Ependyma (ependymocytes) (*ependyma* = upper garment)	Epithelial cells arranged in a single layer and ranging in shape from squamous to columnar; many are ciliated.	Form a continuous epithelial lining for the ventricles of the brain (spaces that form and circulate cerebrospinal fluid) and the central canal of the spinal cord; probably assist in the circulation of cerebrospinal fluid (CSF) in these areas.

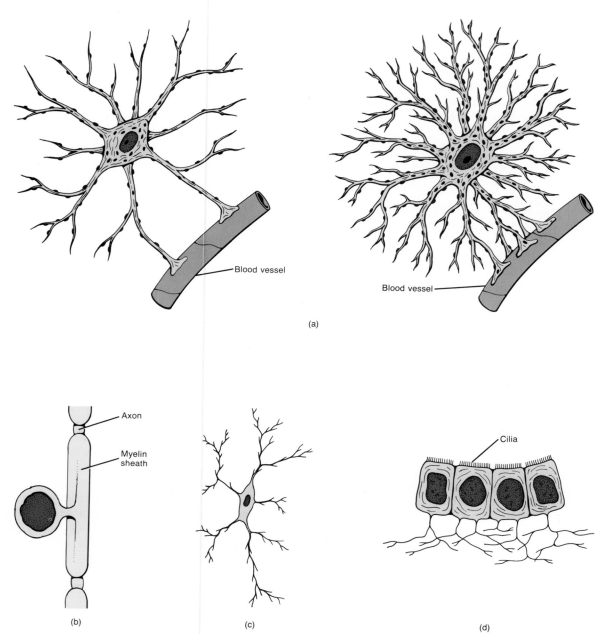

FIGURE 16-2 Histology of neuroglia. (a) Fibrous astrocyte (left) and protoplasmic astrocyte (right) associated with a blood vessel. (b) Oligodendrocyte associated with an axon of a neuron of the central nervous system. (c) Microglial cell. (d) Ependyma. Note the cilia.

function is protein synthesis. Newly synthesized proteins pass from the perikaryon into the neuronal processes, mainly the axon, at the rate of about 1 mm (0.04 in.) per day. These proteins replace those lost during metabolism and are used for growth of neurons and regeneration of peripheral nerve fibers. ***Neurofibrils*** are long, thin fibrils composed of microtubules. They may assume a function in support and the transportation of nutrients. Mature neurons do not contain a mitotic apparatus. The significance of this absence will be noted shortly.

Neurons have two kinds of cytoplasmic processes: dendrites and axons. ***Dendrites*** (*dendro* = tree) are usually highly branched, thick extensions of the cytoplasm of the cell body. They typically contain chromatophilic substance, mitochondria, and other cytoplasmic organelles. A neuron usually has several main dendrites. Their function is to conduct nerve impulses toward the cell body.

The second type of cytoplasmic process, called and ***axon*** (***axis cylinder***), is a single, highly specialized, usually long, thin process that conducts nerve impulses away from the

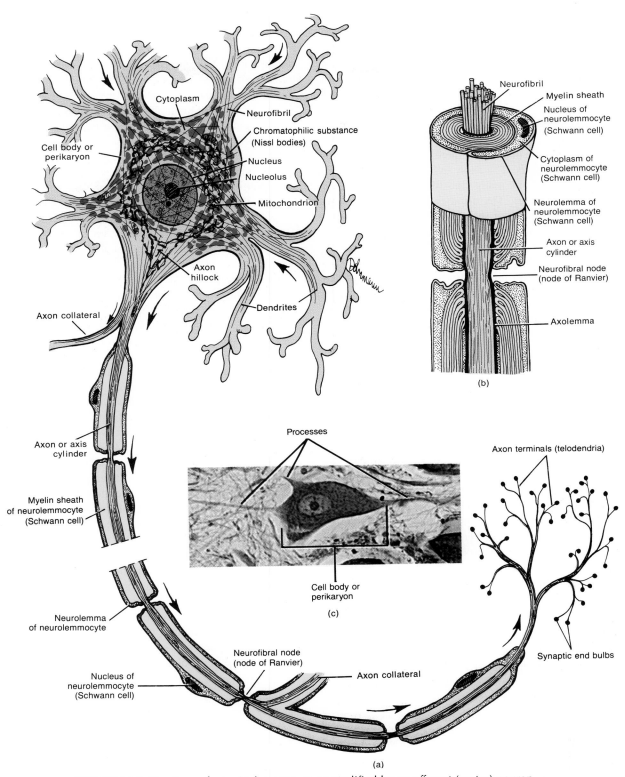

FIGURE 16-3 Structure of a typical neuron as exemplified by an efferent (motor) neuron. (a) An efferent neuron. Arrows indicate the direction in which nerve impulses travel. The break indicates that the process is actually longer than shown. (b) Sectional planes through a myelinated fiber. (c) Photomicrograph of an efferent neuron at a magnification of 640×. (Courtesy of Biophoto Associates/Photo Researchers.)

cell body to another neuron or tissue. It usually originates from the cell body as a small conical elevation called the **axon hillock.** An axon contains mitochondria and neurofibrils but no chromatophilic substance; thus, it does not carry on protein synthesis. Its cytoplasm, called **axoplasm,** is surrounded by a plasma membrane known as the **axolemma** (*lemma* = sheath or husk). Axons vary in length from a few millimeters (1 mm = 0.04 in.) in the brain to a meter (3.28 ft) or more between the spinal cord and toes. Along the length of an axon, there may be side branches called **axon collaterals.** The axon and its collaterals terminate by branching into many fine filaments called **axon terminals** (**telodendria**). The distal ends of axon terminals are expanded into bulblike structures called **synaptic end-bulbs,** which are important in nerve impulse conduction from one neuron to another and from a neuron to muscle or glandular tissue. They contain membrane-enclosed sacs called **synaptic vesicles** that store chemicals called neurotransmitters that determine whether or not nerve impulses pass from one neuron to another or from a neuron to another tissue (muscle or gland).

The cell body of a neuron is essential for the synthesis of many substances that sustain the life of the nerve cell. Neurons have two types of intracellular systems for transporting synthesized materials from the cell body. The slower one, called **axoplasmic flow,** conveys axoplasm in one direction only—from the cell body toward axon terminals. This mechanism may occur by protoplasmic streaming and supplies new axoplasm for developing or regenerating axons and renews axoplasm in growing and mature axons. The faster type of intracellular transport is called **axonal transport.** It conveys materials in both directions—away from the cell body and toward the cell body—possibly along tracks formed by microtubules and filaments. Axonal transport moves various organelles and materials that form the membranes of the axolemma, synaptic end-bulbs, and synaptic vesicles. Materials returning to the cell body are degraded or recycled.

CLINICAL APPLICATION

The route taken by materials back to the cell body by axonal transport is the route by which the **herpes virus** and **rabies virus** make their way back to nerve cell bodies, where they multiply and cause their damage. The toxin produced by the **tetanus bacterium** uses the same route to reach the central nervous system. In fact, the time delay between the release of the toxin and the first appearance of symptoms is in part due to the time required for movement of the toxin by axonal transport.

The term **nerve fiber** may be applied to any process projecting from the cell body. More commonly, it refers to an axon and its sheaths. Figure 16-3b shows two sectional planes of a nerve fiber of the peripheral nervous system.

Many axons, especially large ones outside the CNS, are surrounded by a multilayered, white, phospholipid, segmented covering called the **myelin sheath.** Axons containing such a covering are **myelinated,** while those without it are **unmyelinated** (see Figure 16-4). The function of the myelin sheath is to increase the speed of nerve impulse conduction and to insulate and maintain the axon. Myelin is responsible for the color of the white matter in the nerves, brain, and spinal cord.

The myelin sheath of axons of the peripheral nervous system is produced by flattened cells, called **neurolemmocytes** (**Schwann cells**), located along the axons. In the formation of a sheath, a developing neurolemmocyte encircles the axon until its ends meet and overlap (Figure 16-4). The cell then winds around the axon many times and, as it does so, the cytoplasm and nucleus are pushed to the outside layer. The inner portion, consisting of up to 20–30 layers of neurolemmocyte membrane, is the myelin sheath. The peripheral nucleated cytoplasmic layer of the neurolemmocyte (the outer layer that encloses the sheath) is called the **neurolemma** (**sheath of Schwann**).

The neurolemma is found only around fibers of the peripheral nervous system. Its function is to assist in the regeneration of injured axons by forming a tube in which a regenerating axon grows. Between the segments of the myelin sheath are unmyelinated gaps, called **neurofibral nodes** (**nodes of Ranvier**) (ron-VE-a). Unmyelinated fibers are also enclosed by neurolemmocytes, but without multiple wrappings.

Nerve fibers of the central nervous system may also be myelinated or unmyelinated. Myelination of central nervous system axons is accomplished by oligodendrocytes in somewhat the same manner that neurolemmocytes myelinate peripheral nervous system axons (see Figure 16-2b). Myelinated axons of the central nervous system also contain neurofibral nodes, but they are not so numerous.

Myelin sheaths are first laid down during the later part of fetal development and during the first year of life. The amount of myelin increases from birth to maturity and its presence greatly increases the rate of nerve impulse conduction. Since myelination is still in progress during infancy, an infant's responses to stimuli are not as rapid or coordinated as those of an older child or an adult.

Structural Variation

Although all neurons conform to the general plan described, there are considerable differences in structure. For example, cell bodies range in diameter from 5 μm for the smallest cells to 135 μm for large motor neurons. The pattern of dendritic branching is varied and distinctive for neurons in different parts of the body. The axons of very small neurons are only a fraction of a millimeter in length and lack a myelin sheath, whereas axons of large neurons are over a meter long and are usually enclosed in a myelin sheath.

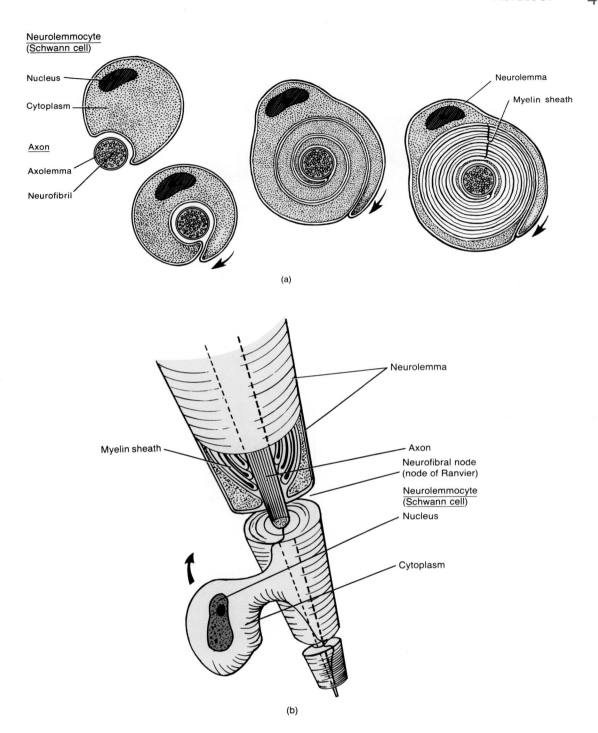

(a)

(b)

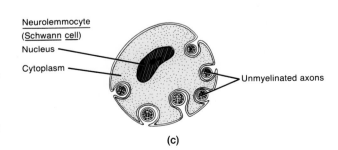

(c)

FIGURE 16-4 Comparison between myelinated and unmyelinated axons. (a) Stages in the formation of a myelin sheath by a neurolemmocyte (Schwann cell) seen in cross section. (b) Myelin sheath formation by a neurolemmocyte (Schwann cell) seen along the length of a portion of an axon. (c) Diagram of an unmyelinated axon.

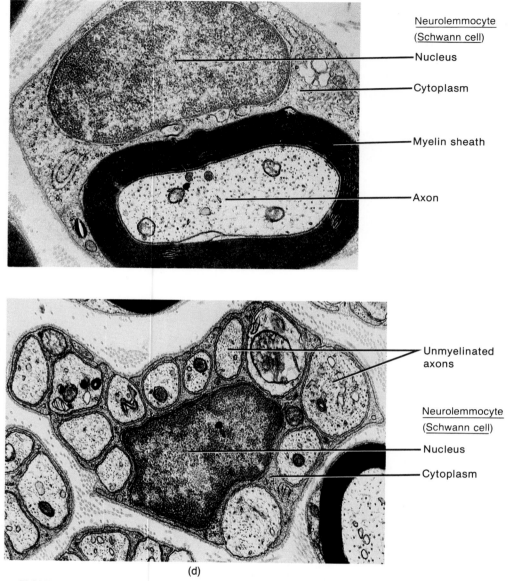

Neurolemmocyte
(Schwann cell)

Nucleus

Cytoplasm

Myelin sheath

Axon

Unmyelinated
axons

Neurolemmocyte
(Schwann cell)

Nucleus

Cytoplasm

(d)

FIGURE 16-4 (*Continued*) Myelin sheath. (d) Electron micrographs of a myelinated axon (above) and several unmyelinated axons (below). (Copyright © 1980, Biology Media, Schultz, Photo Researchers.)

Classification

The different neurons in the body may be classified by structure and function.

The structural classification is based on the number of processes extending from the cell body. *Multipolar neurons* usually have several dendrites and one axon (see Figure 16-3a). Most neurons in the brain and spinal cord are of this type. *Bipolar neurons* have one dendrite and one axon and are found in the retina of the eye, inner ear, and olfactory area. *Unipolar (pseudounipolar) neurons* have only one process extending from the cell body. The single process divides into a central branch, which functions as an axon, and a peripheral branch, which functions as a dendrite.

Unipolar neurons originate in the embryo as bipolar neurons and, during development, the axon and dendrite fuse into a single process. Unipolar neurons are found in posterior (sensory) root ganglia of spinal nerves and the ganglia of cranial nerves that carry general somatic sensations.

The functional classification of neurons is based on the direction in which they transmit impulses. *Sensory (afferent) neurons,* transmit impulses from receptors in the skin, sense organs, and viscera to the brain and spinal cord and from lower to higher centers of the CNS. They are usually unipolar (Figure 16-5). *Motor (efferent) neurons* convey impulses from the brain and spinal cord to effectors, which may be either muscles or glands (see Figure 16-3a) and from higher to lower centers of the CNS. Other neurons,

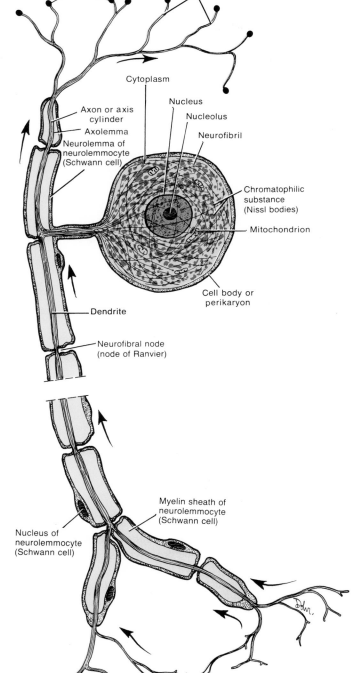

FIGURE 16-5 Structure of a typical afferent (sensory) neuron. Arrows indicate the direction in which the nerve impulse travels. The break indicates that the process is actually longer than shown.

Labels in figure:
Synaptic end bulbs
Axon terminals (telodendria)
Cytoplasm
Nucleus
Nucleolus
Axon or axis cylinder
Neurofibril
Axolemma
Neurolemma of neurolemmocyte (Schwann cell)
Chromatophilic substance (Nissl bodies)
Mitochondrion
Cell body or perikaryon
Dendrite
Neurofibral node (node of Ranvier)
Myelin sheath of neurolemmocyte (Schwann cell)
Nucleus of neurolemmocyte (Schwann cell)
Receptors

called **association (connecting** or **interneuron) neurons,** carry impulses from sensory neurons to motor neurons and are located in the brain and spinal cord. Examples of association neurons are a **stellate cell,** a **cell of Martinotti** (mar'-

ti-NOT-ē), a **horizontal cell of Cajal** (kā-HAL), and a **pyramidal cell** (pi-RAM-i-dal) cell. All are found in the cerebral cortex, the outer layer of the cerebrum. The **granule cell** and **Purkinje** (pur-KIN-jē) **cell** are association neurons in the cortex of the cerebellum. Most neurons in the body, perhaps 90 percent, are association neurons. Several association neurons are shown in Figure 16-6.

The processes of afferent and efferent neurons are arranged into bundles called **nerves** if outside the CNS or **tracts** if inside the CNS. Since nerves lie outside the CNS, they belong to the peripheral nervous system. The functional components of nerves are the nerve fibers, which may be grouped according to the following scheme.

1. **General somatic afferent fibers** conduct nerve impulses from the skin, skeletal muscles, and joints to the central nervous system.
2. **General somatic efferent fibers** conduct nerve impulses from the central nervous system to skeletal muscles. Impulses over these fibers cause the contraction of skeletal muscles.
3. **General visceral afferent fibers** convey nerve impulses from the viscera and blood vessels to the central nervous system.
4. **General visceral efferent fibers** belong to the autonomic nervous system and are also called **autonomic fibers.** They convey nerve impulses from the central nervous system to cause contractions of smooth and cardiac muscle and secretion by glands.

In addition to being grouped as nerves, neural tissue is also organized into other structures such as ganglia, tracts, nuclei, and horns. These are described in Chapter 17.

Two striking features of nervous tissue are (1) its highly developed ability to generate and conduct electrical messages called nerve impulses and (2) its limited ability to regenerate.

NERVE IMPULSE

Although it is beyond the scope of this text to describe the details of a nerve impulse, certain concepts must be understood to know how the nervous system works. Very simply, a **nerve impulse (nerve action potential)** is a wave of negativity that self-propagates along the surface of the membrane of a neuron. Among other things, a nerve impulse depends on the movement of sodium, potassium, and other ions between interstitial fluid and the inside of a neuron. For a nerve impulse to begin, a stimulus of adequate strength must be applied to the neuron. A **stimulus** is a change in the environment of sufficient strength to initiate a nerve impulse. The ability of a neuron to respond to a stimulus and convert it into a nerve impulse is known as **excitability.**

The nerve impulse is the most rapid way that the body can respond to environmental changes. It provides the quick-

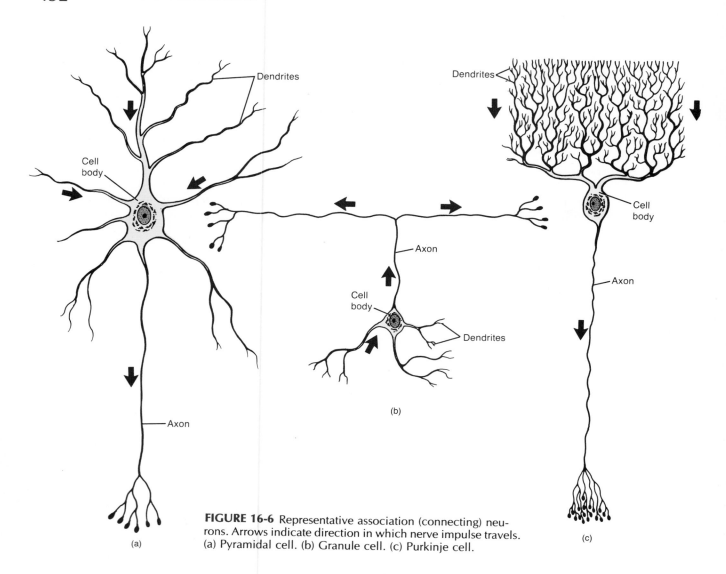

FIGURE 16-6 Representative association (connecting) neurons. Arrows indicate direction in which nerve impulse travels. (a) Pyramidal cell. (b) Granule cell. (c) Purkinje cell.

est means for achieving homeostasis. The speed of a nerve impulse is determined by the size, type, and physiological condition of the nerve fiber. For example, myelinated fibers with the largest diameters can transmit impulses at speeds up to about 100 m (328 ft)/sec. Unmyelinated fibers with the smallest diameters can transmit impulses at the rate of about 0.5 m (1.5 ft)/sec.

In addition to excitability, neurons are also characterized by **conductivity**, the ability to transmit a nerve impulse to another neuron or another tissue, such as a muscle or a gland. The junction between two neurons is called a **synapse** (*synapse* = to join). The synapse is essential for homeostasis because of its ability to transmit certain nerve impulses and inhibit others. Much of an organism's ability to learn will probably be explained in terms of synapses. Moreover, most diseases of the brain and many psychiatric disorders result from a disruption of synaptic communication. And synapses are the sites of action for most drugs that affect the brain, including therapeutic and addictive substances. Figure 16-7 shows that within a synapse is a minute gap,

about 20 nm across, called the **synaptic cleft**. A **presynaptic neuron** is a neuron located before a synapse. A **postsynaptic neuron** is located after a synapse.

Impulses are conducted from a neuron to a muscle fiber (cell) across an area of contact called a **neuromuscular junction (NMJ)**, **myoneural junction**, or **motor end-plate**, the details of which were described in Chapter 9. The area of contact between a neuron and glandular cells is known as a **neuroglandular junction**. Together, neuromuscular and neuroglandular junctions are known as **neuroeffector junctions**.

Axon terminals of neurons end in expanded bulblike structures referred to as **synaptic end bulbs**. The synaptic end-bulbs of a presynaptic neuron commonly synapse with the dendrites, cell body, or axon hillock of a postsynaptic neuron. Accordingly, synapses may be classified as **axodendritic**, **axosomatic**, and **axoaxonic**. The synaptic end-bulbs from a single presynaptic neuron may synapse with several postsynaptic neurons. Such an arrangement, called **divergence**, permits a single presynaptic neuron to influence several postsynaptic neurons or several muscle fibers or gland cells at the same time (see Figure 16-9a). In another arrangement, called ***convergence,*** the synaptic end-bulbs of several presynaptic neurons synapse with a single post-synaptic neuron (see Figure 16-9b). This arrangement permits stimulation or inhibition of the postsynaptic neuron.

At a synapse there is only ***one-way impulse conduction***—from a presynaptic axon to a postsynaptic dendrite, cell body, or axon hillock. Nerve impulses must move forward over their pathways. They cannot back up into another presynaptic neuron. Such a mechanism is crucial in preventing impulse conduction along improper pathways, a situation that would severely disrupt homeostasis.

Whether a nerve impulse is conducted across a synapse, neuromuscular junction, or neuroglandular junction depends on the presence of chemicals called ***neurotransmitters*** (***transmitter substances***). These chemicals are made by the neuron, usually from amino acids. Following its production and transportation to the synaptic end-bulbs, the neurotransmitter is stored in the bulbs in small membrane-enclosed sacs called ***synaptic vesicles*** (Figure 16-7). Each of the thousands of synaptic vesicles present may contain between 10,000 and 100,000 neurotransmitter molecules.

When a nerve impulse arrives at a synaptic end-bulb of a presynaptic neuron, it is believed that a small amount of calcium ions leaks into the bulb, attracts synaptic vesicles to the plasma membrane, and helps liberate the neurotransmitter molecules from the vesicles. Some scientists believe that the synaptic vesicles fuse with the plasma membrane of the presynaptic neuron, form openings, and release the neurotransmitter through the openings into the synaptic cleft. Other scientists think that the neurotransmitter leaves through small channels to enter the synaptic cleft. In either case, the neurotransmitter enters the synaptic cleft and, depending on the chemical nature of the neurotransmitter and the interaction of the neurotransmitter with receptors of the postsynaptic plasma membrane, several things can happen.

An ***excitatory transmitter–receptor interaction*** is one that generates a nerve impulse across a synapse, while an ***inhibitory transmitter–receptor interaction*** is one that inhibits a nerve impulse across a synapse. Many presynaptic neurons synapse with a single postsynaptic neuron. Some presynaptic end-bulbs produce excitation and some produce inhibition. The sum of all the effects, excitatory and inhibitory, determines the effect on the postsynaptic neuron. Thus, the postsynaptic neuron is an ***integrator***. It receives signals, integrates them, and then responds accordingly. The postsynaptic neuron may respond in the following ways:

1. If the excitatory effect is greater than the inhibitory effect but less than the threshold (minimal) level of stimulation, the result is facilitation, that is, near excitation so that subsequent stimuli can generate a nerve impulse.
2. If the excitatory effect is greater than the inhibitory effect but equal to or higher than the threshold level of stimulation, the result is generation of a nerve impulse.
3. If the inhibitory effect is greater than the excitatory effect, the result is inhibition of a nerve impulse.

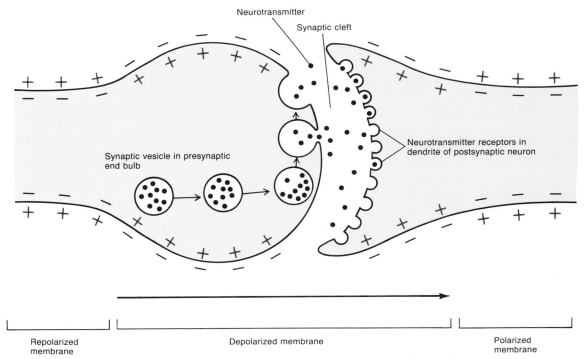

FIGURE 16-7 Impulse conduction at synapses. Shown is impulse conduction from a presynaptic and bulb across a synapse to a postsynaptic dendrite in which synaptic vesicles fuse with the presynaptic membrane and discharge neurotransmitter into the synaptic cleft.

Perhaps the best-studied neurotransmitter substance is *acetylcholine* (as'-ē-til-KŌ-lēn), or *ACh*. It is a neurotransmitter released by many neurons outside the brain and spinal cord and by some neurons inside the brain and cord. This neurotransmitter was discussed in Chapter 9 in relation to nerve impulse conduction from a motor neuron to a muscle fiber across a neuromuscular junction (motor endplate). Following the arrival of a nerve impulse at an axon terminal, calcium (Ca^{2+}) ions enter an axon terminal and cause the release of ACh from synaptic vesicles or the cytoplasm of the terminal. At neuromuscular junctions, ACh binds to receptor sites on the muscle fiber membrane and increases the membrane's permeability to Na^+ ions. The acetylcholine receptor is an integral protein in the plasma membrane of the muscle fiber. When ACh molecules bind to the ACh receptor, a change occurs in the receptor, causing its channel to open. As a result, there is an inward movement of Na^+ ions. In the absence of ACh, the receptor channel remains closed. The inward movement of Na^+ ions leads to a sequence of events that generates a muscle action potential, causing the muscle fiber to contract. As long as ACh is present in the synaptic cleft, it can stimulate a muscle fiber almost indefinitely. The transmission of a continuous succession of impulses by ACh is normally prevented by an enzyme called *acetylcholinesterase (AChE)*, or simply *cholinesterase* (kō'-lin-ES-ter-ās). AChE is found on the surfaces of the subneural clefts of the membrane of muscle fibers. Within 1/500 sec, AChE inactivates ACh. This action permits the membrane of the muscle fiber to discontinue action potential transmission immediately so that another action potential may be generated. When the next nerve impulse comes through, the synaptic vesicles release more ACh, a muscle action potential is generated, and AChE again inactivates ACh. This cycle is repeated over and over again.

ACh is released at some neuromuscular junctions that have cardiac and smooth muscle, as well as those that have skeletal muscle. It is also released at some neuroglandular junctions. Although ACh leads to excitation in many parts of the body, it is inhibitory with respect to the heart (vagus (X) nerve).

CLINICAL APPLICATION

There are many ways that *synaptic conduction can be altered* by disease, drugs, and pressure. In Chapter 9 it was noted that *myasthenia gravis* results from antibodies directed against acetylcholine receptors on muscle fiber membranes at neuromuscular junctions, causing dysfunctions in muscular contractions. *Alkalosis,* an increase in pH above 7.45 up to about 8.0, results in increased excitability of neurons that can cause cerebral convulsions, while *acidosis,* a decrease in pH below 7.35 down to about 6.80, results in a depression of neuronal activity and can produce a comatose state.

Curare, the deadly South American Indian arrowhead poison, also competes for acetylcholine receptor sites. Since the muscles of respiration depend on neuromuscular transmission to initiate their contraction, death results from asphyxiation. *Neostigmine* and *physostigmine* are anticholinesterase agents that combine with acetylcholinesterase to inactivate it for several hours. As a result, acetylcholine accumulates and produces repetitive stimulation of muscles. This causes muscular spasm, and death due to laryngeal spasm can ensue. *Diisopropyl fluorophosphate* is a very powerful nerve gas that inactivates acetylcholinesterase for up to several weeks, making it a particularly lethal drug. The *botulism toxin* inhibits the release of acetylcholine, thus inhibiting muscle contraction. The toxin is exceedingly potent in even small amounts (less than 0.0001 mg) and is the substance involved in one type of food poisoning. *Hypnotics, tranquilizers,* and *anesthetics* depress synaptic conduction, while *caffeine, benzedrine,* and *nicotine* lead to facilitation (near excitation).

Crack, a potent form of cocaine that is smoked rather than sniffed, interferes with the normal functioning of neurotransmitters—dopamine (DA), norepinephrine (NE), and serotonin (5-HT)—that are involved in the regulation of mood and motor functions. Once a neurotransmitter has accomplished its function, it is inactivated and returned to the presynaptic neuron for resynthesis. Initially, crack inhibits the inactivation of the neurotransmitters. The resultant buildup of dopamine, in particular, has been linked to feelings of euphoria. Inhibition of inactivation of neurotransmitters also can cause convulsions, accelerated and abnormal heart rate, vasoconstriction and high blood pressure, weight loss, insomnia, and susceptibility to disease. Repeated use of crack may produce a temporary shortage of neurotransmitters, resulting in depression, anxiety, and craving for more crack. Heavy, prolonged use of crack may eventually deplete neurotransmitters to the point where euphoria no longer occurs and depression is persistent.

Pressure has an effect on nerve impulse conduction also. If excessive or prolonged pressure is applied to a nerve, as when crossing one's legs, nerve impulse conduction is interrupted and part of the body may "go to sleep," producing a tingling sensation. This sensation is caused by an accumulation of waste products and a depressed circulation of blood.

REGENERATION

Unlike the cells of epithelial tissue, neruons have only limited powers for *regeneration,* that is, a natural ability to renew themselves. Around 6 months of age, the cell bodies of most developing nerve cells lose their mitotic apparatus (centrioles and mitotic spindles) and their ability to reproduce. Thus, when a neuron is damaged or destroyed, it cannot be replaced by the daughter cells of other neurons. A neuron destroyed is permanently lost, and only some types of damage may be repaired.

Damage to some types of myelinated axons or other processes often can be repaired if the cell body remains intact, if the cell that performs the myelination remains active, and if a completely severed nerve is surgically apposed. Axons in the peripheral nervous system are myelinated by neurolemmocytes (Schwann cells). They proliferate following axonal damage and their neurolemmas form a tube that assits in regeneration (Chapter 17). Axons in the brain and spinal cord (central nervous system) are myelinated by oligodendroglial cells. These cells do not form neurolemmas to assist in regeneration and do not survive following axonal damage. An added complication in the central nervous system is that following axonal damage, astrocytes appear to stop axons from regenerating by activating a physiological pathway that inhibits axonal regeneration. This is the same pathway that stops axonal growth during development once a target cell has been reached. In addition, following axonal damage, the affected region is rapidly converted into a special form of scar tissue by astroglial proliferation. The scar tissue forms a physical barrier to regeneration. Thus, an injury to the brain or spinal cord is also permanent because axonal regeneration is blocked by a physiological stop pathway and rapid scar tissue formation. An injury to a nerve in the arm (peripheral nervous system) may repair itself before scar tissue forms, and so some nerve function may be restored.

ORGANIZATION OF NEURONS

The central nervous system contains millions of neurons. Their arrangement is not haphazard. They are organized into definite patterns called *neuronal pools.* Each pool differs from all others and has its own role in regulating homeostasis.

A neuronal pool may contain thousands or even millions of neurons. To illustrate the composition of a neuronal pool, a simplified version is given in Figure 16-8. This example contains only five postsynaptic neurons and two incoming presynaptic neurons. The postsynaptic neurons are subject to stimulation by the incoming presynaptic neurons. The postsynaptic neurons in the pool may be stimulated by one or several presynaptic end-bulbs. Moreover, the incoming presynaptic endbulbs may produce facilitation, excitation, or inhibition. A principal feature of a neuronal pool is the location of the presynaptic neurons in relation to the postsynaptic neurons. Compare the location of presynaptic axon 1 with that of postsynaptic neuron B. Since they are aligned, more presynaptic endbulbs of axon number 1 synapse with postsynaptic neuron B than with postsynaptic neuron A or C. We say that postsynaptic neuron B is in the center of the field of presynaptic axon 1. Consequently,

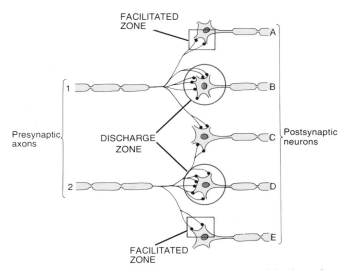

FIGURE 16-8 Relative positions of discharge and facilitated zones in a very simplified version of a neuronal pool.

postsynaptic neuron B usually receives sufficient presynaptic end bulbs from axon 1 to generate a nerve impulse. This region where the neuron in the pool fires is called the *discharge zone.*

Now look outside the field of presynaptic axon 1. Note its relation to postsynaptic neuron A. Here postsynaptic neuron A in the pool is receiving few presynaptic end bulbs from the axon supplying postsynaptic neuron B. Thus, there are insufficient presynaptic end bulbs to fire postsynaptic neuron A, but enough to cause facilitation. We therefore call this area the *facilitated zone.* Presynatpic axon 1, when stimulated, will cause excitation of postsynaptic neuron B and facilitation of postsynaptic neuron A.

Neuronal pools in the central nervous system are arranged in patterns over nerve impulses are conducted. These are termed *circuits. Simple series circuits* are arranged so that a presynaptic neuron stimulates a single neuron in a pool. The single neuron then stimulates another and so on. In other words, the nerve impulse is relayed from one neuron to another in succession as a new nerve impulse is generated at each synapse.

Most circuits, however, are more complex. In a *diverging circuit,* the nerve impulse from a single presynaptic neuron causes the stimulation of increasing numbers of cells along the circuit (Figure 16-9a). An example of such a circuit is a single motor neuron in the brain stimulating numerous other motor neurons in the spinal cord that, in turn, leave the spinal cord where each stimulates many skeletal muscle fibers. Thus, a single nerve impulse may result in the contraction of several skeletal muscle fibers. In another kind of diverging circuit, nerve impulses from one pathway are relayed to other pathways so the same information travels in various directions at the same time. This circuit is common along sensory pathways of the nervous system.

Another kind of circuit is called a *converging circuit*

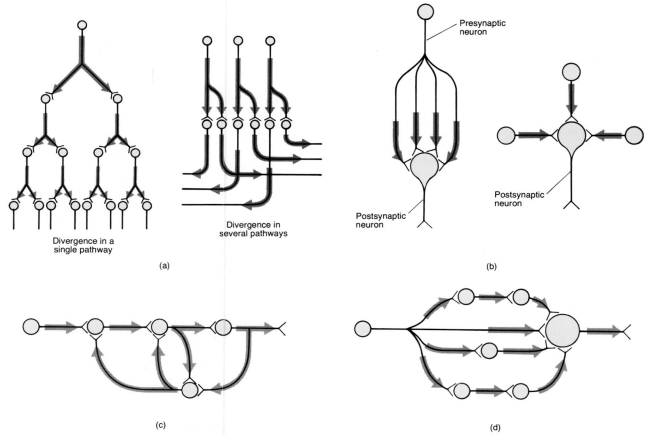

FIGURE 16-9 Circuits of neuronal pools. (a) Diverging. (b) Converging. (c) Reverberating. (d) Parallel after-discharge.

(Figure 16-9b). In one pattern of convergence, the postsynaptic neuron receives nerve impulses from several fibers of the same source. Here there is the possibility of strong excitation or inhibition. In a second pattern, the postsynaptic neuron receives nerve impulses from several different sources. Here there is a possibility of reacting the same way to different stimuli. Suppose your reaction to vomit is distinctly unpleasant. The smell of vomit (one kind of stimulus), the sight of vomit (another kind), or just reading about vomit (still another kind) might all have the same effect on you—an unpleasant one.

Some circuits in your body are constructed so that once the presynaptic cell is stimulated, it will cause the postsynaptic cell to transmit a series of nerve impulses. One such circuit is called a **_reverberating (oscillatory) circuit_** (Figure 16-9c). In this pattern, the incoming nerve impulse stimulates the first neuron, which stimulates the second, which stimulates the third, and so on. Branches from the second and third neurons synapse with the first, however, sending the nerve impulse back through the circuit again and again. A central feature of the reverberating circuit is that once fired, the output signal may last from a few seconds to many hours. The duration depends on the number and ar-

rangement of neurons in the circuit. Among the body responses thought to be the result of output signals from reverberating circuits are the rate of breathing, coordinated muscular activities, waking up, and sleeping (when reverberation stops). Some scientists think reverberating circuits are related to short-term memory. One form of epilepsy (grand mal) is probably caused by abnormal reverberating circuits.

A final circuit worth consideration is the **_parallel after-discharge circuit_** (Figure 16-9d). Like a reverberating circuit, a parallel after-discharge circuit is constructed so the postsynaptic cell transmits a series of nerve impulses. In a parallel after-discharge circuit, a single presynaptic cell stimulates a group of neurons, each of which synapses with a common postsynaptic cell. The advantage of this circuit is that the postsynaptic neuron can send out a stream of nerve impulses in succession as they are received. The nerve impulses leave the postsynaptic neuron once every $1/2000$ sec. This circuit has no feedback system. Once all the neurons in the circuit have transmitted their nerve impulses to the postsynaptic neuron, the circuit is broken. It is thought that the parallel after-discharge circuit is employed for precise activities like mathematical calculations.

STUDY OUTLINE

Organization (p. 444)

1. The nervous system helps control and integrate all body activities by sensing changes (sensory), interpreting them (integrative), and reacting to them (motor).
2. The central nervous system (CNS) consists of the brain and spinal cord.
3. The peripheral nervous system (PNS) is classified into an afferent system and an efferent system.
4. The efferent system is subdivided into a somatic nervous system and an autonomic nervous system.
5. The somatic nervous system (SNS) consists of efferent neurons that conduct nerve impulses for the central nervous system to skeletal muscle tissue.
6. The autonomic nervous system (ANS) contains efferent neurons that convey nerve impulses from the central nervous system to smooth muscle tissue, cardiac muscle tissue, and glands.

Histology (p. 445)

Neuroglia (p. 445)

1. Neuroglia are specialized tissue cells that support neurons, attach neurons to blood vessels, produce the myelin sheath around axons of the CNS, and carry out phagocytosis.
2. Neuroglial cells include astrocytes, oligodendrocytes, microglia, and ependyma.

Neurons (p. 445)

1. Most neurons, or nerve cells, consist of a cell body (perikaryon), dendrites that pick up stimuli and convey nerve impulses to the cell body, and usually a single axon. The axon conducts nerve impulses from the neuron to the dendrites or cell body of another neuron or to an effector organ of the body.
2. On the basis of structure, neurons are multipolar, bipolar, and unipolar.
3. On the basis of function, sensory (afferent) neurons conduct nerve impulses to the central nervous system; association (connecting) neurons conduct nerve impulses to other neurons, including motor neurons; and motor (efferent) neurons conduct nerve impulses to effectors.

Nerve Impulse (p. 451)

1. A nerve impulse (nerve action potential) is a wave of negativity that travels along the surface of the membrane of a neuron.
2. The ability of a neuron to respond to a stimulus and convert it into a nerve impulse is called excitability.
3. The speed of a nerve impulse is determined by the size, type, and physiological condition of the nerve fiber.
4. Conductivity is the ability of a neuron to transmit a nerve impulse to another neuron or another tissue.
5. The junction between neurons is called a synapse. Nerve impulse conduction across a synapse requires the release of neurotransmitters by presynaptic neurons.
6. Impulse conduction at a synapse is one-way conduction.
7. An excitatory transmitter–receptor interaction is one that can generate an action potential across the synapse.
8. An inhibitory transmitter–receptor interaction is one that can inhibit an action potential across a synapse.
9. It is thought that the neurotransmitter that causes excitation in a major portion of the central nervous system is acetylcholine (ACh). An enzyme called acetylcholinesterase (AChE) inactivates acetylcholine.

Regeneration (p. 454)

1. Around the time of birth, the neuronal cell body loses its mitotic apparatus and is no longer able to divide.
2. Nerve fibers (axis cylinders) that have a neurolemma are capable of regeneration.

Organization of Neurons (p. 455)

1. Neurons in the central nervous system are organized into definite patterns called neuronal pools. Each pool differs from all others and has its own role in the functioning nervous system.
2. Neuronal pools are organized into circuits. These include simple series, diverging, converging, reverberating, and parallel afterdischarge circuits.

REVIEW QUESTIONS

1. Describe the three basic functions of the nervous system that are necessary to maintain homeostasis.
2. Distinguish between the central and peripheral nervous systems, and describe the functions of each subdivision.
3. Relate the terms *voluntary* and *involuntary* to the nervous system.
4. What are neuroglia? List the principal types and their functions. Why are they important clinically?
5. Define a neuron. Diagram and label a neuron. Next to each part list its function.
6. What is a myelin sheath? How is it formed?
7. Define the neurolemma. Why is it important?
8. Discuss the structural classification of neurons. Give an example of each.
9. What are the structural differences between a typical afferent and efferent neuron? Give several examples of association neurons.
10. Describe the functional classification of neurons.
11. Distinguish among the following kinds of fibers: general somatic afferent, general somatic efferent, general visceral afferent, and general visceral efferent.
12. Define a nerve impulse.
13. Compare excitability and conductivity.
14. What factors determine the rate of nerve impulse conduction?
15. What is a synapse? How are synapses classified?
16. What is a neurotransmitter? Describe the action of acetylcholine.
17. What events are involved in the conduction of a nerve impulse across a synapse?
18. Distinguish between excitatory and inhibitory transmission.

19. Why does one-way impulse conduction occur?
20. Why is the postsynaptic neuron called an integrator?
21. How do disease, drugs, crack, and pressure affect synaptic transmission?
22. What determines neuron regeneration?

23. Distinguish between the discharge zone and facilitated zone and neuronal pool.
24. What is a neuron circuit? Distinguish among simple series, diverging, converging, reverberating, and parallel after-discharge circuits.

SELF-QUIZ

Choose the one best answer to these questions.

___ 1. Damage to the retina of the eye might involve which kind of neuron?
A. unipolar; B. multipolar; C. bipolar; D. tripolar; E. pseudounipolar.

___ 2. The myelin sheath of central nervous system (CNS) neurons is produced by
A. neurolemmocytes (Schwann cells); B. astrocytes; C. oligodendroctyes; D. microglia; E. ependyma.

___ 3. During an infection of the nervous system, you would expect to find an increase in the number of
A. microglia; B. astrocytes; C. oligodendrocytes; D. association neurons; E. ependyma.

___ 4. Regarding the speed of nerve impulse conduction, which of the following is true?
A. nerve fibers classed as C fibers conduct most rapidly; B. myelinated fibers conduct nerve impulses most rapidly; C. motor impulses are conducted more rapidly than sensory impulses; D. nerve impulses all travel at the same rate; E. none of the above.

___ 5. The point of contact between a nerve fiber and a muscle or a gland is called the
A. synapse; B. exteroceptor; C. neuroeffector junction; D. internuncial neuron; E. ganglion.

___ 6. A neuron cannot undergo regeneration if
(1) its cell body has been destroyed.
(2) its axon has no myelin sheath.
(3) its axon has no neurolemma.
A. (1) only; B. (2) only; C. (3) only; D. any of the above; E. either (1) or (3).

___ 7. Which of the following is true?
A. bipolar neurons are those with two axons; B. a typical motor neuron is a unipolar neuron; C. sensory neurons are usually bipolar; D. most unipolar neurons are found in the central nervous system; E. multipolar neurons possess numerous dendrites and only one axon.

___ 8. Which of the following, regarding neurons, is/are true?
A. the axon of a motor neuron extends from the cell body to an effector; B. a neuron usually has many axons, connected to other neurons; C. sensory and motor neurons have dendrites, while association neurons do not; D. the dendrite of one neuron connects with a dendrite of the next neuron; E. all of the above.

___ 9. There are several types of specialized cells called neuroglia. Which of the following pertain(s) correctly to one or the other of these types?
(1) attach neurons to blood vessels
(2) conduct nerve impulses
(3) produce a myelin sheath around central nervous system (CNS) neurons
(4) engulf and destroy microbes

A. (1) only; B. (2) only; C. (3) only; D. (4) only; E. (1), (3), and (4).

___ 10. Which of the following is/are false?
A. the central nervous system (CNS) is divided into autonomic and peripheral parts; B. the human nervous system consists of the central nervous system (CNS) and the autonomic nervous system; C. the autonomic nervous system consists of sympathetic and parasympathetic parts; D. the central nervous system (CNS) is made up essentially of brain and spinal cord; E. all of the above.

11. Match the following:

___ a. contains nucleus; cannot regenerate since lacks mitotic apparatus
___ b. Yellowish pigment that increases with age; appears to be by-product of lysosomes
___ c. provide energy for neurons
___ d. long, thin fibrils composed of microtubules; may function in transport
___ e. orderly arrangement of rough ER; site of protein synthesis
___ f. conducts nerve impulses toward cell body
___ g. conducts nerve impulses away from cell body; has synaptic end bulbs that contain neurotransmitters
___ h. fine filaments that are branching ends of axon collaterals

A. axon
B. axon terminal
C. cell body
D. chromatophilic substance (Nissl bodies)
E. dendrite
F. lipofuscin
G. mitochondria
H. neurofibrils

Complete the following.

12. The area of contact or junction between two neurons is called a _____.
13. The part of a neuron that conducts nerve impulses toward the perikaryon is the _____.
14. Sacs in the axon terminals that store neurotransmitters are called _____.
15. In a _____ circuit, a nerve impulse from a single presynaptic neuron causes stimulation of several postsynaptic neurons.

Circle T (true) or F (false) for the following.

T F 16. Neurotransmitters are released at synapses and also at neuromuscular junctions.
T F 17. A neuron with several dendrites and one axon is classified as a sensory neuron.
T F 18. The ability of a neuron to transmit a nerve impulse to another neuron or tissue is called conductivity.
T F 19. Neuroglia are a common source of tumors of the nervous system.
T F 20. Acetylcholinesterase (AChE) is an example of a neurotransmitter.

17 The Spinal Cord and the Spinal Nerves

STUDENT OBJECTIVES

1. Describe how neural tissue is grouped.
2. Explain how the spinal cord is protected.
3. Describe the gross anatomical features of the spinal cord.
4. Describe the structure and location of the spinal meninges.
5. Describe the structure of the spinal cord in cross section.
6. Explain the functions of the spinal cord as a conduction pathway and a reflex center.
7. List the location, origin, termination, and function of the principal ascending and descending tracts of the spinal cord.
8. Describe the components of a reflex arc.
9. Describe the composition and coverings of a spinal nerve.
10. Explain how a spinal nerve branches upon leaving an intervertebral foramen.
11. Explain the composition and distribution of the cervical, brachial, lumbar, and sacral plexuses.
12. Define a dermatome and state its clinical importance.
13. Describe spinal cord injury and list the immediate and long-range effects.
14. Identify the effects of peripheral nerve damage and conditions necessary for its regeneration.
15. Explain the causes and symptoms of neuritis, sciatica, and shingles.

CHAPTER OUTLINE

- **Grouping of Neural Tissue**
- **Spinal Cord**

Protection and Coverings
   Vertebral Canal
   Meninges
General Features
Structure in Cross Section
Functions
   Impulse Conduction
   Reflex Center

- **Spinal Nerves**

Names
Composition and Coverings
Distribution
   Branches
   Plexuses
   Intercostal (Thoracic) Nerves
Dermatomes

- **Applications to Health**

Spinal Cord Injury
Peripheral Nerve Damage and Repair
   Axon Reaction
   Wallerian Degeneration
   Retrograde Degeneration
   Regeneration
Neuritis
Sciatica
Shingles

In this chapter, our main concern will be to study the structure and function of the spinal cord and the nerves that originate from it. Keep in mind, however, that the spinal cord is continuous with the brain, and that together they constitute the central nervous system.

GROUPING OF NEURAL TISSUE

The term *white matter* refers to aggregations of myelinated axons from many neurons supported by neuroglia or neurolemmocytes (Schwann cells). The lipid substance myelin has a whitish color that gives white matter its name. The *gray matter* of the nervous system contains either nerve cell bodies and dendrites or bundles of unmyelinated axons and neuroglia. The absence of myelin in these areas accounts for their gray color.

A *nerve* is a bundle of fibers located outside the central nervous system. Since the dendrites of somatic afferent neurons and axons of somatic efferent neurons of the peripheral nervous system are myelinated, most nerves are white matter. Nerve cell bodies that lie outside the central nervous system are generally grouped with other nerve cell bodies to form *ganglia* (GANG-lē-a; *ganglion* = knot). Ganglia, since they are made up principally of nerve cell bodies, are masses of gray matter.

A *tract* is a bundle of fibers in the central nervous system. Tracts may run long distances up or down the spinal cord. Tracts also exist in the brain and connect parts of the brain with each other and with the spinal cord. The chief spinal tracts that conduct impulses up the cord are concerned with sensory impulses and are called *ascending tracts*. Spinal tracts that carry impulses down the cord are motor tracts and are called *descending tracts*. The major tracts consist of myelinated fibers and are therefore white matter. A *nucleus* is a mass of nerve cell bodies and dendrites in the central nervous system. It forms gray matter. *Horns (columns)* are the chief areas of gray matter in the spinal cord. The term *horn* describes the two-dimensional appearance of the organization of gray matter in the spinal cord as seen in cross section. The term *column* describes the three-dimensional appearance of the gray matter in longitudinal columns. Since the white matter of the spinal cord is also arranged in columns, we shall refer to the gray matter as being arranged in horns (see Figure 17-3).

SPINAL CORD

PROTECTION AND COVERINGS

Vertebral Canal

The spinal cord is located in the vertebral (spinal) canal of the vertebral column. The canal is formed by the vertebral foramina of all the vertebrae arranged on top of each other. Since the wall of the vertebral canal is essentially a ring

of bone surrounding the spinal cord, the cord is well protected. A certain degree of protection is also provided by the meninges, cerebrospinal fluid, and the vertebral ligaments.

Meninges

The *meninges* (me-NIN-jēz) are coverings that run continuously around the spinal cord and brain (*meninx* is singular). Those associated with the cord are known as *spinal meninges* (Figure 17-1). The outer spinal meninx is called the *dura mater* (DYOO-ra MĀ-ter), meaning tough mother. It forms a tube from the level of the second sacral vertebra, where it is fused with the filum terminale, to the foramen magnum, where it is continuous with the dura mater of the brain. It is composed of dense, fibrous connective tissue. Between the dura mater and the wall of the vertebral canal is the *epidural space,* which is filled with fat, connective tissue, and blood vessels. It serves as padding around the cord. The epidural space inferior to the second lumbar vertebra is the site for the injection of anesthetics, such as a saddleblock for childbirth.

The middle spinal meninx is called the *arachnoid* (a-RAK-noyd), or spider layer. It is a delicate connective tissue membrane that forms a tube inside the dura mater. It is also continuous with the arachnoid of the brain. Between the dura mater and the arachnoid is the *subdural space,* which contains serous fluid.

The inner meninx is known as the *pia mater* (PĪ-a-MĀ-ter), or delicate mother. It is a transparent fibrous membrane that forms a tube around and adheres to the surface of the spinal cord and brain. It contains numerous blood vessels. Between the arachnoid and the pia mater is the *subarachnoid space,* where the cerebrospinal fluid circulates.

All three spinal meninges cover the spinal nerves up to the point of exit from the spinal column through the intervertebral foramina. The spinal cord is suspended in the middle of its dural sheath by membranous extensions of the pia mater. These extensions, called the *denticulate* (den-TIK-yoo-lāt) *ligaments,* are attached laterally to the dura mater along the length of the cord between the ventral and dorsal nerve roots on either side. The ligaments protect the spinal cord against shock and sudden displacement. Essentially, the spinal cord is fixed in its position in the vertebral canal, since it is anchored to the coccyx inferiorly by the filum terminale, laterally to the dura mater by the denticulate ligaments, and superiorly to the brain.

CLINICAL APPLICATION

Inflammation of the meninges is known as *meningitis.* If only the dura mater becomes inflamed, the condition is *pachymeningitis.* Inflammation of the arachnoid and pia mater is *leptomeningitis,* the most common form of meningitis.

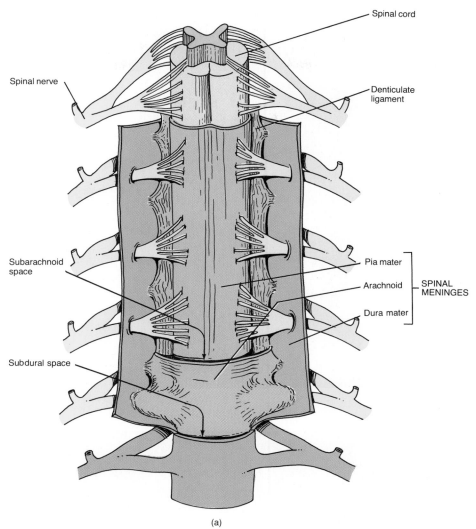

Spinal cord

Spinal nerve

Denticulate ligament

Subarachnoid space

Pia mater

Arachnoid

SPINAL MENINGES

Dura mater

Subdural space

(a)

FIGURE 17-1 Spinal meninges. (a) Location of the spinal meninges as seen in sections of the spinal cord.

Cerebrospinal fluid is removed from the subarachnoid space in the inferior lumbar region of the spinal cord by a *spinal (lumbar) puncture (tap)*. The procedure is normally performed between the third and fourth or fourth and fifth lumbar vertebrae. The spinous process of the fourth lumbar vertebra is easily located. A line drawn across the highest points of the iliac crests will pass through the spinous process. A lumbar puncture is made below the spinal cord and thus poses little danger to it. If the patient lies on one side, drawing the knees and chest together, the vertebrae separate slightly so that a needle can be conveniently inserted. In its course, the needle pierces the skin, superficial fascia, supraspinous ligament, interspinous ligament, epidural space, and arachnoid, to enter the subarachnoid space. Spinal punctures are used to withdraw fluid for diagnostic purposes and to introduce antibiotics (as in the case of meningitis) and contrast media. For example, in a radiologic tech-

nique called *myelography* (mī-e-LOG-ra-fē), a contrast medium (usually metrizamide) is introduced into the spinal subarachnoid space to determine or exlude the presence of lesions within and around the spinal cord.

GENERAL FEATURES

The *spinal cord* is a cylindrical structure that is slightly flattened anteriorly and posteriorly. It begins as a continuation of the medulla oblongata, the inferior part of the brain stem, and extends from the foramen magnum of the occipital bone to the level of the second lumbar vertebra (Figure 17-2). The length of the adult spinal cord ranges from 42 to 45 cm (16 to 18 in.). The diameter of the cord is about 2.54 cm (1 in.) in the midthoracic region but is somewhat larger in the lower cervical and midlumbar regions.

When the cord is viewed externally, two conspicuous

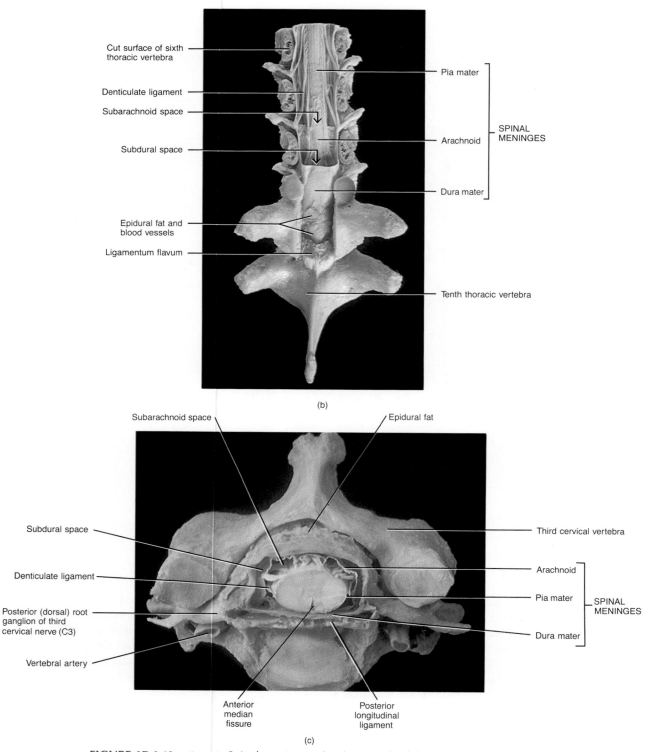

(b)

(c)

FIGURE 17-1 (Continued) Spinal meninges. (b) Photograph of the posterior aspect of the spinal cord. (c) Photograph of a cross section through the spinal cord between the second and third cervical vertebrae. (Courtesy of N. Gluhbegovic and T. H. Williams, *The Human Brain: A Photographic Guide,* Harper & Row, Publishers, Inc., New York, 1980.)

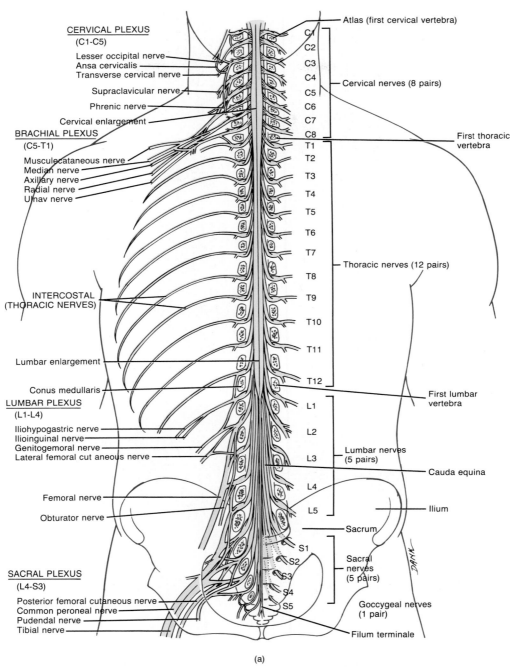

CERVICAL PLEXUS
(C1-C5)

Lesser occipital nerve
Ansa cervicalis
Transverse cervical nerve
Supraclavicular nerve
Phrenic nerve
Cervical enlargement

BRACHIAL PLEXUS
(C5-T1)

Musculocataneous nerve
Median nerve
Axillary nerve
Radial nerve
Ulnav nerve

INTERCOSTAL
(THORACIC NERVES)

Lumbar enlargement

Conus medullaris

LUMBAR PLEXUS
(L1-L4)

Iliohypogastric nerve
Ilioinguinal nerve
Genitogemoral nerve
Lateral femoral cut aneous nerve

Femoral nerve

Obturator nerve

SACRAL PLEXUS
(L4-S3)

Posterior femoral cutaneous nerve
Common peroneal nerve
Pudendal nerve
Tibial nerve

Atlas (first cervical vertebra)

C1
C2
C3
C4
C5
C6
C7
C8

Cervical nerves (8 pairs)

First thoracic
vertebra

T1
T2
T3
T4
T5
T6
T7
T8
T9
T10
T11
T12

Thoracic nerves (12 pairs)

First lumbar
vertebra

L1
L2
L3
L4
L5

Lumbar nerves
(5 pairs)

Cauda equina

Ilium

Sacrum

S1
S2
S3
S4
S5

Sacral
nerves
(5 pairs)

Goccygeal nerves
(1 pair)

Filum terminale

(a)

FIGURE 17-2 Spinal cord and spinal nerves. (a) Diagram of anterior view.

enlargements can be seen. The superior enlargement, the *cervical enlargement,* extends from the fourth cervical to the first thoracic vertebra. Nerves that supply the upper extremities arise from the cervical enlargement. The inferior enlargement, called the *lumbar enlargement,* extends from the ninth to the twelfth thoracic vertebra. Nerves that supply the lower extremities arise from the lumbar enlargement.

Below the lumbar enlargement, the spinal cord tapers to a conical portion known as the *conus medullaris* (KŌ-nus med-yoo-LAR-is). The conus medullaris ends at the

level of the intervertebral disc between the first and second lumbar vertebra. Arising from the conus medullaris is the *filum terminale* (FĪ-lum ter-mi-NAL-ē), a nonnervous fibrous tissue of the spinal cord that extends inferiorly to attach to the coccyx. The filum terminale consists mostly of pia mater, the innermost of three meninges that cover and protect the spinal cord and brain. Some nerves that arise from the lower portion of the cord do not leave the vertebral column immediately. They angle inferiorly in the vertebral canal like wisps of coarse hair flowing from the

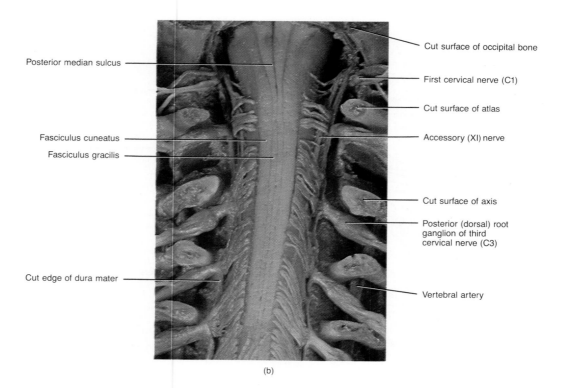

Posterior median sulcus

Fasciculus cuneatus

Fasciculus gracilis

Cut edge of dura mater

Cut surface of occipital bone

First cervical nerve (C1)

Cut surface of atlas

Accessory (XI) nerve

Cut surface of axis

Posterior (dorsal) root ganglion of third cervical nerve (C3)

Vertebral artery

(b)

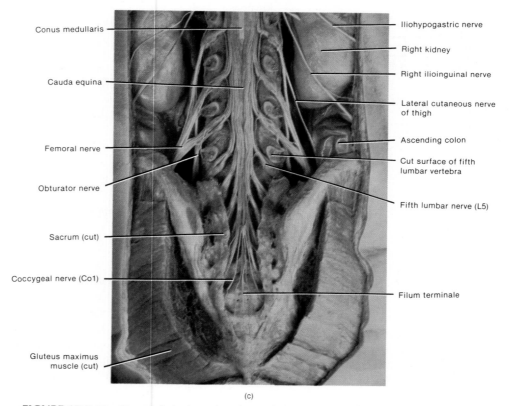

Conus medullaris

Cauda equina

Femoral nerve

Obturator nerve

Sacrum (cut)

Coccygeal nerve (Co1)

Gluteus maximus muscle (cut)

Iliohypogastric nerve

Right kidney

Right ilioinguinal nerve

Lateral cutaneous nerve of thigh

Ascending colon

Cut surface of fifth lumbar vertebra

Fifth lumbar nerve (L5)

Filum terminale

(c)

FIGURE 17-2 (Continued) Spinal cord and spinal nerves. (b) Photograph of the posterior aspect of the inferior portion of the medulla and the superior six segments of the cervical portion of the spinal cord. (c) Photograph of the posterior aspect of the conus medullaris and cauda equina. (Photographs courtesy of N. Gluhbegovic and T. H. Williams, *The Human Brain: A Photographic Guide,* Harper & Row, Publishers, Inc., New York, 1980.)

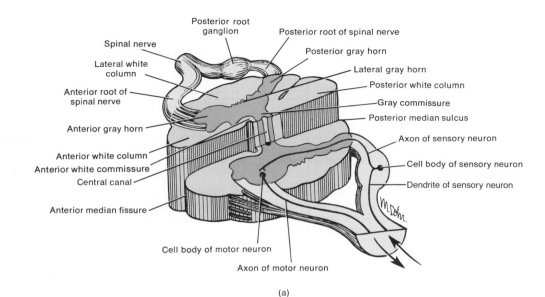

(a)

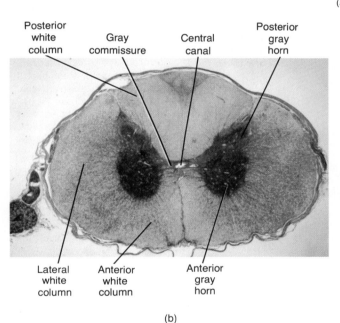

(b)

FIGURE 17-3 Spinal cord. (a) The organization of gray and white matter in the spinal cord as seen in cross section. The front of the figure has been sectioned at a lower level than the back so that you can see what is inside the posterior root ganglion, posterior root of the spinal nerve, anterior root of the spinal nerve, and the spinal nerve. In this and other illustrations of cross sections of the spinal cord, dendrites are not shown in relation to cell bodies of motor or association neurons for purposes of simplicity. (b) Photograph of the spinal cord at the seventh cervical segment. (Courtesy of Victor B. Eichler, Ph.D., Wichita, Kansas.)

end of the cord. They are appropriately named the *cauda equina* (KAW-da ē-KWĪ-na), meaning horse's tail.

The spinal cord is a series of 31 sections called segments, each giving rise to a pair of spinal nerves. *Spinal segment* refers to a region of the spinal cord from which a pair of spinal nerves arises. The cord is divided into right and left sides by two grooves (Figure 17-3). The *anterior median fissure* is a deep, wide groove on the anterior (ventral) surface, and the *posterior median sulcus* is a shallower, narrow groove on the posterior (dorsal) surface.

STRUCTURE IN CROSS SECTION

The spinal cord consists of both gray and white matter. Figure 17-3 shows that the gray matter forms an H-shaped area within the white matter. The gray matter consists pri-

marily of nerve cell bodies and unmyelinated axons and dendrites of association and motor neurons. The white matter consists of bundles of myelinated axons of motor and sensory neurons.

The cross bar of the H is formed by the *gray commissure* (KOM-mi-shur). In the center of the gray commissure is a small space called the *central canal*. This canal runs the length of the spinal cord and is continuous with the fourth ventricle of the medulla. It contains cerebrospinal fluid. Anterior to the gray commissure is the *anterior (ventral) white commissure*, which connects the white matter of the right and left sides of the spinal cord.

The upright portions of the H are further subdivided into regions. Those closer to the front of the cord are called *anterior (ventral) gray horns*. They represent the motor part of the gray matter. The regions closer to the back of the cord are referred to as *posterior (dorsal) gray horns*. They represent the sensory part of the gray matter. The regions between the anterior and posterior gray horns are intermediate *lateral gray horns*. The lateral gray horns are present in the thoracic, upper lumbar, and sacral segments of the cord.

The gray matter of the cord also contains several nuclei that serve as relay stations for nerve impulses and origins for certain nerves. Nuclei are clusters of nerve cell bodies and dendrites in the spinal cord and brain.

The white matter, like the gray matter, is also organized

EXHIBIT 17-1

Selected Ascending and Descending Tracts of Spinal Cord

TRACT	LOCATION (WHITE COLUMN)	ORIGIN	TERMINATION	FUNCTION
ASCENDING TRACTS				
Anterior (ventral) spinothalamic	Anterior (ventral) column.	Posterior (dorsal) gray horn on one side of cord, but crosses to opposite side of brain.	Thalamus; impulses eventually conveyed to cerebral cortex.	Conveys sensations for touch and pressure from one side of body to opposite side of thalamus. Eventually sensations reach cerebral cortex.
Lateral spinothalamic	Lateral column.	Posterior (dorsal) gray horn on one side of cord, but crosses to opposite side of brain.	Thalamus; impulses eventually conveyed to cerebral cortex.	Conveys sensations for pain and temperature from one side of body to opposite side of thalamus. Eventually sensations reach cerebral cortex.
Fasciculus gracilis and fasciculus cuneatus	Posterior (dorsal) column.	Axons of afferent neurons from periphery that enter posterior (dorsal) column on one side of cord and rise to same side of brain.	Nucleus gracilis and nucleus cuneatus of medulla; impulses eventually conveyed to cerebral cortex.	Convey sensations from one side of body to same side of medulla for touch; two-point discrimination (ability to distinguish that two points on skin are touched even though close together); proprioception (awareness of precise position of body parts and their direction of movement); stereognosis (ability to recognize size, shape, and texture of object); weight discrimination (ability to assess weight of an object); and vibration. Eventually sensations may reach cerebral cortex.
Posterior (dorsal) spinocerebellar	Posterior (dorsal) portion of lateral column.	Posterior (dorsal) gray horn on one side of cord and rises to same side of brain.	Cerebellum.	Conveys sensations from one side of body to same side of cerebellum for subconscious proprioception.
Anterior (ventral) spinocerebellar	Anterior (ventral) portion of lateral column.	Posterior (dorsal) gray horn on one side of cord; contains both crossed and uncrossed fibers.	Cerebellum.	Conveys sensations from both sides of body to cerebellum for subconscious proprioception.

into regions. The anterior and posterior gray horns divide the white matter on each side into three broad areas: *anterior (ventral) white columns, posterior (dorsal) white columns,* and *lateral white columns.* Each column in turn consists of distinct bundles of myelinated fibers within the cord. These bundles are called *tracts* or *fasciculi* (fa-SIK-yoo-lī). The longer *ascending tracts* consist of sensory axons that conduct nerve impulses that enter the spinal cord upward to the brain. The longer *descending tracts* consist of motor axons that conduct nerve impulses from the brain downward into the spinal cord, where they synapse with other neurons whose axons pass out to muscles and glands. Thus, the ascending tracts are sensory tracts and the descending tracts are motor tracts.

FUNCTIONS

A major function of the spinal tracts in the spinal cord is to convey sensory impulses from the periphery to the brain and to conduct motor impulses from the brain to the periph-

ery. A second principal function is to provide a means of integrating reflexes. Both functions are essential to maintaining homeostasis.

Impuse Conduction

The vital function of conveying sensory and motor information to and from the brain is carried out by the ascending and descending tracts of the cord. The names of the tracts indicate the white column (funiculus) in which the tract travels, where the cell bodies of the tract originate, and where the axons of the tract terminate. Since the origin and termination are specified, the direction of impulse conduction is also indicated by the name. For example, the anterior spinothalamic tract is located in the *anterior* white column, it originates in the *spinal* cord, and it terminates in the *thalamus* (a region of the brain). It is an ascending (sensory) tract since it conveys nerve impulses from the cord upward to the brain.

EXHIBIT 17-1 (*Continued*)

TRACT	LOCATION (WHITE COLUMN)	ORIGIN	TERMINATION	FUNCTION
DESCENDING TRACTS				
Lateral corticospinal	Lateral column.	Cerebral cortex on one side of brain, but crosses in base of medulla to opposite side of cord	Anterior (ventral) gray horn.	Conveys motor impulses from one side of cortex to anterior gray horn of opposite side. Eventually impulses reach skeletal muscles on opposite side of body that coordinate precise, discrete movements.
Anterior (ventral) corticospinal	Anterior (ventral) column.	Cerebral cortex on both sides of brain, uncrossed in medulla, but crosses to opposite side of cord.	Anterior (ventral) gray horn.	Conveys motor impulses from one side of cortex to anterior gray horn of same side. Impulses cross to opposite side in spinal cord and reach skeletal muscles that coordinate movements of the axial skeleton.
Rubrospinal	Lateral column.	Midbrain (red nucleus) on one side of brain, but crosses to opposite side of cord.	Anterior (ventral) gray horn.	Conveys motor impulses from one side of midbrain to skeletal muscles on opposite side of body that are concerned with precise, discrete movements.
Tectospinal	Anterior (ventral) column.	Midbrain on one side of brain, but crosses to opposite side of cord.	Anterior (ventral) gray horn.	Conveys motor impulses from one side of midbrain to skeletal muscles on opposite side of body that control movements of head in response to auditory, visual, and cutaneous stimuli.
Vestibulospinal	Anterior (ventral) column.	Medulla on one side of brain and descends to same side of cord.	Anterior (ventral) gray horn.	Conveys motor impulses from one side of medulla to skeletal muscles on same side of body that regulate body tone in response to movements of head (equilibrium).
Lateral reticulospinal	Lateral column.	Medulla on one side of brain and descends mainly to same side of cord.	Anterior (ventral) gray horn.	Conveys motor impulses from one side of medulla to axial skeleton and proximal limb muscles that inhibit extensor reflexes and muscle tone.
Anterior (ventral) or medial reticulospinal	Anterior (ventral) column.	Pons on one side of brain and descends mainly to same side of cord.	Anterior (ventral) gray horn.	Conveys motor impulses from one side of pons to axial skeleton and proximal limb muscles that facilitate extensor reflexes and muscle tone.

The principal ascending and descending tracts are listed in Exhibit 17-1 and shown in Figure 17-4.

Reflex Center

The second principal function of the spinal cord is to serve as a center for reflex actions. Spinal nerves are the paths of communication between the spinal cord tracts and the periphery. Figure 17-3 reveals that each pair of spinal nerves is connected to a segment of the cord by two points of attachment called roots. The *posterior* or *dorsal (sensory) root* contains sensory nerve fibers only and conducts nerve impulses from the periphery to the spinal cord. These fibers extend into the posterior (dorsal) gray horn. Each dorsal root also has a swelling, the *posterior* or *dorsal (sensory) root ganglion,* which contains the cell bodies of the sensory neurons from the periphery. The other point of attachment of a spinal nerve to the cord is the *anterior* or *ventral (motor) root.* It contains motor nerve fibers only and conducts nerve impulses from the spinal cord to the periphery.

The cell bodies of the motor neurons are located in the gray matter of the cord. If the motor impulse supplies a skeletal muscle, the cell bodies are located in the anterior (ventral) gray horn. If, however, the motor impulse supplies smooth muscle, cardiac muscle, or a gland through the autonomic nervous system, the cell bodies are located in the lateral gray horn.

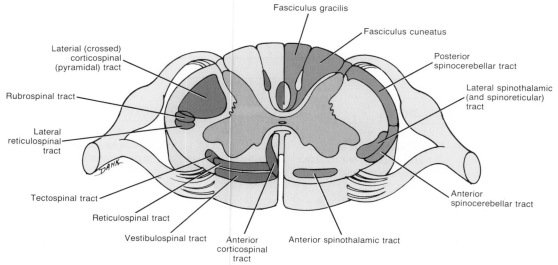

Fasciculus gracilis

Fasciculus cuneatus

Laterial (crossed) corticospinal (pyramidal) tract

Posterior spinocerebellar tract

Rubrospinal tract

Lateral spinothalamic (and spinoreticular) tract

Lateral reticulospinal tract

Tectospinal tract

Reticulospinal tract

Vestibulospinal tract

Anterior corticospinal tract

Anterior spinothalamic tract

Anterior spinocerebellar tract

FIGURE 17-4 Selected tracts of the spinal cord. Ascending (sensory) tracts are shown in pink; descending (motor) tracts are shown in blue.

The path a nerve impulse follows from its origin in the dendrites or cell body of a neuron in one part of the body to its termination elsewhere in the body is called a ***conduction pathway.*** All conduction pathways consist of circuits of neurons. One pathway is known as a ***reflex arc,*** the functional unit of the nervous system. A reflex arc contains two or more neurons over which nerve impulses are conducted from a receptor to the brain or spinal cord and then to an effector. The basic components of a reflex arc are as follows (Figure 17-5).

1. ***Receptor.*** The distal end of a dendrite or a sensory structure associated with the distal end of a dendrite. Its role in the reflex arc is to respond to a change in the internal or external environment by initiating a nerve impulse in a sensory neuron.

2. ***Sensory neuron.*** Passes the nerve impulse from the receptor to its axonal termination in the central nervous system.

3. ***Center.*** A region in the central nervous system where an incoming sensory impulse generates an outgoing motor impulse. In the center, the nerve impulse may be inhibited, transmitted, or rerouted. In the center of some reflex arcs, the sensory neuron directly generates the nerve impulse in the motor neuron. The center may also contain an association neuron between the sensory neuron and the motor neuron leading to a muscle or a gland.

4. ***Motor neuron.*** Transmits the nerve impulse generated by the sensory or association neuron in the center to the organ of the body that will respond.

5. ***Effector.*** The organ of the body, either a muscle or gland, that responds to the motor impulse. This response is called a reflex.

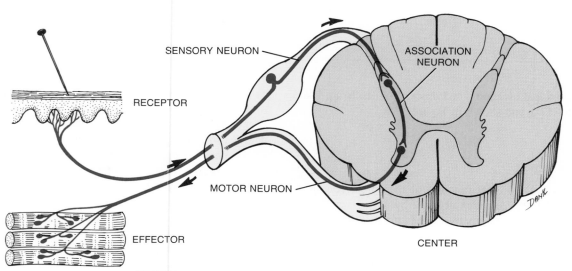

SENSORY NEURON

ASSOCIATION NEURON

RECEPTOR

MOTOR NEURON

EFFECTOR

CENTER

FIGURE 17-5 Components of a generalized reflex arc.

Reflexes are fast responses to changes in the internal or external environment that allow the body to maintain homeostasis. Reflexes are associated not only with skeletal muscle contraction but also with body functions such as heart rate, respiration, digestion, urination, and defecation. Reflexes carried out by the spinal cord alone are called ***spinal reflexes***. Reflexes that result in the contraction of skeletal muscles are known as ***somatic reflexes***. Those that cause the contraction of smooth or cardiac muscle or secretion by glands are ***visceral (autonomic) reflexes***.

SPINAL NERVES

NAMES

The 31 pairs of spinal nerves are named and numbered according to the region and level of the spinal cord from which they emerge (see Figure 17-2). The first cervical pair emerges between the atlas and the occipital bone. All other spinal nerves leave the vertebral column from the intervertebral foramina between adjoining vertebrae. There are 8 pairs of cervical nerves, 12 pairs of thoracic nerves, 5 pairs of lumbar nerves, 5 pairs of sacral nerves, and 1 pair of coccygeal nerves.

During fetal life, the spinal cord and vertebral column grow at different rates, the cord growing more slowly. Thus, not all the spinal cord segments are in line with their corresponding vertebrae. Remember that the spinal cord terminates near the level of the first or second lumbar vertebra. Thus, the lower lumbar, sacral, and coccygeal nerves must descend more and more to reach their foramina before emerging from the vertebral column. This arrangement constitutes the cauda equina.

COMPOSITION AND COVERINGS

A ***spinal nerve*** has two points of attachment to the cord: a posterior root and an anterior root. The posterior and anterior roots unite to form a spinal nerve at the intervertebral foramen. Since the posterior root contains sensory fibers and the anterior root contains motor fibers, a spinal nerve is a ***mixed nerve***. The posterior (dorsal) root ganglion contains cell bodies of sensory neurons.

In Figure 17-6, you can see that a spinal nerve contains many fibers surrounded by different coverings. The individual fibers, whether myelinated or unmyelinated, are wrapped in a connective tissue called the ***endoneurium*** (en'-dō-NYOO-rē-um). Groups of fibers with their endoneurium are arranged in bundles called ***fascicles***, and each bundle is wrapped in connective tissue called the ***perineurium*** (per'-i-NYOO-rē-um). The outermost covering around the entire nerve is the ***epineurium*** (ep'-i-NYOO-rē-um). The spinal meninges fuse with the epineurium as the nerve exits from the vertebral canal.

DISTRIBUTION

Branches

Shortly after a spinal nerve leaves its intervertebral foramen, it divides into several branches (Figure 17-7). These branches are known as ***rami*** (RĀ-mē). The ***dorsal ramus*** (RĀ-mus) innervates the deep muscles and skin of the dorsal

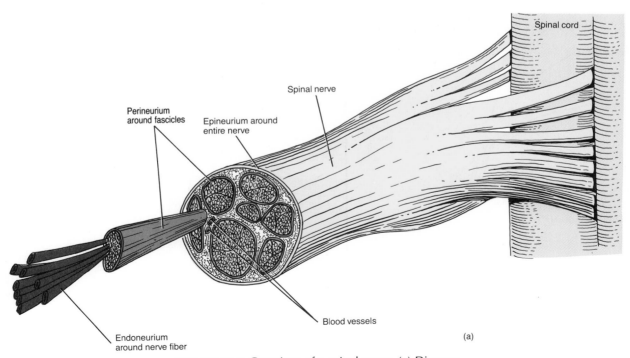

FIGURE 17-6 Coverings of a spinal nerve. (a) Diagram.

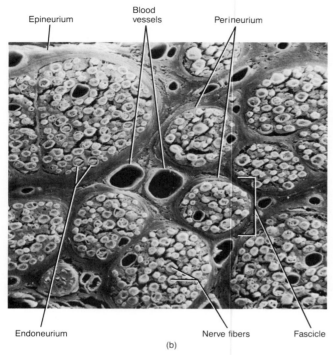

FIGURE 17-6 (*Continued*) (b) Scanning electron micrograph at a magnification of 900×. (Courtesy of Richard G. Kessel and Randy H. Kardon, from *Tissues and Organs: A Text-Atlas of Scanning Electron Microscopy*. Copyright © 1979 by Scientific American, Inc.)

surface of the back. The ***ventral ramus*** of a spinal nerve innervates the muscles and structures of the extremities and the lateral and ventral trunk. In addition to dorsal and ventral rami, spinal nerves also give off a ***meningeal branch.*** This branch reenters the spinal canal through the intervertebral foramen and supplies the vertebrae, vertebral ligaments, blood vessels of the spinal cord, and the meninges. Other branches of a spinal nerve are the ***rami communicantes*** (kō-myoo-ni-KAN-tēz), components of the autonomic nervous system whose structure and function are discussed in Chapter 19.

Plexuses

The ventral rami of spinal nerves, except for thoracic nerves T2–T11, do not go directly to the structures of the body they supply. Instead, they form networks by joining with adjacent nerves on either side of the body. Such a network is called a ***plexus*** (*plexus* = braid). The principal plexuses are the cervical plexus, brachial plexus, lumbar plexus, and sacral plexus (see Figure 17-2). Emerging from the plexuses are nerves bearing names that are often descriptive of the general regions they supply or the course they take. Each of the nerves, in turn, may have several branches named for the specific structures they innervate.

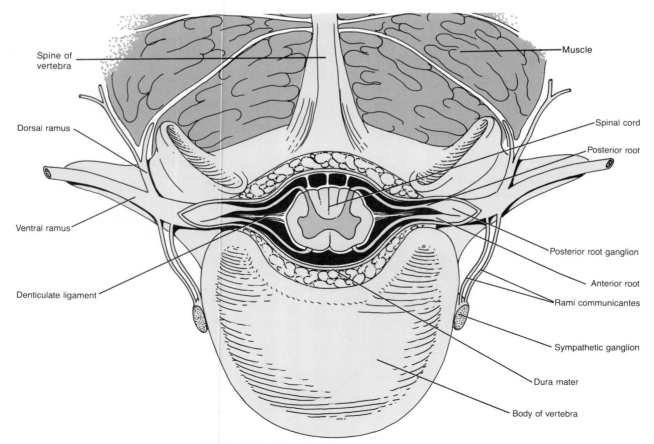

FIGURE 17-7 Branches of a typical spinal nerve.

• **Cervical Plexus** The **cervical plexus** is formed by the ventral rami of the first four cervical nerves (C1–C4) with contributions from C5. There is one on each side of the neck alongside the first four cervical vertebrae (Figure 17-8). The **roots** of the plexus indicated in the diagram are the ventral rami. The cervical plexus supplies the skin and muscles of the head, neck, and upper part of the shoulders. Branches of the cervical plexus also connect with the accessory (XI) and hypoglossal (XII) cranial nerves. The phrenic nerves are a major pair of nerves arising from the cervical plexuses that supply the motor fibers to the diaphragm. Damage to the spinal cord above the origin of the phrenic nerves results in paralysis of the diaphragm, since the phrenic nerves no longer send impulses to the diaphragm. Contractions of the diaphragm are essential for normal breathing.

Exhibit 17-2 summarizes the nerves and distributions of the cervical plexus. The relation of the cervical plexus to the other plexuses is shown in Figure 17-2a.

• **Brachial Plexus** The **brachial plexus** is formed by the ventral rami of spinal nerves C5–C8 and T1. On either side of the last four cervical and first thoracic vertebrae, the brachial plexus extends downward and laterally, passes over the first rib behind the clavicle, and then enters the axilla (Figure 17-9). The brachial plexus constitutes the entire nerve supply for the upper extremities and shoulder region.

The **roots** of the brachial plexus, like those of the cervical plexus, are the ventral rami of the spinal nerves. The roots of C5 and C6 unite to form the **superior trunk,** C7 becomes the **middle trunk,** and C8 and T1 form the **inferior trunk.** Each trunk, in turn, divides into an **anterior division** and a **posterior division.** The divisions then unite to form cords. The **posterior cord** is formed by the union of the posterior divisions of the superior, middle, and inferior trunks. The **medial cord** is formed as a continuation of the anterior division of the inferior trunk. The **lateral cord** is formed by the union of the anterior divisions of the superior and

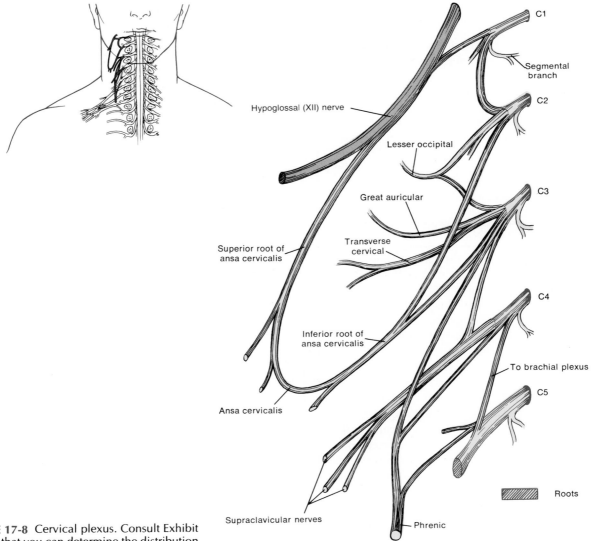

FIGURE 17-8 Cervical plexus. Consult Exhibit 17-2 so that you can determine the distribution of each of the nerves of the plexus.

EXHIBIT 17-2

Cervical Plexus

NERVE	ORIGIN	DISTRIBUTION
SUPERFICIAL OR CUTANEOUS BRANCHES		
Lesser occipital	C2.	Skin of scalp behind and above ear.
Greater auricular	C2–C3.	Skin in front, below, and over ear and over parotid glands.
Transverse cervical	C2–C3.	Skin over anterior aspect of neck.
Supraclavicular	C3–C4.	Skin over upper portion of chest and shoulder.
DEEP OR LARGELY MOTOR BRANCHES		
Ansa cervicalis	This nerve is divided into a superior root and an inferior root.	
Superior root	C1.	Infrahyoid, thyrohyoid, and geniohyoid muscles of neck.
Inferior root	C2–C3.	Omohyoid, sternohyoid, and sternothyroid muscles of neck.
Phrenic	C3–C5.	Diaphragm between thorax and abdomen.
Segmental branches	C1–C5.	Prevertebral (deep) muscles of neck, levator scapulae, and middle scalene muscles.

middle trunk. The peripheral nerves arise from the cords. Thus, the brachial plexus begins as roots that unite to form trunks, the trunks branch into divisions, the divisions form cords, and the cords give rise to the peripheral nerves.

Five important nerves arising from the brachial plexus are the axillary, musculocutaneous, radial, median, and ulnar. The axillary nerve supplies the deltoid and teres minor muscles. The musculocutaneous nerve supplies the flexors of the arm and forearm. The radial nerve supplies the muscles on the posterior aspect of the arm and forearm. The median nerve supplies most of the muscles of the anterior forearm and some of the muscles in the palm. The ulnar nerve supplies the anteromedial muscles of the forearm and most of the muscles of the palm.

CLINICAL APPLICATION

Prolonged use of a crutch that presses into the axilla may result in injury to a portion of the brachial plexus. The usual *crutch palsy* involves the posterior cord of the brachial plexus or, more frequently, just the radial nerve, which, in general, supplies extensors.

Radial nerve damage is indicated by wrist drop: inability to extend the hand at the wrist. Care must be taken not to injure the radial and axillary nerves when deltoid intramuscular injections are given. *Median nerve damage* is indicated by numbness, tingling, and pain in the palm and fingers; weak thumb movements; and inability to pronate the forearm and difficulty in flexing the wrist properly. Compression of the median nerve inside the carpal tunnel, formed anteriorly by the flexor retinaculum (transverse carpal ligament) and posteriorly by the carpal bones, is known as *carpal tunnel syndrome.* It may be caused by any condition that aggravates compression of the contents of the carpal tunnel, such as trauma, edema, and flexion of the wrist. *Ulnar nerve damage* is indicated by an inability to adduct or abduct the fingers (not the thumb) and weakness in flexing and adducting the wrist.

A summary of the nerves and distributions of the brachial plexus is given in Exhibit 17-3. The relation of the brachial plexus to the other plexuses is shown in Figure 17-2a.

● *Lumbar Plexus* The *lumbar plexus* is formed by the ventral rami of spinal nerves L1–L4. It differs from the brachial plexus in that there is no intricate interlacing of fibers. It also consists of *roots* and an *anterior* and *posterior division.* On either side of the first four lumbar vertebrae, the lumbar plexus passes obliquely outward behind the psoas major muscle (posterior divison) and anterior to the quadratus lumborum muscle (anterior division), and then gives rise to its peripheral nerves (Figure 17-10). The lumbar plexus supplies the anterolateral abdominal wall, external genitals, and part of the lower extremity. The largest nerve arising from the lumbar plexus is the femoral nerve.

CLINICAL APPLICATION

Injury to the femoral nerve is indicated by an inability to extend the leg and by loss of sensation in the skin over the anteromedial aspect of the thigh.

A summary of the nerves and distributions of the lumbar plexus is presented in Exhibit 17-4. The relation of the lumbar plexus to the other plexuses is shown in Figure 17-2a.

● *Sacral Plexus* The *sacral plexus* is formed by the ventral rami of spinal nerves L4–L5 and S1–S4. It is situated

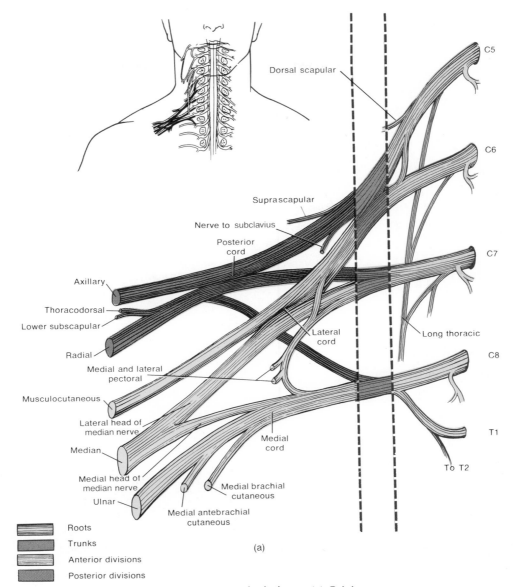

Dorsal scapular

Suprascapular

Nerve to subclavius

Posterior cord

Axillary

Thoracodorsal

Lower subscapular

Radial

Medial and lateral pectoral

Musculocutaneous

Lateral head of median nerve

Median

Medial head of median nerve

Ulnar

Medial antebrachial cutaneous

Medial brachial cutaneous

Lateral cord

Long thoracic

Medial cord

C5

C6

C7

C8

T1

To T2

Roots

Trunks

Anterior divisions

Posterior divisions

(a)

FIGURE 17-9 Brachial plexus. (a) Origin.

largely in front of the sacrum (Figure 17-11). Like the lumbar plexus, it contains *roots* and an *anterior* and *posterior division*. The sacral plexus supplies the buttocks, perineum, and lower extremities. The largest nerve arising from the sacral plexus—and, in fact, the largest nerve in the body—is the sciatic nerve. The sciatic nerve supplies the entire musculature of the leg and foot.

CLINICAL APPLICATION

Damage to the sciatic nerve (common peroneal portion) and its branches results in foot drop, an inability to dorsiflex the foot. This nerve may be injured because of a slipped disc, dislocated hip, pressure from the uterus during pregnancy, or an improperly given gluteal intramuscular injection.

A summary of the nerves and distributions of the sacral plexus is given in Exhibit 17-5. The relation of the sacral plexus to the other plexuses is shown in Figure 17-2a.

Intercostal (Thoracic) Nerves

The ventral rami of spinal nerves T2–T11 do not enter into the formation of plexuses and are known as *intercostal (thoracic) nerves* and are distributed directly to the structures they supply in intercostal spaces (see Figure 17-2). After leaving its intervertebral foramen, the ventral ramus of nerve

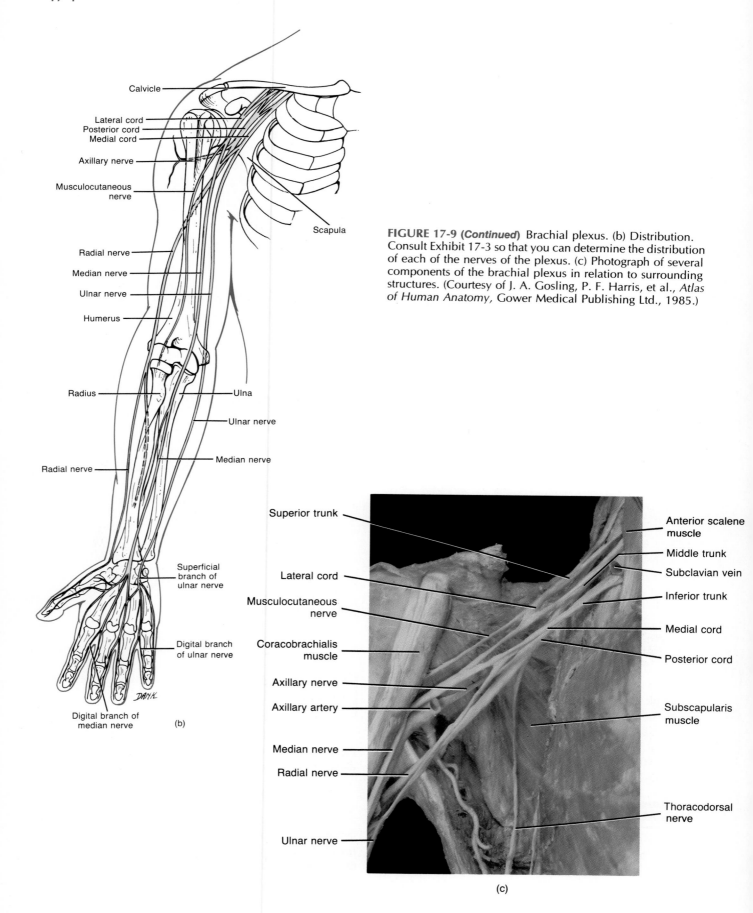

FIGURE 17-9 (Continued) Brachial plexus. (b) Distribution. Consult Exhibit 17-3 so that you can determine the distribution of each of the nerves of the plexus. (c) Photograph of several components of the brachial plexus in relation to surrounding structures. (Courtesy of J. A. Gosling, P. F. Harris, et al., *Atlas of Human Anatomy*, Gower Medical Publishing Ltd., 1985.)

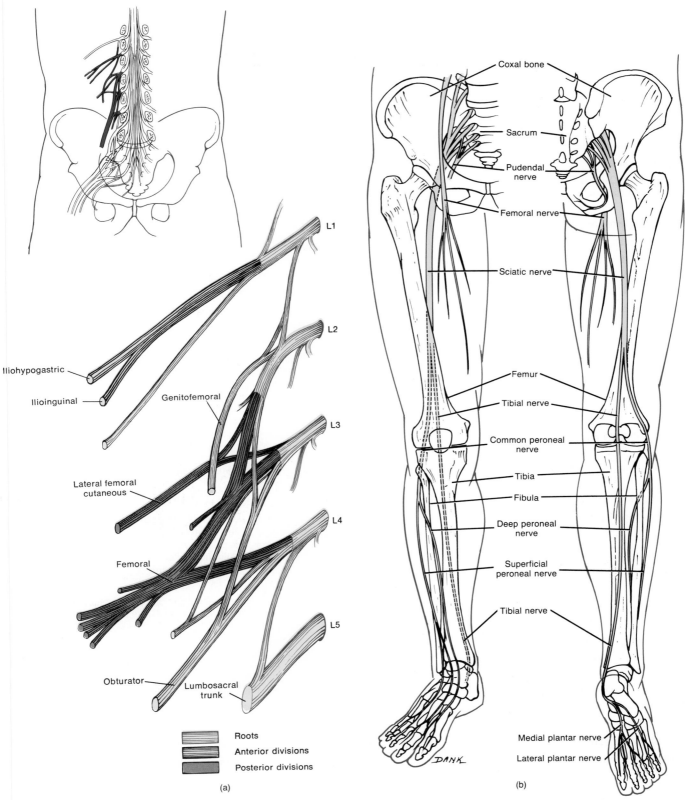

Iliohypogastric

Ilioinguinal

Genitofemoral

L1

L2

L3

L4

L5

Lateral femoral cutaneous

Femoral

Obturator

Lumbosacral trunk

Roots

Anterior divisions

Posterior divisions

(a)

Coxal bone

Sacrum

Pudendal nerve

Femoral nerve

Sciatic nerve

Femur

Tibial nerve

Common peroneal nerve

Tibia

Fibula

Deep peroneal nerve

Superficial peroneal nerve

Tibial nerve

Medial plantar nerve

Lateral plantar nerve

DANK

(b)

FIGURE 17-10 Lumbar plexus. (a) Origin. (b) Distribution of nerves of the lumbar and sacral plexuses in anterior view (left) and posterior view (right). Consult Exhibit 17-4 so that you can determine the distribution of the nerves of the lumbar plexus.

EXHIBIT 17-3

Brachial Plexus

NERVE	ORIGIN	DISTRIBUTION
ROOT NERVES		
Dorsal scapular	C5.	Levator scapulae, rhomboideus major, and rhomboideus minor muscles.
Long thoracic	C5–C7.	Serratus anterior muscle.
TRUNK NERVES		
Nerve to subclavius	C5–C6.	Subclavius muscle.
Suprascapular	C5–C6.	Supraspinatus and infraspinatus muscles.
LATERAL CORD NERVES		
Musculocutaneous	C5–C7.	Coracobrachialis, biceps, brachii, and brachialis muscles.
Median (lateral head)	C5–C7.	See distribution for **Median (medial head)** in this exhibit.
Lateral pectoral	C5–C7.	Pectoralis major muscle.
POSTERIOR CORD NERVES		
Upper subscapular	C5–C6.	Subscapularis muscle.
Thoracodorsal	C6–C8.	Latissimus dorsi muscle.
Lower subscapular	C5–C6.	Subscapularis and teres major muscles.
Axillary (circumflex)	C5–C6.	Deltoid and teres minor muscles; skin over deltoid and upper posterior aspect of arm.
Radial	C5–C8 and T1.	Extensor muscles of arm and forearm (triceps brachii, brachioradialis, extensor carpi radialis longus, extensor digitorum, extensor carpi ulnaris, extensor carpi radialis brevis, extensor indicis); skin of posterior arm and forearm, lateral two-thirds of dorsum of hand, and fingers over proximal and middle phalanges.
MEDIAL CORD NERVES		
Medial pectoral	C8–T1.	Pectoralis major and pectoralis minor muscles.
Medial brachial cutaneous	C8–T1.	Skin of medial and posterior aspects of lower third of arm.
Medial antebrachial cutaneous	C8–T1.	Skin of medial and posterior aspects of forearm.
Median (medial head)	C5–C8 and T1.	Medial and lateral heads of median nerve form median nerve. Distributed to flexors of forearm (pronator teres, flexor carpi radialis, flexor digitorum superficialis, lateral half of flexor digitorum profundus), except flexor carpi ulnaris; skin of lateral two-thirds of palm of hand and fingers.
Ulnar	C8–T1.	Flexor carpi ulnaris and flexor digitorum profundus muscles; skin of medial side of hand, little finger, and medial half of ring finger.
OTHER CUTANEOUS DISTRIBUTIONS		
Intercostobrachial	Second intercostal nerve.	Skin over medial side of arm.
Upper lateral brachial cutaneous	Axillary.	Skin over deltoid muscle and down to elbow.
Posterior brachial cutaneous	Radial.	Skin over posterior aspect of arm.
Lower lateral brachial cutaneous	Radial.	Skin over lateral aspect of elbow.
Lateral antebrachial cutaneous	Musculocutaneous.	Skin over lateral aspect of forearm.
Posterior antebrachial cutaneous	Radial.	Skin over posterior aspect of forearm.

EXHIBIT 17-4

Lumbar Plexus

NERVE	ORIGIN	DISTRIBUTION
Iliohypogastric	T12–L1.	Muscles of anterolateral abdominal wall (external oblique, internal oblique, transversus abdominis); skin of lower abdomen and buttock.
Ilioinguinal	L1.	Muscles of anterolateral abdominal wall as indicated above; skin of upper medial aspect of thigh, root of penis and scrotum in male, and labia majora and mons pubis in female.
Genitofemoral	L1–2.	Cremaster muscle; skin over middle anterior surface of thigh, scrotum in male, and labia majora in female.
Lateral femoral cutaneous	L2–L3.	Skin over lateral, anterior, and posterior aspects of thigh.
Femoral	L2–L4.	Flexor muscles of thigh (iliacus, psoas major, pectineus, rectus femoris, sartorius); extensor muscles of leg (rectus femoris, vastus lateralis, vastus medialis, vastus intermedius); skin on front and over medial aspect of thigh and medial side of leg and foot.
Obturator	L2–L4.	Adductor muscles of leg (obturator externus, pectineus, adductor longus, adductor brevis, adductor magnus, gracilis); skin over medial aspect of thigh.

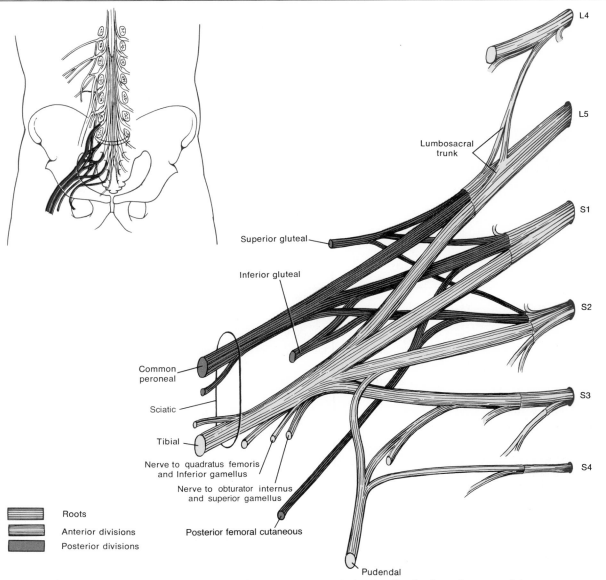

FIGURE 17-11 Sacral plexus. Refer to Figure 17-10b for the distribution of nerves of the sacral plexus. Consult Exhibit 17-5 so that you can determine the distribution of each of the nerves of the plexus.

EXHIBIT 17-5

Sacral Plexus

NERVE	ORIGIN	DISTRIBUTION
Superior gluteal	L4–L5 and S1.	Gluteus minimus and gluteus medius muscles and tensor fasciae latae.
Inferior gluteal	L5–S2.	Gluteus maximus muscle.
Nerve to piriformis	S1–S2.	Piriformis muscle.
Nerve to quadratus femoris	L4–L5 and S1.	Quadratus femoris and inferior gemellus muscles.
Nerve to obturator internus	L5–S2.	Obturator internus and superior gemellus muscles.
Perforating cutaneous	S2–S3.	Skin over lower medial aspect of buttock.
Posterior femoral cutaneous	S1–S3.	Skin over anal region, lower lateral aspect of buttock, upper posterior aspect of thigh, upper part of calf, scrotum in male, and labia majora in female.
Sciatic	L4–S3.	Actually two nerves: tibial and common peroneal, bound together by common sheath of connective tissue. It splits into its two divisions, usually at knee. (See below for distributions.) As sciatic nerve descends through thigh, it sends branches to hamstring muscles (biceps femoris, semitendinosus, semimembranosus) and adductor magnus.
Tibial (medial popliteal)	L4–S3.	Gastrocneumius, plantaris, soleus, popliteus, tibialis posterior, flexor digitorum longus, and flexor hallucis longus muscles. Branches of tibial nerve in foot are medial plantar nerve and lateral plantar nerve.
Medial plantar		Abductor hallucis, flexor digitorum brevis, and flexor hallucis brevis muscles; skin over medial two-thirds of plantar surface of foot.
Lateral plantar		Remaining muscles of foot not supplied by medial plantar nerve; skin over lateral third of plantar surface of foot.
Common peroneal (lateral popliteal)	L4–S2.	Divides into a superficial peroneal and a deep peroneal branch.
Superficial peroneal		Peroneus longus and peroneus brevis muscles; skin over distal third of anterior aspect of leg and dorsum of foot.
Deep peroneal		Tibialis anterior, extensor hallucis longus, peroneus tertius, and extensor digitorum longus and brevis muscles; skin over great and second toes.
Pudendal	S2–S4.	Muscles of perineum; skin of penis and scrotum in male and clitoris, labia majora, labia minora, and lower vagina in female.

T2 supplies the intercostal muscles of the second intercostal space and the skin of the axilla and posteromedial aspect of the arm. Nerves T3 and T6 pass in the costal grooves of the ribs and are distributed to the intercostal muscles and skin of the anterior and lateral chest wall. Nerves T7–T11 supply the intercostal muscles and the abdominal muscles and overlying skin. The dorsal rami of the intercostal nerves supply the deep back muscles and skin of the dorsal aspect of the thorax.

DERMATOMES

The skin over the entire body is supplied segmentally by spinal nerves. This means that the spinal nerves innervate specific, constant segments of the skin. All spinal nerves except C1 supply branches to the skin. The skin segment supplied by the dorsal root of a spinal nerve is a *dermatome* (Figure 17-12).

In the neck and trunk, the dermatomes form consecutive bands of skin. In the trunk, there is an overlap of adjacent dermatome nerve supply. Thus, there is little loss of sensation if only a single nerve supply to a dermatome is interrupted. Most of the skin of the face and scalp is supplied by the trigeminal (V) cranial nerve.

Since physicians know which spinal nerves are associated with each dermatome, it is possible to determine which segment of the spinal cord or spinal nerve is malfunctioning. If a dermatome is stimulated and the sensation is not perceived, it can be assumed that the nerves supplying the dermatome are involved.

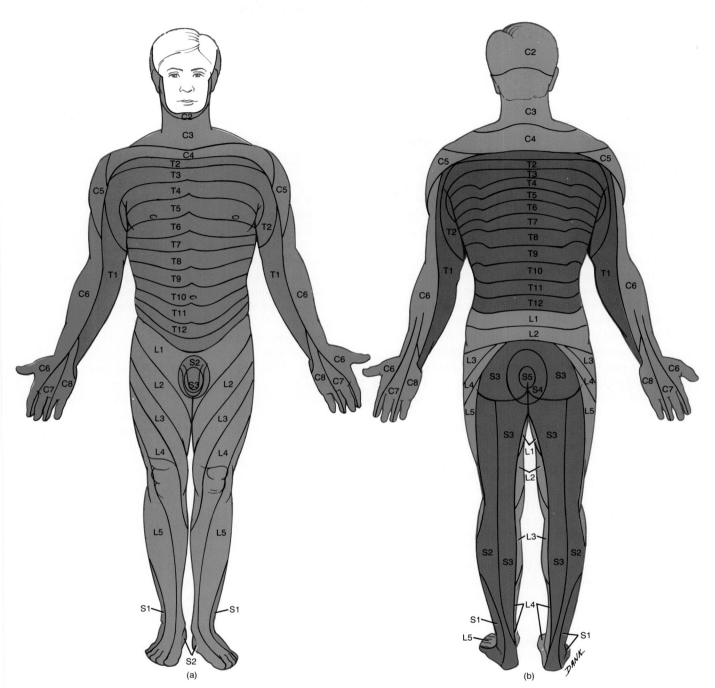

FIGURE 17-12 Distribution of spinal nerves to dermatomes. The lines are not perfectly aligned so that there is often considerable overlap. (a) Anterior view. (b) Posterior view.

APPLICATIONS TO HEALTH

SPINAL CORD INJURY

The spinal cord may be damaged by expanding tumors, herniated intervertebral discs, blood clots, degenerative and demyelinating disorders, fracture or dislocation of the vertebrae enclosing it, penetrating wounds caused by projectile metal fragments, or other traumatic events such as automobile accidents. Depending on the location and extent of the injury, paralysis may occur. The various types of paralysis may be classified as follows: *monoplegia* (*mono* = one; *plege* = stroke), paralysis of one extremity only; *diplegia* (*di* = two), paralysis of both upper extremities or both lower extremities; *paraplegia* (*para* = beyond), paralysis of both lower extremities; *hemiplegia* (*hemi* = half), paralysis of the upper extremity, trunk, and lower extremity on one side of the body; and *quadriplegia* (*quad* = four), paralysis of the two upper and two lower extremities.

Complete transection of the spinal cord means that the cord is cut transversely and severed from one side to the other thus cutting all ascending and descending tracts. It results in a loss of all sensations and voluntary movement below the level of the transection. If the upper cervical cord is transected, quadriplegia results; if the transection is between the cervical and lumbar enlargements, paraplegia results. *Hemisection* of the spinal cord refers to a partial transection. It is characterized, below the hemisection, by a loss of proprioception, tactile discrimination, and feeling of vibration on the same side as the injury; paralysis on the same side; and loss of feelings of pain and temperature on the opposite side. If the hemisection is of the upper cervical cord, hemiplegia results; if the hemisection is of the thoracic cord, paralysis of one lower extremity results (monoplegia).

Following transection, there is an initial period of *spinal shock* that lasts from a few days to several weeks. During this period, all reflex activity is abolished, a condition called *areflexia* (a'-rē-FLEK-sē-a). In time, however, there is a return of reflex activity. The first reflex to return is a stretch reflex, knee jerk. Its reappearance may take several days. Next the flexion reflexes return, over a period of up to several months. Then the crossed extensor reflexes return. Visceral reflexes such as erection and ejaculation are also affected by transection. Moreover, urinary bladder and bowel functions are no longer under voluntary control. An experimental procedure, called *electroejaculation,* has been used with some success in helping males with spinal cord injuries to ejaculate. In the procedure, a probe is inserted into the rectum and attached to a device that delivers an electric current in gradually increasing increments until ejaculation occurs. The husband's sperm is then used to inseminate the wife.

Until recently, severe damage resulting from transection was thought to be irreversible. However, a team of researchers has developed a technique for regenerating severed spinal cords in animals. The technique, called *delayed nerve graft-ing,* involves cutting away the crushed or injured section of the spinal cord and bridging the gap with nerve segments from the arm or leg. The original severed axons in the cord can then grow through the bridge. Delayed nerve grafting offers hope for paraplegics who have lost voluntary movements of the lower extremities, as well as urinary, bowel, and sexual functions.

PERIPHERAL NERVE DAMAGE AND REPAIR

As we have seen, axons that have a neurolemma can be repaired as long as the cell body is intact, fibers are in association with neurolemmocytes (Schwann cells), and scar tissue formation does not occur too rapidly. Most nerves that lie outside the brain and spinal cord consist of processes that are covered with a neurolemma. A person who injures a nerve in the upper extremity, for example, has a good chance of regaining nerve function. Processes in the brain and spinal cord do not have a neurolemma. Injury there is permanent.

When there is damage to an axon (or to dendrites of somatic afferent neurons), there are usually changes in the cell body of the affected neuron called *chromatolysis.* In addition there are always changes that occur in the portion of the axon distal to the site of injury called *Wallerian* (wal-LE-rē-an) *degeneration* and in the portion of the axon proximal to the site of injury called *retrograde degeneration.* Retrograde degeneration occurs in essentially the same way, whether the damaged fiber is in the central or peripheral nervous system. The Wallerian degeneration reaction, however, depends on whether the fiber is central or peripheral.

Chromatolysis

About 24 to 48 hours after injury to a process of a central or peripheral neuron, the chromatophilic substance (Nissl bodies), normally arranged in an orderly fashion in an uninjured cell body, breaks down into finely granular masses. This alteration is called chromatolysis (krō'-ma-TOL-i-sis; *chromo* = color; *lysis* = dissolution). It begins between the axon hillock and nucleus but spreads throughout the cell body. As a result of chromatolysis, the cell body swells and the swelling reaches its maximum between 10 and 20 days after injury (Figure 17-13b). Chromatolysis results in a loss of ribosomes by the rough endoplasmic reticulum and an increase in the number of free ribosomes. Another sign of chromatolysis is the off-center position of the nucleus in the cell body. This change makes it possible to identify the cell bodies of damaged fibers through a microscope.

Wallerian Degeneration

The part of the axon distal to the damage becomes slightly swollen and then breaks up into fragments by the third to fifth day. The myelin sheath around the axon also undergoes degeneration (Figure 17-13c). Degeneration of the distal

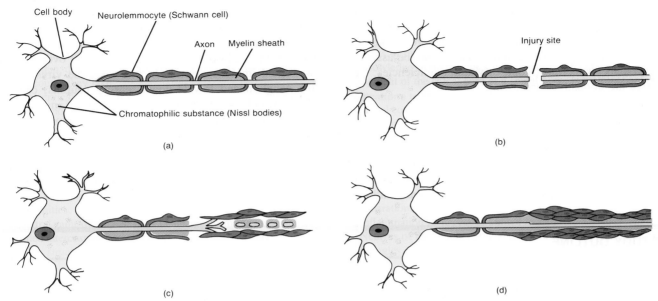

FIGURE 17-13 Peripheral nerve damage and repair. (a) Normal neuron. (b) Chromatolysis. (c) Wallerian degeneration. (d) Regeneration.

portion of the axon and myelin sheath is called Wallerian degeneration. Following degeneration, there is phagocytosis of the remains by macrophages.

Even though there is degeneration of the axon and myelin sheath, the neurolemma of the neurolemmocytes remains. The neurolemmocytes on either side of the site of injury multiply by mitosis and grow toward each other and attempt to form a tube across the injured area. The tube provides a means for new axons to grow from the proximal area across the injured area into the distal area previously occupied by the original nerve fiber (Figure 17-13d). The growth of new axons will not occur if the gap at the site of injury is too large or if the gap becomes filled with dense collagenous fibers.

Retrograde Degeneration

The changes in the proximal portion of the axon, called retrograde degeneration, are similar to those that occur during Wallerian degeneration. The main difference in retrograde degeneration is that the changes occur only as far as the first neurofibral node (node of Ranvier).

Regeneration

Following chromatolysis, there are signs of recovery in the cell body. There is an acceleration of RNA and protein synthesis, which favors regeneration of the axon. Recovery often takes several months and involves the restoration of normal levels of RNA, proteins, and the chromatophilic substance to their usual, uninjured patterns.

Accelerated protein synthesis is required for repair of the damaged axon. The proteins synthesized in the cell body pass into the empty lumen of the tube formed by neurolemmocytes by axoplasmic flow at about the rate of 1 mm (0.04 in.)/day. The proteins assist in regenerating the damaged axon. During the first few days following damage, buds of regenerating axons begin to invade the tube formed by the neurolemmocytes. Axons from the proximal area grow at the rate of about 1.5 mm (0.06 in.)/day across the area of damage, find their way into the distal neurolemmal tubes, and grow toward the distally located receptors and effectors. Thus, sensory and motor connections are reestablished. In time, a new myelin sheath is also produced by the neurolemmocytes. However, function is never completely restored after a nerve is severed.

NEURITIS

Neuritis is inflammation of a single nerve, two or more nerves in separate areas, or many nerves simultaneously. It may result from irritation to the nerve produced by direct blows, bone fractures, contusions, or penetrating injuries. Additional causes include vitamin deficiency (usually thiamine) and poisons such as carbon monoxide, carbon tetrachloride, heavy metals, and some drugs.

SCIATICA

Sciatica (sī-AT-i-ka) is a type of neuritis characterized by severe pain along the path of the sciatic nerve or its branches. The term is commonly applied to a number of disorders affecting this nerve. Because of its length and size, the sciatic nerve is exposed to many kinds of injury. Inflammation of or injury to the nerve causes pain that passes from the back or thigh down its length into the leg, foot, and toes.

18 The Brain and the Cranial Nerves

CHAPTER OUTLINE

■ **Brain**
Principal Parts
Protection and Coverings
Cerebrospinal Fluid (CSF)
Blood Supply
Brain Stem
   Medulla Oblongata
   Pons
   Midbrain
Diencephalon
   Thalamus
   Hypothalamus
Cerebrum
   Lobes
   White Matter
   Basal Ganglia (Cerebral Nuclei)
   Limbic System
   Functional Areas of Cerebral Cortex
   Electroencephalogram (EEG)
Brain Lateralization (Split-Brain Concept)
Cerebellum
   Structure
   Functions
■ **Cranial Nerves**
Olfactory (I)
Optic (II)
Oculomotor (III)
Trochlear (IV)
Trigeminal (V)
Abducens (VI)
Facial (VII)
Vestibulocochlear (VIII)
Glossopharyngeal (IX)
Vagus (X)
Accessory (XI)
Hypoglossal (XII)
■ **Aging and the Nervous System**
■ **Developmental Anatomy of the Nervous System**
■ **Applications to Health**
Cerebrovascular Accident (CVA)
Transient Ischemic Attack (TIA)
Brain Tumors
Poliomyelitis
Cerebral Palsy (CP)
Parkinson's Disease (PD)
Multiple Sclerosis (MS)
Epilepsy
Dyslexia
Tay-Sachs Disease
Headache
Trigeminal Neuralgia (Tic Douloureux)
Reye's Syndrome (RS)
Alzheimer's Disease (AD)
■ **Key Medical Terms Associated with the Central Nervous System**

STUDENT OBJECTIVES

1. Identify the principal parts of the brain and describe how the brain is protected.
2. Explain the formation and circulation of cerebrospinal fluid (CSF).
3. Describe the blood supply to the brain and the concept of the blood–brain barrier (BBB).
4. Compare the components of the brain stem and diencephalon with regard to structure and function.
5. Describe the surface features, lobes, tracts, and basal ganglia of the cerebrum.
6. Describe the structure and functions of the limbic system.
7. Compare the sensory, motor, and association areas of the cerebrum.
8. Describe the principal waves of an electroencephalogram (EEG) and explain its significance in the diagnosis of certain disorders.
9. Explain the concept of brain lateralization.
10. Describe the anatomical characteristics and functions of the cerebellum.
11. Define a cranial nerve and identify the 12 pairs of cranial nerves by name, number, type, location, and function.
12. Describe the effects of aging on the nervous system.
13. Describe the development of the nervous system.
14. List the clinical symptoms of these disorders of the nervous system: cerebrovascular accidents (CVAs), transient ischemic attacks (TIAs), brain tumors, poliomyelitis, cerebral palsy (CP), Parkinson's disease (PD), multiple sclerosis (MS), epilepsy, dyslexia, Tay-Sachs disease, headache, trigeminal neuralgia, Reye's syndrome (RS), and Alzheimer's disease (AD).
15. Define key medical terms associated with the central nervous system.

Now we shall consider the principal parts of the brain, how the brain is protected, and how it is related to the spinal cord and to the 12 pairs of cranial nerves.

The developmental anatomy of the brain is explained in detail at the end of the chapter.

BRAIN

PRINCIPAL PARTS

To understand the terminology used for the principal parts of the brain, it will first be necessary to describe briefly the embryological development of the brain. The developmental anatomy of the brain is explained in more detail at the end of the chapter.

At the end of the fourth week of the embryonic period, the brain develops from three rudimentary regions of the embryo called primary brain vesicles. These are the *prosencephalon* (forebrain), *mesencephalon* (midbrain), and *rhombencephalon* (hindbrain). During the fifth week of development, the prosencephalon develops into two secondary brain vesicles called the *diencephalon* and *telencephalon;* the rhombencephalon also develops into two secondary brain vesicles called the *myelencephalon* and *metencephalon.* Since the mesencephalon remains unchanged, there are five secondary brain vesicles. Ultimately, the diencephalon develops into the thalamus and hypothalamus of the brain, the telencephalon forms the cerebrum, the mesencephalon becomes the midbrain, the myelencephalon develops into the medulla oblongata, and the metencephalon becomes the pons and cerebellum.

These relations are summarized in Exhibit 18-1.

The *brain* of an average adult is one of the largest organs of the body, weighing about 1,300 g (3 lb). Figure 18-1 shows that the brain is mushroom shaped and divided into four principal parts: brain stem, diencephalon, cerebrum, and cerebellum. In some cases, embryological names are retained when distinguishing the various parts of the brain. The *brain stem,* the stalk of the mushroom, consists of the medulla oblongata, pons, and midbrain or mesencephalon (mes-en-SEF-a-lon). The lower end of the brain stem is a continuation of the spinal cord. Above the brain stem is the *diencephalon* (dī-en-SEF-a-lon), consisting primarily of the thalamus and hypothalamus. The *cerebrum* spreads over the diencephalon. The cerebrum constitutes about seven-eighths of the total weight of the brain and occupies most of the cranium. Inferior to the cerebrum and posterior to the brain stem is the *cerebellum.*

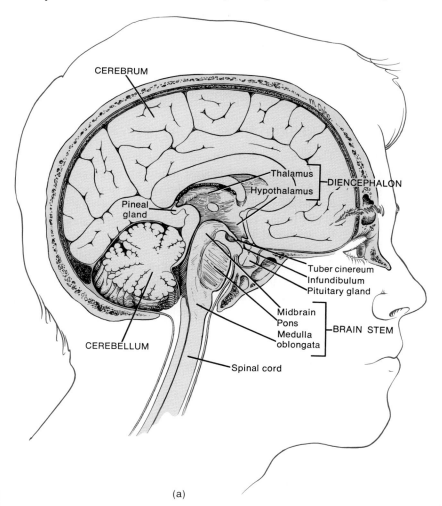

CEREBRUM

Thalamus
Hypothalamus — DIENCEPHALON

Pineal gland

Tuber cinereum
Infundibulum
Pituitary gland

Midbrain
Pons
Medulla oblongata — BRAIN STEM

CEREBELLUM

Spinal cord

(a)

FIGURE 18-1 Brain. (a) Principal parts of the medial aspect of the brain seen in sagittal section. The infundibulum and pituitary gland are discussed in conjunction with the endocrine system in Chapter 21.

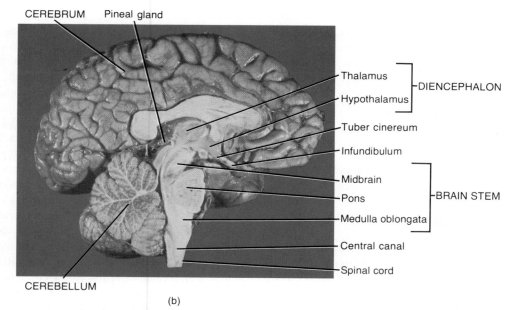

CEREBRUM   Pineal gland

Thalamus — DIENCEPHALON
Hypothalamus
Tuber cinereum
Infundibulum
Midbrain
Pons — BRAIN STEM
Medulla oblongata
Central canal
Spinal cord
CEREBELLUM

(b)

FIGURE 18-1 (Continued) Brain. (b) Photograph of the medial aspect of the brain seen in sagittal section. (Courtesy of C. Yokochi and J. W. Rohen, *Photographic Anatomy of the Human Body,* 2nd ed., 1978, IGAKU-SHOIN, Ltd., Tokyo, New York.)

PROTECTION AND COVERINGS

The brain is protected by the cranial bones (see Figure 6-2). Like the spinal cord, the brain is also protected by meninges. The *cranial meninges* surround the brain, are continuous with the spinal meninges, and have the same basic structure and bear the same names as the spinal meninges: the outermost *dura mater,* middle *arachnoid,* and innermost *pia mater* (Figure 18-2).

The cranial dura mater consists of two layers. The thicker, outer layer (endosteal layer) tightly adheres to the cranial bones and serves as periosteum. The thinner, inner layer (meningeal layer) includes a mesothelial layer on its smooth surface. The spinal dura mater corresponds to the meningeal layer of the cranial dura mater.

CLINICAL APPLICATION

The middle meningeal artery, running on the inner surface of the temporal bone between the dura mater and the skull, is closely adherent to the bones of the skull. It and the meningeal veins may be ruptured by a blow on the temple, especially if the bone is fractured. This rupture produces an *extradural hemorrhage,* which results in gradually increasing cranial pressure, drowsiness, unconsciousness, and death unless there is surgical intervention.

CEREBROSPINAL FLUID (CSF)

The brain, as well as the rest of the central nervous system, is further protected against injury by *cerebrospinal fluid* *(CSF).* This fluid circulates through the subarachnoid space around the brain and spinal cord and through the ventricles of the brain. The subarachnoid space is the area between the arachnoid and pia mater.

The *ventricles* (VEN-tri-kuls) are cavities in the brain that communicate with each other, with the central canal of the spinal cord, and with the subarachnoid space (Figure 18-2). Each of the two *lateral ventricles* is located in a hemisphere (side) of the cerebrum under the corpus callo-

EXHIBIT 18-1

Development of the Brain from Brain Vesicles

PRIMARY BRAIN VESICLE	SECONDARY BRAIN VESICLE	PART OF BRAIN FORMED
Prosencephalon	*Diencephalon*	Thalamus and hypothalamus
	Telencephalon	Cerebrum
Mesencephalon	*Mesencephalon*	Midbrain
Rhombencephalon	*Myelencephalon*	Medulla oblongata
	Metencephalon	Pons and cerebellum

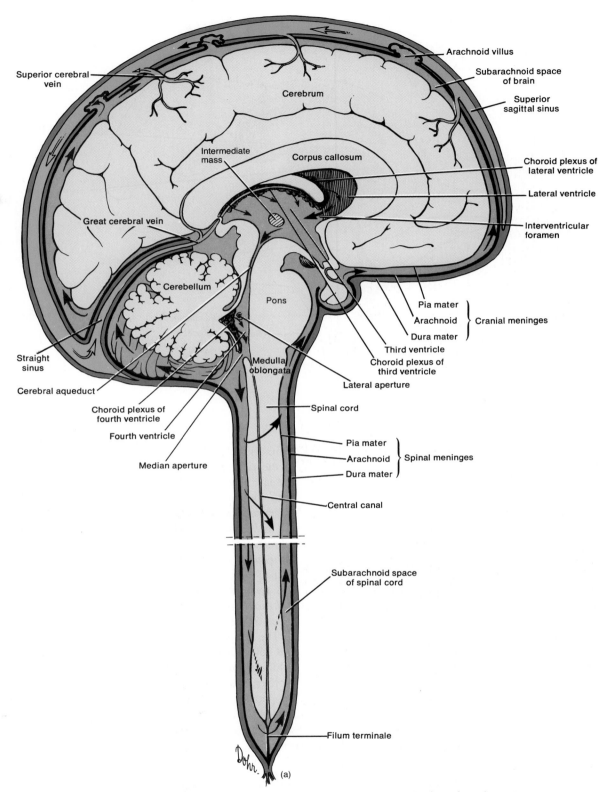

FIGURE 18-2 Meninges and ventricles of the brain. (a) Brain, spinal cord, and meninges seen in sagittal section. Arrows indicate the direction of flow of cerebrospinal fluid.

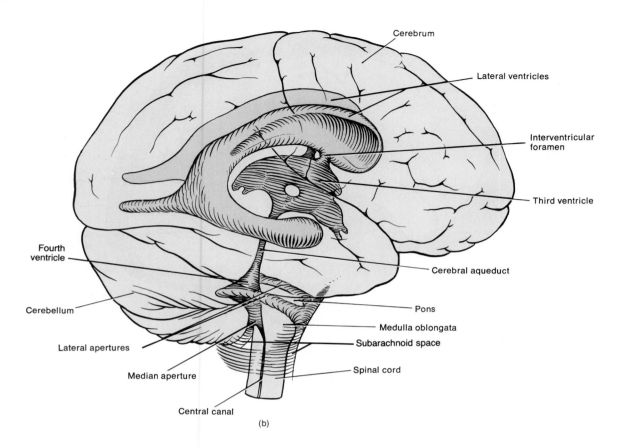

(b)

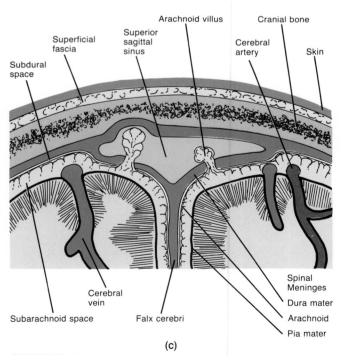

(c)

FIGURE 18-2 (Continued) Meninges and ventricles of the brain. (b) Diagrammatic lateral projection of the ventricles. (c) Frontal section through the superior portion of the brain showing the relation of the superior sagittal sinus to the arachnoid villi.

sum. The ***third ventricle*** is a vertical slit between and inferior to the right and left halves of the thalamus and between the lateral ventricles. Each lateral ventricle communicates with the third ventricle by a narrow, oval opening, the ***interventricular foramen.*** The ***fourth ventricle*** lies between the inferior brain stem and the cerebellum. It communicates with the third ventricle via a canal-like structure, ***cerebral aqueduct,*** which passes through the midbrain. The roof of the fourth ventricle has three openings: a ***median aperture*** and two ***lateral apertures.*** Through these openings, the fourth ventricle also communicates with the subarachnoid space of the brain and cord.

The entire central nervous system contains between 80 and 150 ml (3 to 5 oz) of cerebrospinal fluid. It is a clear, colorless fluid of watery consistency. Chemically, it contains proteins, glucose, urea, and salts. It also contains some lymphocytes. Cerebrospinal fluid has two principal functions related to homeostasis: protection and circulation. The fluid serves as a shock-absorbing medium to protect the brain and spinal cord from jolts that would otherwise cause them to crash against the bony walls of the cranial and vertebral cavities. The fluid also buoys the brain so that it "floats" in the cranial cavity. With regard to its circulatory function, cerebrospinal fluid delivers nutritive substances filtered from blood to the brain and spinal cord and removes wastes and toxic substances produced by brain and spinal cord cells.

Cerebrospinal fluid is formed primarily by filtration and secretion from networks of capillaries in the ventricles called *choroid* (KŌ-royd; *chorion* = delicate) *plexuses* (Figure 18-2a). Various components of the choroid plexuses form a *blood–cerebrospinal fluid barrier* that permits certain substances to enter the fluid but prohibits others. Such a barrier protects the brain and spinal cord from harmful substances. The fluid formed in the choroid plexuses of the lateral ventricles circulates through the interventricular foramina to the third ventricle, where more fluid is added by the choroid plexus of the third ventricle. It then flows through the cerebral aqueduct into the fourth ventricle. Here there are contributions from the choroid plexus of the fourth ventricle. The fluid then circulates through the apertures of the fourth ventricle into the subarachnoid space around the back of the brain. It also passes downward to the subarachnoid space around the posterior surface of the spinal cord, up the anterior surface of the spinal cord, and around the anterior part of the brain. From there it is gradually reabsorbed into veins. Some cerebrospinal fluid may be formed by ependymal (neuroglial) cells lining the central canal of the spinal cord. This small quantity of fluid ascends to reach the fourth ventricle. Most of the fluid is absorbed into a vein called the superior sagittal sinus. The absorption actually occurs through *arachnoid villi*—fingerlike projections of the arachnoid that push into the dural venous sinuses, especially the superior sagittal sinus (Figure 18-2c and 18-13d). Normally, cerebrospinal fluid is absorbed as rapidly as it is formed.

The formation, circulation, and absorption of cerebrospinal fluid are summarized in Figure 18-3.

CLINICAL APPLICATION

If an obstruction, such as a tumor or a congenital blockage, arises in the brain and interferes with the drainage of cerebrospinal fluid from the ventricles into the subarachnoid space, large amounts of fluid accumulate in the ventricles. Fluid pressure inside the brain increases and, if the fontanels have not yet closed, the head bulges to relieve the pressure. This condition is called *internal hydrocephalus* (*hydro* = water; *enkephalos* = brain).

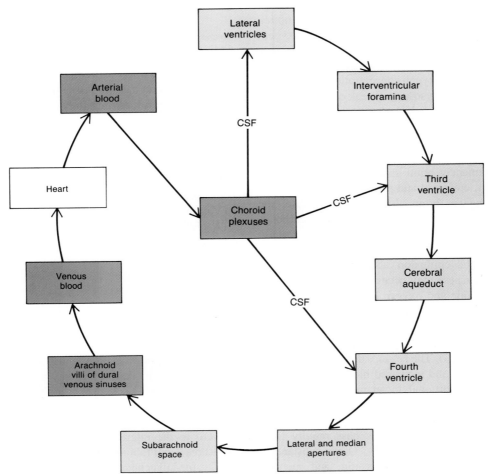

FIGURE 18-3 Summary of the formation, circulation, and absorption of cerebrospinal fluid (CSF).

If an obstruction interferes with drainage somewhere in the subarachnoid space and cerebrospinal fluid accumulates inside the space, the condition is termed *external hydrocephalus.*

BLOOD SUPPLY

The brain is well supplied with oxygen and nutrients by blood vessels that form the cerebral arterial circle (circle of Willis). Cerebral circulation is outlined in Exhibit 14-3 and Figure 14-9d. Blood vessels that enter brain tissue pass along the surface of the brain and, as they penetrate inward, they are surrounded by a loose-fitting layer of pia mater. The space between the penetrating blood vessel and pia mater is called a *perivascular space.*

Although the brain composes only about 2 percent of total body weight, it utilizes about 20 percent of the oxygen used by the entire body. The brain is one of the most metabolically active organs of the body, and the amount of oxygen it uses varies with the degree of mental activity. If the blood flow to the brain is interrupted even briefly, unconsciousness may result. A 1- or 2-minute interruption may weaken the brain cells by starving them of oxygen, and if the cells are totally deprived of oxygen for 4 minutes, many are permanently injured. Lysosomes of brain cells are sensitive to decreased oxygen concentration. If the condition persists long enough, lysosomes break open and release enzymes that bring about self-destruction of brain cells. Occasionally during childbirth, the oxygen supply from the mother's blood is interrupted before the baby leaves the birth canal and can breathe. Often such babies are stillborn or suffer permanent brain damage that may result in mental retardation, epilepsy, and paralysis.

Blood supplying the brain also contains glucose, the principal source of energy for brain cells. Because carbohydrate storage in the brain is limited, the supply of glucose must be continuous. If blood entering the brain has a low glucose level, mental confusion, dizziness, convulsions, and loss of consciousness may occur.

Both carbon dioxide and oxygen have potent effects on cerebral blood flow. Carbon dioxide increases cerebral blood flow by combining with water to form carbonic acid (H_2CO_3), which breaks down into hydrogen ions (H^+) and bicarbonate ions (HCO_3^-). The H^+ ions then cause vasodilation of cerebral vessels and increased blood flow. Dilation is almost directly proportional to an increase in H^+ ion concentration. A decrease in oxygen in the blood also causes vasodilation and increased cerebral blood flow.

Glucose, oxygen, and certain ions pass rapidly from the circulating blood into brain cells. Other substances, such as creatinine, urea, chloride, insulin, and sucrose, enter quite slowly. Still other substances—proteins and most antibiotics—do not pass at all from the blood into brain cells. The differential rates of passage of certain materials from the blood into most parts of the brain are based upon a concept called the *blood–brain barrier* (*BBB*). The barrier is either absent or less selective in the hypothalamus and roof of the fourth ventricle. Electron micrograph studies of the capillaries of the brain reveal that they differ structurally from other capillaries. Brain capillaries are constructed of more densely packed cells and are surrounded by terminations of processes of large numbers of astrocytes (one of the types of neuroglia) and a continuous basement membrane. Current evidence indicates that astrocytes produce a substance that influences the capillaries and confers on them the ability to selectively pass various substances but inhibit others. Substances that cross the barrier are soluble in lipids or water-soluble substances that require the assistance of a carrier molecule to cross by active transport. Some examples of lipid-soluble substances are nicotine, alcohol, and heroin. Water-soluble substances include glucose, certain amino acids, and sodium. The blood–brain barrier functions as a selective barrier to protect brain cells from harmful substances. An injury to the brain due to trauma, inflammation, or toxins causes a breakdown of the blood–brain barrier, permitting the passage of normally restricted substances into brain tissue.

CLINICAL APPLICATION

Various *drugs* differ with respect to their passage through the blood–brain barrier. The antibiotics *chloramphenicol* (Chloromycetin), *tetracycline* (Achromycin V), and *sulfonamides* (Sonilyn) cross easily. *Penicillin* (Bicillin) crosses in only very small amounts.

Thiopental sodium (Pentothal), a general anesthetic, rapidly crosses the blood–brain barrier following intravenous injection. *Atropine* (Atropisol), an antispasmodic, also quickly enters the brain. *Anisotropine methylbromide* (Valpin), another antispasmodic, and *phenylbutazone* (Azolid), an anti-inflammatory drug, cannot cross the blood–brain barrier.

BRAIN STEM

Medulla Oblongata

The *medulla oblongata* (me-DULL-la ob'-long-GA-ta), or simply *medulla,* is a continuation of the upper portion of the spinal cord and forms the inferior part of the brain stem (Figure 18-4). Its position in relation to the other parts of the brain may be noted in Figure 18-1. It lies just superior to the level of the foramen magnum and extends upward to the inferior portion of the pons. The medulla measures 3 cm (about 1 in.) in length.

The medulla contains all ascending and descending tracts that communicate between the spinal cord and various parts of the brain. These tracts constitute the white matter of the medulla. Some tracts cross as they pass through the

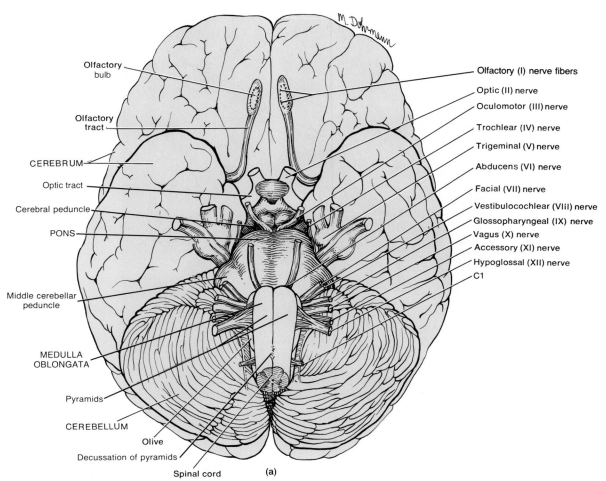

Olfactory bulb
Olfactory tract
CEREBRUM
Optic tract
Cerebral peduncle
PONS
Middle cerebellar peduncle
MEDULLA OBLONGATA
Pyramids
CEREBELLUM
Olive
Decussation of pyramids
Spinal cord   (a)

Olfactory (I) nerve fibers
Optic (II) nerve
Oculomotor (III) nerve
Trochlear (IV) nerve
Trigeminal (V) nerve
Abducens (VI) nerve
Facial (VII) nerve
Vestibulocochlear (VIII) nerve
Glossopharyngeal (IX) nerve
Vagus (X) nerve
Accessory (XI) nerve
Hypoglossal (XII) nerve
C1

FIGURE 18-4 Brain stem. (a) Diagram of the ventral surface of the brain showing the structure of the brain stem in relation to the cranial nerves and associated structures.

medulla. Let us see how this crossing occurs and what it means.

On the ventral side of the medulla are two roughly triangular structure called *pyramids* (Figures 18-4 and 18-5). The pyramids are composed of the largest motor tracts that pass from the outer region of the cerebrum (cerebral cortex) to the spinal cord. Just above the junction of the medulla with the spinal cord, most of the fibers in the left pyramid cross to the right side, and most of the fibers in the right pyramid cross to the left. This crossing is called the *decussation* (dē'-ku-SĀ-shun) *of pyramids.* The adaptive value, if any, of this phenomenon is unknown. Decussation explains why motor areas of one side of the cerebral cortex control muscular movements on the opposite side of the body. The principal motor fibers that undergo decussation belong to the lateral corticospinal tracts. These tracts originate in the cerebral cortex and pass inferiorly to the medulla. The fibers cross in the pyramids and descend in the lateral columns of the spinal cord, terminating in the anterior gray horns. Here synapses occur with motor neurons that terminate in skeletal muscles. As a result of the crossing, fibers that originate in the left cerebral cortex activate muscles

on the right side of the body, and fibers that originate in the right cerebral cortex activate muscles on the left side.

The dorsal side of the medulla contains two pairs of prominent nuclei: the right and left *nucleus gracilis* (gras-I-lis; *gracilis* = slender) and *nucleus cuneatus* (kyoo-nē-Ā-tus; *cuneus* = wedge). These nuclei receive sensory fibers from ascending tracts (right and left fasciculus gracilis and fasciculus cuneatus) of the spinal cord and relay the sensory information to the opposite side of the medulla. The information is conveyed to the thalamus and then to the sensory areas of the cerebral cortex. Nearly all sensory impulses received on one side of the body cross in the medulla or spinal cord and are perceived in the opposite side of the cerebral cortex.

In addition to its function as a conduction pathway for motor and sensory impulses between the brain and spinal cord, the medulla also contains an area of dispersed gray matter containing some white fibers. This region is called the *reticular formation.* Actually, portions of the reticular formation are also located in the spinal cord, pons, midbrain, and diencephalon. The reticular formation functions in consciousness and arousal from sleep.

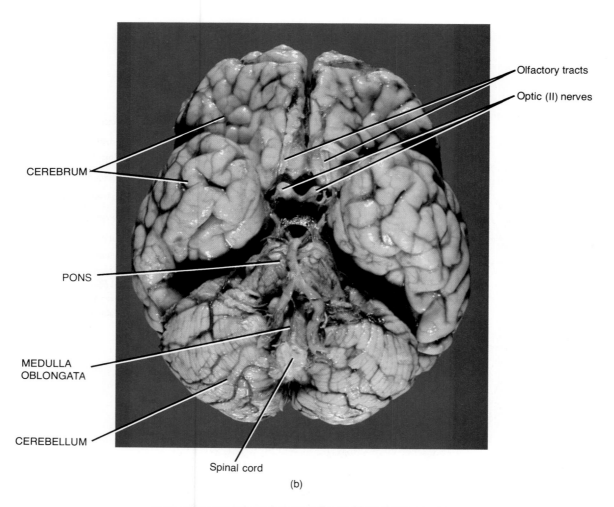

CEREBRUM

Olfactory tracts

Optic (II) nerves

PONS

MEDULLA
OBLONGATA

CEREBELLUM

Spinal cord

(b)

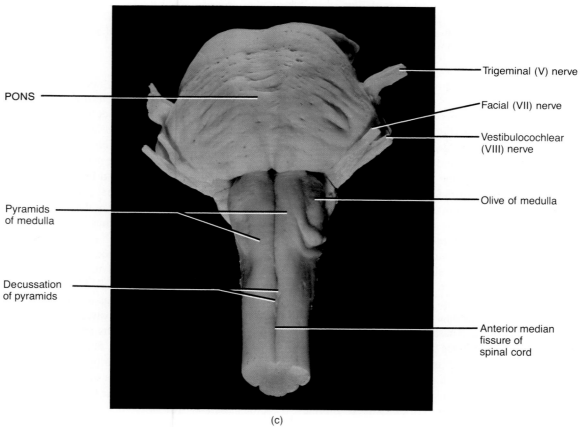

PONS

Pyramids
of medulla

Decussation
of pyramids

Trigeminal (V) nerve

Facial (VII) nerve

Vestibulocochlear
(VIII) nerve

Olive of medulla

Anterior median
fissure of
spinal cord

(c)

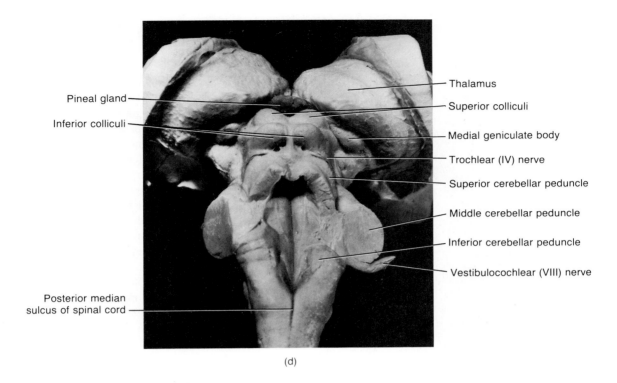

Pineal gland

Inferior colliculi

Thalamus

Superior colliculi

Medial geniculate body

Trochlear (IV) nerve

Superior cerebellar peduncle

Middle cerebellar peduncle

Inferior cerebellar peduncle

Vestibulocochlear (VIII) nerve

Posterior median sulcus of spinal cord

(d)

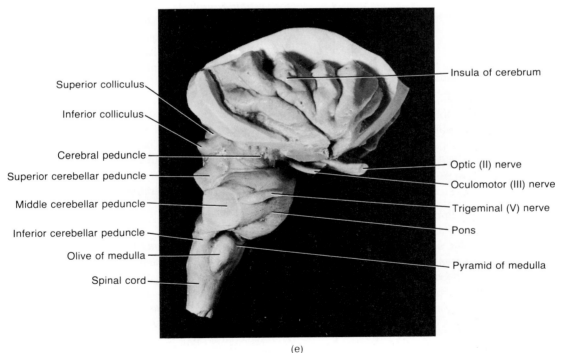

Superior colliculus

Inferior colliculus

Cerebral peduncle

Superior cerebellar peduncle

Middle cerebellar peduncle

Inferior cerebellar peduncle

Olive of medulla

Spinal cord

Insula of cerebrum

Optic (II) nerve

Oculomotor (III) nerve

Trigeminal (V) nerve

Pons

Pyramid of medulla

(e)

FIGURE 18-4 (Continued) (b) Photograph of the ventral surface of the brain showing the brain stem in relation to associated structures. (Courtesy of Martin Rotker, Taurus Photos). (c) Photograph of the ventral surface of the brain stem. (Courtesy of N. Gluhbegovic and T. H. Williams, *The Human Brain: A Photographic Guide,* Harper & Row, Publishers, Inc., New York, 1980.) (d) Photograph of the posterior surface of the brain stem. (Courtesy of N. Gluhbegovic and T. H. Williams, *The Human Brain: A Photographic Guide,* Harper & Row, Publishers, Inc., New York, 1980.) (e) Photograph of the brain stem in right lateral view. (Courtesy of N. Gluhbegovic and T. H. Williams, *The Human Brain: A Photographic Guide,* Harper & Row, Publishers, Inc., New York, 1980.)

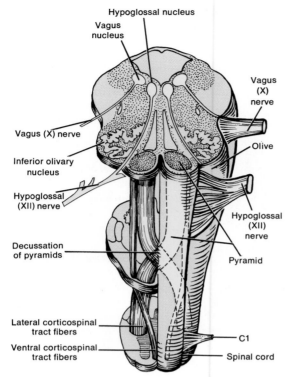

Hypoglossal nucleus

Vagus nucleus

Vagus (X) nerve

Vagus (X) nerve

Olive

Inferior olivary nucleus

Hypoglossal (XII) nerve

Hypoglossal (XII) nerve

Decussation of pyramids

Pyramid

Lateral corticospinal tract fibers

Ventral corticospinal tract fibers

C1

Spinal cord

FIGURE 18-5 Details of the medulla showing the decussation of pyramids.

CLINICAL APPLICATION

The most common knockout blow is one that makes contact with the mandible. Such a blow twists and distorts the brain stem and overwhelms the reticular activating system (RAS) by sending a sudden volley of nerve impulses to the brain, resulting in *unconsciousness.*

Within the medulla are also three vital reflex centers. The *cardiac center* regulates heartbeat and force of contraction, the *medullary rhythmicity area* adjusts the basic rhythm of breathing, and the *vasomotor (vasoconstrictor) center* regulates the diameter of blood vessels. Other centers in the medulla are considered nonvital and coordinate swallowing, vomiting, coughing, sneezing, and hiccuping.

CLINICAL APPLICATION

Neurosurgeons are now using an ultrasonic device called a *cavitron ultrasonic surgical aspirator (CUSA)* that can shatter certain kinds of brain tumors. High-frequency sound waves stimulate the slender tip of the instrument to vibrate 23,000 times per sound. This vibratory action disintegrates the tumors. The instrument also delivers an irrigating saline solution that aspirates the fragmented particles. One major advantage of CUSA is that it decreases the probability of damaging either adjacent nor-

mal tissues such as the brain stem, which could cause abnormal heart rhythms, or large blood vessels, which could cause hemorrhage.

The medulla also contains the nuclei of origin for several pairs of cranial nerves (Figures 18-4 and 18-5). These are the cochlear and vestibular branches of the vestibulocochlear (VIII) nerves, which are concerned with hearing and equilibrium (there is also a nucleus for the vestibular branches in the pons); the glossopharyngeal (IX) nerves, which relay nerve impulses related to swallowing, salivation, and taste; the vagus (X) nerves, which relay nerve impulses to and from many thoracic and abdominal viscera; the spiral portion of the accessory (XI) nerves, which convey nerve impulses related to head and shoulder movements (a part of this nerve, the cranial portion, also arises from the medulla); and the hypoglossal (XII) nerves, which convey nerve impulses that involve tongue movements.

On each lateral surface of the medulla is an oval projection called the *olive* (Figure 18-4), which contains an inferior olivary nucleus and two accessory olivary nuclei. The nuclei are connected to the cerebellum by fibers.

Also associated with the medulla is the greater part of the *vestibular nuclear complex.* This nuclear group consists of the *lateral, medial,* and *inferior vestibular nuclei* in the medulla and the *superior vestibular nucleus* in the pons. As you will see later (Chapter 20), the vestibular nuclei assume an important role in helping the body maintain its sense of equilibrium.

In view of the many vital activities controlled by the medulla, it is not surprising that a hard blow to the base of the skull can be fatal. Nonfatal medullary injury may be indicated by cranial nerve malfunctions on the same side of the body as the area of medullary injury, paralysis and loss of sensation on the opposite side of the body, and irregularities in respiratory control.

Pons

The relation of the *pons* to other parts of the brain can be seen in Figures 18-1 and 18-4. The pons, which means bridge, lies directly above the medulla and anterior to the cerebellum. It measures about 2.5 cm (1 in.) in length. Like the medulla, the pons consists of white fibers scattered throughout with nuclei. As the name implies, the pons is a bridge connecting the spinal cord with the brain and parts of the brain with each other. These connections are provided by fibers that run in two principal directions. The transverse fibers connect with the cerebellum through the *middle cerebellar peduncles.* The longitudinal fibers of the pons belong to the motor and sensory tracts that connect the spinal cord or medulla with the upper parts of the brain stem.

The nuclei for certain paired cranial nerves are also contained in the pons (Figure 18-4a). These include the trigeminal (V) nerves, which relay nerve impulses for chewing and for sensations of the head and face; the abducens (VI) nerves, which regulate certain eyeball movements; the facial (VII) nerves, which conduct nerve impulses related to taste, salivation, and facial expression; and the vestibular branches of the vestibulocochlear (VIII) nerves, which are concerned with equilibrium.

Other important nuclei in the reticular formation of the pons are the *pneumotaxic* (noo-mō-TAK-sik) *area* and the *apneustic* (ap-NOO-stik) *area.* Together with the medullary rhythmicity area in the medulla, they help control respiration.

Midbrain

The *midbrain,* or *mesencephalon* (*meso* = middle; *enke-phalos* = brain), extends from the pons to the lower portion of the diencephalon (Figures 18-1 and 18-4). It is about 2.5 cm (1 in.) in length. The cerebral aqueduct passes through the midbrain and connects the third ventricle above with the fourth ventricle below.

The ventral portion of the midbrain contains a pair of fiber bundles referred to as *cerebral peduncles* (pe-DUNG-kulz). The cerebral peduncles contain some motor fibers that convey nerve impulses from the cerebral cortex to the pons and spinal cord. They also contain sensory fibers that pass from the spinal cord to the thalamus. The cerebral peduncles constitute the main connection for tracts between upper parts of the brain and lower parts of the brain and the spinal cord.

The dorsal portion of the midbrain is called the *tectum* (*tectum* = roof) and contains four rounded eminences: the *corpora quadrigemina* (KOR-po-ra kwad-ri-JEM-ina). Two of the eminences are known as the *superior colliculi* (ko-LIK-yoo-lī). These serve as reflex centers for movements of the eyeballs and head and neck in response to visual and other stimuli. The other two eminences, the *inferior colliculi,* serve as reflex centers for movements of the head and trunk in response to auditory stimuli. The midbrain also contains the *substantia nigra* (sub-STAN-shē-a NĪ-gra), a large, heavily pigmented nucleus near the cerebral peduncles.

A major nucleus in the reticular formation of the midbrain is the *red nucleus.* Fibers from the cerebellum and cerebral cortex terminate in the red nucleus. The red nucleus is also the origin of cell bodies of the descending rubrospinal tract. Other nuclei in the midbrain are associated with cranial nerves (see Figure 18-4). These include the oculomotor (III) nerves, which mediate some movements of the eyeballs and changes in pupil size and lens shape, and the trochlear (IV) nerves, which conduct nerve impulses that move the eyeballs.

A structure called the *medial lemniscus* (*lemniskos* = ribbon or band) is common to the medulla, pons, and midbrain. The medial lemniscus is a band of white fibers containing axons that convey nerve impulses for fine touch, proprioception, pressure, and vibrations from the medulla to the thalamus.

DIENCEPHALON

The *diencephalon* (*dia* = through; *enkephalos* = brain) consists principally of the thalamus and hypothalamus. The relation of these structures to the rest of the brain is shown in Figure 18-1.

Thalamus

The *thalamus* (THAL-a-mus; *thalamos* = inner chamber) is an oval structure above the midbrain that measures about 3 cm (1 in.) in length and constitutes four-fifths of the diencephalon. It consists of two oval masses of mostly gray matter organized into nuclei that form the lateral walls of the third ventricle (Figure 18-6). The masses are joined by a bridge of gray matter called the *intermediate mass.* Each mass is deeply embedded in a cerebral hemisphere and is bounded laterally by the *internal capsule.*

Although the thalamic masses are primarily gray matter, some portions are white matter. Among the white matter portions are the *stratum zonale,* which covers the dorsal surface; the *external medullary lamina,* covering the lateral surface; and the *internal medullary lamina,* which divides the gray matter masses into an anterior nuclear group, a medial nuclear group, and a lateral nuclear group.

Within each group are nuclei that assume various roles. Some nuclei in the thalamus serve as relay stations for all sensory impulses, except smell, to the cerebral cortex. These include the *medial geniculate* (je-NIK-yoo-lāt) *nuclei* (hearing), the *lateral geniculate nuclei* (vision), and the *ventral posterior nuclei* (general sensations and taste). Other nuclei are centers for synapses in the somatic motor system. These include the *ventral lateral nuclei* (voluntary motor actions) and *ventral anterior nuclei* (voluntary motor actions and arousal). The thalamus is the principal relay station for sensory impulses that reach the cerebral cortex from the spinal cord, brain stem, cerebellum, and parts of the cerebrum.

The thalamus also functions as an interpretation center for some sensory impulses, such as pain, temperature, light touch, and pressure. The thalamus also contains a *reticular nucleus* in its reticular formation, which in some way seems to modify neuronal activity in the thalamus, and an *anterior nucleus* in the floor of the lateral ventricle, which is concerned with certain emotions and memory.

Hypothalamus

The *hypothalamus* (*hypo* = under) is a small portion of the diencephalon. Its relation to other parts of the brain is shown in Figure 18-1 and 18-6a. The hypothalamus forms the floor and part of the lateral walls of the third ventricle. It is partially protected by the sella turcica of the sphenoid bone.

Information from the external environment comes to the hypothalamus via afferent pathways originating in the pe-

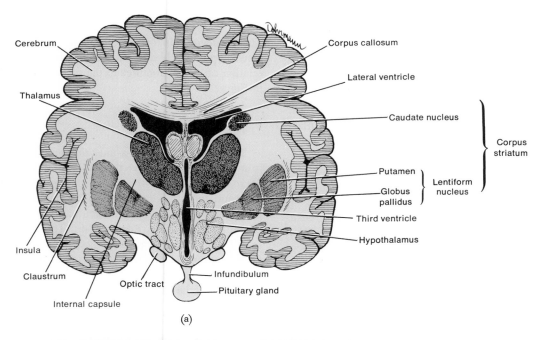

(a)

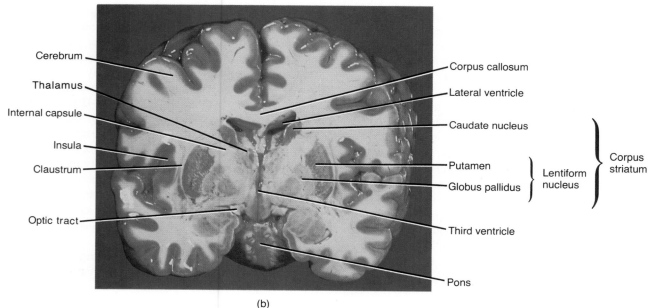

(b)

FIGURE 18-6 Thalamus. (a) Diagram of a frontal section showing the thalamus and associated structures. (b) Photograph of a frontal section of the cerebrum showing the thalamus and associated structures. (Courtesy of C. Yokochi and J. W. Rohen, *Photographic Anatomy of the Human Body,* 2nd ed., 1979, IGAKU-SHOIN, Ltd., Tokyo, New York.)

ripheral sense organs. Sound, taste, smell, and somatic sensations all come to the hypothalamus. Afferent impulses, monitoring the internal environment, arise from the internal viscera and reach the hypothalamus. Other parts of the hypothalamus itself continually monitor the water level, hormone concentrations, and temperature of blood. And, as you will see shortly, the hypothalamus has several very important connections with the pituitary gland.

Despite its small size, nuclei in the hypothalamus control many body activities, most of them related to homeostasis. Although differentiation of the hypothalamic nuclei is far from precise, it is possible to identify certain nuclei. Some of these nuclei are more readily identified in lower animals and are more distinct in fetuses than adults. Also, within a given nucleus there may be several kinds of cells that can be differentiated histologically. The localization of function, with a few exceptions, is not specific to the individual nuclei; certain functions tend to overlap nuclear boundaries.

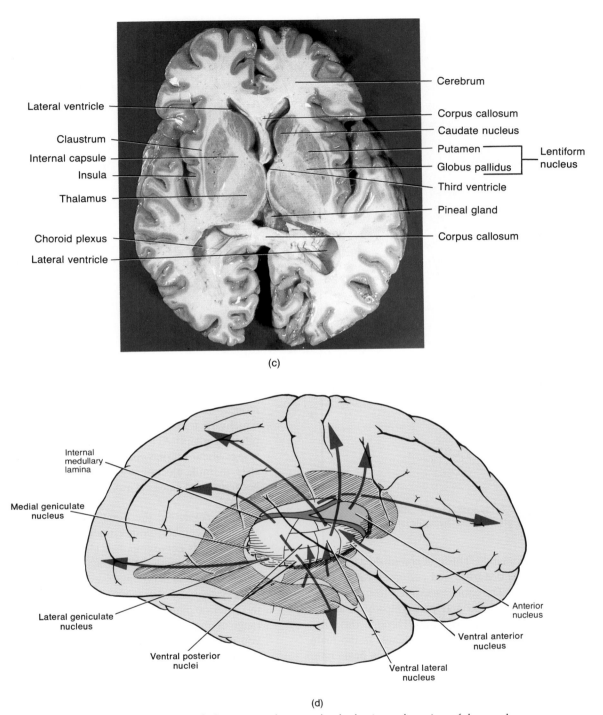

(c)

(d)

FIGURE 18-6 (Continued) Thalamus. (c) Photograph of a horizontal section of the cerebrum showing the thalamus and associated structures. (Courtesy of C. Yokochi and J. W. Rohen, *Photographic Anatomy of the Human Body,* 2nd ed., 1979, IGAKU-SHOIN, Ltd., Tokyo, New York.) (d) Diagram of a right lateral view of the thalamic nuclei.

For this reason, functions are attributed to regions rather than specific nuclei. Since the hypothalamic nuclei are useful landmarks in understanding the subsequent discussion of functions, a three-dimensional view of the hypothalamic nuclei is shown in Figure 18-7.

The chief functions of the hypothalamus are as follows.

1. It controls and integrates the autonomic nervous system, which stimulates smooth muscle, regulates the rate of contraction of cardiac muscle, and controls the se-

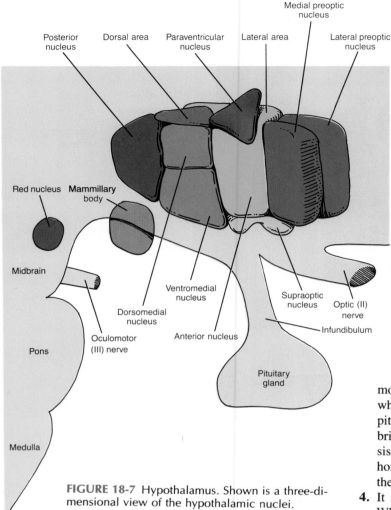

FIGURE 18-7 Hypothalamus. Shown is a three-dimensional view of the hypothalamic nuclei.

cretions of many glands. This is accomplished by axons of neurons whose dendrites and cell bodies are in hypothalamic nuclei. The axons form tracts from the hypothalamus to sympathetic and parasympathetic nuclei in the brain stem and spinal cord. Through the autonomic nervous system, the hypothalamus is the main regulator of visceral activities. It regulates heart rate, movement of food through the gastrointestinal tract, and contraction of the urinary bladder.

2. It is involved in the reception and integration of sensory impulses from the viscera.

3. It is the principal intermediary between the nervous system and the endocrine system—the two major control systems of the body. The hypothalamus lies just above the pituitary, the main endocrine gland. When the hypothalamus detects certain changes in the body, it releases chemicals called regulating hormones (or factors) that stimulate or inhibit the anterior pituitary gland. The anterior pituitary then releases or holds back hormones that regulate various physiological activities of the body. The hypothalamus also produces two hor-

mones, antidiuretic hormone (ADH) and oxytocin (OT), which are transported to and stored in the posterior pituitary gland. ADH decreases urine volume and OT brings about uterine contractions during labor and assists in milk ejection by the mammary glands. The hormones are released from storage when needed by the body.

4. It is the center for the mind-over-body phenomenon. When the cerebral cortex interprets strong emotions, it often sends nerve impulses along the tracts that connect the cortex with the hypothalamus. The hypothalamus then directs nerve impulses via the autonomic nervous system and also releases chemicals that stimulate the anterior pituitary gland. The result can be a wide range of changes in body activities. For instance, when you panic, nerve impulses leave the hypothalamus to stimulate your heart to beat faster. Likewise, continued psychological stress can produce long-term abnormalities in body function that result in serious illness. These so-called psychosomatic disorders are definitely real.

5. It is associated with feelings of rage and aggression.

6. It controls normal body temperature. Certain cells of the hypothalamus serve as a thermostat. If blood flowing through the hypothalamus is above normal temperature, the hypothalamus directs nerve impulses along the autonomic nervous system to stimulate activities that promote heat loss. Heat can be lost through relaxation of the smooth muscle in the blood vessels, causing vasodilation of cutaneous vessels and increased heat loss from the skin. Heat loss also occurs by sweating. Conversely, if the temperature of the blood is below normal, the

hypothalamus generates nerve impulses that promote heat retention. Heat can be retained through the constriction of cutaneous blood vessels, cessation of sweating, and by shivering.

7. It regulates food intake through two centers. The *feeding (hunger) center* is stimulated by hunger sensations from an empty stomach. When sufficient food has been ingested, the *satiety* (sa-TĪ-e-tē) *center* is stimulated and sends out nerve impulses that inhibit the feeding center.

8. It contains a *thirst center.* Certain cells in the hypothalamus are stimulated when the extracellular fluid volume is reduced. The stimulated cells produce the sensation of thirst.

9. It is one of the centers that maintains the waking state and sleep patterns.

10. It exhibits properties of a self-sustained oscillator and, as such, acts as a pacemaker to drive many biological rhythms.

CEREBRUM

Supported on the brain stem and forming the bulk of the brain is the *cerebrum* (see Figure 18-1). The surface of the cerebrum is composed of gray matter 2–4 mm (0.08–0.16 in.) thick and is referred to as the *cerebral cortex* (*cortex* = rind or bark). The cortex, containing billions of cells, consists of six layers of nerve cell bodies in most areas. Beneath the cortex lies the cerebral white matter.

During embryonic development, when there is a rapid increase in brain size, the gray matter of the cortex enlarges out of proportion to the underlying white matter. As a result, the cortical region rolls and folds upon itself. The folds are called *gyri* (JĪ-rī) or *convolutions* (Figure 18-8). The deep grooves between folds are referred to as *fissures;* the shallow grooves between folds are *sulci* (SUL-sī). The most prominent fissure, the *longitudinal fissure,* nearly separates the cerebrum into right and left halves, or *hemispheres.* The hemispheres, however, are connected inter-

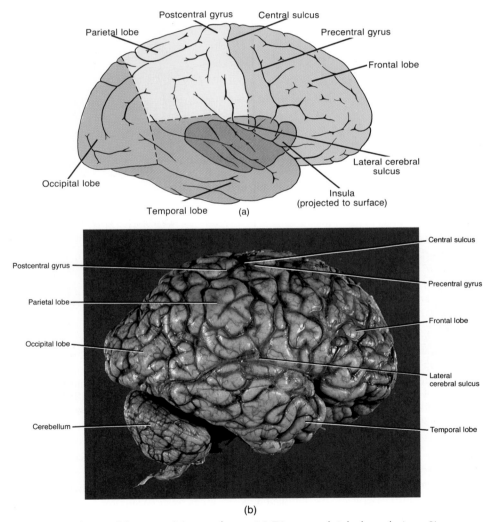

FIGURE 18-8 Lobes and fissures of the cerebrum. (a) Diagram of right lateral view. Since the insula cannot be seen externally, it has been projected to the surface. It can be seen in Figures 18-6a, b and 18-8c. (b) Photograph of the right lateral view. (Courtesy of Martin Rotker, Taurus Photos.)

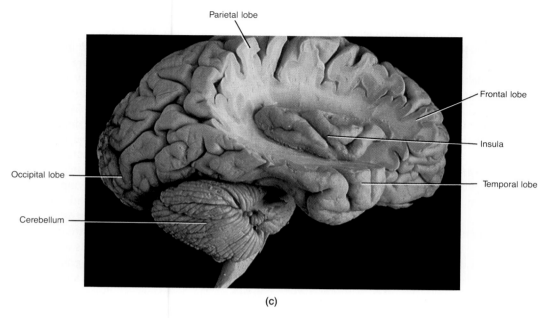

(c)

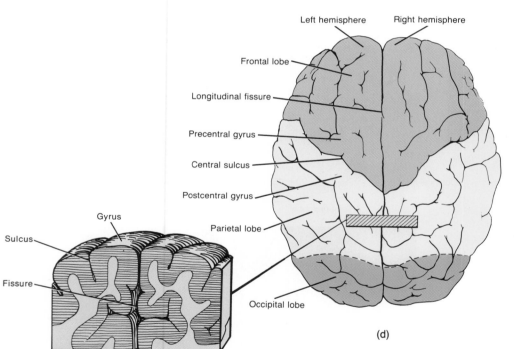

(d)

FIGURE 18-8 (Continued) (c) Photograph of right lateral view showing the insula after removal of a portion of the cerebrum. (Courtesy of N. Gluhbegovic and T. H. Williams, *The Human Brain: A Photographic Guide,* Harper & Row, Publishers, Inc., New York, 1980). (d) Diagram of superior view. The insert to the left indicates the relative differences among a gyrus, sulcus, and fissure.

nally by a large bundle of transverse fibers composed of white matter called the *corpus callosum* (kal-LŌ-sum; *corpus* = body; *callosus* = hard). Between the hemispheres is an extension of the cranial dura mater called the *falx* (FALKS) *cerebri (cerebral fold)*. It encloses the superior and inferior sagittal sinuses (see also Figure 18-13d).

Lobes

Each cerebral hemisphere is further subdivided into four lobes by deep sulci or fissures. The *central sulcus* separates the *frontal lobe* from the *parietal lobe*. A major gyrus, the *precentral gyrus,* is located immediately anterior to the central sulcus. The gyrus is a landmark for the primary

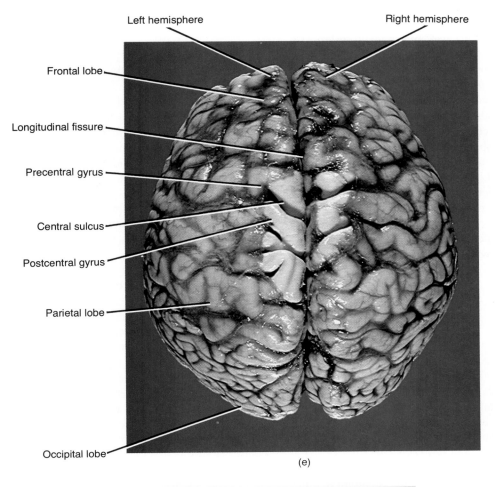

Left hemisphere

Right hemisphere

Frontal lobe

Longitudinal fissure

Precentral gyrus

Central sulcus

Postcentral gyrus

Parietal lobe

Occipital lobe

(e)

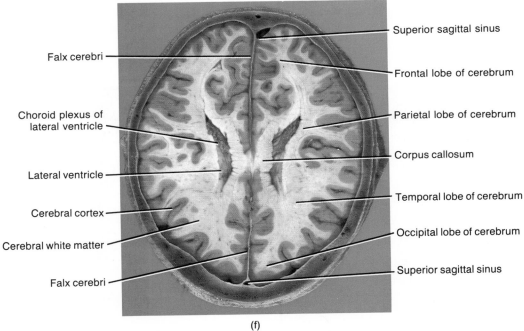

Falx cerebri

Choroid plexus of lateral ventricle

Lateral ventricle

Cerebral cortex

Cerebral white matter

Falx cerebri

Superior sagittal sinus

Frontal lobe of cerebrum

Parietal lobe of cerebrum

Corpus callosum

Temporal lobe of cerebrum

Occipital lobe of cerebrum

Superior sagittal sinus

(f)

FIGURE 18-8 (Continued) Lobes and fissures of the cerebrum. (e) Photograph of superior view. (Courtesy of Martin Rotker, Taurus Photos.) (f) Photograph of a cross section through the cerebrum. (Courtesy of Stephen A. Kieffer and E. Robert Heitzman, *An Atlas of Cross-Sectional Anatomy*, Harper & Row, Publishers, Inc., New York, 1979.)

motor area of the cerebral cortex. Another major gyrus, the *postcentral gyrus,* is located immediately posterior to the central sulcus. This gyrus is a landmark for the general sensory area of the cerebral cortex. The *lateral cerebral sulcus* separates the *frontal lobe* from the *temporal lobe.* The *parieto-occipital sulcus* separates the *parietal lobe* from the *occipital lobe.* Another prominent fissure, the *transverse fissure,* separates the cerebrum from the cerebellum. The frontal lobe, parietal lobe, temporal lobe, and occipital lobe are named after the bones that cover them. A fifth part of the cerebrum, the *insula,* lies deep within the lateral cerebral fissure, under the parietal, frontal, and temporal lobes. It cannot be seen in an external view of the brain (Figure 18-8a, b).

As you will see later, the olfactory (I) and optic (II) nerves are associated with specific lobes of the cerebrum.

White Matter

The white matter underlying the cortex consists of myelinated axons running in three principal directions (Figure 18-9).

1. *Association fibers* connect and transmit nerve impulses between gyri in the same hemisphere.
2. *Commissural fibers* transmit nerve impulses from the gyri in one cerebral hemisphere to the corresponding gyri in the opposite cerebral hemisphere. Three important groups of commissural fibers are the *corpus callosum, anterior commissure,* and *posterior commissure.*
3. *Projection fibers* form ascending and descending tracts that transmit nerve impulses from the cerebrum to other parts of the brain and spinal cord. The internal capsule is an example.

Basal Ganglia (Cerebral Nuclei)

The *basal ganglia (cerebral nuclei)* are paired masses of gray matter in each cerebral hemisphere (Figures 18-6 and 18-10). The largest of the basal ganglia of each hemisphere is the *corpus striatum* (strī-Ā-tum; *corpus* = body; *striatus* = striped). It consists of the *caudate (cauda =* tail) *nucleus* and the *lentiform (lenticula =* shaped like a lentil or lens) *nucleus.* The lentiform nucleus, in turn, is subdivided into a lateral portion called the *putamen* (pu-TĀ-men; *putamen* = shell) and a medial portion called the *globus pallidus (globus* = ball; *pallid* = pale).

The portion of the *internal capsule* passing between the lentiform nucleus and the caudate nucleus and between the lentiform nucleus and thalamus is sometimes considered part of the corpus striatum. The internal capsule is made up of a group of sensory and motor white matter tracts that connect the cerebral cortex with the brain stem and spinal cord.

Other structures frequently considered part of the basal ganglia are the *substantia nigra, subthalamic nucleus,* and *red nucleus.* The substantia nigra is a large nucleus in the midbrain whose axons terminate in the caudate nucleus and putamen. The subthalamic nucleus lies against the internal capsule. Its major connection is with the globus pallidus.

The basal ganglia are interconnected by many fibers. They are also connected to the cerebral cortex, thalamus, and hypothalamus. The caudate nucleus and the putamen control large subconscious movements of the skeletal muscles, such as swinging the arms while walking. Such gross movements are also consciously controlled by the cerebral cortex. The globus pallidus is concerned with the regulation of muscle tone required for specific body movements.

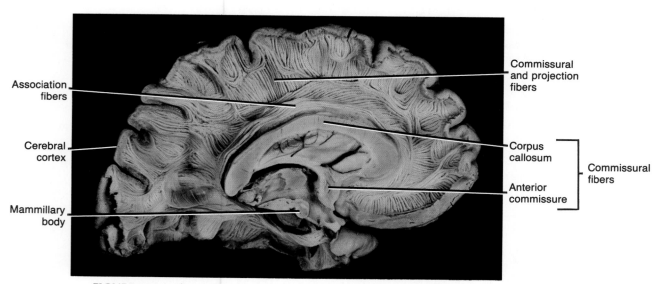

FIGURE 18-9 White matter tracts of the left cerebral hemisphere seen in sagittal section. (Courtesy of N. Gluhbegovic and T. H. Williams, *The Human Brain: A Photographic Guide,* Harper & Row, Publishers, Inc., New York, 1980.)

CLINICAL APPLICATION

Damage to the basal ganglia results in abnormal body movements, such as uncontrollable shaking, called *tremor,* and *involuntary movements of skeletal muscles.* Moreover, destruction of a substantial portion of the caudate nucleus almost totally *paralyzes* the side of the body opposite to the damage. The caudate nucleus is an area often affected by a stroke.

A lesion in the subthalamic nucleus results in a motor disturbance on the opposite side of the body called *hemiballismus* (*hemi* = half; *ballismos* = jumping), which is characterized by involuntary movements occurring suddenly with great force and rapidity. The movements are purposeless and generally of the withdrawal type, although they may be jerky. The spontaneous movements affect the proximal portions of the extremities most severely, especially the arms.

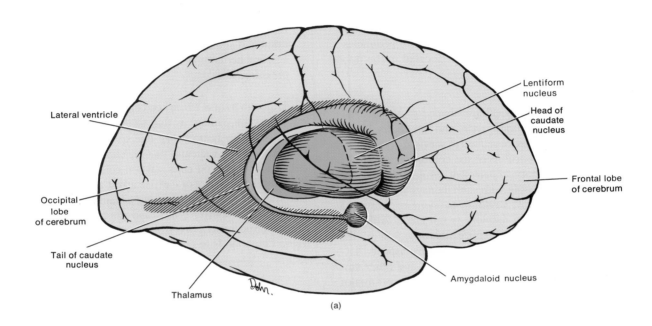

(a)

(b)

FIGURE 18-10 Basal ganglia. (a) In this diagram of the right lateral view of the cerebrum, the basal ganglia have been projected to the surface. Refer to Figure 18-6a, b for the positions of the basal ganglia in the horizontal section of the cerebrum. (b) Photograph of the medial surface of the left cerebral hemisphere showing portions of the basal ganglia. (Courtesy of N. Gluhbegovic and T. H. Williams, *The Human Brain: A Photographic Guide,* Harper & Row, Publishers, Inc., New York, 1980.)

Limbic System

Certain components of the cerebral hemispheres and diencephalon constitute the *limbic* (*limbus* = border) *system.* Among its components are the following regions of gray matter.

1. *Limbic lobe.* Formed by two gyri of the cerebral hemisphere: the cingulate gyrus and the hippocampal gyrus.
2. *Hippocampus.* An extension of the hippocampal gyrus that extends into the floor of the lateral ventricle.
3. *Amygdaloid nucleus.* Located at the tail end of the caudate nucleus.
4. *Mammillary bodies of the hypothalamus.* Two round masses close to the midline near the cerebral peduncles.
5. *Anterior nucleus of the thalamus.* Located in the floor of the lateral ventricle.

The limbic system is a wishbone-shaped group of structures that encircles the brain stem and functions in the emotional aspects of behavior related to survival. The hippocampus, together with portions of the cerebrum, also functions in memory. Memory impairment results from lesions in the limbic system. People with such damage forget recent events and cannot commit anything to memory. How the limbic system functions in memory is not clear. Although behavior is a function of the entire nervous system, the limbic system controls most of its involuntary aspects. Experiments on the limbic system of monkeys and other animals indicate that the amygdaloid nucleus assumes a major role in controlling the overall pattern of behavior.

Other experiments have shown that the limbic system is associated with pleasure and pain. When certain areas of the limbic system of the hypothalamus, thalamus, and midbrain are stimulated in animals, their reactions indicate they are experiencing intense punishment. When other areas are stimulated, the animals' reactions indicate they are experiencing extreme pleasure. In still other studies, stimulation of the perifornical nuclei of the hypothalamus results in a behavioral pattern called rage. The animal assumes a defensive posture—extending its claws, raising its tail, hissing, spitting, growling, and opening its eyes wide. Stimulating other areas of the limbic system results in an opposite behavioral pattern: docility, tameness, and affection. Because the limbic system assumes a primry function in emotions such as pain, pleasure, anger, rage, fear, sorrow, sexual feelings, docility, and affection, it is sometimes called the ''visceral'' or ''emotional'' brain.

CLINICAL APPLICATION

Brain injuries are commonly associated with head injuries and result from displacement and distortion of neuronal tissue at the moment of impact. The various degrees of brain injury are described by the following terms.

1. *Concussion.* An abrupt but temporary loss of consciousness following a blow to the head or a sudden stopping of a moving head. A concussion produces no visible bruising of the brain.
2. *Contusion.* A visible bruising of the brain due to trauma and blood leaking from microscopic vessels. The pia mater is stripped from the brain over the injured area and may be torn, allowing blood to enter the subarachnoid space. A contusion usually results in an extended loss of consciousness, ranging from several minutes to many hours.
3. *Laceration.* Tearing of the brain, usually from a skull fracture or gunshot wound. A laceration results in rupture of large blood vessels with bleeding into the brain and subarachnoid space. Consequences include cerebral hematoma, edema, and increased intracranial pressure.

Although trauma to the head can lead to *brain damage,* not all of the damage is due to the impact alone. Most of the damage is probably due to the release of large numbers of free radicals, that is, charged oxygen molecules from damaged cells. (Brain cells recovering from the effects of a stroke or cardiac arrest also release excess free radicals.) Free radicals cause damage by disrupting cellular DNA and enzymes and altering plasma membrane permeability. Investigations are underway to develop compounds to counter the effects of free radicals.

Functional Areas of Cerebral Cortex

The functions of the cerebrum are numerous and complex. In a general way, the cerebral cortex is divided into sensory, motor, and association areas. the *sensory areas* interpret sensory impulses, the *motor areas* control muscular movement, and the *association areas* are concerned with emotional and intellectual processes.

● *Sensory Areas* The *primary somesthetic* (sō-mes-THET-ik; *soma* = body; *aisthesis* = perception) *area* or *general sensory area* is located directly posterior to the central sulcus of the cerebrum in the postcentral gyrus of the parietal lobe. It extends from the longitudinal fissure on the top of the cerebrum to the lateral cerebral sulcus. In Figure 18-11, the general sensory area is designated by the areas numbered 1, 2, and 3.*

The primary somesthetic area receives sensations from cutaneous, muscular, and visceral receptors in various parts of the body. Each point of the area receives sensations from specific parts of the body, and essentially the entire body is spatially represented in it. The size of the portion of the sensory area receiving stimuli from body parts is not dependent on the size of the part but on the number

* These numbers, as well as most of the others shown, are based on K. Brodmann's cytoarchitectural map of the cerebral cortex. His map, first published in 1909, attempts to correlate structure and function.

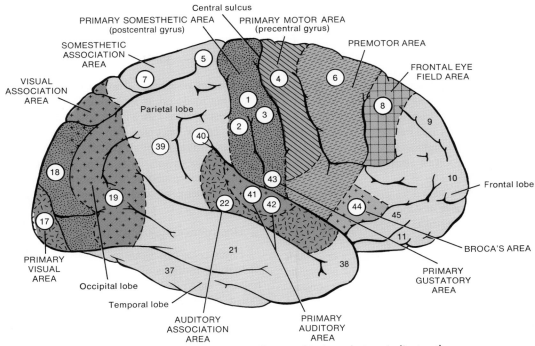

FIGURE 18-11 Functional areas of the cerebrum. This lateral view indicates the sensory and motor areas of the right hemisphere. Although Broca's area is in the left hemisphere of most people, it is shown here to indicate its location.

of receptors the part contains. For example, a larger portion of the sensory area receives impulses from the lips than from the thorax (see Figure 20-4). The major function of the primary somesthetic area is to localize exactly the points of the body where the sensations originate. The thalamus is capable of localizing sensations in a general way; that is, it receives sensations from large areas of the body but cannot distinguish between specific areas of stimulation. This ability is reserved for the primary somesthetic area of the cortex.

The *secondary somesthetic area* is a small region in the posterior wall of the lateral sulcus in line with the postcentral gyrus. It is involved mainly in less discriminative aspects of sensation.

Posterior to the primary somesthetic area is the *somesthetic association area.* It corresponds to the areas numbered 5 and 7 in Figure 18-11. The somesthetic association area receives input from the thalamus, other lower portions of the brain, and the primary somesthetic area. Its role is to integrate and interpret sensations. This area permits you to determine the exact shape and texture of an object without looking at it, to determine the orientation of one object to another as they are felt, and to sense the relation of one body part to another. Another role of the somesthetic association area is the storage of memories of past sensory experiences. Thus, you can compare sensations with previous experiences.

Other sensory areas of the cortex include the following.

1. *Primary visual area* (area 17). Located on the medial surface of the occipital lobe and occasionally extends around to the lateral surface. It receives sensory impulses from the eyes and interprets shape, color, and movement.

2. *Visual association area* (areas 18 and 19). Located in the occipital lobe. It receives sensory impulses from the primary visual area and the thalamus. It relates present to past visual experiences with recognition and evaluation of what is seen.

3. *Primary auditory area* (areas 41 and 42). Located in the superior part of the temporal lobe near the lateral cerebral sulcus. It interprets the basic characteristics of sound such as pitch and rhythm. Whereas the anterolateral portion of the auditory area responds to low pitches, the posterolateral portion responds to high pitches.

4. *Auditory association (Wernicke's) area* (area 22). Inferior to the primary auditory area in the temporal cortex. It determines if a sound is speech, music, or noise. It also interprets the meaning of speech by translating words into thoughts.

5. *Primary gustatory area* (area 43). Located at the base of the postcentral gyrus above the lateral cerebral sulcus in the parietal cortex. It interprets sensations related to taste.

6. *Primary olfactory area.* Located in the temporal lobe on the medial aspect. It interprets sensations related to smell.

7. *Gnostic* (NOS-tik; *gnosis* = knowledge) *area* (areas 5, 7, 39, and 40). This *common integrative area* is located among the somesthetic, visual, and auditory association areas. The gnostic area receives nerve impulses from these areas, as well as from the taste and smell areas, the thalamus, and lower portions of the brain stem. It integrates sensory interpretations from the association areas and nerve impulses from other areas so that a common thought can be formed from the various sensory inputs. It then transmits signals to other parts of the brain to cause the appropriate response to the sensory signal.

• *Motor Areas*   The *primary motor area* (area 4) is located in the precentral gyrus of the frontal lobe (Figure 18-11). Like the primary somesthetic area, the primary motor area consists of regions that control specific muscles or groups of muscles (see Figure 20-20). Stimulation of a specific point of the primary motor area results in a muscular contraction, usually on the opposite side of the body.

The *premotor area* (area 6) is anterior to the primary motor area. It is concerned with learned motor activities of a complex and sequential nature. It generates nerve impulses that cause a specific group of muscles to contract in a specific sequence, for example, writing. Thus, the premotor area controls skilled movements.

The *frontal eye field area* (area 8) in the frontal cortex is sometimes included in the premotor area. This area controls voluntary scanning movements of the eyes—searching for a word in a dictionary, for instance.

The *language areas* are also significant parts of the motor cortex. The translation of speech or written words into thoughts involves sensory areas—primary auditory, auditory association, primary visual, visual association, and gnostic—as we just described. The translation of thoughts into speech involves the *motor speech area* (area 44) or *Broca's* (BRŌ-kaz) *area,* located in the frontal lobe just superior to the lateral cerebral sulcus. From this area, a sequence of nerve impulses is sent to the premotor regions that control the muscles of the larynx, pharynx, and mouth. The nerve impulses from the premotor area to the muscles result in specific, coordinated contractions that enable you to speak. Simultaneously, nerve impulses are sent from Broca's area to the primary motor area. From here, nerve impulses reach your breathing muscles to regulate the proper flow of air past the vocal cords. The coordinated contractions of your speech and breathing muscles enable you to translate your thoughts into speech.

CLINICAL APPLICATION

Broca's area and other language areas are located in the left cerebral hemisphere of most individuals regardless of whether they are left-handed or right-handed. Injury to the sensory or motor speech areas results in *aphasia* (a-FĀ-zē-a; *a* = without; *phasis* = speech), an inability to speak; *agraphia* (*a* = without; *graph* =

write), an inability to write; *word deafness,* an inability to understand spoken words; or *word blindness,* an inability to understand written words.

• *Association Areas*   The *association areas* of the cerebrum are made up of association tracts that connect motor and sensory areas (see Figure 18-9). The association region of the cortex occupies the greater portion of the lateral surfaces of the occipital, parietal, and temporal lobes, and the frontal lobes anterior to the motor areas. The association areas are concerned with memory, emotions, reasoning, will, judgment, personality traits, and intelligence.

Electroencephalogram (EEG)

Brain cells can generate electrical activity as a result of literally millions of nerve impulses (nerve action potentials) of individual neurons. These electrical potentials are called *brain waves* and indicate activity of the cerebral cortex. Brain waves pass through the skull easily and can be detected by sensors called electrodes. A record of such waves is called an *electroencephalogram (EEG).* An EEG is obtained by placing electrodes on the head and amplifying the waves with an electroencephalograph. Distinct EEG patterns appear in certain abnormalities. In fact, the EEG is used clinically in the diagnosis of epilepsy, infectious diseases, tumors, trauma, and hematomas. Electroencephalograms also furnish information regarding sleep and wakefulness.

CLINICAL APPLICATION

In cases of doubt, extinction of an EEG, called a flat EEG, is increasingly being taken as one criterion of *brain death.* Other criteria include unconsciousness, no spontaneous breathing, the absence of a light response and dilation of the pupils, the absence of reflexes, unresponsiveness, and lack of normal muscle tone or strength.

BRAIN LATERALIZATION (SPLIT-BRAIN CONCEPT)

Gross examination of the brain would suggest that it is bilaterally symmetric. However, detailed examination by the use of CT scans reveals certain anatomical differences between the two hemispheres. For example, in left-handed people the parietal and occipital lobes of the right hemisphere are usually narrower than the corresponding lobes of the left hemisphere. In addition, the frontal lobe of the left hemisphere of such individuals is typically narrower than that of the right hemisphere.

In addition to the structural differences between both sides of the brain, there are also several important functional differences (Figure 18-12). It has been shown that the left hemisphere is more important for right-hand control, spoken and written language, numerical and scientific skills, ability

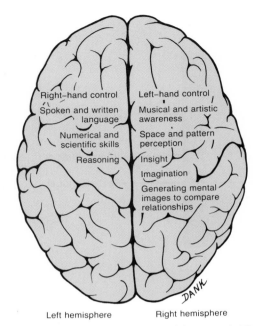

Left hemisphere   Right hemisphere

FIGURE 18-12 Summary of the principal functional differences between the left and right cerebral hemispheres.

to use and understand sign language, and reasoning in most people. Conversely, it has been shown that the right hemisphere is more important for left-hand control; musical and artistic awareness; space and pattern perception; insight; imagination; and generating mental images of sight, sound, touch, taste, and smell in order to compare relations.

CEREBELLUM

The *cerebellum* is the second-largest portion of the brain (almost one-eighth of the brain's mass) and occupies the inferior and posterior aspects of the cranial cavity. Specifically, it is posterior to the medulla and pons and below the occipital lobes of the cerebrum (see Figure 18-1). It is separated from the cerebrum by the *transverse fissure* and by an extension of the cranial dura mater called the *tentorium* (*tentorium* = tent) *cerebelli* (see Figure 18–13d). The tentorium cerebelli partially encloses the transverse sinuses and supports the occipital lobes of the cerebral hemispheres.

Structure

The cerebellum is shaped somewhat like a butterfly. The central constricted area is the *vermis,* which means worm shaped, and the lateral "wings" or lobes are referred to as *hemispheres* (Figure 18-13). Each hemisphere consists of lobes that are separated by deep and distinct fissures. The *anterior lobe* and *posterior lobe* are concerned with subconscious movements of skeletal muscles. The *flocculonodular lobe* is concerned with the sense of equilibrium (see Chapter 20). Between the hemispheres is another extension of the cranial dura mater: the *falx cerebelli.* It passes only a short distance between the cerebellar hemispheres and contains the occipital sinus.

The surface of the cerebellum, called the *cortex,* consists of gray matter in a series of slender, parallel ridges called *folia.* They are less prominent than the convolutions of the cerebral cortex. Beneath the gray matter are *white matter tracts (arbor vitae)* that resemble branches of a tree. Deep within the white matter are masses of gray matter, the *cerebellar nuclei.* The nuclei give rise to nerve fibers that convey information out of the cerebellum to other parts of the nervous system.

The cerebellum is attached to the brain stem by three paired bundles of fibers called *cerebellar penduncles* (see Figure 18-4d,e). *Inferior cerebellar peduncles* connect the cerebellum with the medulla at the base of the brain stem and with the spinal cord. These peduncles contain both afferent and efferent fibers and thus bring information into and out of the cerebellum. *Middle cerebellar peduncles* connect the cerebellum with the pons. These peduncles contain only afferent fibers and thus bring information into the cerebellum. *Superior cerebellar peduncles* connect the cerebellum with the midbrain. These peduncles contain mostly efferent fibers and thus mostly bring information out of the cerebellum.

Functions

Functionally, the cerebellum is a motor area of the brain concerned with coordinating subconscious movements of skeletal muscles. The cerebellar peduncles are the fiber tracts that permit information to pass into and out of the cerebellum. The cerebellum constantly receives input signals from proprioceptors in muscles, tendons, and joints, receptors for equilibrium, and visual receptors of the eyes. Such input permits the cerebellum to collect information on the physical status of the body. In addition, when other motor areas of the brain, such as the motor cortex of the cerebrum and basal ganglia, send signals to skeletal muscles, a duplicate set of signals is also sent to the cerebellum. The cerebellum compares input information regarding the actual status of the body with the intended movement determined by a motor area of the brain. If the intent of the motor area is not being attained by the skeletal muscles, the cerebellum detects the variation and sends feedback signals to the motor area to either stimulate or inhibit the activity of skeletal muscles. This interaction produces smooth, coordinated movements.

The cerebellum also functions in maintaining equilibrium and controlling posture. For example, receptors for equilibrium send nerve impulses to the cerebellum, informing it of body position. When the direction of movement changes, the cerebellum sends corrective signals to the motor cortex of the cerebrum. The motor cortex then sends signals to skeletal muscles to reposition the body.

Another function of the cerebellum is related to predicting the future position of a body part during a particular movement. Just before a moving part of the body reaches its intended position, the cerebellum sends signals to skeletal muscles to slow the moving part and stop it at a specific

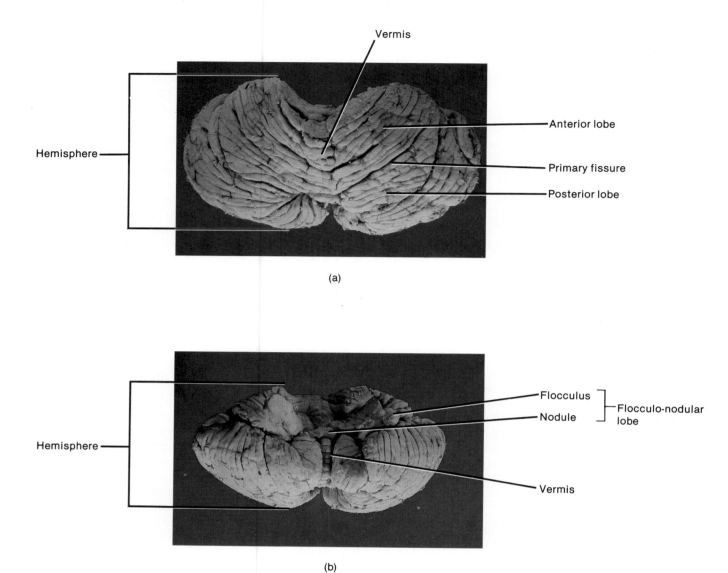

(a)

(b)

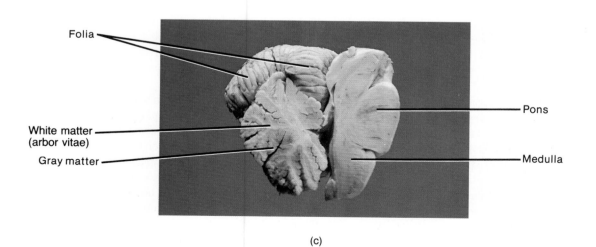

(c)

FIGURE 18-13 Cerebellum. (a) Superior view. (b) Posteroinferior view. (c) Viewed in sagittal section. (Courtesy of Lester V. Bergman & Associates, Inc.)

point. This function of the cerebellum is used in actions such as walking.

There is some evidence that the cerebellum may play a role in a person's emotional development, modulating sensations of anger and pleasure.

CLINICAL APPLICATION

Damage to the cerebellum through trauma or disease is characterized by certain symptoms involving skeletal muscles on the same side of the body as the damage. The effects are ipsilateral because of a double crossing of tracts within the cerebellum. There may be lack of muscle coordination, called ***ataxia*** (*a* = without; *taxis* = order). Blindfolded people with ataxia cannot touch the tip of their nose with a finger because they cannot coordinate movement with their sense of where a body part is located. Another sign of ataxia is a change in the speech pattern due to a lack of coordination of speech muscles. Cerebellar damage may also result in ***distur-***

bances of gait, in which the subject staggers or cannot coordinate normal walking movements, and ***severe dizziness.***

A summary of the functions of the various parts of the brain is presented in Exhibit 18-2.

CRANIAL NERVES

Of the 12 pairs of ***cranial nerves,*** 10 originate from the brain stem, but all leave the skull through foramina of the skull (Figure 18-4a). The cranial nerves are designated with Roman numerals and with names. The Roman numerals indicate the order in which the nerves arise from the brain (front to back). The names indicate the distribution or function.

Some cranial nerves contain only sensory fibers and thus are called ***sensory nerves.*** The remainder contain both sensory and motor fibers and are referred to as ***mixed nerves.***

EXHIBIT 18-2

Summary of Functions of Principal Parts of Brain

PART	FUNCTION
BRAIN STEM **Medulla**	Relays motor and sensory impulses between other parts of the brain and the spinal cord. Recticular formation (also in pons, midbrain, and diencephalon) functions in consciousness and arousal. Vital reflex centers regulate heartbeat, breathing (together with pons), and blood vessel diameter. Nonvital reflex centers coordinate swallowing, vomiting, coughing, sneezing, and hiccuping. Contains nuclei of origin for cranial nerves VIII, IX, X, XI, and XII. Vestibular nuclear complex helps maintain equilibrium.
Pons	Relays impulses within the brain and between parts of the brain and spinal cord. Contains nuclei of origin for cranial nerves V, VI, VII, and VIII. Pneumotaxic area and apneustic area, together with the medulla, help control breathing.
Midbrain	Relays motor impulses from the cerebral cortex to the pons and spinal cord, and relays sensory impulses from the spinal cord to the thalamus. Superior colliculi coordinate movements of the eyeballs in response to visual and other stimuli, and the inferior colliculi coordinate movements of the head and trunk in response to auditory stimuli. Contains nuclei of origin for cranial nerves III and IV.
DIENCEPHALON **Thalamus**	Several nuclei serve as relay stations for all sensory impulses, except smell, to the cerebral cortex. Relays motor impulses from the cerebral cortex to the spinal cord. Interprets pain, temperature, light touch, and pressure sensations. Anterior nucleus functions in emotions and memory.
Hypothalamus	Controls and integrates the autonomic nervous system. Receives sensory impulses from viscera. Articulates with the pituitary gland. Center for mind-over-body phenomena. Functions in rage and aggression. Controls normal body temperature, food intake, and thirst. Helps maintain the waking state and sleep. Functions as a self-sustained oscillator that drives many biological rhythms.
CEREBRUM	Sensory areas interpret sensory impulses, motor areas control muscular movement, and association areas function in emotional and intellectual processes. Basal ganglia control gross muscle movements and regulate muscle tone. Limbic system functions in emotional aspects of behavior related to survival.
CEREBELLUM	Controls subconscious skeletal muscle contractions required for coordination, posture, and balance. Assumes a role in emotional development, modulating sensations of anger and pleasure.

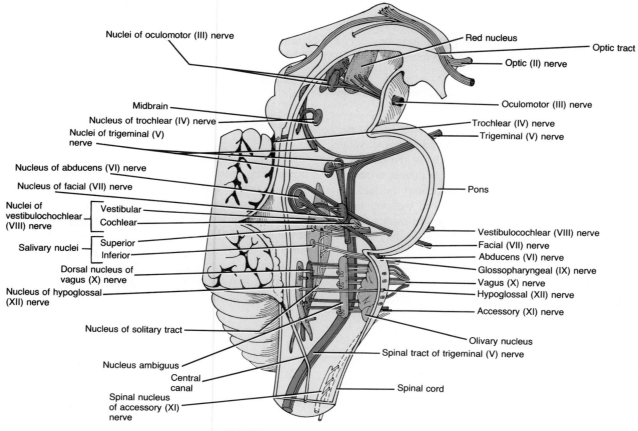

FIGURE 18-14 Cranial nerve nuclei.

At one time, it was believed that some cranial nerves (oculo-motor, trochlear, abducens, accessory, and hypoglossal) were entirely motor. However, it is now known that these cranial nerves also contain some sensory fibers from proprio-ceptors in muscles they innervate. Although these nerves are mixed, they are primarily motor in function, serving to stimulate skeletal muscle contraction. The cell bodies of sensory fibers are found outside the brain, whereas the cell bodies of motor fibers lie in nuclei within the brain (Figure 18-14).

Although the cranial nerves are mentioned singly in the following description of their type, location, and function, remember that they are paired structures.

OLFACTORY (I)

The *olfactory (I) nerve* is entirely sensory and conveys nerve impulses related to smell. It arises as bipolar neurons from the olfactory mucosa of the nasal cavity (see Figure 20-8). The dendrites and cell bodies of these neurons are generally limited to the mucosa covering the superior nasal conchae and the adjacent nasal septum. Axons from the neurons pass through the cribriform plate of the ethmoid bone and synapse with other olfactory neurons in the *olfac-tory bulb,* an extension of the brain lying above the cribri-

form plate. The axons of these neurons make up the *olfactory tract.* The fibers from the tract terminate in the primary olfactory area in the cerebral cortex.

OPTIC (II)

The *optic (II) nerve* is entirely sensory and conveys nerve impulses related to vision. Impulses initiated by rods and cones of the retina are relayed by bipolar neurons to ganglion cells (see Figure 20-11). Axons of the ganglion cells, the optic nerve fibers, exit the optic foramen, after which the two optic nerves unite to form the *optic chiasma* (kī-AZ-ma). Within the chiasma, fibers from the medial half of each retina cross to the opposite side; those from the lateral half remain on the same side. From the chiasma, the re-grouped fibers pass posteriorly to the *optic tracts.* From the optic tracts, the majority of fibers terminate in a nucleus (lateral geniculate) in the thalamus. They then synapse with neurons that pass to the visual areas of the cerebral cortex (see Figure 20-12). Some fibers from the optic chiasma terminate in the superior colliculi of the midbrain. They synapse with neurons whose fibers terminate in the nuclei that convey impulses to the oculomotor (III), trochlear (IV), and abducens (VI) nerves—nerves that control the extrinsic (external) and intrinsic (internal) eye muscles. Through this relay, there are widespread motor responses to light stimuli.

in the carotid sinus, which assumes a major role in blood pressure regulation. The sensory fibers terminate in a nucleus in the thalamus. There are also sensory fibers from proprioceptors in the muscles innervated by this nerve.

VAGUS (X)

The *vagus (X) nerve* is a mixed cranial nerve that is widely distributed from the head and neck into the thorax and abdomen (Figure 18-19). Its motor fibers originate in nuclei of the medulla and terminate in the muscles of the respiratory passageways, lungs, heart, esophagus, stomach, small intestine, most of the large intestine, and gallbladder. Nerve impulses along the motor fibers generate visceral, cardiac, and skeletal muscle movement. Parasympathetic fibers innervate involuntary muscles and glands of the gastrointestinal tract.

Sensory fibers of the vagus nerve supply essentially the same structures as the motor fibers. They convey nerve impulses for various sensations from the larynx, the viscera,

and the ear. The fibers terminate in the medulla and pons. There are also sensory fibers from proprioceptors in the muscles supplied by this nerve.

ACCESSORY (XI)

The *accessory (XI) nerve* (formerly the *spinal accessory nerve*) is a mixed cranial nerve. It differs from all other cranial nerves in that it originates from both the brain stem and the spinal cord (Figure 18-20). The *cranial portion* originates from nuclei in the medulla, passes through the jugular foramen, and supplies the voluntary muscles of the pharynx, larynx, and soft palate that are used in swallowing. The *spinal portion* originates in the anterior gray horn of the first five segments of the cervical portion of the spinal cord. The fibers from the segments join, enter the foramen magnum, and exit through the jugular foramen along with the cranial portion.

The spinal portion conveys motor impulses to the sternocleidomastoid and trapezius muscles to coordinate head

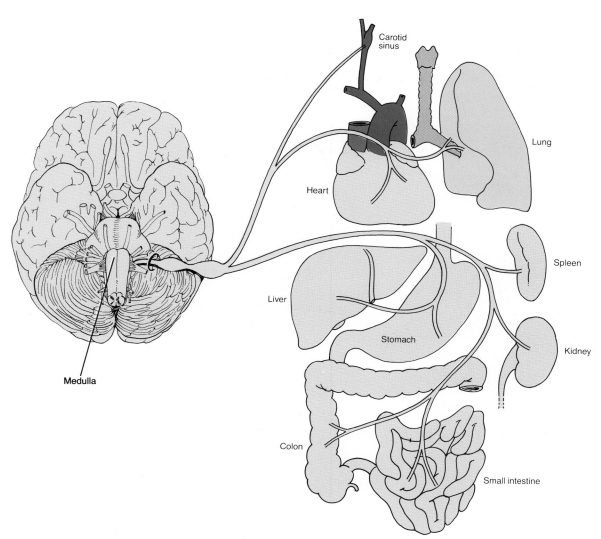

Carotid sinus

Lung

Heart

Spleen

Liver

Kidney

Stomach

Medulla

Colon

Small intestine

FIGURE 18-19 Vagus (X) nerve.

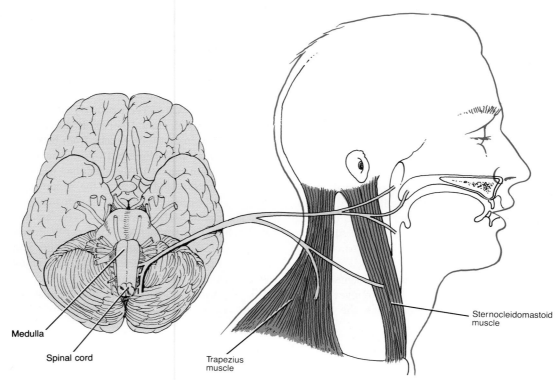

Medulla

Spinal cord

Trapezius muscle

Sternocleidomastoid muscle

FIGURE 18-20 Accessory (XI) nerve.

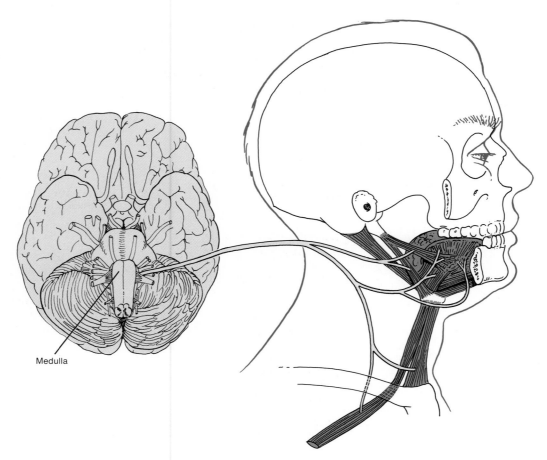

Medulla

FIGURE 18-21 Hypoglossal (XII) nerve.

movements. The sensory fibers originate from proprioceptors in the muscles supplied by its motor neurons and terminate in upper cervical posterior root ganglia.

HYPOGLOSSAL (XII)

The *hypoglossal (XII) nerve* is a mixed cranial nerve. The motor fibers originate in a nucleus in the medulla, pass through the hypoglossal canal, and supply the muscles of the tongue (Figure 18-21). These fibers conduct nerve impulses related to speech and swallowing.

The sensory portion of the hypoglossal nerve consists of fibers originating from proprioceptors in the tongue muscles and terminating in the medulla. The sensory fibers conduct nerve impulses for muscle sense.

A summary of cranial nerves and clinical applications related to dysfunction is presented in Exhibit 18-3.

EXHIBIT 18-3

Summary of Cranial Nerves

NERVE (TYPE)	LOCATION	FUNCTION AND CLINICAL APPLICATION
Olfactory (I) (sensory)	Arises in olfactory mucosa, passes through olfactory bulb and olfactory tract, and terminates in primary olfactory areas of cerebral cortex.	Smell. Loss of the sense of smell, called *anosmia,* may result from head injuries in which the cribriform plate of the ethmoid bone is fractured and from lesions along the olfactory pathway.
Optic (II) (sensory)	Arises in retina of the eye, forms optic chiasma, passes through optic tracts, lateral geniculate nucleus in thalamus, and terminates in visual areas of cerebral cortex.	Vision. Fractures in the orbit, lesions along the visual pathway, and diseases of the nervous system may result in visual field defects and loss of visual acuity. A defect of vision is called *anopsia.*
Oculomotor (III) (mixed, primarily motor)	Motor portion: originates in midbrain and is distributed to levator palpebrae superioris of upper eyelid and four extrinsic eyeball muscles (superior rectus, medial rectus, inferior rectus, and inferior oblique); parasympathetic innervation to ciliary muscles of eyeball and sphincter muscle of iris. Sensory portion: consists of afferent fibers from proprioceptors in eyeball muscles and terminates in midbrain.	Motor: movement of eyelid and eyeball, accommodation of lens for near vision, and constriction of pupil. Sensory: muscle sense (proprioception). A lesion in the nerve causes *strabismus* (squinting) *ptosis* (drooping) of the upper eyelid, pupil dilation, the movement of the eyeball downward and outward on the damaged side, a loss of accommodation for near vision, and double vision (*diplopia*).
Trochlear (IV) (mixed, primarily motor)	Motor portion: originates in midbrain and is distributed to superior oblique muscle, an extrinsic eyeball muscle. Sensory portion: consists of afferent fibers from proprioceptors in superior oblique muscles and terminates in midbrain.	Motor: movement of eyeball. Sensory: muscle sense (proprioception). In trochlear nerve paralysis, the head is tilted to the affected side and diplopia and strabismus occur.
Trigeminal (V) (mixed)	Motor portion: originates in pons and terminates in muscles of mastication, anterior belly of digastric and mylohyoid muscles. Sensory portion: consists of three branches: *ophthalmic*—contains sensory fibers from skin over upper eyelid, eyeball, lacrimal glands, nasal cavity, side of nose, forehead, and anterior half of scalp; *maxillary*—contains sensory fibers from mucosa of nose, palate, parts of pharynx, upper teeth, upper lip, and lower eyelid; *mandibular*—contains sensory fibers from anterior two-thirds of tongue, lower teeth, skin over mandible, cheek and mucosa deep to it, and side of head in front of ear. The three branches terminate in pons. Sensory portion also consists of afferent fibers from proprioceptors in muscles of mastication.	Motor: chewing. Sensory: conveys sensations for touch, pain, and temperature from structures supplied; muscle sense (proprioception). Injury results in paralysis of the msucles of mastication and a loss of sensation of touch and temperature. *Neuralgia* (pain) of one or more branches of trigeminal nerve is called *trigeminal neuralgia (tic douloureux).*

EXHIBIT 18-3 *(Continued)*

NERVE (TYPE)	LOCATION	FUNCTION AND CLINICAL APPLICATION
Abducens (VI) (mixed, primarily motor)	Motor portion: originates in pons and is distributed to lateral rectus muscle, an extrinsic eyeball muscle. Sensory portion: consists of afferent fibers from proprioceptors in lateral rectus muscle and terminates in pons.	Motor: movement of eyeball. Sensory: muscle sense (proprioception). With damage to this nerve, the affected eyeball cannot move laterally beyond the midpoint and the eye is usually directed medially.
Facial (VII) (mixed)	Motor portion: originates in pons and is distributed to facial, scalp, and neck muscles, parasympathetic distribution to lacrimal, sublingual, submandibular, nasal, and palatine glands. Sensory portion: arises from taste buds on anterior two-thirds of tongue, passes through geniculate ganglion, a nucleus in pons that sends fibers to thalamus for relay to gustatory areas of cerebral cortex. Also consists of afferent fibers from proprioceptors in muscles of face and scalp.	Motor: facial expression and secretion of saliva and tears Sensory: taste; muscle sense (proprioception). Injury produces paralysis of the facial muscles, called *Bell's palsy,* loss of taste, and the eyes remain open, even during sleep.
Vestibulocochlear (VIII) (sensory)	Cochlear branch: arises in spiral organ (organ of Corti), forms spiral ganglion passes through nuclei in the medulla, and terminates in thalamus. Fibers synapse with neurons that relay impulses to auditory areas of cerebral cortex. Vestibular branch: arises in semicircular canals, saccule, and utricle and forms vestibular ganglion; fibers terminate in nuclei in thalamus.	Cochlear branch: conveys impulses associated with hearing. Vestibular branch: conveys impulses associated with equilibrium. Injury to the cochlear branch may cause *tinnitus* (ringing) or deafness. Injury to the vestibular branch may cause *vertigo* (a subjective feeling of rotation), *ataxia,* and *nystagmus* (involuntary rapid movement of the eyeball).
Glossopharyngeal (mixed)	Motor portion: originates in medulla and is distributed to stylopharyngeus muscle; parasympathetic distribution to parotid gland. Sensory portion: arises from taste buds on posterior one-third of tongue and from carotid sinus and terminates in thalamus. Also consists of afferent fibers from proprioceptors in swallowing muscles supplied.	Motor: secretion of saliva. Sensory: taste and regulation of blood pressure; muscle sense (proprioception). Injury results in pain during swallowing, reduced secretion of saliva, loss of sensation in the throat, and loss of taste.
Vagus (X) (mixed)	Motor portion: originates in medulla and terminates in muscles of respiratory passageways, lungs, esophagus, heart, stomach, small intestine, most of large intestine, and gallbladder; parasympathetic fibers innervate involuntary muscles and glands of the gastrointestinal tract. Sensory portion: arises from essentially same structures supplied by motor fibers and terminates in medulla and pons. Also consists of afferent fibers from proprioceptors in muscles supplied.	Motor: visceral muscle movement. Sensory: sensations from organs supplied; muscle sense (proprioception). Severing of both nerves in the upper body interferes with swallowing, paralyzes vocal cords, and interrupts sensations from many organs. Injury to both nerves in the abdominal area has little effect, since the abdominal organs are also supplied by autonomic fibers from the spinal cord.
Accessory (XI) (mixed, primarily motor)	Motor portion: consists of a cranial portion and a spinal portion. Cranial portion originates from medulla and supplies voluntary muscles of pharynx, larynx, and soft palate. Spinal portion originates from anterior gray horn of first five cervical segments of spinal cord and supplies sternocleiodomastoid and trapezius muscles. Sensory portion: consists of afferent fibers from proprioceptors in muscles supplied.	Motor: cranial portion mediates swallowing movements; spinal portion mediates movements of head. Sensory: muscle sense (proprioception). If damaged, the sternocleidomastoid and trapezius muscles become paralyzed, with resulting inability to turn the head or raise the shoulders.
Hypoglossal (XII) (mixed, primarily motor)	Motor portion: originates in medulla and supplies muscles of tongue. Sensory portion: consists of fibers from proprioceptors in tongue muscles that terminate in medulla.	Motor: movement of tongue during speech and swallowing. Sensory: muscle sense (proprioception). Injury results in difficulty in chewing, speaking, and swallowing. The tongue, when protruded, curls toward the affected side and the affected side becomes atrophied, shrunken, and deeply furrowed.

AGING AND THE NERVOUS SYSTEM

One of the effects of aging on the nervous system is that nerve cells are lost. Associated with this decline, there is a decreased capacity for sending nerve impulses to and from the brain. Conduction velocity decreases, voluntary motor movements slow down, and the reflex time for skeletal muscles increases. Parkinson's disease is the most common movement disorder involving the central nervous system. Degenerative changes and disease states involving the sense organs can alter vision, hearing, taste, smell, and touch. The disorders that represent the most common visual problems and may be responsible for serious loss of vision are presbyopia (inability to focus on nearby objects), cataracts (cloudiness of the lens), and glaucoma (excessive fluid pressure in the eyeball). Impaired hearing associated with aging, known as presbycusis, is usually the result of changes in important structures of the inner ear.

DEVELOPMENTAL ANATOMY OF THE NERVOUS SYSTEM

The development of the nervous system begins early in the third week of development with a thickening of the *ectoderm* called the **neural plate** (Figure 18-22a–c). The plate folds inward and forms a longitudinal groove, the **neural groove.** The raised edges of the neural plate are called **neural folds.** As development continues the neural folds increase in height, meet, and form a tube, the **neural tube.**

The cells of the wall that encloses the neural tube differentiate into three kinds. The outer or **marginal layer** develops into the *white matter* of the nervous system; the middle or **mantle layer** develops into the *gray matter* of the system; and the inner or **ependymal layer** eventually forms the *lining of the ventricles* of the central nervous system.

The **neural crest** is a mass of tissue between the neural tube and the ectoderm (Figure 18-22). It becomes differentiated and eventually forms the *posterior, (dorsal) root ganglia of spinal nerves, spinal nerves, ganglia of cranial nerves, cranial nerves, ganglia of the automatic nervous system,* and the *adrenal medulla.*

When the neural tube is formed from the neural plate, the anterior portion of the neural tube develops into three enlarged areas called vesicles: (1) **forebrain vesicle (prosencephalon),** (2) **midbrain vesicle (mesencephalon),** and (3) **hindbrain vesicle (rhombencephalon)** (Figure 18-23). The vesicles are fluid-filled enlargements that develop by the fourth week of gestation. Since they are the first vesicles to form, they are called **primary vesicles.** As development progresses, the vesicular region undergoes several flexures (bends), resulting in subdivision of the three primary vesicles, so that by the fifth week of development the embryonic brain consists of five **secondary vesicles.** The forebrain vesicle (prosencephalon) divides into an anterior **telencephalon** and a posterior **diencephalon;** the midbrain vesicle (mesencephalon) remains unchanged; the hindbrain vesicle (rhombencephalon) divides into an anterior **metencephalon** and a posterior **myelencephalon.**

Ultimately, the telencephalon develops into the *cerebral hemispheres* and *basal ganglia;* the diencephalon develops into the *thalamus, hypothalamus,* and *pineal gland;* the midbrain vesicle (mesencephalon) develops into the *midbrain;* the metencephalon develops into the *pons* and *cerebellum;* and the myelencephalon develops into the *medulla oblongata.* The cavities within the vesicles develop into the *ventricles* of the brain, whereas the fluid within them is *cerebrospinal fluid.* The area of the neural tube posterior to the myelencephalon gives rise to the *spinal cord.*

APPLICATIONS TO HEALTH

Many disorders can affect the central nervous system. Some are caused by viruses or bacteria. Others are caused by damage to the nervous system during birth. The origins of many conditions, however, are unknown. Here we discuss the origins and symptoms of some common central nervous system disorders.

CEREBROVASCULAR ACCIDENT (CVA)

The most common brain disorder is a **cerebrovascular accident (CVA),** also called a **stroke** or **cerebral apoplexy.** A CVA is characterized by a relatively abrupt onset of persisting neurological symptoms due to the destruction of brain tissue (infarction), resulting from disorders in the blood vessels that supply the brain. CVAs may be classified into two principal types: (1) *ischemic,* the most common type, due to a decreased blood supply, and (2) *hemorrhagic,* due to a blood vessel in the brain that bursts. Common causes of CVAs are intracerebral hemorrhage from aneurysms, emboli, and atherosclerosis of the cerebral arteries. An **intracerebral hemorrhage** is a rupture of a vessel in the pia mater or brain. Blood seeps into the brain and damages neurons by increasing intracranial fluid pressure. An **embolus** is a blood clot, air bubble, or bit of foreign material, most often debris from an inflammation, that becomes lodged in an artery and blocks circulation. **Atherosclerosis** is the formation of plaques in the artery wall. The plaques may slow down circulation by constricting the vessel. Both emboli and atherosclerosis cause brain damage by reducing the supply of oxygen and glucose needed by brain cells.

Among the risk factors implicated in CVAs are high blood pressure, heart disease, narrowed carotid arteries, transient ischemic attacks (TIAs), diabetes, smoking, obesity, high plasma fibrinogen level, maternal history of CVA (for males), and excessive alcohol intake.

TRANSIENT ISCHEMIC ATTACK (TIA)

A **transient ischemic attack (TIA)** is an episode of temporary, focal, nonconvulsive, cerebral dysfunction caused by an interference of the blood supply to the brain. Symptoms

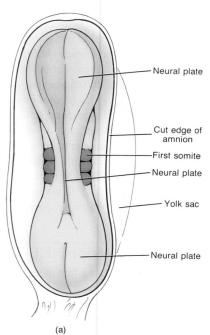

Neural plate

Cut edge of amnion

First somite

Neural plate

Yolk sac

Neural plate

(a)

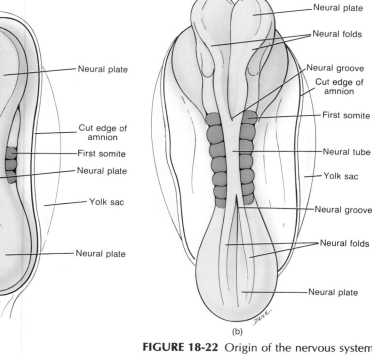

Neural plate

Neural folds

Neural groove

Cut edge of amnion

First somite

Neural tube

Yolk sac

Neural groove

Neural folds

Neural plate

(b)

FIGURE 18-22 Origin of the nervous system. (a) Dorsal view of an embryo with three pairs of somites showing the neural plate. (b) Dorsal view of an embryo with seven pairs of somites in which the neural folds have united just medial to the somites, forming the early neural tube. (c) Cross section through the embryo showing the formation of the neural tube.

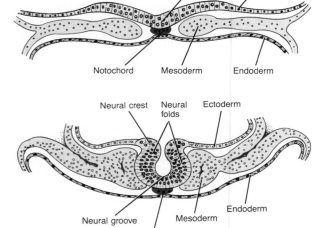

Neural plate   Ectoderm

Notochord   Mesoderm   Endoderm

Neural crest   Neural folds   Ectoderm

Neural groove   Mesoderm   Endoderm

Notochord

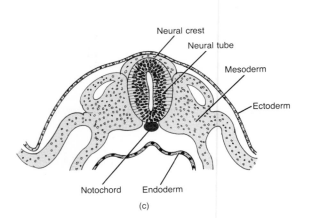

Neural crest

Neural tube

Mesoderm

Ectoderm

Notochord   Endoderm

(c)

include weakness, numbness, or paralysis in a limb or in one-half of the body; drooping of one side of the face, headache; slurred speech, or a partial loss of vision. The onset of symptoms is sudden and reaches maximum intensity almost immediately. A TIA usually persists for 2–15 minutes and only rarely as long as 24 hours. Each TIA leaves no persistent neurologic deficits. The causes of the ischemia that lead to TIAs are emboli, atherosclerosis, impaired blood flow due to hemodynamic disruptions, and hematologic disorders (polycythemia, thrombocytosis, and sickle-cell anemia.)

It is estimated that about one-third of patients with a TIA will have a CVA within 5 years. Therapy for TIAs includes antiplatelet-aggregating agents such as aspirin, anticoagulants, cerebral artery bypass grafting, and carotid endarterectomy (excision of the atheromatous tunica intima of an artery).

BRAIN TUMORS

A **brain tumor** refers to any benign or malignant growth within the cranium. Tumors may arise from neuroglial cells in the cerebrum, brain stem, and cerebellum, or from supporting or neighboring structures such as cranial nerve coverings, meninges, and the pituitary gland. Intracranial pressure from a growing tumor or edema associated with the tumor produces the characteristic signs and symptoms of a brain

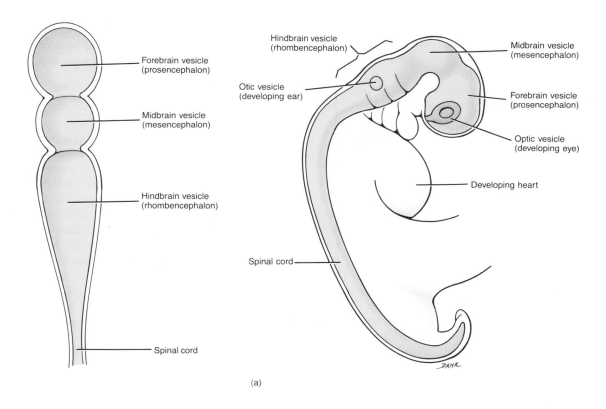

(a)

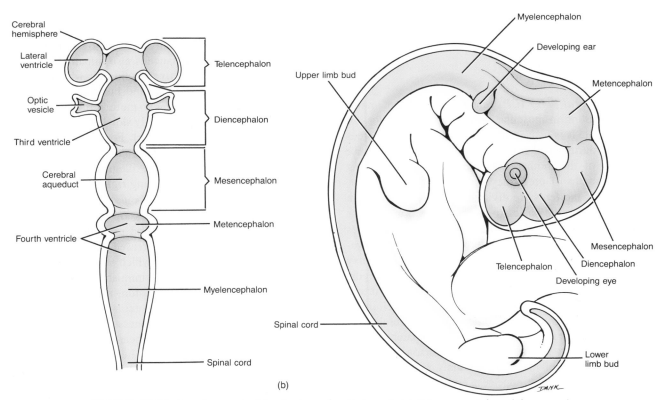

(b)

FIGURE 18-23 Development of the brain and spinal cord. (a) Primary vesicles of the neural tube seen in frontal section (left) and right lateral view (right) at about 3—4 weeks. (b) Secondary vesicles seen in frontal section (left) and right lateral view (right) at about 5 weeks.

tumor. Among these are headache, altered consciousness, edema of the optic disc of the eyeball, and vomiting. Brain tumors can also result in seizures, visual problems, cranial nerve abnormalities, hormonal syndromes, personality changes, dementia, and sensory or motor deficits. Treatment of brain tumors involves surgery, radiation therapy, and chemotherapy.

POLIOMYELITIS

Poliomyelitis (infantile paralysis), or simply **polio,** is most common during childhood and is caused by a virus called poliovirus. The onset of the disease is marked by fever, severe headache, a stiff neck and back, deep muscle pain and weakness, and loss of certain somatic reflexes. The virus can be ingested in drinking water contaminated with feces containing the virus. It may affect almost all parts of the body. In its most serious form, called **bulbar polio,** the virus spreads via blood to the central nervous system, where it destroys the motor nerve cell bodies, specifically those in the anterior horns of the spinal cord and in the nuclei of the cranial nerves. Injury to the spinal gray matter is the basis for the name of this disease (polio = gray matter; myel = spinal cord). Destruction of the anterior horns produces paralysis. The first sign of bulbar polio is difficulty in swallowing, breathing, and speaking. Poliomyelitis can cause death from respiratory or heart failure if the virus invades the brain cells of the vital medullary centers. The incidence of polio in the United States has decreased markedly since the availability of polio vaccines (Salk vaccine and more recently Sabin vaccine).

CEREBRAL PALSY (CP)

The term **cerebral palsy (CP)** refers to a group of motor disorders resulting in muscular uncoordination and loss of muscle control. It is caused by damage to the motor areas of the brain during fetal life, birth, or infancy. One cause is infection of the mother with German measles during the first 3 months of pregnancy. During early pregnancy, certain cells in the fetus are dividing and differentiating in order to lay down the basic structures of the brain. These cells can be abnormally changed by toxin from the measles virus. Radiation during fetal life, temporary oxygen starvation during birth, and hydrocephalus during infancy may also damage brain cells. Cases of cerebral palsy are categorized into three groups, depending on whether the cortex, the basal ganglia of the cerebrum, or the cerebellum is affected most severely. About 70 percent of cerebral palsy victims appear to be mentally retarded. The apparent mental slowness, however, is often due to the person's inability to speak or hear well. Such individuals are often more mentally acute than they appear. Cerebral palsy is not a progressive disease; it does not worsen as time elapses. Once the damage is done, however, it is irreversible. Recently, a surgical procedure called selective posterior rhizot-

omy, has been used on children to reduce muscle spasticity. In the procedure, selected rootlets in posterior roots are severed, allowing the other rootlets to achieve better muscle tone.

PARKINSON'S DISEASE (PD)

Parkinson's disease (PD) or **parkinsonism** is a progressive disorder of the central nervous system that begins inconspicuously and typically affects its victims around age 60. The cause is unknown, but there are indications that an environmental agent might be involved. The disease is related to pathological changes in the substantia nigra and basal ganglia. The substantia nigra contains cell bodies of neurons that produce the neurotransmitter dopamine (DA) in their axon terminals. The axon terminals release DA in the basal ganglia of the cerebrum. Recall that basal ganglia regulate subconscious contractions of skeletal muscles that aid activities also consciously controlled by the motor areas of the cerebral cortex—swinging the arms when walking, for example. In Parkinson's disease there is a degeneration of DA-producing neurons in the substantia nigra, and the severe reduction of DA in the basal ganglia brings about most of the symptoms of Parkinson's disease.

Diminished levels of DA cause unnecessary skeletal muscle movements that often interfere with voluntary movement. For instance, the muscles of the upper extremity may alternately contract and relax, causing the hand to shake. This shaking is called **tremor,** the most common symptom of Parkinson's disease. The tremor may spread to the ipsilateral lower extremity and then to the contralateral extremities. Motor performance is also impaired by **bradykinesia** (brady = slow; kinesis = motion), in which activities such as shaving, cutting food, and buttoning a shirt take longer and become increasingly more difficult. Muscular movements are performed not only slowly but with decreasing range of motion (**hypokinesia).** For example, as handwriting continues, letters get smaller, become poorly formed, and eventually become illegible. Some muscles may contract continuously, causing **rigidity** of the involved body part. Rigidity of the facial muscles gives the face a masklike appearance. The expression is characterized by a wide-eyed, unblinking stare and a slightly open mouth with uncontrolled drooling. Decreased DA production also results in impaired walking, in which steps become shorter and shuffling and arm swing diminishes. There are also changes that lead to stooped posture and loss of postural reflexes, autonomic dysfunction (constipation, retention), sensory complaints (pain, numbness, tingling), and sustained muscle spasms. Vision, hearing, and intelligence are unaffected by the disorder, indicating that Parkinson's disease does not attack the cerebral cortex.

Treatment of the symptoms of Parkinson's disease is directed toward increasing levels of DA. Although people with Parkinson's disease do not manufacture enough DA, injections of it are useless; the blood–brain barrier stops it. However, symptoms are somewhat relieved by a drug developed in the 1960s called levodopa. Administered by itself, levodopa may elevate brain levels of DA, causing undesirable side effects such as low blood pressure, nausea, mental changes, and liver dysfunction. Levodopa has been combined with carbidopa, which inhibits the formation of DA outside the brain. The combined drugs diminish the undesirable side effects. Currently, methods are being devised to help drugs pass through the blood–brain barrier by combining them with fat-soluble chemicals. Once the complex passes through the barrier, the fat-soluble chemical breaks down and is excreted, leaving the drug in the brain where it exerts its effect. In addition, two experimental drugs are being tested: selegiline, that may promote the destuction of DA-producing cells, and tocopherol, that may delay the need for levodopa.

In early 1987, physicians in Mexico City transplanted tissue from the adrenal glands into the brain in order to alleviate the symptoms of Parkinson's disease. The adrenal medulla produces epinephrine, a chemical similar to DA. In the procedure, adrenal medullary tissue (chromaffin cells) was transplanted to the caudate nucleus of the brain. It appears that either the transplanted tissue produces DA or stimulates the brain to produce DA. Preliminary results are encouraging. The first transplant of human fetal cells into the brain of an adult with Parkinson's disease was also performed in 1987. Results are still being evaluated.

MULTIPLE SCLEROSIS (MS)

Multiple sclerosis (*MS*) is often called the "great imitator." It is the progressive destruction of the myelin sheaths of neurons in the central nervous system accompanied by disappearance of oligodendrocytes and the proliferation of astrocytes. The sheaths deteriorate to *scleroses,* which are hardened scars or plaques, in multiple regions. The destruction of myelin sheaths interferes with the transmission of nerve impulses from one neuron to another, literally short-circuiting conduction pathways. Usually the first symptoms occur in early adult life. The average age of onset is 33. The frequency of flare-ups is greatest during the first 3 to 4 years of the disease, but a first attack, which may have been so mild as to escape medical attention, may not be followed by another attack for 10 to 20 years.

Among the first symptoms of MS are muscular weakness of one or more extremities; abnormal sensations, such as burning or pins and needles; visual impairment that includes blurring, double vision, and problems with color and light perception; incoordination; vertigo; and sphincter impairment that results in urinary problems. Following a period of remission during which the symptoms temporarily disappear, a new series of plaques develop and the victim suffers a second attack. One attack follows another over the years, usually every year or two. Each time the plaques form, some neurons are damaged by the hardening of their sheaths, while others are uninjured by their plaques. The result is

a progressive loss of function interspersed with remission periods during which the undamaged neurons regain their ability to transmit impulses.

The symptoms of MS depend on the areas of the central nervous system most heavily laden with plaques. Sclerosis of the white matter of the spinal cord is common. As the sheaths of the neurons in the corticospinal tract deteriorate, the patient loses the ability to contract skeletal muscles. Damage to the ascending tracts produces numbness and short-circuits nerve impulses related to position of body parts and flexion of joints. Damage to either set of tracts also destroys spinal cord reflexes.

The clinical course of MS is unpredictable. During typical episodes, symptoms worsen over a period of a few days to 2–3 weeks and then remit. Relapses occur at an average rate of 0.5 per year during the initial five years, although this rate is highly variable. Some patients experience complete remissions following relapses, while others gradually accumulate neurologic problems. Many patients suffer multiple attacks but are never disabled. About 20 percent of MS patients have symptoms and signs that appear slowly and steadily, without a clear relapsing-remitting pattern. Such a course often occurs in late onset patients (over 40) and is frequently associated with severe disability. About two-thirds of MS patients are ambulatory 25 years after the onset of their disease, half will be working 10 years from the onset, and one-third will have unrestricted functions. MS does not predictably shorten life except in a minority of patients who are bedridden or succumb to a urinary tract infection or pneumonia.

Although the etiology of MS is unclear, there is some evidence that it might result from a viral infection that precipitates an autoimmune response. Viruses may trigger the destruction of oligodendrocytes by the antibodies and killer cells of the body's immune system. Like other demyelinating diseases, MS is incurable. However, in view of the evidence that it might be an autoimmune disease, immunosuppressive therapy is widely used (glucocorticoids such as prednisone). A recent treatment consists of administering cyclophosphamide (a powerful anticancer drug that suppresses the immune system) in combination with adrenocorticotropic hormone (ACTH) (a pituitary gland hormone that stimulates secretion of hormones containing cortisone) to patients with active MS. An experimental therapy for MS is the use of a synthetic protein called copolymer (Cop 1), a decoy that spares destruction of a similar protein in oligodendrocytes. Another treatment is the use of colchicine, a drug used to treat gout. Electrical stimulation of the spinal cord can also improve function in certain patients. Improvement has also been shown in patients who are administered pure oxygen while in a pressure chamber (hyperbaric oxygen). This supports the idea that the destruction of myelin occurs preferentially in parts of the brain that are relatively low in oxygen. Treatment is also directed at management of complications such as spasticity, facial neuralgia and twitching, urinary bladder problems, and constipation.

EPILEPSY

Epilepsy is the second most common neurological disorder after stroke. It is characterized by short, recurrent, periodic attacks of motor, sensory, or psychological malfunction. The attacks, called *epileptic seizures,* are initiated by abnormal and irregular discharges of electricity from millions of neurons in the brain, probably resulting from abnormal reverberating circuits. The discharges stimulate many of the neurons to send impulses over their conduction pathways. As a result, a person undergoing an attack may contract skeletal muscles involuntarily. Lights, noise, or smells may be sensed when the eyes, ears, and nose actually have not been stimulated. The electrical discharges may also inhibit certain brain centers. For instance, the waking center in the brain may be depressed so that the person loses consciousness.

The causes of epilepsy are varied. Many conditions can cause nerve cells to produce periodic bursts of impulses. These causes include metabolic disturbances (hypoglycemia, hypocalcemia, uremia, hypoxia), infections (encephalitis or meningitis), toxins (alcohol, tranquilizers, hallucinogens), vascular disturbances (hemorrhage, hypotension), head injuries, and tumors and abscesses of the brain. Most epileptic seizures, however, are *idiopathic;* that is, they have no demonstrable cause. It should be noted that epilepsy almost never affects intelligence. If frequent severe seizures are allowed to occur over a long period of time, however, some cerebral damage may occasionally result. Damage can be prevented by controlling the seizures with drug therapy.

Epileptic seizures can be eliminated or alleviated by drugs that make neurons more difficult to stimulate. Many of these drugs change the permeability of the neuron cell membrane so that it does not depolarize as easily. One such drug is valproic acid, which increases the quantity of the inhibitory neurotransmitter, gamma aminobutyric acid (GABA).

DYSLEXIA

Dyslexia (dis-LEK-sē-a; *dys* = difficulty; *lexis* = words) is an impairment of the brain's ability to translate images received from the eyes or ears into understandable language. The condition is unrelated to basic intellectual capacity, but it causes a mysterious difficulty in handling words and symbols. Apparently some peculiarity in the brain's organizational pattern distorts the ability to read, write, and count. Letters in words seem transposed, reversed, or upside down—*dog* becomes *god; b* changes identity with *d;* a sign saying "OIL" inverts into "710." Frequently, dyslectics reread portions of paragraphs, skip words, and change the order of letters in a word. Many dyslectics cannot orient themselves in the three dimensions of space and may show bodily awkwardness.

The exact cause of dyslexia is unknown, since it is unaccompanied by outward scars of detectable neurological damage and its symptoms vary from victim to victim. It occurs almost four times as often among boys as among girls. It has been variously attributed to defective vision, brain damage, abnormal brain development, lead in the air, physical trauma, or oxygen deprivation during birth. A recent theory holds that it might be related to a complex language deficiency involving inability to represent and access the sound of a word in order to help remember it, inability to break down words into components, poor vocabulary development, and difficulty discriminating grammatical differences among words and phrases.

A technique used primarily to diagnose dyslexia is called *brain electrical activity mapping* (*BEAM*). It is a noninvasive procedure that measures and displays the electrical activity of the brain on a color television screen and compares the image produced with a normal image (Figure 18-24). BEAM may have other diagnostic applications for learning disabilities, schizophrenia, depression, dementia, epilepsy, and early tumor occurrence and recurrence.

Positron emission tomography (PET) scans are also providing information about dyslexia (see Figure 1–12). For example, PET scans have demonstrated that in persons with dyslexia the left side of the brain is more active than the right side, the language region is less active than normal, and the visual discrimination region is also more active.

TAY-SACHS DISEASE

Tay-Sachs disease is a central nervous system affliction that brings death before age 5. The Tay-Sachs gene is carried mostly by individuals descended from the Ashkenazi Jews of Eastern Europe. Approximately 1 in 3,600 of their offspring will be afflicted with Tay-Sachs disease. The disease involves the neuronal degeneration of the central nervous system because of excessive amounts of a lipid called ganglioside in the nerve cells of the brain. The substance accumulates because of a deficient lysosomal enzyme. The afflicted child develops normally until the age of 4–8 months. Then the symptoms follow a course of progressive degeneration: paralysis, blindness, inability to eat, decubitus ulcers, and death from infection. There is no known cure.

HEADACHE

One of the most common human afflictions is *headache* or *cephalgia* (*enkephalos* = brain, *algia* = painful condition). Based on origin, two general types are distinguished: intracranial and extracranial. Serious headaches of intracranial origin are caused by brain tumors, blood vessel abnormalities, inflammation of the brain or meninges, decrease in oxygen supply to the brain, and damage to brain cells. Extracranial headaches are related to infections of the eyes, ears, nose, and sinuses and are commonly felt as headaches because of the location of these structures.

Most headaches require no special treatment. Analgesic and tranquilizing compounds are generally effective for tension headaches, but not for migraine headaches. Drugs that constrict the blood vessels can be helpful for migraine.

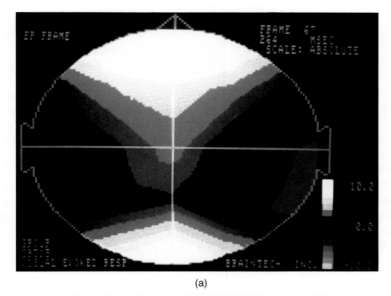

(a)

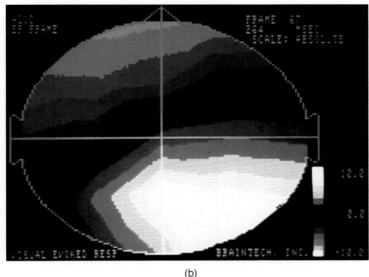

(b)

FIGURE 18-24 Brain electrical activity mapping (BEAM). (a) Image of a visual response (flash stimulus) produced by a normal child. Note the symmetry of the response in the anterior and posterior portions of the brain. (b) Image of a visual response (flash stimulus) produced by a dyslexic child. Note the asymmetry of the response. (Courtesy of Frank H. Duffy, M.D., and Gloria B. McAnulty, Ph.D., The Children's Hospital Medical Center, Boston, Massachusetts.)

Biofeedback training and dietary changes may also be of some help. Taking a careful and complete history is an important procedure in determining the cause or causes of a patient's headaches.

TRIGEMINAL NEURALGIA (TIC DOULOUREUX)

As noted earlier, pain arising from irritation of the trigeminal (V) nerve is known as *trigeminal neuralgia* or *tic douloureux* (doo-loo-ROO). The disorder is characterized by brief but extreme pain in the face and forehead on the affected side. The characteristic pain consists of red-hot, needlelike jabs lasting a few seconds and building to a searing pain like a hot poker being dragged or stuck into the face, lasting 10–15 seconds. Many patients describe sensitive regions around the mouth and nose that can cause an attack when touched. Eating, drinking, washing the face, and exposure to cold may also bring on an attack.

Treatment may be palliative (relieving symptoms without curing the disease) or surgical. One technique involves alcohol injections directly into the semilunar (Gasserian) ganglion, which controls the trigeminal (V) nerve. This method is superior to open surgery because it is safer. Moreover, it can bring lasting pain relief with preservation of touch sensation in the face and release of the patient 24 hours after treatment. Acupuncture can also provide relief in some patients.

REYE'S SYNDROME (RS)

Reye's syndrome (RS), first described in 1963 by the Australian pathologist R. Douglas Reye, seems to occur following a viral infection, particularly chickenpox or influenza. Aspirin at normal doses is believed to be a risk factor in the development of RS. The majority of persons affected are children or teenagers. The disease is characterized by vomiting and brain dysfunction (disorientation, lethargy, and personality changes) and may progress to coma. Also, the liver becomes infiltrated with small lipid droplets and loses some of its ability to detoxify ammonia. The disease runs its course in just a few days.

Brain dysfunction and death are typically caused by swelling of brain cells. The pressure not only kills the cells directly but also results in hypoxia that kills them indirectly. The mortality rate is about 40 percent. Swelling may result in irreversible brain damage, including mental retardation, in children who survive. Therapy is directed at controlling the swelling.

ALZHEIMER'S DISEASE (AD)

Alzheimer's (ALTZ-hī-merz) *disease* or *AD* is a disabling neurological disorder that afflicts about 5 percent (about 1 million people) of the population over age 65. Its causes are unknown, its effects are irreversible, and it has no cure. The disease claims over 120,000 lives a year, making it the fourth leading cause of death among the elderly following heart disease, cancer, and stroke.

Victims of AD initially have trouble remembering recent events. Next, they become more confused and forgetful, often repeating questions or getting lost while traveling to previously familiar places. Disorientation grows, memories of past events disappear, and there may be episodes of paranoia, hallucination, or violent changes in mood. As their minds continue to deteriorate, they lose their ability to read, write, talk, eat, walk, or take care of themselves. Finally, the disease culminates in dementia, the loss of reason. A person with AD usually dies of some complication that affects bedridden patients, such as pneumonia.

At present, there is no diagnostic test for AD. Autopsy findings, however, clearly show several characteristic pathologies in the brains of Alzheimer's victims. Although there is normally gradual loss of neurons in the brain associated with aging, the rate of loss in persons with AD is greater than normal, especially in regions of the brain that are important for memory and other intellectual processes,
such as the cerebral cortex and hippocampus. Another pathological finding is the presence of *neurofibrillary tangles,* bundles of fibrous proteins, in the cell bodies of neurons in the cerebral cortex, hippocampus, and brain stem. Also present is *amyloid,* pathological protein-rich accumulations. In some cases, amyloid surrounds and invades cerebral blood vessels. This form is called *cerebrovascular amyloid,* which collects in the middle muscular layer of the blood vessels and eventually results in hemorrhage. Another form of amyloid is a component of *neuritic (senile) plaques,* which consist of abnormal axons and axon terminals that surround amyloid. Neuritic plaques are abundant in the cerebral cortex, hippocampus, and amygdala.

One hypothesis concerning the possible cause of AD centers on the observation that the enzyme choline acetyltransferase (CAT) is found in very low concentrations in the brains of AD victims. CAT is needed for the synthesis of the neurotransmitter acetylcholine (ACh) in axon terminals. Levels of CAT are quite low in the cerebral cortex and hippocampus of AD patients due to the loss of axon terminals in these regions, and thus acetylcholine levels are also low. Decreased levels of acetylcholine are correlated with reduced transmission of nerve impulses, impulses required for memory.

Other hypotheses suggest that AD may be caused by the loss of neurons due to the inheritance of faulty genes, an abnormal accumulation of proteins in the brain, infectious agents called prions (protein particles), environmental toxins such as aluminum, and decreased blood flow to the brain resulting in inadequate amounts of oxygen and glucose.

Amyloid forms neuritic plaques and another protein called A68 is found in neurofibrillary tangles. Both proteins are found in only two groups of adults; persons with AD and persons with Down's syndrome (DS). DS is a genetic defect that is the leading cause of mental retardation (Chapter 26). Of the two proteins, A68 is emerging as a key factor in the onset of AD. In infants, A68 is believed to cause programmed death of surplus neurons, a process that is completed by age 2, after the A68 disappears from the brain. However, it reappears in the brains of AD patients. It is suspected that A68 may cause AD by triggering the growth of neurofibrillary tangles or mediate programmed cell death as it does in normal infants.

It has recently been learned that a gene on chromosome 21 codes for amyloid, and the extra gene, as found in DS, could lead to excessive amyloid production. Such information suggests that a common genetic defect might be responsible for both AD and DS.

KEY MEDICAL TERMS ASSOCIATED WITH THE CENTRAL NERVOUS SYSTEM

Agnosia (ag-NŌ-zē-a; *a* = without: *gnosis* = knowledge)   Inability to recognize the significance of sensory stimuli such as auditory, visual, olfactory, gustatory, and tactile.

Analgesia (an-al-JĒ-zē-a; *an* = without; *algia* = painful condition)   Insensibility to pain.

Anesthesia (an'-es-THĒ-zē-a; *esthesia* = feeling)   Loss of feeling.

Apraxia (a-PRAK-sē-a; *pratto* = to do)   Inability to carry out purposeful movements in the absence of paralysis.

Coma (KŌ-ma)   Abnormally deep unconsciousness with an absence of voluntary response to stimuli and with varying degrees of reflex activity. It may be due to illness or to an injury.

Electroconvulsive therapy (ECT) (e-lek'-trō-con-VUL-siv THER-a-pē) A form of shock therapy in which convulsions are induced by the passage of an electric current through the brain. A patient undergoing ECT is properly anesthetized and given a muscle relaxant to minimize the convulsions. ECT is regarded as an important therapeutic option in the treatment of severe depression.

Huntington's chorea (HUNT-ing-tunz kō-RĒ-a; *choreia* = dance) A rare hereditary disease characterized by involuntary jerky movements and mental deterioration that terminates in dementia.

Lethargy (LETH-ar-jē) A condition of functional torpor or sluggishness.

Nerve block Loss of sensation in a region, such as in local dental anesthesia.

Neuralgia (noo-RAL-jē-a; *neur* = nerve) Attacks of pain along the entire course or branch of a peripheral sensory nerve.

Paralysis (pa-RAL-a-sis) Diminished or total loss of motor function resulting from damage to nervous tissue or a muscle.

Spastic (SPAS-tik; *spas* = draw or pull) Resembling spasms or convulsions.

Stupor (STOO-por) Condition of unconsciousness, torpor, or lethargy with suppression of sense or feeling.

Torpor (TOR-por) Abnormal inactivity or lack of response to normal stimuli.

Viral encephalitis (VĪ-ral en'-sef-a-LĪ-tis) An acute inflammation of the brain caused by a direct attack by various viruses or by an allergic reaction to any of the many viruses that are normally harmless to the central nervous system. If the virus affects the spinal cord as well, it is called *encephalomyelitis.*

STUDY OUTLINE

Brain (p. 487)
Principal Parts (p. 487)
1. During embryological development, brain vesicles are formed and serve as forerunners of various parts of the brain.
2. The diencephalon develops into the thalamus and hypothalamus, the telencephalon forms the cerebrum, the mesencephalon develops into the midbrain, the myelencephalon forms the medulla, and the metencephalon develops into the pons and cerebellum.
3. The principal parts of the brain are the brain stem, diencephalon, cerebrum, and cerebellum.

Protection and Coverings (p. 488)
1. The brain is protected by cranial bones, meninges, and cerebrospinal fluid.
2. The cranial meninges are continuous with the spinal meninges and are named dura mater, arachnoid, and pia mater.

Cerebrospinal Fluid (CSF) (p. 488)
1. Cerebrospinal fluid is formed in the choroid plexuses and circulates through the subarachnoid space, ventricles, and central canal. Most of the fluid is absorbed by the arachnoid villi of the superior sagittal sinus.
2. Cerebrospinal fluid protects by serving as a shock absorber. It also delivers nutritive substances from the blood and removes wastes.
3. If cerebrospinal fluid accumulates in the ventricles, it is called internal hydrocephalus. If it accumulates in the subarachnoid space, it is called external hydrocephalus.

Blood Supply (p. 492)
1. The blood supply to the brain is via the cerebral arterial circle (circle of Willis).
2. Any interruption of the oxygen supply to the brain can result in weakening, permanent damage, or death of brain cells. Interruption of the mother's blood supply to a child during childbirth before it can breathe may result in paralysis, mental retardation, epilepsy, or death.
3. Glucose deficiency may produce dizziness, convulsions, and unconsciousness.

4. The blood–brain barrier (BBB) is a concept that explains the differential rates of passage of certain material from the blood into the brain.

Brain Stem (p. 492)
1. The medulla oblongata is continuous with the upper part of the spinal cord. It contains nuclei that are reflex centers for regulation of heart rate, respiratory rate, vasoconstriction, swallowing, coughing, vomiting, sneezing, and hiccuping. It also contains the nuclei of origin for cranial nerves VIII (cochlear and vestibular branches) through XII.
2. The pons is superior to the medulla. It connects the spinal cord with the brain and links parts of the brain with one another. It relays nerve impulses related to voluntary skeletal movements from the cerebral cortex to the cerebellum. It contains the nuclei for cranial nerves V through VII and the vestibular branch of VIII. The reticular formation of the pons contains the pneumotaxic center, which helps control respiration.
3. The midbrain connects the pons and diencephalon. It conveys motor impulses from the cerebrum to the cerebellum and cord, sensory impulses from cord to thalamus, and regulates auditory and visual reflexes. It also contains the nuclei of origin for cranial nerves III and IV.

Diencephalon (p. 497)
1. The diencephalon consists primarily of the thalamus and hypothalamus.
2. The thalamus is superior to the midbrain and contains nuclei that serve as relay stations for all sensory impulses, except smell, to the cerebral cortex. It also registers conscious recognition of pain and temperature and some awareness of crude touch and pressure.
3. The hypothalamus is inferior to the thalamus. It controls and integrates the autonomic nervous system, receives sensory impulses from viscera, connects the nervous and endocrine systems, coordinates mind-over-body phenomena, functions in rage and aggression, controls body temperature, regulates food and fluid intake, maintains the waking state and sleep patterns, and acts as a self-sustained oscillator that drives biological rhythms.

Cerebrum (p. 501)

1. The cerebrum is the largest part of the brain. Its cortex contains convolutions, fissures, and sulci.
2. The cerebral lobes are named the frontal, parietal, temporal, and occipital.
3. The white matter is under the cortex and consists of myelinated axons running in three principal directions.
4. The basal ganglia (cerebral nuclei) are paired masses of gray matter in the cerebral hemispheres. They help to control muscular movements.
5. The limbic system is found in the cerebral hemispheres and diencephalon. It functions in emotional aspects of behavior and memory.
6. The sensory areas of the cerebral cortex are concerned with the interpretation of sensory impulses. The motor areas are the regions that govern muscular movement. The association areas are concerned with emotional and intellectual processes.
7. Positron emission tomography (PET) helps identify which parts of the brain are involved in specific sensory and motor activities.
8. Brain waves generated by the cerebral cortex are recorded as an electroencephalogram (EEG). It may be used to diagnose epilepsy, infections, and tumors.

Brain Lateralization (Split-Brain Concept) (p. 508)

1. Recent research indicates that the two hemispheres of the brain are not bilaterally symmetrical, either anatomically or functionally.
2. The left hemisphere is more important for right-handed control, spoken and written language, numerical and scientific skills, and reasoning.
3. The right hemisphere is more important for left-handed control, musical and artistic awareness, space and pattern perception, insight, imagination, and generating mental images of sight, sound, touch, taste, and smell.

Cerebellum (p. 509)

1. The cerebellum occupies the inferior and posterior aspects of the cranial cavity. It consists of two hemispheres and a central, constricted vermis.
2. It is attached to the brain stem by three pairs of cerebellar peduncles.
3. The cerebellum functions in the coordination of skeletal muscles and the maintenance of normal muscle tone and body equilibrium.

Cranial Nerves (p. 511)

1. Twelve pairs of cranial nerves originate from the brain.
2. The pairs are named primarily on the basis of distribution and numbered by order of attachment to the brain. (See Exhibit 18-3 for summary of cranial nerves.)

Aging and the Nervous System (p. 521)

1. Age-related effects involve loss of neurons and decreased capacity for sending nerve impulses.
2. Degenerative changes also affect the sense organs.

Developmental Anatomy of the Nervous System (p. 521)

1. The development of the nervous system begins with a thickening of ectoderm called the neural plate.
2. The parts of the brain develop from primary and secondary vesicles.

Applications to Health (p. 521)

1. Cerebrovascular accidents (CVAs), also called strokes, involve brain tissue destruction due to hemorrhage, thrombosis, or atherosclerosis.
2. A transient ischemic attack (TIA) is an episode of temporary, focal, nonconvulsive, cerebral dysfunction caused by interference of the blood supply to the brain.
3. Brain tumors are neoplasms within the cranium.
4. Poliomyelitis is a viral infection that results in paralysis.
5. Cerebral palsy (CP) refers to a group of motor disorders caused by damage to motor centers of the cerebral cortex, cerebellum, or basal ganglia during fetal development, childbirth, or early infancy.
6. Parkinson's disease (PD) is a progressive degeneration of the dopamine (DA)-producing neurons in the substantia nigra resulting in insufficient DA in the basal ganglia.
7. Multiple sclerosis (MS) is the destruction of meylin sheaths of the neurons of the central nervous system. Impulse transmission is interrupted.
8. Epilepsy results from irregular electrical discharges of brain cells and may be diagnosed by an EEG. Depending on the form of the disease, the victim experiences degrees of motor, sensory, or psychological malfunction.
9. Dyslexia involves an inability of an individual to comprehend written language. It may be diagnosed by using brain electrical activity mapping (BEAM).
10. Tay-Sachs disease is an inherited disorder that involves neurological degeneration of the CNS because of excessive amounts of ganglioside.
11. Headaches are of two types: intracranial and extracranial.
12. Irritation of the trigeminal (V) nerve is known as trigeminal neuralgia.
13. Reye's syndrome (RS) is characterized by vomiting, brain dysfunction, and liver damage.
14. Alzheimer's disease (AD) is a disabling neurological disorder of the elderly that involves widespread intellectual impairment, personality changes, and sometimes delirium.

REVIEW QUESTIONS

1. Identify the four principal parts of the brain and the components of each, where applicable. What is the origin of each of the parts?
2. Describe the location of the cranial meninges. What is an extradural hemorrhage?
3. Where is cerebrospinal fluid (CSF) formed? Describe its circulation. Where is CSF absorbed?
4. What is the blood–brain barrier (BBB)? What is its role?
5. Describe the location and structure of the medulla. Define decussation of pyramids. Why is it important? List the principal functions of the medulla.
6. Describe the location and structure of the pons. What are its functions?
7. Describe the location and structure of the midbrain. What

are some of its functions?

8. Describe the location and structure of the thalamus. List some of its functions.
9. Where is the hypothalamus located? Explain some of its major functions.
10. Where is the cerebrum located? Describe the cortex, convolutions, fissures, and sulci of the cerebrum.
11. List and locate the lobes of the cerebrum. How are they separated from one another? What is the insula?
12. Describe the organization of cerebral white matter. Be sure to indicate the function of each group of fibers.
13. What are basal ganglia? Name the important basal ganglia and list the function of each.
14. Define the limbic system. Explain several of its functions.
15. What is meant by a sensory area of the cerebral cortex? List, locate, and give the function of each sensory area.
16. What is meant by a motor area of the cerebral cortex? List, locate, and give the function of each motor area.
17. What conditions may result from damage to sensory or motor speech areas?
18. What is an association area of the cerebral cortex? What are its functions?
19. Define an electroencephalogram (EEG). What is the diagnostic value of an EEG?

20. Describe brain lateralization (split-brain concept).
21. Describe the location of the cerebellum. List the principal parts of the cerebellum.
22. What are cerebellar peduncles? List and explain the function of each.
23. Explain the functions of the cerebellum. What is ataxia?
24. Define a cranial nerve. How are cranial nerves named and numbered? Distinguish between a mixed and a sensory cranial nerve.
25. For each of the 12 pairs of cranial nerves, list (a) its name, number, and type; (b) its location; and (c) its function. In addition, list the effects of damage, where applicable.
26. Describe the effects of aging on the nervous system.
27. Describe how the nervous system develops.
28. Define each of the following: cerebrovascular accident (CVA), transient ischemic attack (TIA), brain tumors, poliomyelitis, cerebral palsy (CP), Parkinson's disease (PD), multiple sclerosis (MS), epilepsy, dyslexia, Tay-Sachs disease, headache, trigeminal neuralgia, Reye's syndrome (RS), and Alzheimer's disease (AD).
29. What is brain electrical activity mapping (BEAM)? What are its clinical applications?
30. Refer to the glossary of key medical terms associated with the nervous system. Be sure that you can define each term.

SELF-QUIZ

Complete the following.

1. The two medullary nuclei that relay sensory information from the spinal cord to the opposite side of the brain are the nucleus gracilis and nucleus _____.
2. The _____ colliculi of the corpora quadrigemina are reflex centers for head and trunk movements in response to auditory stimuli.
3. The dural extension between the cerebrum and cerebellum is the _____.
4. White matter fibers that transmit nerve impulses between gyri in the same hemisphere are called _____ fibers.
5. The optic (II) nerve exits from the eye and joins its partner from the other eye at the optic _____. Here some fibers cross to the opposite side; all optic fibers proximal to this point are located in the optic _____.
6. The outer layer of the cerebrum is called the _____. It is composed of (white? gray?) matter. This means that it contains mainly (cell bodies? neurons?). The surface of the cerebrum looks much like a view of tightly packed mountains or ridges, called _____. The parts where the cerebral cortex dips down into valleys are called _____ (deep valleys) or _____ (shallow valleys).
7. The caudate and lentiform nuclei, together with several other nuclei, are called _____. They are islands of (gray? white?) matter embedded deep within the cerebrum.
8. The nervous system begins to develop during the third week of gestation when the _____-derm forms a thickening called the _____ plate.
9. Write the name of the correct cranial nerve following the related description.
   a. differs from all other cranial nerves in that it originates from the brain stem and from the spinal cord: _____
   b. eighth (VIII) cranial nerve: _____
   c. is widely distributed into neck, thorax, and abdomen: _____
   d. senses toothache, pain under a contact lens, wind on the face: _____
   e. the largest cranial nerve; has three parts (ophthalmic, maxillary, and mandibular): _____
   f. controls contraction of muscle of the iris, causing constriction of pupil: _____
   g. innervates muscles of facial expression: _____
   h. two nerves that contain taste fibers and autonomic fibers to salivary glands: _____, _____
   i. three purely sensory cranial nerves: _____, _____, _____.

Circle all the correct responses.

10. Choose the function(s) of cerebrospinal fluid (CSF):

    A. serves as a shock absorber for brain and cord; B. contains red blood cells; C. contains white blood cells called lymphocytes; D. contains nutrients.

11. Which statement(s) describes the pathway of cerebrospinal fluid (CSF)?
    A. It circulates around the brain, but not the cord.
    B. It flows below the end of the spinal cord.
    C. It bathes the brain by flowing through the epidural space.
    D. It passes via projections (villi) of the arachnoid into blood vessels (cranial venous sinuses) surrounding the brain.
    E. It is formed initially from blood and finally flows back to blood.

12. Match the following:

___ a. the main regulator of visceral activities since it acts as a liaison between cerebral cortex and autonomic nerves that control viscera

___ b. the site of the red nucleus, the origin of rubrospinal tracts concerned with muscle tone and posture

___ c. cranial nerves III–IV attach to this brain part

___ d. cranial nerves V–VIII attach to this brain part

___ e. cranial nerves VIII–XII attach to this brain part

___ f. feelings of hunger, fullness, and thirst stimulate centers here so that you can respond accordingly

___ g. all sensations except smell are relayed through here

___ h. regulation of heart, blood pressure, and respiration occurs by centers located here

___ i. it constitutes four-fifths of the diencephalon

___ j. it lies under the third ventricle, forming its floor

___ k. it forms most of side walls of the third ventricle

___ l. tumor in this region could compress cerebral aqueduct and cause internal hydrocephalus

___ m. it releases chemicals called regulating hormones (or factors) that control hormones

A. hypothalamus
B. medulla
C. midbrain
D. pons
E. thalamus

Choose the one best answer to these questions.

___ **13.** The meninx that adheres to the surface of the brain and spinal cord and contains blood vessels is the
A. arachnoid; B. dura mater; C. pia mater; D. mia mater; E. perineurium.

___ **14.** The reason that the motor areas of the right cerebral cortex control voluntary movements on the left side of the body is because
A. the cerebrum contains projection fibers; B. the cerebellum controls voluntary movements; C. the medulla contains decussating pyramids; D. the pons connects the spinal cord with the brain; E. the midbrain reroutes all motor impulses.

___ **15.** A patient exhibits the following signs: irregular fluctuations in body temperature, loss of appetite, lack of sensation of thirst, and psychosomatic disorders. What portion of the brain may be malfunctioning?
A. cerebrum; B. midbrain; C. hypothalamus; D. medulla; E. pons.

___ **16.** The vital reflex centers for the control of heart beat, respiration, and blood vessel diameter are located in the
A. pons; B. medulla; C. cerebrum; D. cerebellum; E. midbrain.

___ **17.** An obstruction in the interventricular foramen would interfere with the flow of cerebrospinal fluid into the
A. lateral ventricles; B. third ventricle; C. fourth ventricle; D. subarachnoid space of the spinal cord; E. subdural space of the brain.

___ **18.** The corpus callosum is a bridge of white fibers that connects the
A. midbrain and medulla; B. cerebral hemispheres; C. cerebrum and cerebellum; D. pons and cerebellum; E. cerebellar hemispheres.

___ **19.** The primary motor area is located in the
A. precentral gyrus; B. basal ganglia; C. corpus callosum; D. hypothalamus; E. postcentral gyrus.

___ **20.** Which of the following would you expect to observe in a patient with a tumor of the cerebellum?
(1) loss of general sensation;
(2) inability to execute any voluntary movements;
(3) inability to execute smooth, steady movement.
A. (1) only; B. (2) only; C. (3) only; D. none of the above; E. all of the above.

___ **21.** Which of the following indicates the correct order in which cranial nerves originate from the base of the brain?
A. optic, olfactory, trigeminal, trochlear, ophthalmic; B. olfactory, ophthalmic, oculomotor, trochlear, trigeminal; C. olfactory, optic, trochlear, trigeminal, facial; D. olfactory, optic, oculomotor, trochlear, trigeminal; E. oculomotor, optic, trigeminal, trochlear, facial.

___ **22.** Damage to the occipital lobe of the cerebrum would most likely cause
A. loss of hearing; B. loss of vision; C. loss of ability to smell; D. paralysis; E. loss of muscle sense (proprioception).

23. Arrange the answers in correct sequence.

___ ___ ___ a. From anterior to posterior:
A. fourth ventricle
B. pons and medulla
C. cerebellum

___ ___ ___ b. From superior to inferior:
A. thalamus
B. hypothalamus
C. corpus callosum

Circle T (true) or F (false) for the following.

T F **24.** The thalamus, hypothalamus, and cerebrum are all developed from the forebrain.

T F **25.** The language areas are located in the cerebellar cortex.

T F **26.** The limbic system functions in the control of behavior.

T F **27.** The reticular formation controls arousal and consciousness.

T F **28.** The fourth ventricle communicates with the subarachnoid space of the brain and spinal cord through the cerebral aqueduct.

T F **29.** The dural extension between cerebral hemispheres is known as the falx cerebri.

T F **30.** The cerebral nuclei that control large subconscious movements, such as swinging the arms while walking, are the caudate and putamen.

19 The Autonomic Nervous System

CHAPTER OUTLINE
- ■ **Somatic Efferent and Autonomic Nervous Systems**
- ■ **Structure of the Autonomic Nervous System**

Visceral Efferent Pathways
    Preganglionic Neurons
    Autonomic Ganglia
    Postganglionic Neurons
Sympathetic Division
Parasympathetic Division
- ■ **Physiology of the Autonomic Nervous System**

Neurotransmitters
Receptors
Activities
- ■ **Visceral Autonomic Reflexes**
- ■ **Control by Higher Centers**

Biofeedback
Meditation

STUDENT OBJECTIVES

1. Compare the structural and functional differences between the somatic efferent and autonomic portions of the nervous system.
2. Identify the principal structural features of the autonomic nervous system.
3. Compare the sympathetic and parasympathetic divisions of the autonomic nervous system in terms of structure, physiology, and neurotransmitters released.
4. Describe the various postsynaptic receptors involved in autonomic responses.
5. Explain the role of the hypothalamus and its relation to the sympathetic and parasympathetic divisions.
6. Explain the relation between biofeedback and meditation and the autonomic nervous system.

The portion of the nervous system that regulates the activities of smooth muscle, cardiac muscle, and glands is the *autonomic nervous system (ANS).* Structurally, the system consists of visceral efferent neurons organized into nerves, ganglia, and plexuses. Functionally, it usually operates without conscious control. The system was originally named *autonomic* because physiologists thought it functioned with no control from the central nervous system, that it was autonomous or self-governing. It is now known that the autonomic system is neither structurally nor functionally independent of the central nervous system. It is regulated by centers in the brain, in particular by the cerebral cortex, hypothalamus, and medulla oblongata. However, the old terminology has been retained, and since the autonomic nervous system does differ from the somatic nervous system in some ways, the two are separated for convenience of study.

SOMATIC EFFERENT AND AUTONOMIC NERVOUS SYSTEMS

Whereas the somatic efferent nervous system produces conscious movement in skeletal muscles, the autonomic nervous system (visceral efferent nervous system) regulates visceral activities, and it generally does so involuntarily and automatically. Examples of visceral activities regulated by the autonomic nervous system are changes in the size of the pupil, accommodation for near vision, dilation and constriction of blood vessels, adjustment of the rate and force of the heartbeat, movements of the gastrointestinal tract, and secretion by most glands. These activities usually lie beyond conscious control. They are automatic.

The autonomic nervous system is generally considered to be entirely motor. All its axons are efferent fibers, which transmit impulses from the central nervous system to visceral effectors. Autonomic fibers are called *visceral efferent fibers. Visceral effectors* include cardiac muscle, smooth muscle, and glandular epithelium. This does not mean there are no efferent (sensory) impulses from visceral effectors, however. Nerve impulses that give rise to visceral sensations pass over visceral afferent neurons that have cell bodies located in the posterior (dorsal) root ganglia of spinal nerves. Some functions of these afferent neurons were described in discussing the cranial and spinal nerves. The hypothalamus, which largely controls the autonomic nervous system, also receives nerve impulses from the visceral sensory fibers.

In the motor portion of the neural pathway of the somatic efferent system, an efferent neuron runs from the central nervous system and synapses directly on a skeletal muscle. In the neural pathway of the autonomic nervous system, there are two efferent neurons and a ganglion between them. The first neuron runs from the central nervous system to a ganglion, where it synapses with the second efferent neuron. It is this neuron that ultimately synapses on a visceral effector. Also, whereas fibers of somatic efferent neurons release acetylcholine (ACh) as their neurotransmitter, fibers of autonomic efferent neurons release either ACh or norepinephrine (NE).

The autonomic nervous system consists of two principal divisions: the *sympathetic* and the *parasympathetic.* Many organs innervated by the autonomic nervous system receive visceral efferent neurons from both components of the autonomic system—one set from the sympathetic division, another from the parasympathetic division. In general, nerve impulses transmitted by the fibers of one division stimulate the organ to start or increase activity, whereas nerve impulses from the other division decrease the organ's activity. Organs that receive nerve impulses from both sympathetic and parasympathetic fibers are said to have *dual innervation.* Thus, autonomic innervation may be excitatory or inhibitory. In the somatic efferent nervous system, only one kind of motor neuron innervates an organ, which is always a skeletal muscle. Moreover, innervation is always excitatory. When a somatic neuron stimulates a skeletal muscle, the muscle becomes active. When the neuron ceases to stimulate the muscle, contraction stops altogether.

A summary of the principal differences between the somatic efferent and autonomic nervous systems is presented in Exhibit 19-1.

STRUCTURE OF THE AUTONOMIC NERVOUS SYSTEM

VISCERAL EFFERENT PATHWAYS

Autonomic visceral efferent pathways always consist of two neurons. One extends from the central nervous system to a ganglion. The other extends directly from the ganglion to the effector (muscle or gland).

The first of the visceral efferent neurons in an autonomic pathway is called a *preganglionic neuron* (Figure 19-1). Its cell body is in the brain or spinal cord. Its myelinated axon, called a *preganglionic fiber,* passes out of the central nervous system as part of a cranial or spinal nerve. At some point, the fiber separates from the nerve and travels to an autonomic ganglion, where it synapses with the dendrites or cell body of the postganglionic neuron, the second neuron in the visceral efferent pathway.

The *postganglionic neuron* lies entirely outside the central nervous system. Its cell body and dendrites (if it has dendrites) are located in the autonomic ganglion, where the synapse with one or more preganglionic fibers occurs. The axon of a postganglionic neuron, called a *postganglionic fiber,* is unmyelinated and terminates in a visceral effector.

Thus, preganglionic neurons convey efferent impulses from the central nervous system to autonomic ganglia. Postganglionic neurons relay the efferent impulses from autonomic ganglia to visceral effectors.

EXHIBIT 19-1

Comparison of Somatic Efferent and Autonomic Nervous Systems

	SOMATIC EFFERENT	AUTONOMIC
Effectors	Skeletal muscles.	Cardiac muscle, smooth muscle, glandular epithelium.
Type of control	Voluntary.	Involuntary.
Neural pathway	One efferent neuron extends from CNS and synapses directly on a skeletal muscle.	One efferent neuron extends from the CNS and synapses with another efferent neuron in a ganglion; the second neuron synapses on a visceral effector.
Action on effector	Always excitatory.	May be excitatory or inhibitory, depending on whether stimulation is sympathetic or parasympathetic.
Neurotransmitters	Acetylcholine (ACh).	Acetylcholine (ACh) or norepinephrine (NE).

Preganglionic Neurons

In the sympathetic division, the preganglionic neurons have their cell bodies in the lateral gray horns of the twelve thoracic segments and first two or three lumbar segments of the spinal cord (Figure 19-2). It is for this reason that the sympathetic division is also called the *thoracolumbar* (thō'-ra-kō-LUM-bar) *division* and the fibers of the sympathetic preganglionic neurons are known as the *thoracolumbar outflow*.

The cell bodies of the preganglionic neurons of the parasympathetic division are located in the nuclei of cranial nerves III, VII, IX, and X in the brain stem and in the lateral gray horns of the second through fourth sacral segments of the spinal cord. Hence, the parasympathetic division is also known as the *craniosacral division*, and the fibers of the parasympathetic preganglionic neurons are referred to as the *craniosacral outflow*.

CLINICAL APPLICATION

Raynaud's (rā-NŌZ) *disease* is a disorder characterized by spasms of arteries, especially those in the fingers and toes due to overactivity of the sympathetic nervous system. As a result of the spasmodic contractions, the tissues of the digits receive an inadequate blood supply and the digits exhibit pallor (paleness) or cyanosis and severe pain. The disease is typically bilateral and is provoked by exposure to cold. In extreme cases gangrene may occur. Cutting the preganglionic sympathetic fibers supplying the area abolishes symptoms for patients with severe progression of the disease.

Autonomic Ganglia

Autonomic pathways always include *autonomic ganglia,* where synapses between visceral efferent neurons occur. Autonomic ganglia differ from posterior root ganglia. The latter contain cell bodies of sensory neurons and no synapses occur in them.

The autonomic ganglia may be divided into three general groups. The *sympathetic trunk (vertebral chain) ganglia* are a series of ganglia that lie in a vertical row on either side of the vertebral column, extending from the base of the skull to the coccyx (Figure 19-3). They are also known

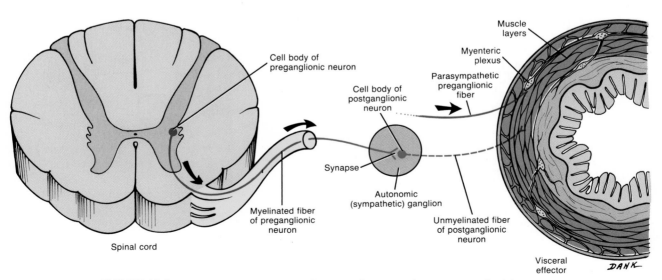

FIGURE 19-1 Relation between preganglionic and postganglionic (sympathetic) neurons.

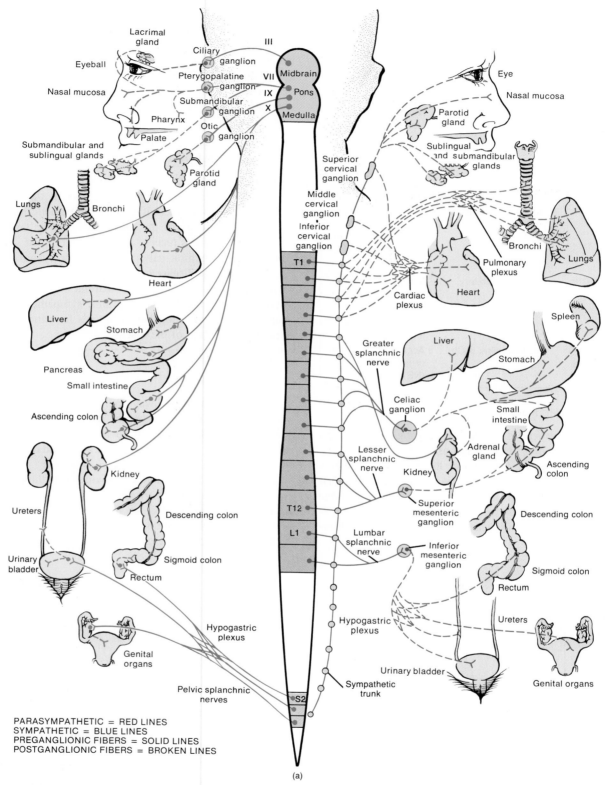

(a)

FIGURE 19-2 Structure of the autonomic nervous system. (a) Diagram. Although the parasympathetic division is shown only on the left side of the figure and the sympathetic division is shown only on the right side, keep in mind that each division is actually on both sides of the body (bilateral symmetry).

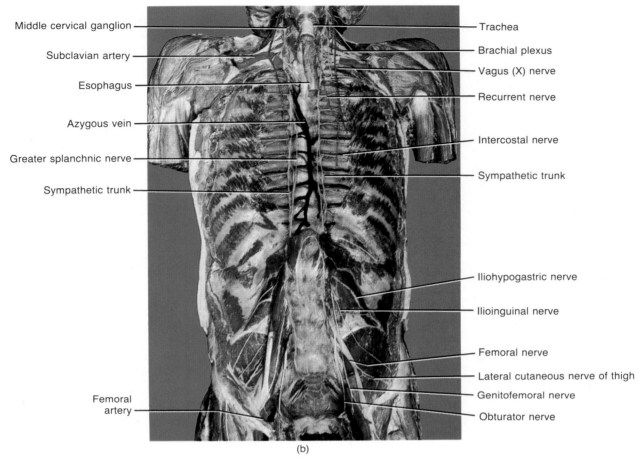

Middle cervical ganglion

Subclavian artery

Esophagus

Azygous vein

Greater splanchnic nerve

Sympathetic trunk

Femoral artery

Trachea

Brachial plexus

Vagus (X) nerve

Recurrent nerve

Intercostal nerve

Sympathetic trunk

Iliohypogastric nerve

Ilioinguinal nerve

Femoral nerve

Lateral cutaneous nerve of thigh

Genitofemoral nerve

Obturator nerve

(b)

FIGURE 19-2 (Continued) Structure of the autonomic nervous system. (b) Photograph. (Courtesy of C. Yokochi and J. W. Rohen, *Photographic Anatomy of the Human Body,* 2nd ed., 1979, IGAKU-SHOIN, Ltd., Tokyo, New York.)

as *paravertebral (lateral) ganglia.* They receive preganglionic fibers only from the sympathetic division (Figure 19-2). Because of this, sympathetic preganglionic fibers tend to be short.

The second kind of autonomic ganglion also belongs to the sympathetic division. It is called a *prevertebral (collateral) ganglion.* (Figure 19-3). The ganglia of this group lie anterior to the spinal column and close to the large abdominal arteries from which their names are derived. Examples of prevertebral ganglia so named are the celiac ganglion, on either side of the celiac artery just below the diaphragm; the superior mesenteric ganglion, near the beginning of the superior mesenteric artery in the upper abdomen; and the inferior mesenteric ganglion, located near the beginning of the inferior mesenteric artery in the middle of the abdomen (Figure 19-2). Prevertebral ganglia receive preganglionic fibers from the sympathetic division.

The third kind of automonic ganglion belongs to the parasympathetic division and is called a *terminal (intramural) ganglion.* The ganglia of this group are located at the end of a visceral efferent pathway very close to visceral effectors or actually within the walls of visceral effectors.

Terminal ganglia receive preganglionic fibers from the parasympathetic division. The preganglionic fibers do not pass through sympathetic trunk ganglia (Figure 19-2). Because of this, parasympathetic preganglionic fibers tend to be long.

In addition to autonomic ganglia, the autonomic nervous system contains *autonomic plexuses.* Slender nerve fibers from ganglia containing postganglionic nerve cell bodies arranged in a branching network constitute an autonomic plexus.

Postganglionic Neurons

Axons from preganglionic neurons of the sympathetic division pass to ganglia of the sympathetic trunk (Figure 19-2). They can either synapse in the sympathetic chain ganglia with postganglionic neurons or they can continue, without synapsing, through the chain ganglia to end at a prevertebral ganglion where synapses with the postganglionic neurons can take place. Each sympathetic preganionic fiber synapses with several postganglionic fibers in the ganglion, and the postganglionic fibers pass to several

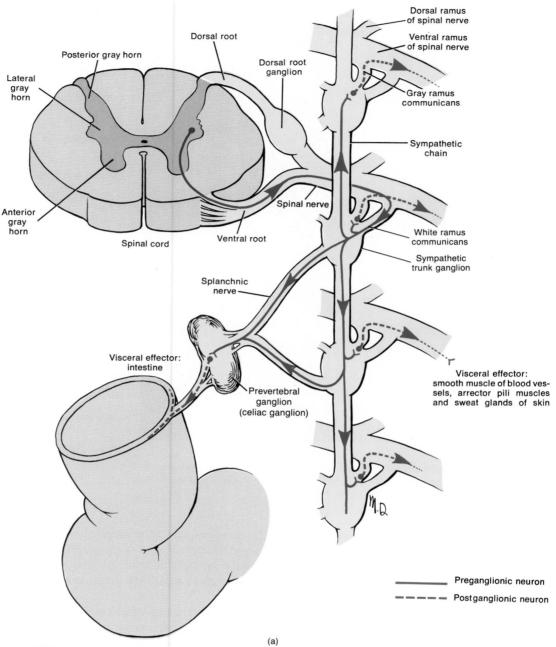

Posterior gray horn

Lateral gray horn

Dorsal root

Dorsal root ganglion

Dorsal ramus of spinal nerve

Ventral ramus of spinal nerve

Gray ramus communicans

Sympathetic chain

Spinal nerve

White ramus communicans

Sympathetic trunk ganglion

Anterior gray horn

Spinal cord

Ventral root

Splanchnic nerve

Visceral effector: intestine

Prevertebral ganglion (celiac ganglion)

Visceral effector: smooth muscle of blood vessels, arrector pili muscles and sweat glands of skin

———— Preganglionic neuron

- - - - - Postganglionic neuron

(a)

FIGURE 19-3 Ganglia and rami communicantes of the sympathetic division of the autonomic nervous system. (a) Diagram.

visceral effectors. After exiting their ganglia, the postsynaptic fibers innervate their visceral effectors.

Axons from preganglionic neurons of the parasympathetic division pass to terminal ganglia near or within a visceral effector. In the ganglion, the presynaptic neuron usually synapses with only four or five postsynaptic neurons to a single visceral effector. After exiting their ganglia, the postsynaptic fibers supply their visceral effectors.

With this background in mind, we can now examine some specific structural features of the sympathetic and parasympathetic divisions of the autonomic nervous system.

SYMPATHETIC DIVISION

The preganglionic fibers of the sympathetic division have their cell bodies located in the lateral gray horns of all the thoracic segments and first two or three lumbar segments of the spinal cord (Figure 19-2). The preganglionic fibers are myelinated and leave the spinal cord through the ventral root of a spinal nerve along with the somatic efferent fibers at the same segmental levels. After exiting through the intervertebral foramina, the preganglionic sympathetic fibers enter a white ramus to pass to the nearest sympathetic trunk

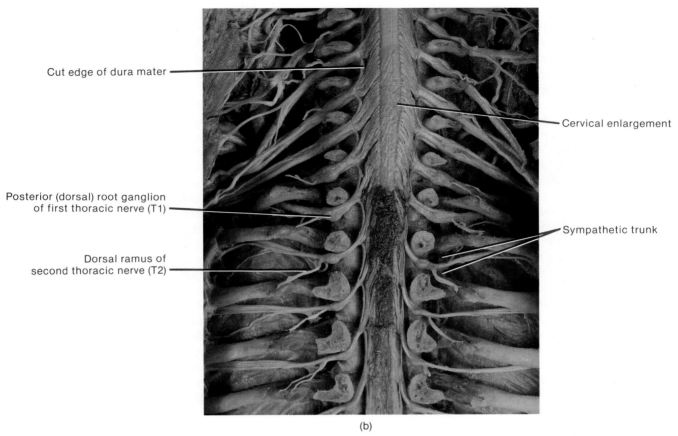

Cut edge of dura mater

Cervical enlargement

Posterior (dorsal) root ganglion of first thoracic nerve (T1)

Sympathetic trunk

Dorsal ramus of second thoracic nerve (T2)

(b)

FIGURE 19-3 (Continued) Ganglia and rami communicantes of the sympathetic division of the autonomic nervous system. (b) Photograph of the posterior aspect of the lower cervical and upper thoracic segments of the spinal cord. (Courtesy of N. Gluhbegovic and T. H. Williams, *The Human Brain: A Photographic Guide*, Harper & Row, Publishers, Inc., New York, 1980.)

ganglion on the same side. Collectively, the white rami are called the *white rami communicantes* (kō-myoo-ni-KAN-tēz). Their name indicates that they contain myelinated fibers. Only thoracic and upper lumbar nerves have white rami communicantes. The white rami communicantes connect the ventral ramus of the spinal nerve with the ganglia of the sympathetic trunk.

The paired sympathetic trunks are situated anterolaterally to the spinal cord, one on either side. Each consists of a series of ganglia arranged more or less segmentally. The divisions of the sympathetic trunk are named on the basis of location. Typically, there are 22 ganglia in each chain: 3 cervical, 11 thoracic, 4 lumbar, and 4 sacral. Although the trunk extends downward from the neck, thorax, and abdomen to the coccyx, it receives preganglionic fibers only from the thoracic and lumbar segments of the spinal cord (Figure 19-2).

The cervical portion of each sympathetic trunk is located in the neck anterior to the prevertebral muscles. It is subdivided into a superior, middle, and inferior ganglion (Figure 19-2). The *superior cervical ganglion* is posterior to the internal carotid artery and anterior to the transverse processes of the second cervical vertebra. Postganglionic fibers leaving the ganglion serve the head, where they are distributed to the sweat glands, the smooth muscle of the eye and blood vessels of the face, the nasal mucosa, and the submandibular, sublingual, and parotid salivary glands. Gray rami communicantes (described shortly) from the ganglion also pass to the upper two to four cervical spinal nerves. The *middle cervical ganglion* is situated near the sixth cervical vertebra at the level of the cricoid cartilage. Postganglionic fibers from it innervate the heart. The *inferior cervical ganglion* is located near the first rib, anterior to the transverse processes of the seventh cervical vertebra. Its postganglionic fibers also supply the heart.

The thoracic portion of each sympathetic trunk usually consists of 11 segmentally arranged ganglia, lying ventral to the necks of the corresponding ribs. This portion of the sympathetic trunk receives most of the sympathetic preganglionic fibers. Postganglionic fibers from the thoracic sympathetic trunk innervate the heart, lungs, bronchi, and other thoracic viscera.

The lumbar portion of each sympathetic trunk is found on either side of the corresponding lumbar vertebrae. The sacral portion of the sympathetic trunk lies in the pelvic cavity on the medial side of the sacral foramina. Postganglionic fibers from the lumbar and sacral sympathetic chain ganglia are distributed with the respective spinal nerves via gray rami, or they may join the hypogastric plexus via direct visceral branches.

When a preganglionic fiber of a white ramus communicans enters the sympathetic trunk, it may terminate (synapse) in several ways. Some fibers synapse in the first ganglion at the level of entry. Others pass up or down the sympathetic trunk for a variable distance to form the fibers on which the ganglia are strung. These fibers, known as *sympathetic chains* (Figure 19-3), may not synapse until they reach a ganglion in the cervical or sacral area. Most rejoin the spinal nerves before supplying peripheral visceral effectors such as sweat glands and the smooth muscle in blood vessels and around hair follicles in the extremities. The *gray ramus communicans* (kō-MYOO-ni-kanz) is the structure containing the postganglionic fibers that connect the ganglion of the sympathetic trunk to the spinal nerve (Figure 19-3). The fibers are unmyelinated. All spinal nerves have gray rami communicantes. Gray rami communicantes outnumber the white rami, since there is a gray ramus leading to each of the 31 pairs of spinal nerves.

In most cases, a sympathetic preganglionic fiber terminates by synapsing with a large number, usually 20 or more, of postganglionic cell bodies in a ganglion. Often the postganglionic fibers then terminate in widely separated organs of the body. Thus, a nerve impulse that starts in a single preganglionic neuron may reach several visceral effectors. For this reason, most sympathetic responses have widespread effects on the body.

Some preganglionic fibers pass through the sympathetic trunk without terminating in the trunk. Beyond the trunk, they form nerves known as *splanchnic* (SPLANK-nik) *nerves* (Figure 19-2). After passing through the trunk of ganglia, the splanchnic nerves from the thoracic area terminate in the *celiac* (SĒ-lē-ak) or *solar plexus.* In the plexus, the preganglionic fibers synapse in ganglia with postganglionic cell bodies. These ganglia are prevertebral ganglia. The greater splanchnic nerve passes to the celiac ganglion of the celiac plexus. From here, postganglionic fibers are distributed to the stomach, spleen, liver, kidney, and small intestine. The lesser splanchnic nerve passes through the celiac plexus to the superior mesenteric ganglion of the superior mesenteric plexus. Postganglionic fibers from this ganglion innervate the small intestine and colon. The lowest splanchnic nerve, not always persent, enters the renal plexus. Postganglionics supply the renal artery and ureter. The lumbar splanchnic nerve enters the inferior mesenteric plexus. In the plexus, the preganglionic fibers synapse with postganglionic fibers in the inferior mesenteric ganglion. These fibers pass through the hypogastric plexus and supply the distal colon and rectum, urinary bladder, and gential organs. As noted earlier, the postganglionic fibers leaving the prevertebral ganglia follow the course of various arteries to abdominal and pelvic visceral effectors.

PARASYMPATHETIC DIVISION

The preganglionic cell bodies of the parasympathetic division are found in nuclei in the brain stem and the lateral gray horn of the second through fourth sacral segments of the spinal cord (Figure 19-2). Their fibers emerge as part of a cranial nerve or as part of the ventral root of a spinal nerve. The *cranial parasympathetic outflow* consists of preganglionic fibers that leave the brain stem by way of the oculomotor (III) nerves, facial (VII) nerves, glossopharyngeal (IX) nerves, and vagus (X) nerves. The *sacral parasympathetic outflow* consist of preganglionic fibers that leave the ventral roots of the second through fourth sacral nerves. The preganglionic fibers of both the cranial and sacral outflows end in terminal ganglia, where they synapse with postganglionic neurons. We shall first look at the cranial outflow.

The cranial outflow has five components: four pairs of ganglia and the plexuses associated with the vagus nerve. The four pairs of cranial parasympathetic ganglia innervate structures in the head and are located close to the organs they innervate. The *ciliary ganglion* is near the back of an orbit lateral to each optic (II) nerve. Preganglionic fibers pass with the oculomotor (III) nerve to the ciliary ganglion. Postganglionic fibers from the ganglion innervate smooth muscle cells in the eyeball. Each *pterygopalatine* (ter'-i-gō-PAL-a-tin) *ganglion* is situated lateral to a sphenopalatine foramen. It receives preganglionic fibers from the facial (VII) nerve and transmits postganglionic fibers to the nasal mucosa, platate, pharynx, and lacrimal gland. Each *submandibular ganglion* is found near the duct of a submandibular salivary gland. It receives preganglionic fibers from the facial (VII) nerve and transmits postganglionic fibers that innervate the submandibular and sublingual salivary glands. The *otic ganglia* are situated just below each foramen ovale. The otic ganglion receives preganglionic fibers form the glossopharyngeal (IX) nerve and transmits postganglionic fibers that innervate the parotid salivary gland. Ganglia associated with the cranial outflow are classified as terminal ganglia. Since the terminal ganglia are close to their visceral effectors, postganglionic parasympathetic fibers are short. Postganglionic sympathetic fibers are relatively long.

The last component of the cranial outflow consists of the preganglionic fibers that leave the brain via the vagus (X) nerves. This component has the most extensive distribution of the parasympathetic fibers, providing about 80 percent of the craniosacral outflow. Each vagus (X) nerve enters into the formation of several plexuses in the thorax and abdomen. As it passes through the thorax, it sends fibers to the *superficial cardiac plexus* in the arch of the aorta and the *deep cardiac plexus* anterior to the branching of the trachea. These plexuses contain terminal ganglia, and the postganglionic parasympathetic fibers emerging from them supply the heart. Also in the thorax is the *pulmo-*

nary plexus, anterior and posterior to the roots of the lungs and within the lungs themselves. It receives preganglionic fibers from the vagus and transmits postganglionic parasympathetic fibers to the lungs and bronchi. Other plexuses associated with the vagus (X) nerve are described in later chapters in conjunction with the appropriate thoracic, abdominal, and pelvic viscera. Postganglionic fibers from these plexuses innervate viscera such as the liver, pancreas, stomach, kidneys, small intestine, and part of the colon.

The sacral parasympathetic outflow consists of preganglionic fibers from the ventral roots of the second through fourth sacral nerves. Collectively, they form the *pelvic splanchnic nerves.* They pass into the hypogastric plexus. From ganglia in the plexus, parasympathetic postganglionic fibers are distributed to the colon, ureters, urinary bladder, and reproductive organs.

The salient structural features of the sympathetic and parasympathetic division are compared in Exhibit 19-2.

PHYSIOLOGY OF THE AUTONOMIC NERVOUS SYSTEM

NEUROTRANSMITTERS

Autonomic fibers, like other axons of the nervous system, release neurotransmitters at synapses as well as at points of contact with visceral effectors. These latter points are called *neuroeffector junctions.* Neuroeffector junctions may be either neuromuscular or neuroglandular junctions. On the basis of the neurotransmitter produced, autonomic fibers may be classified as either cholinergic or adrenergic (Figure 19-4).

Cholinergic (kō'-lin-ER-jik) *fibers* release *acetylcholine*

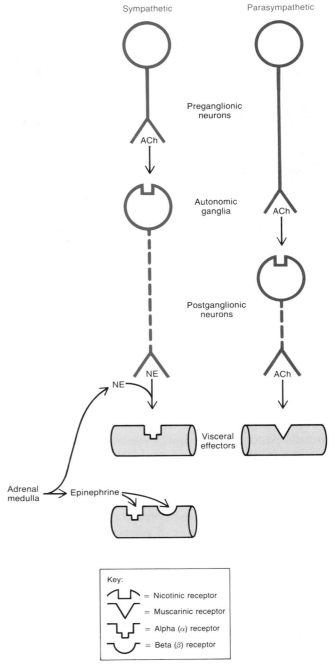

FIGURE 19-4 Neurotransmitters and receptors associated with the autonomic nervous system.

(ACh) and include the following: (1) all sympathetic and parasympathetic preganglionic axons, (2) all parasympathetic postganglionic axons, and (3) a few sympathetic postganglionic axons. The cholinergic sympathetic postganglionic axons include those to sweat glands and blood vessels in skeletal muscles, skin, and external genitalia. Since acetylcholine is quickly inactivated by the enzyme *acetylcholinesterase (AChE),* the effects of cholinergic fibers are short-lived and local.

EXHIBIT 19-2

Structural Features of Sympathetic and Parasympathetic Divisions

SYMPATHETIC	PARASYMPATHETIC
Forms thoracolumbar outflow.	Forms craniosacral outflow.
Contains sympathetic trunk and prevertebral ganglia.	Contains terminal ganglia.
Ganglia are close to the CNS and distant from visceral effectors.	Ganglia are near or within visceral effectors.
Each preganglionic fiber synapses with many postganglionic neurons that pass to many visceral effectors.	Each preganglionic fiber usually synapses with four or five postganglionic neurons that pass to a single visceral effector.
Distributed throughout the body, including the skin.	Distribution limited primarily to head and viscera of thorax, abdomen, and pelvis.

Adrenergic (ad'-ren-ER-jik) *fibers* produce *norepineph-rine (NE)*. Most sympathetic postganglionic axons are ad-renergic. Since norephinephrine is inactivated much more slowly by *catechol-O-methyltransferase (COMT)* or *mo-noamine oxidase (MAO)* than acetylcholine is by acetylcho-linesterase, and since norepinephrine may enter the blood stream, the effects of sympathetic stimulation are longer lasting and more widespread than parasympathetic stimula-tion. The action of NE produced by sympathetic postgan-glionic axons is augmented by both NE and epinephrine secreted by the adrenal medulla. Since both substances are secreted into the blood, their effects are more sustained than NE released by axons. NE and epinephrine released by the adrenal medulla are ultimately destroyed by enzymes in the liver once they have acted on their visceral effectors.

RECEPTORS

The actual effect produced by ACh is determined by the type of postsynaptic receptor with which it interacts (Figure 19-4). The two types of ACh postsynaptic receptors are known as nicotinic receptors and muscarinic receptors. *Nic-otinic receptors* are found on both sympathetic and para-sympathetic postganglionic neurons. These receptors are so named because the actions of ACh on such receptors are similar to those produced by nicotine. *Muscarinic recep-tors* are found on all effectors innervated by parasympathetic postganglionic axons and some effectors innervated by sym-pathetic postganglionic axons. These postsynaptic receptors are so named because the actions of ACh on such receptors are similar to those produced by muscarine, a toxin produced by a mushroom.

NE is released from the most postganglionic sympathetic fibers. It is synthesized and stored in synaptic vesicles lo-cated in the axon terminals of adrenergic fibers. When a nerve impulse reaches the axon terminal, NE is rapidly released into the synaptic cleft. The molecules diffuse across the cleft and combine with specific receptor molecules on the postsynaptic membrane to elicit a specific effector re-sponse.

The effects of NE and epinephrine, like those of ACh, are also determined by the type of postsynaptic receptor with which they interact. Such receptors are found on vis-ceral effectors innervated by most sympathetic postgan-glionic axons and are referred to as *alpha (α) receptors* and *beta (β) receptors* (Figure 19-4). These receptors are distinguished by the specific responses they elicit and by their selective combination with drugs that excite or inhibit them. Alpha receptors are generally excitatory. Beta recep-tors are generally inhibitory.

Although cells of most effectors contain either alpha or beta receptors, some visceral effector cells contain both. NE, in general, stimulates alpha receptors to a greater extent than beta receptors, and epinephrine, in general, stimulates both alpha and beta receptors.

Exhibit 19-3 shows the types of receptors present on the cells of visceral effectors and their response to autonomic stimulation. Note that there are some very important (and perplexing) exceptions to the general rule describing alpha and beta receptors. For example, heart muscle has beta receptors that operate as excitatory receptors and cause the heart muscle to contract more forcibly. Other visceral effec-tor cells possess both types of receptors. In these situations, the majority rules. The beta receptors greatly outnumber the alpha receptors in the smooth muscle fibers (cells) of the blood vessels that flow through skeletal muscles. There-fore, an injection of NE, which binds only to alpha receptors, attaches onto the surface of such smooth muscle fibers and causes a vasoconstriction. An injection of epinephrine, how-ever, would result in a vasodilation because epinephrine acts equally on both types of receptors. But since there is a preponderance of beta receptors, the effect is that of the beta receptor, that is, vasodilation due to relaxation of smooth muscle fibers.

ACTIVITIES

Most visceral effectors have dual innervation; that is, they receive fibers from both the sympathetic and the parasym-pathetic divisions. In these cases, nerve impulses from one division stimulate the organ's activities, whereas nerve im-pulses from the other division inhibit the organ's activities. The stimulating division may be either the sympathetic or the parasympathetic, depending on the organ. For example, sympathetic impulses increase heart activity, whereas para-sympathetic impulses decrease it. On the other hand, para-sympathetic impulses increase digestive activities, whereas sympathetic impulses inhibit them. The actions of the two systems are carefully integrated to help maintain homeosta-sis. A summary of the activities of the autonomic nervous system is presented in Exhibit 19-3.

The parasympathetic division is primarily concerned with activities that restore and conserve body energy. It is a *rest-repose system*. Under normal body conditions, for in-stance, parasympathetic impulses to the digestive glands and the smooth muscle of the gastrointestinal tract dominate over sympathetic impulses. Thus, energy-supplying food can be digested and absorbed by the body.

The sympathetic division, in contrast, is primarily con-cerned with processes involving the expenditure of energy. When the body is in homeostasis, the main function of the sympathetic division is to counteract the parasympathetic effects just enough to carry out normal processes requiring energy. During extreme stress, however, the sympathetic dominates the parasympathetic. When people are confronted with a stress condition, for example, their bodies become alert and they sometimes perform feats of unusual strength. Fear stimulates the sympathetic division.

Activation of the sympathetic division sets into operation

EXHIBIT 19-3

Activities of Autonomic Nervous System

VISCERAL EFFECTOR	RECEPTOR	EFFECT OF SYMPATHETIC STIMULATION	EFFECT OF PARASYMPA-THETIC STIMULATION
Eye			
Radial muscle of iris	α	Contraction that results in dilation of pupil (mydriasis).	No known functional innervation.
Sphincter muscle of iris	α	No known functional innervation.	Contraction that results in constriction of pupil (miosis).
Ciliary muscle	β	Relaxation that results in far vision.	Contraction that results in near vision.
Glands			
Sweat	α	Stimulates local secretion.	Stimulates generalized secretion.
Lacrimal (tear)	—	No known functional innvervation.	Stimulates secretion.
Salivary	α	Vasoconstriction, which decreases secretion.	Stimulates secretion and vasodilation.
Gastric	—	Vasoconstriction, which inhibits secretion.	Stimulates secretion.
Intestinal	—	Vasoconstriction, which inhibits secretion.	Stimulates secretion.
Adrenal medulla	—	Promotes epinephrine and norepinephrine secretion.	No known functional innvervation.
Fat cells	β	Promotes lipolysis.	No known functional innvervation.
Lungs (bronchial muscle)	β	Dilation.	Constriction.
Heart	β	Increases rate and strength of contraction; constricts coronary vessels that supply blood to heart muscle fibers (cells).	Decreases rate and strength of contraction; dilates coronary vessels.
Arterioles			
Skin and mucosa	α	Constriction.	No known functional innervation for most.
Skeletal muscle	α β	Constriction or dilation.	No known functional innervation.
Abdominal viscera	α β	Constriction.	No known functional innervation for most.
Cerebral	α	Slight constriction.	No known functional innervation.
Systemic veins	α β	Constriction and dilation.	No known functional innervation.
Liver	β	Promotes glycogenolysis and gluconeogenesis; decreases bile secretion.	Promotes glycogenesis and gluconeogenesis; increases bile secretion.
Gallbladder and its ducts	—	Relaxation.	Contraction.
Stomach	α β	Decreases motility and tone; contracts sphincters.	Increases motility and tone; relaxes sphincters.
Intestines	α β	Decreases motility and tone; contracts sphincters.	Increases motility and tone; relaxes sphincters.
Kidney	β	Constriction of blood vessels that results in decreased urine volume; secretion of renin.	No known functional innervation.
Ureter	α	Increases motility.	Decreases motility.
Pancreas	α β	Inhibits secretion of enzymes and insulin; promotes secretion of glucagon.	Promotes secretion of enzymes and insulin.
Spleen	—	Contraction and discharge of stored blood into general circulation.	No known functional innervation.

EXHIBIT 19-3 (Continued)

VISCERAL EFFECTOR	RECEPTOR	EFFECT OF SYMPATHETIC STIMULATION	EFFECT OF PARASYMPATHETIC STIMULATION
Urinary bladder	α β	Relaxation of muscular wall; contraction of internal sphincter.	Contraction of muscular wall; relaxation of internal sphincter.
Arrector pili of hair follicles	α	Contraction that results in erection of hairs.	No known functional innervation.
Uterus	α β	Inhibits contraction if nonpregnant; stimulates contraction if pregnant.	Minimal effect.
Sex organs	α	In male, contraction of smooth muscle of ductus (vas) deferens, seminal vesicle, prostate; results in ejaculation. In female, reverse uterine peristalsis.	Vasodilation and erection in both sexes; secretion in females.

a series of physiological responses collectively called the *fight-or-flight-response.* It produces the following effects.

1. The pupils of the eyes dilate.
2. The heart rate increases.
3. The blood vessels of the skin and viscera constrict.
4. The remainder of the blood vessels dilate. This causes a faster flow of blood into the dilated blood vessels of skeletal muscles, cardiac muscle, lungs, and brain—organs involved in fighting off danger.
5. Rapid breathing occurs as the bronchioles dilate to allow faster movement of air in and out of the lungs.
6. Blood sugar level rises as liver glycogen is converted to glucose to supply the body's additional energy needs.
7. The medulla of the adrenal gland is stimulated to produce epinephrine and norepinephrine, hormones that intensify and prolong the sympathetic effects noted previously.
8. Processes that are not essential for meeting the stress situation are inhibited. For example, muscular movements of the gastrointestinal tract and digestive secretions are slowed down or even stopped.

CLINICAL APPLICATION

If the sympathetic trunk is cut on one side, the sympathetic supply to that side of the head is removed, resulting in *Horner's syndrome,* in which the patient exhibits (on the affected side): ptosis (drooping of the upper eyelid), slight elevation of the lower eyelid; narrowing of the palpebral fissure, enophthalmos (the eye appears sunken), miosis (constricted pupil), anhydrosis (lack of sweating), and flushing of the skin.

VISCERAL AUTONOMIC REFLEXES

A *visceral autonomic reflex* adjusts the activity of a visceral effector. In other words, it results in the contraction of smooth or cardiac muscle or secretion by a gland. Such reflexes assume a key role in activities involved in homeostasis such as regulating heart action, blood pressure, respiration, digestion, defecation, and urinary bladder functions.

A visceral autonomic reflex arc consists of the following components.

1. *Receptor.* The receptor is the distal end of an afferent neuron in an exteroceptor or enteroceptor.
2. *Afferent neuron.* This neuron, either a somatic afferent or visceral afferent neuron, conducts the sensory impulse to the spinal cord or brain.
3. *Association neurons.* These neurons are found in the central nervous system.
4. *Visceral efferent preganglionic neuron.* In the thoracic and abdominal regions, this neuron is in the lateral gray horn of the spinal cord. The axon passes through the ventral root of the spinal nerve, the spinal nerve itself, and the white ramus communicans. It then enters a sympathetic trunk or prevertebral ganglion, where it synapses with a postganglionic neuron. In the cranial and sacral regions, the visceral efferent preganglionic axon leaves the central nervous system and passes to a terminal ganglion, where it synapses with a postganglionic neuron. The role of the visceral efferent preganglionic neuron is to convey a motor impulse from the brain or spinal cord to an autonomic ganglion.
5. *Visceral efferent postganglionic neuron.* This neuron conducts a motor impulse from a visceral efferent preganglionic neuron to the visceral effector.
6. *Visceral effector.* A visceral effector is smooth muscle, cardiac muscle, or a gland.

The basic difference between a somatic reflex arc and a visceral autonomic reflex arc is that in a somatic reflex arc only one efferent neuron is involved. In a visceral autonomic reflex arc, two efferent neurons are involved.

Visceral sensations do not always reach the cerebral cortex. Most remain at subconscious levels. Under normal conditions, you are not aware of heartbeat, muscular contractions of the digestive organs, changes in the diameter of blood vessels, and pupil dilation and constriction. Your body adjusts such visceral activities by visceral reflex arcs whose centers are in the spinal cord or lower regions of the brain. Among such centers are the cardiac, respiratory, vasomotor, swallowing, and vomiting centers in the medulla and the temperature control center in the hypothalamus. Stimuli delivered by somatic or visceral afferent neurons synapse in these centers, and the returning motor impulses conducted by visceral efferent neurons bring about an adjustment in the visceral effector without conscious recognition. The nerve impulses are interpreted and acted on subconsciously. Some visceral sensations do give rise to conscious recognition: hunger, nausea, and fullness of the urinary bladder and rectum.

CONTROL BY HIGHER CENTERS

The autonomic nervous system is not a separate nervous system. Although little is known about the specific centers in the brain that regulate specific autonomic functions, it is known that axons from many parts of the central nervous system are connected to both the sympathetic and the parasympathetic divisions of the autonomic nervous system and thus exert considerable control over it. Autonomic centers in the cerebral cortex are connected in autonomic centers of the thalamus, for example. These, in turn, are connected to the hypothalamus. In this hierarchy of command, the thalamus sorts incoming nerve impulses before they reach the cerebral cortex. The cerebral cortex then turns over control and integration of visceral activities to the hypothalamus. It is at the level of the hypothalamus that the major control and integration of the autonomic nervous system are exerted.

The hypothalamus receives input from areas of the nervous system concerned with emotions, visceral functions, olfaction, gustation, as well as changes in temperature, osmolarity, and levels of various substances in blood. Anatomically, the hypothalamus is connected to both the sympathetic and the parasympathetic divisions of the autonomic nervous system by axons of neurons whose dendrites and cell bodies are in various hypothalamic nuclei. The axons form tracts from the hypothalamus to sympathetic and parasympathetic nuclei in the brain stem and spinal cord through relays in the reticular formation. The posterior and lateral portions of the hypothalamus appear to control the sympathetic division. When these areas are stimulated, there is an increase in visceral activities—an increase in heart rate, a rise in blood pressure due to vasoconstriction of blood vessels, an increase in the rate and depth of respiration, dilation of the pupils, and inhibition of the gastrointestinal tract. On the other hand, the anterior and medial portions of the hypothalamus seem to control the parasympathetic division. Stimulation of these areas results in a decrease in heart rate, lowering of blood pressure, constriction of the pupils, and increased motility of the gastrointestinal tract.

Control of the autonomic nervous system by the cerebral cortex occurs primarily during emotional stress. In extreme anxiety, which can result from either conscious or subconscious stimulation in the cerebral cortex, the cortex can stimulate the hypothalamus. This, in turn, stimulates the cardiac and vasomotor centers of the medulla, which increase heart rate and blood pressure. If the cortex is stimulated by an extremely unpleasant sight, the stimulation causes vasodilation of blood vessels, a lowering of blood pressure, and fainting.

Evidence of even more direct control of visceral responses is provided by data gathered from studies of biofeedback and meditation.

BIOFEEDBACK

In the simplest terms, *biofeedback* is a process in which people get constant signals, or feedback, about visceral body functions such as blood pressure, heart rate, and muscle tension. By using special monitoring devices, they can control these visceral functions consciously.

In a study conducted at the Menninger Foundation,[*] subjects suffering from migraine headaches received instructions in the use of a monitor that registers the skin temperature of the right index finger. Subjects were also given a typewritten sheet containing two sets of phrases. The first set was designed to help them relax the entire body. The second set was designed to bring about an increased flow of blood in the hands. The subjects practiced raising their skin temperature at home for 5–15 minutes a day. When skin temperature increased, the monitor emitted a high-pitched sound. In time, the monitor was abandoned.

Once the subjects learned how to vasodilate their blood vessels, the migraine headaches lessened. Since migraine headaches are believed to involve a distension of blood vessels in the head, the shunting of blood from head to hands relieved the distension and thus the pain.

Other experiments have shown that biofeedback can be applied to childbirth. Women were given monitors hooked up to their fingers and arms to measure electrical conductivity of the skin and skeletal muscle tension. Both conductivity and tension increased with nervousness and made labor difficult. Muscle tension was recorded as a sirenlike sound that became louder with nervousness. Skin conductivity was recorded as a crackling noise that also increased with nervousness. The monitors kept the women informed of their nervousness. This was the biofeedback. Having pleasant thoughts reduced the sound levels. The reward was less nervousness. The results of the study indicate that the women in a state of reduced nervousness needed less medication during labor, and labor time was shortened.

There is no way to determine where biofeedback will

[*] Much of the following discussion of the use of biofeedback for the treatment of migraine headaches is based on information provided by Dr. Joseph D. Sargent of the Menninger Foundation, Topeka, Kansas.

lead. Perhaps the outstanding contribution of biofeedback research has been to demonstrate that the autonomic nervous system is not autonomous. Visceral responses can be controlled. Current therapeutic applications of biofeedback include treatment of asthma, Raynaud's disease, hyptertension, gastrointestinal disorders, fecal incontinence, anxiety, pain, and neuromuscular rehabilitation following cerebrovascular accidents (CVAs).

MEDITATION

Yoga, which literally means union, is defined as a higher state of consciousness achieved through a fully rested and relaxed body and a fully awake and relaxed mind. One widely practiced technique for achieving higher consciousness is called *transcendental meditation* (*TM*). One sits in a comfortable position with the eyes closed and concentrates on a suitable sound or thought.

Research indicates that transcendental meditation can alter physiological responses. Oxygen consumption decreases drastically along with carbon dioxide elimination. Subjects have experienced a reduction in metabolic rate and blood pressure. Researchers have also observed a decrease in heart rate, an increase in the intensity of alpha brain waves, a sharp decrease in the amount of lactic acid in the blood, and an increase in the skin's electrical resistance. These last four responses are characteristic of a highly relaxed state of mind. Alpha waves are found in the EEGs of almost all individuals in a resting, but awake, state; they disappear during sleep.

These responses have been called an *integrated response*—essentially, a hypometabolic state due to inactivation of the sympathetic division of the autonomic nervous system. The response is the exact opposite of the fight-or-flight response, which is a hyperactive state of the sympathetic division. The existence of the integrated response suggests that the central nervous system does exert some control over the autonomic nervous system.

STUDY OUTLINE

Somatic Efferent and Autonomic Nervous System (p. 534)
1. The somatic efferent nervous system produces conscious movement in skeletal muscles.
2. The autonomic nervous system, or visceral efferent nervous system, regulates visceral activities, that is, activities of smooth muscle, cardiac muscle, and glands, and it usually operates without conscious control.
3. It is regulated by centers in the brain, in particular by the cerebral cortex, hypothalamus, and medulla oblongata.
4. A single somatic efferent neuron synapses on skeletal muscles; in the autonomic nervous system, there are two efferent neurons—one from the CNS to a ganglion and one from a ganglion to a visceral effector.
5. Somatic efferent neurons release acetylcholine (ACh) and autonomic efferent neurons release either acetylcholine or norepinephrine (NE).

Structure of the Autonomic Nervous System (p. 534)
1. The autonomic nervous system consists of visceral efferent neurons organized into nerves, ganglia, and plexuses.
2. It is entirely motor. All autonomic axons are efferent fibers.
3. Efferent neurons are preganglionic (with myelinated axons) and postganglionic (with unmyelinated axons).
4. The autonomic system consists of two principal divisions: sympathetic (thoracolumbar) and parasympathetic (craniosacral).
5. Autonomic ganglia are classified as sympathetic trunk ganglia (on sides of spinal column), prevertebral ganglia (anterior to spinal column), and terminal ganglia (near or inside visceral effectors).

Physiology of the Autonomic Nervous System (p. 541)
1. Autonomic fibers release neurotransmitters at synapses. On the basis of the neurotransmitter produced, these fibers may be classified as cholinergic or adrenergic.
2. Cholinergic fibers release acetylcholine (ACh). Adrenergic fibers produce norepinephrine (NE).

3. Acetylcholine (ACh) interacts with nicotinic receptors on postganglionic neurons and muscarinic receptors on certain visceral effectors.
4. Norepinephrine (NE) generally interacts with alpha receptors on visceral effectors, and epinephrine generally interacts with alpha and beta receptors on visceral effectors.
5. Sympathetic responses are widespread and, in general, concerned with energy expenditure. Parasympathetic responses are restricted and are typically concerned with energy restoration and conservation.

Visceral Autonomic Reflexes (p. 544)
1. A visceral autonomic reflex adjusts the activity of a visceral effector.
2. A visceral autonomic reflex arc consists of a receptor, afferent neuron, association neuron, visceral efferent preganglionic neuron, visceral efferent postganglionic neuron, and visceral effector.

Control by Higher Centers (p. 545)
1. The hypothalamus controls and integrates the autonomic nervous system. It is connected to both the sympathetic and the parasympathetic divisions.
2. Biofeedback is a process in which people learn to monitor visceral functions and to control them consciously. It has been used to control heart rate, alleviate migraine headaches, and make childbirth easier.
3. Yoga is a higher consciousness achieved through a fully rested and relaxed body and a fully awake and relaxed mind.
4. Transcendental mediation (TM) produces the following physiological responses: decreased oxygen consumption and carbon dioxide elimination, reduced metabolic rate, decrease in heart rate, increase in the intensity of alpha brain waves, a sharp decrease in the amount of lactic acid in the blood, and an increase in the skin's electrical resistance.

REVIEW QUESTIONS

1. What are the principal components of the autonomic nervous system? What is its general function? Why is it called involuntary?
2. What are the principal differences between the voluntary nervous system and the autonomic nervous system?
3. Relate the role of visceral efferent fibers and visceral effectors to the autonomic nervous system.
4. Distinguish between preganglionic neurons and postganglionic neurons with respect to location and function.
5. What is an autonomic ganglion? Describe the location and function of the three types of autonomic ganglia. Define white and gray rami communicantes.
6. On what basis are the sympathetic and parasympathetic divisions of the autonomic nervous system differentiated anatomically and functionally?
7. Discuss the distinction between cholinergic and adrenergic fibers of the autonomic nervous system.
8. How is acetylcholine (ACh) related to nicotinic and muscarinic receptors?
9. How are alpha and beta receptors related to norepinephrine (NE) and epinephrine?
10. Give examples of the antagonistic effects of the sympathetic and parasympathetic divisions of the autonomic nervous system.
11. Summarize the principal functional differences between the voluntary nervous system and the autonomic nervous system.
12. Give the *sympathetic response* in a fear situation for each of the following body parts: hair follicles, iris of eye, lungs, spleen, adrenal medulla, kidneys, urinary bladder, stomach, intestines, gallbladder, liver, heart, arterioles of the abdominal viscera, skeletal muscles, and skin and mucosa.
13. Define a visceral autonomic reflex and give three examples.
14. Describe a complete visceral autonomic reflex in proper sequence.
15. Describe how the hypothalamus controls and integrates the autonomic nervous system.
16. Define biofeedback. Explain how it could be useful.
17. What is transcendental meditation (TM)? How is the integrated response related to the autonomic nervous system?

SELF-QUIZ

1. Write S if the description applies to the sympathetic division of the autonomic nervous system (ANS), P if it applies to the parasympathetic division, and P,S if it applies to both:
   ___ a. also called thoracolumbar outflow
   ___ b. has long preganglionic fibers leading to terminal ganglia and very short postganglionic fibers
   ___ c. has relatively short preganglionic fibers and long postganglionic fibers
   ___ d. sends some preganglionic fibers thorugh cranial nerves
   ___ e. has some preganglionic fibers synapsing in sympathetic trunk ganglia
   ___ f. has more widespread effect in the body, affecting more organs
   ___ g. has some fibers running in gray rami to supply sweat glands, hair muscles, and blood vessels
   ___ h. has fibers in white rami (connecting spinal nerve with sympathetic trunk ganglia)
   ___ i. contains fibers that supply viscera with motor impulses
   ___ j. celiac and superior mesenteric ganglia are sites of postganglionic neuron cell bodies

Choose all correct answers to the following.

___ 2. Which activities are characteristic of the stress response, or fight-or-flight reaction?
   A. the liver breaks down glycogen to glucose; B. the heart rate decreases; C. kidneys increase urine production since blood is shunted to kidneys; D. there is increased blood flow to genitalia, causing erect state; E. hairs stand on end ("goose pimples") due to contraction of arrector pili muscles; F. in general, the sympathetic system is active.
___ 3. Choose all true statements about gray rami:
   A. they contain only sympathetic nerve fibers; B. they contain only postganglionic nerve fibers; C. they carry nerve impulses from sympathetic trunk ganglia to spinal nerves; D. they are located at all levels of the vertebral column (from C1 to coccyx); E. they carry nerve impulses between paravertebral ganglia and prevertebral ganglia; F. they carry preganglionic neurons from anterior ramus of spinal nerve to trunk ganglion.

___ 4. Which are structural features of the parasympathetic system?
   A. ganglia are close to the central nervous system (CNS) and distant from the effector; B. forms the craniosacral outflow; C. distributed throughout the body, including extremities; D. supplies nerves to blood vessels, sweat glands, and adrenal (suprarenal) gland; E. has some of its nerve fibers passing through paravertebral ganglia.

5. Use arrows to show whether parasympathetic (P) or sympathetic (S) fibers stimulate (↑) or inhibit (↓) each of the following activities. Use a dash (—) to indicate that there is no parasympathetic innervation. The first one is done for you.
   a. P ↓ S ↑ dilation of pupil
   b. P ___ S ___ heart rate
   c. P ___ S ___ constriction of skin blood vessels
   d. P ___ S ___ salivation and digestive organ contractions
   e. P ___ S ___ dilation of bronchioles for easier breathing
   f. P ___ S ___ contraction of urinary bladder and relaxation of internal urethral sphincter causing urination
   g. P ___ S ___ contraction of pili of hair follicles causing "goose bumps"
   h. P ___ S ___ contraction of spleen that transfers some of its blood to general circulation, causing increase in blood pressure

i. P ___ S ___   release of epinephrine and norepineph-
                  rine (NE) from adrenal medulla
j. P ___ S ___   coping with stress, "fight-or-flight" re-
                  sponse
k. P ___ S ___   conservation of energy, "rest and re-
                  pose"
l. P ___ S ___   erection of genitalia

Circle T (true) or F (false) for the following.

T F 6. Synapsing occurs in both sympathetic and parasympathetic ganglia.

T F 7. In a visceral autonomic reflex arc only one efferent
neuron is involved.

T F 8. The sympathetic system has a more widespread effect
in the body than the parasympathetic does.

T F 9. In general, the parasympathetic division of the autonomic nervous system (ANS) has long preganglionic
fibers and short postganglionic fibers.

Choose the one best answer to these questions.

___ **10.** Which statement about postganglionic neurons is false?
A. they all lie entirely outside the central nervous system
(CNS); B. their axons are nonmyelinated; C. they terminate in visceral effectors; D. their cell bodies lie in the
lateral gray matter of the cord; E. they are very short
in the parasympathetic system.

___ **11.** All of the following axons are cholinergic *except:*
A. parasympathetic preganglionic; B. parasympathetic
postganglionic; C. sympathetic preganglionic; D. sympathetic postganglionic to sweat glands; E. sympathetic postganglionic to heart muscle.

___ **12.** The autonomic nervous system (ANS) provides the chief
nervous control in which of these activities?
A. following a moving object with the eyes; B. moving
a hand reflexly from a hot object; C. typing; D. digesting
food; E. writing an essay.

___ **13.** Which of the following is not a visceral effector?
A. smooth muscle of the iris; B. pancreas; C. heart; D.
skeletal muscle; E. salivary gland.

___ **14.** The autonomic nervous system (ANS)
A. has two parts—the parasympathetic, which controls
all normal functions, and the sympathetic, which controls

the same functions but to a greater degree; B. is the
part of our nervous system controlling all reflexes; C.
does not function when the body is subjected to stress
situations; D. has two divisions that act antagonistically—
one counteracts the effects of the other; E. is that part
of our anatomy that controls the contraction of skeletal,
smooth, and cardiac muscle tissue.

___ **15.** The preganglionic autonomic nerve fibers that arise from
the thoracic and lumbar parts of the spinal cord
(1) are also called preganglionic sympathetic fibers.
(2) synapse with postganglionic fibers of the parasympathetic nervous system.
(3) form the dorsal roots of the thoracic and lumbar spinal
nerves.
(4) are also called preganglionic parasympathetic fibers.
A. (1) only; B. (2) only; C. (3) only; D. (4) only; E.
(2) and (4).

___ **16.** Which of the following statements is true regarding the
parasympathetic ganglia of the autonomic nervous system
(ANS)?
A. they contain afferent neuron cell bodies only; B. they
are located in or on the walls of the viscera to which
their nerve fibers are going; C. they lie within the spinal
cord; D. they consist of a double chain of structures
along the spinal column; E. none of the above.

___ **17.** Cholinergic fibers are thought to include
A. all preganglionic axons; B. almost all postganglionic
parasympathetic axons; C. a few postganglionic sympathetic axons; D. all axons of somatic motor neurons; E.
all of the above.

___ **18.** Which of the following would indicate increased parasympathetic activity?
A. "cotton mouth" from reduced salivation; B. increased
gastric secretion; C. rise of blood pressure; D. decreased
flow of blood through the skin; E. increased blood glucose
level.

___ **19.** The axons of neurons lying within the central nervous
system (CNS) that connect with the autonomic nervous
system (ANS) are
A. myelinated fibers; B. postganglionic fibers; C. preganglionic fibers; D. cranial nerve fibers; E. none of the
above.

20 Sensory and Motor Systems

STUDENT OBJECTIVES

1. Define a sensation and list the characteristics of sensations.
2. Classify receptors on the basis of location, stimulus detected, and simplicity or complexity.
3. List the location and function of the receptors for tactile sensations (touch, pressure, vibration), thermoreceptive sensations (heat and cold), and pain.
4. Identify the proprioceptive receptors and indicate their functions.
5. Discuss the origin, neuronal components, and destination of the posterior column and spinothalamic pathways.
6. Locate the receptors for olfaction and describe the neural pathway for smell.
7. Identify the gustatory receptors and describe the neural pathway for taste.
8. List and describe the structural divisions of the eye and identify the afferent pathway of light impulses to the brain.
9. Describe the anatomical subdivisions of the ear and the auditory pathway.
10. Identify the receptor organs for static and dynamic equilibrium and describe their neural pathway.
11. Describe how sensory input and motor responses are linked in the central nervous system.
12. Compare the course of the pyramidal and extrapyramidal motor pathways.
13. Describe the development of the eye and ear.
14. Contrast the causes and symptoms of cataracts, glaucoma, conjunctivitis, trachoma, deafness, labyrinthine disease, Ménière's syndrome, vertigo, otitis media, and motion sickness.
15. Define key medical terms associated with sensory structures.

CHAPTER OUTLINE

■ **Sensations**
Definition
Characteristics
Classification of Receptors
  Location
  Stimulus Detected
  Simplicity or Complexity
■ **General Senses**
Cutaneous Sensations
  Tactile Sensations
  Thermoreceptive Sensations
  Pain Sensations
Proprioceptive Sensations
  Receptors
Levels of Sensation
■ **Sensory Pathways**
Somatosensory Cortex
Posterior Column Pathway
Spinothalamic Pathways
Cerebellar Tracts
■ **Special Senses**
Olfactory Sensations
  Structure of Receptors
  Olfactory Pathway
Gustatory Sensations
  Structure of Receptors
  Gustatory Pathway
Visual Sensations
  Accessory Structures of Eye
  Structure of Eyeball
  Visual Pathway
Auditory Sensations and Equilibrium
  External (Outer) Ear
  Middle Ear
  Internal (Inner) Ear
  Auditory Pathway
  Mechanism of Equilibrium
■ **Motor Pathways**
Linkage of Sensory Input and Motor Responses
Motor Cortex
Pyramidal Pathways
Extrapyramidal Pathways
■ **Developmental Anatomy of the Eye and Ear**
■ **Applications to Health**
Cataract
Glaucoma
Conjunctivitis (Pinkeye)
Trachoma
Deafness
Labyrinthine Disease
Ménière's Syndrome
Vertigo
Otitis Media
Motion Sickness
■ **Key Medical Terms Associated with Sensory Structures**

SENSATIONS

The central nervous system requires a continual flow of information to regulate homeostasis and initiate appropriate responses to changes in the internal and external environments. At any given time, our brains receive and respond to many varieties of information. However, we are aware only of the information on which we consciously focus. The central nervous system selects only those bits of information that are important for the moment, and it is only those bits of information that are brought to our conscious level. There is no question that we would collapse into nervous wrecks if our consciousness were forced to deal with all the information arriving at once. The conscious mind is turned off to protect itself from overstimulation.

Your ability to sense stimuli is vital to your survival. If pain could not be sensed, burns would be common. An inflamed appendix or stomach ulcer would progress unnoticed. A lack of sight would increase the risk of injury from unseen obstacles, a loss of smell would allow harmful gas to be inhaled, a loss of hearing would prevent recognition of automobile horns, and a lack of taste would allow toxic substances to be ingested. In short, if you could not ''sense'' your environment and make the necessary homeostatic adjustments, you could not survive very well on your own.

DEFINITION

In its broadest context, *sensation* refers to a state of awareness of external or internal conditions of the body. *Perception* refers to the conscious registration of a sensory stimulus. For a sensation to occur, four prerequisites must be satisfied.

1. A *stimulus,* or change in the environment, capable of initiating a response by the nervous system must be present.
2. A *receptor or sense organ* must pick up the stimulus and transduce (convert) it to a nerve impulse. A receptor or sense organ may be viewed as specialized nervous tissue that is extremely sensitive to internal or external conditions.
3. The nerve impulse must be *conducted* along a neural pathway from the receptor or sense organ to the brain.
4. A region of the brain must *translate* the nerve impulse into a sensation.

Receptors are capable of converting a specific stimulus into a nerve impulse. The stimulus may be light, heat, pressure, mechanical energy, or chemical energy. Each stimulus is capable of causing the membrane of the receptor to depolarize. This depolarization is called a *generator (receptor) potential.*

The generator potential is a graded response within limits, the magnitude increases with stimulus strength and fre-

quency. When the generator potential reaches the threshold level, it initiates a nerve impulse (nerve action potential). Once initiated, the nerve impulse is propagated along the nerve fiber. Whereas a generator potential is a local graded response, a nerve impulse obeys the all-or-none principle. The function of a generator potential is to transduce a stimulus into a nerve impulse.

A receptor may be quite simple. It may consist of the dendrites of a single neuron in the skin that are sensitive to pain stimuli; or it may be contained in a complex organ such as the eye. Regardless of complexity, all sense receptors contain the dendrites of sensory neurons. The dendrites occur either alone or in close association with specialized cells of other tissues.

Once a stimulus is received by a receptor and converted into a nerve impulse, the impulse is conducted along an afferent pathway that enters either the spinal cord or the brain. Many sensory impulses are conducted to the sensory areas of the cerebral cortex. It is in this region that stimuli produce conscious sensations. Sensory impulses that terminate in the spinal cord or brain stem can initiate motor activities but typically do not produce conscious sensation. The thalamus detects pain sensations but cannot distinguish the intensity or location from which they arise. This is a function of the cerebrum.

CHARACTERISTICS

Most conscious sensations or perceptions occur in the cortical regions of the brain. In other words, you see, hear, and feel in the brain. You seem to see with your eyes, hear with your ears, and feel pain in an injured part of your body only because the cortex interprets the sensation as coming from the stimulated sense receptor. The term *projection* describes this process by which the brain refers sensations to their point of stimulation.

A second characteristic of many sensations is *adaptation,* that is, a decrease in sensitivity to continued stimuli. In fact, the perception of a sensation may actually disappear, even though the stimulus is still being applied. For example, when you first get into a tub of hot water, you probably feel a burning sensation, but soon the sensation decreases to one of comfortable warmth, even though the stimulus (hot water) is still present. In time, the sensation of warmth disappears completely. Other examples of adaptation include placing a ring on your finger, putting on your shoes or hat, sitting on a chair, and pushing your glasses up onto the top of your head. Adaptation results from a change in a receptor, a change in a structure associated with a receptor, or inhibitory feedback from the brain. Receptors vary in their ability to adapt. *Rapidly adapting (phasic) receptors,* such as those associated with pressure, touch, and smell, adapt very quickly. Such receptors play a major role in signaling changes in a particular sensation. *Slowly adapting* (tonic) receptors, such as those associated with pain, body position, and detecting chemicals in blood, adapt slowly. These receptors are important in signaling information regarding steady states of the body.

Sensations may also be characterized by *afterimages;* that is, some sensations, persist even though the stimulus has been removed. This phenomenon is the reverse of adaptation. One common example of afterimage occurs when you look at a bright light and then look away or close your eyes. You still see the light for several seconds or minutes afterward.

Another characteristic of sensations is *modality:* the specific type of sensation felt. The sensation may be one of pain, pressure, touch, body position, equilibrium, hearing, vision, smell, or taste. In other words, the distinct property by which one sensation may be distinguished from another is its modality.

CLASSIFICATION OF RECEPTORS

Location

One convenient method of classifying receptors is by their location. *Exteroceptors* (eks'-ter-ō-SEP-tors) provide information about the external environment. They are sensitive to stimuli outside the body and transmit sensations of hearing, sight, smell, taste, touch, pressure, temperature, and pain. Exteroceptors are located near the surface of the body.

Visceroceptors (vis'-er-ō-SEP-tors), or *enteroceptors,* provide information about the internal environment. These sensations arise from within the body and may be felt as pain, pressure, fatigue, hunger, thirst, and nausea. Visceroceptors are located in blood vessels and viscera.

Proprioceptors (prō'-prē-ō-SEP-tors) provide information about body position and movement. Such sensations give us information about muscle tension, the position and tension of our joints, and equilibrium. These receptors are located in muscles, tendons, joints, and the internal ear.

Stimulus Detected

Another method of classifying receptors is by the type of stimuli they detect. *Mechanoreceptors* detect mechanical deformation of the receptor itself or in adjacent cells. Stimuli so detected include those related to touch, pressure, vibration, proprioception, hearing, equilibrium, and blood pressure. *Thermoreceptors* detect changes in temperature. *Nociceptors* detect pain, usually as a result of physical or chemical damage to tissues. *Electromagnetic (photo) receptors* detect light on the retina of the eye. *Chemoreceptors* detect taste in the mouth, smell in the nose, and chemicals in body fluids, such as oxygen, carbon dioxide, water, and glucose.

Simplicity or Complexity

As will be described shortly, receptors may also be classified according to the simplicity or complexity of their structure and the neural pathway involved. *Simple receptors* and neural pathways are associated with *general senses.* The receptors for general sensations are numerous and widespread. Examples include cutaneous sensations such as touch, pressure, vibration, heat, cold, and pain. *Complex receptors* and neural pathways are associated with *special senses.* The receptors for each special sense are found in only one or two specific areas of the body. Among the special senses are smell, taste, sight, and hearing.

GENERAL SENSES

CUTANEOUS SENSATIONS

Cutaneous sensations include tactile sensations (touch, pressures, vibration), thermoreceptive sensations (cold and heat), and pain. The receptors for these sensations are in the skin, connective tissue, and the ends of the gastrointestinal tract.

The cutaneous receptors are distributed over the body surface in such a way that certain parts of the body are densely populated with receptors and other parts contain only a few. This clustering of receptors is called *punctate distribution.* Areas of the body that have few cutaneous receptors are insensitive; those containing many are very sensitive.

Cutaneous receptors have simple structures. They consist of the dendrites of sensory neurons that may or may not be enclosed in a capsule of epithelial or connective tissue. Nerve impulses generated by cutaneous receptors pass along somatic afferent neurons in spinal and cranial nerves, through the thalamus, to the general sensory area of the parietal lobe of the cortex.

Tactile Sensations

Even though the *tactile sensations* are divided into separate sensations of touch, pressure, and vibrations, they are all detected by the same types of receptors—mechanoreceptors—receptors subject to deformation.

● *Touch*   Touch sensations generally result from stimulation of tactile receptors in the skin or tissues immediately beneath the skin. *Light touch* refers to the ability to perceive that something has touched the skin, although its exact location, shape, size, or texture cannot be determined. *Discriminative touch* refers to the ability to recognize exactly what point of the body is touched.

Tactile receptors for touch include root hair plexuses, free nerve endings, tactile discs, corpuscles of touch, and type II cutaneous mechanoreceptors (Figure 20-1). **Root hair plexuses** are dendrites arranged in networks around the roots of hairs. They are not surrounded by supportive or protective structures. If a hair shaft is moved, the dendrites are stimulated. Root hair plexuses detect movements mainly on the surface of the body when hairs are disturbed.

Other receptors that are not surrounded by supportive or protective structures are called *free (naked) nerve endings*. Free nerve endings are found everywhere in the skin and many other tissues. Although they are very important pain receptors, free nerve endings also respond to objects that are in continuous contact with the skin, such as clothing.

Tactile, or **Merkel's** (MER-kelz), **disc** are modified epidermal cells in the stratum basale of hairless skin. Their basal ends are in contact with dendrites of sensory neurons. Tactile discs are distributed in many of the same locations as corpuscles of touch and also function in discriminative touch.

Corpuscles of touch, or **Meissner's** (MĪS-nerz) **corpuscles**, are egg-shaped receptors for discriminative touch containing a mass of dendrites enclosed by connective tissue. They are located in the dermal papillae of the skin and are most numerous in the fingertips, palms of the hands,

and soles of the feet. They are also abundant in the eyelids, tip of the tongue, lips, nipples, clitoris, and tip of penis.

Type II cutaneous mechanoreceptors, or **end-organs of Ruffini**, are embedded deeply in the dermis and in deeper tissues of the body. They detect heavy and continuous touch sensations.

● *Pressure   Pressure sensations* generally result from stimulation of tactile receptors in deeper tissues and are longer lasting and have less variation in intensity than touch sensations. Pressure is really sustained touch. Moreover, pressure is felt over a larger area than touch.

Pressure receptors are free (naked) nerve endings, type II cutaneous mechanoreceptors, and lamellated corpuscles. **Lamellated**, or **Pacinian** (pa-SIN-ē-an), **corpuscles** (Figure 20-1) are oval structures composed of a capsule resembling an onion that consist of connective tissue layers enclosing dendrites. Lamellated corpuscles are located in the subcutaneous tissue under the skin, the deep subcutaneous tissues that lie under mucous membranes, in serous membranes, around joints and tendons, in the perimysium of muscles, in the mammary glands, in the external genitalia of both sexes, and in certain viscera.

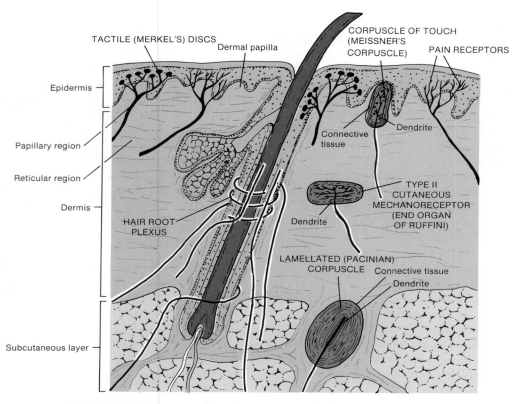

FIGURE 20-1 Structure and location of cutaneous receptors.

● *Vibration*   *Vibration sensations* result from rapidly repetitive sensory signals from tactile receptors.

The receptors for vibration sensations are corpuscles of touch and lamellated corpuscles. Whereas corpuscles of touch detect low-frequency vibration, lamellated corpuscles detect higher-frequency vibration.

Thermoreceptive Sensations

The *thermoreceptive sensations* are heat and cold. The exact nature of *thermoreceptive receptors* is not known, but they might be free (naked) nerve endings.

Pain Sensations

Pain is indispensable for a normal life. It provides us with information about tissue-damaging stimuli and thus often enables us to protect ourselves from greater damage. It is pain that initiates our search for medical assistance, and it is our subjective description and indication of the location of the pain that helps pinpoint the underlying cause of disease.

The receptors for *pain,* called *nociceptors* (nō-sē-SEP-tors), are free (naked) nerve endings, the branching ends of the dendrites of certain sensory neurons (Figure 20-1). Pain receptors are found in practically every tissue of the body. They may respond to any type of stimulus. When stimuli for other sensations, such as touch, pressure, heat, and cold, reach a certain threshold, they stimulate the sensation of pain as well. Excessive stimulation of a sense organ causes pain. Additional stimuli for pain receptors include excessive distension or dilation of a structure, prolonged muscular contractions, muscle spasms, inadequate blood flow to an organ, or the presence of certain chemical substances. Pain receptors, because of their sensitivity to all stimuli, perform a protective function by identifying changes that may endanger the body. Pain receptors adapt only slightly or not at all. Adaptation is the decrease or disappearance of the perception of a sensation even though the stimulus is still present. If there were adaptation to pain, it would cease to be sensed and irreparable damage could result.

Sensory impulses for pain are conducted to the central nervous system along spinal and cranial nerves. The lateral spinothalamic tracts of the spinal cord relay impulses to the thalamus. From here the impulses may be relayed to the postcentral gyrus of the parietal lobe. Recognition of the kind and intensity of most pain is ultimately localized in the cerebral cortex. Some awareness of pain occurs at subcortical levels.

Pain may be classified on the basis of rapidity of onset and duration: acute and slow. *Acute pain* occurs very rapidly, usually within 0.1 sec after a stimulus is applied and is not felt in deeper tissues of the body. This type of pain is also known as sharp, fast, and pricking pain. The pain felt from a needle puncture or knife cut to the skin are examples of acute pain. *Chronic pain,* by contrast, begins after a second or more and then gradually increases over a period of several seconds or minutes. This type of pain may be excruciating and is also referred to as burning, aching, throbbing, and slow pain. Chronic pain can occur both in the skin and deeper tissues or internal organs.

Pain may also be divided into two types on the basis of the location of the stimulated receptors somatic and visceral. *Somatic pain* arises from stimulation of receptors in the skin, in which case it is called *superficial somatic pain,* or from stimulation of receptors in skeletal muscles, joints, tendons, and fascia, then called *deep somatic pain.* *Visceral pain* results from stimulation of receptors in the viscera.

Although receptors for somatic and visceral pain are similar, viscera do not evoke the same pain response as somatic tissue. For example, highly *localized* damage to certain viscera, such as cutting the intestine in two in a patient who is awake, causes very little, if any, pain. But, if stimulation is *diffuse,* involving large areas, visceral pain can be severe. Such stimulation might include distension, spasms, or ischemia. There are even some viscera that are almost entirely insensitive to pain of any type. Examples are the parenchyma of the liver and alveoli of the lungs.

The ability of the cerebral cortex to locate the origin of pain is related to past experience. In most instances of somatic pain and in some instances of visceral pain, the cortex accurately projects the pain back to the stimulated area. If you burn your finger, you feel the pain in your finger. If the pleural membranes are inflamed, you experience pain in the chest. In most instances of visceral pain, however, the sensation is not projected back to the point of stimulation. Rather, the pain may be felt in or just under the skin that overlies the stimulated organ. The pain may also be felt in a surface area far from the stimulated organ. This phenomenon is called *referred pain.* In general, the area to which the pain is referred and the visceral organ involved receive their innervation from the same segment of the spinal cord. Consider the following example. Afferent fibers from the heart as well as from the skin over the heart and along the medial aspect of the left upper extremity enter spinal cord segments T1–T4. Thus, the pain of a heart attack is typically felt in the skin over the heart and along the left arm. Figure 20-2 illustrates cutaneous regions to which visceral pain may be referred.

CLINICAL APPLICATION

A kind of pain frequently experienced by patients who have had a limb amputated is called *phantom pain.* They still experience pain or other sensations in the extremity as if the limb were still there. This probably occurs because the remaining proximal portions of the

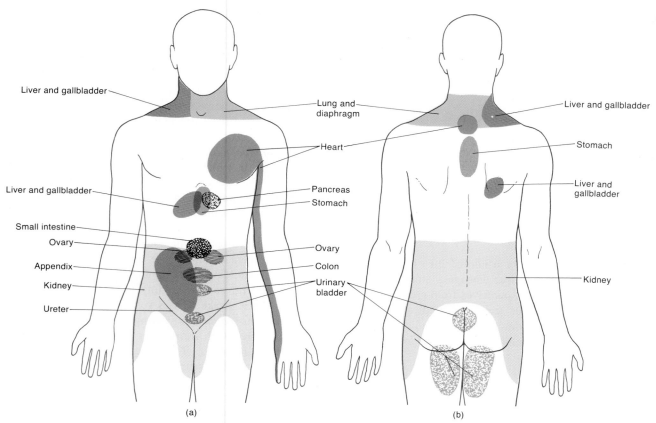

FIGURE 20-2 Referred pain. The colored parts of the diagrams indicate cutaneous areas to which visceral pain is referred. (a) Anterior view. (b) Posterior view.

sensory nerves that previously received nerve impulses from the limb are being stimulated by the trauma of the amputation. Stimuli from these nerves are interpreted by the brain as coming from the nonexistent (phantom) limb.

Pain sensations may be controlled by interrupting the pain impulse between the receptors and the interpretation centers of the brain. This may be done chemically, surgically, or by other means. Most pain sensations respond to pain-reducing drugs, which, in general, act to inhibit nerve impulse conduction at synapses.

Occasionally, however, pain may be controlled only by surgery. The purpose of surgical treatment is to interrupt the pain impulse somewhere between the receptors and the interpretation centers of the brain by severing the sensory nerve, its spinal root, or certain tracts in the spinal cord or brain. *Sympathectomy* is excision of portions of the neural tissue from the autonomic nervous system; *cordotomy* is severing a spinal cord tract, usually the lateral spinothalamic; *rhizotomy* is the cutting of sensory nerve roots; *prefrontal lobotomy* is the destruction of the tracts that connect the thalamus with the prefrontal and frontal lobes of the cerebral cortex. In each instance, the pathway for

pain is severed so that pain impulses are no longer conducted to the cortex.

Another method of inhibiting pain impulses is *acupuncture* (*acus* = needle; *pungere* = sting). Needles are inserted through selected areas of the skin and then twirled by the acupuncturist or by a mechanical device. After 20–30 minutes, pain is deadened for 6–8 hours. The location of needle insertion depends on the part of the body the acupuncturist wishes to anesthetize. To pull a tooth, a needle is inserted in the web between thumb and index finger. For a tonsillectomy, a needle is inserted approximately 5 cm (2 in.) above the wrist. For removal of a lung, a needle is placed in the forearm midway between wrist and elbow.

Acupuncture probably works by taking advantage of the body's natural inhibitory influences that can normally block pain pathways. For example, it has been established that sensory pain fibers in posterior root ganglia release a neurotransmitter called substance P. This substance is required by neurons to produce sensations that result in pain. Some neurons near those in the pain pathway release enkephalin. The enkephalin released by these small neurons blocks the release of substance P from the nerve terminals of the sensory pain fibers and therefore inhibits pain transmission to the brain. Acupuncture enhances the release of these inhibitory substances, such as enkephalin, from various locations, and these substances can then be carried by the circulatory system to the pain fibers. The onset of the effect of acupuncture is delayed until the enkephalin level rises to inhibitory levels. Similarly, the effect of acupuncture lingers after the twirling or vibration of the needles stops.

Currently, acupuncture is used in the United States mostly for childbirth, tic douloureux, arthritis, and other nonsurgical conditions. It is also being used in spinal cord injury rehabilitation to activate useful motion in limbs, provided that there is still some intact neural tissue.

PROPRIOCEPTIVE SENSATIONS

An awareness of the activities of muscles, tendons, and joints and of equilibrium is provided by the *proprioceptive,* or *kinesthetic* (kin'-es-THET-ik), *sense.* It informs us of the degree to which muscles are contracted, the amount of tension created in the tendons, the change of position of a joint, and the orientation of the head relative to the ground and in response to movements (equilibrium). The proprioceptive sense enables us to recognize the location and rate of movement of one body part in relation to others. It also allows us to estimate weight and determine the muscular work necessary to perform a task. With the proprioceptive sense, we can judge the position and movements of our limbs without using our eyes when we walk, type, or dress in the dark.

Receptors

Proprioceptive receptors are located in skeletal muscles, tendons in and around synovial joints, and the internal ear.

● *Muscle Spindles*　*Muscle spindles* are delicate proprioceptive receptors interspersed among skeletal muscle fibers (cells) and oriented parallel to the fibers (Figure 20-3a). The ends of the spindles are anchored to the endomysium and perimysium. Muscle spindles consist of three to ten specialized muscle fibers called *intrafusal fibers,* which are partially enclosed in a connective tissue capsule that is filled with lymph. The spindles are surrounded by skeletal muscle fibers of the muscle called *extrafusal fibers.* The central region of each intrafusal fiber has few or no actin and myosin myofilaments and an accumulation of nuclei. In some intrafusal fibers the nuclei bunch at the center (*nuclear bag fibers*); in others, the nuclei form a chain at the center (*nuclear chain fibers*). The central region of the intrafusal fibers cannot contract and represents the sensory receptor area for a spindle.

Although the central receptor area cannot contract because it lacks myofilaments, it does contain two types of afferent (sensory) fibers. A large sensory fiber, called a *type Ia fiber,* innervates the exact center of the intrafusal fibers. The branches of the Ia fiber, called *primary (annulospiral) endings,* wrap around the center of the intrafusal fibers. When the central part of the spindle is stretched,

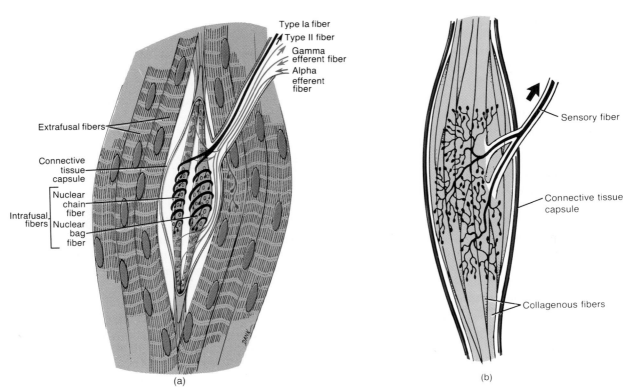

FIGURE 20-3 Proprioceptive receptors. (a) Muscle spindle. (b) Tendon organ.

the primary endings are stimulated and send nerve impulses to the spinal cord at exceedingly great velocities. The central receptor area of most muscle spindles is also innervated by two sensory fibers called *type II fibers.* Their branches, known as *secondary (flower spray) endings,* are located on either side of the primary ending. Secondary endings are also stimulated when the central part of the spindle is stretched, and they too send nerve impulses to the spinal cord.

The ends of the intrafusal fibers contain actin and myosin myofilaments and represent the contractile portions of the fibers. The ends of the fibers contract when stimulated by *gamma efferent (motor) neurons.* These neurons are small motor neurons located in the anterior gray horn of the spinal cord. The fibers of some gamma efferent neurons terminate as motor end plates on the ends of the intrafusal fibers. Extrafusal fibers are innervated by large motor neurons called *alpha efferent neurons.* These neurons are also located in the anterior gray horn of the spinal cord near gamma efferent neurons.

Muscle spindles are stimulated in response to both sudden and prolonged stretch on the central areas of the intrafusal fibers. The muscle spindles monitor changes in the length of a skeletal muscle by responding to the rate and degree of change in length. This information is relayed to the central nervous system to assist in the coordination and efficiency of muscle contraction.

● *Tendon Organs*   Tendon organs *(Golgi tendon organs)* are proprioceptive receptors found at the junction of a tendon with a muscle. They help protect tendons and their associated muscles from damage resulting from excessive tension and also function as receptors for contraction. Each consists of a thin capsule of connective tissue that encloses a few collagenous fibers (Figure 20-3b). The capsule is penetrated by one or more sensory neurons whose terminal branches entwine among and around the collagenous fibers. When tension is applied to a tendon, tendon organs are stimulated and the information is relayed to the central nervous system.

● *Joint Kinesthetic Receptors*   There are several types of *joint kinesthetic receptors* within and around the articular capsules of synovial joints. Encapsulated receptors, similar to type II cutaneous mechanoreceptors (end organs of Ruffini), are present in the capsules of joints and respond to pressure. Small lamellated (Pacinian) corpuscles in the connective tissue outside articular capsules are receptors that respond to acceleration and deceleration. Articular ligaments contain receptors similar to tendon organs that mediate reflex inhibition of the adjacent muscles when excessive strain is placed on the joint.

● *Maculae and Cristae*   The proprioceptors in the internal ear are the macula of the saccule and the utricle and cristae in the semicircular ducts. Their function in equilibrium is discussed later in the chapter.

Proprioceptors adapt only slightly. This feature is advantageous since the brain must be apprised of the status of different parts of the body at all times so that adjustments can be made to ensure coordination.

The afferent pathway for muscle sense consists of impulses generated by proprioceptors via cranial and spinal nerves to the central nervous system (see Figure 20-5). Impulses for conscious proprioception pass along ascending tracts in the cord, where they are relayed to the thalamus and cerebral cortex. The sensation is registered in the general sensory area in the parietal lobe of the cerebral cortex posterior to the central sulcus. Proprioceptive impulses that have resulted in reflex action pass to the cerebellum along spinocerebellar tracts.

LEVELS OF SENSATION

As we have said, a receptor converts a stimulus into a nerve impulse, and only after that impulse has been conducted to a region of the spinal cord or brain can it be translated into a sensation. The nature of the sensation and the type of reaction generated vary with the level of the central nervous system at which the sensation is translated.

Sensory fibers terminating in the spinal cord can generate spinal reflexes without immediate action by the brain. Sensory fibers terminating in the lower brain stem bring about far more complex motor reactions than simply spinal reflexes. When sensory impulses reach the lower brain stem, they cause subconscious motor reactions. Sensory impulses that reach the thalamus can be localized crudely in the body. At the thalamic level sensations are sorted by modality, that is, identified as the *specific* sensation of touch, pressure, pain, position, hearing, vision, smell, or taste. When sensory information reaches the cerebral cortex, we experience precise localization. It is at this level that memories of previous sensory information are stored and the perception of sensation occurs on the basis of past experience.

SENSORY PATHWAYS

SOMATOSENSORY CORTEX

Sensory information from receptors on one side of the body crosses over to the opposite side in the spinal cord or brain stem and then to the *somatosensory cortex (primary somesthetic* or *general sensory area)* of the cerebral cortex where conscious sensations are produced (see Figure 18-11). Areas of the somatosensory cortex have been mapped out that represent the termination of sensory information from all parts of the body. Figure 20-4 shows the location and areas of representation of the somatosensory cortex of the right cerebral hemisphere. The left cerebral hemisphere has a duplicate somatosensory cortex.

Note that some parts of the body are represented by

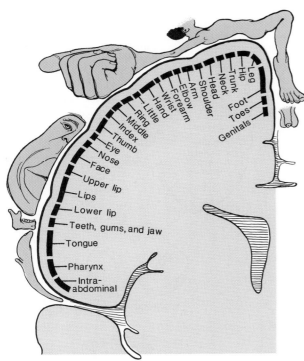

FIGURE 20-4 Somatosensory cortex of the right cerebral hemisphere.

large areas in the somatosensory cortex. These include the lips, face, and thumb. Other parts of the body, such as the trunk and lower extremities, are represented by relatively small areas. The relative sizes of the somatosensory cortex are directly proportional to the number of specialized sensory receptors in each respective part of the body. Thus, there are numerous receptors in the skin of the lips, but relatively few in the skin of the trunk. Essentially, the size of the area for a particular part of the body is determined by the functional importance of the part and its need for sensitivity.

Let us now examine how sensory information is transmitted from receptors to the central nervous system. You will find it helpful to review the principal ascending and descending tracts of the spinal cord (see Exhibit 17-1 and Figure 17-4). Sensory information transmitted from the spinal cord to the brain is conducted along two general pathways: the posterior column pathway and the spinothalamic pathways.

POSTERIOR COLUMN PATHWAY

In the *posterior column pathway* (*fasciculus gracilis* and *fasciculus cuneatus*) to the cerebral cortex there are three separate sensory neurons (Figure 20-5). The *first-order neuron* connects the receptor with the spinal cord and medulla on the same side of the body. The cell body of the first-order neuron is in the posterior root ganglion of a spinal nerve. The first-order neuron synapses with a *second-order neuron,* which passes from the medulla upward to the thalamus. The cell body of the second-order neuron is located

in the nuclei cuneatus or gracilis of the medulla. Before passing into the thalamus, the second-order neuron crosses to the opposite side of the medulla and enters the medial lemniscus, a projection tract that terminates at the thalamus. In the thalamus, the second-order neuron synapses with a *third-order neuron,* which terminates in the somesthetic sensory area of the cerebral cortex.

The posterior column pathway conducts nerve impulses related to proprioception, discriminative touch, two-point discrimination, and vibrations.

SPINOTHALAMIC PATHWAYS

The *spinothalamic pathways* are also composed of three orders of sensory neurons (Figures 20-6 and 20-7) The first-order neuron connects a receptor of the neck, trunk,

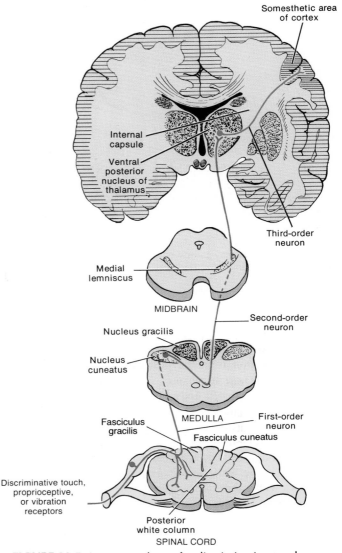

FIGURE 20-5 Sensory pathway for discriminative touch, proprioception, and vibration—the posterior column pathway.

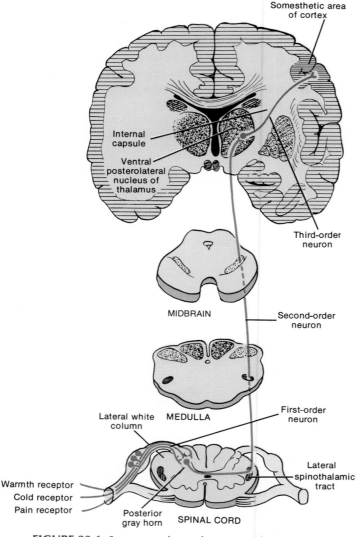

FIGURE 20-6 Sensory pathway for pain and temperature—the lateral spinothalamic pathway.

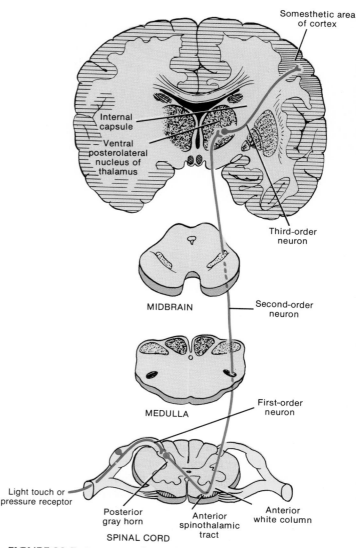

FIGURE 20-7 Sensory pathway for light touch and pressure—the anterior spinothalamic pathway.

and extremities with the spinal cord. The cell body of the first-order neuron is in the posterior root ganglion also. The first-order neuron synapses with the second-order neuron, which has its cell body in the posterior gray horn of the spinal cord. The fiber of the second-order neuron crosses to the opposite side of the spinal cord and passes upward to the brain stem in the lateral spinothalamic tract or anterior spinothalamic tract. The fibers from the second-order neuron terminate in the thalamus. There the second-order neuron synapses with a third-order neuron. The third-order neuron terminates in the somesthetic sensory area of the cerebral cortex. The spinothalamic pathways convey sensory impulses for pain and temperature (lateral spinothalamic) as well as light touch and pressure (anterior spinothalamic).

CEREBELLAR TRACTS

Both the ***posterior spinocerebellar tract*** and ***anterior spinocerebellar tract*** participate in conveying impulses concerned with subconscious proprioception and thus assume a role in posture and muscle tone (see Exhibit 17-1).

SPECIAL SENSES

The special senses—smell, taste, sight, hearing, and equilibrium—have receptor organs that are structurally more complex than receptors for general sensations. The sense of smell is the least specialized, as opposed to the sense of

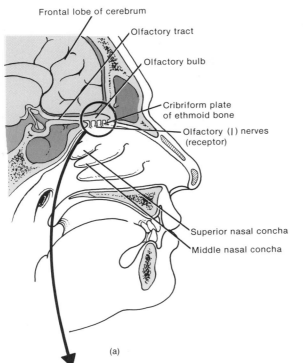

(a)

sight, which is the most specialized. Like the general senses, however, the special senses allow us to detect changes in our environment.

OLFACTORY SENSATIONS

Structure of Receptors

The receptors for the *olfactory* (ol-FAK-tō-rē) *sense,* or sense of smell, are located in the nasal epithelium in the superior portion of the nasal cavity on either side of the nasal septum (Figure 20-8). The nasal epithelium consists of three principal kinds of cells: supporting, olfactory, and basal. The *supporting (sustentacular) cells* are columnar epithelial cells of the mucous membrane lining the nose. The *olfactory cells* are bipolar neurons whose cell bodies lie between the supporting cells. The distal (free) end of each olfactory cell contains a dendrite that terminates in a swelling (*olfactory vesicle*) from which six to eight cilia, called *olfactory hairs,* radiate. The hairs are believed to

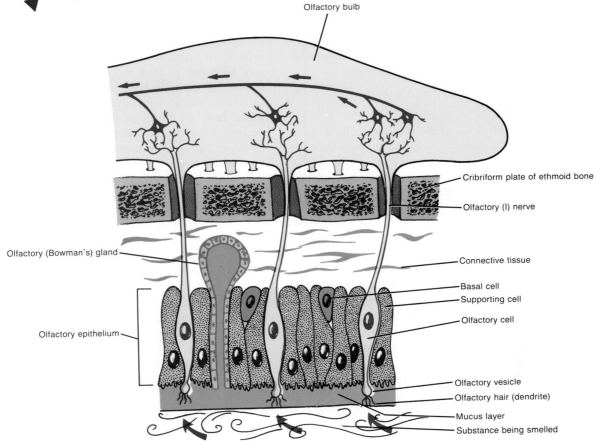

(b)

FIGURE 20-8 Olfactory receptors. (a) Location of receptors in nasal cavity. (b) Enlarged aspect of olfactory receptors.

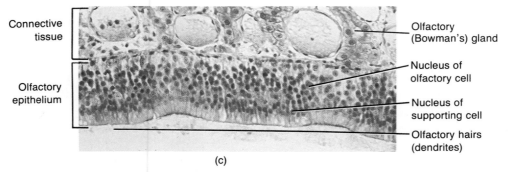

Connective tissue

Olfactory epithelium

Olfactory (Bowman's) gland

Nucleus of olfactory cell

Nucleus of supporting cell

Olfactory hairs (dendrites)

(c)

FIGURE 20-8 (*Continued*) (c) Photomicrograph of the olfactory mucosa at a magnification of 300×. (Copyright © 1985 by Michael H. Ross. Used by permission).

react to odors in the air and then to stimulate the olfactory cells, thus initiating the olfactory response. The proximal (basal) part of each olfactory cell contains a single process that represents its axon. *Basal cells* lie between the bases of the supporting cells and are believed to produce new supporting cells. Within the connective tissue beneath the olfactory epithelium are *olfactory (Bowman's) glands* that produce mucus, which is carried to the surface of the epithelium by ducts. The secretion moistens the surface of the olfactory epithelium and serves as a solvent for odoriferous substances. The continuous secretion of mucus also serves to freshen the surface film of fluid and prevents continuous stimulation of olfactory hairs by the same odor.

Olfactory Pathway

The unmyelinated axons of the olfactory cells unite to form the *olfactory (I) nerves,* which pass through foramina in the cribriform plate of the ethmoid bone (Figure 20-8b). The olfactory (I) nerves terminate in paired masses of gray matter called the *olfactory bulbs.* The olfactory bulbs lie beneath the frontal lobes of the cerebrum on either side of the crista galli of the ethmoid bone. The first synapse of the olfactory neural pathway occurs in the olfactory bulbs between the axons of the olfactory (I) nerves and the dendrites of neurons inside the olfactory bulbs. Axons of these neurons run posteriorly to form the *olfactory tract.* From here, nerve impulses are conveyed to the primary olfactory area of the cerebral cortex. In the cerebral cortex, the impulses are interpreted as odor and give rise to the sensation of smell. Impulses related to olfaction do not pass through the thalamus.

Both the supporting cells of the nasal epithelium and tear glands are innervated by branches of the trigeminal (V) nerve. The nerve receives stimuli of pain, cold, heat, tickling, and pressure. Olfactory stimuli such as pepper, ammonia, and chloroform are irritating and may cause tearing because they stimulate the lacrimal and nasal mucosal receptors of the trigeminal (V) nerve as well as the olfactory neurons.

GUSTATORY SENSATIONS

Structure of Receptors

The receptors for *gustatory* (GUS-ta-tō'rē) *sensations,* or sensations of taste, are located in the taste buds (Figure 20-9). The nearly 2,000 taste buds are most numerous on the tongue, but they are also found on the soft palate and in the throat. The *taste buds* are oval bodies consisting of three kinds of cells: supporting, gustatory, and basal. The *supporting (sustentacular) cells* are a specialized epithelium that forms a capsule, inside each of which are 4–20 *gustatory cells.* Each gustatory cell contains a hairline process (*gustatory hair*) that projects to the external surface through an opening in the taste bud called the *taste pore.* Gustatory cells make contact with taste stimuli through the taste pore. *Basal cells* are found at the periphery of the taste bud near the basal lamina. These cells produce supporting and gustatory cells whose average life span is about 10 days.

Taste buds are found in some elevations on the tongue called *papillae* (pa-PIL-ē). The papillae give the upper surface of the tongue its rough appearance. *Circumvallate* (ser-kum-VAL-āt) or *vallate papillae,* the largest type, are circular and form an inverted V-shaped row at the posterior portion of the tongue. *Fungiform* (FUN-ji-form; meaning mushroom-shaped) *papillae* are knoblike elevations found primarily on the tip and sides of the tongue. All circumvallate and most fungiform papillae contain taste buds. *Filiform* (FIL-i-form) *papillae* are pointed threadlike structures that cover the anterior two-thirds of the tongue. They rarely contain taste buds.

Despite the many substances we seem to taste, there are basically only four primary taste sensations: sour, salt, bitter, and sweet. All other "tastes," such as chocolate, pepper, and coffee, are combinations of these four that are modified by accompanying olfactory sensations.

Each of the four primary tastes is caused by a different response to different chemicals. Certain regions of the tongue react more strongly than others to certain taste sensations. Although the tip of the tongue reacts to all four primary taste sensations, it is highly sensitive to sweet and

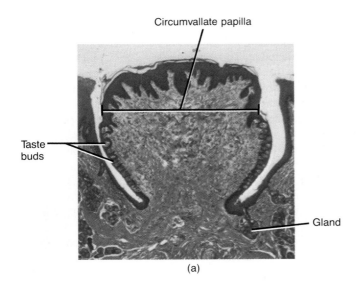

Circumvallate papilla

Taste buds

Gland

(a)

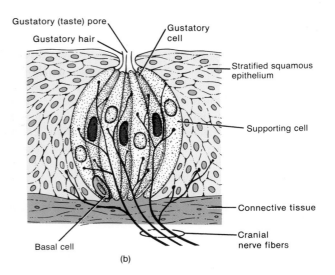

Gustatory (taste) pore
Gustatory hair
Gustatory cell
Stratified squamous epithelium
Supporting cell
Connective tissue
Cranial nerve fibers
Basal cell

(b)

salty substances. The posterior portion of the tongue is highly sensitive to bitter substances. The lateral edges of the tongue are more sensitive to sour substances (Figure 20-9c).

CLINICAL APPLICATION

Persons with colds or allergies sometimes complain that they cannot taste their food. Although their taste sensations may be operating normally, their olfactory sensations are not. This illustrates that much of *what we think of as taste is actually smell.* Odors from foods pass upward into the nasopharynx and stimulate the olfactory system. In fact, a given concentration of a substance will stimulate the olfactory system thousands of times more than it stimulates the gustatory system.

Gustatory Pathway

The cranial nerves that supply afferent fibers to taste buds are the facial (VII), which supplies the anterior two-thirds of the tongue; the glossopharyngeal (IX), which supplies the posterior one-third of the tongue; and the vagus (X), which supplies the throat and epiglottis. Taste impulses are conveyed from the gustatory cells in taste buds along the nerves to the medulla and then to the thalamus. They terminate in the primary gustatory area in the parietal lobe of the cerebral cortex.

VISUAL SENSATIONS

The structures related to vision are the eyeball, the optic (II) nerve, the brain, and a number of accessory structures. The study of the structure, function, and diseases of the eye is known as *ophthalmology* (of'-thal-MOL-ō-jē; *ophthalmo* = eye; *logos* = study of). A physician who

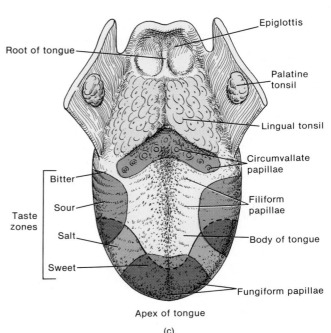

Root of tongue
Epiglottis
Palatine tonsil
Lingual tonsil
Circumvallate papillae
Filiform papillae
Body of tongue
Fungiform papillae
Bitter
Sour
Salt
Sweet
Taste zones
Apex of tongue

(c)

FIGURE 20-9 Gustatory receptors. (a) Photomicrograph of a circumvallate papilla containing taste buds at a magnification of 50×. (Copyright © 1983 by Michael H. Ross. Used by permission). (b) Diagram of the structure of a taste bud. (c) Locations of papillae and four taste zones.

specializes in the diagnosis and treatment of eye disorders with drugs, surgery, and corrective lenses is known as an *ophthalmologist,* whereas an *optometrist* is a specialist with a doctorate in optometry who is licensed to examine and test the eyes and treat visual defects by prescribing corrective lenses. An *optician* is a technician who fits, adjusts, and dispenses corrective lenses on prescription of an ophthalmologist or optometrist.

Accessory Structures of Eye

Among the *accessory structures* are the eyebrows, eyelids, eyelashes, and lacrimal (tearing) apparatus (Figure 20-10). The *eyebrows* form a transverse arch at the junction of the upper eyelid and forehead. Structurally, they resemble the hairy scalp. The skin of the eyebrows is richly supplied with sebaceous (oil) glands. The hairs are generally coarse and directed laterally. Deep to the skin of the eyebrows are the fibers of the orbicularis oculi muscles. The eyebrows help protect the eyeballs from foreign objects, perspiration, and the direct rays of the sun.

The upper and lower *eyelids,* or *palpebrae* (PAL-pe-brē), have several important roles. They shade the eyes during sleep, protect the eyes from excessive light and foreign objects, and spread lubricating secretions over the eyeballs. The upper eyelid is more movable than the lower and contains in its superior region a special levator muscle known as the *levator palpebrae superioris.* The space between the upper and lower eyelids that exposes the eyeball is called the *palpebral fissure.* Its angles are known as the *lateral commissure* (KOM-i-shūr), which is narrower and closer to the temporal bone, and the *medial commissure,* which is broader and nearer the nasal bone. In the medial commissure, there is a small reddish elevation, the *lacrimal caruncle* (KAR-ung-kul), containing sebaceous (oil) and sudoriferous (sweat) glands. A whitish material secreted by the caruncle collects in the medial commissure.

From superficial to deep, each eyelid consists of epidermis, dermis, subcutaneous areolar connective tissue, fibers of the orbicularis oculi muscle, a tarsal plate, tarsal glands, and a conjunctiva. The *tarsal plate* is a thick fold of connective tissue that forms much of the inner wall of each eyelid and gives form and support to the eyelids. Embedded in grooves on the deep surface of each tarsal plate is a row of elongated tarsal glands known as *tarsal* or *Meibomian* (mī-BŌ-mē-an) *glands.* These are modified sebaceous glands, and their oily secretion helps keep the eyelids from adhering to each other. Infection of the tarsal glands produces a tumor or cyst on the eyelid called a *chalazion* (ka-LĀ-zē-on). The *conjunctiva* (kon'-junk-TĪ-va) is a thin mucous membrane. It is called the *palpebral conjunctiva* when it lines the inner aspect of the eyelids. It is called the *bulbar (ocular) conjunctiva* when it is reflected from the eyelids onto the anterior surface of the eyeball. When the blood vessels of the bulbar conjunctiva are dilated and congested due to local irritation or infection, the person has bloodshot eyes.

Projecting from the border of each eyelid, anterior to the tarsal glands, is a row of short, thick hairs, the *eyelashes.* In the upper lid, they are long and turn upward; in the lower lid, they are short and turn downward. Sebaceous glands at the base of the hair follicles of the eyelashes called *sebaceous ciliary glands* or *glands of Zeis* (ZĪS) pour a lubricating fluid into the follicles. Infection of these glands is called a *sty.*

The *lacrimal apparatus* is a term used for a group of structures that manufactures and drains tears. These struc-

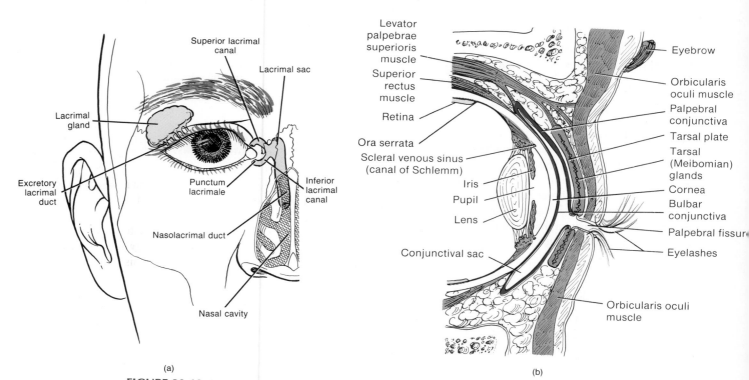

(a)

(b)

FIGURE 20-10 Accessory structures of the eye. (a) Anterior view. (b) Sagittal section of the eyelids and anterior portion of the eyeball.

tures are the lacrimal glands, the excretory lacrimal ducts, lacrimal canals, lacrimal sacs, and nasolacrimal ducts. A *lacrimal gland* is a compound tubuloacinar gland located at the superior lateral portion of each orbit. Each is about the size and shape of an almond. Leading from the lacrimal glands are 6–12 *excretory lacrimal ducts* that empty lacrimal fluid, or tears, onto the surface of the conjunctiva of the upper lid. From here the lacrimal fluid passes medially and enters two small openings called *puncta lacrimalia* that appear as two small pores, one in each papilla of the eyelid, at the medial commissure of the eye. The lacrimal secretion then passes into two ducts, the *lacrimal canals,* and is next conveyed into the lacrimal sac. The lacrimal canals are located in the lacrimal grooves of the lacrimal bones. The *lacrimal sac* is the superior expanded portion of the *nasolacrimal duct,* a canal that transports the lacrimal secretion into the inferior meatus of the nose.

The *lacrimal secretion* is a watery solution containing salts, some mucus, and a bactericidal enzyme called *lyso-zyme.* It cleans, lubricates, and moistens the eyeball. After being secreted by the lacrimal glands, it is spread medially over the surface of the eyeball by the blinking of the eyelids. Usually, 1 ml per day is produced by each gland.

The surface anatomy of the accessory structures of the eye and the eyeball is shown in Figure 11-2.

CLINICAL APPLICATION

Normally, the lacrimal secretion is carried away by evaporation or by passing into the lacrimal canals and then into the nasal cavities as fast as it is produced. If, however, an irritating substance makes contact with the conjunctiva, the lacrimal glands are stimulated to oversecrete. Tears then accumulate more rapidly than they can be carried away. This is a protective mechanism, since the tears dilute and wash away the irritating substance. *"Watery" eyes* also occur when an inflammation of the nasal mucosa, such as a cold, obstructs the nasolacrimal ducts so that drainage of tears is blocked. Humans are unique in that they have the ability to cry to express certain emotions. In response to parasympathetic stimulation, the lacrimal glands produce excessive tears that may spill over the edges of the eyelids and even fill the nasal cavity with fluid. Crying may indicate happiness or sadness.

Structure of Eyeball

The adult *eyeball* measures about 2.5 cm (1 in.) in diameter. Of its total surface area, only the anterior one-sixth is exposed. The remainder is recessed and protected by the orbit into which it fits. Anatomically, the eyeball can be divided into three layers: fibrous tunic, vascular tunic, and retina or nervous tunic (Figure 20-11).

● *Fibrous Tunic* The *fibrous tunic* is the outer coat of the eyeball. It can be divided into two regions: the posterior portion is the sclera and the anterior portion is the cornea. The *sclera* (SKLE-ra), the "white of the eye," is a white coat of dense fibrous tissue that covers all the eyeball except the anterior colored portion (iris). The sclera gives shape to the eyeball and protects its inner parts. Its posterior surface is pierced by the optic (II) nerve. The *cornea* (KOR-nē-a) is a nonvascular, transparent fibrous coat that covers the iris. The cornea's outer surface is covered by an epithelial layer continuous with the epithelium of the bulbar conjunctiva. At the junction of the sclera and cornea is a venous sinus known as the *scleral venous sinus* or *canal of Schlemm.*

CLINICAL APPLICATION

Each year, about 10,000 *corneal transplants (keratoplasty)* are performed in the United States. Corneal transplants are considered to be the most successful type of transplantation. The surgical procedure is performed under an operating microscope at magnifications of between 6× and 40×. A general or local anesthesia is used. The defective cornea, usually 7–8 mm (about 0.3 in.) in diameter, is excised with microcorneal scissors and replaced with a donor cornea of similar diameter. The transplanted cornea is sewn into place with nylon sutures. Patients are later fitted with hard or soft contact lenses or eyeglasses.

In the process of normal vision, light rays pass through the cornea, as well as other structures, to form a clear image on the retina (nervous tunic). In myopia (nearsightedness), the cornea is too curved and images of distant objects fall short of the retina, causing blurred vision. Many people with myopia may be helped by a surgical procedure called *radial keratotomy* (RK). In the procedure, several microscopic incisions, like the spikes of a wheel, are made in the cornea. This flattens the cornea and improves vision without using corrective lenses. Because of various problems associated with the procedure, (vision not fully corrected, overcorrected vision, extra sensitivity to glare at night, and changes in visual acuity during the day), it appears that it is best suited to people with mild to moderate myopia.

An alternative procedure to radial keratotomy for treating myopia is known as *epikeratoplasty* (ep'-ē-KER-a-tō-plas'-tē). In one modification of this procedure, a small circular area of defective cornea is removed and a commercially prepared piece of donor cornea, shaped to provide proper curvature, is sewn into its place.

● *Vascular Tunic* The *vascular tunic* is the middle layer of the eyeball and is composed of three portions: choroid, ciliary body, and iris. Collectively, these three structures

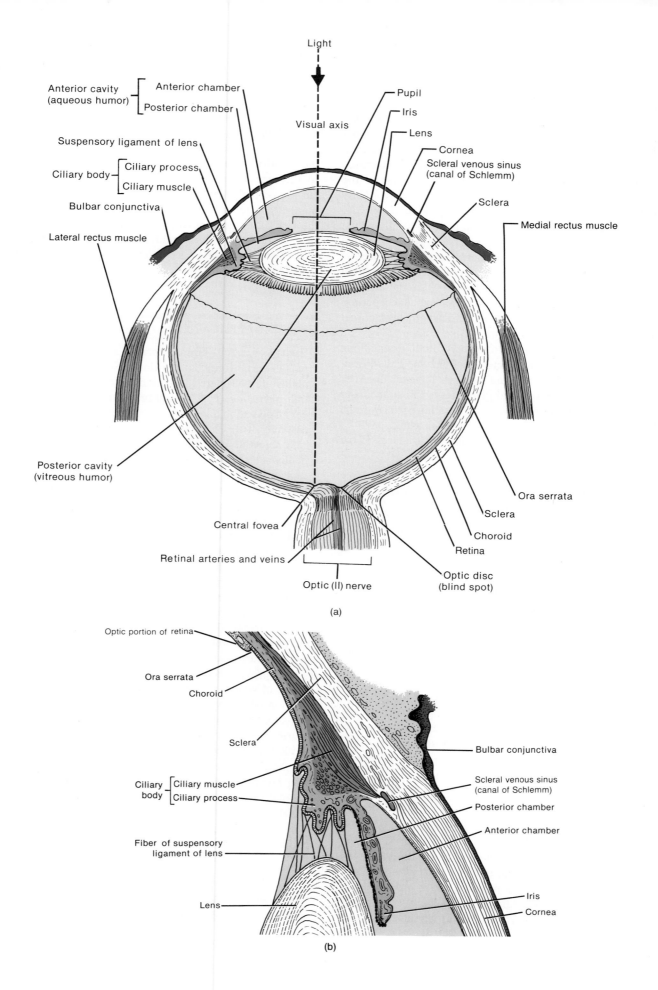

Light

Anterior cavity (aqueous humor) { Anterior chamber / Posterior chamber

Pupil
Iris
Lens

Suspensory ligament of lens

Visual axis

Cornea

Ciliary body { Ciliary process / Ciliary muscle

Scleral venous sinus (canal of Schlemm)

Bulbar conjunctiva

Sclera

Lateral rectus muscle

Medial rectus muscle

Posterior cavity (vitreous humor)

Ora serrata

Sclera

Choroid

Central fovea

Retina

Retinal arteries and veins

Optic disc (blind spot)

Optic (II) nerve

(a)

Optic portion of retina

Ora serrata

Choroid

Sclera

Bulbar conjunctiva

Ciliary body { Ciliary muscle / Ciliary process

Scleral venous sinus (canal of Schlemm)

Posterior chamber

Anterior chamber

Fiber of suspensory ligament of lens

Iris

Lens

Cornea

(b)

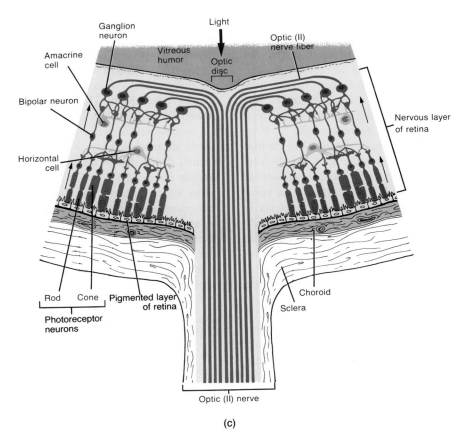

(c)

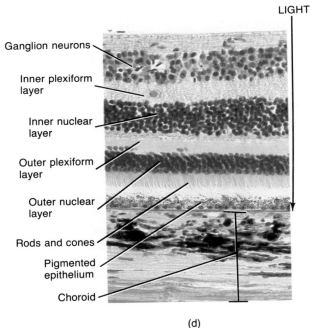

(d)

FIGURE 20-11 Structure of the eyeball. (a) Diagram of gross structure in transverse section. (b) Section through the anterior portion of the eyeball at the sclerocorneal junction. (c) Diagram of the microscopic structure of the retina, exaggerated for emphasis. The arrows pointing upward indicate the direction of the signal passing through the nervous layer of the retina, which ultimately results in a nerve impulse that passes into the optic (II) nerve. (d) Photomicrograph of a portion of the nervous tunic (retina) at a magnification of 300×. The outer nuclear layer contains nuclei of photoreceptor neurons; the outer plexiform layer is where photoreceptor and bipolar neurons synapse; the inner nuclear layer contains the nuclei of bipolar neurons; and the inner plexiform layer is where bipolar and ganglion neurons synapse. (Copyright © 1983 by Michael H. Ross. Used by permission.)

are called the *uvea* (YOO-vē-a). The *choroid* (KŌ-royd), the posterior portion of the vascular tunic, is a thin, dark brown membrane that lines most of the internal surface of the sclera. It contains numerous blood vessels and a large amount of pigment. The choroid absorbs light rays so they are not reflected within the eyeball. Through its blood supply, it nourishes the retina. Like the sclera, it is pierced by the optic (II) nerve at the back of the eyeball.

In the anterior portion of the vascular tunic, the choroid becomes the *ciliary* (SIL-ē-ar'-ē) *body*. It is the thickest portion of the vascular tunic. It extends from the *ora serrata* (Ō-ra ser-RĀ-ta) of the retina (nervous tunic) to a point just behind the sclerocorneal junction. The ora serrata is simply the jagged margin of the retina. The ciliary body consists of the ciliary processes and ciliary muscle. The *ciliary processes* consist of protrusions or folds on the internal surface of the ciliary body that secrete aqueous humor. The *ciliary muscle* is a smooth muscle that alters the shape of the lens for near or far vision.

The *iris* is the third portion of the vascular tunic. It is the colored portion of the eyeball and consists of circular and radial smooth muscle fibers (cells) arranged to form a doughnut-shaped structure. The black hole in the center of the iris is the *pupil,* the area through which light enters the eyeball. The iris is suspended between the cornea and the lens and is attached at its outer margin to the ciliary process. A principal function of the iris is to regulate the amount of light entering the eyeball. When the eye is stimulated by bright light, the circular muscles of the iris contract and decrease the size of the pupil. The pupil constricts. When the eye must adjust to dim light, the radial muscles of the iris contract and increase the pupil's size. The pupil dilates.

● *Retina (Nervous Tunic)*   The third and inner coat of the eye, the *retina (nervous tunic)*, lies only in the posterior portion of the eye. Its primary function is image formation. The retina is one of the few places in the body where blood vessels can be seen directly. This is done by use of a special reflecting light called an ophthalmoscope. It is possible for an ophthalmologist to examine the retina and detect vascular changes associated with hypertension, atherosclerosis, and diabetes. The retina consists of an inner nervous tissue layer (visual portion) and an outer pigmented layer (nonvisual portion). The retina covers the choroid. At the edge of the ciliary body, it ends in a scalloped border called the ora serrata. This is where the nervous layer or visual portion of the retina ends. The pigmented layer extends anteriorly over the back of the ciliary body and the iris as the nonvisual portion of the retina.

The nervous layer of the retina contains three zones of neurons. These three zones, named in the order in which they conduct nerve impulses, are *photoreceptor neurons, bipolar neurons,* and *ganglion neurons.* The dendrites of the photoreceptor neurons are called rods and cones because of their shapes. They are visual receptors highly specialized for stimulation by light rays. Functionally, rods and cones

develop generator potentials. *Rods* are specialized for vision in dim light. They also allow us to discriminate between different shades of dark and light and permit us to see shapes and movement. *Cones* are specialized for color vision and sharpness of vision (*visual acuity*). They are stimulated only by bright light. This is why we cannot see color by moonlight. It is estimated that there are 7 million cones and somewhere between 10 and 20 times as many rods. Cones are most densely concentrated in the **central fovea,** a small depression in the center of the macula lutea. The *macula lutea* (MAK-yoo-la LOO-tē-a), or yellow spot, is in the exact center of the posterior portion of the retina, corresponding to the visual axis of the eye. The fovea is the area of sharpest vision because of the high concentration of cones. Rods are absent from the fovea and macula and increase in density toward the periphery of the retina. It is for this reason that you can see better at night while not looking directly at an object.

CLINICAL APPLICATION

In a disease called *senile macular degeneration (SMD)* new blood vessels grow over the macula lutea. The effect ranges from distorted vision to blindness. SMD accounts for nearly all new cases of blindness in people over 65. Its cause is unknown. Laser beam treatment has been used effectively in arresting blood vessel proliferation and restoring normal vision in some cases. In order to determine if SMD exists, a simple test can be performed without any assistance or special instruments. Stare, one eye at a time (covering the other with your hand), at any long straight line, such as a door frame. If either eye perceives the line as bent or twisted, or if a black spot appears, a physician should be informed immediately.

When information has passed through the photoreceptor neurons, it is conducted across synapses to the bipolar neurons in the intermediate zone of the nervous layer of the retina. From here it is passed to the ganglion neurons. These cells transmit their signals through optic (II) nerve fibers to the brain in the form of nerve impulses. Many rods connect with one bipolar neuron, and many of these bipolar neurons transmit impulses to one ganglion cell. This greatly lowers visual acuity, but it permits summation effects to occur so that low levels of light can stimulate a ganglion cell that would not respond had it been connected directly to a cone. The synaptic connections thus contribute much to the difference in visual acuity and light sensitivity.

The axons of the ganglion neurons extend posteriorly to a small area of the retina called the *optic disc (blind spot).* This region contains openings through which the axons of the ganglion neurons exit as the optic (II) nerve. Since it contains no rods or cones, an image striking it cannot be perceived. Thus, it is called the blind spot.

CLINICAL APPLICATION

Detachment of the retina may occur in trauma, such as a blow to the head. The actual detachment occurs between the inner nervous tissue layer and outer pigmented layer. Fluid accumulates between these layers, forcing the thin, pliable retina to billow out toward the vitreous humor, resulting in distorted vision and blindness in the corresponding field of vision. The retina may be reattached by photocoagulation by laser beam, cryosurgery, or scleral resection. (A laser is an instrument for amplifying light into an intense, highly concentrated beam that is strong enough to burn tissue.)

● *Lens* In addition to the fibrous tunic, vascular tunic, and retina, the eyeball itself contains the lens, just behind the pupil and iris. The *lens* is constructed of three basic parts: lens capsule, lens epithelium, and lens fibers. The *lens capsule* is an elastic basal lamina that surrounds the entire lens. It is formed by the epithelium of the lens and consists of reticular fibers embedded in a matrix of glycoproteins and sulfated glycosaminoglycan. The *lens epithelium* is composed of simple cuboidal epithelium found only on the anterior surface of the lens. At the equator (peripheral edge) of the lens, the lens epithelial cells elongate into columnar cells called *lens fibers.* These cells compose the bulk of the lens and are formed by the division and differentiation of lens epithelial cells. Lens fibers lose their nuclei and consist of a few microtubules and occasional mitochondria in their cytoplasm. As new lens fibers are added, the lens enlarges. Normally, the lens is perfectly transparent. It is enclosed by a clear connective tissue capsule and held in position by *suspensory ligaments.* The lens helps focus light rays for clear vision. A loss of transparency of the lens is known as a *cataract.*

CLINICAL APPLICATION

Contact lenses, rather than conventional lenses, may be worn to improve vision for several reasons: convenience during strenuous physical activity, improvement in peripheral vision, and cosmetic advantages. *Hard lenses* are semiflexible and consist of plastic and silicone that permit oxygen to pass through to the eye. Among their advantages are that their contour effectively corrects for astigmatism, vision is usually clear and stable, they are sturdy, durable, and generally easy to care for. In some cases, they produce distortion and can scratch the corneal surface. *Soft lenses* are constructed of polymers that bind water. They have a soft consistency and conform to the curvature of the cornea. Among the disadvantages of soft lenses are their inability to correct for astigmatism, minor fluctuations in vision, less durability, and care is more complex.

● *Interior* The interior of the eyeball is a large space divided into two cavities by the lens: anterior cavity and posterior cavity. The *anterior cavity,* the division anterior to the lens, is further divided into the *anterior chamber,* which lies behind the cornea and in front of the iris, and the *posterior chamber,* which lies behind the iris and in front of the suspensory ligaments and lens. The anterior cavity is filled with a watery fluid, similar to cerebrospinal fluid, called the *aqueous humor.* The fluid is believed to be secreted into the posterior chamber by choroid plexuses of the ciliary processes of the ciliary bodies behind the iris. Once the fluid is formed, it passes into the posterior chamber and then passes forward between the iris and the lens, through the pupil, into the anterior chamber. From the anterior chamber, the aqueous humor, which is continually produced, is drained into the scleral venous sinus (canal of Schlemm) and then into the blood. The anterior chamber thus serves a function similar to the subarachnoid space around the brain and spinal cord. The scleral venous sinus is analogous to a venous sinus of the dura mater. The pressure in the eye, called *intraocular pressure (IOP),* is produced mainly by the aqueous humor. The intraocular pressure, along with the vitreous humor, maintains the shape of the eyeball and keeps the retina smoothly applied to the choroid so the retina will form clear images. Normal intraocular pressure (about 16 mm Hg) is maintained by drainage of the aqueous humor through the scleral venous sinus. Excessive intraocular pressure, called *glaucoma* (glaw-KŌ-ma), results in degeneration of the retina and blindness. Besides maintaining intraocular pressure, the aqueous humor is also the principal link between the cardiovascular system and the lens and cornea. Neither the lens nor the cornea has blood vessels.

The second, and larger, cavity of the eyeball is the *posterior cavity.* It lies between the lens and the retina and contains a jellylike substance called the *vitreous humor.* This substance contributes to intraocular pressure, helps prevent the eyeball from collapsing, and holds the retina flush against the internal portions of the eyeball. The vitreous humor, unlike the aqueous humor, does not undergo constant replacement. It is formed during embryonic life and is not replaced thereafter.

A summary of structures associated with the eyeball is presented in Exhibit 20-1.

CLINICAL APPLICATION

Eye examinations, in addition to providing information about the status of the eyes, also reveal a good deal of information about general health. A hand-held instrument called an ophthalmoscope (of'-thal-MOS-kōp) is used to examine the interior of the eyeballs. After the pupil is dilated with drops, the instrument which contains a light source and a set of mirrors or prism to reflect light, is used to examine directly the retina (nervous tunic), optic disc, optic nerve (II), macula lutae, and blood vessels.

EXHIBIT 20-1

Summary of Structures Associated with the Eyeball

STRUCTURE	FUNCTION
Fibrous Tunic	
Sclera	Provides shape and protects inner parts.
Cornea	Admits and refracts light.
Vascular Tunic	
Choroid	Provides blood supply and absorbs light.
Ciliary Body	Secretes aqueous humor and alters shape of lens for near or far vision (accommodation).
Iris	Regulates amount of light that enters eyeball.
Retina (Nervous Tunic)	Receives light, converts light into nerve impulses, and transmits impulses to the optic (II) nerve.
Lens	Refracts light.
Anterior Cavity	Contains aqueous humor that helps maintain shape of eyeball and refracts light.
Posterior Cavity	Contains viterous humor that helps maintain shape of eyeball, keeps retina applied to choroid, and refracts light.

Visual Pathway

Before light can reach the rods and cones of the retina to result in image formation, it must pass through the cornea, aqueous humor, pupil, lens, and vitreous humor. For vision to occur, light reaching the rods and cones must form an image on the retina. The resulting nerve impulses must then be conducted to the visual area of the cerebral cortex.

Examination of Figure 20-11c reveals that the retina (nervous tunic) is composed of three principal zones of neurons: photoreceptor (rods and cones), bipolar, and ganglion. For light to stimulate rods and cones to produce generator potentials, light must pass through ganglion and bipolar cells before reaching rods and cones. Note also that there are two special types of cells present in the retina (nervous tunic) called **horizontal cells** and **amacrine cells**.

Once generator potentials are developed by rods and cones, the potentials induce signals in both bipolar neurons and horizontal cells, probably by means of neurotransmitters. In response to the neurotransmitters, bipolar neurons become excited and horizontal cells become inhibited. Functionally, bipolar neurons transmit the excitatory visual signal from rods and cones to ganglion cells. Horizontal cells transmit inhibitory signals to bipolar neurons in the areas lateral to excited rods and cones. This lateral inhibition enhances contrasts in the visual scene between areas of the retina (nervous tunic) that are strongly stimulated and adjacent areas that are weakly stimulated. Horizontal cells also assist in the differentiation of various colors. Amacrine cells, which are also excited by bipolar neurons, synapse with ganglion cells and transmit information to them that signals a change in the level of illumination of the retina (nervous tunic).

When bipolar neurons transmit excitatory visual signals to ganglion cells, the ganglion cells become depolarized and initiate nerve impulses. The cell bodies of the ganglion cells lie in the retina, and their axons leave the eyeball via the **optic (II) nerve** (Figure 20-12). The axons pass through the **optic chiasma** (kī-AZ-ma), a crossing point of the optic (II) nerves. Some fibers cross to the opposite side. Others remain uncrossed. On passing through the optic chiasma, the fibers, now part of the **optic tract,** enter the brain and terminate in the lateral geniculate nucleus of the thalamus. Here the fibers synapse with third-order neurons whose axons pass to the visual areas located in the occipital lobes of the cerebral cortex.

Analysis of the afferent pathway to the brain reveals that the visual field of each eye is divided into two regions: the **nasal (medial) half** and the **temporal (lateral) half.** For each eye, light rays from an object in the nasal half of the visual field fall on the temporal half of the retina. Light rays from an object in the temporal half of the visual field fall on the nasal half of the retina (Figure 20-12). Also, light rays from objects at the top of the visual field of each eye fall on the inferior portion of the retina, and light rays from objects at the bottom of the visual field fall on the superior portion of the retina. In the optic chiasma, nerve fibers from the nasal halves of the retinas cross and continue on to the lateral geniculate nuclei of the thalamus; nerve fibers from the temporal halves of the retinas do not cross, but continue directly on to the lateral geniculate nuclei. As a result, the primary visual area of the cerebral cortex of the right occipital lobe interprets visual sensations from the left side of an object via nerve impulses from the temporal half of the retina of the right eye and the nasal half of the retina of the left eye. The primary visual area of the cerebral cortex of the left occipital lobe interprets visual sensations from the right side of an object via impulses from the nasal half of the right eye and the temporal half of the left eye.

CLINICAL APPLICATION

A **scotoma** (skō-TŌ-ma) or **blind spot** in the field of vision, other than the normal blind spot (optic disc), may indicate a brain tumor along one of the afferent pathways. For instance, a symptom of a tumor in the right optic tract might be an inability to see the left side of a normal field of vision without moving the eyeball.

AUDITORY SENSATIONS AND EQUILIBRIUM

In addition to containing receptors for sound waves, the ear also contains receptors for equilibrium. Anatomically, the ear is divided into three principal regions: the external (outer) ear, middle ear, and internal (inner) ear.

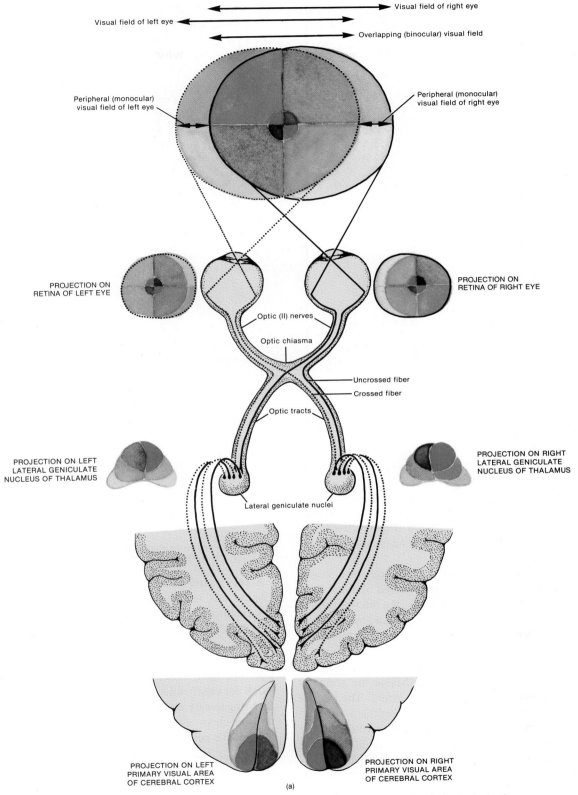

Visual field of right eye

Visual field of left eye

Overlapping (binocular) visual field

Peripheral (monocular) visual field of left eye

Peripheral (monocular) visual field of right eye

PROJECTION ON RETINA OF LEFT EYE

PROJECTION ON RETINA OF RIGHT EYE

Optic (II) nerves

Optic chiasma

Uncrossed fiber

Crossed fiber

Optic tracts

PROJECTION ON LEFT LATERAL GENICULATE NUCLEUS OF THALAMUS

PROJECTION ON RIGHT LATERAL GENICULATE NUCLEUS OF THALAMUS

Lateral geniculate nuclei

PROJECTION ON LEFT PRIMARY VISUAL AREA OF CEREBRAL CORTEX

PROJECTION ON RIGHT PRIMARY VISUAL AREA OF CEREBRAL CORTEX

(a)

FIGURE 20-12 Afferent pathway for visual impulses. (a) Diagram. The dark circle in the center of the visual fields is the macula lutea. The center of the macula lutea is the central fovea, the area of sharpest vision.

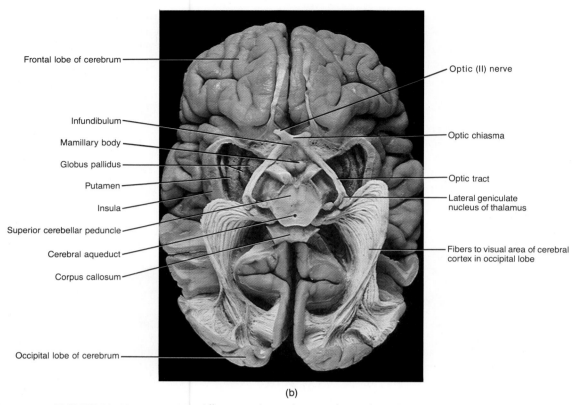

(b)

FIGURE 20-12 (Continued) Afferent pathway for visual impulses. (b) Photograph as seen from the ventral aspect of the brain. (Courtesy of N. Gluhbegovic and T. H. Williams, *The Human Brain: A Photographic Guide,* Harper & Row, Publishers, Inc., New York, 1980.)

External (Outer) Ear

The *external (outer) ear* is structurally designed to collect sound waves and direct them inward (Figure 20-13a). It consists of the pinna, external auditory canal, and tympanic membrane (eardrum).

The *pinna (auricle)* is a flap of elastic cartilage shaped like the flared end of a trumpet and covered by thick skin. The rim of the pinna is called the *helix;* the inferior portion is the *lobule.* The pinna is attached to the head by ligaments and muscles. Additional surface anatomy features of the external ear are shown in Figure 11-3.

The *external auditory canal (meatus)* is a curved tube about 2.5 cm (1 in.) in length that lies in the temporal bone. It leads from the pinna to the eardrum. The wall of the canal consists of bone lined with cartilage that is continuous with the cartilage of the pinna. The cartilage in the external auditory canal is covered with thin, highly sensitive skin. Near the exterior opening, the canal contains a few hairs and specialized sebaceous glands, called *ceruminous* (se-ROO-mi-nus) *glands,* that secrete *cerumen* (earwax). The combination of hairs and cerumen (se-ROO-min) helps prevent foreign objects from entering the ear.

The *tympanic* (tim-PAN-ik) *membrane (eardrum)* is a thin, semitransparent partition of fibrous connective tissue between the external auditory canal and middle ear. Its external surface is concave and covered with skin. Its internal surface is convex and covered with a mucous membrane.

CLINICAL APPLICATION

A **perforated eardrum** is a hole in the tympanic membrane that reduces sound transmission. The condition is characterized by acute pain initially, ringing or roaring in the affected ear, hearing impairment, and sometimes dizziness. Causes of perforated eardrums include compressed air (explosion, scuba diving), trauma from objects (Q tips), or acute middle ear infections. If the condition does not resolve by itself, a surgical procedure called myringoplasty can be performed in which a graft from tissue under the skin is taken from behind the ear to repair the perforation.

Middle Ear

The *middle ear (tympanic cavity)* is a small, epithelial-lined, air-filled cavity hollowed out of the temporal bone (Figure 20-13b–d). It is separated from the external ear by the eardrum and from the internal ear by a thin bony partition that contains two small openings: the oval window and round window.

The posterior wall of the middle ear communicates with the mastoid air cells of the temporal bone through a chamber called the *tympanic antrum.* This anatomical fact explains why a middle ear infection may spread to the temporal bone, causing mastoiditis, or even to the brain.

The anterior wall of the middle ear contains an opening that leads directly into the *auditory (Eustachian) tube.* The auditory tube connects the middle ear with the nasopharynx of the throat. Through this passageway, infections may travel from the throat and nose to the ear. The function

of the tube is to equalize air pressure on both sides of the tympanic membrane. Abrupt changes in external or internal air pressure might otherwise cause the eardrum to rupture. During swallowing and yawning, the tube opens to allow atmospheric air to enter or leave the middle ear until the internal pressure equals the external pressure. If the pressure is not relieved, intense pain, hearing impairment, tinnitus, and vertigo could develop. Any sudden pressure changes against the eardrum may be equalized by deliberately swallowing or pinching the nose closed, closing the mouth, and gently forcing air from the lungs into the nasopharynx.

Extending across the middle ear are three exceedingly

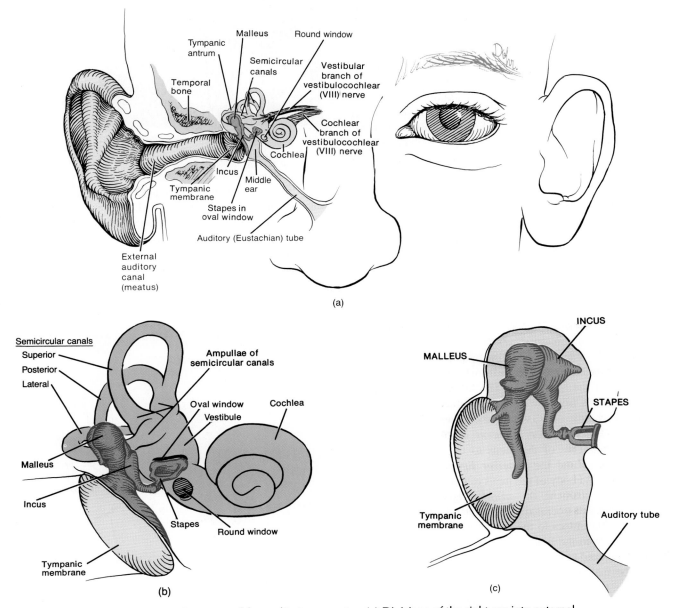

FIGURE 20-13 Structure of the auditory apparatus. (a) Divisions of the right ear into external, middle, and internal portions seen in a frontal section through the right side of the skull. (b) Details of the middle ear and bony labyrinth of the internal ear. (c) Diagram of auditory ossicles of the middle ear.

FIGURE 20-13 (Continued) (d) Photograph of auditory ossicles of the middle ear. (Courtesy of C. Yokochi and J. W. Rohen, *Photographic Anatomy of the Human Body*, 2nd ed., 1978, IGAKU-SHOIN, Ltd., Tokyo, New York.)

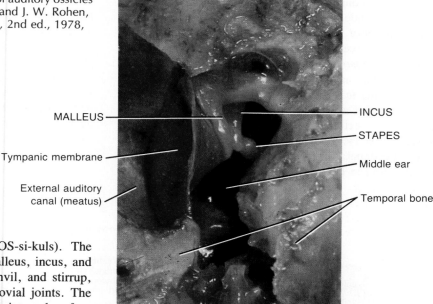

small bones called **auditory ossicles** (OS-si-kuls). The bones, named for their shape, are the malleus, incus, and stapes, commonly called the hammer, anvil, and stirrup, respectively. They are connected by synovial joints. The "handle" of the **malleus** is attached to the internal surface of the tympanic membrane. Its head articulates with the body of the incus. The **incus** is the intermediate bone in the series and articulates with the head of the stapes. The base or footplate of the **stapes** fits into a small opening in the thin bony partition between the middle and inner ear. The opening is called the **oval window**, or **fenestra vestibuli** (fe-NES-tra ves-TIB-yoo-lī). Directly below the oval window is another opening, the **round window**, or **fenestra cochlea** (fe-NES-tra KŌK-lē-a). This opening is enclosed by a membrane called the **secondary tympanic membrane.** Auditory ossicles are attached to the middle ear by means of ligaments.

Three ligaments are associated with the malleus (Figure 20-14). The **anterior ligament of the malleus** is attached by one end to the anterior process of the malleus and by the other to the anterior wall of the middle ear. The **superior ligament of the malleus** extends from the head of the malleus to the roof of the middle ear. The **lateral ligament of the malleus** extends from the neck of the malleus to the lateral wall of the middle ear. Two ligaments are associated with the incus. The **posterior ligament of the incus** extends from the short crus of the incus to the posterior wall of the middle ear, and the **superior ligament of the incus** extends from the body of the incus to the roof of the middle ear. A single ligament, the **annular ligament of the base of the stapes,** is associated with the stapes. It extends from the base of the stapes to the fenestra vestibuli.

In addition to the ligaments, there are also two skeletal muscles attached to the ossicles. The **tensor tympani muscle** draws the malleus medially, thus limiting movement and increasing tension on the tympanic membrane to prevent damage to the inner ear when exposed to loud sounds. Since this response is slow, it only protects the inner ear from prolonged loud noises, not brief ones, such as a gunshot.

The **stapedius muscle** is the smallest of all skeletal muscles. Its action is to draw the stapes posteriorly, thus preventing it from moving too much. Like the tensor tympani muscle, the stapedius muscle has a protective function in that it dampens (checks) large vibrations that result from loud noises. It is for this reason that paralysis of the stapedius muscle is associated with **hyperacusia** (acuteness of hearing).

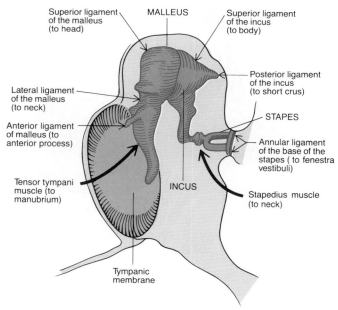

FIGURE 20-14 Areas of attachment of the ligaments and muscles of the auditory ossicles.

Internal (Inner) Ear

The *internal (inner) ear* is also called the *labyrinth* (LAB-i-rinth) because of its complicated series of canals (Figure 20-15). Structurally, it consists of two main divisions: an outer bony labyrinth and an inner membranous labyrinth that fits in the bony labyrinth. The *bony labyrinth* is a series of cavities in the petrous portion of the temporal bone. It can be divided into three areas named on the basis of shape: the vestibule, cochlea, and semicircular canals. The bony labyrinth is lined with periosteum and contains a fluid called *perilymph.* This fluid surrounds the *membranous labyrinth,* a series of sacs and tubes lying inside and having the same general form as the bony labyrinth. The membranous labyrinth is lined with epithelium and contains a fluid called *endolymph.*

The *vestibule* constitutes the oval central portion of the bony labyrinth. The membranous labyrinth in the vestibule consists of two sacs called the *utricle* (YOO-tri-kul) and *saccule* (SAK-yool). These sacs are connected to each other by a small duct.

Projecting upward and posteriorly from the vestibule are the three bony *semicircular canals.* Each is arranged at approximately right angles to the other two. On the basis of their positions, they are called the anterior, posterior, and lateral canals. The anterior and posterior semicircular canals are oriented vertically; the lateral one is oriented horizontally. One end of each canal enlarges into a swelling called the *ampulla* (am-POOL-la). The portions of the membranous labyrinth that lie inside the bony semicircular canals are called the *semicircular ducts (membranous semicircular canals).* These structures are almost identical in shape

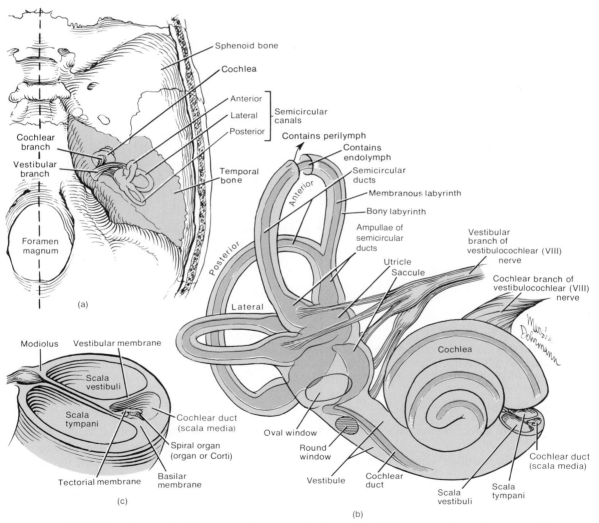

FIGURE 20-15 Details of the internal ear. (a) Relative position of the bony labyrinth projected to the inner surface of the floor of the skull. (b) The outer, blue-colored area belongs to the bony labyrinth. The inner, brown-colored area belongs to the membranous labyrinth. (c) Cross section through the cochlea.

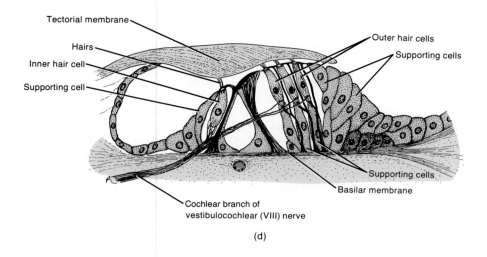

(d)

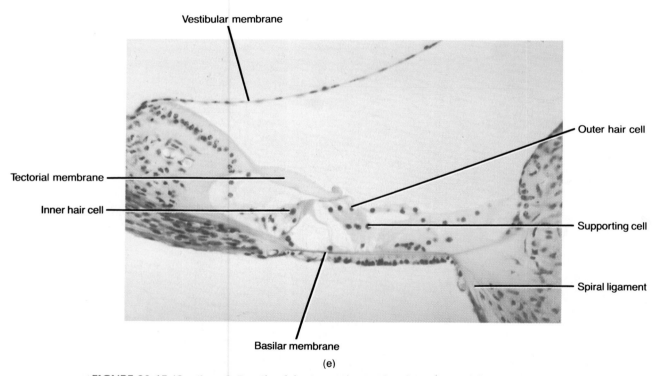

(e)

FIGURE 20-15 (*Continued*) Details of the internal ear. (d) Enlargement of the spiral organ (organ of Corti). (e) Photomicrograph of the spiral organ (organ of Corti) at a magnification of 300×. (Copyright © 1983 by Michael H. Ross. Used by permission.)

to the semicircular canals and communicate with the utricle of the vestibule.

Lying in front of the vestibule is the ***cochlea*** (KŌK-lē-a), so designated because of its resemblance to a snail's shell. The cochlea consists of a bony spiral canal that makes about 2¾ turns around a central bony core called the ***modiolus***. A cross section through the cochlea shows the canal is divided into three separate channels by partitions that together have the shape of the letter Y. The stem of the Y is a bony shelf that protrudes into the canal; the wings of the Y are composed mainly of membranous labyrinth. The

channel above the bony partition is the ***scala vestibuli;*** the channel below is the ***scala tympani.*** The cochlea adjoins the wall of the vestibule, into which the scala vestibuli opens. The scala tympani terminates at the round window. The perilymph of the vestibule is continuous with that of the scala vestibuli. The third channel (between the wings of the Y) is the membranous labyrinth: the ***cochlear duct (scala media).*** The cochlear duct is separated from the scala vestibuli by the ***vestibular membrane***. It is separated from the scala tympani by the ***basilar membrane***.

Resting on the basilar membrane is the ***spiral organ***

(*organ of Corti*), the organ of hearing. The spiral organ is a series of epithelial cells on the inner surface of the basilar membrane. It consists of a number of supporting cells and hair cells, which are the receptors for auditory sensations. The inner hair cells are medially placed in a single row and extend the entire length of the cochlea. The outer hair cells are arranged in several rows throughout the cochlea. The cells have long hairlike processes at their free ends that extend into the endolymph of the cochlear duct. The basal ends of the hair cells are in contact with fibers of the cochlear branch of the vestibulocochlear (VIII) nerve. Projecting over and in contact with the hair cells of the spiral organ is the *tectorial membrane,* a delicate and flexible gelatinous membrane.

CLINICAL APPLICATION

Hair cells of the spiral organ are easily damaged by *exposure to high-intensity noises* such as those produced by engines of jet planes, revved-up motorcycles, and loud music. As a result of the noises, hair cells become arranged in disorganized patterns or they and their supporting cells may degenerate.

Auditory Pathway

Sound waves result from the alternate compression and decompression of air molecules. They originate from a vibrating object, much the same way that waves travel over the surface of water.

The events involved in the physiology of hearing and the conduction of auditory impulses to the brain are as follows (Figure 20-16).

1. Sound waves that reach the ear are directed by the pinna into the external auditory canal.
2. When the waves strike the tympanic membrane, the alternate compression and decompression of the air cause the membrane to vibrate (move forward and backward). The distance the membrane moves is always very small and is relative to the force and velocity of the sound waves. It vibrates slowly in response to low-frequency sounds and rapidly in response to high-frequency sounds.
3. The central area of the tympanic membrane is connected to the malleus, which also starts to vibrate. The vibration is then picked up by the incus, which transmits the vibration to the stapes.

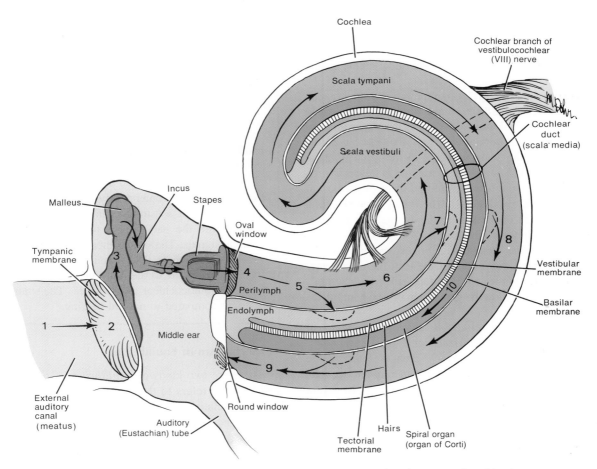

FIGURE 20-16 Physiology of hearing. The numbers correspond to the events listed in the text. The cochlea has been drawn without its normal coiling in order to show the flow of perilymph.

4. As the stapes moves back and forth, it pushes the oval window in and out.

   If sound waves passed directly to the oval window without passing through the tympanic membrane and auditory ossicles, hearing would be inadequate. A minimal amount of sound energy is required to transmit sound waves through the perilymph of the cochlea. Since the tympanic membrane has a surface area about 22 times larger than that of the oval window, it can collect about 22 times more sound energy. This energy is sufficient to transmit sound waves through the perilymph.

5. The movement of the oval window sets up waves in the perilymph.

6. As the oval window bulges inward, it pushes the perilymph of the scala vestibuli and pressure waves are propagated through the scala vestibuli. If the inward movement of the stapes at the oval window is slow, the pressure in the perilymph pushes the perilymph from the scala vestibuli to the scala tympani and eventually to the round window, causing it to bulge outward into the middle ear. See number 9 in the illustration.

7. As the pressure moves through the perilymph of the scala vestibuli, it pushes the vestibular membrane inward and increases the pressure of the endolymph inside the cochlear duct.

8. As a result, the basilar membrane moves slightly and bulges into the scala tympani. Slow movement of the oval window has only a *very slight* effect on the basilar membrane.

9. As the pressure moves through the scala tympani, perilymph moves toward the round window, causing it to bulge outward into the middle ear. If the stapes vibrates rapidly, pressure in the perilymph does not have time to pass from the scala vestibuli to the scala tympani to the round window. Instead, pressure in the perilymph in the scala vestibuli is transmitted through the basilar membrane and eventually to the round window. As a result, the area of the basilar membrane near the oval and round windows vibrates.

10. When the basilar membrane vibrates, the hair cells of the spiral organ move against the tectorial membrane. The movement of the hairs develops generator potentials that ultimately lead to the generation of nerve impulses.

The function of hair cells is to convert a mechanical force (stimulus) into an electrical signal (nerve impulse). It is believed to occur as follows. When the hairs at the top of the cell are moved, the hair cell membrane depolarizes, producing the generator potential. Depolarization spreads through the cell and causes the release of neurotransmitter from the hair cell, which excites a sensory nerve fiber at the base of the hair cell. There is some evidence that the neurotransmitter might be glutamate or gamma aminobutyric acid (GABA).

The nerve impulses are then passed on to the cochlear branch of the vestibulocochlear (VIII) nerve (Figure 20-17) and cochlear nuclei in the medulla. Here most impulses cross to the opposite side and then travel to the midbrain, thalamus, and finally to the auditory area of the temporal lobe of the cerebral cortex.

Differences in pitch are related to differences in width of the basilar membrane and sound waves of various frequencies that cause specific regions of the basilar membrane to vibrate more intensely than others. The membrane is less flexible at the base of the cochlea (portion closer to the oval window) and more flexible near the apex of the cochlea. High-frequency or high-pitched sounds cause the basilar membrane to vibrate near the base of the cochlea and low-frequency or low-pitched sounds cause the basilar membrane to vibrate near the apex of the cochlea. Loudness is determined by the intensity of sound waves. High-intensity sound waves cause greater vibration of the basilar membrane. Thus, more hair cells are stimulated and more impulses reach the brain.

CLINICAL APPLICATION

Artificial ears (*cochlear implants*) are devices that translate sounds into electronic signals that can be interpreted by the brain. They take the place of hair cells of the spiral organ, which normally convert sound waves into electrical signals carried to the brain. The implants are used for individuals with sensorineural deafness, that is, deafness due to disease or injury that has destroyed hair cells of the spiral organ. Such people make up the majority of those who have no hearing in either ear.

The device consists of electrodes implanted in the cochlea that are connected to a plug attached to the outside of the head. A microprocessor pack, about the size of a small transistor radio, is worn on the person's belt or in a shirt pocket and connected by a cord to the plug. Sound waves enter a tiny microphone in the ear, similar to that of a usual hearing aid, and travel down the cord into the pack where they are converted into electrical signals. The signals then travel to the electrodes in the cochlea where they stimulate nerve endings and transmit the signals to the brain via the cochlear branch of the vestibulocochlear (VIII) nerve. The sounds heard are crude compared to normal hearing.

Mechanism of Equilibrium

There are two kinds of *equilibrium*. One, called *static equilibrium*, refers to the orientation of the body (mainly the head) relative to the ground (gravity). The second kind, *dynamic equilibrium*, is the maintenance of body position (mainly the head) in response to sudden movements such as rotation, acceleration, and deceleration. The receptor organs for equilibrium are the maculae of the saccule and the utricle and cristae in the semicircular ducts.

tus by motion. Nerve impulses pass from the internal ear to the vomiting center in the medulla. Visual stimuli and emotional factors like fear and anxiety can also contribute to motion sickness. Ideally, treatment with drugs of susceptible individuals should be instituted prior to their entering into conditions that would produce motion sickness, since prevention is more successful than treatment of symptoms once they have developed. Among the drugs used to treat motion sickness are dimenhydrinate (Dramamine), meclizine (Antivert), and promethazine (Phenegran).

KEY MEDICAL TERMS ASSOCIATED WITH SENSORY STRUCTURES

Achromatopsia (a-krō'-ma-TOP-sē-a; *a* = without; *chrom* = color) Complete color blindness.

Ametropia (am'-e-TRŌ-pē-a; *ametro* = disproportionate; *ops* = eye) Refractive defect of the eye resulting in an inability to focus images properly on the retina.

Anopsia (an-OP-sē-a; *opsia* = vision) A defect of vision.

Astereognosis (a-ster-ē-og-NŌ-sis; *stereos* = solid; *gnosis* = knowledge) Loss of ability to recognize objects or to appreciate their form by touching them.

Audiometer (aw-dē-OM-e-ter; *audire* = to hear; *metron* = to measure) An instrument used to measure hearing by producing acoustic stimuli of known frequency and intensity.

Blepharitis (blef-a-RĪ-tis; *blepharo* = eyelid; *itis* = inflammation of) An inflammation of the eyelid.

Dyskinesia (dis'-ki-NĒ-zē-a; *dys* = difficult; *kinesis* = movement) Abnormality of motor function characterized by involuntary, purposeless movements.

Epiphora (e-PIF-ō-ra; *epi* = above) Abnormal overflow of tears.

Eustachitis (yoo'-stā-KĪ-tis) An inflammation or infection of the auditory (Eustachian) tube.

Exotropia (ek'-sō-TRŌ-pē-a; *ex* = out; *tropia* = turning) Turning outward of the eyes.

Keratitis (ker'-a-TĪ-tis; *kerato* = cornea) An inflammation or infection of the cones.

Kinesthesis (kin'-es-THĒ-sis; *aisthesis* = sensation) The sense of perception of movement.

Labyrinthitis (lab'-i-rin-THĪ-tis) An inflammation of the labyrinth (inner ear).

Mydriasis (mi-DRĒ-a-sis) Dilated pupil.

Myringitis (mir'-in-JĪ-tis; *myringa* = eardrum) An inflammation of the eardrum; also called **tympanitis.**

Nystagmus (nis-TAG-mus; *nystazein* = to nod) A constant, rapid involuntary movement of the eyeballs, possibly caused by a disease of the central nervous system.

Otalgia (o-TAL-jē-a; *oto* = ear; *algia* = pain) Earache.

Otosclerosis (ō'-tō-skle-RŌ-sis; *oto* = ear; *sclerosis* = hardening) Pathological process that may be hereditary in which new bone is deposited around the oval window. The result may be immobilization of the stapes, leading to deafness.

Photophobia (fō'-tō-FŌ-bē-a; *photo* = light; *phobia* = fear) Abnormal visual intolerance to light.

Presbyopia (pres'-bē-Ō-pē-a; *presby* = old) Inability to focus on nearby objects due to loss of elasticity of the crystalline lens. The loss is usually caused by aging.

Ptosis (TŌ-sis; *ptosis* = fall) Falling or drooping of the eyelid. (This term is also used for the slipping of any organ below its normal position.)

Retinoblastoma (ret'-i-nō-blas-TŌ-ma; *blast* = bud; *oma* = tumor) A common tumor arising from immature retinal cells and accounting for 2 percent of childhood malignancies.

Strabismus (stra-BIZ-mus) An eye muscle disorder, commonly called "crossed eyes," in which the eyeballs do not move in unison. It may be caused by lack of coordination of the extrinsic eye muscles.

Tinnitus (ti-NĪ-tus) A ringing, roaring, or clicking in the ears.

STUDY OUTLINE

Sensations (p. 550)
Definition (p. 550)
1. Sensation is a state of awareness of external and internal conditions of the body.
2. The prerequisites for a sensation to occur are reception of a stimulus, conversion of the stimulus into a nerve impulse by a receptor, conduction of the impulse to the brain, and translation of the impulse into a sensation by a region of the brain.
3. Each stimulus is capable of causing the membrane of a receptor to depolarize. This is called the generator potential.

Characteristics (p. 550)
1. Projection occurs when the brain refers a sensation to the point of stimulation.
2. Adaptation is the loss of sensation even though the stimulus is still applied.
3. An afterimage is the persistence of a sensation even though the stimulus is removed.
4. Modality is the property by which one sensation is distinguished from another.

Classification of Receptors (p. 551)
1. According to location, receptors are classified as exteroceptors, visceroceptors, and proprioceptors.
2. On the basis of type of stimulus detected, receptors are classified as mechanoreceptors, thermoreceptors, nociceptors, electromagnetic receptors, and chemoreceptors.
3. In terms of simplicity or complexity, simple receptors are associated with general senses and complex receptors are associated with special senses.

General Senses (p. 551)
Cutaneous Sensations (p. 551)
1. Cutaneous sensations include tactile sensations (touch, pressure, vibration), thermoreceptive sensations (heat and cold), and pain. Receptors for these sensations are located in the skin, connective tissues, and the ends of the gastrointestinal tract.
2. Receptors for touch are root hair plexuses, free nerve endings, tactile (Merkel's) discs, corpuscles of touch (Meissner's corpuscles), and type II cutaneous mechanoreceptors (end-organs of

Ruffini). Receptors for pressure are free nerve endings, type II cutaneous mechanoreceptors, and lamellated (Pacinian) corpuscles. Receptors for vibration are corpuscles of touch and lamellated corpuscles.

3. Pain receptors (nociceptors) are located in nearly every body tissue. Pain may be acute or chronic.

4. Two kinds of pain recognized in the parietal lobe of the cortex are somatic and visceral.

5. Referred pain is felt in the skin near or away from the organ sending pain impulses.

6. Phantom pain is the sensation of pain in a limb that has been amputated.

7. Pain impulses may be inhibited by drugs, surgery, and acupuncture.

8. In acupuncture, enkephalins block the release of substance P and inhibit pain transmission in the brain.

Proprioceptive Sensations (p. 555)

1. Receptors located in skeletal muscles, tendons, in and around joints, and the internal ear convey nerve impulses related to muscle tone, movement of body parts, and body position.

2. The receptors include muscle spindles, tendon organs (Golgi tendon organs), joint kinesthetic receptors, and the maculae and cristae.

Levels of Sensation (p. 556)

1. Sensory fibers terminating in the lower brain stem bring about far more complex motor reactions than simple spinal reflexes.

2. When sensory impulses reach the lower brain stem, they cause subconscious motor reactions.

3. Sensory impulses that reach the thalamus can be localized crudely in the body.

4. When sensory impulses reach the cerebral cortex, we experience precise localization.

Sensory Pathways (p. 556)

1. Sensory information from all parts of the body terminates in a specific area of the somatosensory cortex.

2. In the posterior column pathway and the spinothalamic pathway there are first-order, second-order, and third-order neurons.

3. The neural pathway for pain and temperature is the lateral spinothalamic pathway.

4. The neural pathway for light touch and pressure is the anterior spinothalamic pathway.

5. The neural pathway for discriminative touch, proprioception, and vibration is the posterior column pathway.

6. The pathways to the cerebellum are the anterior and posterior spinocerebellar tracts.

Special Senses (p. 558)

Olfactory Sensations (p. 559)

1. The receptors for olfaction, the olfactory cells, are in the nasal epithelium.

2. Olfactory cells convey impulses to olfactory (I) nerves, olfactory bulbs, olfactory tracts, and cerebral cortex.

Gustatory Sensations (p. 560)

1. The receptors for gustation, the gustatory cells, are located in taste buds.

2. The four primary tastes are salty, sweet, sour, and bitter.

3. Gustatory cells convey impulses to cranial nerves V, VII, IX, and X, the medulla, thalamus, and cerebral cortex.

Visual Sensations (p. 561)

1. Accessory structures of the eyes include the eyebrows, eyelids, eyelashes, and the lacrimal apparatus.

2. The eye is constructed of three coats: (a) fibrous tunic (sclera and cornea), (b) vascular tunic (choroid, ciliary body, and iris), and (c) retina (nervous tunic), which contains rods and cones.

3. The anterior cavity contains aqueous humor; the posterior cavity contains vitreous humor.

4. Rods and cones develop generator potentials and ganglion cells initiate nerve impulses.

5. Impulses from ganglion cells are conveyed through the retina to the optic (II) nerve, the optic chiasma, the optic tract, the thalamus, and the cortex.

Auditory Sensations and Equilibrium (p. 568)

1. The ear consists of three anatomical subdivisions: (a) the external or outer ear (pinna, external auditory canal, and tympanic membrane), (b) the middle ear (auditory or Eustachian tube, ossicles, oval window, and round window), and (c) the internal or inner ear (bony labyrinth and membranous labyrinth). The internal ear contains the spiral organ (organ of Corti), the organ of hearing.

2. Sound waves enter the external auditory canal, strike the tympanic membrane, pass through the ossicles, strike the oval window, set up waves in the perilymph, strike the vestibular membrane and scala tympani, increase pressure in the endolymph, strike the basilar membrane, and stimulate hairs on the spiral organ (organ of Corti). A sound impulse is then initiated.

3. Static equilibrium is the orientation of the body relative to the pull of gravity. The maculae of the utricle and saccule are the sense organs of static equilibrium.

4. Dynamic equilibrium is the maintenance of body position in response to movement. The cristae in the semicircular ducts are the sense organs of dynamic equilibrium.

Motor Pathways (p. 579)

1. Incoming sensory information is added to, subtracted from, or integrated with other information arriving from all other operating sensory receptors.

2. The integration process occurs at many stations along the pathways of the central nervous system such as within the spinal cord, brain stem, cerebellum, and cerebral cortex.

3. A motor response to make a muscle contract or a gland secrete can be initiated at any of these stations or levels.

4. The muscles of all parts of the body are controlled by a specific area of the motor cortex.

5. Voluntary motor impulses are conveyed from the brain through the spinal cord along the pyramidal pathways and the extrapyramidal pathways.

6. Pyramidal pathways include the lateral corticospinal, anterior corticospinal, and corticobulbar tracts.

7. Major extrapyramidal tracts are the rubrospinal, tectospinal, vestibulospinal, and reticulospinal tracts.

Developmental Anatomy of the Eye and Ear (p. 581)

1. The eyes first appear as a depression of the diencephalon called the optic groove.
2. The internal ears develop from the otic placode, the middle ears develop from the first pharyngeal pouch, and the external ears develop from the first branchial groove.

Applications to Health (p. 583)

1. Cataract is the loss of transparency of the lens or capsule.
2. Glaucoma is abnormally high intraocular pressure, which destroys neurons of the retina.
3. Conjunctivitis is an inflammation of the conjunctiva.
4. Trachoma is a chronic, contagious inflammation of the conjunctiva.
5. Deafness is the lack of the sense of hearing or significant hearing loss. It is classified as sensorineural and conduction.
6. Labyrinthine disease is basically a malfunction of the inner ear that has a variety of causes.
7. Ménière's syndrome is the malfunction of the inner ear that may cause deafness and loss of equilibrium.
8. Vertigo is a sensation of motion that is classified as peripheral, central, or psychogenic.
9. Otitis media is an acute infection of the middle ear.
10. Motion sickness is a functional disorder precipitated by repetitive angular, linear, or vertical motion.

REVIEW QUESTIONS

1. Define a sensation and a sense receptor. What prerequisites are necessary for the perception of a sensation?
2. Describe the following characteristics of a sensation: projection, adaptation, afterimage, modality.
3. Classify receptors on the basis of location, stimulus detected, and simplicity or complexity.
4. Distinguish between a general sense and a special sense.
5. What is a cutaneous sensation? Distinguish tactile, thermoreceptive, and pain sensations.
6. For each of the following cutaneous sensations, describe the receptor involved in terms of structure, function, and location: touch, pressure, vibration, and pain.
7. Why are pain receptors important? Differentiate somatic pain, visceral pain, referred pain, and phantom pain.
8. What is the proprioceptive sense? Where are the receptors for this sense located?
9. Describe the structure of muscle spindles, tendon organs (Golgi tendon organs), and joint kinesthetic receptors.
10. Describe the various levels of sensation in the central nervous system.
11. Describe how various parts of the body are represented in the somatosensory cortex.
12. Distinguish between the posterior column and spinothalamic pathways.
13. Identify the receptors for olfaction. Describe the neural pathway for olfaction.
14. Identify the receptors for gustation. Describe the neural pathway for gustation.
15. How are papillae related to taste buds? Describe the structure and location of the papillae.
16. Describe the structure and importance of the following accessory structures of the eye: eyelids, eyelashes, and eyebrows.
17. What is the function of the lacrimal apparatus? Explain how it operates.
18. By means of a labeled diagram, indicate the principal anatomical structures of the eye.
19. Describe the location and contents of the chambers of the eye. What is intraocular pressure (IOP)? How is the scleral venous sinus (canal of Schlemm) related to this pressure?
20. Describe the path of a visual impulse from the optic (II) nerve to the brain.
21. Diagram the principal parts of the outer, middle, and inner ear. Describe the function of each part labeled.
22. Explain the events involved in the transmission of sound from the pinna to the spiral organ (organ of Corti).
23. What is the afferent pathway for sound impulses from the cochlear branch of the vestibulocochlear (VIII) nerve to the brain?
24. Compare the function of the maculae in the saccule and utricle in maintaining static equilibrium with the role of the cristae in the semicircular ducts in maintaining dynamic equilibrium.
25. Describe the path of a nerve impulse that results in static and dynamic equilibrium.
26. Describe how various parts of the body are represented in the motor cortex.
27. Which pathways control voluntary motor impulses from the brain through the spinal cord? How are they distinguished?
28. Describe the development of the eye and ear.
29. Define each of the following: cataract, glaucoma, conjunctivitis, trachoma, deafness, labyrinthine disease, Ménière's syndrome, vertigo, otitis media, and motion sickness.
30. Refer to the glossary of key medical terms associated with the sensory structures. Be sure that you can define each term.

SELF-QUIZ

Complete the following:

1. The final common pathway consist of _____ neurons.
2. The neuron that crosses to the opposite side in sensory pathways is usually the _____ neuron.
3. In general, the extrapyramidal tracts begin in the _____ and end in the _____.
4. Taste buds are located on elevated projections of the tongue called _____.

5. In the macula, the gelatinous membrane is embedded with calcium carbonate crystals called _____. These respond to gravity in such a way that the macula is the receptor for (static? dynamic?) equilibrium.

6. Show the route of nerve impulses along the pyramidal pathway by listing in correct sequence the structures that comprise the pathway.

___ ___ ___ ___ ___ ___ ___ ___

   A. anterior gray horn (lower motor neuron)
   B. midbrain and pons
   C. effector (skeletal muscle)
   D. internal capsule
   E. lateral corticospinal tract
   F. medulla, decussation site
   G. precentral gyrus (upper motor neuron)
   H. anterior root of spinal nerve

7. Match the following:
___ a. sensitive to movement of hair shaft
___ b. egg-shaped masses located in dermal papillae, especially in fingertips, palms, and soles that are receptors for discriminative touch
___ c. onion-shaped structures sensitive to pressure
___ d. modified epidermal cells in the stratum basale of hairless skin

   A. corpuscle of touch (Meissner's corpuscles)
   B. tactile (Merkel's) discs
   C. lamellated (Pacinian) corpuscles
   D. root hair plexuses

Arrange the answers in correct sequence.

___ ___ ___ ___ 8. Levels of sensation, from those causing simplest, least precise reflexes to those causing most complex and precise responses:
   A. thalamus
   B. brain stem
   C. cerebral cortex
   D. spinal cord

___ ___ ___ ___ ___ 9. Pathway for conduction of most nerve impulses for voluntary movement of muscles:
   A. anterior gray horn of the spinal cord
   B. precentral gyrus
   C. internal capsule
   D. location where decussation occurs
   E. lateral corticospinal tract

___ ___ ___ ___ ___ 10. Order of nerve impulses along conduction pathway for smell:
   A. olfactory bulb
   B. olfactory hairs
   C. olfactory (I) nerves
   D. olfactory tract
   E. primary olfactory area of cortex

___ ___ ___ ___ ___ 11. From anterior to posterior:
   A. anterior chamber
   B. iris
   C. lens
   D. posterior cavity
   E. posterior chamber

___ ___ ___ 12. Layers of the eye, from superficial to deep:
   A. sclera
   B. retina
   C. choroid

___ ___ ___ ___ 13. From anterior to posterior:
   A. vitreous humor
   B. optic (II) nerve
   C. cornea
   D. lens

___ ___ ___ ___ 14. Pathway of aqueous humor, from site of formation to destination:
   A. anterior chamber
   B. scleral venous sinus (canal of Schlemm)
   C. ciliary body
   D. posterior chamber

___ ___ ___ ___ ___ 15. Pathway of sound waves and resulting mechanical contraction:
   A. external auditory canal
   B. stapes
   C. malleus and incus
   D. oval window
   E. tympanic membrane

___ ___ ___ ___ ___ 16. Pathway of tears, from site of formation to entrance to nose:
   A. lacrimal gland and ducts
   B. lacrimal sac
   C. nasolacrimal duct
   D. surface of conjunctiva
   E. puncta lacrimalia and lacrimal canals

17. Match the following:
___ a. "white of the eye"
___ b. a clear structure, composed of protein layers arranged like an onion
___ c. blind spot; area in which there are no cones or rods
___ d. area of sharpest vision; area of densest concentration of cones
___ e. nonvascular, transparent, fibrous coat; most anterior eye structure
___ f. layer containing neurons; if detached, causes blindness
___ g. dark brown layer; prevents reflection of light rays; also nourishes eyeball since it is vascular
___ h. a hole; appears black, like a circular doorway leading into a dark room
___ i. regulates the amount of light entering the eye; colored part of the eye
___ j. attaches to the lens by means of radially arranged fibers called the suspensory ligaments
___ k. serrated margin of the retina
___ l. located at the junction of iris and cornea; drains aqueous humor

   A. central fovea
   B. choroid
   C. cornea
   D. ciliary muscle
   E. iris
   F. lens
   G. optic disc
   H. ora serrata
   I. pupil
   J. retina
   K. sclera
   L. scleral venous sinus (canal of Schlemm)

Choose the one best answer to these questions.

___ 18. All of these sensations are conveyed by the posterior column pathway *except*

A. pain and temperature; B. proprioception; C. discriminative touch; D. vibration; E. two-point discrimination.

___ **19.** Receptors for hearing are located in the

A. middle ear; B. cochlea; C. semicircular canals; D. tympanic membrane; E. vestibule.

___ **20.** The incorrect interpretation of pain as having come from regions far from the actual site of pain is known as

A. visceral pain; B. phantom pain; C. referred pain; D. somatosensory pain; E. somatic pain.

___ **21.** Upper motor neurons

A. carry impulses from the cerebellum to the spinal cord; B. carry impulses from the frontal lobe to the cerebellum; C. carry impulses from the cerebral cortex to the spinal cord; D. coordinate reflexes; E. synapse with lower motor neurons in the spinal ganglia.

___ **22.** A layer of tissue that is continuous over the inner surfaces of the eyelids and is reflected over the edge of the outer surface of the cornea is the

A. choroid; B. conjunctiva; C. sclera; D. suspensory ligament; E. palpebra.

___ **23.** Which of the following is/are not part(s) of the vascular tunic of the eyeball?

A. cornea; B. choroid; C. ciliary body; D. iris; E. all of the above.

___ **24.** Rods function in

(1) color vision.
(2) dim light vision.
(3) bright light vision.

A. (1) only; B. (2) only; C. (3) only; D. all of the above; E. none of the above.

___ **25.** Which of these receptors does not belong with the others?

A. muscle spindles; B. tactile (Merkel's) discs; C. tendon (Golgi) organs; D. joint kinesthetic receptors.

26. Match the following. Not all answers will be used.

___ a. tube used to equalize pressure on either side of tympanic membrane

___ b. chamber posterior to middle ear; permits middle ear infection to spread to cause mastoiditis

___ c. eardrum

___ d. structure on which stapes exerts pistonlike action

___ e. flared portion of the outer ear

___ f. ossicle adjacent to eardrum

___ g. anvil-shaped ear bone

A. auditory (Eustachian) tube
B. incus
C. malleus
D. oval window
E. pinna
F. round window
G. stapes
H. tympanic antrum
I. tympanic membrane

21 The Endocrine System

CHAPTER OUTLINE

■ **Endocrine Glands**
Receptors
Feedback Control
■ **Pituitary (Hypophysis)**
Adenohypophysis
Neurohypophysis
■ **Thyroid**
■ **Parathyroids**
■ **Adrenals (Suprarenals)**
Adrenal Cortex
Adrenal Medulla
■ **Pancreas**
■ **Ovaries and Testes**
■ **Pineal (Epiphysis Cerebri)**
■ **Thymus**
■ **Aging and the Endocrine System**
■ **Developmental Anatomy of the Endocrine System**
■ **Other Endocrine Tissues**
■ **Key Medical Terms Associated with the Endocrine System**

STUDENT OBJECTIVES

1. Define an endocrine gland and a exocrine gland, and list the endocrine glands of the body.
2. Describe the structural and functional division of the pituitary gland into the adenohypophysis and the neurohypophysis.
3. Describe the location, histology, and blood and nerve supply of the pituitary gland.
4. Discuss the symptoms of pituitary dwarfism, giantism, acromegaly, and diabetes insipidus as pituitary gland disorders.
5. Describe the location, histology, and blood and nerve supply of the thyroid gland.
6. Discuss the symptoms of cretinism, myxedema, exophthalmic goiter, and simple goiter as thyroid gland disorders.
7. Describe the location, histology, and blood and nerve supply of the parathyroid glands.
8. Discuss the symptoms of tetany and osteitis fibrosa cystica as parathyroid gland disorders.
9. Describe the location, histology, and blood and nerve supply of the adrenal (suprarenal) glands.
10. Explain the subdivisions of the adrenal (suprarenal) glands into cortical and medullary portions.
11. Discuss the symptoms of aldosteronism, Addison's disease, Cushing's syndrome, and adrenogenital syndrome as adrenal cortical disorders.
12. Discuss the symptoms of pheochromocytomas as an adrenal medullary disorder.
13. Describe the location, histology, and blood and nerve supply of the pancreas.
14. Discuss the symptoms of diabetes mellitus and hyperinsulinism as endocrine disorders of the pancreas.
15. Describe the location, histology, and blood and nerve supply of the pineal gland.
16. Describe the location, histology, and blood and nerve supply of the thymus gland.
17. Describe the effects of aging on the endocrine system.
18. Describe the development of the endocrine system.
19. Define key medical terms associated with the endocrine system.

Two regulatory systems are involved in transmitting messages and correlating various body functions: the nervous system and endocrine system. The nervous system controls homeostasis through electrical impulses delivered over neurons. The endocrine system affects bodily activities by releasing chemical messengers, called hormones, into the bloodstream. Whereas the nervous system sends messages to a specific set of cells (muscle fibers, gland cells, or other neurons), the endocrine system as a whole sends messages to cells in virtually any part of the body. The nervous system causes muscles to contract and glands to secrete; the endocrine system brings about changes in the metabolic activities of body tissues. Neurons tend to act within a few milliseconds; hormones can take up to several hours or more to bring about their responses. Also, the effects of nervous system stimulation are generally brief compared to the effects of endocrine stimulation.

Obviously, the body could not function if the two great control systems were to pull in opposite directions. The nervous and endocrine systems coordinate their activities like an interlocking supersystem. Certain parts of the nervous system stimulate or inhibit the release of hormones, and hormones, in turn, are quite capable of stimulating or inhibiting the flow of nerve impulses.

Although the effects of hormones are many and varied, their actions can be categorized into four broad areas:

1. They help to control the internal environment by regulating its chemical composition and volume.
2. They respond to marked changes in environmental conditions to help the body cope with emergency demands such as infection, trauma, emotional stress, dehydration, starvation, hemorrhage, and temperature extremes.
3. They assume a role in the smooth, sequential integration of growth and development.
4. They contribute to the basic processes of reproduction, including gamete (egg and sperm) production, fertilization, nourishment of the embryo and fetus, delivery, and nourishment of the newborn.

The science concerned with the structure and functions of the endocrine glands and the diagnosis and treatment of disorders of the endocrine system is called *endocrinology* (en'-dō-kri-NOL-ō-jē; *endo* = within; *krinein* = to separate; *logos* = study of).

The developmental anatomy of the endocrine system is considered later in the chapter.

ENDOCRINE GLANDS

The body contains two kinds of glands: exocrine and endocrine. *Exocrine glands* secrete their products onto a free surface or into ducts. The ducts carry the secretions into body cavities, into the lumens of various organs, or to the body's surface. Exocrine glands include sudoriferous (sweat), sebaceous (oil), mucous, and digestive glands. *Endocrine glands* by contrast, secrete their products (hormones) into the extracellular space around the secretory cells, rather than into ducts. The secretion then passes into capillaries to be transported in the blood. The endocrine glands make up the *endocrine system.* The endocrine glands of the body include the pituitary (hypophysis), thyroid, parathyroids, adrenals (suprarenals), pineal (epiphysis cerebri), and thymus gland. In addition, there are several organs of the body that contain endocrine tissue but are not exclusively endocrine glands. These include the pancreas, ovaries, testes, kidneys, stomach, small intestine, and placenta. The locations of many organs of the endocrine system and endocrine-containing organs are illustrated in Figure 21-1.

The secretions of endocrine glands are called *hormones* (*hormone* = set in motion). The one thing all hormones have in common is the function of maintaining homeostasis by changing the physiological activities of cells. The amount of hormone released by an endocrine gland or tissue is determined by the body's *need* for the hormone at any given time. This is the basis on which the endocrine system operates. Hormone-producing cells have available to them information from sensing and signaling systems that permit the hormone-producing cells to regulate the amount and duration of hormone release. Secretion is normally regulated so that there is no overproduction or underproduction of a particular hormone.

Once a hormone is released by a secretory cell, it is carried by the blood to *target cells,* cells that respond to the hormone. All cells are target cells for one or more hormones, but not all cells respond to a particular hormone. Target cells contain receptors that bind to the hormone so that it can produce the effect. Once the target cells respond to the hormone, the response must be recognized by the secretory cell by some type of feedback signal. Finally, hormones that have accomplished their goals are degraded by target cells or removed by the liver or kidneys.

RECEPTORS

Cells that produce hormones represent only a limited number of cell types. By contrast, practically all cells of the body are target cells for one hormone or another. As a rule, most of the over 50 or so hormones affect a wide range of target cells. Since all body cells are exposed to equal concentrations of hormones, why is it that some target cells respond to particular hormones and others do not? The answer is *receptors,* large protein molecules found in the plasma membrane (integral proteins), cytoplasm, and nucleus of target cells. Receptors are quite specific in that they recognize only certain hormones. Because of the specific complementary structure of hormone and receptor, only certain hormones bind to certain receptors. It is for this very reason that a hormone influences its target cells but not other cells in the body. In addition, different types of cells can possess receptors for the same hormone but

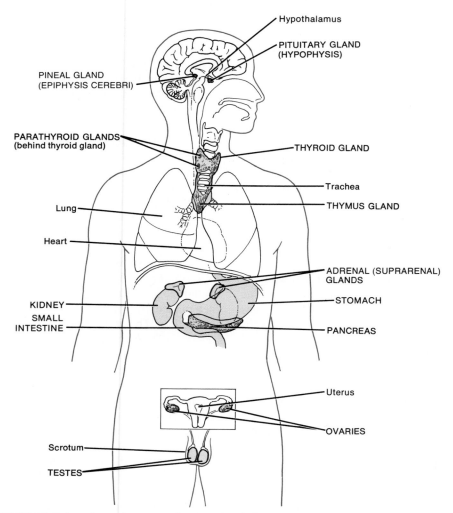

FIGURE 21-1 Location of many endocrine glands, organs containing endocrine tissue, and associated structures.

the responses of the cells are quite different from each other. For example, the hormone insulin acts on fat cells to stimulate glucose transport and lipid synthesis; it acts on liver cells to stimulate amino acid transport and glycogen synthesis; and it acts on pancreatic cells to inhibit their secretion of a hormone called glucagon.

Once a hormone is bound to a specific receptor, the combination activates a chain of events within the target cell in which the physiological effects of the hormone are expressed. Receptors, like other cell proteins, are constantly synthesized and degraded and change in concentration and affinity in response to changes within the body. There are generally 2,000–100,000 receptors per target cell.

FEEDBACK CONTROL

As indicated earlier, the amount of hormone released by an endocrine gland or tissue is determined by the body's need for the hormone at any given time. Most hormones are released in short bursts, with little or no release between bursts. When properly stimulated, an endocrine gland will release hormone in more frequent bursts, and thus blood levels of the hormone increase. Conversely, in the absence of stimulation, bursts are minimal or inhibited, and thus blood levels of the hormone decrease. Secretion is normally regulated so that there is no overproduction or underproduction of a particular hormone. This regulation is one of the very important ways that the body attempts to maintain homeostasis. Unfortunately, there are times when the regulating mechanism does not operate properly and hormonal levels are excessive or deficient. When this happens, disorders result, several of which are discussed later.

The typical way in which hormonal secretions are regulated is by **negative feedback control.** As applied to hormones, information regarding the hormone level or its effect is fed back to the gland, which then responds accordingly.

In one type of negative feedback system, the regulation of hormonal secretion does not involve direct participation by the nervous system. For example, blood calcium level is controlled in part by parathyroid hormone (PTH), produced by the parathyroid glands. If, for some reason, blood calcium level is low, this serves as a stimulus for the parathyroids to release more PTH. PTH then exerts its effects in various parts of the body until the blood calcium level is raised to normal. A high blood calcium level serves as a stimulus for the parathyroids to cease their production of PTH. In the absence of the hormone, other mechanisms take over until blood calcium level is lowered to normal. Note than in negative feedback control the body's response (increased or decreased calcium level) is opposite (negative) to the stimulus (low or high calcium level). Other hormones that are regulated without direct involvement of the nervous system include calcitonin (CT) produced by the thyroid gland that is controlled by blood levels of insulin; insulin produced by the pancreas that is controlled by blood levels of glucose; and aldosterone produced by the adrenal cortex that is controlled by blood volume and blood levels of potassium.

In other negative feedback systems, the hormone is released as a direct result of nerve impulses that stimulate the endocrine gland. Epinephrine and norepinephrine (NE) are released from the adrenal medulla in response to sympathetic nerve impulses. Antidiuretic hormone (ADH) is released from the posterior pituitary in response to nerve impulses from the hypothalamus.

There are also negative feedback systems that involve the nervous system through chemical secretions from the hypothalamus, called *regulating hormones* (or *factors*). Some regulating factors, called *releasing hormones* (or *factors*), stimulate the release of the hormone into the blood, so that it can exert its influence. Other regulating hormones (or factors) called *inhibiting hormones* (or *factors*) prevent the release of the hormone.

One of the few exceptions to the rule of negative feedback control is oxytocin (OT). The regulating system for the release of OT from the pituitary gland in response to nerve impulses from the hypothalamus is a positive feedback cycle; that is, the output intensifies the input. Another exception is the luteinizing hormone (LH) surge that results in ovulation.

PITUITARY (HYPOPHYSIS)

The hormones of the *pituitary gland,* also called the *hypophysis* (hī-POF-i-sis), regulate so many body activities that the pituitary has been nicknamed the "master gland." It is a round structure and surprisingly small, measuring about 1.3 cm (0.5 in.) in diameter. The pituitary gland lies in the sella turcica of the sphenoid bone. Posterior to the optic chiasma is a grayish protuberance, the *tuber cinereum,* which is part of the hypothalamus. The *median eminence* is a raised portion of the tuber cinereum to which is attached the *infundibulum,* a stalklike structure that attaches the pituitary gland to the hypothalamus (see Figure 21-2).

The pituitary gland is divided structurally and functionally into an anterior lobe and a posterior lobe. Both are connected to the hypothalamus. The *anterior lobe* constitutes about 75 percent of the total weight of the gland. It is derived from an outgrowth of ectoderm called the hypophyseal (Rathke's) pouch (see Figure 21-16b). Accordingly, the anterior lobe contains many glandular epithelial cells and forms the glandular part of the pituitary. A system of blood vessels connects the anterior lobe with the hypothalamus.

The *posterior lobe* is also derived from the ectoderm, but from an outgrowth called the neurohypophyseal bud (see Figure 21-16a). Accordingly, the posterior lobe contains axon terminations of neurons whose cell bodies are located in the hypothalamus. The nerve fibers that terminate in the posterior lobe are supported by cells called pituicytes. Other nerve fibers connect the posterior lobe directly with the hypothalamus.

Between the lobes is a small, relatively avascular zone, the *pars intermedia.* Although much larger and more clearly defined in structure and function in some lower animals, its role in humans is obscure.

ADENOHYPOPHYSIS

The anterior lobe of the pituitary is also called the *adenohypophysis* (ad'-e-nō-hī-POF-i-sis). It releases hormones that regulate a whole range of bodily activities from growth to reproduction. The release of these hormones is either stimulated or inhibited by chemical secretions produced by neurosecretory cells in the hypothalamus called *regulating hormones* (or *factors*). These substances constitute an important link between the nervous and endocrine systems.

The hypothalamic regulating hormones (or factors) are delivered to the adenohypophysis through a series of blood vessels. The blood supply to the adenohypophysis and infundibulum is derived principally from several *superior hypophyseal* (hī'-po-FIZ-ē-al) *arteries.* These arteries are branches of the internal carotid and posterior communicating arteries (Figure 21-2). The superior hypophyseal arteries form a network or plexus of capillaries, the *primary plexus,* in the infundibulum near the inferior portion of the hypothalamus diffuse into this plexus. This plexus drains into the *hypophyseal portal veins* that pass down the infundibulum. At the inferior portion of the infundibulum, the veins form a *secondary plexus* in the adenohypophysis. From this plexus, hormones of the adenohypophysis pass into the anterior hypophyseal veins for distribution to tissue cells. Such a delivery system permits regulating hormones (or factors) to act quickly on the adenohypophysis without first circulating through the heart. The short route prevents dilution or destruction of the regulating hormones (or factors).

When the adenohypophysis receives proper stimulation from the hypothalamus via regulating hormones (or factors) its glandular cells secrete any one of seven hormones. Spe-

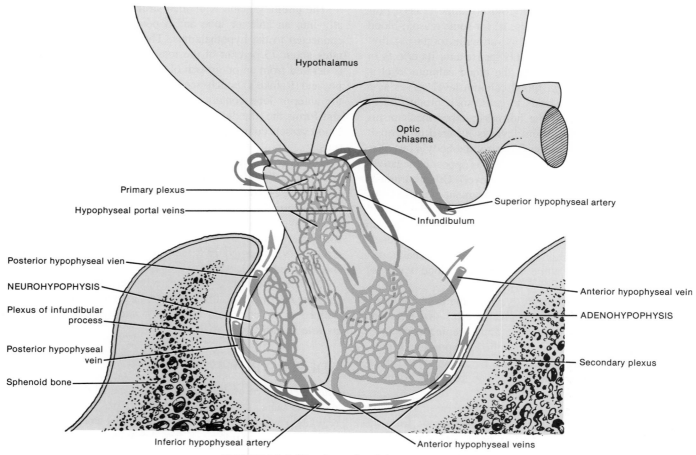

FIGURE 21-2 Blood supply of the pituitary gland.

cial staining techniques have established the division of glandular cells into five principal types (Figure 21-3):

1. *Somatotroph cells* produce *growth hormone (GH)*, which controls general body growth.
2. *Lacotroph cells* synthesize *prolactin (PRL)*, which initiates milk production by the mammary glands.
3. *Corticolipotroph cells* synthesize *adrenocorticotropic hormone (ACTH)*, which stimulates the adrenal cortex to secrete its hormones, and *melanocyte-stimulating hormone (MSH)*, which is related to skin pigmentation.
4. *Thyrotroph cells* manufacture *thyroid-stimulating hormone (TSH)*, which controls the thyroid gland.
5. *Gonadotroph cells* produce *follicle-stimulating hormone (FSH)*, which stimulates the production of eggs and sperm in the ovaries and testes, respectively, and *luetinizing hormone (LH)*, which stimulates other sexual and reproductive activities.

Except for the growth hormone (GH), melanocyte-stimulating hormone (MSH), and prolactin (PRL), all the secretions are referred to as *tropic hormones* (*trop* = turn on), which means that they stimulate other endocrine glands. Follicle-stimulating hormone (FSH) and luteinizing hor-

mone (LH) are also called *gonadotropic* (gō-nad-ō-TRŌ-pik) *hormones* because they regulate the functions of the gonads. The gonads (ovaries and testes) are the endocrine glands that produce sex steroid hormones.

CLINICAL APPLICATION

Disorders of the endocrine system, in general, involve *hyposecretion* (underproduction) of hormones or *hypersecretion* (overproduction).

Among the clinically interesting disorders related to the adenohypophysis are those involving GH. If GH is hyposecreted during the growth years, bone growth is slow and the epiphyseal plates close before normal height is reached. This condition is called *pituitary dwarfism*. Other organs of the body also fail to grow, and the pituitary dwarf is childlike in many physical respects. Treatment requires administration of GH during childhood before the epiphyseal plates close. Dwarfism is also caused by other conditions, in which administration of GH is not corrective.

Hypersecretion of GH during childhood results in *gigantism (gigantism)*, an abnormal increase in the length of long bones. As a result, the person grows to be very

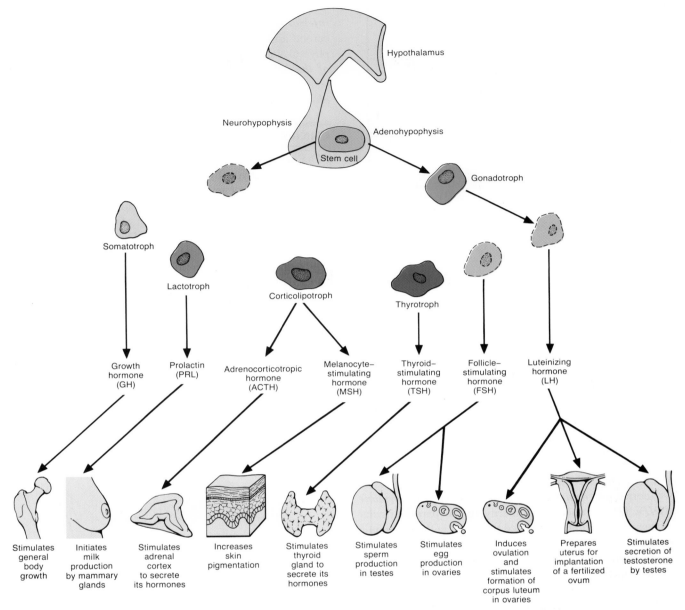

FIGURE 21-3 Cells of the adenohypophysis as revealed by special strains. Most cells that produce growth hormone (somatotrophs) and prolactin (lactotrophs) are separate cells. However, some normal and tumor cells are single cells that produce both growth hormone and prolactin. The corticolipotroph cell produces adrenocorticotropic hormone and melanocyte-stimulating hormone. Thyrotroph cells synthesize thyroid-stimulating hormone. Most gonadotroph cells produce both follicle-stimulating hormone and luteinizing hormone. However, a few separate cells may exist, some producing follicle-stimulating hormone and some producing luteinizing hormone. (Adapted from a slide provided by Calvin Ezrin, M.D., Clinical Professor of Medicine, U.C.L.A., and Adjunct Professor of Pathology, University of Toronto.)

large, but body proportions are about normal. Hypersecretion during adulthood is called *acromegaly* (ak'-rō-MEG-a-lē), which is shown in Figure 21-4. Acromegaly cannot produce further lengthening of the long bones because the epiphyseal plates are already closed. Instead, the bones of the hands, feet, cheeks, and jaws thicken. Other tissues also grow. The eyelids, lips, tongue, and nose enlarge and the skin thickens and furrows, especially on the forehead and soles of the feet.

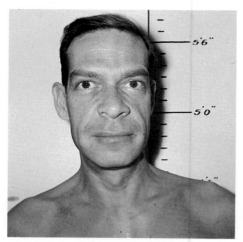

FIGURE 21-4 An individual with acromegaly. (Courtesy of Lester V. Bergman & Associates, Inc.)

NEUROHYPOPHYSIS

In a strict sense, the posterior lobe, or **neurohypophysis,** is not an endocrine gland, since is does not synthesize hormones. Instead, it stores hormones. The posterior lobe consists of cells called **pituicytes** (pi-TOO-i-sītz), which are similar in appearance to the neuroglia of the nervous system. It also contains axon terminals of secretory neurons of the hypothalamus (Figure 21-5). Such neurons are called **neurosecretory cells.** The cell bodies of the neurons originate in nuclei in the hypothalamus. The fibers project from the hypothalamus, form the **hypothalamic–hypophyseal tract,** and terminate on blood capillaries in the neurohypophysis. The cell bodies of the neurosecretory cells produce two hormones: **oxytocin (OT)** and **antidiuretic hormone (ADH).** OT is produced primarily in the paraventricular nucleus and ADH is synthesized primarily in the supraoptic nucleus.

OT stimulates contraction of the smooth muscle of the pregnant uterus during labor and stimulates contractile cells around the ducts of the mammary glands to eject milk. ADH prevents excessive urine production by bringing about water reabsorption and secondarily causes blood pressure to rise by bringing about constriction of arterioles.

Following their production, the hormones are transported through the neuron fibers to the neurohypophysis and are stored in the axon terminals where they become bound to small proteins called **neurophysins.** The transport from the cell bodies to the axon terminations takes about 10 hours. Neurophysins aid in storing the hormones and are important in the release mechanism. The bound hormone is released from the nerve endings in response to nerve impulses reaching the axon terminals. Thus, these fibers perform two tasks. First, they act as conduits for the transport of the hormone molecules from their site of production in the hypothalamic nucleus to their site of secretion in the neurohypophysis. Second, the fibers carry the releasing nerve impulses down to the axon terminals.

The blood supply to the neurohypophysis is from the **inferior hypophyseal arteries,** derived from the internal carotid arteries. In the neurohypophysis, the inferior hypophyseal arteries form a plexus of capillaries called the **plexus of the infundibular process.** From this plexus hormones stored in the neurohypophysis pass into the **posterior hypophyseal veins** for distribution to tissue cells.

CLINICAL APPLICATION

The principal abnormality associated with dysfunction of the neurohypophysis is **diabetes insipidus** (in-SIP-i-dus). Diabetes means overflow and insipidus means tasteless. This disorder should not be confused with diabetes mellitus (*meli* = honey), a disorder of the pancreas characterized by sugar in the urine. Diabetes insipidus is the result of a hyposecretion of ADH, usually caused by damage to the neurohypophysis or the supraoptic nucleus in the hypothalamus. Symptoms include excretion of large amounts of urine and subsequent thirst. Diabetes insipidus is treated by administering ADH.

A summary of pituitary gland hormones, actions, and disorders is presented in Exhibit 21-1.

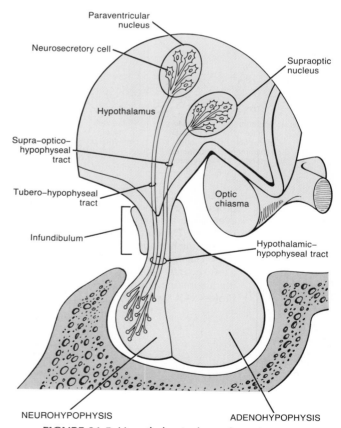

FIGURE 21-5 Hypothalamic–hypophyseal tract.

EXHIBIT 21-1

Summary of Pituitary Gland Hormones, Actions, and Disorders

HORMONE	PRINCIPAL ACTIONS	DISORDERS
Adenohypophyseal Hormones		
Growth Hormone (GH)	Growth of body cells; protein anabolism.	Hyposecretion of GH during the growth years results in pituitary dwarfism; hypersecretion of GH during the growth years results in giantism; hypersecretion of GH during adulthood results in acromegaly.
Thyroid-stimulating Hormone (TSH)	Controls secretion of hormones by thyroid gland.	
Adrenocorticotropic Hormone (ACTH)	Controls secretion of some hormones by adrenal cortex.	
Follicle-stimulating Hormone (FSH)	In female, initiates development of ova and induces ovarian secretion of estrogens. In male, stimulates testes to produce sperm.	
Luteinizing Hormone (LH)	In female, together with estrogens, stimulates ovulation and formation of progesterone-producing corpus luteum, prepares uterus for implantation, and readies mammary glands to secrete milk. In male, stimulates interstitial cells in testes to develop and produce testosterone.	
Prolactin (PRL)	Together with other hormones, initiates and maintains milk secretion by the mammary glands.	
Melanocyte-stimulating Hormone (MSH)	Stimulates dispersion of melanin granules in melanocytes.	
Neurohypophyseal Hormones		
Oxytocin (OT)	Stimulates contraction of smooth muscle cells of pregnant uterus during labor and stimulates contraction of contractile cells of mammary glands for milk ejection.	
Antidiuretic Hormone (ADH)	Principal effect is to decrease urine volume; also to raise blood pressure by constricting arteries during severe hemorrhage.	Hyposecretion of ADH results in diabetes insipidus.

THYROID

The *thyroid gland* is located just below the larynx. The right and left *lateral lobes* lie one on either side of the trachea. The lobes are connected by a mass of tissue called an *isthmus* (IS-mus) that lies in front of the trachea, just below the cricoid cartilage (Figure 21-6). The *pyramidal lobe,* when present, extends upward from the isthmus. The gland has a rich blood supply, receiving about 80–120 ml of blood per minute. Thus, the thyroid gland can deliver high levels of hormones in a short period of time, if necessary.

Histologically, the thyroid gland is composed of spherical sacs called *thyroid follicles* (Figure 21-7). The wall of each follicle consists of two types of cells. Those that reach the surface of the lumen of the follicle are called *follicular cells,* and those that do not reach the lumen are called *parafollicular (C) cells.* When the cells are inactive, they tend to be low cuboidal to squamous, but when actively secreting hormones, they become more columnar. The follicular cells manufacture *thyroxine* (thī-ROK-sēn) or T_4, since it contains four atoms of iodine, and *triiodothyronine* (trī-ī'-ōd-ō-THĪ-rō-nēn) or T_3, since it contains three atoms of iodine. Together these hormones are referred to as the *thyroid hormones.* Thyroxine is normally secreted in greater quantity than triiodothyronine, but triiodothyronine is three to four times more potent. Moreover, in peripheral tissues, especially the liver and lungs, about one-third of triiodothyronine is converted to thyroxine. Both hormones are functionally similar. They control metabolism, regulate growth and development, and increase reactivity of the nervous system. The parafollicular cells produce *calcitonin* (kal-si-TŌ-nin), or *CT.* This hormone decreases blood levels of calcium and phosphate by inhibiting bone breakdown (osteoclastic activity) and accelerating calcium absorption by bones.

One of the thyroid gland's unique features is its ability

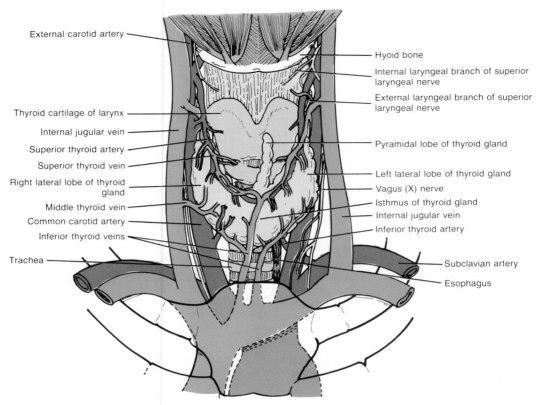

FIGURE 21-6 Location and blood supply of the thyroid gland in anterior view. The pyramidal lobe of the thyroid, when present, may be attached to the hyoid bone by muscle.

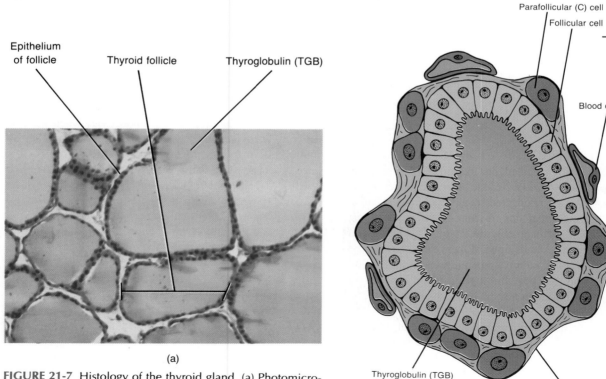

FIGURE 21-7 Histology of the thyroid gland. (a) Photomicrograph at a magnification of 230×. (Copyright © 1983 by Michael H. Ross. Used by permission.) (b) Diagram showing details of a single thyroid follicle.

to store hormones and release them in a steady flow over a long period of time. In the follicle cells, iodide is oxidized to iodine. Through a series of enzymatically controlled reactions, the iodine combines with the amino acid tyrosine to form the thyroid hormones. This combination occurs within a large glycoprotein molecule, called **thyroglobulin (TGB),** which is secreted by the follicle cells into the follicle.

The main blood supply of the thyroid gland is from the superior thyroid artery, a branch of the thyrocervical trunk of the external carotid artery, and the inferior thyroid artery, a branch of the subclavian artery. The thyroid is drained by the superior and middle thyroid veins, which pass into the internal jugular veins, and the inferior thyroid veins, which join the brachiocephalic veins or internal jugular veins.

The nerve supply of the thyroid consists of postganglionic fibers from the superior and middle cervical sympathetic ganglia. Preganglionic fibers from the ganglia are derived from the second through seventh thoracic segments of the spinal cord.

CLINICAL APPLICATION

Hyposecretion of thyroid hormones during the growth years results in *cretinism* (KRĒ-tin-izm), which is shown in Figure 21-8a. Two outstanding clinical symptoms of the cretin are dwarfism and mental retardation. The first is caused by failure of the skeleton to grow and mature; the second is caused by failure of the brain to develop fully. Recall that one function of thyroid hormones is to control tissue growth and development. Cretins also exhibit retarded sexual development and a yellowish skin color. Flat pads of fat develop, giving the cretin a characteristic round face and thick nose; a large, thick, protruding tongue; and protruding abdomen. Because the energy-producing metabolic reactions are slow, the cretin has a low body temperature and suffers from general lethargy. Carbohydrates are stored rather than utilized, and heart rate is also slow. If the condition is diagnosed early, the symptoms can be eliminated by administering thyroid hormones.

Hypothyroidism during the adult years produces *myxedema* (mix-e-DĒ-ma). A hallmark of this disorder is an edema that causes the facial tissues to swell and look puffy. Like the cretin, the person with myxedema suffers from slow heart rate, low body temperature, muscular weakness, general lethargy, and a tendency to gain weight easily. The long-term effect of a slow heart rate may overwork the heart muscle, causing the heart to enlarge. Because the brain has already reached maturity, the person with myxedema does not experience mental retardation. However, in moderately severe cases, nerve reactivity may be dulled so that the person lacks mental alertness.

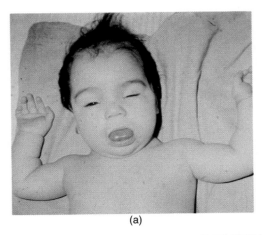

(a)

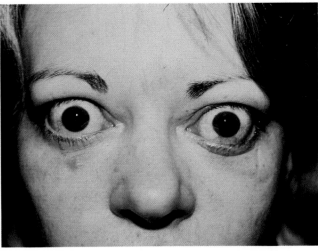

(b)

(c)

FIGURE 21-8 Abnormalities related to the thyroid gland. (a) Photograph of an individual with cretinism. (Courtesy of Lester V. Bergman & Associates, Inc.) (b) Photograph of exophthalmos. (Courtesy of Lester V. Bergman & Associates, Inc.) (c) Simple goiter. (Copyright © Kay, Peter Arnold.)

Myxedema occurs eight times more frequently in females than in males. Its symptoms are alleviated by the administration of thyroid hormones.

Hypersecretion of thyroid hormones gives rise to *exophthalmic* (ek'-sof-THAL-mik) *goiter* (GOY-ter). This disease, like myxedema, is also more frequent in females. One of its primary symptoms is an enlarged thyroid, called a *goiter,* which may be two to three times its original size. Two other symptoms are an edema behind the eye, which causes the eye to protrude (*exophthalmos*), which is shown in Figure 21-8b, and an abnormally high metabolic rate. The high metabolic rate produces a range of effects that are generally opposite to those of myxedema—increased pulse, high body temperature, and moist, flushed skin. The person loses weight and is usually full of ''nervous'' energy. The thyroid hormones also increase the responsiveness of the nervous system, causing the person to become irritable and exhibit tremors of the extended fingers. Hyperthyroidism is usually treated by administering drugs that suppress thyroid hormone synthesis, or by surgically removing part of the gland.

Goiter is a symptom of many thyroid disorders. It may also occur if the gland does not receive enough iodine to produce sufficient thyroxine for the body's needs. The follicular cells then enlarge in a futile attempt to produce more thyroid hormones, and they secrete large quantities of thyroglobulin (TGB). This condition is called *simple goiter* (Figure 21-8c). Simple goiter is most often caused by a lower-than-average amount of iodine in the diet. It may also develop if iodine intake is not increased during certain conditions that put a high demand on the body for thyroxine, such as frequent exposure to cold and high-fat and protein diets.

A summary of thyroid gland hormones, actions, and disorders is presented in Exhibit 21-2.

PARATHYROIDS

Typically embedded on the posterior surfaces of the lateral lobes of the thyroid gland are small, round masses of tissue called the *parathyroid glands.* Usually, two parathyroids, superior and inferior, are attached to each lateral thyroid lobe (Figure 21-9). They measure about 3–6 mm (0.1–0.3 in.) in length. 2–5 mm (0.07–0.2 in.) in width, and 0.5–2 mm (0.02–0.07 in.) in thickness.

Histologically, the parathyroids contain two kinds of epithelial cells (Figure 21-10). The more numerous cells, called *principal (chief) cells,* are believed to be the major synthesizer of *parathyroid hormone (PTH).* Some researchers believe that the other kind of cell, called an *oxyphil cell,* synthesizes a reserve capacity of hormone. Functionally, PTH increases blood calcium level and decreases blood phosphate level by increasing the rate of calcium absorption from the gastrointestinal tract into the blood; increases the number and activity of osteoclasts; increases calcium absorption by the kidneys; increases phosphate excretion by the kidneys; and activates vitamin D.

The parathyroids are abundantly supplied with blood from branches of the superior and inferior thyroid arteries. Blood is drained by the superior, middle, and inferior thyroid veins. The nerve supply of the parathyroids is derived from the thyroid branches of cervical sympathetic ganglia and appear to be vasomotor in function.

CLINICAL APPLICATION

A normal amount of calcium in the extracellular fluid is necessary to maintain the resting state of neurons. A deficiency of calcium caused by hypoparathyroidism (abnormally diminished function of the parathyroid glands) causes neurons to depolarize without the usual stimulus. As a result, nerve impulses increase and result in muscle

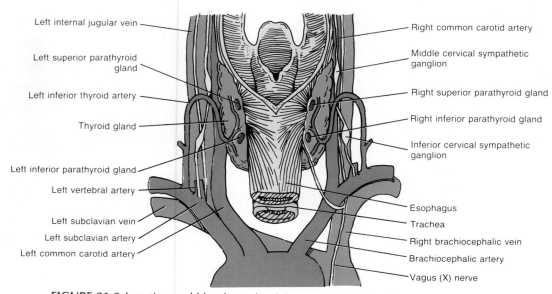

FIGURE 21-9 Location and blood supply of the parathyroid glands in posterior view.

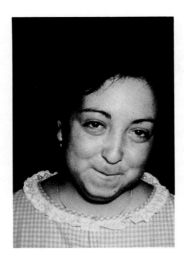

FIGURE 21-13 An individual with Cushing's syndrome. (Courtesy of Lester V. Bergman & Associates, Inc.)

CLINICAL APPLICATIONS

The *adrenogenital syndrome* usually refers to a group of enzyme deficiencies that block the synthesis of glucocorticoids. In an attempt to compensate, the anterior pituitary secretes more ACTH. As a result, excess androgenic (male) hormones are produced, causing *virilism,* or masculinization. For instance, the female develops extremely virile characteristics such as growth of a beard, development of a much deeper voice, occasionally the development of baldness, development of a masculine distribution of hair on the body and on the pubis, growth of the clitoris that resembles a penis, and deposition of proteins in the skin and muscles producing typical masculine characteristics. Such virilism may also result from tumors of the adrenal gland called *virilizing adenomas* (*aden* = gland; *oma* = tumor).

In the prepubertal male, the syndrome causes the same characteristics as in the female, plus rapid development of the male sexual organs and creation of male sexual desires. In the adult male, the virilizing characteristics of the adrenogenital syndrome are usually completely obscured by the normal virilizing characteristics of the testosterone secreted by the testes. As a result, it is often difficult to make a diagnosis of adrenogenital syndrome in the male adult. However, an occasional adrenal tumor secretes sufficient quantities of feminizing hormones (estrogens) that the male patient develops *gynecomastia* (*gyneco* = woman, *mast* = breast), which means excessive growth (benign) of the male mammary glands. Such a tumor is called a *feminizing adenoma.* Gynecomastia is also associated with androgen-deficiency states, pulmonary diseases, chest wall trauma, psychological stress, and certain drugs (alcohol, cimetidine, and digitalis derivatives). As a rule, specific treatment of gynecomastia is not indicated; however, surgery may be undertaken for cosmetic reasons or a chronic condition.

ADRENAL MEDULLA

The adrenal medulla consists of hormone-producing cells, called *chromaffin* (krō-MAF-in) *cells,* which surround large blood-containing sinuses (see Figure 21-12e). Chromaffin cells develop from the same source as the postganglionic cells of the sympathetic division of the autonomic nervous system. They are directly innervated by preganglionic cells of the sympathetic division of the autonomic nervous system and may be regarded as postganglionic cells that are specialized to secrete. In all other visceral effectors, preganglionic sympathetic fibers first synapse with postganglionic neurons before innervating the effector. In the adrenal medulla, however, the preganglionic fibers pass directly into the chromaffin cells of the gland. The secretion of hormones from the chromaffin cells is directly controlled by the autonomic nervous system, and innervation by the preganglionic fibers allows the gland to respond rapidly to a stimulus.

The two principal hormones synthesized by the adrenal medulla are *epinephrine* and *norepinephrine (NE),* also called adrenaline and noradrenaline, respectively. Epinephrine constitutes about 80 percent of the total secretion of the gland and is more potent in its action than norepinephrine. Both hormones are *sympathomimetic* (sim'-pa-thō-mi-MET-ik); that is, they produce effects that mimic those brought about by the sympathetic division of the autonomic nervous system. To a large extent, they are responsible for the fight-or-flight response. Like the glucocorticoids of the adrenal cortices, these hormones help the body resist stress. However, unlike the cortical hormones, the medullary hormones are not essential for life.

The main arteries that supply the adrenal glands are the several superior arteries arising from the inferior phrenic artery, the middle suprarenal artery from the aorta, and the inferior suprarenal arteries from the renal arteries. The suprarenal vein of the right adrenal gland drains into the inferior vena cava, whereas the suprarenal vein of the left adrenal gland empties into the left renal vein.

The principal nerve supply to the adrenal glands is from preganglionic fibers from the thoracic splanchnic nerves, which pass through the celiac and associated sympathetic plexuses. These myelinated fibers end on the secretory cells of the gland found in a region of the medulla.

CLINICAL APPLICATION

Tumors of the chromaffin cells of the adrenal medulla, called *pheochromocytomas* (fē-ō-krō'-mō-sī-TŌ-mas), cause hypersecretion of the medullary hormones. Such tumors are usually benign. The oversecretion causes rapid heart rate, headache, high blood pressure, high levels of sugar in the blood and urine, an elevated basal metabolic rate (BMR), flushing of the face, nervousness, sweating, decreased gastrointestinal motility, and vertigo. Since the medullary hormones create the same effects as sympathetic nervous stimulation, hypersecretion puts the individual into a prolonged version of the fight-

or-flight response. This condition ultimately wears out the body, and the individual eventually suffers from general weakness. The only definitive treatment of pheochromocytomas is surgical removal of the tumor(s).

A summary of adrenal gland hormones, actions, and disorders is presented in Exhibit 21-4.

PANCREAS

The *pancreas* can be classified as both an endocrine and an exocrine gland. Thus, it is referred to as a *heterocrine gland.* We shall treat its endocrine functions at this point; its exocrine functions are discussed in the chapter on the digestive system (Chapter 23). The pancreas is a flattened organ located posterior and slightly inferior to the stomach (Figure 21-14). The adult pancreas consists of a head, body, and tail. Its average length is about 12.5 cm (6 in.), and its average weight is about 85 g (3 oz).

The endocrine portion of the pancreas consists of clusters of cells called *pancreatic islets,* or *islets of Langerhans* (LAHNG-er-hanz) (Figure 21-15). Three kinds of cells are found in these clusters: (1) *alpha cells* that secrete the hormone *glucagon,* which increases blood sugar level; (2) *beta cells* that secrete the hormone *insulin,* which decreases blood sugar level; and (3) *delta cells* that secrete *growth hormone-inhibiting factor* (GHIF), or *somatostatin,* a hormone that inhibits the secretion of insulin and glucagon. The islets are infiltrated by blood capillaries and surrounded by cells (acini) that form the exocrine part of the gland.

The arterial supply of the pancreas is from the superior and inferior pancreaticoduodenal arteries and from the splenic and superior mesenteric arteries. The veins, in general, correspond to the arteries. Venous blood reaches the hepatic portal vein by means of the coronary, splenic, and superior mesenteric veins.

The nerves to the pancreas are autonomic nerves derived from the celiac and superior mesenteric plexuses. Included are preganglionic vagal, postganglionic sympathetic, and afferent fibers. Parasympathetic vagal fibers are said to terminate at both acinar (exocrine) and islet (endocrine) cells. Although the innervation is presumed to influence enzyme formation, pancreatic secretion is controlled largely by the hormones secretin and cholecystokinin (CCK). The sympathetic fibers end on blood vessels and are vasomotor and accompanied by afferent fibers, especially for pain.

CLINICAL APPLICATION

Diabetes mellitus (MEL-i-tus) is not a single hereditary disease but a heterogeneous group of diseases, all of which ultimately lead to an elevation of glucose in the blood (hyperglycemia) and excretion of glucose in the urine as hyperglycemia increases. Diabetes mellitus is also characterized by the three ''polys'': an inability to reabsorb water, resulting in increased urine production (*polyuria*); excessive thirst (*polydipsia*); and excessive eating (*polyphagia*).

Two major types of diabetes mellitus have been distinguished: type I and type II. *Type I diabetes,* which occurs abruptly, is characterized by an absolute deficiency of insulin due to a marked decline in the number of insulin-producing beta cells (perhaps caused by the autoimmune destruction of beta cells), even though target cells contain insulin receptors. Type I diabetes is called *insulin-dependent diabetes* because periodic administration of insulin

EXHIBIT 21-4

Summary of Hormones Produced by the Adrenal Glands, Actions, and Disorders

HORMONE	PRINCIPAL ACTIONS	DISORDERS
Adrenal Cortical Hormones **Mineralocorticoids (Mainly Aldosterone)**	Increase blood levels of sodium and water and decrease blood levels of potassium.	Hypersecretion of aldosterone results in aldosteronism.
Glucocorticoids (Mainly Cortisol)	Help promote normal metabolism, resistance to stress, and counter inflammatory response.	Hyposecretion of glucocorticoids produces Addison's disease. Hypersecretion results in Cushing's syndrome.
Gonadocorticoids	Concentrations secreted by adults are so low that their effects are usually insignificant.	Androgenital syndrome inhibits synthesis of glucocorticoids that results in excess production of ACTH and androgens, causing virilism. The release of sufficient feminizing hormones in males causes gynecomastia.
Adrenal Medullary Hormones **Epinephrine**	Sympathomimetic, that is, produces effects that mimic those of the sympathetic division of the autonomic nervous system (ANS) during stress.	Hypersecretion of medullary hormones results in a prolonged fight-or-flight response.
Norepinephrine (NE)	Same as above.	

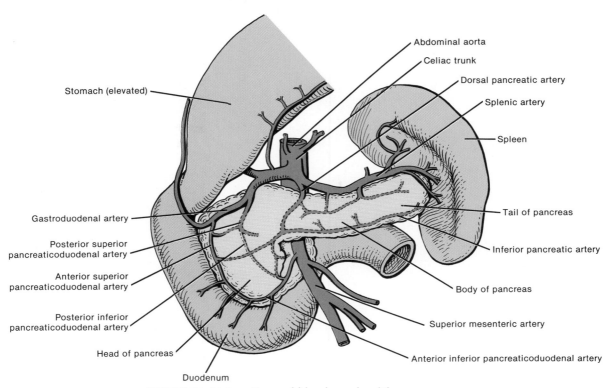

FIGURE 21-14 Location and blood supply of the pancreas.

is required to treat it. It is also known as *juvenile-onset diabetes* because it most commonly develops in people younger than age 20, though it persists throughout life. Although people who develop type I diabetes appear to have certain genes that make them more susceptible, some triggering factor is required. Viral infection seems to be such a factor. The deficiency of insulin accelerates the breakdown of the body's reserve of fat resulting in the production of organic acids called ketones. This causes a form of acidosis called *ketosis,* which lowers the pH of the blood and can result in death. The catabolism of stored fats and proteins also causes weight loss. As lipids are transported by the blood from storage depots to hungry cells, lipid particles are deposited on the walls of blood vessels. The deposition leads to atherosclerosis and a multitude of cardiovascular problems including cerebrovascular insufficiency, ischemic heart disease, peripheral vascular disease, and gangrene. One of the major complications of diabetes is loss of vision due to cataracts (excessive blood sugar chemically attaches to lens proteins, causing cloudiness) or damage to blood vessels of the retina. Severe kidney problems also may result from damage to renal blood vessels.

Type II diabetes is much more common than type I, representing more than 90 percent of all cases. Type II diabetes most often occurs in people who are over 40 and overweight. Since type II diabetes usually occurs later in life, it is called *maturity-onset diabetes.* Clinical symptoms are mild, and the high glucose levels in the blood can usually be controlled by diet alone or with antidiabetic drugs such as *glyburide* (DiaBeta). Many type II diabetics have a sufficient amount or even a surplus of insulin in the blood. For these individuals, diabetes arises not from a shortage of insulin but probably from defects in the molecular machniery that mediates the action of insulin on its target cells. Cells in many parts of the body, especially skeletal muscles and the liver, become less sensitive to insulin because they have fewer insulin receptors. Type II diabetes is therefore called *non-insulin-dependent diabetes.* Many elderly persons with this type of diabetes are required to administer insulin to themselves daily.

Hyperinsulinism is much rarer than hyposecretion and is generally the result of a malignant tumor in an islet. The principal symptom is a decreased blood glucose level, which stimulates the secretion of epinephrine, glucagon, and GH. As a consequence, anxiety, sweating, tremor, increased heart rate, and weakness occur. Moreover, brain cells do not have enough glucose to function efficiently. This condition leads to mental disorientation, convulsions, unconsciousness, shock, and eventually to death, as the vital centers in the medulla are affected.

A summary of pancreatic hormones, actions, and disorders is presented in Exhibit 21-5.

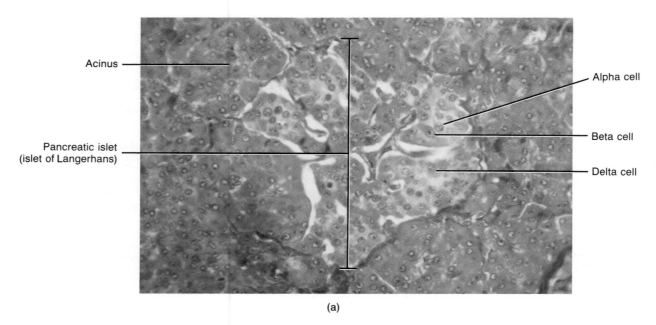

(a)

FIGURE 21-15 Histology of the pancreas. (a) Photomicrograph. (Courtesy of Andrew Kuntzman.) (b) Diagram.

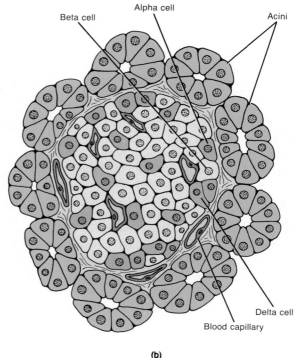

(b)

OVARIES AND TESTES

The female gonads, called the *ovaries,* are paired oval bodies located in the pelvic cavity. The ovaries produce female sex hormones called *estrogens* and *progesterone (PROG).* These hormones are responsible for the development and maintenance of the female sexual characteristics. Along with the gonadotropic hormones of the pituitary gland, the sex hormones also regulate the menstrual cycle, maintain pregnancy, and prepare the mammary glands for lactation. The ovaries (and placenta) also produce a hormone called *relaxin,* which relaxes the symphysis pubis and helps dilate the uterine cervix toward the end of pregnancy and plays a role in increasing sperm motility.

The male has two oval glands, called *testes,* that lie in the scrotum. The testes produce *testosterone,* the primary male sex hormone, that stimulates the development and maintenance of the male sexual characteristics. The testes also produce the hormone *inhibin* that inhibits secretion of FSH to control sperm production. The detailed structure of the ovaries and testes will be discussed in Chapter 25.

A summary of hormones produced by the ovaries and testes and their actions is presented in Exhibit 21-6.

PINEAL (EPIPHYSIS CEREBRI)

The endocrine gland attached to the roof of the third ventricle is known as the *pineal* (pīn-Ē-al) *gland* (because of its resemblance to a pine cone), or *epiphysis cerebri* (see Figure 21-1). The gland is about 5–8 mm (0.2–0.3 in.) long and

5 mm wide. It weighs about 0.2 g. It is covered by a capsule formed by a pia mater and consists of masses of *neuroglial cells* and parenchymal secretory cells called *pinealocytes* (pin-ē-AL-ō-sīts). Around the cells are scattered postganglionic sympathetic fibers. The pineal gland starts to accumulate calcium at about the time of puberty. Such calcium deposits are referred to as *brain sand.* Contrary to a once widely held belief, there is no evidence that the pineal atrophies with age and that the presence of brain sand is an indication of atrophy. In fact, the presence of brain sand may even indicate increased secretory activity.

EXHIBIT 21-5

Summary of Hormones Produced by the Pancreas, Actions, and Disorders

HORMONE	PRINCIPAL ACTIONS	DISORDERS
Glucagon	Raises blood sugar level by accelerating conversion of glucogen into glucose in liver (glycogenolysis) and conversion of other nutrients into glucose in liver (gluconeogenesis) and releasing glucose into blood.	
Insulin	Lowers blood sugar level by accelerating transport of glucose into cells, converting glucose into glycogen (glycogenesis), and decreasing glycogenolysis and gluconeogenesis; also increases lipogenesis and stimulates protein synthesis.	An absolute deficiency of insulin or a defect in the molecular machinery that mediates the action of insulin on target cells produces diabetes mellitus. Hypersecretion of insulin results in hyperinsulinism.
Growth Hormone-inhibiting Factor (GHIF) or Somatostatin	Inhibits secretion of insulin and glucagon.	

Although many anatomical facts concerning the pineal gland have been known for years, its physiology is still somewhat obscure. One hormone secreted by the pineal gland is *melatonin,* which appears to inhibit reproductive activities by inhibiting gonadotropic hormones. Some evidence also exists that the pineal secretes a second hormone called *adrenoglomerulotropin* (a-drē'-nō-glō-mer'-yoo-lō-TRŌ-pin). This hormone may stimulate the adrenal cortex to secrete aldosterone. Other substances found in the pineal gland include norepinephrine (NE), serotonin, histamine, gonadotropin releasing hormone (GnRH) and gammo aminobutyric acid (GABA).

The posterior cerebral artery supplies the pineal with blood, and the great cerebral vein drains it.

A summary of hormones produced by the pineal gland and their actions is presented in Exhibit 21-7.

THYMUS

Usually a bilobed lymphatic organ, the *thymus gland* is located in the superior mediastinum, posterior to the sternum and between the lungs (see Figure 15-7). The anatomy and histology of the thymus gland have already been discussed in Chapter 15. At this point, only its hormonal role in immunity will be discussed.

Lymphoid tissue of the body consists primarily of lymphocytes that may be distinguished into two kinds: B cells and T cells. Both are derived originally in the embryo from lymphocytic stem cells in bone marrow. Before migrating to their positions in lymphoid tissue, the descendants of the stem cells follow two distinct pathways. About half of them migrate to the thymus gland, where they are processed to become thymus-dependent lymphocytes, or *T cells.* The thymus gland confers on some of them the ability to destroy antigens (foreign microbes and substances). These cells, under the influence of hormones produced by the

EXHIBIT 21-6

Summary of Hormones of the Ovaries and Testes and Their Actions

HORMONE	PRINCIPAL ACTIONS
Ovarian Hormones	
Estrogens and Progesterone (PROG)	Development and maintenance of female sexual characteristics. Together with gonadotropic hormones of the adenohypophysis, they also regulate the menstrual cycle, maintain pregnancy, prepare the mammary glands for lactation, and regulate oogenesis.
Relaxin	Relaxes symphysis pubis and helps dilate uterine cervix near the end of pregnancy.
Testicular Hormones	
Testosterone	Development and maintenance of male sexual characteristics, regulation of spermatogenesis, and stimulation of descent of testes before birth.
Inhibin	Inhibits secretion of FSH to control sperm production.

EXHIBIT 21-7

Summary of Hormones of the Pineal Gland and Their Actions

HORMONE	PRINCIPAL ACTIONS
Melatonin	May inhibit reproductive activities by inhibiting gonadotropic hormones.
Adrenoglomerulotropin	May stimulate the adrenal cortex to secrete aldosterone.

thymus gland, called **thymosin, thymic humoral factor (THF), thymic factor (TF),** and **thymopoietin,** promote the maturation of T cells. There is also some evidence that thymic hormones may retard the aging process. The remaining stem cells are processed in some as yet undetermined area of the body, possibly bone marrow, the fetal liver, spleen, or gut-associated lymphoid tissue, and are known as **B cells.** These cells differentiate into plasma cells. Plasma cells, in turn, produce antibodies against antigens.

The arterial supply of the thymus is derived mainly from the internal thoracic and inferior thyroid vessels. The veins that drain the thymus are the internal thoracic, brachiocephalic, and thyroid veins. Postganglionic sympathetic and parasympathetic fibers supply the gland.

A summary of hormones produced by the thymus gland and their actions is presented in Exhibit 21-8.

EXHIBIT 21-8

Summary of Hormones Produced by the Thymus Gland and Their Principal Actions

HORMONES	PRINCIPAL ACTIONS
Thymosin, Thymic Humoral Factor (THF), Thymic Factor (TF), and Thymopoietin	Promote maturation of T cells.

AGING AND THE ENDOCRINE SYSTEM

The endocrine system exhibits a variety of changes, and many researchers look to this system with the hope of finding the key to the aging process. Disorders of the endocrine system are not frequent, but when they do occur, most often they are related to pathologic changes rather than age. Diabetes mellitus and thyroid disorders are common endocrine problems that have a significant effect on health and function.

DEVELOPMENTAL ANATOMY OF THE ENDOCRINE SYSTEM

The development of the endocrine system is not as localized as the development of other systems. The endocrine organs develop in widely separated parts of the embryo.

The *pituitary gland* (hypophysis) originates from two different regions of the **ectoderm.** The *neurohypophysis* (posterior lobe) is derived from an outgrowth of ectoderm called the **neurohypophyseal bud,** located on the floor of the hypothalamus (Figure 21-16a,b). The *infundibulum,* also an outgrowth of the neurohypophyseal bud, connects the neurohypophysis to the hypothalamus. The *adenohypophysis* (anterior lobe) is derived from an outgrowth of **ectoderm**

from the roof of the stomodeum (mouth) called the **hypophyseal (Rathke's) pouch.** The pouch grows toward the neurohypophyseal bud, and the pouch loses its connection with the roof of the mouth.

The *thyroid gland* develops as a midventral outgrowth of **endoderm,** called the **thyroid diverticulum,** from the floor of the pharynx at the level of the second pair of pharyngeal pouches (Figure 21-16b). The outgrowth projects inferiorly and differentiates into the right and left lateral lobes and the isthmus of the gland.

The *parathyroid glands* develop as outgrowths from the third and fourth **pharyngeal pouches** (Figure 21-16a).

The adrenal cortex and adrenal medulla have completely different embryological origins. The *adrenal cortex* is derived from intermediate **mesoderm** from the same region that produces the gonads (see Figure 25-24). The *adrenal medulla* is **ectodermal** in origin and is derived from the **neural crest,** which also produces sympathetic ganglia and other structures of the nervous system (see Figure 18-22c).

The *pancreas* develops from **endoderm** from dorsal and ventral outgrowths of the part of the **foregut** that later becomes the duodenum (see Figure 23-20). The two outgrowths eventually fuse to form the pancreas. The origin of the ovaries and testes is discussed in the section on the reproductive system.

The *pineal gland* arises from **ectoderm** of the **diencephalon** (see Figure 18-23b), as an outgrowth between the thalamus and colliculi.

The *thymus gland* arises from **endoderm** from the third **pharyngeal pouches** (Figure 21-16b).

OTHER ENDOCRINE TISSUES

Before leaving our discussion of hormones, it should be noted that body tissues other than those usually classified as endocrine glands also contain endocrine tissue and thus secrete hormones. The gastrointestinal tract synthesizes several hormones that regulate digestion in the stomach and small intestine. Among these hormones are **stomach gastrin, enteric gastrin, secretin, cholecystokinin (CCK), enterocrinin,** and **gastric inhibitory peptide (GIP).**

The placenta produces **human chorionic gonadotropin (HCG), estrogens, progesterone (PROG), relaxin,** and **human chorionic somatomammotropin (HCS),** all of which are related to pregnancy.

When the kidneys (and liver to a lesser extent) become hypoxic (subject to below normal levels of oxygen), it is believed that they release an enzyme called **renal erythropoietic factor.** It is secreted into the blood where it acts on a plasma protein to bring about the production of a hormone called **erythropoietin** (ē-rith'-rō-POY-ē-tin), which stimulates red blood cell production. The kidneys also help bring about the activation of the hormone **vitamin D.**

The skin produces vitamin D in the presence of sunlight.

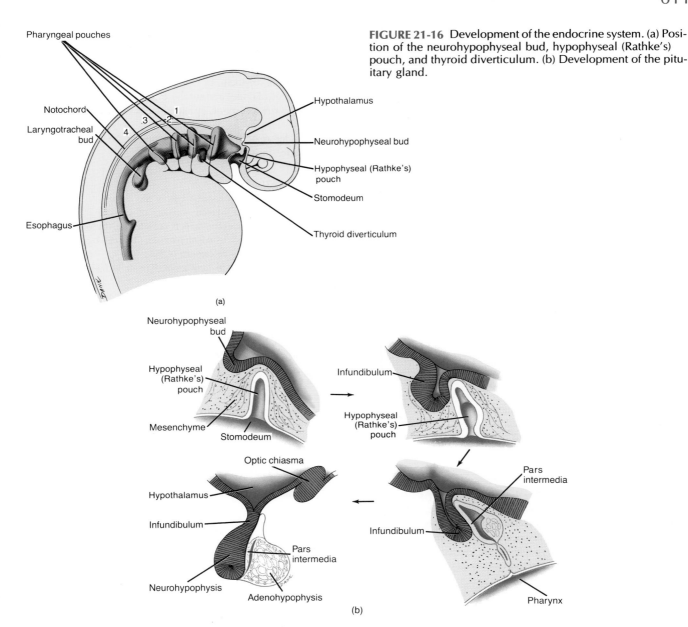

FIGURE 21-16 Development of the endocrine system. (a) Position of the neurohypophyseal bud, hypophyseal (Rathke's) pouch, and thyroid diverticulum. (b) Development of the pituitary gland.

Cardiac muscle fibers (cells) of the atria (upper chambers) of the heart produce a peptide hormone called ***atrial natriuretic factor (ANF)*** that is secreted when they are stretched, as might occur in response to increased blood volume, a factor that increases blood pressure. ANF seems to act as a calcium-channel blocker and may also be secreted by the brain. The general effect of ANF is to increase sodium and water excretion and blood vessel dilatation; the result is a decrease in blood pressure. ANF also acts directly on the kidneys to bring about increases in sodium and water excretion and on the hypothalamus to inhibit the secretion of ADH. Again, these conditions lead to a decrease in blood pressure. The overall function of ANF is therefore to lower blood pressure.

KEY MEDICAL TERMS ASSOCIATED WITH THE ENDOCRINE SYSTEM

Hyperplasia (hī-per-PLĀ-zē-a; *hyper* = over, *plas* = grow)   Increase in the number of cells due to an increase in the frequency of cell division.

Hypoplasia (hi'-pō-PLĀ-zē-a; *hypo* = under)   Defective development of tissue.

Neuroblastoma (noo'-rō-blas-TŌ-ma; *neuro* = nerve)   Malignant tumor arising from the adrenal medulla associated with metastases to bones.

Thyroid (THĪ-roid) **storm**   An aggravation of all symptoms of hyperthyroidism characterized by unregulated hypermetabolism with fever and rapid heart rate; results from trauma, surgery, and unusual emotional stress or labor.

STUDY OUTLINE

Endocrine Glands (p. 591)

1. Both the endocrine and nervous systems assume a role in maintaining homeostasis.
2. Hormones help regulate the internal environment, respond to stress, help regulate growth and development, and contribute to reproductive processes.
3. Exocrine glands (sweat, sebaceous, digestive) secrete their products through ducts into body cavities or onto body surfaces.
4. Endocrine glands secrete hormones into the blood.

Receptors (p. 591)

1. The amount of hormone released is determined by the body's need for the hormone.
2. Cells that respond to the effects of hormones are called target cells.
3. Receptors are found in the plasma membrane, cytoplasm, and nucleus of target cells.
4. The combination of hormone and receptor activates a chain of events in a target cell in which the physiological effects of the hormone are expressed.

Feedback Control (p. 592)

1. A negative feedback control mechanism prevents overproduction or underproduction of a hormone.
2. Hormone secretions are controlled by levels of circulating hormone itself, nerve impulses, and regulating hormones (or factors).

Pituitary (Hypophysis) (p. 593)

1. The pituitary gland is located in the sella turcica and is differentiated into the adenohypophysis (the anterior lobe and glandular portion), neurohypophysis (the posterior lobe and nervous portion), and pars intermedia (avascular zone between lobes).
2. Hormones of the adenohypophysis are released or inhibited by regulating hormones (or factors) produced by the hypothalamus.
3. The blood supply to the adenohypophysis is from the superior hypophyseal arteries. It transports hypothalamic regulating factors.
4. Histologically, the adenohypophysis consists of somatotroph cells that produce growth hormone (GH), which regulates growth; lactotroph cells that produce prolactin (PRL), which helps initiate milk secretion; thyrotroph cells that secrete thyroid-stimulating hormone (TSH), which regulates thyroid gland activities; gonadotroph cells that synthesize follicle-stimulating hormone (FSH), which regulates the activities of the ovaries and testes, and luteinizing hormone (LH), which regulates female and male reproductive activities; and corticolipotroph cells that secrete adrenocorticotropic hormone (ACTH), which regulates the activities of the adrenal cortex, and melanocyte-stimulating hormone (MSH), which increases skin pigmentation.
5. Disorders associated with improper levels of GH are pituitary dwarfism, giantism, and acromegaly.
6. The neural connection between the hypothalamus and neurohypophysis is via the hypothalamic–hypophyseal tract.
7. Hormones made by the hypothalamus and stored in the neurohy-

pophysis are oxytocin, or OT (stimulates contraction of uterus and ejection of milk), and antidiuretic hormone, or ADH (stimulates water reabsorption by the kidneys and arteriole constriction).
8. A disorder associated with dysfunction of the neurohypophysis is diabetes insipidus.

Thyroid (p. 597)

1. The thyroid gland is located below the larynx.
2. Histologically, the thyroid consists of thyroid follicles composed of follicular cells, which secrete the thyroid hormones thyroxine (T_4) and triiodothyronine (T_3), and parafollicular cells, which secrete calcitonin (CT).
3. Thyroid hormones are synthesized from iodine and tyrosine within thyroglobulin (TGB).
4. Thyroid hormones regulate the rate of metabolism, growth and development, and the reactivity of the nervous system.
5. Cretinism, myxedema, exophthalmic goiter, and simple goiter are disorders associated with dysfunction of the thyroid gland.
6. Calcitonin (CT) lowers the blood level of calcium.

Parathyroids (p. 600)

1. The parathyroids are embedded on the posterior surfaces of the lateral lobes of the thyroid.
2. Histologically, the parathyroids consist of principal and oxyphil cells.
3. Parathyroid hormone (PTH) regulates the homeostasis of calcium and phosphate by increasing blood calcium level and decreasing blood phosphate level.
4. Tetany and osteitis fibrosa cystica are disorders associated with the parathyroid glands.

Adrenals (Suprarenals) (p. 601)

1. The adrenal glands are located superior to the kidneys. They consist of an outer cortex and inner medulla.
2. Histologically, the cortex is divided into a zona glomerulosa, zona fasciculata, and zona reticularis; the medulla consists of chromaffin cells.
3. Cortical secretions are mineralocorticoids, glucocorticoids, and gonadocorticoids.
4. Mineralocorticoids (e.g., aldosterone) increase sodium and water reabsorption and decrease potassium reabsorption.
5. A dysfunction related to aldosterone secretion is aldosteronism.
6. Glucocorticoids (e.g., cortisol) promote normal metabolism, help resist stress, and serve as anti-inflammatories.
7. Disorders associated with glucocorticoid secretion are Addison's disease and Cushing's syndrome.
8. Gonadocorticoids secreted by the adrenal cortex usually have minimal effects. Excessive production results in adrenogenital syndrome.
9. Medullary secretions are epinephrine and norepinephrine (NE), which produce effects similar to sympathetic responses. They are released under stress.
10. Tumors of medullary chromaffin cells are called pheochromocytomas.

Pancreas (p. 606)

1. The pancreas is posterior and slightly inferior to the stomach.
2. Histologically, it consists of pancreatic islets, or islets of Langerhans (endocrine cells), and acini (enzyme-producing cells). Three types of cells in the endocrine portion are alpha cells, beta cells, and delta cells.
3. Alpha cells secrete glucagon, beta cells secrete insulin, and delta cells secrete growth hormone-inhibiting factor (GHIF), or somatostatin.
4. Glucagon increases blood sugar level and insulin decreases blood sugar level.
5. Disorders associated with insulin production are diabetes mellitus and hyperinsulinism.

Ovaries and Testes (p. 608)

1. Ovaries are located in the pelvic cavity and produce sex hormones related to development and maintenance of female sexual characteristics, menstrual cycle, pregnancy, lactation, and normal reproductive functions.
2. Testes lie inside the scrotum and produce sex hormones related to the development and maintenance of male sexual characteristics, and normal reproductive functions.

Pineal (Epiphysis Cerebri) (p. 608)

1. The pineal is attached to the roof of the third ventricle.
2. Histologically, it consists of secretory parenchymal cells called pinealocytes, neuroglial cells, and scattered postganglionic sympathetic fibers. Calcium-containing deposits are referred to as brain sand.
3. It secretes melatonin (possibly regulates reproductive activities by inhibiting gonadotropic hormones) and adrenoglomerulotropin (may stimulate adrenal cortex to secrete aldosterone).

Thymus (p. 609)

1. The thymus is a bilobed lymphatic gland located in the superior mediastinum posterior to the sternum and between the lungs.
2. The thymus gland secretes several hormones related to immunity.
3. Thymosin, thymic humoral factor (THF), thymic factor (TF), and thymopoietin promote the maturation of T cells.

Aging and the Endocrine System (p. 610)

1. Most endocrine disorders are related to pathologies rather than age.
2. Diabetes mellitus and thyroid disorders are among the more important endocrine disorders.

Developmental Anatomy of the Endocrine System (p. 610)

1. The development of the endocrine system is not as localized as other systems.
2. The adenohypophysis arises from the hypophyseal (Rathke's) pouch; the neurohypophysis develops from the hypophyseal bud.
3. The thyroid gland develops from the thyroid diverticulum.
4. The parathyroid glands and thymus gland develop from pharyngeal pouches.
5. The adrenal cortex arises from mesoderm; the adrenal medulla develops from ectoderm (neural crest).
6. The pancreas develops from the foregut and the pineal gland develops from the diencephalon.

Other Endocrine Tissues (p. 610)

1. The gastrointestinal tract synthesizes stomach and intestinal gastrin, secretin, cholecystokinin (CCK), enterocrinin, and gastric inhibitory peptide (GIP).
2. The placenta produces human chorionic gonadotropin (HCG), estrogens, progesterone (PROG), relaxin, and human chorionic somatomammotropin (HCS).
3. The kidneys release an enzyme that produces erythropoietin.
4. The skin synthesizes vitamin D.
5. The atria of the heart produce atrial natriuretic factor (ANF).

REVIEW QUESTIONS

1. Distinguish between an endocrine gland and an exocrine gland.
2. What is a hormone? Distinguish between tropic and gonadotropic hormones.
3. Explain how receptors are related to hormones.
4. How are negative feedback systems related to hormonal control? Discuss three models of operation.
5. Describe the histology of the adenohypophysis. Why does the anterior lobe of the gland have such an abundant blood supply?
6. What hormones are produced by the adenohypophysis? What are their functions?
7. Describe the clinical symptoms of pituitary dwarfism, giantism, and acromegaly.
8. Discuss the histology of the neurohypophysis and the functions of its hormones.
9. Describe the structure and importance of the hypothalamic–hypophyseal tract.
10. What are the clinical symptoms of diabetes insipidus?
11. Describe the location and histology of the thyroid gland.
12. Discuss the physiological effects of the thyroid hormones.
13. Discuss the clinical symptoms of cretinism, myxedema, exophthalmic goiter, and simple goiter.
14. Describe the function of calcitonin (CT).
15. Where are the parathyroids located? What is their histology?
16. What are the functions of the parathyroid hormone (PTH)?
17. Discuss the clinical symptoms of tetany and osteitis fibrosa cystica.
18. Compare the adrenal cortex and adrenal medulla with regard to location and histology.
19. Describe the hormones produced by the adrenal cortex in terms of type and function.
20. Describe the clinical symptoms of aldosteronism, Addison's disease, Cushing's syndrome, and adrenogenital syndrome.
21. What relation does the adrenal medulla have to the autonomic nervous system? What is the action of adrenal medullary hormones?

22. What is a pheochromocytoma?
23. Describe the location of the pancreas and the histology of the pancreatic islets (islets of Langerhans).
24. What are the actions of glucagon and insulin?
25. Describe the clinical symptoms of diabetes mellitus and hyperinsulinism. Distinguish the two types of diabetes mellitus.
26. Why are the ovaries and testes considered to be endocrine glands?

27. Where is the pineal gland located? What are its assumed functions?
28. How are hormones of the thymus gland related to immunity?
29. Describe the effects of aging on the endocrine system.
30. Describe the development of the endocrine system.
31. List the hormones secreted by the gastrointestinal tract, placenta, kidneys, skin, and heart.
32. Refer to the glossary of key medical terms associated with the endocrine system. Be sure that you can define each term.

SELF-QUIZ

Complete the following.

1. Glucagon, produced by (alpha? beta? delta?) cells, (increases? decreases?) blood sugar in two ways.
2. Arrange in order the names of vessels that supply blood to the adenohypophysis. Use lines provided. __ __ __ __
   A. Hypophyseal portal veins
   B. Primary plexus
   C. Superior hypophyseal arteries
   D. Secondary plexus

Choose the one best answer to these questions.

___ 3. One of the endocrine glands develops from the sympathetic nervous system during embryological development. Considering your knowledge of function, this endocrine gland is the
   A. pancreas; B. adrenal medulla; C. anterior pituitary; D. thymus; E. posterior pituitary.
___ 4. A generalized anti-inflammatory reaction is most closely associated with
   A. glycocorticoids; B. mineralocorticoids; C. parathyroid hormone (PTH); D. anterior pituitary hormones; E. insulin.
___ 5. A chemical produced by the hypothalamus that causes an endocrine gland to secrete a hormone is called a
   A. gonadotropic hormone; B. tropic hormone; C. regulating hormone (or factor); D. target hormone; E. neurotransmitter.
___ 6. If a person is diagnosed as having a high metabolic rate, which endocrine gland is probably malfunctioning?
   A. parathyroids; B. thymus; C. posterior pituitary; D. thyroid; E. pancreas.
___ 7. A tumor of the beta cells of the pancreatic islets (islets of Langerhans) would probably affect the body's ability to
   A. lower blood sugar level; B. raise blood sugar level; C. lower blood calcium level; D. raise blood calcium level; E. regulate metabolism.
___ 8. The concept by which the level of a hormone is self-regulating is referred to as
   A. hormoregulation; B. hormogenesis; C. negative feedback; D. stimulation; E. estrogenesis.
___ 9. A suspected role of the pineal gland is
   A. regulation of electrolyte balance; B. water retention; C. regulation of the reproductive activities; D. retention of glucose; E. regulation of calcium metabolism.

___ 10. The terms sella turcica and infundibulum are associated with the
   A. pineal; B. thyroid; C. thymus; D. pituitary; E. parathyroids.
___ 11. The stomach, the pancreas, the testes, and the ovaries have something in common. These organs
   A. are all influenced by hormones from the anterior pituitary; B. have tissues that are derived from embryological ectoderm; C. form hormones that influence secondary sex characteristics; D. receive their blood supply from the superior mesenteric artery; E. are considered to be both exocrine and endocrine.
___ 12. A hormone
   (1) is a chemical substance secreted into the body fluids.
   (2) is secreted by one cell or a group of cells.
   (3) exerts a physiological effect on other cells of the body than those that produced it.
   A. (1) only; B. (2) only; C. (3) only; D. all of the above; E. (1) and (2).
___ 13. Which of the following terms is used when referring to hormones that control the activity of other endocrine structures?
   A. cryptic; B. trophic; C. troponic; D. tonic; E. tachyptic.
___ 14. Which endocrine gland consists of lobes made up of follicles that contain the hormone secreted in a colloidal form?
   A. pituitary; B. parathyroid; C. thyroid; D. adrenal medulla; E. pancreatic islets.
___ 15. Which of the endocrine glands are "double" glands; that is each consists of two endocrine tissues having different embryological origins?
   A. testis and ovary; B. pituitary and adrenal (suprarenal); C. adrenal (suprarenal) and thyroid; D. thyroid and parathyroid; E. pancreas and thymus.
___ 16. Which one of these glands is called the "emergency gland" and helps the body meet sudden stress?
   A. pituitary; B. pancreas; C. thyroid; D. thymus; E. adrenal (suprarenal).
___ 17. Whereas an exocrine gland secretes into a duct, an endocrine gland secretes into
   A. open cavities; B. closed cavities; C. blood; D. lymph; E. neural tissue.

Circle T (true) or F (false) for the following.

T F 18. Whereas the anterior pituitary contains axons of neurons, the posterior pituitary contains glandular cells.

T F 19. The middle region of hormone-secreting cells of the adrenal cortex is called the zona reticularis.

T F 20. The amount of a hormone released by an endocrine gland or tissue is determined by the body's need for the hormone at any given time.

T F 21. The gland that assumes a role in the production of antibodies is the pineal.

T F 22. Chromaffin cells are the principal secreting cells of the pancreas.

T F 23. Another term for the anterior pituitary gland is the neurohypophysis.

T F 24. The outer region of the adrenal (suprarenal) gland is called the cortex.

25. Match the following:

___ a. develops from the foregut area that later becomes part of small intestine

___ b. derived from the roof of the stomodeum (mouth) called the hypophyseal (Rathke's) pouch

___ c. originates from the neural crest, which also produces sympathetic ganglia

___ d. derived from tissue from the same region that forms gonads

___ e. arise from the third and fourth pharyngeal pouches (two answers)

A. adrenal cortex
B. adrenal medulla
C. anterior pituitary
D. pancreas
E. parathyroid
F. thymus

22 The Respiratory System

CHAPTER OUTLINE

■ **Organs**
Nose
Pharynx
Larynx
Trachea
Bronchi
Lungs
  Gross Anatomy
  Lobes and Fissures
  Lobules
  Alveolar–Capillary (Respiratory) Membrane
  Blood and Nerve Supply
■ **Nervous Control of Respiration**
Medullary Rhythmicity Area
Pneumotaxic Area
Apneustic Area
Cortical Influences
■ **Aging and the Respiratory System**
■ **Developmental Anatomy of the Respiratory System**
■ **Applications to Health**
Bronchogenic Carcinoma (Lung Cancer)
Bronchial Asthma
Bronchitis
Emphysema
Pneumonia
Tuberculosis (TB)
Respiratory Distress Syndrome (RDS) of the Newborn
Respiratory Failure
Sudden Infant Death Syndrome (SIDS)
Coryza (Common Cold) and Influenza (Flu)
Pulmonary Embolism (PE)
Pulmonary Edema
Carbon Monoxide (CO) Poisoning
Smoke Inhalation Injury
■ **Key Medical Terms Associated with the Respiratory System**

STUDENT OBJECTIVES

1. Identify the organs of the respiratory system.
2. Compare the structure and function of the external and internal nose.
3. Differentiate the three anatomical regions of the pharynx and describe their roles in respiration.
4. Describe the structure of the larynx and explain its function in respiration and voice production.
5. Describe the location and structure of the tubes that form the bronchial tree.
6. Describe the coverings of the lungs, the division of the lungs into lobes, and the composition of a lobule of the lung.
7. Explain the structure of the alveolar–capillary (respiratory) membrane and its function in the diffusion of respiratory gases.
8. Describe the effects of aging on the respiratory system.
9. Describe the development of the respiratory system.
10. Define bronchogenic carcinoma (lung cancer), bronchial asthma, bronchitis, emphysema, pneumonia, tuberculosis, respiratory distress syndrome (RDS) of the newborn, respiratory failure, sudden infant death syndrome (SIDS), coryza (common cold), influenza (flu), pulmonary embolism (PE), pulmonary edema, carbon monoxide (CO) poisoning, and smoke inhalation injury as disorders of the respiratory system.
11. Define key medical terms associated with the respiratory system.

Cells need a continuous supply of oxygen (O_2) for various metabolic reactions that produce energy, some of which is stored in ATP for cellular use. As a result of these reactions, cells also release quantities of carbon dioxide (CO_2). Since an excessive amount of carbon dioxide produces acid conditions that are poisonous to cells, the gas must be eliminated quickly and efficiently. The two systems that supply oxygen and eliminate carbon dioxide are the cardiovascular system and the respiratory system. The *respiratory system* consists of the nose, pharynx, larynx, trachea, bronchi, and lungs (Figure 22-1). The cardiovascular system transports the gases in the blood between the lungs and the cells. The term *upper respiratory system* refers to the nose, throat, and associated structures. The *lower respiratory system* refers to the remainder of the system.

The overall exchange of gases beween the atmosphere, blood, and cells is called *respiration.* Three basic processes are involved. The first process, *pulmonary ventilation,* or breathing, is the inspiration (inflow) and expiration (outflow) of air between the atmosphere and the lungs. The second and third processes involve the exchange of gases within the body. *External respiration* is the exchange of gases between the lungs and blood. *Internal respiration* is the exchange of gases between the blood and cells.

The respiratory and cardiovascular systems participate equally in respiration. Failure of either system has the same effect on the body: disruption of homeostasis and rapid death of cells from oxygen starvation and buildup of waste products.

The developmental anatomy of the respiratory system is considered later in the chapter.

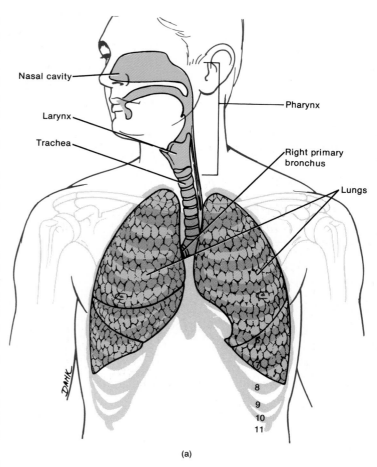

(a)

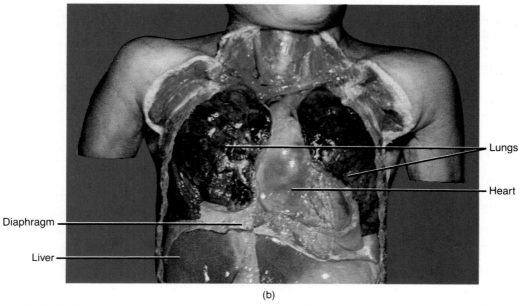

(b)

FIGURE 22-1 Organs of the respiratory system in relation to surrounding structures. (a) Diagram. (b) Photograph. (Courtesy of C. Yokochi and J. W. Rohen, *Photographic Anatomy of the Human Body,* 2nd ed., 1978, IGAKU-SHOIN, Ltd., Tokyo, New York.)

ORGANS

NOSE

The **nose** has an external portion and an internal portion inside the skull (Figure 22-2). The external portion consists of a supporting framework of bone and cartilage covered with skin and lined with mucous membrane. The bridge of the nose is formed by the nasal bones, which hold it in a fixed position. Because it has a framework of pliable cartilage, the rest of the external nose is somewhat flexible. On the undersurface of the external nose are two openings called the **external nares** (NA-rēz; *sing.*, naris) or **nostrils.** The surface anatomy of the nose is shown in Figure 11-4.

The internal portion of the nose is a large cavity in the skull that lies inferior to the cranium and superior to the mouth. Anteriorly, the internal nose merges with the external nose, and posteriorly it communicates with the throat (pharynx) through two openings called the **internal nares (choanae).** Four paranasal sinuses (frontal, sphenoidal, maxillary, and ethmoidal) and the nasolacrimal ducts also open into the internal nose. The lateral walls of the internal nose are formed by the ethmoid, maxillae, lacrimal, palatine, and inferior nasal conchae bones. The ethmoid also forms the roof. The floor is formed by the soft palate and palatine bones and palatine process of the maxilla, which together comprise the hard palate.

The inside of both the external and internal nose consists of a **nasal cavity,** divided into right and left sides by a vertical partition called the **nasal septum.** The anterior por-

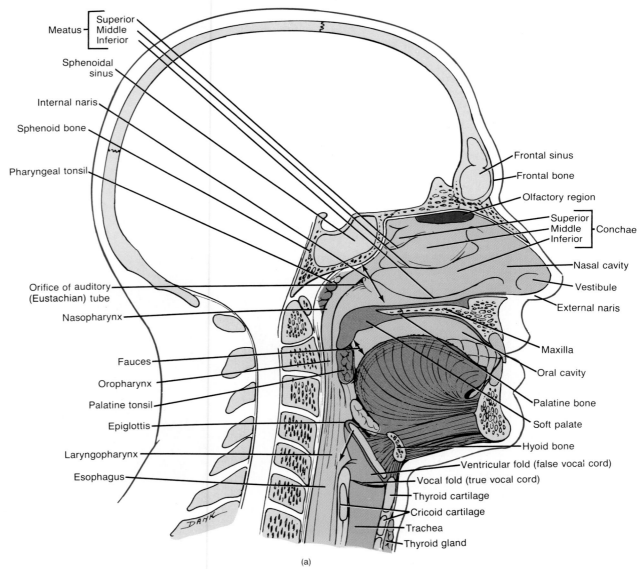

(a)

FIGURE 22-2 Respiratory organs in the head and neck. (a) Diagram of the left side of the head and neck seen in sagittal section with the nasal septum removed.

tion of the septum is made primarily of cartilage. The remainder is formed by the vomer and the perpendicular plate of the ethmoid (see Figure 6-7a). The anterior portion of the nasal cavity, just inside the nostrils, is called the *vestibule.* It is surrounded by cartilage. The upper nasal cavity is surrounded by bone.

CLINICAL APPLICATION

Nasal polyps are protruding growths of the mucous membrane that usually hang down from the nasal septum. The polyps appear as bluish-white tumors and may fill the nasopharynx as they become larger. Nasal polyps usually undergo atrophy if untreated, but they are easily removed by a physician with a nasal snare and cautery.

The interior structures of the nose are specialized for three functions: incoming air is warmed, moistened, and filtered; olfactory stimuli are received; and large hollow resonating chambers are provided for speech sounds.

When air enters the nostrils, it passes first through the vestibule. The vestibule is lined by skin containing coarse hairs that filter out large dust particles. The air then passes into the upper nasal cavity. Three shelves formed by projections of the superior, middle, and inferior nasal conchae

extend out of the lateral wall of the cavity. The conchae, almost reaching the septum, subdivide each side of the nasal cavity into a series of groovelike passageways—the *superior, middle,* and *inferior meatuses.* Mucous membrane lines the cavity and its shelves. The olfactory receptors lie in the membrane lining the area superior to the superior nasal conchae and is also called he *olfactory region.* Below the olfactory region, the membrane contains capillaries and pseudostratified ciliated columnar cells with many goblet cells. As the air whirls around the conchae and meatuses, it is warmed by the capillaries. Mucus secreted by the goblet cells moistens the air and traps dust particles. Drainage from the lacrimal ducts and perhaps secretions from the paranasal sinuses also help moisten the air. The cilia move the mucus–dust packages along the pharynx so they can be eliminated from the body.

CLINICAL APPLICATION

Nosebleed, or *epistaxis* (ep'-i-STAK-sis), is common because of the exposure of the nose to trauma and the extensive blood supply of the nose. In addition to trauma, other causes of nosebleed are intranasal infection, allergy, bleeding disorders, and neoplasms. Bleeding, either arterial or venous, usually occurs on the anterior part of

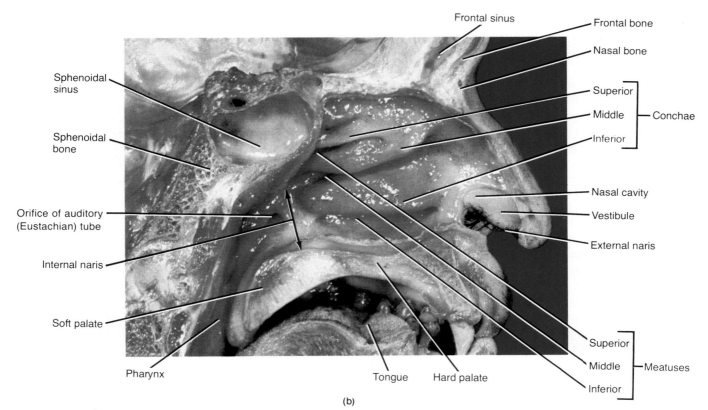

(b)

FIGURE 22-2 (Continued) Respiratory organs in the head and neck. (b) Photograph of the left side of the nasal cavity seen in sagittal section with the nasal septum removed. (Courtesy of C. Yokochi and J. W. Rohen, *Photographic Anatomy of the Human Body,* 2nd ed., 1978, IGAKU-SHOIN, Ltd., Tokyo, New York.)

the septum and can be arrested by cautery with silver nitrate, electrocautery, and firm packing of the external nares. If the point of bleeding is in the posterior region, plugging of both the external and internal nares may be necessary. In extreme emergency, the external carotid artery may have to be ligated (tied) in order to control the hemorrhage.

The arterial supply to the nasal cavity is principally from the sphenopalatine branch of the maxillary artery. The remainder is supplied by the ophthalmic artery. The veins of the nasal cavity drain into the sphenopalatine vein, the facial vein, and the ophthalmic vein.

The nerve supply of the nasal cavity consists of olfactory cells in the olfactory epithelium associated with the olfactory nerve (see Figure 20-8) and the nerves of general sensation. These nerves are branches of the ophthalmic division of the trigeminal (V) nerve and the maxillary division of the trigeminal (V) nerve.

PHARYNX

The *pharynx* (FAR-inks), or throat, is a somewhat funnel-shaped tube about 13 cm (5 in.) long that starts at the internal nares and extends to the level of the cricoid cartilage (Figure 22-3). It lies just posterior to the nasal cavity, oral cavity, and larynx and just anterior to the cervical vertebrae. Its wall is composed of skeletal muscles and lined with mucous membrane. The functions of the pharynx are to serve as a passageway for air and food and to provide a resonating chamber for speech sounds.

The branch of medicine that deals with the diagnosis and treatment of diseases of the ears, nose, and throat is called *otorhinolaryngology* (ō'-tō-rī'-nō-lar'-in-GOL-ō-jē; *otic* = ear; *rhino* = nose).

The uppermost portion of the pharynx, called the *naso-pharynx*, lies posterior to the internal nasal cavity and extends to the plane of the soft palate. There are four openings in its wall: two internal nares and two openings that lead into the auditory (Eustachian) tubes. The posterior wall also contains the pharyngeal tonsil, or adenoid. Through the internal nares the nasopharynx receives air from the nasal cavities and receives the packages of dust-laden mucus. It is lined with pseudostratified ciliated epithelium, and the cilia move the mucus down toward the mouth. The nasopharynx also exchanges small amounts of air with the auditory (Eustachian) tubes so that the air pressure inside the middle ear equals the pressure of the atmospheric air flowing through the nose and pharynx.

The middle portion of the pharynx, the *oropharynx,* lies posterior to the oral cavity and extends from the soft palate inferiorly to the level of the hyoid bone. It has only one opening, the *fauces* (FAW-sēz), the opening from the mouth. It is lined by stratified squamous epithelium. This portion of the pharynx is both respiratory and digestive in function, since it is a common passageway for air, food,

and drink. Two pairs of tonsils, the palatine and lingual tonsils, are found in the oropharynx. The lingual tonsil lies at the base of the tongue (see also Figure 20-9c).

The lowest portion of the pharynx, the *laryngopharynx* (la-rin'-gō-FAR-inks), extends downward from the hyoid bone and becomes continuous with the esophagus (food tube) posteriorly and the larynx (voice box) anteriorly. Like the oropharynx, the laryngopharynx is a respiratory and a digestive pathway and is lined by stratified squamous epithelium.

CLINICAL APPLICATION

Snoring, obstructive breathing during sleep, occurs from time to time in about 45 percent of adults and habitually in about 25 percent. The noise of snoring comes from vibrations of soft tissue between the internal nares and larynx—soft palate uvula, tonsils and their arches, base of the tongue, and pharyngeal muscles. Three circumstances, alone or in combination, contribute to snoring: (1) nasal airway impairment due to injuries, deformities, polyps, infection, tumors, or deviated nasal septum; (2) pharyngeal airway impairment due to enlarged tonsils, cysts, or tumors; and (3) decreased tone of the muscles of the soft palate, tongue, or pharynx.

The arteries of the pharynx are the ascending pharyngeal, the ascending palatine branch of the facial, the descending palatine and pharyngeal branches of the maxillary, and the muscular branches of the superior thyroid artery. The veins of the pharynx drain into the pterygoid plexus and the internal jugular vein.

Most of the muscles of the pharynx (see Figure 10-9) are innervated by the pharyngeal plexus. This plexus is formed by the pharyngeal branches of the glossopharyngeal (IX), vagal (X), and cranial portion of the glossopharyngeal (IX) nerves and the superior cervical sympathetic ganglion.

LARYNX

The *larynx,* or voice box, is a short passageway that connects the pharynx with the trachea. It lies in the midline of the neck anterior to the fourth through sixth cervical vertebrae.

The wall of the larynx is composed of nine pieces of cartilage (Figure 22-4). Three are single and three are paired. The three single pieces are the thyroid cartilage, epiglottic cartilage (epiglottis), and cricoid cartilage. Of the paired cartilages, the arytenoid cartilages are the most important. The paired corniculate and cuneiform cartilages are of lesser significance.

The *thyroid cartilage (Adam's apple)* consists of two fused plates that form the anterior wall of the larynx and give it its triangular shape. It is larger in males than in females.

The *epiglottis* is a large, leaf-shaped piece of cartilage

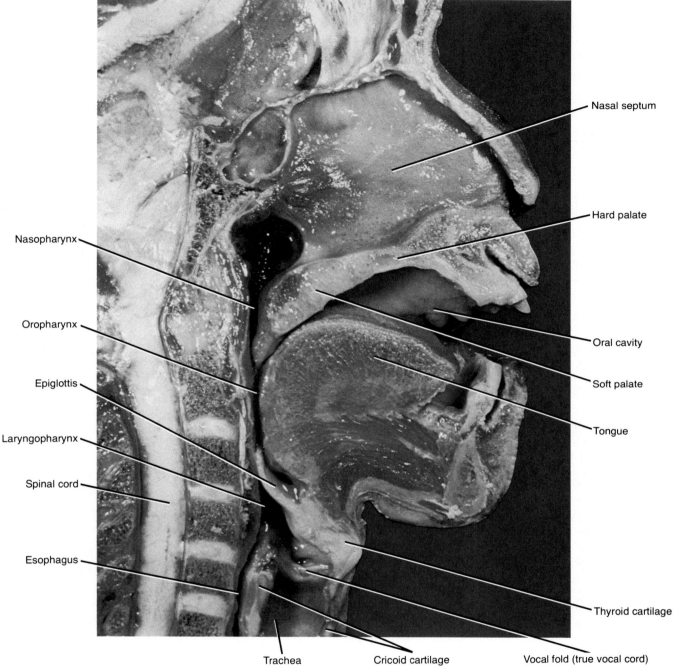

- Nasal septum
- Hard palate
- Oral cavity
- Soft palate
- Tongue
- Thyroid cartilage

Nasopharynx

Oropharynx

Epiglottis

Laryngopharynx

Spinal cord

Esophagus

Trachea · Cricoid cartilage · Vocal fold (true vocal cord)

FIGURE 22-3 Photograph of the head and neck seen in sagittal section. (Courtesy of C. Yokochi and J. W. Rohen, *Photographic Anatomy of the Human Body*, 2nd ed., 1979, IGAKU-SHOIN, Ltd., Tokyo, New York.)

lying on top of the larynx (see also Figure 22-3). The "stem" of the epiglottis is attached to the thyroid cartilage, but the "leaf" portion is unattached and free to move up and down like a trap door. During swallowing, there is elevation of the larynx. This causes, the free edge of the epiglottis to form a lid over the glottis, closing it off. The

glottis is the space between the vocal folds (true vocal cords) in the larynx. In this way, the larynx is closed off and liquids and foods are routed into the esophagus and kept out of the larynx and air passageways below it. When anything but air passes into the larynx, a cough reflex attempts to expel the material.

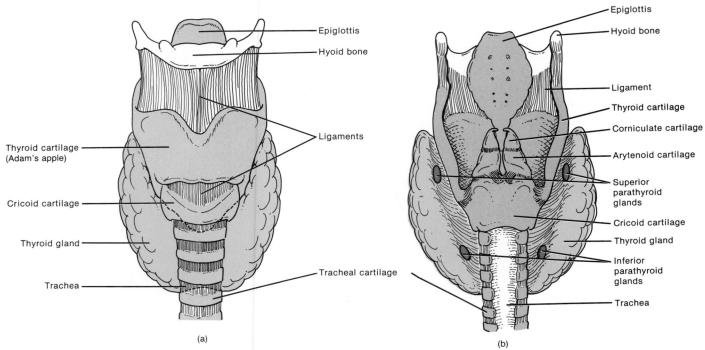

FIGURE 22-4 Larynx. (a) Diagram of anterior view. (b) Diagram of posterior view. (c) Photograph of posterior view. (Courtesy of C. Yokochi and J. W. Rohen, *Photographic Anatomy of the Human Body,* 2nd ed., 1979, IGAKU-SHOIN, Ltd., Tokyo, New York.) (d) Photograph of sagittal section. (Courtesy of C. Yokochi and J. W. Rohen, *Photographic Anatomy of the Human Body,* 2nd ed., 1979, IGAKU-SHOIN, Ltd., Tokyo, New York.) (e) Photograph of the true vocal cords, false vocal cords, and open glottis from above. (Courtesy of C. Yokochi and J. W. Rohen, *Photographic Anatomy of the Human Body,* 2nd ed., 1979, IGAKU-SHOIN, Ltd., Tokyo, New York.)

The *cricoid* (KRĪ-koyd) *cartilage* is a ring of cartilage forming the inferior wall of the larynx. It is attached to the first ring of cartilage of the trachea.

The paired *arytenoid* (ar'-i-TĒ-noyd) *cartilages* are pyramidal in shape and located at the superior border of the cricoid cartilage. They attach to the vocal folds and intrinsic pharyngeal muscles, and by their action can move the vocal folds.

The paired *corniculate* (kor-NIK-yoo-lāt) *cartilages* are cone shaped. One is located at the apex of each arytenoid cartilage. The paired *cuneiform* (kyoo-NĒ-i-form) *cartilages* are club-shaped cartilages anterior to the corniculate cartilages.

The epithelium lining the larynx below the vocal folds is pseudostratified. It consists of ciliated columnar cells, goblet cells, and basal cells, and it helps trap dust not removed in the upper passages.

The mucous membrane of the larynx is arranged into two pairs of folds—an upper pair called the *ventricular folds (false vocal cords)* and a lower pair called simply the *vocal folds (true vocal cords)* (Figure 22-4d,e). When the ventricular folds are brought together, they function in holding the breath against pressure in the thoracic cavity, such as might occur when a person exerts a strain while lifting a heavy weight. The mucous membrane of the vocal folds is lined by nonkeratinized stratified squamous epithe-

lium. Under the membrane lie bands of elastic ligaments stretched between pieces of rigid cartilage like the strings on a guitar. Skeletal muscles of the larynx, called intrinsic muscles, are attached internally to the pieces of rigid cartilage and to the vocal folds themselves (see Figure 10-11). When the muscles contract, they pull the strings of elastic ligaments tight and stretch the vocal folds out into the air passageways so that the glottis is narrowed. If air is directed againt the vocal folds, they vibrate and set up sound waves in the column of air in the pharynx, nose, and mouth. The greater the pressure of air, the louder the sound.

Pitch is controlled by the tension on the vocal folds. If they are pulled taut by the muscles, they vibrate more rapidly and a higher pitch results. Lower sounds are produced by decreasing the muscular tension on the vocal folds. Vocal folds are usually thicker and longer in males than in females, and therefore they vibrate more slowly. Thus, men generally have a lower range of pitch than women.

Sound originates from the vibration of the vocal folds, but other structures are necessary for converting the sound into recognizable speech. The pharynx, mouth, nasal cavity, and parnasal sinuses all act as resonating chambers that give the voice its human and individual quality. By constricting and relaxing the muscles in the wall of the pharynx, we produce the vowel sounds. Muscles of the face, tongue, and lips help us enunciate words.

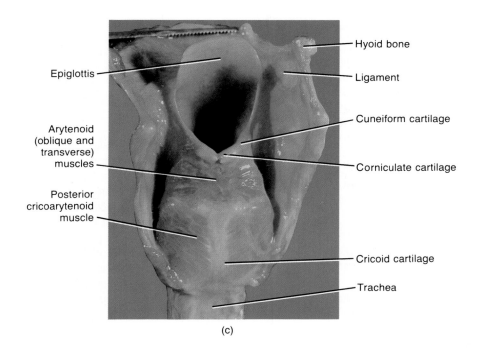

Epiglottis

Arytenoid (oblique and transverse) muscles

Posterior cricoarytenoid muscle

Hyoid bone

Ligament

Cuneiform cartilage

Corniculate cartilage

Cricoid cartilage

Trachea

(c)

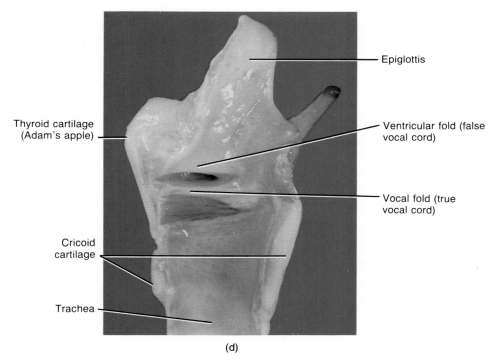

Thyroid cartilage (Adam's apple)

Cricoid cartilage

Trachea

Epiglottis

Ventricular fold (false vocal cord)

Vocal fold (true vocal cord)

(d)

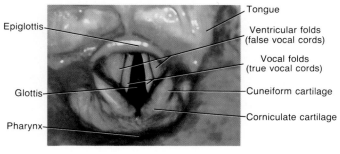

Epiglottis

Glottis

Pharynx

Tongue

Ventricular folds (false vocal cords)

Vocal folds (true vocal cords)

Cuneiform cartilage

Corniculate cartilage

(e)

CLINICAL APPLICATION

Laryngitis is an inflammation of the larynx that is most often caused by a respiratory infection or irritants such as cigarette smoke. Inflammation of the vocal folds causes hoarseness or loss of voice by interfering with the contraction of the folds or by causing them to swell to the point where they cannot vibrate freely. Many long-term smokers acquire a permanent hoarseness from the damage done by chronic inflammation.

Cancer of the larynx is found almost exclusively in individuals who smoke. The condition is characterized by hoarseness, pain on swallowing, or pain radiating to an ear. Treatment is by radiation therapy and/or surgery.

The arteries of the larynx are the superior laryngeal and inferior laryngeal. The superior and inferior laryngeal veins accompany the arteries. The superior laryngeal vein empties into the superior thyroid vein, and the inferior vein empties into the inferior thyroid vein.

The nerves of the larynx are the superior and recurrent (inferior) laryngeal branches of the vagus (X) nerve.

TRACHEA

The *trachea* (TRĀ-kē-a), or windpipe, is a tubular passageway for air about 12 cm (4.5 in.) in length and 2.5 cm (1 in.) in diameter. It is located anterior to the esophagus and extends from the larynx to the fifth thoracic vertebra (T5), where it divides into right and left primary bronchi (see Figure 22-6).

The wall of the trachea consists of a mucosa, submucosa, cartilaginous layer, and adventitia (outer layer of loose connective tissue). The tracheal epithelium of the mucosa is pseudostratified. It consists of ciliated columnar cells that reach the luminal surface, goblet cells, and basal cells that do not reach the luminal surface (Figure 22-5b). The epithelium provides the same protection against dust as the membrane lining the larynx. Seromucous glands and their ducts are present in the submucosa. The cartilaginous layer consists of 16–20 horizontal incomplete rings of hyaline cartilage that look like a series of letter C's stacked one on top of another. The open parts of the C's face the esophagus and permit it to expand slightly into the trachea during swallowing (Figure 22-5c). Transverse smooth muscle fibers, called the *trachealis muscle,* and elastic connective tissue attach the open ends of the cartilage rings. The solid parts of the C's provide a rigid support so the tracheal

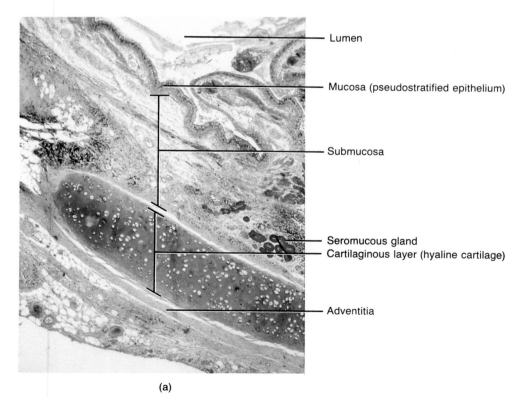

(a)

FIGURE 22-5 Histology of the trachea. (a) Photomicrograph of a portion of the tracheal wall at a magnification of 80×. (Courtesy of Andrew Kuntzman.)

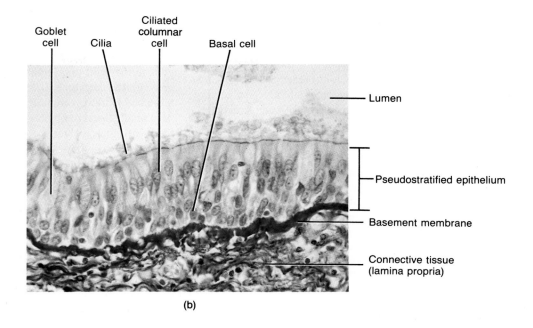

Goblet cell — Cilia — Ciliated columnar cell — Basal cell

Lumen

Pseudostratified epithelium

Basement membrane

Connective tissue (lamina propria)

(b)

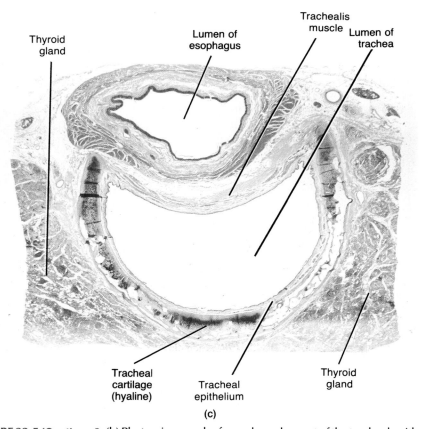

Thyroid gland — Lumen of esophagus — Trachealis muscle — Lumen of trachea

Tracheal cartilage (hyaline) — Tracheal epithelium — Thyroid gland

(c)

FIGURE 22-5 (Continued) (b) Photomicrograph of an enlarged aspect of the tracheal epithelium at a magnification of 600×. (Courtesy of Andrew Kuntzman.) (c) Photomicrograph showing the relation of the trachea to the esophagus in cross section at a magnification of 3×. (Copyright © 1987 by Michael H. Ross. Used by permission.)

wall does not collapse inward and obstruct the air passageway. In some situations, however, such as crushing injuries to the chest, the rings of cartilage may not be strong enough to overcome collapse and obstruction of the trachea.

At the point where the trachea bifurcates into right and left primary bronchi, there is an internal ridge called the *carina* (ka-RĪ-na). It is formed by a posterior and somewhat inferior projection of the last tracheal cartilage. The mucous membrane of the carina is one of the most sensitive areas of the respiratory system and is associated with the cough reflex. Widening and distortion of the carina, which can be seen in an examination by bronchoscopy, is a serious prognostic sign, since it usually indicates a carcinoma of the lymph nodes around the bifurcation of the trachea. *Bronchoscopy* is the visual examination of the bronchi through a *bronchoscope,* an illuminated, tubular instrument that can be passed through the trachea into the bronchi.

CLINICAL APPLICATION

Occasionally, the respiratory passageways are unable to protect themselves from obstruction. The rings of cartilage may accidentally be crushed; the mucous membrane may become inflamed and swell so much that it closes off the air passageways; inflamed membranes secrete a great deal of mucus that may clog the lower respiratory passageways; a large object may be breathed in (aspirated) while the glottis is open; or an aspirated foreign object may cause spasm of the laryngeal muscles. The passageways must be cleared quickly. If the obstruction is above the level of the chest, a *tracheostomy* (trā-kē-OS-tō-mē) may be performed. A midline skin incision is made in the neck from just above the cricoid cartilage to the jugular notch of the sternum. Next, an incision is made in the trachea below the obstructed area. The patient breathes through a metal or plastic tracheal tube inserted through the incision. Another method is *intubation.* A tube is inserted into the mouth or nose and passed down through the larynx and trachea. The firm wall of the tube pushes back any flexible obstruction, and the inside of the tube provides a passageway for air. If mucus is clogging the trachea, it can be suctioned out through the tube.

The arteries of the trachea are branches of the inferior thyroid, internal thoracic, and bronchial arteries. The veins of the trachea terminate in the inferior thyroid veins.

The smooth muscle and glands of the trachea are innervated parasympathetically via the vagus (X), nerve directly and by its recurrent laryngeal branches. Sympathetic innervation is through branches from the sympathetic trunk and its ganglia.

BRONCHI

The trachea terminates in the chest by dividing at the sternal angle into a *right primary bronchus* (BRON-kus), which goes to the right lung, and a *left primary bronchus,* which goes to the left lung (Figure 22-6a). The right primary bronchus is more vertical, shorter, and wider than the left. As a result, foreign objects in the air passageways are more likely to enter it than the left and frequently lodge in it. Like the trachea, the primary bronchi (BRONG-kē) contain incomplete rings of cartilage and are lined by pseudostratified ciliated epithelium.

On entering the lungs, the primary bronchi divide to form smaller bronchi—the *secondary (lobar) bronchi,* one for each lobe of the lung (the right lung has three lobes; the left lung has two). The secondary bronchi continue to branch, forming still smaller bronchi, called *tertiary (segmental) bronchi,* that divide into *bronchioles.* Bronchioles, in turn, branch into even smaller tubes called *terminal bronchioles.* This continuous branching from the trachea resembles a tree trunk with its branches and is commonly referred to as the *bronchial tree.*

As the branching becomes more extensive in the bronchial tree, several structural changes may be noted. First, rings of cartilage are replaced by plates of cartilage that finally disappear in the bronchioles. Second, as the cartilage decreases, the amount of smooth muscle increases. Third, the epithelium changes from pseudostratified ciliated to simple cuboidal in the terminal bronchioles.

CLINICAL APPLICATION

The fact that the walls of the bronchioles contain a great deal of smooth muscle but no cartilage is clinically significant. During an *asthma attack* the muscles go into spasm. Because there is no supporting cartilage, the spasms can close off the air passageways. Movement of air through constricted tubes causes breathing to be loud.

Bronchography (bron-KOG-ra-fē) is a technique for examining the bronchial tree. An intratracheal catheter is passed into the mouth or nose, through the glottis, and into the trachea. Then, an opaque contrast medium, usually containing iodine, is introduced by means of gravity into the trachea and distributed through the bronchial branches. Roentgenograms of the chest in various positions are taken and the developed film, a *bronchogram* (BRONG-kō-gram), provides a picture of the tree (Figure 22-6c).

The blood supply to the bronchi is via the left bronchial and right bronchial arteries. The veins that drain the bronchi are the right bronchial vein, which enters the azygos vein, and the left bronchial vein, which empties into the hemiazygos vein or the left superior intercostal vein.

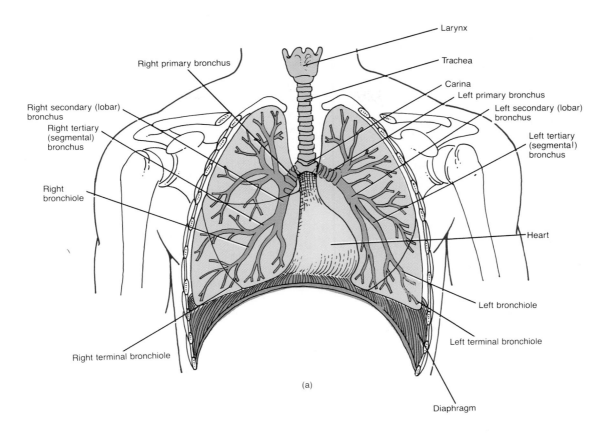

Larynx

Right primary bronchus

Trachea

Carina

Left primary bronchus

Right secondary (lobar) bronchus

Left secondary (lobar) bronchus

Right tertiary (segmental) bronchus

Left tertiary (segmental) bronchus

Right bronchiole

Heart

Left bronchiole

Left terminal bronchiole

Right terminal bronchiole

(a)

Diaphragm

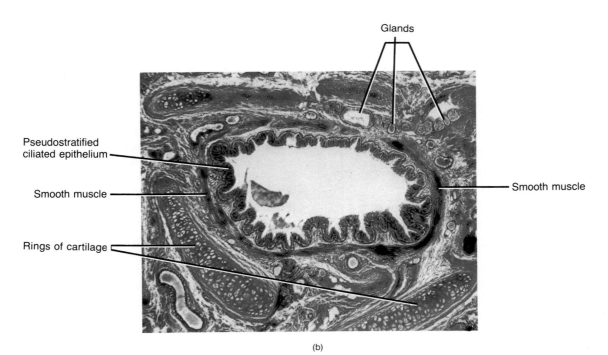

Glands

Pseudostratified ciliated epithelium

Smooth muscle

Smooth muscle

Rings of cartilage

(b)

FIGURE 22-6 Air passageways to the lungs. (a) Diagram of the bronchial tree in relation to the lungs. (b) Photomicrograph of a cross section of a primary bronchus at a magnification of 100×. (Copyright © 1983 by Michael H. Ross. Used by permission.)

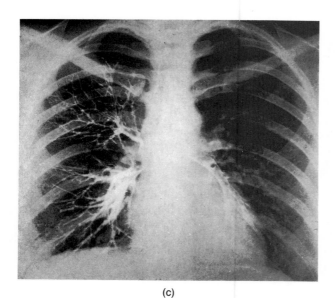

(c)

FIGURE 22-6 (Continued) (c) Anteroposterior bronchogram. (Courtesy of John H. Juhl. From Lester W. Paul and John H. Juhl, *The Essentials of Roentgen Interpretation,* 4th ed., J. B. Lippincott, Philadelphia, 1987.)

LUNGS

The *lungs* are paired, cone-shaped organs lying in the thoracic cavity. They are separated from each other by the heart and other structures in the mediastinum (see Figure 13-1). Two layers of serous membrane, collectively called the *pleural membrane,* enclose and protect each lung. The outer layer is attached to the wall of the thoracic cavity and is called the *parietal pleura.* The inner layer, the *visceral pleura,* covers the lungs themselves. Between the visceral and parietal pleura is a small potential space, the *pleural cavity,* which contains a lubricating fluid secreted by the membranes (see Figure 1-7d). This fluid prevents friction between the membranes and allows them to move easily on one another during breathing.

CLINICAL APPLICATION

In certain conditions, the pleural cavity may fill with air (*pneumothorax*), blood (*hemothorax*), or pus. Air in the pleural cavity, most commonly introduced in a surgical opening of the chest or as a result of a stab or gunshot wound, may cause the lung to collapse. Fluid can be drained from the pleural cavity by inserting a needle, usually posteriorly through the seventh intercostal space. The needle is passed along the superior border of the lower rib to avoid damage to the intercostal nerves and blood vessels. Below the seventh intercostal space there is danger of penetrating the diaphragm.

Inflammation of the pleural membrane, or *pleurisy,* causes friction during breathing that can be quite painful when the swollen membranes rub against each other.

Gross Anatomy

The lungs extend from the diaphragm to a point about 1.5–2.5 cm (0.75–1 in.) superior to the clavicles and lie against the ribs anteriorly and posteriorly. The broad inferior portion of the lung, the *base,* is concave and fits over the convex area of the diaphragm (Figure 22-7). The narrow superior portion of the lung is termed the *apex (cupula).* The surface of the lung lying against the ribs, the *costal surface,* is rounded to match the curvature of the ribs. The *mediastinal (medial) surface* of each lung contains a region, the *hilus,* through which bronchi, pulmonary vessels, lymphatics, and nerves enter and exit. These structures are held together by the pleura and connective tissue, and constitute the *root* of the lung. Medially, the left lung also contains a concavity, the *cardiac notch,* in which the heart lies.

The right lung is thicker and broader than the left. It is also somewhat shorter than the left because the diaphragm is higher on the right side to accommodate the liver that lies below it.

Lobes and Fissures

Each lung is divided into lobes by one or more fissures. Both lungs have an *oblique fissure,* which extends downward and forward. The right lung also has a *horizontal fissure.* The oblique fissure in the left lung separates the *superior lobe* from the *inferior lobe.* The upper part of the oblique fissure of the right lung separates the superior lobe from the inferior lobe, whereas the lower part of the oblique fissure separates the inferior lobe from the *middle lobe.* The horizontal fissure of the right lung subdivides the superior lobe, thus forming a middle lobe.

Each lobe receives its own secondary (lobar) bronchus. Thus, the right primary bronchus gives rise to three secondary (lobar) bronchi called the *superior, middle,* and *inferior secondary (lobar) bronchi.* The left primary bronchus gives rise to a *superior* and an *inferior secondary (lobar) bronchus.* Within the substance of the lung, the secondary bronchi give rise to the *tertiary (segmental) bronchi,* which are constant in both origin and distribution. The segment of lung tissue that each supplies is called a *bronchopulmonary segment* (Figure 22-8). Bronchial and pulmonary disorders, such as tumors or abscesses, may be localized in a bronchopulmonary segment and may be surgically removed without seriously disrupting surrounding lung tissue.

Lobules

Each bronchopulmonary segment of the lungs is broken up into many small compartments called *lobules* (Figure 22-9a). Each lobule is wrapped in elastic connective tissue and contains a lymphatic vessel, an arteriole, a venule, and a branch from a terminal bronchiole. Terminal bronchioles subdivide into microscopic branches called *respira-*

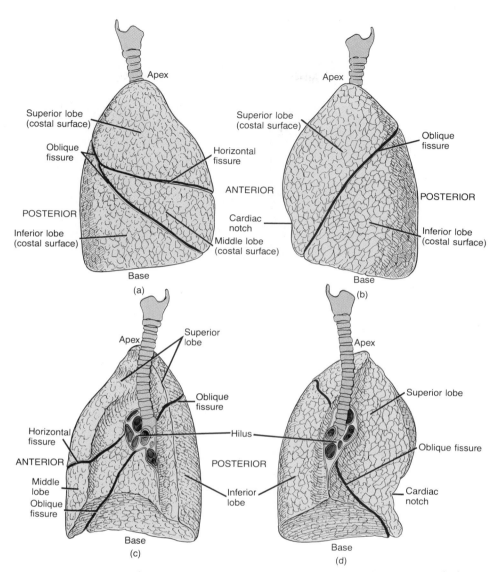

FIGURE 22-7 Lungs. (a) Right lung, lateral view. (b) Left lung, lateral view. (c) Right lung, medial view. (d) Left lung, medial view.

tory bronchioles. As the respiratory bronchioles penetrate more deeply into the lungs, the epithelial lining changes from cuboidal to squamous. Respiratory bronchioles, in turn, subdivide into several (2–11) *alveolar ducts (atria).*

Around the circumference of the alveolar ducts are numerous alveoli and alveolar sacs. An *alveolus* (al-VĒ-ō-lus) is a cup-shaped outpouching lined by epithelium and supported by a thin elastic basement membrane. *Alveolar sacs* are two or more alveoli that share a common opening (Figure 22-9a,b). The alveolar walls consist of several types of epithelial cells: *squamous pulmonary epithelial cells* and *septal cells* (Figure 22-9c,e). The squamous pulmonary epithelial cells are the larger of the two types of cells and form a continuous lining of the alveolar wall, except for occasional septal cells. Septal cells are much smaller, some-

what cuboidal in shape, and are dispersed among the squamous pulmonary epithelial cells. Septal cells produce a phospholipid substance called *surfactant* (sur-FAK-tant), which lowers surface tension (described shortly). Also found within the alveolar wall are free *alveolar macrophages (dust cells),* highly phagocytic cells that serve to remove dust particles or other debris from alveolar spaces; monocytes, white blood cells that become transformed into alveolar macrophages; and fibroblasts. Also present between lining cells of the alveolar wall are reticular and elastic fibers. Deep to the layer of squamous pulmonary epithelial cells is an elastic basement membrane. Over the alveoli, the arteriole and venule disperse into a capillary network. The blood capillaries consist of a single layer of endothelial cells and basement membrane.

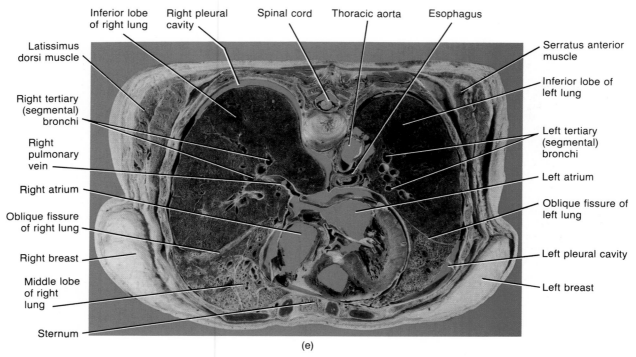

(e)

FIGURE 22-7 (Continued) Lungs. (e) Photograph of a cross section through the inferior portion of the thoracic cavity. (Courtesy of Stephen A. Kieffer and E. Robert Heitzman, *An Atlas of Cross-Sectional Anatomy,* Harper & Row, Publishers, Inc., New York, 1979.)

CLINICAL APPLICATION

Many respiratory disorders are treated by means of ***nebulization*** (neb-yoo-li-ZĀ-shun). This procedure is the administering of medication in the form of droplets that are suspended in air to select areas of the respiratory tract. The patient inhales the medication as a fine mist. The number of droplets suspended in the mist and the area of the respiratory tract that the medication will reach both depend on droplet size. Smaller droplets (approximately 2μm in diameter) can be suspended in greater numbers than can large droplets and will reach the alveolar ducts and sacs. Larger droplets (approximately 7–16 μm in diameter) will be deposited mostly in the bronchi and bronchioles. Droplets of 40 μm and larger will be deposited in the upper respiratory tract—the mouth, pharynx, trachea, and main bronchi. Nebulization therapy can be used with many different types of drugs, such as chemicals that relax the smooth muscle of the respiratory passageways, chemicals that reduce the thickness of mucus, and antibiotics.

Alveolar–Capillary (Respiratory) Membrane

The exchange of respiratory gases between the lungs and blood takes place by diffusion across alveolar and capillary walls. This membrane, through which the respiratory gases move, is collectively known as the ***alveolar–capillary (respiratory) membrane*** (Figure 22-9e). It consists of:

1. A layer of squamous pulmonary epithelial cells with septal cells and free alveolar macrophages that constitute the alveolar (epithelial) wall.
2. An epithelial basement membrane underneath the alveolar wall.
3. A capillary basement membrane that is often fused to the epithelial basement membrane.
4. The endothelial cells of the capillary.

Despite the large number of layers, the alveolar—capillary membrane averages only 0.5 μm in thickness. This is of considerable importance to the efficient diffusion of respiratory gases. Moreover, it has been estimated that the lungs contain 30 million alveoli, providing an immense surface area of 70 m^2 (753 ft^2) for the exchange of gases.

Blood and Nerve Supply

There is a double blood supply to the lungs. Deoxygenated blood passes through the pulmonary trunk, which divides into a left pulmonary artery that enters the left lung and a right pulmonary artery that enters the right lung. The venous return of the oxygenated blood is by way of the pulmonary veins, typically two in number on each side—the right and left superior and inferior pulmonary veins.

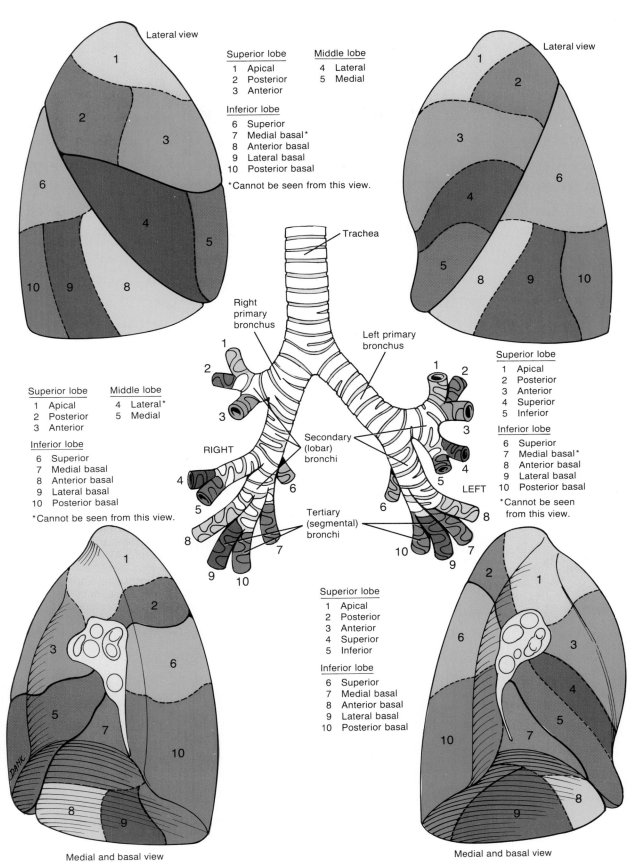

Lateral view

Superior lobe
1 Apical
2 Posterior
3 Anterior

Middle lobe
4 Lateral
5 Medial

Inferior lobe
6 Superior
7 Medial basal*
8 Anterior basal
9 Lateral basal
10 Posterior basal

*Cannot be seen from this view.

Lateral view

Trachea

Right primary bronchus

Left primary bronchus

Secondary (lobar) bronchi

Superior lobe
1 Apical
2 Posterior
3 Anterior
4 Superior
5 Inferior

Inferior lobe
6 Superior
7 Medial basal*
8 Anterior basal
9 Lateral basal
10 Posterior basal

*Cannot be seen from this view.

Superior lobe
1 Apical
2 Posterior
3 Anterior

Middle lobe
4 Lateral*
5 Medial

Inferior lobe
6 Superior
7 Medial basal
8 Anterior basal
9 Lateral basal
10 Posterior basal

*Cannot be seen from this view.

RIGHT

LEFT

Tertiary (segmental) bronchi

Superior lobe
1 Apical
2 Posterior
3 Anterior
4 Superior
5 Inferior

Inferior lobe
6 Superior
7 Medial basal
8 Anterior basal
9 Lateral basal
10 Posterior basal

Medial and basal view

Medial and basal view

FIGURE 22-8 Bronchopulmonary segments of the lungs. The bronchial branches are shown in the center of the figure. The bronchopulmonary segments are numbered and named for convenience.

631

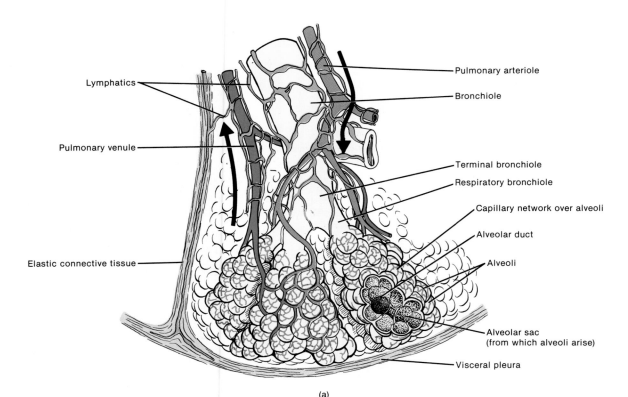

Pulmonary arteriole

Bronchiole

Lymphatics

Pulmonary venule

Terminal bronchiole

Respiratory bronchiole

Capillary network over alveoli

Alveolar duct

Alveoli

Elastic connective tissue

Alveolar sac
(from which alveoli arise)

Visceral pleura

(a)

Alveolar ducts

Alveoli   Alveolar sacs   Surface of lung

(b)

FIGURE 22-9 Histology of the lungs. (a) Diagram of a lobule of the lung. (b) Photomicrograph of alveolar ducts, alveolar sacs, and alveoli at a magnification of 55×. (Courtesy of Michael H. Ross and Edward J. Reith, *Histology: A Text and Atlas*. Copyright © 1985 by Michael H. Ross and Edward J. Reith, Harper & Row, Publishers, Inc., New York.) (c) Details of an alveolus. (d) Scanning electron micrograph of an alveolus showing squamous pulmonary alveolar cells and an alveolar macrophage at a magnification of 3,430×. (Courtesy of Richard K. Kessel and Randy H. Kardon, *Tissues and Organs: A Text-Atlas of Scanning Electron Microscopy*. Copyright © 1979 by Scientific American, Inc.) (e) Diagram of the structure of the alveolar–capillary (respiratory) membrane.

All four veins drain into the left atrium (see Figure 14-17).

Oxygenated blood is delivered through bronchial arteries, direct branches of the aorta. There are communications between the two systems and most blood returns via pulmonary veins. Some blood, however, drains into bronchial veins, branches of the azygos system.

The nerve supply of the lungs is derived from the pulmonary plexus, located anterior and posterior to the roots of the lungs. The pulmonary plexus is formed by branches of the vagus (X) nerves and sympathetic trunks. Efferent parasympathetic fibers arise from the dorsal nucleus of the vagus (X) nerve, whereas efferent sympathetic fibers are postganglionic fibers of the second to fifth thoracic paravertebral ganglia of the sympathetic trunk.

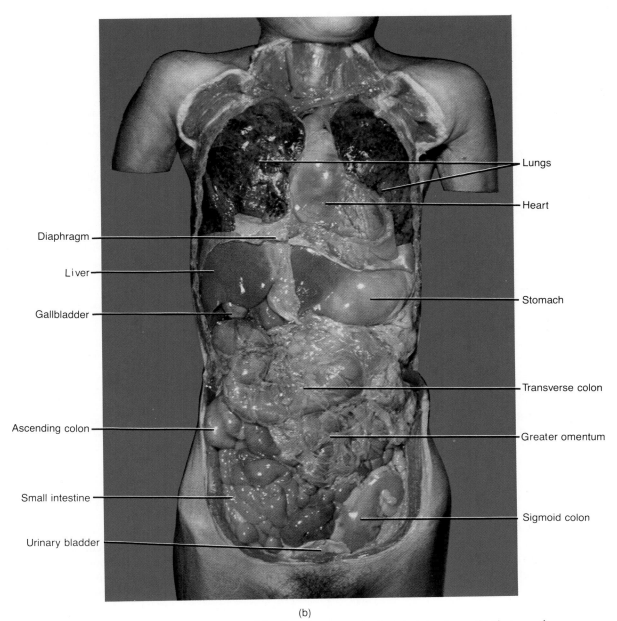

(b)

FIGURE 23-1 (Continued) Organs of the digestive system and related structures. (b) Photograph in relation to surrounding structures. (Courtesy of C. Yokochi and J. W. Rohen, *Photographic Anatomy of the Human Body,* 2nd ed., 1979, IGAKU-SHOIN, Ltd., Tokyo, New York.)

(called mesothelium) and an underlying supporting layer of connective tissue. The *parietal peritoneum* lines the wall of the abdominal cavity. The *visceral peritoneum* covers some of the organs and constitutes their serosa. The potential space between the parietal and visceral portions of the peritoneum is called the *peritoneal cavity* and contains serous fluid. In certain diseases, the peritoneal cavity may become distended by several liters of fluid so that it forms an actual space. Such an accumulation of serous fluid is called *ascites* (a-SĪ-tēz). As you will see later, some organs lie on the posterior abdominal wall and are covered by

peritoneum on their anterior surfaces only. Such organs, including the kidneys and pancreas, are said to be *retroperitoneal.*

Unlike the pericardium and pleurae, the peritoneum contains large folds that weave between the viscera. The folds bind the organs to each other and to the walls of the cavity and contain the blood and lymph vessels and the nerves that supply the abdominal organs. One extension of the peritoneum is called the *mesentery* (MEZ-en-ter'-ē). It is an outward fold of the serous coat of the small intestine (Figure 23-3). The tip of the fold is attached to the posterior

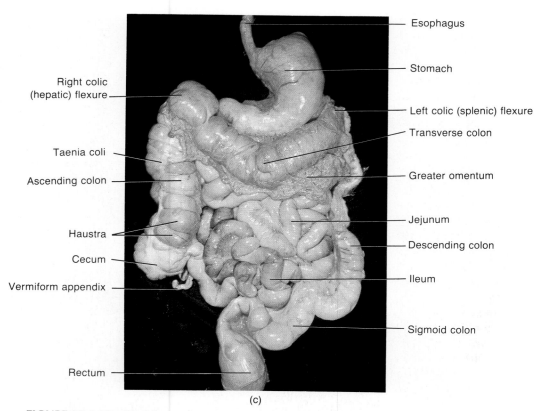

Esophagus

Stomach

Right colic (hepatic) flexure

Left colic (splenic) flexure

Transverse colon

Taenia coli

Ascending colon

Greater omentum

Jejunum

Haustra

Descending colon

Cecum

Ileum

Vermiform appendix

Sigmoid colon

Rectum

(c)

FIGURE 23-1 (Continued) (c) Photograph. (Courtesy of C. Yokochi and J. W. Rohen, *Photographic Anatomy of the Human Body,* 2nd ed., 1978, IGAKU-SHOIN, Ltd., Tokyo, New York.)

abdominal wall. The mesentery binds the small intestine to the wall. A similar fold of parietal peritoneum, called the *mesocolon* (mez'-ō-KŌ-lon), binds the large intestine to the posterior body wall. It also carries blood vessels and lymphatics to the intestines.

Other important peritoneal folds are the falciform ligament, lesser omentum, and greater omentum. The *falciform* (FAL-si-form) *ligament* attaches the liver to the anterior abdominal wall and diaphragm. The *lesser omentum* (ō-MENT-um) arises as two folds in the serosa of the stomach and duodenum suspending the stomach and duodenum from the liver. The *greater omentum* is the largest peritoneal fold that drapes over the transverse colon and coils of the small intestine. It is a double sheet that folds upon itself and thus is a four-layered structure. It is attached along the stomach and duodenum, passes downward over the small intestine for a variable distance, and then turns upward to the transverse colon where it is attached to it. Because the greater omentum contains large quantities of adipose tissue, it commonly is called the "fatty apron." The greater omentum contains numerorus lymph nodes. If an infection occurs in the intestine, plasma cells formed in the lymph nodes combat the infection and help prevent it from spreading to the peritoneum.

MOUTH (ORAL CAVITY)

The *mouth,* also referred to as the *oral* or *buccal* (BUK-al) *cavity,* is formed by the cheeks, hard and soft palates, and tongue (Figure 23-4). Forming the lateral walls of the oral cavity are the *cheeks*—muscular structures covered on the outside by skin and lined by nonkeratinized stratified squamous epithelium. The anterior portions of the cheeks terminate in the superior and inferior lips.

The *lips (labia)* are fleshy folds surrounding the orifice of the mouth. They are covered on the outside by skin and on the inside by a mucous membrane. The transition zone where the two kinds of covering tissue meet is called the *vermilion* (ver-MIL-yon). This portion of the lips is nonkeratinized, and the color of the blood in the underlying blood vessels is visible through the transparent surface layer of the vermilion. The inner surface of each lip is attached to its corresponding gum by a midline fold of mucous membrane called the *labial frenulum* (LĀ-bē-al FREN-yoo-lum).

The orbicularis oris muscle and connective tissue lie between the external integumentary covering and the internal mucosal lining. During chewing, the cheeks and lips help keep food between the upper and lower teeth. They also assist in speech.

(Representative Sections of Gastrointestinal Tract)

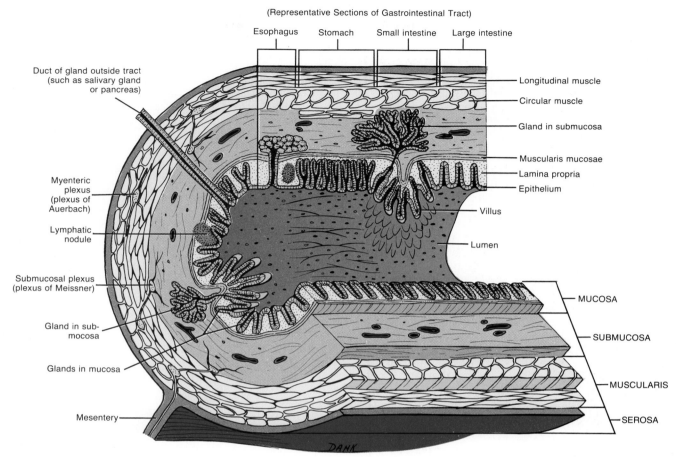

FIGURE 23-2 Portion of the gastrointestinal tract seen in section views showing the various layers and related structures.

The **vestibule** of the oral cavity is bounded externally by the cheeks and lips and internally by the gums and teeth. The **oral cavity proper** extends from the vestibule to the **fauces** (FAW-sēs), the opening between the oral cavity and the pharynx or throat.

The **hard palate,** the anterior portion of the roof of the mouth, is formed by the maxillae and palatine bones, is covered by mucous membrane, and forms a bony partition between the oral and nasal cavities. The **soft palate** forms the posterior portion of the roof of the mouth. It is an arch-shaped muscular partition between the oropharynx and nasopharynx and is lined by mucous membrane.

Hanging from the free border of the soft palate is a conical muscular process called the **uvula** (YOU-vyoo-la). On either side of the base of the uvula are two muscular folds that run down the lateral side of the soft palate. Anteriorly, the **palatoglossal arch (anterior pillar)** extends inferiorly, laterally, and anteriorly to the side of the base of the tongue. Posteriorly, the **palatopharyngeal** (PAL-a-tō-fa-rin'-jē-al) **arch (posterior pillar)** projects inferiorly, laterally, and posteriorly to the side of the pharynx. The palatine tonsils are situated between the arches, and the lingual

tonsil is situated at the base of the tongue. At the posterior border of the soft palate, the mouth opens into the oropharynx through the fauces.

TONGUE

The **tongue,** together with its associated muscles, forms the floor of the oral cavity. It is an accessory structure of the digestive system composed of skeletal muscle covered with mucous membrane (see Figure 20-9c). The tongue is divided into symmetrical lateral halves by a median septum that extends throughout its entire length and is attached inferiorly to the hyoid bone. Each half of the tongue consists of an identical complement of extrinsic and intrinsic muscles.

The **extrinsic muscles** of the tongue originate outside the tongue and insert into it. They include the hyoglossus, genioglossus, and styloglossus (see Figure 10-7). The extrinsic muscles move the tongue from side to side and in and out. These movements maneuver food for chewing, shape the food into a rounded mass, and force the food to the back of the mouth for swallowing. They also form the floor of the mouth and hold the tongue in position. The

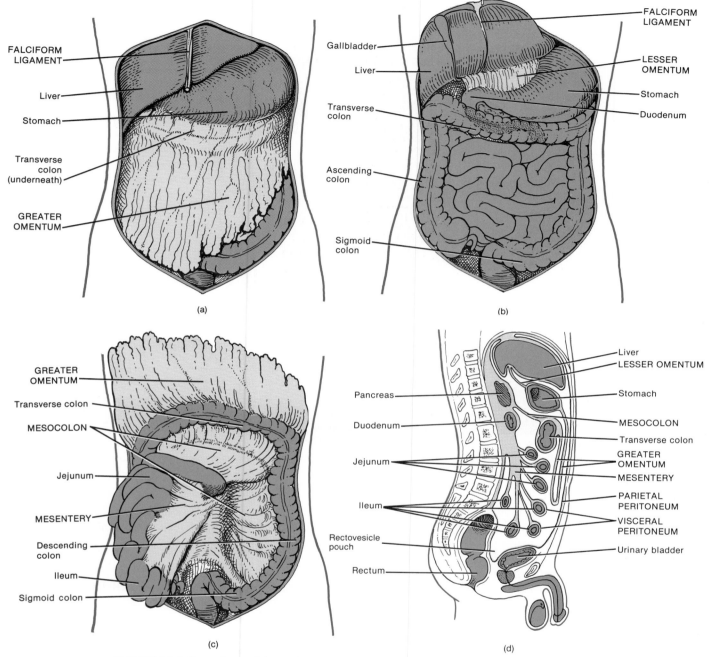

FALCIFORM
LIGAMENT

Liver

Stomach

Transverse
colon
(underneath)

GREATER
OMENTUM

(a)

FALCIFORM
LIGAMENT

Gallblader

Liver

Transverse
colon

LESSER
OMENTUM

Stomach

Duodenum

Ascending
colon

Sigmoid
colon

(b)

GREATER
OMENTUM

Transverse colon

MESOCOLON

Jejunum

MESENTERY

Descending
colon

Ileum

Sigmoid colon

(c)

Pancreas

Duodenum

Jejunum

Ileum

Rectovesicle
pouch

Rectum

Liver
LESSER OMENTUM

Stomach

MESOCOLON

Transverse colon

GREATER
OMENTUM

MESENTERY

PARIETAL
PERITONEUM

VISCERAL
PERITONEUM

Urinary bladder

(d)

FIGURE 23-3 Extensions of the peritoneum. (a) Greater omentum (see also Figure 23-1b). (b) Lesser omentum. The liver and gallbladder have been lifted. (c) Mesentery. The greater omentum has been lifted. (d) Sagittal section through the abdomen and pelvis indicating the relation of the peritoneal extensions to each other.

intrinsic muscles originate and insert within the tongue and alter the shape and size of the tongue for speech and swallowing. The intrinsic muscles include the longitudinalis superior, longitudinalis inferior, transversus linguae, and verticalis linguae (Figure 23-5a). The *lingual frenulum,* a fold of mucous membrane in the midline of the undersurface of the tongue, aids in limiting the movement of the tongue posteriorly.

CLINICAL APPLICATION

If the lingual frenulum is too short, tongue movements are restricted, speech is faulty, and the person is said to be "tongue-tied." This congenital problem is referred to as *ankyloglossia* (ang'-ki-lō-GLOSS-ē-a). It can be corrected by cutting the lingual frenulum.

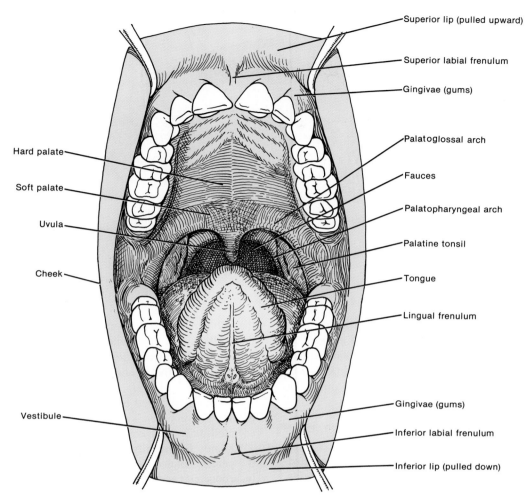

FIGURE 23-4 Mouth (oral cavity).

The upper surface and sides of the tongue are covered with *papillae* (pa-PIL-ē), projections of the lamina propria covered with epithelium (Figure 23-5b,c). *Filiform papillae* are conical projections distributed in parallel rows over the anterior two-thirds of the tongue. They are whitish and contain no taste buds. *Fungiform papillae* are mushroomlike elevations distributed among the filiform papillae and are more numerous near the tip of the tongue. They appear as red dots on the surface of the tongue, and most of them contain taste buds. *Circumvallate papillae,* 10–12 in number are arranged in the form of an inverted V on the posterior surface of the tongue, and all of them contain taste buds. Note the taste zones of the tongue in Figure 20-9c.

SALIVARY GLANDS

Saliva is a fluid that is continuously secreted by glands in or near the mouth. Ordinarily, just enough saliva is secreted to keep the mucous membranes of the mouth moist, but when food enters the mouth, secretion increases so the saliva can lubricate, dissolve, and begin the chemical breakdown of the food. The mucous membrane lining the mouth contains many small glands, the *buccal glands,* that secrete small amounts of saliva. However, the major portion of saliva is secreted by the *salivary glands,* accessory structures that lie outside the mouth and pour their contents into ducts that empty into the oral cavity. There are three pairs of salivary glands: parotid, submandibular (submaxillary), and sublingual glands (Figure 23-6a).

The *parotid glands* are located inferior and anterior to the ears between the skin and the masseter muscle. They are compound tubuloacinar glands. Each secretes into the oral cavity vestibule via a duct, called the *parotid (Stensen's) duct,* that pierces the buccinator muscle to open into the vestibule opposite the upper second molar tooth. The *submandibular glands,* which are compound acinar glands, are found beneath the base of the tongue in the posterior part of the floor of the mouth (Figure 23-6b). Their ducts, the *submandibular (Wharton's) ducts,* run superficially under the mucosa on either side of the midline of the floor

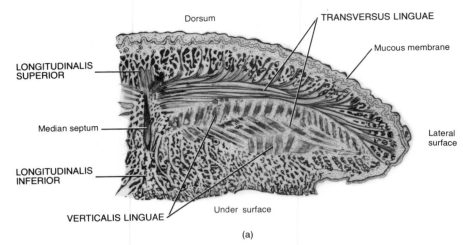

FIGURE 23-5 Tongue. (a) Frontal section through the left side of the tongue showing its intrinsic muscles.

of the mouth and enter the oral cavity proper on either side of the lingual frenulum. The *sublingual glands,* also compound acinar glands, are anterior to the submandibular glands, and their ducts, the *lesser sublingual (Rivinus's) ducts,* open into the floor of the mouth in the oral cavity proper.

CLINICAL APPLICATION

Although any of the salivary glands may become infected as a result of a nasopharyngeal infection, the parotids are typically the target of the mumps virus (myxovirus). *Mumps* is an inflammation and enlargement of the parotid glands accompanied by moderate fever, malaise, and extreme pain in the throat, especially when swallowing sour foods or acid juices. Swelling occurs on one or both sides of the face, just anterior to the ramus of the mandible. In about 20–35 percent of males past puberty, the testes may also become inflamed, and although it rarely occurs, sterility is a possible consequence. In some people, aseptic meningitis, pancreatitis, and hearing loss may also occur as complications.

The parotid gland receives its blood supply from branches of the external carotid artery and is drained by vessels that are tributaries of the external jugular vein. The submandibular gland is supplied by branches of the facial artery and drained by tributaries of the facial vein. The sublingual gland is supplied by the sublingual branch of the lingual artery and the submental branch of the facial artery and is drained by tributaries of the sublingual and submental veins.

The salivary glands receive both sympathetic and parasympathetic innervation. The sympathetic fibers form plexuses on the blood vessels that supply the glands and serve as vasoconstrictors. The parotid gland receives sympathetic fibers from the plexus on the external carotid artery, whereas the submandibular and sublingual glands receive sympathetic fibers that contribute to the sympathetic plexus and accompany the facial artery to the glands. The parasympathetic fibers of the glands consist of secretomotor fibers to the glands.

The fluids secreted by the buccal glands and the three pairs of salivary glands constitute *saliva.* Amounts of saliva secreted daily vary considerably but range from 1,000 to 1,500 ml. Chemically, saliva is 99.5 percent water and 0.5 percent solutes. Among the solutes are salts—chlorides, bicarbonates, and phosphates of sodium and potassium. Some dissolved gases and various organic substances, including urea and uric acid, serum albumin and globulin, mucin, the bacteriolytic enzyme lysozyme, and the digestive enzyme salivary amylase, are also present. *Salivary amylase* initiates the breakdown of starch.

Each saliva-producing gland supplies different ingredients to saliva. The parotids contain cells that secrete a watery serous liquid that includes the enzyme salivary amylase. The submandibular glands contain cells similar to those found in the parotids plus some mucous cells. Therefore, they secrete a fluid that is thickened with mucus but still contains quite a bit of enzyme. The sublingual glands contain mostly mucous cells, so they secrete a much thicker fluid that contributes only a small amount of enzyme to the saliva.

Saliva continues to be secreted heavily some time after food is swallowed. This flow of saliva washes out the mouth and dilutes and buffers the chemical remnants of irritating substances.

TEETH

The *teeth (dentes)* are accessory structures of the digestive system located in sockets of the alveolar processes of the mandible and maxillae. The alveolar processes are covered

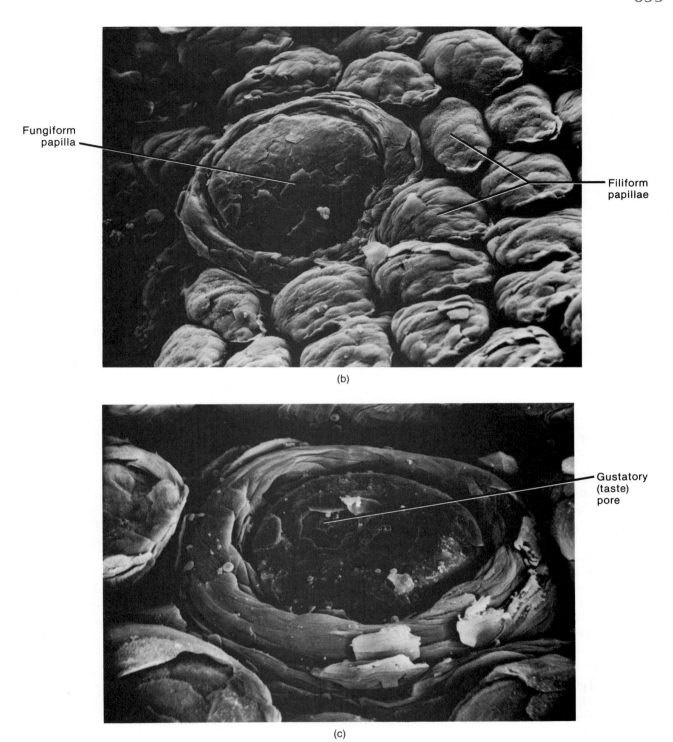

Fungiform papilla

Filiform papillae

(b)

Gustatory (taste) pore

(c)

FIGURE 23-5 (Continued) (b) Scanning electron micrograph of a fungiform papilla surrounded by filiform papillae at a magnification of 500×. (c) Scanning electron micrograph of a circumvallate papilla at a magnification of 500×. (Courtesy of Richard K. Kessel and Randy H. Kardon, *Tissues and Organs: A Text-Atlas of Scanning Electron Microscopy.* Copyright © 1979 by Scientific American, Inc.)

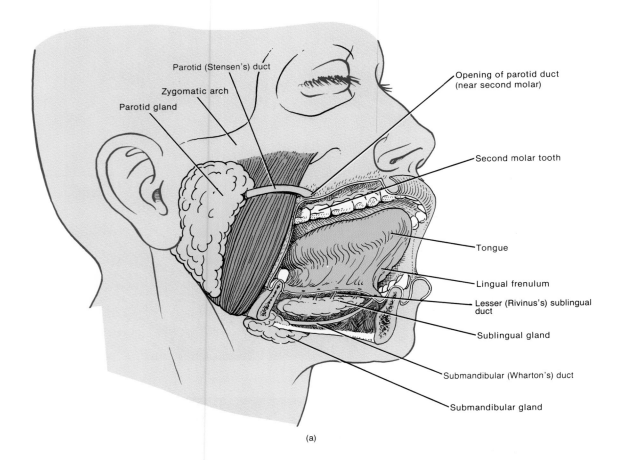

(a)

by the **gingivae** (jin-JI-vē) or gums, which extend slightly into each socket forming the gingival sulcus (Figure 23-7). The sockets are lined by the **periodontal ligament,** which consists of dense fibrous connective tissue and is attached to the socket walls and the cemental surface of the roots. Thus, it anchors the teeth in position and also acts as a shock absorber to dissipate the forces of chewing.

A typical tooth consists of three principal portions. The **crown** is the exposed portion above the level of the gums. The **root** consists of one to three projections embedded in the socket. The **neck** is the constricted junction line of the crown and the root near the gum line.

Teeth are composed primarily of **dentin,** a calcified connective tissue that gives the tooth its basic shape and rigidity. The dentin encloses a cavity. The enlarged part of the cavity, the **pulp cavity,** lies in the crown and is filled with **pulp,** a connective tissue containing blood vessels, nerves, and lymphatics. Narrow extensions of the pulp cavity run through the root of the tooth and are called **root canals.** Each root canal has an opening at its base, the **apical foramen.** Through the foramen enter blood vessels bearing nourishment, lymphatics affording protection, and nerves providing sensation. The dentin of the crown is covered by **enamel** that consists primarily of calcium phosphate and calcium carbonate. Enamel is the hardest substance in the body and protects the tooth from the wear of chewing. It is

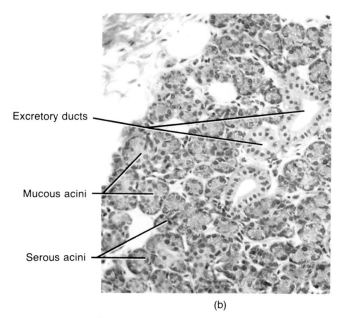

(b)

FIGURE 23-6 Salivary glands. (a) Location of the salivary glands. (b) Photomicrograph of the submandibular gland showing serous and mucous acini at a magnification of 120×. The parotids consist of all serous acini, and the sublinguals consist of mostly mucous acini and a few serous acini. (Courtesy of Andrew Kuntzman.)

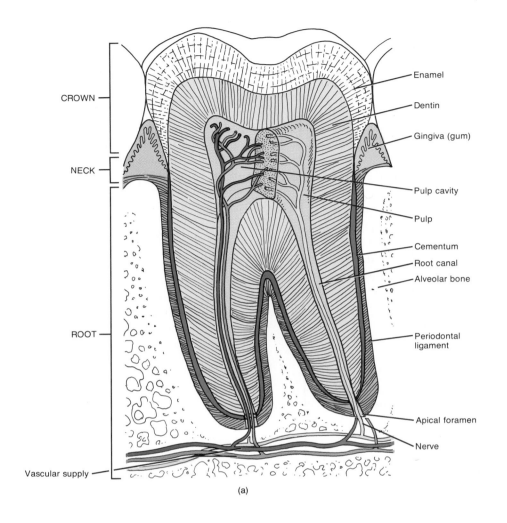

CROWN

NECK

ROOT

Vascular supply

Enamel

Dentin

Gingiva (gum)

Pulp cavity

Pulp

Cementum

Root canal

Alveolar bone

Periodontal ligament

Apical foramen

Nerve

(a)

FIGURE 23-7 Parts of a typical tooth as seen in a section through a molar. (a) Diagram. (b) Photograph. (Courtesy of C. Yokochi and J. W. Rohen, *Photographic Anatomy of the Human Body*, 2nd ed., 1979, IGAKU-SHOIN, Ltd., Tokyo, New York.)

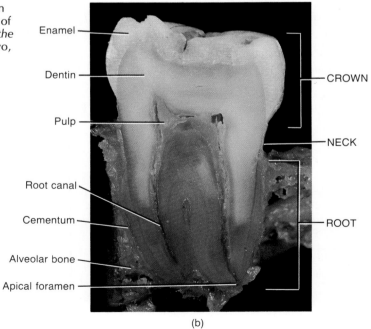

Enamel

Dentin

Pulp

Root canal

Cementum

Alveolar bone

Apical foramen

CROWN

NECK

ROOT

(b)

also a barrier against acids that easily dissolve the dentin. The dentin of the root is covered by *cementum,* another bonelike substance, which attaches the root to the periodontal ligament.

The branch of dentistry that is concerned with the prevention, diagnosis, and treatment of diseases that affect the pulp, root, periodontal ligament, and alveolar bone is known as *endodontics* (en'-dō-DON-tiks; *endo* = within; *odous* = tooth).

CLINICAL APPLICATION

Root canal therapy refers to a procedure, accomplished in several phases, in which all traces of pulp tissue are removed from the pulp cavity and root canal of a badly diseased tooth. After a hole is made in the tooth, the root canal is filed out and irrigated to remove bacteria. Then, the canal is treated with medication and sealed tightly. The damaged crown is then repaired.

Dental Terminology

Because of the curvature of the dental arches, it is necessary to use terms other than anterior, posterior, medial, and lateral in describing the surfaces of the teeth. Accordingly, the following directional terms are used. *Labial* refers to the surface of a tooth in contact with or directed toward the lips. *Buccal* refers to the surface in contact with or directed toward the cheeks. *Lingual* is restricted to the teeth of the lower jaw and refers to the surface directed toward the tongue. *Palatal,* on the other hand, is restricted to the teeth of the upper jaw and refers to the surface directed toward the palate. The term *mesial* designates the anterior or medial side of the tooth relative to its position in the dental arch. *Distal* refers to the posterior or lateral side of the tooth relative to its position in the dental arch. Essentially, mesial and distal refer to the sides of adjacent teeth that are in contact with each other. Finally, *occlusal* refers to the biting surface of a tooth.

Dentitions

Everyone has two *dentitions,* or sets of teeth. The first of these—the *deciduous teeth, milk teeth,* or *baby teeth*—begin to erupt at about 6 months of age, and one pair appears at about each month thereafter until all 20 are present. Figure 23-8a illustrates the deciduous teeth. The incisors, which are closest to the midline, are chisel shaped and adapted for cutting into food. They are referred to as either *central* or *lateral incisors* on the basis of their position. Next to the incisors, moving posteriorly, are the *cuspids* *(canines),* which have a pointed surface called a cusp.

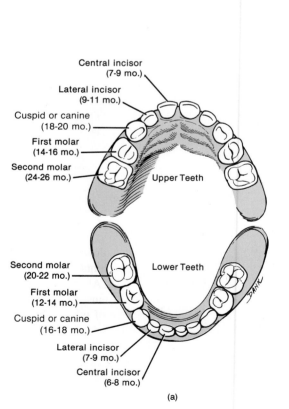

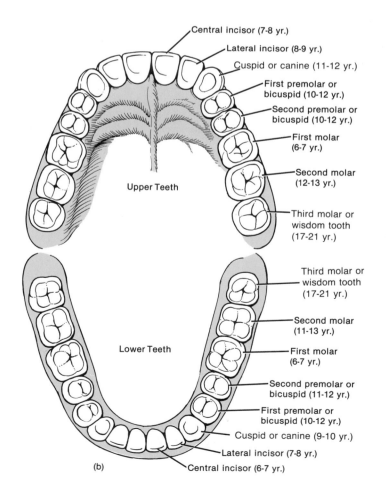

FIGURE 23-8 Dentitions and times of eruptions. The times of eruptions are indicated in parentheses. (a) Deciduous dentition. (b) Permanent dentition.

Cuspids are used to tear and shred food. The incisors and cuspids have only one root apiece. Behind them lie the *first* and *second molars,* which have four cusps. Upper molars have three roots; lower molars have two roots. The molars crush and grind food.

All the deciduous teeth are lost—generally between 6 and 12 years of age—and are replaced by the *permanent dentition* (Figure 23-8b). The permanent dentition contains 32 teeth that appear between the age of 6 and adulthood. It resembles the deciduous dentition with the following exceptions. The deciduous molars are replaced with the *first* and *second premolars (bicuspids),* which have two cusps and one root (upper first bicuspids have two roots) and are used for crushing and grinding. The permanent molars erupt into the mouth behind the bicuspids. They do not replace any deciduous teeth and erupt as the jaw grows to accommodate them—the *first molars* at age 6, the *second molars* at age 12, the *third molars (wisdom teeth)* after age 18. The human jaw has become smaller through time and often does not afford enough room behind the second molars for the eruption of the third molars. In this case, the third molars remain embedded in the alveolar bone and are said to be ''impacted.'' Most often they cause pressure and pain and must be surgically removed. In some individuals, third molars may be dwarfed in size or may not develop at all.

Blood and Nerve Supply

The arteries that supply blood to the teeth are distributed to the pulp cavity and surrounding periodontal ligament. The upper incisors and cuspids are supplied by anterior superior alveolar branches of the maxillary artery; the upper premolars and molars are supplied by the posterior superior alveolar branches of the maxillary artery; the lower incisors and cuspids are supplied by the incisive branches of the inferior alveolar artery; and the lower premolars and molars are supplied by the dental branches of the inferior alveolar artery.

The teeth receive sensory fibers from branches of the maxillary and mandibular divisions of the trigeminal (V) nerve—the upper teeth from branches of the maxillary division and the lower teeth from branches of the mandibular division.

Through chewing, or *mastication,* the teeth grind food and mix it with saliva. As a result, the food is reduced to a soft, flexible mass called a *bolus* that is easily swallowed.

PHARYNX

Swallowing, or *deglutition* (dē-gloo-TISH-un), is a mechanism that moves food from the mouth to the stomach. It is facilitated by saliva and mucus and involves the mouth, pharynx, and esophagus. Swallowing is conveniently divided into three stages: (1) the voluntary stage, in which the bolus is moved into the oropharynx, (2) the pharyngeal stage, the involuntary passage of the bolus through the pharynx into the esophagus, and (3) the esophageal stage, the involuntary passage of the bolus through the esophagus into the stomach.

ESOPHAGUS

The *esophagus* (e-SOF-a-gus), the third organ involved in deglutition, is a muscular, collapsible tube that lies behind the trachea. It is about 23–25 cm (10 in.) long and begins at the end of the laryngopharynx, passes through the mediastinum anterior to the vertebral column, pierces the diaphragm through an opening called *esophageal hiatus,* and terminates in the superior portion of the stomach (see Figure 23-1a).

HISTOLOGY

The *mucosa* of the esophagus consists of nonkeratinized stratified squamous epithelium, lamina propria, and a muscularis mucosae (Figure 23-9c). Near the stomach, the mucosa of the esophagus also contains mucous glands. The *submucosa* contains connective tissue, blood vessels, and mucous glands. The *muscularis* of the upper third is striated, the middle third is striated and smooth, and the lower third is smooth. The outer layer is known as the *adventitia* (adven-TISH-ya) rather than the serosa because the loose connective tissue of the layer is not covered by epithelium (mesothelium) and because the connective tissue merges with the connective tissue of surrounding structures.

ACTIVITIES

The esophagus does not produce digestive enzymes and does not carry on absorption. It secretes mucus and transports food to the stomach. The passage of food from the laryngopharynx into the esophagus is regulated by a sphincter (thick circle of muscle around an opening) at the entrance to the esophagus called the *upper esophageal* (e-sof'-a-JĒ-al) *sphincter.* It consists of the cricopharyngeus muscle attached to the cricoid cartilage. The elevation of the larynx during the pharyngeal stage of swallowing causes the sphincter to relax and the bolus enters the esophagus. The sphincter also relaxes during expiration.

During the esophageal stage of swallowing, food is pushed through the esophagus by involuntary muscular movements called *peristalsis* (per'-is-STAL-sis) (Figure 23-9a,b). Peristalsis is a function of the muscularis and is controlled by the medulla. In the section of the esophagus lying just above and around the top of the bolus, the circular muscle fibers (cells) contract. The contraction constricts the esophageal wall and squeezes the bolus downward. Meanwhile, longitudinal fibers lying around the bottom of and just below the bolus also contract. Contraction of the longitudinal fibers shortens this lower section, pushing its walls outward so it can receive the bolus. The contractions are repeated in a wave that moves down the esophagus,

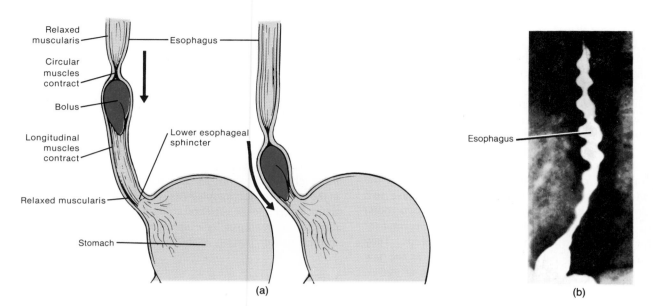

(a)

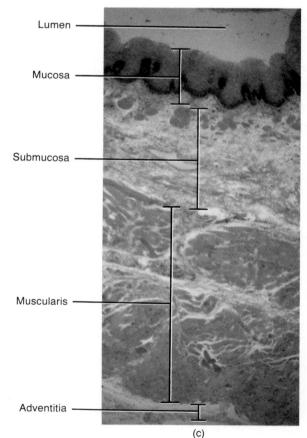

(b)

pushing the food toward the stomach. Passage of the bolus is further facilitated by glands that secrete mucus. The passage of solid or semisolid food from the mouth to the stomach takes 4–8 seconds. Very soft foods and liquids pass through in about 1 second.

Just above the level of the diaphragm, the esophagus is slightly narrowed. This narrowing has been attributed to a physiological sphincter in the inferior part of the esophagus known as the **lower esophageal (gastroesophageal) sphincter.** The lower esophageal sphincter relaxes during swallowing and thus aids the passage of the bolus from the esophagus into the stomach.

CLINICAL APPLICATION

If the lower esophageal sphincter fails to relax normally as food approaches, the condition is called **achalasia** (ak'-a-LĀ-zē-a; *a* = without; *chalasis* = relaxation). As a result, food passage from the esophagus into the stomach is greatly impeded. A whole meal may become lodged in the esophagus, entering the stomach very slowly. Distension of the esophagus results in chest pain that is often confused with pain originating from the heart. The condition is caused by malfunction of the myenteric plexus (plexus of Auerbach).

If, on the other hand, the lower esophageal sphincter fails to close adequately after food has entered the stomach, the stomach contents can enter the lower esophagus. Hydrochloric acid (HCl) from the stomach contents can irritate the esophageal wall, resulting in a burning sensation. The sensation is known as **heartburn** because it is experienced in the region very near the heart, although it is not related to any cardiac problem. Heartburn can be treated by taking antacids (Tums, Gelusil, Rolaids, Maalox) that neutralize the hydrochloric acid and lessen the severity of the burning sensation and discomfort.

(c)

FIGURE 23-9 Esophagus. (a) Diagram of peristalsis. (b) Antero-posterior projection of peristalsis made during fluoroscopic examination while a patient was swallowing barium. (Courtesy of Lester W. Paul and John H. Juhl, *The Essentials of Roentgen Interpretation,* 3rd ed., Harper & Row, Publishers, Inc., New York, 1972.) (c) Histology of the esophagus. Photomicrograph of a portion of the wall of the esophagus at a magnification of 60×. An enlarged aspect of the mucosa of the esophagus is shown in Exhibit 3-1, Stratified squamous. (Courtesy of Andrew Kuntzman.)

BLOOD AND NERVE SUPPLY

The arteries of the esophagus are derived from the arteries along its length: inferior thyroid, thoracic aorta, intercostal arteries, phrenic, and left gastric arteries. It is drained by the adjacent veins. Innervation of the esophagus is by recurrent laryngeal nerves, the cervical sympathetic chain, and vagi.

STOMACH

The *stomach* is a J-shaped enlargement of the GI tract directly under the diaphragm in the epigastric, umbilical, and left hypochondriac regions of the abdomen (see Figure 1-8b). The superior portion of the stomach is a continuation of the esophagus. The inferior portion empties into the duodenum, the first part of the small intestine. Within each individual, the position and size of the stomach vary continually. For instance, the diaphragm pushes the stomach downward with each inspiration and pulls it upward with each expiration. Empty, it is about the size of a large sausage, but it can stretch to accommodate large amounts of food.

ANATOMY

The stomach is divided by gross anatomists into four areas: cardia, fundus, body, and pylorus (Figure 23-10). The *cardia* surrounds the lower esophageal sphincter. The rounded portion above and to the left of the cardia is the *fundus*. Below the fundus is the large central portion of the stomach, called the *body*. The narrow, inferior region is the *pylorus*. The concave medial border of the stomach is called the *lesser curvature,* and the convex lateral border is the *greater curvature.* The pylorus communicates with the duodenum of the small intestine via a sphincter called the *pyloric sphincter (valve).*

CLINICAL APPLICATION

Two abnormalities of the pyloric sphincter can occur in infants. *Pylorospasm* is characterized by failure of the muscle fibers (cells) encircling the opening to relax normally. It can be caused by hypertrophy or continuous spasm of the sphincter and usually occurs between the second and twelfth weeks of life. Ingested food does not pass easily from the stomach to the small intestine, the stomach becomes overly full, and the infant vomits frequently to relieve the pressure. Pylorospasm is treated by adrenergic drugs that relax the muscle fibers of the sphincter. *Pyloric stenosis* is a narrowing of the pyloric sphincter caused by a tumorlike mass that apparently is formed by enlargement of the circular muscle fibers. It must be surgically corrected.

HISTOLOGY

The stomach wall is composed of the same four basic layers as the rest of the GI tract, with certain modifications. When the stomach is empty, the *mucosa* lies in large folds, called *rugae* (ROO-jē), that can be seen with the naked eye (Figure 23-11c). Microscopic inspection of the mucosa reveals a layer of simple columnar epithelium (surface mucous cells) containing many narrow openings that extend down into the lamina propria called *gastric pits* (Figure 23-11b). At the bottoms of the pits are the orifices of *gastric glands.* Each gland consists of four types of secreting cells: zymogenic, parietal, mucous, and enteroendocrine. The *zymogenic (peptic) cells* secrete the principal gastric enzyme precursor, pepsinogen. Hydrochloric acid, involved in the conversion of pepsinogen to the active enzyme pepsin, and intrinsic factor, involved in the absorption of vitamin B_{12} for red blood cell production, are produced by the *parietal (oxyntic) cells.* You may recall from Chapter 12 that inability to produce intrinsic factor can result in pernicious anemia. The *mucous cells* secrete mucus. Secretions of the zymogenic, parietal, and mucous cells are collectively called *gastric juice.* The *enteroendocrine cells* secrete stomach gastrin, a hormone that stimulates secretion of hydrochloric acid and pepsinogen, contracts the lower esophageal sphincter, mildly increases motility of the GI tract, and relaxes the pyloric sphincter.

CLINICAL APPLICATION

The general term *endoscopy* refers to visual inspection of any cavity of the body using an endoscope, an illuminated tube with lenses. Endoscopes can be used to visualize the entire gastrointestinal tract. In fact, endoscopes can be fitted with special devices that also remove foreign objects from the esophagus and stomach, dilate strictures in the esophagus, remove small gallstones, temporarily stop bleeding, biopsy lesions, and remove polyps from the colon.

Endoscopic examination of the stomach is called *gastroscopy.* During the procedure, the patient is anesthetized and the endoscope is passed into the stomach, which is then inflated with air. The gastric mucosa may be examined for any abnormalities and, if necessary, a gastric mucosal biopsy may be performed.

The *submucosa* of the stomach is composed of loose connective tissue, which connects the mucosa to the muscularis.

The *muscularis,* unlike that in other areas of the gastrointestinal tract, has three layers of smooth muscle: an outer longitudinal layer, a middle circular layer, and an inner oblique layer. The oblique layer is limited mostly to the body of the stomach. This arrangement of fibers allows the stomach to contract in a variety of ways to churn food,

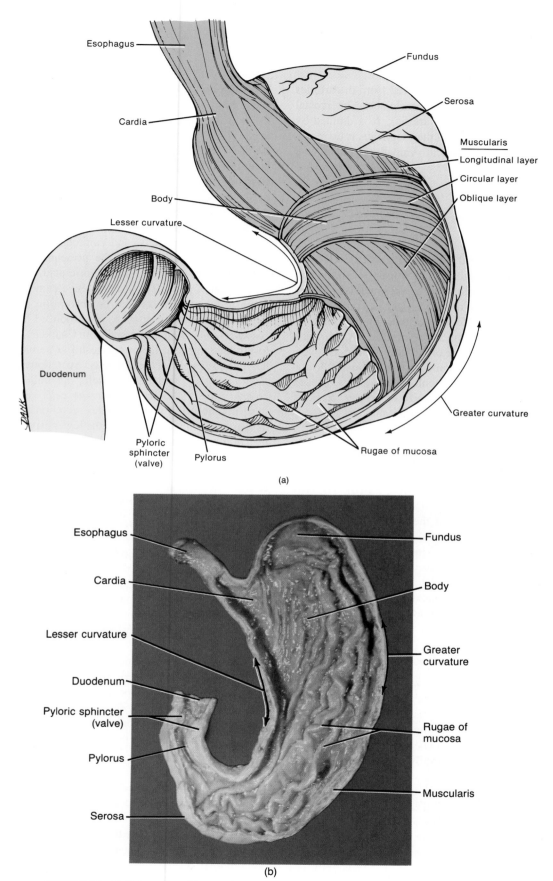

FIGURE 23-10 External and internal anatomy of the stomach. (a) Diagram in anterior view. (b) Photograph of the internal surface in anterior view showing rugae. (Courtesy of C. Yokochi and J. W. Rohen, *Photographic Anatomy of the Human Body,* 2nd ed., 1978, IGAKU-SHOIN, Ltd., Tokyo, New York.)

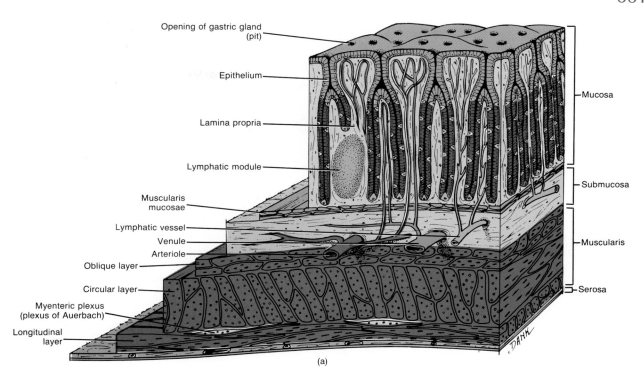

Opening of gastric gland (pit)

Epithelium

Lamina propria

Lymphatic module

Muscularis mucosae

Lymphatic vessel

Venule

Arteriole

Oblique layer

Circular layer

Myenteric plexus (plexus of Auerbach)

Longitudinal layer

Mucosa

Submucosa

Muscularis

Serosa

(a)

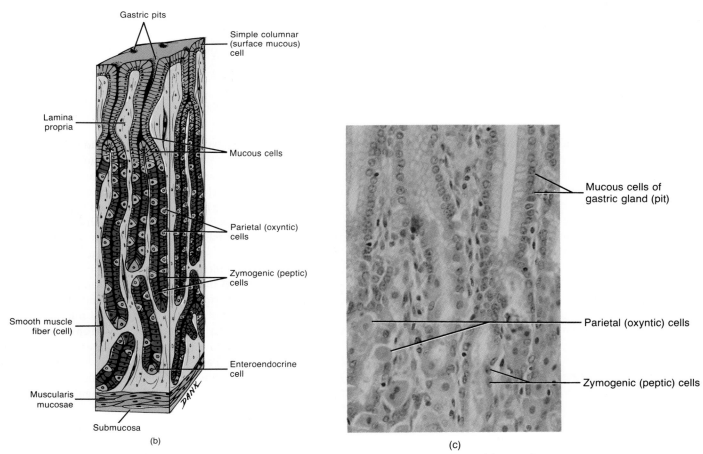

Gastric pits

Simple columnar (surface mucous) cell

Lamina propria

Mucous cells

Parietal (oxyntic) cells

Zymogenic (peptic) cells

Smooth muscle fiber (cell)

Enteroendocrine cell

Muscularis mucosae

Submucosa

(b)

Mucous cells of gastric gland (pit)

Parietal (oxyntic) cells

Zymogenic (peptic) cells

(c)

FIGURE 23-11 Histology of the stomach. (a) Diagram showing four principal layers. (b) Diagram of details of gastric mucosa. (c) Scanning electron micrograph of the gastric mucosa showing surface mucous cells at a magnification of 740×. (Courtesy of Andrew Kuntzman.)

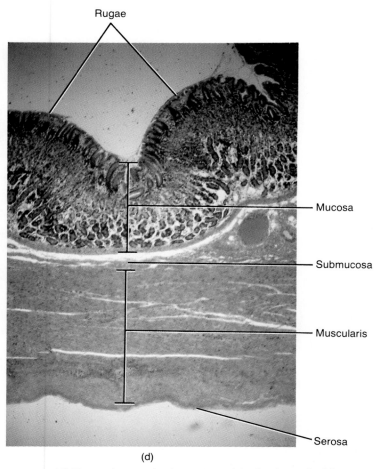

Rugae

Mucosa

Submucosa

Muscularis

Serosa

(d)

FIGURE 23-11 (*Continued*) (d) Photomicrograph of a portion of the fundic wall of the stomach at a magnification of 18×. (Courtesy Andrew Kuntzman.)

break it into small particles, mix it with gastric juice, and pass it to the duodenum.

The *serosa* covering the stomach is part of the visceral peritoneum. At the lesser curvature, the two layers of the visceral peritoneum come together and extend upward to the liver as the lesser omentum. At the greater curvature, the visceral peritoneum continues downward as the greater omentum hanging over the intestines.

ACTIVITIES

Several minutes after food enters the stomach, gentle, rippling, peristaltic movements called *mixing waves* pass over the stomach every 15–25 seconds. These waves macerate food, mix it with the secretions of the gastric glands, and reduce it to a thin liquid called *chyme (kīm)*. Few mixing waves are observed in the fundus, which is primarily a storage area. Foods may remain in the fundus for an hour or more without becoming mixed with gastric juice. During this time, salivary digestion continues.

As digestion proceeds in the stomach, more vigorous mixing waves begin at the body of the stomach and intensify as they reach the pylorus. The pyloric sphincter normally remains almost, but not completely, closed. As food reaches the pylorus, each mixing wave forces a small amount of

the gastric contents into the duodenum through the pyloric sphincter. Most of the food is forced back into the body of the stomach where it is subjected to further mixing. The next wave pushes it forward again and forces a little more into the duodenum. The forward and backward movements of the gastric contents are responsible for almost all the mixing in the stomach.

The principal chemical activity of the stomach is to begin the digestion of proteins. In the adult, digestion is achieved primarily through the enzyme *pepsin*. Another enzyme of the stomach, *gastric lipase,* splits the butterfat molecules found in milk. This enzyme has a limited role in the adult stomach. Adults rely almost exclusively on an enzyme found in the small intestine (pancreatic amylase) to digest fats. The infant stomach also secretes *rennin,* which is important in the digestion of milk. Rennin and calcium act on the casein of milk to produce a curd. The coagulation prevents too rapid a passage of milk from the stomach into the duodenum (first portion of the small intestine). Rennin is absent in the gastric secretions of adults.

The stomach empties all its contents into the duodenum 2–6 hours after ingestion. Food rich in carbohydrate leaves the stomach in a few hours. Protein foods are somewhat slower, and emptying is slowest after a meal containing large amounts of fat.

CLINICAL APPLICATION

Excessive gastric emptying in the wrong direction sometimes occurs. *Vomiting* is the forcible expulsion of the contents of the upper GI tract (stomach and sometimes duodenum) through the mouth. The strongest stimuli for vomiting are irritation and distension of the stomach. Other stimuli include unpleasant sights and dizziness. Nerve impulses are transmitted to the vomiting center in the medulla, and returning impulses to the upper GI tract organs, diaphragm, and abdominal muscles bring about the vomiting act. Basically, vomiting involves squeezing the stomach between the diaphragm and abdominal muscles and expelling of the contents through the open esophageal sphincters. Prolonged vomiting, especially in infants and elderly people, can be serious because the loss of gastric juice and fluids can lead to disturbances in fluid and acid–base balance.

The stomach wall is impermeable to the passage of most materials into the blood, so most substances are not absorbed until they reach the small intestine. However, the stomach does participate in the absorption of some water, electrolytes, certain drugs (especially aspirin), and alcohol.

BLOOD AND NERVE SUPPLY

The arterial supply of the stomach is derived from the celiac artery. The right and left gastric arteries form an anastomosing arch along the lesser curvature, and the right and left gastroepiploic arteries form a similar arch on the greater curvature. Short gastric arteries supply the fundus. The veins of the same name accompany the arteries and drain, directly or indirectly, into the hepatic portal vein.

The vagi convey parasympathetic fibers to the stomach. These fibers form synapses within the submucosal plexus in the submucosa and the myenteric plexus (plexus of Aeurbach) in the muscularis. The sympathetic nerves arise from the celiac ganglia and the nerves reach the stomach along the branches of the celiac artery.

PANCREAS

The next organ of the GI tract involved in the breakdown of food is the small intestine. Chemical digestion in the small intestine depends not only on its own secretions but also on activities of three accessory structures of digestion outside the gastrointestinal tract: the pancreas, liver, and gallbladder.

ANATOMY

The *pancreas* is a soft, oblong tubuloacinar gland about 12.5 cm (6 in.) long and 2.5 cm (1 in.) thick. It lies posterior to the greater curvature of the stomach and is connected to the duodenum, usually by two ducts (Figure 23-12). The pancreas is divided into a head, body, and tail. The *head* is the expanded portion near the C-shaped curve of the duodenum. Moving superiorly and to the left of the head are the centrally located *body* and the terminal tapering *tail*.

Pancreatic secretions pass from the secreting cells in the pancreas to small ducts that unite to form the two ducts that convey the secretions into the small intestine. The larger of the two ducts is called the *pancreatic duct (duct of Wirsung)*. In most people, the pancreatic duct unites with the common bile duct from the liver and gallbladder and enters the duodenum in a common duct, called the *hepatopancreatic ampulla (ampulla of Vater)*. The ampulla opens on an elevation of the duodenal mucosa known as the *duodenal papilla,* about 10 cm (4 in.) below the pylorus of the stomach. The smaller of the two ducts is the *accessory duct (duct of Santorini)*, which leads from the pancreas and empties into the duodenum about 2.5 cm (1 in.) above the hepatopancreatic ampulla.

HISTOLOGY

The pancreas is made up of small clusters of glandular epithelial cells. About 1 percent of the cells, the *pancreatic islets (islets of Langerhans),* form the endocrine portion of the pancreas and consist of alpha, beta, and delta cells that secrete hormones (glucagon, insulin, and somatostatin, respectively). The functions of these hormones may be reviewed in Chapter 21. The remaining 99 percent of the cells, called *acini* (AS-i-nē), constitute the exocrine portions of the organ (see Figure 21-15). Secreting cells of the acini release a mixture of digestive enzymes called *pancreatic juice.*

ACTIVITIES

Each day the pancreas produces 1,200–1,500 ml (about 1.2–1.5 qt) of pancreatic juice, a clear, colorless liquid. It consits mostly of water, some salts, sodium bicarbonate, and enzymes. The sodium bicarbonate gives pancreatic juice a slightly alkaline pH (7.1–8.2) that stops the action of pepsin from the stomach and creates the proper environment for the enzymes in the small intestine. The enzymes in pancreatic juice include a carbohydrate-digesting enzyme called *pancreatic amylase;* several protein-digesting enzymes called *trypsin* (TRIP-sin), *chymotrypsin* (kī'-mō-TRIP-sin), and *carboxypolypeptidase* (kar-bok'-sē-polē'-PEP-ti-dās); the principal fat-digesting enzyme in the adult body called *pancreatic lipase;* and nucleic-acid-digesting enzymes called *ribonuclease* and *deoxyribonuclease.*

BLOOD AND NERVE SUPPLY

The vascular and nerve supply of the pancreas may be reviewed in Chapter 21.

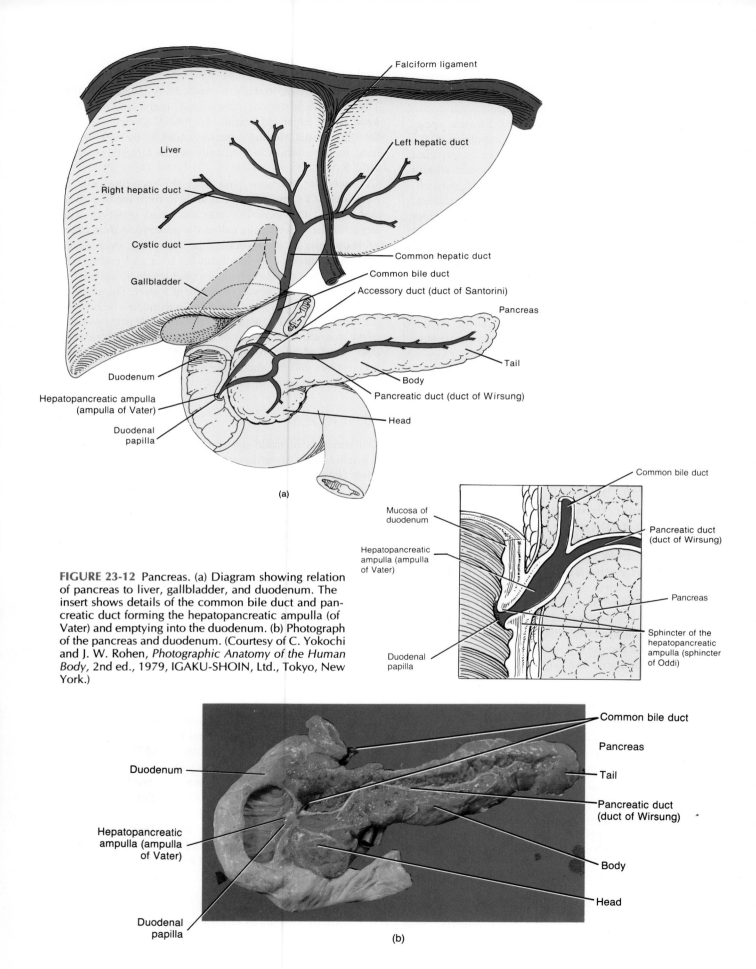

FIGURE 23-12 Pancreas. (a) Diagram showing relation of pancreas to liver, gallbladder, and duodenum. The insert shows details of the common bile duct and pancreatic duct forming the hepatopancreatic ampulla (of Vater) and emptying into the duodenum. (b) Photograph of the pancreas and duodenum. (Courtesy of C. Yokochi and J. W. Rohen, *Photographic Anatomy of the Human Body,* 2nd ed., 1979, IGAKU-SHOIN, Ltd., Tokyo, New York.)

LIVER

The *liver* weighs about 1.4 kg (about 3 lb) in the average adult. It is located under the diaphragm and occupies most of the right hypochondrium and part of the epigastrium of the abdomen (see Figure 1-8c).

ANATOMY

The liver is almost completely covered by peritoneum and completely covered by a dense connective tissue layer that lies beneath the peritoneum. It is divided into two principal lobes— a large *right lobe* and a smaller *left lobe*—separated by the *falciform ligament* (Figure 23-13). The right lobe is considered by many anatomists to consist of an inferior *quadrate lobe* and a posterior *caudate lobe.* However, on the basis of internal morphology, primarily the distribution of blood, the quadrate and caudate lobes more appropriately belong to the left lobe. The falciform ligament is a reflection of the parietal peritoneum, which extends from the undersurface of the diaphragm to the superior surface of the liver, between the two principal lobes of the liver. In the free border of the falciform ligament is the *ligamentum teres* (*round ligament*). It extends from the liver to the umbilicus. The ligamentum teres is a fibrous cord derived from the umbilical vein of the fetus.

Bile, one of the liver's products, enters *bile capillaries* or *canaliculi* (kan'-a-LIK-yoo-lī) that empty into small ducts. These small ducts eventually merge to form the larger *right* and *left hepatic ducts,* which unite to leave the liver as the *common hepatic duct* (Figures 23-12 and 23-13b). Further on, the common hepatic duct joins the *cystic duct* from the gallbladder. The two tubes become the *common bile duct.* The common bile duct and pancreatic duct enter the duodenum in a common duct called the *hepatopancreatic ampulla* (*ampulla of Vater*).

HISTOLOGY

The lobes of the liver are made up of numerous functional units called *lobules,* which may be seen under a microscope (Figure 23-14). A lobule consists of epithelial cells, called *hepatic* (*liver*) *cells,* arranged in irregular, branching, interconnected plates around a *central vein.* These cells secrete bile. Between the plates of cells are endothelial-lined spaces called *sinusoids,* through which blood passes. The sinusoids are also partly lined with phagocytic cells, termed *stellate reticuloendothelial* (*Kupffer's*) *cells,* that destroy worn-out white and red blood cells, bacteria, and toxic substances. The liver contains sinusoids instead of typical capillaries.

Each day the hepatic cells secrete 800–1,000 ml (about 1 qt) of *bile,* a yellow, brownish, or olive-green liquid. It has a pH of 7.6–8.6. Bile consists mostly of water and bile salts, cholesterol, a phospholipid called lecithin, bile pigments, and several ions.

Bile is partially an excretory product and partially a digestive secretion. Bile salts assume a role in *emulsification,* the breakdown of fat globules into a suspension of fat droplets about 1 μm in diameter, and absorption of fats following their digestion. Cholesterol is made soluble in bile by bile salts and lecithin. The principal bile pigment is *bilirubin.* When worn-out red blood cells are broken down, iron, globin, and bilirubin are released. The iron and globin are recycled, but some of the bilirubin is excreted into the bile ducts. Bilirubin eventually is broken down in the intestine, and one of its breakdown products (urobilinogen) gives feces their color.

CLINICAL APPLICATION

If insufficient bile salts or lecithin are present in bile, or if there is excessive cholesterol, the cholesterol precipitates out of solution and crystalizes to form *gallstones* (*biliary calculi*). The problems associated with gallstone formation and the diagnosis and treatment of gallstones are discussed in detail at the end of the chapter.

If the liver is unable to remove bilirubin from the blood because of increased rate of destruction of red blood cells or obstruction of bile ducts, large amounts of bilirubin circulate through the bloodstream and collect in other tissues, giving the skin and eyes a yellow color. This condition is called *jaundice.* If the jaundice is due to damaged red blood cells, it is called *hemolytic jaundice;* if it is due to obstruction in the biliary system, it is known as *obstructive jaundice.* Since the liver of a newborn functions poorly for the first week or so, large amounts of bilirubin are excreted into blood instead of being incorporated into bile in the liver. The result is a type of jaundice called *neonatal* (*physiological*) *jaundice.*

ACTIVITIES

The liver performs many vital functions. Among these are the following:

1. The liver manufactures bile salts, which are used in the small intestine for the emulsification and absorption of fats, cholesterol, phospholipids, and lipoproteins.
2. The liver, together with mast cells, manufactures the anticoagulant heparin. The liver also produces most of the other plasma proteins, such as prothrombin, fibrinogen, and albumin.
3. The stellate reticuloendothelial (Kupffer's) cells of the liver phagocytize worn-out red and white blood cells and some bacteria.
4. Liver cells contain enzymes that either break down poisons or transform them into less harmful compounds. When amino acids are burned for energy, for example, they leave behind toxic nitrogenous wastes (such as ammonia) that are converted to urea by the liver cells.

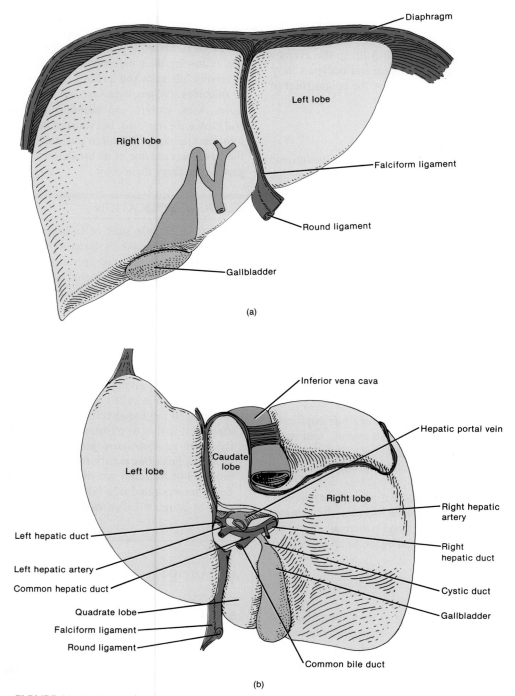

FIGURE 23-13 External anatomy of the liver. (a) Anterior view. (b) Posteroinferior view.

Moderate amounts of urea are harmless to the body and are easily excreted by the kidneys and sweat glands.

5. Newly absorbed nutrients are collected in the liver. Depending on the body's needs, it can change any excess monosaccharides into glycogen or fat, both of which can be stored, or it can transform glycogen, fat, and protein into glucose.

6. The liver stores glycogen, copper, iron, and vitamins A, B_{12}, D, E, and K. It also stores some poisons that cannot be broken down and excreted. (High levels of DDT are found in the livers of animals, including humans, who eat sprayed fruits and vegetables.)

7. The liver and kidneys participate in the activation of vitamin D.

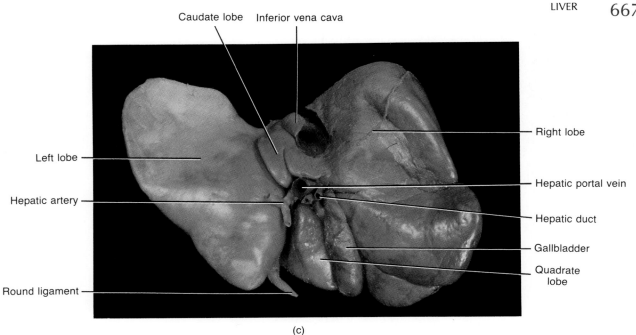

Caudate lobe   Inferior vena cava

Left lobe

Hepatic artery

Round ligament

Right lobe

Hepatic portal vein

Hepatic duct

Gallbladder

Quadrate lobe

(c)

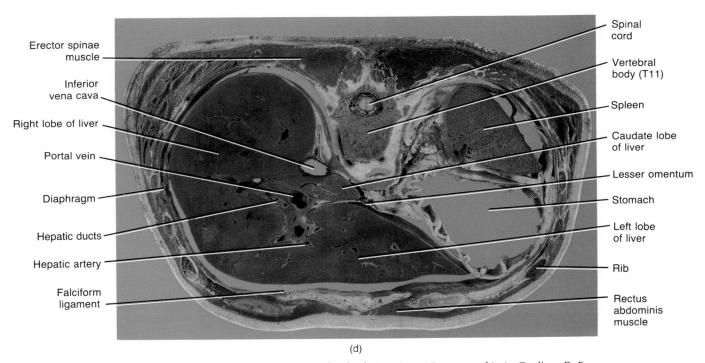

Erector spinae muscle

Inferior vena cava

Right lobe of liver

Portal vein

Diaphragm

Hepatic ducts

Hepatic artery

Falciform ligament

Spinal cord

Vertebral body (T11)

Spleen

Caudate lobe of liver

Lesser omentum

Stomach

Left lobe of liver

Rib

Rectus abdominis muscle

(d)

FIGURE 23-13 (Continued) (c) Photograph of inferior view. (Courtesy of J. A. Gosling, P. F. Harris, et al., *Atlas of Human Anatomy*, Gower Medical Publishing Ltd., 1985.) (d) Photograph of a cross section through the abdomen. (Courtesy of Stephen A. Kieffer and E. Robert Heitzman, *An Atlas of Cross-Sectional Anatomy*, Harper & Row, Publishers, Inc., New York, 1979.)

BLOOD AND NERVE SUPPLY

The liver receives a double supply of blood. From the hepatic artery it obtains oxygenated blood, and from the hepatic portal vein it receives deoxygenated blood containing newly absorbed nutrients (see Figures 14-18 and 23-14).

Branches of both the hepatic artery and the hepatic portal vein carry the blood into the sinusoids of the lobules, where oxygen, most of the nutrients, and certain poisons are extracted by the hepatic cells. Nutrients are stored or used to make new materials. The poisons are stored or detoxified. Products manufactured by the hepatic cells and nutrients

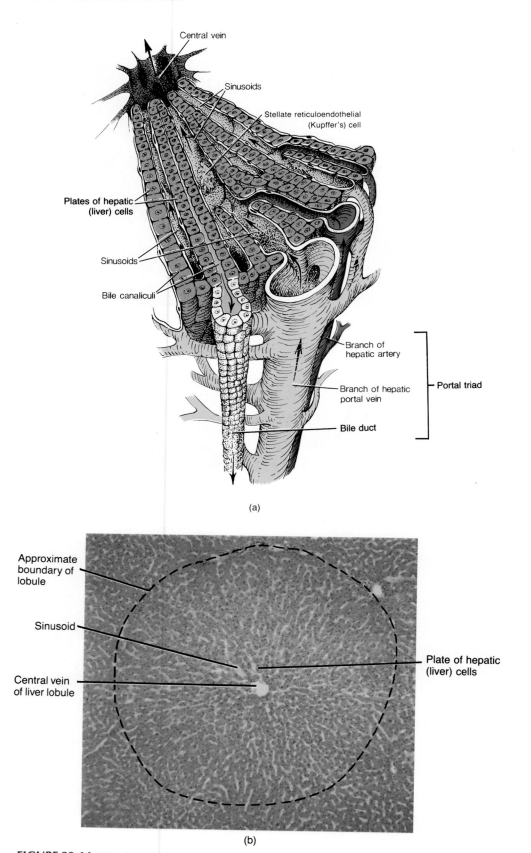

(a)

(b)

FIGURE 23-14 Histology of the liver. (a) Diagram of the microscopic appearance of a portion of a liver lobule. (b) Photomicrograph of a liver lobule at a magnification of 65×. (Copyright © 1983 by Michael H. Ross. Used by permission.)

needed by other cells are secreted back into the blood. The blood then drains into the central vein and eventually passes into a hepatic vein. Unlike the other products of the liver, bile normally is not secreted into the bloodstream.

Branches of the hepatic portal vein, hepatic artery, and bile duct typically accompany each other in their distribution through the liver. Collectively, these three structures are referred to as a *portal triad* (see Figure 23-14a).

The nerve supply to the liver consists of vagal preganglionic parasympathetic fibers and postganglionic sympathetic fibers from the celiac ganglia.

CLINICAL APPLICATION

Liver tissue for diagnostic purposes may be obtained by *liver biopsy*. In the procedure, a needle is inserted through the seventh, eighth, or ninth intercostal space while the patient is holding his or her breath in full expiration. This lessens the possibility of damage to the lung and contamination of the pleural cavity.

GALLBLADDER (GB)

The *gallbladder* (*GB*) is a pear-shaped sac about 7–10 cm (3–4 in.) long. It is located in a fossa of the visceral surface of the liver (see Figures 23-12a and 23-13).

HISTOLOGY

The mucosa of the gallbladder consists of simple columnar epithelium arranged in rugae resembling those of the stomach (Figure 23-15). The gallbladder lacks a submucosa. The middle, muscular coat of the wall consists of smooth muscle fibers (cells). Contraction of these fibers by hormonal stimulation ejects the contents of the gallbladder into the cystic duct. The outer coat is the visceral peritoneum.

ACTIVITIES

The function of the gallbladder is to store and concentrate bile (up to tenfold) until it is needed in the small intestine. In the concentration process, water and many ions are ab-

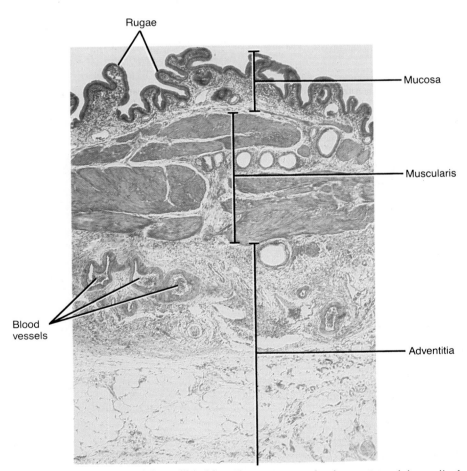

FIGURE 23-15 Histology of the gallbladder. Photomicrograph of a portion of the wall of the gallbladder at a magnification of 40×. (Copyright © 1983 by Michael H. Ross. Used by permission.)

sorbed by the gallbladder mucosa. Bile from the liver enters the small intestine through the common bile duct. When the small intestine is empty, a valve around the hepatopancreatic ampulla (ampulla of Vater), called the **sphincter of the hepatopancreatic ampulla (sphincter of Oddi),** closes, and the backed-up bile overflows into the cystic duct to the gallbladder for storage (see Figure 23-12a).

BLOOD AND NERVE SUPPLY

The gallbladder is supplied by the cystic artery, which usually arises from the right hepatic artery. The cystic veins drain the gallbladder. The nerves to the gallbladder include branches from the celiac plexus and the vagus (X) nerve.

SMALL INTESTINE

The major portions of digestion and absorption occur in a long tube called the **small intestine.** The small intestine begins at the pyloric sphincter of the stomach, coils through the central and lower part of the abdominal cavity, and eventually opens into the large intestine. It averages 2.5 cm (1 in.) in diameter and about 6.35 m (21 ft) in length in the cadaver.

ANATOMY

The small intestine is divided into three segments (see Figure 23-1). The **duodenum** (doo'-ō-DĒ-num), the shortest part, originates at the pyloric sphincter of the stomach and extends about 25 cm (10 in.) until it merges with the jejunum. The **jejunum** (jē-JOO-num) is about 2.5 m (8 ft) long and extends to the ileum. The final portion of the small intestine, the **ileum** (IL-ē-um), measures about 3.6 m (12 ft) and joins the large intestine at the **ileocecal** (il'-ē-ō-SĒ-kal) **sphincter (valve).**

HISTOLOGY

The wall of the small intestine is composed of the same four coats that make up most of the GI tract. However, both the mucosa and the submucosa are modified to allow the small intestine to complete the processes of digestion and absorption (Figure 23-16).

The **mucosa** contains many pits lined with glandular epithelium. These pits—the **intestinal glands (crypts of Lieberkühn)**—secrete intestinal juice. The submucosa of the duodenum contains **duodenal (Brunner's) glands** that secrete an alkaline mucus to protect the wall of the small intestine from the action of the enzymes and to aid in neutralizing acid in the chyme. Some of the epithelial cells in the mucosa and submucosa have been transformed to goblet cells, which secrete additional mucus.

Since almost all the absorption of nutrients occurs in the small intestine, its structure is specially adapted for

this function. Its length alone provides a large surface area for absorption and that area is further increased by modifications in the structure of its wall. The epithelium covering and lining the mucosa consists of simple columnar epithelium. These epithelial cells, except those transformed into goblet cells, contain **microvilli,** fingerlike projections of the plasma membrane. Larger amounts of digested nutrients diffuse into the epithelial cells of the intestinal wall because the microvilli increase the surface area of the plasma membrane. They also increase the surface area for digestion.

The mucosa lies in a series of **villi,** projections 0.5–1 mm high, giving the intestinal mucosa its velvety appearance. The enormous number of villi (10–40 per square millimeter) vastly increases the surface area of the epithelium available for absorption and digestion. Each villus has a core of lamina propria, the connective tissue layer of the mucosa. Embedded in this connective tissue are an arteriole, a venule, a capillary network, and a **lacteal** (LAK-tē-al) or lymphatic vessel. Nutrients that diffuse through the epithelial cells that cover the villus are able to pass through the capillary walls and the lacteal and enter the cardiovascular and lymphatic systems, respectively.

In addition to the microvilli and villi, a third set of projections called **plicae circulares** (PLĪ-kē SER-kyoo-lar-es), or **circular folds,** further increases the surface area for absorption and digestion. The plicae are permanent ridges, about 10 mm (0.4 in.) high, in the mucosa. Some of the folds extend all the way around the intestine, and others extend only partway around. The folds begin near the proximal portion of the duodenum and terminate at about the midportion of the ileum. The plicae circulares enhance absorption by causing the chyme to spiral, rather

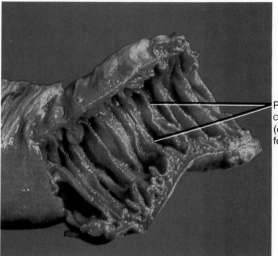

Plicae circulares (circular folds)

(a)

FIGURE 23-16 Small intestine. shown are various structures that adapt the small intestine for digestion and absorption. (a) Photograph of a section of the jejunum cut open to expose the plicae circulares. (Courtesy of C. Yokochi and J. W. Rohen, *Photographic Anatomy of the Human Body,* 2nd ed., 1979, IGAKU-SHOIN, Ltd., Tokyo, New York.)

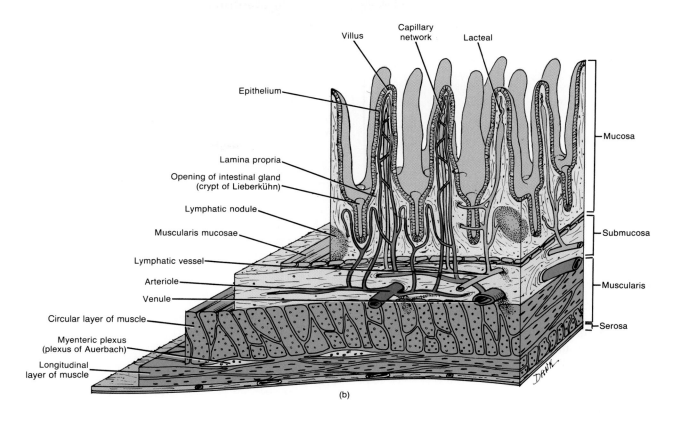

(b)

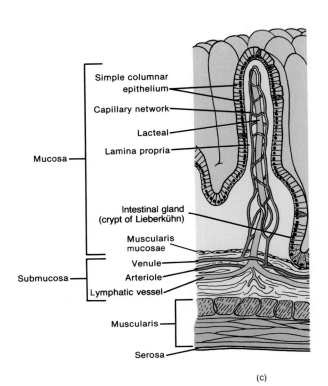

(c)

FIGURE 23-16 (Continued) (b) Diagram showing four principal layers. (c) Enlarged aspect of a single villus.

than to move in a straight line, as it passes through the small intestine. Since the plicae circulares and villi decrease in size in the distal ileum, most absorption occurs in the duodenum and jejunum.

The *muscularis* of the small intestine consists of two layers of smooth muscle. The outer, thinner layer contains longitudinally arranged fibers (cells). The inner, thicker layer contains circularly arranged fibers. Except for a major portion of the duodenum, the serosa (or visceral peritoneum) completely covers the small intestine. Additional histological aspects of the small intestine are shown in Figure 23-17.

There is an abundance of lymphatic tissue in the form of lymphatic nodules, masses of lymphatic tissue not covered by a capsule wall. *Solitary lymphatic nodules* are most numerous in the lower part of the ileum. Groups of lymphatic nodules, referred to as *aggregated lymphatic follicles (Peyer's patches)*, are numerous in the ileum.

ACTIVITIES

Intestinal juice is a clear yellow fluid secreted in amounts of about 2–3 liters (about 2–3 qt) a day. It has a pH of 7.6, which is slightly alkaline, and contains water and mucus. The juice is rapidly reabsorbed by the villi and provides a vehicle for the absorption of substances from chyme as they come in contact with the villi. The intestinal enzymes are formed in the epithelial cells that line the villi, and most digestion by enzymes of the small intestine occurs

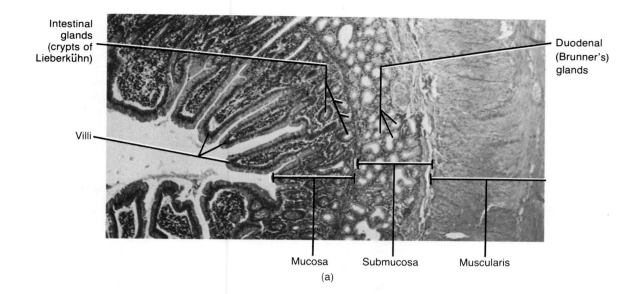

Intestinal glands (crypts of Lieberkühn)

Duodenal (Brunner's) glands

Villi

Mucosa   Submucosa   Muscularis

(a)

within the cells on the surfaces of their microvilli. In addition, as small intestinal cells containing enzymes slough off into the lumen of the intestine, they break apart and release small quantities of enzymes that digest some food in the chyme. Thus, most digestion by enzymes of the small intestine occurs in or on the epithelial cells that line the villi, rather than in the lumen, as in other parts of the gastrointestinal tract. Among the enzymes produced by small intestinal cells are three carbohydrate-digesting enzymes called *maltase, sucrase,* and *lactase;* several protein-digesting enzymes called *peptidases;* and two nucleic-acid-digesting enzymes, *ribonuclease* and *deoxyribonuclease.* Much of the digestion of foods by enzymes produced by the small intestine actually occurs in epithelial cells lining the small intestine rather than in a fluid outside the cells in the lumen of the tube.

The movements of the small intestine are divided into two types: segmentation and peristalsis. *Segmentation* is the major movement of the small intestine. It is strictly a localized contraction in areas containing food. It mixes chyme with the digestive juices and brings the particles of food into contact with the mucosa for absorption. It does not push the intestinal contents along the tract. Segmentation starts with the contractions of circular muscle fibers in a portion of the small intestine, an action that constricts the intestine into segments. Next, muscle fibers that encircle the middle of each segment also contract, dividing each segment again. Finally, the fibers that contracted first relax, and each small segment unites with an adjoining small segment so that large segments are reformed. This sequence of events is repeated 12–16 times a minute, sloshing the chyme back and forth. Segmentation depends mainly on intestinal distension, which initiates nerve impulses to the central nervous system. Returning parasympathetic impulses increase motility. Sympathetic impulses decrease intestinal motility.

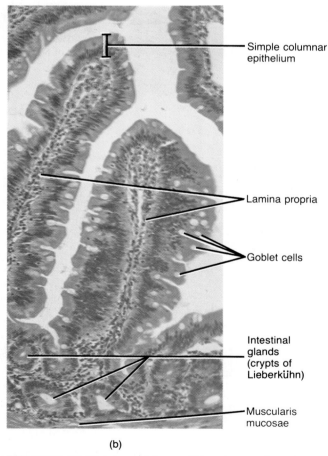

Simple columnar epithelium

Lamina propria

Goblet cells

Intestinal glands (crypts of Lieberkühn)

Muscularis mucosae

(b)

FIGURE 23-17 Histology of the small intestine. (a) Photomicrograph of a portion of the wall of the duodenum at a magnification of 40×. (b) Photomicrograph of an enlarged aspect of two villi from the ileum at a magnification of 120×. (Photomicrographs copyright © 1983 by Michael H. Ross. Used by permission.)

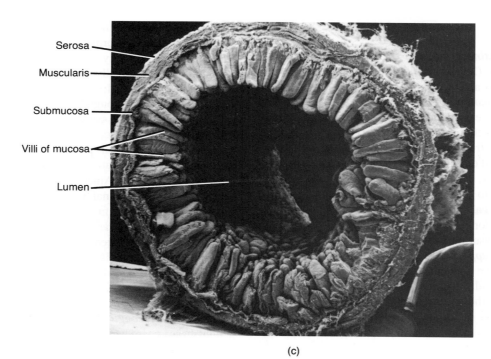

Serosa

Muscularis

Submucosa

Villi of mucosa

Lumen

(c)

FIGURE 23-17 (Continued) (c) Scanning electron micrograph of a cross section of the small intestine at a magnification of 30×. (Courtesy of Richard G. Kessel and Randy H. Kardon. *Tissues and Organs: A Text-Atlas of Scanning Electron Microscopy.* W. H. Freeman and Company. Copyright © 1979.)

Peristalsis propels the chyme onward through the intestinal tract. Peristaltic contractions in the small intestine are normally very weak compared to those in the esophagus or stomach. Chyme moves through the small intestine at a rate of about 1 cm/min. Thus, chyme remains in the small intestine for 3–5 hours. Peristalsis, like segmentation, is initiated by distension and controlled by the autonomic nervous system.

Chyme entering the small intestine contains partially digested carbohydrates, partially digested proteins, and essentially undigested lipids. The completion of the digestion of carbohydrates into monosaccharides, proteins into amino acids, and lipids into fatty acids, glycerol, and glycerides is a collective effort of pancreatic juice, bile, and intestinal juice in the small intestine.

All the chemical and mechanical phases of digestion from the mouth down through the small intestine are directed toward changing food into forms that can pass through the epithelial cells lining the mucosa into the underlying blood and lymph vessels. Passage of these digested nutrients from the GI tract into the blood or lymph is called *absorption*.

About 90 percent of all absorption of nutrients takes place throughout the length of the small intestine. The other 10 percent occurs in the stomach and large intestine. Any undigested or unabsorbed material left in the small intestine is passed on to the large intestine. Absorption of materials in the small intestine occurs specifically through the villi

and depends on diffusion, facilitated diffusion, osmosis, and active transport.

BLOOD AND NERVE SUPPLY

The arterial blood supply of the small intestine is from the superior mesenteric artery and the gastroduodenal artery, coming from the hepatic artery of the celiac trunk. Blood is returned by way of the superior mesenteric vein, which, with the splenic vein, forms the hepatic portal vein.

The nerves to the small intestine are supplied by the superior mesenteric plexus. The branches of the plexus contain postganglionic sympathetic fibers, preganglionic parasympathetic fibers, and afferent fibers. The afferent fibers are both vagal and of spinal nerves. In the wall of the small intestine are two autonomic plexuses: the myenteric plexus between the muscular layers and the submucosal plexus in the submucosa. The nerve fibers are derived chiefly from the sympathetic division of the autonomic nervous system and partly from the vagus (X) nerve.

LARGE INTESTINE

The overall functions of the large intestine are the completion of absorption, the manufacture of certain vitamins, the formation of feces, and the expulsion of feces from the body.

The process of dental caries is initiated when bacteria act on sugars, giving off acids that demineralize the enamel. Microbes that digest sugar into lactic acid are common in the mouth cavity. One that seems to be cariogenic (caries causing) is the bacterium *Streptococcus mutans*. **Dextran,** a sticky polysaccharide produced from sucrose, forms a capsule around the bacteria causing them to stick to the teeth. Masses of bacterial cells, dextran, and other debris adhering to teeth are collectively called **dental plaque.** Saliva cannot reach the tooth surface to buffer the acid because the plaque covers the teeth. Brushing the teeth immediately after eating removes the plaque from flat surfaces before the bacteria have a chance to go to work. Dentists also suggest that the plaque between the teeth be removed every 24 hours with dental floss or by flushing with a water irrigation device or by using an antiplaque dental rinse.

Preventive measures other than brushing, flossing, and irrigation include prenatal diet supplements (chiefly vitamin D, calcium, and phosphorus), fluoride treatments to protect against acids during the period when teeth are being calcified, and dental sealing. In this last procedure, pits and fissures that serve as reservoirs for dental plaque are sealed by the application of a permanent, durable plastic sealant. The sealant is applied to the prepared biting surfaces of the molar teeth. Dental sealants are used primarily for children and adolescents, although some adults could benefit from them.

PERIODONTAL DISEASE

Periodontal disease is a collective term for a variety of conditions characterized by inflammation and degeneration of the gingivae, alveolar bone, periodontal ligament, and cementum. One such condition is called **pyorrhea.** The initial symptoms are enlargement and inflammation of the soft tissue and bleeding gums. Without treatment, the soft tissue may deteriorate and the alveolar bone may be resorbed, causing loosening of the teeth and recession of the gums.

Periodontal diseases are frequently caused by poor oral hygiene; local irritants, such as bacteria, impacted food, and cigarette smoke; or by a poor "bite." The latter may put a strain on the tissues supporting the teeth. Periodontal diseases may also be caused by allergies, vitamin deficiencies (especially vitamin C), and a number of systemic disorders, especially those that affect bone, connective tissue, or circulation.

PERITONITIS

Peritonitis is an acute inflammation of the serous membrane lining the abdominal cavity and covering the abdominal viscera. One possible cause is contamination of the peritoneum by pathogenic bacteria from the external environment. This contamination could result from accidental or surgical wounds in the abdominal wall or from perforation or rupture of organs with consequent exposure to the outside environment. Another possible cause is perforation of the walls of organs that contain bacteria or chemicals beneficial to the organ but toxic to the peritoneum. For example, the large intestine contains colonies of bacteria that live on undigested nutrients and break them down so they can be eliminated. But if the bacteria enter the peritoneal cavity, they attack the cells of the peritoneum for food and produce acute infection. Moreover, the peritoneum has no natural barriers that keep it from being irritated or digested by chemical substances such as bile and digestive enzymes.

Although it contains a great deal of lymphatic tissue and can combat infection fairly well, the peritoneum is in contact with most of the abdominal organs. If infection gets out of hand, it may destroy vital organs and bring on death. For these reasons, perforation of the gastrointestinal tract from an ulcer is considered serious. A surgeon planning to do extensive surgery on the colon may give the patient high doses of antibiotics for several days prior to surgery to kill intestinal bacteria and reduce the risk of peritoneal contamination.

PEPTIC ULCERS

An **ulcer** is a craterlike lesion in a membrane. Ulcers that develop in areas of the gastrointestinal tract exposed to acid gastric juice are called **peptic ulcers.** Peptic ulcers occasionally develop in the lower end of the esophagus, but most occur on the lesser curvature of the stomach, where they are called **gastric ulcers,** or in the first part of the duodenum, where they are called **duodenal ulcers.** Most peptic ulcers are duodenal.

Hypersecretion of acid gastric juice seems to be the immediate cause of duodenal ulcers. In gastric ulcer patients, because the stomach wall is highly adapted to resist gastric juice through the secretion of mucus, the cause may be hyposecretion of mucus. Hypersecretion of pepsin also may contribute to ulcer formation.

Among the factors believed to stimulate an increase in acid secretion are emotions, cigarette smoking, certain foods or medications (alcohol, coffee, aspirin), and overstimulation of the vagus (X) nerve. Normally, the mucous membrane lining the stomach and duodenal walls resists the secretions of hydrochloric acid and pepsin. In some people, however, this resistance breaks down and an ulcer develops. Some evidence suggests that peptic ulcers may be caused by the bacterium *Campylobacter pyloridis*.

The most common complication of peptic ulcers is bleeding. Another is perforation, erosion of the ulcer all the way through the wall of the stomach or duodenum. Perforation allows bacteria and partially digested food to pass into the peritoneal cavity, producing peritonitis. A third complication is obstruction.

APPENDICITIS

Appendicitis is an inflammation of the vermiform appendix. It is preceded by obstruction of the lumen of the appendix by fecal material, inflammation, a foreign body, carcinoma of the cecum, stenosis, or kinking of the organ. The infection that follows may result in edema, ischemia, gangrene, and perforation. Rupture of the appendix develops into peritonitis. Loops of the intestines, the omentum, and the parietal peritoneum may become adherent and form an abscess, either at the site of the appendix or elsewhere in the abdominal cavity.

Typically, appendicitis begins with referred pain in the umbilical region of the abdomen followed by anorexia (lack or loss of appetite for food), nausea, and vomiting. After several hours, the pain localizes in the right lower quadrant (RLQ) and is continuous, dull or severe, and intensified by coughing, sneezing, or body movements.

Early appendectomy (removal of the appendix) is recommended in all suspected cases because it is safer to operate than to risk gangrene, rupture, and peritonitis. Appendectomy may be performed through a muscle-splitting incision in the right lower quadrant (RLQ) in which the cecum is brought into the incision. The base of the appendix is tied, the appendix is excised, and the stump is usually cauterized and then invaginated into the cecum.

TUMORS

Both benign and malignant *tumors* can occur in all parts of the gastrointestinal tract. The benign growths are much more common, but malignant tumors are responsible for 30 percent of all deaths from cancer in the United States. Carcinoma of the colon and rectum is one of the most common malignant diseases, ranking second to that of the lungs in males and lungs and breasts of females. Over 50 percent of colorectal cancers occur in the rectum and sigmoid colon.

Malignant tumors of the rectum may be detected by *digital rectal examination.* The easiest method for detecting cancer of the colon is *fecal occult blood testing,* in which a sample of stool is tested for the presence of occult (hidden) blood, a sign that a malignant tumor might be present. Direct visualization of the rectum and sigmoid colon for the detection of carcinomas is possible by using a *flexible fiberoptic sigmoidoscope.* By using a *flexible fiberoptic endoscope* a physician can visualize the entire gastrointestinal tract. The endoscopic examination of the colon is known as *colonoscopy.* Both the sigmoidoscope and endoscope can also be used to magnify, photograph, biopsy, and remove polyps.

Another test in a routine examination for intestinal disorders is the filling of the gastrointestinal tract with barium, which is either swallowed or given in an enema. Barium, a mineral, shows up on radiographs the same way that calcium appears in bones. Tumors as well as ulcers can

be diagnosed this way. The only definitive treatment of gastrointestinal carcinomas, if they cannot be removed using the endoscope, is surgery.

DIVERTICULITIS

Diverticula are saclike outpouchings of the wall of the colon in places where the muscularis has become weak. The development of diverticula is called *diverticulosis.* Many people who develop diverticulosis are asymptomatic and experience no complications. About 15 percent of people with diverticulosis will eventually develop an inflammation within diverticula, a condition known as *diverticulitis.*

Research indicates that diverticula form because of lack of sufficient bulk in the colon during segmentation. The powerful contractions, working against insufficient bulk, create a pressure so high that it causes the colonic walls to bulge.

The increase in diverticular disease has been attributed to a shift to low-fiber diets. Patients treated for diverticular disease with high-fiber diets show marked relief of symptoms.

Treatment consists of bed rest, cleansing enemas, and drugs to reduce infection. In severe cases, portions of the affected colon may require surgical removal and temporary colostomy.

CIRRHOSIS

Cirrhosis refers to a distorted or scarred liver as a result of chronic inflammation. The parenchymal (functional) liver cells are replaced by fibrous or adipose connective tissue, a process called stromal repair. The liver has a high capacity for parenchymal regeneration, and stromal repair occurs whenever a parenchymal cell is killed or cells are damaged continuously for a long time. The symptoms of cirrhosis include jaundice, edema in the legs, uncontrolled bleeding, and increased sensitivity to drugs. Cirrhosis may be caused by hepatitis (inflammation of the liver), certain chemicals that destroy liver cells, parasites that infect the liver, and alcoholism.

HEPATITIS

Hepatitis refers to inflammation of the liver and can be caused by viruses, drugs, and chemicals, including alcohol. Clinically, several types are recognized.

Hepatitis A (infectious hepatitis) is caused by hepatitis A virus and is spread by fecal contamination of food, clothing, toys, eating utensils, and so forth (fecal–oral route), It is generally a mild disease of children and young adults characterized by anorexia, malaise, nausea, diarrhea, fever, and chills. Eventually, jaundice appears. It does not cause lasting liver damage. Most people recover in 4–6 weeks.

Hepatitis B (serum hepatitis) is caused by hepatitis B

The metabolism of nutrients results in the production of wastes by body cells, including carbon dioxide and excess water and heat. Protein catabolism produces toxic nitrogenous wastes such as ammonia and urea. In addition, many of the essential ions such as sodium, chloride, sulfate, phosphate, and hydrogen tend to accumulate in excess of the body's needs. All the toxic materials and the excess essential materials must be eliminated.

The primary function of the *urinary system* is to help keep the body in homeostasis by controlling the composition and volume of blood. It does so by removing and restoring selected amounts of water and solutes. Two kidneys, two ureters, one urinary bladder, and a single urethra make up the system (Figure 24-1). The kidneys regulate the composition and volume of the blood and remove wastes from the blood in the form of urine. They excrete selected amounts of various wastes, assume a role in erythropoiesis by forming renal erythropoietic factor, help control blood pH, help regulate blood pressure by secreting renin (which activates the renin–angiotensin pathway), and participate in the activation of vitamin D. Urine is excreted from each kidney through its ureter and is stored in the urinary bladder until it is expelled from the body through the urethra. Other systems that aid in waste elimination are the respiratory, integumentary, and digestive systems (Exhibit 24-1).

The specialized branch of medicine that deals with structure, function, and diseases of the male and female urinary systems and the male reproductive system is known as *urology* (yoo-ROL-ō-jē; *uro* = urine or urinary tract; *logos* = study of).

The developmental anatomy of the urinary system is considered at the end of the chapter.

KIDNEYS

The paired *kidneys* are reddish organs that resemble kidney beans in shape. They are found just above the waist between the parietal peritoneum and the posterior wall of the abdomen. Since they are external to the peritoneal lining of the abdominal cavity, their placement is described as *retroperitoneal* (re'-trō-per-i-tō-NĒ-al). Other retroperitoneal structures include the ureters and adrenal (suprarenal) glands. Relative to the vertebral column, the kidneys are located between the levels of the last thoracic and third lumbar vertebrae and are partially protected by the eleventh and twelfth pairs of ribs. The right kidney is slightly lower than the left because of the large area occupied by the liver.

EXTERNAL ANATOMY

The average adult kidney measures about 10–12 cm (4–5 in.) long, 5.0–7.5 cm (2–3 in.) wide, and 2.5 cm (1 in.) thick. Its concave medial border faces the vertebral column.

EXHIBIT 24-1

Excretory Organs and Products Eliminated

EXCRETORY ORGANS	PRODUCTS ELIMINATED	
	PRIMARY	SECONDARY
Kidneys	Water, nitrogenous wastes from protein catabolism, and inorganic salts.	Heat and carbon dioxide.
Lungs	Carbon dioxide.	Heat and water.
Skin (sudoriferous glands)	Heat.	Carbon dioxide, water, salts, and urea.
Gastrointestinal (GI) tract	Solid wastes and secretions.	Carbon dioxide, water, salts, and heat.

Near the center of the concave border is a notch called the *hilus,* through which the ureter leaves the kidney. Blood and lymph vessels and nerves also enter and exit the kidney through the hilus (Figure 24-2). The hilus is the entrance to a cavity in the kidney called the *renal sinus.*

Three layers of tissue surround each kidney. The innermost layer, the *renal capsule,* is a smooth, transparent, fibrous membrane that can easily be stripped off the kidney and is continuous with the outer coat of the ureter at the hilus. It serves as a barrier against trauma and the spread of infection to the kidney. The second layer, the *adipose capsule,* is a mass of fatty tissue surrounding the renal capsule. It also protects the kidney from trauma and holds it firmly in place within the abdominal cavity. The outermost layer, the *renal fascia,* is a thin layer of fibrous connective tissue that anchors the kidneys to its surrounding structures and to the abdominal wall.

CLINICAL APPLICATION

Floating kidney, or *ptosis,* (TŌ-sis) occurs when the kidney is no longer held in place securely by the adjacent organs or its covering of fat and slips from its normal position. Individuals, especially thin people, in whom either the adipose capsule or renal fascia is deficient, may develop ptosis. It is dangerous because it may cause kinking of the ureter with reflux of urine and retrograde pressure. Pain occurs if the ureter is twisted. Also, if the kidneys drop below the rib cage, they become susceptible to blows and penetrating injuries.

INTERNAL ANATOMY

A coronal (frontal) section through a kidney reveals an outer, narrow reddish area called the *cortex* and an inner, wide reddish-brown region called the *medulla* (Figure 24-2b,c). Within the medulla are 5–14 striated, triangular struc-

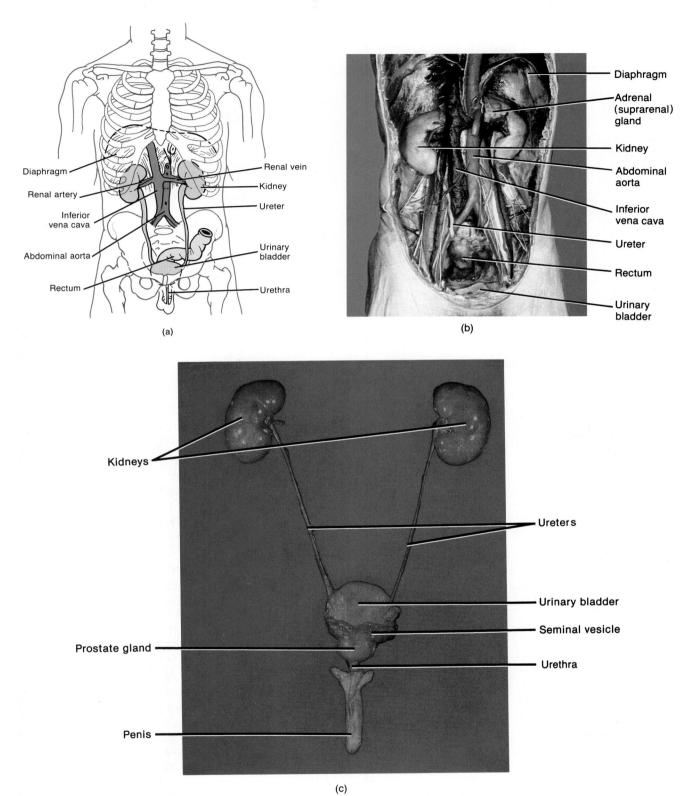

FIGURE 24-1 Organs of the male urinary system in relation to surrounding structures. (a) Diagram. (b) Photograph. (Courtesy of C. Yokochi and J. W. Rohen, *Photographic Anatomy of the Human Body,* 2nd ed., 1978, IGAKU-SHOIN, Ltd., Tokyo, New York.) (c) Photograph of isolated urinary and some reproductive system organs. (Courtesy of C. Yokochi and J. W. Rohen, *Photographic Anatomy of the Human Body,* 2nd ed., 1979, IGAKU-SHOIN, Ltd., Tokyo, New York.)

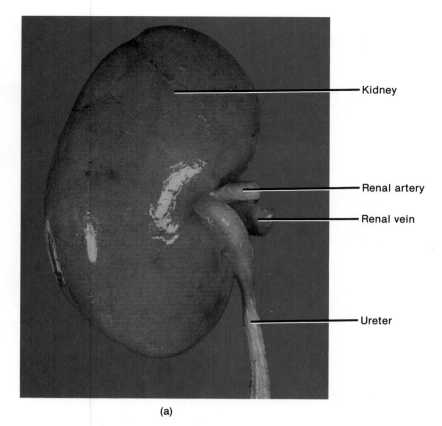

Kidney

Renal artery

Renal vein

Ureter

(a)

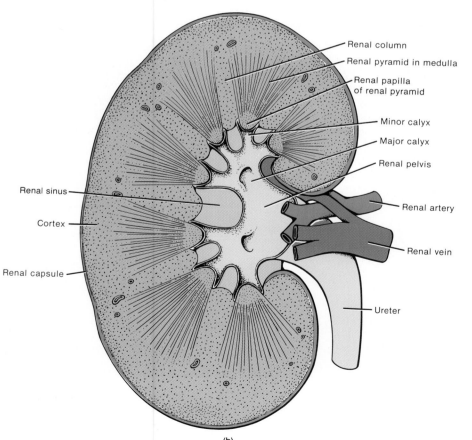

Renal column

Renal pyramid in medulla

Renal papilla
of renal pyramid

Minor calyx

Major calyx

Renal pelvis

Renal artery

Renal vein

Ureter

Renal sinus

Cortex

Renal capsule

(b)

FIGURE 24-2 Kidney. (a) Photograph of the external view of the right kidney. (Courtesy of C. Yokochi and J. W. Rohen, *Photographic Anatomy of the Human Body*, 2nd ed., 1979, IGAKU-SHOIN, Ltd., Tokyo, New York.) (b) Diagram of a coronal section of the right kidney illustrating the internal anatomy.

690

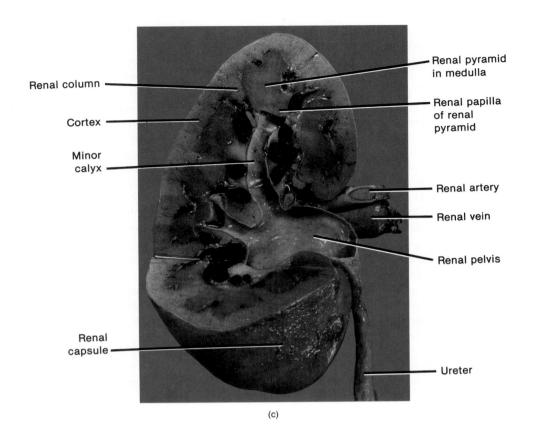

Renal column

Cortex

Minor calyx

Renal capsule

Renal pyramid in medulla

Renal papilla of renal pyramid

Renal artery

Renal vein

Renal pelvis

Ureter

(c)

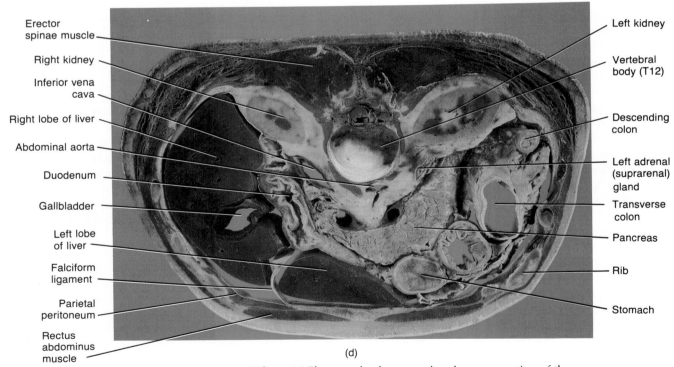

Erector spinae muscle

Right kidney

Inferior vena cava

Right lobe of liver

Abdominal aorta

Duodenum

Gallbladder

Left lobe of liver

Falciform ligament

Parietal peritoneum

Rectus abdominus muscle

Left kidney

Vertebral body (T12)

Descending colon

Left adrenal (suprarenal) gland

Transverse colon

Pancreas

Rib

Stomach

(d)

FIGURE 24-2 (Continued) Kidney. (c) Photograph of a coronal and a cross section of the right kidney illustrating the internal anatomy. (Courtesy of C. Yokochi and J. W. Rohen, *Photographic Anatomy of the Human Body,* 2nd ed., 1978, IGAKU-SHOIN, Ltd., Tokyo, New York.) (d) Photograph of a cross section through the abdomen. (Courtesy of Stephen A. Kieffer and E. Robert Heitzman, *An Atlas of Cross-Sectional Anatomy,* Harper & Row, Publishers, Inc., New York, 1979.)

tures termed *renal (medullary) pyramids.* The striated (striped) appearance is due to the presence of straight tubules and blood vessels. The bases of the pyramids face the cortical area, and their apices, called *renal papillae,* are directed toward the center of the kidney. The cortex is the smooth-textured area extending from the renal capsule to the bases of the pyramids and into the spaces between them. The cortex is divided into an outer cortical zone and an inner juxtamedullary zone. The cortical substance between the renal pyramids forms the *renal columns.*

Together the cortex and renal pyramids constitute the parenchyma (functioning portion) of the kidney. Structurally, the parenchyma of each kidney consists of approximately 1 million microscopic units called nephrons. Nephrons are the functional units of the kidney. They help regulate blood composition and form urine. Associated with nephrons are collecting ducts and a vascular supply.

In the renal sinus of the kidney is a large cavity called the *renal pelvis.* The edge of the pelvis contains cuplike extensions called *major* and *minor calyces* (KĀ-li-sēz). There are 2 or 3 major calyces and 8–18 minor calyces. Each minor calyx collects urine from collecting ducts of the pyramids. From the major calyces, the urine drains into the pelvis and out through the ureter.

NEPHRON

The functional unit of the kidney is the *nephron* (NEF-ron) (Figure 24-3). Nephrons have several functions related to homeostasis. They filter blood; that is, they permit some substances to pass into the kidneys, while keeping others out. As the filtered liquid (filtrate) moves through the nephrons, it is further processed by nephrons by the addition of some substances (wastes and excess substances) and the removal of others (useful materials). As a result of the activities of nephrons, urine is formed.

Essentially, a nephron consists of two portions: a *renal tubule* and a tuft (knot) of capillaries, called the glomerulus. The renal tubule begins as a double-walled epithelial cup, called the *glomerular (Bowman's) capsule,* lying in the cortex of the kidney. The outer wall, or *parietal layer,* is composed of simple squamous epithelium (Figure 24-4). It is separated from the inner wall, known as the *visceral layer,* by the *capsular space.* The visceral layer consists of epithelial cells called podocytes. The capsule surrounds a capillary network called the *glomerulus* (glō-MER-yoo-lus). Collectively, the glomerular capsule and its enclosed glomerulus constitute a *renal corpuscle* (KŌR-pus-sul).

The visceral layer of the glomerular capsule and the endothelium of the glomerulus form an *endothelial-capsular membrane.* This membrane consists of the following parts, listed here in the order in which substances filtered by the kidney must pass through them.

1. *Endothelium of the glomerulus.* The single layer of endothelial cells has completely open pores (fenestrated)

averaging 50–100 mm in diameter. It restricts the passage of blood cells.
2. *Basement membrane of the glomerulus.* This extracellular membrane lies beneath the endothelium and contains no pores. It consists of fibrils in a glycoprotein matrix. It restricts the passage of large-sized proteins.
3. *Epithelium of the visceral layer of the glomerular (Bowman's) capsule.* These epithelial cells, because of their peculiar shape, are called *podocytes.* The podocytes contain footlike structures called *pedicels* (PED-i-sels). The pedicels are arranged parallel to the circumference of the glomerulus and cover the basement membrane except for spaces between them called *filtration slits (slit pores).* Pedicels contain numerous, thin contractile filaments that are believed to regulate the passage of substances through the filtration slits. In addition, another factor that helps regulate the passage of substances through the filtration slits is a thin membrane, the *slit membrane,* that extends between filtration slits. This membrane restricts the passage of intermediate-sized proteins.

The endothelial-capsular membrane filters water and solutes in the blood. Large molecules, such as proteins, and the formed elements in blood do not normally pass through it. The water and solutes that are filtered out of the blood pass into the capsular space between the visceral and parietal layers of the glomerular (Bowman's) capsule and then into the renal tubule.

The glomerular capsule opens into the first section of the renal tubule, called the *proximal convoluted tubule,* which also lies in the cortex. Convoluted means the tubule is coiled rather than straight; proximal signifies that the glomerular (Bowman's) capsule is the origin of the tubule. The wall of the proximal convoluted tubule consists of cuboidal epithelium with microvilli. These surface specializations, like those of the small intestine, increase the surface area for reabsorption and secretion.

Nephrons are frequently classified into two kinds. A *cortical nephron* usually has its glomerulus in the outer cortical zone, and the remainder of the nephron rarely penetrates the medulla. A *juxtamedullary nephron* usually has its glomerulus close to the corticomedullary junction, and other parts of the nephron penetrate deeply into the medulla (see Figure 23-3).

In a juxtamedullary nephron, the proximal convoluted tubule straightens, becomes thinner, and dips into the medulla, where it is called the *descending (thin) limb of the loop of the nephron.* This section consists of squamous epithelium. The tubule then increases in diameter as it bends into a ∪-shaped structure called the *loop of the nephron (loop of Henle).* It then ascends toward the cortex as the *ascending (thick) limb of the loop of the nephron,* which consists of cuboidal and low columnar epithelium.

In the cortex, the tubule again becomes convoluted. Because of its distance from the point of origin at the glomerular (Bowman's) capsule, this section is referred to as the *distal*

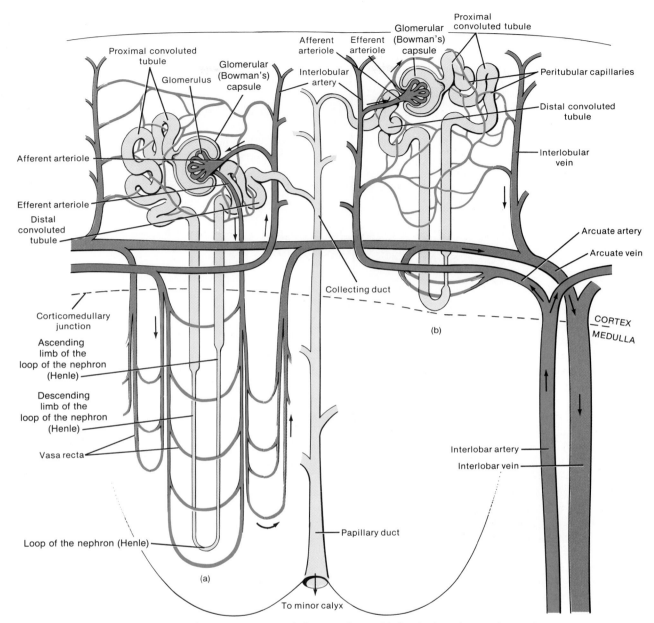

FIGURE 24-3 Nephrons. (a) Juxtamedullary nephron. (b) Cortical nephron. The nephrons are colored pink.

convoluted tubule. The cells of the distal tubule, like those of the proximal tubule, are cuboidal. Unlike the cells of the proximal tubule, however, the cells of the distal tubule have few microvilli. In a cortical nephron, the proximal section runs into the distal tubule without the connecting limbs and loop of the nephron. In both types of nephron, the distal tubule terminates by merging with a straight *collecting duct.*

In the medulla, the collecting ducts receive the distal tubules of several nephrons, pass through the renal pyramids, and open at the renal papillae into the minor calyces through a number of large *papillary ducts.* On the average, there

are 30 papillary ducts per renal papilla. Cells of the collecting ducts are cuboidal; those of the papillary ducts are columnar.

The histology of a nephron and glomerulus is shown in Figure 24-5.

BLOOD AND NERVE SUPPLY

Nephrons are largely responsible for removing wastes from the blood and regulating its fluid and electrolyte content. Thus, they are abundantly supplied with blood vessels. The right and left *renal arteries* transport about one-fourth the total cardiac output to the kidneys (Figure 24-6). Approx-

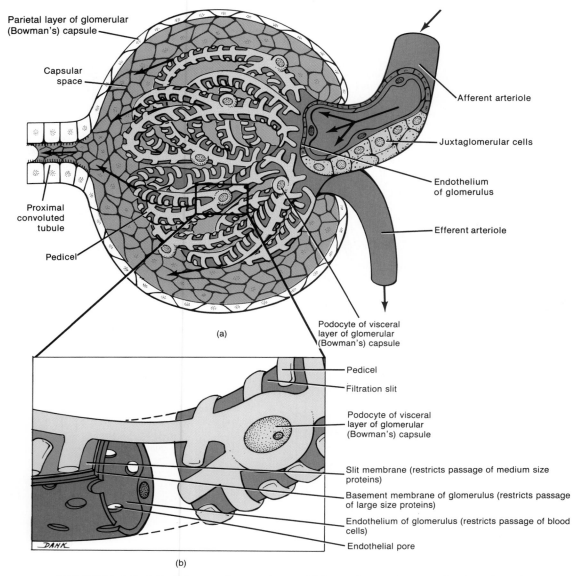

Parietal layer of glomerular (Bowman's) capsule

Capsular space

Proximal convoluted tubule

Pedicel

Afferent arteriole

Juxtaglomerular cells

Endothelium of glomerulus

Efferent arteriole

(a)

Podocyte of visceral layer of glomerular (Bowman's) capsule

Pedicel

Filtration slit

Podocyte of visceral layer of glomerular (Bowman's) capsule

Slit membrane (restricts passage of medium size proteins)

Basement membrane of glomerulus (restricts passage of large size proteins)

Endothelium of glomerulus (restricts passage of blood cells)

Endothelial pore

(b)

FIGURE 24-4 Endothelial–capsular membrane. (a) Parts of a renal corpuscle. (b) Enlarged aspect of a portion of the endothelial–capsular membrane.

imately 1,200 ml passes through the kidneys every minute.

Before or immediately after entering the hilus, the renal artery divides into several branches that enter the parenchyma and pass as the *interlobar arteries* between the renal pyramids in the renal columns. At the bases of the pyramids, the interlobar arteries arch between the medulla and cortex; here they are known as the *arcuate arteries*. Divisions of the arcuate arteries produce a series of *interlobular arteries,* which enter the cortex and divide into *afferent arterioles* (see Figure 24-3).

One afferent arteriole is distributed to each glomerular (Bowman's) capsule, where the arteriole divides into the tangled capillary network called the *glomerulus*. The glomerular capillaries then reunite to form an *efferent arteriole,* which leads away from the capsule and is smaller in diameter

than the afferent arteriole. This variation in diameter helps raise the glomerular pressure. The afferent–efferent arteriole situation is unique because blood usually flows out of capillaries into venules and not into other arterioles.

Each efferent arteriole of a cortical nephron divides to form a network of capillaries, called the *peritubular capillaries,* around the convoluted tubules. The efferent arteriole of a juxtamedullary nephron also forms peritubular capillaries. In addition, it forms long loops of thin-walled vessels called *vasa recta* that dip down alongside the loop of the nephron into the medullary region of the papilla.

The peritubular capillaries eventually reunite to form *interlobular veins*. The blood then drains through the *arcuate veins* to the *interlobar veins* running between the pyramids and leaves the kidney through a single *renal vein*

Parietal layer of glomerular
(Bowman's) capsule

Capsular
space

Podocyte of visceral
layer of glomerular
(Bowman's) capsule

(c)

Podocyte of visceral
layer of glomerular
(Bowman's) capsule

Pedicels

(d)

Filtration slits

FIGURE 24-4 (Continued) Endothelial–capsular membrane. (c) Scanning electron micrograph of a renal corpuscle at a magnification of 2,000×. (Courtesy of CNRI/Science Photo Library/ Photo Researchers.) (d) Scanning electron micrograph of a podocyte at a magnification of 4,500×. (Courtesy of Richard K. Kessel and Randy H. Kardon, *Tissues and Organs: A Text-Atlas of Scanning Electron Microscopy.* Copyright © 1979 by Scientific American, Inc.)

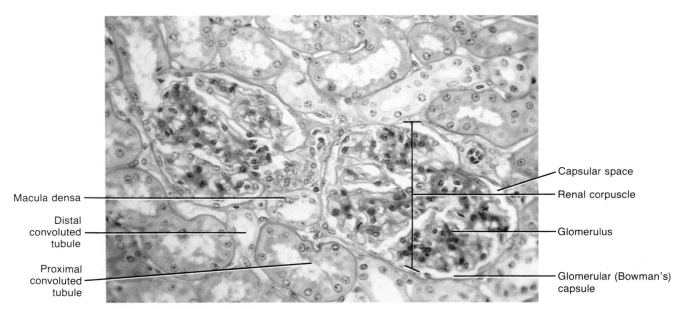

Macula densa

Distal
convoluted
tubule

Proximal
convoluted
tubule

Capsular space

Renal corpuscle

Glomerulus

Glomerular (Bowman's)
capsule

FIGURE 24-5 Histology of a nephron. Photomicrograph of the cortex of the kidney showing a renal corpuscle and surrounding renal tubules at a magnification of 400×. (Courtesy of Andrew Kuntzman.)

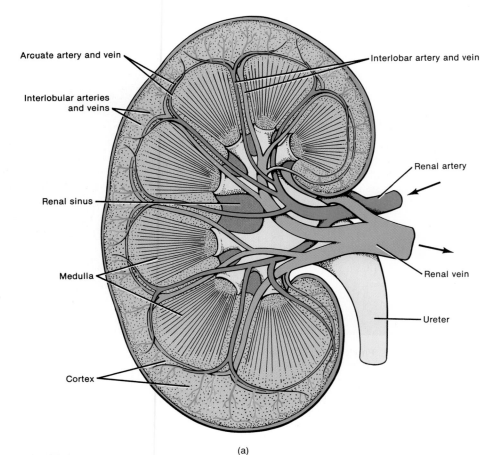

Arcuate artery and vein

Interlobar artery and vein

Interlobular arteries and veins

Renal artery

Renal sinus

Renal vein

Medulla

Ureter

Cortex

(a)

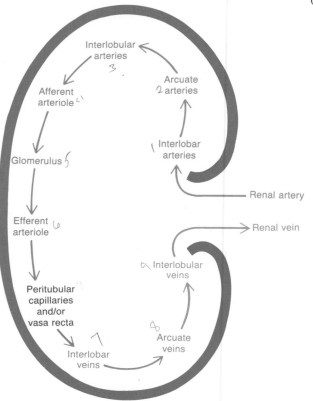

Interlobular arteries

Arcuate arteries

Afferent arteriole

Interlobar arteries

Glomerulus

Renal artery

Efferent arteriole

Renal vein

Peritubular capillaries and/or vasa recta

Interlobular veins

Interlobar veins

Arcuate veins

(b)

FIGURE 24-6 Blood supply of the right kidney. (a) Diagram in coronal section. (b) Scheme of circulation. This view is designed to show the *sequence* of blood flow, not the anatomical location of blood vessels, which is shown in (a).

that exits at the hilus. The vasa recta pass blood into the interlobular veins. From here, it goes to the arcuate veins, the interlobar veins, and then into the renal vein.

The nerve supply to the kidneys is derived from the **renal plexus** of the sympathetic division of the autonomic nervous system. Nerves from the plexus accompany the renal arteries and their branches and are distributed to the vessels. Because the nerves are vasomotor, they regulate the circulation of blood in the kidney by regulating the diameters of the arterioles.

JUXTAGLOMERULAR APPARATUS (JGA)

The smooth muscle fibers (cells) of the tunica media adjacent to the afferent arteriole (and sometimes efferent arteriole) are modified in several ways. Their nuclei are rounded (instead of elongated), and their cytoplasm contains granules (instead of myofibrils). Such modified muscle fibers are called *juxtaglomerular cells.* The cells of the distal convoluted tubule adjacent to the afferent and efferent arterioles are considerably narrower and taller than the other cells. Collectively, these cells are known as the *macula densa.*

Together with the modified cells of the afferent arteriole they constitute the *juxtaglomerular apparatus* or *JGA* (Figure 24-7), which helps regulate renal blood pressure.

PHYSIOLOGY

The major work of the urinary system is done by the nephrons. The other parts of the system are primarily passageways and storage areas. Nephrons carry out three important functions: (1) they control blood concentration and volume by removing selected amounts of water and solutes; (2) they help regulate blood pH; and (3) they also remove toxic wastes from the blood. As the nephrons go about these activities, they remove many materials from the blood, return the ones that the body requires, and eliminate the remainder. The eliminated materials are collectively called *urine*. The entire volume of blood in the body is filtered by the kidneys approximately 60 times a day.

HEMODIALYSIS THERAPY

If the kidneys are so impaired by disease or injury that they are unable to excrete nitrogenous wastes and regulate pH and electrolyte concentration of the plasma, the blood must be filtered by an artificial device. Such filtering of the blood is called *hemodialysis*. Dialysis means the separation of large nondiffusible particles from smaller diffusible ones through a selectively permeable membrane. One of the best-known devices for accomplishing dialysis is the

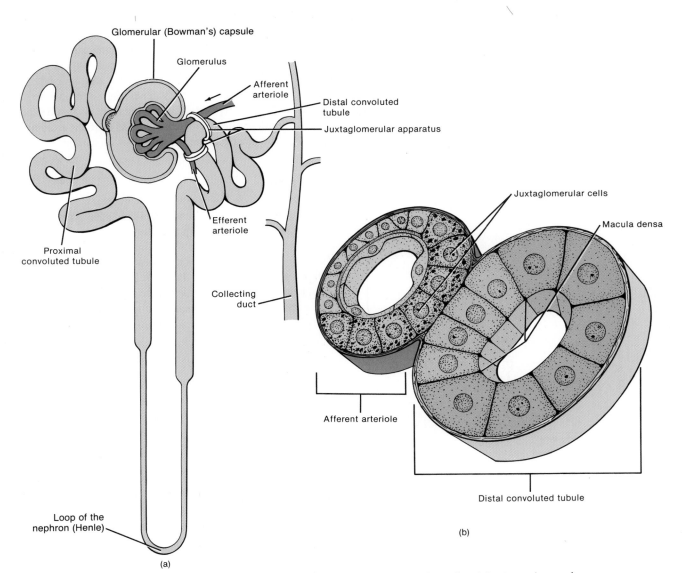

FIGURE 24-7 Juxtaglomerular apparatus. (a) External view. (b) Cells of the juxtaglomerular apparatus seen in cross section. The macula densa adjacent to the efferent arteriole is not illustrated.

artificial kidney machine (Figure 24-8). A tube connects it with the patient's radial artery. The blood is pumped from the artery through the tubes to one side of a selectively permeable dialyzing membrane made of cellulose acetate. The other side of the membrane is continually washed with an artificial solution called the dialysate. The blood that passes through the artificial kidney is treated with an anticoagulant (heparin). Only about 500 ml of the patient's blood is in the machine at a time. This volume is easily compensated for by vasoconstriction and increased cardiac output.

All substances (including wastes) in the blood except protein molecules and blood cells can diffuse back and forth across the selectively permeable membrane. The electrolyte level of the plasma is controlled by keeping the dialysate electrolytes at the same concentration found in normal plasma. Any excess plasma electrolytes move down the concentration gradient and into the dialysate. If the plasma electrolyte level is normal, it is in equilibrium with the dialysate and there is no net gain or loss of electrolytes. Since the dialysate contains no wastes, substances such as urea move down the concentration gradient and into the dialysate. Thus, wastes are removed and normal electrolyte balance is maintained. The usual duration for hemodialysis is 3½–6 hours, depending on factors such as the efficiency of renal function and patient size.

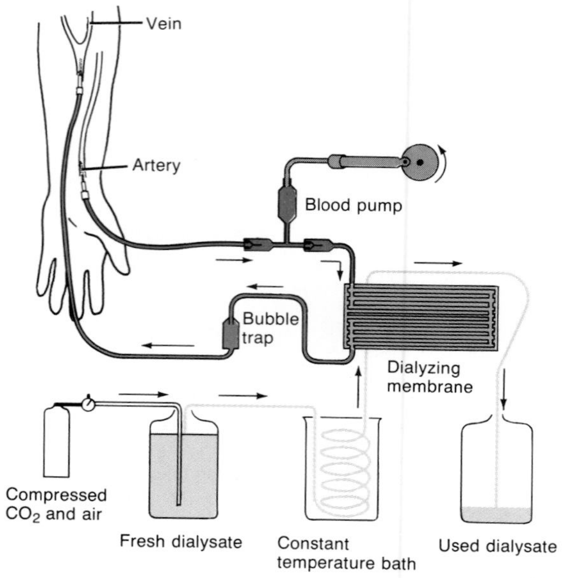

A great advantage of the kidney machine is that nutrition can be bolstered by placing large quantities of glucose in the dialysate. While the blood gives up its wastes, the glucose diffuses into the blood. Thus, the kidney machine beautifully accomplishes the principal function of the fundamental unit of the kidney—the nephron.

There are obvious drawbacks to the artificial kidney, however. Anticoagulants must be added to the blood during dialysis. A large amount of the patient's blood must flow through this apparatus to make the treatment effective, and so the slow rate at which the blood can be processed makes the treatment time consuming. To date, no artificial kidney has been implanted permanently.

Continuous ambulatory peritoneal dialysis (**CAPD**) has made hemodialysis more convenient and less time consuming for many patients. CAPD uses the peritoneum instead of cellulose acetate as the dialyzing membrane. Since the peritoneum is a selectivley permeable membrane, it permits rapid bidirectional transfer of substances. A catheter is placed in the patient's peritoneal cavity and connected to a supply of dialysate. Gravity feeds the solution into the abdominal cavity from its plastic container. When the process is complete, the dialysate is returned from the abdominal cavity to the plastic container and then discarded.

URETERS

Once urine is formed by the nephrons and passed into collecting ducts, it drains through papillary ducts into the calyces surrounding the renal papillae. The minor calyces join to become the major calyces that unite to become the renal pelvis. From the pelvis, the urine drains into the ureters and is carried by peristalsis to the urinary bladder. From the urinary bladder, the urine is discharged from the body through the single urethra.

STRUCTURE

The body has two ***ureters*** (YOO-re-ters)—one for each kidney. Each ureter is an extension of the pelvis of the kidney and extends 25–30 cm (10–12 in.) to the urinary bladder (see Figure 24-1). As the ureters descend, their thick walls increase in diameter, but at their widest point they measure less than 1.7 cm (½ in.) in diameter. Like the kidneys, the ureters are retroperitoneal in placement. The ureters enter the urinary bladder at the superior lateral angle of its base.

Although there are no anatomical valves at the openings of the ureters into the urinary bladder, there is a functional one that is quite effective. Since the ureters pass obliquely through the wall of the urinary bladder, pressure in the urinary bladder compresses the ureters and prevents backflow of urine when pressure builds up in the urinary bladder

FIGURE 24-8 Operation of an artificial kidney. The blood route is indicated in red and blue. The route of the dialysate is indicated in gold.

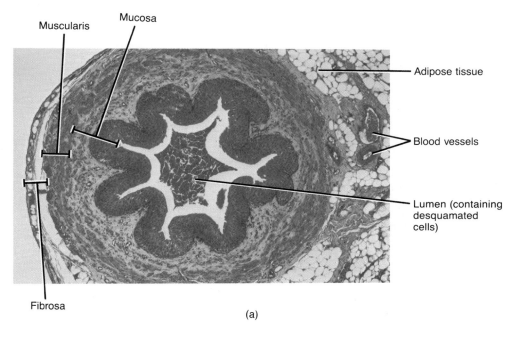

(a)

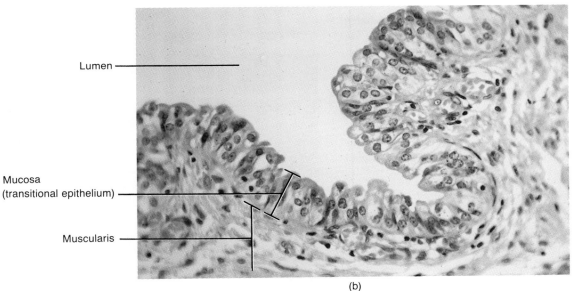

(b)

FIGURE 24-9 Histology of the ureter. (a) Photomicrograph of the ureter seen in cross section at a magnification of 50×. (Copyright © 1983 by Michael H. Ross. Used by permission.) (b) Photomicrograph of an enlarged aspect of the mucosa of the ureter at a magnification of 250×. (Courtesy of Andrew Kuntzman.)

as it fills during urination. When this physiological valve is not operating, it is possible for cystitis (urinary bladder inflammation) to develop into kidney infection.

HISTOLOGY

Three coats of tissue form the wall of the ureters (Figure 24-9). The inner coat, or mucosa, is mucous membrane with transitional epithelium. The solute concentration and pH of urine differ drastically from the internal environment of cells that form the wall of the ureters. Mucus secreted by the mucosa prevents the cells from coming in contact with urine. Throughout most of the length of the ureters, the second or middle coat, the muscularis, is composed of inner longitudinal and outer circular layers of smooth muscle. The muscularis of the proximal third of the ureters also contains a layer of outer longitudinal muscle. Peristalsis is the major function of the muscularis. The third, or external, coat of the ureters is a fibrous coat. Extensions of the fibrous coat anchor the ureters in place.

PHYSIOLOGY

The principal function of the ureters is to transport urine from the renal pelvis into the urinary bladder. Urine is carried through the ureters primarily by peristaltic contractions of the muscular walls of the ureters, but hydrostatic pressure and gravity also contribute. Peristaltic waves pass from the kidney to the urinary bladder, varying in rate from 1 to 5 per minute depending on the amount of urine formation.

BLOOD AND NERVE SUPPLY

The arterial supply of the ureters is from the renal, testicular or ovarian, common iliac, and inferior vesical arteries. The veins terminate in the corresponding trunks.

The ureters are innervated by the renal and vesical plexuses.

URINARY BLADDER

The *urinary bladder* is a hollow muscular organ situated in the pelvic cavity posterior to the symphysis pubis. In the male, it is directly anterior to the rectum. In the female, it is anterior to the vagina and inferior to the uterus. It is a freely movable organ held in position by folds of the peritoneum. The shape of the urinary bladder depends on how much urine it contains. When empty, the size of the lumen is decreased and the wall appears thicker. It becomes spherical when slightly distended. As urine volume increases, it becomes pear shaped and rises into the abdominal cavity.

STRUCTURE

At the base of the urinary bladder is a small triangular area, the *trigone* (TRĪ-gōn), that points anteriorly (Figure 24-10). The opening to the urethra is found in the apex of this triangle. At the two points of the base, the ureters drain into the urinary bladder. It is easily identified because the mucosa is firmly bound to the muscularis so that the trigone is typically smooth.

HISTOLOGY

Four coats make up the wall of the urinary bladder (Figure 24-11). The mucosa, the innermost coat, is a mucous membrane containing transitional epithelium. Transitional epithelium is able to stretch—a marked advantage for an organ that must continually inflate and deflate. Rugae (folds in the mucosa) are also present. The second coat, the submucosa, is a layer of connective tissue that connects the mucosa and muscular coats. The third coat—a muscular one called the *detrusor* (de-TROO-ser) *muscle*—consists of three lay-

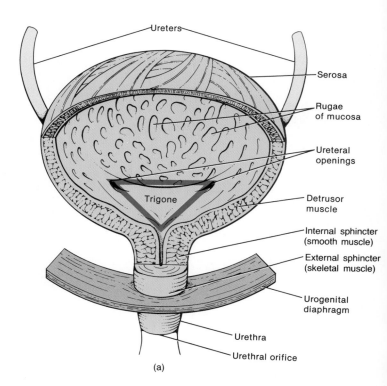

(a)

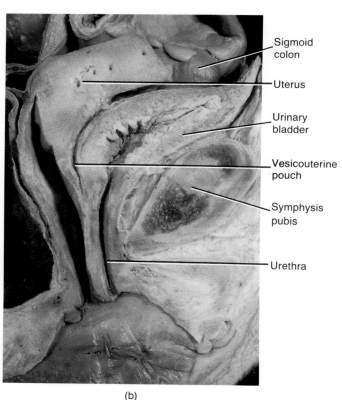

(b)

FIGURE 24-10 Urinary bladder and female urethra. (a) Diagram. (b) Photograph. (Courtesy of J. A. Gosling, P. F. Harris, et al. *Atlas of Human Anatomy,* Gower Medical Publishing Ltd., 1985.)

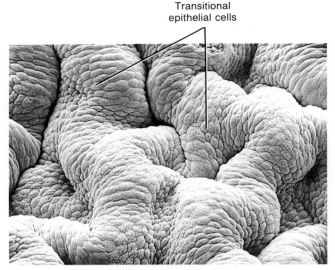

Transitional
epithelial cells

FIGURE 24-11 Histology of the urinary bladder. Photomicrograph of a portion of the wall of the urinary bladder at a magnification of 50×. The details of the mucosa of the urinary bladder are shown in Exhibit 3-1, Stratified transitional. (Copyright © 1983 by Michael H. Ross. Used by permission.)

ers of smooth muscle: inner longitudinal, middle circular, and outer longitudinal muscles. In the area around the opening to the urethra, the circular fibers form an *internal sphincter* muscle. Below the internal sphincter is the *external sphincter,* which is composed of skeletal muscle. The outermost coat, the serous coat, is formed by the peritoneum and covers only the superior surface of the organ.

PHYSIOLOGY

Urine is expelled from the urinary bladder by an act called *micturition* (mik'-too-RISH-un), commonly known as urination or voiding. This response is brought about by a combination of involuntary and voluntary nerve impulses. The average capacity of the urinary bladder is 700–800 ml. When the amount of urine in the urinary bladder exceeds 200–400 ml, stretch receptors in the urinary bladder wall transmit impulses to the lower portion of the spinal cord. These impulses initiate a conscious desire to expel urine and a subconscious reflex referred to as the *micturition reflex.* Parasympathetic impulses transmitted from the sacral area of the spinal cord reach the urinary bladder wall and internal urethral sphincter, bringing about contraction of the detrusor muscle of the urinary bladder and relaxation of the internal sphincter. Then the conscious portion of the brain sends impulses to the external sphincter, the sphincter relaxes, and urination takes place. Although emptying of the urinary bladder is controlled by reflex, it may be initiated voluntarily and stopped at will because of cerebral control of the external sphincter.

CLINICAL APPLICATION

A lack of voluntary control over micturition is referred to as *incontinence.* In infants about 2 years old and under, incontinence is normal because neurons to the external sphincter muscle are not completely developed. Infants void whenever the urinary bladder is sufficiently distended to arouse a reflex stimulus. Proper training overcomes incontinence if the latter is not caused by emotional stress or irritation of the urinary bladder.

Involuntary micturition in the adult may occur as a result of unconsciousness, injury to the spinal nerves controlling the urinary bladder, irritation due to abnormal constituents in urine, disease of the urinary bladder, damage to the external sphincter, and inability of the detrusor muscle to relax due to emotional stress.

Retention, a failure to completely or normally void urine, may be due to an obstruction in the urethra or neck of the urinary bladder, nervous contraction of the urethra, or lack of sensation to urinate.

BLOOD AND NERVE SUPPLY

The arteries of the urinary bladder are the superior vesical, the middle vesical, and the inferior vesical. The veins from the urinary bladder pass to the internal iliac trunk.

The nerves are derived partly from the hypogastric sympathetic plexus and partly from the second and third sacral nerves. The fibers from the sacral nerves constitute the nervi erigentes.

URETHRA

The *urethra* is a small tube leading from the floor of the urinary bladder to the exterior of the body (see Figure 24-10). In females, it lies directly posterior to the symphysis pubis and is anterior to the anterior wall of the vagina. Its undilated diameter is about 6 mm (¼ in.), and its length is approximately 3.8 cm (1½ in.). The female urethra is directed obliquely, inferiorly, and anteriorly. The opening of the urethra to the exterior, the *urethral orifice,* is located between the clitoris and vaginal opening.

In males, the urethra is about 20 cm (8 in.) long. Immediately below the urinary bladder it passes vertically through the prostate gland (prostatic urethra), then pierces the urogenital diaphragm (membranous urethra), and finally pierces the penis (spongy urethra) and takes a curved course through its body (see Figures 25-1 and 25-10).

HISTOLOGY

The wall of the female urethra consists of three coats: an inner mucous coat, an intermediate thin layer of spongy tissue containing a plexus of veins, and an outer muscular coat that is continuous with that of the urinary bladder and consists of circularly arranged fibers (cells) of smooth muscle. The mucosa is usually lined with transitional epithelium near the urinary bladder. The remainder consists of stratified squamous epithelium with areas of stratified columnar or pseudostratified epithelium.

The male urethra is composed of two coats: an inner mucous membrane and an outer submucous tissue that connects the urethra with the structures through which it passes. The mucosa varies in different regions. The mucosa of the prostatic urethra is continuous with that of the urinary bladder and is lined by transitional epithelium. The membranous mucosa is lined by pseudostratified epithelium. The spongy urethra is lined mostly by pseudostratified epithelium. Near its opening to the exterior it is lined by stratified squamous epithelium. In the spongy urethra, especially, there are glands, called urethral (Littré) glands, that produce mucus for lubrication during sexual intercourse.

PHYSIOLOGY

The urethra is the terminal portion of the urinary system. It serves as the passageway for discharging urine from the body. The male urethra also serves as the duct through which reproductive fluid (semen) is discharged from the body.

AGING AND THE URINARY SYSTEM

The effectiveness of kidney function decreases with aging, and by age 70 the filtering mechanism is only about one-half what it was at age 40. Urinary incontinence and urinary tract infections are two major problems associated with aging of the urinary system. Other pathologies include polyuria (excessive urine production), nocturia (excessive urination at night), increased frequency of urination, dysuria (painful urination), retention (failure to produce urine), and hematuria (blood in the urine). Changes and diseases in the kidney include acute and chronic kidney inflammations and renal calculi (kidney stones). The prostate gland is often implicated in various disorders of the urinary tract and cancer of the prostate is the most frequent malignancy in elderly males.

DEVELOPMENTAL ANATOMY OF THE URINARY SYSTEM

As early as the third week of development, a portion of the mesoderm along the posterior half of the dorsal side of the embryo, the *intermediate mesoderm,* differentiates into the kidneys. Three pairs of kidneys form within the intermediate mesoderm in successive time periods: pronephros, mesonephros, and metanephros (Figure 24-12). Only the last one remains as the functional kidneys of the adult.

The first kidney to form, the *pronephros,* is the superior of the three. Associated with its formation is a tube, the *pronephric duct.* This duct empties into the *cloaca,* which is the dilated caudal end of the gut derived from *endoderm.* The pronephros begins to degenerate during the fourth week and is completely gone by the sixth week. The pronephric ducts, however, remain.

The pronephros is replaced by the second kidney, the *mesonephros.* With its appearance, the retained portion of the pronephric duct, which connects to the mesonephros, becomes known as the *mesonephric duct.* The mesonephros begins to degenerate by the sixth week and is just about completely gone by the eighth week.

At about the fifth week, an outgrowth, called a *ureteric bud,* develops from the distal end of the mesonephric duct near the cloaca. This bud is the developing *metanephros.* As it grows toward the head of the embryo, its end widens to form the *pelvis* of the kidney with its *calyces* and associated *collecting tubules.* The unexpanded portion of the bud, the *metanephric duct,* becomes the *ureter.* The *nephrons,* the functional units of the kidney, arise from the intermediate mesoderm around each ureteric bud.

During development, the cloaca divides into a *urogenital sinus,* into which urinary and genital ducts empty, and a *rectum* that discharges into the anal canal. The *urinary bladder* develops from the urogenital sinus. In the female, the *urethra* develops from lengthening of the short duct that extends from the urinary bladder to the urogenital sinus. The *vestibule,* into which the urinary and genital ducts empty, is also derived from the urogenital sinus. In the male, the urethra is considered longer and more complicated but is also derived from the urogenital sinus.

APPLICATIONS TO HEALTH

RENAL CALCULI (KIDNEY STONES)

Occasionally, the crystals of salts found in urine may solidify into insoluble stones called *renal calculi (kidney stones).* They may be formed in any portion of the urinary tract from the kidney tubules to the external opening. Conditions leading to stone formation include the ingestion of excessive mineral salts, a decrease in the amount of water intake, abnormally alkaline or acidic urine, and overactivity of the parathyroid glands. Common constituents of stones are calcium oxalate, uric acid, and calcium phosphate crystals. Calcium oxalate crystals are the most common type. A protein isolated from urine called glycoprotein crystal-growth inhibitor (GCI) inhibits formation of calcium oxalate stones. Individuals who do not synthesize GCI may form calcium oxalate stones. Kidney stones usually form in the pelvis of the kidney, where they cause pain, hematuria, and pyuria.

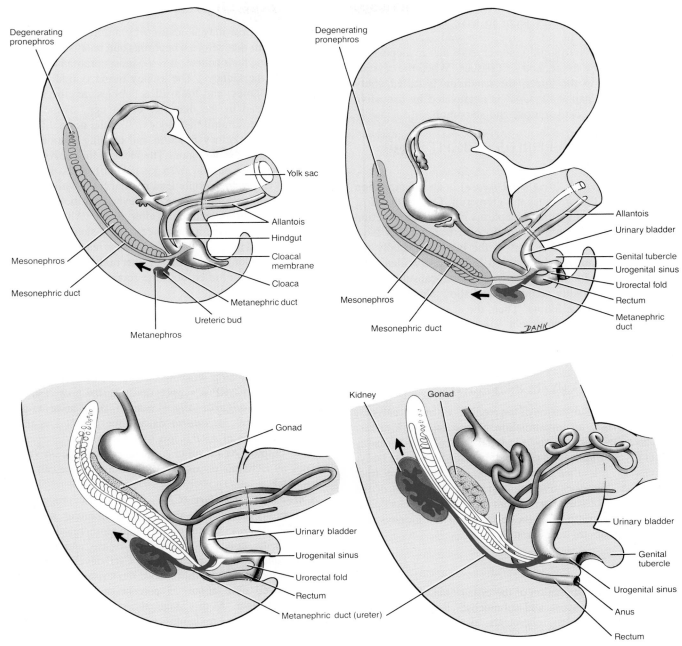

FIGURE 24-12 Development of the urinary system.

For kidney stones that become painful or obstructive, surgical removal is the typical alternative. However, there are now several new methods of treatment that do not involve conventional surgery. One procedure, called ***extracorporeal shock wave lithotripsy (ESWL),*** involves the use of ultrasound waves generated by an instrument called a lithotripter. In this procedure, a patient is placed in a water bath and is subjected to ultrasound waves. Once the stones are shattered, the fragments are eliminated via the urine. In another procedure, a laser fiber is inserted into the urinary tract and the laser devise uses short blasts to shatter the stones. This procedure is used for larger kidney stones that become lodged in the ureters. One other procedure is called ***percutaneous ultrasonic lithotripsy (PUL).*** A tube resembling a cystoscope is inserted into the kidney through a small opening in the back. The surgeon can then pick out small stones whole, or shatter large stones with ultrasound waves, and remove the fragments by suction.

GOUT

Gout is a hereditary condition associated with an excessively high level of uric acid in the blood. When nucleic acids are catabolized, a certain amount of uric acid is produced as a waste. Some people seem to produce excessive amounts of uric acid, and others seem to have trouble excreting normal amounts. In either case, uric acid accumulates in the body and tends to solidify into crystals that are depositied in joints, kidney tissue, and soft tissues. When the crystals are deposited in the joints, the condition is called gouty arthritis (see Chapter 8). Gout is aggravated by excessive use of diuretics, dehydration, and starvation.

GLOMERULONEPHRITIS (BRIGHT'S DISEASE)

Glomerulonephritis (Bright's Disease) is an inflammation of the kidney that involves the glomeruli. One of the most common causes of glomerulonephritis is an allergic reaction to the toxins given off by streptococci bacteria that have recently infected another part of the body, especially the throat. The glomeruli become so inflamed, swollen, and engorged with blood that the endothelial–capsular membranes become highly permeable and allow blood cells and proteins to enter the filtrate. Thus, the urine contains many erythrocytes and much protein. The glomeruli may be permanently changed, leading to chronic renal disease and renal failure.

PYELITIS AND PYELONEPHRITIS

Pyelitis is an inflammation of the renal pelvis and its calyces. *Pyelonephritis,* an inflammation of one or both kidneys, involves the nephrons and the renal pelvis. The disease is generally a complication of infection elsewhere in the body. In females, it is often a complication of lower urinary tract infections. The causative agent in about 75 percent of the cases is the bacterium *Escherichia coli.* Should pyelonephritis become chronic, scar tissue forms in the kidneys and severly impairs their function.

CYSTITIS

Cystitis is an inflammation of the urinary bladder involving principally the mucosa and submucosa. It may be caused by bacterial infection, chemicals, or mechanical injury. Symptoms include burning on urination or painful urination, urgency and frequent urination, and low back pain. Bed wetting may also occur.

NEPHROSIS

Nephrosis is a condition in which the endothelial-capsular membrane leaks, allowing large amounts of protein to escape from the blood into the urine. Water and sodium then accumulate in the body producing edema, especially around the ankles and feet, abdomen, and eyes. Nephrosis is more common in children than adults but occurs in all ages. Although it cannot always be cured, certain synthetic steroid hormones such as cortisone and prednisone, which are similar to the natural hormones secreted by the adrenal (suprarenal) glands, can suppress some forms of it.

POLYCYSTIC DISEASE

Polycystic disease may be caused by a defect in the renal tubular system that deforms nephrons and results in cystlike dilations along their course. It is the most common inherited disorder of the kidneys. The kidney tissue is riddled with cysts, small holes, and fluid-filled bubbles ranging in size from a pinhead to the diameter of an egg. These cysts gradually increase until they squeeze out the normal tissue, interfering with kidney function and causing uremia. The chief symptom is weight gain. The kidneys themselves may enlarge from the normal 0.25 kg (0.5 lb) to as much as 14 kg (30 lb). Many people lead normal lives without ever knowing they have the disease, however, and the condition is sometimes discovered only after death. Although the disease is progressive, its advance can be slowed by diet, drugs, and fluid intake. Kidney failure as a result of polycystic disease seldom occurs before the mid-forties and frequently can be controlled until the sixties.

RENAL FAILURE

Renal failure, a decrease or cessation of glomerular filtration, is classified into two types: acute and chronic. *Acute renal failure (ARF)* is a clinical syndrome in which the kidneys abruptly stop working entirely or almost entirely. The hallmark of ARF is suppression of urine flow, usually categorized by *oliguria* (*olig* = scanty), daily urine output less than 500 ml, or *anuria* daily urine output less than 50 ml. One cause of ARF is low blood volume or decreased cardiac output. Acute tubular necrosis (damage to renal tubules by ischemia or toxins) and a kidney stone can also cause ARF.

Chronic renal failure (CRF) refers to the progressive and generally irreversible decline in glomerular filtration rate (GFR) that may result from chronic glomerulonephritis, pyelonephritis, congenital polycystic disease, and traumatic loss of kidney tissue, among others. CRF develops in three stages. In the first stage, diminished renal reserve, nephrons are destroyed until about 75 percent of the functioning nephrons are lost. At this stage, the person may show no symptoms since the remaining nephrons enlarge and take over the function of those that have been lost. When more nephrons are lost, the balance between glomerular filtration and tubular reabsorption is destroyed, and any change in diet or fluid intake brings on the symptoms. Once 75 percent of the nephrons are lost, the person enters the second stage, called renal insufficiency. In this stage, there is a decrease in GFR and increased blood levels of nitrogenous wastes and creatine. Also, the kidneys cannot effectively concentrate or dilute urine. The final stage, called end stage renal failure (uremia), occurs when about 90 percent of the nephrons have been lost. At this stage, GFR diminishes to about 10 percent normal, and blood levels of nitrogenous wastes and creatine increase further. The low GFR results in oliguria. Individuals with CRF are candidates for hemodialysis therapy and kidney transplantation.

Among the effects of renal failure are edema from salt

Reproduction is the mechanism by which the thread of life is sustained. In one sense, reproduction is the process by which a single cell duplicates its genetic material, allowing an organism to grow and repair itself; thus, reproduction maintains the life of the individual. But reproduction is also the process by which genetic material is passed from generation to generation. In this regard, reproduction maintains the continuation of the species.

The organs of the male and female reproductive systems may be grouped by function. The testes and ovaries, also called *gonads,* function in the production of gametes— sperm cells and ova, respectively. The gonads also secrete hormones. The production of gametes and their discharge into ducts classifies the gonads as exocrine glands, whereas their production of hormones classifies them as endocrine glands. The *ducts* transport, receive, and store gametes. Still other reproductive organs, called *accessory sex glands,* produce materials that support gametes.

The developmental anatomy of the reproductive systems is considered later in the chapter.

MALE REPRODUCTIVE SYSTEM

The organs of the male reproductive system are the testes, or male gonads, that produce sperm; a number of ducts that either store or transport sperm to the exterior; accessory glands that add secretions constituting the semen; and several supporting structures, including the penis (Figure 25-1).

SCROTUM

The *scrotum* is a cutaneous outpouching of the abdomen consisting of loose skin and superficial fascia (Figure 25-1). It is the supporting structure for the testes. Externally, it looks like a single pouch of skin separated into lateral portions by a median ridge called the *raphe* (RĀ-fē). Internally, it is divided by a septum into two sacs, each containing a single testis. The septum consists of superficial fascia and contractile tissue called the *dartos* (DAR-tōs), which consists of bundles of smooth muscle fibers (cells). The dartos is also found in the subcutaneous tissue of the scrotum and is directly continuous with the subcutaneous tissue of the abdominal wall. The dartos causes wrinkling of the skin of the scrotum.

The location of the scrotum and the contraction of its muscle fibers regulate the temperature of the testes. The production and survival of sperm require a temperature that is lower than normal body temperature. Because the scrotum is outside the body cavities, it provides an environment about 3°C below normal body temperature. The *cremaster* (krē-MAS-ter) *muscle* (see Figure 25-8a), a small band of skeletal muscle, elevates the testes during sexual arousal and on exposure to cold, moving them closer to the pelvic cavity where they can absorb body heat. Exposure to warmth reverses the process.

The blood supply of the scrotum is derived from the internal pudendal branch of the internal iliac, the cremasteric branch of the inferior epigastric artery, and the external pudendal artery from the femoral artery. The scrotal veins follow the arteries.

The scrotal nerves are derived from the pudendal, posterior cutaneous of the thigh, and ilioinguinal nerves.

TESTES

The *testes,* or *testicles,* are paired oval glands measuring about 5 cm (2 in.) in length and 2.5 cm (1 in.) in diameter (Figure 25-2). Each weighs between about 10 and 15 g. The testes develop high on the embryo's posterior abdominal wall and usually begin their descent into the scrotum through the inguinal canals during the latter half of the seventh month of fetal development (see Figure 25-8).

CLINICAL APPLICATION

When the testes do not descend, the condition is referred to as *cryptorchidism* (krip-TOR-ki-dizm). The condition occurs in about 3 percent of full-term infants and about 30 percent of premature infants. Cryptorchidism results in sterility because the cells involved in the initial development of sperm cells are destroyed by the higher body temperature of the pelvic cavity. Also, the probability of testicular cancer is 30–50 times greater in cryptorchid testes. Undescended testes can be placed in the scrotum by administering hormones or by surgical means prior to puberty without ill effects. Sometimes, undescended testes undergo spontaneous descent.

The testes are covered by a dense layer of white fibrous tissue, the *tunica albuginea* (al'byoo-JIN-ē-a), that extends inward and divides each testis into a series of internal compartments called *lobules.* Each of the 200–300 lobules contains one to three tightly coiled tubules, the convoluted *seminiferous tubules,* that produce sperm by a process called *spermatogenesis.* This process is considered shortly.

A cross section through a seminiferous tubule reveals that it is lined with spermatogenic cells in various stages of development (Figure 25-3). Spermatogenic cells are successive stages in a continuous process of differentiation of male germ cells. The most immature spermatogenic cells, the *spermatogonia,* are located against the basement membrane. Toward the lumen of the tube, one can see layers of progressively more mature cells. In order of advancing maturity, these are primary spermatocytes, secondary spermatocytes, and spermatids. By the time a *sperm cell,* or *spermatozoon* (sper'-ma-tō-ZŌ-on), has nearly reached maturity, it is in the lumen of the tubule and begins to be moved through a series of ducts. Embedded between the developing sperm cells in the tubules are *sustentacular* (sus'-ten-TAK-yoo-lar), or *Sertoli, cells.* Just internal to

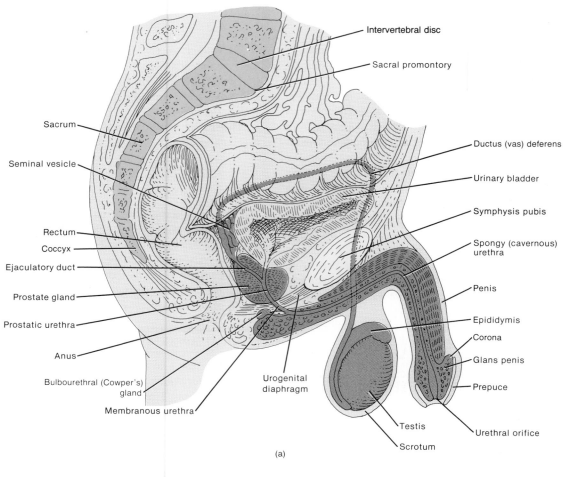

(a)

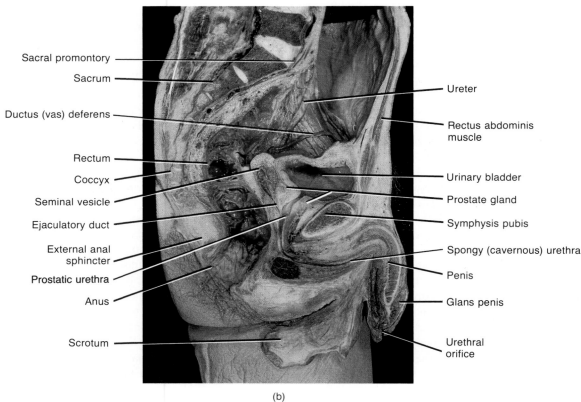

(b)

FIGURE 25-1 Male organs of reproduction and surrounding structures seen in sagittal section. (a) Diagram. (b) Photograph. (Courtesy of C. Yokochi and J. W. Rohen, *Photographic Anatomy of the Human Body*, 2nd ed., 1979, IGAKU-SHOIN, Ltd., Tokyo, New York.)

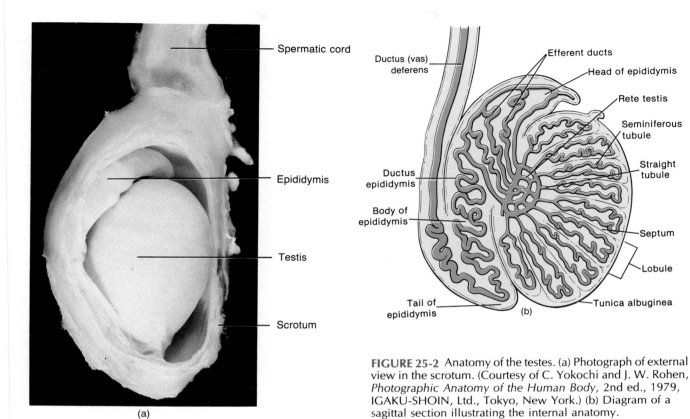

Spermatic cord

Epididymis

Testis

Scrotum

(a)

Ductus (vas) deferens

Ductus epididymis

Body of epididymis

Tail of epididymis

(b)

Efferent ducts

Head of epididymis

Rete testis

Seminiferous tubule

Straight tubule

Septum

Lobule

Tunica albuginea

FIGURE 25-2 Anatomy of the testes. (a) Photograph of external view in the scrotum. (Courtesy of C. Yokochi and J. W. Rohen, *Photographic Anatomy of the Human Body,* 2nd ed., 1979, IGAKU-SHOIN, Ltd., Tokyo, New York.) (b) Diagram of a sagittal section illustrating the internal anatomy.

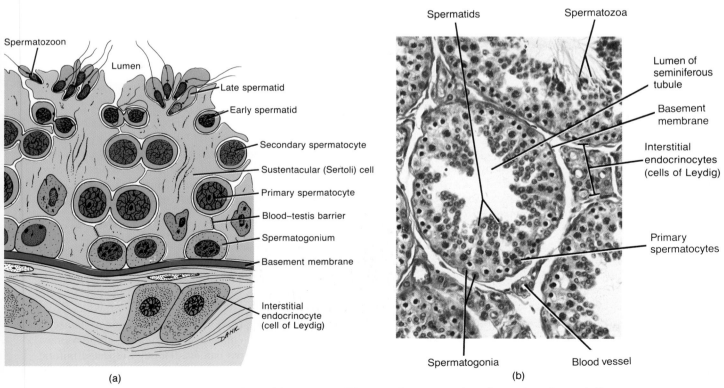

Spermatozoon

Lumen

Late spermatid

Early spermatid

Secondary spermatocyte

Sustentacular (Sertoli) cell

Primary spermatocyte

Blood–testis barrier

Spermatogonium

Basement membrane

Interstitial endocrinocyte (cell of Leydig)

(a)

Spermatids

Spermatozoa

Lumen of seminiferous tubule

Basement membrane

Interstitial endocrinocytes (cells of Leydig)

Primary spermatocytes

Spermatogonia

Blood vessel

(b)

FIGURE 25-3 Histology of the testes. (a) Diagram of a cross section of a portion of a seminiferous tubule showing the stages of spermatogenesis. (b) Photomicrograph of an enlarged aspect of several seminiferous tubules at a magnification of 350×. (Copyright © 1983 by Michael H. Ross. Used by permission.)

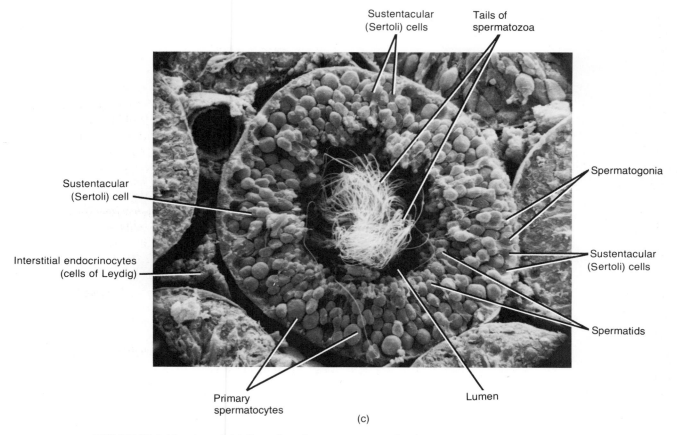

FIGURE 25-3 **(Continued)** (c) Scanning electron micrograph of a seminiferous tubule at a magnification of 400×. (Courtesy of Richard K. Kessel and Randy H. Kardon, *Tissues and Organs: A Text-Atlas of Scanning Electron Microscopy.* Copyright © 1979 by Scientific American, Inc.)

the basement membrane, each sustentacular cell is joined to another by junctional points that form a **blood–testis barrier.** The barrier is important because spermatozoa and developing cells produce surface antigens that are recognized as foreign by the immune system. The barrier prevents an immune response against the antigens by isolating the cells from the blood. Such an immune response is seen following vasectomy (described shortly) in which sperm-specific antibodies, produced in cells of the immune system, are exposed to spermatozoa that no longer remain isolated in the reproductive tract. Sustentacular cells support and protect developing spermatogenic cells; nourish spermatocytes, spermatids, and spermatozoa; phagocytize degenerating spermatogenic cells; control movements of spermatogenic cells and the release of spermatozoa into the lumen of the seminiferous tubule; and secrete the hormone inhibin that helps regulate sperm production and androgen-binding protein, a hormone required for sperm production that concentrates testosterone in the seminiferous tubule. Between the seminiferous tubules are clusters of **interstitial endocrinocytes (interstitial cells of Leydig).** These cells secrete the male hormone testosterone, the most important androgen.

Spermatogenesis

The process by which the testes produce haploid (*n*) spermatozoa involves several phases, including meiosis, and is called **spermatogenesis** (sper'-ma-tō-JEN-e-sis). Before reading the following discussion of spermatogenesis, you should review the details of meiosis in Chapter 2 (see Figure 2-15). At this point, a few key concepts should be kept in mind.

1. In sexual reproduction, a new organism is produced by the union and fusion of sex cells called **gametes.** Male gametes, produced in the testes, are called sperm cells, and female gametes, produced in the ovaries, are called ova.
2. The cell resulting from the union and fusion of gametes, called a **zygote,** contains a mixture of chromosomes (DNA) from the two parents. Through repeated mitotic cell divisions, a zygote develops into a new organism.
3. Gametes differ from all other body cells (somatic cells) in that they contain the **haploid** (one-half) **chromosome number,** symbolized as *n*. In humans, this number is

23, which composes a single set of chromosomes. Uninucleated somatic cells contain the **diploid chromosome number,** symbolized 2n. In humans, this number is 46, which composes two sets of chromosomes.

4. In a diploid cell, two chromosomes that belong to a pair are called **homologous chromosomes (homologues).** In human diploid cells, 22 of the 23 pairs of chromosomes are morphologically similar and are called **autosomes.** The other pair comprises the **sex chromosomes,** designated as X and Y. In the female, the homologous pair of sex chromosomes consists of two X chromosomes; in the male, the pair consists of an X and a Y.

5. If gametes were diploid (2n), like somatic cells, the zygote would contain twice the diploid number (4n), and with every succeeding generation the chromosome number would continue to double.

6. This continual doubling of the chromosome number does not occur because of meiosis, a process of cell division by which gametes produced in the testes and ovaries receive the haploid chromosome number. Thus, when haploid (n) gametes fuse, the zygote contains the diploid chromosome number (2n).

In humans, spermatogenesis takes about 74 days. The seminiferous tubules are lined with immature cells called **spermatogonia** (sper'-ma-tō-GŌ-nē-a), or sperm mother cells (Figures 25-3a and 25-4). Spermatogonia contain the diploid (2n) chromosome number and represent a heterogenous group of cells in which three subtypes can be distinguished. These are referred to as pale type A, dark type A, and type B and are distinguished by the appearance of their nuclear chromatin. Pale type A spermatogonia remain relatively undifferentiated and capable of extensive mitotic division. Following division, some of the daughter cells remain undifferentiated and serve as a reservoir of precursor cells to prevent depletion of the stem cell population. Such

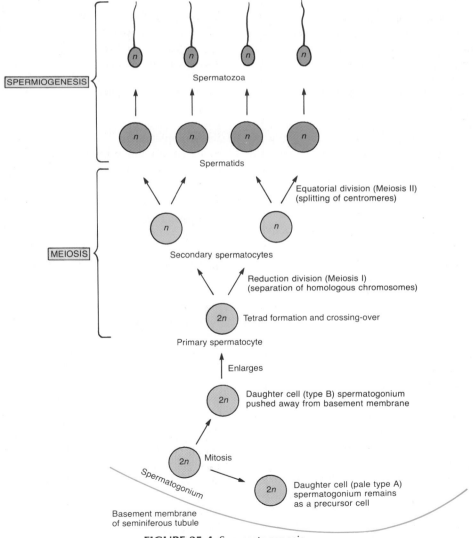

FIGURE 25-4 Spermatogenesis.

cells remain near the basement membrane. The remainder of the daughter cells differentiate into type B spermatogonia. These cells lose contact with the basement membrane of the seminiferous tubule, undergo certain developmental changes, and become known as *primary spermatocytes* (SPER-ma-tō-sītz'). Primary spermatocytes, like spermatogonia, are diploid (2*n*), that is, they have 46 chromosomes. Dark type A spermatogonia are believed to represent reserve stem cells, only becoming activated if pale type A cells become critically depleted.

● *Reduction Division (Meiosis I)* Each primary spermatocyte enlarges before dividing. Then two nuclear divisions take place as part of meiosis. In the first, DNA is replicated and 46 chromosomes (each made up of two chromatids) form and move toward the equatorial plane of the nucleus. There they line up by homologous pairs so that there are 23 pairs of duplicated chromosomes in the center of the nucleus. This pairing of homologous chromosomes is called *synapsis.* The four chromatids of each homologous pair then become associated with each other to form a *tetrad.* In a tetrad, portions of one chromatid may be exchanged with portions of another. This process, called *crossing-over,* permits an exchange of genes among chromatids (see Figure 2-16) that results in the recombination of genes. Thus, the spermatozoa eventually produced are genetically unlike each other and unlike the cell that produced them— one reason for the great variation among humans. Next, the meiotic spindle forms and the chromosomal microtubules produced by the centromeres of the paired chromosomes extend toward the poles of the cell. As the pairs separate, one member of each pair migrates to opposite poles of the dividing nucleus. The random arrangement of chromosome pairs on the spindle is another reason for variation among humans. The cells formed by the first nuclear division (reduction division) are called *secondary spermatocytes.* Each cell has 23 chromosomes—the haploid number. Each chromosome of the secondary spermatocytes, however, is made up of two chromatids. Moreover, the genes of the chromosomes of secondary spermatocytes may be rearranged as a result of crossing-over.

● *Equatorial Division (Meiosis II)* The second nuclear division of meiosis is equatorial division. There is no replication of DNA. The chromosomes (each composed of two chromatids) line up in single file around the equatorial plane, and the chromatids of each chromosome separate from each other. The cells formed from the equatorial division are called *spermatids.* Each contains half the original chromosome number, or 23 chromosomes, and is haploid. Each primary spermatocyte therefore produces four spermatids by meiosis (reduction division and equatorial division). Spermatids lie close to the lumen of the seminiferous tubule.

During spermatogenesis, a very interesting and unique process occurs. As the sperm cells proliferate, they fail to complete cytoplasmic separation (cytokinesis) so that all the daughter cells, except for the least-differentiated sperma-

togonia, remain continuous via cytoplasmic bridges. These cytoplasmic bridges persist until development of the spermatozoa is complete, at which point they float out individually into the lumen of the seminiferous tubule. Thus, the offspring of an original spermatogonium remain in cytoplasmic communication through their entire development. This pattern of development undoubtedly accounts for the synchronized production of spermatozoa in any given area of a seminiferous tubule. This pattern may have survival value in that half the spermatozoa contain an X chromosome and half a Y chromosome. The X chromosome probably carries many essential genes that are lacking on the Y chromosome and, if it were not for the cytoplasmic bridges between the developing sperm, it may be that the Y-bearing spermatozoon could not survive, with the result that no males could be produced in the next generation.

● *Spermiogenesis* The final stage of spermatogenesis, called *spermiogenesis* (sper'-mē-ō-JEN-e-sis), involves the maturation of spermatids into spermatozoa. Each spermatid embeds in a sustentacular (Sertoli) cell and develops a head with an acrosome (described shortly) and a flagellum (tail). Sustentacular cells extend from the basement membrane to the lumen of the seminiferous tubule where they nourish the developing spermatids. Since there is no cell division in spermiogenesis, each spermatid develops into a single *spermatozoon (sperm cell).*

Spermatozoa enter the lumen of the seminiferous tubule and migrate to the ductus epididymis, where in 10–14 days they complete their maturation and become capable of fertilizing an ovum. Spermatozoa are also stored in the ductus (vas) deferens. Here, they can retain their fertility for up to several weeks.

Spermatozoa

Spermatozoa are produced or matured at the rate of about 300 million per day and, once ejaculated, have a life expectancy of about 48 hours within the female reproductive tract. A spermatozoon is highly adapted for reaching and penetrating a female ovum. It is composed of a head, a midpiece, and a tail (Figure 25-5). Within the *head* are the nuclear material and a dense granule called the *acrosome,* which develops from the Golgi complex and contains enzymes (hyaluronidase and proteinases) that facilitate penetration of the sperm cell into the ovum. The acrosome is basically a specialized lysosome. Numerous mitochondria in the *midpiece* carry on the metabolism that provides energy for locomotion. The *tail,* a typical flagellum, propels the sperm along its way.

DUCTS

Ducts of the Testis

Following their production, spermatozoa are moved through the convoluted seminiferous tubules to the *straight tubules* (see Figure 25-2b). The straight tubules lead to a

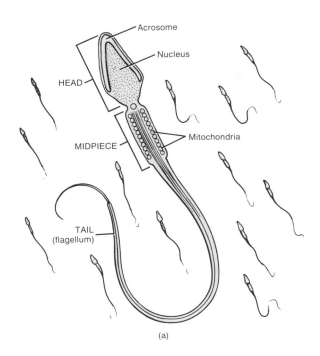

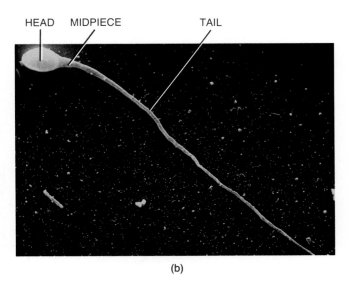

(b)

FIGURE 25-5 Spermatozoa. (a) Diagram of the parts of a spermatozoon. (b) Scanning electron micrograph of a spermatozoon at a magnification of 1100×. (Courtesy of Schatten/Science Photo Library/Photo Researchers.)

network of ducts in the testis called the *rete* (RĒ-tē) *testis.* Some of the cells lining the rete testis possess cilia that probably help move the sperm along. The sperm are next transported out of the testis.

Epididymis

The sperm are transported out of the testis through a series of coiled *efferent ducts* in the epididymis that empty into a single tube called the ductus epididymis. Morphological changes occur in the spermatozoa during their passage through the epididymis.

The *epididymis* (ep'-i-DID-i-mis) is a comma-shaped organ that lies along the posterior border of the testis (see Figures 25-1 and 25-2) and consists mostly of a tightly coiled tube, the *ductus epididymis.* The larger, superior portion of the epididymis is known as the *head.* In the head, the efferent ducts join the ductus epididymis. The *body* is the narrow, midportion of the epididymis. The *tail* is the smaller, inferior portion. At its distal end, the tail of the epididymis continues as the ductus (vas) deferens.

The ductus epididymis is a tightly coiled structure that would measure about 6 m (20 ft) in length and 1 mm in diameter if it were straightened out. The epididymis measures only about 3.8 cm (1.5 in.). The ductus epididymis is lined with pseudostratified columnar epithelium and encircled by layers of smooth muscle. The free surfaces of the columnar cells contain long, branching microvilli called *stereocilia* (Figure 25-6).

Functionally, the ductus epididymis is the site of sperm maturation. They require between 10 and 14 days to complete their maturation, that is, to become capable of fertilizing an ovum. The ductus epididymis also stores spermatozoa and propels them toward the urethra during ejaculation by peristaltic contraction of its smooth muscle. Spermatozoa may remain in storage in the ductus epididymis for up to

4 weeks. After that, they are expelled from the epididymis or reabsorbed.

Ductus (Vas) Deferens

Within the tail of the epididymis, the ductus epididymis becomes less convoluted, its diameter increases, and at this point it is referred to as the *ductus (vas) deferens* or *seminal duct* (see Figure 25-2). The ductus (vas) deferens, about 45 cm (18 in.) long, ascends along the posterior border of the testis, penetrates the inguinal canal, and enters the pelvic cavity, where it loops over the side and down the posterior surface of the urinary bladder (see Figure 25-1a). The dilated terminal portion of the ductus (vas) deferens is known as the *ampulla* (am-POOL-la). The ductus (vas) deferens is lined with pseudostratified epithelium and contains a heavy coat of three layers of muscle (Figure 25-7). Functionally, the ductus (vas) deferens stores sperm and conveys sperm from the epididymis toward the urethra during ejaculation by peristaltic contractions of the muscular coat.

CLINICAL APPLICATION

One method of sterilization of males is called *vasectomy,* a relatively uncomplicated procedure, typically performed under local anesthesia, in which a portion of each ductus (vas) deferens is removed. In the procedure, an incision is made in the scrotum, the ducts are located, each is tied in two places, and the portion between the ties is excised. Although sperm production continues in the testes, the sperm cannot reach the exterior because the ducts are cut, and the sperm degenerate and are destroyed by phagocytosis. Vasectomy has no effect on sexual desire and performance, and if performed correctly, it is virtually 100 percent effective.

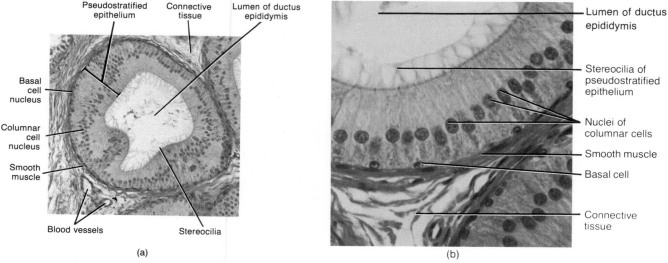

FIGURE 25-6 Histology of the ductus epididymis. (a) Photomicrograph of the ductus epididymis seen in cross section at a magnification of 160×. (b) Photomicrograph of an enlarged aspect of the mucosa of the ductus epididymis at a magnification of 350×. (Copyright © 1983 by Michael H. Ross. Used by permission.)

Traveling with the ductus (vas) deferens as it ascends in the scrotum are the testicular artery, autonomic nerves, veins that drain the testes (pampiniform plexus), lymphatics, and the cremaster muscle. These structures constitute the *spermatic cord*, a supporting structure of the male reproductive system (Figure 25-8a). The cremaster muscle, which also surrounds the testes, elevates the testes during sexual stimulation and exposure to cold. The spermatic cord and ilioinguinal nerve pass through the *inguinal* (IN-gwin-al) *canal* in the male (Figure 25-8b). The canal is an oblique passageway in the anterior abdominal wall just superior and parallel to the medial half of the inguinal ligament. The canal is about 4–5 cm (1.6–2.0 in.) in length. It originates at the *deep (abdominal) inguinal ring*, a slitlike opening in the aponeurosis of the transversus abdominis muscle. The canal terminates at the *superficial (subcutaneous) inguinal ring*, a somewhat triangular opening in the aponeurosis of the external oblique muscle. In the female, the round ligament of the uterus and ilioinguinal nerve pass through the inguinal canal.

CLINICAL APPLICATION

The inguinal region represents a weak area in the abdominal wall. It is frequently the site of an *inguinal hernia*— a rupture or separation of a portion of the abdominal wall resulting in the protrusion of a part of an organ. Inguinal hernias occur much less frequently in females.

Ejaculatory Duct

Posterior to the urinary bladder are the *ejaculatory* (e-JAK-yoo-la-tō'-rē) *ducts* (Figure 25-9). Each duct is about 2 cm (1 in.) long and is formed by the union of the

duct from the seminal vesicle and ductus (vas) deferens. The ejaculatory ducts eject spermatozoa into the prostatic urethra.

Urethra

The *urethra* is the terminal duct of the system, serving as a passageway for spermatozoa or urine. In the male, the urethra passes through the prostate gland, the urogenital diaphragm, and the penis. It measures about 20 cm (8 in.) in length and is subdivided into three parts (see Figures 25-1 and 25-9). The *prostatic urethra* is 2–3 cm (1 in.) long and passes through the prostate gland. It continues inferiorly, and as it passes through the urogenital diaphragm, a muscular partition between the two ischiopubic rami, it is known as the *membranous urethra*. The membranous portion is about 1 cm (0.5 in.) in length. As it passes through the corpus spongiosum of the penis, it is known as the *spongy (cavernous) urethra*. This portion is about 15 cm (6 in.) long. The spongy urethra enters the bulb of the penis and terminates at the external *urethral orifice*. The histology of the male urethra may be reviewed in Chapter 24.

ACCESSORY SEX GLANDS

Whereas the ducts of the male reproductive system store and transport sperm cells, the *accessory sex glands* secrete most of the liquid portion of semen. The paired *seminal vesicles* (VES-i-kuls) are convoluted pouchlike structures, about 5 cm (2 in.) in length, lying posterior to and at the base of the urinary bladder in front of the rectum (Figure 25-9). They secrete an alkaline, viscous fluid, rich in the sugar fructose, and pass it into the ejaculatory duct. This

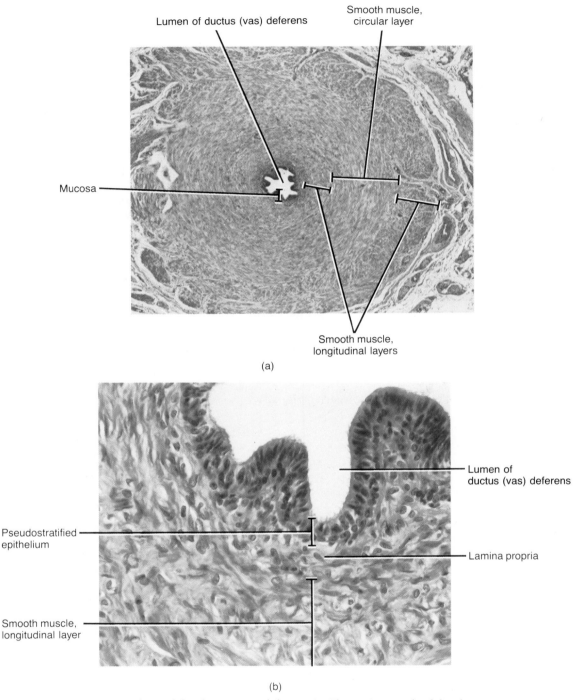

Lumen of ductus (vas) deferens

Smooth muscle, circular layer

Mucosa

Smooth muscle, longitudinal layers

(a)

Pseudostratified epithelium

Lumen of ductus (vas) deferens

Lamina propria

Smooth muscle, longitudinal layer

(b)

FIGURE 25-7 Histology of the ductus (vas) deferens. (a) Photomicrograph of the ductus (vas) deferens seen in cross section at a magnification of 40×. (b) Photomicrograph of an enlarged aspect of the mucosa of the ductus (vas) deferens at a magnification of 160×. (Copyright © 1983 by Michael H. Ross. Used by permission.)

secretion provides a carbohydrate (fructose) that is used as an energy source by sperm. It constitutes about 60 percent of the volume of semen.

The *prostate* (PROS-tāt) *gland* is a single, doughnut-shaped gland about the size of a chestnut (Figure 25-9). It is inferior to the urinary bladder and surrounds the superior portion of the urethra. The prostate secretes a slightly acid fluid rich in citric acid and acid phosphatase into the prostatic urethra through numerous prostatic ducts. The prostatic secretion constitutes 13–33 percent of the volume of semen and contributes to sperm motility and viability.

The paired *bulbourethral* (bul'-bō-yoo-RĒ-thral), or

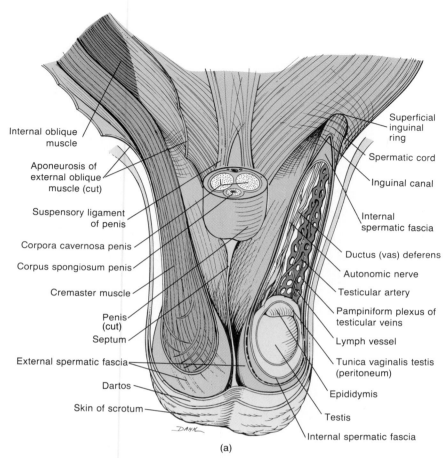

Internal oblique muscle

Aponeurosis of external oblique muscle (cut)

Suspensory ligament of penis

Corpora cavernosa penis

Corpus spongiosum penis

Cremaster muscle

Penis (cut)

Septum

External spermatic fascia

Dartos

Skin of scrotum

Superficial inguinal ring

Spermatic cord

Inguinal canal

Internal spermatic fascia

Ductus (vas) deferens

Autonomic nerve

Testicular artery

Pampiniform plexus of testicular veins

Lymph vessel

Tunica vaginalis testis (peritoneum)

Epididymis

Testis

Internal spermatic fascia

(a)

FIGURE 25-8 Spermatic cord and inguinal canal. (a) The left spermatic cord has been opened to expose its contents.

Cowper's, glands are about the size of peas. They are located beneath the prostate on either side of the membranous urethra within the urogenital diaphragm (Figure 25-9). The bulbourethral glands secrete an alkaline substance that protects sperm by neutralizing the acid environment of the urethra. Their ducts open into the spongy urethra.

SEMEN (SEMINAL FLUID)

Semen (*seminal fluid*) is a mixture of sperm and the secretions of the seminal vesicles, prostate gland, and bulbourethral glands. The average volume of semen for each ejaculation is 2.5–5 ml, and the average range of spermatozoa ejaculated is 50–150 million/ml. When the number of spermatozoa falls below 20 million/ml, the male is likely to be infertile. The very large number is required because only a small percentage eventually reach the ovum. And, although only a single spermatozoon fertilizes an ovum, fertilization seems to require the combined action at the ovum of a larger number of them. The intercellular material of the cells covering the ovum presents a barrier to the sperm. This barrier is digested by the hyaluronidase and proteinases secreted by the acrosomes of sperm, resulting

in the dispersion of the cells surrounding the ovaries. A single sperm does not produce enough of these enzymes to dissolve the barrier. A passageway through which one sperm may enter can be created only by the action of many sperm cells.

Semen has a slightly alkaline pH of 7.20–7.60. The prostatic secretion gives semen a milky appearance, and fluids from the seminal vesicles and bulbourethral glands give it a mucoid consistency. Semen provides spermatozoa with a transportation medium and nutrients. It neutralizes the acid environment of the male urethra and the female vagina. It also contains enzymes that activate sperm after ejaculation.

Semen contains an antibiotic, called *seminalplasmin,* that has the ability to destroy a number of bacteria. Its antimicrobial activity is similar to that of penicillin, streptomycin, and tetracyclines. Since both semen and the lower female reproductive tract contain bacteria, seminalplasmin may keep these bacteria under control to help ensure fertilization.

Once ejaculated into the vagina, liquid semen coagulates rapidly because of a clotting enzyme produced by the prostate that acts on a substance produced by the seminal vesicle.

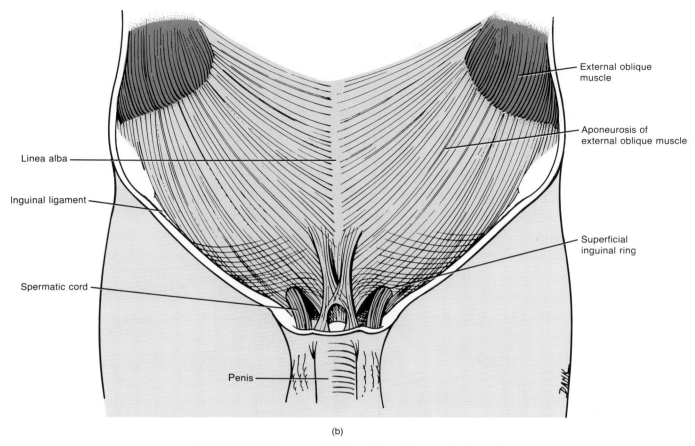

(b)

FIGURE 25-8 (*Continued*) (b) Location of the inguinal canal.

This clot liquefies in about 5–20 minutes because of another enzyme produced by the prostate gland. Abnormal or delayed liquefaction of coagulated semen may cause complete or partial immobilization of spermatozoa, thus inhibiting their movement through the cervix of the uterus.

CLINICAL APPLICATION

Semen analysis is the most valuable test in evaluating sterility. Among the criteria analyzed are the following:

1. *Volume.* A low volume might suggest an anatomical or functional defect or inflammation.
2. *Motility.* This refers to the percentage of motile spermatozoa (40–60 percent) and quality of movement (forward and progressive).
3. *Count.* Sperm counts below 20 million/ml could indicate sterility.
4. *Liquefaction.* Delayed liquefaction of more than 2 hours suggests inflammation of accessory sex glands or enzyme defects in the secretory products of the glands.
5. *Morphology.* No more than about 35 percent of spermatozoa should have abnormal morphology.

6. *Autoagglutination.* Agglutination does not occur normally.
7. *pH.* A rise in pH could indicate prostatitis.
8. *Fructose.* This sugar is present in a normal ejaculate. Its abscence indicates obstruction or congenital absence of the ejaculatory ducts or seminal vesicles.

A normal semen analysis does not guarantee fertility; the absence of spermatozoa or zero motility are the only definitive signs of sterility.

PENIS

The *penis* is used to introduce spermatozoa into the vagina (Figure 25-10). The penis is cylindrical in shape and consists of a body, root, and glans penis. The *body* of the penis is composed of three cylindrical masses of tissue, each bound by fibrous tissue (*tunica albuginea*). The two dorsolateral masses are called the *corpora cavernosa penis.* The smaller midventral mass, the *corpus spongiosum penis,* contains the spongy urethra. All three masses are enclosed by fascia and skin and consist of erectile tissue permeated by blood sinuses. Under the influence of sexual stimulation, the arter-

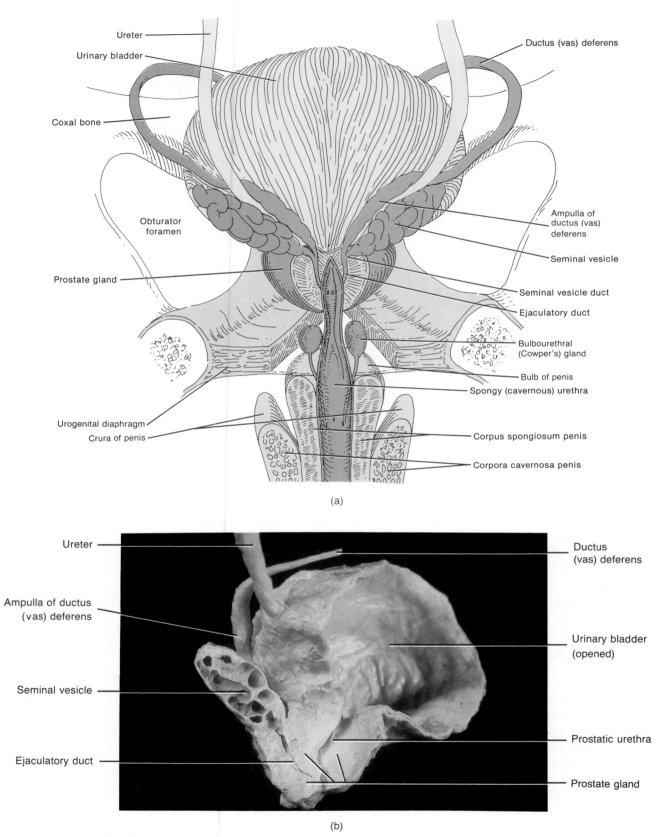

Ureter

Urinary bladder

Coxal bone

Obturator foramen

Prostate gland

Urogenital diaphragm

Crura of penis

Ductus (vas) deferens

Ampulla of ductus (vas) deferens

Seminal vesicle

Seminal vesicle duct

Ejaculatory duct

Bulbourethral (Cowper's) gland

Bulb of penis

Spongy (cavernous) urethra

Corpus spongiosum penis

Corpora cavernosa penis

(a)

Ureter

Ampulla of ductus (vas) deferens

Seminal vesicle

Ejaculatory duct

Ductus (vas) deferens

Urinary bladder (opened)

Prostatic urethra

Prostate gland

(b)

FIGURE 25-9 Male reproductive organs in relation to surrounding structures. (a) Diagram of posterior view. (b) Photograph of parasagittal section. (Courtesy of J. A. Gosling, P. F. Harris, et al., *Atlas of Human Anatomy,* Gower Medical Publishing Ltd., 1985.)

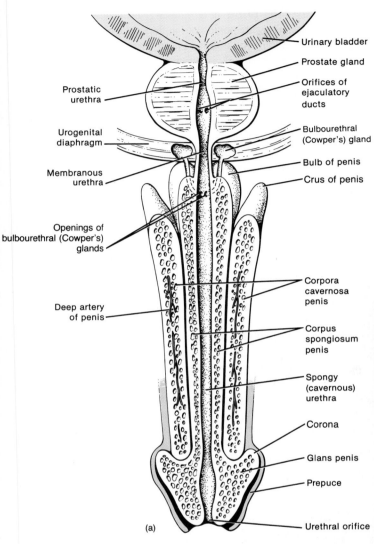

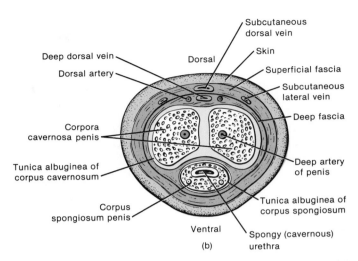

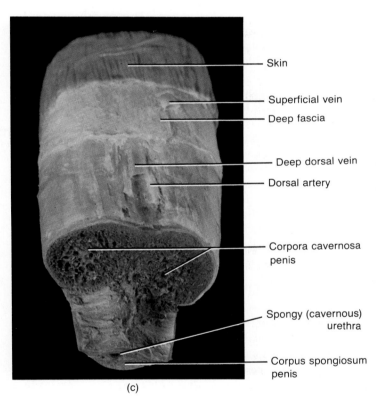

FIGURE 25-10 Internal structure of the penis. (a) Diagram of coronal section. (b) Diagram of cross section. (c) Photograph. (Courtesy of J. A. Gosling, P. F. Harris, et al., *Atlas of Human Anatomy,* Gower Medical Publishing Ltd., 1985.)

ies supplying the penis dilate, and large quantities of blood enter the blood sinuses. Expansion of these spaces compresses the veins draining the penis so most entering blood is retained. These vascular changes result in an *erection,* a parasympathetic reflex. The penis returns to its flaccid state when the arteries constrict and pressure on the veins is relieved. Details of erection are presented later in the chapter. During ejaculation, which is a sympathetic reflex, the smooth muscle sphincter at the base of the urinary bladder is closed. Thus, urine is not expelled during ejaculation, and semen does not enter the urinary bladder.

The *root* of the penis is the attached portion and consists of the *bulb of the penis,* the expanded portion of the base of the corpus spongiosum penis, and the *crura of the penis,* the separated and tapered portion of the corpora cavernosa

penis. The bulb of the penis is attached to the inferior surface of the urogenital diaphragm and enclosed by the bulbocavernosus muscle. Each crus (plural is *crura*) of the penis is attached to the ischial and pubic rami and surrounded by the ischiocavernosus muscle.

The distal end of the corpus spongiosum penis is a slightly enlarged region called the *glans penis,* which means shaped like an acorn. The margin of the glans penis is referred to as the *corona.* Covering the glans is the loosely fitting *prepuce* (PRE-pyoos), or *foreskin.*

CLINICAL APPLICATION

Circumcision (*circumcido* = to cut around) is a surgical procedure in which part or all of the prepuce is removed. It is usually performed in the delivery room or by the third or fourth day after birth (or on the eighth day as part of a Jewish religious rite). Both the American Academy of Pediatrics and the American College of Obstetricians and Gynecologists have concluded that there is no medical justification for circumcision. In recent years, the percentage of males undergoing circumcision is decreasing.

The penis has a very rich blood supply from the internal pudendal artery and the femoral artery. The veins drain into corresponding vessels.

The sensory nerves to the penis are branches from the pudendal and ilioinguinal nerves. The corpora have a sympathetic and parasympathetic supply. As a result of parasympathetic stimulation, the blood vessels dilate, increasing the flow of blood into the erectile tissue. The result is that blood is trapped within the penis and erection is maintained. At ejaculation, sympathetic stimulation causes the smooth muscle located in the walls of the ducts and accessory glands of the reproductive tract to contract and propel the sperm and secretions along their course. The musculature of the penis, which is supplied by the pudendal nerve, also contracts at ejaculation. The muscles include the bulbocavernosus muscle, which overlies the bulb of the penis, the ischiocavernosus muscles on either side of the penis, and the superficial transverse perineus muscles on either side of the bulb of the penis (see Figure 10-15 and 25-20).

FEMALE REPRODUCTIVE SYSTEM

The female organs of reproduction include the ovaries, that produce secondary oocytes (cells that develop into mature ova or eggs following fertilization) and the female sex hormones progesterone, estrogens, and relaxin; the uterine (Fallopian) tubes, that transport ova to the uterus (womb); the vagina; and external organs that constitute the vulva, or pudendum (Figure 25-11). The mammary glands also are considered part of the female reproductive system.

The specialized branch of medicine that deals with the diagnosis and treatment of diseases of the female reproductive system is called *gynecology* (gī'-ne-KOL-ō-jē; *gyneco* = woman).

OVARIES

The *ovaries,* or female gonads, are paired glands resembling unshelled almonds in size and shape. They are homologous to the testes. (*Homologous* means that two organs correspond in structure, position, and origin.) The ovaries descend to the brim of the pelvis during the third month of development. They are positioned in the upper pelvic cavity, one on each side of the uterus. The ovaries are maintained in position by a series of ligaments (Figure 25-12). They are attached to the broad ligament of the uterus, which is itself part of the parietal peritoneum, by a double-layered fold of peritoneum called the *mesovarium.* The ovaries are anchored to the uterus by the *ovarian ligament* and are attached to the pelvic wall by the *suspensory ligament.* Each ovary also contains a *hilus,* the point of entrance for blood vessels and nerves and along which the mesovarium is attached.

The microscope reveals that each ovary consists of the following parts (Figure 25-13).

1. *Germinal epithelium.* A layer of simple epithelium (low cuboidal or squamous) that covers the free surface of the ovary and is continuous with the mesothelium that covers the mesovarium. The term germinal epithelium is a misnomer since it does not give rise to ova, although at one time it was believed that it did. It is now known that the cells that give rise to ova arise from the endoderm of the yolk sac and migrate to the ovaries.
2. *Tunica albuginea.* A capsule of collagenous connective tissue immediately deep to the germinal epithelium.
3. *Stroma.* A region of connective tissue deep to the tunica albuginea and composed of an outer, dense layer called the *cortex* and an inner, loose layer known as the *medulla.* The cortex contains ovarian follicles.
4. *Ovarian follicles.* Oocytes (immature ova) and their surrounding tissues in various stages of development.
5. *Vesicular ovarian (Graafian) follicle.* A relatively large, fluid-filled follicle containing an immature ovum and its surrounding tissues. The follicle secretes hormones called estrogens.
6. *Corpus luteum.* Glandular body that develops from a vesicular ovarian follicle after extrusion of a secondary oocyte (potential mature ovum), a process known as ovulation. The corpus luteum produces the hormones progesterone (PROG), estrogens, and relaxin.

The ovaries produce secondary oocytes, discharge secondary oocytes (ovulation), and secrete the female sex hormones progesterone, estrogens, and relaxin.

The ovarian blood supply is furnished by the ovarian arteries, which anastomose with branches of the uterine arteries. The ovaries are drained by the ovarian veins. On the right side, they drain into the inferior vena cava, and on the left side, they drain into the renal vein.

Sympathetic and parasympathetic nerve fibers to the ovaries are said to terminate on the blood vessels and not enter the substance of the ovaries.

Oogenesis

The formation of haploid (*n*) ova in the ovary involves several phases, including meiosis, and is referred to as

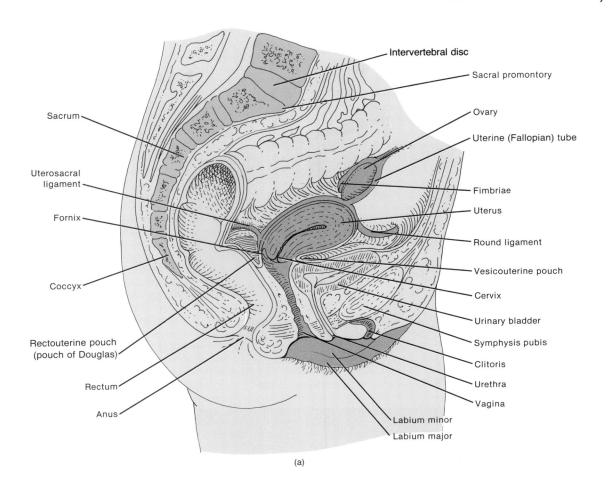

(a)

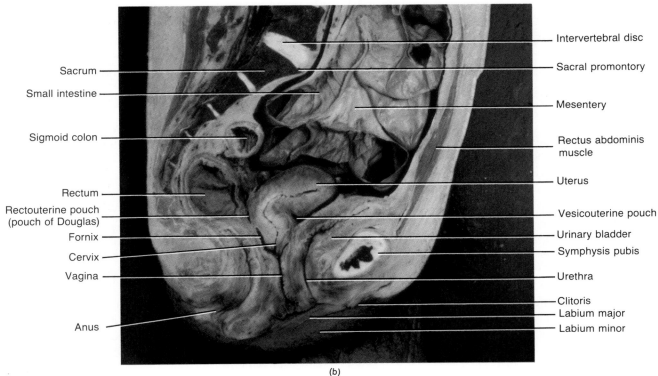

(b)

FIGURE 25-11 Female organs of reproduction and surrounding structures seen in sagittal section. (a) Diagram. (b) Photograph. (Courtesy of J. W. Rohen, C. Yokochi, *Photographic Anatomy of the Human Body*, F. K. Schattauer Verlagsgesellschaft mbH, Stuttgart–New York and Igaku-Shoin Ltd., New York–Tokyo, 1983.)

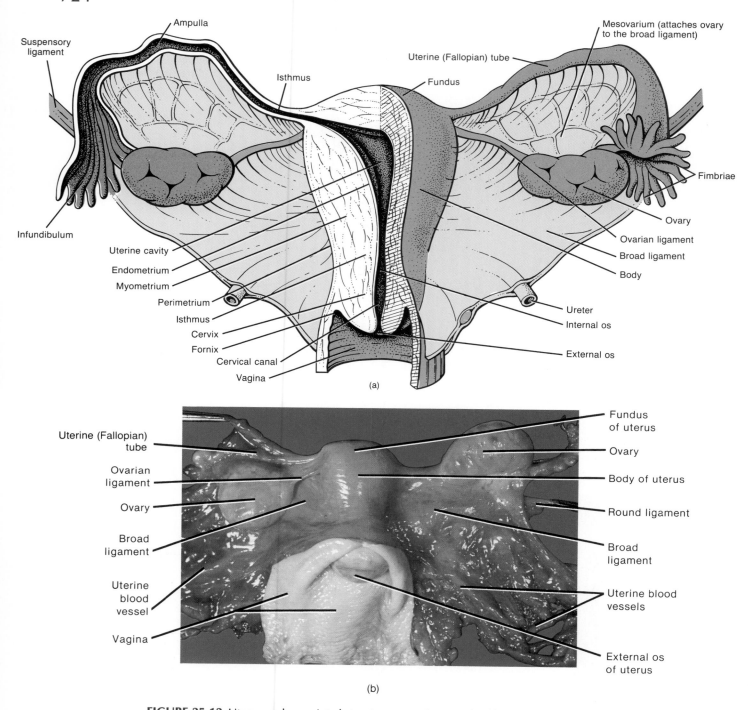

FIGURE 25-12 Uterus and associated structures seen in posterior view. (a) Diagram. The left side of the figure has been sectioned to show internal structures. (b) Photograph. (Courtesy of C. Yokochi, *Photographic Anatomy of the Human Body,* 1969, IGAKU-SHOIN, ltd., Tokyo, New York.)

oogenesis (ō'-ō-JEN-e-sis). With some exceptions, oogenesis occurs in essentially the same manner as spermatogenesis.

● *Reduction Division (Meiosis I)*   During early fetal development, primordial (primitive) germ cells migrate from the endoderm of the yolk sac to the ovaries. There germ cells differentiate within the ovaries into *oogonia* (ō'-o-

GŌ-nē-a), cells that can develop into ova (Figure 25-14). Oogonia are diploid (2*n*) cells that divide mitotically to produce a large population of cells. At about the third month of prenatal development, oogonia divide and develop into larger diploid (2*n*) cells called *primary oocytes* (Ō-o-sitz). These cells enter prophase of reduction division (meiosis I) but do not complete it until after the female reaches puberty. Each primary follicle is surrounded by a single

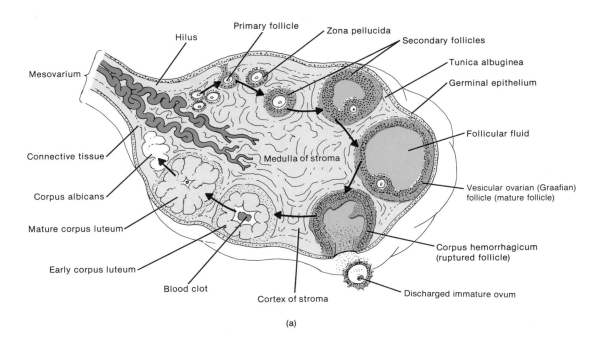

(a)

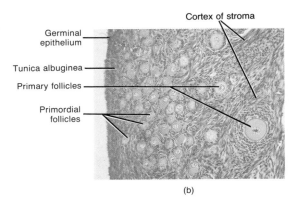

(b)

FIGURE 25-13 Histology of the ovary. (a) Diagram of the parts of an ovary seen in sectional view. The arrows indicate the sequence of developmental stages that occurs as part of the ovarian cycle. (b) Photomicrograph of the cortex of an ovary at a magnification of 60×. (c) Photomicrograph of an enlarged aspect of a secondary follicle at a magnification of 160×. The theca interna and theca externa are connective tissue coverings around the secondary follicle. (Photomicrographs copyright © 1983 by Michael H. Ross. Used by permission.)

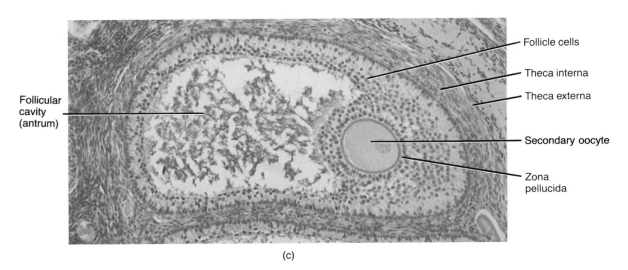

(c)

layer of flattened epithelial cells (follicular) and the entire structure is called a primary follicle. Primary follicles do not begin further development until they are stimulated by follicle-stimulating hormone (FSH) from the anterior pituitary gland, which, in turn, has responded to gonadotropin releasing hormone (GnRH) from the hypothalamus.

Starting with puberty, several primary follicles respond each month to the rising level of FSH. As the preovulatory

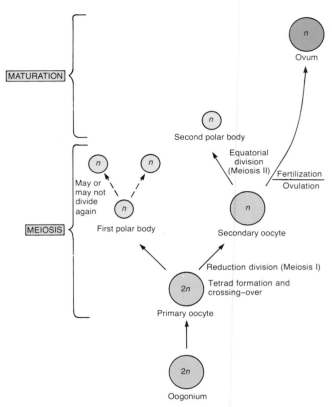

FIGURE 25-14 Oogenesis.

phase of the menstrual cycle proceeds and luteinizing hormone (LH) is secreted from the anterior pituitary, one of the primary follicles reaches a stage in which meiosis resumes and the diploid primary oocyte completes reduction division (meiosis I). Synapsis, tetrad formation, and crossing-over occur, and two cells of unequal size, both with 23 chromosomes (n) of two chromatids each, are produced. The smaller cell, called the *first polar body,* is essentially a packet of discarded nuclear material. The larger cell, known as the *secondary oocyte,* receives most of the cytoplasm. Each secondary oocyte is surrounded by several layers of cuboidal then columnar epithelial cells and the entire structure is called a secondary (growing) follicle. Once a secondary oocyte is formed, it proceeds to the metaphase of equatorial division (meiosis II) and then stops at this stage. The equatorial division (meiosis II) is completed following ovulation and fertilization.

● *Equatorial Division (Meiosis II)* At ovulation, the secondary oocyte with its polar body and some surrounding supporting cells is discharged. The discharged secondary oocyte enters the uterine (Fallopian) tube and, if spermatozoa are present and fertilization occurs, the second division, the equatorial division (meiosis II), is completed.

● *Maturation* The secondary oocyte produces two cells of unequal size, both of them haploid (n). The larger cell

eventually develops into an *ovum,* or mature egg; the smaller is the *second polar body.*

The first polar body may undergo another division to produce two polar bodies. If it does, meiosis of the primary oocyte results in a single haploid (n), secondary oocyte, and three haploid (n) polar bodies. In any event, all polar bodies disintegrate. Thus, each oogonium produces a single secondary oocyte whereas each spermatocyte produces four spermatozoa.

UTERINE (FALLOPIAN) TUBES

The female body contains two *uterine (Fallopian) tubes,* also called *oviducts,* that extend laterally from the uterus and transport the ova from the ovaries to the uterus (see Figure 25-12). Measuring about 10 cm (4 in.) long, the tubes are positioned between the folds of the broad ligaments of the uterus. The funnel-shaped open distal end of each tube, called the *infundibulum,* lies close to the ovary and is surrounded by a fringe of fingerlike projections called *fimbriae* (FIM-brē-ē). One fimbria is attached to the lateral end of the ovary. From the infundibulum, the uterine tube extends medially and inferiorly and attaches to the superior lateral angle of the uterus. The *ampulla* of the uterine tube is the widest, longest portion, making up about two-thirds of its length. The *isthmus* of the uterine tube is the short, narrow, thick-walled portion that joins the uterus.

Histologically, the uterine tubes are composed of three layers. The internal *mucosa* contains ciliated columnar cells and secretory cells, which are believed to aid the movement and nutrition of the ovum. The middle layer, the *muscularis,* is composed of a thick, circular region of smooth muscle and an outer, thin, longitudinal region of smooth muscle. Peristaltic contractions of the muscularis and the ciliated action of the mucosa help move the ovum down into the uterus. The outer layer of the uterine tubes is a serous membrane, the *serosa.*

About once a month a vesicular ovarian (Graafian) follicle (developed from a secondary follicle) ruptures, releasing a secondary oocyte, a process called *ovulation.* The oocyte is swept into the uterine tube by the ciliary action of the epithelium of the infundibulum which becomes associated with the surface of the most mature vesicular ovarian follicle just before ovulation occurs. The oocyte is then moved along the tube by ciliary action that is supplemented by the peristaltic contractions of the muscularis. If the oocyte is fertilized by a sperm cell, it usually occurs in the ampulla of the uterine tube. Fertilization may occur at any time up to about 24 hours following ovulation. With fertilization, the secondary oocyte completes meiosis II in which the oocyte produces a larger cell that develops into an ovum (mature egg) and a smaller second polar body. The fertilized ovum (zygote), now referred to as a blastocyst, descends into the uterus within 7 days. An unfertilized secondary oocyte disintegrates.

CLINICAL APPLICATION

Ectopic (*ektopos* = displaced) *pregnancy* (*EP*) refers to the development of an embryo or fetus outside the uterine cavity. The majority occur in the uterine (Fallopian) tube, usually in the ampullar and infundibular portions. Some occur in the ovaries, abdomen, uterine cervix, and broad ligaments. The basic cause of a tubal pregnancy is impaired passage of the fertilized ovum through the uterine tube related to factors such as pelvic inflammatory disease (PID), previous uterine tube surgery, previous ectopic pregnancy, repeated elective abortions, pelvic tumors, and developmental abnormalities. Ectopic pregnancy may be characterized by one or two missed periods, followed by vaginal bleeding and acute pelvic pain.

Unless removed or discharged from a uterine tube, the developing embryo can rupture the tube, often resulting in death. Standard practice for terminating an ectopic pregnancy in the uterine tube is to remove the tube, occasionally with its associated ovary. An alternative procedure involves injecting prostaglandin F into the uterine tube directly above the improperly implanted embryo. Prostaglandin F induces contractions that expel the embryo.

The uterine tubes are supplied by branches of the uterine and ovarian arteries. Venous return is via the uterine veins.

The uterine tubes are supplied with sympathetic and parasympathetic nerve fibers from the hypogastric plexus and the pelvic splanchnic nerves. The fibers are distributed to the muscular coat of the tubes and their blood vessels.

UTERUS

The site of menstruation, implantation of a fertilized ovum, development of the fetus during pregnancy, and labor is the *uterus.* Situated between the urinary bladder and the rectum, the uterus is shaped like an inverted pear (see Figures 25-11 and 25-12). Before the first pregnancy, the adult uterus measures approximately 7.5 cm (3 in.) long, 5 cm (2 in.) wide, and 2.5 cm (1 in.) thick.

Anatomical subdivisions of the uterus include the dome-shaped portion above the uterine tubes called the *fundus,* the major tapering central portion called the *body,* and the inferior narrow portion opening into the vagina called the *cervix.* Most of the lubrication during sexual intercourse is the result of fluids secreted by the glands of the cervix. Between the body and the cervix is the *isthmus* (IS-mus), a constricted region about 1 cm (0.5 in.) long. The interior of the body of the uterus is called the *uterine cavity,* and the interior of the narrow cervix is called the *cervical canal.* The junction of the isthmus with the cervical canal is the *internal os.* The *external os* is the place where the cervix opens into the vagina.

CLINICAL APPLICATION

Early diagnosis of *cancer of the uterus* is accomplished by the *Papanicolaou* (pap'-a-NIK-ō-la-oo) *test,* or Pap smear. In this generally painless procedure, a few cells from the part of the vagina surrounding the cervix and the cervix itself are removed with a swab and examined microscopically. Malignant cells have a characteristic appearance and indicate an early stage of cancer, even before symptoms occur. Estimates indicate that the Pap smear is more than 90 percent reliable in detecting cancer of the cervix. Females at risk of developing cancer of the uterus should undergo annual gynecological examinations that include a Pap smear. Females with a history of normal Pap smears and no risk factors probably should have a gynecological examination and Pap smear every 2–3 years until age 50 and annually thereafter. In all cases, the decision as to how frequently the test should be performed should be made by a physician in consultation with the patient.

To rule out invasive carcinoma, a *cone biopsy* of the cervix is performed. A cone biopsy is a hospital procedure in which an inverted cone of tissue is excised. It requires an anesthetic and is usually done only when abnormal cells have been detected. In another procedure, *punch biopsy* is combined with an *endocervical curettage* (ku-re-TAZH), or *ECC;* this combination has a high degree of diagnostic accuracy. In a punch biopsy, a disc or segment of tissue is excised. Curettage is a procedure in which the cervix is dilated and the endometrium (lining) of the uterus is scraped with a spoon-shaped instrument called a curette. This procedure is commonly called a *D and C.* If the carcinoma has spread beyond the lining, treatment may involve complete or partial removal of the uterus, called a *hysterectomy,* or radiation treatment.

Normally, the uterus is flexed between the uterine body and the cervix. This is called *anteflexion.* In this position, the body of the uterus projects anteriorly and slightly superiorly over the urinary bladder, and the cervix projects inferiorly and posteriorly and enters the anterior wall of the vagina at nearly a right angle. Several structures that are either extensions of the parietal peritoneum or fibromuscular cords, referred to as ligaments, maintain the position of the uterus. The paired *broad ligaments* are double folds of parietal peritoneum attaching the uterus to either side of the pelvic cavity. Uterine blood vessels and nerves pass through the broad ligaments. The paired *uterosacral ligaments,* also peritoneal extensions, lie on either side of the rectum and connect the uterus to the sacrum. The *cardinal (lateral cervical) ligaments* extend below the bases of the broad ligaments between the pelvic wall and the cervix and vagina. These ligaments contain smooth muscle, uterine blood vessels, and nerves and are the chief ligaments that maintain the position of the uterus and help keep it from dropping down into the vagina. The **round ligaments** are bands of fibrous connective tissue between the layers of the broad ligament. They extend from a point on the uterus just below the uterine (Fallopian) tubes to a portion of the labia majora of the external genitalia. Although the ligaments normally maintain the anteflexed position of the uterus, they also afford the uterine body some movement. As a result, the uterus may become malpositioned. A posterior tilting of the uterus is called **retroflexion.**

Histologically, the uterus consists of three layers of tissue. The outer layer, the **perimetrium (serosa)** is part of the visceral peritoneum. Laterally, it becomes the broad ligament. Anteriorly, it is reflected over the urinary bladder and forms a shallow pouch, the **vesicouterine** (ves'-i-kō-YOO-ter-in) **pouch** (see Figure 25-11). Posteriorly, it is reflected onto the rectum and forms a deep pouch, the **rectouterine** (rek-tō-YOO-ter-in) **pouch (pouch of Douglas)**—the lowest point in the pelvic cavity.

The middle layer of the uterus, the **myometrium,** forms the bulk of the uterine wall (Figure 25-15). This layer consists of three layers of smooth muscle fibers and is thickest in the fundus and thinnest in the cervix. During childbirth, coordinated contractions of the muscles help expel the fetus from the body of the uterus.

The inner layer of the uterus, the **endometrium,** is composed of (1) a surface layer of simple columnar epithelium (ciliated and secretory cells), (2) uterine (endometrial) glands that develop as invaginations of the surface epithelium, and (3) endometrial stroma, a very thick region of lamina propria (connective tissue). The endometrium is divided into two layers. The **stratum functionalis,** the layer closer to the uterine cavity, is shed during menstruation. The second layer, the **stratum basalis** (bā-SAL-is), is permanent. Its function is to produce a new functionalis following menstruation.

CLINICAL APPLICATION

More and more physicians are beginning to use **colposcopy** (kol-POS-ko-pē) to evaluate the status of the mucosa of the vagina and cervix. Colposcopy is the direct examination of the vaginal and cervical mucosa with a magnifying device (a colposcope) similar to a low-power binocular microscope. Various instruments that magnify the mucous membrane from about 6 to 40 times its actual size are commercially available. The application of a 3 percent solution of acetic acid removes mucus and enhances the appearance of mucosal columnar epithelium.

Blood is supplied to the uterus by branches of the internal iliac artery called **uterine arteries** (Figure 25-16). Branches called **arcuate arteries** are arranged in a circular fashion in the myometrium and give off **radial arteries** that penetrate deeply into the myometrium. Just before the branches enter

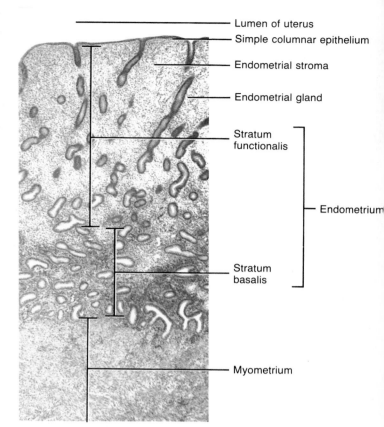

FIGURE 25-15 Histology of the uterus. Photomicrograph of a portion of the uterine wall in which the stratum functionalis is in an early stage of proliferation at a magnification of 25×. (Courtesy of Andrew Kuntzman.)

the endometrium, they divide into two kinds of arterioles. The **straight arterioles** terminate in the basalis and supply it with the materials necessary to regenerate the functionalis. The **spiral arterioles** penetrate the functionalis and change markedly during the menstrual cycle. The uterus is drained by the **uterine veins.**

Sympathetic and parasympathetic fibers are supplied to the uterus via the hypogastric and pelvic plexuses. Both sets of nerves terminate on the uterine vessels. The myometrium is believed to be innervated by the sympathetic fibers alone.

MENSTRUAL CYCLE

The **menstrual cycle** is a series of changes in the endometrium of a nonpregnant female. Each month the endometrium is prepared to receive an already fertilized ovum that eventually normally develops into an embryo and then into a fetus until delivery. If no fertilization occurs, the stratum functionalis portion of the endometrium is shed. The **ovarian cycle** is a monthly series of events associated with the maturation of an ovum.

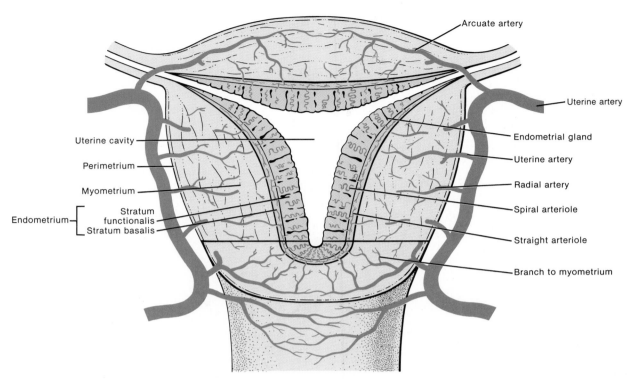

FIGURE 25-16 Blood supply of the uterus.

Hormonal Control

The menstrual cycle, ovarian cycle, and other changes associated with puberty in the female are controlled by a regulating factor from the hypothalamus called gonadotropin releasing hormone (GnRH). It GnRH stimulates the release of follicle-stimulating hormone (FSH) from the anterior pituitary. FSH stimulates the initial development of the ovarian follicles and the secretion of estrogens by the follicles. GnRH also stimulates the release of another anterior pituitary hormone—luteinizing hormone (LH), which stimulates the further development of ovarian follicles, brings about ovulation, and stimulates the production of estrogens, progesterone, and relaxin by ovarian cells.

Estrogens, the hormones of growth, have three main functions. First is the development and maintenance of female reproductive structures, especially the endometrial lining of the uterus, secondary sex characteristics, and the breasts. The secondary sex characteristics include fat distribution to the breasts, abdomen, mons pubis, and hips; voice pitch; broad pelvis; and hair pattern. Second, they control fluid and electrolyte balance. Third, they increase protein anabolism. High levels of estrogens in the blood inhibit the release of GnRH by the hypothalamus, which in turn inhibits the secretion of FSH by the anterior pituitary gland. This inhibition provides the basis for the action of one kind of contraceptive pill.

Progesterone (PROG), the hormone of maturation, works with estrogens to prepare the endometrium for implantation of a fertilized ovum and the mammary glands for milk secretion. High levels of progesterone also inhibit GnRH and prolactin (PRL). Progesterone, like estrogens, is synthesized in the ovaries.

Relaxin goes into operation near the end of pregnancy. It relaxes the symphysis pubis and helps dilate the uterine cervix to facilitate delivery. The hormone also plays a role in increasing sperm motility.

Menstrual Phase (Menstruation)

The duration of the menstrual cycle ranges from 24 to 35 days. For this discussion, we shall assume an average duration of 28 days. Events occurring during the menstrual cycle may be divided into three phases: the menstrual phase, preovulatory phase, and postovulatory phase (Figure 25-17).

The *menstrual phase,* also called *menstruation* or the *menses,* is the periodic discharge of 25–65 ml of blood, tissue fluid, mucus, and epithelial cells. It is caused by a sudden reduction in estrogens and progesterone and lasts for approximately the first 5 days of the cycle. The discharge is associated with endometrial changes in which the stratum functionalis layer degenerates and patchy areas of bleeding develop. Small areas of the stratum functionalis detach one at a time (total detachment would result in hemorrhage), the uterine glands discharge their contents and collapse, and tissue fluid is discharged. The menstrual flow passes

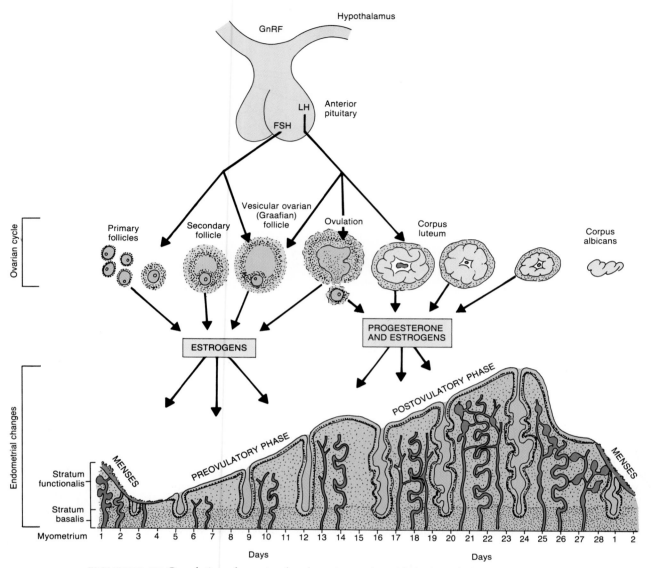

FIGURE 25-17 Correlation of menstrual and ovarian cycles with the hypothalamic and anterior pituitary gland hormones. In the cycle shown, fertilization and implantation have not occurred.

from the uterine cavity to the cervix and through the vagina to the exterior. Generally, the flow terminates by the fifth day of the cycle. At this time the entire stratum functionalis has been shed and the endometrium is very thin because only the stratum basalis remains.

During the menstrual phase, the ovarian cycle is also in operation. Ovarian follicles, called **primary follicles,** begin their development. At birth, each ovary contains about 200,000 such follicles, each consisting of a primary oocyte (potential ovum) surrounded by a single flattened layer of epithelial (follicular) cells. During the early part of each menstrual phase, 20–25 primary follicles start to produce very low levels of estrogens. Toward the end of the menstrual phase (days 4–5) about 20 of the primary follicles develop into **secondary (growing) follicles.** A secondary follicle

consists of a secondary oocyte and several layers of cells formed by division of the single layer of epithelial cells around a primary follicle. The epithelial cells of the secondary follicle are cuboidal and later columnar and are called granular cells. As a secondary follicle continues to grow, it forms a clear glycoprotein layer between the secondary oocyte and granular cells called the **zona pellucida** (pe-LOO-si-da). Also, the granular cells secrete follicular (fō-LIK-yoo-lar) fluid that forces the secondary oocyte to the edge of the secondary follicle and fills the follicular cavity or antrum (see Figure 25-13b). The production of estrogens by the secondary follicles elevates the level of estrogens in the blood slightly. Ovarian follicle development is the result of GnRH secretion by the hypothalamus, which in turn stimulates FSH production by the anterior pituitary.

During this part of the cycle, FSH secretion is relatively high. Although a number of follicles begin development each cycle, only one attains maturity. The others undergo atresia (death).

Preovulatory Phase

The *preovulatory phase,* the second phase of the menstrual cycle, is the time between menstruation and ovulation. This phase of the menstrual cycle is more variable in length than the other phases. It lasts from days 6 to 13 in a 28-day cycle.

FSH and LH stimulate the ovarian follicles to produce more estrogens, and this increase in estrogens stimulates the repair of the endometrium. Cells of the stratum basalis undergo mitosis and produce a new stratum functionalis. As the endometrium thickens, the short, straight endometrial glands develop and the arterioles coil and lengthen as they penetrate the stratum functionalis. The thickness of the endometrium approximately doubles to about 4–6 mm. Because of the proliferation of endometrial cells, the preovulatory phase is also termed the *proliferative phase.* Still another name is the *follicular phase* because of increasing secretion of estrogens by the developing follicle. Functionally, estrogens are the dominant ovarian hormones during this phase of the menstrual cycle.

During the preovulatory phase, one of the secondary follicles in the ovary matures into a *vesicular ovarian (Graafian) follicle* or *mature follicle,* a follicle ready for ovulation. This follicle produces a bulge on the surface of the ovary. During the maturation process, the follicle increases its estrogen production. Early in the preovulatory phase, FSH is the dominant hormone of the anterior pituitary, but close to the time of ovulation, LH is secreted in increasing quantities. Moreover, small amounts of progesterone may be produced by the vesicular ovarian (Graafian) follicle a day or two before ovulation.

Ovulation

Ovulation, the rupture of the vesicular ovarian (Graafian) follicle with release of the secondary oocyte into the pelvic cavity, usually occurs on day 14 in a 28-day cycle. During ovulation, the secondary oocyte remains surrounded by its zona pellucida and a covering of follicle cells directly around it. These cells are referred to as the corona radiata. It generally takes 10–14 days for a primary follicle to develop into a vesicular ovarian (Graafian) follicle and it is during this time that the developing ovum completes reduction division (meiosis I) and reaches metaphase of equatorial division (meiosis II). The developing ovum is in this stage when it is discharged during ovulation.

Just prior to ovulation, the high level of estrogens that developed during the preovulatory phase inhibits GnRH production by the hypothalamus. This, in turn, inhibits FSH secretion by the anterior pituitary via a negative feedback cycle. Concurrently, the high level of estrogens acts in a positive feedback cycle to cause the anterior pituitary to release a surge of LH. Without this surge of LH, ovulation will not occur. (An over-the-counter home test that detects the LH surge associated with ovulation is now available. The test predicts ovulation a day in advance.) Following ovulation, the vesicular ovarian (Graafian) follicle collapses and blood within it forms a clot called the *corpus hemorrhagicum.* The clot is eventually absorbed by the remaining follicular cells. In time, the follicular cells enlarge, change character, and form the *corpus luteum,* or yellow body.

Postovulatory Phase

The *postovulatory phase* of the menstrual cycle is the most constant in duration and lasts from days 15 to 28 in a 28-day cycle. It represents the time between ovulation and the onset of the next menses. Following ovulation, LH secretion stimulates the development of the corpus luteum. The corpus luteum then secretes increasing quantities of estrogens and progesterone. Progesterone is responsible for preparing the endometrium to receive a fertilized ovum. Preparatory activities include secretory activity of the endometrial glands that causes them to appear tortuously coiled, vascularization of the superficial endometrium, thickening of the endometrium, glycogen storage, and an increase in the amount of tissue fluid. These preparatory changes are maximal about 1 week after ovulation, and they correspond to the anticipated arrival of the fertilized ovum. During the postovulatory phase, FSH secretion again gradually increases and LH secretion decreases. The functionally dominant ovarian hormone during this phase is progesterone. The relation of progesterone to prostaglandins in causing painful menstruation will be considered at the end of the chapter.

If fertilization and implantation do not occur, the rising levels of progesterone and estrogens from the corpus luteum inhibit GnRH and LH secretion. As a result, the corpus luteum degenerates and becomes the *corpus albicans,* or white body. The decreased secretion of progesterone and estrogens by the degenerating corpus luteum then initiates another menstrual period. In addition, the decreased levels of progesterone and estrogens in the blood bring about a new output of the anterior pituitary hormones—especially FSH in response to an increased output of GnRH by the hypothalamus. Thus, a new ovarian cycle is initiated.

If, however, fertilization and implantation do occur, the corpus luteum is maintained for 3–4 months into the pregnancy, during which time it secretes estrogens and progesterone. The corpus luteum is maintained by *human chorionic* (kō-rē-ON-ik) *gonadotropin (HCG),* a hormone produced by the developing placenta. The placenta itself secretes estrogens to support pregnancy and progesterone to support pregnancy and breast development for lactation. Once the placenta begins its secretion, the role of the corpus luteum becomes minor.

VAGINA

The *vagina* serves as a passageway for the menstrual flow. It is also the receptacle for the penis during coitus, or sexual intercourse, and the lower portion of the birth canal. It is a muscular, tubular organ lined with mucous membrane and measures about 10 cm (4 in.) in length, extending from the cervix to the vestibule (see Figures 25-11 and 25-12). Situated between the urinary bladder and the rectum, it is directed superiorly and posteriorly, where it attaches to the uterus. A recess, called the *fornix,* surrounds the vaginal attachment to the cervix. The fornix makes possible the use of contraceptive diaphragms.

Histologically, the mucosa of the vagina is continuous with that of the uterus and consists of stratified squamous epithelium and connective tissue that lies in a series of transverse folds, the *rugae* (Figure 25-18). The muscularis is composed of smooth muscle that can stretch considerably. This distension is important because the vagina receives the penis during sexual intercourse and serves as the lower portion of the birth canal. At the lower end of the vaginal opening, the *vaginal orifice,* there may be a thin fold of vascularized mucous membrane called the *hymen,* which forms a border around the orifice, partially closing it (see Figure 25-19).

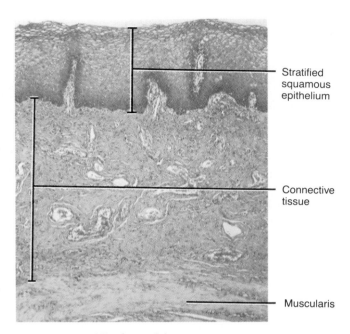

FIGURE 25-18 Histology of the vagina. Photomicrograph of a portion of the wall of the vagina at a magnification of 50×.

Labels on figure:
- Stratified squamous epithelium
- Connective tissue
- Muscularis

CLINICAL APPLICATION

Sometimes the hymen completely covers the orifice, a condition called *imperforate* (im-PER-fō-rāt) *hymen.* Surgery is required to open the orifice to permit the discharge of the menstrual flow.

The mucosa of the vagina contains large amounts of glycogen, which upon decomposition produces organic acids. These acids create a low pH environment that retards microbial growth. However, the acidity is also injurious to sperm cells. Semen neutralizes the acidity of the vagina to ensure survival of the sperm.

VULVA

The term *vulva* (VUL-va), or *pudendum* (pyoo-DEN-dum), is a collective designation for the external genitalia of the female (Figure 25-19). Its components are as follows.

The *mons pubis,* an elevation of adipose tissue covered by skin and coarse pubic hair, is situated over the symphysis pubis. It lies anterior to the vaginal and urethral openings. From the mons pubis, two longitudinal folds of skin, the *labia majora* (LĀ-bē-a ma-JŌ-ra), extend inferiorly and posteriorly. The labia majora are homologous to the scrotum. The labia majora contain an abundance of adipose tissue and sebaceous (oil) and sudoriferous (sweat) glands; they are covered by pubic hair. Medial to the labia majora are two folds of skin called the *labia minor* (MĪ-nō-ra). Unlike the labia majora, the labia minora are devoid of pubic hair

and fat and have few sudoriferous (sweat) glands. They do, however, contain numerous sebaceous (oil) glands.

The *clitoris* (KLI-to-ris) is a richly innervated structure that contains a small, cylindrical mass of erectile tissue and nerves. It is located at the anterior junction of the labia minora. A layer of skin called the *prepuce* (foreskin) is formed at the point where the labia minora unite and covers the body of the clitoris. The exposed portion of the clitoris is the *glans.* The clitoris is homologous to the glans penis of the male. Like the penis, the clitoris is capable of enlargement upon tactile stimulation and assumes a role in sexual excitement of the female.

The cleft between the labia minora is called the *vestibule.* Within the vestibule are the hymen (if present), vaginal orifice, urethral orifice, and the openings of several ducts. The *vaginal orifice,* the opening of the vagina to the exterior, occupies the greater portion of the vestibule and is bordered by the hymen. The *bulb of the vestibule* consists of two elongated masses of erectile tissue just deep to the labia on either side of the vaginal orifice. The bulb becomes engorged with blood during sexual arousal, narrowing the vaginal orifice and placing pressure on the penis during intercourse. The bulb is homologous to the corpus spongiosum penis and bulb of the penis. Anterior to the vaginal orifice and posterior to the clitoris is the *urethral orifice,* the opening of the urethra to the exterior. On either side of the urethral orifice are the openings of the ducts of the *paraurethral (Skene's) glands.* These glands are embedded in the wall of the urethra and secrete mucus. The paraurethral glands are homologous to the male prostate. On either side of the vaginal orifice itself are the *greater vestibular (Bar-*

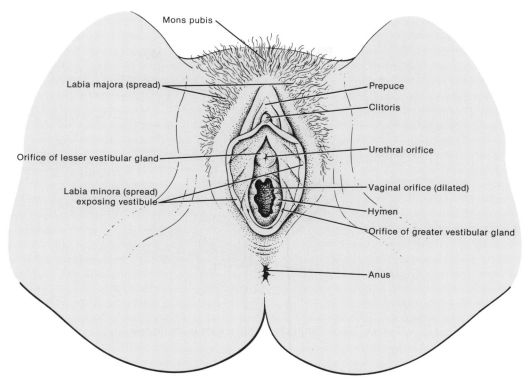

Mons pubis

Labia majora (spread)

Orifice of lesser vestibular gland

Labia minora (spread)
exposing vestibule

Prepuce

Clitoris

Urethral orifice

Vaginal orifice (dilated)

Hymen

Orifice of greater vestibular gland

Anus

FIGURE 25-19 Vulva.

tholin's) glands. These glands open by ducts into a groove between the hymen and labia minora and produce a mucoid secretion that supplements lubrication during sexual intercourse. The greater vestibular glands are homologous to the male bulbourethral (Cowper's) glands. A number of *lesser vestibular glands,* whose orifices are microscopic, open into the vestibule.

CLINICAL APPLICATION

One important sign in the *diagnosis of pregnancy* is a bluish discoloration of the vulva and vagina due to venous congestion. The discoloration appears at about the eighth to twelfth week and increases in intensity as the pregnancy progresses.

PERINEUM

The *perineum* (per'-i-NĒ-um) is the diamond-shaped area between the thighs and buttocks of both males and females. It is bounded anteriorly by the symphysis pubis, laterally by the ischial tuberosities, and posteriorly by the coccyx. A transverse line drawn between the ischial tuberosities divides the perineum into an anterior *urogenital* (yoo'-rō-JEN-i-tal) *triangle* that contains the external genitalia and a posterior *anal triangle* that contains the anus (Figure 25-20).

CLINICAL APPLICATION

In the female, the region between the vagina and anus is known as the *clinical perineum.* If the vagina is too small to accommodate the head of an emerging fetus, the skin, vaginal epithelium, subcutaneous fat, and superficial transverse perineal muscle of the clinical perineum may tear. Moreover, the tissues of the rectum may be damaged. To avoid this, a small incision called an *episiotomy* (e-piz'-ē-OT-ō-mē) is made in the perineal skin and underlying tissues just prior to delivery. After delivery the episiotomy is sutured in layers.

MAMMARY GLANDS

The *mammary glands* are modified sudoriferous (sweat) glands (branched tubuloalveolar) that lie over the pectoralis major and serratus anterior muscles and are attached to them by a layer of connective tissue (Figure 25-21). Internally, each mammary gland consists of 15–20 *lobes,* or compartments, separated by adipose tissue. The amount of adipose tissue determines the size of the breasts. However, breast size has nothing to do with the amount of milk produced. In each lobe are several smaller compartments called *lobules,* composed of connective tissue in which milk-secreting cells referred to as *alveoli* are embedded (Figure 25-22). Alveoli are arranged in grapelike clusters. Between

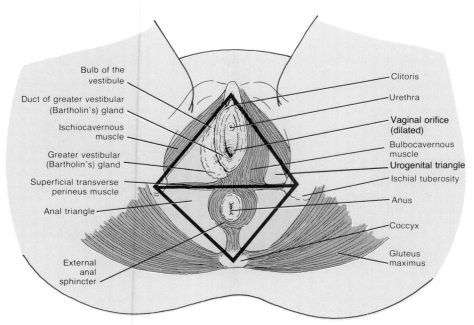

FIGURE 25-20 Perineum. Borders seen in the female.

the lobules are strands of connective tissue called the ***suspensory ligaments of the breast (Cooper's ligaments)***. These ligaments run between the skin and deep fascia and support the breast. Alveoli convey the milk into a series of ***secondary tubules***. From here the milk passes into the ***mammary ducts***. As the mammary ducts approach the nipple, they expand to form sinuses called ***lactiferous sinuses***, where milk may be stored. The sinuses continue as ***lactiferous ducts*** that terminate in the ***nipple***. Each lactiferous duct conveys milk from one of the lobes to the exterior, although some may join before reaching the surface. The circular pigmented area of skin surrounding the nipple is called the ***areola*** (a-RĒ-ō-la). It appears rough because it contains modified sebaceous (oil) glands.

The essential function of the mammary glands is milk secretion and ejection, together called ***lactation***. The secretion of milk is due largely to the hormone prolactin (PRL), with contributions from progesterone and estrogens. The ejection of milk occurs in the presence of oxytocin (OT), which is released from the posterior pituitary gland in response to sucking.

CLINICAL APPLICATION

Early detection—especially by breast self-examination (BSE) and mammography—is still the most promising method to increase the survival rate for ***breast cancer***. It is estimated that 95 percent of breast cancer is first detected by women themselves. Each month after the menstrual period the breasts should be thoroughly examined for lumps, puckering of the skin, nipple retraction, or nipple discharge.

The most effective screening technique for routinely detecting tumors less than 1.27 cm (0.5 in.) in diameter is ***mammography***. One reason that mammography is so useful is that it can detect small calcium deposits, called microcalcifications, in breast tissue. Such calcifications frequently indicate the presence of a tumor. The mammographic image, called a ***mammogram***, is obtained by placing the breasts, one at a time, on a flat surface and using a compressor to smooth the breast for better imaging. Usually, two images are taken of each breast, one from the top and one from the side. There are two basic types of mammography: xeromammography and film-screen mammography. In ***xeromammography***, x rays are beamed onto a specially coated metal plate and the blue-on-white image produced provides the physician with detail of thicker as well as thinner portions of the breast. In ***film-screen mammography***, x rays are beamed onto a fluorescent screen and the image is produced on x-ray film.

One of the most recent breast cancer detecting procedures is ***ultrasound (US)***. The procedure is performed while the patient lies on her stomach in a specially designed hospital bed with her breasts immersed in a tank of water. Ultrasound produces images using a device that first emits a pulse of high-frequency sound and then records the echo on a monitor. Although ultrasound can neither detect microcalcifications or tumors less than 1 cm in diameter, it can be used to determine whether a lump is a benign cyst or a malignant tumor.

Another technique combines computed tomography with mammography (***CT/M***) and appears to overcome some of

Figure labels

Bulb of the vestibule
Duct of greater vestibular (Bartholin's) gland
Ischiocavernous muscle
Greater vestibular (Bartholin's) gland
Superficial transverse perineus muscle
Anal triangle
External anal sphincter
Clitoris
Urethra
Vaginal orifice (dilated)
Bulbocavernous muscle
Urogenital triangle
Ischial tuberosity
Anus
Coccyx
Gluteus maximus

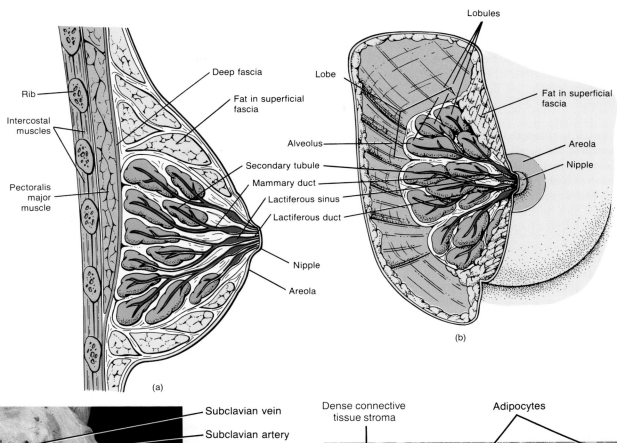

(a)

(b)

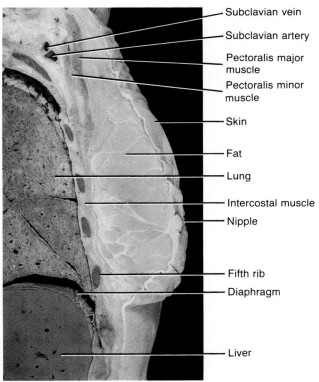

(c)

FIGURE 25-21 Mammary glands. (a) Diagram in sagittal section. (b) Diagram in anterior view, partially sectioned. (c) Photograph in sagittal section. (Courtesy of J. A. Gosling, P. F. Harris, et al., *Atlas of Human Anatomy,* Gower Medical Publishing Ltd., 1985.)

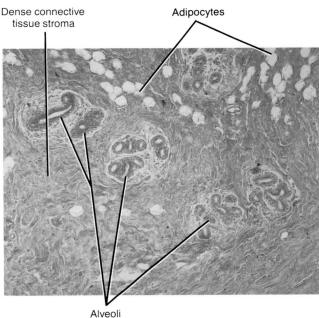

FIGURE 25-22 Histology of the mammary glands. Photomicrograph of several alveoli in a nonlactating mammary gland at a magnification of 60×. (Copyright © 1983 by Michael H. Ross. Used by permission.)

the limitations of mammography. The procedure is based on the fact that breast carcinoma has an abnormal affinity for iodide. CT scans are made before and after the rapid intravenous infusion of an iodide contrast material. Comparison of the initial density of a suspected lesion with the density following infusion of the iodide gives an indication of the status of the tumor. CT/M affords definitive diagnostic help in instances where the mammographic and physical examinations are inconclusive and appears to be a significantly improved method of breast cancer diagnosis.

Long-term studies are now underway to determine the effectiveness of an experimental home device for breast cancer detection. It is called a **breast cancer screening indicator.** It is composed of two plastic discs that contain heat-sensitive chemicals. The discs are worn inside a female's brassiere for 15 minutes each month so that she can observe color changes that could indicate a breast abnormality. The screening device is said to detect tumors as small as 2 mm in diameter, their size about 2–10 years before they become palpable or can be seen by mammography.

SEXUAL INTERCOURSE

Sexual intercourse, or *copulation* (in humans, called *coitus*), is the process by which spermatozoa are deposited in the vagina.

CLINICAL APPLICATION

Artificial insemination (in = into; *seminatus* = sown; *semen* = seed) refers to the deposition of seminal fluid within the vagina or cervix by artificial means. If the husband's seminal fluid is used, the process is known as *homologous insemination.* Such a procedure might be carried out if there is a developmental anomaly that prevents placement of the penis in the vagina or normal ejaculation. A small volume of seminal fluid is also an indication for homologous insemination. If seminal fluid from a donor is used, the process is called *heterologous insemination.* In such cases, an anonymous donor is selected on the basis of race, blood type compatibility, physical appearance, general health, and genetic background.

It has been documented that hepatitis B can be transmitted by semen donors to artificially inseminated females. In order to decrease the risk of passing on sexually transmitted diseases, guidelines have been established for screening semen donors.

MALE SEXUAL ACT

Erection

The male role in the sexual act starts with *erection,* the enlargement and stiffening of the penis. An erection may be initiated in the cerebrum by stimuli such as anticipation, memory, and visual sensation, or it may be a reflex brought on by stimulation of the touch receptors in the penis, especially in the glans. In any case, parasympathetic impulses that pass from the sacral portion of the spinal cord to the penis cause dilation of the arteries of the penis, allowing blood to fill the cavernous spaces of the spongy bodies.

Lubrication

Parasympathetic impulses from the sacral cord also cause the bulbourethral (Cowper's) glands and urethral glands (glands of Littré) to secrete mucus, which affords only a small amount of *lubrication* for intercourse. The mucus flows through the urethra. The major portion of lubricating fluid is produced by the cervical mucosa of the female. Without satisfactory lubrication, the male sexual act is difficult since unlubricated intercourse causes pain impulses that inhibit rather than promote coitus.

Orgasm

Tactile stimulation of the penis brings about emission and ejaculation. When sexual stimulation becomes intense, rhythmic sympathetic nerve impulses leave the spinal cord at the levels of the first and second lumbar vertebrae and pass to the genital organs. These impulses cause peristaltic contractions of the ducts in the testes, epididymides, and ductus (vas) deferens that propel spermatozoa into the urethra—a process called *emission.* Simultaneously, peristaltic contractions of the seminal vesicles and prostate gland expel seminal and prostatic fluid along with the spermatozoa. All these mix with the mucus of the bulbourethral glands, resulting in the fluid called semen. Other rhythmic impulses sent from the spinal cord at the levels of the first and second sacral vertebrae reach the skeletal muscles at the base of the penis, and the penis expels the semen from the urethra to the exterior. The propulsion of semen from the urethra to the exterior constitutes an *ejaculation.* A number of sensory and motor activities accompany ejaculation, including a rapid heart rate, an increase in blood pressure, an increase in respiration, and pleasurable sensations. These activities, together with the muscular events involved in ejaculation, are referred to as an *orgasm.*

During sexual intercourse, ejaculation introduces millions of sperm into the vagina. Next, the sperm move into the cervix where muscular contractions aid their movement into the uterus. Inside the uterus, rhythmic contractions of the muscular wall greatly aid the sperm as they swim toward the uterine (Fallopian) tubes. There is some evidence that suggests the hormone oxytocin (OT) may be released from the posterior pituitary during orgasm and/or that prostaglandins in seminal fluid function to cause the uterine contractions. Of the total number of spermatozoa that enter the vagina, less than 1 percent come in proximity to the ovum.

FEMALE SEXUAL ACT

Erection

The female role in the sex act, like that of the male, also involves erection, lubrication, and orgasm. Stimulation of the female, as in the male, depends on both psychic and tactile responses. Under appropriate conditions, stimulation of the female genitalia, especially the clitoris, results in erection and widespread sexual arousal. This response is controlled by parasympathetic nerve impulses sent from the sacral spinal cord to the external genitalia.

Lubrication

Parasympathetic impulses from the sacral spinal cord also cause the bulk of *lubrication* of the vagina. The impulses result in the secretion of a mucoid fluid from the epithelium of the cervical mucosa. Some mucus is also produced by the greater vestibular (Bartholin's) glands. As was noted earlier, lack of sufficient lubrication results in pain impulses that inhibit rather than promote coitus.

Orgasm (Climax)

When tactile stimulation of the genitalia reaches maximum intensity, reflexes are initiated that cause the female *orgasm* (*climax*). Female orgasm is analogous to male ejaculation. The perineal muscles contract rhythmically from spinal reflexes similar to those that occur in the male ejaculation.

BIRTH CONTROL (BC)

Methods of *birth control* (*BC*) include removal of the gonads and uterus, sterilization, and mechanical and chemical contraception.

REMOVAL OF GONADS AND UTERUS

Castration (removal of the testes), *hysterectomy* (removal of the uterus), and *oophorectomy* (ō'-of-ō-REK-tō-mē; removal of the ovaries) are all absolute preventive methods. Once performed, these operations cannot be reversed, and it is impossible to produce offspring. However, removal of the testes or ovaries has adverse effects because of the importance of these organs in the endocrine system. Generally, these operations are performed only if the organs are diseased. Castration before puberty prevents the development of secondary sex characteristics.

STERILIZATION

One means of *sterilization* of males is *vasectomy* (discussed earlier in the chapter). Sterilization in females generally is achieved by performing a *tubal ligation* (lī-GĀ-shun).

In one procedure, an incision is made into the abdominal cavity, the uterine (Fallopian) tubes are squeezed, and a small loop called a knuckle is made. A suture is tied tightly at the base of the knuckle and the knuckle is then cut. After 4 or 5 days the suture is digested by body fluids and the two severed ends of the tubes separate. The ovum is thus prevented from passing to the uterus, and the sperm cannot reach the ovum. Sterilization normally does not affect sexual performance or enjoyment.

Another procedure for tubal ligation is the *laparoscopic* (lap'-a-rō-SKŌ-pik) *technique*. After a woman receives local or general anesthesia, a harmless gas is introduced into her abdomen to create a gas bubble. The bubble expands the abdominal cavity and pushes the intestines away from the pelvic organs, permitting safe easy access to the uterine tubes. The doctor makes a small incision at the lower rim of the umbilicus and inserts a laparoscope to view the inside of the abdominal cavity and the uterine tubes. The tubes can be closed with this instrument or a second incision can be made at the pubic hairline to insert a cautery forceps. Once the uterine tubes are sealed, the instrument is removed, the gas is released, and the incision is covered with a bandage. After a few hours the patient can usually go home.

CONTRACEPTION

Contraception is the prevention of fertilization by natural, mechanical, or chemical means, without destroying fertility.

Natural

The *natural methods* include complete or periodic abstinence. An example of periodic abstinence is the rhythm method, which takes advantage of the fact that a secondary oocyte is fertilizable for only 24 hours and is available only during a period of 3–5 days in each menstrual cycle. During this time the couple refrains from intercourse. Its effectiveness is limited by the fact that few women have absolutely regular cycles. Moreover, some women occasionally ovulate during the "safe" times of the month, such as during menstruation.

Mechanical

Mechanical methods of contraception include the use of a condom by the male and a diaphragm by the female. The *condom* is a nonporous, elastic (rubber or similar material) covering placed over the penis that prevents deposition of sperm in the female reproductive tract. The *diaphragm* is a dome-shaped structure that fits over the cervix and is generally used in conjunction with a sperm-killing chemical. The diaphragm stops the sperm from passing into the cervix. The chemical kills the sperm cells. Toxic shock syndrome (TSS) and recurrent urinary tract infections are associated with diaphragm use in some females.

Another mechanical method of contraception is an *intrauterine devide (IUD)*. The device is a small object made of plastic, copper, or stainless steel that is inserted into the cavity of the uterus. It is not clear how IUDs operate. Some investigators believe they cause changes in the uterine lining, which, in turn, produce a substance that destroys either the sperm or the fertilized ovum. Some IUDs release contraceptive agents in minute amounts; one secretes progesterone. The dangers associated with the use of IUDs in some females include pelvic inflammatory disease (PID) and infertility. Because of this, IUDs are rapidly declining in popularity. Lawsuits resulting from damage claims have caused manufacturers of copper IUDs in the United States to stop production and sales.

Chemical

Chemical methods of contraception include spermicidal and hormonal methods. Various foams, creams, jellies, suppositories, and douches make the vagina and cervix unfavorable for sperm survival. Several studies have found no increase in the overall frequency of birth defects in association with the use of spermicides. A recent spermicidal development is a *contraceptive sponge,* sold under the trade name Today ®. It is a nonprescription, polyurethane sponge that contains a spermicide called nonoxynol-9, also used in a variety of contraceptive foams and jellies. This spermicide decreases the incidence of chlamydia and gonorrhea but slightly increases the risk of developing vaginal infections caused by *Candida*. The sponge is placed in the vagina where it releases spermicide for up to 24 hours and also acts as a physical barrier to sperm. The sponge is equal to the diaphragm in effectiveness. Some cases of toxic shock syndrome (TSS) have been reported among users of the contraceptive sponge.

Modified human chorionic gonadotropin (HCG), a hormone produced by the placenta, has shown promise in laboratory animals as a chemical method of birth control. In its modified form, HCG prevents implantation of a fertilized egg or terminates an already established pregnancy. Modified versions of gonadotropin releasing hormone (GnRH) are also being tested as contraceptives that inhibit ovulation.

The hormonal method, *oral contraceptive* or *OC* (the pill), has found rapid and widespread use. Although several pills are available, the one most commonly used contains a high concentration of progesterone and a low concentration of estrogens. These two hormones act on the anterior pituitary to decrease the secretion of FSH and LH by inhibiting GnRH by the hypothalamus. The low levels of FSH and LH are not adequate to initiate follicle maturation and ovulation. In the absence of a mature ovum, pregnancy cannot occur.

Women for whom all oral contraceptives are contraindicated include those with a history of thromboembolic disorders (predisposition to blood clotting), cerebral blood vessel damage, hypertension, liver malfunction, heart disease, or cancer of the breast or reproductive system. Recent reports link pill users with an increased risk of infertility. About 40 percent of all pill users experience side effects—generally minor problems such as nausea, weight gain, headache, irregular menses, spotting between periods, and amenorrhea. The statistics on the life-threatening conditions associated with the pill such as blood clots, heart attacks, liver tumors, and gallbladder disease are somewhat more reassuring. For all the problems combined, fewer than 3 deaths occurred per 100,000 users under age 30, 4 among those 30–35, 10 among those 35–39, and 18 among women over 40. The major exception is that women who take the pill and smoke face far higher odds of developing heart attack and stroke than do nonsmoking pill users.

Among the noncontraceptive benefits of oral contraceptives are menstrual regulation, decreased menstrual flow, and prevention of functional ovarian cysts. Some evidence also suggests a protective effect of the pill against endometrial and ovarian cancer.

In late 1986, a team of French physicians reported a new approach to birth control. It involves using a substance called *mifepristone (RU486)* that blocks the action of progesterone (PROG). Progesterone prepares the uterine endometrium for implantation and then maintains the uterine lining after implantation. If progesterone levels fall during pregnancy or if the hormone is inhibited from acting, menstruation occurs and the embryo is sloughed off along with the uterine lining. Mifepristone occupies the endometrial receptor sites for progesterone and, in effect, blocks the access of progesterone to the endometrium and causes a miscarriage. Mifepristone can be taken up to 5 weeks after conception. One side effect of the drug is uterine bleeding, which averages 11 days, and ranges from 5 to 21 days. The drug is not yet available in the United States.

A summary of birth control methods is presented in Exhibit 25-1.

CLINICAL APPLICATION

The quest for an efficient male oral contraceptive has been disappointing until recently. An oral contraceptive, *gossypol,* which is derived from cottonseed oil, has achieved a high efficacy (power to produce effects) rate of 99.9 percent in clinical tests in China. Its action is to inhibit an enzyme required for spermatogenesis. Levels of blood luteinizing hormone (LH) and testosterone remained unchanged and potency was not impaired. The number and morphology of spermatozoa gradually recover and fertility is restored to normal within 3 months after termination of therapy in most individuals. In some cases, fertility is not regained when gossypol is stopped. Gossypol also lowers potassium levels in the blood.

Modified versions of GnRH have also been shown to inhibit testosterone production, with a resultant decrease in sperm count and motility. However, the development of impotence has made this approach unpopular.

EXHIBIT 25-1

Birth Control (BC) Methods

METHOD	COMMENTS
Removal of Gonads and Uterus	Irreversible sterility. Generally performed if organs are diseased rather than as contraceptive method because of importance of hormones produced by gonads.
Sterilization	Procedure involving severing ductus (vas) deferens in males (vasectomy) and uterine (Fallopian) tubes in females (tubal ligation and laparoscopic technique).
Natural Contraception	Abstinence from intercourse during time of month woman is fertile. Under ideal circumstances, effectiveness in women with regular menstrual cycles may approach that of mechanical and chemical contraceptives. Extremely difficult to determine fertile period. Effectiveness can be increased by recording body temperature each morning before getting up; a small rise in temperature indicates that ovulation has occurred 1 or 2 days earlier and ovum can no longer be fertilized.
Mechanical Contraception **Condom**	Thin, strong sheath of rubber or similar material worn by male to prevent sperm from entering vagina. Failures caused by sheath tearing or slipping off after climax or not putting the sheath on soon enough. If used correctly and consistently, effectiveness similar to that of diaphragm.
Diaphragm	Flexible rubber dome inserted into vagina to cover cervix, providing barrier to sperm. Usually used with spermicidal cream or jelly. Must be left in place at least 6 hours after intercourse and may be left in place as long as 24 hours. Must be fitted by physician or other trained personnel and refitted every 2 years and after each pregnancy. Offers high level of protection if used with spermicide. Occasional failures caused by improper insertion or displacement during sexual intercourse.
Intrauterine Device (IUD)	Small object (loop, coil, T, or 7) made of plastic, copper, or stainless steel and inserted into uterus by physician. May be left in place for long periods of time (some must be changed every 2–3 years). Does not require continued attention by user. Some women cannot use them because of expulsion, bleeding, or discomfort. Not recommended for women who have not had children because uterus is too small and cervical canal too narrow. Use of IUDs has diminished because of potential problems such as pelvic inflammatory disease (PID) and infertility. No longer manufactured in the United States.
Chemical Contraception **Foams, Creams, Jellies, Suppositories, Vaginal Douches, Contraceptive Sponge**	Sperm-killing chemicals inserted into vagina to coat vaginal surfaces and cervical opening. Provide protection for about 1 hour. Effective when used alone, but significantly more effective when used with diaphragm or condom. The contraceptive sponge is made of polyurethane and releases a spermicide for up to 24 hours. It rates with the diaphragm in effectiveness. A few cases of toxic shock syndrome (TSS) have been reported among users of the contraceptive sponge.
Oral Contraceptives (OC)	Except for total abstinence or surgical sterilization, most effective contraceptive known. Side effects include nausea, occasional light bleeding between periods, breast tenderness or enlargement, fluid retention, and weight gain. Should not be used by women who have cardiovascular conditions (thromboembolic disorders, cerebrovascular disease, heart disease, hypertension), liver malfunction, cancer or neoplasia of breast or reproductive organs, or by women who smoke. Pill users may have an increased risk of infertility.

AGING AND THE REPRODUCTIVE SYSTEMS

Although there are major age-specific physical changes in structure and function, few age-specific disorders are associated with the reproductive system. In the male, the decreasing production of testosterone produces less muscle strength, fewer viable sperm, and decreased sexual desire. However, abundant spermatozoa may be found even in old age. Most of the age-dependent pathologies do not affect general health, except for prostate problems that could become serious and fatal.

The female reproductive system has a time-limited span of fertility between menarche and menopause. The system demonstrates an age-dependent decline in fertility, possibly as a result of less frequent ovulation and the declining ability of the uterine (Fallopian) tubes and uterus to support the young embryo. There is a decrease in the production of progesterone and estrogens. Menopause is only one of a series of phases that leads to reduced fertility, irregular or absent menstruation, and a variety of physical changes. Uterine cancer peaks at about 65 years of age, but cervical cancer is more common in younger women, and breast cancer is the leading cause of death among women between the ages of 40 and 60. Prolapse (falling down or sinking) of the uterus is possibly the commonest complaint among female geriatric patients.

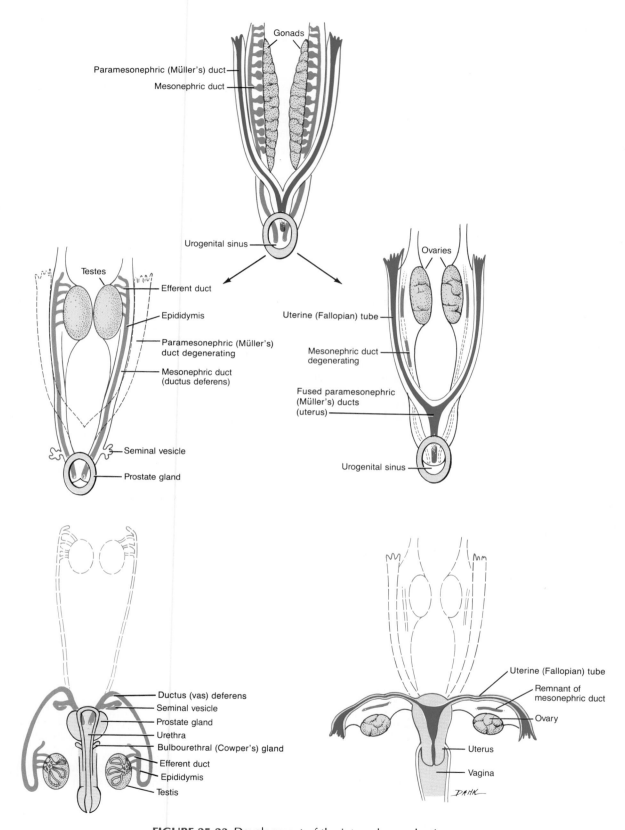

FIGURE 25-23 Development of the internal reproductive systems.

DEVELOPMENTAL ANATOMY OF THE REPRODUCTIVE SYSTEMS

The *gonads* develop from the **intermediate mesoderm.** By the sixth week, they appear as bulges that protrude into the ventral body cavity (Figure 25-23). The gonads develop near the mesonephric ducts. A second pair of ducts, the **paramesonephric (Müller's) ducts,** develop lateral to the mesonephric ducts. Both sets of ducts empty into the urogenital sinus. At about the eighth week, the gonads are clearly differentiated into ovaries or testes.

In the male embryo, the *testes* connect to the mesonephric duct through a series of tubules. These tubules become the *seminiferous tubules.* Continued development of the mesonephric ducts produces the *efferent ducts, ductus epididymis, ductus (vas) deferens, ejaculatory ducts,* and *seminal vesicle.* The *prostate* and *bulbourethral (Cowper's) glands* are **endodermal** outgrowths of the urethra. Shortly after the gonads differentiate into testes, the paramesonephric ducts degenerate without contributing any functional structures to the male reproductive system.

In the female embryo, the gonads develop into *ovaries.* At about the same time, the distal ends of the paramesonephric ducts fuse to form the *uterus* and *vagina.* The unfused portions become the *uterine (Fallopian) tubes.* The *greater (Bartholin's)* and *lesser vestibular glands* develop from **endodermal** outgrowths of the vestibule. The mesonephric ducts in the female degenerate without contributing any functional structures to the female reproductive system.

The *external genitals* of both male and female embryos also remain undifferentiated until about the eighth week. Before differentiation, all embryos have an elevated region, the **genital tubercle,** a point between the tail (future coccyx) and the umbilical cord, where the mesonephric and paramesonephric ducts open to the exterior (Figure 25-24). The tubercle consists of a **urethral groove** (opening into the urogenital sinus), paired **urethral folds,** and paired **labioscrotal swellings.**

In the male embryo, the genital tubercle elongates and develops into a *penis.* Fusion of the urethral folds forms the *spongy (cavernous) urethra* and leaves an opening only at the distal end of the penis, the *urethral orifice.* The labioscrotal swellings develop into the *testes.* In the female, the genital tubercle gives rise to the *clitoris.* The urethral folds remain open as the *labia minora,* and the labioscrotal swellings become the *labia majora.* The urethral groove becomes the *vestibule.*

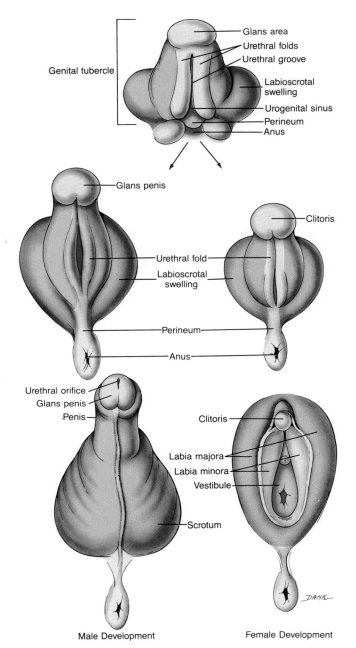

FIGURE 25-24 Development of the external genitals.

APPLICATIONS TO HEALTH

SEXUALLY TRANSMITTED DISEASES (STDs)

The general term **sexually transmitted disease (STD)** is applied to any of the large group of diseases that can be spread by sexual contact. The group includes conditions traditionally specified as **venereal diseases (VD),** from Venus, goddess of love, such as gonorrhea, syphilis, and genital herpes, and several other conditions that are contracted sexually, or may be contracted otherwise, but are then transmitted to a sexual partner.

Gonorrhea

Gonorrhea is an infectious sexually transmitted disease that affects primarily the mucous membrane of the urogenital tract, the rectum, and occasionally the eyes. The disease is caused by the bacterium *Neisseria gonorrhoeae*. Discharges from the involved mucous membranes are the source of infection, and the bacteria are transmitted by direct contact, usually sexual or during passage of a newborn through the birth canal.

Males usually suffer inflammation of the urethra with pus and painful urination. Fibrosis sometimes occurs in an advanced stage, causing stricture of the urethra. There also may be involvement of the epididymis and prostate gland. In females, infection may occur in the urethra, vagina, and cervix, and there may be a discharge of pus. However, infected females often harbor the disease without any symptoms until it has progressed to a more advanced stage. If the uterine (Fallopian) tubes become involved, pelvic inflammation may follow. Peritonitis, or inflammation of the peritoneum, is a very serious disorder. The infection should be treated and controlled immediately because, if neglected, sterility or death may result. Although antibiotics have greatly reduced the mortality rate of acute peritonitis, it is estimated that between 50,000 and 80,000 women are made sterile by gonorrhea every year as a result of scar tissue formation that closes the uterine tubes. If the bacteria are transmitted to the eyes of a newborn in the birth canal, blindness can result.

Administration of a 1 percent silver nitrate solution or penicillin in the infant's eyes prevents infection. Penicillin or tetracycline are the drugs of choice for the treatment of gonorrhea in adults. However, the incidence of antibiotic-resistant gonorrhea has increased substantially since 1984.

Syphilis

Syphilis is a sexually transmitted disease caused by the bacterium *Treponema pallidum*. It is acquired through sexual contact or transmitted through the placenta to a fetus. The disease progresses through several stages: primary, secondary latent, and sometimes tertiary. During the ***primary stage,*** the chief symptom is an open sore, called a ***chancre*** (pronounced SHANKG-ker), at the point of contact. The chancre heals within 1–5 weeks. From 6 to 24 weeks later, symptoms such as a skin rash, fever, and aches in the joints and muscles usher in the ***secondary stage.*** These symptoms also eventually disappear (in about 4–12 weeks), and the disease ceases to be infectious, but a blood test for the presence of the bacteria generally remains positive. During this "symptomless" period, called the ***latent stage,*** the bacteria may invade body organs. When signs of organ degeneration appear, the disease is said to be in the ***tertiary stage.***

If the syphilis bacteria attack the organs of the nervous system, the tertiary stage is called ***neurosyphilis.*** Neurosyphilis may take different forms, depending on the tissue involved. For instance, about 2 years after the onset of the disease, the bacteria may attack the meninges, producing meningitis. The blood vessels that supply the brain may also become infected. In this case, symptoms depend on

the parts of the brain destroyed by oxygen and glucose starvation. Cerebellar damage is manifested by uncoordinated movements in such activities as writing. As the motor areas become extensively damaged, victims may be unable to control urine and bowel movements. Eventually, they may become bedridden, unable even to feed themselves. Damage to the cerebral cortex produces memory loss and personality changes that range from irritability to hallucinations.

Infection of the fetus with syphillis can occur after the fifth month. Infection of the mother is not necessarily followed by fetal infection, provided that the placenta remains intact. But once the bacteria gain access to fetal circulation, there is nothing to impede their growth and multiplication. As many as 80 percent of children born to untreated syphilitic mothers will be infected in the uterus if the fetus is exposed at the onset or in the early stages of the disease. About 25 percent of the fetuses will die within the uterus. Most of the survivors will arrive prematurely, but 30 percent will die shortly after birth. Of the infected and untreated children surviving infancy, about 40 percent will develop symptomatic syphilis during their lifetimes.

Syphilis can be treated with antibiotics (penicillin) during the primary, secondary, and latent periods. Certain forms of neurosyphilis may also be successfully treated, but the prognosis for others is very poor. Noticeable symptoms do not always appear during the first two stages of the disease. Syphilis, however, is usually diagnosed through a blood test whether noticeable symptoms appear or not. The importance of these blood tests and follow-up treatments cannot be overemphasized.

Some evidence suggests that AIDS may alter the course of neurosyphilis by accelerating its progression, possibly by impairing macrophages and antibody production and facilitating penetration of the AIDS virus into the central nervous system.

Genital Herpes

Another sexually transmitted disease, **genital herpes,** is common in the United States. The sexual transmission of the herpes simplex virus is well established. Unlike syphilis and gonorrhea, genital herpes is incurable. Type I herpes simplex virus is the virus that causes the majority of infections above the waist such as cold sores. Type II herpes simplex virus causes most infections below the waist such as painful genital blisters on the prepuce, glans penis, and penile shaft in males and on the vulva or sometimes high up in the vagina in females. The blisters disappear and reappear in most patients, but the virus itself remains in the body.

Genital herpes virus infection causes considerable discomfort, and there is an extraordinarily high rate of recurrence of the symptoms. The infection is usually characterized by fever, chills, flulike symptoms, lymphadenopathy, and numerous clusters of genital blisters. For pregnant women with genital herpes symptoms at the time of delivery, a cesarean section will usually prevent complications in the child. Complications range from a mild asymptomatic infection to central nervous system (CNS) damage to death.

Treatment of the symptoms involves pain medication, saline compresses, sexual abstinence for the duration of the eruption, and use of an oral drug called acyclovir (Zovirax). This drug interferes with viral DNA replication but not with host cell DNA replication. Acyclovir speeds the healing and sometimes reduces the pain of initial genital herpes infections and shortens the duration of lesions in patients with recurrent genital herpes. A topically applied ointment that contains Inter Vir-A (Immuvir), an antiviral substance, is another drug used to treat genital herpes. Inter Vir-A provides rapid relief for the pain, itching, and burning associated with genital herpes. An experimental genital herpes vaccine will involve human testing shortly.

Trichomoniasis

The microorganism *Trichomonas vaginalis,* a flagellated protozoan (one-celled animal), causes **trichomoniasis,** an inflammation of the mucous membrane of the vagina in females and the urethra in males. Symptoms include vaginal discharge and severe vaginal itch in women. Men can have it without symptoms but can transmit it to women nonetheless. Sexual partners must be treated simultaneously.

Chlamydia

Chlamydia is a sexually transmitted disease caused by the bacterium, *Chlamydia trachomatis.* At present, chlamydia is the most prevalent and one of the most damaging of the sexually transmitted diseases. In males, urethritis is the principal result. It is characterized by burning on urination, frequency of urination, painful urination, and low back pain. In females, urethritis may spread through the reproductive tract and develop into inflammation of the uterine (Fallopian) tubes, which increases the risk of ectopic pregnancy and sterility. As in gonorrhea, the organism may be passed from mother to infant during childbirth, infecting the eyes.

MALE DISORDERS

Testicular Cancer

Testicular cancer occurs most often between the ages of 15 and 34 and is one of the most common cancers seen in young males. Although the cause is unknown, the condition is associated with males with a history of undescended testes or late-descended testes. Most testicular cancers arise from the sperm-producing cells. An early sign of testicular cancer is a mass in the testis, often associated with pain or discomfort. Treatment involves removal of the diseased testis.

Prostate

The prostate gland is susceptible to infection, enlargement, and benign and malignant tumors. Because the prostate surrounds the urethra, any of these disorders can obstruct the flow of urine. Prolonged obstruction may result in serious changes in the urinary bladder, ureters, and kidneys and may perpetuate urinary tract infections. Therefore, if the obstruction cannot be relieved by other means, surgical removal of part of or the entire gland is indicated. The surgical procedure is called **prostatectomy** (pros'-ta-TEK-tō-mē).

Acute and chronic infections of the prostate gland are common in postpubescent males, often in association with inflammation of the urethra. In *acute prostatitis,* the prostate gland becomes swollen and tender. Appropriate antibiotic therapy, bed rest, and above-normal fluid intake are effective treatment.

Chronic prostatitis is one of the most common chronic infections in men of the middle and later years. On examination, the prostate gland feels enlarged, soft, and extremely tender and its surface outline is irregular. This disease frequently produces no symptoms, but the prostate is believed to harbor infectious microorganisms responsible for some allergic conditions, arthritis, and inflammation of nerves (neuritis), muscles (myositis), and the iris (iritis).

An *enlarged prostate* gland, increasing to two to four times larger than normal, occurs in approximately one-third of all males over age 60. The enlarged condition usually can be detected by rectal examination.

Tumors of the male reproductive system usually involve the prostate gland. Carcinoma of the prostate is the second leading cause of death from cancer in men in the United States, and it is responsible for approximately 19,000 deaths annually. Its incidence is related to age, race, occupation, geography, and ethnic origin. Both benign and malignant growths are common in elderly men. Both types of tumors put pressure on the urethra, making urination painful and difficult. At times, the excessive back pressure destroys kidney tissue and gives rise to an increased susceptibility to infection. Therefore, even when the tumor is benign, surgery is indicated.

Sexual Functional Abnormalities

Impotence (*impotenia* = lack of strength) is the inability of an adult male to attain or hold an erection long enough for normal intercourse. Impotence could be the result of physical abnormalities of the penis, systemic disorders such as syphilis, vascular disturbances, neurological disorders, testosterone deficiency, or psychic factors such as fear of causing pregnancy, fear of sexually transmitted diseases, religious inhibitions, and emotional immaturity. In selected individuals, penile prostheses may be indicated. Penile injections of papaverine and phentolamine mesylate can produce excellent effects in overcoming both physical and psychological impotence.

Infertility (sterility) is an inability to fertilize the ovum. It does not imply impotence. Male fertility requires production of adequate amounts of viable, normal spermatozoa by the testes, unobstructed transportation of sperm through the seminal tract, and satisfactory deposition in the vagina. The tubules of the testes are sensitive to many factors—x rays, infections, toxins, malnutrition—that may cause degenerative changes and produce male sterility. If inadequate spermatozoa production is suspected, a sperm analysis should be performed. At least one type of fertility may be improved by administration of vitamin C.

FEMALE DISORDERS

Menstrual Abnormalities

Because menstruation reflects not only the health of the uterus but also the health of the endocrine glands that control it, the ovaries and the pituitary gland, disorders of the female reproductive system frequently involve menstrual disorders.

Amenorrhea is the absence of menstruation. If a woman has never menstruated, the condition is called *primary amenorrhea.* Primary amenorrhea can be caused by endocrine disorders, most often in the pituitary gland and hypothalamus, or by a genetically caused abnormal development of the ovaries or uterus. *Secondary amenorrhea,* the skipping of one or more periods, is commonly experienced by women at some time during their lives. Changes in body weight, either gains or losses, often cause amenorrhea. Obesity may disturb ovarian function, and similarly, the extreme weight loss that characterizes anorexia nervosa often leads to a suspension of menstrual flow. When amenorrhea is unrelated to weight, analysis of levels of estrogens often reveals deficiencies of pituitary and ovarian hormones. Amenorrhea may also be caused by continuous involvement in rigorous athletic training.

Dysmenorrhea is painful menstruation caused by forceful contraction of the uterus. It is often accompanied by nausea, vomiting, diarrhea, headache, fatigue, and nervousness. Some cases are caused by pathological conditions such as uterine tumors, ovarian cysts, endometriosis, and pelvic inflammatory disease (PID). However, other cases of dysmenorrhea are not related to any pathologies. Although the cause of these cases is unknown, it appears that they may be triggered by an overproduction of prostaglandins by the uterus. Prostaglandins are known to stimulate uterine contractions, but they cannot do so in the presence of high levels of progesterone. As we have noted earlier, progesterone levels are high during the last half of the menstrual cycle. During this time, prostaglandins are apparently inhibited by progesterone from producing uterine contractions. However, if pregnancy does not occur, progesterone levels drop rapidly and prostaglandin production increases. This causes the uterus to contract and slough off its lining and may result in dysmenorrhea. The other symptoms of dysmenorrhea—nausea, vomiting, diarrhea, and headache—may be due to prostaglandin-stimulated contractions of the smooth muscle of the stomach, intestines, and blood vessels in the brain. Drugs that inhibit prostaglandin synthesis (naproxen and ibuprofen) are used to treat dysmenorrhea.

Abnormal uterine bleeding includes menstruation of excessive duration or excessive amount, too frequent menstruation, intermenstrual bleeding, and postmenopausal bleeding. These abnormalities may be caused by disordered hormonal regulation, emotional factors, fibroid tumors of the uterus, and systemic diseases.

Premenstrual syndrome (PMS) is a term usually reserved for severe physical and emotional distress occurring late in the postovulatory phase of the menstrual cycle and sometimes overlapping with menstruation. Signs and symptoms include edema, weight gain, breast swelling and tenderness, abdominal distension, backache, joint pain, constipation, skin eruptions, fatigue and lethargy, greater need for sleep, depression or anxiety, irritability, mood swings, headache, poor coordination and clumsiness, and cravings for sweet or salty foods. The basic cause of PMS is unknown. Although PMS is related to the cyclic production of ovarian hormones, the symptoms are not directly due to changes in the levels of these hormones. Treatment is individualized, depending on the type and severity of symptoms, and may include dietary changes, exercise, diuretics, tranquilizers, hormone therapy (progesterone), and vitamin B_6.

Toxic Shock Syndrome (TSS)

Toxic shock syndrome (TSS), first described in 1978, is primarily a disease of previously healthy, young, menstruating females who use tampons. It is also recognized in males, children, and nonmenstruating females. Clinically, TSS is characterized by high fever up to 40.6°C (105°F), sore throat or very tender mouth, headache, fatigue, irritability, muscle soreness and tenderness, conjunctivitis, diarrhea and vomiting, abdominal pain, vaginal irritation, and erythematous rash. Other symptoms include lethargy, unresponsiveness, memory loss, hypotension, peripheral vasoconstriction, respiratory distress syndrome, intravascular coagulation, decreased platelet count, renal failure, circulatory shock, and liver involvement.

It is now clear that toxin-producing strains of the bacterium *Staphylococcus aureus* are necessary for development of the disease. Actually, it appears that a virus has become incorporated into *S. aureus,* causing the bacterium to produce the toxins. Although all tampon users are at some risk for developing TSS, the risk is increased considerably by females who use highly absorbent tampons, regardless of the chemical composition of the tampon. Apparently, high-absorbency tampons provide a substrate on which the bacteria grow and produce toxins. TSS can also occur as a complication of influenza and influenzalike illness and use of contraceptive sponges. Initial therapy is directed at correcting all homeostatic imbalances as quickly as possible.

Antistaphylococcal antibiotics, such as penicillin or clindamycin, are also administered. In severe cases, high doses of corticosteroids are also administered.

Ovarian Cysts

Ovarian cysts are fluid-containing tumors of the ovary. Follicular cysts may occur in the ovaries of elderly women, in ovaries that have inflammatory diseases, and in menstruating females. They have thin walls and contain a serous albuminous material. Cysts may also arise from the corpus luteum or the endometrium.

Endometriosis

Endometriosis (*endo* = within; *metri* = uterus; *osis* = condition) is a benign condition characterized by the growth of endometrial tissue outside the uterus. The tissue enters the pelvic cavity via the open uterine (Fallopian) tubes and may be found in any of several sites—on the ovaries, rectouterine pouch, surface of the uterus, sigmoid colon, pelvic and abdominal lymph nodes, cervix, abdominal wall, kidneys, and urinary bladder. One theory for the development of endometriosis is that there is regurgitation of menstrual flow through the uterine tubes. Endometriosis is common in women 25 to 40 years of age who have not had children. Symptoms include premenstrual pain or unusual menstrual pain. The unusual pain is caused by the displaced tissue sloughing off at the same time the normal uterine endometrium is being shed during menstruation. Infertility can be a consequence. Treatment usually consists of hormone therapy, videolaseroscopy (laparoscope with camera and laser), or conventional surgery. Endometriosis disappears at menopause or when the ovaries are removed.

Infertility

Female infertility, or the inability to conceive, occurs in about 10 percent of married females in the United States. Once it is established that ovulation occurs regularly, the reproductive tract is examined for functional and anatomical disorders to determine the possibility of union of the sperm and the ovum in the uterine tube. Some research suggests that an autoimmune disease might underlie many cases of infertility.

Disorders Involving the Breasts

The breasts of females are highly susceptible to cysts and tumors. Men are also susceptible to breast tumors, but certain breast cancers are 100 times more common in women.

In the female, the benign *fibroadenoma* is a common tumor of the breast. It occurs most frequently in young women. Fibroadenomas have a firm rubbery consistency and are easily moved about within the mammary tissue.

The usual treatment is excision of the growth. The breast itself is not removed.

Breast cancer has one of the highest fatality rates of all cancers affecting women, but it is rare in men. In the female, breast cancer is rarely seen before age 30, and its occurrence rises rapidly after menopause. Breast cancer is generally not painful until it becomes quite advanced, so often it is not discovered early or, if noted, is ignored. Any lump, no matter how small, should be reported to a doctor at once. Treatment for breast cancer may involve hormone therapy, chemotherapy, *lumpectomy* (removal of just the tumor and immediate surrounding tissue), a modified or radical mastectomy, or a combination of these. A *radical mastectomy* involves removal of the affected breast along with the underlying pectoral muscles and the axillary lymph nodes. Metastasis of cancerous cells is usually through the lymphatics or blood. Radiation treatment and chemotherapy may follow the surgery to ensure the destruction of any stray cancer cells.

Among the factors that clearly increase the risk of breast cancer development are (1) a family history of breast cancer, especially in a mother or sister; (2) never having a child or having a first child after age 34; (3) previous cancer in one breast; (4) exposure to ionizing radiation; (5) an abnormal mammogram; and (6) excessive fat and alcohol intake. Females who take birth control pills do not have a higher risk of developing breast cancer than females who do not.

The American Cancer Society recommends the following steps in order to help diagnose breast cancer as early as possible:

1. A mammogram should be taken between the ages of 35 and 39, to be used later for comparison.
2. A physician should examine the breasts every 3 years when a female is between the ages of 20 and 40, and every year after 40.
3. Females with no symptoms should have a mammogram every year or two between ages 40 and 49, and every year after 50.
4. Females of any age with a history of breast cancer, a strong family history of the disease, or other risk factors should consult a physician to determine a schedule for mammography.
5. All females over 20 should develop the habit of breast self-examination (BSE).

Cervical Cancer

Another common disorder of the female reproductive tract is *cervical cancer,* carcinoma of the cervix of the uterus. The condition starts with *cervical dysplasia* (dis-PLĀ-sē-a), a change in the shape, growth, and number of the cervical cells. If the condition is minimal, the cells may regress to normal. If it is severe, it may progress to cancer. Cervical cancer may be detected in most cases in its earliest stages by a Pap smear. There is some evidence linking cervical cancer to penile virus (papillomavirus) infections of male sexual partners. Depending on the progress of the disease, treatment may consist of excision of lesions, radiotherapy, chemotherapy, and hysterectomy.

Pelvic Inflammatory Disease (PID)

Pelvic inflammatory disease (*PID*) is a collective term for any extensive bacterial infection (primarily involving *Chlamydia trachomatis, Neisseria gonorrhoeae, Bacteroides, Peptostreptococcus,* and *Gardnerella vaginalis.*) of the pelvic organs, especially the uterus, uterine (Fallopian) tubes, or ovaries. A vaginal or uterine infection may spread into the uterine tube (*salpingitis*) or even farther into the abdominal cavity, where it infects the peritoneum (*peritonitis*). Diagnosis of PID depends on three findings: abdominal tenderness, cervical tenderness, and ovarian, uterine tube, and uterine ligament tenderness. In addition, diagnosis is based on at least one of the following: fever, leucocytosis, pelvic abscess or inflammation, purulent cervical discharge, and the presence of certain bacteria in smears. Early treatment with bed rest and antibiotics (cefoxitin, penicillin, tetracycline, dioxycycline) can stop the spread of PID.

KEY MEDICAL TERMS ASSOCIATED WITH THE REPRODUCTIVE SYSTEMS

Colpotomy (kol-POT-ō-mē; *colp* = vagina; *tome* = cutting)   Incision of the vagina.

Culdoscopy (Kul-DOS-kō-pē; *skopein* = to examine)   A procedure in which a culdoscope (endoscope) is used to view the pelvic cavity. The approach is through the vagina.

Hermaphroditism (her-MAF-rō-di-tizm')   Presence of both male and female sex organs in one individual.

Leukorrhea (loo'-kō-RĒ-a; *leuco* = white; *rrhea* = discharge)   A nonbloody vaginal discharge that may occur at any age and affects most women at some time.

Salpingectomy (sal'pin-JEK-tō-mē; *salpingo* = tube)   Excision of a uterine (Fallopian) tube.

Smegma (SMEG-ma; *smegma* = soap)   The secretion, consisting principally of desquamated epithelial cells, found chiefly about the external genitalia and especially under the foreskin of the male.

Vaginitis (vaj'-i'NĪ-tis)   Inflammation of the vagina.

STUDY OUTLINE

Male Reproductive System (p. 709)

1. Reproduction is the process by which genetic material is passed on from one generation to the next.
2. The organs of reproduction are grouped as gonads (produce gametes), ducts (transport and store gametes), and accessory glands (produce materials that support gametes).
3. The male structures of reproduction include the testes, ductus epididymis, ductus (vas) deferens, ejaculatory duct, urethra, seminal vesicles, prostate gland, bulbourethral (Cowper's) glands, and penis.

Scrotum (p. 709)

1. The scrotum is a cutaneous outpouching of the abdomen that supports the testes.
2. It regulates the temperature of the testes by contraction of the cremaster muscle to elevate them closer to the pelvic cavity.

Testes (p. 709)

1. The testes are oval-shaped glands (gonads) in the scrotum containing seminiferous tubules, in which sperm cells are made; sustentacular (Sertoli) cells, which nourish sperm cells; and interstitial endocrinocytes (cells of Leydig), which produce the male sex hormone testosterone.
2. Failure of the testes to descend is called cryptorchidism.
3. Ova and sperm are collectively called gametes, or sex cells, and are produced in gonads.
4. Uninucleated somatic cells divide by mitosis, the process in which each daughter cell receives the full complement of 23 chromosome pairs (46 chromosomes). Somatic cells are said to be diploid ($2n$).
5. Immature gametes divide by meiosis, in which the pairs of chromosomes are split so that the mature gamete has only 23 chromosomes. It is said to be haploid (n).
6. Spermatogenesis occurs in the testes. It results in the formation of four haploid spermatozoa from each primary spermatocyte.
7. Spermatogenesis is a process in which immature spermatogonia develop into mature spermatozoa. The spermatogenesis sequence includes reduction division (meiosis I), equatorial division (meiosis II), and spermiogenesis.
8. Mature spermatozoa consist of a head, midpiece, and tail. Their function is to fertilize an ovum.

Ducts (p. 714)

1. The duct system of the testes includes the seminiferous tubules, straight tubules, and rete testis.
2. Sperm are transported out of the testes through the efferent ducts.
3. The ductus epididymis is lined by stereocilia and is the site of sperm maturation and storage.
4. The ductus (vas) deferens stores sperm and propels them toward the urethra during ejaculation.
5. Alteration of the ductus (vas) deferens to prevent fertilization is called vasectomy.
6. The ejaculatory ducts are formed by the union of the ducts from the seminal vesicles and ductus (vas) deferens and eject spermatozoa into the prostatic urethra.

7. The male urethra is subdivided into three portions: prostatic, membranous, and spongy (cavernous).

Accessory Sex Glands (p. 716)

1. The seminal vesicles secret an alkaline, viscous fluid that constitutes about 60 percent of the volume of semen and contributes to sperm viability.

2. The prostate gland secretes an alkaline fluid that constitutes about 13–33 percent of the volume of semen and contributes to sperm motility.

3. The bulbourethral (Cowper's) gland secretes mucus for lubrication and a substance that neutralizes acid.

4. Semen (seminal fluid) is a mixture of spermatozoa and accessory sex gland secretions that provides the fluid in which spermatozoa are transported, provides nutrients, and neutralizes the acidity of the male urethra and female vagina.

Penis (p. 719)

1. The penis is the male organ of copulation that consists of a root, body, and glans penis.

2. Expansion of its blood sinuses under the influence of sexual excitation is called erection.

Female Reproductive System (p. 722)

1. The female organs of reproduction include the ovaries (gonads), uterine (Fallopian) tubes, uterus, vagina, and vulva.

2. The mammary glands are considered part of the reproductive system.

Ovaries (p. 722)

1. The ovaries are female gonads located in the upper pelvic cavity, on either side of the uterus.

2. They produce secondary oocytes, discharge secondary oocytes (ovulation), and secrete estrogens, progesterone, and relaxin.

3. Oogenesis occurs in the ovaries. It results in the formation of a single haploid secondary oocyte.

4. Oogenesis is a process in which immature oogonia develop into mature ova. The oogenesis sequence includes reduction division (meiosis I), equatorial division (meiosis II), and maturation.

Uterine (Fallopian) Tubes (p. 726)

1. The uterine (Fallopian) tubes transport ova from the ovaries to the uterus and are the normal sites of fertilization.

2. Implantation outside the uterus (pelvic or tubular) is called an ectopic pregnancy.

Uterus (p. 727)

1. The uterus is an inverted, pear-shaped organ that functions in menstruation, implantation of a fertilized ovum, development of a fetus during pregnancy, and labor.

2. The uterus is normally held in position by a series of ligaments.

3. Histologically, the uterus consists of an outer perimetrium, middle myometrium, and inner endometrium.

Menstrual Cycle (p. 728)

1. The function of the menstrual cycle is to prepare the endometrium each month for the reception of a fertilized egg.

2. The menstrual and ovarian cycles are controlled by GnRH, which stimulates the release of FSH and LH.

3. FSH stimulates the initial development of ovarian follicles and secretion of estrogens by the ovaries. LH stimulates further development of ovarian follicles, ovulation, and the secretion of estrogens and progesterone by the ovaries.

4. Estrogens stimulate the growth, development, and maintenance of female reproductive structures; stimulate the development of secondary sex characteristics; regulate fluid and electrolyte balance; and stimulate protein anabolism.

5. Progesterone works with estrogens to prepare the endometrium for implantation and the mammary glands for milk secretion.

6. Relaxin relaxes the symphysis publis and helps dilate the uterine cervix to facilitate delivery, and increases sperm motility.

7. During the menstrual phase, the stratum functionalis layer of the endometrium is shed with a discharge of blood, tissue fluid, mucus, and epithelial cells. Primary follicles develop into secondary follicles.

8. During the preovulatory phase, endometrial repair occurs. A secondary follicle develops into a vesicular (Graafian) follicle. Estrogens are the dominant ovarian hormones.

9. Ovulation is the rupture of a vesicular (Graafian) follicle and the release of an immature ovum into the pelvic cavity brought about by inhibition of FSH and a surge of LH.

10. During the postovulatory phase, the endometrium thickens in anticipation of implantation. Progesterone is the dominant ovarian hormone.

11. If fertilization and implantation do not occur, the corpus luteum degenerates and low levels of estrogens and progesterone initiate another menstrual and ovarian cycle.

12. If fertilization and implantation do occur, the corpus luteum is maintained by placental HCG, and the corpus luteum and placenta secrete estrogens and progesterone to support pregnancy and breast development for lactation.

Vagina (p. 732)

1. The vagina is a passageway for the menstrual flow, the receptacle of the penis during sexual intercourse, and the lower portion of the birth canal.

2. It is capable of considerable distension to accomplish its functions.

Vulva (p. 732)

1. The vulva is a collective term for the external genitals of the female.

2. It consists of the mons pubis, labia majora, labia minora, clitoris, vestibule, vaginal and urethral orifices, hymen, bulb of the vestibule, and the paraurethral (Skene's), greater vestibular (Bartholin's), and lesser vestibular glands.

Perineum (p. 733)

1. The perineum is a diamond-shaped area at the inferior end of the trunk between the thighs and buttocks.

2. An incision in the perineal skin prior to delivery is called an episiotomy.

Mammary Glands (p. 733)

1. The mammary glands are modified sweat glands (branched tubuloalveolar) over the pectoralis major muscles. Their function is to secrete and eject milk (lactation).

2. Milk secretion is mainly due to PRL and milk ejection is stimulated by OT.

Sexual Intercourse (p. 736)

1. The role of the male in the sex act involves erection, minimal lubrication, and orgasm.
2. The female role also involves erection, lubrication, and orgasm (climax).

Birth Control (BC) (p. 737)

1. Methods of birth control include removal of gonads and uterus, sterilization (vasectomy, tubal ligation, laparoscopic technique), and contraception (natural, mechanical, and chemical).
2. Contraceptive pills of the combination type contain estrogens and progesterone in concentrations that decrease the secretion of FHS and LH and thereby inhibit ovulation.

Aging and the Reproductive Systems (p. 739)

1. In the male, decreased levels of testosterone decrease muscle strength, sexual desire, and viable sperm; prostate disorders are common.
2. In the female, levels of progesterone and estrogens decrease resulting in changes in menstruation; uterine and breast cancer increase in incidence.

Developmental Anatomy of the Reproductive Systems (p. 741)

1. The gonads develop from intermediate mesoderm and are differentiated into ovaries or testes by about the eighth week.
2. The external genitals develop from the genital tubercle.

Applications to Health (p. 741)

1. Sexually transmitted diseases (STDs) are diseases spread by sexual contact and include gonorrhea, syphilis, genital herpes, trichomoniasis, and nongonococcal urethritis (NGU).
2. Testicular cancer originates in sperm-producing cells.
3. Conditions that affect the prostate gland are prostatitis, enlarged prostate, and tumors.
4. Impotence is the inability of the male to attain or hold an erection long enough for intercourse.
5. Infertility is the inability of a male's sperm to fertilize an ovum.
6. Menstrual disorders include amenorrhea, dysmenorrhea, abnormal bleeding, and premenstrual syndrome (PMS).
7. Toxic shock syndrome (TSS) includes widespread homeostatic imbalances and is a reaction to toxins produced by *Staphylococcus aureus*.
8. Ovarian cysts are tumors that contain fluid.
9. Endometriosis refers to the growth of uterine tissue outside the uterus.
10. Female infertility is the inability of the female to conceive.
11. The mammary glands are susceptible to benign fibroadenomas and malignant tumors. The removal of a malignant breast, pectoral muscles, and lymph nodes is called a radical mastectomy.
12. Cervical cancer starts with dysplasia and can be diagnosed by a Pap test.
13. Pelvic inflammatory disease (PID) refers to bacterial infection of pelvic organs.

REVIEW QUESTIONS

1. Define reproduction. Describe how the reproductive organs are classified and list the male and female organs of reproduction.
2. Describe the function of the scrotum in protecting the testes from temperature fluctuations. What is cryptorchidism?
3. Describe the internal structure of a testis. Where are the sperm cells made?
4. Describe the principal events of spermatogenesis. Why is meiosis important? Distinguish between haploid (*n*) and diploid cells (2*n*).
5. Identify the principal parts of a spermatozoon. List the function of each.
6. Which ducts are involved in transporting sperm within the testes?
7. Describe the location, structure, and functions of the ductus epididymis, ductus (vas) deferens, and ejaculatory duct.
8. What is the spermatic cord? What is an inguinal hernia?
9. Give the location of the three subdivisions of the male urethra.
10. Trace the course of spermatozoa through the male system of ducts from the seminiferous tubules through the urethra.
11. Briefly explain the locations and functions of the seminal vesicles, prostate gland, and bulbourethral (Cowper's) glands.
12. How is the penis structurally adapted as an organ of copulation? What is circumcision?
13. How are the ovaries held in position in the pelvic cavity?
14. Describe the microscopic structure of an ovary. What are the functions of the ovaries?
15. Describe the principal events of oogenesis.
16. Where are the uterine (Fallopian) tubes located? What is their function? What is an ectopic pregnancy?
17. Diagram the principal parts of the uterus.
18. Describe the arrangement of ligaments that hold the uterus in its normal position. What is retroflexion?
19. Discuss the blood supply to the uterus. Why is an abundant blood supply important?
20. Describe the histology of the uterus.
21. Define the menstrual cycle. Describe its principal events.
22. What is the function of the vagina? Describe its histology.
23. List the parts of the vulva and the functions of each part.
24. What is the perineum? Define episiotomy.
25. Describe the structure of the mammary glands. How are they supported?
26. Describe the passage of milk from the alveolar cells of the mammary gland to the nipple. Define lactation.
27. Explain the role of the male's erection, lubrication, and orgasm in the sex act. How do the female's erection, lubrication, and orgasm (climax) contribute to the sex act?
28. Briefly describe the various methods of birth control (BC) and the effectiveness of each.
29. Explain the effects of aging on the reproductive systems.
30. Describe the development of the reproductive systems.
31. Define a sexually transmitted disease (STD). Describe the cause, clinical symptoms, and treatment of gonorrhea, syphilis, genital herpes, trichomoniasis, and chlamydia.
32. What is testicular cancer?
33. Describe several disorders that affect the prostate gland.

34. What are some of the causes of amenorrhea, dysmenorrhea, and abnormal uterine bleeding?
35. Describe the clinical symptoms of premenstrual syndrome (PMS) and toxic shock syndrome (TSS).
36. What are ovarian cysts? Define endometriosis.

37. What is a radical mastectomy?
38. Define pelvic inflammatory disease (PID).
39. Refer to the glossary of key medical terms associated with the reproductive systems. Be sure that you can define each term.

SELF-QUIZ

Complete the following:

1. Identify male homologues for each of the following:
   a. labia majora: _____
   b. clitoris: _____
   c. paraurethral (Skene's) glands: _____
   d. greater vestibular (Bartholin's) glands: _____
2. Gametes are (haploid? diploid?); human gametes contain _____ chromosomes. Fusion of gametes produces a cell called a _____. This cell, and all cells of the organism derived from it, contain the (haploid? diploid?) number of chromosomes, written as (n? $2n$?). These chromosomes, received from both sperm and ovum, are said to exist in _____ pairs.
3. Write E (endoderm) or M (mesoderm) next to the structures below to indicate their embryonic origins.
   ___ a. ovaries and testes
   ___ b. prostate and bulbourethral (Cowper's) glands
   ___ c. greater (Bartholin's) vestibular and lesser vestibular glands
4. The process by which the testes produce spermatozoa involves meiosis and is referred to as _____.
5. Which of the following cells is *least* mature?
   (primary spermatocyte, spermatid, spermatogonium, secondary spermatocyte)
6. The most important ligaments in prevention of drooping (prolapse) of the uterus are the _____ ligaments, which are attached to the base of the uterus. Most of the uterus consists of _____-metrium. This layer is (smooth muscle? epithelium?). The innermost layer of the uterus is the _____-metrium. Which portion of it is shed during menstruation? Stratum (basalis? functionalis?). Which arteries supply the stratum functionalis? (spiral? straight?)
7. The two dorsolaterally located masses of tissue within the penis are called the _____ penis.
8. The ovary is attached to the uterus by the _____ ligament.
9. The rupture of the vesicular ovarian (Graafian) follicle with release of a secondary oocyte into the pelvic cavity is called _____.
10. If fertilization and implantation do not occur, the corpus luteum degenerates and becomes the _____.
11. The organ that serves as a passageway for the menstrual flow, the receptacle for the penis during copulation, and the lower portion of the birth canal is the _____.
12. Arrange the answers in correct sequence.
   ___ ___ ___ a. From most external to deepest:
      A. endometrium
      B. serosa
      C. myometrium

   ___ ___ ___ b. Portions of the urethra, from proximal (closest to urinary bladder) to distal (closest to outside of the body):
      A. prostatic
      B. membranous
      C. spongy
   ___ ___ ___ c. Thickness of the endometrium, from thinnest to thickest:
      A. on day 5 of cycle
      B. on day 13 of cycle
      C. on day 23 of cycle
   ___ ___ ___ d. From anterior to posterior:
      A. uterus and vagina
      B. urinary bladder and urethra
      C. rectum and anus
   ___ ___ ___ ___ e. Order of events in the female monthly cycle, beginning with day 1:
      A. ovulation
      B. formation of follicle
      C. menstruation
      D. formation of corpus luteum
   ___ ___ ___ ___ f. From anterior to posterior:
      A. anus
      B. vaginal orifice
      C. clitoris
      D. urethral orifice
   ___ ___ ___ ___ ___ g. Pathway of sperm:
      A. ejaculatory duct
      B. testis
      C. urethra
      D. ductus (vas) deferens
      E. epididymis
   ___ ___ ___ ___ ___ h. Pathway of milk in breasts:
      A. lactiferous sinuses
      B. alveoli
      C. secondary tubules
      D. mammary ducts
      E. lactiferous ducts
   ___ ___ ___ ___ ___ ___ i. Pathway of sperm entering female reproductive system:
      A. uterine cavity
      B. cervical canal
      C. external os
      D. internal os
      E. uterine (Fallopian) tube
      F. vagina

— — — j. Sequence involved in spermatogenesis:
  A. spermiogenesis
  B. reduction division
  C. maturation division

13. Match the following:
  ___ a. removal of testes
  ___ b. removal of uterus
  ___ c. removal of a portion of the ductus (vas) deferens
  ___ d. tying off of the uterine (Fallopian) tubes
  ___ e. a natural method of birth control involving abstinence during the period when ovulation is most likely
  ___ f. a mechanical method of birth control in which a dome-shaped structure is placed over the cervix
  ___ g. small object, an intrauterine device, placed in the uterus by physician
  ___ h. a rubber sheath over the penis
  ___ i. spermicidal foams, jellies, and creams
  ___ j. "the pill," combination of progesterone and estrogens, causing decrease in FSH and LH so ovulation does not occur

  A. castration
  B. chemical methods
  C. condom
  D. diaphragm
  E. hysterectomy
  F. IUD
  G. oral contraceptive
  H. rhythm
  I. tubal ligation
  J. vasectomy

Choose the one best answer to these questions.

___ 14. The ovaries
  (1) are female gonads.
  (2) produce secondary oocytes and female hormones.
  (3) are homologous to the testes.
  (4) are attached to the pelvic wall by the suspensory ligament.
  A. (1), (3), (4); B. (2), (3), (4); C. (1), (2), (4); D. (1) and (4) only; E. all of the above.

___ 15. Which is *not* a component of the vulva?
  A. labia majora; B. clitoris; C. greater vestibular (Bartholin's) glands; D. ovaries; E. labia minora.

___ 16. Which of these structures is the male gonad?
  A. seminal vesicle; B. scrotum; C. testis; D. prostate gland; E. penis.

___ 17. Which of the following produce(s) a secretion that helps maintain the motility and viability of spermatozoa?
  A. prostate; B. penis; C. bulbourethral (Cowper's) glands; D. ejaculatory duct; E. all of the above.

___ 18. In the human male, the highly coiled duct in which sperm are stored for maturation before being released is called the
  A. epididymis; B. urethra; C. seminal vesicle; D. ductus (vas) deferens; E. ejaculatory duct.

___ 19. The site where fertilization normally occurs is
  A. about one-third the way down the uterine (Fallopian) tube; B. the uterine wall, somewhere in the fundus; C. the cervix of the uterus; D. the abdominal cavity; E. the vagina.

___ 20. The vagina
  (1) is the anterior part of the uterus, sometimes called the cervix.
  (2) is a thin-walled organ whose external opening lies between two folds of skin, the labia minora.
  (3) is the site of the growth of the fetus; it is also called the womb.
  (4) has its inner mucosa slough off each month as the menstrual flow.
  A. (1) only; B. (2) only; C. (3) only; D. (4) only; E. (3) and (4).

___ 21. As part of oogenesis,
  (1) oogonia develop into primary oocytes.
  (2) four haploid (*n*) secondary oocytes are produced from each oogonium.
  (3) further development of primary follicles requires stimulation by FSH.
  (4) equatorial division of a secondary oocyte is completed only if ovulation and fertilization occur.
  A. (1), (2), (3); B. (2), (3), (4); C. (1), (3), (4); D. (1), (2), (4); E. all of the above.

___ 22. Which of these is true concerning the menstrual cycle?
  (1) During the menstrual phase primary follicles develop into secondary follicles.
  (2) A portion of the endometrium is shed during the preovulatory phase.
  (3) A surge of FSH is required for ovulation.
  (4) Rising levels of progesterone and estrogens bring about another menstrual cycle.
  A. (1), (2), (4); B. (1), (2), (3); C. (1) only; D. (3) only; E. (2) only.

23. Match the following:
  ___ a. 200–300 of these per testis
  ___ b. tightly coiled tubes (1–3 per lobule) composed of cells that develop into sperm
  ___ c. cells between developing sperm cells that form the blood–testis barrier and provide nourishment
  ___ d. cells located between seminiferous tubules that secrete testosterone

  A. interstitial endocrinocytes (cells of Leydig)
  B. lobule
  C. sustentacular (Sertoli) cells
  D. seminiferous tubules

26 Developmental Anatomy

STUDENT OBJECTIVES

1. Explain the activities associated with fertilization, morula formation, blastocyst development, and implantation.
2. Describe how external human fertilization, embryo transfer, gamete intrafallopian transfer (GIFT), and transvaginal oocyte retrieval are accomplished.
3. Discuss the formation of the primary germ layers, embryonic membranes, placenta, and umbilical cord as the principal events of the embryonic period.
4. List representative body structures produced by the primary germ layers.
5. Discuss the principal body changes associated with fetal growth.
6. Describe some of the anatomical and physiological changes associated with gestation.
7. Explain amniocentesis and chorionic villi sampling (CVS) as procedures for diagnosing diseases in the newborn.
8. Explain the events associated with the three stages of labor.
9. Define key medical terms associated with developmental anatomy.

CHAPTER OUTLINE

■ **Development During Pregnancy**
Fertilization and Implantation
 Fertilization
 Formation of the Morula
 Development of the Blastocyst
 Implantation
External Human Fertilization
Embryo Transfer
Gamete Intrafallopian Transfer
 (*GIFT*) *and Transvaginal Oocyte Retrieval*
■ **Embryonic Development**
Beginnings of Organ Systems
 Embryonic Membranes
 Placenta and Umbilical Cord
■ **Fetal Growth**
■ **Gestation**
■ **Prenatal Diagnostic Techniques**
Amniocentesis
Chorionic Villi Sampling (CVS)
■ **Parturition and Labor**
■ **Key Medical Terms Associated with Developmental Anatomy**

Developmental anatomy is the study of the sequence of events from the fertilization of a secondary oocyte to the formation of an adult organism. As we look at the sequence from fertilization to birth, we shall consider fertilization, implantation, embryonic development, fetal growth, gestation, parturition, and labor.

DEVELOPMENT DURING PREGNANCY

Pregnancy may occur following the deposition of mature spermatozoa into the vagina. *Pregnancy* is a sequence of events that normally includes fertilization, implantation, embryonic growth, and fetal development that terminates in birth.

FERTILIZATION AND IMPLANTATION

Fertilization

The term *fertilization* refers to the penetration of a secondary oocyte by a spermatozoon and the subsequent union of the sperm nucleus and the nucleus of the oocyte. Of the hundreds of millions of sperm cells introduced into the vagina, very few, perhaps only several hundred to several thousand, arrive in the vicinity of the oocyte. Fertilization normally occurs in the uterine (Fallopian) tube when the oocyte is about one-third of the way down the tube, usually within 24 hours after ovulation. Peristaltic contractions and the action of cilia transport the oocyte through the uterine tube. The mechanism by which sperm reach the uterine tube is apparently related to several factors. Sperm probably swim up the female tract by means of whiplike movements

of their flagella. In addition, the acrosome of sperm produces an enzyme called *acrosin* that stimulates sperm motility and migration within the female reproductive tract. Finally, sperm are probably transported by muscular contractions of the uterus.

In addition to assisting in the transport of sperm, the female reproductive tract also confers on sperm the capacity to fertilize a secondary oocyte. Although sperm undergo maturation in the epididymis, they are still not able to fertilize an oocyte until they have remained in the female reproductive tract for about 10 hours. The functional changes that sperm undergo in the female reproductive tract that allow them to fertilize a secondary oocyte are referred to as *capacitation* (ka'-pas'i'TĀ-shun). During this process, it is believed that the enzymes hyaluronidase and proteinases are secreted by the acrosomes. The enzymes help dissolve the intercellular materials covering the secondary oocyte, the gelatinous glycoprotein layer called the *zona pellucida* (pe-LOO-si-da) and several layers of cells, the innermost of which are follicle cells, known as the *corona radiata* (Figure 26-1a). Once this is accomplished, normally only one spermatozoon enters and fertilizes a secondary oocyte because once union is achieved, the electrical changes in the surface of the oocyte block the entry of other sperm, and enzymes produced by the fertilized ovum (egg) alter receptor sites so that sperm already bound are detached and others are prevented from binding. In this way, polyspermy, fertilization by more than one spermatozoon, is prevented.

When a spermatozoon has entered a secondary oocyte, the tail is shed and the nucleus in the head develops into a structure called the *male pronucleus.* The nucleus of the oocyte develops into a *female pronucleus* (Figure 26-

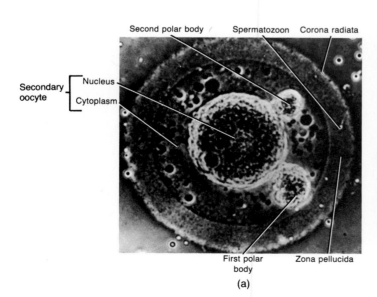

Second polar body / Spermatozoon Corona radiata

Secondary oocyte { Nucleus / Cytoplasm }

First polar body   Zona pellucida

(a)

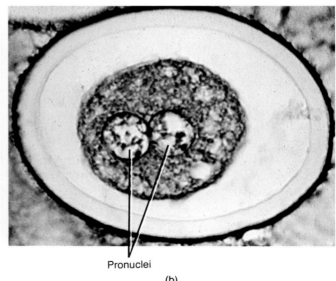

Pronuclei

(b)

FIGURE 26-1 Fertilization and implantation. (a) Photomicrograph of a spermatozoon moving through the zona pellucida on its way to reach the nucleus of the secondary oocyte. (Courtesy of Popperfoto.) (b) Photomicrograph showing male and female pronuclei. (Courtesy of Carolina Biological Supply Company.)

1b). After the pronuclei are formed, they fuse to produce a *segmentation nucleus.* The segmentation nucleus contains 23 chromosomes (*n*) from the male pronucleus and 23 chromosomes (*n*) from the female pronucleus. Thus, the fusion of the haploid (*n*) pronuclei restores the diploid number (2*n*). Once a spermatozoon has entered a secondary oocyte, the oocyte completes equatorial division (meiosis II). The secondary oocyte divides into a larger ovum (mature egg) and a smaller second polar body that fragments and disintegrates. The fertilized ovum, consisting of a segmentation nucleus, cytoplasm, and enveloping membrane, is called a *zygote.*

CLINICAL APPLICATION

Dizygotic (fraternal) twins are produced from the independent release of two ova and the subsequent fertilization of each by different spermatozoa. They are the same age and are in the uterus at the same time, but they are genetically as dissimilar as any other siblings. They may or may not be the same sex. *Monozygotic (identical) twins* are derived from a single fertilized ovum that splits at an early stage in development. They contain the same genetic material and are always the same sex.

In September 1987, a surgical team at Johns Hopkins University performed a unique operation in which 7-month-old *Siamese twins* were successfully separated after being joined at the head since birth. The 22-hour procedure, performed by a 70-member team, was exceedingly complex and involved a total stoppage of circulation and hypothermia for about an hour. This was necessary to prevent hemorrhage while separating a shared sagittal sinus between the infants and to reduce brain activity to near zero to reduce brain swelling. In all known previous attempts to separate Siamese twins joined at their heads, one or both infants died or suffered serious neurological impairment.

Formation of the Morula

Immediately after fertilization, rapid cell division of the zygote takes place. This early division of the zygote is called *cleavage.* During this time, the dividing cells are contained by the zona pellucida. Although cleavage increases the number of cells, it does not result in an increase in the size of the developing organism.

The first cleavage is completed after about 36 hours, and each succeeding division takes slightly less time (Figure 26-2). By the second day after conception, the second cleavage is completed. By the end of the third day there are 16 cells. The progressively smaller cells produced by cleavage are called *blastomeres* (BLAS-tō-mērz). A few days after fertilization the successive cleavages have produced a solid mass of cells, the *morula* (MOR-yoo-la) or mulberry, which is about the same size as the original zygote.

Development of the Blastocyst

As the number of cells in the morula increases, it moves from the original site of fertilization down through the ciliated uterine (Fallopian) tube toward the uterus and enters the uterine cavity. By this time, the dense cluster of cells is altered to form a hollow ball of cells. The mass is now referred to as a *blastocyst* (Figure 26-3).

The blastocyst is differentiated into an outer covering of cells called the *trophoblast* (TRŌF-ō-blast), an *inner cell mass* (*embryoblast*), and an internal fluid-filled cavity called the *blastocoele* (BLAS-tō-sēl). The trophoblast ultimately forms part of the membranes composing the fetal portion of the placenta; the inner cell mass develops into the embryo.

Implantation

The blastocyst remains free within the cavity of the uterus from 2 to 4 days before it actually attaches to the uterine wall. During this time, nourishment is provided by secretions

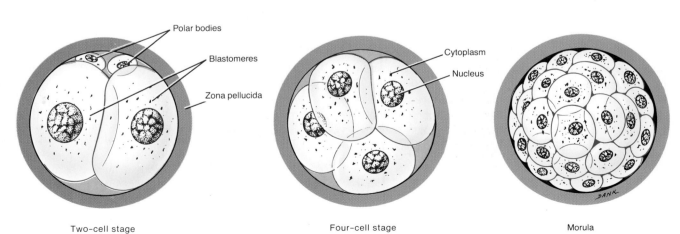

Two-cell stage   Four-cell stage   Morula

FIGURE 26-2 Formation of the morula.

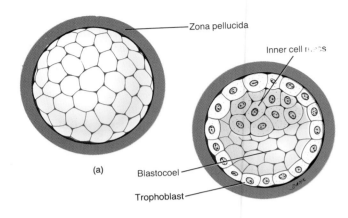

FIGURE 26-3 Blastocyst. (a) External view. (b) Internal view.

of the endometrium, sometimes called uterine milk. The attachment of the blastocyst to the endometrium occurs 7–8 days after fertilization and is called *implantation* (Figure 26-4). At this time, the endometrium is in its postovulatory phase. As the blastocyst becomes implanted, the trophoblast separates into two layers in the region of contact between the blastocyst and endometrium. These layers are an outer *syncytiotrophoblast* (sin-sīt'-ē-ō-TRŌF-ō-blast) that contains no cell boundaries and an inner *cytotrophoblast* (sī-tō-TRŌF-ō-blast) that is composed of distinct cells. During implantation, the syncytiotrophoblast secretes enzymes that enable the blastocyst to penetrate the uterine lining. The enzymes digest and liquefy the endometrial cells. The fluid and nutrients further nourish the burrowing blastocyst for about a week after implantation. Eventually, the blastocyst becomes buried in the endometrium, usually on the posterior wall of the fundus or body of the uterus. The blastocyst becomes oriented so that the inner cell mass is toward the endometrium. Eventually, nutrients are delivered through the placenta for the subsequent growth and development of the embryo and fetus.

A summary of the principal events associated with fertilization and implantation is shown in Figure 26-5.

CLINICAL APPLICATION

In the early months of pregnancy, some females develop *hyperemesis gravidarum* (*morning sickness*), characterized by nausea and vomiting. Although the exact cause is unknown, it is possible that the degenerative products of digested portions of the endometrium during implantation may be responsible. Another possible cause is high levels of human chorionic gonadotropin (HCG) secreted by the placenta.

EXTERNAL HUMAN FERTILIZATION

On July 12, 1978, Louise Joy Brown was born near Manchester, England. Her birth was the first recorded case of *external human fertilization* (*in vitro fertilization*)—fertili-

zation in a glass dish. The procedure developed for external human fertilization is carried out as follows. The female is given follicle-stimulating hormone (FSH) soon after menstruation, so that several secondary oocytes, rather than the typical single one, will be produced. Administration of leuteinizing hormone (LH) may also ensure the maturation of the secondary oocytes. Next, a small incision is made near the umbilicus, and the secondary oocytes are aspirated from the follicles and placed in a medium that simulates the fluids in the female reproductive tract. The secondary oocytes are then transferred to a solution of the male's sperm. Once fertilization has taken place, the fertilized ovum is put in another medium and is observed for cleavage. When the fertilized ovum reaches the 8-cell or 16-cell stage, it is introduced into the uterus for implantation and subsequent growth. The growth and developmental sequences that occur are similar to those in internal fertilization. It is also possible to freeze unused embryos (cryopreservation) to permit parents a successive pregnancy several years later or allow a second attempt at implantation if the first attempt is unsuccessful.

EMBRYO TRANSFER

Embryo transfer is an alternate to external human fertilization. It is a procedure in which a husband's seminal fluid is used to artificially inseminate a fertile sondary oocyte donor and, following fertilization, the morula or blastocyst is transferred from the donor to the infertile wife who carries it to term. Embryo transfer is indicated for females who are infertile, who have surgically untreatable blocked uterine (Fallopian) tubes, or who are afraid of passing on their own genes because they are carriers of a serious genetic disorder.

In the procedure, the donor is monitored to ascertain the time of ovulation by checking her blood levels of luteinizing hormone (LH) and by ultrasound. The wife is also monitored to make sure that her ovarian cycle is synchronized with that of the donor. Once ovulation occurs in the donor, she is artificially inseminated with the husband's seminal fluid. Four days later, a morula or blastocyst is flushed from the donor's uterus through a soft plastic catheter and transferred to the uterus of the wife, where it grows and develops until the time of birth. Embryo transfer is an office procedure that requires no anesthesia and may be performed in a few minutes.

GAMETE INTRAFALLOPIAN TRANSFER (GIFT) AND TRANSVAGINAL OOCYTE RETRIEVAL

Recently, two alternatives to external human fertilization and embryo transfer have been developed. Both are designed to improve the low success rates of external human fertilization. The first is called *gamete intrafallopian transfer* (*GIFT*). The technique is essentially an attempt to mimic the normal process of conception by uniting sperm and secondary oocyte in the prospective mother's uterine (Fallo-

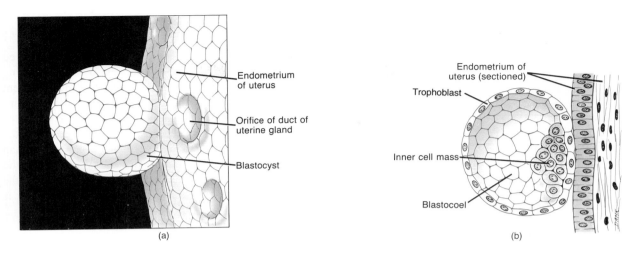

(a)

(b)

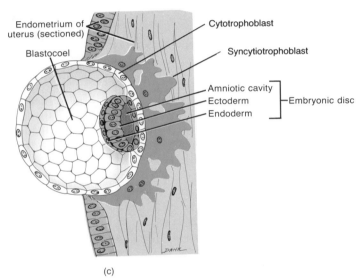

(c)

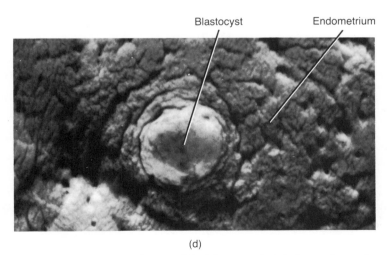

(d)

FIGURE 26-4 Implantation. (a) External view of the blastocyst in relation to the endometrium of the uterus about 5 days after fertilization. (b) Internal view of the blastocyst in relation to the endometrium about 6 days after fertilization. (c) Internal view of the blastocyst at implantation about 7 days after fertilization. (d) Photomicrograph of Implantation. (From *From Conception to Birth: The Drama of Life's Beginnings* by Roberts Rugh, Landrum B. Shettles with Richard Einhorn. Copyright © 1971 by Roberts Rugh and Landrum B. Shettles. By permission of Harper & Row, Publishers, Inc.)

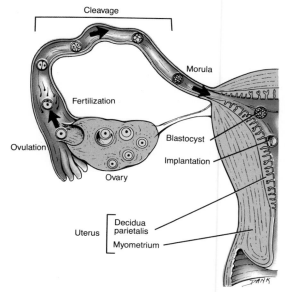

FIGURE 26-5 Summary of events associated with fertilization and implantation.

pian) tubes. In the procedure, the female is given FSH and LH to stimulate the production of several secondary oocytes. The secondary oocytes are aspirated with a laparoscope fitted with a suction device, mixed with a solution of the male's sperm outside the body, and then immediately inserted into the uterine (Fallopian) tubes.

In the second technique, called *transvaginal oocyte retrieval,* the female is given hormones to produce several secondary oocytes, a needle is placed through the vaginal wall and guided to the ovaries by ultrasound (US), suction is applied to the needle, and the secondary oocytes are removed and placed in a solution outside the body. The male's sperm are added to the solution, and the fertilized ova are then implanted in the uterus.

EMBRYONIC DEVELOPMENT

The first 2 months of development are generally considered the *embryonic period.* During this period the developing human is called an *embryo.* The study of development from the fertilized egg through the eighth week in utero is referred to as *embryology* (em-brē-OL-ō-jē). The months of development after the second month are considered the *fetal period,* and during this time the developing human is called a *fetus.* By the end of the embryonic period the rudiments of all the principal adult organs are present, the embryonic membranes are developed, and the placenta is functioning.

BEGINNINGS OF ORGAN SYSTEMS

Following implantation, the inner cell mass of the blastocyst begins to differentiate into the three *primary germ layers:* ectoderm, endoderm, and mesoderm. They are the embryonic tissues from which all tissues and organs of the body will develop. The various movements of cell groups leading to the establishment of the primary germ layers are referred to as *gastrulation.*

In the human, the germ layers form so quickly that it is difficult to determine the exact sequence of events. Within 8 days after fertilization, the top layer of cells of the inner cell mass proliferates and forms the amnion (a fetal membrane) and a space, the *amniotic (amnionic) cavity,* over the inner cell mass. The upper layer of cells of the inner cell mass that is closer to the amniotic cavity develops into the *ectoderm.* The bottom layer of cells of the inner cell mass that borders the blastocoel develops into the *endoderm.*

About the twelfth day after fertilization, striking changes appear (Figure 26-6a). The cells below the amniotic cavity are called the *embryonic disc.* They will form the embryo. At this stage, the embryonic disc contains ectodermal and

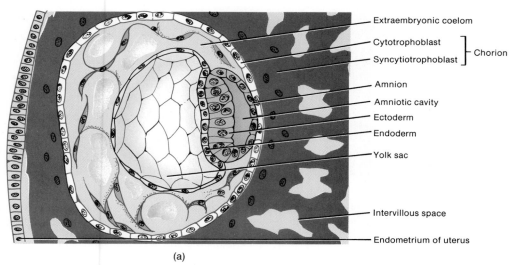

(a)

FIGURE 26-6 Formation of the primary germ layers and associated structures. (a) Internal view of the developing embryo about 12 days after fertilization.

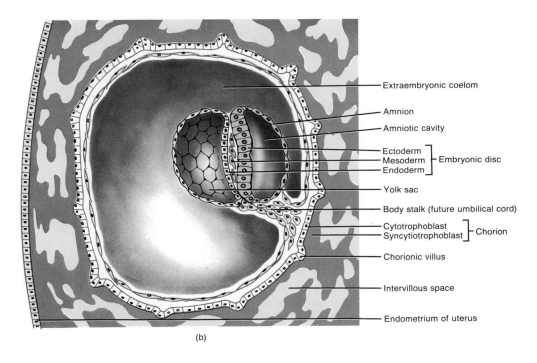

(b)

Extraembryonic coelom

Amnion

Amniotic cavity

Ectoderm
Mesoderm — Embryonic disc
Endoderm

Yolk sac

Body stalk (future umbilical cord)

Cytotrophoblast — Chorion
Syncytiotrophoblast

Chorionic villus

Intervillous space

Endometrium of uterus

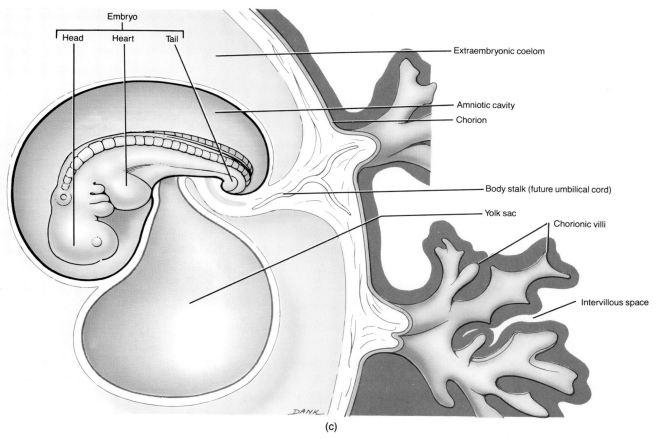

(c)

Embryo
Head Heart Tail

Extraembryonic coelom

Amniotic cavity
Chorion

Body stalk (future umbilical cord)

Yolk sac

Chorionic villi

Intervillous space

FIGURE 26-6 (Continued) (b) Internal view of the developing embryo about 14 days after fertilization. (c) External view of the developing embryo about 25 days after fertilization.

endodermal cells; the mesodermal cells are scattered external to the disc. The cells of the endodermal layer have been dividing rapidly, so that groups of them now extend downward in a circle, forming the yolk sac, another fetal membrane. The cells of the *mesoderm,* which develop between the ectodermal and endodermal layers, also have been dividing, and many have left the area of the embryonic disc and can be seen around the structures that are becoming fetal membranes.

About the fourteenth day, the cells of the embryonic disc differentiate into three distinct layers: the upper ectoderm, the middle mesoderm, and the lower endoderm (Figure 26-6b). The mesoderm in the disc soon splits into two layers, and the space between the layers becomes the *extraembryonic coelom.*

As the embryo develops (Figure 26-6c), the endoderm becomes the epithelial lining of the gastrointestinal tract, respiratory tract, and a number of other organs. The mesoderm forms the peritoneum, muscle, bone, and other connective tissue. The ectoderm develops into the skin and nervous system. Exhibit 26-1 provides more details about the fates of these primary germ layers.

Embryonic Membranes

During the embryonic period, the *embryonic membranes* form (Figure 26-7). These membranes lie outside the embryo and protect and nourish the embryo and, later, the fetus. The membranes are the yolk sac, amnion, chorion, and allantois.

The *yolk sac* is an endoderm-lined membrane that, in many species, provides the primary or exclusive nutrient for the embryo (Figure 26-8; see also Figures 26-6c and 26-7). However, the human embryo receives its nourishment from the endometrium, and the yolk sac remains small. During an early stage of development it becomes a nonfunctional part of the umbilical cord.

The *amnion* is a thin, protective membrane that initially overlies the embryonic disc and is formed by the eighth day following fertilization. As the embryo grows, the amnion entirely surrounds the embryo and becomes filled with

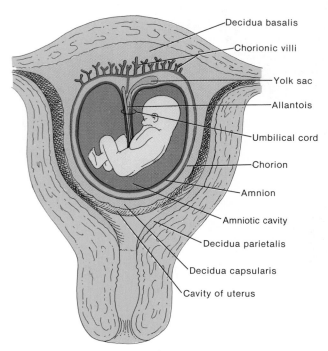

Decidua basalis
Chorionic villi
Yolk sac
Allantois
Umbilical cord
Chorion
Amnion
Amniotic cavity
Decidua parietalis
Decidua capsularis
Cavity of uterus

FIGURE 26-7 Embryonic membranes.

EXHIBIT 26-1

Structures Produced by the Three Primary Germ Layers

ENDODERM	MESODERM	ECTODERM
Epithelium of gastrointestinal tract (except the oral cavity and anal canal) and the epithelium of its glands.	All skeletal, most smooth, and all cardiac muscle.	All nervous tissue.
	Cartilage, bone, and other connective tissues.	Epidermis of skin.
Epithelium of urinary bladder, gallbladder, and liver.		Hair follicles, arrector pili muscles, nails, and epithelium of skin glands (sebaceous and sudoriferous).
	Blood, bone marrow, and lymphoid tissue.	
Epithelium of pharynx, auditory (Eustachian) tube, tonsils, larynx, trachea, bronchi, and lungs.	Endothelium of blood vessels and lymphatics.	Lens, cornea, and internal eye muscles.
	Dermis of skin.	Internal and external ear.
Epithelium of thyroid, parathyroid, pancreas, and thymus glands.	Fibrous tunic and vascular tunic of eye.	Neuroepithelium of sense organs.
	Middle ear.	Epithelium of oral cavity, nasal cavity, paranasal sinuses, salivary glands, and anal canal.
Epithelium of prostate and bulbourethral (Cowper's) glands, vagina, vestibule, urethra, and associated glands such as the greater (Bartholin's) vestibular and lesser vestibular glands.	Mesothelium of ventral body and joint cavities.	
	Epithelium of kidneys and ureters.	Epithelium of pineal gland, pituitary gland, and adrenal medulla.
	Epithelium of adrenal cortex.	
	Epithelium of gonads and genital ducts.	

of pregnancy. The *placenta* has the shape of a flat cake when fully developed and is formed by the chorion of the embryo and a portion of the endometrium (decidua basalis) of the mother (Figure 26-9). It allows the fetus and mother to change nutrients and wastes and secrets hormones necessary to maintain pregnancy.

If implantation occurs, a portion of the endometrium becomes modified and is known as the *decidua* (dē-SID-yoo-a). The decidua includes all but the deepest layer of the endometrium and is shed when the fetus is delivered. Different regions of the decidua, all areas of the stratum functionalis, are named on the basis of their positions relative

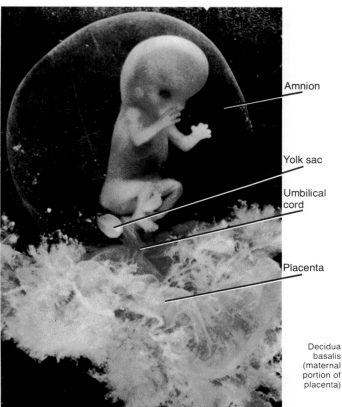

FIGURE 26-8 Ten-week fetus in which the amnion, yolk sac, umbilical cord, and placenta are clearly visible. (From *From Conception to Birth: The Drama of Life's Beginnings* by Roberts Rugh, Landrum B. Shettles with Richard Einhorn. Copyright © 1971 by Roberts Rugh and Landrum B. Shettles. By permission of Harper & Row, Publishers, Inc.)

amniotic fluid or *AF* (Figure 26-8). Amniotic fluid is actually fetal urine and serves as a shock absorber for the fetus. The amnion usually ruptures just before birth and with its fluid constitutes the "bag of waters (BOW)."

The *chorion* (KŌ-rē-on) is derived from the trophoblast of the blastocyst and the mesoderm that lines the trophoblast. It surrounds the embryo and, later, the fetus. Eventually, the chorion becomes the principal embryonic part of the placenta, the structure through which materials are exchanged between mother and fetus. The amnion also surrounds the fetus and eventually fuses to the inner layer of the chorion.

The *allantois* (a-LAN-tō-is) is a small vascularized membrane. Later its blood vessels serve as connections in the placenta between mother and fetus. This connection is the umbilical cord.

Placenta and Umbilical Cord

Development of the placenta, the third major event of the embryonic period, is accomplished by the third month

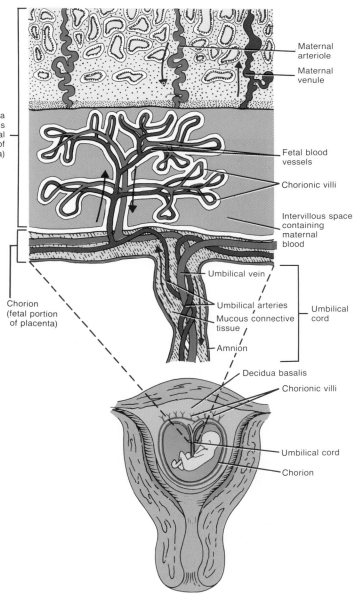

(a)

FIGURE 26-9 Placenta and umbilical cord. (a) Diagram of the structure of the placenta and umbilical cord.

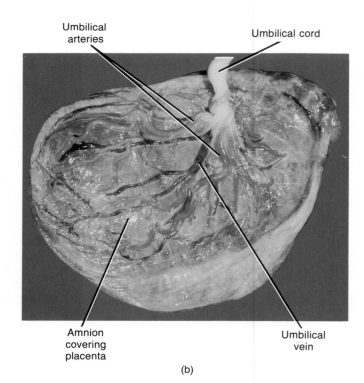

Umbilical
arteries

Umbilical cord

Amnion
covering
placenta

Umbilical
vein

(b)

to the site of the implanted, fertilized ovum (Figure 26-10). The *decidua parietalis* (pa-rī-e-TAL-is) is the portion of the modified endometrium that lines the entire pregnant uterus, except for the area where the placenta is forming. The *decidua capsularis* is the portion of the endometrium between the embryo and the uterine cavity. The *decidua basalis* is the portion of the endometrium between the chorion and the stratum basalis of the uterus. The decidua basalis becomes the maternal part of the placenta.

During embryonic life, fingerlike projections of the chorion, called *chorionic villi* (kō'-rē-ON-ik VIL-ē), grow into the decidua basalis of the endometrium (see also Figures 26-6 and 26-7). These will contain fetal blood vessels of the allantois. They continue growing until they are bathed in maternal blood sinuses called *intervillous* (in-ter-VIL-us) *spaces*. Thus, maternal and fetal blood vessels are brought into proximity. It should be noted, however, that maternal and fetal blood do not normally mix. Oxygen and nutrients from the mother's blood diffuse into the capillaries of the villi. From the capillaries the nutrients circulate

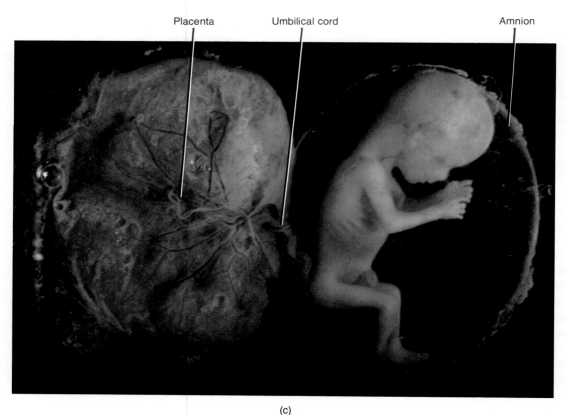

Placenta

Umbilical cord

Amnion

(c)

FIGURE 26-9 (*Continued*) Placenta and umbilical cord. (b) Photograph of the fetal aspect of the placenta. Although the umbilical arteries are red, they carry deoxygenated blood. Similarly, although the umbilical vein is blue, it carries oxygenated blood. (Photograph courtesy of C. Yokochi and J. W. Rohen, *Photographic Anatomy of the Human Body*, 2nd ed., 1979, IGAKU-SHOIN, Ltd., Tokyo, New York.) (c) Twelve-week fetus showing the relation of the placenta to the umbilical cord. (From *From Conception to Birth: The Drama of Life's Beginnings* by Roberts Rugh, Landrum B. Shettles with Richard Einhorn. Copyright © 1971 by Roberts Rugh and Landrum B. Shettles. By permission of Harper & Row, Publishers, Inc.)

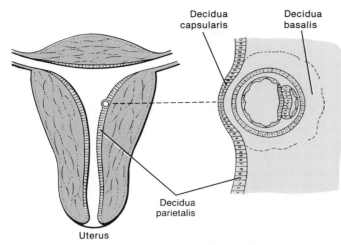

Decidua
capsularis
Decidua
basalis

Decidua
parietalis

Uterus

FIGURE 26-10 Regions of the decidua.

into the umbilical vein. Wastes leave the fetus through the umbilical arteries, pass into the capillaries of the villi, and diffuse into the maternal blood. The **umbilical cord** consists of an outer layer of amnion containing the umbilical arteries and umbilical vein, supported internally by mucous connective tissue (Wharton's jelly) from the allantois.

CLINICAL APPLICATION

At delivery, the placenta detaches from the uterus and is referred to as the **afterbirth.** At this time, the umbilical cord is severed, leaving the baby on its own. The scar that marks the site of the entry of the fetal umbilical cord into the abdomen is the **umbilicus (navel).** Pharmaceutical houses use human placenta as a source of hormones, drugs, and blood. Portions of placentas are also used for burn coverage. The placental and umbilical cord veins are used in blood vessel grafts.

Fetomaternal hemorrhage refers to the entrance of blood into maternal circulation brought on by a dysfunction in placental circulation. Although the condition occurs in at least 50 percent of all pregnancies, in most instances blood loss is so small that the pregnancy is not adversely affected. In some situations, however, the hemorrhage can compromise the fetus and lead to serious complications later in the same pregnancy or a future pregnancy. Among the causes of fetomaternal hemorrhage are trauma, rapid deceleration, placental and umbilical cord abnormalities, amniocentesis, intrauterine fetal surgery, umbilical vein thrombosis, and operative delivery (e.g., cesarean section). As a result of fetomaternal hemorrhage, certain complications may result. Examples include intrauterine fetal death, hypovolemic shock and anemia in the newborn, edema of the newborn, fetal cardiac arrhythmia, and anaphylactic shock in the mother.

FETAL GROWTH

During the **fetal period,** organs established by the primary germ layers grow rapidly. The organism takes on a human appearance. A summary of changes associated with the embryonic and fetal period is presented in Exhibit 26-2.

CLINICAL APPLICATION

Fetal surgery is a new medical field that had its beginnings in 1985. In a pioneering operation, a team of surgeons removed a 23-week-old fetus from its mother's uterus, operated to correct a blocked urinary tract, and then returned the fetus to the uterus. Nine weeks later, a healthy baby was delivered. Surgeons are now experimenting on animals with fetal surgical procedures that could repair diaphragmatic hernias and spina bifida and correct hydrocephalus.

Another relatively new therapy for treating certain diseases is known as **fetal-cell surgery.** In the procedure, tissue is used from aborted fetuses in order to correct certain defects. For example, surgeons in China have been transplanting fetal pancreatic islet (islet of Langerhans) cells to treat type I diabetes since 1982. Swedish researchers hope to transplant fetal brain cells into the brains of patients with Parkinson's disease. It is hoped that fetal liver tissue may be transplanted to cure hereditary blood disorders such as thalassemia.

Fetal cells have the advantage of being immunologically naive; that is, they have not yet developed all the antigens that allow a recipient's immune system to reject the cells. In addition, fetal cells are usually not mature enough to cause graft-versus-host disease, in which the tissues of a transplant recipient are attacked by implanted adult cells. Moreover, fetal nerve cells have the ability to regenerate and thus have the potential to repair damaged brain or spinal cord tissue.

GESTATION

The time the zygote, embryo, or fetus is carried in the female reproductive tract is called **gestation** (jes-TĀ-shun) The total human gestation period is about 280 days from the beginning of the last menstrual period. The specialized branch of medicine that deals with pregnancy, labor, and the period of time immediately following delivery is called **obstetrics** (ob-STET-riks; *obstetrix* = midwife).

By about the end of the third month of gestation, the uterus occupies most of the pelvic cavity, and as the fetus continues to grow, the uterus extends higher and higher into the abdominal cavity. In fact, toward the end of a full-term pregnancy, the uterus occupies practically all of

EXHIBIT 26-2

Changes Associated with Embryonic and Fetal Growth

END OF MONTH	APPROXIMATE SIZE AND WEIGHT	REPRESENTATIVE CHANGES
1	0.6 cm (3/16 in.)	Eyes, nose, and ears not yet visible. Backbone and vertebral canal form. Small buds that will develop into upper and lower extremities form. Heart forms and starts beating. Body systems begin to form.
2	3 cm (1¼ in.) 1 g (1/30 oz)	Eyes far apart, eyelids fused, nose flat. Ossification begins. Limbs become distinct as upper and lower extremities. Digits are well formed. Major blood vessels form. Many internal organs continue to develop.
3	7.5 cm (3 in.) 28 g (1 oz)	Eyes almost fully developed but eyelids still fused, nose develops bridge, and external ears are present. Ossification continues. Appendages are fully formed and nails develop. Heartbeat can be detected. Body systems continue to develop.
4	18 cm (6½–7 in.) 113 g (4 oz)	Head large in proportion to rest of body. Face takes on human features and hair appears on head. Skin bright pink. Many bones ossified, and joints begin to form. Continued development of body systems.
5	25–30 cm (10–12 in.) 227–454 g (½–1 lb)	Head less disproportionate to rest of body. Fine hair (lanugo) covers body. Skin still bright pink. Rapid development of body systems.
6	27–35 cm (11–14 in.) 567–781 g (1¼–1½ lb)	Head becomes even less disproportionate to rest of body. Eyelids separate and eyelashes form. Skin wrinkled and pink.
7	32–42 cm (13–17 in.) 1,135–1,362 g (2½–3 lb)	Head and body become more proportionate. Skin wrinkled and pink. Seven-month fetus (premature baby) is capable of survival.
8	41–45 cm (16½–18 in.) 2,043–2,270 g (4½–5 lb)	Subcutaneous fat deposited. Skin less wrinkled. Testes descend into scrotum. Bones of head are soft. Chances of survival much greater at end of eighth month.
9	50 cm (20 in.) 3,178–3,405 g (7–7½ lb)	Additional subcutaneous fat accumulates. Lanugo shed. Nails extend to tips of fingers and maybe even beyond.

the abdominal cavity, reaching above the costal margin nearly to the xiphoid process of the sternum (Figure 26-11), causing displacement of the maternal intestines, liver, and stomach upward, elevation of the diaphragm, and widening of the thoracic cavity. In the pelvic cavity, there is compression of the ureters and urinary bladder.

In addition to the anatomical changes associated with pregnancy, there are also certain pregnancy-induced physiological changes. General changes include weight gain due to the fetus, amniotic fluid, placenta, uterine enlargement, and increased total body water; increased proteins, fat, and mineral storage; marked breast enlargement in anticipation of lactation; and lower back pain due to lordosis. With respect to the cardiovascular system, there is an increase in stroke volume by about 30 percent; a rise in cardiac output (CO) by 20–30 percent by the twenty-seventh week due to increased maternal blood flow to the placenta and increased metabolism; an increase in heart rate by about 10–15 percent; and an increase in blood volume up to 30–50 percent, mostly during the latter half of pregnancy. In the supine position, the enlarged uterus may compress the aorta, resulting in diminished blood flow to the uterus. Hormonal changes associated with pregnancy and compression of the inferior vena cava can also produce varicose veins.

Pulmonary function is also altered during pregnancy in that tidal volume can increase 30–40 percent, expiratory reserve volume can decrease up to 40 percent, functional

residual capacity can decrease up to 25 percent, minute volume of respiration (MVR) can increase up to 40 percent, and airway resistance in the bronchial tree can decrease up to 36 percent. Dyspnea also occurs.

With regard to the gastrointestinal tract, there is an increase in appetite and a general decrease in motility that can result in constipation and a delay in gastric emptying time. Nausea, vomiting, and heartburn also occur.

Pressure on the urinary bladder by the enlarging uterus can produce urinary symptoms, such as frequency, urgency, and stress incontinence. Other conditions related to the urinary system include an increase in renal plasma flow up to 35 percent, an increase in glomerular filtration rate (GFR) up to 40 percent, and a decrease in ureteral muscle tone.

Changes in the skin during pregnancy are more apparent in some patients than others. Included are increased pigmentation around the eyes and cheekbones in a masklike pattern (chloasma), in the areolae of the breasts, and in the linea alba of the lower abdomen (linea nigra). Striae (stretch marks) over the abdomen occur as the uterus enlarges and hair loss also increases.

Changes in the reproductive system include edema and increased vascularity of the vulva and increased pliability and vascularity of the vagina. The uterus increases in weight from its nonpregnant state of 60–80 g to 900–1,200 g at term. This increase is due to hyperplasia of muscle fibers (cells) in the myometrium in early pregnancy and hypertrophy of muscle fibers during the second and third trimesters.

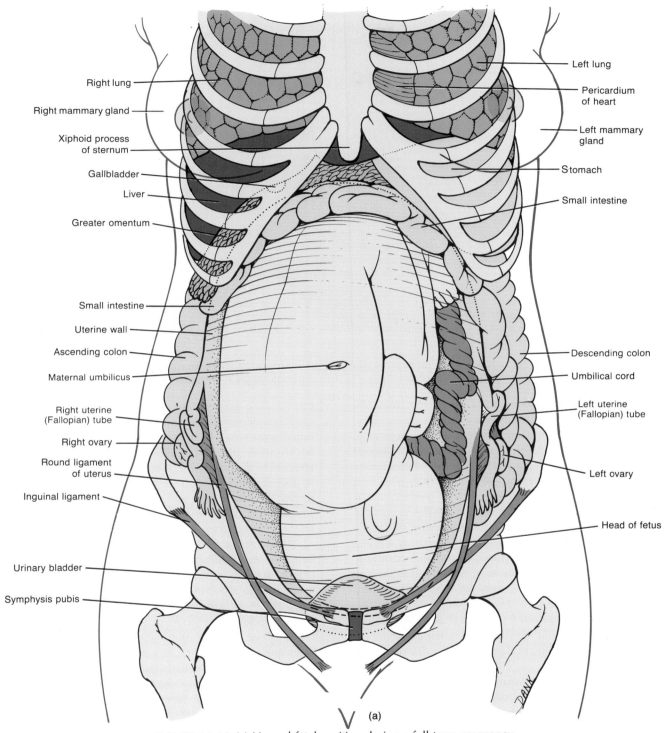

Right lung

Right mammary gland

Xiphoid process
of sternum

Gallbladder

Liver

Greater omentum

Small intestine

Uterine wall

Ascending colon

Maternal umbilicus

Right uterine
(Fallopian) tube

Right ovary

Round ligament
of uterus

Inguinal ligament

Urinary bladder

Symphysis pubis

Left lung

Pericardium
of heart

Left mammary
gland

Stomach

Small intestine

Descending colon

Umbilical cord

Left uterine
(Fallopian) tube

Left ovary

Head of fetus

(a)

FIGURE 26-11 (a) Normal fetal position during a full-term pregnancy.

CLINICAL APPLICATION

Physicians can see into the uterus of a pregnant woman without exposing her to the known dangers of x rays and without pain or intrusion. One such technique, called **_fetal ultrasound,_** uses high-frequency, inaudible sound waves, which are directed into the abdomen of the mother-to-be and then reflected back to a receiver. The reflected waves give a visual "echo" of what's inside the uterus. This echo is transformed electronically into an image on a screen. The potential risks of ultrasonography, such as decreased immune response, changes in plasma membrane functions, and breakdown of macromolecules, are currently under investigation.

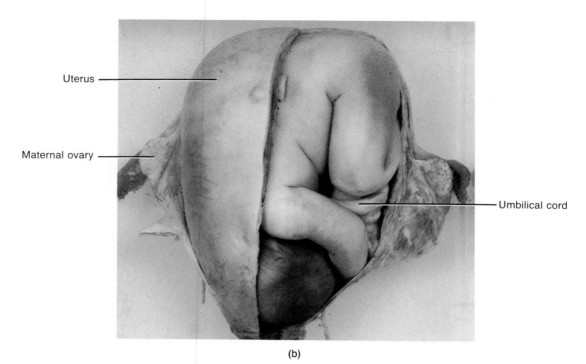

Uterus

Maternal ovary

Umbilical cord

(b)

FIGURE 26-11 (Continued) (b) Normal fetal position during a full-term pregnancy. (Courtesy M. England, *A Colour Atlas of Life Before Birth*, Year Book Medical Publishers.)

In the United States today, most obstetricians prescribe ultrasound examinations only when there is some clinical question about the normal progress of the pregnancy. By far the most common use of diagnostic ultrasound is to determine true fetal age when the date of conception is unknown or mistaken by the mother. It is also used to evaluate fetal growth, determine fetal position, determine the time of ovulation in infertile females to enhance the chances of pregnancy, ascertain the cause of vaginal bleeding, diagnose ectopic and multiple pregnancies, and as an adjunct to special procedures such as amniocentesis. Ultrasound is not used routinely to determine the sex of a fetus; it is performed only for a specific medical indication.

Ultrasound has gained a wide application beyond obstetrics, including the detection of tumors, gallstones, and other abnormal internal masses; heart disorders; and brain hemorrhages.

Electronic fetal monitoring **(EFM)** records fetal heart rate and maternal uterine contractions and is used to evaluate fetal well-being during labor and to detect any early signs of potential problems. EFM is also used to monitor the fetus during special tests that may be given prior to labor (nonstress test and oxytocin [OT] challenge test). Generally, EFM is used to monitor high-risk pregnancies.

In external fetal monitoring, two rubber straps are placed around the abdomen. Attached to the straps are sensors that detect fetal heart rate using ultrasound. The other measures uterine contractions or fetal movements. Measurements are recorded by a small machine that traces them on moving graph paper. External fetal monitoring can be done at any time, including early labor before the cervix dilates and the amniotic sac ruptures. In internal fetal monitoring, fetal heart rate is measured through an electrode placed through the mother's vagina and attached to the fetal scalp. A sensor attached to a catheter is also inserted through the vagina and placed in the uterus. Measurements are also made on graph paper. Internal fetal monitoring can be used only if cervical dilation has occurred and the amniotic sac has ruptured.

The nonstress test uses EFM to check a fetus' well-being before labor begins. After an external monitoring belt is applied to the abdomen, fetal movements are noted on the recording of the fetal heart rate. In this way, fetal movements and heart rate are timed simultaneously. A normal test shows that the fetus moves at least two to three times during a 20-minute period and the heart rate increases with each movement.

The oxytocin (OT) challenge or stress test uses EFM to check fetal well-being also before labor begins. After two external monitoring belts are applied to the abdomen, fetal heart rate and maternal uterine contractions are measured. A small amount of oxytocin (OT) is given intravenously to bring about uterine contractions. Fetal response to the contractions is observed. A normal fetus can adjust to the decreased amount of oxygen that accompanies a contraction, as evidenced by a heart rate that remains the same or increases. This suggests fetal well-being when natural contractions occur during labor. If, instead, the test shows a decreased heart rate during contractions, this suggests that fetal distress may occur during delivery and a cesarean section may be indicated.

PRENATAL DIAGNOSTIC TECHNIQUES

AMNIOCENTESIS

Amniocentesis is a technique of withdrawing some of the amniotic fluid that bathes the developing fetus to diagnose genetic disorders or to determine fetal maturity or well-being. Using ultrasound (US) and palpation, the position of the fetus and placenta are first determined. After a local anesthetic is given, a needle is inserted into the amniotic cavity. About 10–20 ml of fluid is removed by hypodermic needle puncture of the uterus, usually 16–20 weeks after conception (Figure 26-12). Cells and fluid are subjected to microscopic examination and biochemical testing to determine abnormalities in chromosome number or structure and biochemical defects. Close to 300 chromosomal disorders and over 50 inheritable biochemical defects can be detected through amniocentesis, including hemophilia, certain muscular dystrophies, Tay-Sachs disease, myelocytic leukemia, Klinefelter's and Turner's syndromes, sickle-cell anemia, thalassemia, and cystic fibrosis. When both parents are known or suspected to be genetic carriers of any one of these disorders, or when maternal age is over 30, amniocentesis is advised.

CLINICAL APPLICATION

One chromosome disorder that may be diagnosed through amniocentesis is **Down's syndrome** (**DS**). This disorder is the commonest cause of mental retardation in the United States and is characterized by retarded physical development (short stature, muscular flaccidity, and stubby fingers), distinctive facial structures (large tongue, flat profile, broad skull, slanting eyes, epicanthic folds of the eyes, and round head), and malformation of the heart, ears, hands, and feet. DS individuals are also at increased risk of developing cataracts or other vision problems and they have elevated levels of purines that can lead to neurological impairment and immune system deficiencies. Other complications include increased susceptibility to infection and a greater than normal risk for developing leukemia. Sexual maturity is rarely attained.

Individuals with the disorder usually have 47 chromosomes instead of the normal 46 (an extra chromosome in the twenty-first pair). The genes believed to be responsible for many of the abnormalities associated with DS are being identified and mapped on specific sites on the extra twenty-first chromosome. All persons with DS over 35 develop the same kind of abnormal microscopic senile plaques and neurofibrillary tangles in the brain as individuals who die from Alzheimer's disease (AD). Individuals with DS also seem to have a significantly higher risk of developing the neurological symptoms associated with AD. It is possible that DS and AD share a common genetic defect.

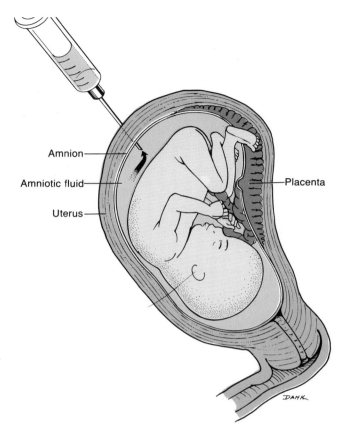

FIGURE 26-12   Amniocentesis.

CHORIONIC VILLI SAMPLING (CVS)

A new test used to detect prenatal genetic defects is now available and is referred to as *chorionic villi sampling* (*CVS*). Although the test picks up the same defects as amniocentesis, it has several advantages. First of all, it can be performed during the first trimester of pregnancy, usually at 8–10 weeks gestation. Also, results are available within 24 hours. Moreover, the procedure does not require penetration of the abdominal wall, uterine wall, or amniotic cavity.

CVS is performed as follows. A catheter is placed through the vagina into the uterus and then to the chorionic villi under ultrasound guidance. About 30 mg of tissue is suctioned out and prepared for chromosomal analysis. Chorion cells and fetal cells contain identical genetic information. The safety of the procedure is believed to be comparable to that for amniocentesis.

PARTURITION AND LABOR

The term *parturition* (par'-too-RISH-un) refers to birth. Parturition is accompanied by a sequence of events commonly called *labor*. The onset of labor is apparently related to a complex interaction of many factors. Just prior to birth, the muscles of the uterus contract rhythmically and forcefully. Both placental and ovarian hormones seem to play a role in these contractions. Since progesterone inhibits uterine contractions, labor cannot take place until its effects are diminished. At the end of gestation, the level of estrogens in the mother's blood is sufficient to overcome the inhibiting effects of progesterone and labor commences. It has been suggested that some factor released by the placenta, fetus, or mother rather suddenly overcomes the inhibiting effects of progesterone so that estrogens can exert their effect.

Prostaglandins may also play a role in labor. Oxytocin (OT) from the posterior pituitary gland also stimulates uterine contractions and relaxin assists by relaxing the symphysis pubis and helping to dilate the uterine cervix.

Uterine contractions occur in waves, quite similar to peristaltic waves, that start at the top of the uterus and move downward. These waves expel the fetus. *True labor* begins when pains occur at regular intervals. The pains correspond to uterine contractions. As the interval between

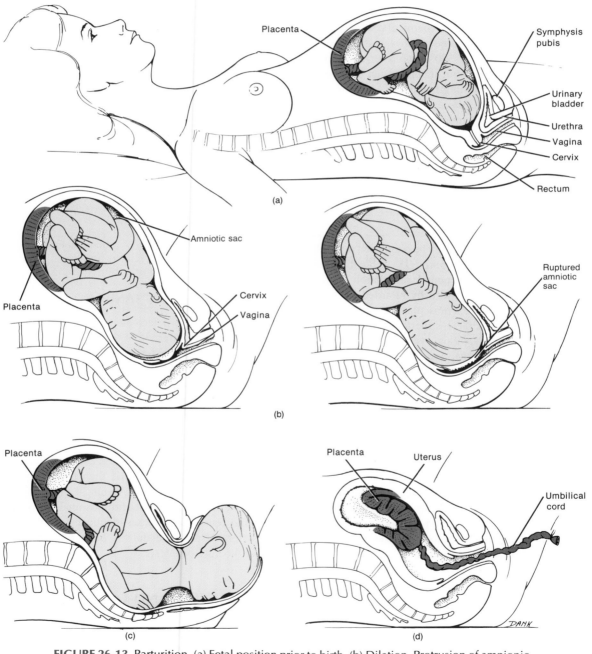

FIGURE 26-13 Parturition. (a) Fetal position prior to birth. (b) Dilation. Protrusion of amnionic sac through partly dilated cervic (left). Amnionic sac ruptured and complete dilation of cervix (right). (c) Stage of expulsion. (d) Placental stage.

contractions shortens, the contractions intensify. Another sign of true labor in some females is localization of pain in the back, which is intensified by walking. A reliable indication of true labor is the "show" and dilation of the cervix. The "show" is a discharge of a blood-containing mucus that accumulates in the cervical canal during labor. In *false labor,* pain is felt in the abdomen at irregular intervals. The pain does not intensify and is not altered significantly by walking. There is no "show" and no cervical dilation.

Labor can be divided into three stages (Figure 26-13).

1. The *stage of dilation* is the time from the onset of labor to the complete dilation of the cervix. During this stage there are regular contractions of the uterus, usually a rupturing of the amniotic sac, and complete dilation (10 cm) of the cervix. If the amniotic sac does not rupture spontaneously, it is done artificially.
2. The *stage of expulsion* is the time from complete cervical dilation to delivery.
3. The *placental stage* is the time after delivery until the placenta or "afterbirth" is expelled by powerful uterine contractions. These contractions also constrict blood vessels that were torn during delivery. In this way, the possibility of hemorrhage is reduced.

During labor, the fetus may be squeezed through the birth canal for up to several hours. As a result, the fetal head is compressed and there is some degree of intermittent hypoxia due to compression of the umbilical cord and placenta during uterine contractions. In response to this compression, the adrenal medulla of a fetus secrets very high levels of epinephrine and norepinephrine (NE), the "fight-or-flight" hormones. Much of the protection afforded against the stresses of the birth process and preparation of the infant to survive extrauterine life are provided by the adrenal medullary hormones. Among other functions, the hormones clear the lungs and alter their physiology for breathing outside the uterus, mobilize readily usable nutrients for cellular metabolism, and promote a rich vascular supply to the brain and heart.

CLINICAL APPLICATION

Pudendal (pyoo-DEN-dal) *nerve block* is used for procedures such as episiotomy. The primary innervation to the skin and muscles of the perineum is the pudendal nerve. In the transvaginal approach, the needle is passed through the lateral vaginal wall to a point just medial to the ischial spine. The anesthesia results in loss of the anal reflex, relaxation of the muscles of the floor of the pelvis, and loss of sensation to the vulva and lower one-third of the vagina.

Various deformities of the female pelvis may be responsible for *dystocia* (dis-TŌ-sē-a), that is, difficult labor. Pelvic deformities may be congenital or acquired from disease, fractures, or poor posture. Among other conditions associated with difficult labor are malposition of the fetus, malpresentation of the fetus, and premature rupture of the fetal membranes.

CLINICAL APPLICATION

If either dystocia or prolonged labor is indicated, it may be necessary to deliver the baby via a *cesarean* (*caedere* = to cut) *section* (*C-section*). In this procedure, a low, horizontal incision is made through the abdominal wall and uterus, through which the baby and placenta are removed.

KEY MEDICAL TERMS ASSOCIATED WITH DEVELOPMENTAL ANATOMY

Abortion (a-BOR-shun) Premature expulsion from the uterus of the products of conception—embryo or nonviable fetus. Nonspontaneous methods include vacuum aspiration (suction curettage), up to the twelfth week of pregnancy; dilation and evacuation (D&E), commonly used between the thirteenth and fifteenth weeks of pregnancy and sometimes up to twenty weeks; and use of saline or prostaglandin preparations to induce labor and delivery, usually after the fifteenth week of pregnancy.

Karyotype (KAR-ē-ō-tīp; *karyon* = nucleus) The chromosomal elements typical of a cell, drawn in their true proportions, based on the average of measurements determined in a number of cells. Useful in judging whether or not chromosomes are normal in number and structure.

Lethal gene (LĒ-thal jēn; *lethum* = death) A gene that, when expressed, results in death either in the embryonic state or shortly after birth.

Lochia (LŌ-kē-a) The discharge from the birth canal consisting initially of blood and later of serous fluid occurring after childbirth. The discharge is derived from the former placental site and may last up to about 2–4 weeks.

Mutation (myoo-TĀ-shun; *mutare* = change) A permanent heritable change in a gene that causes it to have a different effect than it had previously.

Preeclampsia (prē'-e-KLAMP-sē-a) A syndrome characterized by sudden hypertension, large amounts of protein in urine, and generalized edema; might be related to autoimmune or allergic reaction due to the presence of a fetus; when the condition is also associated with convulsions and coma, it is referred to as **eclampsia.**

Puerperal (pyoo-ER-per-al; *puer* = child, *parere* = to bring forth) **fever** Infectious disease of childbirth, also called puerperal sepsis and childbed fever. The disease results from an infection originating in the birth canal and affects the endometrium. It may spread to other pelvic structures and lead to septicemia.

STUDY OUTLINE

Development During Pregnancy (p. 752)
1. Pregnancy is a sequence of events that includes fertilization.
2. Its various events are hormonally controlled.

Fertilization and Implantation (*p. 752*)
1. Fertilization refers to the penetration of a secondary oocyte by a sperm cell and the subsequent union of the sperm and oocyte nuclei to form a zygote.
2. Penetration is facilitated by hyaluronidase and proteinases produced by sperm.
3. Normally only one sperm fertilizes a secondary oocyte.
4. Early rapid cell division of a zygote is called cleavage, and the cells produced by cleavage are called blastomeres.
5. The solid mass of cells produced by cleavage is a morula.
6. The morula develops into a blastocyst, a hollow ball of cells differentiated into a trophoblast (future embryonic membranes) and inner cell mass (future embryo).
7. The attachment of a blastocyst to the endometrium is called implantation; it occurs by enzymatic degradation of the endometrium.
8. External human fertilization refers to the fertilization of a secondary oocyte outside the body and the subsequent implantation of the zygote.
9. In embryo transfer, a husband's seminal fluid is used to artificially inseminate a fertile ovum donor, and following fertilization, the morula or blastocyst is transferred from the donor to the infertile wife.
10. Alternatives to external human fertilization and embryo transfer include gamete intrafallopian transfer (GIFT) and transvaginal oocyte retrieval.

Embryonic Development (p. 756)
1. During embryonic growth, the primary germ layers and embryonic membranes are formed and the placenta is functioning.
2. The primary germ layers—ectoderm, mesoderm, and endoderm—form all tissues of the developing organism.
3. Embryonic membranes include the yolk sac, amnion, chorion, and allantois.
4. Fetal and maternal materials are exchanged through the placenta.

Fetal Growth (p. 761)
1. During the fetal period, organs established by the primary germ layers grow rapidly.
2. The principal changes associated with fetal growth are summarized in Exhibit 26-2.

Gestation (p. 761)
1. The time an embryo or fetus is carried in the uterus is called gestation.
2. Human gestation lasts about 280 days from the beginning of the last menstrual period.
3. During gestation, several anatomical and physiological changes occur.

Prenatal Diagnostic Techniques (p. 765)
1. Amniocentesis is the withdrawal of amniotic fluid. It can be used to diagnose inherited biochemical defects and chromosomal disorders, such as hemophilia, Tay-Sachs disease, sickle-cell anemia, and Down's syndrome.
2. Down's syndrome is a chromosomal abnormality characterized by mental retardation and retarded physical development.
3. Chorionic villi sampling (CVS) involves withdrawal of chorionic villi for chromosomal analysis.
4. CVS can be done sooner than amniocentesis, and the results are available sooner.

Parturition and Labor (p. 765)
1. Parturition refers to birth and is accompanied by a sequence of events called labor.
2. The birth of a baby involves dilation of the cervix, expulsion of the fetus, and delivery of the placenta.

REVIEW QUESTIONS

1. Define developmental anatomy.
2. Define fertilization. Where does it normally occur? How is a morula formed?
3. Explain how dizygotic (fraternal) and monozygotic (identical) twins are produced.
4. Describe the components of a blastocyst.
5. What is implantation? How does the fertilized ovum implant itself? What causes morning sickness?
6. Describe the procedure for external human fertilization, embryo transfer, gamete intrafallopian transfer (GIFT), and transvaginal oocyte retrieval.
7. Define the embryonic period and the fetal period.
8. List several body structures formed by the endoderm, mesoderm, and ectoderm.
9. What is an embryonic membrane? Describe the functions of the four embryonic membranes.
10. Explain the importance of the placenta and umbilical cord to fetal growth.
11. Outline some of the major developmental changes during fetal growth.
12. Define gestation and parturition.
13. Describe several anatomical and physiological changes that occur during gestation.
14. What is amniocentesis? What is its value?
15. Describe chorionic villi sampling (CVS). What are its advantages over amniocentesis?
16. Distinguish between false and true labor. Describe what happens during the stage of dilation, the stage of expulsion, and the placental stage of delivery.
17. Refer to the glossary of key medical terms associated with developmental anatomy. Be sure that you can define each term.

SELF-QUIZ

Choose the one best answer to these questions.

___ 1. Which of these statements is true?
   (1) Capacitation occurs in the male reproductive tract.
   (2) The acrosome produces enzymes that help a sperm cell penetrate a secondary oocyte.
   (3) A segmentation nucleus contains the diploid (2*n*) number of chromosomes.
   (4) The zona pellucida is a gelatinous covering around the acrosome of a sperm cell.
   A. (1), (2), (3); B. (2), (3), (4); C. (2), (3); D. (1), (3); E. all are true.

___ 2. Implantation of a developing individual (blastocyst stage) usually occurs about _____ after fertilization.
   A. 3 weeks; B. 1 week; C. 1 day; D. 7 hours; E. 7 minutes.

___ 3. During pregnancy
   (1) the uterus extends upward into the abdominal cavity.
   (2) there is an increased pulse rate.
   (3) there is a decrease in tidal volume.
   (4) frequency and urgency of urination may occur.
   A. (1), (2), (4); B. (1), (2), (3); C. (2), (3), (4); D. (2) and (3) only; E. all are true.

___ 4. Which procedures involve use of the mother's *own* secondary oocytes for fertilization?
   (1) External human fertilization.
   (2) Embryo transfer.
   (3) Gamete intrafallopian transfer (GIFT).
   (4) Transvaginal oocyte retrieval.
   A. (1), (3), (4); B. (1), (2), (3); C. (1), (2), (4); D. (2), (3), (4); E. all are correct.

___ 5. Which of the following is *not* an embryonic membrane? A. amnion; B. membranous labyrinth; C. chorion; D. allantois; E. yolk sac.

___ 6. The placenta is formed by the union of the decidua basalis of the endometrium with the
   A. yolk sac; B. amnion; C. chorion; D. umbilicus; E. allantois.

___ 7. The "bag of waters" that ruptures just before birth is the
   A. placenta; B. allantois; C. chorion; D. amnion; E. umbilical cord.

Complete the following.

8. Which portion of the decidua is located between the embryo and uterine cavity? (basalis, capsularis, parietalis)

9. The embryonic tissues from which all tissues and organs develop are called _____ layers.

10. In a developing embryo, the upper layer of cells of the inner cell mass near the amniotic cavity develop into _____, whereas the bottom layer of the inner cell mass that borders the blastocoel develops into _____.

11. Identify descriptions of the three phases of labor (first, second, or third).
   ___ a. stage of expulsion: from complete cervical dilation through delivery of the baby
   ___ b. time after the delivery of the baby until the placenta ("afterbirth") is expelled; the placental stage
   ___ c. time from onset of labor to complete dilation of the cervix; the stage of dilation

12. Amniocentesis involves withdrawal of _____ fluid, usually at about (2–4? 8–10? 16–20?) weeks after conception. CVS (meaning _____) is a procedure most often performed at about (2–4? 8–10? 16–20?) weeks. It (does? does not?) involve penetration of the uterine cavity.

13. Write the name of each fetal membrane next to its description.
   a. Originally formed from ectoderm, this membrane encloses fluid that acts as a shock absorber for the developing baby: _____.
   b. Derived from mesoderm and trophoblast, it becomes the principal embryonic part of the placenta: _____.
   c. Endoderm-lined membrane serving as exclusive nutrient supply for embryos of some species: _____.
   d. A small membrane that forms umbilical blood vessels: _____.

14. Identify what structures will ultimately form from these two parts of the implanted blastocyst. Trophoblast: _____; inner cell mass: _____.

15. The early rapid division of a zygote that results in the production of smaller cells called blastomers is known as _____.

16. The normal gestation period is about _____ days from the beginning of the last menstrual period (LMP).

17. Successive divisions of a zygote produce a solid mass of cells called the _____.

18. Match the following:
   ___ a. epithelial lining of all of gastrointestinal, respiratory, and genitourinary tracts except near openings to the exterior of the body
   ___ b. epidermis of skin, epithelial lining of entrances to the body (such as mouth, nose, and anus), hair, nails
   ___ c. all of the skeletal system (bone, cartilage, joint cavities)
   ___ d. muscle (skeletal, smooth, and cardiac)
   ___ e. blood and all blood and lymphatic vessels
   ___ f. entire nervous system
   ___ g. thyroid, parathyroid, thymus, and pancreas

   A. ectoderm
   B. endoderm
   C. mesoderm

19. Arrange the answers in correct sequence.
   ___ ___ ___ ___ a. From most superficial to deepest (closest to embryo):
      A. amnion
      B. amniotic cavity
      C. chorion
      D. decidua
   ___ ___ ___ ___ ___ b. Stages in development:
      A. morula
      B. blastocyst
      C. zygote
      D. fetus
      E. embryo

Appendix A
Symbols and Abbreviations

Many terms, especially medical terms, are commonly expressed in abbreviated form. In order to familiarize you with some of these abbreviations, an alphabetical list has been prepared. Most of the terms listed have been used in the book; some terms not referred to in the book have been included because of their frequent use.

SYMBOLS

♀, ○	female
♂, □	male
*	birth
†	death
∞	infinity

ABBREVIATIONS

Ab	antibody; abortion
ABC	airway, breathing, circulation
ABX	antibiotics
ACh	acetylcholine
ACS	acute confusional state; delirium
ACTH	adrenocorticotropic hormone
ACU	acute care unit
AD	Alzheimer's disease
ADD	attention deficit disorder
ADH	antidiuretic hormone
ADR	adverse drug reaction
AEM	ambulatory electrocardiogram monitoring
AF	amniotic fluid
Ag	antigen
AID	automatic implantable defibrillation; artificial insemination by donor
AIDS	acquired immune deficiency syndrome
ALS	amyotrophic lateral sclerosis
AMI	acute myocardial infarction
ANG	angiogram
ANS	autonomic nervous system
APPY	appendectomy
ARD	acute respiratory disease
ARF	acute renal failure
ASC	altered state of consciousness
ASX	asymptomatic

ATP	adenosine triphosphate
AV	atrioventricular
BBB	blood–brain barrier; bundle branch block
BBT	basal body temperature
BC	birth control
BE	barium enema; bacterial endocarditis; base excess
BEAM	brain electrical activity mapping
BMR	basal metabolic rate
BOW	bag of waters
BP	blood pressure
BPM	beats per minute
BS	blood sugar
BSE	breast self-examiniation
BX	biopsy
C	Celcius
CA	cancer; carcinoma
CABG	coronary artery bypass grafting
CAC	cardioacceleratory center
CAD	coronary artery disease
CAPD	continuous ambulatory peritoneal dialysis
CAVS	calcific aortic valve stenosis
CBC	complete blood count
CBE	clinical breast examination
CCCC	closed-chest cardiac compression
CCI	chronic coronary insufficiency
CCU	cardiac care unit; coronary care unit
CF	cystic fibrosis; cardiac failure
CH	cholesterol
CHF	congestive heart failure
CHR	chronic
CIBD	chronic inflammatory bowel disease
CIC	cardioinhibitory center
CNS	central nervous system
CO	cardiac output; carbon monoxide
COAD	chronic obstructive airways disease
COPD	chronic obstructive pulmonary disease
CP	cor pulmonale; cerebral palsy
CPR	cardiopulmonary resuscitation
CRC	colorectal carcinoma
CRF	chronic renal failure
C-section	cesarean section
CSF	cerebrospinal fluid
CT (CAT)	computed tomography (computed axial tomography)

CT	calcitonin		FOBT	fecal occult blood testing
CUSA	cavitron ultrasonic surgical aspirator		FP	false positive
CVA	cerebrovascular accident		FSH	follicle-stimulating hormone
CVD	cardiovascular disease		FT	full term
CVP	central venous pressure		FUO	fever of unknown origin
CVS	chronic villi sampling		Fx	fracture
CXR	chest x ray			
			GA	gastric analysis; general anesthesia
D & C	dilation and curettage		GB	gallbladder
D & E	dilation and evacuation		GE	gastroenterology
D & V	diarrhea and vomiting		GH	growth hormone
DES	diethylstilbestrol		GI	gastrointestinal
DFU	dead fetus in utero		GIFT	gamete intrafallopian transfer
DH	delayed hypersensitivity; dental hygienist		GIS	gastrointestinal series
DIF	differential blood count		GP	general practitioner
DJD	degenerative joint disease		GUS	genitourinary system
DMD	Duchenne muscular dystrophy		GYN	gynecology; gynecologist
DMSO	dimethyl sulfoxide			
DNA	deoxyribonucleic acid		H & P	history and physical
DNR	do not resuscitate		Hb	hemoglobin
DNS	deviated nasal septum		HBP	high blood pressure
DOA	dead on arrival		HBV	hepatitis B virus
DRG	diagnostic related group		HCG	human chorionic gonadotropin
DS	Down's syndrome		HCS	human chorionic somatotropin
DSA	digital subtraction angiography		Hct	hematocrit
DSR	dynamic spatial reconstructor		HCW	health care worker
DVT	deep-venous thrombosis		HD	Huntington's disease; Hodgkin's disease
Dx	diagnosis		HDL	high-density lipoprotein
			HDN	hemolytic disease of newborn
EBV	Epstein-Barr virus		HF	heart failure
ECF	extracellular fluid		HIVD	herniated intervertebral disc
ECG (EKG)	electrocardiogram		HLA antigen	human leucocyte-associated antigen
ECT	electroconvulsive therapy		HPI	history of present illness
EEG	electroencephalogram		HR	heart rate
EENT	eye, ear, nose, and throat		HSV	herpes simplex virus
EFM	electronic fetal monitoring		HTLV	human T-cell leukemia-lymphoma virus
EM	electron micrograph		HTN	hypertension
EMG	electromyogram; electromyography		Hx	history
EMT	emergency medical technician			
EP	ectopic pregnancy		I & D	incision and drainage
EPSDT	early and periodic screening, diagnosis, and treatment		I & O	intake and output
			IBD	inflammatory bowel disease
EPSP	excitatory postsynaptic potential		IBS	irritable bowel syndrome
ER	endoplasmic reticulum; emergency room		ICC	intensive coronary care
ERT	estrogen replacement therapy		ICF	intracellular fluid
ESR	erythrocyte sedimentation rate		ICU	intensive care unit
ESRD	end-stage renal disease		ID	intradermal
ESWL	extracorporeal shock wave lithotripsy		IF	intrinsic factor
ET	endotracheal; endotracheal tube		IFN	interferon
EW	emergency ward		IM	intramuscular; infectious mononucleosis
EX	examination		IN	internist
			IOP	intraocular pressure
F	Farenheit		IPPA	inspection, palpation, percussion, auscultation
FAS	fetal alcohol syndrome			
FHx	family history		IPSP	inhibitory postsynaptic potential
MHH	family medical history		IUD	intrauterine device

IV	intravenous
IVC	inferior vena cava
IVF	in vitro fertilization
IVP	intravenous pyelogram
IVT	intravenous transfusion
JGA	juxtaglomerular apparatus
KUB	kidneys, ureters, bladder
LBB	left breast biopsy
LDL	low-density lipoprotein
LG	laryngectomy
LH	luteinizing hormone
LLQ	left lower quadrant
LMP	last menstrual period
LOC	loss of consciousness
LP	lumbar puncture
LPN	licensed practical nurse
LRI	lower respiratory infection
LUQ	left upper quadrant
MG	myasthenia gravis
MI	myocardial infarction
MLT	medical laboratory technologist
mm^3	cubic millimeter
mm Hg	millimeters of mercury
MOA	medical office assistant
MRI	magnetic resonance imaging
MS	multiple sclerosis
MSH	melanocyte-stimulating hormone
MVP	mitral valve prolapse
ND	natural death
NE	norepinephrine
NGU	nongonococcal urethritis
NLMC	nocturnal leg muscle cramping
NMJ	neuromuscular junction
NSAID	nonsteroidal anti-inflammatory drug
NSU	nonspecific urethritis
NTG	nitroglycerin
NTP	normal temperature and pressure
OB/GYN	obstetrician–gynecologist; obstetrics–gynecology
OC	oral contraceptive
OD	overdose; right eye
OHS	open heart surgery
OI	opportunistic infection
OR	operating room
ORT	operating room technician
OT	oxytocin
OTC	over-the-counter
OV	office visit
P	pressure
PABA	para-aminobenzoic acid

PCP	*Pneumocystis carinii* pneumonia
PCV	packed cell volume
PD	Parkinson's disease
PE	pulmonary embolism; physical examination
PED	pediatrics; pediatrician
PEG	pneumoencephalogram
PEMF	pulsating electromagnetic field
PET	positron emission tomography
PG	prostaglandin
pH	hydrogen-ion concentration
PID	pelvic inflammatory disease
PKU	phenylketonuria
PMH	past medical history
PMP	plasma membrane protein
PMN	polymorphonuclear leucocyte
PMS	premenstrual syndrome
PNS	peripheral nervous system
PRL	prolactin
PROG	progesterone
PT	prothrombin time; physical therapist
PTCA	percutaneous transluminal coronary angioplasty
PTD	permanent and total disability
PTH	parathyroid hormone
PTT	partial thromboplastin time
PTX	pneumothorax
PUL	percutaneous ultrasonic lithotripsy
Px	prognosis; pneumothorax
PX	physical examination
R	roentgen (unit of x radiation)
RA	rheumatoid arthritis
RAS	reticular activating system
RBB	right breast biopsy
RBC	red blood cell; red blood count
RBOW	rupture of bag of waters
RDA	recommended daily allowance
RDS	respiratory distress syndrome
REM	rapid eye movement
Rh	*Rhesus*
RHC	respirations have ceased
RK	radial keratotomy
RLQ	right lower quandrant
RLX	relaxin
RM	radical mastectomy
RN	registered nurse
RNA	ribonucleic acid
ROS	review of symptoms
RR	respiratory rate
RRR	regular rate and rhythm (heart)
RS	Reye's syndrome
RT	radiotherapy; radiologic technologist
RUQ	right upper quadrant
Rx	prescription

SA	sinoatrial (sinuatrial)	**t-PA**	tissue plasminogen activator
SC	subcutaneous	**TPE**	therapeutic plasma exchange
SCA	sickle-cell anemia	**TPR**	temperature, pulse, and respiration
SCD	sudden cardiac death	**TSH**	thyroid-stimulating hormone
SCID	severe combined immunodeficiency syndrome	**TSS**	toxic shock syndrome
		Tx	treatment
SDS	same-day surgery		
SF	synovial fluid	**UA**	urinalysis
SG	skin graft; specific gravity	**UDO**	undertermined origin
SH	social history	**UG**	urogenital
SIDS	sudden infant death syndrome	**URI**	upper respiratory infection
SIG	sigmoidoscopy; sigmoidoscope	**US**	ultrasound; ultrasonography
SIW	self-inflicted wound	**UTI**	urinary tract infection
SLE	systemic lupus erythematosus	**UV**	ultraviolet
SMD	senile macular degeneration		
SNS	somatic nervous system	**VD**	venereal disease
SPF	sun protection factor	**VDRL**	venereal disease research laboratory test (blood test for syphilis)
S/S (Sx)	signs and symptoms		
STD	sexually transmitted disease	**VF**	ventricular fibrillation
SubQ or **SQ**	subcutaneous	**VPC**	ventricular premature contraction
SVC	superior vena cava	**VS**	vital signs
		VV	varicose veins; vulva and vagina
T	temperature		
TB	tuberculosis	**WBC**	white blood cell; white blood count
TIA	transient ischemic attack	**WNL**	within normal limits
TM	transcendental meditation		
TMJ	temporomandibular joint	**X match**	cross match
TOP	termination of pregnancy	**XRT**	x-ray therapy

Appendix B
Eponyms Used in This Text

Eponymous terms are those named after a person. In general, eponyms should be avoided where possible, since they are totally nondescriptive, often vague, and do not necessarily indicate that the person whose name is used actually contributed anything very original. However, since eponyms are still in frequent use, this glossary has been prepared to indicate which current terms have been used to replace eponyms in this book. In the body of the text eponyms are cited in parentheses, immediately following the current terms where they are used for the first time in a chapter or later in the book. In addition, although eponyms are included in the index, they have been cross-referenced to their current terminology.

EPONYM	CURRENT TERMINOLOGY
Achilles tendon	calcaneal tendon
Adam's apple	thyroid cartilage
ampulla of Vater (VA-ter)	hepatopancreatic ampulla
Bartholin's (BAR-tō-linz) gland	greater vestibular gland
Billroth's (BIL-rōtz) cord	splenic cord
Bowman's (BŌ-manz) capsule	glomerular capsule
Bowman's (BŌ-manz) gland	olfactory gland
Broca's (BRŌ-kaz) area	motor speech area
Brunner's (BRUN-erz) gland	duodenal gland
bundle of His (HISS)	atrioventricular (AV) bundle
canal of Schlemm (SHLEM)	scleral venous sinus
circle of Willis (WIL-is)	cerebral arterial circle
Cooper's (KOO-perz) ligament	suspensory ligament of the breast
Cowper's (KOW-perz) gland	bulbourethral gland
crypt of Lieberkühn (LĒ-ber-kyoon)	intestinal gland
duct of Rivinus (re-VĒ-nus)	lesser sublingual duct
duct of Santorini (san'-tō-RĒ-nē)	accessory duct
duct of Wirsung (VĒR-sung)	pancreatic duct
end-organ of Ruffini (roo-FĒ-nē)	type II cutaneous mechanoreceptor
Eustachian (yoo-STĀ-kē-an) tube	auditory tube
Fallopian (fal-LŌ-pē-an) tube	uterine tube
gland of Littré (LĒ-tra)	urethral gland
gland of Zeis (ZĪS)	sebaceous ciliary gland
Golgi (GOL-jē) tendon organ	tendon organ
Graafian (GRAF-ē-an) follicle	vesicular ovarian follicle
Granstein (GRAN-stēn) cell	nonpigmented granular dendrocyte
Hassall's (HAS-alz) corpuscle	thymic corpuscle
Haversian (ha-VĒR-shun) canal	central canal
Haversian (ha-VĒR-shun) system	osteon
interstitial cell of Leydig (LĪ-dig)	interstitial endocrinocyte
islet of Langerhans (LANG-er-hanz)	pancreatic islet
Kupffer's (KOOP-ferz) cells	stellate reticuloendothelial cell
Langerhans (LANG-er-hanz) cell	nonpigmented granular dendrocyte
loop of Henle (HEN-lē)	loop of the nephron
Malpighian (mal-PIG-ē-an) corpuscle	splenic nodule
Meibomian (mi-BŌ-mē-an) gland	tarsal gland
Meissner's (MĪS-nerz) corpuscle	corpuscle of touch
Merkel's (MER-kelz) disc	tactile disc
Müller's (MIL-erz) duct	paramesonephric duct
Nissl (NIS-l) bodies	chromatophilic substance
node of Ranvier (ron-VĒ-ā)	neurofibral node
organ of Corti (KOR-tē)	spiral organ
Pacinian (pa-SIN-ē-an) corpuscle	lamellated corpuscle
Peyer's (PĪ-erz) patches	aggregated lymphatic follicles
plexus of Auerbach (OW-er-bak)	myenteric plexus
plexus of Meissner (MĪS-ner)	submucosal plexus
pouch of Douglas	rectouterine pouch
Purkinje (pur-KIN-jē) fiber	conduction myofiber
Rathke's (rath-KĒZ) pouch	hypophyseal pouch
Schwann (SCHVON) cell	neurolemmocyte
Sertoli (ser-TŌ-lē) cell	sustentacular cell
Skene's (SKĒNZ) gland	paraurethral gland
sphincter of Oddi (OD-dē)	sphincter of the hepatopancreatic ampulla

Stensen's (STEN-senz) duct	parotid duct
Volkmann's (FŌLK-manz) canal	perforating canal
Wernicke's (VER-ni-kēz) area	auditory association area
Wharton's (HWAR-tunz) duct	submandibular duct
Wharton's (HWAR-tunz) jelly	mucous connective tissue
Wormian (WER-mē-an) bone	sutural bone

Appendix C
Answers to
Self-Quizzes

CHAPTER 1
1. a. armpit b. brachium c. head d. thorax e. neck
2. D **3.** D **4.** B **5.** B **6.** E **7.** C **8.** E **9.** A **10.** B
11. E **12.** C **13.** D **14.** B **15.** E **16.** C **17.** right iliac
18. anterior **19.** ipsi **20.** midsagittal

CHAPTER 2
1. C **2.** C **3.** A **4.** C **5.** A **6.** A **7.** D **8.** E **9.** a. J
b. L c. C d. E e. F f. K g. I h. G i. H j. A k. D
l. B **10.** eating **11.** phospholipids **12.** meiosis **13.** more,
fewer **14.** 23, n **15.** a. A b. D c. E d. C e. B

CHAPTER 3
1. C **2.** A **3.** C **4.** B **5.** D **6.** A **7.** E **8.** D **9.** C
10. E **11.** E **12.** B **13.** C **14.** a. F b. B c. H d. J
e. L f. D g. K h. G i. C j. M k. A l. N m. E n.
I
15. osseous, lamellae, osteocytes, lacunae **16.** visceral, parietal
17. mesenchyme **18.** basement membrane **19.** vascular,
plasma **20.** exocrine

CHAPTER 4
1. C **2.** A **3.** C **4.** B **5.** A **6.** a. A b. E c. C d. D
e. B **7.** melanocyte **8.** ecto, fourth, meso **9.** matrix **10.** papillae, touch **11.** epidermis, epithelium **12.** dermis, connective
tissue **13.** A, C, B **14.** B, C, A **15.** A. sudoriferous
B. sebaceous C. ceruminous **16.** T **17.** T **18.** F **19.** F
20. F

CHAPTER 5
1. a. A b. D c. C d. E e. B **2.** C **3.** D **4.** B **5.** E
6. B **7.** E **8.** C **9.** E **10.** intercellular material **11.** meso-,
mesenchyme **12.** hyaline cartilage and fibrous membranes
13. a. lacunae b. osteocytes c. lamellae d. osteons (Haversian
canals) e. medullary f. canaliculi g. perforating (Volkmann's) canals **14.** a. B, C, A b. A, C, B **15.** T **16.** F
17. T **18.** T **19.** T **20.** F

CHAPTER 6
1. a. E b. N c. D d. G e. L f. H g. J h. F i. I
j. B k. A l. K m. M n. C **2.** D **3.** C **4.** B **5.** A
6. D **7.** C **8.** E **9.** E **10.** B **11.** D **12.** C **13.** A
14. A **15.** C **16.** E **17.** B **18.** C **19.** a. A, B, C b. C,
B, A c. A, C, B **20.** a. A b. A c. C d. E e. C f. E
g. B h. D

CHAPTER 7
1. D **2.** B **3.** E **4.** B **5.** C **6.** B **7.** D **8.** C **9.** B
10. E **11.** C **12.** A, C, B **13.** C, B, A **14.** carpals, 8
15. metacarpals, 5, I

CHAPTER 8
1. a. F b. G c. E d. A e. J f. D **2.** a. knee b. knee
c. hip d. shoulder e. knee f. shoulder **3.** lack **4.** more,
syndesmoses **5.** hyaline, epiphyseal **6.** a. E b. F c. A
d. D e. D f. B, F **7.** D **8.** A **9.** D **10.** E **11.** D
12. A **13.** D **14.** a. D b. B c. G d. C e. F f. A
g. E **15.** B **16.** A **17.** F **18.** F **19.** F **20.** T **21.** F
22. B, A, C

CHAPTER 9
1. a. S b. D c. S d. S e. D **2.** myofiber, sarcolemma,
sarcoplasm **3.** triad, myofilaments **4.** shortens, stays the same
length, stays the same length, shortens, shortens or disappears
5. spindle, one nucleus, do, nonstriated or smooth, slowly, longer
6. skeletal, somites **7.** a. C b. A c. B **8.** a. B b. C
c. A d. D **9.** a. A, C, B b. C, A, B c. C, A, B **10.** B
11. C **12.** B **13.** B **14.** E **15.** C **16.** B **17.** C

CHAPTER 10
1. lever, fulcrum **2.** insertion **3.** R, E, occipital, first **4.** prime
mover (agonist), antagonists **5.** a. frontalis b. buccinator
6. facial, VII **7.** elevating the mandible, temporalis, masseter
8. the same as, superiorly **9.** increase, inspiration
10. urogenital **11.** latissimus dorsi **12.** femur **13.** thenar,
hypothenar **14.** flexor retinaculum **15.** anterior, flex
16. segmental **17.** anterior, 4, tibia, extension, antagonists,
posterior **18.** tibial **19.** ischium, extension, flexion
20. plantar **21.** B **22.** A **23.** B **24.** E **25.** D **26.** a. B,
C b. C, F c. D, C d. E e. A, F **27.** a. D b. E c. C
d. B e. A

CHAPTER 11
1. a. head b. cranium c. side of skull d. supercilia e. eyelids f. mental g. buccal h. ear i. ankle j. antebrachium
k. hand l. phalanges (digits) m. gluteal region n. leg **2.**
cranium, face **3.** sternocleidomastoid, mandible, cervical midline, clavicle, trapezius **4.** a. deltoid b. brachioradialis
c. acromioclavicular joint d. site of the ulnar nerve e. cubital
fossa f. depression between tendons of two muscles that move
the thumb g. distal ends of second through fifth metacarpals
h. styloid process i. iliac crest j. vastus lateralis k. tibial
tuberosity **5.** a. B, C, A b. A, C, B c. C, A, B d. B, A,
C e. B, C, A f. C, A, B g. B, A, C h. A, B, C
6. T **7.** F **8.** T **9.** T **10.** F **11.** T **12.** T

CHAPTER 12
1. a. E b. C c. C d. A e. B f. D g. D, E h. A, B,
E **2.** 55, 45, plasma, hemopoiesis **3.** C, A, B, D **4.** in, 15,000,
leucocytosis, differential blood (DIF) **5.** C **6.** A **7.** E
8. D **9.** B **10.** A **11.** C **12.** C **13.** A **14.** B **15.** C

16. A   **17.** C   **18.** E   **19.** T   **20.** F   **21.** T   **22.** F   **23.** F
24. T

CHAPTER 13
1. a. B   b. D   c. C   d. B   e. A, C   f. B, D   **2.** meso, third, primitive heart   **3.** lubb, closing, atrioventricular   **4.** veins, left atrium   **5.** apex, fifth, second rib   **6.** atria, ventricles   **7.** right atrium, superior vena cava, coronary sinus   **8.** E   **9.** e
10. C   **11.** D   **12.** D   **13.** A   **14.** B   **15.** A   **16.** B   **17.** A
18. C, E, D, A, B   **19.** C, D, E, A, B

CHAPTER 14
1. A. umbilical artery;   B. ductus venosus;   C. femoral artery;   D. pulmonary artery;   E. thoracic portion of inferior vena cava   **2.** hepatic portal   **3.** fourth lumbar, common iliac arteries   **4.** vertebral, basilar   **5.** aorta, coronary   **6.** A. ↓
B. ↓   C. ↑   D. ↑   **7.** A. superior vena cava;   B. inferior vena cava;   C. coronary sinus (answers can be in any sequence)
8. internal jugular   **9.** the heart and a lung   **10.** Yes, Yes, Yes
11. A, C, E, F, G   **12.** A, D, G   **13.** D   **14.** A   **15.** B
16. C   **17.** D   **18.** B   **19.** C   **20.** D   **21.** C   **22.** D   **23.** D
24. C   **25.** D   **26.** D   **27.** D   **28.** E   **29.** A, B, C, D
30. A, B, C, D, E

CHAPTER 15
1. afferent   **2.** cisterna chyli   **3.** meso, fifth   **4.** T   **5.** T
6. F   **7.** F   **8.** T   **9.** D, C, B, A, E   **10.** A, C, B, D
11. a. F   b. I   c. H   d. C   e. B   f. D   g. E   h. A   i. G
j. J   **12.** D   **13.** E   **14.** D   **15.** D   **16.** D   **17.** E   **18.** C
19. C

CHAPTER 16
1. C   **2.** C   **3.** A   **4.** B   **5.** C   **6.** E   **7.** E   **8.** A   **9.** E
10. A   **11.** a. C   b. F   c. G   d. H   e. D   f. E   g. A   h. B
12. synapse   **13.** dendrite   **14.** synaptic vesicles   **15.** parallel after-discharge   **16.** T   **17.** F   **18.** T   **19.** T   **20.** F

CHAPTER 17
1. a. B   b. C   c. D   d. A   **2.** endo, peri, epi   **3.** sensory, anterior, lateral   **4.** white columns, myelinated, sensory, brain   **5.** a. A   b. B   c. A   d. C   e. B   f. D   g. E   **6.** C
7. C   **8.** D   **9.** B   **10.** B   **11.** A   **12.** C   **13.** D   **14.** C
15. B   **16.** C   **17.** a. D   b. C   c. A   d. B   e. F   f. E
18. F   **19.** T   **20.** T   **21.** F   **22.** T   **23.** F   **24.** a. B, C, A,
b. B, C, A,   c. C, B, A, D

CHAPTER 18
1. cuneatus   **2.** inferior   **3.** tentorium cerebelli   **4.** association
5. chiasma, tracts   **6.** cortex, gray, cell bodies, gyri, fissures, sulci   **7.** basal ganglia, gray   **8.** ecto, neural   **9.** a. accessory;   b. vestibulocochlear;   c. vagus;   d. trigeminal;   e. trigeminal;   f. oculomotor;   g. facial;   h. facial, glossopharyngeal;   i. olfactory, optic, vestibulocochlear (answers can be in any sequence)   **10.** A, C, D   **11.** B, D, E   **12.** a. A   b. C
c. C   d. D   e. B   f. A   g. E   h. B   i. E   j. A   k. E   l. C
m. A   **13.** C   **14.** C   **15.** C   **16.** B   **17.** B   **18.** B   **19.** A
20. C   **21.** D   **22.** B   **23.** a. B, A, C   b. C, A, B   **24.** T
25. F   **26.** T   **27.** T   **28.** F   **29.** T   **30.** T

CHAPTER 19
1. a. S   b. P   c. S   d. P   e. S   f. S   g. S   h. S   i. P, S
j. S   **2.** A, E, F   **3.** A, C, D   **4.** B   **5.** a. P ↓   S ↑
b. P ↓   S ↑   c. P =   S ↑   d. P ↑   S ↓
e. P ↓   S ↑   f. P ↑   S ↓   g. P =   S ↑
h. P =   S ↓   i. P ↑   S =   j. P ↓   S ↑
k. P ↑   S ↓   l. P ↑   S ↓   **6.** T   **7.** F   **8.** T
9. T   **10.** D   **11.** E   **12.** D   **13.** D   **14.** D   **15.** A   **16.** B
17. E   **18.** B   **19.** C

CHAPTER 20
1. upper motor   **2.** second-order   **3.** basal ganglia, reticular formation   **4.** papillae   **5.** otoliths, static   **6.** G, D, B, F, E, A, H, C   **7.** a. D   b. A   c. C   d. B   **8.** D, B, A, C   **9.** B, C, D, E, A   **10.** B, C, A, D, E   **11.** A, B, E, C, D   **12.** A, C, B   **13.** C, D, A, B   **14.** C, D, A, B   **15.** A, E, C, B, D   **16.** A, D, E, B, C   **17.** a. K   b. F   c. G   d. A   e. C
f. J   g. B   h. I   i. E   j. D   k. H   l. L   **18.** A   **19.** B
20. C   **21.** C   **22.** B   **23.** A   **24.** B   **25.** B   **26.** a. A
b. H   c. I   d. D   e. E   f. C   g. B

CHAPTER 21
1. alpha, increases   **2.** SHA, PP, HPV, SP   **3.** B   **4.** A
5. C   **6.** D   **7.** A   **8.** C   **9.** C   **10.** D   **11.** E   **12.** D
13. B   **14.** C   **15.** B   **16.** E   **17.** C   **18.** F   **19.** F   **20.** T
21. F   **22.** F   **23.** F   **24.** T   **25.** a. D   b. C   c. B   d. A
e. E, F

CHAPTER 22
1. base, apex, ribs   **2.** C, B, A   **3.** right, left, left, right
4. endo   **5.** a. D   b. E   c. F   d. B   e. A   f. A   g. D, E
h. C, F   **6.** a. A, C, B   b. D, C, E, B, A   c. A, G, C, B, D
d. D, C, E, B, A, F   **7.** A   **8.** E   **9.** D   **10.** B   **11.** E
12. C   **13.** D   **14.** E   **15.** D   **16.** C   **17.** T   **18.** T   **19.** F
20. F   **21.** F

CHAPTER 23
1. pepsin   **2.** a. E   b. E   c. M   **3.** 32, 20, incisors, cuspids, premolars, molars   **4.** defecation   **5.** cystic   **6.** mouth   **7.** circular folds   **8.** anal canal   **9.** lateral   **10.** a. A, D, B, C   b. E, D, B, A, C   c. A, D, C, B, E   d. A, B, D, C, E
11. a. A   b. D   c. E   d. B   e. C   **12.** D   **13.** C   **14.** B
15. B   **16.** E   **17.** B   **18.** A   **19.** C   **20.** A   **21.** D
22. B   **23.** B   **24.** C   **25.** D   **26.** B   **27.** A

CHAPTER 24
1. podocytes   **2.** detrusor   **3.** hemodialysis   **4.** renal corpuscle   **5.** nephron   **6.** renal pyramids   **7.** ureter   **8.** a. A, B, C   b. A, C, B, D   c. A, C, E, B, D   d. E, B, A, D, C
e. C, E, A, B, D   **9.** A   **10.** C   **11.** D   **12.** E   **13.** D
14. B   **15.** B   **16.** C   **17.** C   **18.** B   **19.** C   **20.** A

CHAPTER 25
1. a. scrotum;   b. penis;   c. prostate gland;   d. bulbourethral (Cowper's) glands   **2.** haploid, 23, zygote, diploid, $2n$, homologous   **3.** a. M;   b. E;   c. E   **4.** spermatogenesis
5. spermatogonium   **6.** cardinal, myo, smooth muscle, endo, functionalis, spiral   **7.** corpora cavernosa   **8.** ovarian
9. ovulation   **10.** corpus albicans   **11.** vagina   **12.** a. B, C, A
b. A, B, C   c. A, B, C   d. B, A, C   e. C, B, A, D   f. C, D, B, A   g. B, E, D, A, C   h. B, C, D, A, E   i. F, C, B,

D, A, E   j. B, C, A   **13.** a. A   b. E   c. J   d. I   e. H   f. D
g. F   h. C   i. B   j. G   **14.** E   **15.** E   **16.** C   **17.** A   **18.** A
19. A   **20.** B   **21.** C   **22.** C   **23.** a. B   b. D   c. C   d. A

CHAPTER 26

1. C   **2.** B   **3.** A   **4.** A   **5.** B   **6.** C   **7.** D   **8.** capsularis
9. primary germ   **10.** ectoderm, endoderm   **11.** a. second;
b. third;   c. first   **12.** amniotic, 16–20;   chorionic villi sampling,
8–10, does not   **13.** a. amnion, b. chorion, c. yolk sac,
d. allantois   **14.** fetal portion of placenta, embryo
15. cleavage   **16.** 280   **17.** morula   **18.** a. B   b. A   c. C
d. C   e. C   f. A   g. B   **19.** a. D, C, A, B   b. C, A, B,
E, D

Selected Readings

Anderson, J. E., *Grant's Atlas of Anatomy,* 8th ed. Baltimore: Williams & Wilkins, 1983.

Barr, M. L., and J. A. Kiernan, *The Human Nervous System,* 4th ed. New York: Harper & Row, 1983.

Basmajian, J. V., *Grant's Method of Anatomy,* 10th ed. Baltimore: Williams & Wilkins, 1980.

Basmajian, J. V., *Primary Anatomy,* 8th ed. Baltimore: Williams & Wilkins, 1982.

Basmajian, J. V., *Surface Anatomy,* Baltimore: Williams & Wilkins, 1977.

Borysenko, M., and T. Beringer, *Functional Histology,* 2nd ed. Boston: Little, Brown & Co., 1984.

Carlson, B. M., *Patten's Foundations of Human Embryology,* 4th ed. New York: McGraw-Hill, 1981.

Carpenter, M. B., *Care Text of Neuroanatomy,* 3rd ed. Baltimore: Williams & Wilkins, 1985.

Clemente, C. D., *Anatomy: A Regional Atlas of the Human Body,* 3rd ed. Baltimore: Urban and Schwarzenberg, 1987.

Crafts, R. C., *A Textbook of Human Anatomy,* 3rd ed. New York: Wiley, 1985.

Cunningham's Textbook of Anatomy, 12th ed. Edited by G. J. Romanes. London: Oxford University Press, 1981.

Danforth, D. N., and J. R. Scott, *Obstetrics and Gynecology,* 5th ed. Philadelphia: Lippincott, 1986.

DiFiore, M. S. H., *An Atlas of Human Histology,* 5th ed. Philadelphia: Lea & Febiger, 1981.

Ezrin, C., J. O. Godden, and R. Volpé, *Systematic Endocrinology,* 2d ed. New York: Harper & Row, 1979.

Fawcett, D. W., *A Textbook of Histology,* 11th ed. Philadelphia: Saunders, 1986.

Geneser, F., *Color Atlas of Histology,* Philadelphia: Lea & Febiger, 1985.

Ger, R., and P. Abrahams, *Essentials of Clinical Anatomy,* London: Pitman, 1986.

Gluhbegovic, N., and T. H. Williams, *The Human Brain: A Photographic Guide,* New York: Harper & Row, 1980.

Gray, H., *Anatomy of the Human Body,* 30th ed. Edited by Carmine D. Clemente, Philadelphia: Lea & Febiger, 1985.

Hafez, E. S. E., *Human Reproduction,* 2d ed. New York: Harper & Row, 1980.

Ham, A. W., and D. H. Cormack, *Histology,* 8th ed., Philadelphia: Lippincott, 1979.

Hamilton, W. J., G. Simon, and S. G. I. Hamilton, *Surface and Radiological Anatomy,* 5th ed. Cambridge: Heffer & Sons, 1971.

Hollinshead, W. H., and C. Rosse, *Textbook of Anatomy,* 4th ed. New York: Harper & Row, 1985.

Kelly, D. E., R. L. Wood, and A. C. Enders, *Bailey's Textbook of Microscopic Anatomy,* 18th ed. Baltimore: Williams & Wilkins, 1984.

Keogh, B., and S. Ebbs, *Normal Surface Anatomy with Practical Applications,* Philadelphia: Lippincott, 1984.

Kieffer, S. A., and E. R. Heitzman, *An Atlas of Cross-Sectional Anatomy.* New York: Harper & Row, 1979.

Langman, J., *Medical Embryology,* 5th ed. Baltimore: Williams & Wilkins, 1985.

Langman, J., and M W. Woerdeman, *Atlas of Medical Anatomy.* Philadelphia: Saunders, 1982.

Lockhart, R. D., *Living Anatomy,* 7th ed. London: Faber & Faber, 1974.

Martin, A. H., *Introduction to Human Anatomy,* New York: Thieme-Stratton, 1985.

McMinn, R. M. H., and R. T. Hutchings, *Color Atlas of Human Anatomy.* Chicago: Year Book Medical Publishers, 1977.

Moore, K. L., *Clinically Oriented Anatomy,* 2nd ed. Baltimore: Williams & Wilkins, 1985.

Moyer, K. E., *Neuroanatomy.* New York: Harper & Row, 1980.

Netter, F. H., *Ciba Collection of Medical Illustrations,* Complete collection. Summit, NJ: CIBA, 1962–1983.

Noback, C. R., and R. J. Demarest, *The Nervous System: Introduction and Review,* 3d ed. New York: McGraw-Hill, 1986.

O'Rahilly, R., *Basic Human Anatomy.* Philadelphia: Saunders, 1983.

Pernkopf, E., *Atlas of Topographical and Applied Human Anatomy.* Edited by H. Ferner; translated by H. Monsen. Philadelphia: Saunders, Vols. 1 and 2, 1981.

Reith, E. J., and M. N. Ross, *Histology: A Text and Atlas.* New York: Harper & Row, 1985.

Rohen, J. W., and C. Yokochi, *Color Atlas of Anatomy.* Tokyo, New York: Igaku-Shoin, Ltd., 1983.

Royce, J., *Surface Anatomy.* Philadelphia: Davis, 1973.

Snell, R. S., *Clinical Anatomy for Medical Students,* 3rd ed. Boston: Little, Brown & Co., 1986.

Snell, R. S., *Clinical and Functional Histology for Medical Students.* Boston: Little, Brown & Co., 1984.

Sobotta Atlas of Human Anatomy, Vols. 1 and 2, 10th English ed. Edited by H. Ferner and J. Staubesand. Baltimore: Urban & Schwarzenberg, 1983.

Sobotta, J., and F. Hammersen, *Histology.* Baltimore: Urban & Schwarzenberg, 1985.

Williams, R. H., *Textbook of Endocrinology,* 7th ed. Philadelphia: Saunders, 1985.

Woodburne, R. T., *Essentials of Human Anatomy,* 7th ed. New York: Oxford University Press, 1983.

Glossary of Terms

PRONUNCIATION KEY

1. The strongest accented syllable appears in capital letters, for example, bilateral (bī-LAT-er-al) and diagnosis (dī-ag-NŌ-sis).
2. If there is a secondary accent, it is noted by a single quote mark ('), for example, constitution (kon'-sti-TOO-shun) and physiology (fiz'-ē-OL-ō-jē). Any additional secondary accents are also noted by a single quote mark, for example, decarboxylation (dē'-kar-bok'-si-LĀ-shun).
3. Vowels marked with a line above the letter are pronounced with the long sound, as in the following common words:

   ā as in *māke*
   ē as in *bē*
   ī as in *īvy*
   ō as in *pōle*

4. Vowels not so marked are pronounced with the short sound, as in the following words:

   e as in *bet*
   i as in *sip*
   o as in *not*
   u as in *bud*

5. Other phonetic symbols are used to indicate the following sounds:

   a as in *above*
   oo as in *sue*
   yoo as in *cute*
   oy as in *oil*

Abatement (a-BĀT-ment) A decrease in the seriousness of a disorder or in the severity of pain or other symptoms.

Abdomen (ab-DŌ-men or AB-dō-men) The area between the diaphragm and pelvis.

Abdominal (ab-DŌM-i-nal) *cavity* Superior portion of the abdominopelvic cavity that contains the stomach, spleen, liver, gallbladder, pancreas, small intestine, and most of the large intestine.

Abdominal thrust maneuver A first-aid procedure for choking. Employs a quick, upward thrust against the diaphragm that forces air out of the lungs with sufficient force to eject any lodged material. Also called the *Heimlich* (HĪM-lik) *maneuver*.

Abdominopelvic (ab-dom'-i-nō-PEL-vic) *cavity* Inferior component of the ventral body cavity that is subdivided into an upper abdominal cavity and a lower pelvic cavity.

Abduction (ab-DUK-shun) Movement away from the axis or midline of the body or one of its parts.

Abnormal Curvature (KUR-va-tūr) A nonangular deviation of a straight line, as in the greater and lesser curvatures of the stomach. Abnormal curvatures of the vertebral column include kyphosis, lordosis, and scoliosis.

Abortion (a-BOR-shun) The premature loss or removal of the embryo or nonviable fetus; any failure in the normal process of developing or maturing.

Abscess (AB-ses) A localized collection of pus and liquefied tissue in a cavity.

Absorption (ab-SORP-shun) The taking up of liquids by solids or of gases by solids or liquids; intake of fluids or other substances by cells of the skin or mucous membranes; the passage of digested foods from the gastrointestinal tract into blood or lymph.

Accessory duct A duct of the pancreas that empties into the duodenum about 2.5 cm (1 in.) superior to the ampulla of Vater (hepatopancreatic ampulla). Also called the *duct of Santorini* (san'-tō-RE-ne).

Accretion (a-KRĒ-shun) A mass of material that has accumulated in a space or cavity; the adhesion of parts.

Acetabulum (as'-e-TAB-yoo-lum) The rounded cavity on the external surface of the coxal (hip) bone that receives the head of the femur.

Acetylcholine (as'-ē-til-KŌ-lēn) *(ACh)* A neurotransmitter, liberated at synapses in the central nervous system, that stimulates skeletal muscle contraction.

Achille's tendon *See Calcaneal tendon.*

Achlorhydria (ā-klōr-HĪ-drē-a) Absence of hydrochloric acid in the gastric juice.

Acid (AS-id) A proton donor, or substance that dissociates into hydrogen ions (H^+) and anions, characterized by an excess of hydrogen ions and a pH less than 7.

Acinar (AS-i-nar) Flasklike.

Acini (AS-i-nē) Masses of cells in the pancreas that secrete digestive enzymes.

Acne (AK-nē) Inflammation of sebaceous (oil) glands that usually begins at puberty; the basic acne lesions in order of increasing severity are comedones, papules, pustules, and cysts.

Acoustic (a-KOOS-tik) Pertaining to sound or the sense of hearing.

Acquired immune deficiency syndrome *(AIDS)* A deficiency of helper T cells and a reversed ratio of helper T cells to suppressor T cells that results in fever or night sweats, coughing, sore throat, fatigue, body aches, weight loss, and enlarged lymph nodes. Caused by a virus called human immunodeficiency virus (HIV).

Acromegaly (ak'-rō-MEG-a-lē) Condition caused by hypersecretion of growth hormone (GH) during adulthood characterized by thickened bones and enlargement of other tissues.

Acrosome (AK-rō-sōm)   A dense granule in the head of a spermatozoon that contains enzymes that facilitate the penetration of a spermatazoon into a secondary oocyte.

Actin (AK-tin)   The contractile protein that makes up thin myofilaments in muscle fiber (cell).

Action potential   A wave of negativity that self-propagates along the outside surface of the membrane of a neuron or muscle fiber (cell). Also called a *nerve impulse* (*nerve action potential*) as it relates to a neuron and a *muscle action potential* as it relates to a muscle fiber.

Active transport   The movement of substances, usually ions, across cell membranes, against a concentration gradient, requiring the expenditure of energy (ATP).

Acuity (a-KYOO-i-tē)   Clearness or sharpness, usually of vision.

Acupuncture (AK-ū-punk'-chur)   The insertion of a needle into a tissue for the purpose of drawing fluid or relieving pain. It is also an ancient Chinese practice employed to cure illnesses by inserting needles into specific locations of the skin.

Acute (a-KYOOT)   Having rapid onset, severe symptoms, and a short course; not chronic.

Adam's apple   *See Thyroid cartilage.*

Adaptation (ad'-ap-TĀ-shun)   The adjustment of the pupil of the eye to light variations. The property by which a neuron relays a decreased frequency of action potentials from a receptor even though the strength of the stimulus remains constant. The decrease in perception of a sensation over time while the stimulus is still present.

Addison's (AD-i-sonz) *disease*   Disorder caused by hyposecretion of glucocorticoids characterized by muscular weakness, mental lethargy, weight loss, low blood pressure, and dehydration.

Adduction (ad-DUK-shun)   Movement toward the axis or midline of the body or one of its parts.

Adenohypophysis (ad'-e-nō-hī-POF-i-sis)   The anterior portion of the pituitary gland.

Adenoids (AD-e-nyods)   The pharyngeal tonsils.

Adenosine triphosphate (a-DEN-ō-sēn trī-FOS-fāt (*ATP*)   The universal energy-carrying molecule manufactured in all living cells as a means of capturing and storing energy. It consists of the purine base *adenine* and the five-carbon sugar *ribose,* to which are added, in linear array, three *phosphate* molecules.

Adherence (ad-HĒR-ens)   Firm contact between the plasma membrane of a phagocyte and an antigen.

Adhesion (ad-HĒ-zhun)   Abnormal joining of parts to each other.

Adipocyte (AD-i-pō-sīt)   Fat cell, derived from a fibroblast.

Adrenal cortex (a-DRĒ-nal KOR-teks)   The outer portion of an adrenal gland, divided into three zones, each of which has a different cellular arrangement and secretes different hormones.

Adrenal (a-DRĒ-nal) *glands*   Two glands located superior to each kidney. Also called the *suprarenal* (soo'-pra-RĒ-nal) *glands.*

Adrenal medulla (me-DUL-a)   The inner portion of an adrenal gland, consisting of cells that secrete epinephrine and norepinephrine (NE) in response to the stimulation of preganglionic sympathetic neurons.

Adrenergic (ad'-ren-ER-jik) *fiber*   A nerve fiber that, when stimulated releases norepinephrine (noradrenaline) at a synapse.

Adrenocorticotropic (ad-rē'-nō-kor-ti-kō-TRŌP-ik) *hormone* (*ACTH*)   A hormone produced by the adenohypophysis (anterior lobe) of the pituitary gland that influences the production and secretion of certain hormones of the adrenal cortex.

Adrenoglomerulotropin (a-drē'-nō-glō-mer'-yoo-lō-TRŌ-pin)   A hormone secreted by the pineal gland that may stimulate aldosterone secretion.

Adventitia (ad-ven-TISH-ya)   The outermost covering of a structure or organ.

Aerobic (air-Ō-bik)   Requiring molecular oxygen.

Afferent arteriole (AF-er-ent ar-TĒ-rē-ōl)   A blood vessel of a kidney that breaks up into the capillary network called a glomerulus; there is one afferent arteriole for each glomerulus.

Afferent neuron (NOO-ron)   A neuron that carries a nerve impulse toward the central nervous system. Also called a *sensory neuron.*

Afterimage   Persistence of a sensation even though the stimulus has been removed.

Agglutination (a-gloo'-ti-NĀ-shun)   Clumping of microorganisms or blood corpuscles; typically an antigen–antibody reaction.

Agglutinin (a-GLOO-ti-nin)   A specific principle or antibody in blood serum capable of causing the clumping of bacteria, blood corpuscles, or particles. Also called an *isoantibody.*

Agglutinogen (ag'-loo-TIN-ō-gen)   A genetically determined antigen located on the surface of erythrocytes; basis for the ABO grouping and Rh system of blood classification. Also called an *isoantigen.*

Aggregated lymphatic follicles   Aggregated lymph nodules that are most numerous in the ileum. Also called *Peyer's* (PĪ-erz) *patches.*

Aging   Progressive failure of the body's homeostatic adaptive responses.

Agnosia (ag-NŌ-zē-a)   A loss of the ability to recognize the meaning of stimuli from the various senses (visual, auditory, touch).

Agraphia (a-GRAF-ē-a)   An inability to write.

Albinism (AL-bin-izm)   Abnormal, nonpathological, partial or total absence of pigment in skin, hair, and eyes.

Albumin (al-BYOO-min)   The most abundant (60 percent) and smallest of the plasma proteins, which functions primarily to regulate osmotic pressure of plasma.

Albuminuria (al-byoo'-min-UR-ēa)   Presence of albumin in the urine.

Aldosterone (al-do-STĒR-ōn)   A mineralocorticoid produced by the adrenal cortex that brings about sodium and water reabsorption and potassium excretion.

Aldosteronism (al'-do-STER-ōn-izm')   Condition caused by hypersecretion of aldosterone characterized by muscular paralysis, high blood pressure, and edema.

Alimentary (al-i-MEN-ta-rē)   Pertaining to nutrition.

Alkaline (AL-ka-līn)   Containing more hydroxyl ions than hydrogen ions to produce a pH of more than 7.

Alkalosis (al-ka-LŌ-sis)   A condition in which blood pH is between 7.45 and 8.00.

Allantois (a-LAN-tō-is)   A small, vascularized membrane between the chorion and amnion of the fetus.

Allele (a-LĒL)   One of two or more alternative forms of a gene occupying corresponding sites (loci) on homologous chromosomes.

Allergen (AL-er-jen)   An antigen that evokes a hypersensitivity reaction.

Allergic (a-LER-jik)   Pertaining to or sensitive to an allergen.

All-or-none principle   In muscle physiology, muscle fibers (cells) of a motor unit contract to their fullest extent or not at all.

In neuron physiology, if a stimulus is strong enough to initiate an action potential, an impulse is transmitted along the entire neuron at a constant and minimum strength.

Alpha (AL-fa) *cell* A cell in the pancreatic islets (islets of Langerhans) in the pancreas that secretes glucagon.

Alpha receptor Receptor found on visceral effectors innervated by most sympathetic postganglionic axons; in general, stimulation of alpha receptors leads to excitation.

Alveolar–capillary (al-VĒ-ō-lar) *membrane* Structure in the lungs consisting of the alveolar wall and basement membrane and a capillary endothelium and basement membrane through which the diffusion of respiratory gases occurs. Also called the *respiratory membrane.*

Alveolar duct Branch of a respiratory bronchiole around which alveoli and alveolar sacs are arranged.

Alveolar macrophage (MAK-rō-fāj) Cell found in the alveolar walls of the lungs that is highly phagocytic. Also called a *dust cell.*

Alveolar sac A collection or cluster of alveoli that share a common opening.

Alveolus (al-VĒ-ō-lus) A small hollow or cavity; an air sac in the lungs; milk-secreting portion of a mammary gland.

Alzheimer's (ALTZ-hī-merz) *disease* (*AD*) Disabling neurological disorder characterized by dysfunction and death of specific cerebral neurons resulting in widespread intellectual impairment, personality changes, and fluctuations in alertness.

Ambulatory (AM-byoo-la-tō'-rē) Capable of walking.

Amenorrhea (ā-men-ō-RĒ-a) Absence of menstruation.

Amnesia (am-NĒ-zē-a) A lack or loss of memory.

Amniocentesis (am'-nē-ō-sen-TĒ-sis) Removal of amniotic fluid by inserting a needle transabdominally into the amniotic cavity.

Amnion (AM-nē-on) The innermost fetal membrane; a thin transparent sac that holds the fetus suspended in amniotic fluid. Also called the *"bag of waters."*

Amniotic (am'-nē-OT-ik) *fluid* Fluid in the amniotic cavity, the space between the developing embryo (or fetus) and amnion.

Amorphous (a-MOR-fus) Without definite shape or differentiation in structure; pertains to solids without crystalline structure.

Amphiarthrosis (am'-fē-ar-THRŌ-sis) Articulation midway between diarthrosis and synarthrosis, in which the articulating bony surfaces are separated by an elastic substance to which both are attached, so that the mobility is slight but may be exerted in all directions.

Ampulla (am-POOL-la) A saclike dilation of a canal.

Ampulla of Vater See Hepatopancreatic ampulla.

Amyotrophic (a-mē-ō-TROF-ik) *lateral sclerosis* (*ALS*) Progressive neuromuscular disease characterized by degeneration of motor cells in the spinal cord that leads to muscular weakness. Also called *Lou Gehrig's disease.*

Anaerobic (an-AIR-ō-bik) Not requiring molecular oxygen.

Anal (Ā-nal) *canal* The terminal 2 or 3 cm (1 in.) of the rectum; opens to the exterior of the anus.

Anal column A longitudinal fold in the mucous membrane of the anal canal that contains a network of arteries and veins.

Analgesia (an-al-JĒ-zē-a) Absence of normal sense of pain.

Anal triangle The subdivision of the male or female perineum that contains the anus.

Anaphase (AN-a-fāz) The third stage of mitosis in which the chromatids that have separated at the centromeres move to opposite poles of the cell.

Anaphylaxis (an'-a-fi-LAK-sis) A hypersensitivity (allergic) reaction in which IgE antibodies attach to mast cells and basophils, causing them to produce mediators of anaphylaxis (histamine and prostaglandins) that bring about increased blood permeability, increased smooth muscle contraction, and increased mucus production. Examples are hayfever, bronchial asthma, hives, and anaphylactic shock.

Anastomosis (a-nas-tō-MŌ-sis) An end-to-end union or joining together of blood vessels, lymphatics, or nerves.

Anatomical (an'-a-TOM-i-kal) *position* A position of the body universally used in anatomical descriptions in which the body is erect, facing the observer, the upper extremities are at the sides, the palms of the hands are facing forward, and the feet are on the floor.

Anatomy (a-NAT-ō-mē) The structure or study of structure of the body and the relation of its parts to each other.

Androgen (AN-drō-jen) Substance producing or stimulating male characteristics, such as the male hormone testosterone.

Anemia (a-NĒ-mē-a) Condition of the blood in which the number of functional red blood cells or their hemoglobin content is below normal.

Anesthesia (an'-es-THĒ-zē-a) A total or partial loss of feeling or sensation, usually defined with respect to loss of pain sensation.

Aneurysm (AN-yoo-rizm) A saclike enlargement of a blood vessel caused by a weakening of its wall.

Angina pectoris (an-JĪ-na *or* AN-ji-na PEK-tō-ris) A pain in the chest related to reduced coronary circulation that may or may not involve heart or artery disease.

Angiography (an-jē-OG-ra-fē) X-ray examination of blood vessels after injection of a radiopaque substance into the common carotid or vertebral artery; used to demonstrate cerebral blood vessels and may detect brain tumors with specific vascular patterns.

Ankyloglossia (ang'-ki-lō-GLOSS-ē-a) "Tongue-tied"; restriction of tongue movements by a short lingual frenulum.

Ankylosis (ang'-ki-LŌ-sus) Severe or complete loss of movement at a joint.

Anomaly (a-NOM-a-lē) An abnormality that may be a developmental (congenital) defect; a variant from the usual standard.

Anopsia (an-OP-sē-a) A defect of vision.

Anorexia nervosa (an-ō-REK-sē-a ner-VŌ-sa) A disorder characterized by loss of appetite and bizarre patterns of eating.

Anosmia (an-OZ-mē-a) Loss of the sense of smell.

Anoxia (an-OK-sē-a) Deficiency of oxygen.

Antagonist (an-TAG-ō-nist) A muscle that has an action opposite that of the agonist and yields to the movement of the prime mover (agonist).

Antepartum (an-tē-PAR-tum) Before delivery of the child; occurring (to the mother) before childbirth.

Anterior (an-TĒR-ē-or) Nearer to or at the front of the body. Also called *ventral.*

Anterior root The structure composed of axons of motor or efferent fibers that emerges from the anterior aspect of the spinal cord and extends laterally to join a posterior root, forming a spinal nerve. Also called a *ventral root.*

Antibiotic (an'-ti-bī-OT-ik) Literally, "antilife"; a chemical produced by a microorganism that is able to inhibit the growth of or kill other microorganisms.

Antibody (AN-ti-bod'-ē) A substance produced by certain cells in the presence of a specific antigen that combines with that antigen to neutralize, inhibit, or destroy it.

Anticoagulant (an-tī-cō-AG-yoo-lant)   A substance that is able to delay, suppress, or prevent the clotting of blood.

Antidiuretic (an'ti-dī-yoo-RET-ik)   Substance that inhibits urine formation.

Antidiuretic hormone (**ADH**)   Hormone produced by neurosecretory cells in the paraventricular nucleus of the hypothalamus that stimulates water reabsorption from kidney cells into the blood and vasoconstriction of arterioles.

Antigen (AN-ti-jen)   Any substance that when introduced into the tissues or blood induces the formation of antibodies or reacts with them.

Antrum (AN-trum)   Any nearly closed cavity or chamber, especially one within a bone, such as a sinus.

Anulus fibrosus (AN-yoo-lus fī-BRŌ-sus)   A ring of fibrous tissue and fibrocartilage that encircles the pulpy substance (nucleus pulposus) of an intervertebral disc.

Anuria (a-NOO-rē-a)   A daily urine output of less than 50 ml.

Anus (Ā-nus)   The distal end and outlet of the rectum.

Aorta (ā-OR-ta)   The main systemic trunk of the arterial system of the body; emerges from the left ventricle.

Aortic (ā-OR-tik) *body*   Receptor on or near the arch of the aorta that responds to alterations in blood levels of oxygen, carbon dioxide, and hydrogen ions.

Aperture (AP-er-chur)   An opening or orifice.

Apex (Ā-peks)   The pointed end of a conical structure.

Aphasia (a-FĀ-zē-a)   Loss of ability to express oneself properly through speech or loss of verbal comprehension.

Apheresis (a-FER-ē-sis)   Procedure in which blood is removed from the body, components are selectively separated, the undesired component is removed, and the remainder is returned to the body.

Apnea (ap-NĒ-a)   Temporary cessation of breathing.

Apneustic (ap-NOO-stik) *area*   Portion of the respiratory center in the pons that sends stimulatory nerve impulses to the inspiratory area that activate and prolong inspiration and inhibit expiration.

Apocrine (AP-ō-krin) *gland*   A type of gland in which the secretory products gather at the free end of the secreting cell and are pinched off, along with some of the cytoplasm, to become the secretion, as in mammary glands.

Aponeurosis (ap'-ō-noo-RŌ-sis)   A sheetlike layer of dense, regularly arranged connective tissue joining one muscle with another or with bone.

Appendage (a-PEN-dij)   A structure attached to the body.

Appendicitis (a-pen-di-SĪ-tis)   Inflammation of the vermiform appendix.

Appositional (ap'-ō-ZISH-o-nal) *growth*   Growth due to surface deposition of material, as in the growth in diameter of cartilage and bone. Also called *exogenous* (eks-OJ-e-nus) *growth*.

Aqueduct (AK-we-duct)   A canal or passage, especially for the conduction of a liquid.

Aqueous humor (AK-wē-us HYOO-mor)   The watery fluid that fills the anterior cavity of the eye.

Arachnoid (a-RAK-noyd)   The middle of the three coverings (meninges) of the brain.

Arachnoid villus (VIL-us)   Berrylike tuft of arachnoid that protrudes into the superior sagittal sinus and through which the cerebrospinal fluid enters the bloodstream.

Arbor vitae (AR-bōr VĒ-tē)   The treelike appearance of the white matter tracts of the cerebellum when seen in midsagittal section.

A series of branching ridges within the cervix of the uterus.

Arch of the aorta (ā-OR-ta)   The most superior portion of the aorta, lying between the ascending and descending segments of the aorta.

Areflexia (a'-rē-FLEK-sē-a)   Absence of reflexes.

Areola (a-RĒ-ō-la)   Any tiny space in a tissue. The pigmented ring around the nipple of the breast.

Arm   The portion of the upper extremity from the shoulder to the elbow.

Arrector pili (a-REK-tor PI-lē)   Smooth muscles attached to hairs; contraction pulls the hairs into a more vertical position, resulting in ''goose bumps.''

Arrhythmia (a-RITH-mē-a)   Irregular heart rhythm. Also called a *dysrhythmia.*

Arteriogram (ar-TĒR-ē-ō-gram)   A roentgenogram of an artery after injection of a radiopaque substance into the blood.

Arteriole (ar-TĒ-rē-ōl)   A small arterial branch that delivers blood to a capillary.

Artery (AR-ter-ē)   A blood vessel that carries blood away from the heart.

Arthritis (ar-THRĪ-tis)   Inflammation of a joint.

Arthrology (ar-THROL-ō-jē)   The study or description of joints.

Arthroscopy (ar-THROS-co-pē)   Surgical technique in which an arthroscope is inserted into a small incision, usually in the knee, to repair torn cartilage.

Arthrosis (ar-THRŌ-sis)   A joint or articulation.

Articular (ar-TIK-yoo-lar) *capsule*   Sleevelike structure around a synovial joint composed of a fibrous capsule and a synovial membrane.

Articular cartilage (KAR-ti-lij)   Hyaline cartilage attached to articular bone surfaces.

Articular disc   Fibrocartilage pad between articular surfaces of bones of some synovial joints. Also called a *meniscus* (men-IS-cus).

Articulate (ar-TIK-yoo-lāt)   To join together as a joint to permit motion between parts.

Articulation (ar-tik'-yoo-LĀ-shun)   A joint.

Artificial insemination (in-sem'-i-NĀ-shun)   The deposition of seminal fluid within the vagina or cervix by artificial means. It may be homologous (using the husband's semen) or heterologous (using a donor's semen).

Artificial pacemaker   A device that generates and delivers electrical signals to the heart to maintain a regular heart rhythm.

Arytenoid (ar'-i-TĒ-noyd)   Ladle shaped.

Arytenoid (ar'-i-TĒ-noyd) *cartilages*   A pair of small cartilages of the larynx that articulates with the cricoid cartilage.

Ascending colon (KŌ-lon)   The portion of the large intestine that passes upward from the cecum to the lower edge of the liver where it bends at the right colic (hepatic) flexure to become the transverse colon.

Ascites (as-SĪ-tēz)   Serous fluid in the peritoneal cavity.

Aseptic (ā-SEP-tik)   Free from any infectious or septic material.

Asphyxia (as-FIX-ē-a)   Unconsciousness due to interference with the oxygen supply of the blood.

Aspirate (AS-pir-āt)   To remove by suction.

Association area   A portion of the cerebral cortex connected by many motor and sensory fibers to other parts of the cortex. The association areas are concerned with motor patterns, memory, concepts of word-hearing and word-seeing, reasoning, will, judgment, and personality traits.

Association neuron (NOO-ron) A nerve cell lying completely within the central nervous system that carries nerve impulses from sensory neurons to motor neurons. Also called a **connecting neuron.**

Astereognosis (as-ter'-ē-ōg-NŌ-sis) Inability to recognize objects or forms by touch.

Asthenia (as-THĒ-nē-a) Lack or loss of strength; debility.

Astigmatism (a-STIG-ma-tizm) An irregularity of the lens or cornea of the eye causing the image to be out of focus and producing faulty vision.

Astrocyte (AS-trō-sīt) A neuroglial cell having a star shape that supports neurons in the brain and spinal cord and attaches the neurons to blood vessels.

Ataxia (a-TAK-sē-a) A lack of muscular coordination, lack of precision.

Atelectasis (at'-ē-LEK-ta-sis) A collapsed or airless state of all or part of the lung, which may be acute or chronic.

Atherosclerosis (ath'-er-ō-skle-RŌ-sis) A process in which fatty substances (cholesterol and triglycerides) are deposited in the walls of medium and large arteries in response to certain stimuli (hypertension, carbon monoxide, dietary cholesterol). Following endothelial damage, monocytes stick to the tunica interna, develop into macrophages, and take up cholesterol and low-density lipoproteins. Smooth muscle fibers (cells) in the tunica media ingest cholesterol. This results in the formation of an atherosclerotic plaque that decreases the size of the arterial lumen.

Atresia (a-TRĒ-zē-a) Abnormal closure of a passage, or absence of a normal body opening.

Atrial fibrillation (Ā-trē-al fib-ri-LĀ-shun) Asynchronous contraction of the atria that results in the cessation of atrial pumping.

Atrioventricular (AV) (ā'-trē-ō-ven-TRIK-yoo-lar) **bundle** The portion of the conduction system of the heart that begins at the atrioventricular (AV) node, passes through the cardiac skeleton separating the atria and the ventricles, then runs a short distance down the interventricular septum before splitting into right and left bundle branches. Also called the **bundle of His (HISS).**

Atrioventricular (AV) node The portion of the conduction system of the heart made up of a compact mass of conducting cells located near the orifice of the coronary sinus in the right atrial wall.

Atrioventricular (AV) valve A structure made up of membranous flaps or cusps that allows blood to flow in one direction only, from an atrium into a ventricle.

Atrium (Ā-trē-um) A superior chamber of the heart.

Atrophy (AT-rō-fē) Wasting away or decrease in size of a part, due to a failure, abnormality of nutrition, or lack of use.

Auditory ossicle (AW-di-tō-rē OS-si-kul) One of the three small bones of the middle ear called the malleus, incus, and stapes.

Auditory tube The tube that connects the middle ear with the nose and nasopharynx of the throat. Also called the **Eustachian** (yoo-STĀ-kē-an) **tube.**

Aura (OR-a) A feeling or sensation that precedes an epileptic seizure or any paroxysmal attack (like those of bronchial asthma).

Auscultation (aws-kul-TĀ-shun) Examination by listening to sounds in the body.

Autoimmunity An immunologic response against a person's own tissue antigens.

Autolysis (aw-TOL-i-sis) Spontaneous self-destruction of cells by their own digestive enzymes upon death or a pathological process.

Autonomic ganglion (aw'-tō-NOM-ik GANG-lē-on) A cluster of sympathetic or parasympathetic cell bodies located outside the central nervous system.

Autonomic nervous system (ANS) Visceral efferent neurons, both sympathetic and parasympathetic, that transmit nerve impulses from the central nervous system to smooth muscle, cardiac muscle, and glands; so named because this portion of the nervous system was thought to be self-governing or spontaneous.

Autonomic plexus (PLEK-sus) An extensive network of sympathetic and parasympathetic fibers; the cardiac, celiac, and pelvic plexuses are located in the thorax, abdomen, and pelvis, respectively.

Autophagy (aw-TOF-a-jē) Process by which worn-out organelles are digested within lysosomes.

Autopsy (AW-top-sē) The examination of the body after death.

Autosome (AW-tō-sōm) Any chromosome other than the pair of sex chromosomes.

Axilla (ak-SIL-a) The small hollow beneath the arm where it joins the body at the shoulders. Also called the **armpit.**

Axon (AK-son) The process of a nerve cell that carries a nerve impulse away from the cell body.

Axon terminal Terminal branch of an axon and its collateral. Also called a **telodendrium** (tel-ō-DEN-drē-um).

Azygos (AZ-i-gos) An anatomical structure that is not paired; occurring singly.

Babinski (ba-BIN-skē) **sign** Extension of the great toe, with or without fanning of the other toes, in response to stimulation of the outer margin of the sole of the foot; normal up to 1½ years of age.

Back The posterior part of the body; the dorsum.

Ball-and-socket joint A synovial joint in which the rounded surface of one bone moves within a cup-shaped depression or fossa of another bone, as in the shoulder or hip joint. Also called a **spheroid** (SFĒ-roid) **joint.**

Baroreceptor (bar'-ō-re-SEP-tor) Nerve cell capable of responding to changes in blood pressure. Also called a **pressoreceptor.**

Bartholin's glands See **Greater vestibular glands.**

Basal ganglia (GANG-glē-a) Paired clusters of cell bodies that make up the central gray matter in each cerebral hemisphere, including the caudate nucleus, lentiform nucleus, claustrum, and amygdaloid body. Also called **cerebral nuclei** (SER-e-bral NOO-klē-ī).

Basal metabolic (BĀ-sal met'-a-BOL-ik) **rate (BMR)** The rate of metabolism measured under standard or basal conditions.

Base The broadest part of a pyramidal structure. A nonacid or a proton acceptor, characterized by excess of hydroxide ions and a pH greater than 7. A ring-shaped, nitrogen-containing organic molecule that is one of the components of a nucleotide, for example, adenine, guanine, cytosine, thymine, and uracil.

Basement membrane Thin, extracellular layer consisting of basal lamina secreted by epithelial cells and reticular lamina secreted by connective tissue cells.

Basilar (BAS-i-lar) **membrane** A membrane in the cochlea of the inner ear that separates the cochlear duct from the scala tympani and on which the spiral organ (organ of Corti) rests.

Basophil (BĀ-sō-fil) A type of white blood cell characterized by a pale nucleus and large granules that stain readily with basic dyes.

B cell   A lymphocyte that develops into a plasma cell that produces antibodies or a memory cell.

Belly   The abdomen. The gaster or prominent, fleshy part of a skeletal muscle.

Benign (be-NĪN)   Not malignant; favorable for recovery; a mild disease.

Beta (BĀ-ta) *cell*   A cell in the pancreatic islets (islets of Langerhans) in the pancreas that secretes insulin.

Beta receptor   Receptor found on visceral effectors innervated by most sympathetic postganglionic axons; in general, stimulation of beta receptors leads to inhibition.

Bicuspid (bī-KUS-pid) *valve*   Atrioventricular (AV) valve on the left side of the heart. Also called the *mitral valve.*

Bifurcate (bī-FUR-kāt)   Having two branches or divisions; forked.

Bilateral (bī-LAT-er-al)   Pertaining to two sides of the body.

Bile (bīl)   A secretion of the liver.

Biliary (BIL-ē-er-ē)   Relating to bile, the gallbladder, or the bile ducts.

Biliary calculi (BIL-ē-er-ē CAL-kyoo-lē)   Gallstones formed by the crystallization of cholesterol in bile.

Bilirubin (bil-ē-ROO-bin)   A red pigment that is one of the end products of hemoglobin breakdown in the liver cells and is excreted as a waste material in the bile.

Bilirubinuria (bil-ē-roo-bi-NOO-rē-a)   The presence of above-normal levels of bilirubin in urine.

Biliverdin (bil-ē-VER-din)   A green pigment that is one of the first products of hemoglobin breakdown in the liver cells and is converted to bilirubin or excreted as a waste material in bile.

Biofeedback   Process by which an individual gets constant signals (feedback) about various visceral body functions.

Biopsy (BĪ-op-sē)   Removal of tissue or other material from the living body for examination, usually microscopic.

Blastocoel (BLAS-tō-sēl)   The fluid-filled cavity within the blastocyst.

Blastocyst (BLAS-tō-sist)   In the development of an embryo, a hollow ball of cells that consists of a blastocoele (the internal cavity), trophoblast (outer cells), and inner cell mass.

Blastomere (BLAS-tō-mēr)   One of the cells resulting from the cleavage of a fertilized ovum.

Blastula (BLAS-tyoo-la)   An early stage in the development of a zygote.

Blepharism (BLEF-a-rizm)   Spasm of the eyelids; continuous blinking.

Blind spot   Area in the retina at the end of the optic (II) nerve in which there are no light receptor cells.

Blood   The fluid that circulates through the heart, arteries, capillaries, and veins and that constitutes the chief means of transport within the body.

Blood–brain barrier (BBB)   A special mechanism that prevens the passage of materials from the blood to the cerebrospinal fluid and brain.

Blood reservoir (REZ-er-vwar)   Systemic veins that contain large amounts of blood that can be moved quickly to parts of the body requiring the blood.

Blood–testis barrier   A barrier formed by sustentacular (Sertoli) cells that prevents an immune response against antigens produced by spermatozoa and developing cells by isolating the cells from the blood.

Body cavity   A space within the body that contains various internal organs.

Bolus (BŌ-lus)   A soft, rounded mass, usually food, that is swallowed.

Bony labyrinth (LAB-i-rinth)   A series of cavities within the petrous portion of the temporal bone forming the vestibule, cochlea, and semicircular canals of the inner ear.

Bowman's capsule   See Glomerular capsule.

Brachial plexus (BRĀ-kē-al PLEK-sus)   A network of nerve fibers of the anterior rami of spinal nerves C5, C6, C7, C8, and T1. The nerves that emerge from the brachial plexus supply the upper extremity.

Brain   A mass of nervous tissue located in the cranial cavity.

Brain electrical activity mapping (BEAM)   Noninvasive procedure that measures and displays the electrical activity of the brain; used primarily to diagnose epilepsy.

Brain sand   Calcium deposits in the pineal gland that are laid down starting at puberty.

Brain stem   The portion of the brain immediately superior to the spinal cord, made up of the medulla oblongata, pons, and midbrain.

Bright's disease   See Glomerulonephritis.

Broad ligament   A double fold of parietal peritoneum attaching the uterus to the side of the pelvic cavity.

Broca's (BRŌ-kaz) *area*   Motor area of the brain in the frontal lobe that translates thoughts into speech. Also called the *motor speech area.*

Bronchi (BRONG-kē)   Branches of the respiratory passageway including primary bronchi (the two divisions of the trachea), secondary or lobar bronchi (divisions of the primary that are distributed to the lobes of the lung), and tertiary or segmental bronchi (divisions of the secondary that are distributed to bronchopulmonary segments of the lung).

Bronchial asthma (BRONG-kē-al AZ-ma)   Usually allergic reaction characterized by smooth muscle spasms in bronchi resulting in wheezing and difficult breathing.

Bronchial tree   The trachea, bronchi, and their branching structures.

Bronchiectasis (brong'-kē-EK-ta-sis)   A chronic disorder in which there is a loss of the normal tissue and expansion of lung air passages; characterized by difficult breathing, coughing, expectoration of pus, and foul breath.

Bronchiole (BRONG-kē-ol)   Branch of a tertiary bronchus further dividing into terminal bronchioles (distributed to lobules of the lung), which divide into respiratory bronchioles (distributed to alveolar sacs).

Bronchitis (brong-KĪ-tis)   Inflammation of the bronchi characterized by hypertrophy and hyperplasia of seromucous glands and goblet cells that line the bronchi and resulting in a productive cough.

Bronchogenic carcinoma (brong'-kō-JEN-ik kar'-si-NŌ-ma)   Cancer originating in the bronchi.

Bronchogram (BRONG-kō-gram)   A roentgenogram of the lungs and bronchi.

Bronchopulmonary (brong'-kō-PUL-mō-ner-ē) *segment*   One of the smaller divisions of a lobe of a lung supplied by its own branches of a bronchus.

Bronchoscope (BRONG-kō-skōp)   An instrument used to examine the interior of the bronchi of the lungs.

Bronchus (BRONG-kus)   One of the two large branches of the trachea. *Plural,* **bronchi** (BRONG-kē).

Brunner's gland See Duodenal gland.

Buccal (BUK-al) Pertaining to the cheek or mouth.

Bulb of penis Expanded portion of the base of the corpus spongiosum penis.

Bulbourethral (bul'-bō-yoo-RĒ-thral) *gland* One of a pair of glands located inferior to the prostate gland on either side of the urethra that secretes an alkaline fluid into the cavernous urethra. Also called a *Cowper's* (KOW-perz) *gland.*

Bulimia (boo-LIM-ē-a) A disorder characterized by uncontrollable overeating followed by forced vomiting or overdoses of laxatives.

Bullae (BYOOL-ē) Blisters beneath or within the epidermis.

Bundle branch One of the two branches of the atrioventricular (AV) bundle made up of specialized muscle fibers (cells) that transmit electrical impulses to the ventricles.

Bundle of His See Atrioventricular (AV) bundle.

Bunion (BUN-yun) Lateral deviation of the great toe that produces inflammation and thickening of the bursa, bone spurs, and calluses.

Burn An injury caused by heat (fire, steam), chemicals, electricity, or the ultraviolet rays of the sun.

Bursa (BUR-sa) A sac or pouch of synovial fluid located at friction points, especially about joints.

Bursitis (bur-SĪ-tis) Inflammation of a bursa.

Buttocks (BUT-oks) The two fleshy masses on the posterior aspect of the lower trunk, formed by the gluteal muscles.

Cachexia (kah-KEK-sē-ah) A state of ill health, malnutrition, and wasting.

Calcaneal tendon The tendon of the soleus, gastrocnemius, and plantaris muscles at the back of the heel. Also called the *Achilles* (a-KIL-ēz) *tendon.*

Calcify (KAL-si-fī) To harden by deposits of calcium salts.

Calcitonin (kal-si-TŌ-nin) *(CT)* A hormone produced by the thyroid gland that lowers the calcium and phosphate levels of the blood by inhibiting bone breakdown and accelerating calcium absorption by bones.

Calculus (KAL-kyoo-lus) A stone, or insoluble mass of crystallized salts or other material, formed within the body, as in the gallbladder, kidney, or urinary bladder.

Callus (KAL-lus) A growth of new bone tissue in and around a fractured area, ultimately replaced by mature bone. An acquired, localized thickening.

Calyx (KĀL-iks) Any cuplike division of the kidney pelvis. *Plural, calyces* (KĀ-li-sēz).

Canal (ka-NAL) A narrow tube, channel, or passageway.

Canaliculus (kan'-a-LIK-yoo-lus) A small channel or canal, as in bones, where they connect the lacunae. *Plural, canaliculi* (kan'-a-LIK-yoo-lī).

Canal of Schlemm See Scleral venous sinus.

Cancellous (KAN-sel-us) Having a reticular or latticework structure, as in spongy tissue of bone.

Cancer (KAN-ser) A malignant tumor of epithelial origin tending to infiltrate and give rise to new growths or metastases. Also called *carcinoma* (kar'-si-NŌ-ma).

Capacitation (ka'-pas-i-TĀ-shun) The functional changes that sperm undergo in the female reproductive tract that allow them to fertilize a secondary oocyte.

Capillary (KAP-i-lar'-ē) A microscopic blood vessel located between an arteriole and venule through which materials are exchanged between blood and body cells.

Carbohydrate (kar'-bō-HĪ-drāt) An organic compound containing carbon, hydrogen, and oxygen in a particular amount and arrangement and comprised of sugar subunits; usually has the formula $(CH_2O)_n$.

Carbon monoxide (CO) poisoning Hypoxia due to increased levels of carbon monoxide as a result of its preferential and tenacious combination with hemoglobin compared to oxygen.

Carcinogen (kar-SIN-ō-jen) Any substance that causes cancer.

Carcinoma (kar'-si-NŌ-ma) A malignant tumor consisting of epithelial cells.

Cardiac arrest Complete stoppage of heartbeat.

Cardiac catheterization (kath'-e-ter-i-ZĀ-shun) An invasive procedure in which a catheter is introduced into a vein or artery and threaded, via fluoroscopy, through blood vessels to reach the heart.

Cardiac muscle An organ specialized for contraction, composed of striated muscle fibers (cells), forming the wall of the heart, and stimulated by an intrinsic conduction system and visceral efferent neurons.

Cardiac notch An angular notch in the anterior border of the left lung.

Cardiac tamponade (tam'-pon-ĀD) Compression of the heart due to excessive fluid or blood in the pericardial sac that could result in cardiac failure.

Cardinal ligament A ligament of the uterus, extending laterally from the cervix and vagina as a continuation of the broad ligament.

Cardioacceleratory (kar-dē-ō-ak-SEL-er-a-tō-rē) *center (CAC)* A group of neurons in the medulla from which cardiac nerves (sympathetic) arise; nerve impulses along the nerves release epinephrine that increases the rate and force of heartbeat.

Cardioinhibitory (kar-dē-ō-in-HIB-i-tō-rē) *center (CIC)* A group of neurons in the medulla from which parasympathetic fibers that reach the heart via the vagus (X) nerve arise; nerve impulses along the nerves release acetylcholine that decreases the rate and force of heartbeat.

Cardiology (kar-de-OL-ō-jē) The study of the heart and diseases associated with it.

Cardiopulmonary resuscitation (rē-sus-i-TĀ-shun) *(CPR)* A technique employed to restore life or consciousness to a person apparently dead or dying; includes external respiration (exhaled air respiration) and external cardiac massage.

Carina (ka-RĪ-na) A ridge on the inside of the division of the right and left primary bronchi.

Carotid (ka-ROT-id) *body* Receptor on or near the carotid sinus that responds to alterations in blood levels of oxygen, carbon dioxide, and hydrogen ions.

Carotid sinus A dilated region of the internal carotid artery immediately above the bifurcation of the common carotid artery that contains receptors that monitor blood pressure.

Carotid sinus reflex A reflex concerned with maintaining normal blood pressure in the brain.

Carpus (KAR-pus) A collective term for the eight bones of the wrist.

Cartilage (KAR-ti-lij) A type of connective tissue consisting of chondrocytes in lacunae embedded in a dense network of collagenous and elastic fibers and a matrix of chondroitin sulfate.

Cartilaginous (kar'-ti-LAJ-i-nus) *joint* A joint without a synovial (joint) cavity where the articulating bones are held tightly together by cartilage, allowing little or no movement.

Caruncle (KAR-ung-kul)   A small fleshy eminence, often abnormal.

Cast   A small mass of hardened material formed within a cavity in the body and then discharged from the body; can originate in different areas and be composed of various materials.

Castration (kas-TRĀ-shun)   The removal of the testes.

Catabolism (ka-TAB-ō-lizm)   Chemical reactions that break down complex organic compounds into simple ones with the release of energy.

Cataract (KAT-a-rakt)   Loss of transparency of the lens of the eye or its capsule or both.

Catheter (KATH-i-ter)   A tube that can be inserted into a body cavity through a canal or into a blood vessel; used to remove fluids, such as urine and blood, and to introduce diagnostic materials or medication.

Cauda equina (KAW-da ē-KWĪ-na)   A taillike collection of roots of spinal nerves at the inferior end of the spinal canal.

Caudal (KAW-dal)   Pertaining to any taillike structure; inferior in position.

Cecum (SĒ-kum)   A blind pouch at the proximal end of the large intestine to which the ileum is attached.

Celiac (SĒ-lē-ak)   Pertaining to the abdomen.

Celiac plexus (PLEK-sus)   A large mass of ganglia and nerve fibers located at the level of the upper part of the first lumbar vertebra. Also called the *solar plexus.*

Cell   The basic structural and functional unit of all organisms; the smallest structure capable of performing all the activities vital to life.

Cell division   Process by which a cell reproduces itself that consists of a nuclear division and a cytoplasmic division; types include somatic and reproductive cell division.

Cell inclusion   A lifeless, often temporary, constituent in the cytoplasm of a cell as opposed to an organelle.

Cementum (se-MEN-tum)   Calcified tissue covering the root of a tooth.

Center   An area in the brain where a particular function is localized.

Center of ossification (os'-i-fi-KĀ-shun)   An area in the cartilage model of a future bone where the cartilage cells hypertrophy and then secrete enzymes that result in the calcification of their matrix, resulting in the death of the cartilage cells, followed by the invasion of the area by osteoblasts that then lay down bone.

Central canal   A circular channel running longitudinally in the center of an osteon (Haversian system) of mature compact bone, containing blood and lymph vessels and nerves. Also called a *Haversian* (ha-VĒR-shun) *canal.* A microscopic tube running the length of the spinal cord in the gray commissure.

Central fovea (FŌ-vē-a)   A cuplike depression in the center of the macula lutea of the retina, containing cones only; the area of clearest vision.

Central nervous system (CNS)   Brain and spinal cord.

Centrioles (SEN-trē-ōlz)   Paired, cylindrical organelles within a centrosome, each consisting of a ring of microtubules and arranged at right angles to each other; function in cell division.

Centromere (SEN-trō-mēr)   The clear, constricted portion of a chromosome where the two chromatids are joined; serves as the point of attachment for the chromosomal microtubules.

Centrosome (SEN-trō-sōm)   A rather dense area of cytoplasm, near the nucleus of a cell, containing a pair of centrioles.

Cephalic (se-FAL-ik)   Pertaining to the head; superior in position.

Cerebellar peduncle (ser-e-BEL-ar pe-DUNG-kul)   A bundle of nerve fibers connecting the cerebellum with the brain stem.

Cerebellum (ser-e-BEL-um)   The portion of the brain lying posterior to the medulla and pons, concerned with coordination of movements.

Cerebral aqueduct (SER-e-bral AK-we-dukt)   A channel through the midbrain connecting the third and fourth ventricles and containing cerebrospinal fluid.

Cerebral arterial circle   A ring of arteries forming an anastomosis at the base of the brain between the internal carotid and vertebral arteries and arteries supplying the brain. Also called the *circle of Willis.*

Cerebral cortex   The surface of the cerebral hemispheres, 2–4 mm thick, consisting of six layers of nerve cell bodies (gray matter) in most areas.

Cerebral palsy (PAL-zē)   A group of nonprogressive motor disorders caused by damage to motor areas of the brain (cerebral cortex, basal ganglia, and cerebellum) during fetal life, birth, or infancy.

Cerebral peduncle (pe-DUNG-kul)   One of a pair of nerve fiber bundles located on the ventral surface of the midbrain, conducting nerve impulses between the pons and the cerebral hemispheres.

Cerebrospinal (se-rē'-brō-SPĪ-nal) *fluid (CSF)*   A fluid produced in the choroid plexuses of the ventricles of the brain that circulates in the ventricles and the subarachnoid space around the brain and spinal cord.

Cerebrovascular (se-rē'-brō-VAS-kyoo-lar) *accident (CVA)*   Destruction of brain tissue (infarction) resulting from disorders of blood vessels that supply the brain. Also called a *stroke.*

Cerebrum (SER-ē-brum)   The two hemispheres of the forebrain, making up the largest part of the brain.

Ceruminous (se-ROO-mi-nus) *gland*   A modified sudoriferous (sweat) gland in the external auditory meatus that secretes cerumen (ear wax).

Cervical dysplasia (dis-PLĀ-sē-a)   A change in the shape, growth, and number of cervical cells of the uterus that, if severe, may progress to cancer.

Cervical ganglion (SER-vi-kul GANG-glē-on)   A cluster of nerve cell bodies of postganglionic sympathetic neurons located in the neck, near the vertebral column.

Cervical plexus (PLEK-sus)   A network of neuron fibers formed by the anterior rami of the first four cervical nerves.

Cervix (SER-viks)   Neck; any constricted portion of an organ, especially the lower cylindrical part of the uterus.

Cesarean (se-SA-rē-an) *section*   Procedure in which a low, horizontal incision is made through the abdominal wall and uterus for removal of the baby and placenta. Also called a *C-section.*

Chalazion (ka-LĀ-zē-on)   A small tumor of the eyelid.

Chemonucleolysis (kē'-mō-noo'-klē-OL-i-sis)   Dissolution of the nucleus pulposus of an intervertebral disc by injection of a proteolytic enzyme (chymopapain) to relieve the pressure and pain associated with a herniated (slipped) disc.

Chemoreceptor (kē'-mō-rē-SEP-tor)   Receptor outside the central nervous system on or near the carotid and aortic bodies that detects the presence of chemicals.

Chemotaxis (kē-mō-TAK-sis)   Attraction of phagocytes to microbes by a chemical stimulus.

Chemotherapy (kē'-mō-THER-a-pē) The treatment of illness or disease by chemicals.

Chiasma (kī-AZ-ma) A crossing; especially the crossing of the optic (II) nerve fibers.

Chiropractic (kī'-rō-PRAK-tik) A system of treating disease by using one's hands to manipulate body parts, mostly the vertebral column.

Chlamydia (kla-MID-e-a) A sexually transmitted disease characterized by burning on urination, frequent and painful urination, and low back pain; may spread to uterine (Fallopian) tubes in females.

Choana (KŌ-a-na) A funnel-shaped structure; the posterior opening of the nasal fossa, or internal naris.

Cholecystectomy (kō'-lē-sis-TEK-tō-mē) Surgical removal of the gallbladder.

Cholesterol (kō-LES-te-rol) Classified as a lipid, the most abundant steroid in animal tissues; located in cell membranes and used for the synthesis of steroid hormones and bile salts.

Cholinergic (kō'-lin-ER-jik) *fiber* A nerve ending that liberates acetylcholine at a synapse.

Cholinesterase (kō'-lin-ES-ter-ās) An enzyme that hydrolyzes acetylcholine.

Chondrocyte (KON-drō-sīt) Cell of mature cartilage.

Chondroitin (kon-DROY-tin) *sulfate* An amorphous matrix material found outside the cell.

Chordae tendineae (KOR-dē TEN-di-nē) Cords that connect the heart valves with the papillary muscles.

Chorion (KŌ-rē-on) The outermost fetal membrane; serves a protective and nutritive function.

Chorionic villi sampling (CVS) The removal of a sample of chorionic villi tissue by means of a catheter to analyze the tissue for prenatal genetic defects.

Chorionic villus (kō'-rē-ON-ik VIL-lus) Fingerlike projection of the chorion that grows into the decidua basalis of the endometrium and contains fetal blood vessels.

Choroid (KŌ-royd) One of the vascular coats of the eyeball.

Choroid plexus (PLEK-sus) A vascular structure located in the roof of each of the four ventricles of the brain; produces cerebrospinal fluid.

Chromaffin (krō-MAF-in) *cell* Cell that has an affinity for chrome salts, due in part to the presence of the precursors of the neurotransmitter epinephrine; found, among other places, in the adrenal medulla.

Chromatid (KRŌ-ma-tid) One of a pair of identical connected nucleoprotein strands that are joined at the centromere and separate during cell division, each becoming a chromosome of one of the two daughter cells.

Chromatin (KRŌ-ma-tin) The threadlike mass of the genetic material consisting principally of DNA, which is present in the nucleus of a nondividing or interplasm cell.

Chromatolysis (krō'-ma-TOL-i-sis) The breakdown of chromatophilic substance (Nissl bodies) into finely granular masses in the cell body of a neuron whose axon has been damaged.

Chromatophilic substance Rough endoplasmic reticulum in the cell bodies of neurons that functions in protein synthesis. Also called *Nissl bodies*.

Chromosomal microtubule (mī-krō-TOOB-yool) Microtubule formed during prophase of mitosis that originates from centromeres, extends from a centromere to a pole of the cell, and assists in chromosomal movement; constitutes a part of the mitotic spindle.

Chromosome (KRŌ-mō-sōm) One of the 46 small, dark-staining bodies that appear in the nucleus of a human diploid (2 *n*) cell during cell division.

Chronic (KRON-ik) Long-term or frequently recurring; applied to a disease that is not acute.

Chyle (kīl) The milky fluid found in the lacteals of the small intestine after digestion.

Chyme (kīm) The semifluid mixture of partly digested food and digestive secretions found in the stomach and small intestine during digestion of a meal.

Cicatrix (SIK-a-triks) A scar left by a healed wound.

Ciliary (SIL-ē-ar'-ē) *body* One of the three portions of the vascular tunic of the eyeball, the others being the choroid and the iris; includes the ciliary muscle and the ciliary processes.

Ciliary ganglion (GANG-glē-on) A very small parasympathetic ganglion whose preganglionic fibers come from the oculomotor (III) nerve and whose postganglionic fibers carry nerve impulses to the ciliary muscle and the sphincter muscle of the iris.

Cilium (SIL-ē-um) A hair or hairlike process projecting from a cell that may be used to move the entire cell or to move substances along the surface of the cell.

Circadian (ser-KĀ-dē-an) *rhythm* A cycle of active and nonactive periods in organisms determined by internal mechanisms and repeating about every 24 hours.

Circle of Willis *See Cerebral Arterial Circle.*

Circumcision (ser'-kum-SIZH-un) Removal of the foreskin (prepuce), the fold over the glans penis.

Circumduction (ser'-kum-DUK-shun) A movement at a synovial joint in which the distal end of a bone moves in a circle while the proximal end remains relatively stable.

Circumvallate papilla (ser'-kum-VAL-āt pa-PIL-a) One of the circular projections that is arranged in an inverted V-shaped row at the posterior portion of the tongue; the largest of the elevations on the upper surface of the tongue containing taste buds.

Cirrhosis (si-RŌ-sis) A liver disorder in which the parenchymal cells are destroyed and replaced by connective tissue.

Cisterna chyli (sis-TER-na KĪ-lē) The origin of the thoracic duct.

Cleavage The rapid mitotic divisions following the fertilization of a secondary oocyte, resulting in an increased number of progressively smaller cells, called blastomeres, so that the overall size of the zygote remains the same.

Cleft palate Condition in which the palatine processes of the maxillae do not unite before birth; cleft lip, a split in the upper lip, is often associated with cleft palate.

Climacteric (klī-mak-TER-ik) Cessation of the reproductive function in the female or diminution of testicular activity in the male.

Climax The peak period or moments of greatest intensity during sexual excitement.

Clitoris (KLI-to-ris) An erectile organ of the female that is homologous to the male penis.

Clot The end result of a series of biochemical reactions that changes liquid plasma into a gelatinous mass; specifically, the conversion of fibrinogen into a tangle of polymerized fibrin molecules.

Coarctation (kō'-ark-TĀ-shun) *of the aorta* Congenital condition in which the aorta is too narrow and results in reduced blood supply, increased ventricular pumping, and high blood pressure.

Coccyx (KOK-six)   The fused bones at the end of the vertebral column.

Cochlea (KŌK-lē-a)   A winding, cone-shaped tube forming a portion of the inner ear and containing the spiral organ (organ of Corti).

Cochlear duct   The membranous cochlea consisting of a spirally arranged tube enclosed in the bony cochlea and lying along its outer wall. Also called the *scala media* (SCA-la MĒ-dē-a).

Coitus (KŌ-i-tus)   Sexual intercourse. Also called *copulation* (cop-yoo-LĀ-shun).

Collagen (KOL-a-jen)   A protein that is the main organic constituent of connective tissue.

Collateral circulation   The alternate route taken by blood through an anastomosis.

Colliculus (ko-LIK-yoo-lus)   A small elevation.

Colon   The division of the large intestine consisting of ascending, transverse, descending, and sigmoid portions.

Color blindness   Any deviation in the normal perception of colors, resulting from the lack of one or more of the photopigments of the cones.

Colostomy (kō-LOS-tō-mē)   The surgical creation of a new opening from the colon to the body surface.

Colostrum (kō-LOS-trum)   A thin, cloudy fluid secreted by the mammary glands a few days prior to or after delivery before true milk is secreted.

Colposcopy (kol-POS-kō-pē)   Direct examination of the vaginal and cervical mucosa using a magnifying device.

Coma (KŌ-ma)   Profound unconsciousness from which one cannot be roused.

Commissure (KOM-i-shūr)   The angular junction of the eyelids at either corner of the eyes.

Common bile duct   A tube formed by the union of the common hepatic duct and the cystic duct that empties bile into the duodenum at the hepatopancreatic ampulla (ampulla of Vater).

Compact (dense) bone   Bone tissue with no apparent spaces in which the layers of lamellae are fitted tightly together. Compact bone is found immediately deep to the periosteum and external to spongy bone.

Complete blood count (CBC)   Hematology test that usually includes hemoglobin determination, hematocrit, red and white blood cell count, differential blood count, and comments about blood cell morphology.

Computed tomography (tō-MOG-ra-fē) *(CT)*   X-ray technique that provides a cross-sectional picture of any area of the body. Also called *computed axial tomography (CAT)*.

Concha (KONG-ka)   A scroll-like bone found in the skull. *Plural, conchae* (KONG-kē).

Concussion (kon-KUSH-un)   Traumatic injury to the brain that produces no visible bruising but may result in abrupt, temporary loss of consciousness.

Conduction myofiber   Muscle fiber in the subendocardial tissue of the heart specialized for conducting an action potential to the myocardium; part of the conduction system of the heart. Also called a *Purkinje* (pur-KIN-jē) *fiber.*

Conduction system   An intrinsic regulating system composed of specialized muscle tissue that generates and distributes electrical impulses that stimulate cardiac muscle fibers (cells) to contract.

Conductivity (kon'-duk-TIV-i-tē)   The ability to carry the effect of a stimulus from one part of a cell to another; highly developed in nerve and muscle fibers (cells).

Cone   The light-sensitive receptor in the retina concerned with color vision.

Congenital (kon-JEN-i-tal)   Present at the time of birth.

Congestive heart failure (CHF)   Chronic or actue state that results when the heart is not capable of supplying the oxygen demands of the body.

Conjunctiva (kon'-junk-TĪ-va)   The delicate membrane covering the eyeball and lining the eyes.

Conjunctivitis (kon-junk'-ti-VĪ-tis)   Inflammation of the conjunctiva, the delicate membrane covering the eyeball and lining the eyelids.

Connective tissue   The most abundant of the four basis tissue types in the body, performing the functions of binding and supporting; consists of relatively few cells in a great deal of intercellular substance.

Constipation (con-sti-PĀ-shun)   Infrequent or difficult defecation caused by decreased motility of the intestines.

Contact inhibition   Phenomenon by which migration of a growing cell is stopped when it makes contact with another cell of its own kind.

Continuous microtubules (mī-krō-TOOB-yoolz)   Microtubules formed during prophase of mitosis that originate from the vicinity of the centrioles, grow toward each other, extend from one pole of the cell to another, and assist in chromosomal movement; constitute a part of the mitotic spindle.

Contraception (kon'-tra-SEP-shun)   The prevention of conception or impregnation without destroying fertility.

Contractility (kon'-trak-TIL-i-tē)   The ability of muscle tissue to shorten when an adequate stimulus is applied.

Contralateral (kon'-tra-LAT-er-al)   On the opposite side; affecting the opposite side of the body.

Conus medullaris (KŌ-nus med-yoo-LAR-is)   The tapered portion of the spinal cord below the lumbar enlargement.

Convergence (con-VER-jens)   An anatomical arrangement in which the synaptic end-bulbs of several presynaptic neurons terminate on one postsynaptic neuron. The medial movement of the two eyeballs so that both are directed toward a near object being viewed in order to produce a single image.

Convulsion (con-VUL-shun)   Violent, involuntary, tetanic contractions of an entire group of muscles.

Cornea (KOR-nē-a)   The transparent fibrous coat that covers the iris of the eye.

Corona (kō-RŌ-na)   Margin of the glans penis.

Coronal (kō-RŌ-nal) *plane*   A plane that runs vertical to the ground and divides the body into anterior and posterior portions. Also called *frontal plane.*

Corona radiata   Innermost layer of follicle cells surrounding a secondary oocyte.

Coronary (KOR-ō-na-rē) *artery disease (CAD)*   A condition in which the heart muscle receives inadequate blood due to an interruption of its blood supply.

Coronary artery spasm   A condition in which the smooth muscle of a coronary artery undergoes a sudden contraction, resulting in vasoconstriction.

Coronary circulation   The pathway followed by the blood from the ascending aorta through the blood vessels supplying the heart and returning to the right atrium. Also called *cardiac circulation.*

Coronary sinus (SĪ-nus)   A wide venous channel on the posterior surface of the heart that collects the blood from the coronary circulation and returns it to the right atrium.

Corpora quadrigemina (KOR-por-a kwad-ri-JEM-in-a) Four small elevations (superior and inferior colliculi) on the dorsal region of the midbrain concerned with visual and auditory functions.

Cor pulmonale (kor pul-mōn-ALE) **(CP)** Right ventricular hypertrophy from disorders that bring about hypertension in pulmonary circulation.

Corpus (KOR-pus) The principal part of any organ; any mass or body.

Corpus albicans (KOR-pus AL-bi-kanz) A white fibrous patch in the ovary that forms after the corpus luteum regresses.

Corpus callosum (kal-LŌ-sum) The great commissure of the brain between the cerebral hemispheres.

Corpuscle of touch The sensory receptor for the sensation of touch; found in the dermal papillae, especially in palms and soles. Also called a **Meissner's** (MĪS-nerz) **corpuscle.**

Corpus luteum (LOO-tē-um) A yellow endocrine gland in the ovary formed when a follicle has discharged its secondary oocyte; secretes estrogens, progesterone, and relaxin.

Corpus striatum (strī-Ā-tum) An area in the interior of each cerebral hemisphere composed of the caudate and lentiform nuclei of the basal ganglia and white matter of the internal capsule, arranged in a striated manner.

Cortex (KOR-teks) An outer layer of an organ. The convoluted layer of gray matter covering each cerebral hemisphere.

Costal (KOS-tal) Pertaining to a rib.

Costal cartilage (KOS-tal KAR-ti-lij) Hyaline cartilage that attaches a rib to the sternum.

Cowper's gland See **Bulbourethral gland.**

Cramp A spasmodic, especially a tonic, contraction of one of many muscles, usually painful.

Cranial (KRĀ-nē-al) **cavity** A subdivision of the dorsal body cavity formed by the cranial bones and containing the brain.

Cranial nerve One of 12 pairs of nerves that leave the brain, pass through foramina in the skull, and supply the head, neck, and part of the trunk; each is designated by a Roman numeral and a name.

Craniosacral (krā-nē-ō-SĀ-kral) **outflow** The fibers of parasympathetic preganglionic neurons, which have their cell bodies located in nuclei in the brain stem and in the lateral gray matter of the sacral portion of the spinal cord.

Craniotomy (krā'-nē-OT-ō-mē) Any operation on the skull, as for surgery on the brain or decompression of the fetal head in difficult labor.

Cranium (KRĀ-nē-um) The skeleton of the skull that protects the brain and the organs of sight, hearing, and balance; includes the frontal, parietal, temporal, occipital, sphenoid, and ethmoid bones.

Crenation (krē-NĀ-shun) The shrinkage of red blood cells into knobbed, starry forms when placed in a hypertonic solution.

Cretinism (KRĒ-tin-izm) Severe congenital thyroid deficiency during childhood leading to physical and mental retardation.

Crista (KRIS-ta) A crest or ridged structure. A small elevation in the ampulla of each semicircular duct that serves as a receptor for dynamic equilibrium.

Crossing-over The exchange of a portion of one chromatid with another in a tetrad during meiosis. It permits an exchange of genes among chromatids and is one factor that results in genetic variation.

Crus (krus) **of penis** Separated, tapered portion of the corpora cavernosa penis. *Plural,* **crura** (KROO-ra).

Cryosurgery (KRĪ-ō-ser-jer-ē) The destruction of tissue by application of extreme cold.

Crypt of Lieberkühn See **Intestinal gland.**

Cryptorchidism (krip-TOR-ki-dizm) The conditon of undescended testes.

Cupula (KUP-yoo-la) A mass of gelatinous material covering the hair cells of a crista, a receptor in the ampulla of a semicircular canal stimulated when the head moves.

Cushing's syndrome Condition caused by a hypersecretion of glucocorticoids characterized by spindly legs, ''moon face,'' ''buffalo hump,'' pendulous abdomen, flushed facial skin, and poor wound healing.

Cutaneous (kyoo-TĀ-nē-us) Pertaining to the skin.

Cyanosis (sī'-a-NŌ-sis) Slightly bluish or dark purple discoloration of the skin and the mucous membrane due to an oxygen deficiency.

Cyst (SIST) A sac with a distinct connective tissue wall, containing a fluid or other material.

Cystic (SIS-tik) **duct** The duct that transports bile from the gallbladder to the common bile duct.

Cystitis (sis-TĪ-tis) Inflammation of the urinary bladder.

Cystoscope (SIS-to-skōp) An instrument used to examine the inside of the urinary bladder.

Cytokinesis (sī'-tō-ki-NĒ-sis) Division of the cytoplasm.

Cytology (sī-TOL-ō-jē) The study of cells.

Cytoplasm (SĪ-tō-plazm) Substance within a cell's plasma membrane and external to its nucleus. Also called **protoplasm.**

Cytoskeleton Complex internal structure of cytoplasm consisting of microfilaments, microtubules, and intermediate filaments.

Dartos (DAR-tōs) The contractile tissue under the skin of the scrotum.

Deafness Lack of the sense of hearing or a significant hearing loss.

Debility (dē-BIL-i-tē) Weakness of tonicity in functions or organs of the body.

Decidua (dē-SID-yoo-a) That portion of the endometrium of the uterus (all but the deepest layer) that is modified for pregnancy and shed after childbirth.

Deciduous (dē-SID-yoo-us) Falling off or being shed seasonally or at a particular stage of development. In the body, referring to the first set of teeth.

Decubitus (dē-KYOO-bi-tus) **ulcer** Tissue destruction due to a constant deficiency of blood to tissues overlying a bony projection that has been subjected to prolonged pressure against an object such as a bed, cast, or splint. Also called **bedsore, pressure sore,** or **trophic ulcer.**

Decussation (dē-ku-SĀ-shun) A crossing-over; usually refers to the crossing of most of the fibers in the large motor tracts to opposite sides in the medullary pyramids.

Deep Away from the surface of the body.

Deep fascia (FASH-ē-a) A sheet of connective tissue wrapped around a muscle to hold it in place.

Deep inguinal (IN-gwi-nal) **ring** A slitlike opening in the aponeurosis of the transversus abdominis muscle that represents the origin of the inguinal canal.

Deep-venous thrombosis **(DVT)** The presence of a thrombus in a vein, usually a deep vein of the lower extremities.

Defecation (def-e-KĀ-shun) The discharge of feces from the rectum.

Defibrillation (dē-fib-ri-LĀ-shun) Delivery of a very strong electrical current to the heart in an attempt to stop ventricular fibrillation.

Degeneration (dē-jen-er-Ā-shun) A change from a higher to a lower state; a breakdown in structure.

Deglutition (dē-gloo-TISH-un) The act of swallowing.

Dehydration (dē-hī-DRĀ-shun) Excessive loss of water from the body or its parts.

Delta cell A cell in the pancreatic islets (islets of Langerhans) in the pancreas that secretes somatostatin.

Demineralization (de-min'-er-al-i-ZĀ-shun) Loss of calcium and phosphorus from bones.

Dendrite (DEN-drīt) A nerve cell process that carries a nerve impulse toward the cell body.

Dens (denz) Tooth.

Dental caries (KA-rēz) Gradual demineralization of the enamel and dentin of a tooth that may invade the pulp and alveolar bone. Also called **tooth decay.**

Denticulate (den-TIK-yoo-lāt) Finely toothed or serrated; characterized by a series of small, pointed projections.

Dentin (DEN-tin) The osseous tissues of a tooth enclosing the pulp cavity.

Dentition (den-TI-shun) The eruption of teeth. The number, shape, and arrangement of teeth.

Deoxyribonucleic (dē-ok'-sē-ri'-bō-nyoo-KLĒ-ik) **acid (DNA)** A nucleic acid in the shape of a double helix constructed of nucleotides consisting of one of four nitrogen bases (adenine, cytosine, guanine, or thymine), deoxyribose, and a phosphate group; encoded in the nucleotides is genetic information.

Depression (dē-PRESS-shun) Movement in which a part of the body moves downward.

Dermal papilla (pa-PILL-a) Fingerlike projection of the papillary region of the dermis that may contain blood capillaries or corpuscles of touch (Meissner's corpuscles).

Dermatology (der-ma-TOL-ō-jē) The medical specialty dealing with diseases of the skin.

Dermatome (DER-ma-tōm) An instrument for incising the skin or cutting thin transplants of skin. The cutaneous area developed from one embryonic spinal cord segment and receiving most of its innervation from one spinal nerve.

Dermis (DER-mis) A layer of dense connective tissue lying deep to the epidermis; the true skin or corium.

Descending colon (KŌ-lon) The part of the large intestine descending from the left colic (splenic) flexure to the level of the left iliac crest.

Detritus (de-TRI-tus) Any broken-down or degenerative tissue or carious matter.

Detrusor (de-TROO-ser) **muscle** Muscle in the wall of the urinary bladder.

Developmental anatomy The study of development from the fertilized egg to the adult form. The branch of anatomy called embryology is generally restricted to the study of development from the fertilized egg through the eighth week in utero.

Diabetes insipidus (dī-a-BĒ-tēz in-SIP-i-dus) Condition caused by hyposecretion of antidiuretic hormone (ADH) and characterized by excretion of large amounts of urine and thirst.

Diabetes mellitus (MEL-i-tus) Hereditary condition caused by hyposecretion of insulin and characterized by hyperglycemia, increased urine production, excessive thirst, and excessive eating.

Diagnosis (dī-ag-NŌ-sis) Recognition of disease states from signs and symptoms by inspection, palpation, laboratory tests, and other means.

Dialysis (dī-AL-i-sis) The process of separating crystalloids (smaller particles) from colloids (larger particles) by the difference in their rates of diffusion through a selectively permeable membrane.

Diapedesis (dī-a-pe-DĒ-sis) The passage of white blood cells though intact blood vessel walls.

Diaphragm (DĪ-a-fram) Any partition that separates one area from another, especially the dome-shaped skeletal muscle between the thoracic and abdominal cavities.

Diaphysis (dī-AF-i-sis) The shaft of a long bone.

Diarrhea (dī-a-RĒ-a) Frequent defecation of liquid feces caused by increased motility of the intestines.

Diarthrosis (dī-'ar-THRŌ-sis) Articulation in which opposing bones move freely, as in a hinge joint.

Diastole (dī-AS-tō-lē) In the cardiac cycle, the phase of relaxation or dilation of the heart muscle, especially of the ventricles.

Diencephalon (dī-'en-SEF-a-lon) A part of the brain consisting primarily of the thalamus and the hypothalamus.

Differential (dif-fer-EN-shal) **blood count (DIF)** Determination of the number of each kind of white blood cell in a sample of 100 cells for diagnostic purposes.

Differentiation (dif'-e-ren'-shē-Ā-shun) Acquisition of specific functions different from those of the original general type.

Diffusion (dif-YOO-zhun) A passive process in which there is a net or greater movement of molecules or ions from a region of high concentration to a region of low concentration until equilibrium is reached.

Digestion (di-JES-chun) The mechanical and chemical breakdown of food to simple molecules that can be absorbed by the body.

Dilate (DĪ-lāte) To expand or swell.

Diploid (DIP-loyd) Having the number of chromosomes characteristically found in the somatic cells of an organism. Symbolized $2n$.

Diplopia (di-PLŌ-pē-a) Double vision.

Disease Any change from a state of health.

Dislocation (dis-lō-KĀ-shun) Displacement of a bone from a joint with tearing of ligaments, tendons, and articular capsules. Also called **luxation** (luks-Ā-shun).

Dissect (DIS-sekt) To separate tissues and parts of a cadaver (corpse) or an organ for anatomical study.

Distal (DIS-tal) Farther from the attachment of an extremity to the trunk or a structure; farther from the point of origin.

Diuretic (dī-yoo-RET-ik) A chemical that increases urine volume by inhibiting facultative reabsorption of water.

Diurnal (dī-UR-nal) Daily.

Divergence (di-VER-jens) An anatomical arrangement in which the synaptic end bulbs of one presynaptic neuron terminate on several postsynaptic neurons.

Diverticulitis (dī-ver-tik-yoo-LĪ-tis) Inflammation of diverticula, saclike outpouchings of the colonic wall, when the muscularis becomes weak.

Diverticulum (dī-ver-TIK-yoo-lum) A sac or pouch in the wall of a canal or organ, especially in the colon.

Dorsal body cavity Cavity near the dorsal surface of the body that consists of a cranial cavity and vertebral canal.

Dorsal ramus (RĀ-mus) A branch of a spinal nerve containing motor and sensory fibers supplying the muscles, skin, and bones of the posterior part of the head, neck and trunk.

Dorsiflexion (dor'-si-FLEK-shun) Bending the foot in the direction of the dorsum (upper surface).

Down's syndrome (DS) An inherited defect due to an extra copy of chromosome 21. Symptoms include mental retardation; a small skull, flattened front to back; a short, flat nose; short fingers; and a widened space between the first two digits of the hand and foot. Also called **trisomy 21.**

Dropsy (DROP-sē) A condition in which there is an abnormal accumulation of water in the tissues and cavities.

Duct of Santorini *See Accessory duct.*

Duct of Wirsung *See Pancreatic duct.*

Ductus arteriosus (DUK-tus ar-tē-rē-Ō-sus) A small vessel connecting the pulmonary trunk with the aorta; found only in the fetus.

Ductus (vas) deferens (DEF-er-ens) The duct that conducts spermatozoa from the epididymis to the ejaculatory duct. Also called the **seminal duct.**

Ductus epididymis (ep'-i-DID-i-mis) A tightly coiled tube inside the epididymis, distinguished into a head, body, and tail, in which spermatozoa undergo maturation.

Ductus venosus (ve-NŌ-sus) A small vessel in the fetus that helps the circulation bypass the liver.

Duodenal (doo-ō-DĒ-nal) **gland** Gland in the submucosa of the duodenum that secretes an alkaline mucus to protect the lining of the small intestine from the action of enzymes and to help neutralize the acid in chyme. Also called **Brunner's** (BRUN-erz) **gland.**

Duodenal papilla (pa-PILL-a) An elevation on the duodenal mucosa that receives the hepatopancreatic ampulla (ampulla of Vater).

Duodenum (doo'-ō-DĒ-num) The first portion of the small intestine.

Dura mater (DYOO-ra MĀ-ter) The outer membrane (meninx) covering the brain and spinal cord.

Dynamic equilibrium (ē-kwi-LIB-rē-um) The maintenance of body position, mainly the head, in response to sudden movements such as rotation, acceleration, and deceleration.

Dynamic spatial reconstructor (DSR) An x-ray machine that has the ability to construct moving, three-dimensional, life-size images of all or part of an internal organ from any view desired.

Dysfunction (dis-FUNK-shun) Absence of complete normal function.

Dyslexia (dis-LEX-sē-a) Impairment of ability to comprehend written language.

Dysmenorrhea (dis'-men-ō-RĒ-a) Painful menstruation.

Dysphagia (dis-FĀ-jē-a) Difficulty in swallowing.

Dysplasia (dis-PLĀ-zē-a) Change in the size, shape, and organization of cells due to chronic irritation or inflammation; may revert to normal if stress is removed or progress to neoplasia.

Dyspnea (DISP-nē-a) Labored breathing.

Dystocia (dis-TŌ-sē-a) Difficult labor due to factors such as pelvic deformities, malpositioned fetus, and premature rupture of fetal membranes.

Dystrophia (dis-TRŌ-fē-a) Progressive weakening of a muscle.

Dysuria (dis-SOO-rē-a) Painful urination.

Ectoderm The outermost of the three primary germ layers that gives rise to the nervous system and the epidermis of skin and its derivatives.

Ectopic (ek-TOP-ik) Out of the normal location, as in ectopic pregnancy.

Eczema (EK-ze-ma) A skin rash characterized by itching, swelling, blistering, oozing, and scaling of the skin.

Edema (e-DĒ-ma) An abnormal accumulation of fluid in body tissues.

Effector (e-FEK-tor) The organ of the body, either a muscle or a gland, that responds to a motor neuron impulse.

Efferent arteriole (EF-er-ent ar-TĒ-rē-ōl) A vessel of the renal vascular system that transports blood from the glomerulus to the peritubular capillary.

Efferent (EF-er-ent) **ducts** A series of coiled tubes that transport spermatozoa from the rete testis to the epididymis.

Efferent neuron (NOO-ron) A neuron that conveys nerve impulses from the brain and spinal cord to effectors that may be either muscles or glands. Also called a **motor neuron.**

Effusion (e-FYOO-zhun) The escape of fluid from the lymphatics or blood vessels into a cavity or into tissues.

Ejaculation (e-jak-yoo-LĀ-shun) The reflex ejection or expulsion of semen from the penis.

Ejaculatory (e-JAK-yoo-la-tō'-rē) **duct** A tube that transports spermatozoa from the ductus (vas) deferens to the prostatic urethra.

Elasticity (e-las-TIS-i-tē) The ability of tissue to return to its original shape after contraction or extension.

Electrocardiogram (e-lek'-trō-KAR-dē-ō-gram) **(ECG** or **EKG)** A recording of the electrical changes that accompany the cardiac cycle.

Electroencephalogram (e-lek'-trō-en-SEF-a-lō-gram) **(EEG)** A recording of the electrical impulses of the brain.

Eleidin (el-Ē-i-din) A translucent substance found in the skin.

Elevation (el-e-VĀ-shun) Movement in which a part of the body moves upward.

Ellipsoidal (e-lip-SOY-dal) **joint** A synovial joint structured so that an oval-shaped condyle of one bone fits into an elliptical cavity of another bone, permitting side-to-side and back-and-forth movements, as at the joint at the wrist between the radius and carpals. Also called a **condyloid** (KON-di-loid) **joint.**

Embolism (EM-bō-lizm) Obstruction or closure of a vessel by an embolus.

Embolus (EM-bō-lus) A blood clot, bubble of air, fat from broken bones, mass of bacteria, or other debris or foreign material transported by the blood.

Embryo (EM-brē-ō) The young of any organism in an early stage of development; in humans, the developing organism from fertilization to the end of the eighth week in utero.

Embryology (em'-brē-OL-ō-jē) The study of development from the fertilized egg to the end of the eighth week in utero.

Embryo transfer A procedure in which a husband's semen is used to artificially inseminate a fertile secondary oocyte donor and the morula or blastocyst is then transferred from the donor to the infertile wife who carries it to term.

Emesis (EM-e-sis) Vomiting.

Emmetropia (em'-e-TRŌ-pē-a) The ideal optical condition of the eyes.

Emphysema (em'-fi-SĒ-ma) A swelling or inflation of air passages due to loss of elasticity in the alveoli.

Emulsification (ē-mul'-si-fi-KĀ-shun) The dispersion of large fat globules to smaller uniformly distributed particles.

Enamel (e-NAM-el) The hard, white substance covering the crown of a tooth.

Endocardium (en-dō-KAR-dē-um)   The layer of the heart wall, composed of endothelium and smooth muscle, that lines the inside of the heart and covers the valves and tendons that hold the valves open.

Endochondral ossification (en'-dō-KON-dral os'-i-fi-KĀ-shun)   The replacement of cartilage by bone. Also called *intra-cartilaginous* (in'-tra-kar'-ti-LAJ-i-nus) *ossification.*

Endocrine (EN-dō-krin) *gland*   A gland that secretes hormones into the blood; a ductless gland.

Endocrinology (en'-dō-kri-NOL-ō-jē)   The science concerned with the structure and functions of endocrine glands and the diagnosis and treatment of disorders of the endocrine system.

Endocytosis (en'-dō-sī-TŌ-sis)   The uptake into a cell of large molecules and particles in which a segment of plasma membrane surrounds the substance, encloses it, and brings it in; includes phagocytosis, pinocytosis, and receptor-mediated endocytosis.

Endoderm   The innermost of the three primary germ layers of the developing embryo that gives rise to the gastrointestinal tract, urinary bladder and urethra, and respiratory tract.

Endodontics (en'-dō-DON-tiks)   The branch of dentistry concerned with the prevention, diagnosis, and treatment of diseases that affect the pulp, root, periodontal ligament, and alveolar bone.

Endogenous (en-DOJ-e-nus)   Growing from or beginning within the organism.

Endolymph (EN-dō-lymf')   The fluid within the membranous labyrinth of the inner ear.

Endometriosis (en'-dō-MĒ-trē-ō'-sis)   The growth of endometrial tissue outside the uterus.

Endometrium (en'-dō-MĒ-trē-um)   The mucous membrane lining the uterus.

Endomysium (em'-dō-MĪZ-ē-um)   Invagination of the perimysium separating each individual muscle fiber (cell).

Endoneurium (en'-dō-NYOO-rē-um)   Connective tissue wrapping around individual nerve fibers (cells).

Endoplasmic reticulum (en'-dō-PLAZ-mik re-TIK-yoo-lum) **(ER)**   A network of channels running through the cytoplasm of a cell that serves in intracellular transportation, support, storage, synthesis, and packaging of molecules. Portions of ER where ribosomes are attached to the outer surface are called granular or rough reticulum; portions that have no ribosomes are called agranular or smooth reticulum.

End organ of Ruffini   *See Type II cutaneous mechanoreceptor.*

Endorphin (en-DOR-fin)   A neuropeptide in the central nervous system that acts as a painkiller.

Endoscope (EN-dō-skōp')   An illuminated tube with lenses used to look inside hollow organs such as the stomach (gastroscope) or urinary bladder (cystoscope).

Endoscopy (en-DOS-kō-pē)   The visual examination of any cavity of the body using an endoscope, an illuminated tube with lenses.

Endosteum (en-DOS-tē-um)   The membrane that lines the medullary cavity of bones, consisting of osteoprogenitor cells and scattered osteoclasts.

Endothelial–capsular (en-dō-THĒ-lē-al) *membrane*   A filtration membrane in a nephron of a kidney consisting of the endothelium and basement membrane of the glomerulus and the epithelium of the visceral layer of the glomerular (Bowman's) capsule.

Endothelium (en'-dō-THĒ-lē-um)   The layer of simple squamous epithelium that lines the cavities of the heart and blood and lymphatic vessels.

Enkephalin (en-KEF-a-lin)   A peptide found in the central nervous sytem that acts as a painkiller.

Enteroendocrine (en-ter-ō-EN-dō-krin) *cell*   A stomach cell that secretes the hormone stomach gastrin.

Enuresis (en'-yoo-RĒ-sis)   Involuntary discharge of urine, complete or partial, after age 3.

Enzyme (EN-zīm)   A substance that affects the speed of chemical changes; an organic catalyst, usually a protein.

Eosinophil (ē'-ō-SIN-ō-fil)   A type of white blood cell characterized by granular cytoplasm readily stained by eosin.

Ependyma (e-PEN-de-ma)   Neuroglial cells that line ventricles of the brain and probably assist in the circulation of cerebrospinal fluid (CSF). Also called *ependymocytes* (e-PEN-di-mō-sītz).

Epicardium (ep-i-KAR-dē-um)   The thin outer layer of the heart wall, composed of serous tissue and mesothelium. Also called the *visceral pericardium.*

Epidemic (ep'-i-DEM-ik)   A disease that occurs above the expected level among individuals in a population.

Epidemiology (ep'-i-dē-mē-OL-ō-jē)   Medical science concerned with the occurrence and distribution of disease in human populations.

Epidermis (ep'-i-DERM-is)   The outermost layer of skin, composed of stratified squamous epithelium.

Epididymis (ep'-i-DID-i-mis)   A comma-shaped organ that lies along the posterior border of the testis and contains the ductus epididymis, in which sperm undergo maturation. *Plural, epididymides* (ep'-i-DID-i-mi-dēz).

Epidural space (ep'-i-DOO-ral)   A space between the spinal dura mater and the vertebral canal, containing loose connective tissue and a plexus of veins.

Epiglottis (ep'-i-GLOT-is)   A large, leaf-shaped piece of cartilage lying on top of the larynx, with its ''stem'' attached to the thyroid cartilage and its ''leaf'' portion unattached and free to move up and down to cover the glottis.

Epilepsy (EP-i-lep'-sē)   Neurological disorder characterized by short, periodic attacks of motor, sensory, or psychological malfunction.

Epimysium (ep'-i-MĪZ-ē-um)   Fibrous connective tissue around muscles.

Epinephrine (ep-ē-NEF-rin)   Hormone secreted by the adrenal medulla that produces actions similar to those that result from sympathetic stimulation. Also called *adrenaline* (a-DREN-a-lin).

Epineurium (ep'-i-NYOO-rē-um)   The outermost covering around the entire nerve.

Epiphyseal (ep'-i-FIZ-ē-al) *line*   The remnant of the epiphyseal plate in a long bone.

Epiphyseal (ep'-i-FIZ-ē-al) *plate*   The cartilaginous plate between the epiphysis and diaphysis.

Epiphysis (ē-PIF-i-sis)   The end of a long bone, usually larger in diameter than the shaft (the diaphysis).

Epiphysis cerebri (se-RĒ-brē)   Pineal gland.

Episiotomy (e-piz'-ē-OT-ō-mē)   Incision made to avoid tearing of the clinical perineum at the end of second stage of labor.

Epistaxis (ep'-i-STAK-sis)   Hemorrhage from the nose; nosebleed.

Epithelial (ep'-i-THĒ-lē-al) *tissue*   The tissue that forms glands or the outer part of the skin and lines blood vessels, hollow organs, and passages that lead externally from the body.

Eponychium (ep'-ō-NIK-ē-um)   Narrow band of stratum corneum at the proximal border of a nail that extends from the margin of the nail wall. Also called the *cuticle.*

Erection (ē-REK-shun)   The enlarged and stiff state of the penis (or clitoris) resulting from the engorgement of the spongy erectile tissue with blood.

Eructation (e-ruk'-TĀ-shun)   The forceful expulsion of gas from the stomach. Also called *belching.*

Erythema (er'-e-THĒ-ma)   Skin redness usually caused by engorgement of the capillaries in the lower layers of the skin.

Erythematosus (er-i'-them-a-TŌ-sus)   Pertaining to redness.

Erythrocyte (e-RITH-rō-sīt)   Red blood cell.

Erythropoiesis (e-rith'-rō-poy-Ē-sis)   The process by which erythrocytes (red blood cells) are formed.

Erythropoietin (e-rith'-rō-POY-ē-tin)   A hormone formed from a plasma protein that stimulates erythrocyte (red blood cell) production.

Esophagus (e-SOF-a-gus)   A hollow muscular tube connecting the pharynx and the stomach.

Estrogens (ES-tro-jens)   Female sex hormones produced by the ovaries concerned with the development and maintenance of female reproductive structures and secondary sex characteristics, fluid and electrolyte balance, and protein anabolism. Examples are β-estradiol, estrone, and estriol.

Etiology (ē'-tē-OL-ō-jē)   The study of the causes of disease, including theories of origin and the organisms, if any, involved.

Euphoria (yoo-FŌR-ē-a)   A subjectively pleasant feeling of well-being marked by confidence and assurance.

Eupnea (yoop-NĒ-a)   Normal quiet breathing.

Eustachian tube   *See Auditory tube.*

Euthanasia (yoo'-tha-NĀ-zē-a)   The practice of ending a life in case of incurable disease.

Eversion (ē-VER-zhun)   The movement of the sole outward at the ankle joint.

Exacerbation (eg-zas'-er-BĀ-shun)   An increase in the severity of symptoms or of disease.

Excitability (ek-sīt'-a-BIL-i-tē)   The ability of muscle tissue to receive and respond to stimuli; the ability of nerve cells to respond to stimuli and convert them into nerve impulses.

Excrement (EKS-kre-ment)   Material cast out from the body as waste, especially fecal matter.

Excretion (eks-KRĒ-shun)   The process of eliminating waste products from a cell, tissue, or the entire body; or the products excreted.

Exocrine (EK-sō-krin) *gland*   A gland that secretes substances into ducts that empty at covering or lining epithelium or directly onto a free surface.

Exocytosis (ex'-ō-sī-TŌ-sis)   A process of discharging cellular products too big to go through the membrane. Particles for export are enclosed by Golgi membranes when they are synthesized. Vesicles pinch off from the Golgi complex and carry the enclosed particles to the interior surface of the cell membrane, where the vesicle membrane and plasma membrane fuse and the contents of the vesicle are discharged.

Exogenous (ex-SOJ-e-nus)   Originating outside an organ or part.

Exophthalmic goiter (ek'-sof-THAL-mik GOY-ter)   Condition caused by hypersecretion of thyroid hormones characterized by protrusion of the eyeballs (exophthalmos) and an enlarged thyroid.

Exophthalmos (ek'-sof-THAL-mus)   An abnormal protrusion or bulging of the eyeball.

Expiration (ek-spi-RĀ-shun)   Breathing out; expelling air from the lungs into the atmosphere. Also called *exhalation.*

Expiratory (eks-PĪ-ra-tō-rē) *reserve volume*   The volume of air in excess of tidal volume that can be exhaled forcibly; about 1,200 ml.

Extensibility (ek-sten'-si-BIL-i-tē)   The ability of muscle tissue to be stretched when pulled.

Extension (ek-STEN-shun)   An increase in the angle between two bones; restoring a body part to its anatomical position after flexion.

External   Located on or near the surface.

External auditory (AW-di-tōr-ē) *canal* or *meatus* (mē-Ā-tus)   A canal in the temporal bone that leads to the middle ear.

External ear   The outer ear, consisting of the pinna, external auditory canal, and tympanic membrane or eardrum.

External nares (NA-rēz)   The external nostrils, or the openings into the nasal cavity on the exterior of the body.

External respiration   The exchange of respiratory gases between the lungs and blood.

Exteroceptor (eks'-ter-ō-SEP-tor)   A receptor adapted for the reception of stimuli from outside the body.

Extracellular fluid (ECF)   Fluid outside body cells, such as interstitial fluid and plasma.

Extracorporeal (eks'-tra-kor-PŌ-rē-al)   The circulation of blood outside the body.

Extravasation (eks-trav-a-SĀ-shun)   The escape of fluid, especially blood, lymph, or serum, from a vessel into the tissues.

Exudate (EKS-yoo-dāt)   Escaping fluid or semifluid material that oozes from a space that may contain serum, pus, and cellular debris.

Eyebrow   The hairy ridge above the eye.

Face   The anterior aspect of the head.

Facilitated diffusion (fa-SIL-i-tā-ted dif-YOO-zhun)   Diffusion in which a substance not soluble by itself in lipids is transported across a selectively permeable membrane by combining with a carrier substance.

Falciform ligament (FAL-si-form LIG-a-ment)   A sheet of parietal peritoneum between the two principal lobes of the liver. The ligamentum teres, or remnant of the umbilical vein, lies within its fold.

Fallopian tube   *See Uterine tube.*

Falx cerebelli (FALKS ser'-e-BEL-lē)   A small triangular process of the dura mater attached to the occipital bone in the posterior cranial fossa and projecting inward between the two cerebellar hemispheres.

Falx cerebri (SER-e-brē)   A fold of the dura mater extending down into the longitudinal fissure between the two cerebral hemispheres.

Fascia (FASH-ē-a)   A fibrous membrane covering, supporting, and separating muscles.

Fascicle (FAS-i-kul)   A small bundle or cluster, especially of nerve or muscle fibers (cells). Also called a *fasciculus* (fa-SIK-yoo-lus); *plural, fasciculi* (fa-SIK-yoo-lī).

Fat   A lipid compound formed from one molecule of glycerol and three molecules of fatty acids; the body's most highly concentrated source of energy. Adipose tissue, composed of adipocytes specialized for fat storage and present in the form of soft pads between various organs for support, protection, and insulation.

Fauces (FAW-sēz)   The opening from the mouth into the pharynx.

Febrile (FĒ-bril)   Feverish; pertaining to a fever.

Feces (FĒ-sēz)   Material discharged from the rectum and made up of bacteria, excretions, and food residue. Also called **stool.**

Feeding (hunger) center   A cluster of neurons in the lateral nuclei of the hypothalamus that, when stimulated, brings about feeding.

Fenestration (fen-e-STRĀ-shun)   Surgical opening made into the labyrinth of the ear for some conditions of deafness.

Fertilization (fer'-ti-li-ZĀ-shun)   Penetration of a secondary oocyte by a spermatozoon and subsequent union of the nuclei of the cells.

Fetal (FĒ-tal) **alcohol syndrome (FAS)**   Term applied to the effects of intrauterine exposure to alcohol, such as slow growth, defective organs, and mental retardation.

Fetal circulation   The cardiovascular system of the fetus, including the placenta and special blood vessels involved in the exchange of materials between fetus and mother.

Fetus (FĒ-tus)   The latter stages of the developing young of an animal; in humans, the developing organism in utero from the beginning of the third month to birth.

Fever   An elevation in body temperature above its normal temperature of 37°C (98.6°F).

Fibrillation (fi-bri-LĀ-shun)   Irregular twitching of individual muscle fibers (cells) or small groups of muscle fibers preventing effective action by an organ or muscle.

Fibrin (FĪ-brin)   An insoluble protein that is essential to blood clotting; formed from fibrinogen by action of thrombin.

Fibroblast (FĪ-brō-blast)   A large, flat cell that forms collagenous and elastic fibers and the viscous ground substance of loose connective tissue.

Fibrocyte (FĪ-brō-sīt)   A mature fibroblast that no longer produces fibers or matrix in connective tissue.

Fibromyositis (fī'-brō-mī-ō'-SĪ-tis)   A group of symptoms including pain, tenderness, and stiffness of joints, muscles, or adjacent structures. Called **"charley horse"** when it invovles the thigh.

Fibrosis (fī-BRŌ-sis)   Abnormal formation of fibrous tissue.

Fibrous (FĪ-brus) **joint**   A joint that allows little or no movement, such as a suture and syndesmosis.

Fibrous tunic (TOO-nik)   The outer coat of the eyeball, made up of the posterior sclera and the anterior cornea.

Fight-or-flight response   The effect of the stimulation of the sympathetic division of the autonomic nervous system.

Filiform papilla (FIL-i-form pa-PIL-a)   One of the conical projections that are distributed in parallel rows over the anterior two-thirds of the tongue and contain no taste buds.

Filum terminale (FĪ-lum ter-mi-NAL-ē)   Nonnervous fibrous tissue of the spinal cord that extends inferiorly from the conus medullaris to the coccyx.

Fimbriae (FIM-brē-ē)   Fingerlike structures, especially the lateral ends of the uterine (Fallopian) tubes.

Fissure (FISH-ur)   A groove, fold, or slit that may be normal or abnormal.

Fistula (FIS-choo-la)   An abnormal passage between two organs or between an organ cavity and the outside.

Fixator   A muscle that stabilizes the origin of the prime mover so that the prime mover can act more efficiently.

Fixed macrophage (MAK-rō-fāj)   Stationary phagocytic cell found in the liver, lungs, brain, spleen, lymph nodes, subcutaneous tissue, and bone marrow. Also called a **histiocyte** (HIS-tē-ō-sīt).

Flaccid (FLAS-sid)   Relaxed, flabby, or soft; lacking muscle tone.

Flagellum (fla-JEL-um)   A hairlike, motile process on the extremity of a bacterium or protozoan. *Plural,* **flagella** (fla-JEL-a).

Flatfoot   A condition in which the ligaments and tendons of the arches of the foot are weakened and the height of the longitudinal arch decreases.

Flatus (FLĀ-tus)   Gas or air in the gastrointestinal tract, commonly used to denote passage of gas rectally.

Flexion (FLEK-shun)   A folding movement in which there is a decrease in the angle between two bones.

Fluoroscope (FLOOR-ō-skōp)   An instrument for visual observation of the body by means of x ray.

Follicle (FOL-i-kul)   A small secretory sac or cavity.

Follicle-stimulating (FOL-i-kul) **hormone (FSH)**   Hormone secreted by the adenohypophysis (anterior lobe) of the pituitary gland that initiates development of ova and stimulates the ovaries to secrete estrogens in females and initiates sperm production in males.

Fontanel (fon'-ta-NEL)   A membrane-covered spot where bone formation is not yet complete, especially between the cranial bones of an infant's skull.

Foot   The terminal part of the lower extremity.

Foramen (fo-RĀ-men)   A passage or opening; a communication between two cavities of an organ or a hole in a bone for passage of vessels or nerves.

Foramen ovale (ō-VAL-ē)   An opening in the fetal heart in the septum between the right and left atria. A hole in the greater wing of the sphenoid bone that transmits the mandibular branch of the trigeminal (V) nerve.

Forearm (FOR-arm)   The part of the upper extremity between the elbow and the wrist.

Fornix (FOR-niks)   An arch or fold; a tract in the brain made up of association fibers, connecting the hippocampus with the mammillary bodies; a recess around the cervix of the uterus where it protrudes into the vagina.

Fossa (FOS-a)   A furrow or shallow depression.

Fourth ventricle (VEN-tri-kul)   A cavity within the brain lying between the cerebellum and the medulla and pons.

Fracture (FRAK-chur)   Any break in a bone.

Frenulum (FREN-yoo-lum)   A small fold of mucous membrane that connects two parts and limits movement.

Frontal plane   A plane at a right angle to a midsagittal plane that divides the body or organs into anterior and posterior portions. Also called a **coronal** (kō-RŌ-nal) **plane.**

Fulminate (FUL-mi-nāt')   To occur suddenly with great intensity.

Fundus (FUN-dus)   The part of a hollow organ farthest from the opening.

Fungiform papilla (FUN-ji-form pa-PIL-a)   A mushroomlike elevation on the upper surface of the tongue appearing as a red dot; most contain taste buds.

Furuncle (FYOOR-ung-kul)   A boil; painful nodule caused by bacterial infection and inflammation of a hair follicle or sebaceous (oil) gland.

Gallbladder   A small pouch that stores bile, located under the liver, which is filled with bile and emptied via the cystic duct.

Gallstone   A concretion, usually consisting of cholesterol, formed anywhere between bile canaliculi in the liver and the hepatopancreatic ampulla (ampulla of Vater), where bile enters the duodenum. Also called a **biliary calculus.**

Gamete (GAM-ēt)   A male or female reproductive cell; the spermatozoon or ovum.

Gamete intrafallopian transfer (GIFT)   Procedure in which aspirated secondary oocytes are combined with a solution containing sperm outside the body and then the secondary oocytes are inserted into the uterine (Fallopian) tubes.

Ganglion (GANG-glē-on)   A group of nerve cell bodies that lie outside the central nervous system. *Plural, ganglia* (GANG-glē-a).

Gangrene (GANG-rēn)   Death and rotting of a considerable mass of tissue that usually is caused by interruption of blood supply followed by bacterial (*Clostridium*) invasion.

Gastroenterology (gas'-trō-en'-ter-OL-ō-jē)   The medical specialty that deals with the structure, function, diagnosis, and treatment of diseases of the stomach and intestines.

Gastrointestinal (gas-trō-in-TES-ti-nal) **(GI) tract**   A continuous tube running through the ventral body cavity extending from the mouth to the anus. Also called the *alimentary* (al'-i-MEN-tar-ē) *canal.*

Gastrulation (gas'-troo-LĀ-shun)   The various movements of groups of cells that lead to the establishment of the primary germ layers.

Gavage (ga-VAZH)   Feeding through a tube passed through the esophagus and into the stomach.

Gene (jēn)   Biological unit of heredity; an ultramicroscopic, self-reproducing DNA particle located in a definite position on a particular chromosome.

Genetic engineering   The manufacture and manipulation of genetic material.

Genetics   The study of heredity.

Genital herpes (JEN-i-tal HER-pēz)   A sexually transmitted disease caused by type II herpes simplex virus.

Genitalia (jen'-i-TĀL-ya)   Reproductive organs.

Genotype (JĒ-nō-tīp)   The total hereditary information carried by an individual; the genetic makeup of an organism.

Geriatrics (jer'-ē-AT-riks)   The branch of medicine devoted to the medical problems and care of elderly persons.

Germinal (JER-mi-nal) *epithelium*   A layer of epithelial cells that covers the ovaries and lines the seminiferous tubules of the testes.

Germinativum (jer'-mi-na-TĒ-vum)   Skin layers where new cells are germinated.

Gestation (jes-TĀ-shun)   The period of intrauterine fetal development.

Giantism (GĪ-an-tizm)   Condition caused by hypersecretion of growth hormone (GH) during childhood characterized by excessive bone growth and body size. Also called *gigantism.*

Gingivae (jin-JI-vē)   Gums. They cover the alveolar processes of the mandible and maxilla and extend slightly into each socket.

Gingivitis (jin'-je-VĪ-tis)   Inflammation of the gums.

Gland   Single or group of specialized epithelial cells that secrete substances.

Glans penis (glanz PĒ-nis)   The slightly enlarged region at the distal end of the penis.

Glaucoma (glaw-KŌ-ma)   An eye disorder in which there is increased pressure due to an excess of intraocular fluid.

Gliding joint   A synovial joint having articulating surfaces that are usually flat, permitting only side-to-side and back-and-forth movements, as between carpal bones, tarsal bones, and the scapula and clavicle. Also called an *arthrodial* (ar-THRŌ-dē-al) *joint.*

Glomerular (glō-MER-yoo-lar) *capsule*   A double-walled globe at the proximal end of a nephron that encloses the glomerulus. Also called *Bowman's* (BŌ-manz) *capsule.*

Glomerulonephritis (glō-mer-yoo-lō-nef-RĪ-tis)   Inflammation of the glomeruli of the kidney. Also called *Bright's disease.*

Glomerulus (glō-MER-yoo-lus)   A rounded mass of nerves or blood vessels, especially the microscopic tuft of capillaries that is surrounded by the glomerular (Bowman's) capsule of each kidney tubule.

Glottis (GLOT-is)   The air passageway between the vocal folds in the larynx.

Glucagon (GLOO-ka-gon)   A hormone produced by the pancreas that increases the blood glucose level.

Glucocorticoids (gloo-kō-KOR-ti-koyds)   A group of hormones of the adrenal cortex.

Gluconeogenesis (gloo'-kō-nē'-ō-JEN-e-sis)   The conversion of a substance other than carbohydrate into glucose.

Glucose (GLOO-kōs)   A six-carbon sugar, $C_6H_{12}O_6$; the major energy source for every cell type in the body. Its metabolism is possible by every known living cell for the production of ATP.

Glycogen (GLĪ-ko-jen)   A highly branched polymer of glucose containing thousands of subunits; functions as a compact store of glucose molecules in liver and muscle fibers (cells).

Glycosuria (glī'-kō-SOO-rē-a)   The presence of glucose in the urine; may be temporary or pathological.

Gnostic (NOS-tik)   Pertaining to the faculties of perceiving and recognizing.

Gnostic area   Sensory area of the cerebral cortex that receives and integrates sensory input from various parts of the brain so that a common thought can be formed.

Goblet cell   A goblet-shaped unicellular gland that secretes mucus. Also called a *mucus cell.*

Goiter (GOY-ter)   An enlargement of the thyroid gland.

Golgi (GOL-jē) *complex*   An organelle in the cytoplasm of cells consisting of four to eight flattened channels, stacked upon one another, with expanded areas at their ends; functions in packaging secreted proteins, lipid secretion, and carbohydrate synthesis.

Golgi tendon organ   *See Tendon organ.*

Gomphosis (gom-FŌ-sis)   A fibrous joint in which a cone-shaped peg fits into a socket.

Gonad (GŌ-nad)   A gland that produces gametes and hormones; the ovary in the female and the testis in the male.

Gonadocorticoids (gō-na-dō-KOR-ti-koydz)   Sex hormones secreted by the adrenal cortex.

Gonadotropic (gō'-nad-ō-TRŌ-pik) *hormone*   A hormone that regulates the functions of the gonads.

Gonorrhea (gon'-ō-RĒ-a)   Infectious, sexually transmitted disease caused by the bacterium *Neisseria gonorrhoeae* and characterized by inflammation of the urogenital mucosa, discharge of pus, and painful urination.

Gout (gowt)   Hereditary condition associated with excessive uric acid in the blood; the acid crystallizes and deposits in joints, kidneys, and soft tissue.

Graafian follicle   *See Vesicular ovarian follicle.*

Gray commissure (KOM-i-shur)   A narrow strip of gray matter connecting the two lateral gray masses within the spinal cord.

Gray matter   Area in the central nervous system and ganglia consisting of nonmyelinated nerve tissue.

Gray ramus communicans (RĀ-mus kō-MYOO-ni-kans)   A short nerve containing postganglionic sympathetic fibers; the cell bodies of the fibers are in a sympathetic chain ganglion, and the nonmyelinated axons run by way of the gray ramus to a spinal nerve and then to the periphery to supply smooth muscle in blood vessels, arrector pili muscles, and sweat glands. *Plural*, *rami communicantes* (RĀ-mē kō-myoo-ni-KAN-tēz).

Greater omentum (ō-MEN-tum)   A large fold in the serosa of the stomach that hangs down like an apron over the front of the intestines.

Greater vestibular (ves-TIB-yoo-lar) *glands*   A pair of glands on either side of the vaginal orifice that open by a duct into the space between the hymen and the labia minora. Also called **Bartholin's** (BAR-to-linz) *glands*.

Groin (groyn)   The depression between the thigh and the trunk; the inguinal region.

Gross anatomy   The branch of anatomy that deals with structures that can be studied without using a microscope. Also called **macroscopic anatomy**.

Growth   An increase in size due to an increase in the number of cells or an increase in the size of existing cells.

Growth hormone (GH)   Hormone secreted by the adenohypophysis (anterior lobe) of the pituitary that brings about growth of body tissues, especially skeletal and muscular. Also known as **somatotropin** and **somatotropic hormone (STH)**.

Gustatory (GUS-ta-tō'-rē)   Pertaining to taste.

Gynecology (gī'-ne-KOL-ō-jē)   The branch of medicine dealing with the study and treatment of disorders of the female reproductive system.

Gyrus (JĪ-rus)   One of the folds of the cerebral cortex of the brain. *Plural*, **gyri** (JĪ-rī). Also called a **convolution**.

Hair   A threadlike structure produced by hair follicles that develops in the dermis. Also called **pilus** (PI-lus).

Hair follicle (FOL-li-kul)   Structure composed of epithelium surrounding the root of a hair from which hair develops.

Hallucination (ha-loo'-si-NĀ-shun)   A sensory perception of something that does not really exist in the world, that is, a sensory experience created from within the brain.

Hand   The terminal portion of an upper extremity, including the carpus, metacarpus, and phalanges.

Haploid (HAP-loyd)   Having half the number of chromosomes characteristically found in the somatic cells of an organism; characteristic of mature gametes. Symbolized *n*.

Hard palate (PAL-at)   The anterior portion of the roof of the mouth, formed by the maxillae and palatine bones and lined by mucous membrane.

Haustra (HAWS-tra)   The sacculated elevations of the colon.

Haversian canal   *See* **Central canal.**

Haversian system   *See* **Osteon.**

Head   The superior part of a human, cephalic to the neck. The superior or proximal part of a structure.

Heart   A hollow muscular organ lying slightly to the left of the midline of the chest that pumps the blood through the cardiovascular system.

Heart block   An arrhythmia (dysrhythmia) of the heart in which the atria and ventricles contract independently because of a blocking of electrical impulses through the heart at a critical point in the conduction system.

Heartburn   Burning sensation in the esophagus due to reflux of hydrochloric acid (HCl) from the stomach.

Heart–lung machine   A device that pumps blood, functioning as a heart, and removes carbon dioxide from blood and oxygenates it, functioning as lungs; used during heart transplantation, open-heart surgery, and coronary artery bypass grafting.

Heimlich maneuver   *See* **Abdominal thrust maneuver.**

Hematocrit (hē-MAT-ō-krit) **(Hct)**   The percentage of blood made up of red blood cells. Usually calculated by centrifuging a blood sample in a graduated tube and then reading off the volume of red blood cells and total blood.

Hematology (hē'-ma-TOL-ō-jē)   The study of blood.

Hematoma (hē'-ma-TŌ-ma)   A tumor or swelling filled with blood.

Hematopoiesis (hem'-a-tō-poy-Ē-sis)   Blood cell production occurring in the red marrow of bones. Also called **hemopoiesis** (hē-mō-poy-Ē-sis).

Hematuria (hē'-ma-TOOR-ē-a)   Blood in the urine.

Hemiballismus (hem'-i-ba-LIZ-mus)   Violent muscular restlessness of half of the body, especially of the upper extremity.

Hemiplegia (hem-i-PLĒ-jē-a)   Paralysis of the upper extremity, trunk, and lower extremity on one side of the body.

Hemocytoblast (hē'-mō-SĪ-tō-blast)   Immature stem cell in bone marrow that develops along different lines into all the different mature blood cells.

Hemodialysis (hē'-mō-dī-AL-i-sis)   Filtering of the blood by means of an artificial device so that certain substances are removed from the blood as a result of the difference in rates of their diffusion through a selectively permeable membrane while the blood is being circulated outside the body.

Hemodynamics (hē-mō-dī-NA-miks)   The study of factors and forces that govern the flow of blood through blood vessels.

Hemoglobin (hē'-mō-GLŌ-bin) **(Hb)**   A substance in erythrocytes (red blood cells) consisting of the protein globin and the iron-containing red pigment heme and constituting about 33 percent of the cell volume; involved in the transport of oxygen and carbon dioxide.

Hemolysis (hē-MOL-i-sis)   The escape of hemoglobin from the interior of the red blood cell into the surrounding medium; results from disruption of the integrity of the cell membrane by toxins or drugs, freezing or thawing, or hypotonic solutions.

Hemolytic disease of the newborn   A hemolytic anemia of a newborn child that results from the destruction of the infant's red blood cells by antibodies produced by the mother; usually the antibodies are due to an Rh blood type incompatibility. Also called **erythroblastosis fetalis** (e-rith'-rō-blas-TŌ-sis fe-TAL-is).

Hemophilia (hē'-mō-FĒL-ē-a)   A hereditary blood disorder where there is a deficient production of certain factors involved in blood clotting, resulting in excessive bleeding into joints, deep tissues, and elsewhere.

Hemorrhage (HEM-or-rij)   Bleeding; the escape of blood from blood vessels, especially when it is profuse.

Hemorrhoids (HEM-ō-royds)   Dilated or varicosed blood vessels (usually veins) in the anal region. Also called **piles**.

Hemostasis (hē-MOS-tā-sis)   The stoppage of bleeding.

Hemostat (HĒ-mō-stat)   An agent or instrument used to prevent the flow or escape of blood.

Hepatic (he-PAT-ik)   Refers to the liver.

Hepatic duct   A duct that receives bile from the bile capillaries. Small hepatic ducts merge to form the larger right and left hepatic ducts that unite to leave the liver as the common hepatic duct.

Hepatic portal circulation   The flow of blood from the gastrointestinal organs to the liver before returning to the heart.

Hepatitis (hep-a-TĪ-tis)   Inflammation of the liver due to a virus, drugs, and chemicals.

Hepatopancreatic (hep'-a-tō-pan'-krē-A-tik) *ampulla* A small, raised area in the duodenum where the combined common bile duct and main pancreatic duct empty into the duodenum. Also called the *ampulla of Vater* (VA-ter).

Hering-Breuer reflex *See Inflation reflex.*

Hernia (HER-nē-a) The protrusion or projection of an organ or part of an organ through the wall of the cavity containing it.

Herniated (her'-nē-A-ted) *disc* A rupture of an intervertebral disc so that the nucleus pulposus protrudes into the vertebral cavity. Also called a *slipped disc.*

Hiatus (hī-A-tus) An opening; a foramen.

Hilus (HI-lus) An area, depression, or pit where blood vessels and nerves enter or leave an organ. Also called a *hilum.*

Hinge joint A synovial joint in which a convex surface of one bone fits into a concave surface of another bone, such as the elbow, knee, ankle, and interphalangeal joints. Also called a *ginglymus* (JIN-gli-mus) *joint.*

Hirsutism (HER-soot-izm) An excessive growth of hair in females and children, with a distribution similar to that in adult males, due to the conversion of vellus hairs into large terminal hairs in response to higher-than-normal levels of androgens.

Histamine (HISS-ta-mēn) Substance found in many cells, especially mast cells, basophils, and platelets, released when the cells are injured; results in vasodilation, increased permeability of blood vessels, and bronchiole constriction.

Histology (hiss-TOL-ō-jē) Microscopic study of the structure of tissues.

Hodgkin's disease (HD) A malignant disorder, usually arising in lymph nodes.

Holocrine (HŌL-ō-krin) *gland* A type of gland in which the entire secreting cell, along with its accumulated secretions, makes up the secretory product of the gland, as in the sebaceous (oil) glands.

Holter monitor Electrocardiograph worn by a person while going about everyday routines.

Homeostasis (hō'-mē-ō-STA-sis) The condition in which the body's internal environment remains relatively constant, within physiological limits.

Homologous (hō-MOL-ō-gus) Correspondence of two organs in structure, position, and origin.

Homologous chromosomes Two chromosomes that belong to a pair.

Horizontal plane A plane that runs parallel to the ground and divides the body or organs into superior and inferior portions. Also called a *transverse plane.*

Hormone (HOR-mōn) A secretion of endocrine tissue that alters the physiological activity of target cells of the body.

Horn Principal area of gray matter in the spinal cord.

Human chorionic gonadotropin (kō-rē-ON-ik gō-nad-ō-TRŌ-pin) A hormone produced by the developing placenta that maintains the corpus luteum.

Human chorionic somatomammotropin (sō-mat-ō-mam-ō-TRŌ-pin) *(HCS)* A hormone produced by the chorion of the placenta that may stimulate breast tissue for lactation, enhance body growth, and regulate metabolism.

Human leucocyte associated (HLA) antigens Surface proteins on white blood cells and other nucleated cells that are unique for each person (except for identical twins) and are used to type tissues and help prevent rejection.

Hyaluronic (hī'-a-loo-RON-ik) *acid* An amorphous matrix material found outside the cell.

Hyaluronidase (hī'-a-loo-RON-i-dās) An enzyme that breaks down hyaluronic acid, increasing the permeability of connective tissues by dissolving the substances that hold body cells together.

Hydrocele (HĪ-drō-sēl) A fluid-containing sac or tumor. Specifically, a collection of fluid formed in the space along the spermatic cord and in the scrotum.

Hydrocephalus (hī-drō-SEF-a-lus) Abnormal accumulation of cerebrospinal fluid on the brain.

Hydrophobia (hī'-drō-FŌ-bē-a) Rabies; a condition characterized by severe muscle spasms when attempting to drink water. Also, an abnormal fear of water.

Hymen (HĪ-men) A thin fold of vascularized mucous membrane at the vaginal orifice.

Hyperemia (hī'-per-Ē-mē-a) An excess of blood in an area or part of the body.

Hyperextension (hī'-per-ek-STEN-shun) Continuation of extension beyond the anatomical position, as in bending the head backward.

Hyperglycemia (hī'-per-glī-SĒ-mē-a) An elevated blood sugar level.

Hypermetropia (hī'-per-mē-TRŌ-pē-a) A condition in which visual images are focused behind the retina with resulting defective vision of near objects; farsightedness.

Hyperplasia (hī'-per-PLA-zē-a) An abnormal increase in the number of normal cells in a tissue or organ, increasing its size.

Hypersecretion (hī'-per-se-KRĒ-shun) Overactivity of glands resulting in excessive secretion.

Hypersensitivity (hī'-per-sen-si-TI-vi-tē) Overreaction to an allergen that results in pathological changes in tissues. Also called *allergy.*

Hypertension (hī'-per-TEN-shun) High blood pressure.

Hyperthermia (hī'-per-THERM-ē-a) An elevated body temperature.

Hypertonic (hī'-per-TON-ik) Having an osmotic pressure greater than that of a solution with which it is compared.

Hypertrophy (hī-PER-trō-fē) An excessive enlargement or overgrowth of tissue without cell division.

Hyperventilation (hī'-per-ven-ti-LA-shun) A rate of respiration higher than that required to maintain a normal level of plasma PCO_2.

Hyponychium (hī'-pō-NIK-ē-um) Free edge of the fingernail.

Hypophyseal (hī'-pō-FIZ-ē-al) *pouch* An outgrowth of ectoderm from the roof of the stomodeum (mouth) from which the adenohypophysis (anterior lobe) of the pituitary gland develops.

Hypophysis (hī-POF-i-sis) Pituitary gland.

Hypoplasia (hī-pō-PLA-zē-a) Defective development of tissue.

Hyposecretion (hī'-pō-se-KRĒ-shun) Underactivity of glands resulting in diminished secretion.

Hypothalamic–hypophyseal (hī'-pō-thal-AM-ik hī'-po-FIZ-ē-al) *tract* A bundle of nerve processes made up of fibers that have their cell bodies in the hypothalamus but release their neurosecretions in the posterior pituitary gland or neurohypophysis.

Hypothalamus (hī'-pō-THAL-a-mus) A portion of the diencephalon, lying beneath the thalamus and forming the floor and part of the wall of the third ventricle.

Hypothermia (hī-pō-THER-mē-a) Low body temperature; in surgical procedures, it refers to deliberate cooling of the body to slow down metabolism and reduce oxygen needs of tissues.

Hypotonic (hī'-pō-TON-ik) Having an osmotic pressure lower than that of a solution with which it is compared.

Hypoxia (hī-POKS-ē-a) Lack of adequate oxygen at the tissue level.

Hysterectomy (his-te-REK-to-mē) The surgical removal of the uterus.

Ileocecal (il'-ē-ō-SĒ-kal) *sphincter* A fold of mucous membrane that guards the opening from the ileum into the large intestine. Also called the *ileocecal valve.*

Ileum (IL-ē-um) The terminal portion of the small intestine.

Immunity (i-MYOON-i-tē) The state of being resistant to injury, particularly by poisons, foreign proteins, and invading parasites, due to the presence of antibodies.

Immunoglobulin (im-yoo-nō-GLOB-yoo-lin) *(Ig)* An antibody synthesized by plasma cells derived from B lymphocytes in response to the introduction of antigen. Immunoglobulins are divided into five kinds (IgG, IgM, IgA, IgD, IgE) based primarily on the larger protein component present in the immunoglobulin.

Immunology (im'-yoo-NOL-ō-jē) The branch of science that deals with the responses of the body when challenged by antigens.

Immunosuppression (im'-yoo-nō-su-PRESH-un) Inhibition of the immune response.

Imperforate (im-PER-fō-rāt) Abnormally closed.

Impetigo (im'-pe-TĪ-go) A contagious skin disorder characterized by pustular eruptions.

Implantation (im-plan-TĀ-shun) The insertion of a tissue or a part into the body. The attachment of the blastocyst to the lining of the uterus 7–8 days after fertilization.

Impotence (IM-pō-tens) Weakness; inability to copulate; failure to maintain an erection.

Incontinence (in-KON-ti-nens) Inability to retain urine, semen, or feces, through loss of sphincter control.

Infant respiratory distress syndrome (RDS) A disease of newborn infants, especially premature ones, in which insufficient amounts of surfactant are produced and breathing is labored. Also called *hyaline* (HĪ-a-lin) *membrane disease (HMD).*

Infarction (in-FARK-shun) The presence of a localized area of necrotic tissue, produced by inadequate oxygenation of the tissue.

Infection (in-FEK-shun) Invasion and multiplication of microorganisms in body tissues, which may be inapparent or characterized by cellular injury.

Infectious mononucleosis (mon-ō-nook'-lē-Ō-sis) *(IM)* Contagious disease caused by the Epstein-Barr virus (EBV) and characterized by an elevated mononucleocyte and lymphocyte count, fever, sore throat, stiff neck, cough, and malaise.

Inferior (in-FĒR-ē-or) Away from the head or toward the lower part of a structure. Also called *caudad* (KAW-dad).

Inferior vena cava (VĒ-na CĀ-va) *(IVC)* Large vein that collects blood from parts of the body inferior to the heart and returns it to the right atrium.

Infertility Inability to conceive or to cause conception. Also called *sterility.*

Inflammation (in'-fla-MĀ-shun) Localized, protective response to tissue injury designed to destroy, dilute, or wall off the infecting agent or injured tissue; characterized by redness, pain, heat, swelling, and sometimes loss of function.

Inflation reflex Reflex that prevents overinflation of the lungs. Also called *Hering-Breuer reflex.*

Infraspinatous (in'-fra-SPĪ-na-tus) Bony process found below the spine of the scapula used for muscle attachment.

Infundibulum (in'-fun-DIB-yoo-lum) The stalklike structure that attaches the pituitary gland (hypophysis) to the hypothalamus of the brain. The funnel-shaped, open, distal end of the uterine (Fallopian) tube.

Ingestion (in-JES-chun) The taking in of food, liquids, or drugs, by mouth.

Inguinal (IN-gwi-nal) Pertaining to the groin.

Inguinal canal An oblique passageway in the anterior abdominal wall just superior and parallel to the medial half of the inguinal ligament that transmits the spermatic cord and ilioinguinal nerve in the male and round ligament of the uterus and ilioinguinal nerve in the female.

Inheritance The acquisition of body characteristics and qualities by transmission of genetic information from parents to offspring.

Inhibin A male sex hormone secreted by sustentacular (Sertoli) cells that inhibits FSH release by the adenohypophysis (anterior pituitary) and thus spermatogenesis.

Inner cell mass A region of cells of a blastocyst that differentiates into the three primary germ layers—ectoderm, mesoderm, and endoderm—from which all tissues and organs develop; also called an *embryoblast.*

Insertion (in-SER-shun) The manner or place of attachment of a muscle to the bone that it moves.

Insomnia (in-SOM-nē-a) Difficulty in falling asleep and, usually, frequent awakening.

Inspiration (in-spi-RĀ-shun) The act of drawing air into the lungs.

Insula (IN-su-la) A triangular area of cerebral cortex that lies deep within the lateral cerebral fissure, under the parietal, frontal, and temporal lobes, and cannot be seen in an external view of the brain. Also called the *island* or *isle of Reil* (RĪL).

Insulin (IN-su-lin) A hormone produced by the pancreas that decreases the blood glucose level.

Integumentary (in-teg'-yoo-MEN-tar-ē) Relating to the skin.

Intercalated (in-TER-ka-lāt-ed) *disc* An irregular transverse thickening of sarcolemma that separates cardiac muscle fibers (cells) from each other.

Intercellular fluid That portion of extracellular fluid that bathes the cells of the body; the internal environment of the body. Also called *interstitial* (in'-ter-STISH-al) *fluid.*

Intercostal (in'-ter-KOS-tal) *nerve* A nerve supplying a muscle located between the ribs.

Interferon (in'-ter-FĒR-on) Three principal types of protein (alpha, beta, gamma) naturally produced by virus-infected host cells that inhibit intracellular viral replication in uninfected host cells; artificially synthesized through recombinant DNA techniques.

Intermediate Between two structures, one of which is medial and one of which is lateral.

Intermediate filament Cytoplasmic structures, ranging from 8 to 12 nm in diameter, that may provide structural reinforcement and assist in contraction.

Internal Away from the surface of the body.

Internal capsule A thick sheet of white matter made up of myelinated fibers connecting various parts of the cerebral cortex and lying between the thalamus and the caudate and lentiform nuclei of the basal ganglia.

Internal ear   The inner ear or labyrinth, lying inside the temporal bone, containing the organs of hearing and balance.

Internal nares (NA-rēz)   The two openings posterior to the nasal cavities opening into the nasopharynx. Also called the **choanae** (kō-A-nē).

Internal respiration   The exchange of respiratory gases between the blood and body cells.

Interphase (IN-ter-fāz)   The period during its life cycle when a cell is carrying on every life process except division; the stage between two mitotic divisions. Also called **metabolic phase.**

Interstitial cell of Leydig   See **Interstitial endocrinocyte.**

Interstitial (in'-ter-STISH-al) **endocrinocyte**   A cell located in the connective tissue between seminiferous tubules in a mature testis that secretes testosterone. Also called an **interstitial cell of Leydig** (LĪ-dig).

Interstitial fluid   The fluid filling the microscopic spaces between the cells of tissues.

Interstitial growth   Growth from within, as in the growth of cartilage. Also called **endogenous** (en-DOJ-e-nus) **growth.**

Interventricular foramen (in'-ter-ven-TRIK-yoo-lar)   A narrow, oval opening through which the lateral ventricles of the brain communicate with the third ventricle. Also called the **foramen of Monro.**

Intervertebral (in'-ter-VER-te-bral) **disc**   A pad of fibrocartilage located between the bodies of two vertebrae.

Intestinal gland   Simple tubular gland that opens onto the surface of the intestinal mucosa and secretes digestive enzymes. Also called a **crypt of Lieberkühn** (LE-ber-kyoon).

Intracellular (in'-tra-SEL-yoo-lar) **fluid** (**ICF**)   Fluid located within cells.

Intrafusal (in'-tra-FYOO-zal) **fibers**   Three to ten specialized muscle fibers (cells), partially enclosed in a connective tissue capsule that is filled with lymph; the fibers compose muscle spindles.

Intramembranous ossification (in'-tra-MEM-bra-nus os'-i'-fī-KĀ-shun)   The method of bone formation in which the bone is formed directly in membranous tissue.

Intraocular (in-tra-OC-yoo-lar) **pressure** (**IOP**)   Pressure in the eyeball, produced mainly by aqueous humor.

Intrauterine device (**IUD**)   A small metal or plastic object inserted into the uterus for the purpose of preventing pregnancy.

Intrinsic (in-TRIN-sik)   Of internal origin; for example, the intrinsic factor, a mucoprotein formed by the gastric mucosa that is necessary for the absorption of vitamin B_{12}.

Intrinsic clotting pathway   Sequence of reactions leading to blood clotting that is initiated by the release of a substance contained *within* blood itself.

Intrinsic factor   A glycoprotein synthesized and secreted by the gastric mucosa that facilitates vitamin B_{12} absorption.

Intubation (in'-too-BĀ-shun)   Insertion of a tube through the nose or mouth into the larynx and trachea for entrance of air or to dilate a stricture.

Intussusception (in'-ta-sa-SEP-shun)   The infolding (invagination) of one part of the intestine within another segment.

In utero (YOO-ter-ō)   Within the uterus.

Invagination (in-vaj'-i-NĀ-shun)   The pushing of the wall of a cavity into the cavity itself.

Inversion (in-VER-zhun)   The movement of the sole inward at the ankle joint.

In vitro (VE-trō)   Literally, in glass; outside the living body and in an artificial environment such as a laboratory test tube.

In vivo (VĒ-vō)   In the living body.

Ipsilateral (ip'-si-LAT-er-al)   On the same side, affecting the same side of the body.

Iris   The colored portion of the eyeball that consists of circular and radial smooth muscle; the hole in the center of the iris is the pupil.

Ischemia (is-KĒ-mē-a)   A lack of sufficient blood to a part due to obstruction of circulation.

Island of Reil   See **Insula.**

Islet of Langerhans   See **Pancreatic islet.**

Isotonic (ī-sō-TON-ik)   Having equal tension or tone. Having equal osmotic pressure between two different solutions or between two elements in a solution.

Isthmus (IS-mus)   A narrow strip of tissue or narrow passage connecting two larger parts.

Jaundice (JAWN-dis)   A condition characterized by yellowness of skin, white of eyes, mucous membranes, and body fluids.

Jejunum (jē-JOO-num)   The middle portion of the small intestine.

Joint kinesthetic (kin'-es-THET-ik) **receptor**   A proprioceptive receptor located in a joint, stimulated by joint movement.

Juxtaglomerular (juks-ta-glō-MER-yoo-lar) **apparatus** (**JPA**)   Consists of the macula densa (cells of the distal convoluted tubule adjacent to the afferent and efferent arteriole) and juxtaglomerular cells (modified cells of the afferent and sometimes efferent arteriole); secretes renin when blood pressure starts to fall.

Karyotype (KAR-ē-ō-tīp)   Chromosome characteristics of an individual or a group of cells.

Keratin (KER-a-tin)   An insoluble protein found in the hair, nails, and other keratinized tissues of the epidermis.

Keratinocyte (ker-A-tin'-ō-sīt)   The most numerous of the epidermal cells that function in the production of keratin.

Keratohyalin (ker'-a-tō-HĪ-a-lin)   A compound involved in the formation of keratin.

Keratosis (ker'-a-TŌ-sis)   Formation of a hardened growth of tissue.

Kidney (KID-nē)   One of the paired reddish organs located in the lumbar region that regulates the composition and volume of blood and produces urine.

Kidney stone   A concretion, usually consisting of calcium oxalate, uric acid, and calcium phosphate crystals, that may form in any portion of the urinary tract. Also called a **renal calculus.**

Kinesiology (ki-nē'-sē-OL-ō-jē)   The study of the movement of body parts.

Kinesthesia (kin-is-THĒ-szē-a)   Ability to perceive extent, direction, or weight of movement; muscle sense.

Korotkoff (kō-ROT-kof) **sounds**   The various sounds that are heard while taking blood pressure.

Kupffer's cell   See **Stellate reticuloendothelial cell.**

Kyphosis (kī-FŌ-sis)   An exaggeration of the thoracic curve of the vertebral column, resulting in a "round-shouldered" or hunchback appearance.

Labial frenulum (LĀ-bē-al FREN-yoo-lum)   A medial fold of mucous membrane between the inner surface of the lip and the gums.

Labia majora (LĀ-bē-a ma-JO-ra)   Two longitudinal folds of skin extending downward and backward from the mons pubis of the female.

Labia minora (min-OR-a)   Two small folds of mucous membrane lying medial to the labia majora of the female.

Labium (LĀ-bē-um)   A lip. A liplike structure. *Plural,* **labia** (LĀ-bē-a).

Labor   The process by which the product of conception is expelled from the uterus through the vagina.

Labyrinth (LAB-i-rinth)   Intricate communicating passageway, especially in the internal ear.

Labyrinthine (lab-i-RIN-thēn) *disease*   Malfunction of the internal ear characterized by deafness, tinnitus, vertigo, nausea, and vomiting.

Laceration (las'-er-Ā-shun)   A torn, ragged, or mangled wound due to trauma.

Lacrimal (LAK-ri-mal)   Pertaining to tears.

Lacrimal (LAK-ri-mal) *canal*   A duct, one on each eyelid, commencing at the punctum at the medial margin of an eyelid and conveying the tears medially into the nasolacrimal sac.

Lacrimal gland   Secretory cells located at the superior lateral portion of each orbit that secrete tears into excretory ducts that open onto the surface of the conjunctiva.

Lacrimal sac   The superior expanded portion of the nasolacrimal duct that receives the tears from a lacrimal canal.

Lactation (lak-TĀ-shun)   The secretion and ejection of milk by the mammary glands.

Lacteal (LAK-tē-al)   One of many intestinal lymph vessels in villi that absorb fat from digested food.

Lacuna (la-KOO-na)   A small, hollow space, such as that found in bones in which the osteoblasts lie. *Plural,* **lacunae** (la-KOO-nē).

Lambdoidal suture (lam-DOY-dal)   The line of union in the skull between the parietal bones and the occipital bone.

Lamellae (la-MEL-ē)   Concentric rings found in compact bone.

Lamellated corpuscle   Oval pressure receptor located in subcutaneous tissue and consisting of concentric layers of connective tissue wrapped around an afferent nerve fiber. Also called a **Pacinian** (pa-SIN-ē-an) *corpuscle.*

Lamina (LAM-i-na)   A thin, flat layer or membrane, as the flattened part of either side of the arch of a vertebra. *Plural,* **laminae** (LAM-i-nē).

Lamina propria (PRO-prē-a)   The connective tissue layer of a mucous membrane.

Lanugo (lan-YOO-gō)   Fine downy hairs that cover the fetus.

Large intestine   The portion of the gastrointestinal tract extending from the ileum of the small intestine to the anus, divided structurally into the cecum, colon, rectum, and anal canal.

Laryngitis (la-rin-JĪ-tis)   Inflammation of the mucous membrane lining the larynx.

Laryngopharynx (la-rin'-gō-FAR-inks)   The inferior portion of the pharynx, extending downward from the level of the hyoid bone to divide posteriorly into the esophagus and anteriorly into the larynx.

Laryngoscope (la-RINJ-ō-skōp)   An instrument for examining the larynx.

Larynx (LAR-inks)   The voice box, a short passageway that connects the pharynx with the trachea.

Lateral (LAT-er-al)   Farther from the midline of the body or a structure.

Lateral ventricle (VEN-tri-kul)   A cavity within a cerebral hemisphere that communicates with the lateral ventricle in the other cerebral hemisphere and with the third ventricle by way of the interventricular foramen.

Leg   The part of the lower extremity between the knee and the ankle.

Lens   A transparent organ lying posterior to the pupil and iris of the eyeball and anterior to the vitreous humor.

Lesion (LĒ-zhun)   Any localized, abnormal change in tissue formation.

Lesser omentum (ō-MEN-tum)   A fold of the peritoneum that extends from the liver to the lesser curvature of the stomach and the commencement of the duodenum.

Lesser vestibular (ves-TIB-yoo-lar) *gland*   One of the paired mucus-secreting glands that have ducts that open on either side of the urethral orifice in the vestibule of the female.

Lethargy (LETH-ar-jē)   A condition of drowsiness or indifference.

Leucocyte (LOO-kō-sīt)   A white blood cell.

Leucocytosis (loo'-kō-sī-TŌ-sis)   An increase in the number of white blood cells, characteristic of many infections and other disorders.

Leucopenia (loo-kō-PĒ-nē-a)   A decrease of the number of white blood cells below 5,000/mm³.

Leukemia (loo-KĒ-mē-a)   A malignant disease of the blood-forming tissues characterized by either uncontrolled production and accumulation of immature leucocytes in which many cells fail to reach maturity (acute) or an accumulation of mature leucocytes in the blood because they do not die at the end of their normal life span (chronic).

Leukoplakia (loo-kō-PLĀ-kē-a)   A disorder in which there are white patches in the mucous membranes of the tongue, gums, and cheeks.

Libido (li-BĒ-dō)   The sexual drive, conscious or unconscious.

Ligament (LIG-a-ment)   Dense, regularly arranged connective tissue that attaches bone to bone.

Limbic system   A portion of the forebrain, sometimes termed the visceral brain, concerned with various aspects of emotion and behavior, that includes the limbic lobe (hippocampus and associated areas of gray matter plus the cingulate gyrus), certain parts of the temporal and frontal cortex, some thalamic and hypothalamic nuclei, and parts of the basal ganglia.

Lingual frenulum (LIN-gwal FREN-yoo-lum)   A fold of mucous membrane that connects the tongue to the floor of the mouth.

Lipase (LĪ-pās)   A fat-splitting enzyme.

Lipid   An organic compound composed of carbon, hydrogen, and oxygen that is usually insoluble in water, but soluble in alcohol, ether, and chloroform; examples include fats, phospholipids, steroids, and prostaglandins.

Lipoma (li-PŌ-ma)   A fatty tissue tumor, usually benign.

Liver   Large gland under the diaphragm that occupies most of the right hypochondriac region and part of the epigastric region; functionally, it produces bile salts, heparin, and plasma proteins; converts one nutrient into another; detoxifies substances; stores glycogen, minerals, and vitamins; carries on phagocytosis of blood cells and bacteria; and helps activate vitamin D.

Lobe (lōb)   A curved or rounded projection.

Local   Pertaining to or restricted to one spot or part.

Locus coeruleus (LŌ-kus sē-ROO-lē-us)   A group of neurons in the brain stem where norepinephrine (NE) is concentrated.

Lordosis (lor-DŌ-sis)   An exaggeration of the lumbar curve of the vertebral column.

Lou Gehrig's disease   *See* **Amyotrophic lateral sclerosis.**

Lower extremity   The appendage attached at the pelvic (hip) girdle, consisting of the thigh, knee, leg, ankle, foot, and toes.

Lumbar (LUM-bar)   Region of the back and side between the ribs and pelvis; loins.

Lumbar plexus (PLEK-sus)   A network formed by the anterior branches of spinal nerves L1 through L4.

Lumen (LOO-men)   The space within an artery, vein, intestine, or tube.

Lung   One of the two main organs of respiration, lying on either side of the heart in the thoracic cavity.

Lunula (LOO-nyoo-la)   The moon-shaped white area at the base of a nail.

Luteinizing (LOO-tē-in'-īz-ing) *hormone* (*LH*)   A hormone secreted by the adenohypophysis (anterior lobe) of the pituitary gland that stimulates ovulation, progesterone secretion by the corpus luteum, and readies the mammary glands for milk secretion in females and stimulates testosterone secretion by the testes in males.

Lymph (limf)   Fluid confined in lymphatic vessels and flowing through the lymphatic system to be returned to the blood.

Lymphangiography (lim-fan'-jē-OG-ra-fē)   A procedure by which lymphatic vessels and lymph organs are filled with a radiopaque substance in order to be x-rayed.

Lymphatic (lim-FAT-ik)   Pertaining to lymph. A large vessel that collects lymph from lymph capillaries and converges with other lymphatics to form the thoracic and right lymphatic ducts.

Lymphatic tissue   A specialized form of reticular tissue that contains large numbers of lymphocytes.

Lymph capillary   Blind-ended microscopic lymph vessel that begins in spaces between cells and converges with other lymph capillaries to form lymphatics.

Lymph node   An oval or bean-shaped structure located along lymphatic vessels.

Lymphocyte (LIM-fō-sīt)   A type of white blood cell, found in lymph nodes, associated with the immune system.

Lysosome (LĪ-sō-sōm)   An organelle in the cytoplasm of a cell, enclosed by a single membrane and containing powerful digestive enzymes.

Macrophage (MAK-rō-fāj)   Phagocytic cell derived from a monocyte. May be fixed or wandering.

Macula (MAK-yoo-la)   A discolored spot or a colored area. A small, flat region on the wall of the utricle and saccule that serves as a receptor for static equilibrium.

Macula lutea (LOO-tē-a)   The yellow spot in the center of the retina.

Magnetic resonance imaging (*MRI*)   A diagnostic procedure that focuses on the nuclei of atoms of a single element in a tissue, usually hydrogen, to determine if they behave normally in the presence of an external magnetic force; used to indicate the biochemical activity of a tissue. Formerly called *nuclear magnetic resonance* (*NMR*).

Malaise (ma-LĀYZ)   Discomfort, uneasiness, and indisposition, often indicative of infection.

Malignant (ma-LIG-nant)   Referring to diseases that tend to become worse and cause death; especially the invasion and spreading of cancer.

Mammary (MAM-ar-ē) *gland*   Modified sudoriferous (sweat) gland of the female that secretes milk for the nourishment of the young.

Marfan (MAR-fan) *syndrome*   Inherited disorder that results in abnormalities of connective tissue, especially in the skeleton, eyes, and cardiovascular system.

Marrow (MAR-ō)   Soft, spongelike material in the cavities of bone. Red marrow produces blood cells; yellow marrow, formed mainly of fatty tissue, has no blood-producing function.

Mast cell   A cell found in loose connective tissue along blood vessels that produces heparin, an anticoagulant. The name given to a basophil after it has left the bloodstream and entered the tissues.

Mastectomy (mas-TEK-tō-mē)   Surgical removal of breast tissue.

Mastication (mas'-ti-KĀ-shun)   Chewing.

Meatus (mē-Ā-tus)   A passage or opening, especially the external portion of a canal.

Mechanoreceptor (me-KAN-ō-rē'-sep-tor)   Receptor that detects mechanical deformation of the receptor itself or adjacent cells; stimuli so detected include touch, pressure, vibration, proprioception, hearing, equilibrium, and blood pressure.

Medial (MĒ-dē-al)   Nearer the midline of the body or a structure.

Medial lemniscus (lem-NIS-kus)   A flat band of myelinated nerve fibers extending through the medulla, pons, and midbrain and terminating in the thalamus on the same side. Sensory neurons in this tract transmit impulses for proprioception, fine touch, pressure, and vibration sensations.

Median aperture (AP-er-choor)   One of the three openings in the roof of the fourth ventricle through which cerebrospinal fluid enters the subarachnoid space of the brain and cord. Also called the *foramen of Magendie.*

Mediastinum (mē'-dē-as-TĪ-num)   A broad, median partition, actually a mass of tissue found between the pleurae of the lungs that extends from the sternum to the vertebral column.

Medulla (me-DULL-la)   An inner layer of an organ, such as the medulla of the kidneys.

Medulla oblongata (ob'-long-GA-ta)   The most inferior part of the brain stem.

Medullary (MED-yoo-lar'-ē) *cavity*   The space within the diaphysis of a bone that contains yellow marrow.

Medullary rhythmicity (rith-MIS-i-tē) *area*   Portion of the respiratory center in the medulla that controls the basic rhythm of respiration.

Meibomian gland   *See* **Tarsal gland.**

Meiosis (mē-Ō-sis)   A type of cell division restricted to sex-cell production involving two successive nuclear divisions that result in daughter cells with the haploid (*n*) number of chromosomes.

Meissner's corpuscle   *See* **Corpuscle of touch.**

Melanin (MEL-a-nin)   A dark black, brown, or yellow pigment found in some parts of the body such as the skin.

Melanoblast (mel'-a-NŌ-blast)   Precursor cell in the epidermis that gives rise to melanocytes, cells that produce melanin.

Melanocyte (MEL-a-nō-sīt')   A pigmented cell located between or beneath cells of the deepest layer of the epidermis that synthesizes melanin.

Melanocyte-stimulating hormone (*MSH*)   A hormone secreted by the adenohypophysis (anterior lobe) of the pituitary gland that stimulates the dispersion of melanin granules in melanocytes.

Melanoma (mel'-a-NŌ-ma)   A usually dark, malignant tumor of the skin containing melanin.

Melatonin (mel-a-TŌN-in)   A hormone secreted by the pineal gland that may inhibit reproductive activities.

Membrane   A thin, flexible sheet of tissue composed of an epithelial layer and an underlying connective tissue layer, as in an epithelial membrane, or of loose connective tissue only, as in a synovial membrane.

Membranous labyrinth (mem-BRA-nus LAB-i-rinth)   The portion of the labyrinth of the inner ear that is located inside the bony labyrinth and separated from it by the perilymph; made up of the membranous semicircular canals, the saccule and utricle, and the cochlear duct.

Memory   The ability to recall thoughts; commonly classified as activated (short-term) and long-term.

Menarche (me-NAR-kē)   Beginning of the menstrual function.

Ménière's (men-YAIRZ) **syndrome**   A type of labyrinthine disease characterized by fluctuating loss of hearing, vertigo, and tinnitus.

Meninges (me-NIN-jēz)   Three membranes covering the brain and spinal cord, called the dura mater, arachnoid, and pia mater. *Singular, meninx* (MEN-inks).

Meningitis (men-in-JĪ-tis)   Inflammation of the meninges, most commonly the pia mater and arachnoid.

Menopause (MEN-ō-pawz)   The termination of the menstrual cycles.

Menstrual (MEN-stroo-al) **cycle**   A series of changes in the endometrium of a nonpregnant female that prepares the lining of the uterus to receive a fertilized ovum.

Menstruation (men'-stroo-Ā-shun)   Periodic discharge of blood, tissue fluid, mucus, and epithelial cells that usually lasts for 5 days; caused by a sudden reduction in estrogens and progesterone. Also called the **menstrual phase** or **menses**.

Merocrine (MER-ō-krin) **gland**   A secretory cell that remains intact throughout the process of formation and discharge of the secretory product, as in the salivary and pancreatic glands.

Mesenchyme (MEZ-en-kīm)   An embryonic connective tissue from which all other connective tissues arise.

Mesentery (MEZ-en-ter'-ē)   A fold of peritoneum attaching the small intestine to the posterior abdominal wall.

Mesocolon (mez'-ō-KŌ-lon)   A fold of peritoneum attaching the colon to the posterior abdominal wall.

Mesoderm   The middle of the three primary germ layers that gives rise to connective tissues, blood and blood vessels, and muscles.

Mesothelium (mez'-ō-THĒ-lē-um)   The layer of simple squamous epithelium that lines serous cavities.

Mesovarium (mez'-ō-VAR-ēum)   A short fold of peritoneum that attaches an ovary to the broad ligament of the uterus.

Metabolism (me-TAB-ō-lizm)   The sum of all the biochemical reactions that occur within an organism, including the synthetic (anabolic) reactions and decomposition (catabolic) reactions.

Metacarpus (met'-a-KAR-pus)   A collective term for the five bones that make up the palm of the hand.

Metaphase (MET-a-phāz)   The second stage of mitosis in which chromatid pairs line up on the equatorial plane of the cell.

Metaphysis (me-TAF-i-sis)   Growing portion of a bone.

Metaplasia (met'-a-PLĀ-zē-a)   The transformation of one cell into another.

Metarteriole (met'-ar-TĒ-rē-ōl)   A blood vessel that emerges from an arteriole, traverses a capillary network, and empties into a venule.

Metastasis (me-TAS-ta-sis)   The transfer of disease from one organ or part of the body to another.

Metatarsus (met'-a-TAR-sus)   A collective term for the five bones located in the foot between the tarsals and the phalanges.

Microcephalus (mi-krō-SEF-a-lus)   An abnormally small head; premature closing of the anterior fontanel so that the brain has insufficient room for growth, resulting in mental retardation.

Microfilament (mī-krō-FIL-a-ment)   Rodlike cytoplasmic structure about 6 nm in diameter; comprises contractile units in muscle fibers (cells) and provides support, shape, and movement in nonmuscle cells.

Microglia (mi-krō-GLĒ-a)   Neuroglial cells that carry on phagocytosis. Also called **brain macrophages** (MAK-rō-fāj-ez).

Microphage (MĪK-rō-fāj)   Granular leucocyte that carries on phagocytosis, especially neutrophils and eosinophils.

Microtomography (mī-krō-tō-MOG-ra-fē)   A procedure that combines the principles of electron microscopy and computed tomography to produce highly magnified, three-dimensional images of living cells.

Microtrabeculae (mī-krō-tra-BEK-yoo-lē)   Three-dimensional meshwork of fine filaments, about 10–15 nm in diameter, that hold together microfilaments, microtubules, and intermediate filaments that together constitute the microtrabecular lattice.

Microtrabecular (mī-krō-tra-BEK-yoo-lar) **lattice** (LAT-is)   Collective term for microfilaments, microtubules, and intermediate filaments held together by microtrabeculae in cytoplasm.

Microtubule (mī-krō-TOOB-yool')   Cylindrical cytoplasmic structure, ranging in diameter from 18 to 30 nm, consisting of the protein tubulin; provides support, structure, and transportation.

Microvilli (mī'-krō-VIL-ē)   Microscopic, fingerlike projections of the cell membranes of small intestinal cells that increase surface area for absorption.

Micturition (mik'-too-RISH-un)   The act of expelling urine from the urinary bladder. Also called **urination** (yoo-ri-NĀ-shun).

Midbrain   The part of the brain between the pons and the diencephalon. Also called the **mesencephalon** (mes'-en-SEF-a-lon).

Middle ear   A small, epithelial-lined cavity hollowed out of the temporal bone, separated from the external ear by the eardrum and from the internal ear by a thin bony partition containing the oval and round windows; extending across the middle ear are the three auditory ossicles. Also called the **tympanic** (tim-PAN-ik) **cavity**.

Midline   An imaginary vertical line that divides the body into equal left and right sides.

Midsagittal plane   A vertical plane through the midline of the body that divides the body or organs into *equal* right and left sides. Also called a **median plane**.

Milk letdown   Contraction of alveolar cells to force milk into ducts of mammary glands, stimulated by oxytocin (OT), which is released from the posterior pituitary in response to suckling action.

Mitochondrion (mī'-tō-KON-drē-on)   A double-membraned organelle that plays a central role in the production of ATP; known as the ''powerhouse'' of the cell.

Mitosis (mī-TŌ-sis)   The orderly division of the nucleus of a cell that ensures that each new daughter nucleus has the same number and kind of chromosomes as the original parent nucleus. The process includes the replication of chromosomes and the distribution of the two sets of chromosomes into two separate and equal nuclei.

Mitotic apparatus   Collective term for continuous and chromosomal microtubules and centrioles; involved in cell division.

Mitotic spindle   The combination of continuous and chromosomal microtubules, involved in chromosomal movement during mitosis.

Mittelschmerz (MIT-el-shmerz) Abdominopelvic pain that supposedly indicates the release of a secondary oocyte from the ovary.

Modality (mō-DAL-i-tē) Any of the specific sensory entities, such as vision, smell, or taste.

Modiolus (mō-DĪ-ō'-lus) The central pillar or column of the cochlea.

Monoclonal antibody Antibody produced by in vitro clones of B cells hybridized with cancerous cells.

Monocyte (MON-ō-sīt') A type of white blood cell characterized by agranular cytoplasm; the largest of the leucocytes.

Mons pubis (monz PYOO-bis) The rounded, fatty prominence over the symphysis pubis, covered by coarse pubic hair.

Morbid (MOR-bid) Diseased; pertaining to disease.

Morula (MOR-yoo-la) A solid mass of cells produced by successive cleavages of a fertilized ovum a few days after fertilization.

Motor area The region of the cerebral cortex that governs muscular movement, particularly the precentral gyrus of the frontal lobe.

Motor unit A motor neuron together with the muscle fibers (cells) it stimulates.

Mucin (MYOO-sin) A protein found in mucus.

Mucous (MYOO-kus) *cell* A unicellular gland that secretes mucus. Also called a *goblet cell.*

Mucous membrane A membrane that lines a body cavity that opens to the exterior. Also called the *mucosa* (myoo-KŌ-sa).

Mucus The thick fluid secretion of the mucous glands and mucous membranes.

Multiple sclerosis (skler-Ō-sis) Progressive destruction of myelin sheaths of neurons in the central nervous system, short-circuiting conduction pathways.

Mumps Inflammation and enlargement of the parotid glands accompanied by fever and extreme pain during swallowing.

Murmur An unusual heart sound; may indicate a disorder such as a malfunctioning bicuspid (mitral) valve or may have no clinical significance.

Muscarinic (mus'-ka-RIN-ik) *receptor* Receptor found on all effectors innervated by parasympathetic postganglionic axons and some effectors innervated by sympathetic postganglionic axons; so named because the actions of acetylcholine (ACh) on such receptors are similar to those produced by muscarine.

Muscle An organ composed of one of three types of muscle tissue (skeletal, cardiac, or visceral), specialized for contraction to produce voluntary or involuntary movement of parts of the body.

Muscle spindle An encapsulated receptor in a skeletal muscle, consisting of specialized muscle fiber (cell) and nerve endings, stimulated by changes in length or tension of muscle fibers; a proprioceptor. Also called a *neuromuscular* (noo-rō-MUS-kyoo-lar) *spindle.*

Muscle tissue A tissue specialized to produce motion in response to muscle action potentials by its qualities of contractility, extensibility, elasticity, and excitability.

Muscle tone A sustained, partial contraction of portions of a skeletal muscle in response to activation of stretch receptors.

Muscular dystrophy (DIS-trō-fē') Inherited myopathy characterized by degeneration of muscle fibers (cells) that leads to progressive atrophy.

Muscularis (MUS-kyoo-la'-ris) A muscular layer or tunic of an organ.

Muscularis mucosae (myoo-KŌ-sē) A thin layer of smooth muscle fibers (cells) located in the outermost layer of the mucosa of the gastrointestinal tract, underlying the lamina propria of the mucosa.

Myasthenia (mī-as-THĒ-nē-a) *gravis* Weakness of skeletal muscles caused by antibodies directed against acetylcholine receptors that inhibit muscle contraction.

Myelin (MĪ-e-lin) *sheath* A white, phospholipid, segmented covering, formed by neurolemmocytes (Schwann cells), around the axons and dendrites of many peripheral neurons.

Myenteric plexus A network of nerve fibers from both autonomic divisions located in the muscularis coat or tunic of the small intestine. Also called the *plexus of Auerbach* (OW-er-bak).

Myocardial infarction (mī'-ō-KAR-dē-al in-FARK-shun) *(MI)* Gross necrosis of myocardial tissue due to interrupted blood supply. Also called a *heart attack.*

Myocardium (mī'-ō-KAR-dē-um) The middle layer of the heart wall, made up of cardiac muscle, comprising the bulk of the heart, and lying between the epicardium and the endocardium.

Myofibril (mī'-ō-FĪ-bril) A threadlike structure, running longitudinally through a muscle fiber (cell) consisting mainly of thick myofilaments (myosin) and thin myofilaments (actin).

Myoglobin (mī-ō-GLŌ-bin) The oxygen-binding, iron-containing conjugated protein complex present in the sarcoplasm of muscle fibers (cells); contributes the red color to muscle.

Myogram (MĪ-ō-gram) The record or tracing produced by the myograph, the apparatus that measures and records the effects of muscular contractions.

Myology (mī-OL-ō-jē) The study of the muscles.

Myometrium (mī'-ō-MĒ-trē-um) The smooth muscle coat or tunic of the uterus.

Myopia (mī-Ō-pē-a) Defect in vision so that objects can be seen distinctly only when very close to the eyes; nearsightedness.

Myosin (MĪ-ō-sin) The contractile protein that makes up the thick myofilaments of muscle fibers (cells).

Myotonia (mī-ō-TŌ-nē-a) A continuous spasm of muscle; increased muscular irritability and tendency to contract, and less ability to relax.

Myxedema (mix-e-DĒ-ma) Condition caused by hypothyroidism during the adult years characterized by swelling of facial tissues.

Nail A hard plate, composed largely of keratin, that develops from the epidermis of the skin to form a protective covering on the dorsal surface of the distal phalanges of the fingers and toes.

Nail matrix (MĀ-triks) The part of the nail beneath the body and root from which the nail is produced.

Narcosis (nar-KŌ-sis) Unconscious state due to narcotics.

Nasal (NĀ-zal) *cavity* A mucosa-lined cavity on either side of the nasal septum that opens onto the face at an external naris and into the nasopharynx at an internal naris.

Nasal septum (SEP-tum) A vertical partition composed of bone and cartilage, covered with a mucous membrane, separating the nasal cavity into left and right sides.

Nasolacrimal (nā'-zō-LAK-ri-mal) *duct* A canal that transports the lacrimal secretion from the nasolacrimal sac into the nose.

Nasopharynx (nā'-zō-FAR-inks) The uppermost portion of the pharynx, lying posterior to the nose and extending down to the soft palate.

Nausea (NAW-sē-a) Discomfort preceding vomiting.

Nebulization (neb'-yoo-li-ZĀ-shun)   Treatment with medication by spray method.

Neck   The part of the body connecting the head and the trunk. A constricted portion of an organ such as the neck of the femur or uterus.

Necrosis (ne-KRŌ-sis)   Death of a cell or group of cells as a result of disease or injury.

Neonatal (nē'-ō-NĀ-tal)   Pertaining to the first 4 weeks after birth.

Neoplasm (NĒ-ō-plazm)   A mass of new, abnormal tissue; a tumor.

Nephritis (ne-FRĪT-is)   Inflammation of the kidney.

Nephron (NEF-ron)   The functional unit of the kidney.

Nephrosis (nef-RŌ-sis)   A degenerative disease of the kidney in which glomerular damage results in leakage of protein into urine.

Nerve   A cordlike bundle of nerve fibers and their associated connective tissue coursing together outside the central nervous system.

Nervous tissue   Tissue that initiates and transmits nerve impulses to coordinate homeostasis.

Neuralgia (noo-RAL-jē-a)   Attacks of pain along the entire course or branch of a peripheral sensory nerve.

Neuritis (noo-RĪ-tis)   Inflammation of a nerve.

Neuroeffector (noo-rō-e-FEK-tor) *junction*   Collective term for neuromuscular and neuroglandular junctions.

Neurofibral node (noo-rō-FĪ-bral)   A space, along a myelinated nerve fiber, between the individual neurolemmocytes (Schwann cells) that form the myelin sheath and the neurolemma. Also called *node of Ranvier* (ron-VĒ-ā).

Neurofibril (noo-rō-FĪ-bril)   One of the delicate threads that forms a complicated network in the cytoplasm of the cell body and processes of a neuron.

Neuroglandular (noo-rō-GLAND-yoo-lar) *junction*   Area of contact between a motor neuron and a gland.

Neuroglia (noo-RŌG-lē-a)   Cells of the nervous system that are specialized to perform the functions of connective tissue. The neuroglia of the central nervous system are the astrocytes, oligodendrocytes, microglia, and ependyma; neuroglia of the peripheral nervous system include the neurolemmocytes (Schwann cells) and the ganglion satellite cells. Also called *glial* (GLĒ-al) *cells.*

Neurohypophysis (noo-rō-hī-POF-i-sis)   The posterior lobe of the pituitary gland.

Neurolemma (noo-rō-LEM-ma)   The peripheral, nucleated cytoplasmic layer of the neurolemmocyte (Schwann cell). Also called *sheath of Schwann* (SCHVON).

Neurolemmocyte   A neuroglial cell of the peripheral nervous system that forms the myelin sheath and neurolemma of a nerve fiber by wrapping around a nerve fiber in a jelly-roll fashion. Also called a *Schwann* (SCHVON) *cell.*

Neurology (noo-ROL-ō-jē)   The branch of science that deals with the normal functioning and disorders of the nervous system.

Neuromuscular (noo-rō-MUS-kyoo-lar) *junction*   The area of contact between the axon terminal of a motor neuron and a portion of the sarcolemma of a muscle fiber (cell). Also called a *myoneural* (mi-o-NOO-ral) *junction* or *motor end-plate.*

Neuron (NOO-ron)   A nerve cell, consisting of a cell body, dendrites, and an axon.

Neurosecretory (noo-rō-SĒC-re-tō-rē) *cell*   A cell in a nucleus in the hypothalamus that produces oxytocin (OT) or antidiuretic

hormone (ADH), hormones stored in the neurohypophysis of the pituitary gland.

Neurosyphilis (noo-rō-SIF-i-lis)   A form of the tertiary stage of syphilis in which various types of nervous tissue are attacked by bacteria and degenerate.

Neurotransmitter   One of a variety of molecules synthesized within the nerve axon terminals, released into the synaptic cleft in response to a nerve impulse, and affecting the membrane potential of the postsynaptic neuron. Also called a *transmitter substance.*

Neutrophil (NOO-trō-fil)   A type of white blood cell characterized by granular cytoplasm that stains as readily with acid or basic dyes. Also called a *polymorph* (POL-ē-morf).

Nicotinic (nik'-ō-TIN-ik) *receptor*   Receptor found on both sympathetic and parasympathetic postganglionic neurons so named because the actions of acetylcholine (ACh) in such receptors are similar to those produced by nicotine.

Night blindness   Poor or no vision in dim light or at night, although good vision is present during bright illumination; frequently caused by a deficiency of vitamin A. Also referred to as *nyctalopia* (nik'-ta-LŌ-pē-a).

Nipple   A pigmented, wrinkled projection on the surface of the mammary gland that is the location of the openings of the lactiferous ducts for milk release.

Nissl bodies   See *Chromatophilic substance.*

Nociceptor (nō'-sē-SEP-tor)   A receptor that detects pain.

Node of Ranvier   See *Neurofibral node.*

Nonpigmented granular dendrocytes   Two distinct cell types found in the epidermis, formerly known as *Langerhans cells* and *Granstein cells,* that differ in their sensitivity to damage by ultraviolet (UV) radiation and their functions in immunity.

Norepinephrine (nor'-ep-ē-NEF-rin) *(NE)*   A hormone secreted by the adrenal medulla that produces actions similar to those that result from sympathetic stimulation. Also called *noradrenaline* (nor-a-DREN-a-lin).

Nuclear medicine   The branch of medicine concerned with the use of radioisotopes in the diagnosis of disease and therapy.

Nuclease (NOO-klē-ās)   An enzyme that breaks nucleotides into pentoses and nitrogen bases; examples are ribonuclease and deoxyribonuclease.

Nucleic (noo-KLĒ-ic) *acid*   An organic compound that is a long polymer of nucleotides, with each nucleotide containing a pentose sugar, a phosphate group, and one of four possible nitrogen bases (adenine, cytosine, guanine, and thymine or uracil).

Nucleolus (noo-KLĒ-ō-lus)   Nonmembranous spherical body within the nucelus composed of protein, DNA, and RNA that functions in the synthesis and storage of ribosomal RNA.

Nucleosome (NOO-klē-ō-sōm)   Elementary structural subunit of a chromosome consisting of histones and DNA.

Nucleus (NOO-klē-us)   A spherical or oval organelle of a cell that contains the hereditary factors of the cell, called genes. A cluster of nerve cell bodies in the central nervous system. The central portion of an atom made up of protons and neutrons.

Nucleus cuneatus (kyoo-nē-Ā-tus)   A group of nerve cells in the inferior portion of the medulla in which fibers of the fasciculus cuneatus terminate.

Nucleus gracilis (gras-I-lis)   A group of nerve cells in the inferior portion of the medulla in which fibers of the fasciculus gracilis terminate.

Nucleus pulposus (pul-PŌ-sus)   A soft, pulpy, highly elastic substance in the center of an intervertebral disc, a remnant of the notochord.

Nutrient A chemical substance in food that provides energy, forms new body components, or assists in the functioning of various body processes.

Nystagmus (nis-TAG-mus) Constant, involuntary, rhythmic movement of the eyeballs; horizontal, rotary, or vertical.

Obesity (ō-BĒS-i-tē) Body weight 10–20 percent over a desirable standard as a result of excessive accumulation of fat. Types of obesity are hypertrophic (adult-onset) and hyperplastic (lifelong).

Obstetrics (ob-STET-riks) The specialized branch of medicine that deals with pregnancy, labor, and the period of time immediately following delivery.

Obturator (OB-tyoo-rā'-ter) Anything that obstructs or closes a cavity or opening.

Occlusion (ō-KLOO-zhun) The act of closure or state of being closed.

Olfactory (ōl-FAK-tō-rē) Pertaining to smell.

Olfactory bulb A mass of gray matter at the termination of an olfactory (I) nerve, lying beneath the frontal lobe of the cerebrum on either side of the crista galli of the ethmoid bone.

Olfactory cell A bipolar neuron with its cell body lying between supporting cells located in the mucous membrane lining the upper portion of each nasal cavity.

Olfactory tract A bundle of axons that extends from the olfactory bulb posteriorly to the olfactory portion of the cortex.

Oligodendroctye (o-lig-ō-DEN-drō-sīt) A neuroglial cell that supports neurons and produces a phospholipid myelin sheath around axons of neurons of the central nervous system.

Oligospermia (ol'-i-gō-SPER-mē-a) A deficiency of spermatozoa in the semen.

Olive A prominent oval mass on each lateral surface of the superior part of the medulla.

Oncogene (ONG-kō-jēn) Gene that has the ability to transform a normal cell into a cancerous cell.

Oncology (ong-KOL-ō-jē) The study of tumors.

Oogenesis (ō'-ō-JEN-e-sis) Formation and development of the ovum.

Oophorectomy (ō'-of-ō-REK-tō-mē) The surgical removal of the ovaries.

Ophthalmic (of-THAL-mik) Pertaining to the eye.

Ophthalmologist (of'-thal-MOL-ō-jist) A physician who specializes in the diagnosis and treatment of eye disorders with drugs, surgery, and corrective lenses.

Ophthalmology (of'-thal-MOL-ō-jē) The study of the structure, function, and diseases of the eye.

Optic (OP-tik) Refers to the eye, vision, or properties of light.

Optic chiasma (kī-AZ-ma) A crossing point of the optic (II) nerves, anterior to the pituitary gland.

Optic disc A small area of the retina containing openings through which the fibers of the ganglion neurons emerge as the optic (II) nerve. Also called the *blind spot.*

Optician (op-TISH-an) A technician who fits, adjusts, and dispenses corrective lenses on prescription of an ophthalmologist or optometrist.

Optic tract A bundle of axons that transmits nerve impulses from the retina of the eye between the optic chiasma and the thalamus.

Optometrist (op-TOM-e-trist) Specialist with a doctorate degree in optometry who is licensed to examine and test the eyes and treat visual defects by prescribing corrective lenses.

Oral contraceptive (*OC*) A hormonal compound that is swallowed and prevents ovulation, and thus pregnancy. Also called *"the pill."*

Ora serrata (Ō-ra ser-RĀ-ta) The irregular margin of the retina lying internal and slightly posterior to the junction of the choroid and ciliary body.

Orbit (OR-bit) The bony, pyramid-shaped cavity of the skull that holds the eyeball.

Organ A structure of definite form and function composed of two or more different kinds of tissues.

Organelle (or-gan-EL) A permanent structure within a cell with characteristic morphology that is specialized to serve a specific function in cellular activities.

Organic (or-GAN-ik) *compound* Compound that always contains carbon and hydrogen and is held together by covalent bonds. Examples include carbohydrates, lipids, protein, and nucleic acids (DNA and RNA).

Organism (OR-ga-nizm) A total living form; one individual.

Orgasm (OR-gazm) Sensory and motor events involved in ejaculation for the male and involuntary contraction of the perineal muscles in the female at the climax of sexual intercourse.

Orifice (OR-i-fis) Any aperture or opening.

Origin (OR-i-jin) The place of attachment of a muscle to the more stationary bone, or the end opposite the insertion.

Oropharynx (or'-ō-FAR-inks) The second portion of the pharynx, lying posterior to the mouth and extending from the soft palate down to the hyoid bone.

Orthopedics (or'-thō-PĒ-diks) The branch of medicine that deals with the preservation and restoration of the skeletal system, articulations, and associated structures.

Orthopnea (or'-thop-NĒ-a) Inability to breath in horizontal position.

Osmosis (os-MŌ-sis) The net movement of water molecules through a selectively permeable membrane from an area of high water concentration to an area of lower water concentration until an equilibrium is reached.

Osmotic pressure The pressure required to prevent the movement of pure water into a solution containing solutes when the solutions are separated by a selectively permeable membrane.

Osseous (OS-ē-us) Bony.

Ossicle (OS-si-kul) Small bone, as in the middle ear (malleus, incus, stapes).

Ossification (os'-i-fi-KĀ-shun) Formation of bone. Also called *osteogenesis.*

Osteoblast (OS-tē-ō-blast') Cell formed from an osteoprogenitor cell that participates in bone formation by secreting some organic components and inorganic salts.

Osteoclast (OS-tē-ō-clast') A large multinuclear cell that develops from a monocyte and destroys or resorbs bone tissue.

Osteocyte (OS-tē-ō-sīt') A mature bone cell that maintains the daily activities of bone tissue.

Osteogenic (os'-tē-ō-JEN-ik) *layer* The inner layer of the periosteum that contains cells responsible for forming new bone during growth and repair.

Osteology (os'-tē-OL-ō-jē) The study of bones.

Osteomalacia (os'-tē-ō-ma-LĀ-shē-a) A deficiency of vitamin D in adults causing demineralization and softening of bone.

Osteomyelitis (os'-tē-ō-mī-i-LĪ-tis) Inflammation of bone marrow or of the bone and marrow.

Osteon The basic unit of structure in adult compact bone, consisting of a central (Haversian) canal with its concentrically arranged lamellae, lacunae, osteocytes, and canaliculi. Also called a *Haversian* (ha-VER-shun) *system.*

Osteoporosis (os'-tē-ō-pō-RŌ-sis)   Age-related disorder characterized by decreased bone mass and increased susceptibility to fracture.

Osteoprogenitor (os'-tē-ō-prō-JEN-i-tor) *cell*   Stem cell derived from mesenchyme that has mitotic potential and the ability to differentiate into an osteoblast.

Otalgia (ō-TAL-jē-a)   Pain in the ear; earache.

Otic (Ō-tik)   Pertaining to the ear.

Otitis media (ō-TĪ-tus ME-dē-a)   Acute infection of the middle ear cavity characterized by an inflamed tympanic membrane, subject to rupture.

Otolith (Ō-tō-lith)   A particle of calcium carbonate embedded in the otolithic membrane that functions in maintaining static equilibrium.

Otolithic (ō-tō-LITH-ik) *membrane*   Thick, gelatinous, glycoprotein layer located directly over hair cells of the macula in the saccule and utricle of the inner ear.

Otorhinolaryngology (ō'-tō-rī-nō-lar'-in-GOL-ō-jē)   The branch of medicine that deals with the diagnosis and treatment of diseases of the ears, nose, and throat.

Oval window   A small opening between the middle ear and inner ear into which the footplate of the stapes fits. Also called the *fenestra vestibuli* (fe-NES-tra ves-TIB-yoo-lē).

Ovarian (ō-VAR-ē-an) *cycle*   A monthly series of events in the ovary associated with the maturation of an ovum.

Ovarian follicle (FOL-i-kul)   A general name for oocytes (immature ova) in any stage of development, along with their surrounding epithelial cells.

Ovarian ligament (LIG-a-ment)   A rounded cord of connective tissue that attaches the ovary to the uterus.

Ovary (Ō-var-ē)   Female gonad that produces ova and the hormones estrogens, progesterone, and relaxin.

Ovulation (ō-vyoo-LĀ-shun)   The rupture of a vesicular ovarian (Graafian) follicle with discharge of a secondary oocyte into the pelvic cavity.

Ovum (Ō-vum)   The female reproductive or germ cell; an egg cell.

Oxyhemoglobin (ok'-sē-HĒ-mō-glō-bin) (*HbO₂*)   Hemoglobin combined with oxygen.

Oxyphil cell   A cell found in the parathyroid gland that secretes parathyroid hormone (PTH).

Oxytocin (ok'-sē-TŌ-sin) (*OT*)   A hormone secreted by neurosecretory cells in the paraventricular nucleus of the hypothalamus that stimulates contraction of the smooth muscle fibers (cells) in the pregnant uterus and contractile cells around the ducts of mammary glands.

Pacinian corpuscle   *See Lamellated corpuscle.*

Paget's (PAJ-ets) *disease*   A disorder characterized by a greatly accelerated remodeling process in which osteoclastic resorption is massive and new bone formation by osteoblasts is extensive. As a result, there is an irregular thickening and softening of the bones.

Palate (PAL-at)   The horizontal structure separating the oral and the nasal cavities; the roof of the mouth.

Palliative (PAL-ē-a-tiv)   Serving to relieve or alleviate without curing.

Palpate (PAL-pāt)   To examine by touch; to feel.

Palpitation (pal'-pi-TĀ-shun)   A fluttering of the heart or abnormal rate or rhythm of the heart.

Pancreas (PAN-krē-as)   A soft, oblong organ lying along the greater curvature of the stomach and connected by a duct to the duodenum. It is both exocrine (secreting pancreatic juice) and endocrine (secreting insulin, glucagon, and somatostatin).

Pancreatic (pan'-krē-AT-ik) *duct*   A single, large tube that unites with the common bile duct from the liver and gallbladder and drains pancreatic juice into the duodenum at the hepatopancreatic ampulla (ampulla of Vater). Also called the *duct of Wirsung.*

Pancreatic islet   A cluster of endocrine gland cells in the pancreas that secretes insulin, glucagon, and somatostatin. Also called an *islet of Langerhans* (LANG-er-hanz).

Papanicolaou (pap'-a-NIK-ō-la-oo) *test*   A cytological staining test for the detection and diagnosis of premalignant and malignant conditions of the female genital tract. Cells scraped from the genital epithelium are smeared, fixed, stained, and examined microscopically. Also called a *Pap smear.*

Papilla (pa-PIL-a)   A small nipple-shaped projection or elevation.

Paralysis (pa-RAL-a-sis)   Loss or impairment of motor function due to a lesion of nervous or muscular origin.

Paranasal sinus (par'-a-NĀ-zal SĪ-nus)   A mucus-lined air cavity in a skull bone that communicates with the nasal cavity. Paranasal sinuses are located in the frontal, maxillary, ethmoid, and sphenoid bones.

Paraplegia (par-a-PLĒ-jē-a)   Paralysis of both lower extremities.

Parasagittal plane   A vertical plane that does not pass through the midline and that divides the body or organs into *unequal* left and right portions.

Parasympathetic (par'-a-sim-pa-THET-ik) *division*   One of the two subdivisions of the autonomic nervous system, having cell bodies of preganglionic neurons in nuclei in the brain stem and in the lateral gray matter of the sacral portion of the spinal cord; primarily concerned with activities that restore and conserve body energy. Also called the *craniosacral* (krā-nē-ō-SĀ-kral) *division.*

Parathyroid (par'-a-THĪ-royd) *gland*   One of four small endocrine glands embedded on the posterior surfaces of the lateral lobes of the thyroid gland.

Parathyroid hormone (PTH)   A hormone secreted by the parathyroid glands that decreases blood phosphate level and increases blood calcium level.

Paraurethral (par'-a-yoo-RĒ-thral) *gland*   Gland embedded in the wall of the urethra whose duct opens on either side of the urethral orifice and secretes mucus. Also called *Skene's* (SKĒNZ) *gland.*

Parenchyma (par-EN-ki-ma)   The functional parts of any organ, as opposed to tissue that forms its stroma or framework.

Parenteral (par-EN-ter-al)   Situated or occurring outside the intestines; referring to introduction of substances into the body other than by way of the intestines such as intradermal, subcutaneous, intramuscular, intravenous, or intraspinal.

Parietal (pa-RĪ-e-tal)   Pertaining to or forming the outer wall of a body cavity.

Parietal cell   The secreting cell of the gastric glands that produces hydrochloric acid and intrinsic factor. Also called an *oxyntic cell.*

Parietal pleura (PLOO-ra)   The outer layer of the serous pleural membrane that encloses and protects the lungs; the layer that is attached to the wall of the pleural cavity.

Parkinson's disease Progressive degeneration of the basal ganglia and substantia nigra of the cerebrum resulting in decreased production of dopamine (DA) that leads to tremor, slowing of voluntary movements, and muscle weakness.

Parotid (pa-ROT-id) **gland** One of the paired salivary glands located inferior and anterior to the ears connected to the oral cavity via a duct (Stensen's) that opens into the inside of the cheek opposite the upper second molar tooth.

Paroxysm (PAR-ok-sizm) A sudden periodic attack or recurrence of symptoms of a disease.

Pars intermedia A small avascular zone between the adenohypophysis and neurohypophysis of the pituitary gland.

Parturition (par'-too-RISH-un) Act of giving birth to young; childbirth, delivery.

Patent ductus arteriosus Congenital anatomical heart defect in which the fetal connection between the aorta and pulmonary trunk remains open instead of closing completely after birth.

Pathogen (PATH-ō-jen) A disease-producing organism.

Pathogenesis (path'-ō-JEN-e-sis) The development of disease or a morbid or pathological state.

Pathological (path'-ō-LOJ-i-kal) Pertaining to or caused by disease.

Pathological (path'-ō-LOJ-i-kal) **anatomy** The study of structural changes caused by disease.

Pectinate (PEK-ti-nāt) **muscles** Projecting muscle bundles of the anterior atrial walls and the lining of the auricles.

Pectoral (PEK-tō-ral) Pertaining to the chest or breast.

Pediatrician (pē'-dē-a-TRISH-un) A physician who specializes in the care and treatment of children and their illnesses.

Pedicel (PED-i-sel) Footlike structure, as on podocytes of a glomerulus.

Pedicle (PED-i-kul) A short, thick process found on vertebrae.

Pelvic (PEL-vik) **cavity** Inferior portion of the abdominopelvic cavity that contains the urinary bladder, sigmoid colon, rectum, and internal female and male reproductive structures.

Pelvic inflammatory disease (PID) Collective term for any extensive bacterial infection of the pelvic organs, especially the uterus, uterine (Fallopian) tubes, and ovaries.

Pelvic splanchnic (PEL-vik SPLANGK-nik) **nerves** Preganglionic parasympathetic fibers from the levels of S2, S3, and S4 that supply the urinary bladder, reproductive organs, and the descending and sigmoid colon and rectum.

Pelvimetry (pel-VIM-e-trē) Measurement of the size of the inlet and outlet of the birth canal.

Pelvis The basinlike structure formed by the two pelvic (hip) bones, the sacrum, and the coccyx. The expanded, proximal portion of the ureter, lying within the kidney and into which the major calyces open.

Penis (PĒ-nis) The male copulatory organ, used to introduce spermatozoa into the female vagina.

Pepsin Protein-digesting enzyme secreted by zymogenic (chief) cells of the stomach as the inactive form pepsinogen, which is converted to active pepsin by hydrochloric acid.

Peptic ulcer An ulcer that develops in areas of the gastrointestinal tract exposed to hydrochloric acid; classified as a gastric ulcer if in the lesser curvature of the stomach and as a duodenal ulcer if in the first part of the duodenum.

Percussion (per-KUSH-un) The act of striking (percussing) an underlying part of the body with short, sharp blows as an aid in diagnosing the part by the quality of the sound produced.

Perforating canal A minute passageway by means of which blood vessels and nerves from the periosteum penetrate into compact bone. Also called **Volkmann's** (FŌLK-manz) **canal.**

Pericardial (per'-i-KAR-dē-al) **cavity** Small potential space between the visceral and parietal layers of the serous pericardium.

Pericardium (per'-i-KAR-dē-um) A loose-fitting membrane that encloses the heart, consisting of an outer fibrous layer and an inner serous layer.

Perichondrium (per'-i-KON-drē-um) The membrane that covers cartilage.

Perikaryon (per'-i-KAR-ē-on) The nerve cell body that contains the nucleus and other organelles.

Perilymph (PER-i-lymf') The fluid contained between the bony and membranous labyrinths of the inner ear.

Perimetrium (per-i-MĒ-trē-um) The serosa of the uterus.

Perimysium (per'-i-MĪZ-ē-um) Invagination of the epimysium that divides muscles into bundles.

Perineum (per'-i-NĒ-um) The pelvic floor; the space between the anus and the scrotum in the male and between the anus and the vulva in the female.

Perineurium (per'-i-NYOO-rē-um) Connective tissue wrapping around fascicles in a nerve.

Periodontal (per-ē-ō-DON-tal) **disease** A collective term for conditions characterized by degeneration of gingivae, alveolar bone, periodontal ligament, and cementum.

Periodontal membrane The periosteum lining the alveoli (sockets) for the teeth in the alveolar processes of the mandible and maxillae.

Periosteum (per'-ē-OS-tē-um) The membrane that covers bone and consists of connective tissue, osteoprogenitor cells, and osteoblasts and is essential for bone growth, repair, and nutrition.

Peripheral (pe-RIF-er-al) Located on the outer part or a surface of the body.

Peripheral nervous system (PNS) The part of the nervous system that lies outside the central nervous system—nerves and ganglia.

Periphery (pe-RIF-er-ē) Outer part or a surface of the body; part away from the center.

Peristalsis (per'-i-STAL-sis) Successive muscular contractions along the wall of a hollow muscular structure.

Peritoneum (per'-i-tō-NĒ-um) The largest serous membrane of the body that lines the abdominal cavity and covers the viscera.

Peritonitis (per'-i-tō-NĪ-tis) Inflammation of the peritoneum.

Pernicious (per-NISH-us) Fatal.

Peroxisome (pe-ROKS-ī-sōm) Organelle similar in structure to a lysosome that contains enzymes related to hydrogen peroxide metabolism; abundant in liver cells.

Perspiration Substance produced by sudoriferous (sweat) glands containing water, salts, urea, uric acid, amino acids, ammonia, sugar, lactic acid, and ascorbic acid; helps maintain body temperature and eliminate wastes.

Peyer's patches See **Aggregated lymphatic follicles.**

pH A symbol of the measure of the concentration of hydrogen ions in a solution. The pH scale extends from 0 to 14, with a value of 7 expressing neutrality, values lower than 7 expressing increasing acidity, and values higher than 7 expressing increasing alkalinity.

Phagocytosis (fag'-ō-sī-TŌ-sis) The process by which cells (phagocytes) ingest particulate matter; especially the ingestion and destruction of microbes, cell debris, and other foreign matter.

Phalanx (FĀ-lanks)   The bone of a finger or toe. *Plural,* ***phalanges*** (fa-LAN-jēz).

Phantom pain   A sensation of pain as originating in a limb that has been amputated.

Pharmacology (far'-ma-KOL-ō-jē)   The science that deals with the effects and uses of drugs in the treatment of disease.

Pharynx (FAR-inks)   The throat; a tube that starts at the internal nares and runs partway down the neck where it opens into the esophagus posteriorly and into the larynx anteriorly.

Phenotype (FĒ-nō-tīp)   The observable expression of genotype; physical characteristics of an organism determined by genetic makeup and influenced by interaction between genes and internal and external environmental factors.

Phenylketonuria (fen'-il-kē'-tō-NOO-rē-a) (***PKU***)   A disorder characterized by an elevation of the amino acid phenylalanine in the blood.

Pheochromocytoma (fē-ō-krō'-mō-sī-TŌ-ma)   Tumor of the chromaffin cells of the adrenal medulla that results in hypersecretion of medullary hormones.

Phlebotomy (fle-BOT-ō-me)   The cutting of a vein to allow the escape of blood.

Photoreceptor   Receptor that detects light on the retina of the eye. Also called *electromagnetic receptor.*

Physiology (fiz'-ē-OL-ō-jē)   Science that deals with the functions of an organism or its parts.

Pia mater (PĪ-a MĀ-ter)   The inner membrane (meninx) covering the brain and spinal cord.

Pilonidal (pī-lō-NĪ-dal)   Containing hairs resembling a tuft inside a cyst or sinus.

Pineal (PIN-ē-al) *gland*   The cone-shaped gland located in the roof of the third ventricle. Also called the ***epiphysis cerebri*** (ē-PIF-i-sis se-RĒ-brē).

Pinealocyte (pin-ē-AL-ō-sīt)   Secretory cell of the pineal gland that produces hormones.

Pinna (PIN-na)   The projecting part of the external ear composed of elastic cartilage and covered by skin and shaped like the flared end of a trumpet. Also called the ***auricle*** (OR-i-kul).

Pinocytosis (pi'-nō-sī-TŌ-sis)   The process by which cells ingest liquid.

Pituicyte (pi-TOO-i-sīt)   Supporting cell of the posterior lobe of the pituitary gland.

Pituitary (pi-TOO-i-tar'-ē) *dwarfism*   Condition caused by hyposecretion of growth hormone (GH) during the growth years and characterized by childlike physical traits in an adult.

Pituitary gland   A small endocrine gland lying in the sella turcica of the sphenoid bone and attached to the hypothalamus by the infundibulum; nicknamed the ''master gland.'' Also called the ***hypophysis*** (hī-POF-i-sis).

Pivot joint   A synovial joint in which a rounded, pointed, or conical surface of one bone articulates with a ring formed partly by another bone and partly by a ligament, as in the joint between the atlas and axis and between the proximal ends of the radius and ulna. Also called a ***trochoid*** (TRŌ-koid) *joint.*

Placenta (pla-SEN-ta)   The special structure through which the exchange of materials between fetal and maternal circulations occurs. Also called the ***afterbirth.***

Plantar flexion (PLAN-tar FLEK-shun)   Bending the foot in the direction of the plantar surface (sole).

Plaque (plak)   A cholesterol-containing mass in the tunica media of arteries. A mass of bacterial cells, dextran (polysaccharide), and other debris that adheres to teeth.

Plasma (PLAZ-ma)   The extracellular fluid found in blood vessels; blood minus the formed elements.

Plasma cell   Cell that produces antibodies and develops from a B cell (lymphocyte).

Plasma (*cell*) membrane   Outer, limiting membrane that separates the cell's internal parts from extracellular fluid and the external environment.

Pleura (PLOOR-a)   The serous membrane that enfolds the lungs and lines the walls of the chest and diaphragm.

Pleural cavity   Small potential space between the visceral and parietal pleurae.

Plexus (PLEK-sus)   A network of nerves, veins, or lymphatic vessels.

Plexus of Auerbach   *See* **Myenteric plexus.**

Plexus of Meissner   *See* **Submucosal plexus.**

Plicae circulares (PLĪ-kē SER-kyoo-lar-es)   Permanent, deep, transverse folds in the mucosa and submucosa of the small intestine that increase the surface area for absorption. Also called ***circular folds.***

Pneumonia (noo-MŌ-nē-a)   Acute infection or inflammation of the alveoli of the lungs.

Pneumotaxic (noo-mō-TAK-sik) *area*   Portion of the respiratory center in the pons that continually sends inhibitory nerve impulses to the inspiratory area that limit inspiration and facilitate expiration.

Podiatry (pō-DĪ-a-trē)   The diagnosis and treatment of foot disorders.

Polar body   The smaller cell resulting from the unequal division of cytoplasm during the meiotic divisions of an oocyte. The polar body has no function and is resorbed.

Poliomyelitis (pō'-lē-ō-mī-e-LĪ-tis)   Viral infection marked by fever, headache, stiff neck and back, deep muscle pain and weakness, and loss of certain somatic reflexes; a serious form of the disease, ***bulbar polio,*** results in destruction of motor neurons in anterior horns of spinal nerves that leads to paralysis.

Polycythemia (pol'-ē-sī-THĒ-mē-a)   An abnormal increase in the number of red blood cells.

Polyp (POL-ip)   A tumor on a stem found especially on a mucous membrane.

Polysaccharides (pol'-ē-SAK-a-rīds)   Three or more monosaccharides joined chemically.

Polyunsaturated fat   A fat that contains two or more double covalent bonds between its carbon atoms; examples are corn oil, safflower oil, and cottonseed oil.

Polyuria (pol'-ē-YOO-rē-a)   An excessive production of urine.

Pons (ponz)   The portion of the brain stem that forms a ''bridge'' between the medulla and the midbrain, anterior to the cerebellum.

Positron emission tomography (*PET*)   A type of radioactive scanning based on the release of gamma rays when positrons collide with negatively charged electrons in body tissues; it indicates where radioisotopes are used in the body.

Postcentral gyrus   A gyrus immediately posterior to the central sulcus that contains the general sensory area of the cerebral cortex.

Posterior (pos-TĒR-ē-or)   Nearer to or at the back of the body. Also called *dorsal.*

Posterior root   The structure composed of afferent (sensory) fibers lying between a spinal nerve and the dorsolateral aspect of the spinal cord. Also called the ***dorsal (sensory) root.***

Posterior root ganglion   A group of cell bodies of sensory (afferent) neurons and their supporting cells located along the posterior root of a spinal nerve. Also called a **dorsal (sensory) root ganglion** (GANG-glē-on).

Postganglionic neuron (pōst'-gang-lē-ON-ik NOO-ron) The second visceral efferent neuron in an autonomic pathway, having its cell body and dendrites located in an autonomic ganglion and its unmyelinated axon ending at cardiac muscle, smooth muscle, or a gland.

Postpartum (pōst-PAR-tum) After parturition; occurring after the delivery of a baby.

Postsynaptic (pōst-sin-AP-tik) **neuron** The nerve cell that is activated by the release of a neurotransmitter substance from another neuron and carries nerve impulses away from the synapse.

Pouch of Douglas See **Rectouterine pouch.**

Precentral gyrus A gyrus immediately anterior to the central sulcus that contains the primary motor area of the cerebral cortex.

Preeclampsia (pre'-e-KLAMP-sē-a) A syndrome characterized by sudden hypertension, large amounts of protein in urine, and generalized edema; it might be related to an autoimmune or allergic reaction due to the presence of a fetus.

Preganglionic (prē'-gang-lē-ON-ik) **neuron** The first visceral efferent neuron in an autonomic pathway, with its cell body and dendrites in the brain or spinal cord and its myelinated axon ending at an autonomic ganglion, where it synapses with a postganglionic neuron.

Pregnancy Sequence of events that normally includes fertilization, implantation, embryonic growth, and fetal growth that terminates in birth.

Premenstrual syndrome (PMS) Severe physical and emotional stress occurring late in the postovulatory phase of the menstrual cycle and sometimes overlapping with menstruation.

Premonitory (prē-MON-i-tō-rē) Giving previous warning; as premonitory symptoms.

Prepuce (PRĒ-pyoos) The loose-fitting skin covering the glans of the penis and clitoris. Also called the **foreskin.**

Presynaptic (prē-sin-AP-tik) **neuron** A nerve cell that carries nerve impulses toward a synapse.

Prevertebral ganglion (prē-VERT-e-bral GANG-lē-on) A cluster of cell bodies of postganglionic sympathetic neurons anterior to the spinal column and close to large abdominal arteries. Also called a **collateral ganglion** (GANG-lē-on).

Primary germ layer One of three layers of embryonic tissue, called ectoderm, mesoderm, and endoderm, that give rise to all tissues and organs of the organism.

Primary motor area A region of the cerebral cortex in the precentral gyrus of the frontal lobe of the cerebrum that controls specific muscles or groups of muscles.

Primary somesthetic (sō-mes-THET-ik) **area** A region of the cerebral cortex posterior to the central sulcus in the postcentral gyrus of the parietal lobe of the cerebrum that localizes exactly the points of the body where sensations originate.

Prime mover The muscle directly responsible for producing the motion in question. Also called an **agonist** (AG-ō-nist).

Primigravida (prī-mi-GRAV-i-da) A woman pregnant for the first time.

Primitive gut Embryonic structure composed of endoderm and mesoderm that gives rise to most of the gastrointestinal (GI) tract.

Primordial (prī-MŌR-dē-al) Existing first; especially primordial egg cells in the ovary.

Principal cell Cell found in the parathyroid glands that secretes parathyroid hormone (PTH). Also called a **chief cell.**

Proctology (prok-TOL-ō-jē) The branch of medicine that treats the rectum and its disorders.

Progeny (PROJ-e-nē) Refers to offspring or descendants.

Progesterone (prō-JES-te-rōn) **(PROG)** A female sex hormone produced by the ovaries that helps prepare the endometrium for implantation of a fertilized ovum and the mammary glands for milk secretion.

Prognosis (prog-NŌ-sis) A forecast of the probable results of a disorder; the outlook for recovery.

Projection (prō-JEK-shun) The process by which the brain refers sensations to their point of stimulation.

Prolactin (prō-LAK-tin) **(PRL)** A hormone secreted by the adenohypophysis (anterior lobe) of the pituitary gland that initiates and maintains milk secretion by the mammary glands.

Prolapse (PRŌ-laps) A dropping or falling down of an organ, especially the uterus or rectum.

Proliferation (pro-lif'-er-Ā-shun) Rapid and repeated reproduction of new parts, especially cells.

Pronation (prō-NĀ-shun) A movement of the forearm in which the palm of the hand is turned posteriorly.

Prophase (PRŌ-fāz) The first stage in mitosis during which chromatid pairs are formed and aggregate around the equatorial plane region of the cell.

Proprioception (prō-prē-ō-SEP-shun) The receipt of information from muscles, tendons, and the labyrinth that enables the brain to determine movements and position of the body and its parts. Also called **kinesthesia** (kin'-es-THĒ-zē-a).

Proprioceptor (prō-'prē-ō-SEP-tor) A receptor located in muscles, tendons, or joints that provides information about body position and movements.

Prostaglandin (pros'-ta-GLAN-din) **(PG)** A membrane-associated lipid composed of 20-carbon fatty acids with 5 carbon atoms joined to form a cyclopentane ring; synthesized in small quantities and basically mimics hormones in activities.

Prostatectomy (pros'-ta-TEK-tō-mē) The surgical removal of part of or the entire prostate gland.

Prostate (PROS-tāt) **gland** A muscular, doughnut-shaped gland inferior to the urinary bladder that surrounds the superior portion of the male urethra and secretes a slightly acid solution that contributes to sperm motility and viability.

Prosthesis (pros-THĒ-sis) An artificial device to replace a missing body part.

Protein An organic compound consisting of carbon, hydrogen, oxygen, nitrogen, and sometimes sulfur and phosphorus, and made up of amino acids linked by peptide bonds.

Proto-oncogene (prō'-tō-ONG-kō-jēn) Gene responsible for some aspect of normal growth and development; it may transform into an oncogene, a gene capable of causing cancer.

Protraction (prō-TRAK-shun) The movement of the mandible or shoulder girdle forward on a plane parallel with the ground.

Proximal (PROK-si-mal) Nearer the attachment of an extremity to the trunk or a structure; nearer to the point of origin.

Pruritus (proo'-RĪ-tus) Itching.

Pseudopodia (soo'-dō-PŌ-dē-a) Temporary, protruding projections of cytoplasm.

Psoriasis (sō-RĪ-a-sis) Chronic skin disease characterized by reddish plaques or papules covered with scales.

Psychosomatic (sī'-kō-sō-MAT-ik) Pertaining to the relation between mind and body. Commonly used to refer to those physiological disorders thought to be caused entirely or partly by emotional disturbances.

Pterygopalatine ganglion (ter'-i-gō-PAL-a-tīn GANG-glē-on) A cluster of cell bodies of parasympathetic postganglionic neurons ending at the lacrimal and nasal glands.

Ptosis (TŌ-sis)   Drooping, as of the eyelid or the kidney.

Puberty (PYOO-ber-tē)   The time of life during which the secondary sex characteristics begin to appear and the capability for sexual reproduction is possible; usually between the ages of 10 and 15.

Pudendum (pyoo-DEN-dum)   A collective designation for the external genitalia of the female.

Puerperium (pyoo'-er-PER-ē-um)   The state immediately after childbirth, usually 4–6 weeks.

Pulmonary (PUL-mo-ner'-ē)   Concerning or affected by the lungs.

Pulmonary circulation   The flow of deoxygenated blood from the right ventricle to the lungs and the return of oxygenated blood from the lungs to the left atrium.

Pulmonary edema (e-DĒ-ma)   An abnormal accumulation of interstitial fluid in the tissue spaces and alveoli of the lungs due to increased pulmonary capillary permeability or increased pulmonary capillary pressure.

Pulmonary embolism (EM-bō-lizm) **(PE)**   The presence of a blood clot or other foreign substance in a pulmonary arterial blood vessel that obstructs circulation to lung tissue.

Pulmonary ventilation   The inflow (inspiration) and outflow (expiration) of air between the atmosphere and the lungs. Also called **breathing**.

Pulp cavity   A cavity within the crown and neck of a tooth, filled with pulp, a connective tissue containing blood vessels, nerves, and lymphatics.

Pulsating electromagnetic fields (PEMFs)   A procedure that uses electrotherapy to treat improperly healing fractures.

Pupil   The hole in the center of the iris, the area through which light enters the posterior cavity of the eyeball.

Purkinje fiber   *See* **Conduction myofiber.**

Pus   The liquid product of inflammation containing leucocytes or their remains and debris of dead cells.

Pyelitis (pī'-e-LĪ-tis)   Inflammation of the kidney pelvis and its calyces.

Pyemia (pī-Ē-mē-a)   Infection of the blood, with multiple abscesses, caused by pus-forming microorganisms.

Pyloric (pī-LOR-ik) **sphincter**   A thickened ring of smooth muscle through which the pylorus of the stomach communicates with the duodenum. Also called the **pyloric valve.**

Pyogenesis (pi'-ō-JEN-e-sis)   Formation of pus.

Pyorrhea (pī-ō-RĒ-a)   A discharge or flow of pus, especially in the alveoli (sockets) and the tissues of the gums.

Pyramid (PIR-a-mid)   A pointed or cone-shaped structure; one of two roughly triangular structures on the ventral side of the medulla composed of the largest motor tracts that run from the cerebral cortex to the spinal cord; a triangular-shaped structure in the renal medulla composed of the straight segments of renal tubules.

Pyramidal (pi-RAM-i-dal) **pathways**   Collections of motor nerve fibers arising in the brain and passing down through the spinal cord to motor cells in the anterior horns.

Pyrexia (pī-REK-sē-a)   A condition in which the temperature is above normal.

Pyuria (pī-YOO-rē-a)   The presence of leucocytes and other components of pus in urine.

Quadrant (KWOD-rant)   One of four parts.

Quadriplegia (kwod'-ri-PLĒ-jē-a)   Paralysis of the two upper and two lower extremities.

Radiographic (rā'-dē-ō-GRAF-ic) **anatomy**   Diagnostic branch of anatomy that includes the use of x rays.

Rami communicantes (RĀ-mē ko-myoo-ni-KAN-tēz)   Branches of a spinal nerve. *Singular,* **ramus communicans** (RĀ-mus ko-MYOO-ni-kans).

Rathke's pouch   *See* **Hypophyseal pouch.**

Receptor   A specialized cell or a nerve cell terminal modified to respond to some specific sensory modality, such as touch, pressure, cold, light, or sound. A specific molecule or arrangement of molecules organized to accept only molecules with a complementary shape.

Receptor-mediated endocytosis   A highly selective process in which cells take up large molecules or particles (ligands). In the process, successive compartments called vesicles, endosomes, and CURLs form. Ligands are eventually broken down by enzymes in lysosomes.

Reciprocal innervation (re-SIP-rō-kal in-ner-VĀ-shun)   The phenomenon by which nerve impulses stimulate contraction of one muscle and simultaneously inhibit contraction of antagonistic muscles.

Recombinant DNA   Synthetic DNA, formed by joining a fragment of DNA from one source to a portion of DNA from another.

Recruitment (rē-KROOT-ment)   The process of increasing the number of active motor units.

Rectouterine pouch   A pocket formed by the parietal peritoneum as it moves posteriorly from the surface of the uterus and is reflected onto the rectum; the lowest point in the pelvic cavity. Also called the **pouch** or **cul de sac of Douglas.**

Rectum (REK-tum)   The last 20 cm (7 in.) of the gastrointestinal tract, from the sigmoid colon to the anus.

Recumbent (re-KUM-bent)   Lying down.

Red nucleus   A cluster of cell bodies in the midbrain, occupying a large portion of the tegmentum and sending fibers into the rubroreticular and rubrospinal tracts.

Red pulp   That portion of the spleen that consists of venous sinuses filled with blood and cords of splenic tissue called splenic (Billroth's) cords.

Referred pain   Pain that is felt at a site remote from the place of origin.

Reflex   Fast response to a change in the internal or external environment that attempts to restore homeostasis; passes over a reflex arc.

Reflex arc   The most basic conduction pathway through the nervous system, connecting a receptor and an effector and consisting of a receptor, a sensory neuron, a center in the central nervous system for a synapse, a motor neuron, and an effector.

Regeneration (rē-jen'-er-Ā-shun)   The natural renewal of a structure.

Regimen (REJ-i-men)   A strictly regulated scheme of diet, exercise, or activity designed to achieve certain ends.

Regional anatomy   The division of anatomy dealing with a specific region of the body, such as the head, neck, chest, or abdomen.

Regulating factor   Chemical secretion of the hypothalamus that can either stimulate or inhibit secretion of hormones of the adenohypophysis (anterior pituitary).

Regulating hormone   Chemical secretion of the hypothalamus that can either stimulate or inhibit secretion of hormones of the adenohypophysis (anterior pituitary).

Regurgitation (rē-gur'-ji-TA-shun)   Return of solids or fluids to the mouth from the stomach; flowing backward of blood through incompletely closed heart valves.

Relapse (RĒ-laps)   The return of a disease weeks or months after its apparent cessation.

Relaxin (RLX)   A female hormone produced by the ovaries that relaxes the symphysis pubis, helps dilate the uterine cervix to facilitate delivery, and plays a role in increasing sperm motility.

Remodeling   Replacement of old bone by new bone tissue.

Renal (RĒ-nal)   Pertaining to the kidney.

Renal corpuscle (KOR-pus'-l)   A glomerular (Bowman's) capsule and its enclosed glomerulus.

Renal erythropoietic (ē-rith'-rō-poy-Ē-tik) *factor*   An enzyme released by the kidneys and liver in response to hypoxia that acts on a plasma protein to bring about the production of erythropoietin, which stimulates red blood cell production.

Renal failure   Inability of the kidneys to function properly, due to abrupt failure (acute) or progressive failure (chronic).

Renal pelvis   A cavity in the center of the kidney formed by the expanded, proximal portion of the ureter, lying within the kidney, and into which the major calyces open.

Renal pyramid   A triangular structure in the renal medulla composed of the straight segments of renal tubules.

Reproduction (rē'-prō-DUK-shun)   Either the formation of new cells for growth, repair, or replacement, or the production of a new individual.

Reproductive cell division   Type of cell division in which sperm and egg cells are produced; consists of meiosis and cytokinesis.

Resistance   Ability to ward off disease. The hindrance encountered by an electrical charge as it moves through a substance from one point to another. The hindrance encountered by blood as it flows through the vascular system or by air through respiratory passageways.

Respiration (res-pi-RA-shun)   Overall exchange of gases between the atmosphere, blood, and body cells consisting of pulmonary ventilation, external respiration, and internal respiration.

Respiratory center   Neurons in the reticular formation of the brain stem that regulate the rate of respiration.

Respiratory distress syndrome (RDS) of the newborn   A disease of newborn infants, especially premature ones, in which insufficient amounts of surfactant are produced and breathing is labored. Also called *hyaline* (HĪ-a-lin) *membrane disease (HMD)*.

Respiratory failure   Condition in which the respiratory system cannot supply sufficient oxygen to maintain metabolism or eliminate enough carbon dioxide to prevent respiratory acidosis.

Resuscitation (rē-sus'-i-TA-shun)   Act of bringing a person back to full consciousness.

Retention (rē-TEN-shun)   A failure to void urine due to obstruction, nervous contraction of the urethra, or absence of sensation of desire to urinate.

Rete (RĒ-tē) *testis*   The network of ducts in the testes.

Reticular (re-TIK-yoo-lar) *activating system (RAS)*   An extensive network of branched nerve cells running through the core of the brain stem. When these cells are activated, a generalized alert or arousal behavior results.

Reticular formation   A network of small groups of nerve cells scattered among bundles of fibers beginning in the medulla

as a continuation of the spinal cord and extending upward through the central part of the brain stem.

Reticulocyte (rē-TIK-yoo-lō-sīt)   An immature red blood cell.

Reticulum (re-TIK-yoo-lum)   A network.

Retina (RET-i-na)   The inner coat of the eyeball, lying only in the posterior portion of the eye and consisting of nervous tissue and a pigmented layer comprised of epithelial cells lying in contact with the choroid. Also called the *nervous tunic* (TOO-nik).

Retinal (RE'-ti-nal)   The pigment portion of the photopigment rhodopsin. Also called *visual yellow*.

Retraction (rē-TRAK-shun)   The movement of a protracted part of the body backward on a plane parallel to the ground, as in pulling the lower jaw back in line with the upper jaw.

Retroflexion (re-trō-FLEK-shun)   A malposition of the uterus in which it is tilted posteriorly.

Retrograde degeneration (RE-trō-grād dē-jen-er-A-shun)   Degeneration of the portion of the axon and myelin sheath of a neuron proximal to the site of injury.

Retroperitoneal (re'-trō-per-i-tō-NĒ-al)   External to the peritoneal lining of the abdominal cavity.

Rheumatism (ROO-ma-tizm')   Any painful state of the supporting structures of the body—bones, ligaments, joints, tendons, or muscles.

Rhinology (rī-NOL-ō-jē)   The study of the nose and its disorders.

Ribonucleic (rī'-bō-nyoo-KLĒ-ik) *acid (RNA)*   A single-stranded nucleic acid constructed of nucleotides consisting of one of four possible nitrogen bases (adenine, cytosine, guanine, or uracil), ribose, and a phosphate group; three types are messenger RNA (mRNA), transfer RNA (tRNA), and ribosomal RNA (rRNA), each of which cooperates with DNA for protein synthesis.

Ribosome (RĪ-bō-sōm)   An organelle in the cytoplasm of cells, composed of ribosomal RNA and ribosomal proteins, that synthesizes proteins; nicknamed the ''protein factory.''

Rickets (RIK-ets)   Condition affecting children characterized by soft and deformed bones resulting from inadequate calcium metabolism due to a vitamin D deficiency.

Right lymphatic (lim-FAT-ik) *duct*   A vessel of the lymphatic system that drains lymph from the upper right side of the body and empties it into the right subclavian vein.

Rigor mortis   State of partial contraction of muscles following death due to lack of ATP that causes cross bridges of thick myofilaments to remain attached to thin myofilaments, thus preventing relaxation.

Rod   A visual receptor in the retina of the eye that is specialized for vision in dim light.

Roentgen (RENT-gen)   The international unit of radiation; a standard quantity of x or gamma radiation.

Roentgenogram (RENT-gen-ō-gram)   A photographic image.

Root canal   A narrow extension of the pulp cavity lying within the root of a tooth.

Root hair plexus (PLEK-sus)   A network of dendrites arranged around the root of a hair as free or naked nerve endings that are stimulated when a hair shaft is moved.

Root of penis   Attached portion of penis that consists of the bulb and crura.

Rotation (rō-TA-shun)   Moving a bone around its own axis, with no other movement.

Round ligament (LIG-a-ment)   A band of fibrous connective tissue enclosed between the folds of the broad ligament of the uterus, emerging from a point on the uterus just below the uterine (Fallopian) tube, extending laterally along the pelvic wall, and penetrating the abdominal wall through the deep inguinal ring to end in the labia majora.

Round window A small opening between the middle and inner ear, directly below the oval window, covered by the secondary tympanic membrane. Also called the *fenestra cochlea* (fe-NES-tra KŌK-lē-a).

Rugae (ROO-jē) Large folds in the mucosa of an empty hollow organ, such as the stomach and vagina.

Saccule (SAK-yool) The lower and smaller of the two chambers in the membranous labyrinth inside the vestibule of the inner ear containing a receptor organ for static equilibrium.

Sacral hiatus (hi-Ā-tus) Inferior entrance to the vertebral canal formed when the laminae of the fifth sacral vertebra (and sometimes fourth) fail to meet.

Sacral plexus (PLEK-sus) A network formed by the anterior branches of spinal nerves L4 through S3.

Sacral promontory (PROM-on-tor'-ē) The superior surface of the body of the first sacral vertebra that projects anteriorly into the pelvic cavity; a line from the sacral promontory to the superior border of the symphysis pubis divides the abdominal and pelvic cavities.

Saddle joint A synovial joint in which the articular surface of one bone is saddle shaped and the articular surface of the other bone is shaped like a rider sitting in the saddle, as in the joint between the trapezium and the metacarpal of the thumb. Also called a *sellaris* (sel-LA-ris) *joint*.

Sagittal (SAJ-i-tal) *plane* A vertical plane that divides the body or organs into left and right portions. Such a plane may be *midsagittal (median)*, in which the divisions are equal, or *parasagittal*, in which the divisions are unequal.

Saliva (sa-LĪ-va) A clear, alkaline, somewhat viscous secretion produced by the three pairs of salivary glands; contains various salts, mucin, lysozyme, and salivary amylase.

Salivary amylase (SAL-i-ver-ē AM-i-lās) An enzyme in saliva that initiates the chemical breakdown of starch, mostly in the mouth.

Salivary gland One of three pairs of glands that lie outside the mouth and pour their secretory product (called saliva) into ducts that empty into the oral cavity; the parotid, submandibular, and sublingual glands.

Salpingitis (sal'-pin-JĪ-tis) Inflammation of the uterine (Fallopian) or auditory (Eustachian) tube.

Sarcolemma (sar'-kō-LEM-ma) The cell membrane of a muscle fiber (cell), especially of a skeletal muscle fiber.

Sarcoma (sar-KŌ-ma) A connective tissue tumor, often highly malignant.

Sarcomere (SAR-kō-mēr) A contractile unit in a striated muscle fiber (cell) extending from one Z line to the next Z line.

Sarcoplasm (SAR-kō-plazm) The cytoplasm of a muscle fiber (cell).

Sarcoplasmic reticulum (sar'-kō-PLAZ-mik re-TIK-yoo-lum) A network of saccules and tubes surrounding myofibrils of a muscle fiber (cell), comparable to endoplasmic reticulum; functions to reabsorb calcium ions during relaxation and to release them to cause contraction.

Satiety (sa-TĪ-e-tē) Fullness or gratification, as of hunger or thirst.

Satiety center A collection of nerve cells located in the ven-

tromedial nuclei of the hypothalamus that, when stimulated, brings about the cessation of eating.

Saturated fat A fat that contains no double bonds between any of its carbon atoms, all are single bonds and all carbon atoms are bonded to the maximum number of hydrogen atoms; found naturally in animal foods such as meat, milk, milk products, and eggs.

Scala tympani (SKA-la TIM-pan-ē) The lower spiral-shaped channel of the bony cochlea, filled with perilymph.

Scala vestibuli (ves-TIB-yoo-lē) The upper spiral-shaped channel of the bony cochlea, filled with perilymph.

Schwann cell *See Neurolemmocyte.*

Sciatica (sī-AT-i-ka) Inflammation and pain along the sciatic nerve; felt at the back of the thigh running down the inside of the leg.

Sclera (SKLE-ra) The white coat of fibrous tissue that forms the outer protective covering over the eyeball except in the area of the anterior cornea; the posterior portion of the fibrous tunic.

Scleral venous sinus A circular venous sinus located at the junction of the sclera and the cornea through which aqueous humor drains from the anterior chamber of the eyeball into the blood. Also called the *canal of Schlemm* (SHLEM).

Sclerosis (skle-RŌ-sis) A hardening with loss of elasticity of the tissues.

Scoliosis (skō'-lē-Ō-sis) An abnormal lateral curvature from the normal vertical line of the spine.

Scotoma (skō-TŌ-ma) A blind spot or area of depressed vision within the visual field.

Scrotum (SKRŌ-tum) A skin-covered pouch that contains the testes and their accessory structures.

Sebaceous (se-BĀ-shus) Secreting oil.

Sebaceous (se-BĀ-shus) *gland* An exocrine gland in the dermis of the skin, almost always associated with a hair follicle, that secretes sebum. Also called an *oil gland*.

Sebum (SĒ-bum) Secretion of sebaceous (oil) glands.

Secondary sex characteristic A feature characteristic of the male or female body that develops at puberty under the stimulation of sex hormones but is not directly involved in sexual reproduction, such as distribution of body hair, voice pitch, body shape, and muscle development.

Secretion (se-KRĒ-shun) Production and release from a gland cell of a fluid, especially a functionally useful product as opposed to a waste product.

Selectively permeable membrane A membrane that permits the passage of certain substances, but restricts the passage of others. Also called a *semipermeable* (sem'-ē-PER-mē-a-bl) *membrane*.

Sella turcica (SEL-a TUR-si-ka) A depression on the superior surface of the sphenoid bone that houses the pituitary gland.

Semen (SĒ-men) A fluid discharged at ejaculation by a male that consists of a mixture of spermatozoa and the secretions of the seminal vesicles, the prostate gland, and the bulbourethral (Cowper's) glands. Also called *seminal* (SEM-i-nal) *fluid*.

Semicircular canals Three bony channels projecting superiorly and posteriorly from the vestibule of the inner ear, filled with perilymph, in which lie the membranous semicircular canals filled with endolymph. They contain receptors for equilibrium.

Semicircular ducts The membranous semicircular canals filled with endolymph and floating in the perilymph of the bony semicircular canals. They contain cristae that are concerned with dynamic equilibrium.

Semilunar (sem'-ē-LOO-nar) *valve*   A valve guarding the entrance into the aorta or the pulmonary trunk from a ventricle of the heart.

Seminal vesicle (SEM-i-nal VES-i-kul)   One of a pair of convoluted, pouchlike, structures, lying posterior and inferior to the urinary bladder and anterior to the rectum, that secrete a component of semen into the ejaculatory ducts.

Seminiferous tubule (sem'-i-NI-fer-us TOO-byool)   A tightly coiled duct, located in a lobule of the testis, where spermatozoa are produced.

Senescence (se-NES-ens)   The process of growing old; the period of old age.

Senile macular (MAK-yoo-lar) *degeneration* **(SMD)**   A disease in which blood vessels grow over the macula lutea.

Senility (se-NIL-i-tē)   A loss of mental or physical ability due to old age.

Sensation   A state of awareness of external or internal conditions of the body.

Sensory area   A region of the cerebral cortex concerned with the interpretation of sensory impulses.

Sepsis (SEP-sis)   A morbid condition that results from the presence in the blood or other body tissues of pathogenic bacteria and their products.

Septal defect   An opening in the septum (interatrial or interventricular) between the left and right sides of the heart.

Septicemia (sep'-ti-SĒ-mē-a)   Toxins or disease-causing bacteria in blood. Also called *"blood poisoning."*

Septum (SEP-tum)   A wall dividing two cavities.

Serosa (ser-Ō-sa)   Any serous membrane. The outermost layer or tunic of an organ formed by a serous membrane. The membrane that lines the pleural, pericardial, and peritoneal cavities.

Serous (SIR-us) *membrane*   A membrane that lines body cavities that does not open to the exterior. Also called the *serosa* (se-RŌ-sa).

Serum   Plasma minus its clotting proteins.

Sesamoid bones (SES-a-moyd)   Small bones usually found in tendons.

Sex chromosomes   The twenty-third pair of chromosomes, designated X and Y, which determine the genetic sex of an individual; in males, the pair is XY; in females, XX.

Sexual intercourse   The insertion of the erect penis of a male into the vagina of a female. Also called *coitus* (KŌ-i-tus) or *copulation.*

Sexually transmitted disease (STD)   General term for any of a large number of diseases spread by sexual contact. Also called a *venereal disease (VD).*

Sheath of Schwann   *See* **Neurolemma.**

Shingles   Acute infection of the peripheral nervous system caused by a virus.

Shinsplints   Soreness or pain along the tibia probably caused by inflammation of the periosteum brought on by repeated tugging of the muscles and tendons attached to the periosteum. Also called *tibia stress syndrome.*

Shoulder   A synovial or diarthrotic joint where the humerus joins the scapula.

Sigmoid colon (SIG-moyd KŌ-lon)   The S-shaped portion of the large intestine that begins at the level of the left iliac crest, projects inward to the midline, and terminates at the rectum at about the level of the third sacral vertebra.

Sign   Any objective evidence of disease such as a lesion, swelling, or fever.

Sinoatrial (si-nō-Ā-trē-al) **(SA)** *node*   A compact mass of cardiac muscle fibers (cells) specialized for conduction, located in the right atrium beneath the opening of the superior vena cava. Also called the *sinuatrial node* or *pacemaker.*

Sinus (SĪ-nus)   A hollow in a bone (paranasal sinus) or other tissue; a channel for blood (vascular sinus); any cavity having a narrow opening.

Sinusitis (sīn-yoo-SĪT-is)   Inflammation of the mucous membrane of a paranasal sinus.

Sinusoid (SĪN-yoo-soyd)   A microscopic space or passage for blood in certain organs such as the liver or spleen.

Skeletal muscle   An organ specialized for contraction, composed of striated muscle fibers (cells), supported by connective tissue, attached to a bone by a tendon or an aponeurosis, and stimulated by somatic efferent neurons.

Skene's gland   *See* **Paraurethral gland.**

Skull   The skeleton of the head consisting of the cranial and facial bones.

Sliding-filament theory   The most commonly accepted explanation for muscle contraction in which actin and myosin myofilaments move into interdigitation with each other, decreasing the length of the sarcomeres.

Small intestine   A long tube of the gastrointestinal tract that begins at the pyloric sphincter of the stomach, coils through the central and lower part of the abdominal cavity, and ends at the large intestine; divided into three segments: duodenum, jejunum, and ileum.

Smooth muscle   An organ specialized for contraction, composed of smooth muscle fibers (cells), located in the walls of hollow internal structures, and innervated by a visceral efferent neuron.

Soft palate (PAL-at)   The posterior portion of the roof of the mouth, extending posteriorly from the palatine bones and ending at the uvula. It is a muscular partition lined with mucous membrane.

Somatic cell division   Type of cell division in which a single starting cell (parent cell) duplicates itself to produce two identical cells (daughter cells); consists of mitosis and cytokinesis.

Somatic (sō-MAT-ik) *nervous system* **(SNS)**   The portion of the peripheral nervous system made up of the somatic efferent fibers that run between the central nervous system and the skeletal muscles and skin.

Somesthetic (sō'-mes-THET-ik)   Pertaining to sensations and sensory structures of the body.

Spasm (spazm)   An involuntary, convulsive, muscular contraction.

S period   Period of interphase during which chromosomes are replicated preceded by a G_1 period and followed by a G_2 period, when cells grow, metabolize, and produce substances required for division.

Spermatic (sper-MAT-ik) *cord*   A supporting structure of the male reproductive system, extending from a testis to the deep inguinal ring, that includes the ductus (vas) deferens, arteries, veins, lymphatics, nerves, cremaster muscle, and connective tissue.

Spermatogenesis (sper'-ma-tō-JEN-e-sis)   The formation and development of spermatozoa.

Spermatozoon (sper'-ma-tō-ZŌ-on)   A mature sperm cell.

Spermicide (SPER-mi-sīd')   An agent that kills spermatozoa.

Spermiogenesis (sper'-mē-ō-JEN-e-sis)   The maturation of spermatids into spermatozoa.

Sphincter (SFINGK-ter)   A circular muscle constricting an orifice.

Sphincter of Oddi   *See* **Sphincter of the hepatopancreatic ampulla.**

Sphincter of the hepatopancreatic ampulla   A circular muscle at the opening of the common bile and main pancreatic ducts in the duodenum. Also called the *sphincter of Oddi* (OD-ē).

Sphygmomanometer (sfig'-mō-ma-NOM-e-ter)   An instrument for measuring arterial blood pressure.

Spina bifida (SPĪ-na BIF-i-da)   A congenital defect of the vertebral column in which the halves of the neural arch of a vertebra fail to fuse in midline.

Spinal (SPĪ-nal) *cord*   A mass of nerve tissue located in the vertebral canal from which 31 pairs of spinal nerves originate.

Spinal (lumbar) puncture   Withdrawal of some of the cerebrospinal fluid from the subarachnoid space in the lumbar region.

Spinal nerve   One of the 31 pairs of nerves that originate on the spinal cord from posterior and anterior roots.

Spinal shock   A period of time, from several days to several weeks, following transection of the spinal cord and characterized by the abolition of all reflex activity.

Spinous (SPĪ-nus) *process*   A sharp or thornlike process or projection. Also called a *spine*. A sharp ridge running diagonally across the posterior surface of the scapula.

Spiral organ   The organ of hearing, consisting of supporting cells and hair cells that rest on the basilar membrane and extend into the endolymph of the cochlear duct. Also called the *organ of Corti* (KOR-tē).

Spirometer (spī-ROM-e-ter)   An apparatus used to measure air capacity of the lungs.

Splanchnic (SPLANK-nik)   Pertaining to the viscera.

Spleen (SPLĒN)   Large mass of lymphatic tissue between the fundus of the stomach and the diaphragm that functions in phagocytosis, production of lymphocytes, and blood storage.

Sprain   Forcible wrenching or twisting of a joint with partial rupture or other injury to its attachments without dislocation.

Sputum (SPYOO-tum)   Substance ejected from the mouth containing saliva and mucus.

Squamous (SKWĀ-mus)   Scalelike.

Starvation (star-VĀ-shun)   The loss of energy stores in the form of glycogen, fats, and proteins due to inadequate intake of nutrients or inability to digest, absorb, or metabolize ingested nutrients.

Stasis (STĀ-sis)   Stagnation or halt of normal flow of fluids, as blood, urine, or of the intestinal mechanism.

Static equilibrium (ē-kwi-LIB-rē-um)   The maintenance of posture in response to changes in the orientation of the body, mainly the head, relative to the ground.

Stellate reticuloendothelial (STEL-āte re-tik'-yoo-lō-en'-dō-THĒ-lē-al) *cell*   Phagocytic cell that lines a sinusoid of the liver. Also called a *Kupffer's* (KOOP-ferz) *cell.*

Stenosis (sten-Ō-sis)   An abnormal narrowing or constriction of a duct or opening.

Stereocilia (ste'-rē-ō-SIL-ē-a)   Groups of extremely long, slender, nonmotile microvilli projecting from epithelial cells lining the epididymis.

Stereognosis (ste'-rē-og-NŌ-sis)   The ability to recognize the size, shape, and texture of an object by touch.

Sterile (STE-ril)   Free from any living microorganisms. Unable to conceive or produce offspring.

Sterilization (ster'-i-li-ZĀ-shun)   Elimination of all living microorganisms. The rendering of an individual incapable of reproduction (e.g., castration, vasectomy, hysterectomy).

Sternal puncture   Introduction of a wide-bore needle into the marrow cavity of the sternum for aspiration of a sample of red bone marrow.

Stimulus   Any change in the environment capable of altering the membrane potential.

Stomach   The J-shaped enlargement of the gastrointestinal tract directly under the diaphragm in the epigastric, umbilical, and left hypochondriac regions of the abdomen, between the esophagus and small intestine.

Strabismus (stra-BIZ-mus)   A condition in which the visual axes of the two eyes differ, so that they do not fix on the same object.

Straight tubule (TOO-byool)   A duct in a testis leading from a convoluted seminiferous tubule to the rete testis.

Stratum (STRĀ-tum)   A layer.

Stratum basalis (STRĀ-tum ba-SAL-is)   The outer layer of the endometrium, next to the myometrium, that is maintained during menstruation and gestation and produces a new functionalis following menstruation or parturition.

Stratum functionalis (funk'-shun-AL-is)   The inner layer of the endometrium, the layer next to the uterine cavity, that is shed during menstruation and that forms the maternal portion of the placenta during gestation.

Stricture (STRIK-cher)   A local contraction of a tubular structure.

Stroke volume   The volume of blood ejected by either ventricle in one systole; about 70 ml.

Stroma (STRŌ-ma)   The tissue that forms the ground substance, foundation, or framework of an organ, as opposed to its functional parts.

Stupor (STOO-por)   Condition of unconsciousness, torpor, or lethargy with suppression of sense of feeling.

Subarachnoid (sub'-a-RAK-noyd) *space*   A space between the arachnoid and the pia mater that surrounds the brain and spinal cord and through which cerebrospinal fluid circulates.

Subcutaneous (sub'-kyoo-TĀ-nē-us)   Beneath the skin. Also called *hypodermic* (hī-po-DER-mik).

Subcutaneous layer   A continuous sheet of loose connective tissue and adipose tissue between the dermis of the skin and the deep fascia of the muscles. Also called the *superficial fascia* (FASH-ē-a).

Subdural space (sub-DOO-ral)   A space between the dura mater and the arachnoid of the brain and spinal cord that contains a small amount of fluid.

Sublingual (sub-LING-gwal) *gland*   One of a pair of salivary glands situated in the floor of the mouth under the mucous membrane and to the side of the lingual frenulum, with a duct (Rivinus's) that opens into the floor of the mouth.

Submandibular (sub'-man-DIB-yoo-lar) *gland*   One of a pair of salivary glands found beneath the base of the tongue under the mucous membrane in the posterior part of the floor of the mouth, posterior to the sublingual glands, with a duct (Wharton's) situated to the side of the lingual frenulum. Also called the *submaxillary* (sub'-MAK-si-ler-ē) *gland.*

Submucosa (sub-myoo-KŌ-sa)   A layer of connective tissue located beneath a mucous membrane, as in the gastrointestinal tract or the urinary bladder, where a submucosa connects the mucosa to the muscularis tunic.

Submucosal plexus   A network of autonomic nerve fibers located in the outer portion of the submucous layer of the small intestine. Also called the **plexus of Meissner** (MĪS-ner).

Subserous fascia (sub-SE-rus FASH-ē-a)   A layer of connective tissue internal to the deep fascia, lying between the deep fascia and the serous membrane that lines the body cavities.

Sudoriferous (soo'-dor-IF-er-us) **gland**   An apocrine or eccrine exocrine gland in the dermis or subcutaneous layer that produces perspiration. Also called a **sweat gland.**

Sulcus (SUL-kus)   A groove or depression between parts, especially between the convolutions of the brain. *Plural,* **sulci** (SUL-sē).

Superficial (soo'-per-FISH-al)   Located on or near the surface of the body.

Superficial fascia (FASH-ē-a)   A continuous sheet of fibrous connective tissue between the dermis of the skin and the deep fascia of the muscles. Also called **subcutaneous** (sub'-kyoo-TĀ-nē-us) **layer.**

Superficial inguinal (IN-gwi-nal) **ring**   A triangular opening in the aponeurosis of the external oblique muscle that represents the termination of the inguinal canal.

Superior (soo-PĒR-ē-or)   Toward the head or upper part of a structure. Also called **cephalad** (SEF-a-lad) or **craniad.**

Superior vena cava (VĒ-na CĀ-va) **(SVC)**   Large vein that collects blood from parts of the body superior to the heart and returns it to the right atrium.

Supination (soo-pī-NĀ-shun)   A movement of the forearm in which the palm of the hand is turned anteriorly.

Suppuration (sup'-yoo-RĀ-shun)   Pus formation and discharge.

Surface anatomy   The study of the structures that can be identified from the outside of the body.

Surfactant (sur-FAK-tant)   A phospholipid substance produced by the lungs that decreases surface tension.

Susceptibility (sus-sep'-ti-BIL-i-tē)   Lack of resistance of a body to the deleterious or other effects of an agent such as pathogenic microorganisms.

Suspensory ligament (sus-PEN-so-rē LIG-a-ment)   A fold of peritoneum extending laterally from the surface of the ovary to the pelvic wall.

Sustentacular (sus'-ten-TAK-yoo-lar) **cell**   A supporting cell of seminiferous tubules that produces secretions for supplying nutrients to spermatozoa and the hormone inhibin. Also called a **Sertoli** (ser-TŌ-lē) **cell.**

Sutural bone (SOO-cher-al)   A small bone located within a suture between certain cranial bones. Also called **Wormian** (WER-mē-an) **bone.**

Suture (SOO-cher)   A fibrous joint in the skull where bone surfaces are closely united.

Sympathetic (sim'-pa-THET-ik) **division**   One of the two subdivisions of the autonomic nervous system, having cell bodies of preganglionic neurons in the lateral gray columns of the thoracic segment and first two or three lumbar segments of the spinal cord; primarily concerned with processes involving the expenditure of energy. Also called the **thoracolumbar** (thō'-ra-kō-LUM-bar) **division.**

Sympathetic trunk ganglion (GANG-glē-on)   A cluster of cell bodies of postganglionic sympathetic neurons lateral to the vertebral column, close to the body of a vertebra. These ganglia extend downward through the neck, thorax, and abdomen to the coccyx on both sides of the vertebral column and are connected to one another to form a chain on each side of the vertebral column. Also called **lateral,** or **sympathetic, chain** or **vertebral chain ganglia.**

Sympathomimetic (sim'-pa-thō-mi-MET-ik)   Producing effects that mimic those brought about by the sympathetic division of the autonomic nervous system.

Symphysis (SIM-fi-sis)   A line of union. A slightly movable cartilaginous joint such as the symphysis pubis between the anterior surfaces of the coxal (hip) bones.

Symphysis pubis (PYOO-bis)   A slightly movable cartilaginous joint between the anterior surfaces of the coxal (hip) bones.

Symptom (SIMP-tum)   An observable abnormality that indicates the presence of a disease or disorder of the body.

Synapse (SIN-aps)   The junction between the process of two adjacent neurons; the place where the activity of one neuron affects the activity of another.

Synapsis (sin-AP-sis)   The pairing of homologous chromosomes during prophase I of meiosis.

Synaptic (sin-AP-tik) **cleft**   The narrow gap that separates the axon terminal of one nerve cell from another nerve cell or muscle fiber (cell) and across which a neurotransmitter diffuses to affect the postsynaptic cell.

Synaptic end bulb   Expanded distal end of an axon terminal that contains synaptic vesicles. Also called a **synaptic knob** or **end foot.**

Synaptic gutter   Invaginated portion of a sarcolemma under an axon terminal. Also called a **synaptic trough** (TROF).

Synaptic vesicle   Membrane-enclosed sac in a synaptic end-bulb that stores neurotransmitters.

Synarthrosis (sin'-ar-THRŌ-sis)   An immovable joint.

Synchondrosis (sin'-kon-DRŌ-sis)   A cartilaginous joint in which the connecting material is hyaline cartilage.

Syncope (SIN-kō-pē)   A temporary cessation of consciousness, most commonly due to cerebral ischemia; a faint.

Syndesmosis (sin'-dez-MŌ-sis)   A fibrous joint in which articulating bones are united by dense fibrous tissue.

Syndrome (SIN-drōm)   A group of signs and symptoms that occur together in a pattern characteristic of a particular disease or abnormal condition.

Syneresis (si-NER-e-sis)   The process of clot retraction.

Synergist (SIN-er-jist)   A muscle that assists the prime mover by reducing undesired action or unnecessary movement.

Synostosis (sin'-os-TŌ-sis)   A joint in which the dense fibrous connective tissue that unites bones at a suture has been replaced by bone, resulting in a complete fusion across the suture line.

Synovial (si-NŌ-vē-al) **cavity**   The space between the articulating bones of a synovial (diarthrotic) joint, filled with synovial fluid. Also called a **joint cavity.**

Synovial fluid   Secretion of synovial membranes that lubricates joints and nourishes articular cartilage.

Synovial joint   A fully movable or diarthrotic joint in which a synovial (joint) cavity is present between the two articulating bones.

Synovial membrane   The inner of the two layers of the articular capsule of a synovial joint, composed of loose connective tissue that secretes synovial fluid into the synovial (joint) cavity.

Syphilis (SIF-i-lis)   A sexually transmitted disease caused by the bacterium *Treponema pallidum.*

System   An association of organs that have a common function.

Systemic (sis-TEM-ik)   Affecting the whole body; generalized.

Systemic anatomy   The study of particular systems of the body, such as the skeletal, muscular, nervous, cardiovascular, or urinary systems.

Systemic circulation   The routes through which oxygenated blood flows from the left ventricle through the aorta to all the organs of the body and deoxygenated blood returns to the right atrium.

Systemic lupus erythematosus (er-i-them-a-TŌ-sus) **(SLE)** An autoimmune, inflammatory disease that may affect every tissue of the body.

Systole (SIS-tō-lē) In the cardiac cycle, the phase of contraction of the heart muscle, especially of the ventricles.

Tachycardia (tak'-i-KAR-dē-a) A rapid heartbeat or pulse rate.

Tactile (TAK-tīl) Pertaining to the sense of touch.

Tactile disc An encapsulated, cutaneous receptor for touch located in deeper layers of epidermal cells. Also called a **Merkel's** (MER-kelz) **disc**.

Taenia coli (TĒ-nē-a KŌ-lī) One of three flat bands of thickened, longitudinal muscles running the length of the large intestine.

Target cell A cell whose activity is affected by a particular hormone.

Tarsal gland Sebaceous (oil) gland that opens on the edge of each eyelid. Also called a **Meibomian** (mī-BŌ-mē-an) **gland**.

Tarsal plate A thin, elongated sheet of connective tissue, one in each eyelid, giving the eyelid form and support. The aponeurosis of the levator palpebrae superioris is attached to the tarsal plate of the superior eyelid.

Tarsus (TAR-sus) A collective term for the seven bones of the ankle.

Tay-Sachs disease Inherited, progressive neuronal degeneration of the central nervous system due to a deficient lysosomal enzyme that causes excessive accumulations of a lipid called ganglioside.

T cell A lymphocyte that can differentiate into one of six kinds of cells—killer, helper, suppressor, memory, amplifier, or delayed hypersensitivity—all of which function in cellular immunity.

Tectorial (tek-TŌ-rē-al) **membrane** A gelatinous membrane projecting over and in contact with the hair cells of the spiral organ (organ of Corti) in the cochlear duct.

Telophase (TEL-ō-fāz) The final stage of mitosis in which the daughter nuclei become established.

Temporomandibular joint (TMJ) syndrome A disorder of the temporomandibular joint (TMJ) characterized by dull pain around the ear, tenderness of jaw muscles, a clicking or popping noise when opening or closing the mouth, limited or abnormal opening of the mouth, headache, tooth sensitivity, and abnormal wearing of the teeth.

Tendinitis (ten'-din-Ī-tis) Inflammation of a tendon and synovial membrane at a joint. Also called **tenosynovitis** (ten'-o-sin'-ō-VĪ-tis).

Tendon (TEN-don) A white fibrous cord of dense, regularly arranged connective tissue that attaches muscle to bone.

Tendon organ A proprioceptive receptor, sensitive to changes in muscle tension, found chiefly near the junction of tendons and muscles. Also called a **Golgi** (GOL-jē) **tendon organ**.

Tentorium cerebelli (ten-TŌ-rē-um ser'-e-BEL-ē) A transverse shelf of dura mater that forms a partition between the occipital lobe of the cerebral hemispheres and the cerebellum and that covers the cerebellum.

Teratogen (TER-a-tō-jen) Any agent or factor that causes physical defects in a developing embryo.

Terminal ganglion (TER-min-al GANG-lē-on) A cluster of cell bodies of postganglionic parasympathetic neurons either lying very close to the visceral effectors or located within the walls of the visceral effectors supplied by the postganglionic fibers.

Testis (TES-tis) Male gonad that produces sperm and the hormones testosterone and inhibin. Also called a **testicle**.

Testosterone (tes-TOS-te-rōn) A male sex hormone (androgen) secreted by interstitial endocrinocytes (cells of Leydig) of a mature testis; controls the growth and development of male sex organs, secondary sex characteristics, spermatozoa, and body growth.

Tetanus (TET-a-nus) An infectious disease caused by the toxin of *Clostridium tetani,* characterized by tonic muscle spasms and exaggerated reflexes, lockjaw, and arching of the back. A smooth, sustained contraction produced by a series of very rapid stimuli to a muscle.

Tetany (TET-a-nē) A nervous condition caused by hypoparathyroidism and characterized by intermittent or continuous tonic muscular contractions of the extremities.

Tetralogy of Fallot (tet-RAL-ō-jē of fal-Ō) A combination of four congenital heart defects: (1) constricted pulmonary semilunar valve, (2) interventricular septal opening, (3) emergence of aorta from both ventricles instead of from the left only, and (4) enlarged right ventricle.

Thalamus (THAL-a-mus) A large, oval structure located above the midbrain, consisting of two masses of gray matter covered by a thin layer of white matter.

Thalassemia (thal'-a-SĒ-mē-a) A group of hereditary hemolytic anemias.

Therapy (THER-a-pē) The treatment of a disease or disorder.

Thermoreceptor (THER-mō-rē-sep-tor) Receptor that detects changes in temperature.

Thigh The portion of the lower extremity between the hip and the knee.

Third ventricle (VEN-tri-kul) A slitlike cavity between the right and left halves of the thalamus and between the lateral ventricles.

Thoracic (thō-RAS-ik) **cavity** Superior component of the ventral body cavity that contains two pleural cavities, the mediastinum, and the pericardial cavity.

Thoracic duct A lymphatic vessel that begins as a dilation called the cisterna chyli, receives lymph from the left side of the head, neck, and chest, the left arm, and the entire body below the ribs, and empties into the left subclavian vein. Also called the **left lymphatic** (lim-FAT-ik) **duct**.

Thoracolumbar (thō'-ra-kō-LUM-bar) **outflow** The fibers of the sympathetic preganglionic neurons, which have their cell bodies in the lateral gray columns of the thoracic segment and first two or three lumbar segments of the spinal cord.

Thorax (THŌ-raks) The chest.

Thrombocyte (THROM-bō-sīt) A fragment of cytoplasm enclosed in a cell membrane and lacking a nucleus; found in the circulating blood; plays a role in blood clotting. Also called a **platelet** (PLĀT-let).

Thombophlebitis (throm'-bo-fle-BĪ-tis) A disorder in which inflammation of a vein wall is followed by the formation of a blood clot (thrombus).

Thrombosis (throm-BŌ-sis) The formation of a clot in an unbroken blood vessel.

Thrombus A clot formed in an unbroken blood vessel.

Thymectomy (thī-MEK-tō-mē) Surgical removal of the thymus.

Thymus (THĪ-mus) **gland** A bilobed organ, located in the upper mediastinum posterior to the sternum and between the lungs, that plays a role in the immune mechanism of the body.

Thyroglobulin (thī-rō-GLŌ-byoo-lin) **(TGB)** A large glycoprotein molecule secreted by follicle cells of the thyroid gland in which iodine is combined with tyrosine to form thyroid hormones.

Thyroid cartilage (THĪ-royd KAR-ti-lij) The largest single cartilage of the larynx, consisting of two fused plates that form the anterior wall of the larynx. Also called the **Adam's apple.**

Thyroid colloid (KOL-loyd) A complex in thyroid follicles consisting of thyroglobulin and stored thyroid hormones.

Thyroid follicle (FOL-i-kul) Spherical sac that forms the parenchyma of the thyroid gland and consists of follicular cells that produce thyroxine (T_4) and triiodothyronine (T_3) and parafollicular cells that produce calcitonin (CT).

Thyroid gland An endocrine gland with right and left lateral lobes on either side of the trachea connected by an isthmus located in front of the trachea just below the cricoid cartilage.

Thyroid-stimulating hormone (TSH) A hormone secreted by the adenohypophysis (anterior lobe) of the pituitary gland that stimulates the synthesis and secretion of hormones produced by the thyroid gland.

Thyroxine (thī-ROK-sēn) **(T_4)** A hormone secreted by the thyroid that regulates metabolism, growth and development, and the activity of the nervous system.

Tic Spasmodic, involuntary twitching by muscles that are ordinarily under voluntary control.

Tinnitus (ti-NĪ-tus) A ringing or tingling sound in the ears.

Tissue A group of similar cells and their intercellular substance joined together to perform a specific function.

Tissue plasminogen activator (t-PA) An enzyme that dissolves small blood clots by initiating a process that converts plasminogen to plasmin, which degrades the fibrin of a clot.

Tissue rejection Phenomenon by which the body recognizes the proteins (HLA antigens) in transplanted tissues or organs as foreign and produces antibodies against them.

Tongue A large skeletal muscle on the floor of the oral cavity.

Tonsil (TON-sil) A multiple aggregation of large lymphatic nodules embedded in mucous membrane.

Topical (TOP-i-kal) Applied to the surface rather than ingested or injected.

Torn cartilage A tearing of an articular disk in the knee.

Torpor (TOR-por) Abnormal inactivity or lack of response to normal stimuli.

Toxic (TOK-sik) Pertaining to poison; poisonous.

Toxic shock syndrome (TSS) A disease caused by the bacterium *Staphylococcus aureus,* occurring among menstruating females who use tampons and characterized by high fever, sore throat, headache, fatigue, irritability, and abdominal pain.

Trabecula (tra-BEK-yoo-la) Irregular latticework of thin plate of spongy bone. Fibrous cord of connective tissue serving as supporting fiber by forming a septum extending into an organ from its wall or capsule. *Plural, **trabeculae*** (tra-BEK-yoo-lē).

Trabeculae carnae (tra-BEK-yoo-lē KAR-nē) Ridges and folds of the myocardium in the ventricles.

Trachea (TRĀ-kē-a) Tubular air passageway extending from the larynx to fifth thoracic vertebra. Also called the **windpipe.**

Tracheostomy (trā-kē-OS-tō-mē) Creation of an opening into the trachea through the neck, with insertion of a tube to facilitate passage of air or evacuation of secretions.

Trachoma (tra-KŌ-ma) A chronic infectious disease of the conjunctiva and cornea of the eye caused by the TRIC agent.

Tract A bundle of nerve fibers in the central nervous system.

Transfusion (trans-FYOO-shun) Transfer of whole blood, blood components, or bone marrow directly into the bloodstream.

Transplantation (trans-plan-TĀ-shun) The replacement of injured or diseased tissues or organs with natural ones.

Transvaginal oocyte retrieval Procedure in which aspirated ova are combined with a solution containing sperm outside the body and then the fertilized ova are implanted in the uterus.

Transverse colon (trans-VERS KŌ-lon) The portion of the large intestine extending across the abdomen from the right colic (hepatic) flexure to the left colic (splenic) flexure.

Transverse fissure (FISH-er) The deep cleft that separates the cerebrum from the cerebellum.

Transverse tubules (TOO-byools) **(T tubules)** Minute, cylindrical invaginations of the muscle fiber (cell) membrane that carry the muscle action potentials deep into the muscle fiber (cell).

Trauma (TRAW-ma) An injury, either a physical wound or psychic disorder, caused by an external agent or force, such as a physical blow or emotional shock; the agent or force that causes the injury.

Triad (TRĪ-ad) A complex of three units in a muscle fiber (cell) composed of a transverse tubule and the segments of sarcoplasmic reticulum on both sides of it.

Tricuspid (trī-KUS-pid) *valve* Artrioventricular (AV) valve on the right side of the heart.

Trigeminal neuralgia (trī-JEM-i-nal noo-RAL-jē-a) Pain in one or more of the branches of the trigeminal (V) nerve. Also called **tic douloureux** (doo-loo-ROO).

Trigone (TRĪ-gon) A triangular area at the base of the urinary bladder.

Triiodothyronine (trī-ī-od-ō-THĪ-rō-nēn) **(T_3)** A hormone produced by the thyroid gland that regulates metabolism, growth and development, and the activity of the nervous system.

Trochlea (TROK-lē-a) A pulleylike surface.

Trophoblast (TRŌF-ō-blast) The outer covering of cells of the blastocyst.

Tropic (TRŌ-pik) *hormone* A hormone whose target is another endocrine gland.

Trunk The part of the body to which the upper and lower extremities are attached.

Tubal ligation (lī-GĀ-shun) A sterilization procedure in which the uterine (Fallopian) tubes are tied and cut.

Tuberculosis (too-berk-yoo-LŌ-sis) An inflammation of the lungs and pleurae caused by *Mycobacterium tuberculosis* resulting in destruction of lung tissue and its replacement by fibrous connective tissue.

Tumor (TOO-mor) A growth of excess tissue due to an unusually rapid division of cells.

Tunica albuginea (TOO-ni-ka al'-byoo-JIN-ē-a) A dense layer of white fibrous tissue covering a testis or deep to the surface of an ovary.

Tunica externa (eks-TER-na) The outer coat of an artery or vein, composed mostly of elastic and collagenous fibers. Also called the **adventitia.**

Tunica interna (in-TER-na) The inner coat of an artery or vein, consisting of a lining of endothelium and its supporting layer of connective tissue. Also called the **tunica intima** (IN-ti-ma).

Tunica media (MĒ-dē-a) The middle coat of an artery or vein, composed of smooth muscle and elastic fibers.

T wave The deflection wave of an electrocardiogram that records ventricular repolarization.

Tympanic antrum (tim-PAN-ik AN-trum) An air space in the posterior wall of the middle ear that leads into the mastoid air cells or sinus.

Tympanic (tim-PAN-ik) **membrane** A thin, semitransparent partition of fibrous connective tissue between the external auditory meatus and the middle ear. Also called the **eardrum.**

Type II cutaneous mechanoreceptor A receptor embedded deeply in the dermis and deeper tissues that detects heavy and continuous touch sensations. Also called an **end organ of Ruffini.**

Ulcer (UL-ser) An open lesion of the skin or a mucous membrane of the body with loss of substance and necrosis of the tissue.

Ultrasound (**US**) Medical imaging technique that utilizes high-frequency sound waves to produce an image called a **sonogram.**

Umbilical (um-BIL-i-kal) Pertaining to the umbilicus or navel.

Umbilical (um-BIL-i-kal) **cord** The long, ropelike structure, containing the umbilical arteries and vein that connect the fetus to the placenta.

Umbilicus (um-BIL-i-kus) A small scar on the abdomen that marks the former attachment of the umbilical cord to the fetus. Also called the **navel.**

Unsaturated fat A fat that contains one or more double covalent bonds between its carbon atoms; examples are olive oil and peanut oil.

Upper extremity The appendage attached at the shoulder girdle, consisting of the arm, forearm, wrist, hand, and fingers.

Uremia (yoo-RĒ-mē-a) Accumulation of toxic levels of urea and other nitrogenous waste products in the blood, usually resulting from severe kidney malfunction.

Ureter (YOO-re-ter) One of two tubes that connect the kidney with the urinary bladder.

Urethra (yoo-RĒ-thra) The duct from the urinary bladder to the exterior of the body that conveys urine in females and urine and semen in males.

Urinalysis The physical, chemical, and microscopic analysis or examination of urine.

Urinary (YOO-ri-ner-ē) **bladder** A hollow, muscular organ situated in the pelvic cavity posterior to the symphysis pubis.

Urinary tract infection (**UTI**) An infection of a part of the urinary tract or the presence of large numbers of microbes in urine.

Urine The fluid produced by the kidneys that contains wastes or excess materials and excreted from the body through the urethra.

Urobilinogenuria The presence of urobilinogen in urine.

Urogenital (yoo'-rō-JEN-i-tal) **triangle** The region of the pelvic floor below the symphysis pubis, bounded by the symphysis pubis and the ischial tuberosities and containing the external genitalia.

Urology (yoo-ROL-ō-jē) The specialized branch of medicine that deals with the structure, function, and diseases of the male and female urinary systems and the male reproductive system.

Urticaria (ur'-ti-KĀ-rē-a) A skin reaction to certain foods, drugs, or other substances to which a person may be allergic; hives.

Uterine (YOO-ter-in) **tube** Duct that transports ova from the ovary to the uterus. Also called the **Fallopian** (fal-LŌ-pē-an) **tube** or **oviduct.**

Uterosacral ligament (yoo'-ter-ō-SĀ-kral LIG-a-ment) A fibrous band of tissue extending from the cervix of the uterus laterally to attach to the sacrum.

Uterovesical (yoo'-ter-ō-VES-ī-kal) **pouch** A shallow pouch formed by the reflection of the peritoneum from the anterior surface of the uterus, at the junction of the cervix and the body, to the posterior surface of the urinary bladder.

Uterus (YOO-te-rus) The hollow, muscular organ in females that is the site of menstruation, implantation, development of the fetus, and labor. Also called the **womb.**

Utricle (YOO-tri-kul) The larger of the two divisions of the membranous labyrinth located inside the vestibule of the inner ear, containing a receptor organ for static equilibrium.

Uvea (OO-vē-a) The three structures that make up the vascular tunic of the eye.

Uvula (OO-vyoo-la) A soft, fleshy mass, especially the V-shaped pendant part, descending from the soft palate.

Vagina (va-JĪ-na) A muscular, tubular organ that leads from the uterus to the vestibule, situated between the urinary bladder and the rectum of the female.

Valvular stenosis (VAL-vyoo-lar STEN-Ō-sis) A narrowing of heart valve, usually the bicuspid (mitral) valve.

Varicocele (VAR-i-kō-sēl) A twisted vein; especially, the accumulation of blood in the veins of the spermatic cord.

Varicose (VAR-i-kōs) Pertaining to an unnatural swelling, as in the case of a varicose vein.

Vas A vessel or duct.

Vasa recta (REK-ta) Extensions of the efferent arteriole of a juxtaglomerular nephron that run alongside the loop of the nephron (Henle) in the medullary region.

Vasa vasorum (VĀ-sa va-SŌ-rum) Blood vessels supplying nutrients to the larger arteries and veins.

Vascular (VAS-kyoo-lar) Pertaining to or containing many blood vessels.

Vascular tunic (TOO-nik) The middle layer of the eyeball, composed of the choroid, ciliary body, and iris.

Vascular (**venous**) **sinus** A vein with a thin endothelial wall that lacks a tunica media and externa and is supported by surrounding tissue.

Vasectomy (va-SEK-tō-mē) A means of sterilization of males in which a portion of each ductus (vas) deferens is removed.

Vasoconstriction (vāz-ō-kon-STRIK-shun) A decrease in the size of the lumen of a blood vessel caused by contraction of the smooth muscle in the wall of the vessel.

Vasodilation (vās'-ō-DĪ-lā-shun) An increase in the size of the lumen of a blood vessel caused by relaxation of the smooth muscle in the wall of the vessel.

Vasomotion (vāz-ō-MŌ-shun) Intermittent contraction and relaxation of the smooth muscle of the metarterioles and precapillary sphincters that result in an intermittent blood flow.

Vasomotor (vā-sō-MŌ-tor) **center** A cluster of neurons in the medulla that controls the diameter of blood vessels, especially arteries.

Vein A blood vessel that conveys blood from tissues back to the heart.

Vena cava (VĒ-na KĀ-va) One of two large veins that open into the right atrium, returning to the heart all of the deoxygenated blood from the systemic circulation except from the coronary circulation.

Venesection (vēn'-e-SEK-shun) Opening of a vein for withdrawal of blood.

Ventral (VEN-tral) Pertaining to the anterior or front side of the body; opposite of dorsal.

Ventral body cavity Cavity near the ventral aspect of the body that contains viscera and consists of a superior thoracic cavity and an inferior abdominopelvic cavity.

Ventral ramus (RĀ-mus) The anterior branch of a spinal nerve, containing sensory and motor fibers to the muscles and skin of the anterior surface of the head, neck, trunk, and the extremities.

Ventricle (VEN-tri-kul) A cavity in the brain or an inferior chamber of the heart.

Ventricular fibrillation (ven-TRIK-yoo-lar fib-ri-LĀ-shun) Asynchronous ventricular contractions that result in circulatory failure.

Venule (VEN-yool) A small vein that collects blood from capillaries and delivers it to a vein.

Vermiform appendix (VER-mi-form a-PEN-diks) A twisted, coiled tube attached to the cecum.

Vermillion (ver-MIL-yon) The area of the mouth where the skin on the outside meets the mucous membrane on the inside.

Vermis (VER-mis) The central constricted area of the cerebellum that separates the two cerebellar hemispheres.

Vertebral (VER-te-bral) **canal** A cavity within the vertebral column formed by the vertebral foramina of all the vertebrae and containing the spinal cord. Also called the **spinal canal.**

Vertebral column The 26 vertebrae; encloses and protects the spinal cord and serves as a point of attachment for the ribs and back muscles. Also called the **spine, spinal column,** or **backbone.**

Vertigo (VER-ti-go) Sensation of dizziness.

Vesicle (VES-i-kul) A small bladder or sac containing liquid.

Vesicular ovarian follicle A relatively large, fluid-filled follicle containing an immature ovum and its surrounding tissues that secretes estrogens. Also called a **Graafian** (GRAF-ē-an) **follicle.**

Vestibular (ves-TIB-yoo-lar) **membrane** The membrane that separates the cochlear duct from the scala vestibuli.

Vestibule (VES-ti-byool) A small space or cavity at the beginning of a canal, especially the inner ear, larynx, mouth, nose, and vagina.

Villus (VIL-lus) A projection of the intestinal mucosal cells containing connective tissue, blood vessels, and a lymphatic vessel; functions in the absorption of food. *Plural,* **villi** (VIL-ē).

Viscera (VIS-er-a) The organs inside the ventral body cavity. *Singular,* **viscus** (VIS-kus).

Visceral (VIS-er-al) Pertaining to the organs or to the covering of an organ.

Visceral effector (e-FEK-tor) Cardiac muscle, smooth muscle, and glandular epithelium.

Visceral muscle An organ specialized for contraction, composed of smooth muscle fibers (cells), located in the walls of hollow internal structures, and stimulated by visceral efferent neurons.

Visceral pleura (PLOO-ra) The inner layer of the serous membrane that covers the lungs.

Visceroceptor (vis'-er-ō-SEP-tor) Receptor that provides information about the body's internal environment.

Viscosity (vis-KOS-i-tē) The state of being sticky or thick.

Vital signs Signs necessary to life that include temperature (T), pulse (P), respiratory rate (RR), and blood pressure (BP).

Vitamin An organic molecule necessary in trace amounts that acts as a catalyst in normal metabolic processes in the body.

Vitiligo (vit-i-LĪ-go) Patchy, white spots on the skin due to partial or complete loss of melanocytes.

Vitreous humor (VIT-rē-us HYOO-mor) A soft, jellylike substance that fills the posterior cavity of the eyeball, lying between the lens and the retina.

Vocal folds Pair of mucous membrane folds below the ventricular folds that function in voice production. Also called **true vocal cords.**

Volkmann's canal *See* **Perforating canal.**

Vomiting Forcible expulsion of the contents of the upper gastrointestinal tract through the mouth.

Vulva (VUL-va) Collective designation for the external genitalia of the female. Also called the **pudendum** (poo-DEN-dum).

Wallerian (wal-LE-rē-an) **degeneration** Degeneration of the portion of the axon and myelin sheath of a neuron distal to the site of injury.

Wandering macrophage (MAK-rō-fāj) Phagocytic cell that develops from a monocyte, leaves the blood, and migrates to infected tissues.

Wart Generally benign tumor of epithelial skin cells caused by a virus.

Wheal (hwēl) Elevated lesion of the skin.

White matter Aggregations or bundles of myelinated axons located in the brain and spinal cord.

White matter tract The treelike appearance of the white matter of the cerebellum when seen in midsagittal section. Also called **arbor vitae** (AR-bōr VĒ-te). A series of branching ridges within the cervix of the uterus.

White pulp The portion of the spleen composed of lymphatic tissue, mostly lymphocytes, arranged around central arteries; in some areas of white pulp, lymphocytes are thickened into lymphatic nodules called splenic nodules (Malpighian corpuscles).

White ramus communicans (RĀ-mus co-MYOO-ni-kans) The portion of a preganglionic sympathetic nerve fiber that branches away from the anterior ramus of a spinal nerve to enter the nearest sympathetic trunk ganglion.

Wormian bone *See* **Sutural bone.**

Xiphoid (ZĪ-foyd) Sword-shaped. The lowest portion of the sternum.

Yolk sac An extraembryonic membrane that connects with the midgut during early embryonic development, but is nonfunctional in humans.

Zona fasciculata (ZŌ-na fa-sik'-yoo-LA-ta) The middle zone of the adrenal cortex that consists of cells arranged in long, straight cords and that secretes glucocorticoid hormones.

Zona glomerulosa (glo-mer'-yoo-LŌ-sa) The outer zone of the adrenal cortex, directly under the connective tissue covering, that consists of cells arranged in arched loops or round balls and that secretes mineralocorticoid hormones.

Zona pellucida (pe-LOO-si-da) Gelatinous glycoprotein covering that surrounds a secondary oocyte.

Zona reticularis (ret-ik'-yoo-LAR-is) The inner zone of the adrenal cortex, consisting of cords of branching cells that secrete sex hormones, chiefly androgens.

Zygote (ZĪ-gōt) The single cell resulting from the union of a male and female gamete; the fertilized ovum.

Zymogenic (zī'-mō-JEN-ik) **cell** One of the cells of a gastric gland that secretes the principal gastric enzyme precursor, pepsinogen. Also called a **peptic cell.**

Index

Page numbers followed by the letter E indicate term to be found in exhibits.

A band, 229
Abdomen
   lymph nodes of, 432E, 433E
   muscles that act on walls of, 267E
   surface anatomy of, 312, 315, 317E
   veins of, 399E, 400, 401
Abdominal aorta, 347, 376, 377E, 385E, 388, 387, 388
Abdominopelvic cavity, 16
Abdominopelvic quadrants, 16
Abdominopelvic regions, 16
   representative structures found in, 16E
Abduction, 189, 193E
Abductor muscles, 249
Abnormal cell division, 54–57
   definition, 54–55
   possible causes of, 56–57
   spread of, 55–56
   treatment, 57
   types of, 56
Abnormal contraction of a muscle, 240
Abnormal uterine bleeding, 744
Abortion, 767
Abrasion, 109
Absorption, 4, 67, 645
Accessory ligaments, 188
Accessory sex glands, 709, 716–718
Accessory spleens, 421
Accutane, 106
Acetylcholine (ACh), 231, 454, 535, 541
Acetylcholinesterase (AChE), 454, 541
Achalasia, 658
Achondroplasia, 127
Achromatopsia, 585
Acidosis, 454
Acinar gland, 74
Acne, 106
Acquired immune deficiency syndrome (AIDS), 435–438
   classification system for, 437
   drugs and vaccines against HIV, 437–438
   HIV virus and, 56, 435–436, 437
   prevention of transmission, 438
   symptoms, 437
   transmission of, 437
Acromegaly, 595
Acromion, 168
Acrosome, 714
Actin, 46
Active processes, 35, 36–38
Active transport, 36–37
Acupuncture, 554–555
Acute pain, 553
Acute prostatitis, 743
Acute renal failure (ARF), 704
Adam's apple, 312, 620
Adaptation, 550
Addison's disease, 603

Adduction, 189, 193E
Adductor muscle, 249
Adenine nucleotide, 49
Adenitis, 440
Adenohypophysis, 593–594
Adenoid, see Pharyngeal tonsil
Adenosarcoma, 56
Adhesion proteins, 64
Adipocytes, 83
Adipose tissue, 78E, 83
Adrenal cortex, 602–605
Adrenal gland, 591, 601–606
   adrenal cortex, 602–605
   adrenal medulla, 543E, 602, 605–606
   development of, 610
   disorders of, 602, 603, 605
   hormones of, 602, 603, 605
Adrenal medulla, 543E, 602, 605–606
Adrenergic fibers, 542
Adrenocorticotropic hormones (ACTH), 594, 597E
Adrenogenital syndrome, 605
Adrenoglomerulotropin, 609
Adult connective tissue, 76–85
   cartilage, 84
   connective tissue proper, 76–84
   osseous, 84
   vascular, 84–85
Afferent nervous system, 444
Afferent neurons, 444
"Afterbirth," 409, 761
Afterimages, 551
Age spots, see Liver spots
Aging
   blood vessels and, 409
   cardiovascular system and, 409
   cells and, 57–58
   digestive system and, 678
   endocrine system and, 610
   free radical theory of, 58
   integumentary system and, 105
   muscle tissue and, 238
   nervous system and, 521
   reproductive system and, 739
   respiratory system and, 635
   skeletal system and, 123
   urinary system and, 702
Agnosia, 528
Agonist muscle, 248
Agranular ER, 40
Agranular leucocytes (agranulocytes), 329, 330, 334, 337E
Agraphia, 508
Albininism, 97
Albino, 97
Albumins, 337
Aldosterone, 602, 606E
Aldosteronism, 602

Alkalosis, 454
Alimentary canal, *see* Gastrointestinal (GI) tract
Allantois, 759
Allergens, 439
Allergy, 439
All-or-none principle, 233
Alpha receptors, 542
Alveolar-capillary membrane, 630
Alveolar ducts, 629
Alveolar macrophage, 629
Alzheimer's disease (AD), 528
Amacrine cells, 568
Amenorrhea, 744
Ametropia, 585
Amniocentesis, 765
Amnion, 758–759
Amniotic fluid (AF), 759
Amorphous matrix materials, 48
Amphiarthroses, 186, 197
Amygdaloid nucleus, 506
Amyloid, 528
Anabolic steroids, 229–230
Anabolism, 4
Anal canal, 674
Analgesia, 528
Anaphase stage, 51
    activity of, 52E
Anaphylaxis, 439
Anastomosis, 358, 370–371
Anatomical position, 4
Anatomy
    branches of, 2, 312
    characteristics, 4
    definition, 2
    See also Developmental anatomy; Surface anatomy
Anchoring filaments, 416
Anemia, 338–339
Anesthesia, 528
Aneurysm, 410
Angina pectoris, 358
Angiocardiography, 365
Angiogenin, 56
Angular movements, 189, 193E
Anhidrosis, 109
Anisotropine methylbromide, 492
Ankle, 216–217, 216E
    edema of, 392
    surface anatomy of, 312, 324E, 325
Ankle sprain, 216
Ankyloglossia, 650
Ankylosis, 219
Annulus fibrosus, 148
Anopsia, 585
Anorexia nervosa, 682
Antagonist muscle, 248
Anterior corticospinal tract, 467E
Anterior reticulospinal tract, 467E
Anterior spinocerebellar tract, 466E
Anterior spinothalamic tract, 466E
Antibodies, 335
Anticholinesterase drugs, 240
Antidiuretic hormone (ADH), 596, 597E
Antigens, 335
Anti-oncogenes, 57
Antiperspirant, 109
Anus, 674

Aorta, 375, 376
    abdominal, 347, 376, 377E, 385E, 386, 387, 388
    arch of, 347, 376, 377E, 381E, 382, 383, 384
    ascending, 347, 375–376, 377E, 380E
    coarction of, 352
    descending, 376
    thoracic, 347, 376, 377E, 385E, 387, 388
Aortic insufficiency, 365
Aortography, 411
Aphasia, 508
Apnea, 640
Apneustic area, 635
Apocrine glands, 75
Apocrine sweat glands, 104
Aponeuroses, 83, 226
Appendages, 133
Appendicitis, 681
Appendicular skeleton, 132, 164–184
    divisions of, 135E
Apraxia, 528
Arachnoid, 460, 488
Arches of the foot, 180
Arch of the aorta, 347, 376, 377E, 381E, 382, 383, 384
Areflexia, 480
Areola, 734
Arm
    muscles that move, 278E
    surface anatomy of, 312, 315, 318E
Arrector pili, 103
Arrhythmias, 364–365
Arterial stick, 328
Arteries, 370–371
    anastomosis, 370–371
    arcuate, 694, 728
    arterial branches of the aorta, 377E, 378
    arterioles, 370, 371, 372
    articular, 122
    basilar, 381E
    brachocephalic, 381E
    bronchial, 385E
    celiac, 385E
    common hepatic, 385E
    common iliac, 376, 389E
    coronary, 347, 357, 380E
    definition, 370
    elastic, 370
    end, 371
    esophageal, 385E
    external iliac, 389E
    gonadal, 385E
    inferior mesenteric, 385E
    inferior phrenic, 385E
    interlobar, 694
    internal iliac, 389E
    left common carotid, 381E
    left gastric, 385E
    left pulmonary, 405
    left subclavian, 381E
    lumbar, 385E
    mediastinal, 385E
    middle sacral, 385E
    muscular, 370
    nutrient, 122
    pericardial, 385E
    periosteal, 122
    posterior intercostal, 385E
    radial, 315, 320E, 728

renal, 385E
right pulmonary, 405
splenic, 385E
subcostal, 385E
superior mesenteric, 385E
superior phrenic, 385E
suprarenal, 385E
umbilical, 407–408, 409
uterine, 728
Arterioles, 370, 371, 372
Arteritis, 411
Arthralgia, 219
Arthritis, 218
Arthrodia, *see* Gliding joints
Arthrology, 186
Arthroscopy, 188
Arthrosis, 219
Articular cartilage, 84, 114, 119, 187
Articular discs, 188
Articular tubercle, 139
Articulations, 185–222
    atlanto-occipital, 199E
    ball-and-socket, 196, 196E
    cartilaginous, 186, 187, 196E
    coxal (hip), 210–211, 210E
    cracking sound and, 188
    definition, 186
    disorders of, 218–219
    elbow, 206–207, 206E
    ellipsoidal, 193, 196E
    fibrous, 186, 196E
    functional classification of, 186
    glenohumeral (shoulder), 204–205, 204E
    gliding, 193, 196E
    gomphosis, 186, 196E
    hinge, 193, 196E
    humeroscapular (shoulder), 204E
    intervertebral, 200–201, 200E
    key medical terms, 219
    lumbosacral, 202–203, 202E
    pivot, 193, 196E
    radiocarpal (wrist), 208–209, 208E
    saddle, 193, 196E
    structural classification of, 186
    symphysis, 187, 196E
    synchondrosis, 187, 196E
    syndesmosis, 186, 196E
    synovial, 186, 187–196, 196E
    talocrural (ankle), 216–217, 216E
    temporomandibular, 145–146, 197E, 198
    tibiofemoral (knee), 212E, 213–215
Artificial ear, 576
Artificial heart, 358–369
Artificial insemination, 736
Artificial kidney, 697–698
Artificial pacemaker, 356
Arytenoid cartilage, 622
Ascending aorta, 347, 375–376, 377E, 380E
Ascending colon, 675
Ascending tracts of the spinal cord, 466E
Asphyxia, 640
Aspiration, 640
Assimilation, 4
Association neurons, 451
Astereognosis, 585
Asthma attack, 626
Astrocytes, 445E

Ataxia, 511
Atelectasis, 640
Atherosclerosis, 360, 520
Atherosclerotic plaque, 360
Atlanto-occipital joints, 199E
Atria, 347, 348
Atrial fibrillation, 365
Atrial natriuretic factor (ANF), 611
Atrioventricular (AV) block, 364–365
Atrioventricular (AV) node, 354
Atrioventricular (AV) valves, 348–351
Atrophy, 58, 234
Atropine, 492
Audiometer, 585
Auditory ossicles, 572
Auditory sensations, 568–578
    auditory pathways, 575–576
    external (outer) ear, 570, 578E
    internal (inner) ear, 573–575, 578E
    mechanism of equilibrium, 576–578
    middle ear, 570–572, 578E
Auditory tube, 571
Auricle, 347
Auscultation, 356
Autocrine motility factor (AMF), 55
Autoimmune diseases, 106, 348
Autologous transfusion, 340
Autolysis, 45
Autonomic ganglia, 535–537
Autonomic nervous system (ANS), 444, 533–548
    activities of, 542–544
    biofeedback and, 545–546
    control of, 545–546
    definition, 534
    meditation and, 546
    neurotransmitters of, 541–542
    parasympathetic division, 540–541, 541E
    receptors of, 542
    somatic efferent nervous system compared to, 534, 535E
    structure of, 534–541
    sympathetic division, 538–540
    visceral autonomic reflexes of, 544–545
    visceral efferent neurons of, 534–538
Autophagy, 45
Autopsy, 19
Autoregulation, 372
Autorhythmicity, 235
Autosomes, 53, 713
Avascular blood vessels, 64
Axial skeleton, 131–163
    divisions of, 135E
Axonal transport, 448
Axon collaterals, 448
Axon hillock, 448
Axons, 446–448
Axon terminals, 230
Axoplasma, 448
Axoplasmic flow, 448
Azidothymidine (AZT), 438
Azotemia, 705

Babinski sign, 581
Baby teeth, 656
Back, 312–314
Backbone, *see* Vertebral column
Baldness, 104
Ball-and-socket joint, 196, 196E

Baroreceptors, 357
Basal cell carcinoma (BCC), 107
Basal cells, 560
Basal ganglia, 504–505
Basal lamina, 64
Basement membrane, 64
Basophils, 334–335, 337E
Belly (gaster), 246
Benign growth, 55
Beta blockers, 361
Beta receptors, 542
Bilaterally symmetrical, 4
Bile capillaries, 665
Biliary colic, 682
Bilirubin, 665
Binge-purge syndrome, 682
Biofeedback, 545–546
Biopsy, 58
Bipennate muscle, 248
Birth control (BC), 737–738, 739E
Black eye, 136
Blackheads, *see* Comedo
Blastocyst, 753
Blastomers, 753
Blepharitis, 585
Blind spot, *see* Scotoma
Blood, 327–343
    components, 329
    definition, 328
    development of, 409
    disorders of, 338
    formed elements in, 329–337, 337E
    functions, 328–329
    key medical terms, 340–341
    physical characteristics of, 328
    plasma, 337–338
Blood-brain barrier (BBB), 492
Blood-cerebrospinal fluid barrier, 491
Blood reservoirs, 374–375
Blood samples, 328
Blood substitute, 333
Blood supply
    of the bone, 122
    of the brain, 492
    of the esophagus, 659
    of the heart, 357–358
    of the kidneys, 693–696
    of the large intestine, 678
    of the liver, 667–669
    of the lungs, 630–632
    of muscle tissue, 226
    of the skin, 98
    of the small intestine, 673
    of the stomach, 663
    of the teeth, 657
    of the ureter, 700
    of the urinary bladder, 701
Blood-testis barrier, 712
Blood transfusion, *see* Transfusion
Blood vessels, 369–414
    aging and, 409
    arteries, 370–371
    arterioles, 370, 371, 372
    capillaries, 370, 371–373
    definition, 370
    development of, 409
    disorders of, 409–411

of fetal circulation, 406–409
of hepatic portal circulation, 405–406
key medical terms, 411
of pulmonary circulation, 405
of systemic circulation, 375–376, 377E, 380E, 381E, 385E, 389E, 392E, 394E, 396E, 399E, 402E
veins, 370, 373–374
venules, 370, 373
Body cavities, 13–16
Boils, 104
Bolus, 657
Bone
    arch of the foot, 180–183
    arm, 169–170
    blood supply of, 122
    carpals, 170
    cervical vertebral, 148, 151
    clavicle, 166
    coccyx, 148, 156–157
    compact, 115, 116
    coxal, 174
    cranial, 133–143, 146
    disorders of, 123–127
    ethmoid, 142–143
    facial, 135, 143–145, 146
    of the female skeleton, 183
    femur, 175
    fibula, 179
    of the finger, 170
    flat, 132
    of the foot, 180
    of the forearm, 170
    formation of, 123
    frontal, 135–136
    growth of, 119–121
    head, 133E
    heel, 325
    hip, 170–171
    histology, 114–118
    humerus, 169–170
    hyoid, 146–148, 315E
    inferior nasal conchae, 146
    irregular, 132
    kneecap, 175–176
    lacrimal, 146
    long, 114–115, 132
    of lower extremity, 174–183
    lumbar vertebrae, 148, 151–156
    of the male skeleton, 183
    mandible, 145
    maxillary, 144
    metacarpals, 170
    metatarsals, 180
    nasal, 143
    nerve supply of, 122
    occipital, 140
    osseous tissue, 82E, 84
    palatine, 146
    parietal, 136
    patella, 175–176
    pectoral shoulder girdle, 165–168
    pelvic girdle, 170–174
    phalanges of the finger, 170
    phalanges of the foot, 180
    pisiform, 135, 320E
    radius, 170
    replacement of, 121–122

ribs, 158
sacral vertebrae, 148, 156–157
sacrum, 156–157
scapula, 166–168
sesamoid, 132
shinbone, 176–179
short, 132
of the skull, 133–146
spheroid, 140–142
spongy, 115–117
sternum, 157–158
surface markings, 132, 133E
sutural, 132
tarsals, 179–180
temporal, 139–140
thoracic vertebrae, 148, 151
thorax, 157–158
thighbone, 175
tibia, 176–177
types of, 132
ulna, 169
of upper extremity, 169–170
vertebral column, 148–149
vomer, 146
wormian, 132
zygomatic, 145
Bone marrow, yellow, 115
Bone marrow transplantation, 336
Borborygmus, 682
Botulism, 682
Botulism toxin, 454
Bowman's gland, *see* Olfactory gland
Brachial plexus, 471–476, 476E
nerves of, 476E
Bradykinesia, 524
Brain, 487–511
blood supply of, 492
brain stem, 487, 492–497, 511E
cerebellum, 487, 509–511, 511E
cerebrospinal fluid, 488–492
cerebrum, 487, 501–508, 511E
coverings, 488
development from brain vesicles, 487, 488E
diencephalon, 487, 488E, 497–501, 511E
injuries to, 506
principal part of, 487
protection of, 488
split-brain concept, 508–509
Brain death, 508
Brain damage, 506
Brain electrical activity mapping (BEAM), 526
Brain lateralization, *see* Split-brain concept
Brain sand, 608
Brain stem, 478, 492–497, 511E
medulla oblongata, 488E, 492–496, 511E
mesencephalon (midbrain), 487, 488E, 497, 511E
pons, 488E, 496–497, 511E
Brain tumors, 522–524
Brain vesicles, 487, 488E
Brain waves, 508
Breast cancer, 734, 745
Breast cancer screening indicator, 735
Breathing
costal, 634
diaphragmatic, 634
muscles used in, 270E
Bright's disease, 704

Brim of the pelvis, 171
Bronchi, 617, 626
Bronchial asthma, 636
Bronchial tree, 626
Bronchiectasis, 640
Bronchitis, 636
Bronchogenic carcinoma, 636
Bronchogram, 626
Bronchography, 626
Bronchoscope, 626
Bronchoscopy, 626
Buccal glands, 651
Bulbourethral glands, 717–718
Bulimia, 682
Bullae, 108
Bunion, 183
Burkitt's lymphoma, 56
Burns, 107–109
classification of, 107–109
Lund-Browder method for, 108
treatment of, 109
Bursae, 188
Bursectomy, 219
Bursitis, 188, 219
Buttocks, surface anatomy of, 312, 317, 322E

Calcaneal tendon, 302, 325
Calcification, 118
Calcified cartilage, 118
Calcitonin, 597, 601E
Callus, 109, 126–127
Canaliculi, 84, 116
Cancer (CA)
breast, 734, 745
cell division and, 54–57
cervical, 745–746
definition, 54–55
of the larynx, 624
of the lung, 636
possible causes of, 56–57
skin, 97, 107
spread of, 55–56
testicular, 743
treatment, 57
types of, 56
of the uterus, 727
Capillaries, 370, 371–373
definition, 370
Carbon monoxide poisoning, 639
Carboxypolypeptidase, 663
Capacitation, 752
Carbuncle, 109
Carcinogens, 56
Carcinoma, 56
basal cell, 107
bronchogenic, 636
nasopharyngeal, 56
squamous cell, 107
Cardiac arrest, 365
Cardiac center, 496
Cardiac cycle, 354
Cardiac muscle tissue, 85–87, 86E, 224, 235–237, 238E
Cardioacceleratory center (CAC), 356
Cardioinhibitory center (CIC), 356
Cardiology, 345
Cardiomegaly, 365–366
Cardiopulmonary resuscitation (CPR), 640

Cardiovascular system 6E
    aging and, 409
    blood, 327–343
    blood vessels, 369–414
    the heart, 344–368
    relationship of lymphatic system to, 417
Carotene, 97
Carotid foramen, 139
Carpal tunnel syndrome, 472
Carpals, 170
Carrier molecules, 35
Cartilage, 84
    articular, 84, 114, 119, 187
    arytenoid, 622
    calcified, 118
    corniculate, 622
    costal, 84, 158
    cricoid, 622
    cuneiform, 622
    elastic, 81E, 84
    hyaline, 80E, 84
Cartilaginous joints, 186, 187, 196E
Castration, 737
Catabolism, 4
Cataract, 567, 583
Catechol-O-methyltransferase (COMT), 542
Cauda equina, 465
Caudal anesthesia, 156
Caveolae, 238
Cavities of the body, 13–16
Cavitron ultrasonic surgical aspirator (CUSA), 496
Cell body, 87
Cell division
    abnormal, 54–57
    cancer, 54–57
    definition, 48
    events associated with, 52E
    reproductive division, 49, 52–54
    somatic division, 48, 49–52
Cell junctions, 64, 65–67
Cells, 31–62
    aging and, 57–58
    definition, 2, 32
    division of, 48–54
    functions of cell parts, 48E
    inclusions, 47, 48E
    key medical terms, 58
    organelles, 32, 38–47
    parts of, 32, 48
    plasma membrane of, 32–38
Cell shapes, 65
Cellular level of structural organization, 2
Cementum, 656
Center, 468
Central canal, 84, 116
Central nervous system (CNS), 7E, 444
    aging and, 521
    development of, 521
    disorders of, 521–528
    neuroglia of, 445E
    neuronal pools in, 455–456
Centrioles, 47
    function of, 48E
Centromere, 50
Centrosome, 47
Cerebellar peduncles, 509–511
Cerebellar tract, 558

Cerebellum, 487, 509–511, 511E
    damage to, 511
    functions of, 511
    structure, 509–511
Cerebral apoplexy, 521
Cerebral aqueduct, 490
Cerebral cortex, 501
    functional areas of, 506–508
Cerebral palsy (CP), 524
Cerebral peduncles, 497
Cerebrospinal fluid (CSF), 488–492
Cerebrovascular accident (CVA), 521
Cerebrum, 487, 501–508, 511E
    basal ganglia (cerebral nuclei), 504–505
    functional areas of cerebral cortex, 506–508
    limbic system, 506
    lobes, 502–504
    white matter, 460, 465–466, 504
Cerumen, 104
Ceruminous glands, 104
Cervical canal, 727
Cervical cancer, 745–746
Cervical enlargement, 463
Cervical plexus, 471
    nerves of, 472E
Cervical vertebrae, 148, 151
Cesarean section (C-section), 767
Chancre, 742
"Charley horse," 240
Chemical contraception, 738, 739E
Chemical digestion, 645
Chemical exfoliation, see Skin peel
Chemical level of structural organization, 2
Chemonucleolysis, 159
Chemoreceptors, 551
Chemotaxis, 334
Chest, surface anatomy of, 312, 316E
Cheyne-Stokes respiration, 640
Chickenpox, 109
Chlamydia, 743
Chloramphenicol, 492
Cholecystitis, 682
Cholecystokinin (CCK), 610
Cholelithiasis, 682
Cholinergic fibers, 541
Chondritis, 219
Chondroblasts, 118
Chondrocytes, 84
Chondroitin sulfate, 48
Chondrosacromas, 56
Chordae tendinea, 348
Chorion, 759
Chorionic villi, 760
Chorionic villi sampling (CVS), 765
Choroid plexuses, 491
Chromaffin, 605
Chromatids, 50
Chromatin, 40, 49
Chromosomal microtubules, 50
Chromosomes, 40, 49
Chrondoblasts, 123
Chronic Epstein-Barr virus (EBV) syndrome, 340
Chronic obstructive pulmonary disease (COPD), 637
Chronic pain, 553
Chronic prostatitis, 743
Chronic renal failure (CRF), 704
Chvostek sign, 601

Chyme, 662, 677
Chymotrypsin, 663
Cilia, 47, 65
    function of, 48E
Ciliary body, 566
Ciliary ganglion, 513, 540
Cine CT, 22
Circular pattern of fasciculi, 248
Circumcision, 722
Circumduction, 189, 193E
Cirrhosis, 681
Citrated whole blood, 340
Claudication, 411
Clavicle, 166, 316E
Clawfoot, 183
Cleavage, 753
    lines of, 96
Cleft lip, 144
Cleft palate, 144
Clinical perineum, 733
Clitoris, 732
Cloaca, 702
Cloacal membrane, 678
Closed fracture, 125, 126
Closed reduction, 125
Coarctation of the aorta, 362
Coccygeal vertebrae, 148
Coccyx, 148, 156–157
Cochlea, 574
Cochlear duct, 574
Cochlear implants, 576
Coitus, see Sexual intercourse
Cold sore, 109
Colitis, 682
Collagen implant, 98
Collagenous fibers, 48, 76
Collarbones, see Clavicles
Colles' fracture, 125, 126, 170
Colon, 674
Colonoscopy, 681
Colostomy, 682
Colposcopy, 728
Colpotomy, 746
Column, 460
Columnar cells, 65
Coma, 528–529
Comedo (blackheads), 104, 109
Comminuted fracture, 125, 126
Commissurotomy, 366
Common cold, 639
Compensation, 411
Complete blood count (CBC), 337
Complete fracture, 125
Compliance, 366
Compound gland, 75
Computed tomography (CT) scanning, 19–22
Computed tomography with mammography (CT/M), 734–735
Computerized axial tomography (CAT) scanning, 19–22
Concentric lamellae, 116
Concussion, 506
Condom, 737, 739E
Conduction deafness, 584
Conduction myofibers, 354
Conduction pathway, 469
Conductivity, 4, 452
Condylar process, 145
Condyle, 133E

Cone biopsy, 727
Cones, 566
Congenital defects, 362–364
Congestive heart failure (CHF), 365
Conjunctivitis, 583–584
Connective tissue, 64, 75–85
    adult, 76–85
    classification of, 76
    embryonic, 76, 77E
    skeletal muscle tissue and, 224–226
Constipation, 678
Constrictive pericarditis, 366
Contact inhibition, 56, 98–99
Contact lenses, 567
Continuous ambulatory peritoneal dialysis (CAPD), 698
Continuous capillaries, 372
Continuous microtubules, 50
Contraception, 737–738, 739E
Contraceptive sponge, 738, 739E
Contractility, 4
Contusion, 109, 506
Conus medullaris, 463
Conus tendon, 353
Convergence, 453
Converging circuit, 455–456
Convulsions, 240
Copulation, see Sexual intercourse
Coracoid process, 168
Cordotomy, 554
Corn, 109
Cornea, 563
Corneal transplants, 563
Corniculate cartilage, 622
Coronal suture, 135
Coronary artery bypass grafting (CABG), 361–362
Coronary artery disease (CAD), 360–362, 410
Coronary artery spasm, 362
Coronary enderterectomy, 411
Coronary sinus, 347, 376
Coronary sulcus, 347
Coronoid process, 145
Corpora quadrigemina, 497
Cor pulmonale (CP), 365
Corpus albicans, 731
Corpus callosum, 502
Corpuscles of touch, 96, 552
Corpus striatum, 504
Cortical nephron, 692
Corticolipotroph cells, 594
Corticosterone, 603
Cortisol, 603, 606E
Cortisone, 603, 606E
Corzyo, see Common cold
Costal breathing, 634
Costal cartilage, 84, 158
Covergent pattern of fasciculi, 248
Covering and lining epithelium, 64, 65–74
    classification of, 67
Coxal joint, 210–211, 210E
Crack, 454
Cracking sound of synvovial joint, 188
Cramp, 240
Cranial bones, 135–144, 146
    ethmoid bone, 142–143
    frontal bone, 135–136
    occipital bone, 140
    parietal bone, 136

Cranial bones (*Continued*)
  spheroid bone, 140–142
  temporal bone, 139–140
Cranial cavity, 13
Cranial fossae, 143
Cranial meninges, 488
Cranial nerves, 511–520
  abducens, 515–516, 520E
  accessory, 517–519, 520E
  clinical application of, 519E, 520E
  facial, 516, 520E
  function of, 519E, 520E
  glossopharyngeal, 516–517, 520E
  hypoglossal, 519, 520E
  oculomotor, 513, 519E
  olfactory, 511, 519E
  optic, 512, 519E
  trigeminal, 514–515, 519E
  trochlear, 513–514, 519E
  vagus, 517, 520E
  vestibulocochlear, 516, 520E
Cranial parasympathetic outflow, 540
Craniosacral outflow, 535
Craniotomy, 127
Cremaster muscle, 709
Crenation, 36
Crest, 133E
Cretinism, 599
Cricoid cartilage, 622
Cristae, 556
Crista galli, 143
Crossing-over, 53
Cross section, 13
Crutch palsy, 472
Cryptorchidism, 709
CT scan, *see* Computed tomography (CT) scanning
Cuboidal cells, 65
Culdoscopy, 746
Cuneiform cartilage, 622
Curare, 454
CURL (compartment of uncoupling of receptor and ligand), 38
Cushing's syndrome, 603
Cuspids, 656–657
Cutaneous membrane, 85
Cutaneous plexus, 98
Cutaneous sensations, 551–555
Cuticle, 105
Cuticle of the hair, 100
Cyanosis, 340
Cyst, 109
Cystitis, 704, 705
Cystocele, 705
Cytokinesis, 49, 52
  activity of, 52E
  definition, 52
Cytology, 2, 32
Cytoplasm, 32
  composition of, 38
  definition, 38
  function of, 38, 48E
Cytoskeleton, 46–47
Cytotrophoblast, 754

D and C (dilation and curettage), 722
Dartos, 709
Daughter cells, 48
Deafness, 584

Decidua, 759–760, 761
Deciduous teeth, 656
Decubitus ulcers, 107
Decussation of pyramids, 493
Deep fascia, 225
Deep-venous thrombosis (DVT), 411
Deep wound healing, 99
Defecation, 645, 677
Defensins, 334
Defibrillation, 365
Deflection waves, 354
Deglutition, 657
Delayed nerve grafting, 480
Demineralization, 125
Dendrites, 87, 446
Dense bodies, 237
Dense connective tissue, 79E, 83
Dental caries, 678–680
Dental plaque, 680
Dental terminology, 656
Dentes, *see* Teeth
Denticulate ligaments, 461
Dentin, 654
Dentitions, 656–657
Deodorant, 109
Deoxyribonuclease, 663, 672
Deoxyribose, 49
Depilatory, 103
Depression, 192, 193E
Depressor muscle, 249
Dermabrasion, 109
Dermal papillae, 96
Dermatan sulfate, 48
Dermatome, 109, 239, 478–479
Dermis, 93, 96
  development of, 106
Descending aorta, 376
Descending colon, 674
Descending tract of the spinal cord, 467E
Desmosome, 65–66
Deterioration, 58
Detritus, 109
Detrusor muscle, 700–701
Developmental anatomy, 2, 751–769
  of blood, 409
  of blood vessels, 409
  of the digestive system, 678
  of the ear, 581
  embryonic development, 756–761, 762E
  of the endocrine system, 610
  of the epidermis, 105–106
  of the eye, 581
  fetal growth, 761, 762E
  gestation, 761–762
  of the heart, 360
  of the integumentary system, 105–106
  key medical terms, 767
  of the lymphatic system, 435
  of the muscular system, 238–239
  of the nervous system, 521
  parturition and labor, 765–767
  during pregnancy, 752–756
  prenatal diagnostic techniques, 765
  of the reproductive systems, 741
  of the respiratory system, 635–636
  of the skeletal system, 123
  of the urinary system, 702

Deviated nasal septum (DNS), 146
Dextran, 680
Diabetes insipidus, 596
Diabetes mellitus, 606–607
Diapedesis, 334
Diaphragm, 737, 739E
Diaphragmatic breathing, 634
Diaphysis, 114
Diarrhea, 677–678
Diarthroses, 186, 197
Diastole, 354
Diencephalon, 487, 488E, 497–501, 511E
    hypothalamus, 488E, 497–501, 511E
    thalamus, 488E, 497, 511E
Differential blood count (DIF), 336
Differentiation, 4
Diffuse lymphatic tissue, 416
Diffusion, 35
Digestion, 4, 645
Digestive system, 9E, 644–686
    aging and, 678
    basic activities of, 645
    definition, 645
    development of, 678
    disorders of, 678–682
    general histology, 645–646
    key medical terms, 682–683
    organization of the GI tract, 645
    organs of, 648–678
    peritoneum of, 85, 646–648
Digital rectal examination, 681
Digital subtraction angiography (DSA), 19, 25
Diisopropyl fluorophosphate, 454
Dimethyl sulfoxide (DMSO), 219
Diphtheria, 640
Diplegia, 480
Diploid cells, 53
Diploid chromosome number, 713
Directional terms, 4–12
Direct (immediate) transfusion, 340
Dislocation, 219
    of the jaw, 145
    of the shoulder joint, 204
Displaced fracture, 125
Disturbance of gait, 511
Divergence, 452
Diverging circuits, 455
Diverticulitis, 681
Diverticulosis, 681
Dizygotic twins, 753
Dizziness, 511
DNA molecule, 49, 50
Dorsal body cavity, 13
Dorsal ramus, 469–470
Dorsiflexion, 189–192, 193E
Double helix, 49
Down's syndrome (DS), 765
Drugs
    passage through blood-brain barrier of, 492
    See also names of drugs
Dual inversion, 535
Duchenne muscular dystrophy (DMD), 240
Ducts, 709
Ductus arteriosus, 408–409
Ductus (vas) deferens, 715–716
Ductus venosus, 408, 409
Duodenal glands, 670

Duodenal ulcers, 680
Duodenum, 670
Dura mater, 460, 488
Dynamic equilibrium, 576, 578
Dynamic spatial reconstructor (DSR), 19, 22–23
Dyskinesia, 585
Dyslexia, 526
Dysmenorrhea, 744
Dysphagia, 682
Dysplasia, 58
Dyspnea, 640
Dysrhythmia, 364
Dystocia, 767
Dysuria, 705

Ear
    artificial, 576
    development of, 581
    disorders of, 583–585
    external (outer), 570, 578E
    internal (inner), 573–575, 578E
    middle (tympanic cavity), 570–572, 578E
    surface anatomy of, 314E
Eardrum, 570
    perforated, 570
Eccrine sweat glands, 104
Ectoderm, 105–106, 123, 678, 756
    structures produced by, 758E
Ectopic pregnancy (EP), 727
Eczema, 109
Edema, 427
Effector, 468
Efferent nervous system, 444
Efferent neurons, 444
Effort, 247
Ejaculation, 736
Ejaculatory duct, 716
Elastic arteries, 370
Elastic cartilage, 81E, 84
Elastic connective tissue, 79E, 84
Elastic fibers, 48, 76
Elbow, surface anatomy of, 312, 315, 318E
Elbow joint, 206–207, 206E
Electrocardiogram (ECG), 354–356
Electrocardiograph, 354
Electroconvulsive therapy (ECT), 529
Electroejaculation, 480
Electroencephalogram (EEG), 508
Electrolysis, 103
Electrolytes, 48, 76
Electromagnetic receptors, 551
Electromyogram, 235
Electromyography (EMG), 235
Electronic fetal monitoring (EFM), 764
Electron micrograph (EM), 32n
Eleidin, 95
Elephantiasis, 440
Elevation, 192, 193E
Ellipsoidal joint, 193, 196E
Embolus, 521
Embolus, 521
Embryo, 76
    definition, 756
    development of, 756–761, 762E
Embryology, 2, 756
Embryonic connective tissue, 76, 77E
Embryonic development, 756–761, 762E
Embryonic membranes, 758–759
Embryo transfer, 754

Emission, 736
Emphysema, 637
Enamel, 654–656
Encephalomyelitis, 529
End arteries, 371
Endocarditis, 347
Endocardium, 347
Endocervical curettage (ECC), 727
Endochrondial ossification, 118–119, 123
Endocrine glands, 73E, 74, 591–593
Endocrine system, 8E, 590–615
    aging and, 610
    development of, 610
    disorders of, 594–595, 596, 599–600, 600–601, 602, 603, 605–
        606, 606–607
    feedback control, 592–593
    glands of, 591–610
    hormones of, 591–610
    key medical terms, 611
    other endocrine tissues, 610–611
    receptors, 591–592
Endocytosis, 37–38
Endoderm, 756
    structures produced by, 758E
Endolymph, 573
Endometriosis, 745
Endomysium, 225
Endoneurium, 469
Endoplasmic reticulum (ER), 40
    function of, 48E
Endoscopy, 659
Endosome, 38
Endosteum, 115
Endothelial-capsule membrane, 692
Endoethelial tubes, 360
Enlarged prostate, 743
Enteric gastrin, 610
Enteritis, 683
Enteroceptors, 551
Enterocrinin, 610
Enteroendocrine cells, 659
Enuresis, 705
Eosinophils, 334, 337E
Ependymocytes, 445E
Epicarditis, 347
Epicardium, 345
Epicondyle, 133E
Epidemiology, 58
Epidermal grooves, 97
Epidermal ridges, 97
Epidermal wound healing, 98–99
Epidermis, 93–95
    development of, 105–106
    epidermal derivatives, 99–105
Epididymis, 715
Epidural space, 460
Epigastric region, 16E
Epiglottis, 620–621
Epikeratoplasty, 563
Epilepsy, 526
Epileptic seizures, 526
Epimysium, 225
Epineurium, 469
Epinephrine, 605, 606E
Epiphora, 585
Epiphyseal line, 121
Epiphyseal plate, 119

Epiphyses, 114
Episiotomy, 733
Epithelial tissue, 64–75, 645
    covering and lining, 64, 65–74
    glandular, 64, 72E, 73E, 74–75
Epithelium, see Epithelial tissue
Eponychium, see Cuticle
Epstein-Barr virus (EBV), 56, 340
Equatorial plane region of the cell, 50
Equilibrium, 351, 576–578
    dynamic, 576, 578
    static, 576, 577–578
    structures of the ear related to, 578E
Erection, 721
    female, 737
    male, 736
Eructation, 683
Erythema, 109
Erythrocytes (red blood cells), 84–85, 329, 330, 332–334, 337E
    functions of, 332
    life span of, 333
    number, 333
    production of, 333
    structure, 332
Erythropoiesis, 333
Esophagus, 657–659
    blood supply of, 659
    nerve supply of, 659
Estrogen replacement therapy (ERT), 123
Estrogens, 608, 609E, 610, 729
Ethmoidal sinuses, 143
Etiology, 125
Eupnea, 634
Eustachian tube, 571
Eustachitis, 585
Eversion, 189, 193E
Exchange transfusion, 340
Excitability, 4, 451
Excitatory transmitter-receptor interaction, 453
Excretion, 4
Exocrine glands, 72E, 591
    definition, 74
    functional classification of, 75
    structural classification of, 74–75
Exocytosis, 37
Exophthalmic goiter, 600
Exotropia, 585
Extension, 189, 193E
Extensor muscle, 249
Extensor retinaculum, 283
External human fertilization, 754
External hydrocephalus, 491–492
External occipital protuberance, 140
External respiration, 617
Exteroceptors, 551
Extracapsular ligament, 188
Extracellular materials, 32, 48
Extracorporeal shock wave lithotripsy (ESWL), 703
Extradural hemorrhage, 488
Extraembryonic coelom, 758
Extrafusal fibers, 555
Extrapyramidal pathways, 581
Extrinsic muscles, 225E, 257E
Eye
    accessory structures of, 562–563
    development of, 581
    disorders of, 583–585

effect of sympathetic and parasympathetic stimulation on, 543E
surface anatomy of, 313E
Eyeball
fibrous tunic of, 563
interior of, 567
lens, 567
muscles that move, 255E
retina, 566
structure, 563–567
structures associated with, 568E
vascular tunic of, 463–566
Eyebrows, 313E, 562
Eye examination, 567
Eyelashes, 313E
Eyelids, 562

Face, surface anatomy of, 312, 313E, 314E
Facet, 133E
Facial bones, 143–146
Facial expression muscles, 252E, 253E
Facilitated diffusion, 35
Falciform ligament, 648
Fallopian tubes, *see* Uterine tubes
False labor, 767
False ribs, 158
Falx cerebri, 502
Fascia, 224–225
Fascia lata, 294E
Fascicles, 469
Fasciculi, 225
arrangement of, 248
Fasciculus cuneatus tract, 466E
Fasciculus gracilis tract, 466E
Fast-twitch red muscle fibers, 234
Fast-twitch white muscle fibers, 234
Fecal occult-blood testing, 681
Feces, 677
Feedback control, 592–593
Feeding center, 501
Female gonads, *see* Ovaries
Female infertility, 745
Female reproductive system, 722–737
Female skeleton, 183
Feminizing adenoma, 605
Femur, 175
Fenestrated capillaries, 372
Fertilization, 53, 752–754
external human, 754
Fetal-cell surgery, 761
Fetal circulation, 406–409
Fetal growth, 761, 762E
Fetal period, 756, 761, 762E
Fetal surgery, 761
Fetal ultrasound, 763–764
Fetomaternal hemorrhage, 761
Fetus, 756
growth of, 761, 762E
Fibrillation, 240, 365
Fibrinogen, 337
Fibroadenoma, 745
Fibroblasts, 76
Fibrocartilage, 81E, 84
Fibrocyte, 76
Fibromyositis, 240
Fibrosis, 239
Fibrositis, 239
Fibrous astrocytes, 445E

Fibrous joints, 186, 196E
Fibrous layer, 114
Fibrous matrix materials, 48
Fibrous pericardium, 345
Fibrous ring, 351–353
Fibrous trigone, 351–353
Fibula, 179
muscles that act on, 297E, 298E
Fight-or-flight response, 544
Film-screen mammography, 734
Filtration, 36
Filum terminale, 463
Fingers, muscles that move, 283E, 284E
Fingerstick, 328
First-class levers, 247
First-degree burns, 107–108
Fissure, 133E, 628
Fixators, 248
Flaccid muscles, 234
Flaccid paralysis, 581
Flagella, 47
function of, 48E
Flatfoot, 180
Flatus, 683
Flexible fiberoptic sigmoidoscope, 681
Flexion, 189, 193E
Flexor muscle, 249
Flexor retinaculum, 283
Floating kidney, 688
Floating rib, 158
Fluid mosaic model, 32
Fluosol-DA (blood substitute), 333
Flutter, 365
Follicle-stimulating hormone (FSH), 594, 597E, 729
Fontanels, 135
Foot
arch of, 180–183
intrinsic muscles of, 305E
muscles that move, 301E, 302E
surface anatomy of, 312, 324E, 325
Foot plates, 123
Foramen, 133E
Foramen ovale, 408, 409
Foramina, 146, 147E
Forearm
muscles that move, 281E
surface anatomy of, 312, 315, 319E
Foreskin, *see* Prepuce
Fornix, 732
Fossa, 133E
Fossa ovalis, 347
Fracture (Fx), 125
Colles', 125, 126, 170
Pott's, 125, 126, 179
repair of, 125–127
types of, 125
of the vertebral column, 160
Fractured clavicle, 166
Fracture hematoma, 126
Fraternal twins, *see* Dizygotic twins
Freckles, 97
Free nerve endings, 552
Free radical theory of aging, 58
Free ribosomes, 40
Frontal plane, 12
Frontal section, 13
Frontal sinuses, 136

Frontal squama, 136
Fulcrum, 246–247
Functional classification of joints, 186
Furuncle, 109
Fusiform, 248

G_1 period, 49
  activity of, 52E
G_2 period, 49
  activity of, 52E
Gallbladder (GB), 669–670
  effect of sympathetic and parasympathetic stimulation on, 543E
Gallstones, 665, 682
Gamete intrafallopian transfer (GIFT), 754–756
Gametes, 52–53, 709, 712–714
Gamma globulin, 340
Ganglia
  basal, 504–505
  definition, 461
Gangrene, 240
Gap junction, 66, 235
Gastrectomy, 683
Gastric gland, 543E, 659
Gastric inhibitory peptide (GIP), 610
Gastric lipase, 662
Gastric pits, 659
Gastroenterology, 645
Gastroileal reflex, 676
Gastrointestinal (GI) tract, 645
  products eliminated from, 688E
Gastroscopy, 659
Gastrulation, 756
Generalized animal cell, 32
General senses, 551–556
Generator potential, 550
Genetic material, 40
Geniculate ganglion, 516
Genital herpes, 742–743
Geriatrics, 57–58
German measles, 109–110
Gestation, 761–762
Giantism (gigantism), 594–595
Gingivae, 654
Girdles, 133
Glabella, 136
Glands
  acinar, 74
  adrenal, 591, 601–602
  apocrine, 75
  buccal, 651
  bulbourethral, 717–718
  ceruminous, 104
  duodenal, 670
  effect of sympathetic and parasympathetic stimulation on, 543E
  endocrine, 73E, 74, 591–593
  exocrine, 72E, 591
  gastric, 543E, 659
  intestinal, 543E, 670
  lacrimal, 543E
  mammary, 316E, 733–735
  mandibular, 651
  merocrine, 75
  multicellular, 74–75
  parotid, 651
  prostate, 717
  of the skin, 104
  parathyroid, 591, 600–601, 602E
  pineal, 608–609
  pituitary, 591, 593–596, 597E
  salivary, 543E, 651–652
  sebaceous, 104, 106
  sebaceous ciliary, 562
  sudoriferous, 104, 106
  sweat, 543E
  tarsal, 562
  thymus, 416, 424, 609–610
  thyroid, 597–600
  tubular, 74
  tubuloacinar, 75
Glandular epithelium, 64, 72E, 73E, 74–75
Glans penis, 721
Glaucoma, 567, 583
Glenohumeral joint, 204–205, 204E
Glenoid cavity, 168
Glial cells, *see* Neuroglia cells
Gliding joint, 193, 196E
Gliding movement, 189, 193E
Globulins, 337
Glomerulonephritis, 704
Glomerulus, 694
Glottis, 621
Glucagon, 606
Glucocorticoids, 603, 606E
Gluteus medius muscle, 307
Glycogen, 47
Goblet cells, 67
Goiter, 600
Golgi complex, 40–42
  function of, 48E
Gomphosis, 186, 196E
Gonadocorticoids, 603, 606E
Gonadotroph cells, 594
Gonadotropic hormones, 594
Gonadotropin releasing hormone (GnHR), 729
Gonads, 709
  removal of, 737
Gonorrhea, 741–742
Gossypol, 738
Gout, 703–704
Gouty arthritis, 218
Grading malignant tumors, 56
Granstein cells, 93
Granular ER, 40
Granular leucocytes (granulocytes), 329, 330, 334, 337E
Granulation tissue, 99
Greenstick, 125, 126
Gravity inversion, 148
Gray commisure, 465
Gray matter, 460, 465
Gray ramus communicans, 540
Groin, pulled, 295
Groove, 133E
Gross anatomy, 2
Growth, 4, 55
Growth hormone (GH), 594, 597E
Growth hormone-inhibiting factor (GHIF), 606
Gustatory sensations, 560–561
  gustatory pathway, 561
  receptors, 550–551
Gynecology, 722
Gynecomastia, 605
Gyri (convolutions), 501

Hair, 99–104
  color, 103
  definition, 99
  development of, 106
  development of distribution, 99–100
  growth and replacement, 103–104
  structure, 100–103
Hair cells, 577
Hair follicles, 100
Hand
  intrinsic muscles of, 288E
  muscles that move, 283E, 294E
  surface anatomy of, 312, 317, 321E
Hand plates, 123
Haploid cells, 53
Haploid chromosome number, 712–713
Hard palate, 649
Haustral churning, 677
Head
  lymph nodes of, 428E
  muscles of, 266E
  surface anatomy of, 312, 313E, 314E
  veins of, 394E, 395
Headache, 526–527
Heart, 344–368
  artificial, 358–359
  autonomic control of, 356–357
  blood supply of, 357–358
  cardiac cycle, 356
  chambers of, 347
  conduction system, 354
  definition, 345
  development of, 360
  disorders of, 360–365
  effect of sympathetic and parasympathetic stimulation on, 543E
  electrocardiogram (EGC), 354–356
  great vessels of, 347–348
  heart wall, 345–347
  key medical terms, 365–366
  location, 345
  pericardium, 85, 345
  skeleton of, 351–352
  surface projection of, 353–354
  valves of, 348–354
Heart block, 364–365
Heart disease, risk factors in, 359–360
Heart sound, 354
Heart transplant, three-way, 359
Heimlich maneuver, 640
Hematocrit (Hct), 334
Hematology, 328
Hematopoiesis, 332
Hemiballismus, 505
Hemidesmosomes, 66
Hemiplegia, 480
Hemochromatosis, 340
Hemocytoblasts, 332
Hemodialysis therapy, 697–698
Hemoglobin, 332
Hemolysis, 36
Hemolytic anemia, 338–339
Hemolytic jaundice, 665
Hemopoiesis, 332
Hemoptysis, 640
Hemorrhage, 340
Hemorrhagic anemia, 338
Hemorrhoids, 674

Hemothorax, 628
Hepatic cells, 665
Hepatic portal circulation, 405–406
Hepatitis, 681–682
Hepatitis A, 681
Hepatitis B, 681–682
Hepatitis B virus (HBV), 57
Hermaphroditism, 746
Hernia, 683
Herniated disc, 159–160
Herpes, genital, 742–743
Herpes virus, 448
Herpes simplex virus, type 2, 57
Heterologous insemination, 736
Hinge joint, 193, 196E
Hip, 210–211, 210E
  dislocation of, 210
Hippocampus, 506
Hirsutism, 100
Histology, 2, 64
Histones, 40
Hives, 110
Hodgkin's disease (HD), 56, 439–440
Holocrine glands, 75
Holter monitor, 356
Homologous chromosomes, 53, 713
Homologous insemination, 736
Homologous organs, 722
Horizontal plane, 12
Hormones, 591
  inhibiting, 593
  regulating, 593
  releasing, 593
  tropic, 594
Horn, 460
Horner's syndrome, 544
Human chorionic gonadotropin (HCG), 610, 731
Human chorionic somatomammotropin (HCS), 610
Human immunodeficiency virus (HIV), 56, 435
  drugs and vaccines against, 437–438
  outside the body, 437
  structure and pathogenesis, 435–436
  transmission of, 437
Human leucocyte associated (HLA) antigens, 334
Human T-cell leukemia-lymphoma virus-1 (HTLV-1), 56, 340
Humeroscapular joint, 204E
Humerus, 169–170
Huntington's chorea, 529
Hyaline cartilage, 80E, 84
Hyaluronic acid, 48
Hyaluronidase, 76
Hymen, 732
Hyperacusia, 572
Hypercholesteremia, 411
Hyperemesis gravidarum, 754
Hyperextension, 189, 193E
Hyperinsulinism, 607
Hyperparathyroidism, 601, 602E
Hyperplasia, 58, 611
Hypersecretion, 594
Hypersensitivity, 439
Hypersplenism, 440
Hypertension, 409–410
Hypertonic solution, 36
Hypertrophy, 58
Hypochondriac regions, 16E
Hypodermic, 110

Hypodermis, 110
Hypogastric region, 16E
Hyponychium, 105
Hypoparathyroidism, 600, 602E
Hypoplasia, 611
Hyposecretion, 594
Hypotension, 411
Hypothalamus, 488E, 497–501, 511E
  mammillary bodies of, 506
Hypotonic solution, 35–36
Hypoxia, 640
Hysterectomy, 727, 737
H zone, 229

I band, 229
Identical twins, see Monozygotic twins
Idiopathic seizures, 526
Ileum, 670
Iliotribal tract, 294E
Ilium, 174
Immovable joints, see Synarthroses
Immunologic tolerance, 438
Immunology, 336
Immunosuppressive drugs, 439
Impacted cerumen, 104
Impacted fracture, 125, 126
Imperforate hymen, 732
Impetigo, 110
Implantation, 753–754
Impotence, 743
Incisors, 656
Inclusions, 32
Incompetent valve, 366
Incontinence, 701
Indirect (mediate) transfusion, 340–341
Infarction, 358
Infectious hepatitis, see Hepatitis A
Infectious mononucleosis (IM), 339–340
Inferior cervical ganglion, 539
Inferior vena cava, 392
Infertility
  female, 745
  male, 744
Inflammatory bowel disease (IBD), 683
Inflammatory phase of deep wound healing, 99
Influenza (flue), 639
Infrahyoid muscle, 264
Infrared photocoagulation, 674
Infundibulum, 726
Ingestion, 4, 645
Inguinal canal, 276, 716
Inguinal hernia, 716
Inguinal ligament, 267
Inhibin, 608, 609E
Inhibiting hormones, 593
Insertion, 246
Insidious, 58
Insulin, 606
Insulin-dependent diabetes, 606–607
Integral proteins, 32–33
Integrans, 64
Integrated response, 546
Integrator, 453
Integumentary system, 5E, 92–112
  aging and, 105
  definition, 93
  development of, 105–106

disorders of, 106–109
  key medical terms, 109–110
  skin, 93–105
  skin wound healing, 98–99
Interatrial septum, 347
Intercalated discs, 86, 235
Intermediate filaments, 46–47, 237
  function of, 48E
Internal hydrocephalus, 491–492
Internal respiration, 617
Interstitial fluid, 328
Interstitial lamellae, 116
Interventricular foramen, 490
Interventricular septum, 347
Intervertebral discs, 148
Intervertebral joints, 200–201, 200E
Intestinal gland, 543E, 670
Intestinal juice, 761–762
Intestines
  effect of sympathetic and parasympathetic stimulation on, 543E
  large, 673–678
  small, 670–673
Intracapusular ligament, 188
Intracerebral hemorrhage, 521
Intradermal, 110
Intrafusal fibers, 555
Intramembranous ossification, 118, 123
Intramuscular (IM) injection, 307
Intraorbital foramen, 144
Intrauterine device (IUD), 738, 739E
Intravenous pyelogram, 706
Intrinsic muscles, 255E, 257E
Intubation, 626
Inversion, 189, 193E
In vitro fertilization, 754
Involuntary muscle tissue, 224
Iris, 313E
Irritable bowel syndrome (IBS), 683
Ischemia, 358
Ischium, 174
Isotonic solution, 35
Islets of Langerhans, 606, 663

Jarvik-7 (artificial heart), 359
Jaundice, 665
Jaw
  dislocation of, 145
  lower jaw muscles, 254E
Jejunum, 670
Joint kinesthetic receptors, 556
Jugular foramen, 139
Juxtaglomerular apparatus (JGA), 696–697
Juxtamedullary nephron, 692

Kaposi's sarcoma (KS), 435, 436
Karyolymph, 38
Karyotype, 767
Keratan sulfate, 48
Keratin, 73, 95
Keratinization, 95
Keratinocyte, 93–94
Keratitis, 585
Keratosis, 110
Ketosis, 607
Kidneys, 688–697
  anatomy of, 688–691

artificial, 697–698
blood supply of, 693–696
effect of sympathetic and parasympathetic stimulation on, 543E
hemodialysis therapy, 697–698
juxtaglomerular apparatus (JGA), 696–697
nephrons, 692–693
nerve supply of, 693–696
products eliminated from, 688E
Kidney stones, 702–703
Killer T cells, 335
Kinesthesis, 585
Knee, 212E, 213–215
dislocated, 213
injury to, 213
runner's knee, 213
surface anatomy of, 312, 317–325, 323E
swollen, 213
transplant, 213
Kyphosis, 160

Labia, *see* Lips
Labia major, 732
Labia minor, 732
Labile cells, 65
Labor, 765–767
three stages of, 767
Labyrinthine disease, 584
Labyrinthitis, 585
Laceration, 110, 506
Lacotroph cells, 594
Lacrimal apparatus, 562–563
Lacrimal caruncle, 562
Lacrimal gland, 543E
Lacrimal secretion, 563
Lactase, 672
Lactation, 734
Lactiferous sinus, 734
Lacunae, 84, 116
Laetrile, 57
Labdoidal suture, 135
Lamellae, 84
Lamellated corpuscles, 96, 552
Laminae, 151
Lamina propria, 85, 645
Langerhans cells, 93
Lanugo, 100, 106
Laparoscopic technique, 737
Large intestine, 673–678
blood supply of, 678
nerve supply of, 678
Laryngitis, 624
Laryngopharynx, 620
Larynx, 617, 620–621
cancer of, 624
muscles of, 264E
Lateral corticospinal tract, 467E
Lateral reticulospinal tract, 467E
Lateral spinothalamic tract, 466E
Lateral ventricles, 488–490
Left iliac region, 16E
Left lumbar region, 16E
Left lymphatic duct, 425
Leg
muscles that act on, 297E, 298E
surface anatomy of, 312, 324E, 325
Lens, 567
Lens capsule, 567

Lens epithelium, 567
Lens fibers, 567
Leptomeningitis, 460
Lesser cornu, 148
Lesser pelvis, 171–174
Lethal gene, 767
Lethargy, 529
Leucocytes (white blood cells), 83, 85, 329, 330, 334–336, 337E
functions of, 334–336
life span of, 336
number, 336
production of, 336
structure, 334
types, 334
Leukemia, 56, 340
Leukorrhea, 746
Levator muscle, 249
Lever, 246, 247
Leverage, 247
Life processes, 4
Ligaments, 83, 187
denticulate, 461
extracapsular, 188
falciform, 648
intracapsular, 188
periodontal, 654
reconstruction of, 188
Ligands, 38
Limb buds, 123
Limbic system, 506
Lines of cleavage, 96
Lingual frenulum, 650
Lingual tonsils, 421
Lipids, 47
Lips, 648
surface anatomy of, 314E
Little-league elbow, 206
Liver, 665–669
blood supply of, 667–669
effect of sympathetic and parasympathetic stimulation on, 543E
nerve supply of, 667–669
Liver biopsy, 669
Liver spots, 97
Lobes
of the cerebrum, 502–504
of the lungs, 628
Lobules, 628–629
Local disease, 58
Lochia, 767
Long bone, structure of, 114–115
Longitudinal arch of the foot, 180
Longitudinal axis of the human body, 132–133
Loose connective tissue, 76–83
Lordosis, 160
Lower extremities, 174–183
arteries of, 389E, 390, 391
lymph nodes of, 431E
surface anatomy of, 312, 317–325, 322E, 323E, 324E
veins of, 402E, 403, 404
Lower jaw, muscles of, 254E
Lower respiratory system, 617
Lubb, 356
Lubrication for sexual intercourse, 736
Lubrication of the vagina, 737
Luetinizing hormone (LH), 594, 597E
Lumbago, 239
Lumbar enlargement, 463

Lumbar plexus, 475
  nerves of, 477E
Lumbar vertebrae, 148, 151–156
Lumpectomy, 745
Lumbosacral joint, 202–203, 202E
Lumen, 370
Lund-Browder method, 108
Lung cancer, 636
Lungs, 617, 628–633
  alveolar-capillary membrane, 630
  blood supply of, 630–632
  effect of sympathetic and parasympathetic stimulation on, 543E
  fissures, 628
  gross anatomy of, 628
  lobes, 628
  lobules, 628–629
  nerve supply of, 632
  products eliminated from, 688E
  surgical access to, 158
Luxation, 219
Lymphadenectomy, 440
Lymphadenopathy, 440
Lymphangiography, 416
Lymphangioma, 440
Lymphangitis, 440
Lymphatic nodules, 416
Lymphatic organs, 416, 418–424
Lymphatics, 416
Lymphatic system, 7E, 328, 415–442
  definition, 416
  development of, 435
  disorders of, 435–440
  key medical terms, 440
  lymph circulation, 424–427
  lymph nodes, 416, 418–419, 427–434
  lymphatic organs, 416, 418–424
  lymphatic tissue, 416, 418–424
  lymphatic vessels, 416
  relationship of cardiovascular system to, 417
Lymphatic tissue, 416, 418–424
  lymph nodes, 416, 418–419, 427–434
  spleen, 416, 421–424
  thymus gland, 416, 424
  tonsils, 416, 421
Lymphatic vessels, 416
Lymph capillaries, 416
Lymph circulation, 424–427
Lymphedema, 440
Lymph nodes, 416, 418–419, 427–434
  anterior mediastinal, 434E
  anterior phrenic, 434E
  axillary, 429E
  bronchial, 434E
  bronchopulmonary, 434E
  buccal, 428E
  celiac, 433E
  central, 429E
  common iliac, 432E
  deep cervical, 428E
  deep inguinal, 431E
  deltopectoral, 429E
  external iliac, 432E
  facial, 428E
  gastric, 433E
  hepatic, 433E
  ileocolic, 433E
  inferior deep cerival, 428E

  inferior mesenteric, 433E
  infraorbital, 428E
  intercostal, 434E
  internal iliac, 432E
  lateral, 429E
  lumbar, 432E
  mandibular, 428E
  mesenteric, 433E
  mesocolic, 433E
  middle phrenic, 434E
  occiptal, 428E
  pancreaticosplenic, 433E
  parotid, 428E
  pectoral, 429E
  phrenic, 434E
  popliteal, 431E
  posterior mediastinal, 434E
  posterior phrenic, 434E
  preauricular, 428E
  pulmonary, 434E
  retroauricular, 428E
  sacral, 432E
  sternal, 434E
  subclavicular, 429E
  submandibular, 428E
  submental, 428E
  subscapular, 429E
  superficial cervical, 428E
  superficial inguinal, 431E
  superior deep cervical, 428E
  superior mesenteric, 433E
  supratrochlear, 429E
  tracheal, 434E
  tracheobronchial, 434E
Lymphoblasts, 332
Lymphocytes, 334, 335, 337E
Lymphoma, 440
Lymphostasis, 440
Lymph sacs, 435
Lysosomes, 44–45
  function of, 48E

Macrophages, 76
  alveolar, 629
  wandering, 334
Macroscopic anatomy, 2
Macula adherens, 65–66
Maculae, 556
Macula lutea, 566
Magnetic resonance imaging (MRI), 19, 23–24
Male infertility, 744
Male-pattern baldness, 104
Male reproductive system, 709–722
Male skeleton, 183
Malignancy, 55
Malignant melanoma, 97, 107
Malignant tumors, 56
Malocclusion, 683
Maltase, 672
Mammary ducts, 734
Mammary glands, 316E, 733–735
Mammogram, 734
Mammography, 734
Mandible, 145
Mandibular fossa, 139
Mandibular ganglion, 540
Mandibular glands, 651

Manubrium, 157–158
Marfan syndrome, 83
Marrow, *see* Bone marrow
Marrow biopsy, 158
Marrow cavity, 115
Mast cell, 83
Mastectomy, 745
Mastication, 657
Mastoiditis, 139
Mastoid process, 139–140
Matrix of the mitochondria, 42–44
Maturation phase of deep wound healing, 99
Maturity-onset diabetes, 607
Maxillae, 144
Measles, 110
Measurement of the human body, 25
Meatus, 133E
Mechanical digestion, 645
Mechanical methods of contraception, 737, 739E
Mechanoreceptors, 551
Medial lemniscus, 497
Medial malleolus, 177
Median nerve damage, 472
Mediastinum, 16
Medical imaging techniques, 19–25
Meditation, 546
Medulla, 100
Medulla oblongata, 488E, 492–496, 511E
Medullary cavity, 115
Medullary rhythmicity area, 634–635
Megakaryoblasts, 332
Meiosis, 49, 53–54
   comparison between mitosis and, 55
Meiosis I, 53
Meiosis II, 53
Meissner's corpuscles, 96
Melanin, 47
Melanoblasts, 97
Melanocytes, 93, 97
Melanoma, malignant, 97, 107
Melatonin, 609
Membrane capsules, 83
Membranes, 85
Ménière's syndrome, 584
Meninges, 460
Meningitis, 461
Menstrual abnormalities, 744
Menstrual cycle, 729–731
Menstrual phase, 729–731
Menstruation, 729–731
Mental foramen, 145
Merocrine glands, 75
Mesencephalon, 487, 488E, 497, 511E
Mesenchyme, 76, 77E
Mesentery, 647–648
Mesoappendix, 674
Mesocolon, 648, 674
Mesoderm, 123, 238–239, 360, 435, 758
   structures produced by, 758E
Mesodermal cells, 239
Mesonephros, 702
Metabolism, 4
Metacarpals, 170
Metanocyte-stimulating hormone (MSH), 594, 597E
Metaphase, 50
   activity of, 52E
Metaphysis, 112, 114, 120–121

Metaplasia, 58
Metarteriole, 372
Metastasis, 55–56, 58, 419
Metatarsus, 180
Metencephalon, 487, 488E
Metopic suture, 136
Metric units
   of length, 26E
   of mass, 26E
   of volume, 26E
Microfilaments, 46
   function of, 48E
Microglia, 445E
Microtomography, 32
Microtrabeculae, 46–47
Microtrabecular lattice, 47
Microtubules, 46
   function of, 48E
Microvilli, 67, 670
Micturition, 701
Micturition reflex, 701
Midbrain, *see* Mesencephalon
Middle cervical ganglion, 539
Midline, 12
Midsagittal plane, 12
Midsagittal section, 13
Mifepristone (RU486), 738
Migratory phase of deep wound healing, 99
Milk teeth, 656
Mineralocorticoids, 602, 606E
Mitochondria, 42–44
   function of, 48E
Mitosis, 48, 49–52
   comparison between meiosis and, 55
   time required for, 52
Mitral valve prolapse (MVP), 354
M line, 229
Molars, 657
Moles, 97
Monoamine oxidase (MAO), 542
Monoblasts, 332
Monocytes, 334
Monoplegia, 480
Monozygotic twins, 753
Mons pubis, 732
Morning sickness, 754
Morula, 753
Motion sickness, 584–585
Motor area of the cerebral cortex, 508
Motor cortex, 580
Motor neuron, 230
Motor pathways, 579–581
   extrapyramidal pathways, 581
   linkage of sensory input and motor responses, 579
   motor cortex, 580
   pyramidal pathways, 580
Motor unit, 231
Mouth, 648–657
   muscles of the floor, 262E
Movable joints, *see* Amphiarthroses; Diarthroses
Movement (of food), 648
Mucosa, *see* Mucous membrane
Mucous connective tissue, 76, 77E
Mucous membrane, 85, 645
Multicellular glands, 74–75
Multiple myeloma, 335
Multiple sclerosis (MS), 525

Multiunit smooth muscle tissue, 237–238
Mumps, 652
Murmurs of the heart, 354
Muscle-building anabolic steroids, 229–230
Muscle fibers, 226
Muscle spindles, 234, 555–556
Muscle tissue, 64, 85–87, 223–244, 246
    aging and, 238
    blood supply of, 226
    cardiac muscle tissue, 84–87, 86E, 224, 235–237, 238E
    characteristics of, 224
    classification of, 224
    connective tissue components, 224–226
    contraction of, 230–235
    development of, 238–239
    disorders of, 239–240
    functions, 224
    histology, 226–230
    key medical terms, 240–241
    nerve supply of, 226
    skeletal muscle tissue, 85, 86E, 224–230, 238E
    smooth muscle tissue, 86–87, 86E, 224, 237–238, 238E
    types of, 224, 234–235
Muscle tone, 233–234
Muscarinic receptors, 542
Muscular arteries, 370
Muscular dystrophies, 240
Muscular hypertrophy, 234
Muscularis, 646
Muscular mucosae, 85
Muscular system, see Skeletal muscles
Mutation, 767
Myalgia, 240
Myastenia gravis (MG), 240
Mydriasis, 585
Myelencephalon, 487, 488E
Myelin sheath, 448
Myeloblasts, 332
Myelography, 461
Myeloma, 56
    multiple, 335
Myocardial infarction (MI), 358
Myocarditis, 347
Myocardium, 345–347
Myofibers, 226
Myofibrils, 228–229
Myofilaments, 85, 229
Myoglobin, 234
Myology, 224
Myoma, 240
Myomalacia, 241
Myoneural function, 230
Myopathy, 241
Myosclerosis, 241
Myosin-binding site, 229
Myositis, 241
Myospasm, 241
Myotome, 239
Myotonia, 241
Myringitis, 585
Myxedema, 599

Nail matrix, 105
Nails, 105
    development of, 106
Nasal cavity, 618
Nasal polyps, 619

Nasal septum, 146, 618
Nasopharyngeal carcinoma, 56
Nasopharynx, 620
Natural method of contraception, 737, 739E
Nausea, 683
Navel, 761
Nebulization, 630
Neck
    lymph nodes of, 428E
    surface anatomy of, 312, 315E
    veins of, 394E, 395
Necrosis, 58, 127
Negative foodback control, 592
Neonatal jaundice, 665
Neoplasm, 55, 58
Nephroblastoma, 705
Nephrocele, 705
Nephron, 692–693
Nephrosis, 704
Nerve, 451
    abducens, 515–516, 520E
    accessory, 517–519, 520E
    ansa cervicalis, 472E
    axiliary, 476E
    cardiac, 356
    common peroneal, 478E
    cranial, 511–520
    cutaneous branch, 472E
    deep peroneal, 478E
    definition, 460
    doral scapular, 476E
    facial, 516, 520E
    femoral, 474
    genitofemoral, 478E
    glossopharyngeal, 516–517, 520E
    greater auricular, 474E
    hypoglossal, 519, 520E
    iliohypogastric, 478E
    ilioinguinal, 478E
    inferior gluteal, 478E
    inferior root, 472E
    intercostal, 474–477
    intercostobrachial, 476E
    lateral antibrachial cutaneous, 476E
    lateral cord, 476E
    lateral femoral cutaneous, 478E
    lateral pectoral, 476E
    lateral plantar, 478E
    lesser occipital, 472E
    long thoracic, 476E
    lower lateral brachial cutaneous, 476E
    lower motor, 580, 581
    lower subscapular, 476E
    medial antebrachial cutaneous, 476E
    medial brachial cutaneous, 476E
    medial cord, 476E
    medial pectoral, 476E
    medial plantar, 478E
    median, 476E
    mixed, 469, 511
    motor branch, 472E
    musculocutaneous, 476E
    obturator, 478E
    oculomotor, 513, 519E
    olfactory, 512, 519E
    optic, 512, 519E
    pelvic splanchnic, 541

perforating cutaneous, 478E
phrenic, 472E
piriformis, 478E
posterior antebrachial cutaneous, 476E
posterior brachial cutaneous, 476E
posterior cord, 476E
posterior femoral cutaneous, 478E
pudendal, 478E
quadratus femoris, 478E
radial, 476E
root, 476E
sciatic, 474, 478E
segmental branches, 472E
sensory, 511
splanchnic, 540, 541
subclavius, 476E
superficial branch, 472E
superficial peroneal, 478E
superior gluteal, 478E
superior root, 472E
supraclavicular, 472E
suprascapular, 476E
thoracodorsal, 476E
tibial, 478E
transverse cervical, 472E
trigeminal, 514–515, 519E
trochlear, 513–514, 519E
ulnar, 315, 476E
upper lateral brachial cutaneous, 476E
upper subscapular, 476E
vagus, 356, 517, 520E
vestibulocochlear, 516, 520E
Nerve block, 529
Nerve fiber, 448, 451
Nerve impulse, 451–454
Nerve optic, 568
Nerve supply
of the bone, 122
of the esophagus, 659
of the kidneys, 693–696
of the large intestine, 678
of the liver, 667–669
of the lungs, 632
of muscle tissue, 226
of the small intestine, 673
of the stomach, 663
of the teeth, 657
of the ureter, 700
of the urinary bladder, 701
Nervous tissue, 64, 87, 87E, 443–458
classification of, 450–451
conductivity of, 445–451
definition, 444
histology, 445–451
nerve impulse, 451–454
organization of, 444–445
organization of neurons, 455–456
regeneration of neurons, 454–455
Neural deafness, 584
Neuralgia, 529
Neural tissue, 460
Neuritic plaques, 528
Neuritis, 481
Neuroblastoma, 611
Neurofibral nodes, 448
Neurofibrillary tangles, 528
Neurofibrils, 446

Neuroglandular junction, 452
Neuroglia, 87
Neuroglia cells, 444
Neurohypophysis, 596
Neurolemmocytes, 448
Neurology, 444
Neuromuscular function, 230–231
Neuromuscular junction (NMJ), 452
Neuronal pools, 455–456
Neurons, 87, 230
afferent, 444, 450, 544
alpha efferent, 556
association, 451, 544
bipolar, 450, 566
efferent, 444, 450
first-order, 557
gamma efferent, 556
ganglion, 566
motor, 450, 468
multipolar, 450
organization of, 455–456
photoreceptor, 566
postganglionic, 534, 537–538
postsynaptic, 452, 453
preganglionic, 534, 535
presynaptic, 452–453
regeneration of, 454–455
second-order, 557
sensory, 450, 468
structural variation of, 448
structure of, 445–448
third-order, 557
unipolar, 450
upper motor, 580, 581
visceral efferent, 534, 544
visceral efferent postganglionic, 544
visceral efferent preganglionic, 544
Neurophysins, 596
Neurosyphilis, 742
Neurotransmitters, 231, 453, 541–542
Neutrophils, 334, 337E
Nevus, 110
Nexus, 66
Nicotinic receptors, 542
Nipples, 316E, 734
Nitrogen bases, 49
Nitroglycerin, 361
Nociceptors, 553
Nodes of Ranvier, 448
Nodules, 110
Non-A, non-B (NANB) hepatitis, 682
Nonaxial joints, 193
Nondisplaced fracture, 125
Non-insulin-dependent diabetes, 607
Nonpigmented granular dendrocytes, 93
Nonstriated muscle tissue, 224
Norepinephrine (NE), 542, 605, 606E
Normal curves of the vertebrae column, 148
Normotensive, 411
Nose, 617, 618–620
surface anatomy of, 314E
Nosebleed, 619–620
Nostrils, 618
Notochord, 123
Nuclear magnetic resonance (NMR), 23
Nuclear medicine, 25
Nuclear membrane, 38

Nucleosomes, 40
Nucleotides, 49
Nucleus
    function of, 48E
    of organelles, 38–40
Nucleus pulposus, 148
Nutrient canal, 122
Nutrient foramen, 122
Nutritional anemia, 338
Nystagmus, 585

Obstetrics, 761
Obstructive jaundice, 665
Occlusion, 411
Occult blood, 677
Oil glands, see Sebaceous glands
Olfactory bulb, 512, 560
Olfactory foramina, 143
Olfactory gland, 560
Olfactory region, 619
Olfactory sense, 559–560
    olfactory pathway, 560
    receptors, 559–560
Olfactory tract, 512, 560
Oligodendrocytes, 445E
Oncogenes, 57
Oogenesis, 722–726
Oogonia, 724
Oophorectomy, 737
Open fracture, 125, 126
Open reduction, 125
Ophthalmologist, 561
Ophthalmology, 561
Optic chiasma, 512, 568
Optic disc, 566
Optician, 561
Optic tracts, 512, 568
Optometrist, 561
Oral cavity, see Mouth
Oral contraceptive (OC), 738, 739E
Ora serrata, 566
Orbits, 135
Organ, 2, 93
Organelles, 32, 38–47
    centrioles, 47
    dentrosome, 47
    cilia, 47, 65
    cytoskeleton, 46–47
    definition, 38
    endoplasmic reticulum, 40
    flagella, 47
    functions of, 48E
    Golgi complex, 40–42
    lysosomes, 44–45
    mitochondria, 42–44
    nucleus, 38–40
    peroxisomes, 45–46
    ribosomes, 40
Organism, 4
Organismic level of structural organization, 4
Organ level of structural organization, 2
Orgasm
    female, 737
    male, 736
Origin, 246
Oropharynx, 620
Orthopedics, 114

Orthopnea, 640
Osmosis, 35
Osseous tissue, 82E, 84, 113–130
    disorders of, 123–127
    functions of, 114
    histology, 114–118
    key medical terms, 127
    ossification, 118–119
Ossification, 118–119
    endochondral, 118–119, 123
    intramembranous, 118, 123
Osteitis, 127
Osteitis fibrosa cystica, 601
Osteoarthritis, 127, 218
Osteoblastoma, 127
Osteoblasts, 114, 123
Osteochondroma, 127
Osteoclasts, 114
Osteocytes, 114, 116
Osteogenesis, 118
Osteogenic layer, 114
Osteogenic sarcoma, 56, 121
Osteoma, 127
Osteomalacia, 125
Osteomyelitis, 125
Osteon, 84, 116
Osteoporosis, 123
Osteoprogenitor cells, 114
Osteosarcoma, 127
Otalgia, 585
Otic ganglion, 540
Otitis media, 584
Otolithic membrane, 577
Otorhinolaryngology, 620
Otosclerosis, 585
Ovarian cycle, 728
Ovarian cysts, 745
Ovaries, 608, 609E, 722–726
Ovulation, 726, 731
Oxytocin (OT), 596, 597E

Pacemaker
    artificial, 356
    SA node, 354
Pachymeningitis, 460
Pacinian corpuscles, 96
Paget's disease, 125
Pain sensations, 553–555
Palatine tonsils, 421
Palpation, 312
Palpebral fissure, 562
Palpitation, 361
Pancarditis, 366
Pancreas, 606–607, 663
    development of, 610
    disorders of, 606–607, 608E
    effect of sympathetic and parasympathetic stimulation on, 543E
    hormones of, 606, 608E
Pancreatic amylase, 663
Pancreatic duct, 663
Pancreatic islets, 606, 663
Pancreatic juice, 663
Pancreatitis, 683
Pannus, 218
Papanicolaou (Pap) test, 727
Papillae, 560
Papilla of the hair, 103

Papillary plexus, 98
Papillary region, 96
Papule, 110
Parallel after-discharge circuit, 456
Parallel pattern of fasciculi, 248
Paralysis, 241, 529
Paranasal sinus, 133E, 144–145
Paraplegia, 480
Parasagittal plane, 12
Parasympathetic nervous system, 535
    structural features of, 541E
Parathyroid glands, 591, 600–601, 602E
    disorders of, 600–601, 602E
    hormone of, 600, 602E
Parathyroid hormone (PTH), 600, 602E
Paravertebral ganglia, 537
Parent cell, 48
Parietal layer, 226
Parietal peritoneum, 647
Parkinson's disease (PD), 524–525
Parotid glands, 651
Paroxysmal tachycardia, 366
Partial fracture, 125
Parturition, 765–767
Passive processes, 35–36
Patella, 175–176, 323E
Patellar ligament, 325
Patent ductus arteriosus, 364
Pathological anatomy, 2
Pathologic fracture, 125
Pectinate muscles, 347
Pectoral girdle, 165–168
    muscles that move, 275E
Pedicles, 151
Pelvic cavity, 16
Pelvic diaphragm, 272E
Pelvic girdle, 170–174
Pelvic inflammatory disease (PID), 746
Pelvic outlet, 174
Pelvimetry, 174
Pelvic
    arteries of, 389E, 390, 391
    comparison of female and male, 183
    lymph nodes of, 432E, 433E
    muscles of pelvic floor, 272E
    surface anatomy of, 312, 315, 317E
    veins of, 399E, 400, 401
Penicillin, 492
Penis, 719–722
Pennate, 248
Pepsin, 662
Peptic ulcers, 680
Peptidases, 672
Perception, 550
Percussion, 411
Percutaneous balloon valvuloplasty, 362
Percutaneous lumbar discectomy with aspiration probe, 159–160
Percutaneous transluminal coronary angioplasty (PTCA), 361, 362
Percutaneous ultrasonic lithotripsy (PUL), 703
Perforating canals, 116
Pericardial cavity, 15
Pericardial fluid, 345
Pericarditis, 345
Pericardium, 85, 345
Perichondrium, 84, 118
Periderm, 106
Perimysium, 225

Perineum, 733
    muscles of, 273E
Perineurium, 469
Perinuclear cisterna, 38
Periodontal disease, 680
Periodontal ligament, 654
Periosteum, 114–115, 118
Peripheral nerve damage, 480
Peripheral nervous system (PNS), 444
Peripheral proteins, 33
Peristalsis, 657–658, 673, 677
Peritoneum, 85, 646–648
Peritonitis, 680
Perivascular space, 492
Permanent cells, 65
Permanent dentition, 657
Pernicious anemia, 338
Peroxisomes, 45–46
    function of, 48E
Perspiration, 104
Phagocytic vesicle, 37
Phagocytosis, 37
Phagocytotic, 334
Phalanges
    of the finger, 170
    of the foot, 180
Phantom pain, 553–554
Pharmacology, 58
Pharyngeal tonsil, 421
Pharynx, 617, 620, 657
    muscles of, 260E
Phenylbutazone, 492
Pheochromocytoma, 410, 605
Phlebitis, 411
Phlebotomy, 341
Phosphate group, 49
Phospholipid bilayer, 32
Photophobia, 585
Physiology, 2
Pia mater, 470, 488
Piles, see Hemorrhoids
Pimples, 104
Pineal gland, 608–609
Pinealocytes, 608
Pinkeye, see Conjunctivitis
Pinocytosis, 37–38
Pituitary dwarfism, 594
Pituitary gland, 591, 593–596, 597E
    adenohypophysis, 593–594
    development of, 610
    disorders of, 594, 597E
    hormone of, 593, 597E
    neurohypophysis, 596
Pivot joint, 193, 196E
Placenta, 406–407, 409, 759–761
Planes, 12–13
Plane suture, 186
Plantar aponeurosis, 305E
Plantarflexion, 192, 193E
Plasma, 84, 337–338
    chemical composition and description of substances in, 338E
Plasma cells, 83
Plasma membrane, 32–38
    chemistry and structure of, 32–33
    cytoplasm of, 38
    functions of, 33–35, 48E
    movement of materials across, 35–38

Plasma membrane proteins (PMPs), 32–33
Plasmapheresis, 240, 337–338
Plasma proteins, 337
Platelet concentration, 341
Platelet-derived growth factor (PDGF), 360
Platelets, *see* Thrombocytes
Pleural cavity, 16, 628
Pleural membrane, 628
Pleurisy, 628
Plexus, 470–474
    brachial, 471–474, 474E
    cervical, 471
    lumbar, 474
    pulmonary, 540–541
    renal, 696
    sacral, 474
Pneumocystis carinii pneumonia (PCP), 435, 436
Pneumonectomy, 640
Pneumonia, 637
Pneumotaxic area, 635
Pneumothorax, 628
Poliomyelitis (polio), 524
Polycystic disease, 704
Polycythemia, 339
Polyp, 110
Polyuria, 705
Pons, 488E, 496–497, 511E
Popliteal fossa, 323E, 325
Portal triad, 669
Positron emission tomography (PET), 19, 25
Positron emission tomography (PET) scans, 25, 526
Positrons, 25
Postcentral gyrus, 504
Posterior column pathway, 557
Posterior spinocerebellar tract, 466E
Postganglionic fiber, 538
Postovulatory phase of the menstrual cycle, 731
Postural hypotension, 411
Pott's disease, 127
Pott's fracture, 125, 126, 179
Preeclampsia, 767
Prefrontal lobotomy, 554
Preganglionic fiber, 538
Pregnancy
    definition, 752
    diagnosis of, 733
    development during, 752–756
    ectopic, 727
Premenstrual syndrome (PMS), 744
Prenatal diagnostic techniques, 765
Preovulatory phase of the menstrual cycle, 731
Prepuce, 721
Presbyopia, 585
Pressure, 454
Pressure sensations, 552
Prevertebral ganglion, 537
Primary follicle, 725, 730
Primary hypertension, 410
Primary lysosome, 45
Primary ossification center, 118
Primary somesthetic area, 506–507
Primary tumor, 55
Prime mover muscles, 248
Primitive gut, 678
Primitive heart tube, 360
Proctodeum, 678
Proctology, 674

Proerythroblasts, 332, 333
Progeny, 58
Progesterone (PROG), 608, 609E, 610, 729
Projection, 550
Prolactin (PRL), 594, 597E
Pronation, 192, 193E
Pronator muscle, 249
Pronephros, 702
Prone position, 12
Prophase stage, 49–50
    activity of, 52E
Proprioceptive sensations, 555–556
Proprioceptors, 551
Prosencephalon, 487, 488E
Prostatectomy, 743
Prostate gland, 717
    enlarged, 743
Prostatic urethra, 716
Prostatitis
    acute, 743
    chronic, 743
Proto-oncogenes, 57
Protoplasmic astrocytes, 445E
Protraction, 192, 193E
Pruritus, 110
Pseudopodia, 37
Pseudostratified epithelium, 65
Psoriasis, 106
Pterygoid processes, 141–142
Pterygopalantine ganglion, 540
Ptosis, 585, 688
Pudendal nerve block, 767
Pudendum, *see* Vulva
Puerperal fever, 767
Pulled groin, 295
''Pulled hamstrings,'' 298
Pulmonary circulation, 405
Pulmonary edema, 639
Pulmonary embolism (PE), 639
Pulmonary plexus, 540–541
Pulmonary trunk, 405
Pulmonary ventilation, 617
Pulsing electromagnetic fields (PEMFs), 127
Punch biopsy, 727
Punctate distribution, 551
Pupil, 313E, 566
Purkinje cell, 451
Purkinje fibers, 354
Pustule, 110
P wave, 354
Pyelitis, 704
Pyelonephritis, 704, 705
Pyloric sphincter, 659
Pyloric stenosis, 659
Pylorospasm, 659
Pylorus, 659
Pyorrhea, 680
Pyramidal pathways, 580
Pyramids, 493

Quadriplegia, 480
QRS wave, 356

Rabies virus, 448
Radial keratotomy (RK), 563
Radial nerve damage, 472
Radical mastectomy, 745

Radiocarpal joint, 208–209, 208E
Radiographic anatomy, 2, 19
Radioisotopic scanning, 25
Radiology, 19
Rales, 640
Rami, 145, 469–470
Raphe, 709
Rapidly adapting receptors, 550
Raynaud's disease, 411, 535
Raynaud's phenomenon, 411
Receptor-mediated endocytosis, 38
Receptors, 468
    of the autonomic nervous system, 542, 544
    of the endocrine system, 591–592
    as sense organs, 550–551, 555–556, 559–560
Reciprocal transfusion, 341
Recruitment, 231
Rectum, 674
Red blood cells, see Erythrocytes
Red muscle fibers, 234
Red nucleus, 497, 504
Referred pain, 553
Reflex arc, 468
Reflexes, 469
Regeneration, 454–455
Regional anatomy, 2
Regulating hormones, 593
Relaxin, 608, 609E, 610, 729
Releasing hormones, 593
Remodeling, 121–122
    of the calli, 127
Renal calculi, 702–703
Renal capsule, 688
Renal failure, 704–705
    acute, 704
    chronic, 704
Renal fascia, 688
Renal pelvis, 692
Renal plexus, 696
Renal pyramids, 692
Renal sinus, 688
Rennin, 662
Reproduction, 4
    definition, 709
Reproductive cell division, 49, 52–54
    meiosis, 49, 53–54
Reproductive system, 10E, 708–750
    aging and, 739
    birth control, 737–738, 739E
    development of, 741
    disorders of, 741–746
    female organs, 722–737
    gamete production, 709, 712–714
    key medical terms, 746
    male organs, 709–722
    sexual intercourse, 735, 736–737
Resistance, 247
Respiration, 4, 617
    nervous control of, 634–635
Respirator, 640
Respiratory bronchioles, 628–629
Respiratory center, 634
Respiratory distress syndrome (RDS) of the newborn, 637–638
Respiratory failure, 638
Respiratory system, 8E, 616–643
    aging and, 635
    development of, 635–636

disorders of, 636–640
    key medical terms, 640
    nervous control of, 634–635
    organs of, 617, 618–633
Rest-repose system, 542
Retention, 701
Rete testis, 715
Reticular connective tissue, 80E, 84
Reticular fibers, 48, 76
Reticular lamina, 64
Reticular region, 96
Reticulocyte count, 333–334
Retina (nervous tunic), 566
    detachment of, 567
Retinoblastoma, 585
Retraction, 192, 193E
Retrograde degeneration, 480, 481
Reverberating circuit, 456
Reye's syndrome, 528
Rheumatic fever, 354
Rheumatism, 218
Rheumatoid arthritis (RA), 218
Rheumatology, 219
Rhinitis, 640
Rhizotomy, 554
Rhombencephalon, 487, 488E
Ribonuclease, 663, 672
Ribosomes, 40
    function of, 48E
Rib fractures, 158
Ribs, 158
Rickets, 124
Right iliac region, 16E
Right lumbar region, 16E
Right lymphatic duct, 425–427
Rigidity, 524
Rigor mortis, 233
Rods, 566
Roentgenogram, 19
Root canal therapy, 656
Root canals, 654
Root hair plexuses, 103, 552
Root of the hair, 103
Rotation, 189, 193E
Rotator cuff, 278E
Rotator muscle, 249
Rubber band ligation, 674
Rubrospinal tract, 467E
Runner's knee, 213
Rupture of the spleen, 424

Sacral parasympathetic outflow, 540
Sacral plexus, 472
    nerves of, 478E
Sacral vertebrae, 148
Sacrum, 148, 156–157
Saddle joint, 193, 196E
Sagittal suture, 135
Saliva, 652
Salivary amylase, 651
Salivary glands, 543E, 651–652
Salpingectomy, 746
SA node, see Sinoatrial node
Sarcolemma, 85, 226–228
Sarcoma, 56
    osteogenic, 56, 121
Sarcomeres, 229

Sarcoplasm, 85, 228
Sarcoplasmic reticulum, 228
Satiety center, 501
Scala vestibuli, 574
Scapulae, 166–168
Scapular muscles, 278E
Schwann cells, 448
Sciatica, 481–482
Sclera ("white of the eye"), 563
Scleral venous sinus, 563
Scleroses, 525
Sclerotome, 239
Scoliosis, 160
Scotoma, 568
Scrotum, 709
Sebaceous ciliary glands, 562
Sebaceous glands, 104
    development of, 106
Sebum, 104
Secondary follicle, 730
Secondary hypertension, 410
Secondary lysosome, 45
Secondary ossification center, 119
Secondary somesthetic area, 506–507
Secondary spermatocytes, 714
Secondary tumor, 56
Second-class levers, 247
Second-degree burns, 108
Secretin, 610
Secretion, 4, 67
Secretory granules, 40
Section, 12–13
Segmental muscles, 290E
Segmentation, 672
Selective permeability, 33–34
Semen (seminal fluid), 718–719
Semen analysis, 719
Semicircular canals, 573–574
Semilunar ganglion, 514
Semilunar valves, 351
Seminalplasmin, 718
Seminal vesicles, 716–717
Senescence, 58
Senile macular degeneration (SMD), 566
Sensations, 550–551
    characteristics of, 550–551
    complexity of, 551
    cutaneous, 551–555
    definition, 550
    gustatory, 560–561
    levels of, 556
    location of, 551
    olfactory, 559–560
    pain, 553–555
    proprioceptive, 555–556
    receptors, 555–556
    simplicity of, 551
    stimulus detected, 551
    tactile, 551–553
    thermoreceptive, 553
Sense organs, 550
Sensory areas of the cerebral cortex, 506–508
Sensory pathways, 556–558
    cerebellar tract, 558
    posterior column pathway, 557
    somatosensory cortex, 556–557
    spinothalamic pathway, 557–558

Separation of the shoulder, 204
Septal defect, 364
Septicemia, 341
Serosa, 646
Serous membranes, 85
Serous pericardium, 345
Serrated suture, 186
Serum hepatitis, see Hepatitis B
Severe combined immunodeficiency disease (SCID), 336, 438–439
Severe dizziness, 511
Sex chromosomes, 53, 713
Sex hormones, 603
Sex organs, effect of sympathetic and parasympathetic stimulation, on, 544E
Sexual intercourse, 735, 736–737
Sexually transmitted diseases (STDs), 741–743
Shaft of hair, 100
Shingles, 482
Shinsplints, 177
Shoulder, 204–205, 204E
    bones of, 165, 166, 167
    muscles that move, 275E
    surface anatomy of, 312, 315, 318E
Shoulder blade, see Scapulae
Shunt, 411
Siamese twins, 753
Sickle-cell anemia, 339
Simple epithelium, 67, 68E, 69E
Simple gland, 75
Simple goiter, 600
Simple-series circuits, 455
Sinoatrial node (SA node), 354
Sinusitis, 145
Sinus node dysfunction, 354
Sinusoids, 372–373
Skeletal materials, 132
Skeletal muscles, 6E
    abductor digiti minimi, 288E, 305E
    abductor hallucis, 305E
    abductor pollicis brevis, 288E
    abductor pollicis longus, 284E
    adductor brevis, 294E, 297E
    adductor hallucis, 305E
    adductor longus, 294E, 297E
    adductor magnus, 295E, 297E
    adductor pollicis, 288E
    anconeus, 281E
    anterior scalene, 291E
    arytenoid, 264E
    biceps brachii, 281E
    biceps femoris, 298E
    brachialis, 281E
    brahcioradialis, 281E
    buccinator, 253E
    bulbocavernosus, 273E
    cocygeus, 272E
    coracobrachialis, 278E
    corrugator supercilli, 253E
    cricothyroid, 264E
    deep transverse perineus, 273E
    deltoid, 278E
    depressor labil inferioris, 253E
    development of, 238–239
    diagastric, 262E
    diaphragm, 270E
    dorsal interossei, 288E, 305E

effect of sympathetic and parasympathetic stimulation on, 543E
epicranius, 252E
external anal sphincter, 273E
extensor carpi radialis brevis, 284E
extensor carpi radialis longus, 284E
extensor carpi ulnarius, 284E
extensor digiti minimi, 284E
extensor digitorum, 284E
extensor digitorum brevis, 305E
extensor digitorum longus, 301E
extensor hallucis longus, 301E
extensor indicis, 284E
extensor pollicis brevis, 284E
extensor pollicis longus, 284E
external intercostal, 270E
external oblique, 267E
facial expression muscles, 252E, 253E
fasciculi arrangement, 248
flexor carpi radialis, 283E
flexor carpi ulnaris, 283E
flexor digiti mimini brevis, 288E, 305E
flexor digitorum brevis, 305E
flexor digitorum longus, 301E
flexor digitorum profundus, 283E
flexor digitorum superficialis, 283E
flexor hallucis brevis, 305E
flexor hallucis longus, 301E
flexor pollicis brevis, 288E
flexor pollicis longus, 283E
frontalis, 252E
gastrochemius, 301E
genioglossus, 257E
geniohyoid, 262E
gluteus maximus, 294E
gluteus medius, 294E
gluteus minimus, 294E
gracillis, 297E
group actions, 248
hypoglossus, 257E
iliacus, 294E
iliococcygeus, 272E
iliocostalis lumborum, 290E
iliocostalis thoracis, 290E
inferior constrictor, 260E, 264E
inferior gemelius, 294E
inferior oblique, 255E
inferior rectus, 255E
infraspinatus, 278E
internal intercostals, 270E
internal oblique, 267E
interspinales, 291E
intertransversarii, 291E
ischiocavernosus, 273E
latissimus dorsi, 278E
lateral cricoarytenoid, 264E
lateral pterygoid, 254E
lateral rectur, 255E
levator ani, 272E
levator labil superioris, 252E
levator palpebrae superioris, 253E
levator scapulae, 275E
levator veil palatini, 258E
lever systems and leverage, 246–247
longissimus capitis, 266E, 290E
longissimus cervicis, 290E
longissimus thoracis, 290E
lumbricales, 288E, 305E

masseter, 254E
medial pterygoid, 254E
medial rectus, 255E
mentalis, 253E
middle constrictor, 260E, 264E
middle scalene, 291E
multifidus, 291E
musculus uvulae, 258E
mylohyoid, 262E
naming, 248–249
obturator internus, 294E
obturator externus, 294E
occipitalis, 252E
omohyoid, 264E
opponens digiti minimi, 288E
opponens pollicis, 288E
orbicularis oculi, 253E
orbicularis oris, 252E
origin and insertion of, 246
palatoglossus, 257E, 258E
palatopharyngeus, 258E, 260E, 264E
palmar interossei, 288E
palmaris brevis, 288E
palmaris longus, 283E
pectineus, 295E
pectoralis major, 278E
pectoralis minor, 275E
peroneus brevis, 301E
peroneus longus, 301E
peroneus tertius, 301E
piriformis, 294E
plantar interossei, 305E
plantaris, 301E
platysma, 253E
popliteus, 301E
posterior cricoarytenoid, 264E
posterior scalene, 291E
principal actions of, 249E
pronator quadratus, 283E
pronator teres, 283E
psoas major, 294E
pubococcygeus, 272E
quadratus femoris, 294E
quadratus lumborum, 267E
quadratus plantae, 305E
quadriceps femoris, 297E
rectus abdominis, 267E
rectus femoris, 297E
rhomboideus major, 275E
rhomboideus minor, 275E
risorius, 253E
rotatores, 291E
salpingopharyngeus, 260E
sartorius, 297E
semimembranosus, 298E
semispinalis capitis, 266E, 291E
semispinalis cervicis, 291E
semispinalis thoracis, 291E
semitendinosus, 298E
serratus anterior, 275E
soleus, 301E
spinalis capitis, 291E
spinalis cervicis, 290E
spinalis thoracis, 290E
splenius capitis, 266E, 290E
splenius cervicis, 290E
sternocleidomastoid, 266E

Skeletal muscles (*Continued*)
  sternohyoid, 264E
  sternothyroid, 264E
  styloglossus, 257E
  stylohyoid, 262E
  stylopharyngeus, 260E, 264E
  subclavius, 275E
  subscapularis, 278E
  superficial transverse perineus, 273E
  superior constrictor, 260E
  superior gemelius, 294E
  superior oblique, 255E
  superior rectus, 255E
  supinator, 284E
  supraspinatus, 278E
  temporalis, 254E
  tensor fasciae latae, 294E
  tensor veil palatini, 258E
  teres major, 278E
  teres minor, 278E
  thyroarytenoid, 264E
  thyrohyoid, 264E
  tibialis anterior, 301E
  tibialis posterior, 302E
  transversus abdominis, 267E
  trapezius, 275E
  triceps brachii, 281E
  urethral sphincter, 274E
  vastus intermedius, 297E
  vastus lateralis, 297E
  vastus medialis, 297E
  zygomaticus major, 252E
Skeletal muscle tissue, 85, 86E, 224–230, 238E
  blood supply of, 226
  connective tissue components, 224–226
  contraction of, 233E
  definition, 224
  fascia, 224–225
  histology, 226–229
  nerve supply of, 226
  relaxation of, 233E
  types of, 234–235
Skeletal system, 5E
  aging and, 123
  appendicular skeleton, 132, 164–184
  axial skeleton, 131–163
  definition, 114
  development of, 123
  divisions of, 132–133, 134, 135E
  functions of, 114
  histology, 114–118
  ossification, 118–119
Skeleton of the heart, 351–352
Skin, 93–105
  blood supply of, 98
  color of, 96, 97
  dermis, 93, 96
  effect of sympathetic and parasympathetic stimulation on, 543E
  epidermal ridges and grooves, 97
  epidermis, 93–95
  functions of, 93
  products eliminated from, 688E
  structure, 93
  temperature regulation of, 98
Skin cancer, 97, 107
  risk factors for, 107
Skin grafts (SGs), 99

Skin peel, 98
Skin wound healing, 98–99
  deep wounds, 99
  epidermal wounds, 98–99
Skull, 133–146
  cranial bones, 135–144, 146
  cranial fossae, 143
  facial bones, 143–146
  fontanels, 135
  foramina, 146, 147E
  sutures, 135
Sliding-filament theory, 230
Slipped discs, 159–160
Slowly adapting receptors, 550–551
Slow-twitch red muscle fibers, 234
Small intestine, 670–673
  blood supply of, 673
  nerve supply of, 673
Smegma, 746
Smoke inhalation injury, 639–640
Smooth muscle tissue, 86–87, 86E, 224, 237–238, 238E
Snoring, 620
Soft palate, 649
  muscles of, 258E
Solar plexus, 540
Solenoid, 40
Somatic cell division, 48, 49–52
  cytokinesis, 49, 52
  mitosis, 48, 49–52
Somatic nervous system (SNS), 444
  autonomic nervous system compared with, 534, 535E
Somatic pain, 553
Somatic reflexes, 469
Somatosensory cortex, 556–557
Somatostatin, 606
Somatotroph cells, 594
Somites, 238–239
  mesoderms of, 238–239
Songram, 24
Sound valves, 575–576
Spasm, 240
Spastic, 529
Spasticity, 581
Special senses, 558–578
  auditory sensations, 568–578
  equilibrium, 576–578
  gustatory sensations, 560–561
  olfactory sensations, 559–560
  visual sensations, 561–568
S period, 49
  activity of, 52E
Spermatic cord, 716
Spermatids, 714
Spermatogenesis, 709, 712–714
Spermatogonia, 709, 713–714
Spermatozoa, 714
Spermatozoon, 709
Sperm cell, 709
Spermiogenesis, 714
Sphincter
  external, 701
  of the hepatopancreatic ampulla, 769
  ileocecal, 670, 674
  internal, 701
  lower esophageal, 658
  precapillary, 372

pyloric, 659
    upper esophageal, 657
Sphincter muscle, 249
Spider burst veins, 374
Spina bifida, 160
Spinal cord, 460–469
    ascending tract, 466E
    coverings, 460–461
    descending tract, 467E
    functions of, 466–469
    general features, 460–465
    injury to, 480
    protection, 460–461
    structure, 465–466
Spinal meninges, 460
Spinal nerves, 460–479
    branches, 469–470
    coverings, 469
    dermatomes, 477–479
    intercostal, 474–477
    names, 469
    plexus, 470–472
Spinal puncture (tap), 461
Spinal reflexes, 469
Spinal segment, 465
Spinal shock, 480
Spine, see Vertebral column
Spinothalamic pathway, 557–558
Spiral fracture, 125
Spiral ganglion, 516
Spleen, 416, 421–424
    effect or sympathetic and parasympathetic stimulation on, 543E
    rupture of, 424
Splenectomy, 424
Splenius muscles, 290E
Splenomegaly, 440
Split-brain concept, 508–509
Sprain, 219
Spurs, 218
Squamosal suture, 135
Squamous cell carcinoma (SCC), 107
Squamous cells, 65
Squamous pulmonary epithelial cells, 629
Squamous suture, 186
Stable cells, 65
Staging, 56
Stapedius muscle, 572
Static equilibrium, 576, 577–578
Stereocilia, 715
Sterility, see Infertility
Sterilization, 737, 739E
Sternal puncture, 158
Sternocleidomastoid muscles, 312
Sternum, 157–158, 316E
Steroid hormones, 45
Steroid ointments, 106
Steroids, 229–230
Stimulus, 451, 550
Stokes-Adams syndrome, 366
Stomach, 659–663
    bloody supply of, 663
    effect of sympathetic and parasympathetic stimulation on, 543E
    nerve supply of, 663
Stomach gastrin, 610
Strabismus, 585
Strain, 219
Stratified epithelium, 67–74, 70E, 71E

Stratum basale, 95
Stratum corneum, 95
Stratum granulosum, 95
Stratum lucidum, 95
Stratum spinosum, 95
Stress fracture, 125
Stress-relaxation, 238
Striae, 96
Striated muscle tissue, 85, 224
Stricture, 705
Stroke, 521
Stroma, 76
Structural organization, levels of, 2–4
Stupor, 529
Stylomastoid foramen, 140
Subcutaneous, 110
Subcutaneous (SC) layer, 83, 93
Sublaxation, 219
Submandibular ducts, 651–652
Submucosa, 646
Subneural clefts, 231
Subserous fascia, 225
Substantial nigra, 504
Subthalamic nucleus, 504
Subthreshold stimulus, 233
Sucrase, 672
Suction lipectomy, 83
Sudden infant death syndrome (SIDS), 638–639
Sudoriferous glands, 104
    development of, 106
Sulfonamides, 492
Sunburn, 107
Suntanning salons, 107
Superficial fascia, 93, 224
Superior vena cava (SVC), 347, 376
Supinator muscle, 249
Supination, 192, 193E
Supine position, 12
Suprahyoid muscles, 263
Suprarenal gland, see Adrenal gland
Surface anatomy, 2, 311–326
    of the abdomen, 312, 315, 317E
    of the ankle, 312, 324E, 325
    of the arm, 312, 315, 318E
    of the buttocks, 312, 317, 322E
    of the chest, 312, 316E
    definition, 312
    of the ear, 314E
    of the elbow, 312, 315, 318E
    of the eye, 313E
    of the face, 312, 313E, 314E
    of the foot, 312, 324E, 325
    of the forearm, 312, 315, 319E
    of the hand, 312, 317, 321E
    of the head, 312, 313E, 314E
    of the knee, 312, 317–325, 323E
    of the leg, 312, 324E, 325
    of the lips, 314E
    of the lower extremity, 312, 317–325, 322E, 323E, 324E
    of the neck, 312, 315E
    of the nose, 314E
    of the pelvis, 312, 315, 317E
    of the shoulder, 312, 315, 318E
    of the thigh, 312, 317, 322E
    of the thorax, 314
    of the trunk, 312–315, 316E, 317E

Surface anatomy (*Continued*)
    of the upper extremity, 312, 315–316, 318E, 319E, 320E, 321E
    of the wrist, 312, 315–317, 320E
Surfactant, 629
Surgical access to the lungs, 158
Surgical incision, 225
Sutures, 135, 186
Swallowing, 657
Sweat, 104
Sweat gland, 543E
Sympathectomy, 554
Sympathetic nervous system, 535
    structural features of, 541E
Sympathetic trunk ganglia, 535–537
Sympathomimetic hormones, 605
Symphysis, 187, 196E
Symphysis pubis, 174
Synapse, 452
Synapsis, 53
Synaptic cleft, 231, 452
Synaptic conduction, 454
Synaptic end-bulbs, 231, 452
Synaptic gutter, 231
Synaptic vesicles, 448, 453
Synarthroses, 186, 196
Synchondrosis, 187, 196E
Syncope, 411
Syncytiotrophoblast, 754
Syndesmosis, 186, 196E
Synergists, 248
Synostoses, 186
Synovial fluid, 85, 188
Synovial joints, 186, 187–196, 196E
    ball-and-socket, 196, 196E
    ellipsoidal, 193, 196E
    gliding, 193, 196E
    hinge, 193, 196E
    movements of, 189–192, 193E
    pivot, 193, 196E
    saddle, 193, 196E
    structure, 187–189
    types, 192–196
Synovial membranes, 85, 188
Synovitis, 219
Syphilis, 742
System, 2–4, 93
Systemic anatomy, 2
Systemic circulation of blood, 375–376, 377E, 380E, 381E, 385E, 389E, 392E, 394E, 396E, 399E, 402E
Systemic disease, 58
Systemic lupus erythematosus (SLE), 106
System level of structural organization, 2
Systole, 354

T_3 (triiodothyronine), 597, 601E
T_4 (thyroxine), 597, 601E
Tachypnea, 640
Tactile discs, 95
Tactile sensations, 551–553
Taeniae coli, 674
Talocrural joint, 216–217, 216E
Target cells, 591
Tarsal glands, 562
Tarsals, 179–180
Taste pore, 560
Tay-Sachs disease, 526

T cells, 335, 609–610
Tectorial membrane, 575
Tectospinal tract, 467E
Tectum, 497
Teeth, 652–657
    blood supply of, 657
    nerve supply of, 657
Telencephalon, 487, 488E
Telophase, 51
    activity of, 52E
Temperature regulation of the skin, 98
Temporomandibular joint (TMJ), 145–146, 197E, 198
TMJ syndrome, 145–146
Tendinitis, 226
Tendon organs, 556
Tendons, 83, 236
    reconstructing, 83
Tendon sheaths, 226
Tennis elbow, 206
Tenosynovitis, 226
Tension lines, 96
Tensor muscle, 249
Terminal cisterns, 228
Terminal ganglion, 537
Terminal web, 65
Tertiary bronchus, 626
Testes (testicles), 608, 609E, 709–714
    ducts of, 714–716
Testicular cancer, 743
Testosterone, 608, 609E
Tetanus bacterium, 448
Tetany, 601
Tetracycline, 492
Tetralogy of Fallot, 364
Thalamus, 448E, 497, 511E
    anterior nucleus of, 506
Thalassemia, 339
Therapeutic plasma exchange (TPE), 337–338
Thermoreceptive sensations, 553
Thermoreceptors, 551
Thigh
    muscles that move, 294E, 295E
    surface anatomy of, 312, 317, 322E
Thiopental sodium, 492
Third-class levers, 247
Third-degree burns, 108
Thirst center, 501
Thoracic aorta, 347, 376, 377E, 385E, 387, 388
Thoracic cavity, 16
Thoracic duct, 524
Thoracic vertebrae, 148, 151
Thoracolumbar outflow, 535
Thorax, 157–158
    lymph nodes of, 434E
    surface anatomy of, 314
    veins of, 399E, 400, 401
Threshold stimulus, 233
Throat, *see* Pharynx
Thrombectomy, 411
Thrombocytes (platelets), 85, 329, 330, 336, 337E
Thrombocytopenia, 341
Thrombophlebitis, 411
Thymic corpuscles, 424
Thymic factor (TF), 610
Thymic humoral factor (THF), 610
Thymine nucleotide, 49
Thymopoietin, 610

Thymosin, 610
Thymus gland, 416, 424, 609–610
Thyroglobulin (TGB), 599
Thyroid cartilage, *see* Adam's apple
Thyroid gland, 597–600
   development of, 610
   disorders of, 599–600, 601E
   hormone of, 597, 601E
Thyroid-stimulating hormone (THS), 594, 597E
Thyroid storm, 611
Thyrotroph cells, 594
Thyroxine (T$_4$), 597, 601E
Tibia, 176–179
   muscles that act on, 297E, 298E
Tibia stress syndrome, 177
Tibiofemoral joint, 212E, 213–215
Tic, 240
Tic douloureux, 527
Tinnitus, 585
Tissue, 63–91
   connective, 64, 75–85
   definition, 2, 64
   epithelial, 64–75, 645
   muscular, 64, 85–87
   nervous, 64, 87, 443–458
   osseous, 82E, 84, 113–130
   types, 64
   vascular, 82E, 84–85
Tissue level of structural organization, 2
Tissue rejection, 439
Toes, muscles that move, 301E, 302E
Tongue, 649–651
   muscles that move, 257E
Tonofilaments, 66
Tonsils, 416, 421
Topical, 110
Torn cartilage, 188
Torpor, 529
Torticollis, 241
Touch sensations, 551–552
Toxic shock syndrome (TSS), 744–745
Trabeculae, 118
Trabeculae carnea, 348
Trachea, 617, 624–626
Trachealis muscle, 624
Tracheostomy, 626
Trachoma, 584
Tracts, 451
   definition, 460
Transcendental meditation (TM), 546
Transducer, 24
Transfusion, 341
   direct, 340
   exchange, 340
   indirect, 340–341
   reciprocal, 341
Transient ischemic attack (TIA), 521–522
Transitional cells, 65
Transplantation (transplants), 439
   bone marrow, 336
   corneal, 563
   heart, 359
   knee, 213
Transvaginal oocyte retrieval, 756
Transverse arch of the foot, 180
Transverse fracture, 125
Transverse tubules, 228

Transversospinalis muscles, 290E
Trapezius muscle, 312
Tremor, 505
Trichinosis, 241
Trichomoniasis, 743
Trigeminal neuralgia, 527
Trigone, 700
Triiodothyronine (T$_3$), 597, 601E
Trochanter, 133E
Trophoblast, 753
Tropic hormones, 594
Tropomyosin, 229
Tropomyosin-troponin complex, 229
Troponin, 229
Trousseau sign, 601
True capillaries, 372
True labor, 766–767
True ribs, 158
Trunk, surface anatomy of, 312–315, 316E, 317E
Trypsin, 663
Tubal ligation, 737
Tubercle, 133E
Tuberculosis (TB), 637
Tubular gland, 74
Tubulin, 46
Tubuloacinar gland, 75
Tumor, 55
   brain, 522–524
   of the GI tract, 681
   grading, 56
   of the male reproductive system, 743
   malignant, 56
   primary, 55
   secondary, 56
Tunica externa, 370
Tunica interna, 370
Tunica media, 370
T wave, 354
Tympanic membrane, *see* Ear drum
Type I diabetes, 606–607
Type I cutaneout mechanoreceptors, 552
Type II diabetes, 607
Type 2 herpes simplex virus, 57
Tyrosinase, 97

Ulcer
   decubitus, 107
   duodenal, 680
   peptic, 680
Ulna, 170
Ultrasound (US), 19, 24, 732
   fetal, 763–764
Umbilical cord, 407, 761
Umbilical region, 16E
Umbilicus, *see* Navel
Unconsciousness, 496
Unicellular glands, 74, 75
Unipennate muscle, 248
Upper extremities, 169–170
   lymph nodes of, 429E, 430
   surface anatomy of, 312, 315–316, 318E, 319E, 320E, 321E
   veins of, 396E, 397, 398
Upper respiratory system, 617
Uremia, 705
Ureter, 698–700
   blood supply of, 700

Ureter (*Continued*)
  effect of sympathetic and parasympathetic stimulation on, 543E
  nerve supply of, 700
Ureteric bud, 702
Urethra, 708, 716
Urethral orifice, 708
Urethritis, 705
Urinary bladder, 700–701
  blood supply of, 701
  effect of sympathetic and parasympathetic stimulation on, 544E
  nerve supply of, 701
Urinary system, 9E, 687–707
  aging and, 702
  development of, 702
  disorders of, 702–705
  key medical terms, 705
  organs of, 688–702
Urinary tract infections (UTIs), 705
Urine, 697
Urogenital diaphragm, 273
Urogenital sinus, 702
Urogenital triangle, 273
Urology, 688
Uterine (Fallopian) tubes, 726–727
Uterus, 727–278
  cancer of, 727
  effect of sympathetic and parasympathetic stimulation on, 544E
  removal of, 737
Uvula, 649

Vacuole, 45
Vagina, 732
  lubrication of, 737
Vaginal orifice, 732
Vaginitis, 746
Valves of the heart, 348
Valvular heart disease, *see* Rheumatic fever
Valvular stenosis, 364
Varicose veins (VVs), 373–374, 392
Vasa recta, 694
Vasa vasorum, 370
Vascular tissue, 82E, 84–85
Vascular tunic, 563–566
Vasectomy, 715, 737
Vasoconstriction, 370
Vasodilation, 370
Vasodilator substances, 372
Vasomotor center, 496
Vastus lateralis muscle, 307
Veins, 370, 373–374
  accessory hemiazygos, 399E
  anterior tibial, 402E
  arculate, 694
  axiliary, 396E
  azygos, 399E
  basilic, 396E
  as blood reservoirs, 374–375
  brachial, 396E
  brachiocephalic, 399E
  cephalic, 396E
  common iliac, 399E
  definition, 370
  external iliac, 399E
  external jugular, 312, 394E
  femoral, 402E
  gastric, 406
  gastroepiploic, 406

  gonadal, 399E
  great cardiac, 358
  great saphenous, 402E
  hemiazygos, 399E
  hepatic, 399E, 406
  hepatic portal, 405
  inferior phrenic, 399E
  inferior vena cava, 392E, 399E
  interlobular, 694
  internal iliac, 399E
  internal jugular, 394E
  lumbar, 399E
  median antibrachial, 396E
  middle cardiac, 358
  pancreatic, 406
  popliteal, 402E
  posterior tibial, 402E
  pulmonary, 347, 405
  pyloric, 406
  radial, 396E
  renal, 399E, 694–696
  small saphenous, 402E
  spider burst, 374
  splenic, 406
  subclavian, 396E
  superior mesenteric, 405–406
  superior vena cava, 392E
  suprarenal, 399E
  of systemic circulation, 392E, 393
  ulnar, 396E
  umbilical, 408, 409
  uterine, 728
  varicose, 373–374, 392
  vascular sinus, 374
Vellus, 100
Venereal diseases (VD), 741
Venesection, 341
Venipuncture, 328
Ventral body cavity, 16
Ventricles, 347, 488–490
Ventricular fibrillation (VF), 365
Ventricular premature contraction (VPC), 365
Venules, 370, 373
  definition, 370
Vermiform appendix, 574
Vermilion, 648
Vermis, 509
Vertebrae, 148
Vertebral canal, 13, 460
Vertebral column, 4, 148–157
  cervical region, 148, 151
  disorders of, 159–160
  divisions of, 148
  fracture of, 160
  lumbar region, 148, 151–156
  muscles that move, 290E, 291E
  sacrum, 148, 156–157
  structure, 148–149
  thoracic region, 148, 151
Vertebrates, 4
Vertigo, 584
Vesicles
  brain, 487, 488E
  phagocytic, 37
  seminal, 716–717
  synaptic, 448, 453
Vestibular ganglion, 516

Vestibular nuclear complex, 496
Vestibule, 619
  of the oral cavity, 649
Vestibulospinal tract, 467E
Vibration sensations, 553
Viral encephalitis, 529
Virilism, 605
Virilizing adenomas, 605
Viscera, 16
Visceral autonomic reflexes, 544–545
Visceral effector, 544
Visceral efferent fibers, 534
Visceral efferent neurons, 534–538
Visceral muscle tissue, 238
Visceral peritoneum, 647
Visceral reflexes, 469
Visceroceptors, 551
Visual sensations, 561–568
  accessory structures of the eye, 562–563
  structure of the eyeball, 563–567
  visual pathway, 558
Vitamin D, 93, 610–611
  deficiency of, 124, 125
Vitiligo, 97
Vitreous humor, 567
Vocal folds (vocal cords), 622
Volkmann's contracture, 241
Voluntary muscle tissue, 224
Vomiting, 663
Vulva, 732–733

Wallerian degeneration, 480–481
Wandering macrophage, 334
Wart, 110
"Watery" eye, 563
Whiplash injury, 151
White blood cells (WBCs), see Leucocytes
White commissure, 465
White matter, 460, 465–466, 504
White muscle fibers, 234

White rami communicantes, 539
Whole blood, 341
Windpipe, see Trachea
Word blindness, 508
Word deafness, 508
Wound healing, 98–99
  deep, 99
  epidermal, 98–99
Wrist, 208–209, 208E
  muscles that move, 283E, 284E
  surface anatomy of, 312, 315–317, 320E
Wrist crease, 317, 320
Wryneck, 241

Xanthinoxidase, 358
X chromosomes, 53, 713
Xeromammography, 734
Xiphoid process, 158

Y chromosomes, 53, 713
Yellow marrow, 115
Yoga, 546
Yolk sac, 758

Z lines, 229
Zona fasciculata, 603
Zona glomerulosa, 602
Zona pellucida, 703, 752
Zona radiata, 752
Zona reticularis, 603
Zone of calcified matrix, 120–121
Zone of hypertrophic cartilage, 120
Zone of proliferating cartilage, 119–120
Zone of reserved cartilage, 119
Zonula adherens, 65
Zonula occludens, 65
Zygomatic arch, 139
Zygomaticofacial foramen, 145
Zygote, 53, 712
Zymogenic (peptic) cells, 659